Precalculus
With Limits

A GRAPHING APPROACH

Roland E. Larson
Robert P. Hostetler
THE PENNSYLVANIA STATE UNIVERSITY
THE BEHREND COLLEGE

Bruce H. Edwards
UNIVERSITY OF FLORIDA

WITH THE ASSISTANCE OF
David E. Heyd
THE PENNSYLVANIA STATE UNIVERSITY
THE BEHREND COLLEGE

D. C. Heath and Company
Lexington, Massachusetts   Toronto

Address editorial correspondence to:

D. C. Heath and Company
125 Spring Street
Lexington, MA 02173

Acquisitions Editor: Ann Marie Jones
Developmental Editor: Cathy Cantin
Production Editor: Andrea Cava
Designer and Art Editor: Sally Steele
Production Coordinators: Lisa Merrill, Richard Tonachel
Technical Art: Folium
Cover Designer: Linda Wade
Cover Photographer: Gabrielle Keller

Trademark Acknowledgments: TI is a registered trademark of Texas Instruments Incorporated. Casio is a registered trademark of Casio, Inc. Sharp is a registered trademark of Sharp Electronics Corporation.

Art and Photo Credits: p. 163, from *The Fractal Geometry of Nature* by Benoit B. Mandelbrot, by permission; p.164 (top), from *Chaos and Fractals: The Mathematics Behind the Computer Graphics,* edited by Robert L. Devaney and Linda Keen. Copyright © 1989 by the American Mathematical Society. All rights reserved. Used by permission of the publisher; p. 164 (bottom), from *Technology in the Classroom: Fractal Programs for the TI-81 Graphing Calculator* by L. Charles Biehl, Leadership Program in Discrete Mathematics, Rutgers University, New Brunswick, N.J. Copyright © 1991 END Products/Publications.

We have included examples and exercises that use real-life data. This would not have been possible without the help of many people and organizations. Our wholehearted thanks goes to all for their time and effort.

PREFACE

Precalculus with Limits: A Graphing Approach has two basic goals. The first is to help students develop a good understanding of algebra and trigonometry. The other goal is to show students how algebra can be used as a modeling language for real-life problems.

FEATURES

The text has several key features designed to help students develop their problem-solving skills, as well as acquire an understanding of mathematical concepts.

GRAPHICS

The ability to visualize a problem is a critical part of a student's ability to solve the problem. To encourage the development of this skill, the text has many figures in examples and exercise sets and in answers to odd-numbered exercises in the back of the text. Various types of graphics show geometric representations, including graphs of functions, geometric figures, symmetry, displays of statistical information, and numerous screen outputs from graphing technology. All graphs of functions, computer- or calculator-generated for accuracy, are designed to resemble students' actual screen outputs as closely as possible.

APPLICATIONS

Numerous pertinent applications are integrated throughout every section of the text, both as solved examples and as exercises. This encourages students to use and review their problem-solving skills. The text applications are current and often involve multiple parts. Students learn to apply the process of mathematical modeling to real-world situations in many areas, such as business, economics, biology, engineering, chemistry, and physics. Many applications in the text use real data, and source lines are included to help motivate student interest. We tried to use the data accurately—to give honest and unbiased portrayals of real-life situations. In the cases in which models were fit to data, we used the least-squares method. In all cases the square of the correlation coefficient r^2 was at least 0.95. In most cases it was 0.99 or greater.

EXPLORING DATA

Modeling real-life problems requires the ability to organize, represent, and interpret real data. The *Exploring Data* sections throughout the text help students learn basic statistical skills to represent data graphically, to find linear and nonlinear models for data, and to describe sets of data. The exercises in the sections allow students to practice these skills using real data.

EXAMPLES

Each example was carefully chosen to illustrate a particular concept or problem-solving technique. Examples are titled for quick reference, and many include color side comments to justify or explain the solution. We have included problems solved graphically, analytically, numerically, or by a combination of these strategies, and the text helps students choose appropriate approaches to the problems. Several examples also preview ideas from calculus, building an intuitive foundation for future study.

DISCOVERY

Discovery boxes encourage students to strengthen their intuition and understanding by exploring the relationships between functions and the behaviors of functions. The powerful features of graphing utilities enhance the study of functions because a graph of each step in a solution can be generated quickly and easily for use in the problem-solving process.

INTUITIVE FOUNDATION FOR CALCULUS

Throughout the text, many examples discuss algebraic techniques or graphically illustrate concepts that are used in calculus. These help students develop a natural and intuitive foundation for later work.

EXERCISE SETS

Exercise sets, including warm-up exercises, appear at the end of each text section. Many sets include a group of exercises that provide the graphs of functions involved. Review exercises are included at the end of each chapter, and cumulative tests are included to review what students have learned from the preceding chapters. The opportunity to use calculators is provided with topics that allow students to see patterns, experiment, calculate, or create graphic models.

DISCUSSION PROBLEMS

The discussion problems offer students the opportunity to think, reason, and communicate about mathematics in different ways. Individually or in teams, for in-class discussion, writing assignments, or class presentations, students are encouraged to draw new conclusions about the concepts presented. The problem might ask for further explanation, synthesis, experimentation, or extension of the section concepts. Discussion problems appear at the end of each text section.

TECHNOLOGY NOTES

Technology notes to students appear in the margins throughout the text. These notes offer additional insights, help students avoid common errors, and provide opportunities for problem solving using technology.

WARM-UP EXERCISES AND CUMULATIVE TESTS

We have found that students can benefit greatly from reinforcement of previously learned concepts. Most sections in the text contain a set of ten warm-up exercises that efficiently give students practice using techniques studied earlier in the course that are necessary to master the new ideas presented in the section.

Cumulative tests are included after Chapters 2, 5, 8, and 11. These tests help students assess their level of success and help them maintain the knowledge base they have been building throughout the text—preparing them for other exams and future courses.

These and other features of the text are described in greater detail on the following pages.

FEATURES OF THE TEXT

CHAPTER OPENER

Each chapter begins with a list of the topics to be covered. Each section begins with a list of important topics covered in that section.

FUNCTIONS AND GRAPHS

1.1 GRAPHS AND GRAPHING UTILITIES

The Graph of an Equation / Using a Graphing Utility / Determining a Viewing Rectangle / Applications

The Graph of an Equation

News magazines often show graphs comparing the rate of inflation, the federal deficit, wholesale prices, or the unemployment rate to the time of year. Industrial firms and businesses use graphs to report their monthly production and sales statistics. Such graphs provide geometric pictures of the way one quantity changes with respect to another.

Frequently, the relationship between two quantities is expressed as an equation. This section introduces the basic procedure for determining the geometric picture associated with an equation.

For an equation in variables x and y, a point (a, b) is a **solution point** if the substitution of $x = a$ and $y = b$ satisfies the equation. Most equations have *infinitely* many solution points. For example, the equation $3x + y = 5$ has solution points $(0, 5), (1, 2), (2, -1), (3, -4)$, and so on. The set of all solution points of an equation is the **graph** of the equation.

DEFINITIONS

All of the important formulas and definitions are boxed for emphasis. Each is also titled for easy reference.

INTUITIVE FOUNDATION FOR CALCULUS

Special emphasis has been given to skills that are needed in calculus. Many examples include algebraic techniques or graphically show concepts that are used in calculus, providing an intuitive foundation for future work.

Technology Note _____

When you use a graphing utility to estimate the x- and y-values of a relative minimum or relative maximum, the automatic zoom feature will often produce graphs that are nearly flat. To overcome this problem, you can manually change the vertical setting of the viewing rectangle. You can stretch the graph vertically by making the values of Y_{min} and Y_{max} closer together.

EXAMPLE 4   **Approximating a Relative Minimum**

Use a graphing utility to approximate the relative minimum of the function $f(x) = 3x^2 - 4x - 2$.

SOLUTION

The graph of f is shown in Figure 1.35. By using the zoom and trace features of a graphing utility, you can estimate that the function has a relative minimum at the point

$$(0.67, -3.33).$$ *Relative minimum*

Later, in Section 3.1, you will be able to determine that the exact point at which the relative minimum occurs is $(\frac{2}{3}, -\frac{10}{3})$.

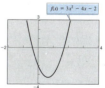

$f(x) = 3x^2 - 4x - 2$

FIGURE 1.35

Technology Note _____

Some graphing utilities, such as a TI-85, can automatically determine the maximum and minimum value of a function defined on a closed interval. If your graphing utility has this feature, use it to **graph** $y = -x^3 + x$ on the interval $-1 \le x \le 1$ and verify the results of Example 5. What happens if you use the interval $-10 \le x \le 10$?

EXAMPLE 5   **Approximating Relative Minimums and Maximums**

Use a graphing utility to approximate the relative minimum and relative maximum of the function $f(x) = -x^3 + x$.

SOLUTION

A sketch of the graph of f is shown in Figure 1.36. By using the zoom and trace features of the graphing utility, you can estimate that the function has a relative minimum at the point

$$(-0.58, -0.38).$$ *Relative minimum*

and a relative maximum at the point

$$(0.58, 0.38).$$ *Relative maximum*

If you go on to take a course in calculus, you will learn a technique for finding the exact points at which this function has a relative minimum and a relative maximum.

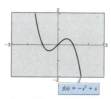

$f(x) = -x^3 + x$

FIGURE 1.36

...ORS AND ... GRAPHING

...he range of teaching and learning ...otes for working with calculators ... many places. Students with access to ...ics calculators or graphing utilities can ...ve exercises both graphically and analytically beginning with Chapter 1. Additionally, many exercises require a graphing utility.

TECHNOLOGY NOTES

These notes appear in the margins. They provide additional insight and help students avoid common errors.

Throughout the text, when solving equations, be sure to check your solutions—either *algebraically*, by substituting in the original equation, or *graphically*. For instance, the graphs shown in Figure 2.22 visually reinforce the solutions obtained in Example 1.

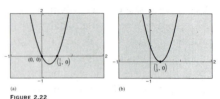

(a)   (b)

FIGURE 2.22

EXAMPLE 2   Extracting Square Roots

Solve the quadratic equations.

A. $4x^2 = 12$ **B.** $(x - 3)^2 = 7$

SOLUTION

A. $4x^2 = 12$ *Original equation*
$\quad x^2 = 3$ *Divide both sides by 4*
$\quad x = \pm\sqrt{3}$ *Extract square roots*
B. $(x - 3)^2 = 7$ *Original equation*
$\quad x - 3 = \pm\sqrt{7}$ *Extract square roots*
$\quad x = 3 \pm \sqrt{7}$ *Add 3 to both sides*

The graphs of $y = 4x^2 - 12$ and $y = (x - 3)^2 - 7$, shown in Figure 2.23, reinforce these solutions.

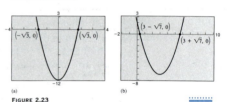

(a)   (b)

FIGURE 2.23

Technology Note

Most graphing utilities have two log keys: *ln* is the natural logarithmic function and *log* is the logarithm to base 10. You can graph logarithms to other bases, $y = \log_a(x)$, by using the change of base formula.

$$y = \log_a(x) = \frac{\ln x}{\ln a} = \frac{\log x}{\log a}$$

EXAMPLE 1   Changing Bases

Use *common logarithms* to evaluate the following.

A. $\log_4 30$ **B.** $\log_2 14$

SOLUTION

A. Using the change of base formula with $a = 4$, $b = 10$, and $x = 30$, convert to common logarithms and obtain

$$\log_4 30 = \frac{\log_{10} 30}{\log_{10} 4} \approx \frac{1.47712}{0.60206} \approx 2.4534.$$

B. Using the change of base formula with $a = 2$, $b = 10$, and $x = 14$, convert to common logarithms and obtain

$$\log_2 14 = \frac{\log_{10} 14}{\log_{10} 2} \approx \frac{1.14613}{0.30103} \approx 3.8074.$$

EXAMPLE 2   Changing Bases

Use *natural logarithms* to evaluate the following.

A. $\log_4 30$ **B.** $\log_2 14$

SOLUTION

A. Using the change of base formula with $a = 4$, $b = e$, and $x = 30$, convert to natural logarithms and obtain

$$\log_4 30 = \frac{\ln 30}{\ln 4} \approx \frac{3.40120}{1.38629} \approx 2.4534.$$

B. Using the change of base formula with $a = 2$, $b = e$, and $x = 14$, convert to natural logarithms and obtain

$$\log_2 14 = \frac{\ln 14}{\ln 2} \approx \frac{2.63906}{0.693147} \approx 3.8074.$$

Note that the results agree with those obtained in Example 1, using common logarithms.

Properties of Logarithms

You know from the previous section that the logarithmic function with base a is the *inverse* of the exponential function with base a. Thus, it makes sense that the properties of exponents should have corresponding properties involving logarithms. For instance, the exponential property $a^0 = 1$ corresponds to the logarithmic property $\log_a 1 = 0$.

DISCOVERY

Use a graphing utility to graph $y = \ln x$ and $y = \ln x/\ln a = \log_a x$ with $a = 2, 3,$ and 5 on the same viewing rectangle. (Use a viewing rectangle in which $0 \le x \le 10$ and $-4 \le y \le 4$.) On the interval $(0, 1)$, which graph is on top? On the interval $(1, \infty)$, which graph is on top? Which is on the bottom?

DISCOVERY

Throughout the text, the discovery boxes take advantage of the power of graphing utilities to explore and examine the behavior of complicated functions.

PROBLEM SOLVING

A consistent strategy for solving problems is emphasized throughout: analyze the problem, create a verbal model, construct an algebraic model, solve the problem, and check the answer in the statement of the original problem. This problem-solving process has wide applicability and can be used with analytical, graphical, and numerical approaches to problem solving.

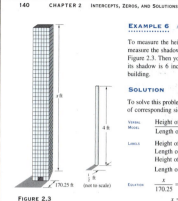

FIGURE 2.3

EXAMPLE 6   An Application Involving Similar Triangles

To measure the height of the World Trade Center (in New York City), you measure the shadow cast by the building to be 170.25 feet long, as shown in Figure 2.3. Then you measure the shadow cast by a 4-foot post and find that its shadow is 6 inches long. Use this information to find the height of the building.

SOLUTION

To solve this problem, use a theorem from geometry that states that the ratios of corresponding sides of similar triangles are equal.

VERBAL MODEL

$$\frac{\text{Height of building}}{\text{Length of shadow}} = \frac{\text{Height of post}}{\text{Length of shadow}}$$

LABELS

Height of building $= x$	(feet)
Length of building's shadow $= 170.25$	(feet)
Height of post $= 4$	(feet)
Length of post's shadow $= \frac{1}{2}$	(feet)

EQUATION

$$\frac{x}{170.25} = \frac{4}{\frac{1}{2}}$$

$$x = 1362 \text{ feet}$$

The World Trade Center is about 1362 feet high.

EXAMPLE 7   An Inventory Problem

A store has $30,000 of inventory in 12-inch and 19-inch television sets. The profit on a 12-inch set is 22%. The profit on a 19-inch set is 40%. If the profit for the entire stock is 35%, how much was invested in each type of television set?

SOLUTION

VERBAL MODEL

$$\frac{\text{Profit from}}{\text{12-inch sets}} + \frac{\text{Profit from}}{\text{19-inch sets}} = \frac{\text{Total}}{\text{profit}}$$

LABELS

Inventory of 12-inch sets $= x$	(dollars)
Inventory of 19-inch sets $= 30{,}000 - x$	(dollars)
Profit from 12-inch sets $= 0.22x$	(dollars)
Profit from 19-inch sets $= 0.40(30{,}000 - x)$	(dollars)
Total profit $= 0.35(30{,}000) = 10{,}500$	(dollars)

EQUATION

$$0.22(x) + 0.40(30{,}000 - x) = 0.35(30{,}000)$$
$$0.22x + 12{,}000 - 0.4x = 10{,}500$$
$$-0.18x = -1500$$
$$x \approx \$8333.33 \text{ 12-inch sets}$$
$$30{,}000 - x \approx \$21{,}666.67 \text{ 19-inch sets}$$

EXAMPLE 14   The Cost of a New Car

Between 1970 and 1990, the average cost of a new car increased according to the model

$$C = 30.5t^2 + 4192, \qquad 0 \le t \le 20,$$

as shown in Figure 2.34. In this model the cost is measured in dollars and the time t represents the year with $t = 0$ corresponding to 1970. If the average cost of a new car continued to increase according to this model, when would the average cost reach $20,000? (*Source:* Commerce Department, American Auto-datum)

FIGURE 2.34

SOLUTION

To solve this problem, let the cost be $20,000 and solve the equation $20{,}000 = 30.5t^2 + 4192$ for t.

$30.5t^2 + 4192 = 20{,}000$	*Set cost equal to 20,000*
$30.5t^2 = 15{,}808$	*Subtract 4192 from both sides*
$t^2 \approx 518.295$	*Divide both sides by 30.5*
$t \approx \sqrt{518.295}$	*Extract positive square root*

Thus, the solution is $t \approx 23$. Because $t = 0$ represents 1970, you can conclude that the average cost of a new car reached $20,000 in 1993. You could use a graphing utility to solve this problem. By finding the positive zero of $30.5t^2 + 4192 - 20{,}000 = 0$, you can see that $t \approx 23$.

EXAMPLES

The text contains over 650 examples. They are titled for easy reference, and many include side comments that explain or justify steps in the solution. Students are encouraged to check their solutions.

EXPLORING DATA

Students work with real data to develop basic statistical skills in these sections, learning to graph, model, and describe sets of data in the examples and exercises.

2.6 EXPLORING DATA: LINEAR MODELS AND SCATTER PLOTS

Scatter Plots / Fitting a Line to Data

Technology Note

Most graphing utilities have built-in statistical programs that can create scatter plots. Use your graphing utility to plot the points given in Table 2.2.

Scatter Plots

Many real-life situations involve finding relationships between two variables, such as the year and the number of people in the labor force. In a typical situation, data are collected and written as a set of ordered pairs. The graph of such a set is called a **scatter plot.**

EXAMPLE 1   Constructing a Scatter Plot

The data in Table 2.2 show the number of people P (in millions) in the United States who were part of the labor force from 1980 through 1990. In the table, t represents the year, with $t = 0$ corresponding to 1980. Sketch a scatter plot of the data. (*Source:* U.S. Bureau of Labor Statistics)

TABLE 2.2

t	0	1	2	3	4	5	6	7	8	9	10
P	109	110	112	113	115	117	120	122	123	126	126

SOLUTION

Begin by representing the data with a set of ordered pairs.

(0, 109), (1, 110), (2, 112), (3, 113), (4, 115), (5, 117), (6, 120), (7, 122), (8, 123), (9, 126), (10, 126)

Then plot each point in a coordinate plane, as shown in Figure 2.48.

FIGURE 2.48

DISCUSSION PROBLEM

AN APPLICATION OF SLOPE

In 1982, a college had an enrollment of 5000 students. By 1992, the enrollment had increased to 7000 students.

(a) What is the average annual change in enrollment from 1982 to 1992?

(b) Use the average annual change in enrollment to estimate the enrollment in 1986, 1990, and 1994.

Year	1982	1986	1990	1992	1994
Enrollment	5000			7000	

(c) Graph the line represented by the data given in the table in part (b). What is the slope of this line?

(d) Write a short paragraph that compares the concepts of *slope* and *average rate of change.*

DISCUSSION PROBLEMS

A discussion problem appears at the end of each section. Each one encourages students to think, reason, and write about mathematics, individually or in groups. Presenting the mathematics in a different way from in the section, these problems emphasize synthesis and experimentation.

WARM-UPS

Each section (except those in Prerequisites) contains a set of 10 warm-up exercises for students to review and practice the previously learned skills that are necessary to master the new skills and concepts presented in the section. All warm-up exercises are answered in the back of the text.

WARM-UP

The following warm-up exercises involve skills that were covered in earlier sections. You will use these skills in the exercise set for this section.

In Exercises 1 and 2, simplify the expression.

1. $\dfrac{4 - (-5)}{-3 - (-1)}$

2. $\dfrac{-5 - 8}{0 - (-3)}$

3. Find $-1/m$ for $m = 4/5$.

4. Find $-1/m$ for $m = -2$.

In Exercises 5–10, solve for y in terms of x.

5. $2x - 3y = 5$

6. $4x + 2y = 0$

7. $y - (-4) = 3[x - (-1)]$

8. $y - 7 = \frac{2}{3}(x - 3)$

9. $y - (-1) = \dfrac{3 - (-1)}{2 - 4}(x - 4)$

10. $y - 5 = \dfrac{3 - 5}{0 - 2}(x - 2)$

SECTION 3.2   POLYNOMIAL FUNCTIONS OF HIGHER DEGREE   229

SECTION 3.2 · EXERCISES

1. Compare the graph of f with the graph of $y = x^3$.
   (a) $f(x) = (x - 2)^3$
   (b) $f(x) = x^3 - 2$
   (c) $f(x) = (x - 2)^3 - 2$
   (d) $f(x) = -\frac{1}{2}x^3$

2. Compare the graph of f with the graph of $y = x^4$.
   (a) $f(x) = (x + 3)^4$
   (b) $f(x) = x^4 - 3$
   (c) $f(x) = 4 - x^4$
   (d) $f(x) = \frac{1}{2}(x - 1)^4$

In Exercises 3–10, match the polynomial function with its graph and describe the viewing rectangle. [The graphs are labeled (a), (b), (c), (d), (e), (f), (g), and (h).]

3. $f(x) = -3x + 5$
4. $f(x) = x^2 - 2x$
5. $f(x) = -2x^2 - 8x - 9$
6. $f(x) = 3x^3 - 9x + 1$
7. $f(x) = -\frac{1}{3}x^3 + x - \frac{2}{3}$
8. $f(x) = -\frac{1}{4}x^4 + 2x^2$
9. $f(x) = 3x^4 + 4x^3$
10. $f(x) = x^5 - 5x^3 + 4x$

(a)

(b)

(c)

(d)

(e)

(f)

(g)

(h)

In Exercises 11–14, use a graphing utility to graph the functions f and g in the same viewing rectangle. Zoom out sufficiently far to show that the right-hand and left-hand behavior of f and g are identical.

11. $f(x) = x^3 - 9x + 1$ $g(x) = x^3$
12. $f(x) = -\frac{1}{3}(x^3 - 3x + 2)$ $g(x) = -\frac{1}{3}x^3$
13. $f(x) = -(x^4 - 4x^3 + 16x)$ $g(x) = -x^4$
14. $f(x) = 3x^4 - 6x^2$ $g(x) = 3x^4$

In Exercises 15–20, determine the right-hand and left-hand behavior of the graph of the polynomial function based on the Leading Coefficient Test. Use a graphing utility to verify your result.

15. $f(x) = 2x^2 - 3x + 1$
16. $g(x) = 5 - \frac{7}{2}x - 3x^2$
17. $h(t) = -\frac{2}{3}(t^2 - 5t + 3)$
18. $f(s) = -\frac{7}{8}(s^3 + 5s^2 - 7s + 1)$
19. $f(x) = 6 - 2x + 4x^2 - 5x^3$
20. $f(x) = \dfrac{3x^4 - 2x + 5}{4}$

In Exercises 21–36, find all the real zeros of the polynomial function. In each case, state whether you solved the problem algebraically or graphically and give reasons for your choice.

21. $f(x) = x^2 - 25$
22. $f(x) = 49 - x^2$
23. $h(t) = t^2 - 6t + 9$
24. $f(x) = x^2 + 10x + 25$
25. $f(x) = x^2 + x - 2$
26. $f(x)$
27. $f(x) = 3x^2 - 12x + 3$
28. $g(x)$
29. $f(t) = t^3 - 4t^2 + 4t$
30. $f(x)$
31. $g(t) = \frac{1}{2}t^4 - \frac{1}{2}$
32. $f(x)$
33. $f(x) = 2x^4 - 2x^2 - 40$
34. $g(t) = t^5 - 6t^3 + 9t$
35. $f(x) = 5x^4 + 15x^2 + 10$
36. $f(x) = x^3 - 4x^2 - 25x + 100$

EXERCISES

The approximately 7000 exercises—computational, conceptual, exploratory, and applied problems—are designed to build competence, skill, and understanding. Each exercise set is graded in difficulty to allow students to gain confidence as they progress. Many exercises require the use of a graphing utility. All odd-numbered exercises are solved in detail in the *Study and Solutions Guide,* with answers appearing in the back of the text.

GEOMETRY

Geometric formulas and concepts are reviewed throughout the text. For easy reference, common formulas are given inside the back cover.

APPLICATIONS

Real-world applications are integrated throughout the text in examples and exercises. This offers students insight about the usefulness of algebra and trigonometry, develops strategies for solving problems, and emphasizes the relevance of the mathematics. Titled for reference, many of the applications involve multiple parts, use current real data, and include source lines.

GRAPHICS

Students must be able to visualize problems in order to solve them. To develop this skill and reinforce concepts, the text has over 1100 figures.

SECTION 6.7   LAW OF COSINES   507

40. *Navigation* On a certain map, Minneapolis is 6.5 inches due west of Albany, Phoenix is 8.5 inches from Minneapolis, and Phoenix is 14.5 inches from Albany (see figure).
   (a) Find the bearing of Minneapolis from Phoenix.
   (b) Find the bearing of Albany from Phoenix.

FIGURE FOR 40

41. *Baseball* In a (square) baseball diamond with 90-foot sides, the pitcher's mound is 60 feet from home plate.
   (a) How far is it from the pitcher's mound to third base?
   (b) When a runner is halfway from second to third, how far is the runner from the pitcher's mound?

42. *Baseball* The baseball player in center field is playing approximately 330 feet from the television camera that is behind home plate. A batter hits a fly ball that goes to the wall 420 feet from the camera (see figure). Approximate the number of feet that the center fielder had to run to make the catch if the camera turned 9° in following the play.

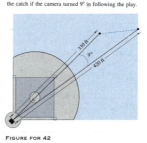

FIGURE FOR 42

43. *Awning Design* A retractable awning lowers at an angle of 50° from the top of a patio door that is 7 feet high (see figure). Find the length x of the awning if no direct sunlight is to enter the door when the angle of elevation of the sun is greater than 65°.

44. *Circumscribed and Inscribed Circles* Let R and r be the radii of the circumscribed and inscribed circles of a triangle ABC, respectively, and let $s = (a + b + c)/2$ (see figure). Prove the following.
   (a) $2R = \dfrac{a}{\sin A} = \dfrac{b}{\sin B} = \dfrac{c}{\sin C}$
   (b) $r = \sqrt{\dfrac{(s - a)(s - b)(s - c)}{s}}$

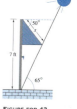

FIGURE FOR 43

FIGURE FOR 44

Circumscribed and Inscribed Circles In Exercises 45 and 46, use the results of Exercise 44.

45. Given the triangle with $a = 25$, $b = 55$, and $c = 72$, find the areas of (a) the triangle, (b) the circumscribed circle, and (c) the inscribed circle.

46. Find the length of the largest circular track that can be built on a triangular piece of property whose sides measure 200 feet, 250 feet, and 325 feet.

47. Use the Law of Cosines to prove that
$$\frac{1}{2}bc(1 + \cos A) = \frac{a + b + c}{2} \cdot \frac{-a + b + c}{2}.$$

48. Use the Law of Cosines to prove that
$$\frac{1}{2}bc(1 - \cos A) = \frac{a - b + c}{2} \cdot \frac{a + b - c}{2}.$$

REVIEW EXERCISES

A set of review exercises at the end of each chapter gives students an opportunity for additional practice. The review exercises include computational, conceptual, and applied problems covering a wide range of topics.

CUMULATIVE TESTS

Cumulative tests appear after Chapters 2, 5, 8, and 11. These tests help students judge their mastery of previously covered concepts. They also help students maintain the knowledge base they have been building throughout the text, preparing them for other exams and future courses.

---

204   CHAPTER 2   INTERCEPTS, ZEROS, AND SOLUTIONS

20. *School Enrollment* The table gives the preprimary school enrollments y (in millions) for the years 1985 through 1991 where $t = 5$ corresponds to 1985. (*Source: U.S. Bureau of the Census*).

5	6	7	8	9	10	11
10.73	10.87	10.87	11.00	11.04	11.21	11.37

(a) Use a computer or calculator to find the least squares regression line. Use the equation to estimate enrollment in 1992.
(b) Make a scatter plot of the data and sketch the graph of the regression line.
(c) Use the computer or calculator to determine the correlation coefficient.

CHAPTER 2 · REVIEW EXERCISES

In Exercises 1 and 2, determine whether the equation is an identity or a conditional equation.

1. $6 - (x - 2)^2 = 2 + 4x - x^2$
2. $3(x - 2) + 2x = 2(x + 3)$

In Exercises 3 and 4, determine whether the values of x are solutions of the equation.

Equation	Values
3. $3x^2 + 7x + 5 = x^2 + 9$	(a) $x = 0$ (b) $x = -4$ (c) $x = \frac{1}{2}$ (d) $x = -1$
4. $6 + \dfrac{3}{x - 4} = 5$	(a) $x = 4$ (b) $x = 0$ (c) $x = -2$ (d) $x = 1$

In Exercises 5–28, solve the equation (if possible) and check your answer either algebraically or graphically.

5. $3x - 2(x + 5) = 10$
6. $4(x + 3) - 3 = 2(4 - 3x) - 4$
7. $3\left(1 - \dfrac{1}{5t}\right) = 0$
8. $\dfrac{1}{x - 2} = 3$
9. $6x^2 = 5x + 4$
10. $15 + x - 2x^2 = 0$
11. $(x + 4)^2 = 18$
12. $16x^2 = 25$
13. $x^2 - 12x + 30 = 0$
14. $5x^4 - 12x^3 = 0$
15. $4t^3 - 12t^2 + 8t = 0$
16. $2 - x^{-2} = 0$
17. $\dfrac{4}{(x - 4)^2} = 1$
18. $\dfrac{4}{x - 3} - \dfrac{4}{x} = 1$
19. $\sqrt{x + 4} = 3$
20. $\sqrt{x - 2} - 8 = 0$

21. $2\sqrt{x} - 5 = 0$
22. $\sqrt{2x + 3} + \sqrt{x - 2} = 2$
23. $(x - 1)^{2/3} - 25 = 0$
24. $(x + 2)^{3/4} = 27$
25. $(x + 4)^{1/2} + 5x(x + 4)^{3/2} = 0$
26. $8x^2(x^2 - 4)^{1/3} + (x^2 - 4)^{4/3} = 0$
27. $|x - 5| = 10$
28. $|x^2 - 6| = x$

In Exercises 29–36, use a graphing utility to solve the equation (if possible).

29. $x^2 + 6x - 3 = 0$
30. $12t^3 - 84t^2 + 120t = 0$
31. $5\sqrt{x} - \sqrt{x - 1} = 6$
32. $\sqrt{3x - 2} = 4 - x$
33. $\dfrac{1}{x} + \dfrac{1}{x + 1} = 2$
34. $\dfrac{1}{(t + 1)^2} = 1$
35. $|x^2 - 3| = 2x$
36. $|2x + 3| = 7$

In Exercises 37–40, solve the equation for the indicated variable.

37. Solve for r: $V = \frac{1}{3}\pi r^2 h$
38. Solve for X: $Z = \sqrt{R^2 - X^2}$
39. Solve for p: $L = \dfrac{k}{3\pi r^2 p}$
40. Solve for v: $E = 2kw\left(\dfrac{v}{2}\right)^2$

---

438   CHAPTER 5   TRIGONOMETRIC FUNCTIONS

CUMULATIVE TEST FOR CHAPTERS 3–5

Take this test as you would take a test in class. After you are done, check your work against the answers in the back of the book.

1. Use a graphing utility to graph the quadratic function
$f(x) = \frac{1}{4}(4x^2 - 12x + 17)$.
Find the coordinates of the vertex of the parabola.
2. Find a quadratic function whose graph is a parabola with vertex at $(0, 6)$ and passes through the point $(2, 5)$.
3. Describe the right-hand and left-hand behavior of the polynomial function
$f(x) = -\frac{1}{3}x^3 + 3x^2 - 2x + 1$.
4. Find a polynomial function with integer coefficients whose zeros are $-4, \frac{1}{3}$, and 2.
5. Sketch a graph of the function $f(t) = \frac{1}{4}t(t - 2)^2$ without the aid of a graphing utility.
6. Perform the division: $\dfrac{6x^3 - 4x^2}{2x^2 + 1}$
7. Use synthetic division to perform the division: $\dfrac{3x^3 - 5x + 4}{x - 2}$
8. Find the rational zeros of the function
$f(x) = 6x^3 - 25x^2 - 8x + 48$.
(*Hint:* Use a graphing utility to eliminate some of the possible rational zeros.)
9. Use a graphing utility to approximate (accurate to one decimal place) the zero of the function $g(t) = t^3 - 5t - 2$ in the interval $[2, 3]$.
10. Sketch a graph of each of the following.
(a) $g(s) = \dfrac{2s}{s - 3}$ (b) $g(s) = \dfrac{2s^2}{s - 3}$
11. Sketch a graph of each of the following.
(a) $f(x) = 6(2^{-x})$ (b) $g(x) = \log_2 x$
12. Evaluate without the aid of a calculator: $\log_5 125$
13. Use the properties of a logarithm to write the expression
$2 \ln x - \frac{1}{2}\ln(x + 5)$
as the logarithm of a single quantity.
14. Solve each of the following, giving your answers accurate to two decimal places.
(a) $6e^{2x} = 72$ (b) $\log_2 x + \log_2 5 = 6$
15. On the day a grandchild is born, a grandparent deposits $2500 into a fund earning 7.5%, compounded continuously. Determine the balance in the account at the time of the grandchild's 25th birthday.
16. Express the angle $4\pi/9$ in degree measure and sketch the angle in standard position.

CUMULATIVE TEST FOR CHAPTERS 3–5   439

17. Express the angle $-120°$ in radian measure as a multiple of π and sketch the angle in standard position.
18. The terminal side of an angle θ in standard position passes through the point $(12, 5)$. Evaluate the six trigonometric functions of the angle.
19. If $\cos t = -\frac{4}{5}$ where $\pi/2 < t < \pi$, find $\sin t$ and $\tan t$.
20. Use a calculator to approximate $\sin(-1.25)$. Round your answer to four decimal places.
21. Use a calculator to approximate two values of $\theta(0° \le \theta < 360°)$ such that $\sec \theta = 2.125$. Round your answers to two decimal places.
22. Sketch the graph of each of the following functions through two periods.
(a) $y = -3 \sin 2x$ (b) $f(x) = 2 \cos\left(x - \dfrac{\pi}{2}\right)$
(c) $g(x) = \tan\left(\dfrac{\pi x}{2}\right)$ (d) $h(t) = \sec t$
23. Write a sentence describing the relationship between the graphs of the functions $f(x) = \sin x$ and g.
(a) $g(x) = 10 + \sin x$ (b) $g(x) = \sin \dfrac{\pi x}{2}$
(c) $g(x) = \sin\left(x + \dfrac{\pi}{4}\right)$ (d) $g(x) = -\sin x$
24. Find a, b, and c so that the graph of $f(x) = a \sin(bx + c)$ matches the graph in the figure.
25. Consider the function $f(x) = \sin 3x - 2 \cos x$.
(a) Use a graphing utility to graph the function.
(b) Determine the period of the function.
(c) Approximate (accurate to one decimal place) the zero of the function in the interval $[0, 3]$.
(d) Approximate (accurate to one decimal place) the maximum value of the function in the interval $[0, 3]$.
26. Consider the function
$f(t) = 2^{-t/2} \cos\left(\dfrac{\pi t}{2}\right)$
where t is the time in seconds.
(a) Use a graphing utility to graph the function.
(b) Is the function periodic?
(c) Beyond what time t is the maximum value of the function less than 0.3?
27. Evaluate the expression *without* the aid of a calculator.
(a) $\arcsin(\frac{1}{2})$ (b) $\arctan \sqrt{3}$
28. Write an algebraic expression that is equivalent to $\sin(\arccos 2x)$.
29. From a point on the ground 600 feet from the foot of a cliff, the angle of elevation of the top of the cliff is 32° 30′. How high is the cliff?

FIGURE FOR 24

SUPPLEMENTS

This text is accompanied by a comprehensive supplements package for maximum teaching effectiveness and efficiency.

Instructor's Guide by Roland E. Larson and Robert P. Hostetler, *The Pennsylvania State University*, and Bruce H. Edwards, *University of Florida*

Study and Solutions Guide by Bruce H. Edwards, *University of Florida*, and Dianna L. Zook*, Indiana University—Purdue University at Fort Wayne*

Test Item File and Resource Guide

Graphing Technology Guide by Benjamin N. Levy

Transparency Package

Precalculus Videotapes by Dana Mosely

Test-Generating Software (IBM, Macintosh)

The Algebra of Calculus by Eric J. Braude

BestGrapher Software (IBM, Macintosh) by George Best

This complete supplements package offers ancillary materials for students, for instructors, and for classroom resources. Most items are keyed directly to the textbook for easy use. For the convenience of software users, a technical support telephone number is available with all D. C. Heath software products. The components of this comprehensive teaching and learning package are outlined on the following pages.

	PRINTED ANCILLARIES	SOFTWARE AND VIDEOS
INSTRUCTORS	**Instructor's Guide** • Solutions to all even-numbered text exercises, all discussion problems, and all cumulative tests **Test Item File and Resource Guide** • Printed test bank • Over 2000 test items • Open-ended and multiple-choice test items • Available as test-generating software • Sample tests	**BestGrapher** • Function grapher • Screen simultaneously displays equation, graph, and table of values • Some features anticipate calculus • Includes zoom and print features for use on assignments **Computerized Testing Software** • Test-generating software • Over 2000 test items • Also available as a printed test item file **Derive** • Computer algebra system • Discount available to adopters
STUDENTS	**Study and Solutions Guide** • Solutions to all odd-numbered text exercises • Solutions match methods of text • Summaries of key concepts in each text chapter • Study strategies **Graphing Technology Guide** • Keystroke instructions for graphics calculators • Examples **The Algebra of Calculus** • Reviews the algebra, trigonometry, and analytic geometry that students will encounter in calculus • Over 200 examples • Pretests and exercise sets	**BestGrapher** • Function grapher • Screen simultaneously displays equation, graph, and table of values • Some features anticipate calculus • Includes zoom and print features for use on assignments
CLASSROOM RESOURCES	**Transparency Package** • 50 color transparencies • Color-coded by text topic	**BestGrapher** • Function grapher • Screen simultaneously displays equation, graph, and table of values • Some features anticipate calculus • Includes zoom and print features for use on assignments **Videotapes** • Text-specific videos

INTEGRATED LEARNING PACKAGE

Instructor's Guide

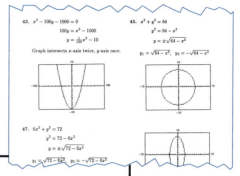

40. $y = 4x^3 - x^4$

42. $y = 100x\sqrt{25 - x^2}$

Graph intersects x-axis twice, y-axis once.

Graph intersects x-axis three times, y-axis once.

44. $2x^3 - 100x - 15,625 + 250y = 0$

$250y = -2x^3 + 100x + 15,625$

$y = -\frac{1}{125}x^3 + \frac{2}{5}x + \frac{125}{2}$

46. $x^2 + y^2 = 49$

$y^2 = 49 - x^2$

$y = \pm\sqrt{49 - x^2}$

$y_1 = \sqrt{49 - x^2}, \quad y_2 = -\sqrt{49 - x^2}$

Study and Solutions Guide

43. $x^2 - 100y - 1000 = 0$

$100y = x^2 - 1000$

$y = \frac{1}{100}x^2 - 10$

Graph intersects x-axis twice, y-axis once.

45. $x^2 + y^2 = 64$

$y^2 = 64 - x^2$

$y = \pm\sqrt{64 - x^2}$

$y_1 = \sqrt{64 - x^2}, \quad y_2 = -\sqrt{64 - x^2}$

47. $6x^2 + y^2 = 72$

$y^2 = 72 - 6x^2$

$y = \pm\sqrt{72 - 6x^2}$

$y_1 = \sqrt{72 - 6x^2}, \quad y_2 = -\sqrt{72 - 6x^2}$

BestGrapher

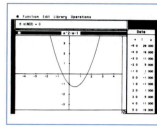

CHAPTER 1

1.1 GRAPHS AND GRAPHING UTILITIES

1.2 LINES IN THE PLANE

1.3 FUNCTIONS

1.4 GRAPHS OF FUNCTIONS

1.5 SHIFTING, REFLECTING, AND STRETCHING GRAPHS

1.6 COMBINATIONS OF FUNCTIONS

1.7 INVERSE FUNCTIONS

FUNCTIONS AND GRAPHS

Test Item File and Resource Guide

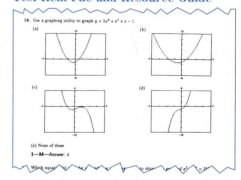

18. Use a graphing utility to graph $y = 2x^3 + x^3 + x - 1$.

(a)   (b)

(c)   (d)

(e) None of these

1—M—Answer: d

Which equation ... to obt... of $x^2 ... = 2...$

1.1 GRAPHS AND GRAPHING UTILITIES

The Graph of an Equation / Using a Graphing Utility / Determining a Viewing Rectangle / Applications

The Graph of an Equation

News magazines often show graphs comparing the rate of inflation, the federal deficit, wholesale prices, or the unemployment rate to the time of year. Industrial firms and businesses use graphs to report their monthly production and sales statistics. Such graphs provide geometric pictures of the way one quantity changes with respect to another.

Frequently, the relationship between two quantities is expressed as an equation. This section introduces the basic procedure for determining the geometric picture associated with an equation.

For an equation in variables x and y, a point (a, b) is a **solution point** if the substitution of $x = a$ and $y = b$ satisfies the equation. Most equations have *infinitely* many solution points. For example, the equation $3x + y = 5$ has solution points $(0, 5)$, $(1, 2)$, $(2, -1)$, $(3, -4)$, and so on. The set of all solution points of an equation is the **graph** of the equation.

THE POINT-PLOTTING METHOD OF GRAPHING

To sketch the graph of an equation by point plotting, use the following steps.

1. If possible, rewrite the equation so that one of the variables is isolated on the left side of the equation.
2. Make up a table of several solution points.
3. Plot these points in the coordinate plane.
4. Connect the points with a smooth curve.

57

Transparency Package

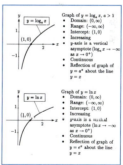

Graph of $y = \log_a x, a > 1$
- Domain: $(0, \infty)$
- Range: $(-\infty, \infty)$
- Intercept: $(1, 0)$
- Increasing
- y-axis is a vertical asymptote ($\log_a x \to -\infty$ as $x \to 0^+$)
- Continuous
- Reflection of graph of $y = a^x$ about the line $y = x$

Graph of $y = \ln x$
- Domain: $(0, \infty)$
- Range: $(-\infty, \infty)$
- Intercept: $(1, 0)$
- Increasing
- y-axis is a vertical asymptote ($\ln x \to -\infty$ as $x \to 0^+$)
- Continuous
- Reflection of graph of $y = e^x$ about the line $y = x$

Videotapes

Computerized Testing Software

Open Select Edit View Print Quit | F1=HELP

Select Questions by Review
Chapter 8&2 - Question 20 of 27
Find the x-intercept of the function: $f(x) = \ln(x + 1)$

(a) (0, 0)   (b) (-1, 0)   (c) (0, -1)

(d) (1, 0)   (e) None of these

Answer

F5 Chap   F3 Prev   F4 Next   F2 Attr   F8 Select   <Esc>

Scroll - PgUp/PgDn for Question - <Ctrl>PgUp/PgDn for Answer   <Esc> - Done

The Algebra of Calculus with Trigonometry and Analytic Geometry

8   Functions II: Combinations of Functions. Difficult Graphs

REVIEW OF FUNDAMENTALS

EXAMPLE 1

XVII

ACKNOWLEDGMENTS

We would like to thank the many people who have helped us prepare the text and supplements package. Their encouragement, criticisms, and suggestions have been invaluable to us.

REVIEWERS

Marilyn Carlson, University of Kansas
John Dersh, Grand Rapids Community College
Patricia J. Ernst, St. Cloud State University
Eunice F. Everett, Seminole Community College
James R. Fryxell, College of Lake County
Bernard Greenspan, University of Akron
Lynda Hollingsworth, Northwest Missouri State University
Spencer Hurd, The Citadel
Luella Johnson, State University of New York, College at Buffalo
Peter A. Lappan, Michigan State University
Marilyn McCollum, North Carolina State University
David R. Peterson, University of Central Arkansas
Antonio Quesada, University of Akron
Stephen Slack, Kenyon College
Judith Smalling, St. Petersburg Junior College
Howard L. Wilson, Oregon State University

FOCUS GROUP

John Dersh, Grand Rapids Community College
Donald Shriner, Frostburg State University

TELEPHONE FOCUS GROUP

John Dersh, Grand Rapids Community College
Ruth Pruitt, Fort Hays State University
Sharon Sledge, San Jacinto College
Fredric W. Tufte, University of Wisconsin—Platteville
Darlene Whitkanack, Northern Illinois University

SURVEY RESPONDENTS

William C. Allgyer, Mountain Empire Community College
Gloria Child, Rollins College
Allan C. Cochran, University of Arkansas at Fayetteville
Ronald Dalla, Eastern Washington University
Ann Dinkheller, Xavier University
Gloria Dion, Pennsylvania State University—Ogontz Campus
Iris B. Fetta, Clemson University
Spencer P. Hurd, The Citadel
Marvin L. Johnson, College of Lake County
Donald A. Josephson, Wheaton College (Ill.)
Thomas J. Kearns, Northern Kentucky University
N. J. Kuenzi, University of Wisconsin—Oshkosh
Edward Laughbaum, Columbus State Community College
M. S. McCollum, North Carolina State University
Carolyn Meitler, Concordia University
Harold M. Ness, University of Wisconsin Centers—Fond du Lac
Michele Olson, College of the Redwoods
John Savige, St. Petersburg Junior College—Clearwater
Stephen Slack, Kenyon College
Marjie Vittum-Jones, South Seattle Community College
Charles Vander Embse, Central Michigan University

Our thanks to the staffs of Texas Instruments Incorporated; Casio, Inc.; and Sharp Electronics Corporation for their help and cooperation with the programs.

A special thanks to all the people at D. C. Heath and Company who worked with us in the development and production of the text, especially Ann Marie Jones, Mathematics Acquisitions Editor; Cathy Cantin, Senior Developmental Editor; Andrea Cava, Senior Production Editor; Sally Steele, Designer; Elinor Stapleton, Art Editor; Carolyn Johnson, Editorial Associate; Mike O'Dea, Production Manager; Lisa Merrill, Production Supervisor; and Richard Tonachel, Production Coordinator.

We would also like to thank the staff at Larson Texts, Inc., who assisted with proofreading the manuscript, preparing and proofreading the art package, and checking and typesetting the supplements.

On a personal level, we are grateful to our wives, Deanna Gilbert Larson, Eloise Hostetler, and Consuelo Edwards, for their love, patience, and support. Also, a special thanks goes to R. Scott O'Neil.

If you have suggestions for improving the text, please feel free to write to us. Over the past two decades, we have received many useful comments from both instructors and students, and we value these very much.

Roland E. Larson
Robert P. Hostetler
Bruce H. Edwards

CONTENTS

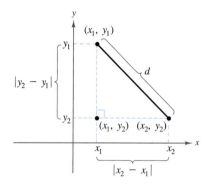
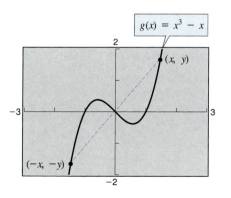

Chapter 2 INTERCEPTS, ZEROS, AND SOLUTIONS   135

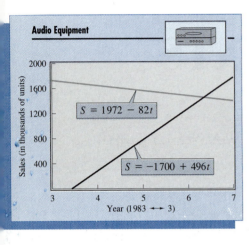

Chapter 3 POLYNOMIAL AND RATIONAL FUNCTIONS   209

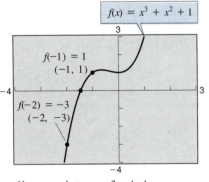

f has a zero between -2 and -1.

Chapter 4   EXPONENTIAL AND LOGARITHMIC FUNCTIONS   282

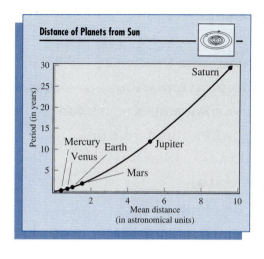

Distance of Planets from Sun

Chapter 5   TRIGONOMETRIC FUNCTIONS   341

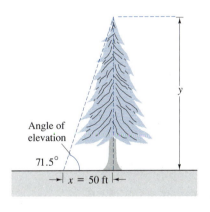

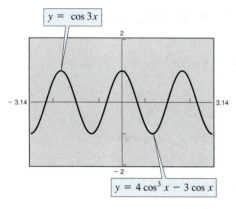
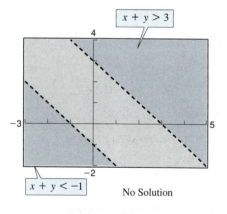

Chapter 8   MATRICES AND DETERMINANTS

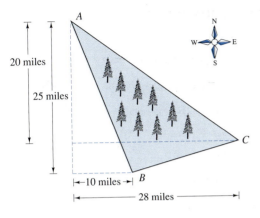

Chapter 9   SEQUENCES, PROBABILITY, AND STATISTICS   653

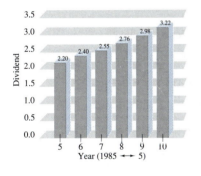

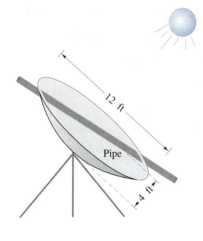
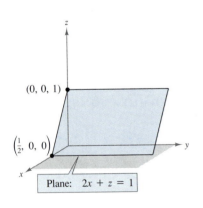

Chapter 12   LIMITS AND AN INTRODUCTION TO CALCULUS   883

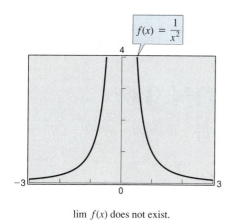

$f(x) = \dfrac{1}{x^2}$

$\lim\limits_{x \to 0} f(x)$ does not exist.

APPENDIXES

ANSWERS

INDEXES

INTRODUCTION TO CALCULATORS

Many examples and exercises in the text require the use of a calculator. Using this tool allows you to examine math in the context of real-life applications.

GRAPHING CALCULATORS

One of the basic differences in calculators is their order of operations. You should practice entering and evaluating expressions on your calculator to make sure you understand its order of operations. Some calculators use an order of operations called RPN (for Reverse Polish Notation). In this text, however, all calculator steps are given using algebraic order of operations. For example, the calculation

$$4.69[5 + 2(6.87 - 3.042)]$$

can be performed with the following steps using an algebraic order of operations.

4.69 ⟨ × ⟩ ⟨ (⟩ 5 ⟨ + ⟩ 2 ⟨ × ⟩ ⟨ (⟩ 6.87 ⟨ − ⟩ 3.042 ⟨) ⟩ ⟨) ⟩ ⟨ ENTER ⟩

This yields the value of 59.35664. Without parentheses, you could enter the expression from the inside out with the sequence

6.87 ⟨ − ⟩ 3.042 ⟨ ENTER ⟩ ⟨ × ⟩ 2 ⟨ + ⟩ 5 ⟨ ENTER ⟩ ⟨ × ⟩ 4.69 ⟨ ENTER ⟩

to obtain the same result.

Using a calculator with RPN, the calculation can be performed with the following steps.

6.87 ⟨ ENTER ⟩ 3.042 ⟨ − ⟩ 2 ⟨ × ⟩ 5 ⟨ + ⟩ 4.69 ⟨ × ⟩

You should also practice using your calculator's graphing features. To help you become familiar with the graphing features, we have included Appendix A,

Graphing Utilities. Be sure to read the examples in Appendix A and work the exercises. Answers to odd-numbered exercises in Appendix A are given in the back of the book.

You can program your calculator to perform various procedures. Try entering and using the selection of optional programs included in this text. For convenience, Appendix B, Programs, contains translations of the text programs for several types of graphing calculators.

ROUNDING NUMBERS

For all their usefulness, calculators do have a problem representing numbers because they are limited to a finite number of digits. For instance, what does your calculator display when you compute 2 ÷ 3? Some calculators simply truncate (drop) the digits that exceed their display range and display .66666666. Others will round the number and display .66666667. Although the second display is more accurate, both of these decimal representations of 2/3 contain a rounding error.

When rounding decimals, we suggest the following guidelines.

1. Determine the number of digits of accuracy you want to keep. The digit in the last position you keep is the **rounding digit,** and the digit in the first position you discard is the **decision digit.**

2. If the decision digit is 5 or greater, round up by adding 1 to the rounding digit.

3. If the decision digit is 4 or less, round down by leaving the rounding digit unchanged.

Here are some examples. Note that you round down in the first example because the decision digit is 4 or less, and you round up in the other two examples because the decision digit is 5 or greater.

Number	Rounded to Three Decimal Places	
a. $\sqrt{2} = 1.4142136$	1.414	*Round down*
b. $\pi = 3.1415927$	3.142	*Round up*
c. $\frac{7}{9} = 0.7777777$	0.778	*Round up*

One of the best ways to minimize error due to rounding is to leave numbers in your calculator until your calculations are complete. If you want to save a number for future use, store it in your calculator's memory.

Remember that once you (or your calculator) have rounded a number, a round-off error has been introduced. For instance, if a number is rounded to

$x = 27.3$, then the actual value of x can lie anywhere between 27.25 and 27.35, or at 27.25 exactly. That is, $27.25 \leq x < 27.35$.

PROBLEM SOLVING USING A CALCULATOR

Here are some guidelines to consider when using any type of calculator in problem solving.

1. Be sure you understand the operation of your own calculator. You need to be skilled at entering expressions in a way that will guarantee that your calculator is performing the operations correctly.

2. Focus first on analyzing the problem. After you have developed a strategy, you may be able to use your calculator to help implement the strategy. Write down your steps in an organized way to clearly outline the strategy used and the results.

3. Most problems can be solved in a variety of ways. If you choose to solve a problem using a table, try checking the solution with an analytic (or algebraic) approach. If you choose to solve a problem using algebra, try checking the solution with a graphing approach. Or, if you choose to solve a problem using a graphing approach, try checking the solution with an algebraic approach.

4. After obtaining a solution with a calculator, be sure to ask yourself if the solution is reasonable (within the context of the problem).

5. To lessen the chance of errors, clear the calculator display (and check the settings) before beginning a new problem.

PREREQUISITES: REVIEW OF BASIC ALGEBRA

1   THE REAL NUMBER SYSTEM

The Real Number System / The Real Number Line / Ordering the Real Numbers / The Absolute Value of a Real Number / The Distance Between Two Real Numbers

The Real Number System

Real numbers are used in everyday life to describe quantities such as age, miles per gallon, container size, population, and so on. To represent real numbers we use symbols such as

$$9, \ -5, \ 0, \ \frac{4}{3}, \ 0.6666\ldots, \ 28.21, \ \sqrt{2}, \ \pi, \quad \text{and} \quad \sqrt[3]{-32}.$$

The set of real numbers contains some important subsets with which you should be familiar:

$\{1, 2, 3, 4, \ldots\}$ *Set of **natural** numbers*

$\{0, 1, 2, 3, 4, \ldots\}$ *Set of **whole** numbers*

$\{\ldots, -3, -2, -1, 0, 1, 2, 3, \ldots\}$ *Set of **integers***

A real number is **rational** if it can be written as the ratio p/q of two integers, where $q \neq 0$. For instance, the numbers

$$\frac{1}{3} = 0.3333\ldots, \quad \frac{1}{8} = 0.125, \quad \text{and} \quad \frac{125}{111} = 1.126126\ldots$$

1

are rational. The decimal representation of a rational number either repeats (as in 3.1454545 . . .) or terminates (as in $\frac{1}{2} = 0.5$). A real number that cannot be written as the ratio of two integers is called **irrational.** Irrational numbers have infinite *nonrepeating* decimal representations. For instance, the numbers

$$\sqrt{2} \approx 1.4142135 \ldots \quad \text{and} \quad \pi \approx 3.1415926 \ldots$$

are irrational. (The symbol $\approx$ means "is approximately equal to.")

EXAMPLE 1   Identifying Real Numbers

Consider the following subset of real numbers:

$$\left\{-8, \; -\sqrt{5}, \; 1, \; \frac{2}{3}, \; 0, \; -\frac{1}{7}, \; \sqrt{3}, \; \pi, \; 9\right\}.$$

List the numbers in this set that are

A. Natural numbers   **B.** Integers
C. Rational numbers   **D.** Irrational numbers

SOLUTION

A. Natural numbers: 1, 9
B. Integers: -8, 1, 0, 9
C. Rational numbers: -8, 1, $\frac{2}{3}$, 0, $-\frac{1}{7}$, 9

D. Irrational numbers: $-\sqrt{5}$, $\sqrt{3}$, π

The Real Number Line

The model used to represent the real number system is called the **real number line.** It consists of a horizontal line with a point (the **origin**) labeled 0. Numbers to the right of 0 are positive, and numbers to the left of 0 are negative, as shown in Figure 1. We use the term **nonnegative** to describe a number that is either positive or zero.

The Real Number Line

FIGURE 1

One-to-One Correspondence

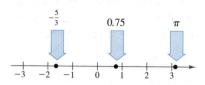

(a) Every real number corresponds to exactly one point on the real number line.

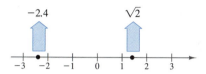

(b) Every point on the real number line corresponds to exactly one real number.

FIGURE 2

Each point on the real number line corresponds to one and only one real number and each real number corresponds to one and only one point on the real number line. This type of relationship is called a **one-to-one correspondence,** as shown in Figure 2.

The number associated with a point on the real number line is called the **coordinate** of the point. For example, in Figure 2(a), $-\frac{5}{3}$ is the coordinate of the leftmost point and π is the coordinate of the rightmost point.

Ordering the Real Numbers

One important property of real numbers is that they are **ordered.**

DEFINITION OF ORDER ON THE REAL NUMBER LINE

If a and b are real numbers, then a is **less than** b if $b - a$ is positive. We denote this order by the **inequality**

$$a < b.$$

This relationship can also be described by saying that b is **greater than** a and writing $b > a$. The inequality $a \leq b$ means that a is **less than or equal to** b and the inequality $b \geq a$ means that b is **greater than or equal to** a. The symbols $<$, $>$, $\leq$, and $\geq$ are called **inequality symbols.**

$a < b$ if and only if a lies to the left of b.

FIGURE 3

Geometrically, this definition implies that $a < b$ if and only if a lies to the *left* of b on the real number line, as shown in Figure 3. For example, $1 < 2$ because 1 lies to the left of 2 on the real number line.

Inequalities are useful in denoting subsets of real numbers, as shown in Examples 2 and 3.

(a)

(b)

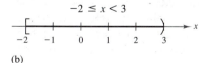

(c)

FIGURE 4

EXAMPLE 2   Interpreting Inequalities

A. The inequality $x \leq 2$ denotes all real numbers less than or equal to 2.
B. The inequality $-2 \leq x < 3$ means that $x \geq -2$ *and* $x < 3$. This "double" inequality denotes all real numbers between -2 and 3, including -2 but *not* including 3.
C. The inequality $x > -5$ denotes all real numbers greater than -5.

The graphs of all three inequalities are shown in Figure 4. In the graphs, note that the brackets correspond to $\geq$ and $\leq$ and the parentheses correspond to $>$ and $<$.

Each of the subsets of real numbers in Example 2 is an **interval.**

BOUNDED INTERVALS ON THE REAL NUMBER LINE

Let a and b be real numbers such that $a < b$. The following intervals on the real number line are **bounded intervals.** The numbers a and b are the **endpoints** of each interval.

Notation	Interval Type	Inequality	Graph
$[a, b]$	Closed	$a \leq x \leq b$	
(a, b)	Open	$a < x < b$	
$[a, b)$	Half-open	$a \leq x < b$	
$(a, b]$	Half-open	$a < x \leq b$	

UNBOUNDED INTERVALS ON THE REAL NUMBER LINE

Let a and b be real numbers. The following intervals on the real number line are **unbounded intervals.**

Notation	Interval Type	Inequality	Graph
$[a, \infty)$	Half-open	$x \geq a$	
(a, ∞)	Open	$x > a$	
$(-\infty, b]$	Half-open	$x \leq b$	
$(-\infty, b)$	Open	$x < b$	
$(-\infty, \infty)$	Entire real line		

The symbols ∞ (**positive infinity**) and $-\infty$ (**negative infinity**) do not represent real numbers. They are simply convenient symbols used to describe the unboundedness of an interval such as $(1, \infty)$.

EXAMPLE 3   Intervals and Inequalities

Write an inequality to represent each of the intervals and state whether the interval is bounded or unbounded.

A. $(-3, 5]$ **B.** $(-3, \infty)$ **C.** $[0, 2]$

SOLUTION

A. $(-3, 5]$ corresponds to $-3 < x \le 5$.   *Bounded*
B. $(-3, \infty)$ corresponds to $-3 < x$.   *Unbounded*
C. $[0, 2]$ corresponds to $0 \le x \le 2$.   *Bounded*

EXAMPLE 4   Using Inequalities to Represent Sets of Real Numbers

Use inequality and interval notation to describe each of the following.

A. c is nonnegative.   **B.** b is at most 5.
C. d is negative and greater than -3.
D. x is positive or x is less than -6.

SOLUTION

A. "c is nonnegative" means that c is greater than or equal to zero. This can be written as $c \ge 0$. The interval notation is $[0, \infty)$.
B. "b is at most 5" can be written as $b \le 5$. The interval notation is $(-\infty, 5]$.
C. "d is negative" can be written as $d < 0$, and "d is greater than -3" can be written as $-3 < d$. Combining these two inequalities produces the double inequality $-3 < d < 0$. The interval notation is $(-3, 0)$.
D. "x is positive" can be written as $0 < x$, and "x is less than -6" can be written as $x < -6$. Combining these two inequalities produces $0 < x$ or $x < -6$. This yields *two* intervals: $(-\infty, -6), (0, \infty)$.

The Absolute Value of a Real Number

The **absolute value** of a real number is its *magnitude.*

DEFINITION OF ABSOLUTE VALUE

If a is a real number, then the **absolute value** of a is given by

$$|a| = \begin{cases} a, & \text{if } a \ge 0 \\ -a, & \text{if } a < 0. \end{cases}$$

Notice from this definition that the absolute value of a real number is never negative. For instance, if $a = -5$, then $|-5| = -(-5) = 5$. Zero is the only real number whose absolute value is zero. That is, $|0| = 0$.

The following list gives four useful properties of absolute value.

PROPERTIES OF ABSOLUTE VALUES

Let a and b be real numbers.

1. $|a| \geq 0$ 2. $|-a| = |a|$

3. $|ab| = |a||b|$ 4. $\left|\dfrac{a}{b}\right| = \dfrac{|a|}{|b|}, \quad b \neq 0$

The Distance Between Two Real Numbers

Absolute value is used to define the distance between two numbers on the real number line.

DISTANCE BETWEEN TWO POINTS ON THE REAL NUMBER LINE

Let a and b be real numbers. The **distance between a and b** is

$$d(a, b) = |b - a| = |a - b|.$$

EXAMPLE 5   **Distance and Absolute Value**

A. The distance between $\sqrt{7}$ and 4 is given by

$$d\left(\sqrt{7}, 4\right) = \left|4 - \sqrt{7}\right| = 4 - \sqrt{7}.$$

B. The statement "the distance between c and -2 is at least 7" is written as

$$d(c, -2) = |c - (-2)| = |c + 2| \geq 7.$$

C. The distance between -4 and the origin is given by

$$d(-4, 0) = |-4 - 0| = |-4| = 4.$$

See Figure 5.

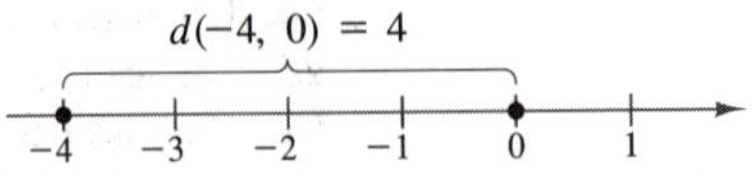

FIGURE 5

SECTION 1 · EXERCISES

In Exercises 1–6, determine which numbers in the set are (a) natural numbers, (b) integers, (c) rational numbers, and (d) irrational numbers.

*1. $-9, -\frac{7}{2}, 5, \frac{2}{3}, \sqrt{2}, 0, 1$

2. $\sqrt{5}, -7, -\frac{7}{3}, 0, 3.12, \frac{5}{4}$

3. $12, -13, 1, \sqrt{4}, \sqrt{6}, \frac{3}{2}$

4. $\frac{8}{2}, -\frac{8}{3}, \sqrt{10}, -4, 9, 14.2$

5. $-\pi, -\frac{1}{3}, \frac{6}{3}, \frac{1}{2}\sqrt{2}, -7.5$

6. $25, -17, -\frac{12}{5}, \sqrt{9}, 3.12, \frac{1}{2}\pi$

In Exercises 7–12, plot the two real numbers on the real number line, and then write them with the appropriate inequality sign ($<$ or $>$) between them.

7. $\frac{3}{2}, 7$

8. $-3.5, 1$

9. $-4, -8$

10. $1, \frac{16}{3}$

11. $\frac{5}{6}, \frac{2}{3}$

12. $-\frac{8}{7}, -\frac{3}{7}$

In Exercises 13–16, write an inequality to represent the given interval and state whether the interval is bounded or unbounded.

13. $[-1, 3]$

14. $(4, 10]$

15. $(10, \infty)$

16. $(-6, \infty)$

In Exercises 17–26, use interval notation to describe the subset of real numbers that is represented by the inequality. Then sketch the subset on the real number line.

17. $x \le 5$

18. $x \ge -2$

19. $x < 0$

20. $x > 3$

21. $x \ge 4$

22. $x < 2$

23. $-2 < x < 2$

24. $0 \le x \le 5$

25. $-1 \le x < 0$

26. $0 < x \le 6$

In Exercises 27–34, use inequality and interval notation to describe the set of real numbers.

27. x is negative.

28. z is at least 10.

29. y is no more than 25.

30. y is greater than 5 and less than or equal to 12.

31. The person's age A is at least 30.

32. The yield Y is no more than 45 bushels per acre.

33. The annual rate of inflation r is expected to be at least 3.5%, but no more than 6%.

34. The price p of unleaded gasoline is not expected to go above \$1.35 per gallon during the coming year.

In Exercises 35–40, write the expression without using absolute value signs.

35. $\dfrac{-5}{|-5|}$

36. $|4 - \pi|$

37. $-3|-3|$

38. $|-1| - |-2|$

39. $-|16.25| + 20$

40. $2|33|$

In Exercises 41–46, fill in the blank with $<$, $>$, or $=$.

41. $|-3| \; \rule{0.5cm}{0.4pt} \; -|-3|$

42. $|-4| \; \rule{0.5cm}{0.4pt} \; |4|$

43. $-5 \; \rule{0.5cm}{0.4pt} \; -|5|$

44. $-|-6| \; \rule{0.5cm}{0.4pt} \; |-6|$

45. $-|-2| \; \rule{0.5cm}{0.4pt} \; -|2|$

46. $-(-2) \; \rule{0.5cm}{0.4pt} \; -2$

In Exercises 47–52, find the distance between a and b.

47.

$a = -\frac{5}{2}$ $b = 0$

$-3 \quad -2 \quad -1 \quad 0$

48.

$a = \frac{1}{4}$ $b = \frac{11}{4}$

$0 \quad 1 \quad 2 \quad 3$

49. $a = 126, b = 75$

50. $a = -126, b = -75$

51. $a = 9.34, b = -5.65$

52. $a = \frac{16}{5}, b = \frac{112}{75}$

In Exercises 53–58, use absolute value notation to describe the expression.

53. The distance between x and 5 is no more than 3.

54. The distance between x and -10 is at least 6.

55. The distance between z and $\frac{3}{2}$ is greater than 1.

56. The distance between z and 0 is less than 8.

57. y is at least 6 units from 0.

58. y is at most 2 units from a.

* Detailed solutions to all odd-numbered exercises can be found in the *Study and Solutions Guide*.

In Exercises 59 and 60, use a calculator to order the numbers from smallest to largest.

59. $\frac{7071}{5000}$, $\frac{584}{413}$, $\sqrt{2}$, $\frac{47}{33}$, $\frac{127}{90}$

60. $\frac{26}{15}$, $\sqrt{3}$, 1.7320, $\frac{381}{220}$, $\sqrt{10} - \sqrt{2}$

In Exercises 61–64, use a calculator to find the decimal form of the rational number. If it is a nonterminating decimal, write the repeating pattern.

61. $\frac{5}{8}$

62. $\frac{1}{3}$

63. $\frac{41}{333}$

64. $\frac{6}{11}$

In Exercises 65–68, determine whether the statement is true or false. Explain your reasoning.

65. The reciprocal of every nonzero integer is an integer.

66. Every integer is a rational number.

67. Every real number is either rational or irrational.

68. The absolute value of every real number is positive.

2   PROPERTIES OF REAL NUMBERS AND THE BASIC RULES OF ALGEBRA

Algebraic Expressions / Basic Rules of Algebra / Equations / Exponents / Scientific Notation

Algebraic Expressions

One characteristic of algebra is the use of letters or combinations of letters to represent numbers. The letters are called **variables** and combinations of letters and numbers are called **algebraic expressions.** Here are a few examples of algebraic expressions:

$$5x, \qquad 2x - 3, \qquad \frac{4}{x^2 + 2}, \qquad 7x + y.$$

DEFINITION OF AN ALGEBRAIC EXPRESSION

An **algebraic expression** is a collection of letters (**variables**) and real numbers (**constants**) combined using the operations of addition, subtraction, multiplication, division, and exponentiation.

The **terms** of an algebraic expression are those parts that are separated by *addition.* For example,

$$x^2 - 5x + 8 = x^2 + (-5x) + 8$$

has three terms: x^2 and $-5x$ are the **variable terms** and 8 is the **constant term.** The numerical factor of a variable term is the **coefficient** of the variable term. For instance, the coefficient of $-5x$ is -5, and the coefficient of x^2 is 1.

To **evaluate** an algebraic expression, substitute numerical values for each of the variables in the expression. Here are two examples.

Expression	Value of Variable	Substitute	Value of Expression
$-3x + 5$	$x = 3$	$-3(3) + 5$	$-9 + 5 = -4$
$3x^2 + 2x - 1$	$x = -1$	$3(-1)^2 + 2(-1) - 1$	$3 - 2 - 1 = 0$

Basic Rules of Algebra

There are four arithmetic operations with real numbers: **addition, multiplication, subtraction,** and **division,** denoted by the symbols $+$, $\times$ or $\cdot$, $-$, and $\div$. Of these, addition and multiplication are the two primary operations. Subtraction and division are defined as the inverse operations of addition and multiplication.

Subtraction	Division
$a - b = a + (-b)$	If $b \neq 0$, then $\dfrac{a}{b} = a\left(\dfrac{1}{b}\right)$.

In these definitions, $-b$ is the **additive inverse** (or opposite) of b, and $1/b$ is the **multiplicative inverse** (or reciprocal) of b. In the fractional form a/b, a is the **numerator** of the fraction and b is the **denominator.**

Be sure you see that the following **basic rules of algebra** are true for variables and algebraic expressions as well as for real numbers.

BASIC RULES OF ALGEBRA

Let a, b, and c be real numbers, variables, or algebraic expressions.

Property		Example
Commutative Property of Addition:	$a + b = b + a$	$4x + x^2 = x^2 + 4x$
Commutative Property of Multiplication:	$ab = ba$	$(4 - x)x^2 = x^2(4 - x)$
Associative Property of Addition:	$(a + b) + c = a + (b + c)$	$(-x + 5) + 2x^2 = -x + (5 + 2x^2)$
Associative Property of Multiplication:	$(ab)c = a(bc)$	$(2x \cdot 3y)(8) = (2x)(3y \cdot 8)$
Distributive Property:	$a(b + c) = ab + ac$	$x(5 + 2x) = x \cdot 5 + x \cdot 2x$
	$(a + b)c = ac + bc$	$(y + 8)y = y \cdot y + 8 \cdot y$
Additive Identity Property:	$a + 0 = a$	$5y^2 + 0 = 5y^2$
Multiplicative Identity Property:	$a \cdot 1 = 1 \cdot a = a$	$(4x^2)(1) = (1)(4x^2) = 4x^2$
Additive Inverse Property:	$a + (-a) = 0$	$5x^3 + (-5x^3) = 0$
Multiplicative Inverse Property:	$a \cdot \dfrac{1}{a} = 1, \quad a \neq 0$	$(x^2 + 4)\left(\dfrac{1}{x^2 + 4}\right) = 1$

REMARK   The Distributive Properties are also true for subtraction. For instance, the "subtraction form" of $a(b + c) = ab + ac$ is

$$a(b - c) = a[b + (-c)] = ab + a(-c) = ab - ac.$$

■■■■■

PROPERTIES OF NEGATION

Let a and b be real numbers, variables, or algebraic expressions.

Property	*Example*
1. $(-1)a = -a$	$(-1)7 = -7$
2. $-(-a) = a$	$-(-6) = 6$
3. $(-a)b = -(ab) = a(-b)$	$(-5)3 = -(5 \cdot 3) = 5(-3) = -15$
4. $(-a)(-b) = ab$	$(-2)(-6) = 2 \cdot 6 = 12$
5. $-(a + b) = (-a) + (-b)$	$-(3 + 8) = (-3) + (-8) = -11$

Be sure you see the difference between the *additive inverse of a number* and a *negative number.* If a is already negative, then its additive inverse, $-a$, is positive. For instance, if $a = -5$, then $-a = -(-5) = 5$.

PROPERTIES OF ZERO

Let a and b be real numbers, variables, or algebraic expressions.

1. $a + 0 = a$ and $a - 0 = a$
2. $a \cdot 0 = 0$
3. $\dfrac{0}{a} = 0, \quad a \neq 0$
4. $\dfrac{a}{0}$ is undefined.
5. Zero-Factor Property: If $ab = 0$, then $a = 0$ or $b = 0$.

The "or" in the Zero-Factor Property includes the possibility that both factors may be zero. This is an **inclusive or,** and it is the way the word "or" is always used in mathematics.

PROPERTIES OF FRACTIONS

Let a, b, c, and d be real numbers, variables, or algebraic expressions such that $b \neq 0$ and $d \neq 0$.

1. Equivalent Fractions: $\dfrac{a}{b} = \dfrac{c}{d}$ if and only if $ad = bc$.

2. Rules of Signs: $-\dfrac{a}{b} = \dfrac{-a}{b} = \dfrac{a}{-b}$ and $\dfrac{-a}{-b} = \dfrac{a}{b}$

3. Generate Equivalent Fractions: $\dfrac{a}{b} = \dfrac{ac}{bc}$, $c \neq 0$

4. Add or Subtract with Like Denominators: $\dfrac{a}{b} \pm \dfrac{c}{b} = \dfrac{a \pm c}{b}$

5. Add or Subtract with Unlike Denominators: $\dfrac{a}{b} \pm \dfrac{c}{d} = \dfrac{ad \pm bc}{bd}$

6. Multiply Fractions: $\dfrac{a}{b} \cdot \dfrac{c}{d} = \dfrac{ac}{bd}$

7. Divide Fractions: $\dfrac{a}{b} \div \dfrac{c}{d} = \dfrac{a}{b} \cdot \dfrac{d}{c} = \dfrac{ad}{bc}$, $c \neq 0$

In Property 1 (equivalent fractions) the phrase "if and only if" implies two statements. One statement is: If $a/b = c/d$, then $ad = bc$. The other statement is: If $ad = bc$, where $b \neq 0$ and $d \neq 0$, then $a/b = c/d$.

EXAMPLE 1   Properties of Zero and Properties of Fractions

A. $x - \dfrac{0}{5} = x - 0 = x$ *Properties 3 and 1 of zero*

B. $\dfrac{x}{5} = \dfrac{3 \cdot x}{3 \cdot 5} = \dfrac{3x}{15}$ *Generate equivalent fractions*

C. $\dfrac{x}{3} + \dfrac{2x}{5} = \dfrac{5 \cdot x + 3 \cdot 2x}{15}$ *Add fractions with unlike denominators*

D. $\dfrac{7}{x} \div \dfrac{3}{2} = \dfrac{7}{x} \cdot \dfrac{2}{3} = \dfrac{14}{3x}$ *Divide fractions*

If a, b, and c are integers such that $ab = c$, then a and b are **factors** or **divisors** of c. For example, 2 and 3 are factors of 6. A **prime number** is a positive integer that has exactly two factors: itself and 1. For example, 2, 3, 5, 7, and 11 are prime numbers. The numbers 4, 6, 8, 9, and 10 are **composite** because they can be written as the product of two or more prime numbers. The number 1 is neither prime nor composite. The **Fundamental Theorem of Arithmetic** states that every positive integer greater than 1 can be written as the product of prime numbers in precisely one way (disregarding order). For instance, the prime factorization of 24 is $24 = 2 \cdot 2 \cdot 2 \cdot 3$.

When adding or subtracting fractions with unlike denominators, you have two options. You can use Property 5 of fractions as in Example 1(C). Or you can use the **least common denominator** (LCD) method—rewrite both fractions so that they have the same denominator. For adding or subtracting *two* fractions, Property 5 is often more convenient. For *three or more* fractions, the LCD method is usually preferred.

EXAMPLE 2   The LCD Method of Adding or Subtracting Fractions

Evaluate the following as a single fraction.

$$\frac{2}{15} - \frac{5}{9} + \frac{4}{5}$$

SOLUTION

By factoring the denominators ($15 = 3 \cdot 5$, $9 = 3 \cdot 3$, and $5 = 5$), you can see that the least common denominator is $3 \cdot 3 \cdot 5 = 45$. Therefore, it follows that

$$\frac{2}{15} - \frac{5}{9} + \frac{4}{5} = \frac{2(3)}{15(3)} - \frac{5(5)}{9(5)} + \frac{4(9)}{5(9)} = \frac{6 - 25 + 36}{45} = \frac{17}{45}.$$

Equations

An **equation** is a statement of equality between two expressions. Thus, the statement $a + b = c + d$ means that the expressions $a + b$ and $c + d$ represent the same number.

PROPERTIES OF EQUALITY

Let a, b, and c be real numbers, variables, or algebraic expressions.

1. Reflexive: $a = a$
2. Symmetric: If $a = b$, then $b = a$.
3. Transitive: If $a = b$ and $b = c$, then $a = c$.
4. Substitution Principle: If $a = b$, then a can be replaced by b in any expression involving a.

Two important consequences of the Substitution Principle are as follows.

1. If $a = b$, then $a + c = b + c$.   *Add c to both sides*
2. If $a = b$, then $ac = bc$.   *Multiply both sides by c*

The first rule allows you to add the same number to both sides of an equation. The second allows you to multiply both sides of an equation by the same number. The converses of these two rules are the **Cancellation Laws** for addition and multiplication. Note that $c \neq 0$ in the second rule.

1. If $a + c = b + c$, then $a = b$.   *Subtract c from both sides*
2. If $ac = bc$ and $c \neq 0$, then $a = b$.   *Divide both sides by nonzero c*

Exponents

Repeated factors can be written in **exponential form.**

Repeated Factors	Exponential Form
$7 \cdot 7$	7^2
$(-4)(-4)(-4)$	$(-4)^3$
$(2x)(2x)(2x)(2x)$	$(2x)^4$

EXPONENTIAL NOTATION

Let a be a real number, a variable, or an algebraic expression, and let n be a positive integer. Then

$$a^n = \underbrace{a \cdot a \cdot a \cdots a}_{n \text{ factors}},$$

where n is the **exponent** and a is the **base.** The expression a^n is read as "a to the nth **power.**"

It is important to recognize the difference between expressions such as $(-2)^4$ and -2^4. In $(-2)^4$, the parentheses indicate that the exponent applies to the negative sign as well as to the 2, but in $-2^4 = -(2^4)$, the exponent applies only to the 2. Hence, $(-2)^4 = 16$, whereas $-2^4 = -16$.

When multiplying exponential expressions with the same base, you *add* exponents, $a^m \cdot a^n = a^{m+n}$. When dividing exponential expressions with the same base, you *subtract* exponents,

$$\frac{a^m}{a^n} = a^{m-n}, \qquad a \neq 0.$$

PROPERTIES OF EXPONENTS

Let a and b be real numbers, variables, or algebraic expressions, and let m and n be integers. (Assume all denominators and bases are nonzero.)

Property	*Example*
1. $a^m a^n = a^{m+n}$	$3^2 \cdot 3^4 = 3^{2+4} = 3^6$
2. $\dfrac{a^m}{a^n} = a^{m-n}$	$\dfrac{x^7}{x^4} = x^{7-4} = x^3$
3. $a^{-n} = \dfrac{1}{a^n} = \left(\dfrac{1}{a}\right)^n$	$y^{-4} = \dfrac{1}{y^4} = \left(\dfrac{1}{y}\right)^4$
4. $a^0 = 1, \quad a \neq 0$	$(x^2 + 1)^0 = 1$
5. $(ab)^m = a^m b^m$	$(5x)^3 = 5^3 x^3 = 125x^3$
6. $(a^m)^n = a^{mn}$	$(y^3)^{-4} = y^{3(-4)} = y^{-12} = \dfrac{1}{y^{12}}$
7. $\left(\dfrac{a}{b}\right)^m = \dfrac{a^m}{b^m}$	$\left(\dfrac{2}{x}\right)^3 = \dfrac{2^3}{x^3} = \dfrac{8}{x^3}$
8. $\lvert a^2 \rvert = \lvert a \rvert^2 = a^2$	$\lvert (-2)^2 \rvert = \lvert -2 \rvert^2 = (-2)^2 = 4$

EXAMPLE 3 Using Properties of Exponents to Simplify

A. $(-3ab^4)(4ab^{-3}) = -12(a)(a)(b^4)(b^{-3}) = -12a^2 b$

B. $(2xy^2)^3 = 2^3(x)^3(y^2)^3 = 8x^3 y^6$

C. $3a(-4a^2)^0 = 3a(1) = 3a, \quad a \neq 0$

D. $\left(\dfrac{5x^3}{y}\right)^2 = \dfrac{5^2(x^3)^2}{y^2} = \dfrac{25x^6}{y^2}$

EXAMPLE 4 Rewriting with Positive Exponents

A. $x^{-1} = \dfrac{1}{x}$ *Property 3: $a^{-n} = \dfrac{1}{a^n}$*

B. $\dfrac{1}{3x^{-2}} = \dfrac{1(x^2)}{3} = \dfrac{x^2}{3}$ *-2 exponent does not apply to 3*

C. $\dfrac{12a^3 b^{-4}}{4a^{-2}b} = \dfrac{12a^3 \cdot a^2}{4b \cdot b^4} = \dfrac{3a^5}{b^5}$

D. $\left(\dfrac{3x^2}{y}\right)^{-2} = \dfrac{3^{-2}(x^2)^{-2}}{y^{-2}} = \dfrac{3^{-2}x^{-4}}{y^{-2}} = \dfrac{y^2}{3^2 x^4} = \dfrac{y^2}{9x^4}$

Rarely in algebra is there only one way to solve a problem. Don't be concerned if the steps you use to solve a problem are not exactly the same as the steps presented in this text. The important thing is to use steps that you understand *and*, of course, that are justified by the rules of algebra. For instance, you might prefer the following solution to Example 4(D).

$$\left(\frac{3x^2}{y}\right)^{-2} = \left(\frac{y}{3x^2}\right)^2 = \frac{y^2}{9x^4}$$

EXAMPLE 5 Using a Calculator to Raise a Number to a Power

Use a calculator to evaluate the following expressions. Round your answers to two decimal places.

REMARK In Example 5, note that the number of displayed digits depends on the particular type of calculator being used.

Expression	Display	Rounded Answer
A. $13^4 + 5$	28566	28,566.00
B. $3^{-2} + 4^{-1}$	.3611111111	0.36
C. $\dfrac{3^5 + 1}{3^5 - 1}$	1.008264463	1.01

Scientific Notation

Exponents provide an efficient way of writing and computing with very large (or very small) numbers. For instance, a drop of water contains more than 33 billion billion molecules—that is, 33 followed by 18 zeros:

33,000,000,000,000,000,000.

It is convenient to write such numbers in **scientific notation.** This notation has the form $c \times 10^n$, where $1 \le c < 10$ and n is an integer. Thus, the number of molecules in a drop of water can be written in scientific notation as

$$3.3 \times 10,000,000,000,000,000,000 = 3.3 \times 10^{19}.$$

The *positive* exponent 19 indicates that the number is large (10 or more) and that the decimal point has been moved 19 places. A *negative* exponent in scientific notation indicates that the number is *small* (less than 1). For instance, the mass (in grams) of one electron is approximately

$$9.0 \times 10^{-28} = 0.00000000000000000000000000009.$$

28 decimal places

EXAMPLE 6 Scientific Notation

A. $1.345 \times 10^2 = 134.5$ **B.** $9.36 \times 10^{-6} = 0.00000936$
C. $0.0000782 = 7.82 \times 10^{-5}$ **D.** $836,100,000.0 = 8.361 \times 10^8$

Most calculators automatically switch to scientific notation when they are showing large (or small) numbers that exceed the display range. Try multiplying $86,500,000 \times 6000$. If your calculator follows standard conventions, its display should be $\boxed{5.19 \quad 11}$ or $\boxed{5.19 \quad E \quad 11}$. This means that $c = 5.19$ and the exponent of 10 is $n = 11$, which implies that the number is 5.19×10^{11}.

EXAMPLE 7   Using Scientific Notation with a Calculator

Use a calculator to evaluate $65,000 \times 3,400,000,000$.

SOLUTION

Because $65,000 = 6.5 \times 10^4$ and $3,400,000,000 = 3.4 \times 10^9$, you can multiply the two numbers as follows:

$$65,000 \times 3,400,000,000 = (6.5 \times 10^4)(3.4 \times 10^9).$$

Display: $\boxed{2.21 \quad E \quad 14}$

The product of the two numbers is

$$(6.5 \times 10^4)(3.4 \times 10^9) = 2.21 \times 10^{14} = 221,000,000,000,000.$$

SECTION 2 · EXERCISES

In Exercises 1–6, identify the terms of the expression.

1. $7x + 4$
2. $-5 + 3x$
3. $x^2 - 4x + 8$
4. $3x^2 - 8x - 11$
5. $4x^3 + x - 5$
6. $3x^4 + 3x^3$

In Exercises 7–12, evaluate the expression for the given values of x. If not possible, state the reason.

Expression	Values	
7. $4x - 6$	(a) $x = -1$	(b) $x = 0$
8. $9 - 7x$	(a) $x = -3$	(b) $x = 3$
9. $x^2 - 3x + 4$	(a) $x = -2$	(b) $x = 2$
10. $-x^2 + 5x - 4$	(a) $x = -1$	(b) $x = 1$
11. $\dfrac{x + 1}{x - 1}$	(a) $x = 1$	(b) $x = -1$
12. $\dfrac{x}{x + 2}$	(a) $x = 2$	(b) $x = -2$

In Exercises 13–22, identify the rule(s) of algebra illustrated by the equation.

13. $x + 9 = 9 + x$
14. $(x + 3) - (x + 3) = 0$
15. $\dfrac{1}{h + 6}(h + 6) = 1, h \neq -6$
16. $2\left(\dfrac{1}{2}\right) = 1$
17. $2(x + 3) = 2x + 6$
18. $(z - 2) + 0 = z - 2$
19. $1 \cdot (1 + x) = 1 + x$
20. $x + (y + 10) = (x + y) + 10$
21. $x(3y) = (x \cdot 3)y = (3x)y$
22. $\frac{1}{7}(7 \cdot 12) = (\frac{1}{7} \cdot 7)12 = 1 \cdot 12 = 12$

In Exercises 23–26, evaluate the expression. If not possible, state the reason.

23. $\dfrac{81 - (90 - 9)}{5}$
24. $10(23 - 30 + 7)$
25. $\dfrac{8 - 8}{-9 + (6 + 3)}$
26. $15 - \dfrac{3 - 3}{5}$

In Exercises 27–40, perform the indicated operations. (Write fractional answers in reduced form.)

27. $10 - 6 - 2$

28. $-3(5 - 2)$

29. $(4 - 7)(-2)$

30. $\dfrac{27 - 35}{4}$

31. $\frac{3}{16} + \frac{5}{16}$

32. $\frac{6}{7} - \frac{4}{7}$

33. $\frac{5}{8} - \frac{5}{12} + \frac{1}{6}$

34. $\frac{10}{11} + \frac{6}{33} - \frac{13}{66}$

35. $\frac{4}{5} \cdot \frac{1}{2} \cdot \frac{3}{4}$

36. $\frac{1}{3} \cdot \frac{5}{8} \cdot \frac{5}{16} \cdot \frac{3}{4}$

37. $\frac{2}{3} \div 8$

38. $\frac{11}{16} \div \frac{3}{4}$

39. $12 \div \frac{1}{4}$

40. $\left(\frac{3}{5} \div 3\right) - \left(6 \cdot \frac{4}{8}\right)$

In Exercises 41–48, use a calculator to evaluate the expression. (Round the result to two decimal places.)

41. $-3 + \frac{3}{7}$

42. $3\left(-\frac{5}{12} + \frac{3}{8}\right)$

43. $\dfrac{11.46 - 5.37}{3.91}$

44. $\dfrac{\frac{1}{5}(-8 - 9)}{-\frac{1}{3}}$

45. $5^{-1} + 2^{-3}$

46. $\left(\frac{3}{2}\right)^{-2} - \left(\frac{3}{4}\right)^3$

47. $\dfrac{7^4 + 32}{7^4 - 200}$

48. $\dfrac{1 - 3^{-5}}{0.5 + (0.3)^4}$

In Exercises 49–56, evaluate the expression.

49. $\dfrac{5^5}{5^2}$

50. $3 \cdot 3^3$

51. $(3^3)^2$

52. -3^2

53. $(2^3 \cdot 3^2)^2$

54. $\left(-\frac{3}{5}\right)^3\left(\frac{5}{3}\right)^2$

55. $\dfrac{4 \cdot 3^{-2}}{2^{-2} \cdot 3^{-1}}$

56. $(-2)^0$

In Exercises 57–60, evaluate the expression for the given value of x.

Expression	Value
57. $-3x^3$	2
58. $7x^{-2}$	4
59. $6x^0 - (6x)^0$	10
60. $5(-x)^3$	3

In Exercises 61–78, simplify the expression. (Use only positive exponents.)

61. $(-5z)^3$

62. $(3x)^2$

63. $6y^2(2y^4)^2$

64. $(-z)^3(3z^4)$

65. $\dfrac{3x^5}{x^3}$

66. $\dfrac{25y^8}{10y^4}$

67. $\dfrac{7x^2}{x^3}$

68. $\dfrac{r^4}{r^6}$

69. $\dfrac{12(x + y)^3}{9(x + y)}$

70. $\left(\dfrac{4}{y}\right)^3\left(\dfrac{3}{y}\right)^4$

71. $(x + 5)^0, \quad x \neq -5$

72. $(2x^5)^0, \quad x \neq 0$

73. $(-2x^2)^3(4x^3)^{-1}$

74. $(4y^{-2})(8y^4)$

75. $\left(\dfrac{x}{10}\right)^{-1}$

76. $\left(\dfrac{x^{-3}y^4}{5}\right)^{-3}$

77. $3^n \cdot 3^{2n}$

78. $\dfrac{x^2 \cdot x^n}{x^3 \cdot x^n}$

In Exercises 79–82, write the number in scientific notation.

79. Land area of earth: 57,500,000 square miles

80. Light year: 9,461,000,000,000,000 kilometers

81. Relative density of hydrogen: 0.0000899 gram per cm^3

82. One micron (millionth of a meter): 0.00003937 inch

In Exercises 83–86, write the number in decimal form.

83. U.S. daily Coca-Cola consumption: 5.24×10^8 servings

84. Interior temperature of the sun: 1.3×10^7 degrees Celsius

85. Charge of electron: 4.8×10^{-10} electrostatic units

86. Width of human hair: 9.0×10^{-4} meters

In Exercises 87 and 88, use a calculator to evaluate the expression. (Round the result to three decimal places.)

87. (a) $750\left(1 + \dfrac{0.11}{365}\right)^{800}$

(b) $\dfrac{67,000,000 + 93,000,000}{0.0052}$

88. (a) $(9.3 \times 10^6)^3(6.1 \times 10^{-4})$

(b) $\dfrac{(2.414 \times 10^4)^6}{(1.68 \times 10^5)^5}$

3   RADICALS AND RATIONAL EXPONENTS

Radicals and Properties of Radicals / Simplifying Radicals / Rationalizing Denominators and Numerators / Rational Exponents / Radicals and Calculators

Radicals and Properties of Radicals

You already know how to square a number—raise it to the second power by using the number *twice* as a factor. For instance, 5 squared is $5^2 = 5 \cdot 5 = 25$. Conversely, the **square root of a number** is one of its two equal factors. For example, 5 is a square root of 25 because 5 is one of the two equal factors of 25. In a similar way, the **cube root** of a number is one of its three equal factors. Consider the following examples.

Number	Equal Factors	Root
$16 = 4^2$	$4 \cdot 4$	4 (square root)
$-64 = (-4)^3$	$(-4)(-4)(-4)$	-4 (cube root)
$81 = 3^4$	$3 \cdot 3 \cdot 3 \cdot 3$	3 (fourth root)

DEFINITION OF *N*TH ROOT OF A NUMBER

Let a and b be real numbers, and let $n \geq 2$ be a positive integer. If

$$a = b^n,$$

then b is the **nth root of a.** If $n = 2$, then the root is a **square root.** If $n = 3$, then the root is a **cube root.**

Some numbers have more than one *n*th root. For example, both 5 and -5 are square roots of 25. The **principal *n*th root** of a number is defined as follows.

PRINCIPAL *N*TH ROOT OF A NUMBER

Let a be a real number that has at least one *n*th root. The **principal *n*th root of a** is the *n*th root that has the same sign as a. It is denoted by a **radical symbol:**

$$\sqrt[n]{a}.$$ *Principal nth root*

The positive integer n is the **index** of the radical, and the number a is the **radicand.** If $n = 2$, we omit the index and write $\sqrt{a}$ rather than $\sqrt[2]{a}$.

EXAMPLE 1   **Evaluating Expressions Involving Radicals**

A. $\sqrt{49} = 7$ because $7^2 = 49$.

B. $-\sqrt{49} = -7$ because $-(\sqrt{49}) = -(7) = -7$.

C. $\sqrt[3]{\dfrac{125}{64}} = \dfrac{5}{4}$ because $\left(\dfrac{5}{4}\right)^3 = \dfrac{5^3}{4^3} = \dfrac{125}{64}$.

D. $\sqrt[5]{-32} = -2$ because $(-2)^5 = -32$.

E. $\sqrt[4]{-81}$ is not a real number because there is no real number that can be raised to the fourth power to produce -81.

Here are some generalizations about the *n*th roots of a real number.

1. If a is a positive real number and n is a positive *even* integer, then a has exactly two (real) *n*th roots. We denote these roots by $\sqrt[n]{a}$ and $-\sqrt[n]{a}$. See Example 1(A) and (B).
2. If a is any real number and n is an *odd* integer, then a has only one (real) *n*th root, which is denoted by $\sqrt[n]{a}$. See Example 1(C) and (D).
3. If a is a negative real number and n is an *even* integer, then a has no (real) *n*th root. See Example 1(E).
4. $\sqrt[n]{0} = 0$.

Integers such as 1, 4, 9, 16, 25, and 49 are called **perfect squares** because they have integer square roots. Similarly, integers such as 1, 8, 27, 64, and 125 are called **perfect cubes** because they have integer cube roots.

PROPERTIES OF RADICALS

Let a and b be real numbers, variables, or algebraic expressions such that the indicated roots are real numbers, and let m and n be positive integers.

Property	*Example*
1. $\sqrt[n]{a^m} = \left(\sqrt[n]{a}\right)^m$	$\sqrt[3]{8^2} = \left(\sqrt[3]{8}\right)^2 = (2)^2 = 4$
2. $\sqrt[n]{a} \cdot \sqrt[n]{b} = \sqrt[n]{ab}$	$\sqrt{3} \cdot \sqrt{12} = \sqrt{3 \cdot 12} = \sqrt{36} = 6$
3. $\dfrac{\sqrt[n]{a}}{\sqrt[n]{b}} = \sqrt[n]{\dfrac{a}{b}}, b \neq 0$	$\dfrac{\sqrt[4]{27}}{\sqrt[4]{9}} = \sqrt[4]{\dfrac{27}{9}} = \sqrt[4]{3}$
4. $\sqrt[m]{\sqrt[n]{a}} = \sqrt[mn]{a}$	$\sqrt[3]{\sqrt{10}} = \sqrt[6]{10}$
5. $\left(\sqrt[n]{a}\right)^n = a$	$\left(\sqrt{3}\right)^2 = 3$
6. For n even, $\sqrt[n]{a^n} = \|a\|$.	$\sqrt{(-12)^2} = \|-12\| = 12$
For n odd, $\sqrt[n]{a^n} = a$.	$\sqrt[3]{(-12)^3} = -12$

REMARK   A common special case of Property 6 is $\sqrt{a^2} = |a|$. For instance, $\sqrt{(-2)^2} = |-2| = 2$.

Simplifying Radicals

An expression involving radicals is in **simplest form** when the following conditions are satisfied.

1. All possible factors have been removed from the radical.
2. All fractions have radical-free denominators (accomplished by a process called *rationalizing the denominator*).
3. The index of the radical is reduced.

To simplify a radical, we factor the radicand into factors whose exponents are multiples of the index. The roots of these factors are written outside the radical, and the "leftover" factors make up the new radicand.

EXAMPLE 2   Simplifying Even Roots

Perfect   Leftover
4th power   factor

A. $\sqrt[4]{48} = \sqrt[4]{16 \cdot 3} = \sqrt[4]{2^4 \cdot 3} = \sqrt[4]{2^4} \cdot \sqrt[4]{3} = 2\sqrt[4]{3}$

Perfect   Leftover
square   factor

B. $\sqrt{75x^3} = \sqrt{25x^2 \cdot 3x}$ *Find largest square factor*

$= \sqrt{(5x)^2 \cdot 3x}$

$= 5|x|\sqrt{3x}, \quad x \geq 0$ *Find root of perfect square*

C. $\sqrt[4]{(5x)^4} = |5x| = 5|x|$

EXAMPLE 3   Simplifying Odd Roots

Perfect   Leftover
cube   factor

A. $\sqrt[3]{24} = \sqrt[3]{8 \cdot 3} = \sqrt[3]{2^3 \cdot 3} = 2\sqrt[3]{3}$

Perfect   Leftover
cube   factor

B. $\sqrt[3]{24a^4} = \sqrt[3]{8a^3 \cdot 3a}$ *Find largest cube factor*

$= \sqrt[3]{(2a)^3 \cdot 3a}$

$= 2a\sqrt[3]{3a}$ *Find root of perfect cube*

C. $\sqrt[3]{-40x^6} = \sqrt[3]{(-8x^6) \cdot 5} = \sqrt[3]{(-2x^2)^3 \cdot 5} = -2x^2\sqrt[3]{5}$

Rationalizing Denominators and Numerators

To rationalize a denominator or numerator of the form $a - b\sqrt{m}$, or $a + b\sqrt{m}$, multiply both numerator and denominator by a **conjugate** factor: $a + b\sqrt{m}$ and $a - b\sqrt{m}$ are conjugates of each other. If $a = 0$, then the rationalizing factor for $\sqrt{m}$ is itself, $\sqrt{m}$.

EXAMPLE 4   Rationalizing Single-Term Denominators

A. $\dfrac{5}{2\sqrt{3}} = \dfrac{5}{2\sqrt{3}} \cdot \dfrac{\sqrt{3}}{\sqrt{3}} = \dfrac{5\sqrt{3}}{2(3)} = \dfrac{5\sqrt{3}}{6}$

B. $\dfrac{2}{\sqrt[3]{5}} = \dfrac{2}{\sqrt[3]{5}} \cdot \dfrac{\sqrt[3]{5^2}}{\sqrt[3]{5^2}}$

Multiply numerator and denominator by $\sqrt[3]{5^2}$ to produce a perfect cube radicand.

$\qquad = \dfrac{2\sqrt[3]{(5)^2}}{\sqrt[3]{5^3}} = \dfrac{2\sqrt[3]{25}}{5}$

EXAMPLE 5   Rationalizing a Denominator with Two Terms

$\dfrac{2}{3 + \sqrt{7}} = \dfrac{2}{3 + \sqrt{7}} \cdot \dfrac{3 - \sqrt{7}}{3 - \sqrt{7}}$

Multiply numerator and denominator by conjugate.

$\qquad = \dfrac{2(3 - \sqrt{7})}{(3)^2 - (\sqrt{7})^2}$

$\qquad = \dfrac{2(3 - \sqrt{7})}{9 - 7}$

$\qquad = \dfrac{2(3 - \sqrt{7})}{2}$

Cancel like factors.

$\qquad = 3 - \sqrt{7}$

In some applications, it helps to rationalize the numerator of a fraction.

EXAMPLE 6   Rationalizing the Numerator

$\dfrac{\sqrt{5} - \sqrt{7}}{2} = \dfrac{\sqrt{5} - \sqrt{7}}{2} \cdot \dfrac{\sqrt{5} + \sqrt{7}}{\sqrt{5} + \sqrt{7}}$

Multiply numerator and denominator by conjugate.

$\qquad = \dfrac{5 - 7}{2(\sqrt{5} + \sqrt{7})}$

$\qquad = \dfrac{-2}{2(\sqrt{5} + \sqrt{7})}$

$\qquad = \dfrac{-1}{\sqrt{5} + \sqrt{7}}$

Do not confuse an expression such as $\sqrt{5} + \sqrt{7}$ with the expression $\sqrt{5 + 7}$. In general, $\sqrt{x + y}$ does not equal $\sqrt{x} + \sqrt{y}$. Similarly, $\sqrt{x^2 + y^2}$ does not equal $x + y$. For instance, $\sqrt{3^2 + 4^2} \neq 3 + 4$.

Rational Exponents

In the following definition, radicals are used to define **rational exponents.**

DEFINITION OF RATIONAL EXPONENTS

If a is a real number and n is a positive integer such that the principal nth root of a exists, then we define $a^{1/n}$ to be

$$a^{1/n} = \sqrt[n]{a}.$$

Moreover, if m is a positive integer that has no common factor with n, then

$$a^{m/n} = (a^{1/n})^m = \left(\sqrt[n]{a}\right)^m \quad \text{and} \quad a^{m/n} = (a^m)^{1/n} = \sqrt[n]{a^m}.$$

The numerator of a rational exponent denotes the *power* to which the base is raised, and the denominator denotes the *index* or the *root* to be taken, as shown below.

$$b^{m/n} = \left(\sqrt[n]{b}\right)^m = \sqrt[n]{b^m}$$

(Power, Index)

EXAMPLE 7   Changing from Radical to Exponential Form

A. $\sqrt{3} = 3^{1/2}$

B. $\sqrt{(3xy)^5} = \sqrt[2]{(3xy)^5} = (3xy)^{(5/2)}$

C. $2x\sqrt[4]{x^3} = (2x)(x^{3/4}) = 2x^{1+(3/4)} = 2x^{7/4}$

EXAMPLE 8   Changing from Exponential to Radical Form

A. $(x^2 + y^2)^{3/2} = \left(\sqrt{x^2 + y^2}\right)^3 = \sqrt{(x^2 + y^2)^3}$

B. $2y^{3/4}z^{1/4} = 2(y^3z)^{1/4} = 2\sqrt[4]{y^3z}$

C. $a^{-3/2} = \dfrac{1}{a^{3/2}} = \dfrac{1}{\sqrt{a^3}}$

D. $x^{0.2} = x^{1/5} = \sqrt[5]{x}$

REMARK Rational exponents can be tricky, and you must remember that the expression $b^{m/n}$ is not defined unless $\sqrt[n]{b}$ is a real number. For instance, $(-8)^{1/2}$ is not defined because $\sqrt{-8}$ is not a real number, whereas $(-8)^{1/3}$ is defined because $\sqrt[3]{-8} = -2$.

Rational exponents are particularly useful for evaluating roots of numbers on a calculator, for reducing the index of a radical, and for simplifying expressions encountered in calculus.

EXAMPLE 9   Simplifying with Rational Exponents

A. $(27)^{2/6} = (27)^{1/3} = \sqrt[3]{27} = 3$

B. $(-32)^{-4/5} = \left(\sqrt[5]{(-32)}\right)^{-4} = (-2)^{-4} = \dfrac{1}{(-2)^4} = \dfrac{1}{16}$

C. $(-5x^{5/3})(3x^{-3/4}) = -15x^{(5/3)-(3/4)} = -15x^{11/12}, \quad x \neq 0$

EXAMPLE 10   Reducing the Index of a Radical

$$\sqrt[9]{a^3} = a^{3/9} \qquad \textit{Rewrite with rational exponents}$$
$$= a^{1/3} \qquad \textit{Reduce exponent}$$
$$= \sqrt[3]{a} \qquad \textit{Rewrite in radical form}$$

B. $\sqrt[3]{\sqrt{125}} = \sqrt[6]{125} = \sqrt[6]{(5)^3} = 5^{3/6} = 5^{1/2} = \sqrt{5}$

EXAMPLE 11   Simplifying Algebraic Expressions Involving Exponents

A. $(2x - 1)^{4/3}(2x - 1)^{-1/3} = (2x - 1)^{(4/3)-(1/3)} \qquad \textit{Add exponents}$
$$= (2x - 1)^1$$
$$- 2x - 1, \quad x \neq \dfrac{1}{2}$$

B. $\dfrac{x - 1}{(x - 1)^{-1/2}} = (x - 1)^{1+(1/2)} \qquad \textit{Subtract exponents}$
$$= (x - 1)^{3/2}$$

Radical expressions can be combined (added or subtracted) if they are **like radicals**—that is, if they have the same index and radicand. For instance, $2\sqrt{3x}$, $-\sqrt{3x}$, and $\sqrt{3x}/2$ are like radicals but $\sqrt[3]{3x}$ and $2\sqrt{3x}$ are not like radicals. To determine whether two radicals are like radicals, you should first simplify each radical.

EXAMPLE 12 **Combining Radicals**

A. $2\sqrt{48} - 3\sqrt{27} = 2\sqrt{16 \cdot 3} - 3\sqrt{9 \cdot 3}$ *Find square factors*
$= 8\sqrt{3} - 9\sqrt{3}$ *Find square roots*
$= (8 - 9)\sqrt{3}$ *Distributive Property*
$= -\sqrt{3}$

B. $\sqrt[3]{16x} - \sqrt[3]{54x^4} = \sqrt[3]{8 \cdot 2x} - \sqrt[3]{27 \cdot x^3 \cdot 2x}$
$= 2\sqrt[3]{2x} - 3x\sqrt[3]{2x}$
$= (2 - 3x)\sqrt[3]{2x}$

Radicals and Calculators

There are two basic methods of evaluating radicals on most calculators. For square roots, use the *square root key*. For other roots, first convert the radical to exponential form and then use the *exponential key*.

EXAMPLE 13 **Evaluating Radicals with a Calculator**

A. $\dfrac{1 + \sqrt{5}}{2} = 1.618033989 \approx 1.618$

B. $\sqrt[3]{56} = 56^{1/3} = 3.825862366 \approx 3.826$

C. $\sqrt[3]{-4} = \sqrt[3]{(-1)(4)} = \sqrt[3]{-1}\sqrt[3]{4} = -\sqrt[3]{4} \approx -1.587$

D. $(1.2)^{-1/6} \approx 0.970$

SECTION 3 · EXERCISES

In Exercises 1–10, fill in the missing description.

Radical Form	Rational Exponent Form
1. $\sqrt{9} = 3$	
2. $\sqrt[3]{64} = 4$	
3.	$196^{1/2} = 14$
4.	$-(144^{1/2}) = -12$
5. $\sqrt[3]{-216} = -6$	
6. $\sqrt[3]{614.125} = 8.5$	
7.	$27^{2/3} = 9$
8.	$(-243)^{1/5} = -3$
9. $\sqrt[4]{81^3} = 27$	
10. $\left(\sqrt[4]{81}\right)^3 = 27$	

In Exercises 11–30, evaluate the expression. (Do each exercise with and without a calculator.)

11. $\sqrt{9}$ **12.** $\sqrt{49}$

13. $\sqrt[3]{8}$ **14.** $\sqrt[3]{64}$

15. $\sqrt{36}$ **16.** $\sqrt[3]{\frac{27}{8}}$

17. $-\sqrt[3]{-27}$ **18.** $\sqrt[3]{0}$

19. $\dfrac{4}{\sqrt{64}}$ **20.** $\dfrac{\sqrt[4]{81}}{3}$

21. $\left(\sqrt[3]{-125}\right)^3$ **22.** $\sqrt[4]{562^4}$

23. $16^{1/2}$ **24.** $27^{1/3}$

25. $32^{-3/5}$ **26.** $100^{-3/2}$

27. $\left(\frac{16}{81}\right)^{-3/4}$ **28.** $\left(\frac{9}{4}\right)^{-1/2}$

29. $\left(-\frac{1}{64}\right)^{-1/3}$ **30.** $\left(-\frac{125}{27}\right)^{-1/3}$

In Exercises 31–40, simplify by removing all possible factors from the radical.

31. $\sqrt{8}$

32. $\sqrt[3]{\frac{16}{27}}$

33. $\sqrt{9 \times 10^{-4}}$

34. $\sqrt{4.5 \times 10^9}$

35. $\sqrt{72x^3}$

36. $\sqrt{54xy^4}$

37. $\sqrt[3]{16x^5}$

38. $\sqrt[4]{(3x^2)^4}$

39. $\sqrt{75x^2y^{-4}}$

40. $\sqrt[5]{96x^5}$

In Exercises 41–48, rationalize the denominator. Then, simplify the result.

41. $\dfrac{1}{\sqrt{3}}$

42. $\dfrac{5}{\sqrt{10}}$

43. $\dfrac{8}{\sqrt[3]{2}}$

44. $\dfrac{5}{\sqrt[3]{(5x)^2}}$

45. $\dfrac{2x}{5 - \sqrt{3}}$

46. $\dfrac{5}{\sqrt{14} - 2}$

47. $\dfrac{3}{\sqrt{5} + \sqrt{6}}$

48. $\dfrac{5}{2\sqrt{10} - 5}$

In Exercises 49–54, rationalize the numerator. Then, simplify the result.

49. $\dfrac{\sqrt{8}}{2}$

50. $\dfrac{\sqrt{2}}{3}$

51. $\dfrac{\sqrt{5} + \sqrt{3}}{3}$

52. $\dfrac{\sqrt{3} - \sqrt{2}}{2}$

53. $\dfrac{\sqrt{7} - 3}{4}$

54. $\dfrac{2\sqrt{3} + \sqrt{3}}{3}$

In Exercises 55–58, reduce the index of the radical.

55. $\sqrt[4]{3^2}$

56. $\sqrt[6]{x^3}$

57. $\sqrt[6]{(x + 1)^4}$

58. $\sqrt[4]{(3x^2)^4}$

In Exercises 59–62, write as a single radical. Then, simplify the result.

59. $\sqrt{\sqrt{32}}$

60. $\sqrt{\sqrt{243(x + 1)}}$

61. $\sqrt{\sqrt[4]{2x}}$

62. $\sqrt{\sqrt[3]{10a^7b}}$

In Exercises 63–68, simplify the expression.

63. $5\sqrt{x} - 3\sqrt{x}$

64. $3\sqrt{x + 1} + 10\sqrt{x + 1}$

65. $2\sqrt{50} + 12\sqrt{8}$

66. $4\sqrt{27} - \sqrt{75}$

67. $-2\sqrt{9y} + 10\sqrt{y}$

68. $7\sqrt{80x} - 2\sqrt{125x}$

In Exercises 69–78, perform the indicated operations and simplify. Assume that all variables are positive real numbers.

69. $\dfrac{x^{-3} \cdot x^{1/2}}{x^{3/2} \cdot x^{-1}}$

70. $\dfrac{5^{-1/2} \cdot 5x^{5/2}}{(5x)^{3/2}}$

71. $(3x^{-1/3}y^{3/4})^2$

72. $(-2u^{3/5}v^{-1/5})^3$

73. $\dfrac{18y^{4/3}z^{-1/3}}{24y^{-2/3}z}$

74. $\dfrac{a^{3/4} \cdot a^{1/2}}{a^{5/2}}$

75. $\left(\dfrac{x^{1/4}}{x^{1/6}}\right)^3$

76. $\left(\dfrac{3m^{1/6}n^{1/3}}{4n^{-2/3}}\right)^2$

77. $(c^{3/2})^{1/3}$

78. $(k^{-1/3})^{3/2}$

In Exercises 79–84, use a calculator to approximate the number. (Round the result to three decimal places.)

79. $\sqrt{57}$

80. $\sqrt[3]{45^2}$

81. $\sqrt[6]{125}$

82. $(15.25)^{-1.4}$

83. $\sqrt{75 + 3\sqrt{8}}$

84. $(2.65 \times 10^{-4})^{1/3}$

In Exercises 85–90, fill in the blank with $<$, $=$, or $>$.

85. $\sqrt{5} + \sqrt{3}$ ____ $\sqrt{5 + 3}$

86. $\sqrt{3} - \sqrt{2}$ ____ $\sqrt{3 - 2}$

87. 5 ____ $\sqrt{3^2 + 2^2}$ **88.** 5 ____ $\sqrt{3^2 + 4^2}$

89. $\sqrt{3} \cdot \sqrt[4]{3}$ ____ $\sqrt[8]{3}$ **90.** $\sqrt{\dfrac{3}{11}}$ ____ $\dfrac{\sqrt{3}}{\sqrt{11}}$

91. *Calculator Experiment* Enter any positive real number in your calculator and repeatedly take the square root. What real number does the display appear to be approaching?

92. Square the real number $2/\sqrt{5}$ and note that the radical is eliminated from the denominator. Is this equivalent to rationalizing the denominator? Why or why not?

4   POLYNOMIALS AND SPECIAL PRODUCTS

Polynomials   /   **Operations with Polynomials**   /   **Special Products**

Polynomials

The most common kind of algebraic expression is the **polynomial.** Some examples are

$$2x + 5, \quad 3x^4 - 7x^2 + 2x + 4, \quad \text{and} \quad 5x^2y^2 - xy + 3.$$

The first two are *polynomials in x* and the third one is a *polynomial in x and y.* The terms of a polynomial in x have the form ax^k, where a is called the **coefficient** and k the **degree** of the term. For instance, the third-degree polynomial

$$2x^3 - 5x^2 + 1 = 2x^3 + (-5)x^2 + (0)x + 1$$

has coefficients 2, −5, 0, and 1.

<table>
<tr><td colspan="2">**DEFINITION OF A POLYNOMIAL IN X**</td></tr>
<tr><td>

REMARK   Polynomials with one, two, and three terms are called **monomials, binomials,** and **trinomials,** respectively.

</td><td>

Let $a_0, a_1, a_2, \ldots, a_n$ be *real numbers*, and let n be a *nonnegative integer.* A **polynomial in x** is an expression of the form

$$a_nx^n + a_{n-1}x^{n-1} + \cdots + a_1x + a_0$$

where $a \neq 0$. The polynomial is of **degree n**, a_n is the **leading coefficient,** and a_0 is the **constant term.**

</td></tr>
</table>

In **standard form,** a polynomial is written with descending powers of x.

EXAMPLE 1   Writing Polynomials in Standard Form

Polynomial	*Standard Form*	*Degree*
A. $4x^2 - 5x^7 - 2 + 3x$	$-5x^7 + 4x^2 + 3x - 2$	7
B. $4 - 9x^2$	$-9x^2 + 4$	2
C. 8	$8 \quad (8 = 8x^0)$	0

A polynomial that has all zero coefficients is called the **zero polynomial,** denoted by 0. We do not assign a degree to the zero polynomial. Expressions such as $\sqrt{x^2 - 3x}$ and $x^2 + 5x^{-1}$ are not polynomials.

For polynomials in more than one variable, the degree of a *term* is the sum of the exponents of the variables in the term. The degree of the *polynomial* is the highest degree of its terms. For instance, the polynomial $5x^3y - x^2y^2 + 2xy - 5$ has two terms of degree 4, one term of degree 2, and one term of degree 0. The degree of the polynomial is 4.

Operations with Polynomials

You can **add** and **subtract** polynomials in much the same way you add and subtract real numbers. Simply add or subtract the *like terms* (terms having the same variables to the same powers) by adding their coefficients. For instance, $-3xy^2$ and $5xy^2$ are like terms and their sum is

$$-3xy^2 + 5xy^2 = (-3 + 5)xy^2 = 2xy^2.$$

EXAMPLE 2   Sums and Differences of Polynomials

A. $(5x^3 - 7x^2 - 3) + (x^3 + 2x^2 - x + 8)$

$$= (5x^3 + x^3) + (2x^2 - 7x^2) - x + (8 - 3) \qquad \textit{Group like terms}$$

$$= 6x^3 - 5x^2 - x + 5 \qquad \textit{Combine like terms}$$

B. $(7x^4 - x^2 - 4x + 2) - (3x^4 - 4x^2 + 3x)$

$$= 7x^4 - x^2 - 4x + 2 - 3x^4 + 4x^2 - 3x$$

$$= (7x^4 - 3x^4) + (4x^2 - x^2) + (-3x - 4x) + 2 \qquad \textit{Group like terms}$$

$$= 4x^4 + 3x^2 - 7x + 2 \qquad \textit{Combine like terms}$$

REMARK   A common mistake is to fail to change the sign of *each* term inside parentheses preceded by a negative sign. For instance, note that

$$-(3x^4 - 4x^2 + 3x) = -3x^4 + 4x^2 - 3x \quad \text{and}$$

$$-(3x^4 - 4x^2 + 3x) \quad \text{does not equal} \quad -3x^4 - 4x^2 + 3x.$$

To find the **product** of two polynomials, use the left and right Distributive Properties. For example, if you treat $(5x + 7)$ as a single quantity, you can multiply $(3x - 2)$ by $(5x + 7)$ as follows.

$$(3x - 2)(5x + 7) = 3x(5x + 7) - 2(5x + 7)$$

$$= (3x)(5x) + (3x)(7) - (2)(5x) - (2)(7)$$

$$= 15x^2 + 21x - 10x - 14$$

Product of	Product of	Product of	Product of
First terms	Outer terms	Inner terms	Last terms

$$= 15x^2 + 11x - 14$$

Note in this **FOIL method** that for binomials the outer (O) and inner (I) terms are alike and can be combined into one term.

When multiplying two polynomials, be sure to multiply *each* term of one polynomial by *each* term of the other. The vertical pattern shown in Example 3 is a convenient way to multiply two polynomials.

EXAMPLE 3   **Using a Vertical Arrangement to Multiply Polynomials**

Multiply $(x^2 - 2x + 2)$ by $(x^2 + 2x + 2)$.

SOLUTION

$$
\begin{array}{l}
x^2 - 2x + 2 \qquad\qquad\qquad \textit{Standard form} \\
\underline{x^2 + 2x + 2} \qquad\qquad\qquad \textit{Standard form} \\
x^4 - 2x^3 + 2x^2 \qquad\qquad \longleftarrow \;\; x^2(x^2 - 2x + 2) \\
\quad\;\; 2x^3 - 4x^2 + 4x \qquad \longleftarrow \;\; 2x(x^2 - 2x + 2) \\
\underline{\qquad\quad\;\; 2x^2 - 4x + 4 \qquad \longleftarrow \;\; 2(x^2 - 2x + 2)} \\
x^4 + 0x^3 + 0x^2 - 0x + 4 = x^4 + 4
\end{array}
$$

Thus,

$$(x^2 - 2x + 2)(x^2 + 2x + 2) = x^4 + 4.$$

Special Products

SPECIAL PRODUCTS

Let u and v be real numbers, variables, or algebraic expressions.

Special Product	*Example*
Sum and Difference of Two Terms	
$(u + v)(u - v) = u^2 - v^2$	$(x + 4)(x - 4) = x^2 - 4^2$
	$= x^2 - 16$
Square of a Binomial	
$(u + v)^2 = u^2 + 2uv + v^2$	$(x + 3)^2 = x^2 + 2(x)(3) + 3^2$
	$= x^2 + 6x + 9$
$(u - v)^2 = u^2 - 2uv + v^2$	$(3x - 2)^2$
	$= (3x)^2 - 2(3x)(2) + 2^2$
	$= 9x^2 - 12x + 4$
Cube of a Binomial	
$(u + v)^3 = u^3 + 3u^2v + 3uv^2 + v^3$	$(x + 2)^3$
	$= x^3 + 3x^2(2) + 3x(2^2) + 2^3$
	$= x^3 + 6x^2 + 12x + 8$
$(u - v)^3 = u^3 - 3u^2v + 3uv^2 - v^3$	$(x - 1)^3$
	$= x^3 - 3x^2(1) + 3x(1^2) - 1^3$
	$= x^3 - 3x^2 + 3x - 1$

EXAMPLE 4   Sum and Difference of Two Terms

Find the product of $(5x + 9)$ and $(5x - 9)$.

SOLUTION

The product of a sum and a difference of the *same* two terms has no middle term and it takes the form $(u + v)(u - v) = u^2 - v^2$.

$$(5x + 9)(5x - 9) = (5x)^2 - 9^2 = 25x^2 - 81$$

EXAMPLE 5   Square and Cube of a Binomial

A. $(6x - 5)^2 = (6x)^2 - 2(6x)(5) + 5^2$ *Square of a binomial*
$$= 36x^2 - 60x + 25$$
B. $(3x + 2)^3 = (3x)^3 + 3(3x)^2(2) + 3(3x)(2)^2 + 2^3$ *Cube of a binomial*
$$= 27x^3 + 54x^2 + 36x + 8$$

Occasionally the formulas for special products can be extended to cover products of two trinomials, as demonstrated in the following example.

EXAMPLE 6   The Product of Two Trinomials

$$(x + y - 2)(x + y + 2)$$
$$= [(x + y) - 2][(x + y) + 2]$$ *Group $(x + y)$*
$$= (x + y)^2 - 2^2$$ *Sum and difference of terms*
$$= x^2 + 2xy + y^2 - 4$$

SECTION 4 · EXERCISES

In Exercises 1–6, find the degree and leading coefficient of the polynomial.

1. $2x^2 - x + 1$ **2.** $-3x^4 + 2x^2 - 5$
3. $x^5 - 1$ **4.** 3
5. $4x^5 + 6x^4 - x - 1$ **6.** $2x$

In Exercises 7–12, determine whether the expression is a polynomial. If it is, write the polynomial in standard form.

7. $2x - 3x^3 + 8$ **8.** $2x^3 + x - 3x^{-1}$
9. $\dfrac{3x + 4}{x}$ **10.** $\dfrac{x^2 + 2x - 3}{2}$
11. $y^2 - y^4 + y^3$ **12.** $\sqrt{y^2 - y^4}$

In Exercises 13–26, perform the indicated operations. Write the resulting polynomial in standard form.

13. $(6x + 5) - (8x + 15)$
14. $(2x^2 + 1) - (x^2 - 2x + 1)$
15. $-(x^3 - 2) + (4x^3 - 2x)$
16. $-(5x^2 - 1) - (-3x^2 + 5)$
17. $(15x^2 - 6) - (-8x^3 - 14x^2 - 17)$
18. $(15x^4 - 18x - 19) - (13x^4 - 5x + 15)$
19. $5z - [3z - (10z + 8)]$
20. $(y^3 + 1) - [(y^2 + 1) + (3y - 7)]$
21. $3x(x^2 - 2x + 1)$ **22.** $y^2(4y^2 + 2y - 3)$
23. $-5z(3z - 1)$ **24.** $-4x(3 - x^3)$
25. $(-2x)(-3x)(5x + 2)$ **26.** $(1 - x^3)(4x)$

In Exercises 27–60, find the product. Write the resulting polynomial in standard form.

27. $(x + 3)(x + 4)$

28. $(x - 5)(x + 10)$

29. $(3x - 5)(2x + 1)$

30. $(7x - 2)(4x - 3)$

31. $(x + 6)^2$

32. $(3x - 2)^2$

33. $(2x - 5y)^2$

34. $(5 - 8x)^2$

35. $[(x - 3) + y]^2$

36. $[(x + 1) - y]^2$

37. $(x + 10)(x - 10)$

38. $(2x + 3)(2x - 3)$

39. $(x + 2y)(x - 2y)$

40. $(2x + 3y)(2x - 3y)$

41. $(m - 3 + n)(m - 3 - n)$

42. $(x + y + 1)(x + y - 1)$

43. $(2r^2 - 5)(2r^2 + 5)$

44. $(3a^3 - 4b^2)(3a^3 + 4b^2)$

45. $(x + 1)^3$

46. $(x - 2)^3$

47. $(2x - y)^3$

48. $(3x + 2y)^3$

49. $(\sqrt{x} + \sqrt{y})(\sqrt{x} - \sqrt{y})$

50. $(5 + \sqrt{x})(5 - \sqrt{x})$

51. $(4x^3 - 3)^2$

52. $(8x + 3)^2$

53. $(x^2 + 9)(x^2 - x - 4)$

54. $(x - 2)(x^2 + 2x + 4)$

55. $(x^2 - x + 1)(x^2 + x + 1)$

56. $(x^2 + 3x - 2)(x^2 - 3x - 2)$

57. $5x(x + 1) - 3x(x + 1)$

58. $(2x - 1)(x + 3) + 3(x + 3)$

59. $(x + \sqrt{5})(x - \sqrt{5})(x + 4)$

60. $(x + y)(x - y)(x^2 + y^2)$

61. Determine the degree of the product of two polynomials of degrees m and n.

62. Determine the degree of the sum of two polynomials of degrees m and n if $m < n$.

5   FACTORING

Introduction / Factoring Special Polynomial Forms / Trinomials with Binomial Factors / Factoring by Grouping

Introduction

The process of writing a polynomial as a product is called **factoring.** It is an important tool for solving equations and for reducing fractional expressions.

Unless noted otherwise, we will limit our discussion of factoring to polynomials whose factors have integer coefficients. If a polynomial cannot be factored using integer coefficients, then it is called **prime** or **irreducible over the integers.** For instance, the polynomial $x^2 - 3$ is irreducible over the integers. [Over the *real numbers,* this polynomial can be factored as $x^2 - 3 = (x + \sqrt{3})(x - \sqrt{3})$.]

A polynomial is said to be **completely factored** when each of its factors is prime. For instance,

$$x^3 - x^2 + 4x - 4 = (x - 1)(x^2 + 4)$$

is completely factored, but

$$x^3 - x^2 - 4x + 4 = (x - 1)(x^2 - 4)$$

is not completely factored. Its complete factorization is

$$(x - 1)(x + 2)(x - 2).$$

Some polynomials can be written as the product of a monomial and another polynomial. The technique used is the Distributive Property, $a(b + c) = ab + ac$, in the *reverse* direction:

$$ab + ac = a(b + c).$$ *a is common factor*

EXAMPLE 1 Removing Common Factors

A. $6x^3 - 4x = 2x(3x^2) - 2x(2)$ *2x is common factor*

$\qquad\qquad = 2x(3x^2 - 2)$

B. $(x - 2)(2x) + (x - 2)(3) = (x - 2)(2x + 3)$ *x − 2 is common factor*

Factoring Special Polynomial Forms

FACTORING SPECIAL POLYNOMIAL FORMS

Factored Form	Example
Difference of Two Squares	
$u^2 - v^2 = (u + v)(u - v)$	$9x^2 - 4 = (3x)^2 - 2^2$
	$\qquad\quad = (3x + 2)(3x - 2)$
Perfect Square Trinomial	
$u^2 + 2uv + v^2 = (u + v)^2$	$x^2 + 6x + 9$
	$\qquad = x^2 + 2(x)(3) + 3^2$
	$\qquad = (x + 3)^2$
$u^2 - 2uv + v^2 = (u - v)^2$	$x^2 - 6x + 9$
	$\qquad = x^2 - 2(x)(3) + 3^2$
	$\qquad = (x - 3)^2$
Sum or Difference of Two Cubes	
$u^3 + v^3 = (u + v)(u^2 - uv + v^2)$	$x^3 + 8 = x^3 + 2^3$
	$\qquad = (x + 2)(x^2 - 2x + 4)$
$u^3 - v^3 = (u - v)(u^2 + uv + v^2)$	$27x^3 - 1$
	$\qquad = (3x)^3 - 1^3$
	$\qquad = (3x - 1)(9x^2 + 3x + 1)$

One of the easiest special polynomial forms to factor is the difference of two squares. Think of the form as follows:

$$u^2 \ominus v^2 = (u + v)(u - v).$$

Difference   Opposite signs

To recognize perfect square terms, look for coefficients that are squares of integers and variables raised to *even powers*.

REMARK In Example 2, note that the first step in factoring a polynomial is to check for common factors. Once the common factor has been removed, it is often possible to recognize patterns that were not immediately obvious.

EXAMPLE 2   **Removing a Common Factor First**

$$3 - 12x^2 = 3(1 - 4x^2) \qquad\qquad \textit{Common factor}$$
$$= 3[1^2 - (2x)^2] \qquad\qquad \textit{Difference of squares}$$
$$= 3(1 + 2x)(1 - 2x)$$

EXAMPLE 3   **Factoring the Difference of Two Squares**

A. $(x + 2)^2 - y^2 = [(x + 2) + y][(x + 2) - y] \qquad \textit{Difference of two squares}$
$$= (x + 2 + y)(x + 2 - y)$$
$$= (x + y + 2)(x - y + 2)$$
B. $16x^4 - 81 = (4x^2)^2 - 9^2$
$$= (4x^2 + 9)(4x^2 - 9) \qquad\qquad \textit{Difference of two squares}$$
$$= (4x^2 + 9)[(2x)^2 - 3^2]$$
$$= (4x^2 + 9)(2x + 3)(2x - 3) \qquad \textit{Difference of two squares}$$

A perfect square trinomial is the square of a binomial, and it has the following form:

$$u^2 + 2uv + v^2 = (u + v)^2 \quad \text{or} \quad u^2 - 2uv + v^2 = (u - v)^2.$$

Same sign   Same sign

Note that the first and last terms are squares and the middle term is twice the product of u and v.

EXAMPLE 4   **Factoring Perfect Square Trinomials**

A. $16x^2 + 8x + 1 = (4x)^2 + 2(4x)(1) + 1^2$
$$= (4x + 1)^2$$
B. $x^2 - 10x + 25 = x^2 - 2(x)(5) + 5^2$
$$= (x - 5)^2$$

The next two formulas show sums and differences of cubes. Pay special attention to the signs of the terms.

Like signs   Like signs

$$u^3 + v^3 = (u + v)(u^2 - uv + v^2) \qquad u^3 - v^3 = (u - v)(u^2 + uv + v^2)$$

Unlike signs   Unlike signs

EXAMPLE 5   Factoring the Difference of Cubes

$$x^3 - 27 = x^3 - 3^3 = (x - 3)(x^2 + 3x + 9)$$

Notice that the last factor is prime.

EXAMPLE 6   Factoring the Sum of Cubes

$$\begin{aligned}
3x^3 + 192 &= 3(x^3 + 64) && \textit{Common factor} \\
&= 3(x^3 + 4^3) \\
&= 3(x + 4)(x^2 - 4x + 16) && \textit{Sum of two cubes}
\end{aligned}$$

Trinomials with Binomial Factors

To factor a trinomial of the form $ax^2 + bx + c$, use the following pattern:

Factors of a

$$ax^2 + bx + c = (\blacksquare x + \blacksquare)(\blacksquare x + \blacksquare).$$

Factors of c

The goal is to find a combination of factors of a and c so that the outer and inner products add up to the middle term bx. For instance, in the trinomial $6x^2 + 17x + 5$, you can write

$$\begin{array}{cccc} \mathbf{F} & \mathbf{O} & \mathbf{I} & \mathbf{L} \\ \downarrow & \downarrow & \downarrow & \downarrow \end{array}$$

$$(2x + 5)(3x + 1) = 6x^2 + 2x + 15x + 5 = 6x^2 + 17x + 5.$$

Note that the outer (**O**) and inner (**I**) products add up to $17x$.

EXAMPLE 7   Factoring a Trinomial: Leading Coefficient Is 1

Factor the trinomial $x^2 - 7x + 12$.

SOLUTION

The possible factorizations are

$$(x - 2)(x - 6), \quad (x - 1)(x - 12), \quad \text{and} \quad (x - 3)(x - 4).$$

Testing the middle term in each, you will find the correct factorization to be

$$x^2 - 7x + 12 = (x - 3)(x - 4).$$

EXAMPLE 8 **Factoring a Trinomial: Leading Coefficient Is Not 1**

Factor the trinomial $2x^2 + x - 15$.

SOLUTION

The eight possible factorizations are as follows:

$$(2x - 1)(x + 15) \qquad (2x + 1)(x - 15)$$
$$(2x - 3)(x + 5) \qquad (2x + 3)(x - 5)$$
$$(2x - 5)(x + 3) \qquad (2x + 5)(x - 3)$$
$$(2x - 15)(x + 1) \qquad (2x + 15)(x - 1).$$

Testing the middle term in each, you will find the correct factorization to be

$$2x^2 + x - 15 = (2x - 5)(x + 3).$$

Factoring by Grouping

Sometimes polynomials with more than three terms can be factored by a method called **factoring by grouping.** It is not always obvious which terms to group, and sometimes several different groupings will work.

EXAMPLE 9 **Factoring by Grouping**

$$\begin{aligned} x^3 - 2x^2 - 3x + 6 &= (x^3 - 2x^2) - (3x - 6) && \textit{Group terms} \\ &= x^2(x - 2) - 3(x - 2) && \textit{Factor groups} \\ &= (x - 2)(x^2 - 3) && \textit{Common factor} \end{aligned}$$

As general guidelines for completely factoring polynomials, consider the following factorizations, in order.

GUIDELINES FOR FACTORING POLYNOMIALS

1. Factor out any common factors.
2. Factor according to one of the special polynomial forms.
3. Factor $ax^2 + bx + c$ using the FOIL method.
4. Factor by grouping.

SECTION 5 · EXERCISES

In Exercises 1–6, factor out the common factor.

1. $3x + 6$

2. $5y - 30$

3. $2x^3 - 6x$

4. $4x^3 - 6x^2 + 12x$

5. $(x - 1)^2 + 6(x - 1)$

6. $3x(x + 2) - 4(x + 2)$

In Exercises 7–12, factor the difference of two squares.

7. $x^2 - 36$

8. $x^2 - \frac{1}{4}$

9. $16y^2 - 9$

10. $49 - 9y^2$

11. $(x - 1)^2 - 4$

12. $25 - (z + 5)^2$

In Exercises 13–18, factor the perfect square trinomial.

13. $x^2 - 4x + 4$

14. $x^2 + 10x + 25$

15. $4t^2 + 4t + 1$

16. $9x^2 - 12x + 4$

17. $25y^2 - 10y + 1$

18. $z^2 + z + \frac{1}{4}$

In Exercises 19–32, factor the trinomial.

19. $x^2 + x - 2$

20. $x^2 + 5x + 6$

21. $s^2 - 5s + 6$

22. $t^2 - t - 6$

23. $y^2 + y - 20$

24. $z^2 - 5z - 24$

25. $x^2 - 30x + 200$

26. $x^2 - 13x + 42$

27. $3x^2 - 5x + 2$

28. $2x^2 - x - 1$

29. $9z^2 - 3z - 2$

30. $12x^2 + 7x + 1$

31. $5x^2 + 26x + 5$

32. $5u^2 + 13u - 6$

In Exercises 33–38, factor the sum or difference of cubes.

33. $x^3 - 8$

34. $x^3 - 27$

35. $y^3 + 64$

36. $z^3 + 125$

37. $8t^3 - 1$

38. $27x^3 + 8$

In Exercises 39–44, factor by grouping.

39. $x^3 - x^2 + 2x - 2$

40. $x^3 + 5x^2 - 5x - 25$

41. $2x^3 - x^2 - 6x + 3$

42. $5x^3 - 10x^2 + 3x - 6$

43. $6 + 2x - 3x^3 - x^4$

44. $x^5 + 2x^3 + x^2 + 2$

In Exercises 45–50, factor the trinomial by grouping. For instance, to factor the polynomial in Exercise 45, you can group its terms as $(3x^2 + 6x) + (4x + 8)$.

45. $3x^2 + 10x + 8$

46. $2x^2 + 9x + 9$

47. $6x^2 + x - 2$

48. $6x^2 - x - 15$

49. $15x^2 - 11x + 2$

50. $12x^2 - 13x + 1$

In Exercises 51–54, make a "geometric factoring model" to represent the given factorization. For instance, a factoring model for $2x^2 + 5x + 2 = (2x + 1)(x + 2)$ is shown in the accompanying figure.

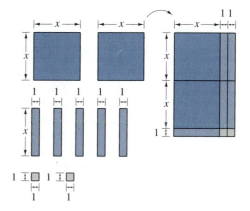

FIGURE FOR 51–54

51. $3x^2 + 7x + 2 = (3x + 1)(x + 2)$

52. $x^2 + 4x + 3 = (x + 3)(x + 1)$

53. $2x^2 + 7x + 3 = (2x + 1)(x + 3)$

54. $x^2 + 3x + 2 = (x + 2)(x + 1)$

In Exercises 55–80, completely factor the expression.

55. $x^3 - 9x$

56. $12x^2 - 48$

57. $x^3 - 4x^2$

58. $6x^2 - 54$

59. $x^2 - 2x + 1$

60. $16 + 6x - x^2$

61. $1 - 4x + 4x^2$

62. $9x^2 - 6x + 1$

63. $-2x^2 - 4x + 2x^3$

64. $2y^3 - 7y^2 - 15y$

65. $9x^2 + 10x + 1$

66. $13x + 6 + 5x^2$

67. $3x^3 + x^2 + 15x + 5$

68. $5 - x + 5x^2 - x^3$

69. $x^4 - 4x^3 + x^2 - 4x$

70. $3u - 2u^2 + 6 - u^3$

71. $25 - (x + 5)^2$

72. $(t - 1)^2 - 49$

73. $(x^2 + 1)^2 - 4x^2$

74. $(x^2 + 8)^2 - 36x^2$

75. $2t^3 - 16$

76. $5x^3 + 40$

77. $4x(2x - 1) + (2x - 1)^2$

78. $5(3 - 4x)^2 - 8(3 - 4x)(5x - 1)$

79. $2(x + 1)(x - 3)^2 - 3(x + 1)^2(x - 3)$

80. $7(3x + 2)^2(1 - x)^2 + (3x + 2)(1 - x)^3$

---

6   FRACTIONAL EXPRESSIONS

**Domain of an Algebraic Expression / Simplifying Rational Expressions /
Operations with Rational Expressions / Compound Fractions**

Domain of an Algebraic Expression

The set of real numbers for which an expression is defined is the **domain** of
the expression. Two algebraic expressions are said to be **equivalent** if they
yield the same value for all numbers in their domain. For instance, the expressions $[(x + 1) + (x + 2)]$ and $2x + 3$ are equivalent.

EXAMPLE 1   **Finding the Domain of an Algebraic Expression**

A. The domain of the polynomial $2x^3 + 3x + 4$ is the set of all real numbers.
In fact, the domain of any polynomial is the set of all real numbers (unless
the domain is specifically restricted).

B. The domain of the polynomial $x^2 + 5x + 2$, $x > 0$, is the set of positive
real numbers, because the polynomial is specifically restricted to that set.

C. The domain of the radical expression $\sqrt{x}$ is the set of nonnegative real
numbers, because the square root of a negative number is not a real
number.

D. The domain of the expression $(x + 2)/(x - 3)$ is the set of all real numbers except $x = 3$, which would produce an undefined division by zero.

The quotient of two algebraic expressions is a **fractional expression.**
Moreover, the quotient of two *polynomials* such as

$$\frac{1}{x}, \quad \frac{2x - 1}{x + 1}, \quad \text{or} \quad \frac{x^2 - 1}{x^2 + 1}$$

is a **rational expression.** Recall that a fraction is in reduced form if its
numerator and denominator have no factors in common aside from ± 1. To
write a fraction in reduced form, apply the following rule.

$$\frac{a \cdot \cancel{c}}{b \cdot \cancel{c}} = \frac{a}{b}, \qquad b \neq 0 \quad \text{and} \quad c \neq 0$$

The key to success in simplifying rational expressions lies in your ability to *factor* polynomials.

EXAMPLE 2   Reducing a Rational Expression

Write the following rational expression in reduced form.

$$\frac{x^2 + 4x - 12}{3x - 6}$$

SOLUTION

Factoring both numerator and denominator, and then reducing, produces the following:

$$\frac{x^2 + 4x - 12}{3x - 6} = \frac{(x + 6)\cancel{(x - 2)}}{3\cancel{(x - 2)}} \qquad \textit{Factor completely}$$

$$= \frac{x + 6}{3}, \qquad x \neq -2. \qquad \textit{Reduce}$$

Note that the original expression is undefined when $x = 2$ (because division by zero is undefined). To make sure that the reduced expression is *equivalent* to the original expression, you must restrict the domain of the reduced expression by excluding the value $x = 2$.

Simplifying Rational Expressions

When simplifying rational expressions, be sure to factor each polynomial completely before concluding that the numerator and denominator have no factors in common. Moreover, changing the sign of a factor may allow further reduction, as seen in part (B) of the next example.

EXAMPLE 3   Reducing Rational Expressions

A. $\dfrac{x^3 - 4x}{x^2 + x - 2} = \dfrac{x(x^2 - 4)}{(x + 2)(x - 1)}$

$$= \frac{x\cancel{(x + 2)}(x - 2)}{\cancel{(x + 2)}(x - 1)} \qquad \textit{Factor completely}$$

$$= \frac{x(x - 2)}{x - 1}, \qquad x \neq -2 \qquad \textit{Reduce}$$

B. $\dfrac{12 + x - x^2}{2x^2 - 9x + 4} = \dfrac{(4 - x)(3 + x)}{(2x - 1)(x - 4)}$ *Factor completely*

$\qquad\qquad = \dfrac{-(x - 4)(3 + x)}{(2x - 1)(x - 4)}$ $(4 - x) = -(x - 4)$

$\qquad\qquad = -\dfrac{3 + x}{2x - 1}, \quad x \neq 4$ *Reduce*

Operations with Rational Expressions

To multiply or divide rational expressions, use the properties of fractions discussed in Section 2.

EXAMPLE 4   **Multiplying and Dividing Rational Expressions**

A. $\dfrac{2x^2 + x - 6}{x^2 + 4x - 5} \cdot \dfrac{x^3 - 3x^2 + 2x}{4x^2 - 6x}$

$\qquad = \dfrac{(2x - 3)(x + 2)}{(x + 5)(x - 1)} \cdot \dfrac{x(x - 2)(x - 1)}{2x(2x - 3)}$ *Factor and reduce*

$\qquad = \dfrac{(x + 2)(x - 2)}{2(x + 5)}, \quad x \neq 0, x \neq 1, x \neq \dfrac{3}{2}$

B. $\dfrac{x^3 - 8}{x^2 - 4} \div \dfrac{x^2 + 2x + 4}{x^3 + 8} = \dfrac{x^3 - 8}{x^2 - 4} \cdot \dfrac{x^3 + 8}{x^2 + 2x + 4}$ *Invert and multiply*

$\qquad = \dfrac{(x - 2)(x^2 + 2x + 4)}{(x + 2)(x - 2)} \cdot \dfrac{(x + 2)(x^2 - 2x + 4)}{x^2 + 2x + 4}$

$\qquad = x^2 - 2x + 4, \quad x \neq \pm 2$

To add or subtract rational expressions, use the least common denominator (LCD) method or the basic definition

$$\frac{a}{b} \pm \frac{c}{d} = \frac{ad \pm bc}{bd}, \qquad b \neq 0 \quad \text{and} \quad d \neq 0.$$

This definition provides an efficient way of adding or subtracting *two* fractional expressions that have no common factors in their denominators.

EXAMPLE 5   **Subtracting Rational Expressions Using the Basic Definition**

$\dfrac{x}{x - 3} - \dfrac{2}{3x + 4} = \dfrac{x(3x + 4) - 2(x - 3)}{(x - 3)(3x + 4)}$ *Definition of subtraction*

$\qquad\qquad = \dfrac{3x^2 + 4x - 2x + 6}{(x - 3)(3x + 4)}$ *Remove parentheses*

$\qquad\qquad = \dfrac{3x^2 + 2x + 6}{(x - 3)(3x + 4)}$ *Combine like terms*

For three or more fractions, or for fractions with a repeated factor in the denominator, the LCD method works well. Recall that the LCD of several fractions consists of the product of all prime factors in the denominators, with each factor given the highest power of its occurrence in any denominator.

EXAMPLE 6   **The LCD Method for Combining Rational Expressions**

Perform the given operations and simplify.

$$\frac{3}{x - 1} - \frac{2}{x} + \frac{x + 3}{x^2 - 1}$$

SOLUTION

Using the factored denominators $(x - 1)$, x, and $(x + 1)(x - 1)$, you can see that the LCD is $x(x + 1)(x - 1)$.

$$\frac{3}{x - 1} - \frac{2}{x} + \frac{x + 3}{(x + 1)(x - 1)}$$

$$= \frac{3(x)(x + 1)}{x(x + 1)(x - 1)} - \frac{2(x + 1)(x - 1)}{x(x + 1)(x - 1)} + \frac{(x + 3)(x)}{x(x + 1)(x - 1)}$$

$$= \frac{3(x)(x + 1) - 2(x + 1)(x - 1) + (x + 3)(x)}{x(x + 1)(x - 1)}$$

$$= \frac{3x^2 + 3x - 2x^2 + 2 + x^2 + 3x}{x(x + 1)(x - 1)}$$

$$= \frac{2x^2 + 6x + 2}{x(x + 1)(x - 1)}$$

$$= \frac{2(x^2 + 3x + 1)}{x(x + 1)(x - 1)}$$

Compound Fractions

Fractional expressions with fractions in the numerator or denominator, or both, are called **compound fractions** (or complex fractions).

A compound fraction can be simplified by first simplifying both its numerator and its denominator into single fractions, then inverting the denominator and multiplying.

EXAMPLE 7 **Simplifying a Compound Fraction**

$$\frac{\left(\frac{2}{x} - 3\right)}{\left(1 - \frac{1}{x-1}\right)} = \frac{\left(\frac{2 - 3(x)}{x}\right)}{\left(\frac{1(x-1) - 1}{x-1}\right)} \qquad \textit{Combine fractions}$$

$$= \frac{\left(\frac{2 - 3x}{x}\right)}{\left(\frac{x-2}{x-1}\right)} \qquad \textit{Simplify}$$

$$= \frac{2 - 3x}{x} \cdot \frac{x-1}{x-2} \qquad \textit{Invert and multiply}$$

$$= \frac{(2 - 3x)(x-1)}{x(x-2)}, \qquad x \neq 1$$

Another way to simplify a compound fraction is to multiply each term in its numerator and denominator by the LCD of all fractions in both its numerator and denominator. Each product is then reduced to obtain a single fraction.

EXAMPLE 8 **Simplifying a Compound Fraction by Multiplying by the LCD**

Use the LCD to simplify the following compound fraction.

$$\frac{\left(\frac{1}{x^2} - \frac{1}{y^2}\right)}{\left(\frac{1}{x} + \frac{1}{y}\right)}$$

SOLUTION

For the four fractions in the numerator and denominator, the LCD is $x^2 y^2$. Multiplying each term of the numerator and denominator by this LCD yields the following.

$$\frac{\left(\frac{1}{x^2} - \frac{1}{y^2}\right)x^2 y^2}{\left(\frac{1}{x} + \frac{1}{y}\right)x^2 y^2} = \frac{\left(\frac{1}{x^2}\right)x^2 y^2 - \left(\frac{1}{y^2}\right)x^2 y^2}{\left(\frac{1}{x}\right)x^2 y^2 + \left(\frac{1}{y}\right)x^2 y^2}$$

$$= \frac{y^2 - x^2}{xy^2 + x^2 y}$$

$$= \frac{(y - x)(y + x)}{xy(y + x)}$$

$$= \frac{y - x}{xy}, \qquad x \neq -y$$

The next two examples illustrate some methods for simplifying expressions involving radicals and negative exponents. (These types of expressions occur frequently in calculus.)

EXAMPLE 9   Simplifying an Expression with Negative Exponents

Simplify the expression $x(1 - 2x)^{-3/2} + (1 - 2x)^{-1/2}$.

SOLUTION

By rewriting the given expression with positive exponents, we obtain

$$\frac{x}{(1 - 2x)^{3/2}} + \frac{1}{(1 - 2x)^{1/2}},$$

which could then be combined by the LCD method. However, if we first remove the common factor with the *smaller exponent,* the process is simpler.

$$x(1 - 2x)^{-3/2} + (1 - 2x)^{-1/2} = (1 - 2x)^{-3/2}\left[x + (1 - 2x)^{(-1/2)-(-3/2)}\right]$$
$$= (1 - 2x)^{-3/2}\left[x + (1 - 2x)^{1}\right]$$
$$= \frac{1 - x}{(1 - 2x)^{3/2}}$$

Note that when factoring, you subtract exponents.

EXAMPLE 10   Simplifying a Compound Fraction: LCD Method

$$\frac{\sqrt{4 - x^2} + \dfrac{x^2}{\sqrt{4 - x^2}}}{4 - x^2} = \frac{\sqrt{4 - x^2} + \dfrac{x^2}{\sqrt{4 - x^2}}}{4 - x^2} \cdot \frac{\sqrt{4 - x^2}}{\sqrt{4 - x^2}}$$
$$= \frac{(4 - x^2) + x^2}{(4 - x^2)^{3/2}}$$
$$= \frac{4}{\sqrt{(4 - x^2)^3}}$$

SECTION 6 · EXERCISES

In Exercises 1–10, find the domain of the expression.

1. $3x^2 - 4x + 7$

2. $2x^2 + 5x - 2$

3. $4x^3 + 5x + 3, \quad x \geq 0$

4. $6x^2 + 7x - 9, \quad x > 0$

5. $\dfrac{1}{x - 2}$

6. $\dfrac{x + 1}{2x + 1}$

7. $\dfrac{x - 1}{x^2 - 4x}$

8. $\dfrac{2x + 1}{x^2 - 9}$

9. $\sqrt{x + 1}$

10. $\dfrac{1}{\sqrt{x + 1}}$

In Exercises 11–16, find the factor that makes the two fractions equivalent.

11. $\dfrac{5}{2x} = \dfrac{5()}{6x^2}$ **12.** $\dfrac{3}{4} = \dfrac{3()}{4(x+1)}$

13. $\dfrac{x+1}{x} = \dfrac{(x+1)()}{x(x-2)}$

14. $\dfrac{3y-4}{y+1} = \dfrac{(3y-4)()}{y^2-1}$

15. $\dfrac{3x}{x-3} = \dfrac{3x()}{x^2-x-6}$ **16.** $\dfrac{1-z}{z^2} = \dfrac{(1-z)()}{z^3+z^2}$

In Exercises 17–30, write the expression in reduced form.

17. $\dfrac{15x^2}{10x}$ **18.** $\dfrac{18y^2}{60y^5}$

19. $\dfrac{3xy}{xy+x}$ **20.** $\dfrac{9x^2+9x}{2x+2}$

21. $\dfrac{x-5}{10-2x}$ **22.** $\dfrac{x^2-25}{5-x}$

23. $\dfrac{x^3+5x^2+6x}{x^2-4}$ **24.** $\dfrac{x^2+8x-20}{x^2+11x+10}$

25. $\dfrac{y^2-7y+12}{y^2+3y-18}$ **26.** $\dfrac{10+x}{x^2+11x+10}$

27. $\dfrac{2-x+2x^2-x^3}{x-2}$ **28.** $\dfrac{x^2-9}{x^3+x^2-9x-9}$

29. $\dfrac{z^3-8}{z^2+2z+4}$ **30.** $\dfrac{y^3-2y^2-3y}{y^3+1}$

In Exercises 31–54, perform the indicated operations and simplify.

31. $\dfrac{5}{x-1} \cdot \dfrac{x-1}{25(x-2)}$

32. $\dfrac{(x+5)(x-3)}{x+2} \cdot \dfrac{1}{(x+5)(x+2)}$

33. $\dfrac{r}{r-1} \cdot \dfrac{r^2-1}{r^2}$ **34.** $\dfrac{4y-16}{5y+15} \cdot \dfrac{2y+6}{4-y}$

35. $\dfrac{y^3-8}{2y^3} \cdot \dfrac{4y}{y^2-5y+6}$

36. $\dfrac{x^3-1}{x+1} \cdot \dfrac{x^2+1}{x^2-1}$ **37.** $\dfrac{3(x+y)}{4} \div \dfrac{x+y}{2}$

38. $\dfrac{x+2}{5(x-3)} \div \dfrac{x-2}{5(x-3)}$

39. $\dfrac{\left(\dfrac{x^2}{(x+1)^2}\right)}{\left(\dfrac{x}{(x+1)^3}\right)}$ **40.** $\dfrac{\left(\dfrac{x^2-1}{x}\right)}{\left(\dfrac{(x-1)^2}{x}\right)}$

41. $\dfrac{5}{x-1} + \dfrac{x}{x-1}$ **42.** $\dfrac{2x-1}{x+3} + \dfrac{1-x}{x+3}$

43. $6 - \dfrac{5}{x+3}$ **44.** $\dfrac{3}{x-1} - 5$

45. $\dfrac{3}{x-2} + \dfrac{5}{2-x}$ **46.** $\dfrac{2x}{x-5} - \dfrac{5}{5-x}$

47. $\dfrac{2}{x^2-4} - \dfrac{1}{x^2-3x+2}$

48. $\dfrac{x}{x^2+x-2} - \dfrac{1}{x+2}$ **49.** $-\dfrac{1}{x} + \dfrac{2}{x^2+1} + \dfrac{1}{x^3+x}$

50. $\dfrac{2}{x+1} + \dfrac{2}{x-1} + \dfrac{1}{x^2-1}$

51. $x^2(x^2-1)^{-1/2} + (x^2-1)^{1/2}$

52. $(1-x^3)^{1/3} - x^3(1-x^3)^{-2/3}$

53. $3(x-2)^{-1/3} - x(x-2)^{-4/3}$

54. $2(2+x)^{-1/2} + (1-x)(2+x)^{-3/2}$

In Exercises 55–66, simplify the compound fraction.

55. $\dfrac{\left(\dfrac{x}{2}-1\right)}{x-2}$ **56.** $\dfrac{x-4}{\left(\dfrac{x}{4}-\dfrac{4}{x}\right)}$

57. $\dfrac{\left(\dfrac{1}{x}-\dfrac{1}{x+1}\right)}{\left(\dfrac{1}{x+1}\right)}$ **58.** $\dfrac{\left(\dfrac{5}{y}-\dfrac{6}{2y+1}\right)}{\left(\dfrac{5}{y}+4\right)}$

59. $\dfrac{\left(\dfrac{1}{(x+h)^2}-\dfrac{1}{x^2}\right)}{h}$ **60.** $\dfrac{\left(\dfrac{x+h}{x+h+1}-\dfrac{x}{x+1}\right)}{h}$

61. $\dfrac{2a^{-1}+b^{-1}}{ab}$ **62.** $\dfrac{z^{-1}-2(z+2)^{-1}}{z^{-2}}$

63. $\dfrac{\left(\sqrt{x}-\dfrac{1}{2\sqrt{x}}\right)}{\sqrt{x}}$ **64.** $\dfrac{3x^{1/3}-x^{-2/3}}{3x^{-2/3}}$

65. $\dfrac{\left(\dfrac{t^2}{\sqrt{t^2+1}}-\sqrt{t^2+1}\right)}{t^2}$

66. $\dfrac{-x^3(1-x^2)^{-1/2}-2x(1-x^2)^{1/2}}{x^4}$

In Exercises 67 and 68, rationalize the numerator of the expression.

67. $\dfrac{\sqrt{x+2}-\sqrt{x}}{2}$ **68.** $\dfrac{\sqrt{z-3}-\sqrt{z}}{3}$

7   THE CARTESIAN PLANE

**The Cartesian Plane   /   The Distance Between Two Points in the Plane   /
The Midpoint Formula   /   The Equation of a Circle**

The Cartesian Plane

Just as real numbers are represented by points on the real number line, ordered pairs of real numbers are represented by points in a plane. This plane is called a **rectangular coordinate system** or the **Cartesian plane,** after the French mathematician René Descartes (1596–1650).

The Cartesian plane is formed by two real number lines intersecting at right angles, as shown in Figure 6(a). The horizontal number line is usually called the *x*-**axis** and the vertical number line is usually called the *y*-**axis.** (The plural of axis is *axes.*) The point of intersection of the two axes is the **origin.** The axes separate the plane into four regions called **quadrants.**

REMARK   It is customary to use the notation (x, y) to denote both a point in the plane and an open interval on the real number line. The nature of a specific problem will show which of the two we are talking about.

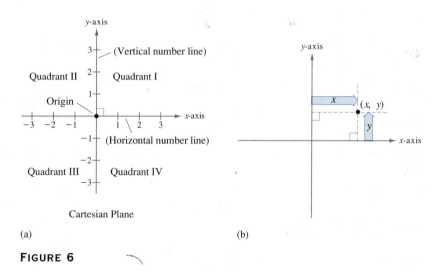

Cartesian Plane

(a)   (b)

FIGURE 6

Each point in the plane corresponds to an **ordered pair** (x, y) of real numbers x and y, called **coordinates** of the point. The first number (*x*-**coordinate**) tells how far to the left or right the point is from the vertical axis, and the second number (*y*-**coordinate**) tells how far up or down the point is from the horizontal axis, as shown in Figure 6(b). Figure 7 shows several points that have been plotted in the plane.

The rectangular coordinate system allows you to visualize relationships between variables x and y. Today, Descartes's ideas are commonly used in every scientific and business-related field.

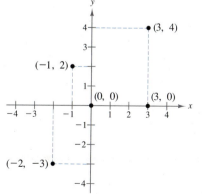

FIGURE 7

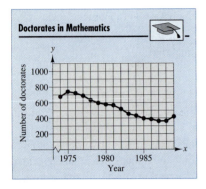

Doctorates in Mathematics

FIGURE 8

EXAMPLE 1   Number of Doctorates in Mathematics

The numbers of doctorates in mathematics granted to United States citizens by universities in the United States in the years 1974 to 1989 are given in Table 1. Plot these points in a rectangular coordinate system. (*Source*: American Mathematical Society)

TABLE 1

Year	1974	1975	1976	1977	1978	1979	1980	1981
Degrees	677	741	722	689	634	596	578	567
Year	1982	1983	1984	1985	1986	1987	1988	1989
Degrees	519	455	433	396	386	362	363	419

SOLUTION

The points are shown in Figure 8. Note that the break in the x-axis indicates that the numbers for years prior to 1974 have been omitted.

The Distance Between Two Points in the Plane

Let (x_1, y_1) and (x_2, y_2) represent two points in the plane that do not lie on the same horizontal or vertical line. With these two points, a right triangle can be formed, as shown in Figure 9. Note that the third vertex of the triangle is (x_1, y_2). Because (x_1, y_1) and (x_1, y_2) lie on the same vertical line, the length of the vertical side of the triangle is $|y_2 - y_1|$. Similarly, the length of the horizontal side is $|x_2 - x_1|$. Thus, by the Pythagorean Theorem, the square of the distance d between (x_1, y_1) and that (x_2, y_2) is

$$d^2 = |x_2 - x_1|^2 + |y_2 - y_1|^2.$$

Because the distance d must be positive, choose the positive square root and write

$$d = \sqrt{|x_2 - x_1|^2 + |y_2 - y_1|^2}.$$

Finally, replacing $|x_2 - x_1|^2$ and $|y_2 - y_1|^2$ by the equivalent expressions $(x_2 - x_1)^2$ and $(y_2 - y_1)^2$ gives the following formula for the distance between two points in a rectangular coordinate plane.

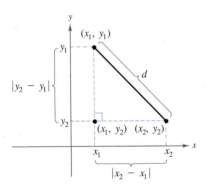

Distance Between Two Points

FIGURE 9

THE DISTANCE FORMULA

The distance d between the two points (x_1, y_1) and (x_2, y_2) in the coordinate plane is

$$d = \sqrt{(x_2 - x_1)^2 + (y_2 - y_1)^2}.$$

EXAMPLE 2   Finding the Distance Between Two Points

Find the distance between the points $(-2, 1)$ and $(3, 4)$.

SOLUTION

Letting $(x_1, y_1) = (-2, 1)$ and $(x_2, y_2) = (3, 4)$, apply the Distance Formula to obtain

$$d = \sqrt{[3 - (-2)]^2 + (4 - 1)^2}$$
$$= \sqrt{5^2 + 3^2}$$
$$= \sqrt{25 + 9}$$
$$= \sqrt{34}$$
$$\approx 5.83.$$

See Figure 10.

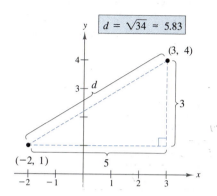

$d = \sqrt{34} \approx 5.83$

FIGURE 10

EXAMPLE 3   An Application of the Distance Formula

Show that the points $(2, 1)$, $(4, 0)$, and $(5, 7)$ are vertices of a right triangle.

SOLUTION

The three points are plotted in Figure 11. Using the Distance Formula, you can find the lengths of the three sides of the triangle:

$$d_1 = \sqrt{(5 - 2)^2 + (7 - 1)^2} = \sqrt{9 + 36} = \sqrt{45}$$
$$d_2 = \sqrt{(4 - 2)^2 + (0 - 1)^2} = \sqrt{4 + 1} = \sqrt{5}$$
$$d_3 = \sqrt{(5 - 4)^2 + (7 - 0)^2} = \sqrt{1 + 49} = \sqrt{50}.$$

Because $d_1^2 + d_2^2 = 45 + 5 = 50 = d_3^2$, you can conclude from the Pythagorean Theorem that the triangle is a right triangle.

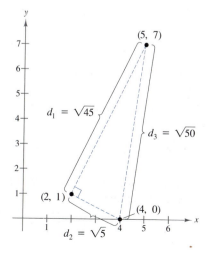

FIGURE 11

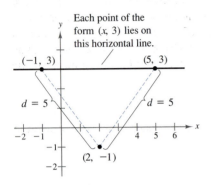

Each point of the form $(x, 3)$ lies on this horizontal line.

$(-1, 3)$ $(5, 3)$

$d = 5$ $d = 5$

$(2, -1)$

FIGURE 12

EXAMPLE 4 **Finding Points at a Specified Distance from a Given Point**

Find x so that the distance between $(x, 3)$ and $(2, -1)$ is 5.

SOLUTION

$$\sqrt{(x-2)^2 + (3+1)^2} = 5 \qquad \textit{Distance Formula}$$
$$(x^2 - 4x + 4) + 16 = 25 \qquad \textit{Square both sides}$$
$$x^2 - 4x - 5 = 0 \qquad \textit{Standard form}$$
$$(x - 5)(x + 1) = 0 \qquad \textit{Factor}$$
$$x - 5 = 0 \;\longrightarrow\; x = 5 \qquad \textit{Set 1st factor equal to 0}$$
$$x + 1 = 0 \;\longrightarrow\; x = -1 \qquad \textit{Set 2nd factor equal to 0}$$

There are two solutions: $(5, 3)$ and $(-1, 3)$. Note that each of the points $(5, 3)$ and $(-1, 3)$ lies five units from the point $(2, -1)$, as shown in Figure 12.

The Midpoint Formula

The coordinates of the midpoint of a line segment are the average values of the corresponding coordinates of the two endpoints.

THE MIDPOINT FORMULA

The **midpoint** of the line segment joining the points (x_1, y_1) and (x_2, y_2) in the coordinate plane is

$$\left(\frac{x_1 + x_2}{2}, \frac{y_1 + y_2}{2} \right).$$

EXAMPLE 5 **Finding the Midpoint of a Line Segment**

Find the midpoint of the line segment joining the points $(-5, -3)$ and $(9, 3)$.

SOLUTION

Figure 13 shows the two given points and their midpoint. Using the Midpoint Formula, you can write

$$\text{Midpoint} = \left(\frac{-5 + 9}{2}, \frac{-3 + 3}{2} \right)$$
$$= (2, 0).$$

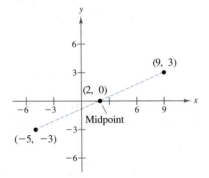

$(9, 3)$

$(2, 0)$

Midpoint

$(-5, -3)$

FIGURE 13

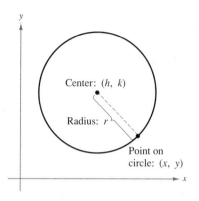

FIGURE 14

The Equation of a Circle

The Distance Formula provides a convenient way to define circles. A **circle of radius** r with **center** at the point (h, k) is shown in Figure 14. The point (x, y) is on this circle if and only if its distance from the center (h, k) is r. This means that a **circle** in the plane consists of all points (x, y) that are a given positive distance r from a fixed point (h, k). Using the Distance Formula, you can express this relationship by saying that the point (x, y) lies on the circle if and only if

$$\sqrt{(x - h)^2 + (y - k)^2} = r.$$

By squaring both sides of this equation, you can obtain the **standard form of the equation of a circle.**

STANDARD FORM OF THE EQUATION OF A CIRCLE

The **standard form of the equation of a circle** is

$$(x - h)^2 + (y - k)^2 = r^2.$$

The point (h, k) is the **center** of the circle, and the positive number r is the **radius** of the circle. The standard form of the equation of a circle whose center is the *origin* is $x^2 + y^2 = r^2$.

EXAMPLE 6   Finding an Equation of a Circle

The point $(3, 4)$ lies on a circle whose center is at $(-1, 2)$, as shown in Figure 15. Find an equation for the circle.

SOLUTION

The radius r of the circle is the distance between $(-1, 2)$ and $(3, 4)$.

$$r = \sqrt{[3 - (-1)]^2 + (4 - 2)^2}$$
$$= \sqrt{16 + 4}$$
$$= \sqrt{20}$$

Thus, the center of the circle is $(h, k) = (-1, 2)$ and the radius is $r = \sqrt{20}$, and you can write the standard form of the equation of the circle as follows.

$$(x - h)^2 + (y - k)^2 = r^2 \qquad \textit{Standard form}$$
$$[x - (-1)]^2 + (y - 2)^2 = (\sqrt{20})^2 \qquad \textit{Let } h = -1, k = 2, \text{ and } r = \sqrt{20}$$
$$(x + 1)^2 + (y - 2)^2 = 20 \qquad \textit{Equation of circle}$$

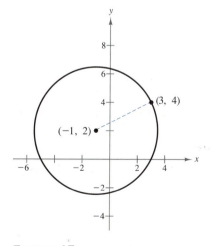

FIGURE 15

SECTION 7 · EXERCISES

In Exercises 1–4, sketch the polygon with the indicated vertices.

1. Triangle: $(-1, 1)$, $(2, -1)$, $(3, 4)$
2. Triangle: $(0, 3)$, $(-1, -2)$, $(4, 8)$
3. Square: $(2, 4)$, $(5, 1)$, $(2, -2)$, $(-1, 1)$
4. Parallelogram: $(5, 2)$, $(7, 0)$, $(1, -2)$, $(-1, 0)$

In Exercises 5–8, find the distance between the points (*Note:* In each case, the two points lie on the same horizontal or vertical line.)

5. $(6, -3)$, $(6, 5)$
6. $(1, 4)$, $(8, 4)$
7. $(-3, -1)$, $(2, -1)$
8. $(-3, -4)$, $(-3, 6)$

In Exercises 9–12, (a) find the lengths of the two perpendicular sides of the right triangle and use the Pythagorean Theorem to find the length of the hypotenuse, and (b) use the Distance Formula to find the length of the hypotenuse of the triangle.

9.
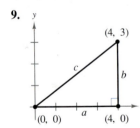

10.
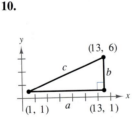

11.
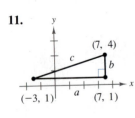

12.
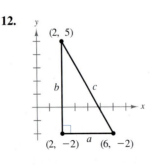

In Exercises 13–24, (a) plot the points, (b) find the distance between the points, and (c) find the midpoint of the line segment joining the points.

13. $(1, 1)$, $(9, 7)$
14. $(1, 12)$, $(6, 0)$
15. $(-4, 10)$, $(4, -5)$
16. $(-7, -4)$, $(2, 8)$
17. $(-1, 2)$, $(5, 4)$
18. $(2, 10)$, $(10, 2)$
19. $(\frac{1}{2}, 1)$, $(-\frac{5}{2}, \frac{4}{3})$
20. $(-\frac{1}{3}, -\frac{1}{3})$, $(-\frac{1}{6}, -\frac{1}{2})$
21. $(6.2, 5.4)$, $(-3.7, 1.8)$
22. $(-16.8, 12.3)$, $(5.6, 4.9)$
23. $(-36, -18)$, $(48, -72)$
24. $(1.451, 3.051)$, $(5.906, 11.360)$

In Exercises 25 and 26, use the Midpoint Formula to estimate the sales of a company for 1991, given the sales in 1989 and 1993. Assume the sales followed a linear pattern.

25.

Year	1989	1993
Sales	$520,000	$740,000

26.

Year	1989	1993
Sales	$4,200,000	$5,650,000

In Exercises 27–30, show that the points form the vertices of the indicated polygon. (A rhombus is a parallelogram whose sides are all of the same length.)

27. Right triangle: $(4, 0)$, $(2, 1)$, $(-1, -5)$
28. Isosceles triangle: $(1, -3)$, $(3, 2)$, $(-2, 4)$
29. Rhombus: $(0, 0)$, $(1, 2)$, $(2, 1)$, $(3, 3)$
30. Parallelogram: $(0, 1)$, $(3, 7)$, $(4, 4)$, $(1, -2)$

In Exercises 31 and 32, find x so that the distance between the points is 13.

31. $(1, 2)$, $(x, -10)$
32. $(-8, 0)$, $(x, 5)$

In Exercises 33 and 34, find y so that the distance between the points is 17.

33. $(0, 0)$, $(8, y)$
34. $(-8, 4)$, $(7, y)$

In Exercises 35 and 36, find an equation that relates x and y so that (x, y) is equidistant from the two given points.

35. $(4, -1)$, $(-2, 3)$
36. $(3, \frac{5}{2})$, $(-7, 1)$

In Exercises 37–48, determine the quadrant(s) in which (x, y) is located so that the given conditions are satisfied.

37. $x > 0$ and $y < 0$
38. $x < 0$ and $y < 0$
39. $x > 0$ and $y > 0$
40. $x < 0$ and $y > 0$
41. $x = -4$ and $y > 0$
42. $x > 2$ and $y = 3$
43. $y < -5$
44. $x > 4$
45. $xy > 0$
46. $xy < 0$
47. $(x, -y)$ is in the second quadrant.
48. $(-x, y)$ is in the fourth quadrant.

49. Plot the points $(2, 1)$, $(-3, 5)$, and $(7, -3)$ on the rectangular coordinate system. Now plot the corresponding points when the sign of the x-coordinate is negated. What inference can you make about the result of the location of a point when the sign of the x-coordinate is changed?

50. Plot the points $(2, 1)$, $(-3, 5)$, and $(7, -3)$ on the rectangular coordinate system. Now plot the corresponding points when the sign of the y-coordinate is negated. What inference can you make about the result of the location of a point when the sign of the y-coordinate is changed?

In Exercises 51–58, find the standard form of the equation of the specified circle.

51. Center: $(0, 0)$; radius: 3
52. Center: $(0, 0)$; radius: 5
53. Center: $(2, -1)$; radius: 4
54. Center: $(0, \frac{1}{3})$; radius: $\frac{1}{3}$
55. Center: $(-1, 2)$; solution point: $(0, 0)$
56. Center: $(3, -2)$; solution point: $(-1, 1)$
57. Endpoints of a diameter: $(0, 0)$, $(6, 8)$
58. Endpoints of a diameter: $(-4, -1)$, $(4, 1)$

8   EXPLORING DATA: REPRESENTING DATA GRAPHICALLY

Line Plots / Stem and Leaf Plots / Histograms and Frequency Distributions / Line Graphs

Line Plots

Statistics is the branch of mathematics that studies techniques for collecting organizing, and interpreting data. In this section, you will study several ways to organize data. The first is a **line plot,** which uses a portion of a real number line to order numbers. Line plots are especially useful for ordering small sets of numbers (about 50 or less) by hand.

EXAMPLE 1   Constructing a Line Plot

Use a line plot to organize the following test scores. Which score occurred with the greatest frequency?

93, 70, 76, 58, 86, 93, 82, 78, 83, 86, 64, 78, 76, 66, 83, 83, 96, 74, 69, 76, 64, 74, 79, 76, 88, 76, 81, 82, 74, 70

SOLUTION

Begin by scanning the data to find the smallest and largest numbers. For these data, the smallest number is 58 and the largest is 96. Next, draw a portion of a real number line that includes the interval [58, 96]. To create the line plot, start with the first number, 93, and enter an $\times$ above 93 on the number line.

Technology Note _____

Many computer programs and calculators will sort data. Try using a computer or calculator to sort the data in Example 1.

Continue recording ×'s for each number in the list until you obtain the line plot shown in Figure 16. From the line plot, you can see that 76 had the greatest frequency.

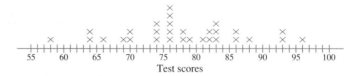

FIGURE 16

Stem and Leaf Plots

Another type of plot that can be used to organize sets of numbers by hand is a **stem-and-leaf plot.** A stem-and-leaf plot for the test scores in Example 1 is shown below.

Stems	Leaves
5	8
6	4 4 6 9
7	0 0 4 4 4 6 6 6 6 6 8 8 9
8	1 2 2 3 3 3 6 6 8
9	3 3 6

Note that the *leaves* represent the units digits of the numbers and the *stems* represent the tens digits. Stem-and-leaf plots can also be used to compare two sets of data, as shown in the next example.

EXAMPLE 2 Comparing Two sets of Data

Use a stem-and-leaf plot to compare the test scores in Example 1 with the following test scores. Which set of test scores is better?

90, 81, 70, 62, 64, 73, 81, 92, 73, 81, 92, 93, 83, 75, 76, 83, 94, 96, 86, 77, 77, 86, 96, 86, 77, 86, 87, 87, 79, 88

SOLUTION

Begin by ordering the second set of scores. (You could use a line plot, as shown in Example 1.)

62, 64, 70, 73, 73, 75, 76, 77, 77, 77, 79, 81, 81, 81, 83, 83, 86, 86, 86, 86, 87, 87, 88, 90, 92, 92, 93, 94, 96, 96

Now that the data have been ordered, you can construct a *double* stem-and-leaf plot by letting the leaves to the right of the stems represent the units digits for the first group of test scores and letting the leaves to the left of the stems represent the units digits for the second group of test scores.

Leaves (2nd Group)	Stems	Leaves (1st Group)
	5	8
4 2	6	4 4 6 9
9 7 7 7 6 5 3 3 0	7	0 0 4 4 4 6 6 6 6 6 8 8 9
8 7 7 6 6 6 6 3 3 1 1 1	8	1 2 2 3 3 3 6 6 8
6 6 4 3 2 2 0	9	3 3 6

From the leaves on the left, you can see that the second group of test scores is better than the first group.

EXAMPLE 3   Using a Stem-and-Leaf Plot

Table 2 shows the percent of the population of each state and the District of Columbia that was 65 or older in 1989. Use a stem-and-leaf plot to organize the data. (*Source:* U.S. Bureau of Census)

TABLE 2

AK	4.1	AL	12.7	AR	14.8	AZ	13.1	CA	10.6
CO	9.8	CT	13.6	DC	12.5	DE	11.8	FL	18.0
GA	10.1	HI	10.7	IA	15.1	ID	11.9	IL	12.3
IN	12.4	KS	13.7	KY	12.7	LA	11.1	MA	13.8
MD	10.8	ME	13.4	MI	11.9	MN	12.6	MO	13.9
MS	12.4	MT	13.2	NC	12.1	ND	13.9	NE	13.9
NH	11.4	NJ	13.2	NM	10.5	NV	10.9	NY	13.0
OH	12.8	OK	13.3	OR	13.9	PA	15.1	RI	14.8
SC	11.1	SD	14.4	TN	12.6	TX	10.1	UT	8.6
VA	10.8	VT	11.9	WA	11.9	WI	13.4	WV	14.6
WY	9.8								

SOLUTION

Begin by ordering the numbers, as shown below.

4.1, 8.6, 9.8, 9.8, 10.1, 10.1, 10.5, 10.6, 10.7, 10.8, 10.8,
10.9, 11.1, 11.1, 11.4, 11.8, 11.9, 11.9, 11.9, 11.9, 12.1, 12.3,
12.4, 12.4, 12.5, 12.6, 12.6, 12.7, 12.7, 12.8, 13.0, 13.1, 13.2,
13.2, 13.3, 13.4, 13.4, 13.6, 13.7, 13.8, 13.9, 13.9, 13.9, 13.9,
14.4, 14.6, 14.8, 14.8, 15.1, 15.1, 18.0

Next construct the stem-and-leaf plot using the leaves to represent the digits to the right of the decimal points.

Stems	Leaves
4.	1 *Alaska has the lowest percent.*
5.	
6.	
7.	
8.	6
9.	8 8
10.	1 1 5 6 7 8 8 9
11.	1 1 4 8 9 9 9 9
12.	1 3 4 4 5 6 6 7 7 8
13.	0 1 2 2 3 4 4 6 7 8 9 9 9 9
14.	4 6 8 8
15.	1 1
16.	
17.	
18.	0 *Florida has the highest percent.*

Histograms and Frequency Distributions

With data such as that given in Example 3, it is useful to group the numbers into intervals and plot the frequency of the data in each interval. For instance, the **frequency distribution** and **histogram** shown in Figure 17 represent the data given in Example 3.

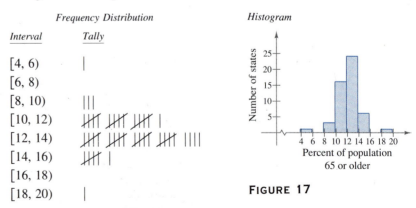

Interval	Tally
[4, 6)	|
[6, 8)	
[8, 10)	|||
[10, 12)	JHT JHT JHT |
[12, 14)	JHT JHT JHT JHT ||||
[14, 16)	JHT |
[16, 18)	
[18, 20)	|

FIGURE 17

Technology Note

Try using a computer or graphing calculator to create a histogram for the data at the right. How does the histogram change when the intervals change?

A histogram has a portion of a real number line as its horizontal axis. A **bar graph** is similar to a histogram, except that the rectangles (bars) can be either horizontal or vertical and the labels of the bars are not necessarily numbers.

Another difference between a bar graph and a histogram is that the bars in a bar graph are usually separated by spaces, whereas the bars in a histogram are not separated by spaces.

EXAMPLE 4 Constructing a Bar Graph

The data below show the average monthly precipitation (in inches) in Houston, Texas. Construct a bar graph for these data. What can you conclude? (*Source: PC USA*)

January	3.2	February	3.3	March	2.7
April	4.2	May	4.7	June	4.1
July	3.3	August	3.7	September	4.9
October	3.7	November	3.4	December	3.7

SOLUTION

To create a bar graph, begin by drawing a vertical axis to represent the precipitation and a horizontal axis to represent the months. The bar graph is shown in Figure 18. From the graph, you can see that Houston receives a fairly consistent amount of rain throughout the year—the driest month tends to be March and the wettest month tends to be September.

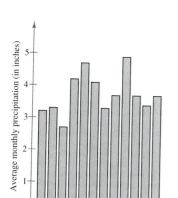

FIGURE 18

The next example shows how to use a *double* bar graph to compare two sets of data.

EXAMPLE 5 Constructing a Double Bar Graph

Table 3 shows the percent of bachelor's degrees awarded to males and females for selected majors in the United States in 1988. Construct a double bar graph for these data.

TABLE 3

Field of study	% female	% male
Agriculture and natural resources	31.5	68.5
Area and ethnic studies	59.8	40.2
Education	76.9	23.1
Engineering	13.7	86.3
Home economics	91.7	8.3
Liberal/general studies	56.4	43.6
Mathematics	46.4	53.6
Military sciences	5.2	94.8
Physical sciences	30.4	69.6
Social sciences	43.9	56.1

SOLUTION

For these data, a horizontal bar graph seems to be appropriate. Such a graph is shown in Figure 19.

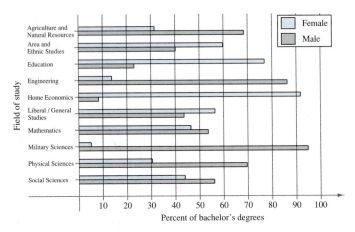

FIGURE 19

Line Graphs

A **line graph** is similar to a standard coordinate graph. Line graphs are usually used to show trends over periods of time.

EXAMPLE 6   Constructing a Line Graph

The following data show the number of immigrants (in thousands) entering the United States by decade. Construct a line graph of these data. What can you conclude? (*Source:* U.S. Bureau of Census)

Decade	Number	Decade	Number
1851–1860	2598	1861–1870	2315
1871–1880	2812	1881–1890	5247
1891–1900	3688	1901–1910	8795
1911–1920	5736	1921–1930	4107
1931–1940	528	1941–1950	1035
1951–1960	2515	1961–1970	3322
1971–1980	4493	1981–1990	6447

SOLUTION

Begin by drawing a vertical axis to represent the number of immigrants in thousands. Then label the horizontal axis with decades and plot the points shown in the table. Finally, connect the points with line segments, as shown in Figure 20. From the line graph, you can see that the number of immigrants hits a low point during the depression of the 1930s. Since then the numbers have steadily increased.

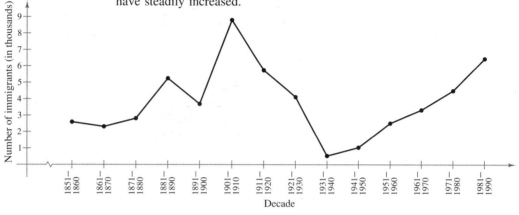

FIGURE 20

SECTION 8 · EXERCISES

Quiz and Exam Scores In Exercises 1–6, use the following scores from a math class with 30 students. The scores are given for two 25-point quizzes and two 100-point exams.

Quiz #1 20, 15, 14, 20, 16, 19, 10, 21, 24, 15, 15, 14, 15, 21, 19, 15, 20, 18, 18, 22, 18, 16, 18, 19, 21, 19, 16, 20, 14, 12

Quiz #2 22, 22, 23, 22, 21, 24, 22, 19, 21, 23, 23, 25, 24, 22, 22, 23, 23, 23, 23, 22, 24, 23, 22, 24, 21, 24, 16, 21, 16, 14

Exam #1 77, 100, 77, 70, 83, 89, 87, 85, 81, 84, 81, 78, 89, 78, 88, 85, 90, 92, 75, 81, 85, 100, 98, 81, 78, 75, 85, 89, 82, 75

Exam #2 76, 78, 73, 59, 70, 81, 71, 66, 66, 73, 68, 67, 63, 67, 77, 84, 87, 71, 78, 78, 90, 80, 77, 70, 80, 64, 74, 68, 68, 68

1. Construct a line plot for Quiz #1. Which score occurred with the greatest frequency?

2. Construct a line plot for Quiz #2. Which score occurred with the greatest frequency?

3. Construct a line plot for Exam #1. Which score occurred with the greatest frequency?

4. Construct a line plot for Exam #2. Which score occurred with the greatest frequency?

5. Construct a stem-and-leaf plot for Exam #1.

6. Construct a double stem-and-leaf plot to compare the scores for Exam #1 and Exam #2. Which set of test scores is higher?

7. *Education Expenses* The following table shows the per capita expenditures for public elementary and secondary education in the 50 states and the District of Columbia in 1991. Use a stem-and-leaf plot to organize the data. (*Source:* National Education Association)

AK	1626	AL	694	AR	668	AZ	892	CA	918
CO	841	CT	1151	DC	1010	DE	891	FL	862
GA	859	HI	784	IA	846	ID	725	IL	788
IN	925	KS	906	KY	725	LA	758	MA	866
MD	944	ME	1062	MI	926	MN	990	MO	742
MS	671	MT	983	NC	813	ND	719	NE	757
NH	881	NJ	1223	NM	915	NV	1004	NY	1186
OH	861	OK	776	OR	925	PA	889	RI	892
SC	835	SD	716	TN	618	TX	905	UT	828
VA	941	VT	992	WA	1095	WI	928	WV	883
WY	1178								

8. *Snowfall* The data below show the seasonal snowfall (in inches) at Erie, Pennsylvania for the years 1960 through 1989 (the amounts are listed in order by year). How would you organize these data? Explain your reasoning. (*Source: National Oceanic and Atmospheric Administration*)

69.6, 42.5, 75.9, 115.9, 92.9, 84.8, 68.6, 107.9, 79.7, 85.6, 120.0, 92.3, 53.7, 68.6, 66.7, 66.0, 111.5, 142.8, 76.5, 55.2, 89.4, 71.3, 41.2, 110.0, 106.3, 124.9, 68.2, 103.5, 76.5, 114.9

9. *Fruit Crops* The data below show the cash receipts (in millions of dollars) from fruit crops for farmers in 1990. Construct a bar graph for these data. (*Source: U.S. Department of Agriculture*)

Apples	1159	Peaches	365
Grapefruit	317	Pears	266
Grapes	1668	Plums and Prunes	293
Lemons	278	Strawberries	560
Oranges	1707		

10. *Travel to the United States* The data below give the places of origin and the numbers of travelers (in millions) to the United States in 1991. Construct a horizontal bar graph for these data. (*Source: U.S. Travel and Tourism Administration*)

Canada	18.9	Mexico	7.0
Europe	7.4	Latin America	2.0
Other	6.8		

11. *Sports Participants* The following table shows the numbers of males and the numbers of females (in millions) over the age of seven that participated in popular sports activities in the year 1990 in the United States. Construct a double bar graph for these data. (*Source: National Sporting Goods Association*)

Activity	Male	Female
Exercise walking	25.1	46.3
Swimming	31.5	35.9
Bicycling	28.2	27.1
Camping	24.4	21.8
Bowling	20.9	19.2
Basketball	18.9	7.4
Running	13.1	10.8
Aerobic exercising	3.7	19.5

12. *College Attendance* The following table shows the enrollment in a liberal arts college. Construct a line graph for these data.

Year	1985	1986	1987	1988
Enrollment	1675	1704	1710	1768

Year	1989	1990	1991	1992
Enrollment	1833	1918	1967	1972

13. *Oil Imports* The following table shows the amounts of crude oil imported into the United States (in millions of barrels) for the years 1982 through 1990. Construct a line graph for these data and state what information they reveal. (*Source: U.S. Energy Information Administration*)

Year	1982	1983	1984	1985	1986
Oil Imports	1273	1215	1254	1168	1525

Year	1987	1988	1989	1990	
Oil Imports	1706	1869	2133	2145	

14. *Federal Income* The following data show the receipts (in billions of dollars) for the federal government of the United States. Construct a line graph for these data. What can you conclude? (*Source: U.S. Office of Management and Budget*)

Year	Receipts	Year	Receipts
1972	207.3	1983	600.6
1973	230.8	1984	666.5
1974	263.2	1985	734.1
1975	279.1	1986	769.1
1976	298.1	1987	854.1
1977	355.6	1988	909.0
1978	399.6	1989	990.7
1979	463.3	1990	1031.3
1980	517.1	1991	1054.3
1981	599.3	1992	1075.7
1982	617.8		

FUNCTIONS AND GRAPHS

1.1 GRAPHS AND GRAPHING UTILITIES

The Graph of an Equation / Using a Graphing Utility / Determining a Viewing Rectangle / Applications

The Graph of an Equation

News magazines often show graphs comparing the rate of inflation, the federal deficit, wholesale prices, or the unemployment rate to the time of year. Industrial firms and businesses use graphs to report their monthly production and sales statistics. Such graphs provide geometric pictures of the way one quantity changes with respect to another.

Frequently, the relationship between two quantities is expressed as an equation. This section introduces the basic procedure for determining the geometric picture associated with an equation.

For an equation in variables x and y, a point (a, b) is a **solution point** if the substitution of $x = a$ and $y = b$ satisfies the equation. Most equations have *infinitely* many solution points. For example, the equation $3x + y = 5$ has solution points $(0, 5)$, $(1, 2)$, $(2, -1)$, $(3, -4)$, and so on. The set of all solution points of an equation is the **graph** of the equation.

THE POINT-PLOTTING METHOD OF GRAPHING

To sketch the graph of an equation by point plotting, use the following steps.

1. If possible, rewrite the equation so that one of the variables is isolated on the left side of the equation.
2. Make up a table of several solution points.
3. Plot these points in the coordinate plane.
4. Connect the points with a smooth curve.

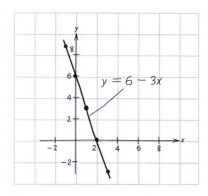

FIGURE 1.1

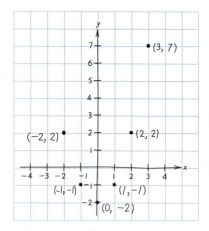

Plot several points.

FIGURE 1.2

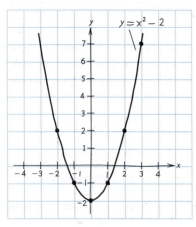

Connect points with a smooth curve.

FIGURE 1.3

EXAMPLE 1   **Sketching the Graph of an Equation by Point Plotting**

Use point plotting and graph paper to sketch the graph of $3x + y = 6$.

SOLUTION

In this case you can isolate the variable y to obtain

$y = 6 - 3x.$

Using negative, zero, and positive values for x, you can obtain the following table of values (solution points).

x	-1	0	1	2	3
$y = 6 - 3x$	9	6	3	0	-3

Next, plot these points and connect them as shown in Figure 1.1. It appears that the graph is a straight line. (We will discuss lines extensively in Section 1.2.)

━━━━━━━

The points at which a graph touches or crosses an axis are the **intercepts** of the graph. For instance, in Example 1 the point $(0, 6)$ is the y-intercept of the graph because the graph crosses the y-axis at that point. The point $(2, 0)$ is the x-intercept of the graph because the graph crosses the x-axis at that point.

EXAMPLE 2   **Sketching the Graph of an Equation by Point Plotting**

Use point plotting and graph paper to sketch the graph of $y = x^2 - 2$.

SOLUTION

First, make a table of values by choosing several convenient values of x and calculating the corresponding values of y.

x	-2	-1	0	1	2	3
$y = x^2 - 2$	2	-1	-2	-1	2	7

Next, plot the corresponding solution points, as shown in Figure 1.2. Finally, connect the points with a smooth curve, as shown in Figure 1.3. This graph is called a **parabola.**

━━━━━━━

Using a Graphing Utility

One of the disadvantages of the point-plotting method is that to get a good idea about the shape of a graph you need to plot *many* points. With only a few points, you could badly misrepresent the graph. For instance, consider the equation

$$y = \frac{1}{30}x(x^4 - 10x^2 + 39).$$

Suppose you plotted only five points: $(-3, -3)$, $(-1, -1)$, $(0, 0)$, $(1, 1)$ and $(3, 3)$, as shown in Figure 1.4. From these five points, a person might assume that the graph of the equation is a straight line. That, however, is not correct. By plotting several more points, you can see that the actual graph is not straight at all! (See Figure 1.5.)

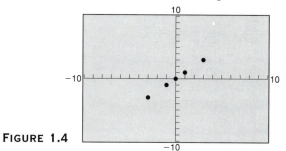

FIGURE 1.4

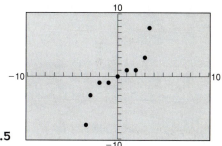

FIGURE 1.5

Thus, the point-plotting method leaves us with a dilemma. On the one hand, the method can be very inaccurate if only a few points are plotted. But on the other hand, it is very time consuming to plot a dozen (or more) points. Technology can help solve this dilemma. Plotting several (even several hundred) points in a rectangular coordinate system is something that a computer or calculator can do easily.

The point-plotting method is the method used by *all* graphing utilities. Each computer or calculator screen is made up of a grid of hundreds or thousands of small areas called **pixels.** Screens that have many pixels per inch are said to have a higher **resolution** than screens that don't have as many. Screens on most graphing calculators have 48 pixels per inch, whereas screens on computer monitors typically have between 32 and 100 pixels per inch.

With most graphing utilities, you can graph an equation by using the following steps.

USING A GRAPHING UTILITY TO GRAPH AN EQUATION

To graph an equation in x and y on a graphing utility, use the following steps.

1. Rewrite the equation so that y is isolated on the left side of the equation.
2. Enter the equation into a graphing utility.
3. Determine a **viewing window** for the graph. For some graphing utilities, the standard viewing window ranges between -10 and 10 for both the x and y values.
4. Activate the graphing utility.

EXAMPLE 3 **Using a Graphing Utility**

Use a graphing utility to graph $2y + x^3 = 4x$.

SOLUTION

To begin, solve the equation for y in terms of x.

$$2y + x^3 = 4x \qquad \textit{Given equation}$$
$$2y = -x^3 + 4x \qquad \textit{Subtract } x^3 \textit{ from both sides}$$
$$y = -\frac{1}{2}x^3 + 2x \qquad \textit{Divide both sides by 2}$$

Now, by entering this equation into a graphing utility (using a standard viewing rectangle), you can obtain the graph shown in Figure 1.6.

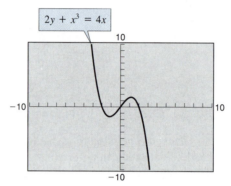

$2y + x^3 = 4x$

FIGURE 1.6

Determining a Viewing Rectangle

A **viewing rectangle** for a graph is a rectangular portion of the Cartesian plane. A viewing rectangle is determined by six values: the minimum x-value, the maximum x-value, the x-scale, the minimum y-value, the maximum y-value, and the y-scale. When you enter these six values into the graphing utility, you are setting the **range** of the viewing rectangle. The **standard** viewing rectangle for some graphing utilities is shown in Figure 1.7.

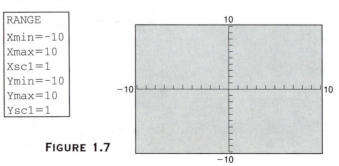

```
RANGE
Xmin=-10
Xmax=10
Xscl=1
Ymin=-10
Ymax=10
Yscl=1
```

FIGURE 1.7

By choosing different viewing rectangles for a graph, it is possible to obtain very different impressions of the graph's shape. For instance, Figure 1.8 shows four different viewing rectangles for the graph of

$$y = 0.1x^4 - x^3 + 2x^2.$$

Of these, the view shown in Figure 1.8(a) is the most complete because it shows more of the distinguishing portions of the graph.

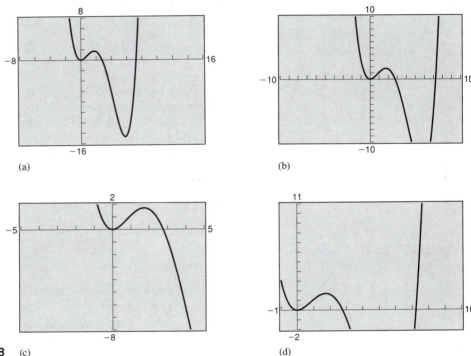

(a)

(b)

FIGURE 1.8 (c)

(d)

EXAMPLE 4 Sketching a Circle with a Graphing Utility

Use a graphing utility to graph $x^2 + y^2 = 9$.

SOLUTION

Technology Note

The standard viewing rectangle on many graphing utilities does not give a true geometric perspective. That is, perpendicular lines will not appear to be perpendicular and circles will not appear to be circular. To overcome this, you can use a square setting, as demonstrated in Example 4.

The graph of $x^2 + y^2 = 9$ is a circle whose center is the origin and whose radius is 3. (See Section 7 in preceding chapter on prerequisites.) To graph the equation, begin by solving the equation for y.

$$x^2 + y^2 = 9$$
$$y^2 = 9 - x^2$$
$$y = \pm\sqrt{9 - x^2}$$

The graph of $y = \sqrt{9 - x^2}$ is the upper semicircle. The graph of $y = -\sqrt{9 - x^2}$ is the lower semicircle. Enter *both* the equations in your

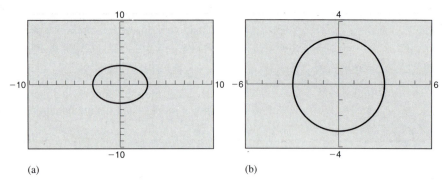

(a)   (b)

FIGURE 1.9

graphing utility and generate the resulting graphs. In Figure 1.9(a), note that if you use a standard viewing rectangle, the two graphs do not appear to form a circle. You can overcome this problem by using a **square setting,** in which the horizontal and vertical tick marks have equal spacing, as shown in Figure 1.9(b). On many graphing utilities, a square setting can be obtained by using the ratio $(Y_{max} - Y_{min})/(X_{max} - X_{min}) = 2/3$.

Applications

The following two applications show how to develop mathematical models to represent real-world situations. You will see that both a graphing utility and algebra can be used to understand and resolve the problems posed.

Once an appropriate viewing rectangle is chosen for a particular graph, the *zoom* and *trace* features of a graphing utility are useful for approximating specific values from the graph.

EXAMPLE 5   Using the Zoom and Trace Features of a Graphing Utility

Technology Note

In applications, it is convenient to use variable names that suggest real-life quantities; d for distance, t for time, and so on. Most graphing utilities, however, require the variable names to be x and y.

A runner runs at a constant rate of 4.8 miles per hour. The verbal model and algebraic equation relating distance run in terms of time are given by

VERBAL
MODEL   Distance = Rate · Time

EQUATION $d = 4.8t$

A. Use a graphing utility and an appropriate viewing rectangle to graph the equation $d = 4.8t$. (Represent d by y and t by x.)

B. Use the zoom and trace features to estimate how far the runner can run in 3.2 hours.

C. Use the zoom and trace features to estimate how long it will take to run a 26-mile marathon.

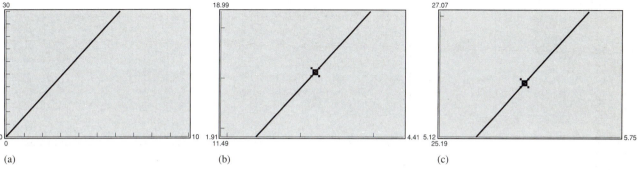

(a)   (b)   (c)

FIGURE 1.10

SOLUTION

A. An appropriate viewing rectangle and the associated graph are shown in Figure 1.10(a).

REMARK The viewing rectangle on your graphing utility may differ from those shown in parts (b) and (c) of Figure 1.10.

B. Figure 1.10(b) shows the viewing rectangle after zooming in (near $x = 3.2$) once by a factor of 4. Using the trace feature, you can determine that for $x = 3.2$, the distance is $y \approx 15.36$ miles.

C. Figure 1.10(c) shows the viewing rectangle after zooming in (near $y = 26$) twice by a factor of 4. Using the trace feature, you can determine that for $y = 26$, the time is $x \approx 5.42$ hours.

■■■■■■■■

EXAMPLE 6   An Application: Monthly Wages

You receive a monthly salary of $2000 plus a commission of 10% of sales.

A. Find an equation expressing the monthly wages y in terms of sales x.

B. If sales are $x = 1480$ in August, what are your wages for that month?

C. If you receive $2225 in wages for September, what were your sales for that month?

SOLUTION

A. The monthly wages are the sum of the fixed $2000 salary and the 10% commission on sales x.

VERBAL
MODEL   Wages = Salary + Commission on sales

EQUATION $y = 2000 + 0.1x$

B. If $x = 1480$, then the corresponding wages would be

$$y = 2000 + 0.1(1480) = 2000 + 148 = \$2148.$$

You can confirm this result using a graphing utility. Because $x \geq 0$ and the monthly wages are at least $2000, a reasonable viewing rectangle is shown below. [See Figure 1.11(a).] Using the zoom and trace features near $x = 1480$ shows that the wages are about $2148.

```
RANGE
Xmin=0
Xmax=3000
Xscl=300
Ymin=2000
Ymax=2300
Yscl=30
```

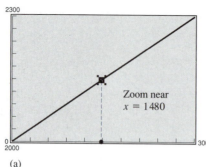

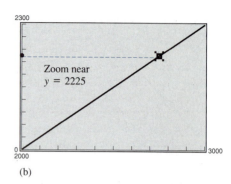

FIGURE 1.11   (a)   (b)

c. To answer the third question, you can use the graphing utility to find the value along the *x*-axis (sales) that corresponds to a *y*-value of 2225 (wages). Beginning with Figure 1.11(b) and using the zoom and trace features, you can estimate *x* to be about 2250. You can verify this answer by observing that

$$\text{Wages} = 2000 + 0.1(2250)$$
$$= 2000 + 225$$
$$= 2225.$$

DISCUSSION PROBLEM

COMPARISON OF WAGES

Michael receives a monthly salary of $3100 plus commission of 7% of sales, whereas Janet receives a salary of $3400 plus a 6% commission. Discuss their relative wages depending on sales. Can there be a month in which they receive exactly the same amount of wages? Be sure to take advantage of your graphing utility.

WARM-UP

The following warm-up exercises involve skills that were covered in earlier sections. You will use these skills in the exercise set for this section.

In Exercises 1–6, simplify the expression.

1. $4(x - 3) - 5(6 - 2x)$ **2.** $-s(5s + 2) + 5(s^2 - 3s)$

3. $3y(-2y^2)^3$ **4.** $\dfrac{18a^2b^{-3}}{27ab^2}$

5. $\sqrt{150x^4}$ **6.** $8^{-1/3}$

In Exercises 7–10, completely factor the expression.

7. $2x^3 - 6x^2$ **8.** $2t(t + 3)^2 - 4(t + 3)$

9. $6z^2 + 5z - 50$ **10.** $4s^2 - 25$

SECTION 1.1 · EXERCISES

In Exercises 1–6, determine whether the points lie on the graph of the equation.

Equation	Points	
*1. $y = \sqrt{x} + 4$	(a) $(0, 2)$	(b) $(5, 3)$
2. $y = x^2 - 3x + 2$	(a) $(2, 0)$	(b) $(-2, 8)$
3. $2x - y - 3 = 0$	(a) $(1, 2)$	(b) $(1, -1)$
4. $x^2 + y^2 = 20$	(a) $(3, -2)$	(b) $(-4, 2)$
5. $x^2y - x^2 + 4y = 0$	(a) $(1, \frac{1}{5})$	(b) $(2, \frac{1}{2})$
6. $y = \dfrac{1}{x^2 + 1}$	(a) $(0, 0)$	(b) $(3, 0.1)$

In Exercises 7–10, find the constant C such that the ordered pair is a solution point of the equation.

7. $y = x^2 + C$, $(2, 6)$ **8.** $y = Cx^3$, $(-4, 8)$

9. $y = C\sqrt{x + 1}$, $(3, 8)$ **10.** $x + C(y + 2) = 0$, $(4, 3)$

In Exercises 11 and 12, complete the table and use the resulting solution points to graph the equation.

11. $2x + y = 3$

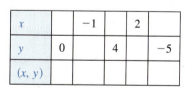

12. $y = 4 - x^2$

x		-1		2	
y	0		4		-5
(x, y)					

In Exercises 13–18, match the equation with its graph and describe the given viewing rectangle. [The graphs are labeled (a), (b), (c), (d), (e), and (f).]

13. $y = 4 - x$ **14.** $y = x^2 + 2x$

15. $y = \sqrt{4 - x^2}$ **16.** $y = \sqrt{x}$

17. $y = x^3 - x$ **18.** $y = |x| - 2$

*Detailed solutions to all odd-numbered exercises can be found in the *Study and Solutions Guide.*

(a)
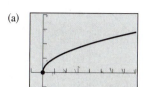

(b)
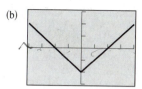

(c)
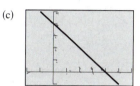

(d)
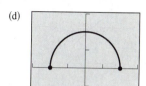

(e)
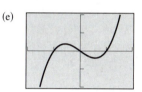

(f)
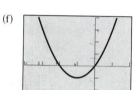

In Exercises 19–30, use the point-plotting method to graph the equation. Use a graphing utility to verify your graph. (Use the standard viewing rectangle.)

19. $y = -3x + 2$ **20.** $y = 2x - 3$

21. $y = 1 - x^2$ **22.** $y = x^2 - 1$

23. $y = x^3 + 2$ **24.** $y = x^3 - 1$

25. $y = (x - 3)(x + 2)$ **26.** $y = x(x - 5)$

27. $y = \sqrt{x - 3}$ **28.** $y = \sqrt{1 - x}$

29. $y = |x - 2|$ **30.** $y = 4 - |x|$

In Exercises 31–38, use a graphing utility to graph the equation. (Use the standard viewing rectangle.) Determine the number of times (if any) the graph intersects each coordinate axis.

31. $y = x^2 - 4x + 3$ **32.** $y = 4 - 4x - x^2$

33. $y = 3x^4 - 6x^2$ **34.** $y = \frac{1}{27}(x^4 + 4x^3)$

35. $y = x\sqrt{4 - x}$ **36.** $y = x\sqrt{4 - x^2}$

37. $y = \dfrac{10x}{x^2 + 1}$ **38.** $y = \dfrac{10}{x^2 + 1}$

In Exercises 39–44, use a graphing utility to graph the equation. Use the specified viewing rectangle. Determine the number of times the graph intersects each coordinate axis.

39. $y = x^4 - 4x^3 + 16x$ **40.** $y = 4x^3 - x^4$

RANGE
Xmin=-5
Xmax=5
Xscl=1
Ymin=-15
Ymax=30
Yscl=5

RANGE
Xmin=-2
Xmax=6
Xscl=1
Ymin=-2
Ymax=30
Yscl=2

41. $y = 100x\sqrt{25 - x}$ **42.** $y = 100x\sqrt{25 - x^2}$

RANGE
Xmin=-30
Xmax=30
Xscl=5
Ymin=-5000
Ymax=5000
Yscl=1000

RANGE
Xmin=-8
Xmax=8
Xscl=1
Ymin=-2000
Ymax=2000
Yscl=500

43. $x^2 - 100y - 1000 = 0$

RANGE
Xmin=-100
Xmax=100
Xscl=10
Ymin=-10
Ymax=10
Yscl=1

44. $2x^3 - 100x - 15,625 + 250y = 0$

RANGE
Xmin=-20
Xmax=25
Xscl=2
Ymin=-2
Ymax=100
Yscl=5

In Exercises 45–48, solve for y and use a graphing utility to graph each of the resulting equations on the same viewing rectangle. Adjust the viewing rectangle so a circle really does appear circular. (Your graphing utility may have a *square* setting that does this automatically.)

45. Circle: $x^2 + y^2 = 64$ **46.** Circle: $x^2 + y^2 = 49$

47. Ellipse: $6x^2 + y^2 = 72$ **48.** Ellipse: $x^2 + 9y^2 = 81$

In Exercises 49–52, describe the given viewing rectangle.

49. $9x + 27y - 1000 = 0$ **50.** $0.75x - 3y + 200 = 0$

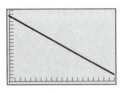

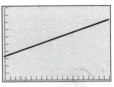

51. $y = -(x - 5)^2(x - 15)$ **52.** $y = \sqrt{x^3 + 8}$

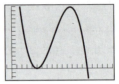

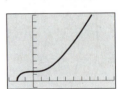

In Exercises 53 and 54, use a graphing utility to graph the equation using each of the suggested viewing rectangles. Assume that the equation gives the profit y when x units of a product are sold. Note that a graph can distort the information presented simply by changing the viewing rectangle. Which viewing rectangle would be selected by a person who wishes to argue that profits will increase dramatically with increased sales?

53. $y = 0.25x - 50$

RANGE
Xmin=-3
Xmax=800
Xscl=50
Ymin=-20
Ymax=100
Yscl=10

RANGE
Xmin=-3
Xmax=1000
Xscl=100
Ymin=-100
Ymax=500
Yscl=40

54. $y = 2.44x - \dfrac{x^2}{20,000} - 5000$

RANGE
Xmin=-5000
Xmax=22000
Xscl=5000
Ymin=-20000
Ymax=60000
Yscl=10000

RANGE
Xmin=-5000
Xmax=22000
Xscl=5000
Ymin=-5000
Ymax=24000
Yscl=5000

In Exercises 55 and 56, use the equation to complete each table. Note the importance of increasing the number of solution points to increase the accuracy of the graph.

(a) Use the table to sketch a graph of the equation.

x	-1	0	1
y			

(b) Use the table to sketch a graph of the equation.

x	-1	$-\frac{3}{4}$	$-\frac{1}{2}$	$-\frac{1}{4}$	0	$\frac{1}{4}$	$\frac{1}{2}$	$\frac{3}{4}$	1
y									

(c) Use a graphing utility to graph the equation for $-1 \le x \le 1$.

55. $y = \sqrt[3]{x}$ **56.** $y = x^3(3x + 4)$

In Exercises 57–60, use a graphing utility to graph the equation. Move the cursor along the curve to approximate the unknown coordinate(s) of the given solution point accurate to two decimal places. (*Hint:* You may need to use the zoom feature of the calculator to obtain the required accuracy.)

57. $y = \sqrt{5 - x}$
   (a) $(2, y)$
   (b) $(x, 3)$
58. $y = x^3(x - 3)$
   (a) $(2.25, y)$
   (b) $(x, 20)$
59. $y = x^5 - 5x$
   (a) $(-0.5, y)$
   (b) $(x, -4)$
60. $y = |x^2 - 6x + 5|$
   (a) $(2, y)$
   (b) $(x, 1.5)$

61. *Depreciation*   A manufacturing plant purchases a new molding machine for $225,000. The depreciated value y after x years is given by $y = 225{,}000 - 20{,}000x$, $0 \le x \le 8$.
   (a) Use the constraints of the model to determine an appropriate viewing rectangle.
   (b) Graph the equation.
62. *Dimensions of a Rectangle*   A rectangle of length x and width w has a perimeter of 12 meters.
   (a) Show that the area of the rectangle is $y = x(6 - x)$.
   (b) Use the physical constraints of the problem to determine the horizontal component of the viewing rectangle.
   (c) Use a graphing utility to graph the equation for the area. (*Note:* You may have to sketch the graph more than once in order to determine an appropriate vertical component of the viewing rectangle.)
   (d) From the graph in part (c), estimate the dimensions of the rectangle that yield a maximum area.

In Exercises 63 and 64, (a) graph the model and compare it with the data, and (b) use the model to predict y for the year 1994.

63. *Federal Debt*   The table gives the per capita federal debt for the United States for selected years from 1950 to 1990. (*Source:* U.S. Treasury Department)

Year	1950	1960	1970	1980	1985	1990
Per capita debt	$1688	$1572	$1807	$3981	$7614	$12,848

A model for the per capita debt during this period is

$$y = 0.40x^3 - 9.42x^2 + 1053.24,$$

where y represents the per capita debt and x is the year, with $x = 0$ corresponding to 1950.

64. *Life Expectancy*   The table gives the life expectancy of a child (at birth) for selected years from 1920 to 1989. (*Source:* Department of Health and Human Services)

Year	1920	1930	1940	1950
Life expectancy	54.1	59.7	62.9	68.2

Year	1960	1970	1980	1989
Life expectancy	69.7	70.8	73.7	75.2

A model for the life expectancy during this period is

$$y = \frac{x + 66.94}{0.01x + 1},$$

where y represents the life expectancy and x represents the year, with $x = 0$ corresponding to 1950.

65. *Earnings Per Share*   The earnings per share for Eli Lilly Corporation from 1980 to 1986 can be approximated by the model

$$y = 1.097x + 0.15, \qquad 0 \le x \le 6,$$

where y is the earnings and x represents the year, with $x = 0$ corresponding to 1980. Graph this equation. (*Source:* NYSE Stock Reports)

66. *Copper Wire*   The resistance y in ohms of 1000 feet of solid copper wire at 77 degrees Fahrenheit can be approximated by the model

$$y = \frac{10{,}770}{x^2} - 0.37, \qquad 5 \le x \le 100,$$

where x is the diameter of the wire in mils (0.001 in.). Use the model to estimate the resistance when $x = 50$.

67. Find a and b if the x-intercept of the graph of $y = \sqrt{ax + b}$ is $(5, 0)$. (There is more than one correct answer.)

1.2   LINES IN THE PLANE

The Slope of a Line / The Point-Slope Form of the Equation of a Line /
Sketching Graphs of Lines / Changing the Viewing Rectangle /
Parallel and Perpendicular Lines

The Slope of a Line

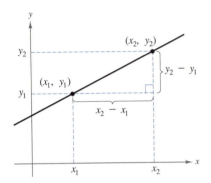

FIGURE 1.12

In this section, you will study lines and their equations. Throughout this text, the term **line** always means a *straight* line.

The **slope** of a nonvertical line represents the number of units a line rises or falls vertically for each unit of horizontal change from left to right. For instance, consider the two points (x_1, y_1) and (x_2, y_2) on the line shown in Figure 1.12. As you move from left to right along this line, a change of $(y_2 - y_1)$ units in the vertical direction corresponds to a change of $(x_2 - x_1)$ units in the horizontal direction. That is,

$$y_2 - y_1 = \text{the change in } y$$

and

$$x_2 - x_1 = \text{the change in } x.$$

The slope of the line is given by the ratio of these two changes.

DEFINITION OF THE SLOPE OF A LINE

The **slope** m of the nonvertical line passing through the points (x_1, y_1) and (x_2, y_2) is

$$m = \frac{y_2 - y_1}{x_2 - x_1} = \frac{\text{change in } y}{\text{change in } x},$$

where $x_1 \neq x_2$.

DISCOVERY

Use a graphing utility to compare the slopes of the lines given by $y = ax$ with $a = 0.5, 1, 2,$ and 4. What do you observe about the slopes of the lines? Repeat the experiment with $a = -0.5, -1, -2,$ and -4. What do you observe about the slopes of these lines? (*Hint:* Use a square setting to guarantee a true geometric perspective.)

When this formula is used, the *order of subtraction* is important. Given two points on a line, you are free to label either one of them as (x_1, y_1), and the other as (x_2, y_2). However, once this is done, you must form the numerator and denominator using the same order of subtraction.

$$m = \frac{y_2 - y_1}{x_2 - x_1} \qquad m = \frac{y_1 - y_2}{x_1 - x_2} \qquad m = \frac{y_2 - y_1}{x_1 - x_2}$$

Correct   Correct   Incorrect

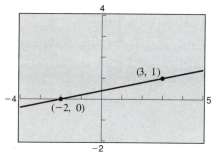

(a) The slope is $\frac{1}{5}$.

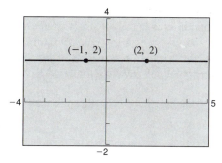

(b) The slope is zero.

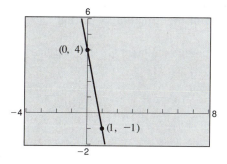

(c) The slope is -5.

FIGURE 1.13

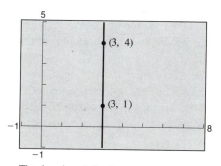

The slope is undefined.

FIGURE 1.14

EXAMPLE 1   Finding the Slope of a Line Passing Through Two Points

Find the slopes of the lines passing through the following pairs of points.

A. $(-2, 0)$ and $(3, 1)$
B. $(-1, 2)$ and $(2, 2)$
C. $(0, 4)$ and $(1, -1)$

SOLUTION

Difference in y-values

A. $m = \dfrac{\overbrace{y_2 - y_1}}{\underbrace{x_2 - x_1}} = \dfrac{1 - 0}{3 - (-2)} = \dfrac{1}{3 + 2} = \dfrac{1}{5}$

Difference in x-values

B. $m = \dfrac{2 - 2}{2 - (-1)} = \dfrac{0}{3} = 0$

C. $m = \dfrac{-1 - 4}{1 - 0} = \dfrac{-5}{1} = -5$

The graphs of the three lines are shown in Figure 1.13. In parts (a) and (c), note that the "square" setting gives the correct "steepness" of the line.

The definition of slope does not apply to vertical lines. For instance, consider the points $(3, 4)$ and $(3, 1)$ on the vertical line shown in Figure 1.14. Applying the formula for slope, you obtain

$$m = \frac{4 - 1}{3 - 3}. \qquad \textit{Undefined division by zero}$$

Because division by zero is undefined, the slope of a vertical line is undefined.

From the slopes of the lines shown in Figures 1.13 and 1.14, you can make the following generalizations about the slope of a line.

1. A line with positive slope $(m > 0)$ *rises* from left to right.
2. A line with negative slope $(m < 0)$ *falls* from left to right.
3. A line with zero slope $(m = 0)$ is *horizontal*.
4. A line with undefined slope is *vertical*.

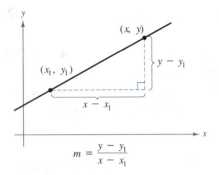

$$m = \frac{y - y_1}{x - x_1}$$

FIGURE 1.15

The Point-Slope Form of the Equation of a Line

If you know the slope of a line *and* you also know the coordinates of one point on the line, then you can find an equation for the line. For instance, in Figure 1.15, let (x_1, y_1) be a given point on the line whose slope is m. If (x, y) is any *other* point on the line, then it follows that

$$\frac{y - y_1}{x - x_1} = m.$$

This equation in the variables x and y can be rewritten in the form

$$y - y_1 = m(x - x_1),$$

which is the **point-slope form** of the equation of a line.

> **POINT-SLOPE FORM OF THE EQUATION OF A LINE**
>
> The **point-slope** form of the equation of the line that passes through the point (x_1, y_1) and has a slope of m is
>
> $$y - y_1 = m(x - x_1).$$

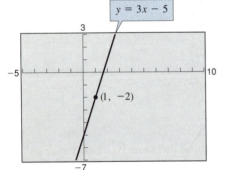

FIGURE 1.16

EXAMPLE 2   The Point-Slope Form of the Equation of a Line

Find an equation of the line that passes through the point $(1, -2)$ and has a slope of 3.

SOLUTION

$$\begin{aligned}
y - y_1 &= m(x - x_1) &&\textit{Point-slope form} \\
y - (-2) &= 3(x - 1) &&\textit{Substitute } y_1 = -2,\ x_1 = 1,\ \textit{and } m = 3 \\
y + 2 &= 3x - 3 \\
y &= 3x - 5 &&\textit{Equation of line}
\end{aligned}$$

This line is shown in Figure 1.16.

The point-slope form can be used to find the equation of a nonvertical line passing through two points (x_1, y_1) and (x_2, y_2). First, use the formula for the slope of the line passing through two points.

$$m = \frac{y_2 - y_1}{x_2 - x_1}$$

Then, once you know the slope, use the point-slope form to obtain the equation

$$y - y_1 = \frac{y_2 - y_1}{x_2 - x_1}(x - x_1).$$

This is sometimes called the **two-point form** of the equation of a line.

EXAMPLE 3   A Linear Model for Sales Prediction

During the first two quarters of the year, a company had total sales of $3.4 million and $3.7 million, respectively.

A. Write a linear equation giving the total sales y in terms of the quarter x.
B. Use the equation to predict the total sales during the fourth quarter.

SOLUTION

A. In Figure 1.17, let $(1, 3.4)$ and $(2, 3.7)$ be two points on the line representing the total sales. The slope of the line passing through these two points is

$$m = \frac{3.7 - 3.4}{2 - 1} = 0.3.$$

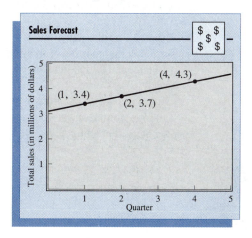

FIGURE 1.17

By the point-slope form, the equation of the line is as follows.

$$y - y_1 = m(x - x_1)$$
$$y - 3.4 = 0.3(x - 1)$$
$$y = 0.3x - 0.3 + 3.4$$
$$y = 0.3x + 3.1$$

B. Using the equation from part (A), estimate the fourth-quarter sales ($x = 4$) to be $y = 0.3(4) + 3.1 = 1.2 + 3.1 = \$4.3$ million.

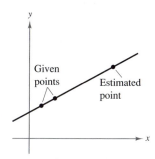

Linear Extrapolation

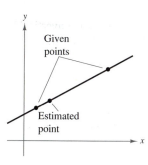

Linear Interpolation

FIGURE 1.18

The approximation method illustrated in Example 3 is **linear extrapolation.** Note in Figure 1.18 that for linear extrapolation, the estimated point lies *outside* of the given points. When the estimated point lies *between* two given points, the procedure is called **linear interpolation.**

Sketching Graphs of Lines

Many problems in coordinate (or analytic) geometry can be classified in two basic categories.

1. Given a graph (or parts of it), find its equation.
2. Given an equation, find its graph.

For lines, the first problem is solved easily by using the point-slope form. This formula, however, is not particularly useful for solving the second type of problem. The form that is better suited to graphing linear equations is the **slope-intercept form** $y = mx + b$ of the equation of a line.

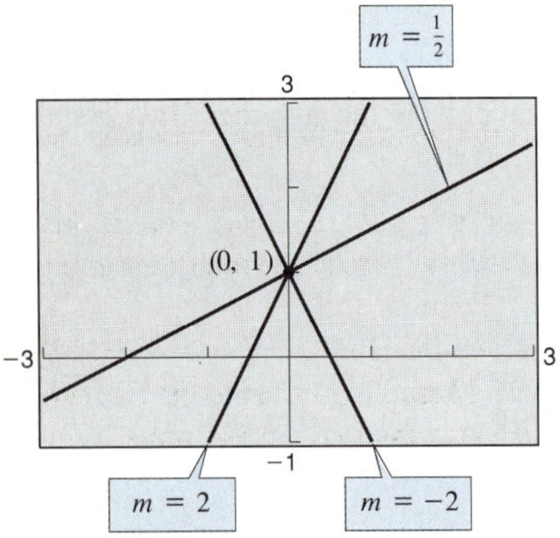

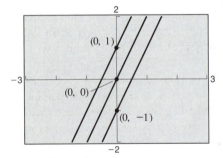

$m = \frac{1}{2}$

(0, 1)

$m = 2$ $m = -2$

FIGURE 1.19

EXAMPLE 4   Determining the Slope and y-Intercepts

A. By graphing the equations $y = 2x + 1$, $y = \frac{1}{2}x + 1$, and $y = -2x + 1$ on the same coordinate axes, you can see that they each have the same y-intercept $(0, 1)$, but slopes of 2, $\frac{1}{2}$, and -2, respectively (see Figure 1.19).

B. By graphing the equations $y = 2x + 1$, $y = 2x$, and $y = 2x - 1$ on the same coordinate axes, you can see that they each have the same slope of 2, but y-intercepts of $(0, 1)$, $(0, 0)$, and $(0, -1)$, respectively (see Figure 1.20).

(0, 1)

(0, 0)

(0, -1)

FIGURE 1.20

SLOPE-INTERCEPT FORM OF THE EQUATION OF A LINE

The graph of the equation

$$y = mx + b$$

is a line whose slope is m and whose y-intercept is $(0, b)$.

To derive algebraically the slope-intercept form, write the following.

$$y - y_1 = m(x - x_1) \qquad \textit{Point-slope form}$$
$$y = mx - mx_1 + y_1$$
$$y = mx + b \qquad \textit{Slope-intercept form}$$

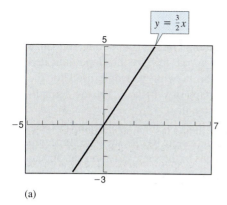

(a)

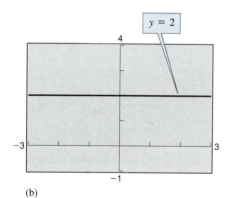

(b)

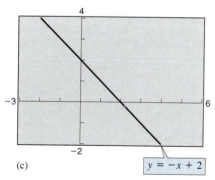

(c)

FIGURE 1.21

EXAMPLE 5   **Using the Slope-Intercept Form**

Describe the graphs of the linear equations.

A. $y = \dfrac{3}{2}x$

B. $y = 2$

C. $x + y = 2$

SOLUTION

A. Because $b = 0$, the y-intercept is $(0, 0)$. Moreover, because the slope is $m = \frac{3}{2}$, this line *rises* three units for every two units the line moves to the right, as shown in Figure 1.21(a).

B. By writing the equation $y = 2$ in the form $y = (0)x + 2$, you can see that the y-intercept is $(0, 2)$ and the slope is zero. A zero slope implies that the line is horizontal, as shown in Figure 1.21(b).

C. By writing the equation $x + y = 2$ in slope-intercept form, $y = -x + 2$, you can see that the y-intercept is $(0, 2)$. Moreover, because the slope is $m = -1$, this line *falls* one unit for each unit the line moves to the right, as shown in Figure 1.21(c).

From the slope-intercept form of the equation of a line, you can see that a horizontal line ($m = 0$) has an equation of the form $y = b$. This is consistent with the fact that each point on a horizontal line through $(0, b)$ has a y-coordinate of b.

Similarly, each point on a vertical line through $(a, 0)$ has an x-coordinate of a. Hence, a vertical line has an equation of the form $x = a$. This equation cannot be written in the slope-intercept form, because the slope of a vertical line is undefined. However, *every* line has an equation that can be written in the **general form** $Ax + By + C = 0$, where A and B are not *both* zero.

SUMMARY OF EQUATIONS OF LINES

1. General form: $Ax + By + C = 0$
2. Vertical line: $x = a$
3. Horizontal line: $y = b$
4. Slope-intercept form: $y = mx + b$
5. Point-slope form: $y - y_1 = m(x - x_1)$

Changing the Viewing Rectangle

When using a graphing utility to sketch a straight line, it is important to realize that the graph of the line may not visually appear to have the slope indicated by its equation. This occurs because of the viewing rectangle used for the graph. For instance, Figure 1.22 shows a graph of $y = 2x + 1$ using a graphing utility with three different viewing rectangles.

Notice that the slopes of the first two lines do not visually appear to be equal to 2. If you use the "square" viewing rectangle (right), then the slope will visually appear to be 2. In general, two graphs of the same equation can appear to be quite different depending upon the viewing rectangle selected.

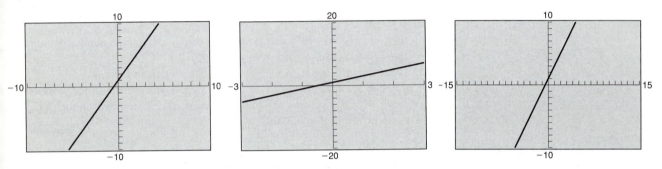

FIGURE 1.22

Effects of Different Viewing Rectangles on Graph of $y = 2x + 1$

EXAMPLE 6   Different Viewing Rectangles

The graphs of the two lines $y = -x - 1$ and $y = -10x - 1$ are shown in Figure 1.23. Even though the slopes of these lines are different (-1 and -10, respectively), the graphs seem similar because the viewing rectangles are different.

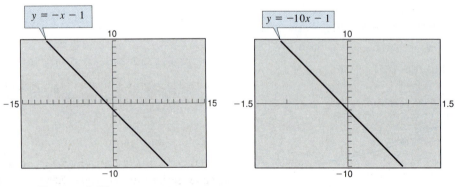

FIGURE 1.23

Parallel and Perpendicular Lines

The slope of a line is a convenient tool for determining whether two lines are parallel or perpendicular. Example 4(B) suggests the following property of parallel lines.

PARALLEL LINES

Two distinct nonvertical lines are **parallel** if and only if their slopes are equal.

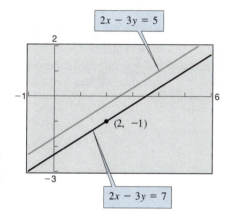

FIGURE 1.24

DISCOVERY

Use a graphing utility with the viewing rectangle $-15 \leq x \leq 15$ and $-10y \leq y \leq 10$ to graph the following linear equations.

$y_1 = -\frac{1}{3}x + 2$

$y_2 = 3x - 6$

$y_3 = 3x + 3$

What do you observe? How are the slopes of parallel lines related? How are the slopes of perpendicular lines related? Verify your conclusions using the lines $y_1 = -2x - 5$, $y_2 = -2x + 6$, and $y_3 = \frac{1}{2}x + 1$.

EXAMPLE 7 **Equations of Parallel Lines**

Find an equation of the line that passes through the point $(2, -1)$ and is parallel to the line $2x - 3y = 5$, as shown in Figure 1.24.

SOLUTION

Write the given equation in slope-intercept form.

$$2x - 3y = 5 \qquad \text{\textit{Given equation}}$$
$$3y = 2x - 5$$
$$y = \frac{2}{3}x - \frac{5}{3} \qquad \text{\textit{Slope-intercept form}}$$

Therefore, the given line has a slope of $m = \frac{2}{3}$. Because any line parallel to the given line must also have a slope of $\frac{2}{3}$, the required line through $(2, -1)$ has the following equation.

$$y - (-1) = \frac{2}{3}(x - 2) \qquad \text{\textit{Point-slope form}}$$
$$y = \frac{2}{3}x - \frac{4}{3} - 1$$
$$y = \frac{2}{3}x - \frac{7}{3} \qquad \text{\textit{Slope-intercept form}}$$

Notice the similarity between the slope-intercept form of the original equation and the slope-intercept form of the parallel equation.

Two nonvertical lines are *perpendicular* if and only if their slopes are negative reciprocals of each other. For instance, the lines $y = 2x$ and $y = -\frac{1}{2}x$ are perpendicular because one has a slope of 2 and the other has a slope of $-\frac{1}{2}$.

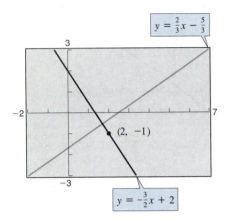

$y = \frac{2}{3}x - \frac{5}{3}$

$(2, -1)$

$y = -\frac{3}{2}x + 2$

FIGURE 1.25

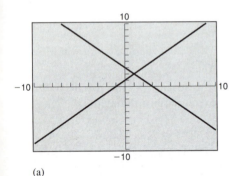

(a)

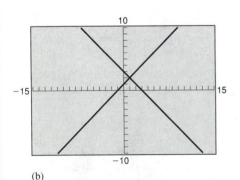

(b)

FIGURE 1.26

PERPENDICULAR LINES

Two nonvertical lines are **perpendicular** if and only if their slopes are negative reciprocals of each other. That is,

$$m_1 = -\frac{1}{m_2}.$$

EXAMPLE 8   **Equations of Perpendicular Lines**

Find an equation of the line that passes through the point $(2, -1)$ and is perpendicular to the line $2x - 3y = 5$.

SOLUTION

By writing the given line in the form $y = \frac{2}{3}x - \frac{5}{3}$ you can see that the line has a slope of $\frac{2}{3}$. Hence, any line that is perpendicular to this line must have a slope of $-\frac{3}{2}$ (because $-\frac{3}{2}$ is the negative reciprocal of $\frac{2}{3}$). Therefore, the required line through the point $(2, -1)$ has the following equation.

$$y - (-1) = -\frac{3}{2}(x - 2) \qquad \textit{Point-slope form}$$

$$y = -\frac{3}{2}x + 3 - 1$$

$$y = -\frac{3}{2}x + 2 \qquad \textit{Slope-intercept form}$$

The graphs of both equations are shown in Figure 1.25.

EXAMPLE 9   **Graphs of Perpendicular Lines**

Use a graphing utility to graph the lines given by $y = x + 1$ and $y = -x + 3$. Display *both* graphs on the same screen. The lines are supposed to be perpendicular (they have slopes of $m_1 = 1$ and $m_2 = -1$). Do they appear to be perpendicular on the display?

SOLUTION

If the viewing window has the *standard* range settings, then the tick marks on both the x-axis and the y-axis will vary between -10 and 10, as in Figure 1.26(a). However, because most display screens are not square, the lines might *not* appear to be perpendicular. That is, the graphs of two perpendicular lines will appear to be perpendicular only if the tick marks on the x-axis have the same spacing as the tick marks on the y-axis, as in Figure 1.26(b). This can be done by using a "square" setting for the viewing rectangle.

DISCUSSION PROBLEM

....

AN
APPLICATION
OF SLOPE

In 1982, a college had an enrollment of 5000 students. By 1992, the enrollment had increased to 7000 students.

(a) What is the average annual change in enrollment from 1982 to 1992?
(b) Use the average annual change in enrollment to estimate the enrollment in 1986, 1990, and 1994.

Year	1982	1986	1990	1992	1994
Enrollment	5000			7000	

(c) Graph the line represented by the data given in the table in part (b). What is the slope of this line?
(d) Write a short paragraph that compares the concepts of *slope* and *average rate of change*.

WARM-UP

....

The following warm-up exercises involve skills that were covered in earlier sections. You will use these skills in the exercise set for this section.

In Exercises 1 and 2, simplify the expression.

1. $\dfrac{4 - (-5)}{-3 - (-1)}$

2. $\dfrac{-5 - 8}{0 - (-3)}$

3. Find $-1/m$ for $m = 4/5$.

4. Find $-1/m$ for $m = -2$.

In Exercises 5–10, solve for y in terms of x.

5. $2x - 3y = 5$

6. $4x + 2y = 0$

7. $y - (-4) = 3[x - (-1)]$

8. $y - 7 = \frac{2}{3}(x - 3)$

9. $y - (-1) = \dfrac{3 - (-1)}{2 - 4}(x - 4)$

10. $y - 5 = \dfrac{3 - 5}{0 - 2}(x - 2)$

SECTION 1.2 · EXERCISES

In Exercises 1–6, estimate the slope of the line.

1.
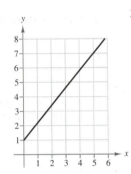

2.
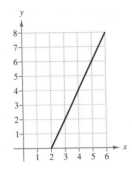

3.
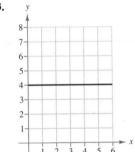

4.
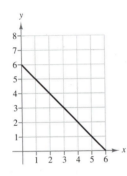

5.
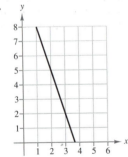

6.
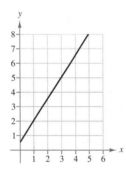

In Exercises 7 and 8, graph the lines through the given point with the indicated slopes. Graph the lines on the same set of coordinate axes.

Point	Slopes			
7. (2, 3)	(a) 0	(b) 1	(c) 2	(d) −3
8. (−4, 1)	(a) 3	(b) −3	(c) $\frac{1}{2}$	(d) Undefined

In Exercises 9–14, plot the points and find the slope of the line passing through the points.

9. $(-3, -2)$, $(1, 6)$ **10.** $(2, 4)$, $(4, -4)$

11. $(-6, -1)$, $(-6, 4)$ **12.** $(0, -10)$, $(-4, 0)$

13. $(1, 2)$, $(-2, -2)$ **14.** $\left(\frac{7}{8}, \frac{3}{4}\right)$, $\left(\frac{5}{4}, -\frac{1}{4}\right)$

In Exercises 15–18, use the given point on the line and the slope of the line to find three additional points through which the line passes. (There are many correct answers.)

Point	Slope		Point	Slope
15. (2, 1)	$m = 0$		**16.** (−4, 1)	m is undefined
17. (5, −6)	$m = 1$		**18.** (10, −6)	$m = -1$

In Exercises 19–22, determine whether the lines L_1 and L_2 passing through the given pairs of points are parallel, perpendicular, or neither.

19. L_1: $(0, -1)$, $(5, 9)$ **20.** L_1: $(-2, -1)$, $(1, 5)$
  L_2: $(0, 3)$, $(4, 1)$ L_2: $(1, 3)$, $(5, -5)$

21. L_1: $(3, 6)$, $(-6, 0)$ **22.** L_1: $(4, 8)$, $(-4, 2)$
  L_2: $(0, -1)$, $\left(5, \frac{7}{3}\right)$ L_2: $(3, -5)$, $\left(-1, \frac{1}{3}\right)$

In Exercises 23 and 24, use the concept of slope to determine whether the three points are collinear.

23. $(0, -4)$, $(2, 0)$, $(3, 2)$ **24.** $(-2, 1)$, $(-1, 0)$, $(2, -2)$

25. *Mountain Driving*   While driving down a mountain road, you notice a "12% grade" warning sign. This means that the slope of the road is $-\frac{12}{100}$. Determine the amount of horizontal change in your position if you note from elevation markers that you have descended 2000 feet vertically.

26. *Attic Height*   The "rise to run" in determining the steepness of the roof on the house in the figure is 3 to 4. Determine the maximum height in the attic of the house if the house is 30 feet wide.

FIGURE FOR 26

In Exercises 27–30, find the slope and y-intercept (if possible) of the line specified by the equation. Graph the line.

27. $5x - y + 3 = 0$ **28.** $2x + 3y - 9 = 0$

29. $5x - 2 = 0$ **30.** $3y + 5 = 0$

In Exercises 31–36, find an equation for the line passing through the points.

31. $(5, -1), (-5, 5)$ **32.** $(4, 3), (-4, -4)$

33. $(2, \frac{1}{2}), (\frac{1}{2}, \frac{5}{4})$ **34.** $(-1, 4), (6, 4)$

35. $(-8, 1), (-8, 7)$ **36.** $(-8, 0.6), (2, -2.4)$

In Exercises 37–42, find an equation of the line that passes through the point and has the indicated slope.

Point	Slope		Point	Slope
37. $(0, -2)$	$m = 3$	**38.**	$(0, 0)$	$m = 4$
39. $(4, \frac{5}{2})$	$m = -\frac{4}{3}$	**40.**	$(-2, -5)$	$m = \frac{3}{4}$
41. $(6, -1)$	m is undefined	**42.**	$(-10, 4)$	$m = 0$

In Exercises 43 and 44, use a graphing utility to graph the equation using each of the suggested viewing rectangles. Note that the viewing rectangle selected will alter the appearance of the slope.

43. $y = 0.5x - 3$

```
RANGE
Xmin=-5
Xmax=10
Xscl=1
Ymin=-10
Ymax=5
Yscl=1
```

```
RANGE
Xmin=-2
Xmax=10
Xscl=1
Ymin=-4
Ymax=1
Yscl=1
```

44. $y = -8x + 5$

```
RANGE
Xmin=-5
Xmax=5
Xscl=1
Ymin=-10
Ymax=10
Yscl=1
```

```
RANGE
Xmin=-5
Xmax=10
Xscl=1
Ymin=-80
Ymax=80
Yscl=20
```

In Exercises 45–48, use a graphing utility to graph the three equations on the same viewing rectangle. Adjust the viewing rectangle so the slope appears visually correct. (Your calculator may have a *square* setting that does this automatically.)

45. $y = 2x$ $y = -2x$ $y = \frac{1}{2}x$

46. $y = \frac{2}{3}x$ $y = -\frac{3}{2}x$ $y = \frac{2}{3}x + 2$

47. $y = -\frac{1}{2}x$ $y = -\frac{1}{2}x + 3$ $y = 2x - 4$

48. $y = x - 8$ $y = x + 1$ $y = -x + 3$

In Exercises 49 and 50, use the given values of a and b and a graphing utility to graph the equation of the line given by

$$\frac{x}{a} + \frac{y}{b} = 1, \qquad a \neq 0, b \neq 0.$$

Use the graphs to determine the meaning of the constants a and b.

49. $a = 5, \quad b = -3$ **50.** $a = -6, \quad b = 2$

In Exercises 51 and 52, use the results of Exercises 49 and 50 to write an equation of the line that passes through the given points.

51. x-intercept: $(2, 0)$ **52.** x-intercept: $(-\frac{1}{6}, 0)$
 y-intercept: $(0, 3)$ y-intercept: $(0, -\frac{2}{3})$

In Exercises 53–56, write an equation of the line through the point (a) parallel to the given line and (b) perpendicular to the given line.

Point	Line		Point	Line
53. $(2, 1)$	$4x - 2y = 3$	**54.**	$(\frac{7}{8}, \frac{3}{4})$	$5x + 3y = 0$
55. $(-1, 0)$	$y = -3$	**56.**	$(2, 5)$	$x = 4$

In Exercises 57–60, you are given the dollar value of a product in 1990 *and* the rate at which the value of the item is expected to increase during the next five years. Use this information to write a linear equation that gives the dollar value V of the product in terms of the year t. (Let $t = 0$ represent 1990.)

	1990 Value	Rate
57.	$2540	$125 per year
58.	$156	$4.50 per year
59.	$20,400	$2000 per year
60.	$245,000	$5600 per year

In Exercises 61–64, match the description with a graph. Determine the slope and how it is interpreted in the situation. [The graphs are labeled (a), (b), (c), and (d).]

61. A person is paying $10 per week to a friend to repay a $100 loan.

62. An employee is paid $12.50 per hour plus $1.50 for each unit produced per hour.

63. A sales representative receives $20 per day for food plus $0.25 for each mile traveled.

64. A typewriter purchased for $600 depreciates $100 per year.

(a)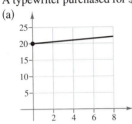

(b)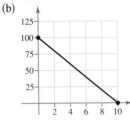

(c)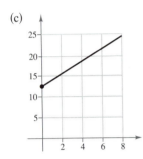

(d)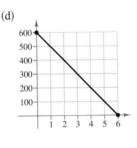

65. *Temperature*   Find an equation of the line giving the relationship between the temperature in degrees Celsius, C, and degrees Fahrenheit, F. Remember that water freezes at 0° Celsius (32° Fahrenheit) and boils at 100° Celsius (212° Fahrenheit).

66. *Temperature*   Use the result of Exercise 65 to complete the table.

C		−10°	10°			177°
F	0°			68°	90°	

67. *Annual Salary*   Suppose your salary was $28,500 in 1990 and $32,900 in 1992. Assume your salary follows a linear growth pattern.
   (a) Write a linear equation giving the salary S in terms of the year t where $t = 0$ corresponds to the year 1990.
   (b) Use the linear equation to predict your salary in 1995.

68. *College Enrollment*   A small college had 2546 students in 1990 and 2702 students in 1992. Assume the enrollment follows a linear growth pattern.
   (a) Write a linear equation giving the enrollment E in terms of the year t where $t = 0$ corresponds to the year 1990.
   (b) Use the linear equation to predict the enrollment in 1997.

69. *Straight-Line Depreciation*   A small business purchases a piece of equipment for $875. After 5 years the equipment will be outdated and have no value.
   (a) Write a linear equation giving the value y of the equipment in terms of the time x, $0 \leq x \leq 5$.
   (b) Use a graphing utility to graph the equation.
   (c) Move the cursor along the graph and estimate (to two decimal place accuracy) the value of the equipment when $x = 2$.
   (d) Move the cursor along the graph and estimate (to two decimal place accuracy) the time when the value of the equipment is $200.

70. *Hourly Wages*   A manufacturer pays its assembly line workers $11.50 per hour plus $0.75 per unit produced.
   (a) Write the linear equation for the hourly wages y in terms of the number of units x produced per hour.
   (b) Use a graphing utility to graph the equation.
   (c) Move the cursor along the graph and estimate (to two decimal place accuracy) the hourly wage when a worker produces $x = 6$ units per hour.
   (d) Move the cursor along the graph and determine the number of units that must be produced per hour to have an hourly wage of $18.25.

71. *Sales Commission*   A salesperson receives a monthly salary of $2500 plus a commission of 7% of sales. Write a linear equation for the salesperson's monthly wage W in terms of monthly sales S.

72. *Daily Cost*   A sales representative of a company receives $120 per day for lodging and meals plus $0.26 per mile driven. Write a linear equation giving the daily cost C to the company in terms of x, the number of miles driven.

73. *Contracting Purchase*   A contractor purchases a piece of equipment for $36,500. The equipment has an average expense of $5.25 per hour for fuel and maintenance, and the operator is paid $11.50 per hour.
   (a) Write a linear equation for the total cost C of operating the equipment for t hours. (Include the purchase cost.)
   (b) Customers are charged $37 per hour of machine use. Write an equation for the revenue R derived from t hours of use.
   (c) *Break-Even Point*   Use a graphing utility to graph the cost and revenue equations on the same viewing rectangle. Move the cursor to the point of intersection of the graphs to estimate (to the nearest hour) the number of hours the equipment must be used to break even.

74. *Real Estate*   A real estate office handles an apartment complex with 50 units. When the rent per unit is $380 per month, all 50 units are occupied. However, when the rent is $425 per month, the average number of occupied units drops to 47. Assume that the relationship between the monthly rent p and the demand x is linear.

(a) Write an equation of the line giving the demand x in terms of the rent p.

(b) Use a graphing utility to graph the demand equation.

(c) Move the cursor along the graph to predict the number of units occupied if the rent is raised to $455.

(d) Move the cursor along the graph to predict the number of units occupied if the rent is lowered to $395.

In Exercises 75 and 76, create a realistic problem that is modeled by the graph.

75.

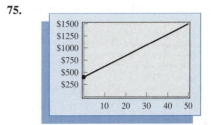

76.

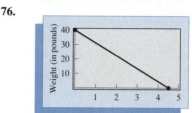

77. Use the theorem that the ratios of corresponding sides of similar triangles are equal to verify that any two points on a line can be used to calculate its slope.

1.3   FUNCTIONS

Introduction to Functions   /   Function Notation   /   The Domain of a Function   /   Function Keys on a Graphing Utility   /   Applications

Introduction to Functions

Many everyday phenomena involve two quantities that are related to each other by some rule of correspondence. Such a rule of correspondence is called a **function.** Here are two examples.

1. The simple interest I earned on $1000 for one year is related to the annual percentage rate r by the formula $I = 1000r$.
2. The area A of a circle is related to its radius r by the formula $A = \pi r^2$.

DEFINITION OF A FUNCTION

A **function** f from a set A to a set B is a rule of correspondence that assigns to each element x in the set A exactly one element y in the set B. The set A is the **domain** (or set of inputs) of the function f, and the set B contains the **range** (or set of outputs).

REMARK   Note that this use of the word *range* is not the same as the use of *range* relating to the viewing window for a graph.

To help understand this definition, look at the function illustrated in Figure 1.27. This function can be represented by the following set of ordered pairs.

$$\{(1, 9°), (2, 13°), (3, 15°), (4, 15°), (5, 12°), (6, 4°)\}$$

In each ordered pair, the first coordinate is the input and the second coordinate is the output.

The following characteristics are true of a function from a set A to a set B.

1. Each element in A must be matched with an element in B.
2. Some elements in B may not be matched with any element in A.
3. Two or more elements of A may be matched with the same element of B.

The converse of the third statement is not true. That is, an element of A (the domain) cannot be matched with two different elements of B.

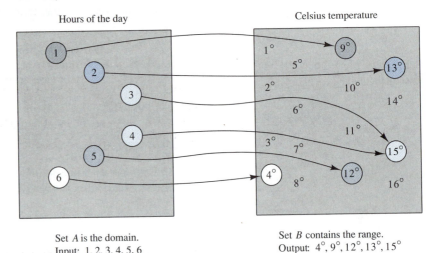

Hours of the day   Celsius temperature

Set A is the domain.   Set B contains the range.
Input: 1, 2, 3, 4, 5, 6   Output: 4°, 9°, 12°, 13°, 15°

FIGURE 1.27   Function from Set A to Set B

EXAMPLE 1   Testing for Functions

Let $A = \{a, b, c\}$ and $B = \{1, 2, 3, 4, 5\}$. Does the set of ordered pairs or figures represent a function from set A to set B?

A. $\{(a, 2), (b, 3), (c, 4)\}$ **B.** $\{(a, 4), (b, 5)\}$

C. 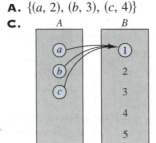   **D.**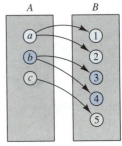

SOLUTION

A. Yes, because each element of A is matched with exactly one element of B.

B. No, because not all elements of A are matched with an element of B.

C. Yes. It does not matter that each element of A is matched with the same element of B.

D. No, because the element a in A is matched with *two* elements, 1 and 2, in B. This is also true of the element b.

Representing functions by sets of ordered pairs is a common practice in *discrete mathematics*. In algebra, however, it is more common to represent functions by equations or formulas involving two variables. For instance, the equation

$$y = x^2$$

represents the variable y as a function of the variable x. Here, x is the **independent variable** and y is the **dependent variable.** The domain of the function is the set of all values (inputs) taken on by the independent variable x, and the range of the function is the set of all values (outputs) taken on by the dependent variable y.

EXAMPLE 2   Testing for Functions Represented by Equations

Determine whether the equations represent y as a function of x.

A. $x^2 + y = 1$ **B.** $-x + y^2 = 1$

SOLUTION

In each case, to determine whether y is a function of x, it is helpful to solve for y in terms of x.

A. $x^2 + y = 1$ *Given equation*

   $\quad\quad y = 1 - x^2$ *Solve for y*

To each value of x there corresponds one value of y. Therefore, y is a function of x.

B. $-x + y^2 = 1$ *Given equation*

   $\quad\quad\quad y^2 = 1 + x$ *Add x to both sides*

   $\quad\quad\quad y = \pm\sqrt{1 + x}$ *Solve for y*

The $\pm$ indicates that to a given value of x there correspond two values of y. Therefore, y is *not* a function of x. For instance, if $x = 3$, then y could be either 2 or -2.

DISCOVERY

Use a graphing utility to graph $x^2 + y = 1$. Then use the graph to write a convincing argument that each x-value has at most one y-value.

Use a graphing utility to graph $-x + y^2 = 1$. (*Hint:* You will need to use two equations.) Then use the graph to find an x-value that corresponds to two y-values. Why does the graph not represent y as a function of x?

Function Notation

When using an equation to represent a function, it is convenient to name the function so that it can be referenced easily. For example, the equation $y = 1 - x^2$, Example 2(A), describes y as a function of x. Suppose you give this function the name "f." Then you can use the following **function notation.**

Input	Output	Equation
x	$f(x)$	$f(x) = 1 - x^2$

The symbol $f(x)$ is read as the **value of f at x** or simply "f of x." This corresponds to the y-value for a given x. Thus, you can write $y = f(x)$.

Keep in mind that f is the *name* of the function, whereas $f(x)$ is the *value* of the function at x. For instance, the function

$$f(x) = 3 - 2x$$

has *function values* denoted by $f(-1)$, $f(0)$, $f(2)$, and so on. To find these values, substitute the specified input values into the equation.

For $x = -1$, $f(-1) = 3 - 2(-1) = 3 + 2 = 5$.
For $x = 0$, $f(0) = 3 - 2(0) = 3 + 0 = 3$.
For $x = 2$, $f(2) = 3 - 2(2) = 3 - 4 = -1$.

Although it is convenient to use f as a function name and x as the independent variable, you can use other letters. For instance, $f(x) = x^2 - 4x + 7$, $f(t) = t^2 - 4t + 7$, and $g(s) = s^2 - 4s + 7$ all define the same function. In fact, the role of the independent variable in a function is simply that of a "placeholder." Consequently, the above function could be described by the form

$$f(\ ) = (\ )^2 - 4(\ ) + 7,$$

where the parentheses are used in place of a letter. To evaluate $f(-2)$, simply place -2 in each set of parentheses.

$f(\ ) = (\ )^2 - 4(\ ) + 7$
$f(-2) = (-2)^2 - 4(-2) + 7$ *Place -2 in each set of parentheses*
$\quad = 4 + 8 + 7$ *Evaluate each term*
$\quad = 19$ *Simplify*

Similarly, the value of $f(3x)$ is obtained as follows.

$f(\ ) = (\ )^2 - 4(\ ) + 7$
$f(3x) = (3x)^2 - 4(3x) + 7$ *Place $3x$ in each set of parentheses*
$\quad = 9x^2 - 12x + 7$ *Simplify*

EXAMPLE 3   Evaluating a Function

Let $g(x) = -x^2 + 4x + 1$ and find the following.

A. $g(2)$ **B.** $g(t)$ **C.** $g(x + 2)$

Technology Note _____

Most graphing utilities can be used to evaluate a function at a real value of x.

The technique depends on the graphing utility, but here is a sample program.

```
Prgm1: EVALUATE
Lbl 1
:Disp "ENTER X"
:Input X
:Disp Y₁
:Goto 1
```

To use this program, enter a function in y_1. Then run the program—it will allow you to evaluate the function at several values of x.

REMARK Example 3 shows that
$g(x + 2) \neq g(x) + g(2)$ because
$-x^2 + 5 \neq (-x^2 + 4x + 1) + 5$. In
general, $g(u + v) \neq g(u) + g(v)$.

SOLUTION

A. Replacing x with 2 in $g(x) = -x^2 + 4x + 1$ yields

$$g(2) = -(2)^2 + 4(2) + 1 = -4 + 8 + 1 = 5.$$

B. Replacing x with t yields

$$g(t) = -(t)^2 + 4(t) + 1 = -t^2 + 4t + 1.$$

C. Replacing x with $x + 2$ yields

$$
\begin{aligned}
g(x + 2) &= -(x + 2)^2 + 4(x + 2) + 1 \\
&= -(x^2 + 4x + 4) + 4x + 8 + 1 \\
&= -x^2 - 4x - 4 + 4x + 8 + 1 \\
&= -x^2 + 5.
\end{aligned}
$$

Sometimes a function is defined using more than one equation.

EXAMPLE 4   A Function Defined by Two Equations

Evaluate the function

$$f(x) = \begin{cases} x^2 + 1, & x < 0 \\ x - 1, & x \geq 0 \end{cases}$$

at $x = -1$, 0, and 1.

Technology Note

Most graphing utilities can graph
functions that are defined using more
than one equation. For example, on
the TI-81, TI-82, or TI-85, you can
obtain the graph of the function in
Example 4 as follows:

$Y_1 = (x^2 + 1)(x < 0)$
$\qquad + (x - 1)(x \geq 0)$

SOLUTION

Because $x = -1$ is less than 0, use $f(x) = x^2 + 1$ to obtain

$$f(-1) = (-1)^2 + 1 = 2.$$

For $x = 0$, use $f(x) = x - 1$ to obtain

$$f(0) = (0) - 1 = -1.$$

For $x = 1$, use $f(x) = x - 1$ to obtain

$$f(1) = (1) - 1 = 0.$$

The Domain of a Function

The domain of a function may be explicitly described along with the function,
or it may be *implied* by the expression used to define the function. The **implied
domain** is the set of all real numbers for which the expression is defined. For
instance, the function

$$f(x) = \frac{1}{x^2 - 4}$$

has an implied domain that consists of all real x other than $x = \pm 2$. These two
values are excluded from the domain because division by zero is undefined.
Another common type of implied domain is used to avoid even roots of
negative numbers. For example, the function

$$f(x) = \sqrt{x}$$

is defined only for $x \geq 0$. Hence, its implied domain is the interval $[0, \infty)$. In
general, the domain of a function *excludes* values that would cause division by
zero *or* result in the even root of a negative number.
    The *range* of a function is more difficult to find, and can best be obtained
from the graph of the function (see Section 1.4).

EXAMPLE 5   **Finding the Domain of a Function**

Find the domain of each of the following functions.

A. f: $\{(-3, 0), (-1, 4), (0, 2), (2, 2), (4, -1)\}$

B. Volume of a sphere: $V = \dfrac{4}{3}\pi r^3$

C. $g(x) = \dfrac{1}{x + 5}$

D. $h(x) = \sqrt{4 - x}$

SOLUTION

A. The domain of f consists of all first coordinates in the set of ordered pairs,
and is therefore the set

Domain = $\{-3, -1, 0, 2, 4\}$.

REMARK In Example 5(B), note that
the domain of a function may be implied by the physical context. For instance, from the equation $V = \frac{4}{3}\pi r^3$,
you would have no reason to restrict r
to nonnegative values, but the physical
context tells you that a sphere cannot
have a negative radius.

B. For the volume of a sphere you must choose nonnegative values for the
radius r. Thus, the domain is the set of all real numbers r such that $r \geq 0$.

C. Excluding x-values that yield zero in the denominator, the domain of g is
the set of all real numbers $x \neq -5$.

D. Choose x-values for which $4 - x \geq 0$. The domain is all real numbers
that are less than or equal to 4.

Function Keys on a Graphing Utility

Calculators and computers have many built-in functions that can be evaluated
with a simple keystroke. For instance, you can use the square root key to
calculate square roots of nonnegative numbers. If you attempt to take the
square root of a negative number, you will get an error message. Here are some
more examples of built-in functions that you can verify on your calculator or
computer.

Technology Note _____

You can evaluate the square root of a number in more than one way. For instance, the square root of 10 can be obtained in any of the following ways.

$$\sqrt{10} \quad 10 \wedge \left(\tfrac{1}{2}\right) \quad 10 \wedge .5$$

Notice that you need parentheses around the fraction $\tfrac{1}{2}$.

EXAMPLE 6   Function Keys

A. $\sqrt{10} = 3.16227766$
B. $64^{1/3} = 4$
C. $|-3| = \text{abs}(-3) = 3$
D. $10^3 = 1000$
E. $\log 2 = 0.3010299957$ (You will study logarithms in Chapter 4.)

Applications

EXAMPLE 7   The Dimensions of a Container

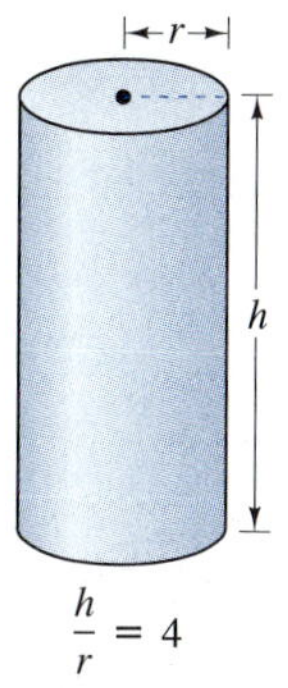

A standard soft-drink can has a height of about 4.75 inches and a radius of about 1.3 inches. For this standard can, the ratio of the height to the radius is about 3.65. Suppose you work in the marketing department of a soft-drink company, and are experimenting with a new soft-drink can that is slightly narrower and taller. For your experimental can, the ratio of the height to the radius is 4, as shown in Figure 1.28.

A. Express the volume of the can as a function of the radius r.
B. Express the volume of the can as a function of the height h.

$$\frac{h}{r} = 4$$

FIGURE 1.28

SOLUTION

The volume of a right circular cylinder is given by the formula

$$V = \pi(\text{radius})^2(\text{height}) = \pi r^2 h.$$

Because the ratio of the height to the radius is 4, you can write $h = 4r$.

A. To write the volume as a function of the radius, use the fact that $h = 4r$.

$$V = \pi r^2 h = \pi r^2(4r) = 4\pi r^3$$

B. To write the volume as a function of the height, use the fact that $r = h/4$.

$$V = \pi\left(\frac{h}{4}\right)^2 h = \frac{\pi h^3}{16}$$

EXAMPLE 8 **The Path of a Baseball**

A baseball is hit three feet above the ground at a velocity of 100 feet per second and at an angle of 45° with respect to the ground. The path of the baseball is given by the function

$$y = -0.0032x^2 + x + 3,$$

where the height y and the horizontal distance x are measured in feet, as shown in Figure 1.29. (From this equation, note that the height of the baseball is a function of the horizontal distance from home plate.) Will the baseball clear a 10-foot fence located 300 feet from home plate?

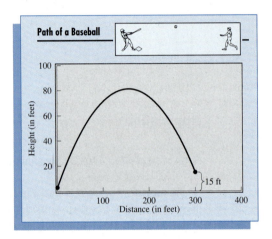

Path of a Baseball

FIGURE 1.29

DISCOVERY

Use a graphing utility to graph $y = -0.0032x^2 + x + 3$ and $y = 10$ on the same viewing rectangle. (Use a viewing rectangle in which $0 \le x \le 300$ and $0 \le y \le 90$.) Explain how the graphs can be used to answer the question asked in Example 8.

Suppose the fence in Example 8 is raised to 20 feet and the distance from home plate is 295 feet. Will the ball clear this fence?

SOLUTION

When $x = 300$, the height of the baseball is

$$y = -0.0032(300^2) + 300 + 3 = 15 \text{ feet.}$$

Thus, the ball will clear the fence.

One of the basic definitions in calculus employs a ratio called a **difference quotient.**

$$\frac{f(x + h) - f(x)}{h}, \qquad h \neq 0$$

EXAMPLE 9 Evaluating a Difference Quotient

For the function given by $f(x) = x^2 - 4x + 7$, find

$$\frac{f(x + h) - f(x)}{h}.$$

SOLUTION

$$\frac{f(x + h) - f(x)}{h} = \frac{[(x + h)^2 - 4(x + h) + 7] - [x^2 - 4x + 7]}{h}$$

$$= \frac{x^2 + 2xh + h^2 - 4x - 4h + 7 - x^2 + 4x - 7}{h}$$

$$= \frac{2xh + h^2 - 4h}{h}$$

$$= \frac{h(2x + h - 4)}{h}$$

$$= 2x + h - 4, \quad (h \neq 0)$$

SUMMARY OF FUNCTION TERMINOLOGY

Function

A **function** is a relationship between two variables such that to each value of the independent variable there corresponds exactly one value of the dependent variable.

Function Notation: $y = f(x)$

    f is the **name** of the function.
    y is the **dependent variable.**
    x is the **independent variable.**
    $f(x)$ is the **value of the function at x.**

Domain

The **domain** of a function is the set of all values (inputs) of the independent variable for which the function is defined. If x is in the domain of f, then f is **defined** at x. If x is not in the domain of f, then f is **undefined** at x.

Range

The **range** of a function is the set of all values (outputs) assumed by the dependent variable (that is, the set of all function values).

Implied Domain

If f is defined by an algebraic expression and the domain is not specified, then the **implied domain** consists of all real numbers for which the expression is defined.

DISCUSSION
PROBLEM
....

DETERMINING
RELATIONSHIPS
THAT ARE
FUNCTIONS

Write two statements describing relationships in everyday life that *are* functions and two that are *not* functions. Here are two examples.

(a) The statement "The sale price of an item is a function of the amount of sales tax on the item" is *not* a correct mathematical use of the word "function." The problem is that the sales tax does not determine the sale price. For instance, suppose the sales tax is 6%. Knowing that the sales tax on a particular item is $0.12 is not sufficient information to determine the sale price. (It could be any price between $1.92 and $2.08.)

(b) The statement "Your federal income tax is a function of your adjusted gross income" *is* a correct mathematical use of the word "function." Once you have determined your adjusted gross income, then your income tax can be determined.

WARM-UP
....

The following warm-up exercises involve skills that were covered in earlier sections. You will use these skills in the exercise set for this section.

In Exercises 1–4, simplify the expression.

1. $2(-3)^3 + 4(-3) - 7$

2. $4(-1)^2 - 5(-1) + 4$

3. $(x + 1)^2 + 3(x + 1) - (x^2 + 3x)$

4. $(x - 2)^2 - 4(x - 2) - (x^2 - 4)$

In Exercises 5 and 6, solve for y in terms of x.

5. $2x + 5y - 7 = 0$

6. $y^2 = x^2$

In Exercises 7–10, simplify the expression. Assume the variables are positive.

7. $\sqrt{50x^3y^2}$

8. $\sqrt{18z^4 - 9z^2}$

9. $\dfrac{(x - y)^2}{x^2 - y^2}$

10. $\dfrac{x^2}{x + 2} \cdot \dfrac{x^2 - x - 6}{2x}$

SECTION 1.3 · EXERCISES
....

In Exercises 1 and 2, determine which of the sets of ordered pairs represents a function from A to B. Give reasons for your answers.

1. $A = \{0, 1, 2, 3\}$ and $B = \{-2, -1, 0, 1, 2\}$
  (a) $\{(0, 1), (1, -2), (2, 0), (3, 2)\}$
  (b) $\{(2, 2), (1, -2), (3, 0), (1, 1)\}$
  (c) $\{(0, 0), (1, 0), (2, 0), (3, 0)\}$
  (d) $\{(0, 2), (3, 0), (1, 1)\}$

2. $A = \{a, b, c\}$ and $B = \{0, 1, 2, 3\}$
  (a) $\{(a, 1), (c, 2), (c, 3), (b, 3)\}$
  (b) $\{(a, 1), (b, 2), (c, 3)\}$
  (c) $\{(1, a), (0, a), (2, c), (3, b)\}$
  (d) $\{(c, 0), (b, 0), (a, 3)\}$

In Exercises 3–10, state whether the equation determines y as a function of x.

3. $x^2 + y^2 = 4$

4. $x = y^2$

5. $x^2 + y = 4$

6. $x + y^2 = 4$

7. $x^2y - x^2 + 4y = 0$

8. $x^2 + y^2 - 2x - 4y + 1 = 0$

9. $y^2 = x^2 - 1$

10. $y = \sqrt{x + 5}$

In Exercises 11–14, fill in the blanks and simplify the result.

11. $f(x) = 6 - 4x$
   (a) $f(3) = 6 - 4(\quad)$
   (b) $f(-7) = 6 - 4(\quad)$
   (c) $f(t) = 6 - 4(\quad)$
   (d) $f(c + 1) = 6 - 4(\quad)$

12. $g(x) = x^2 - 2x$
   (a) $g(2) = (\quad)^2 - 2(\quad)$
   (b) $g(-3) = (\quad)^2 - 2(\quad)$
   (c) $g(t + 1) = (\quad)^2 - 2(\quad)$
   (d) $g(x + h) = (\quad)^2 - 2(\quad)$

13. $f(s) = \dfrac{1}{s + 1}$
   (a) $f(4) = \dfrac{1}{(\quad) + 1}$
   (b) $f(0) = \dfrac{1}{(\quad) + 1}$
   (c) $f(4x) = \dfrac{1}{(\quad) + 1}$
   (d) $f(x + h) = \dfrac{1}{(\quad) + 1}$

14. $f(t) = \sqrt{25 - t^2}$
   (a) $f(3) - \sqrt{25 - (\quad)^2}$
   (b) $f(5) = \sqrt{25 - (\quad)^2}$
   (c) $f(x + 5) = \sqrt{25 - (\quad)^2}$
   (d) $f(2 + h) = \sqrt{25 - (\quad)^2}$

In Exercises 15–22, evaluate (if possible) the function at the specified value of the independent variable and simplify the result.

15. $f(x) = 2x - 3$
   (a) $f(1)$
   (b) $f(-3)$
   (c) $f(x - 1)$
   (d) $f(\frac{1}{4})$

16. $V(r) = \frac{4}{3}\pi r^3$
   (a) $V(3)$
   (b) $V(0)$
   (c) $V(\frac{3}{2})$
   (d) $V(2r)$

17. $h(t) = t^2 - 2t$
   (a) $h(2)$
   (b) $h(-1)$
   (c) $h(x + 2)$
   (d) $h(1.5)$

18. $f(x) = \sqrt{x + 8} + 2$
   (a) $f(-8)$
   (b) $f(1)$
   (c) $f(x - 8)$
   (d) $f(h + 8)$

19. $f(x) = \dfrac{|x|}{x}$
   (a) $f(2)$
   (b) $f(-2)$
   (c) $f(x^2)$
   (d) $f(x - 1)$

20. $q(t) = \dfrac{2t^2 + 3}{t^2}$
   (a) $q(2)$
   (b) $q(0)$
   (c) $q(x)$
   (d) $q(-x)$

21. $f(x) = \begin{cases} 2x + 1, & x < 0 \\ 2x + 2, & x \geq 0 \end{cases}$
   (a) $f(-1)$
   (b) $f(0)$
   (c) $f(1)$
   (d) $f(2)$

22. $f(x) = \begin{cases} x^2 + 2, & x \leq 1 \\ 2x^2 + 2, & x > 1 \end{cases}$
   (a) $f(-2)$
   (b) $f(0)$
   (c) $f(1)$
   (d) $f(2)$

In Exercises 23–30, find the domain of the function.

23. $f(x) = 5x^2 + 2x - 1$

24. $f(t) = \sqrt[3]{t + 4}$

25. $g(y) = \sqrt{y - 10}$

26. $f(x) = \sqrt[4]{1 - x^2}$

27. $h(t) = \dfrac{4}{t}$

28. $h(x) = \dfrac{10}{x^2 - 2x}$

29. $g(x) = \dfrac{1}{x} - \dfrac{3}{x + 2}$

30. $f(s) = \dfrac{\sqrt{s - 1}}{s - 4}$

In Exercises 31–34, assume that the domain of f is the set $A = \{-2, -1, 0, 1, 2\}$. Determine the set of ordered pairs representing the function f.

31. $f(x) = x^2$

32. $f(x) = \dfrac{2x}{x^2 + 1}$

33. $f(x) = \sqrt{x + 2}$

34. $f(x) = |x + 1|$

In Exercises 35–38, use a function key on a calculator to evaluate each expression. (Round the result to three decimal places.)

35. (a) 32.5^2 (b) 4.3^5

36. (a) $\sqrt{232}$ (b) $\sqrt[3]{2500}$

37. (a) $\dfrac{1}{8.5}$ (b) $\dfrac{1}{0.047}$

38. (a) $|-326.8|$ (b) $10^{3.8}$

In Exercises 39–42, select a function from (a) $f(x) = cx$, (b) $g(x) = cx^2$, (c) $h(x) = c\sqrt{|x|}$, or (d) $r(x) = c/x$ and determine the value of the constant c so that the function fits the data given in the table.

39.

x	-4	-1	0	1	4
y	-32	-2	0	-2	-32

40.

x	-4	-1	0	1	4
y	-1	$-\frac{1}{4}$	0	$\frac{1}{4}$	1

41.

x	-4	-1	0	1	4
y	-8	-32	undef.	32	8

42.

x	-4	-1	0	1	4
y	6	3	0	3	6

In Exercises 43–46, evaluate the difference quotient. Then simplify the result.

43. $f(x) = 2x$

$$\frac{f(x + h) - f(x)}{h}$$

44. $f(x) = 5x - x^2$

$$\frac{f(5 + h) - f(5)}{h}$$

45. $f(x) = x^3$

$$\frac{f(x + h) - f(x)}{h}$$

46. $f(t) = \dfrac{1}{t}$

$$\frac{f(t) - f(1)}{t - 1}$$

47. *Area of a Circle*   Express the area A of a circle as a function of its circumference C.

48. *Area of a Triangle*   Express the area A of an equilateral triangle as a function of the length s of its sides.

49. *Area of a Triangle*   A right triangle is formed in the first quadrant by the x- and y-axes and a line through the point $(1, 2)$ (see figure). Express the area of the triangle as a function of x, and determine the domain of the function.

50. *Area of a Rectangle*   A rectangle is bounded by the x-axis and the semicircle $y = \sqrt{25 - x^2}$ (see figure). Express the area of the rectangle as a function of x, and determine the domain of the function.

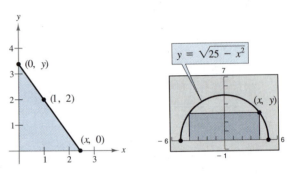

FIGURE FOR 49   FIGURE FOR 50

51. *Volume of a Package*   A rectangular package to be sent by a postal service can have a maximum combined length and girth (perimeter of a cross section) of 108 inches (see figure). Express the volume of the package as a function of x. What is the domain of the function?

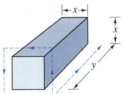

FIGURE FOR 51

52. *Volume of a Box*   An open box is to be made from a square piece of material 12 inches on a side by cutting equal squares from each corner and turning up the sides (see figure). Express the volume V of the box as a function of x. What is the domain of this function?

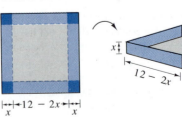

FIGURE FOR 52

53. *Height of a Balloon*   A transmitting balloon ascends vertically from a point 2000 feet from the receiving station (see figure). Let d be the distance between the balloon and the receiving station. Express the height of the balloon as a function of d. What is the domain of the function?

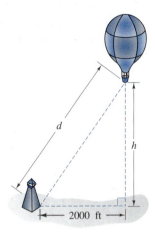

d

h

2000 ft
FIGURE FOR 53

54. *Price of Mobile Homes*   The average price p of a new mobile home in the United States from 1974 to 1988 can be modeled by the function

$$p(t) = \begin{cases} 19{,}503.6 + 1753.6t, & -6 \le t \le -1 \\ 19{,}838.8 + 81.11t^2, & 0 \le t \le 8 \end{cases}$$

where t is the year with $t = 0$ corresponding to 1980. Use this model to find the average price of a mobile home in 1978 and in 1988. (*Source:* U.S. Bureau of Census, Construction Reports)

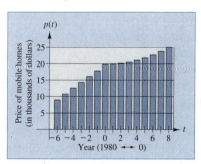

FIGURE FOR 54

55. *Cost, Revenue, and Profit*   A company invests $98,000 for equipment to produce a product. Each unit of the product costs $12.30 and is sold for $17.98. Let x be the number of units produced and sold.
(a) Write the total cost C as a function of x.
(b) Write the revenue R as a function of x.
(c) Write the profit P as a function of x.
   (*Note:* $P = R - C$)

56. *Charter Bus Fares*   For groups of 80 or more people, a charter bus company determines the rate per person according to the formula

Rate $= 8 - 0.05(n - 80)$, $n \ge 80$,

where the rate is given in dollars and n is the number of people.
(a) Express the revenue R for the bus company as a function of n.
(b) Use the function from part (a) to complete the following table.

n	90	100	110	120	130	140	150
$R(n)$							

(c) Graph R and determine the number of people that will produce a maximum revenue.

1.4 GRAPHS OF FUNCTIONS

The Graph of a Function / Increasing and Decreasing Functions / Relative Minimum and Maximum Values / Step Functions / Even and Odd Functions

The Graph of a Function

In Section 1.3 you studied functions from an algebraic point of view. In this section, you will study functions from a geometric perspective. The **graph of a function** f is the collection of ordered pairs $(x, f(x))$ such that x is in the domain of f. As you study this section, remember the following geometrical interpretation of x and $f(x)$.

$$x = \text{the directed distance from the } y\text{-axis}$$

$$f(x) = \text{the directed distance from the } x\text{-axis}$$

EXAMPLE 1 **Finding Domain and Range from the Graph of a Function**

Use the graph of the function f, shown in Figure 1.30.

A. Find the domain of f.
B. Find the function values $f(-1)$ and $f(2)$.
C. Find the range of f.

SOLUTION

A. The closed dot (on the left) indicates that $x = -1$ is in the domain of f, whereas the open dot (on the right) indicates $x = 4$ is not in the domain. Thus, the domain of f is all x in the interval $[-1, 4)$.
B. Because $(-1, -5)$ is a point on the graph of f, it follows that $f(-1) = -5$. Similarly, because $(2, 4)$ is a point on the graph of f, it follows that $f(2) = 4$.
C. Because the graph does not extend below $f(-1) = -5$ or above $f(2) = 4$, the range of f is the interval $[-5, 4]$. ⬛⬛⬛⬛⬛

By the definition of a function, at most one y-value corresponds to a given x-value. It follows, then, that a vertical line can intersect the graph of a function at most once. This observation provides a convenient visual test for functions.

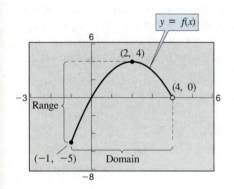

FIGURE 1.30

REMARK In Figure 1.30, the solid dot representing the point $(-1, -5)$ indicates that $(-1, -5)$ is a point on the graph. The open dot indicates that $(4, 0)$ is not a point on the graph.

VERTICAL LINE TEST FOR FUNCTIONS

A set of points in a coordinate plane is the graph of y as a function of x if and only if no vertical line intersects the graph at more than one point.

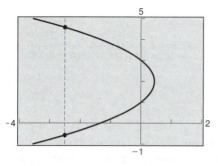

(a)

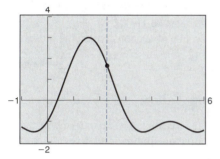

(b)

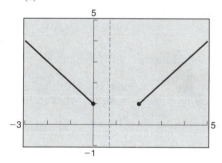

(c)

FIGURE 1.31

EXAMPLE 2 **Vertical Line Test for Functions**

Which of the graphs in Figure 1.31 represent y as a function of x?

SOLUTION

A. This is *not* a graph of y as a function of x because you can find a vertical line that intersects the graph twice.

B. This *is* a graph of y as a function of x because every vertical line intersects the graph at most once.

C. This *is* a graph of y as a function of x. (Note that if a vertical line does not intersect the graph, it simply means that the function is undefined for this particular value of x.)

Increasing and Decreasing Functions

The more you know about the graph of a function, the more you know about the function itself. Consider the graph shown in Figure 1.32. Moving from *left to right,* this graph falls from $x = -2$ to $x = 0$, is constant from $x = 0$ to $x = 2$, and rises from $x = 2$ to $x = 4$. These observations indicate that the function has the following characteristics.

1. The function is **decreasing** on the interval $(-2, 0)$.
2. The function is **constant** on the interval $(0, 2)$.
3. The function is **increasing** on the interval $(2, 4)$.

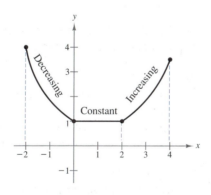

FIGURE 1.32

INCREASING, DECREASING, AND CONSTANT FUNCTIONS

A function f is **increasing** on an interval if, for any x_1 and x_2 in the interval, $x_1 < x_2$ implies $f(x_1) < f(x_2)$.

A function f is **decreasing** on an interval if, for any x_1 and x_2 in the interval, $x_1 < x_2$ implies $f(x_1) > f(x_2)$.

A function f is **constant** on an interval if, for any x_1 and x_2 in the interval, $f(x_1) = f(x_2)$.

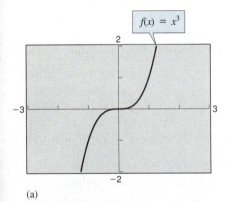

(a)

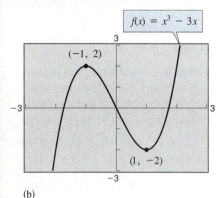

(b)

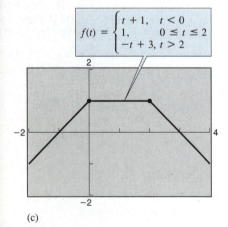

(c)

FIGURE 1.33

EXAMPLE 3 **Increasing and Decreasing Functions**

In Figure 1.33, determine the open intervals on which each function is increasing, decreasing, or constant.

SOLUTION

A. Although it might appear that there is an interval about zero over which this function is constant, you can see that if $x_1 < x_2$, then $f(x_1) = x_1^3 < x_2^3 = f(x_2)$. Thus, you can conclude that the function is increasing over the entire real line.

B. This function is increasing on the interval $(-\infty, -1)$, decreasing on the interval $(-1, 1)$, and increasing on the interval $(1, \infty)$.

C. This function is increasing on the interval $(-\infty, 0)$, constant on the interval $(0, 2)$, and decreasing on the interval $(2, \infty)$.

Relative Minimum and Maximum Values

The points at which a function changes its increasing, decreasing, or constant behavior are helpful in determining the relative maximum or relative minimum values of the function.

DEFINITION OF RELATIVE MINIMUM AND RELATIVE MAXIMUM

A function value $f(a)$ is called a **relative minimum** of f if there exists an interval (x_1, x_2) that contains a such that

$$x_1 < x < x_2 \quad \text{implies} \quad f(a) \leq f(x).$$

A function $f(a)$ is called a **relative maximum** of f if there exists an interval (x_1, x_2) that contains a such that

$$x_1 < x < x_2 \quad \text{implies} \quad f(a) \geq f(x).$$

Figure 1.34 shows two different examples of relative minimums and two of relative maximums.

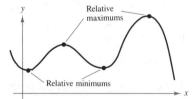

FIGURE 1.34

In Section 3.1, you will study a technique for finding the *exact points* at which a second-degree polynomial function has a relative minimum or relative maximum. For the time being, however, you can use a graphing utility to find reasonable approximations for these points.

Technology Note

When you use a graphing utility to estimate the x- and y-values of a relative minimum or relative maximum, the automatic zoom feature will often produce graphs that are nearly flat. To overcome this problem, you can manually change the vertical setting of the viewing rectangle. You can stretch the graph vertically by making the values of Y_{min} and Y_{max} closer together.

Technology Note

Some graphing utilities, such as a TI-85, can automatically determine the maximum and minimum value of a function defined on a closed interval. If your graphing utility has this feature, use it to graph $y = -x^3 + x$ on the interval $-1 \le x \le 1$ and verify the results of Example 5. What happens if you use the interval $-10 \le x \le 10$?

EXAMPLE 4   Approximating a Relative Minimum

Use a graphing utility to approximate the relative minimum of the function $f(x) = 3x^2 - 4x - 2$.

SOLUTION

The graph of f is shown in Figure 1.35. By using the zoom and trace features of a graphing utility, you can estimate that the function has a relative minimum at the point

$(0.67, -3.33)$.   *Relative minimum*

Later, in Section 3.1, you will be able to determine that the exact point at which the relative minimum occurs is $\left(\frac{2}{3}, -\frac{10}{3}\right)$.

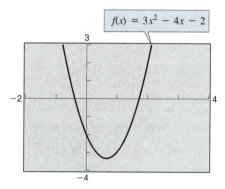

$f(x) = 3x^2 - 4x - 2$

FIGURE 1.35

EXAMPLE 5   Approximating Relative Minimums and Maximums

Use a graphing utility to approximate the relative minimum and relative maximum of the function $f(x) = -x^3 + x$.

SOLUTION

A sketch of the graph of f is shown in Figure 1.36. By using the zoom and trace features of the graphing utility, you can estimate that the function has a relative minimum at the point

$(-0.58, -0.38)$ *Relative minimum*

and a relative maximum at the point

$(0.58, 0.38)$.   *Relative maximum*

If you go on to take a course in calculus, you will learn a technique for finding the exact points at which this function has a relative minimum and a relative maximum.

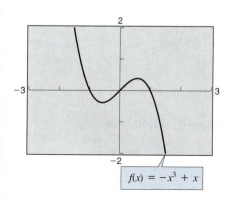

$f(x) = -x^3 + x$

FIGURE 1.36

EXAMPLE 6   The Price of Diamonds

During the 1980s the average price of a 1-carat polished diamond decreased and then increased according to the model

$$C = -0.7t^3 + 16.25t^2 - 106t + 388, \qquad 2 \le t \le 10,$$

where C is the average price in dollars (on the Antwerp Index) and t represents the calendar year with $t = 2$ corresponding to January 1, 1982. (*Source: Diamond High Council*) According to this model, during which years was the price of diamonds decreasing? During which years was the price of diamonds increasing? Approximate the minimum price of a 1-carat diamond between 1982 and 1990.

SOLUTION

To solve this problem, sketch an accurate graph of the function, as shown in Figure 1.37. From the graph, you can see that the price of diamonds decreased from 1982 until late 1984. Then from late 1984 to 1990 the price increased. The minimum price during the 8-year period was approximately $175.

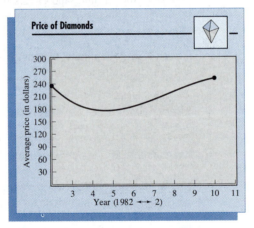

Price of Diamonds

Average price (in dollars) vs. Year (1982 ↔ 2)

FIGURE 1.37

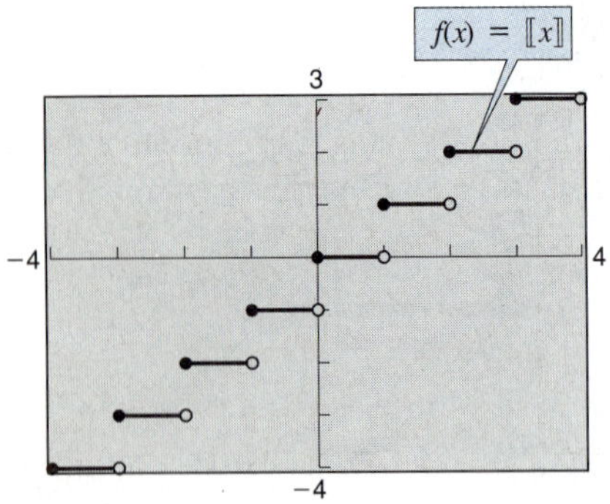

$f(x) = [[x]]$

Greatest Integer Function

FIGURE 1.38

Step Functions

EXAMPLE 7   The Greatest Integer Function

The **greatest integer function** is denoted by $[[x]]$ and is defined by

$$f(x) = [[x]] = \text{the greatest integer less than or equal to } x.$$

The graph of this function is shown in Figure 1.38. Note that the graph of the greatest integer function jumps vertically one unit at each integer and is constant (a horizontal line segment) between each pair of consecutive integers.

Because of the jumps in its graph, the greatest integer function is an example of a **step function.** Some values of the greatest integer function are as follows.

$$[[-1]] = -1 \qquad [[-0.5]] = -1$$
$$[[0]] = 0 \qquad [[0.5]] = 0$$
$$[[1]] = 1 \qquad [[1.5]] = 1$$

The range of the greatest integer function is the set of all integers.

Technology Note

Most graphing utilities display graphs in what is called a *connected mode,* which means that the graph has no breaks. Thus, when you are sketching graphs that do have breaks, it is better to change the graphing utility to *dot mode*. Try sketching the graph of the greatest integer function [often called Int(x)] on a graphing utility in connected and dot modes, and compare the two results.

EXAMPLE 8   The Cost of a Telephone Call

Suppose the cost of a telephone call between Los Angeles and San Francisco is $0.50 for the first minute and $0.36 for each additional minute. The greatest integer function can be used to create a model for the cost of this call.

$$C = 0.50 + 0.36[[t]], \qquad 0 < t,$$

where C is the total cost of the call in dollars and t is the length of the call in minutes. Sketch the graph of this function.

SOLUTION

For calls up to 1 minute, the cost is $0.50. For calls between 1 and 2 minutes, the cost is $0.86, and so on.

Length of call: $0 < t < 1$ $1 \le t < 2$ $2 \le t < 3$ $3 \le t < 4$ $4 \le t < 5$
Cost of call: $0.50 $0.86 $1.22 $1.58 $1.94

Using these values, you can sketch the graph shown in Figure 1.39.

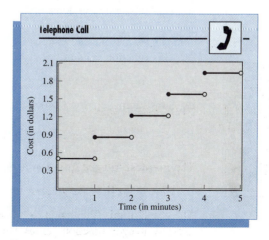

FIGURE 1.39

Even and Odd Functions

A graph has symmetry with respect to the y-axis if, whenever (x, y) is on the graph, so is the point $(-x, y)$. A function whose graph is symmetric with respect to the y-axis is an **even** function.

A graph has symmetry with respect to the origin if, whenever (x, y) is on the graph, so is the point $(-x, -y)$. A function whose graph is symmetric with respect to the origin is an **odd** function.

A graph has symmetry with respect to the x-axis if, whenever (x, y) is on the graph, so is the point $(x, -y)$. The graph of a (nonzero) function cannot be symmetric with respect to the x-axis.

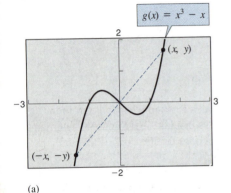

$g(x) = x^3 - x$

(x, y)

$(-x, -y)$

(a)

TEST FOR EVEN AND ODD FUNCTIONS

A function f is **even** if, for each x in the domain of f, $f(-x) = f(x)$.
A function f is **odd** if, for each x in the domain of f, $f(-x) = -f(x)$.

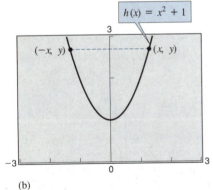

$h(x) = x^2 + 1$

$(-x, y)$ (x, y)

(b)

EXAMPLE 9   Even and Odd Functions

Determine whether the following functions are even, odd, or neither.

A. $g(x) = x^3 - x$
B. $h(x) = x^2 + 1$
C. $f(x) = x^3 - 1$

SOLUTION

A. This function is odd because

$$g(-x) = (-x)^3 - (-x) = -x^3 + x = -(x^3 - x) = -g(x).$$

B. This function is even because

$$h(-x) = (-x)^2 + 1 = x^2 + 1 = h(x).$$

C. Substituting $-x$ for x produces

$$f(-x) = (-x)^3 - 1 = -x^3 - 1.$$

Because $f(x) = x^3 - 1$ and $-f(x) = -x^3 + 1$, you can conclude that $f(-x) \neq f(x)$ and $f(-x) \neq -f(x)$. Hence, the function is neither even nor odd.

The graphs of the three functions are shown in Figure 1.40.

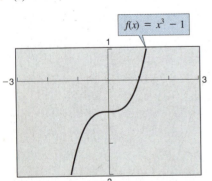

$f(x) = x^3 - 1$

(c)

FIGURE 1.40

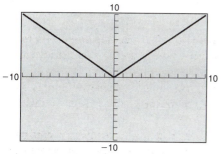

FIGURE 1.41

EXAMPLE 10   The Absolute Value Function

Figure 1.41 shows the V-shape of the absolute value function, $f(x) = |x|$. This was produced using the "abs" key and the standard viewing rectangle. Because the graph is symmetric about the y-axis, you can see that it is an even function. You can verify this by observing that $f(-x) = |-x| = |x| = f(x)$.

DISCUSSION PROBLEM

INCREASING AND DECREASING FUNCTIONS

Use your school's library or some other reference source to find examples of three different functions that represent quantities between 1980 and 1990. Find one that decreased during the decade, one that increased, and one that was constant. For instance, the value of the dollar decreased, the population of the United States increased, and the land size of the United States remained constant. Can you find three other examples? Present your results graphically.

WARM-UP

The following warm-up exercises involve skills that were covered in earlier sections. You will use these skills in the exercise set for this section.

1. Find $f(2)$ for $f(x) = -x^3 + 5x$.

2. Find $f(6)$ for $f(x) = x^2 - 6x$.

3. Find $f(-x)$ for $f(x) = \dfrac{3}{x}$.

4. Find $f(-x)$ for $f(x) = x^2 + 3$.

In Exercises 5 and 6, find the x- and y-intercepts of the graph of the line.

5. $5x - 2y + 12 = 0$

6. $-0.3x - 0.5y + 1.5 = 0$

In Exercises 7–10, find the domain of the function.

7. $g(x) = \dfrac{4}{x - 4}$

8. $f(x) = \dfrac{2x}{x^2 - 9x + 20}$

9. $h(t) = \sqrt[4]{5 - 3t}$

10. $f(t) = t^3 + 3t - 5$

SECTION 1.4 · EXERCISES

In Exercises 1–6, use a graphing utility to graph the function and find its domain and range.

1. $f(x) = \sqrt{x - 1}$ **2.** $f(x) = 4 - x^2$

3. $f(x) = \sqrt{x^2 - 4}$ **4.** $f(x) = |x - 2|$

5. $f(x) = \sqrt{25 - x^2}$ **6.** $f(x) = \dfrac{|x - 2|}{x - 2}$

In Exercises 7–12, use the vertical line test to determine whether y is a function of x. Describe how your graphing utility can be used to produce the given graph.

7. $y = x^2$ **8.** $y = x^3 - 1$

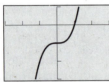

9. $x - y^2 = 0$ **10.** $x^2 + y^2 = 9$

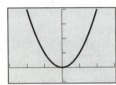

11. $x^2 = xy - 1$ **12.** $x = |y|$

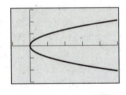

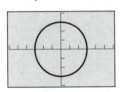

In Exercises 13–16, select the viewing rectangle on a graphing utility that shows the most complete graph of the function.

13. $f(x) = -0.2x^2 + 3x + 32$

(a)
```
RANGE
Xmin=-2
Xmax=20
Xscl=1
Ymin=-10
Ymax=30
Yscl=4
```
(b)
```
RANGE
Xmin=-10
Xmax=30
Xscl=5
Ymin=-5
Ymax=50
Yscl=5
```
(c)
```
RANGE
Xmin=0
Xmax=10
Xscl=0.5
Ymin=0
Ymax=200
Yscl=25
```

14. $f(x) = 6[x - (0.1x)^5]$

(a)
```
RANGE
Xmin=-500
Xmax=500
Xscl=50
Ymin=-500
Ymax=500
Yscl=50
```
(b)
```
RANGE
Xmin=-25
Xmax=25
Xscl=5
Ymin=-25
Ymax=25
Yscl=5
```
(c)
```
RANGE
Xmin=-20
Xmax=20
Xscl=5
Ymin=-100
Ymax=100
Yscl=20
```

15. $f(x) = 4x^3 - x^4$

(a)
```
RANGE
Xmin=-2
Xmax=6
Xscl=1
Ymin=-10
Ymax=30
Yscl=4
```
(b)
```
RANGE
Xmin=-50
Xmax=50
Xscl=5
Ymin=-50
Ymax=50
Yscl=5
```
(c)
```
RANGE
Xmin=0
Xmax=2
Xscl=0.2
Ymin=-2
Ymax=2
Yscl=0.5
```

16. $f(x) = 10x\sqrt{400 - x^2}$

(a)
```
RANGE
Xmin=-5
Xmax=50
Xscl=5
Ymin=-5000
Ymax=5000
Yscl=500
```
(b)
```
RANGE
Xmin=-20
Xmax=20
Xscl=2
Ymin=-500
Ymax=500
Yscl=50
```
(c)
```
RANGE
Xmin=-25
Xmax=25
Xscl=5
Ymin=-2000
Ymax=2000
Yscl=200
```

In Exercises 17–24, use a graphing utility to graph the function. Approximate the intervals over which the function is increasing, decreasing, or constant.

17. $f(x) = 2x$ **18.** $f(x) = x^2 - 2x$

19. $f(x) = x^3 - 3x^2$ **20.** $f(x) = \sqrt{x^2 - 4}$

21. $f(x) = 3x^4 - 6x^2$ **22.** $f(x) = x^{2/3} = (x^2)^{1/3}$

23. $f(x) = x\sqrt{x + 3}$ **24.** $f(x) = |x + 1| + |x - 1|$

In Exercises 25–30, use a graphing utility to approximate (to two decimal place accuracy) any relative minimum or maximum values of the function.

25. $f(x) = x^2 - 6x$ **26.** $f(x) = (x - 1)^2(x + 2)$

27. $g(x) = 2x^3 + 3x^2 - 12x$ **28.** $g(x) = x^3 - 6x^2 + 15$

29. $h(x) = (x - 1)\sqrt{x}$ **30.** $h(x) = x\sqrt{4 - x}$

31. *Maximum Area*   The perimeter of a rectangle is 100 feet (see figure).

(a) Show that the area of the rectangle is given by $A = x(50 - x)$, where x is its length.

(b) Use a graphing utility to graph the area function.

(c) Move the cursor along the graph to approximate (to two decimal place accuracy) the maximum of the area function. Approximate the dimensions of the rectangle that yield the maximum area.

FIGURE FOR 31

32. *Minimum Cost*   A power station is on one side of a river that is $\frac{1}{2}$ mile wide, and a factory is 6 miles downstream on the other side (see figure). It costs $6 per foot to run power lines overland and $8 per foot to run them underwater.

(a) Show that the total cost for the power line is given by
$$T = 8(5280)\sqrt{x^2 + \tfrac{1}{4}} + 6(5280)(6 - x).$$

(b) Use a graphing utility to graph the cost function.

(c) To determine the most economical path for the power line, approximate (to two decimal place accuracy) the value of x that minimizes the cost function.

Factory

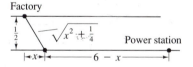

Power station

$\leftarrow x \rightarrow$ 6 − x   **FIGURE FOR 32**

33. *Maximum Profit*   The marketing department of a company estimates that the demand for a product is given by $p = 100 - 0.0001x$, where p is the price per unit and x is the number of units. The cost of producing x units is given by $C = 350,000 + 30x$, and the profit for producing and selling x units is given by

$$P = R - C = xp - C.$$

Graph the profit function and estimate the number of units that would produce a maximum profit.

34. *Maximum Revenue*   When a wholesaler sold a certain product at $25 per unit, sales were 800 units per week. After a price increase of $5, the average number of units sold dropped to 775 per week. Assume that the price p is a linear function of the demand x. Graph the revenue function $R = xp$ and estimate (to two decimal place accuracy) the price that will maximize total revenue.

In Exercises 35–38, use a graphing utility to obtain the graph of the compound function.

35. $f(x) = \begin{cases} 2x + 3, & x < 0 \\ 3 - x, & x \geq 0 \end{cases}$

36. $f(x) = \begin{cases} \sqrt{4 + x}, & x < 0 \\ \sqrt{4 - x}, & x \geq 0 \end{cases}$

37. $f(x) = \begin{cases} x^2 + 5, & x \leq 1 \\ -x^2 + 4x + 3, & x > 1 \end{cases}$

38. $f(x) = \begin{cases} 1 - (x - 1)^3, & x \leq 2 \\ \sqrt{x - 2}, & x > 2 \end{cases}$

In Exercises 39 and 40, use a graphing utility to graph the step function. State the domain and range of the function.

39. $s(x) = 2[[x - 1]]$ **40.** $g(x) = 6 - [[x]]$

41. *Price of a Telephone Call*   The cost of a telephone call between two cities is $0.65 for the first minute and $0.42 for each additional minute (or portion thereof).

(a) Use the greatest integer function to create a model for the cost C of a telephone call between the two cities lasting t minutes.

(b) Use a graphing utility to graph the cost model.

(c) Move the cursor along the graph to determine the maximum length of a call that cannot cost more than $6.

42. *Cost of Overnight Delivery*   Suppose the cost of sending an overnight package from New York to Atlanta is $9.80 for the first pound and $2.50 for each additional pound (or portion thereof).

(a) Use the greatest integer function to create a model for the cost C of overnight delivery of a package weighing x pounds.

(b) Use a graphing utility to graph the cost function.

(c) Move the cursor along the graph to determine the cost of overnight delivery of a package weighing 12 pounds.

In Exercises 43–46, find the coordinates of a second point on the graph of a function f if the given point is on the graph and the function is (a) even and (b) odd.

43. $(5, 6)$ **44.** $(-3, 7)$

45. $\left(-\frac{3}{2}, -2\right)$ **46.** $\left(\frac{3}{4}, -\frac{7}{8}\right)$

In Exercises 47–56, use a graphing utility to graph the function and determine whether the function is even, odd, or neither.

47. $f(x) = 3$ **48.** $g(x) = x$

49. $f(x) = 5 - 3x$ **50.** $h(x) = x^2 - 4$

51. $g(s) = \dfrac{s^3}{4}$ **52.** $f(t) = -t^4$

53. $f(x) = \sqrt{1-x}$ **54.** $f(x) = x^2\sqrt[3]{x}$

55. $g(x) = \dfrac{5x}{x^2+1}$ **56.** $f(x) = |x+2|$

In Exercises 57–62, use an algebraic test to determine whether the function is even, odd, or neither. Verify your result graphically.

57. $f(x) = \frac{1}{3}x^6 - 2x^2$ **58.** $h(x) = x^3 - 5$

59. $g(x) = x^3 - 5x$ **60.** $f(x) = x\sqrt{1-x^2}$

61. $f(t) = t^2 + 2t - 3$ **62.** $g(s) = 4s^{2/3}$

In Exercises 63–68, graph the function and determine the interval(s) (if any) on the real axis for which $f(x) \geq 0$.

63. $f(x) = 4 - x$ **64.** $f(x) = x^2 - 4x$

65. $f(x) = 1 - x^4$ **66.** $f(x) = \sqrt{x+2}$

67. $f(x) = 5$ **68.** $f(x) = -(1 + |x|)$

In Exercises 69–72, express the height h of the rectangle as a function of x.

69.

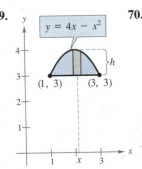

70.

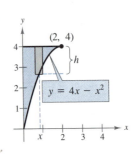

71.

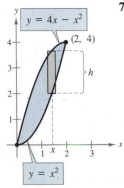

72.

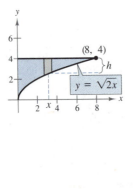

In Exercises 73 and 74, express the length L of the rectangle as a function of y.

73.

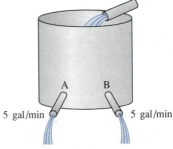

74.

75. *Water Intake*   When the intake pipe of a 100-gallon tank is turned on, water flows into the tank at the rate of 10 gallons per minute. When either of the two drain pipes is turned on, water flows out of the tank at the rate of 5 gallons per minute. The graph below shows the volume of water in the tank over a 60-minute period. Which pipes are turned on during the various periods of time indicated by the graph? (There is more than one correct answer.)

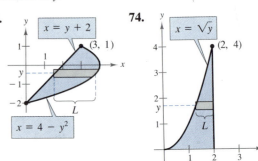

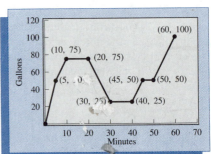

FIGURE FOR 75

76. Prove that the following function is odd.

$$f(x) = a_{2n+1}x^{2n+1} + a_{2n-1}x^{2n-1} + \cdots + a_3x^3 + a_1x$$

77. Prove that the following function is even.

$$f(x) = a_{2n}x^{2n} + a_{2n-2}x^{2n-2} + \cdots + a_2x^2 + a_0$$

1.5   SHIFTING, REFLECTING, AND STRETCHING GRAPHS

Summary of Graphs of Common Functions / Vertical and Horizontal
Shifts / Reflections / Nonrigid Transformations

Summary of Graphs of Common Functions

One of the goals of this text is to enable you to build your intuition for the basic
shapes of the graphs of different types of functions. For instance, from your
study of lines in Section 1.2, you can determine the basic shape of the graph
of the linear function

$$f(x) = ax + b.$$

Specifically, you know that the graph of this function is a line whose slope is
a and whose y-intercept is b.

The six graphs shown in Figure 1.42 represent the most commonly used
functions in algebra. Familiarity with the basic characteristics of these simple
graphs will help you analyze the shapes of more complicated graphs. Try using
a graphing utility to verify these graphs.

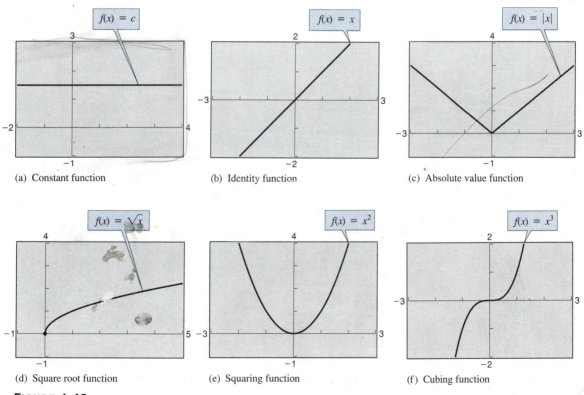

(a) Constant function

(b) Identity function

(c) Absolute value function

(d) Square root function

(e) Squaring function

(f) Cubing function

FIGURE 1.42

Vertical and Horizontal Shifts

Many functions have graphs that are simple transformations of the common graphs summarized in Figure 1.42. For example, you can obtain the graph of $h(x) = x^2 + 2$ by shifting the graph of $f(x) = x^2$ *up* two units, as shown in Figure 1.43. In function notation, h and f are related as follows.

$$h(x) = x^2 + 2 = f(x) + 2 \qquad \textit{Upward shift of 2}$$

Similarly, you can obtain the graph of $g(x) = (x - 2)^2$ by shifting the graph of $f(x) = x^2$ to the *right* two units, as shown in Figure 1.44. In this case, the functions g and f have the following relationship.

$$g(x) = (x - 2)^2 = f(x - 2) \qquad \textit{Right shift of 2}$$

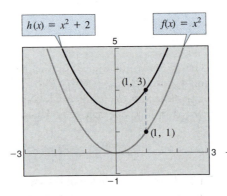

Vertical shift upward: 2 units

FIGURE 1.43

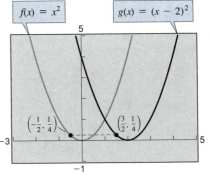

Horizontal shift to the right: 2 units

FIGURE 1.44

VERTICAL AND HORIZONTAL SHIFTS

Let c be a positive real number. **Vertical** and **horizontal shifts** in the graph of $y = f(x)$ are represented as follows.

1. Vertical shift c units **upward:** $h(x) = f(x) + c$
2. Vertical shift c units **downward:** $h(x) = f(x) - c$
3. Horizontal shift c units to the **right:** $h(x) = f(x - c)$
4. Horizontal shift c units to the **left:** $h(x) = f(x + c)$

EXAMPLE 1   **Shifts in the Graph of a Function**

Compare the graph of each function with the graph of $f(x) = x^3$.

A. $g(x) = x^3 + 1$ **B.** $h(x) = (x - 1)^3$ **C.** $k(x) = (x + 2)^3 + 1$

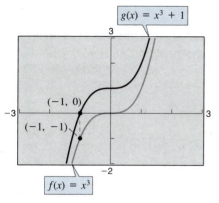

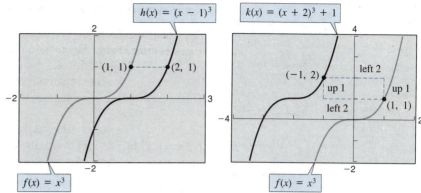

(a) Vertical shift: 1 unit up

(b) Horizontal shift: 1 unit right

(c) Horizontal shift: 2 units left
    Vertical shift: 1 unit up

FIGURE 1.45

SOLUTION

Relative to the graph of $f(x) = x^3$, the graph of $g(x) = x^3 + 1$ is an upward shift of one unit. The graph of $h(x) = (x - 1)^3$ is a right shift of one unit. The graph of $k(x) = (x + 2)^3 + 1$ involves a left shift of two units *followed* by an upward shift of one unit. (See Figure 1.45.)

EXAMPLE 2   Finding Equations from Graphs

Each of the graphs shown in Figure 1.46 is a vertical or horizontal shift of the graph of $f(x) = x^2$. Find an equation for each function.

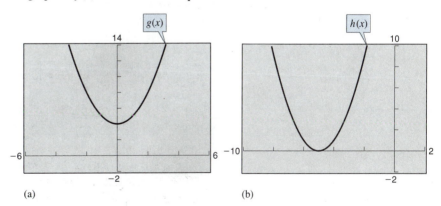

FIGURE 1.46   (a)   (b)

SOLUTION

A. The graph of g is a vertical shift of four units upward of the graph of $f(x) = x^2$. Thus, the equation for g is $g(x) = x^2 + 4$.

B. The graph of h is a horizontal shift of five units to the left of the graph of $f(x) = x^2$. Thus, the equation for h is $h(x) = (x + 5)^2$.

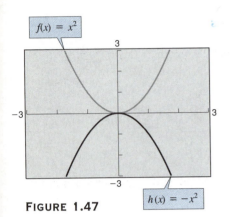

FIGURE 1.47

Reflections

The second common type of transformation is called a **reflection.** For instance, if you consider the x-axis to be a mirror, the graph of

$$h(x) = -x^2$$

is the mirror image (or reflection) of the graph of

$$f(x) = x^2,$$

as shown in Figure 1.47.

REFLECTIONS IN THE COORDINATE AXES

Reflections in the coordinate axes of the graph of $y = f(x)$ are represented as follows.

1. Reflection in the x-axis: $h(x) = -f(x)$
2. Reflection in the y-axis: $h(x) = f(-x)$

EXAMPLE 3   Finding Equations from Graphs

Each of the graphs shown in Figure 1.48(a) and 1.48(b) is a transformation of the graph of $f(x) = x^4$ (shown at left in Figure 1.48). Find an equation for each function.

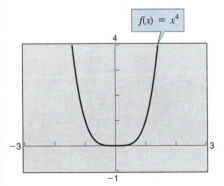

FIGURE 1.48

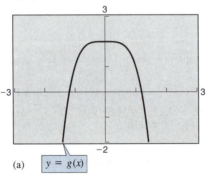

(a) $y = g(x)$

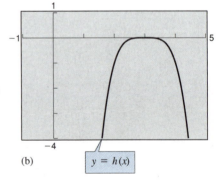

(b) $y = h(x)$

SOLUTION

A. The graph of g is a reflection in the x-axis *followed* by an upward shift of two units of the graph of $f(x) = x^4$. Thus, the equation for g is $g(x) = -x^4 + 2$.

B. The graph of h is a horizontal shift of three units to the right followed by a reflection in the x-axis of the graph of $f(x) = x^4$. Thus, the equation for h is $h(x) = -(x - 3)^4$.

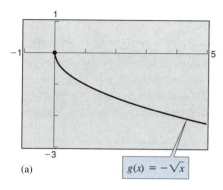

(a)

$$g(x) = -\sqrt{x}$$

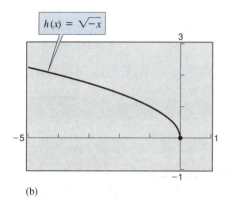

$$h(x) = \sqrt{-x}$$

(b)

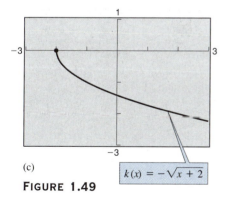

(c)

FIGURE 1.49

$$k(x) = -\sqrt{x + 2}$$

EXAMPLE 4   Reflections and Shifts

Compare the graph of each function with the graph of $f(x) = \sqrt{x}$.

A. $g(x) = -\sqrt{x}$

B. $h(x) = \sqrt{-x}$

C. $k(x) = -\sqrt{x + 2}$

SOLUTION

A. Relative to the graph of $f(x) = \sqrt{x}$, the graph of g is a reflection in the x-axis because

$$g(x) = -\sqrt{x} = -f(x).$$

B. The graph of h is a reflection of the graph of $f(x) = \sqrt{x}$ in the y-axis because

$$h(x) = \sqrt{-x} = f(-x).$$

C. From the equation

$$k(x) = -\sqrt{x + 2} = -f(x + 2)$$

you can conclude that the graph of k is a left shift of two units, followed by a reflection in the x-axis.

The graphs of all three functions are shown in Figure 1.49.

EXAMPLE 5   A Program for Practice

If you have a programmable calculator, try entering the following program. (These program steps are for a Texas Instruments TI-81. Programs for other calculators are listed in the appendix.)

```
:Rand→H                    :-6→Ymin
:-6+Int (12H)→H            :6→Ymax
:Rand→V                    :1→Yscl
:-3+ Int (6V)→V            :DispGraph
:Rand→R                    :Pause
:If R<.5                   :Disp "Y=R(X+H)²+V"
:-1→R                      :Disp "R="
:If R>.49                  :Disp R
:1→R                       :Disp "H="
:"R(X+H)²+V"→Y₁            :Disp H
:-9→Xmin                   :Disp "V="
:9→Xmax                    :Disp V
:1→Xscl                    :End
```

This program will sketch a graph of the function

$$y = R(x + H)^2 + V,$$

where $R = \pm 1$, H is an integer between -6 and 6, and V is an integer between -3 and 3. (Each time you run the program, different values of R, H, and V are possible.) From the graph, you should be able to determine the values of R, H, and V. After you have determined the values, press ENTER to see the answer. (To look at the graph again, press GRAPH.)

For example, in the graph shown in Figure 1.50, you can make the following conclusions. Because the graph of $f(x) = x^2$ has been reflected about the x-axis, you know that $R = -1$. Because the graph of $f(x) = x^2$ has been shifted four units to the left, you know that $H = 4$. Because the graph of $f(x) = x^2$ has been shifted three units up, you know that $V = 3$. Thus, the equation of the graph shown in Figure 1.50 must be

$$y = -(x + 4)^2 + 3.$$

Try running this program several times. It will give you practice in working with reflections, horizontal shifts, and vertical shifts.

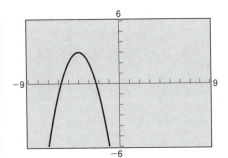

FIGURE 1.50

Nonrigid Transformations

Horizontal shifts, vertical shifts, and reflections are called **rigid** transformations because the basic shape of the graph is unchanged. These transformations change only the *position* of the graph in the xy-plane. **Nonrigid** transformations are those that cause a *distortion*—a change in the shape of the original graph. For instance, a nonrigid transformation of the graph of $y = f(x)$ is represented by $y = cf(x)$, where the transformation is a **vertical stretch** if $c > 1$ and a **vertical shrink** if $0 < c < 1$.

EXAMPLE 6   Nonrigid Transformations

Compare the graph of each function with the graph of $f(x) = |x|$.

A. $h(x) = 3|x|$ **B.** $g(x) = \dfrac{1}{3}|x|$

SOLUTION

A. Relative to the graph of $f(x) = |x|$, the graph of

$$h(x) = 3|x| = 3f(x)$$

is a vertical stretch (multiply each y-value by 3) of the graph of f.

B. Similarly, the equation

$$g(x) = \frac{1}{3}|x| = \frac{1}{3}f(x)$$

indicates that the graph of g is a vertical shrink of the graph of f.

The graphs of all three functions are shown in Figure 1.51.

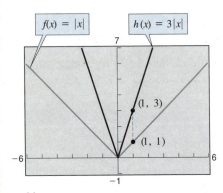

(a)

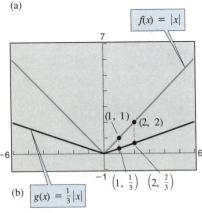

(b)

FIGURE 1.51

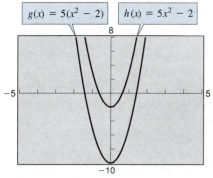

FIGURE 1.52

EXAMPLE 7   Sequences of Transformations

Use a graphing utility to graph the two functions $g(x) = 5(x^2 - 2)$ and $h(x) = 5x^2 - 2$ on the same screen. Describe how each function was obtained from $f(x) = x^2$ as a sequence of shifts and stretches.

SOLUTION

Notice that the two graphs in Figure 1.52 are different. The graph of g is a downward shift of two units followed by a vertical stretch, whereas h is a vertical stretch followed by a downward shift of two units. Hence, the order of applying the transformations is important.

DISCUSSION PROBLEM

COMPARING MATHEMATICAL MODELS

The following two second-degree polynomial functions are models of the United States population from 1800 to 1990. In the first model, $f(t)$ represents the population and t represents the year.

Model 1

$$f(t) = 6722.5t^2 - 24,201,000t + 21,787,790,000, \qquad 1800 \le t \le 1990$$

In the second model, $g(t)$ represents the population in millions and t represents the year, with $t = 0$ corresponding to 1800, $t = 1$ corresponding to 1810, and so on. (See figure.)

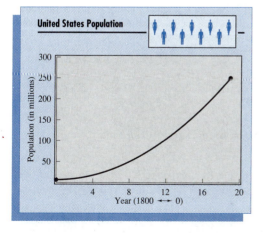

Model 2

$$g(t) = 0.67225t^2 + 6.89, \qquad 0 \le t \le 19$$

These two models are related as follows.

$$f(t) = 1,000,000g\left(\frac{t - 1800}{10}\right)$$

Which model do you think is better? Write a short paragraph explaining your reasons.

WARM-UP

The following warm-up exercises involve skills that were covered in earlier sections. You will use these skills in the exercise set for this section.

In Exercises 1–6, graph the function. Determine its domain and range.

1. $f(x) = \sqrt{16 - x^2}$
2. $h(x) = \frac{1}{2}|x - 3|$
3. $g(x) = \frac{1}{4}x^2 - 1$
4. $f(x) = 1 - \frac{1}{3}x^3$
5. $f(s) = \dfrac{|s - 2|}{s - 2}$
6. $g(t) = \sqrt{t - 2}$

In Exercises 7–10, determine whether the function is even, odd, or neither.

7. $p(x) = -0.25x$
8. $h(x) = x^2 - 3$
9. $g(t) = \frac{3}{8}t^4 - t^2$
10. $f(s) = s^{4/3} + 3$

SECTION 1.5 · EXERCISES

In Exercises 1–8, sketch the graphs of the three functions *by hand* on the same coordinate plane. Verify your result with a graphing utility.

1. $f(x) = x$
  $g(x) = x - 4$
  $h(x) = 3x$

2. $f(x) = \frac{1}{2}x$
  $g(x) = \frac{1}{2}x + 2$
  $h(x) = \frac{1}{2}(x - 2)$

3. $f(x) = x^2$
  $g(x) = x^2 + 2$
  $h(x) = (x - 2)^2$

4. $f(x) = x^2$
  $g(x) = x^2 - 4$
  $h(x) = (x + 2)^2 + 1$

5. $f(x) = -x^2$
  $g(x) = -x^2 + 1$
  $h(x) = -(x - 2)^2$

6. $f(x) = (x - 2)^2$
  $g(x) = (x - 2)^2 + 2$
  $h(x) = -(x - 2)^2 + 4$

7. $f(x) = x^2$
  $g(x) = (\frac{1}{2}x)^2$
  $h(x) = (2x)^2$

8. $f(x) = x^2$
  $g(x) = (\frac{1}{4}x)^2 + 2$
  $h(x) = -(\frac{1}{4}x)^2$

11.
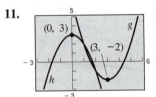

12.
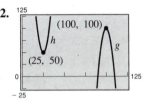

In Exercises 13–16, use a graphing utility to graph the three functions on the same set of coordinate axes. Describe the graphs of g and h relative to the graph of f.

13. $f(x) = x^3 - 3x^2$
  $g(x) = f(x + 2)$
  $h(x) = f(\frac{1}{2}x)$

14. $f(x) = x^3 - 3x^2 + 2$
  $g(x) = f(x - 1)$
  $h(x) = f(2x)$

15. $f(x) = x^3 - 3x^2$
  $g(x) = -\frac{1}{3}f(x)$
  $h(x) = f(-x)$

16. $f(x) = x^3 - 3x^2 + 2$
  $g(x) = -f(x)$
  $h(x) = f(-x)$

In Exercises 9–12, use the graph of $f(x) = x^2$ to write formulas for the functions g and h shown in the given graphs.

In Exercises 17 and 18, use the graph of $f(x) = x^3 - 3x^2$ (see Exercise 13) to write a formula for the function g shown in the given graph.

9.
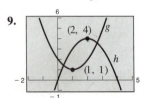

10.
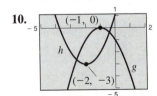

17.
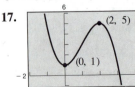

18.
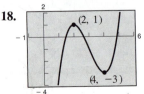

In Exercises 19–24, compare the graph of the given function with the graph of $f(x) = \sqrt{x}$.

19. $y = \sqrt{x} + 2$

20. $y = -\sqrt{x}$

21. $y = \sqrt{x} - 2$

22. $y = \sqrt{x} + 3$

23. $y = \sqrt{2x}$

24. $y = \sqrt{-x}$

In Exercises 25–30, compare the graph of the given function with the graph of $f(x) = \sqrt[3]{x}$.

25. $y = \sqrt[3]{x} - 1$

26. $y = \sqrt[3]{x + 1}$

27. $y = \sqrt[3]{x - 1}$

28. $y = -\sqrt[3]{x} - 2$

29. $y = \sqrt[3]{-x}$

30. $y = \frac{1}{2}\sqrt[3]{x}$

In Exercises 31–36, use the graph of $f(x)$ (see figure) to sketch the graph of the specified function.

31. $f(x - 4)$

32. $f(x + 2)$

33. $f(x) + 4$

34. $f(x) - 1$

35. $2f(x)$

36. $\frac{1}{2}f(x)$

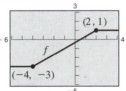

FIGURE FOR 31–36

In Exercises 37–42, specify a sequence of transformations that will yield the graph of the given function from the graph of the function $f(x) = x^3$.

37. $g(x) = 4 - x^3$

38. $g(x) = (x - 4)^3$

39. $h(x) = \frac{1}{4}(x + 2)^3$

40. $h(x) = -2(x - 1)^3 + 3$

41. $p(x) = (\frac{1}{3}x)^3 + 2$

42. $p(x) = [3(x - 2)]^3$

43. *Profit*   The profit P per week on a certain product is given by the model

$$P(x) = 80 + 20x - 0.5x^2, \qquad 0 \le x \le 20,$$

where x is the amount spent on advertising. In this model, x and P are both measured in hundreds of dollars.

(a) Use a graphing utility to graph the profit function.

(b) The business estimates that taxes and operating costs will increase by an average of $2500 per week during the next year. Rewrite the profit equation to reflect this expected decrease in profits. Identify the type of transformation made to the graph of the equation.

(c) Rewrite the original profit equation so that x measures advertising expenditures in dollars. [Find $P(x/100)$.] Describe the transformation made to the graph of the profit function.

44. *Automobile Aerodynamics*   The number of horsepower H required to overcome wind drag on a certain automobile is approximated by

$$H(x) = 0.002x^2 + 0.005x - 0.029, \qquad 10 \le x \le 100,$$

where x is the speed of the car in miles per hour.

(a) Use a graphing utility to graph the power function.

(b) Rewrite the power function so that x represents the speed in kilometers per hour. [Find $H(1.6x)$.] Describe the transformation made to the graph of the power function.

45. Use a graphing utility to graph $f(x) = x^2$, $g(x) = x^4$, and $h(x) = x^6$ in the same viewing rectangle. Describe any similarities and differences you observe among the graphs.

46. Use a graphing utility to graph $f(x) = x$, $g(x) = x^3$, and $h(x) = x^5$ in the same viewing rectangle. Describe any similarities and differences you observe among the graphs.

47. Use the results of Exercises 45 and 46 to sketch the graphs of $f(x) = x^{10}$ and $g(x) = x^{11}$, without the aid of a graphing utility.

48. Use the results of Exercises 45 and 46 to sketch the graphs of $f(x) = (x - 3)^3$, $g(x) = (x + 1)^2$, and $h(x) = (x - 4)^5$, without the aid of a graphing utility.

In Exercises 49–52, use the results of Exercises 45–48 and your imagination to guess the shape of the graph of the function. Then use a graphing utility to graph the function, and compare the result with the graph drawn by hand.

49. $f(x) = x^2(x - 6)^2$

50. $f(x) = x^3(x - 6)^2$

51. $f(x) = x^2(x - 6)^3$

52. $f(x) = x^3(x - 6)^3$

1.6 COMBINATIONS OF FUNCTIONS
....

Arithmetic Combinations of Functions / **Compositions of Functions** /
Applications

Arithmetic Combinations of Functions

Just as two real numbers can be combined by the operations of addition, subtraction, multiplication, and division to form other real numbers, two *functions* can be combined to create new functions. For example, if

$$f(x) = 2x - 3 \quad \text{and} \quad g(x) = x^2 - 1,$$

you can form the sum, difference, product, and quotient of f and g as follows.

$$f(x) + g(x) = (2x - 3) + (x^2 - 1) = x^2 + 2x - 4 \qquad \textit{Sum}$$
$$f(x) - g(x) = (2x - 3) - (x^2 - 1) = -x^2 + 2x - 2 \qquad \textit{Difference}$$
$$f(x)g(x) = (2x - 3)(x^2 - 1) = 2x^3 - 3x^2 - 2x + 3 \qquad \textit{Product}$$
$$\frac{f(x)}{g(x)} = \frac{2x - 3}{x^2 - 1}, \qquad x \neq \pm 1 \qquad \textit{Quotient}$$

The domain of an arithmetic combination of functions f and g consists of all real numbers that are common to the domains of f and g. In the case of the quotient $f(x)/g(x)$, there is the further restriction that $g(x) \neq 0$.

DEFINITIONS OF SUM, DIFFERENCE, PRODUCT, AND QUOTIENT OF FUNCTIONS

Let f and g be two functions with overlapping domains. Then, for all x common to both domains, the **sum, difference, product,** and **quotient** of f and g are defined as follows.

1. Sum: $\quad (f + g)(x) = f(x) + g(x)$
2. Difference: $\quad (f - g)(x) = f(x) - g(x)$
3. Product: $\quad (fg)(x) = f(x) \cdot g(x)$
4. Quotient: $\quad \left(\dfrac{f}{g}\right)(x) = \dfrac{f(x)}{g(x)}, \quad g(x) \neq 0$

EXAMPLE 1 **Finding the Sum of Two Functions**
....

Given $f(x) = 2x + 1$ and $g(x) = x^2 + 2x - 1$, find $(f + g)(x)$. Then evaluate this sum when $x = 2$.

SOLUTION

The sum of the functions f and g is given by

$$(f + g)(x) = f(x) + g(x)$$
$$= (2x + 1) + (x^2 + 2x - 1)$$
$$= x^2 + 4x.$$

When $x = 2$, the value of this sum is

$$(f + g)(2) = 2^2 + 4(2) = 12.$$

EXAMPLE 2   Finding the Difference of Two Functions

Given $f(x) = 2x + 1$ and $g(x) = x^2 + 2x - 1$, find $(f - g)(x)$. Then evaluate this difference when $x = 2$.

SOLUTION

The difference of the functions f and g is given by

$$(f - g)(x) = f(x) - g(x)$$
$$= (2x + 1) - (x^2 + 2x - 1)$$
$$= -x^2 + 2.$$

When $x = 2$, the value of this difference is

$$(f - g)(2) = -(2)^2 + 2 = -2.$$

In Examples 1 and 2, both f and g have domains that consist of all real numbers. Thus, the domain of both $(f + g)$ and $(f - g)$ is also the set of all real numbers. Remember that any restrictions on the domains of f or g must be taken into account when forming the sum, difference, product, or quotient of f and g. For instance, the domain of $f(x) = 1/x$ is all $x \neq 0$, and the domain of $g(x) = \sqrt{x}$ is $[0, \infty)$. This implies that the domain of $f + g$ is $(0, \infty)$.

EXAMPLE 3   The Quotient of Two Functions

Find $(f/g)(x)$ and $(g/f)(x)$ for the functions $f(x) = \sqrt{x}$ and $g(x) = \sqrt{4 - x^2}$. Then find the domains of f/g and g/f.

SOLUTION

The quotient of f and g is given by

$$\left(\frac{f}{g}\right)(x) = \frac{f(x)}{g(x)} = \frac{\sqrt{x}}{\sqrt{4 - x^2}},$$

and the quotient of g and f is given by

$$\left(\frac{g}{f}\right)(x) = \frac{g(x)}{f(x)} = \frac{\sqrt{4 - x^2}}{\sqrt{x}}.$$

The domain of f is $[0, \infty)$ and the domain of g is $[-2, 2]$. The intersection of these two domains is $[0, 2]$. Thus, the domains for f/g and g/f are as follows.

$$\text{Domain of } \frac{f}{g}: [0, 2) \qquad \text{Domain of } \frac{g}{f}: (0, 2]$$

Can you see why these two domains differ slightly?

Compositions of Functions

Another way of combining two functions is to form the **composition** of one with the other. For instance, if $f(x) = x^2$ and $g(x) = x + 1$, then the composition of f with g is

$$f(g(x)) = f(x + 1) = (x + 1)^2.$$

This composition is denoted as $f \circ g$.

DEFINITION OF COMPOSITION OF TWO FUNCTIONS

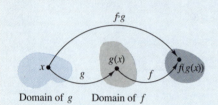

The **composition** of the functions f and g is

$$(f \circ g)(x) = f(g(x)).$$

The domain of $f \circ g$ is the set of all x in the domain of g such that $g(x)$ is in the domain of f. (See figure.)

EXAMPLE 4   Forming the Composition of f with g

Find $(f \circ g)(x)$ for $f(x) = \sqrt{x}, x \geq 0$, and $g(x) = x - 1, x \geq 1$. If possible, find $(f \circ g)(2)$ and $(f \circ g)(0)$.

SOLUTION

$$
\begin{aligned}
(f \circ g)(x) &= f(g(x)) & & \textit{Definition of } f \circ g \\
&= f(x - 1) & & \textit{Definition of } g(x) \\
&= \sqrt{x - 1}, & x \geq 1 & \textit{Definition of } f(x)
\end{aligned}
$$

The domain of $f \circ g$ is $[1, \infty)$. Thus,

$$(f \circ g)(2) = \sqrt{2 - 1} = 1$$

is defined, but $(f \circ g)(0)$ is not defined because 0 is not in the domain of $f \circ g$.

The composition of f with g is generally *not* the same as the composition of g with f. This is illustrated in Example 5.

EXAMPLE 5 **Compositions of Functions**

Given $f(x) = x + 2$ and $g(x) = 4 - x^2$, find the following.

A. $(f \circ g)(x)$
B. $(g \circ f)(x)$

SOLUTION

A. $(f \circ g)(x) = f(g(x))$ *Definition of $f \circ g$*

$\qquad\qquad = f(4 - x^2)$ *Definition of $g(x)$*

$\qquad\qquad = (4 - x^2) + 2$ *Definition of $f(x)$*

$\qquad\qquad = -x^2 + 6$

B. $(g \circ f)(x) = g(f(x))$ *Definition of $g \circ f$*

$\qquad\qquad = g(x + 2)$ *Definition of $f(x)$*

$\qquad\qquad = 4 - (x + 2)^2$ *Definition of $g(x)$*

$\qquad\qquad = 4 - (x^2 + 4x + 4)$

$\qquad\qquad = -x^2 - 4x$

Note in this case that $(f \circ g)(x) \neq (g \circ f)(x)$.

EXAMPLE 6 **Finding the Domain of a Composite Function**

Given $f(x) = x^2 - 9$ and $g(x) = \sqrt{9 - x^2}$, find $(f \circ g)(x)$. Then find the domain of $f \circ g$.

SOLUTION

To begin, notice that the domain of g is $-3 \leq x \leq 3$.

$(f \circ g)(x) = f(g(x)),$ $-3 \leq x \leq 3$

$\qquad\qquad = f(\sqrt{9 - x^2}),$ $-3 \leq x \leq 3$

$\qquad\qquad = (\sqrt{9 - x^2})^2 - 9,$ $-3 \leq x \leq 3$

$\qquad\qquad = 9 - x^2 - 9,$ $-3 \leq x \leq 3$

$\qquad\qquad = -x^2,$ $-3 \leq x \leq 3$

Thus, the domain of $f \circ g$ is $-3 \leq x \leq 3$. To convince yourself of this, use a graphing utility to graph

$$y = (\sqrt{9 - x^2})^2 - 9,$$

as shown in Figure 1.53. Notice that the graphing utility does not extend the graph to the left of $x = -3$ or to the right of $x = 3$.

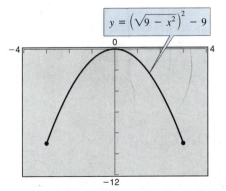

$y = (\sqrt{9 - x^2})^2 - 9$

FIGURE 1.53

EXAMPLE 7   A Case in Which $f \circ g = g \circ f$

Given $f(x) = 2x + 3$ and $g(x) = \frac{1}{2}(x - 3)$, find the following.

A. $(f \circ g)(x)$
B. $(g \circ f)(x)$

SOLUTION

REMARK   In Example 7, note that the two composite functions $f \circ g$ and $g \circ f$ are equal, and both represent the identity function. That is, $(f \circ g)(x) = (g \circ f)(x) = x$. We will discuss this special case in the next section.

A. $(f \circ g)(x) = f(g(x)) = f\left(\frac{1}{2}(x - 3)\right) = 2\left[\frac{1}{2}(x - 3)\right] + 3$

$= x - 3 + 3 = x$

B. $(g \circ f)(x) = g(f(x)) = g(2x + 3) = \frac{1}{2}[(2x + 3) - 3] = \frac{1}{2}(2x) = x$

In Examples 5, 6, and 7 we formed the composition of two given functions. In calculus, it is also important to be able to identify two functions that *make up a given* composite function. For instance, the function h given by $h(x) = (3x - 5)^3$ is the composition of f with g, where $f(x) = x^3$ and $g(x) = 3x - 5$. That is,

$$h(x) = (3x - 5)^3 = [g(x)]^3 = f(g(x)).$$

Basically, to "decompose" a composite function, look for an "inner" and an "outer" function. In the function h above, $g(x) = 3x - 5$ is the inner function and $f(x) = x^3$ is the outer function.

EXAMPLE 8   Identifying a Composite Function

Express the function

$$h(x) = \frac{1}{(x - 2)^2}$$

as a composition of two functions.

SOLUTION

One way to write h as a composite of two functions is to take the inner function to be $g(x) = x - 2$ and the outer function to be

$$f(x) = \frac{1}{x^2} = x^{-2}.$$

Then you can write

$$h(x) = \frac{1}{(x - 2)^2} = (x - 2)^{-2} = f(x - 2) = f(g(x)).$$

This function can be decomposed in other ways. For instance, let $g(x) = 1/(x - 2)$ and $f(x) = x^2$. Then $h(x) = f(g(x))$.

Applications

EXAMPLE 9   Bacteria Count

The number of bacteria in a refrigerated food is given by

$$N(T) = 20T^2 - 80T + 500, \qquad 2 \le T \le 14,$$

where T is the Celsius temperature of the food. When the food is removed from refrigeration, the temperature is given by

$$T(t) = 4t + 2, \qquad 0 \le t \le 3,$$

where t is the time in hours. Find the following.

A. The composite function $N(T(t))$. What does this function represent?

B. The number of bacteria in the food when $t = 2$ hours

C. The time when the bacteria count reaches 2000

SOLUTION

A. $N(T(t)) = 20(4t + 2)^2 - 80(4t + 2) + 500$
$= 20(16t^2 + 16t + 4) - 320t - 160 + 500$
$= 320t^2 + 320t + 80 - 320t - 160 + 500$
$= 320t^2 + 420$

This composite function represents the number of bacteria as a function of the amount of time the food has been out of refrigeration.

B. When $t = 2$, the number of bacteria is

$$N - 320(2)^2 + 420 = 1280 + 420 = 1700.$$

C. The bacteria count will reach $N = 2000$ when $320t^2 + 420 = 2000$. You can solve this equation for t algebraically as follows.

$$320t^2 + 420 = 2000$$
$$320t^2 = 1580$$
$$t^2 = \frac{1580}{320} = \frac{79}{16}$$
$$t = \frac{\sqrt{79}}{4} \approx 2.2 \text{ hours}$$

Or you can use a graphing utility to approximate the solution, as shown in Figure 1.54.

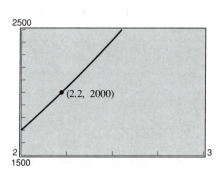

FIGURE 1.54

DISCUSSION PROBLEM

THE COMPOSITION OF TWO FUNCTIONS

You are considering buying a new car for which the regular price is $15,800. The dealership has advertised a factory rebate of $1500 *and* a 12% discount. Using function notation, the rebate and the discount can be represented as

$$f(x) = x - 1500 \qquad \text{\textit{Rebate of \$1500}}$$

and

$$g(x) = 0.88x, \qquad \text{\textit{Discount of 12\%}}$$

where x is the price of the car, $f(x)$ is the price after subtracting the rebate, and $g(x)$ is the price after taking the 12% discount. Compare the sale price obtained by subtracting the rebate first and then taking the discount with the sale price obtained by taking the discount first and then subtracting the rebate.

WARM-UP

The following warm-up exercises involve skills that were covered in earlier sections. You will use these skills in the exercise set for this section.

In Exercises 1–10, perform the indicated operations and simplify the result.

1. $\dfrac{1}{x} + \dfrac{1}{1 - x}$

2. $\dfrac{2}{x + 3} - \dfrac{2}{x - 3}$

3. $\dfrac{3}{x - 2} - \dfrac{2}{x(x - 2)}$

4. $\dfrac{x}{x - 5} + \dfrac{1}{3}$

5. $(x - 1)\left(\dfrac{1}{\sqrt{x^2 - 1}}\right)$

6. $\left(\dfrac{x}{x^2 - 4}\right)\left(\dfrac{x^2 - x - 2}{x^2}\right)$

7. $(x^2 - 4) \div \left(\dfrac{x + 2}{5}\right)$

8. $\left(\dfrac{x}{x^2 + 3x - 10}\right) \div \left(\dfrac{x^2 + 3x}{x^2 + 6x + 5}\right)$

9. $\dfrac{(1/x) + 5}{3 - (1/x)}$

10. $\dfrac{(x/4) - (4/x)}{x - 4}$

SECTION 1.6 · EXERCISES

In Exercises 1–6, find (a) $(f + g)(x)$, (b) $(f - g)(x)$, (c) $(fg)(x)$, and (d) $(f/g)(x)$. What is the domain of f/g?

1. $f(x) = x + 1, \quad g(x) = x - 1$

2. $f(x) = 2x - 5, \quad g(x) = 5$

3. $f(x) = x^2 + 5, \quad g(x) = \sqrt{1 - x}$

4. $f(x) = \sqrt{x^2 - 4}, \quad g(x) = \dfrac{x^2}{x^2 + 1}$

5. $f(x) = \dfrac{1}{x}, \quad g(x) = \dfrac{1}{x^2}$

6. $f(x) = \dfrac{x}{x + 1}, \quad g(x) = x^3$

In Exercises 7–16, evaluate the indicated function for $f(x) = x^2 + 1$ and $g(x) = x - 4$.

7. $(f + g)(3)$

8. $(f - g)(-2)$

9. $(f - g)(2t)$

10. $(f + g)(t - 1)$

11. $(fg)(4)$

12. $(fg)(-6)$

13. $\left(\dfrac{f}{g}\right)(5)$

14. $\left(\dfrac{f}{g}\right)(0)$

15. $\left(\dfrac{f}{g}\right)(-1) - g(3)$

16. $(2f)(5)$

In Exercises 17–20, graph the functions f, g, and $f + g$ on the same set of coordinate axes.

17. $f(x) = \frac{1}{2}x, \quad g(x) = x - 1$

18. $f(x) = \frac{1}{3}x, \quad g(x) = -x + 4$

19. $f(x) = x^2, \quad g(x) = -2x$

20. $f(x) = 4 - x^2, \quad g(x) = x$

In Exercises 21 and 22, use a graphing utility to graph f, g, and $f + g$ in the same viewing rectangle. Which function contributes most to the magnitude of the sum when $0 \le x \le 2$? Which function contributes most to the sum when $x > 5$?

21. $f(x) = 3x, \quad g(x) = -\dfrac{x^3}{10}$

22. $f(x) = \dfrac{x}{2}, \quad g(x) = \sqrt{x}$

23. *Stopping Distance*   While traveling in a car at x miles per hour, you are required to stop quickly to avoid an accident. The distance the car travels during your reaction time is given by $R(x) = \frac{3}{4}x$. The distance traveled while braking is given by $B(x) = \frac{1}{15}x^2$. Find the function giving total stopping distance T. Graph the functions R, B, and T on the same set of coordinate axes for $0 \le x \le 60$.

24. *Comparing Sales*   Suppose you own two fast-food restaurants in town. From 1985 to 1990, the sales for one restaurant have been decreasing according to the function

$$R_1 = 500 - 0.8t^2, \qquad t = 5, 6, 7, 8, 9, 10,$$

where R_1 represents the sales for the first restaurant (in thousands of dollars) and t represents the calendar year with $t = 5$ corresponding to 1985. During the same 6-year period, the sales for the second restaurant have been increasing according to the function

$$R_2 = 250 + 0.78t, \qquad t = 5, 6, 7, 8, 9, 10.$$

Write a function that represents the total sales for the two restaurants. Use the *stacked bar graph* in the figure, which represents the total sales during the 6-year period, to determine whether the total sales have been increasing or decreasing.

FIGURE FOR 24

In Exercises 25–28, find (a) $f \circ g$, (b) $g \circ f$, and (c) $f \circ f$.

25. $f(x) = x^2, \quad g(x) = x - 1$

26. $f(x) = \sqrt[3]{x - 1}, \quad g(x) = x^2 + 1$

27. $f(x) = 3x + 5, \quad g(x) = 5 - x$

28. $f(x) = x^3, \quad g(x) = \dfrac{1}{x}$

In Exercises 29–34, find (a) $f \circ g$ and (b) $g \circ f$.

29. $f(x) = \sqrt{x + 4}, \quad g(x) = x^2$

30. $f(x) = \sqrt[5]{x + 1}, \quad g(x) = x^5 - 2$

31. $f(x) = \frac{1}{3}x - 3, \quad g(x) = 3x + 1$

32. $f(x) = \sqrt{x}, \quad g(x) = \sqrt{x}$

33. $f(x) = |x|, \quad g(x) = x + 6$

34. $f(x) = x^{2/3}, \quad g(x) = x^6$

In Exercises 35–38, use the graphs of f and g to evaluate the indicated functions.

35. (a) $(f + g)(3)$ (b) $\left(\dfrac{f}{g}\right)(2)$

36. (a) $(f - g)(1)$ (b) $(fg)(4)$
37. (a) $(f \circ g)(2)$ (b) $(g \circ f)(2)$
38. (a) $(f \circ g)(1)$ (b) $(g \circ f)(3)$

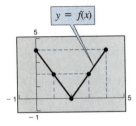
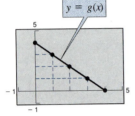

FIGURES FOR 35–38

In Exercises 39–46, find two functions f and g such that $(f \circ g)(x) = h(x)$. (There are many correct answers to these exercises.)

39. $h(x) = (2x + 1)^2$ **40.** $h(x) = (1 - x)^3$
41. $h(x) = \sqrt[3]{x^2 - 4}$ **42.** $h(x) = \sqrt{9 - x}$

43. $h(x) = \dfrac{1}{x + 2}$ **44.** $h(x) = \dfrac{4}{(5x + 2)^2}$

45. $h(x) = (x + 4)^2 + 2(x + 4)$ **46.** $h(x) = (x + 3)^{3/2}$

In Exercises 47–50, determine the domain of (a) f, (b) g, and (c) $f \circ g$.

47. $f(x) = \sqrt{x}, \quad g(x) = x^2 + 1$

48. $f(x) = \dfrac{1}{x}, \quad g(x) = x + 3$

49. $f(x) = \dfrac{3}{x^2 - 1}, \quad g(x) = x + 1$

50. $f(x) = 2x + 3, \quad g(x) = \dfrac{x}{2}$

51. *Ripples*   A pebble is dropped into a calm pond, causing ripples in the form of concentric circles (see figure). The radius (in feet) of the outer ripple is given by $r(t) = 0.6t$, where t is the time in seconds after the pebble strikes the water. The area of the circle is given by the function $A(r) = \pi r^2$.
(a) Find and interpret $(A \circ r)(t)$.

(b) Use a graphing utility to graph the area as a function of t. Move the cursor along the graph to estimate (to two decimal place accuracy) the time required for the area enclosed by a ripple to increase to 20 square feet.

FIGURE FOR 51

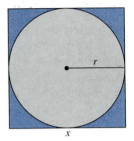

FIGURE FOR 52

52. *Area*   A square concrete foundation was prepared as a base for a large cylindrical gasoline tank (see figure).
(a) Express the radius r of the tank as a function of the length x of the sides of the square.
(b) Express the area A of the circular base of the tank as a function of the radius r.
(c) Find and interpret $(A \circ r)(x)$.

53. *Cost*   The weekly cost (in dollars) of producing x units in a manufacturing process is given by the function $C(x) = 60x + 750$. The number of units produced in t hours is given by $x(t) = 50t$.
(a) Find and interpret $(C \circ x)(t)$.
(b) Use a graphing utility to graph the cost as a function of time. Move the cursor along the curve to estimate (to two decimal place accuracy) the time that must elapse until the cost increases to \$15,000.

54. *Air Traffic Control*   An air traffic controller spots two planes at the same altitude flying toward the same point (see figure). Their flight paths form a right angle at point P. One plane is 150 miles from point P and is moving at 450 miles per hour. The second plane is 200 miles from point P and is moving at 450 miles per hour. Write the distance s between the planes as a function of time t.

FIGURE FOR 54

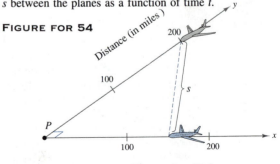

Distance (in miles)

55. *Rebate and Discount* A car dealership is offering a rebate of $1200 *and* a discount of 15% on a car for which the regular price is $18,400. Write a function that represents each type of discount. Then use composition of functions to compare the sale price obtained by subtracting the rebate first with the sale price obtained by taking the discount first.

56. Prove that the product of two odd functions is an even function, and that the product of two even functions is an even function.

57. Use examples to hypothesize whether the product of an odd function and an even function is even or odd. Then prove your hypothesis.

58. Given a function f, prove that $g(x)$ is even and $h(x)$ is odd where

$$g(x) = \tfrac{1}{2}[f(x) + f(-x)] \quad \text{and} \quad h(x) = \tfrac{1}{2}[f(x) - f(-x)].$$

59. Use the result of Exercise 58 to prove that any function can be written as a sum of even and odd functions. (*Hint:* Add the two equations in Exercise 58.)

60. Use the result of Exercise 59 to write each of the following functions as a sum of even and odd functions.

(a) $f(x) = x^2 - 2x + 1$ (b) $f(x) = \dfrac{1}{x + 1}$

1.7 INVERSE FUNCTIONS

**The Inverse of a Function / The Graph of the Inverse of a Function /
The Existence of an Inverse Function / Finding the Inverse of a Function**

The Inverse of a Function

In Section 1.3 you saw that a function can be represented by a set of ordered pairs. For instance, the function $f(x) = x + 4$ from the set $A = \{1, 2, 3, 4\}$ to the set $B = \{5, 6, 7, 8\}$ can be written as follows.

$$f(x) = x + 4: \{(1, 5), (2, 6), (3, 7), (4, 8)\}$$

By interchanging the first and second coordinates of each of these ordered pairs, you can form the **inverse function** of f, denoted by f^{-1}. This is a function from the set B to the set A, and can be written as follows.

$$f^{-1}(x) = x - 4: \{(5, 1), (6, 2), (7, 3), (8, 4)\}$$

Note that the domain of f is equal to the range of f^{-1}, and vice versa, as shown in Figure 1.55. Also note that the functions f and f^{-1} have the effect of "undoing" each other. In other words, when you form the composition of f with f^{-1} or the composition of f^{-1} with f, you obtain the identity function.

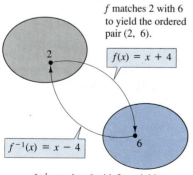

f matches 2 with 6 to yield the ordered pair (2, 6).

$f(x) = x + 4$

$f^{-1}(x) = x - 4$

f^{-1} matches 6 with 2 to yield the ordered pair (6, 2).

FIGURE 1.55

REMARK Don't be confused by the use of -1 to denote the inverse function f^{-1}. In this book f^{-1} will *always* refer to the inverse of the function f and *not* to the reciprocal of $f(x)$.

EXAMPLE 1   Finding Inverse Functions Informally

Find the inverse of the following.

A. $f(x) = 4x$
B. $f(x) = x - 6$

Technology Note

On most graphing utilities you can graph $y = x^{1/3}$ in two ways:

$y = x \wedge (\frac{1}{3})$ or $y = \sqrt[3]{x}$. On most graphing utilities, however, you cannot obtain the complete graph of $y = x^{2/3}$ by entering $y = x \wedge (\frac{2}{3})$. Instead, you need to enter $y = (x \wedge (\frac{1}{3})) \wedge 2$ or $y = \sqrt[3]{x^2}$.

SOLUTION

A. The given function is a function that *multiplies* each input by 4. To "undo" this function, you need to *divide* each input by 4. Thus, the inverse function of $f(x) = 4x$ is

$$f^{-1}(x) = \frac{x}{4}.$$

B. The given function is a function that *subtracts* 6 from each input. To "undo" this function, you need to *add* 6 to each input. Thus, the inverse function of $f(x) = x - 6$ is

$$f^{-1}(x) = x + 6.$$

DEFINITION OF THE INVERSE OF A FUNCTION

Let f and g be two functions such that $f(g(x)) = x$ for every x in the domain of g and $g(f(x)) = x$ for every x in the domain of f. Then the function g is the **inverse** of the function f. The inverse function is denoted by f^{-1} (read "f-inverse"). Thus,

$$f(f^{-1}(x)) = x \quad \text{and} \quad f^{-1}(f(x)) = x.$$

The domain of f must be equal to the range of f^{-1}, and the range of f must be equal to the domain of f^{-1}.

Note from this definition that if the function g is the inverse of the function f, then it must also be true that the function f is the inverse of the function g.

EXAMPLE 2   Verifying Inverse Functions

Show that $f(x) = x^3 - 1$ and $g(x) = \sqrt[3]{x + 1}$ are inverses of each other.

D I S C O V E R Y

Sketch the graphs of the functions given in Example 2, along with the graph of $y = x$.

$f(x) = x^3 - 1$

$g(x) = \sqrt[3]{x + 1}$

(Use the viewing rectangle $-6 \le x \le 6$ and $-4 \le y \le 4$.) Describe how the two graphs are related.

SOLUTION

Begin by noting that the domain (and the range) of both functions is the entire set of real numbers. To show that f and g are inverses of each other, you need to show that $f(g(x)) = x$ and $g(f(x)) = x$.

$$f(g(x)) = f\left(\sqrt[3]{x + 1}\right) = \left(\sqrt[3]{x + 1}\right)^3 - 1 = (x + 1) - 1 = x$$

$$g(f(x)) = g(x^3 - 1) = \sqrt[3]{(x^3 - 1) + 1} = \sqrt[3]{x^3} = x$$

You can see that the two functions f and g "undo" each other: the function f first cubes the input x and then subtracts 1, whereas the function g first adds 1 and then takes the cube root of the result.

EXAMPLE 3   Verifying Inverse Functions

Which of the functions

$$g(x) = \frac{x - 2}{5} \quad \text{and} \quad h(x) = \frac{5}{x} + 2$$

is the inverse of the function $f(x) = \dfrac{5}{x - 2}$?

SOLUTION

$$f(g(x)) = f\left(\frac{x - 2}{5}\right) = \frac{5}{[(x - 2)/5] - 2} = \frac{25}{(x - 2) - 10} = \frac{25}{x - 12} \neq x$$

Because this composition is not equal to the identity function x, you can conclude that g is *not* the inverse of f.

$$f(h(x)) = f\left(\frac{5}{x} + 2\right) = \frac{5}{(5/x) + 2 - 2} = \frac{5}{5/x} = x$$

Thus, it appears that h is the inverse of f. You can confirm this by showing that the composition of h with f is also equal to the identity function.

The Graph of the Inverse of a Function

The graphs of f and f^{-1} are related to each other in the following way. If the point (a, b) lies on the graph of f, then the point (b, a) lies on the graph of f^{-1} and vice versa. This means that the graph of f^{-1} is a reflection of the graph of f in the line $y = x$, as shown in Figure 1.56.

In Examples 2 and 3, inverse functions were verified algebraically. A graphing utility can also be helpful in checking to see whether one function is the inverse of another.

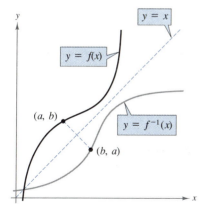

The graph of f^{-1} is a reflection of the graph of f in the line $y = x$.

FIGURE 1.56

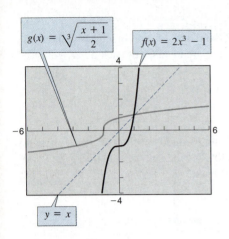

$g(x) = \sqrt[3]{\dfrac{x+1}{2}}$ $f(x) = 2x^3 - 1$

$y = x$

FIGURE 1.57

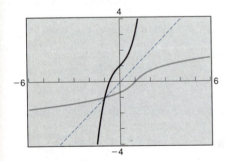

FIGURE 1.58

EXAMPLE 4   Graphical Check for Inverse Functions

Sketch the graphs of

$$f(x) = 2x^3 - 1 \quad \text{and} \quad g(x) = \sqrt[3]{\dfrac{x+1}{2}}$$

on the same coordinate plane and show that the graph of g is the reflection of the graph of f in the line $y = x$.

SOLUTION

Using a graphing utility, you can obtain the graphs shown in Figure 1.57. From this figure you can see that the graph of g is the reflection of the graph of f in the line $y = x$, which graphically confirms that g is the inverse of f.

EXAMPLE 5   A Graph Reflecting Program

The following program will graph a function f *and* its reflection in the line $y = x$. (The program steps listed are for a Texas Instruments TI-81. Program steps for other programmable calculators are given in the appendix.)

```
:2Xmin/3→Ymin          :Lbl 1
:2Xmax/3→Ymax          :PT-On(Y₁,X)
:Xscl→Yscl             :X+I→X
:"X"→Y₂                :If X>Xmax
:DispGraph             :End
:(Xmax-Xmin)/95→I      :Goto 1
:Xmin→X
```

Use this program to graph the inverse of $f(x) = x^3 + x + 1$.

SOLUTION

To run the program, enter the function f into the calculator. Then set a viewing rectangle. (For the program to run, the viewing rectangle must not contain x-values outside the domain of f.) The resulting display is shown in Figure 1.58.

The Existence of an Inverse Function

A function need not have an inverse function. For instance, the function $f(x) = x^2$ has no inverse [assuming a domain of $(-\infty, \infty)$]. To have an inverse, a function must be **one-to-one,** which means that no two elements in the domain of f correspond to the same element in the range of f.

DEFINITION OF A ONE-TO-ONE FUNCTION

A function f is **one-to-one** if, for a and b in its domain,

$f(a) = f(b)$ implies that $a = b$.

The function $f(x) = x + 1$ *is* one-to-one because $a + 1 = b + 1$ implies that a and b must be equal. However, the function $f(x) = x^2$ is *not* one-to-one because $a^2 = b^2$ does not imply that $a = b$. For instance, $(-1)^2 = 1^2$ and yet $-1 \neq 1$.

EXISTENCE OF AN INVERSE FUNCTION

A function f has an inverse function f^{-1} if and only if f is one-to-one.

EXAMPLE 6   Testing for One-to-One Functions

Which functions are one-to-one and have inverse functions?

A. $f(x) = x^3 + 1$ **B.** $g(x) = x^2 - x$ **C.** $h(x) = \sqrt{x}$

SOLUTION

A. Let a and b be real numbers with $f(a) = f(b)$.

$$a^3 + 1 = b^3 + 1 \qquad \textit{Set } f(a) = f(b)$$
$$a^3 = b^3$$
$$a = b$$

Therefore, $f(a) = f(b)$ implies that $a = b$. From this, it follows that f *is* one-to-one and has an inverse function.

B. Because $g(-1) = (-1)^2 - (-1) = 2$ and $g(2) = 2^2 - 2 = 2$, you have two distinct inputs matched with the same output. Thus, g *is not* a one-to-one function and has no inverse.

C. Let a and b be nonnegative real numbers with $h(a) = h(b)$.

$$\sqrt{a} = \sqrt{b} \qquad \textit{Set } h(a) = h(b)$$
$$a = b$$

Therefore, $h(a) = h(b)$ implies that $a = b$. Thus, h *is* one-to-one and has an inverse.

From its graph, it is easy to tell whether a function of x is one-to-one. Simply check to see that every *horizontal* line intersects the graph of the function at most once. For instance, Figure 1.59 shows the graphs of the three functions given in Example 6. On the graph of $g(x) = x^2 - x$, you can find a horizontal line that intersects the graph twice.

Two special types of functions that pass the **horizontal line test** are those that are increasing or decreasing on their entire domains.

1. If f is *increasing* on its entire domain, then f is one-to-one.
2. If f is *decreasing* on its entire domain, then f is one-to-one.

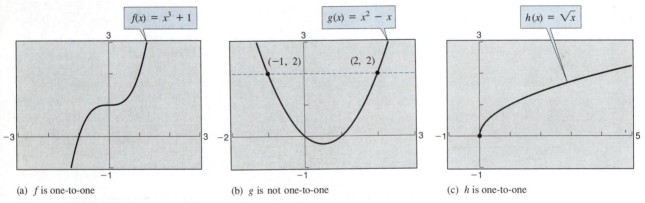

$f(x) = x^3 + 1$ $g(x) = x^2 - x$ $h(x) = \sqrt{x}$

$(-1, 2)$ $(2, 2)$

(a) f is one-to-one   (b) g is not one-to-one   (c) h is one-to-one

FIGURE 1.59

Finding the Inverse of a Function

For simple functions (such as the ones in Example 1) you can find inverse functions by inspection. For instance, the inverse of $f(x) = 8x$ is equal to $f^{-1}(x) = x/8$. For more complicated functions, however, it is best to use the following procedure for finding the inverse of a function.

FINDING THE INVERSE OF A FUNCTION

To find the inverse of f, use the following steps.

1. In the equation for $f(x)$, replace $f(x)$ by y.
2. Interchange the roles of x and y.
3. If the new equation does not represent y as a function of x, then the function f does not have an inverse function. If the new equation does represent y as a function of x, then solve the new equation for y.
4. Replace y by $f^{-1}(x)$.
5. Verify that f and f^{-1} are inverses of each other by showing that $f(f^{-1}(x)) = x$ and $f^{-1}(f(x)) = x$.

EXAMPLE 7   Finding the Inverse of a Function

Find the inverse (if it exists) of

$$f(x) = \frac{5 - 3x}{2}.$$

SOLUTION

From Figure 1.60, you can see that f is one-to-one, and therefore has an inverse.

$$y = \frac{5 - 3x}{2} \qquad \textit{Write in form } y = f(x)$$

$$x = \frac{5 - 3y}{2} \qquad \textit{Interchange x and y}$$

$$2x = 5 - 3y$$

$$3y = 5 - 2x$$

$$y = \frac{5 - 2x}{3} \qquad \textit{Solve for y}$$

$$f^{-1}(x) = \frac{5 - 2x}{3} \qquad \textit{Replace y by } f^{-1}(x)$$

The domain and range of both f and f^{-1} consist of all real numbers.

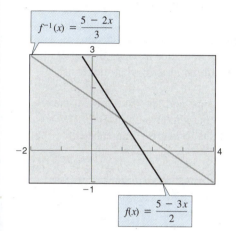

FIGURE 1.60

EXAMPLE 8   Finding the Inverse of a Function

Find the inverse of the function $f(x) = \sqrt{2x - 3}$ and sketch the graphs of f and f^{-1}.

SOLUTION

$$y = \sqrt{2x - 3}$$

$$x = \sqrt{2y - 3} \qquad \textit{Interchange x and y}$$

$$x^2 = 2y - 3$$

$$2y = x^2 + 3 \qquad \textit{Solve for y}$$

$$y = \frac{x^2 + 3}{2} \qquad x \geq 0$$

$$f^{-1}(x) = \frac{x^2 + 3}{2}, \quad x \geq 0 \qquad \textit{Replace y by } f^{-1}(x)$$

The graph of f^{-1} is the reflection of the graph of f in the line $y = x$, as shown in Figure 1.61. Note that the domain of f is the interval $[\frac{3}{2}, \infty)$ and the range of f is the interval $[0, \infty)$. Moreover, the domain of f^{-1} is the interval $[0, \infty)$ and the range of f^{-1} is the interval $[\frac{3}{2}, \infty)$.

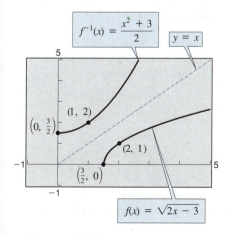

FIGURE 1.61

The problem of finding the inverse of a function can be difficult (or even impossible) for two reasons. First, given $y = f(x)$, it may be algebraically difficult to solve for x in terms of y. Second, if f is not one-to-one, then f^{-1} does not exist.

DISCUSSION PROBLEM

THE EXISTENCE OF AN INVERSE FUNCTION

Write a short paragraph describing why the following functions do or do not have inverse functions.

(a) Let x represent the retail price of an item (in dollars), and let $f(x)$ represent the sales tax on the item. Assume that the sales tax is 7% of the retail price *and* that the sales tax is rounded to the nearest cent. Does this function have an inverse? (*Hint:* Can you undo this function? For instance, if you know that the sales tax is $0.14, can you determine *exactly* what the retail price is?)

(b) Let x represent the temperature in degrees Celsius, and let $f(x)$ represent the temperature in degrees Fahrenheit. Does this function have an inverse? (*Hint:* The formula for converting from degrees Celsius to degrees Fahrenheit is $F = \frac{9}{5}C + 32$.)

WARM-UP

The following warm-up exercises involve skills that were covered in earlier sections. You will use these skills in the exercise set for this section.

In Exercises 1–4, find the domain of the function.

1. $f(x) = \sqrt[3]{x + 1}$ 2. $f(x) = \sqrt{x + 1}$

3. $g(x) = \dfrac{2}{x^2 - 2x}$ 4. $h(x) = \dfrac{x}{3x + 5}$

In Exercises 5–8, simplify the expression.

5. $2\left(\dfrac{x + 5}{2}\right) - 5$ 6. $7 - 10\left(\dfrac{7 - x}{10}\right)$

7. $\sqrt[3]{2\left(\dfrac{x^3}{2} - 2\right) + 4}$ 8. $\left(\sqrt[5]{x + 2}\right)^5 - 2$

In Exercises 9 and 10, solve for x in terms of y.

9. $y = \dfrac{2x - 6}{3}$ 10. $y = \sqrt[3]{2x - 4}$

SECTION 1.7 · EXERCISES

In Exercises 1–6, find the inverse f^{-1} of the function f informally. Verify that $f(f^{-1}(x))$ and $f^{-1}(f(x))$ are equal to the identity function.

1. $f(x) = 8x$
2. $f(x) = \frac{1}{5}x$
3. $f(x) = x + 10$
4. $f(x) = x - 5$
5. $f(x) = \sqrt[3]{x}$
6. $f(x) = x^5$

In Exercises 7–16, show that f and g are inverse functions (a) algebraically and (b) graphically.

7. $f(x) = 2x, \quad g(x) = \frac{x}{2}$

8. $f(x) = x - 5, \quad g(x) = x + 5$

9. $f(x) = 5x + 1, \quad g(x) = \frac{x - 1}{5}$

10. $f(x) = 3 - 4x, \quad g(x) = \frac{3 - x}{4}$

11. $f(x) = x^3, \quad g(x) = \sqrt[3]{x}$

12. $f(x) = \frac{1}{x}, \quad g(x) = \frac{1}{x}$

13. $f(x) = \sqrt{x - 4}$
    $g(x) = x^2 + 4, \quad x \geq 0$

14. $f(x) = 1 - x^3$
    $g(x) = \sqrt[3]{1 - x}$

15. $f(x) = 9 - x^2, \quad x \geq 0$
    $g(x) = \sqrt{9 - x}, \quad x \leq 9$

16. $f(x) = \frac{1}{1 + x}, \quad x \geq 0$

    $g(x) = \frac{1 - x}{x}, \quad 0 < x \leq 1$

In Exercises 17–22, use a graphing utility to graph the function and use the graph to determine whether the function is one-to-one.

17. $g(x) = \frac{4 - x}{6}$

18. $f(x) = 10$

19. $h(x) = |x + 4| - |x - 4|$

20. $g(x) = (x + 5)^3$

21. $f(x) = -2x\sqrt{16 - x^2}$

22. $f(x) = \frac{1}{8}(x + 2)^2 - 1$

In Exercises 23–32, find the inverse of the one-to-one function f. Then use a graphing utility to graph both f and f^{-1} in the same viewing rectangle.

23. $f(x) = 2x - 3$
24. $f(x) = 3x$
25. $f(x) = x^5$
26. $f(x) = x^3 + 1$
27. $f(x) = \sqrt{x}$
28. $f(x) = x^2, \quad x \geq 0$
29. $f(x) = \sqrt{4 - x^2}, \quad 0 \leq x \leq 2$

30. $f(x) = \frac{4}{x}$

31. $f(x) = \sqrt[3]{x - 1}$
32. $f(x) = x^{3/5}$

In Exercises 33–46, determine whether the given function is one-to-one. If it is, find its inverse.

33. $f(x) = x^4$
34. $f(x) = \frac{1}{x^2}$

35. $g(x) = \frac{x}{8}$
36. $f(x) = -3$

37. $f(x) = (x + 3)^2, \quad x \geq -3$
38. $q(x) = (x - 5)^2$

39. $h(x) = \frac{1}{x}$
40. $f(x) = |x - 2|, \quad x \leq 2$

41. $f(x) = \sqrt{2x + 3}$
42. $f(x) = \sqrt{x - 2}$

43. $g(x) = x^2 - x^4$
44. $f(x) = \frac{x^2}{x^2 + 1}$

45. $f(x) = 25 - x^2, \quad x \leq 0$
46. $f(x) = ax + b, \quad a \neq 0$

In Exercises 47–50, delete part of the graph of the function so that the part that remains is one-to-one. Find the inverse of the remaining part and give the domain of the inverse. (There is more than one correct answer.)

47. $f(x) = (x - 3)^2$
48. $f(x) = 16 - x^4$

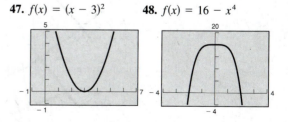

49. $f(x) = |x + 3|$ **50.** $f(x) = |x - 3|$

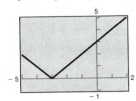
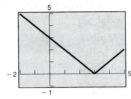

In Exercises 51 and 52, use the graph of the function f to complete the table and to graph f^{-1}.

51.

x	0	1	2	3	4
$f^{-1}(x)$					

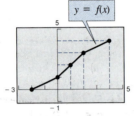

52.

x	0	2	4	6
$f^{-1}(x)$				

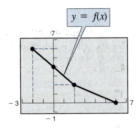

In Exercises 53–56, use the functions $f(x) = \frac{1}{8}x - 3$ and $g(x) = x^3$ to find the indicated value.

53. $(f^{-1} \circ g^{-1})(1)$ **54.** $(g^{-1} \circ f^{-1})(-3)$

55. $(f^{-1} \circ f^{-1})(6)$ **56.** $(g^{-1} \circ g^{-1})(-4)$

In Exercises 57–60, use the functions $f(x) = x + 4$ and $g(x) = 2x - 5$ to find the specified functions.

57. $g^{-1} \circ f^{-1}$ **58.** $f^{-1} \circ g^{-1}$

59. $(f \circ g)^{-1}$ **60.** $(g \circ f)^{-1}$

Diesel Engine   In Exercises 61 and 62, use the following function, which approximates the exhaust temperature y in degrees Fahrenheit

$$y = 0.03x^2 + 254.50, \quad 0 < x < 100,$$

where x is the percentage load for a diesel engine (see figure).

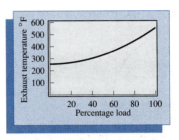

FIGURE
FOR 61 AND 62

61. (a) Find the inverse of the function and state what the variables x and y represent in the inverse function.
   (b) Graph the inverse function.

62. Determine the percentage load interval if the exhaust temperature of the engine must not exceed 500 degrees Fahrenheit.

In Exercises 63–66, determine whether the statement is true or false. Explain your reasoning.

63. If f is an even function, then f^{-1} exists.

64. If the inverse of f exists, then the y-intercept of f is an x-intercept of f^{-1}.

65. If $f(x) = x^n$ where n is odd, then f^{-1} exists.

66. There exists no function f such that $f = f^{-1}$.

CHAPTER 1 · REVIEW EXERCISES

In Exercises 1–10, sketch the graph of the equation *by hand*. Verify your result with a graphing utility.

1. $y - 2x - 3 = 0$ **2.** $3x + 2y + 6 = 0$

3. $x - 5 = 0$ **4.** $y = 8 - |x|$

5. $y = \sqrt{5 - x}$ **6.** $y = \sqrt{x + 2}$

7. $y + 2x^2 = 0$ **8.** $y = x^2 - 4x$

9. $y = \sqrt{25 - x^2}$ **10.** $x^2 + y^2 = 10$

In Exercises 11–18, use a graphing utility to graph the equation. Determine the number of times (if any) the graph intersects the coordinate axes.

11. $y = \frac{1}{4}(x + 1)^3$ **12.** $y = 4 - (x - 4)^2$

13. $y = \frac{1}{4}x^4 - 2x^2$ **14.** $y = \frac{1}{4}x^3 - 3x$

15. $y = x\sqrt{9 - x^2}$ **16.** $y = x\sqrt{x + 3}$

17. $y = |x - 4| - 4$ **18.** $y = |x + 2| + |3 - x|$

In Exercises 19 and 20, solve for y and use a graphing utility to graph each of the resulting equations on the same viewing rectangle.

19. $2y^2 = x^3$

20. $(x + 2)^2 + y^2 = 16$

In Exercises 21–24, use the concept of slope to find t so that the three points are collinear.

21. $(-2, 5), (0, t), (1, 1)$ **22.** $(-6, 1), (1, t), (10, 5)$

23. $(1, -4), (t, 3), (5, 10)$ **24.** $(-3, 3), (t, -1), (8, 6)$

In Exercises 25–30, find an equation of the line through the two points.

25. $(0, 0), (0, 10)$ **26.** $(-1, 4), (2, 0)$

27. $(2, 1), (14, 6)$ **28.** $(-2, 2), (3, -10)$

29. $(-1, 0), (6, 2)$ **30.** $(1, 6), (4, 2)$

In Exercises 31–34, find an equation of the line that passes through the given point and has the specified slope. Sketch the graph of the line *by hand*.

Point	Slope	Point	Slope
31. $(0, -5)$	$m = \frac{3}{2}$	**32.** $(-2, 6)$	$m = 0$
33. $(3, 0)$	$m = -\frac{2}{3}$	**34.** $(5, 4)$	m is undefined

In Exercises 35 and 36, write an equation of the line through the point (a) parallel to the given line and (b) perpendicular to the given line. Verify your result with a graphing utility.

Point	Line	Point	Line
35. $(3, 2)$	$5x - 4y = 8$	**36.** $(-8, 3)$	$2x + 3y = 5$

37. *Fourth Quarter Sales*   During the second and third quarters of the year, a business had sales of $160,000 and $185,000. If the growth of sales follows a linear pattern, estimate sales during the fourth quarter.

38. *Dollar Value*   The dollar value of a product in 1990 is $85, and the item will increase in value at an expected rate of $3.75 per year.
(a) Write a linear equation that gives the dollar value V of the product in terms of the year t. (Let $t = 0$ represent 1990.)
(b) Use a graphing utility to graph the sales equation.
(c) Move the cursor along the graph of the sales model to estimate the dollar value of the product in 1995.

In Exercises 39–42, evaluate the function at the specified values of the independent variable. Simplify your results.

39. $f(x) = x^2 + 1$
(a) $f(2)$ (b) $f(-4)$
(c) $f(t^2)$ (d) $-f(x)$

40. $g(x) = x^{4/3}$
(a) $g(8)$ (b) $g(t + 1)$
(c) $\dfrac{g(8) - g(1)}{8 - 1}$ (d) $g(-x)$

41. $h(x) = 6 - 5x^2$
(a) $h(2)$
(b) $h(x + 3)$
(c) $\dfrac{h(4) - h(2)}{4 - 2}$
(d) $\dfrac{h(x + t) - h(x)}{t}$

42. $f(t) = \sqrt[4]{t}$
(a) $f(16)$
(b) $f(t + 5)$
(c) $\dfrac{f(16) - f(0)}{16}$
(d) $f(t + h)$

43. Sketch (on the same set of coordinate axes) a graph of f for $c = -2, 0,$ and 2.
(a) $f(x) = \frac{1}{2}x + c$ (b) $f(x) = \frac{1}{2}(x - c)$
(c) $f(x) = \frac{1}{2}(cx)$

44. Sketch (on the same set of coordinate axes) a graph of f for $c = -2, 0,$ and 2.
(a) $f(x) = x^3 + c$ (b) $f(x) = (x - c)^3$
(c) $f(x) = (x - 2)^3 + c$

In Exercises 45–50, determine the domain of the function. Verify your result with a graphing utility.

45. $f(x) = \sqrt{25 - x^2}$ **46.** $f(x) = 3x + 4$

47. $g(s) = \dfrac{5}{3s - 9}$ **48.** $f(x) = \sqrt{x^2 + 8x}$

49. $h(x) = \dfrac{x}{x^2 - x - 6}$ **50.** $h(t) = |t + 1|$

51. *Vertical Motion*   The velocity of a ball thrown vertically upward from ground level is given by $v(t) = -32t + 48$, where t is the time in seconds and v is the velocity in feet per second.
(a) Find the velocity when $t = 1$.
(b) Find the time when the ball reaches its maximum height. [*Hint:* Find the time when $v(t) = 0$.]
(c) Find the velocity when $t = 2$.

52. *Cost and Profit*   A company produces a product for which the variable cost is $5.35 per unit and the fixed costs are $16,000. The company sells the product for $8.20, and can sell all that it produces.
(a) Find the total cost as a function of x, the number of units produced.
(b) Find the profit as a function of x.

53. *Dimensions of a Rectangle*   A wire 24 inches long is to be cut into four pieces to form a rectangle whose shortest side has a length of x. Express the area A of the rectangle as a function of x. Determine the domain of the function and use a graphing utility to graph the function over that domain.

54. *Cost of a Phone Call*   Suppose the cost of a telephone call between Dallas and Philadelphia is $0.70 for the first minute and $0.38 for each additional minute (or portion thereof). A model for the total cost of the phone call is

$$C = 0.70 + 0.38[[x]],$$

where C is the total cost of the call in dollars and x is the length of the call in minutes. Sketch the graph of this function.

In Exercises 55–58, use a graphing utility to graph the function and use the graph to (a) approximate the intervals in which the function is increasing, decreasing, or constant; (b) approximate (to two decimal place accuracy) any relative maximum or minimum values of the function; and (c) determine if the function is even, odd, or neither.

55. $g(x) = |x + 2| - |x - 2|$ **56.** $f(x) = (x^2 - 4)^2$

57. $h(x) = 4x^3 - x^4$ **58.** $g(x) = \sqrt[3]{x(x + 3)^2}$

59. *Navigation*   At noon, ship A was 100 miles due east of ship B. Ship A is sailing west at 12 miles per hour, and ship B is sailing south at 10 miles per hour (see figure).
(a) Verify that the distance between the ships is given by
$$d = \sqrt{(100 - 12t)^2 + (10t)^2},$$
where t is the time in hours with $t = 0$ corresponding to noon.
(b) Use a graphing utility to graph the distance function. Move the cursor along the function to estimate (to two decimal place accuracy) the minimum distance between the ships. At what time will this occur?

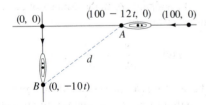

FIGURE FOR 59

60. *Maximum Revenue*   For groups of 80 or more, a charter bus company determines the rate per person according to the formula

Rate $= \$8.00 - \$0.05(n - 80)$, $n \geq 80$.

(a) Determine the revenue as a function of n.
(b) Use a graphing utility to graph the revenue function. Move the cursor along the function to estimate the number of passengers that will maximize the revenue.

In Exercises 61–66, (a) find f^{-1}, (b) use a graphing utility to graph both f and f^{-1} in the same viewing rectangle, and (c) verify that $f^{-1}(f(x)) = x$ and $f(f^{-1}(x)) = x$.

61. $f(x) = \frac{1}{2}x - 3$ **62.** $f(x) = 5x - 7$

63. $f(x) = \sqrt{x + 1}$ **64.** $f(x) = x^3 + 2$

65. $f(x) = x^2 - 5, \quad x \geq 0$

66. $f(x) = \sqrt[3]{x + 1}$

In Exercises 67–70, restrict the domain of the function f to an interval where the function is increasing and determine f^{-1} over that interval. Then use a graphing utility to graph both f and f^{-1} in the same viewing rectangle.

67. $f(x) = 2(x - 4)^2$ **68.** $f(x) = |x - 2|$

69. $f(x) = \sqrt{x^2 - 4}$ **70.** $f(x) = x^{4/3}$

In Exercises 71–78, let
$$f(x) = 3 - 2x, \quad g(x) = \sqrt{x}, \quad \text{and} \quad h(x) = 3x^2 + 2$$
and find the indicated value.

71. $(f - g)(4)$ **72.** $(f + h)(5)$

73. $(fh)(1)$ **74.** $\left(\dfrac{g}{h}\right)(1)$

75. $(h \circ g)(7)$ **76.** $(g \circ f)(-2)$

77. $g^{-1}(3)$ **78.** $(h \circ f^{-1})(1)$

INTERCEPTS, ZEROS, AND SOLUTIONS

2.1 LINEAR EQUATIONS AND MODELING

Equations and Solutions of Equations / Linear Equations / Using Mathematical Models to Solve Problems / Common Formulas

Equations and Solutions of Equations

An **equation** is a statement that two algebraic expressions are equal. Some examples of equations in x are $3x - 5 = 7$, $x^2 - x - 6 = 0$, and $\sqrt{2x} = 4$. To **solve** an equation in x means to find all values of x for which the equation is true. Such values are **solutions.** For instance, $x = 4$ is a solution of the equation $3x - 5 = 7$, because $3(4) - 5 = 7$ is a true statement.

The solutions of an equation depend upon the kinds of numbers being considered. For instance, in the set of rational numbers, the equation $x^2 = 10$ has no solution because there is no rational number whose square is 10. However, in the set of real numbers this equation has two solutions, $\sqrt{10}$ and $-\sqrt{10}$, because $\left(\sqrt{10}\right)^2 = 10$ and $\left(-\sqrt{10}\right)^2 = 10$.

An equation that is true for *every* real number in the domain of the variable is an **identity.** Two examples of identities are

$$x^2 - 9 = (x + 3)(x - 3) \qquad \text{and} \qquad \frac{x}{3x^2} = \frac{1}{3x}, \quad x \neq 0.$$

The first equation is an identity because it is a true statement for any real value of x. The second equation is an identity because it is true for any nonzero real value of x.

135

An equation that is true for just *some* (or even none) of the real numbers in the domain of the variable is a **conditional equation.** For example, the equation $x^2 - 9 = 0$ is conditional because $x = 3$ and $x = -3$ are the only values in the domain that satisfy the equation. To solve a conditional equation in x, isolate x on one side of the equation by a sequence of **equivalent** (and usually simpler) equations, each having the same solutions(s) as the original equation.

GENERATING EQUIVALENT EQUATIONS

An equation can be transformed into an *equivalent equation* by one or more of the following steps.

	Original Equation	*Equivalent Equation*
1. Remove symbols of grouping, combine like terms, or reduce fractions on one or both sides of the equation.	$2x - x = 4$	$x = 4$
2. Add (or subtract) the same quantity to *both* sides of the equation.	$x + 1 = 6$	$x = 5$
3. Multiply (or divide) *both* sides of the equation by the same *nonzero* quantity.	$2x = 6$	$x = 3$
4. Interchange the two sides of the equation.	$2 = x$	$x = 2$

Linear Equations

The most common type of conditional equation is a linear equation. A **linear equation** in one variable x is an equation that can be written in the standard form

$$ax + b = 0,$$

where a and b are real numbers with $a \neq 0$.

DISCOVERY

Use a graphing utility to graph $y = 6(x - 1) + 4$ and $y = 3(7x + 1)$ on the same viewing rectangle. Explain how the result can be used to approximate the solution of

$6(x - 1) + 4 = 3(7x + 1)$

or to perform a graphic check for a solution that was obtained algebraically.

EXAMPLE 1   Solving a Linear Equation

$$
\begin{aligned}
6(x - 1) + 4 &= 3(7x + 1) && \text{\textit{Original equation}} \\
6x - 6 + 4 &= 21x + 3 && \text{\textit{Remove parentheses}} \\
6x - 2 &= 21x + 3 && \text{\textit{Simplify}} \\
-15x &= 5 && \text{\textit{Add 2 and subtract 21x}} \\
x &= -\frac{1}{3} && \text{\textit{Divide by }} -15
\end{aligned}
$$

CHECK: Check this solution by substituting in the original equation.

$$6(x - 1) + 4 = 3(7x + 1) \qquad \textit{Original equation}$$

$$6\left(-\frac{1}{3} - 1\right) + 4 \stackrel{?}{=} 21\left(-\frac{1}{3}\right) + 3 \qquad \textit{Replace x by } -\frac{1}{3}$$

$$6\left(-\frac{4}{3}\right) + 4 \stackrel{?}{=} -7 + 3 \qquad \textit{Simplify}$$

$$-8 + 4 \stackrel{?}{=} -7 + 3$$
$$-4 = -4 \qquad \textit{Solution checks}$$

To solve an equation involving fractional expressions, find the least common denominator of all terms in the equation and multiply every term by this LCD. This procedure clears the equation of fractions, as shown in the following example.

EXAMPLE 2 Solving an Equation Involving Fractional Expressions

$$\frac{x}{3} + \frac{3x}{4} = 2 \qquad \textit{Original equation}$$

$$(12)\frac{x}{3} + (12)\frac{3x}{4} = (12)2 \qquad \textit{Multiply by the LCD}$$

$$4x + 9x = 24 \qquad \textit{Reduce and multiply}$$

$$13x = 24 \qquad \textit{Combine like terms}$$

$$x = \frac{24}{13} \qquad \textit{Divide by 13}$$

Thus, the equation has one solution: $\frac{24}{13}$. Try checking this solution in the original equation.

When multiplying or dividing an equation by a *variable* quantity, it is possible to introduce an **extraneous** solution—one that does not satisfy the original equation. The next example demonstrates the importance of checking your solution when you have multiplied or divided by a variable.

EXAMPLE 3 An Equation with an Extraneous Solution

Solve the equation for x.

$$\frac{1}{x - 2} = \frac{3}{x + 2} - \frac{6x}{x^2 - 4}$$

SOLUTION

In this case, the LCD is $x^2 - 4 = (x + 2)(x - 2)$. Multiplying every term by this LCD and reducing produces the following.

$$\frac{1}{x-2}(x+2)(x-2) = \frac{3}{x+2}(x+2)(x-2) - \frac{6x}{x^2-4}(x+2)(x-2)$$

$$x + 2 = 3(x-2) - 6x, \qquad x \neq \pm 2$$
$$x + 2 = 3x - 6 - 6x$$
$$4x = -8$$
$$x = -2$$

By checking $x = -2$ in the original equation, you can see that it yields a denominator of zero. Therefore, $x = -2$ is an extraneous solution, and the equation has *no solution*. ▬▬▬

Using Mathematical Models to Solve Problems

One of the primary goals of this text is to demonstrate how algebra can be used to solve problems that occur in real-life situations. This procedure is called **mathematical modeling.**

A good approach to mathematical modeling is to use two stages. Begin by using the verbal description of the problem to form a *verbal model.* Then, after assigning labels to the unknown quantities in the verbal model, form a *mathematical model* or *algebraic equation.*

$$\boxed{\begin{array}{c} \textit{Verbal} \\ \textit{description} \end{array}} \rightarrow \boxed{\begin{array}{c} \textit{Verbal} \\ \textit{model} \end{array}} \rightarrow \boxed{\begin{array}{c} \textit{Algebraic} \\ \textit{equation} \end{array}}$$

When you are trying to construct a verbal model, it is helpful to look for a *hidden equality*—a statement that two algebraic expressions are equal. These two expressions might be explicitly stated as being equal, or they might be known to be equal (based on prior knowledge or experience).

EXAMPLE 4   Finding the Dimensions of a Room

A rectangular family room is twice as long as it is wide, and its perimeter is 84 feet. Find the dimensions and area of the family room.

SOLUTION

For this problem, it helps to sketch a picture, as shown in Figure 2.1.

VERBAL
MODEL $2 \cdot \text{Width} + 2 \cdot \text{Length} = \text{Perimeter}$

LABELS   Width $= w$ (*feet*)
           Length $= l = 2w$ (*feet*)
           Perimeter $= 84$ (*feet*)

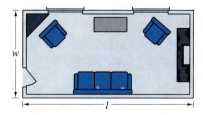

FIGURE 2.1

EQUATION
$$2w + 2(2w) = 84$$
$$6w = 84$$
$$w = 14 \text{ feet}$$
$$l = 2w = 28 \text{ feet}$$

The dimensions of the room are 14 feet by 28 feet. The area is 392 square feet.

EXAMPLE 5   A Distance Problem

A plane is flying at a constant speed from New York to San Francisco, a distance of about 2700 miles, as shown in Figure 2.2. After $1\frac{1}{2}$ hours in the air, the plane flies over Chicago (800 miles from New York). How long will it take the plane to fly from New York to San Francisco?

San
Francisco

New
York

FIGURE 2.2

SOLUTION

VERBAL
MODEL
Distance = Rate · Time

LABELS
Distance = 2700   *(miles)*

Time = t *(hours)*

Rate = $\dfrac{800}{1.5} \approx 533.33$ *(miles per hour)*

EQUATION
$$2700 = 533.33t$$
$$t \approx 5.06 \text{ hours}$$

The trip will take about 5 hours and 4 minutes.

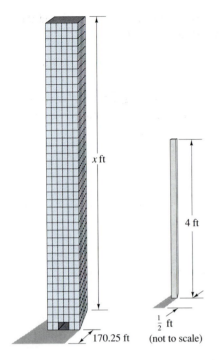

x ft

4 ft

$\frac{1}{2}$ ft

170.25 ft   (not to scale)

FIGURE 2.3

EXAMPLE 6   An Application Involving Similar Triangles

To measure the height of the World Trade Center (in New York City), you measure the shadow cast by the building to be 170.25 feet long, as shown in Figure 2.3. Then you measure the shadow cast by a 4-foot post and find that its shadow is 6 inches long. Use this information to find the height of the building.

SOLUTION

To solve this problem, use a theorem from geometry that states that the ratios of corresponding sides of similar triangles are equal.

VERBAL
MODEL
$$\frac{\text{Height of building}}{\text{Length of shadow}} = \frac{\text{Height of post}}{\text{Length of shadow}}$$

LABELS
Height of building $= x$ (*feet*)
Length of building's shadow $= 170.25$ (*feet*)
Height of post $= 4$ (*feet*)
Length of post's shadow $= \frac{1}{2}$ (*feet*)

EQUATION
$$\frac{x}{170.25} = \frac{4}{\frac{1}{2}}$$

$$x = 1362 \text{ feet}$$

The World Trade Center is about 1362 feet high.

EXAMPLE 7   An Inventory Problem

A store has $30,000 of inventory in 12-inch and 19-inch television sets. The profit on a 12-inch set is 22%. The profit on a 19-inch set is 40%. If the profit for the entire stock is 35%, how much was invested in each type of television set?

SOLUTION

VERBAL
MODEL
$$\frac{\text{Profit from}}{\text{12-inch sets}} + \frac{\text{Profit from}}{\text{19-inch sets}} = \frac{\text{Total}}{\text{profit}}$$

LABELS
Inventory of 12-inch sets $= x$ (*dollars*)
Inventory of 19-inch sets $= 30,000 - x$ (*dollars*)
Profit from 12-inch sets $= 0.22x$ (*dollars*)
Profit from 19-inch sets $= 0.40(30,000 - x)$ (*dollars*)
Total profit $= 0.35(30,000) = 10,500$ (*dollars*)

EQUATION
$$0.22(x) + 0.40(30,000 - x) = 0.35(30,000)$$
$$0.22x + 12,000 - 0.4x = 10,500$$
$$-0.18x = -1500$$
$$x \approx \$8333.33 \text{ 12-inch sets}$$
$$30,000 - x \approx \$21,666.67 \text{ 19-inch sets}$$

Thus, the total inventory for 12-inch sets is approximately $8333.33 and the total inventory for 19-inch sets is approximately $21,666.67.

Common Formulas

Many common types of geometric, scientific, and investment problems use ready-made equations, called **formulas.** Knowing these formulas will help you translate and solve a wide variety of real-life problems.

COMMON FORMULAS

These formulas give the area A, perimeter P (or circumference C), and volume V.

Square	*Rectangle*	*Circle*	*Triangle*	*Trapezoid*
$A = s^2$	$A = lw$	$A = \pi r^2$	$A = \dfrac{1}{2}bh$	$A = \dfrac{1}{2}h(b_1 + b_2)$
$P = 4s$	$P = 2l + 2w$	$C = 2\pi r$		

  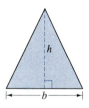 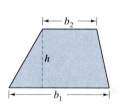

Cube	*Rectangular Solid*	*Circular Cylinder*	*Sphere*
$V = s^3$	$V = lwh$	$V = \pi r^2 h$	$V = \dfrac{4}{3}\pi r^3$

Temperature

$$F = \frac{9}{5}C + 32$$

F = degrees Fahrenheit
C = degrees Celsius

Simple Interest

$I = Prt$
I = interest
P = principal
r = annual interest rate
t = time in years

Compound Interest

$$A = P\left(1 + \frac{r}{n}\right)^{nt}$$

A = balance
P = principal
r = annual interest rate
n = compoundings per year
t = time in years

Distance

$d = rt$
d = distance traveled
r = rate
t = time

|←3 cm→|

h

FIGURE 2.4

When working with applied problems, you often need to rewrite one of the common formulas. For instance, the formula $P = 2l + 2w$, for the perimeter of a rectangle, can be rewritten or solved for w to produce $w = (P - 2l)/2$.

EXAMPLE 8   Using a Formula

A cylindrical can has a volume of 300 cubic centimeters (cm^3) and a radius of 3 centimeters (cm), as shown in Figure 2.4. Find the height of the can.

SOLUTION

The formula for the *volume of a cylinder* is $V = \pi r^2 h$. To find the height of the can, solve for h.

$$h = \frac{V}{\pi r^2}$$

Then, using $V = 300 \ cm^3$ and $r = 3$ cm, find the height.

$$h = \frac{300 \ cm^3}{\pi (3 \ cm)^2}$$

$$= \frac{300 \ cm^3}{9\pi \ cm^2} \qquad\qquad \pi \approx 3.14159$$

$$\approx 10.61 \ cm \qquad\qquad \textit{Round to two decimal places}$$

DISCUSSION PROBLEM

RED HERRINGS

Applied problems in textbooks often give precisely the right amount of information that is necessary to solve a given problem. In real life, however, you often must sort through the given information and discard information that is irrelevant to the problem. Such irrelevant information is called a **red herring.** Find the red herrings in the following problems.

(a) You have accepted a job for which your annual salary will be $18,600. You will be paid once a month, which implies that your monthly salary will be $1550. How much of your annual salary will be deducted for Social Security tax? (Assume that the Social Security tax rate is 7.65%.)

(b) You are driving 240 miles to attend a concert. After traveling for 1 hour, you stop for a snack, which takes 20 minutes. You then continue driving for 4 more hours until you reach the theater where the concert is being held. At this point you notice that you have used $10\frac{1}{2}$ gallons of gasoline (for the entire trip). How many miles per gallon did your car average on the trip?

WARM-UP

The following warm-up exercises involve skills that were covered in earlier sections. You will use these skills in the exercise set for this section.

In Exercises 1–10, perform the indicated operations and simplify your result.

1. $(2x - 4) - (5x + 6)$

2. $(3x - 5) + (2x - 7)$

3. $2(x + 1) - (x + 2)$

4. $-3(2x - 4) + 7(x + 2)$

5. $\dfrac{x}{3} + \dfrac{x}{5}$

6. $x - \dfrac{x}{4}$

7. $\dfrac{1}{x + 1} - \dfrac{1}{x}$

8. $\dfrac{2}{x} + \dfrac{3}{x}$

9. $\dfrac{4}{x} + \dfrac{3}{x - 2}$

10. $\dfrac{1}{x + 1} - \dfrac{1}{x - 1}$

SECTION 2.1 · EXERCISES

In Exercises 1–10, determine whether the equation is an identity or a conditional equation.

1. $2(x - 1) = 2x - 2$

2. $3(x + 2) = 3x + 4$

3. $-2(x - 3) + 5 = -2x + 10$

4. $3(x + 2) - 5 = 3x + 1$

5. $4(x + 1) - 2x = 2(x + 2)$

6. $-7(x - 3) + 4x = 3(7 - x)$

7. $x^2 - 8x + 5 = (x - 4)^2 - 11$

8. $x^2 + 2(3x - 2) = x^2 + 6x - 4$

9. $3 + \dfrac{1}{x + 1} = \dfrac{4x}{x + 1}$

10. $\dfrac{5}{x} + \dfrac{3}{x} = 24$

In Exercises 11–14, determine whether the value of x is a solution of the equation.

Equation	Values	
11. $5x - 3 = 3x + 5$	(a) $x = 0$	(b) $x = -5$
	(c) $x = 4$	(d) $x = 10$
12. $7 - 3x = 5x - 17$	(a) $x = -3$	(b) $x = 0$
	(c) $x = 8$	(d) $x = 3$
13. $(x + 5)(x - 3) = 20$	(a) $x = 3$	(b) $x = -5$
	(c) $x = 5$	(d) $x = -7$
14. $\sqrt[3]{x - 8} = 3$	(a) $x = 2$	(b) $x = -2$
	(c) $x = 35$	(d) $x = 8$

In Exercises 15–52, solve the equation (if possible). Check your result algebraically or graphically.

15. $x + 10 = 15$

16. $7 - x = 18$

17. $7 - 2x = 15$

18. $7x + 2 = 16$

19. $8x - 5 = 3x + 10$

20. $7x + 3 = 3x - 13$

21. $2(x + 5) - 7 = 3(x - 2)$

22. $2(13t - 15) + 3(t - 19) = 0$

23. $6[x - (2x + 3)] = 8 - 5x$

24. $8(x + 2) - 3(2x + 1) = 2(x + 5)$

25. $\dfrac{5x}{4} + \dfrac{1}{2} = x - \dfrac{1}{2}$

26. $\dfrac{x}{5} - \dfrac{x}{2} = 3$

27. $\frac{3}{2}(z + 5) - \frac{1}{4}(z + 24) = 0$

28. $0.25x + 0.75(10 - x) = 3$

29. $x + 8 = 2(x - 2) - x$

30. $3(x + 3) = 5(1 - x) - 1$

31. $\dfrac{100 - 4u}{3} = \dfrac{5u + 6}{4} + 6$

32. $\dfrac{17 + y}{y} + \dfrac{32 + y}{y} = 100$

33. $\dfrac{5x - 4}{5x + 4} = \dfrac{2}{3}$

34. $\dfrac{10x + 3}{5x + 6} = \dfrac{1}{2}$

35. $10 - \dfrac{13}{x} = 4 + \dfrac{5}{x}$

36. $\dfrac{15}{x} - 4 = \dfrac{6}{x} + 3$

37. $\dfrac{1}{x-3} + \dfrac{1}{x+3} = \dfrac{10}{x^2-9}$

38. $\dfrac{1}{x-2} + \dfrac{3}{x+3} = \dfrac{4}{x^2+x-6}$

39. $\dfrac{x}{x+4} + \dfrac{4}{x+4} + 2 = 0$

40. $\dfrac{2}{(x-4)(x-2)} = \dfrac{1}{x-4} + \dfrac{2}{x-2}$

41. $\dfrac{7}{2x+1} - \dfrac{8x}{2x-1} = -4$

42. $\dfrac{4}{u-1} + \dfrac{6}{3u+1} = \dfrac{15}{3u+1}$

43. $\dfrac{1}{x} + \dfrac{2}{x-5} = 0$

44. $\dfrac{6}{x} - \dfrac{2}{x+3} = \dfrac{3(x+5)}{x(x+3)}$

45. $\dfrac{3}{x(x-3)} + \dfrac{4}{x} = \dfrac{1}{x-3}$

46. $3 = 2 + \dfrac{2}{z+2}$

47. $(x+2)^2 - x^2 = 4(x+1)$

48. $(2x+1)^2 = 4(x^2+x+1)$

49. $4(x+1) - ax = x+5$

50. $6x + ax = 2x + 5$

51. $4 - 2(x - 2b) = ax + 3$

52. $5 + ax = 12 - bx$

In Exercises 53–56, solve the equation. (Round the result to three decimal places.)

53. $0.275x + 0.725(500 - x) = 300$

54. $2.763 - 4.5(2.1x - 5.1432) = 6.32x + 5$

55. $\dfrac{x}{0.6321} + \dfrac{x}{0.0692} = 1000$

56. $\dfrac{2}{7.398} - \dfrac{4.405}{x} = \dfrac{1}{x}$

In Exercises 57 and 58, find an equation of the form $ax + b = cx$ that has the specified solution. (There are many correct answers.)

57. $x = 2$ **58.** $x = \frac{1}{3}$

59. *Dimensions of a Room*   A room is 1.5 times as long as it is wide, and its perimeter is 75 feet (see figure). Find the dimensions of the room.

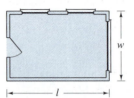

FIGURE FOR 59

60. *Dimensions of a Picture*   A picture frame has a total perimeter of 3 feet (see figure). The width of the frame is 0.62 times its height. Find the dimensions of the frame.

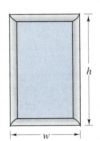

FIGURE FOR 60

61. *Travel Time*   Suppose you are driving on a freeway to a town 150 miles from your home. After 30 minutes you pass a freeway exit that you know is 25 miles from your home. Assuming that you continue at the same constant speed, how long will the entire trip take?

62. *Travel Time*   On the first part of a 317-mile trip, a salesman averaged 58 miles per hour. He averaged only 52 miles per hour on the last part of the trip because of the increased volume of traffic. Find the amount of time at each of the speeds if the total time was 5 hours and 45 minutes.

63. *Travel Time*   Two cars start at one point and travel in the same direction at average speeds of 40 and 55 miles per hour. How much time must elapse before the two cars are 5 miles apart?

64. *Catch-Up Time*   Students are traveling in two cars to a football game 135 miles away. The first car travels at an average speed of 45 miles per hour. The second car starts one-half hour after the first and travels at an average speed of 55 miles per hour. How long will it take the second car to catch up to the first car? Will the second car catch up to the first car before the first car arrives at the game?

65. *Travel Time*   Two families meet at a park for a picnic. At the end of the day one family travels east at an average speed of 42 miles per hour and the other family travels west at an average speed of 50 miles per hour. Both families have approximately 160 miles to travel.
(a) Find the time it takes each family to get home.
(b) Find the time that will have elapsed when they are 100 miles apart.
(c) Find the distance the eastbound family has to travel after the westbound family has arrived home.

66. *Average Speed*   A truck driver traveled at an average speed of 55 miles per hour on a 200-mile trip to pick up a load of freight. On the return trip (with the truck fully loaded), the average speed was 40 miles per hour. Find the average speed for the round trip.

67. *Wind Speed*   An executive flew in the corporate jet to a meeting in a city 1500 miles away (see figure). After traveling the same amount of time on the return flight, the pilot mentioned that they still had 300 miles to go. If the air speed of the plane was 600 miles per hour, how fast was the wind blowing? (Assume that the wind direction was parallel to the flight path and constant all day.)

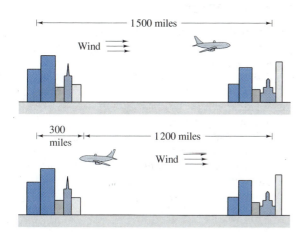

FIGURE FOR 67

68. *Speed of Light*   Light travels at the speed of 3.0×10^8 meters per second. Find the time in minutes required for light to travel from the sun to the earth (a distance of 1.5×10^{11} meters).

69. *Height of a Building*   Suppose you want to measure the height of a building (see figure). To do this, you measure the building's shadow and find that it is 50 feet long. You also measure the shadow of a 4-foot stake and find that its shadow is $3\frac{1}{2}$ feet long. How tall is the building?

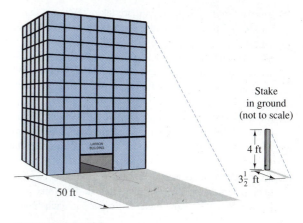

FIGURE FOR 69

70. *Height of a Tree*   Suppose you want to measure the height of a tree. To do this, you measure the tree's shadow and find that it is 25 feet long. You also measure the shadow of a 5-foot lamp post and find that its shadow is 2 feet long (see figure). How tall is the tree?

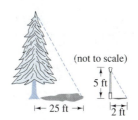

FIGURE FOR 70

71. *Height of a Flagpole*   A person who is 6 feet tall walks away from a flagpole toward the tip of the shadow of the pole. When the person is 30 feet from the pole, the tip of the person's shadow and the shadow cast by the pole coincide at a point 5 feet in front of the person. Find the height of the pole.

72. *Walking Distance*   A person who is 6 feet tall walks away from a 50-foot silo toward the tip of the silo's shadow. At a distance of 32 feet from the silo, the person's shadow begins to emerge beyond the silo's shadow. How much farther must the person walk to be completely out of the silo's shadow?

73. *Course Grade*   To get an A in a course you must have an average of at least 90 on four tests of 100 points each. Your scores on the first three tests were 87, 92, and 84. What must you score on the fourth test to get an A in the course?

74. *Course Grade*   Suppose you are taking a course that has four tests. The first three tests have 100 points each and the fourth test has 200 points. To get an A in the course, you must have an average of at least 90% on the four tests. Your scores on the first three tests were 87, 92, and 84. What must you score on the fourth test to get an A in the course?

75. *Investment Mix*   Suppose you invest $12,000 in two funds paying $10\frac{1}{2}\%$ and 13% simple interest. The total annual interest is $1447.50. How much is invested in each fund?

76. *Investment Mix*   Suppose you invest $25,000 in two funds paying 11% and $12\frac{1}{2}\%$ simple interest. The total annual interest is $2975. How much is invested in each fund?

77. *Comparing Investment Returns*   A person invested $12,000 in a fund paying $9\frac{1}{2}\%$ simple interest and $8000 in a fund with a variable interest rate. At the end of the year, the person received notification that the total interest for both funds was $2054.40. Find the equivalent simple interest rate on the variable rate fund.

78. *Comparing Investment Returns*   A person has $10,000 on deposit earning simple interest with the interest rate linked to the *prime rate*. Because of a drop in the prime rate, the rate on the person's investment dropped by $1\frac{1}{2}\%$ for the last quarter of the year. The annual earnings on the fund are $1112.50. Find the interest rate for the first three quarters of the year and for the last quarter.

79. *Mixture Problem*   A 55-gallon barrel contains a mixture with a concentration of 40% (see figure). How much of this mixture must be withdrawn and replaced by 100% concentrate to bring the mixture up to 75% concentration?

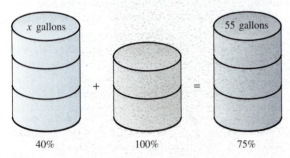

x gallons   +   100%   =   55 gallons

40%   100%   75%

FIGURE FOR 79

80. *Mixture Problem*   A farmer mixed gasoline and oil in order to have 2 gallons of mixture for his two-cycle chain saw engine. This mixture was 32 parts gasoline and 1 part two-cycle oil. How much gasoline must be added to bring the mixture to 40 parts gasoline and 1 part oil?

81. *Production Limit*   A company has fixed costs of $10,000 per month and variable costs of $8.50 per unit manufactured. The company has $85,000 available to cover the monthly costs. How many units can the company manufacture? (*Fixed costs* occur regardless of the level of production. *Variable costs* depend on the level of production.)

82. *Production Limit*   A company has fixed costs of $10,000 per month and variable costs of $9.30 per unit manufactured. The company has $85,000 available to cover the monthly costs. How many units can the company manufacture?

In Exercises 83–94, solve for the indicated variable.

83. *Area of a Triangle*
    Solve for h: $A = \frac{1}{2}bh$

84. *Volume of a Right Circular Cylinder*
    Solve for h: $V = \pi r^2 h$

85. *Investment at Simple Interest*
    Solve for r: $I = Prt$

86. *Investment at Compound Interest*
    Solve for P: $A = P\left(1 + \frac{r}{n}\right)^{nt}$

87. *Area of a Trapezoid*
    Solve for b: $A = \frac{1}{2}(a + b)h$

88. *Area of a Sector of a Circle*
    Solve for θ: $A = \dfrac{\pi r^2 \theta}{360}$

89. *Volume of a Spherical Segment*
    Solve for r: $V = \frac{1}{3}\pi h^2(3r - h)$

90. *Free-Falling Body*
    Solve for a: $h = v_0 t + \frac{1}{2}at^2$

91. *Newton's Law of Universal Gravitation*
    Solve for m_2: $F = \alpha\dfrac{m_1 m_2}{r^2}$

92. *Capacitance in Series Circuits*
    Solve for C_1: $C = \dfrac{1}{\dfrac{1}{C_1} + \dfrac{1}{C_2}}$

93. *Geometric Progression*
    Solve for r: $S = \dfrac{rL - a}{r - 1}$

94. *Prismoidal Formula*
    Solve for S_1: $V = \frac{1}{6}H(S_0 + 4S_1 + S_2)$

Static Problems   In Exercises 95 and 96, suppose you have a uniform beam of length L with a fulcrum x feet from the one end (see figure). If objects with weights W_1 and W_2 are placed at opposite ends of the beam, the beam will balance if

$$W_1x = W_2(L - x).$$

Find x so the beam will balance.

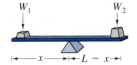

FIGURE FOR 95 AND 96

95. Two children weighing 50 and 75 pounds are going to play on a seesaw that is 10 feet long.

96. A person weighing 200 pounds is attempting to move a 550-pound rock with a bar that is 5 feet long.

2.2   SOLVING EQUATIONS GRAPHICALLY

Intercepts, Zeros, and Solutions / Finding Solutions Graphically / Viewing Rectangles, Scale, and Accuracy / Points of Intersection of Two Graphs

Intercepts, Zeros, and Solutions

In Section 1.1, you learned that the **intercepts** of a graph are the points at which the graph intersects the x- or y-axis.

DEFINITION OF INTERCEPTS

1. The point $(a, 0)$ is called an **x-intercept** of the graph of an equation if it is a solution point of the equation. To find the x-intercepts, let y be zero and solve the equation for x.
2. The point $(0, b)$ is called a **y-intercept** of the graph of an equation if it is a solution point of the equation. To find the y-intercepts, let x be zero and solve the equation for y.

REMARK   Sometimes it is convenient to denote the x-intercept as simply the x-coordinate of the point $(a, 0)$ rather than the point itself. Unless it is necessary to make a distinction, we will use "intercept" to mean either the point or the coordinate.

It is possible that a particular graph will have no intercepts or several intercepts. For instance, consider the three graphs in Figure 2.5.

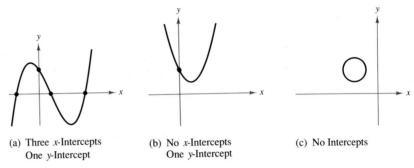

(a) Three x-Intercepts
    One y-Intercept

(b) No x-Intercepts
    One y-Intercept

(c) No Intercepts

FIGURE 2.5

EXAMPLE 1   Finding x- and y-Intercepts

Find the x- and y-intercepts of the graph of $2x + 3y = 5$.

SOLUTION

To find the x-intercept, let $y = 0$. This produces

$$2x = 5$$
$$x = \frac{5}{2},$$

which implies that the graph has one x-intercept: $(\frac{5}{2}, 0)$. To find the y-intercept, let $x = 0$. This produces

$$3y = 5$$
$$y = \frac{5}{3},$$

which implies that the graph has one y-intercept: $(0, \frac{5}{3})$. See Figure 2.6.

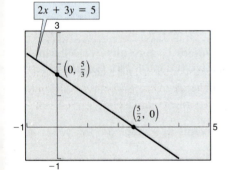

FIGURE 2.6

A **zero** of a function $y = f(x)$ is a number a such that $f(a) = 0$. Thus, to find the zeros of a function, you must solve the equation $f(x) = 0$.

EXAMPLE 2   Verifying Zeros of Functions

A. The real number 3 is a zero of the function $f(x) = 4x - 12$. Check that $f(3) = 0$.

B. The real numbers -2 and 1 are zeros of the function $f(x) = x^2 + x - 2$. Check that $f(-2) = 0$ and $f(1) = 0$.

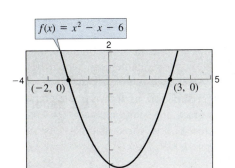

$f(x) = x^2 - x - 6$

$(-2, 0)$ $(3, 0)$

FIGURE 2.7

The concepts of x-intercepts, zeros of functions, and solutions of equations are closely related. In fact, the following statements are equivalent.

1. The point $(a, 0)$ is an x-intercept of the graph of $y = f(x)$.
2. The number a is a *zero* of the function f.
3. The number a is a *solution* of the equation $f(x) = 0$.

Figure 2.7 shows the graph of the function $y = f(x) = x^2 - x - 6$. Notice that the graph has *x-intercepts* of $(3, 0)$ and $(-2, 0)$ because the equation

$$0 = x^2 - x - 6$$

has the *solutions* $x = 3$ and $x = -2$.

This close connection among x-intercepts, zeros, and solutions is crucial to our study of algebra, and you can take advantage of this connection in two basic ways. You can use your algebraic "equation solving skills" to find the x-intercepts of a graph, and you can use your "graphing skills" to approximate the solutions of an equation.

REMARK   In Chapter 3 you will learn techniques for determining the number of solutions of a polynomial equation. For now, you should know that a polynomial equation of degree n cannot have more than n different solutions.

Finding Solutions Graphically

Polynomial equations of degree one or two can be solved in relatively straightforward ways. Polynomial equations of higher degrees can, however, be quite difficult, especially if you rely on only algebraic techniques. For such equations, a graphing utility can be very helpful.

GRAPHICAL APPROXIMATIONS OF SOLUTIONS OF AN EQUATION

To use a graphing utility to approximate the solutions of an equation in the variable x, use the following steps.

1. Write the equation in *standard form, $f(x) = 0$*, with all of the nonzero terms on one side of the equation and zero on the other side.
2. Use a graphing utility to graph the function $y = f(x)$. Be sure the viewing rectangle shows all the relevant features of the graph.
3. Use the zoom and trace features of the graphing utility to approximate each of the x-intercepts of the graph of f. Remember that a graph can have more than one x-intercept, so you may need to change the viewing rectangle a few times.

EXAMPLE 3   **Finding Solutions of an Equation Graphically**

Use a graphing utility to approximate the solutions of $2x^3 - 3x + 2 = 0$.

SOLUTION

Begin by graphing the function $y = 2x^3 - 3x + 2$, as shown in Figure 2.8(a). You can see from the graph that there is only one x-intercept, which lies between -1 and -2 and is approximately -1.5. By using the zoom feature of a graphing utility you can improve the approximation, as shown in the graph in Figure 2.8(b). To three decimal place accuracy, the solution is $x \approx -1.476$. Check this approximation on your calculator. You will find that the value of y is $y = 2(-1.476)^3 - 3(-1.476) + 2 \approx -0.003$.

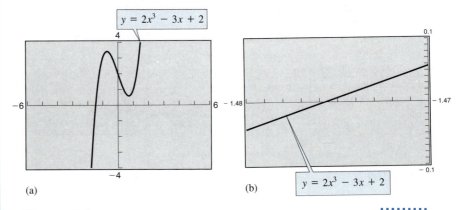

FIGURE 2.8

DISCOVERY

Use a graphing utility to sketch the graph of $y = 24x^3 - 36x + 17$. Describe a viewing rectangle that allows you to determine the number of real solutions of the equation

$$24x^3 - 36x + 17 = 0.$$

Use the same technique to determine the number of real solutions of

$$97x^3 - 102x^2 - 200x - 63 = 0.$$

Viewing Rectangles, Scale, and Accuracy

Here are some suggestions for using the *zoom-in* feature of a graphing utility.

1. With each successive zoom-in, adjust the x-scale (if necessary) so that the resulting viewing rectangle shows at least the two scale marks between which the solution lies.
2. The accuracy of the approximation will always be such that the error is less than the distance between two scale marks.
3. If you have a *trace* feature on your graphing utility, you can generally add one more decimal place of accuracy without changing the viewing rectangle.

Unless stated otherwise, this book will approximate all real solutions with an error of *at most* 0.01.

EXAMPLE 4   Rewriting in Standard Form First

Use a graphing utility to approximate the solutions of $x^2 + 3 = 5x$.

SOLUTION

In standard form, this equation is $x^2 - 5x + 3 = 0$. Thus, you can begin by graphing $y = x^2 - 5x + 3$, as shown in Figure 2.9(a). This graph has two x-intercepts, and by using the zoom and trace features you can approximate the corresponding solutions to be $x \approx 0.697$ and $x \approx 4.303$, as shown in Figure 2.9(b) and (c).

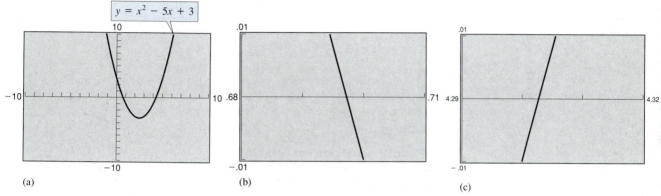

(a)   (b)   (c)

FIGURE 2.9

EXAMPLE 5   An Equation with Three Solutions

Use a graphing utility to approximate the solutions of $2x^3 - 3x - 0.75 = 0$.

SOLUTION

Begin by graphing the function $y = 2x^3 - 3x - 0.75$, as shown in Figure 2.10. From the viewing rectangle shown, you can see that the graph has three x-intercepts. Using the zoom and trace features, approximate the corresponding solutions to be

$$x \approx -1.07, \quad x \approx -0.26, \quad \text{and} \quad x \approx 1.33,$$

using an x-scale of 0.01.

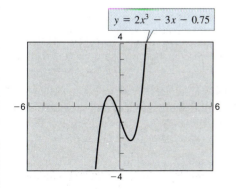

FIGURE 2.10

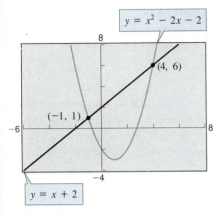

FIGURE 2.11

Points of Intersection of Two Graphs

An ordered pair that is a solution of two different equations is called a **point of intersection** of the graphs of the two equations. For instance, from Figure 2.11 you can see that the graphs of the equations

$$y = x + 2 \quad \text{and} \quad y = x^2 - 2x - 2$$

have two points of intersection. The point $(-1, 1)$ is a solution of both equations, and the point $(4, 6)$ is a solution of both equations.

To find the points of intersection of the graphs of two equations, solve each equation for y (or x) and set the two results equal to each other. The resulting equation will be an equation in one variable, which can be solved using standard procedures, as shown in Example 6.

EXAMPLE 6   Finding Points of Intersection

Find the points of intersection of the graphs of $2x - 3y = -2$ and $4x - y = 6$.

SOLUTION

To begin, solve each equation for y.

$$y = \frac{2}{3}x + \frac{2}{3} \quad \text{and} \quad y = 4x - 6$$

Next, set the two expressions for y equal to each other and solve the resulting equation for x, as follows.

$$\frac{2}{3}x + \frac{2}{3} = 4x - 6$$
$$2x + 2 = 12x - 18$$
$$-10x = -20$$
$$x = 2$$

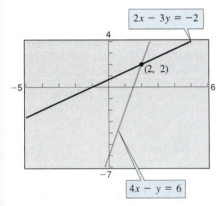

FIGURE 2.12

When $x = 2$, the y-value of each of the given equations is 2. Thus, the graphs of the given equations have one point of intersection, $(2, 2)$, as shown in Figure 2.12.

Another way to approximate points of intersection of two graphs is to graph both equations and use the zoom and trace features to find the point or points at which the two graphs intersect.

EXAMPLE 7   Approximating Points of Intersection

Approximate the point(s) of intersection of the graphs of

$$y = x^2 - 3x - 4 \quad \text{and} \quad y = x^3 + 3x^2 - 2x - 1.$$

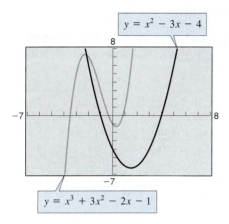

$y = x^2 - 3x - 4$

$y = x^3 + 3x^2 - 2x - 1$

FIGURE 2.13

SOLUTION

Begin by using a graphing utility to graph both functions, as shown in Figure 2.13. From the display, you can see that the two graphs have only one point of intersection. Then, using the zoom and trace features, approximate the point of intersection to be $(-2.17, 7.25)$. To test the reasonableness of this approximation, you can evaluate both functions when $x = -2.17$.

Quadratic function: $y = (-2.17)^2 - 3(-2.17) - 4 \approx 7.22$

Cubic function: $y = (-2.17)^3 + 3(-2.17)^2 - 2(-2.17) - 1$
$$\approx 7.25$$

Because both functions yield approximately the same y-value, you can conclude that the approximate coordinates of the point of intersection are $x \approx -2.17$ and $y \approx 7.25$.

Technology Note

You can obtain a more accurate solution of the equation $x^3 + 2x^2 + x + 3 = 0$ by continuing to use the zoom feature. For instance, if you change the *Xscl* value to 0.00001 and zoom repeatedly, you will find that the curve crosses the x-axis between the two scale marks corresponding to $x = -2.17456$ and $x = -2.17455$.

If your graphing utility has an automatic root-finding feature, use it to verify that an even better approximation to the solution is $x = -2.17455941$.

The method shown in Example 7 gives a nice graphical picture of points of intersection of two graphs. However, for actual approximation purposes, it is better to use the procedure described in Example 6. That is, the point of intersection of $y = x^2 - 3x - 4$ and $y = x^3 + 3x^2 - 2x - 1$ coincides with the solution of the equation

$$x^3 + 3x^2 - 2x - 1 = x^2 - 3x - 4$$
$$x^3 + 2x^2 + x + 3 = 0.$$

By graphing $y = x^3 + 2x^2 + x + 3$ on a graphing utility and using the zoom feature, you can approximate the solution of this equation to be $x \approx -2.175$. The corresponding y-value for *both* of the functions given in Example 7 is $y \approx 7.25$.

EXAMPLE 8   Compact Disc Players and Turntables

Between 1983 and 1987 the number of compact disc players sold each year in the United States was *increasing* and the number of turntables was *decreasing*. Two models that approximate these sales are

$$S = -1700 + 469t \qquad \textit{Compact disc players}$$

and

$$S = 1972 - 82t, \qquad \textit{Turntables}$$

where S represents the annual sales in thousands of units and t represents the year, with $t = 3$ corresponding to 1983. According to these two models, when would you expect the sales of compact disc players to have exceeded the sales of turntables? (*Source:* Dealerscope Merchandising)

SOLUTION

Because the first equation has already been solved for S in terms of t, substitute this value into the second equation and solve for t, as follows.

$$-1700 + 496t = 1972 - 82t$$
$$496t + 82t = 1972 + 1700$$
$$578t = 3672$$
$$t \approx 6.35$$

Thus, from the given models, you would expect that the sales of compact disc players exceeded the sales of turntables sometime during 1986. The graphs of $y_1 = -1700 + 496t$ and $y_2 = 1972 - 82t$ are shown in Figure 2.14, confirming that they intersect at approximately $t = 6.35$.

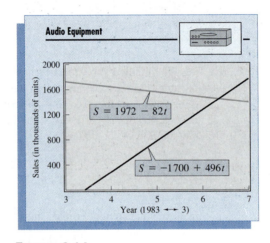

FIGURE 2.14

DISCUSSION PROBLEM

JUDGING THE ACCURACY OF AN APPROXIMATE SOLUTION

Suppose you are solving the equation

$$\frac{x}{x-1} - \frac{99}{100} = 0$$

for x, and you obtain $x = -99.1$ as your solution. Substituting this value back into the equation produces

$$\frac{-99.1}{-99.1 - 1} - \frac{99}{100} = 0.00000999 = 9.99 \times 10^{-6} \approx 0.$$

Does this mean that -99.1 is a good approximation to the solution? Explain your reasoning.

WARM-UP

The following warm-up exercises involve skills that were covered in earlier sections. You will use these skills in the exercise set for this section.

In Exercises 1–6, solve the equation.

1. $9 - 2(x - 3) = 12$

2. $5(t + 5) - 2(t - 1) = 0$

3. $\dfrac{z}{4} - \dfrac{z}{10} = 2$

4. $\dfrac{y + 2}{3} + \dfrac{y - 6}{4} = 20$

5. $\dfrac{8}{x} - \dfrac{6}{x} = 1$

6. $\dfrac{3}{x - 2} + \dfrac{3}{2} = 1$

In Exercises 7–10, use a graphing utility to graph the equation.

7. $y = 0.5x^2 - 2x - 1.2$

8. $y = 5 - 0.3(x - 4)^2$

9. $y = \frac{1}{9}x^3 - 6x + 3$

10. $y = \sqrt{225 - x^2}$

SECTION 2.2 · EXERCISES

In Exercises 1–8, find the *x*- and *y*-intercepts of the graph of the equation.

1. $y = x - 5$

2. $y = (x - 1)(x - 3)$

3. $y = x^2 + x - 2$

4. $y = 4 - x^2$

5. $y = x\sqrt{x + 2}$

6. $xy = 4$

7. $xy - 2y - x + 1 = 0$

8. $x^2y - x^2 + 4y = 0$

In Exercises 9–14, use a graphing utility to graph the function and verify its zero(s).

Function	Zeros
9. $f(x) = 12 - 4x$	$x = 3$
10. $f(x) = 3(x - 5) + 9$	$x = 2$
11. $f(x) = x^2 - 2.5x - 6$	$x = -1.5, \quad x = 4$
12. $f(x) = x^3 - 9x^2 + 18x$	$x = 0, \quad x = 3, \quad x = 6$
13. $f(x) = \dfrac{x + 2}{3} - \dfrac{x - 1}{5} - 1$	$x = 1$
14. $f(x) = x - 3 - \dfrac{10}{x}$	$x = -2, \quad x = 5$

In Exercises 15–20, write the equation in the form $f(x) = 0$. There are many correct answers.

15. $25(x - 3) = 12(x + 2) - 10$

16. $1200 = 300 + 2(x - 500)$

17. $\dfrac{2x}{3} = 10 - \dfrac{1}{x}$

18. $\dfrac{x - 3}{25} = \dfrac{x - 5}{12}$

19. $\dfrac{3}{x + 2} - \dfrac{4}{x - 2} = 5$

20. $\dfrac{6}{x} + \dfrac{8}{x + 5} = 10$

In Exercises 21–26, solve the equation algebraically. Then write the equation in the form $f(x) = 0$ and use a graphing utility to verify the algebraic solution.

21. $27 - 4x = 12$

22. $3.5x - 8 = 0.5x$

23. $\dfrac{3x}{2} + \dfrac{1}{4}(x - 2) = 10$

24. $0.60x + 0.40(100 - x) = 50$

25. $3(x + 3) = 5(1 - x) - 1$

26. $(x + 1)^2 + 2(x - 2) = (x + 1)(x - 2)$

In Exercises 27–36, use a graphing utility to approximate any solutions (accurate to three decimal places) of the equation. [Remember to write the equation in the form $f(x) = 0$.]

27. $\frac{1}{4}(x^2 - 10x + 17) = 0$

28. $-2(x^2 - 6x + 6) = 0$

29. $x^3 + x + 4 = 0$

30. $\frac{1}{9}x^3 + x + 4 = 0$

31. $2x^3 - x^2 - 18x + 9 = 0$

32. $4x^3 + 12x^2 - 26x - 24 = 0$

33. $x^4 = 2x^3 + 1$

34. $x^5 = 3 + 2x^3$

35. $\dfrac{2}{z + 2} = 3$

36. $\dfrac{5}{x} = 1 + \dfrac{3}{x + 2}$

In Exercises 37–46, use a graphing utility to approximate any points of intersection (accurate to three decimal places) of the graphs of the equations.

37. $y = 2 - x$
$y = 2x - 1$

38. $y = 7 - x$
$y = \frac{3}{2} - \frac{11}{2}x$

39. $y = 4 - x^2$
$y = 2x - 1$

40. $y = x^3 - 3$
$y = 5 - 2x$

41. $y = 8$
$y = 3x^2 + 2x$

42. $y = 32$
$y = x^5 - x^2$

43. $y = 3(x + 1)$
$y = x^2 + 2x + 1$

44. $y = x + 2$
$y = -x^2 + 4x + 2$

45. $y = 2x^2$
$y = x^4 - 2x^2$

46. $y = -x$
$y = 2x - x^2$

In Exercises 47–50, evaluate the expression in two ways. (a) Calculate entirely on your calculator by storing intermediate results, and then round the answer to two decimal places. (b) Round both the numerator and the denominator to two decimal places before dividing, and then round the final answer to two decimal places. Does the second method decrease the accuracy?

47. $\dfrac{1 + 0.73205}{1 - 0.73205}$

48. $\dfrac{1 + 0.86603}{1 - 0.86603}$

49. $\dfrac{3.33 + \frac{1.98}{0.74}}{4 + \frac{6.25}{3.15}}$

50. $\dfrac{1.73205 - 1.19195}{3 - (1.73205)(1.19195)}$

51. *Travel Time* On the first part of a 280-mile trip, a salesman averaged 63 miles per hour. He averaged only 54 miles per hour on the last part of the trip because of the increased volume of traffic.
(a) Express the total time for the trip as a function of the distance x traveled at an average speed of 63 miles per hour.
(b) Use a graphing utility to graph the time function. What is the domain of the function?
(c) Approximate the number of miles traveled at 63 miles per hour if the total time was 4 hours and 45 minutes.

52. *Production Limit* A company has fixed costs of $25,000 per month and a variable cost of $18.65 per unit manufactured. (*Fixed costs* are those that occur regardless of the level of production.)
(a) Write the total monthly costs C as a function of the number of units x produced.
(b) Approximate the number of units that can be produced per month if total costs cannot exceed $200,000. Is this problem better solved algebraically or graphically? Explain your reasoning.

53. *Mixture Problem* A 55-gallon barrel contains a mixture with a concentration of 33%. You are asked to remove x gallons of this mixture and replace it with 100% concentrate.
(a) Write the amount of concentrate in the final mixture as a function of x.
(b) Use a graphing utility to graph the concentration function. What is the domain of the function?
(c) Approximate (accurate to one decimal place) the value of x if the final mixture is 60% concentrate.

54. *Dimensions of a Region* A rectangular region with perimeter 230 feet has length x.
(a) Express the area of the rectangle as a function of x.
(b) Use a graphing utility to graph the area function. Because area is nonnegative, what is the domain of the function?
(c) Approximate (accurate to one decimal place) the dimensions of the region if its area is 2000 square feet. (Note that there are two values of x such that the area of the region is 2000 square feet.) Which x-value is appropriate in this problem? Explain.

Negative Income Tax In Exercises 55–57, use the following information about a possible negative income tax for a family consisting of two adults and two children. The plan would guarantee the poor a minimum income while encouraging the family to increase their private income.

Family's earned income: $I = x$

Government payment: $G = 8000 - \dfrac{1}{2}x,$ $0 \le x \le 16{,}000$

Spendable income: $S = I + G$

55. Express the spendable income S in terms of x. Use a graphing utility to graph the functions I, G, and S. What effect does an increase in the family's earned income have on (a) the size of the government payment and (b) the family's spendable income?

56. Find the earned income x if the government payment is $4600.

57. Find the earned income x if the spendable income is $11,800.

58. Four feet of wire is to be cut into two pieces in order to form a square and a circle.
(a) Express the sum A of the areas of the square and the circle in terms of x, the length of the side of the square, and r, the radius of the circle.
(b) Use the fact that there are four feet of wire to express A as a function of x alone.
(c) What is the domain of A?
(d) Use a graphing utility to graph A.
(e) Find the value of x that gives the maximum total area.
(f) Find the value of x that gives the minimum total area.

2.3   COMPLEX NUMBERS

The Imaginary Unit *i* / **Operations with Complex Numbers** / **Complex Conjugates and Division** / **Applications**

The Imaginary Unit *i*

Some quadratic equations have no real solutions. For instance, the quadratic equation $x^2 + 1 = 0$ has no real solution because there is no real number x that can be squared to produce -1. To overcome this deficiency, mathematicians created an expanded system of numbers using the **imaginary unit *i*,** defined as

$$i = \sqrt{-1},$$

where $i^2 = -1$. Adding real numbers to real multiples of this imaginary unit produces the set of **complex numbers.** Each complex number can be written in the **standard form,** $a + bi$.

DEFINITION OF A COMPLEX NUMBER

For real numbers a and b, the number

$a + bi$

is a **complex number.** If $a = 0$ and $b \neq 0$, then the complex number bi is an **imaginary number.**

The set of real numbers is a subset of the set of complex numbers because every real number a can be written as a complex number using $b = 0$. That is, for every real number a, you can write $a = a + 0i$.

Two complex numbers $a + bi$ and $c + di$, written in standard form, are **equal** to each other

$$a + bi = c + di \qquad \qquad \textit{Equality of two complex numbers}$$

if and only if $a = c$ and $b = d$.

Operations with Complex Numbers

To add (or subtract) two complex numbers, you add (or subtract) the real and imaginary parts of the numbers separately.

> ### ADDITION AND SUBTRACTION OF COMPLEX NUMBERS
>
> If $a + bi$ and $c + di$ are two complex numbers written in standard form, then their sum and difference are defined as follows.
>
> $Sum:$ $\quad\quad (a + bi) + (c + di) = (a + c) + (b + d)i$
> $Difference:$ $(a + bi) - (c + di) = (a - c) + (b - d)i$

Technology Note _____

Some graphing utilities can perform operations with complex numbers. Check your user's manual to see if your graphing utility has this capability.

The **additive identity** in the complex number system is zero (the same as in the real number system). Furthermore, the **additive inverse** of the complex number $a + bi$ is

$$-(a + bi) = -a - bi. \qquad \textit{Additive inverse}$$

Thus, $(a + bi) + (-a - bi) = 0 + 0i = 0$.

EXAMPLE 1 Adding and Subtracting Complex Numbers

Write the sums and differences in standard form.

A. $(3 - i) + (2 + 3i)$
B. $2i + (-4 - 2i)$
C. $3 - (-2 + 3i) + (-5 + i)$

SOLUTION

A. $(3 - i) + (2 + 3i) = 3 - i + 2 + 3i$ $\qquad$ *Remove parentheses*
$\qquad\qquad\qquad\quad = 3 + 2 - i + 3i$ $\qquad$ *Group real and imaginary terms*
$\qquad\qquad\qquad\quad = (3 + 2) + (-1 + 3)i$
$\qquad\qquad\qquad\quad = 5 + 2i$ $\qquad\qquad$ *Standard form*

B. $2i + (-4 - 2i) = 2i - 4 - 2i$ $\qquad$ *Remove parentheses*
$\qquad\qquad\qquad\; = -4 + 2i - 2i$ $\qquad$ *Group real and imaginary terms*
$\qquad\qquad\qquad\; = -4$ $\qquad\qquad\qquad$ *Standard form*

C. $3 - (-2 + 3i) + (-5 + i) = 3 + 2 - 3i - 5 + i$
$\qquad\qquad\qquad\qquad\qquad = 3 + 2 - 5 - 3i + i$
$\qquad\qquad\qquad\qquad\qquad = 0 - 2i$
$\qquad\qquad\qquad\qquad\qquad = -2i$

REMARK   Note in Example 1 (B) that the sum of two complex numbers can be a real number.

REMARK   When performing operations with complex numbers, it is sometimes necessary to evaluate powers of the imaginary unit i. The first several powers of i are as follows.

$i^1 = i$

$i^2 = -1$

$i^3 = i(i^2) = i(-1) = -i$

$i^4 = (i^2)(i^2) = (-1)(-1) = 1$

$i^5 = i(i^4) = i(1) = i$

$i^6 = (i^2)(i^4) = (-1)(1) = -1$

$i^7 = (i^3)(i^4) = (-i)(1) = -i$

$i^8 = (i^4)(i^4) = (1)(1) = 1$

Note how the pattern of values i, -1, $-i$, and 1 repeats for powers greater than 4.

Many of the properties of real numbers are valid for complex numbers as well. Here are some.

Associative Property of Addition and Multiplication
Commutative Property of Addition and Multiplication
Distributive Property of Multiplication over Addition

Notice how these properties are used when two complex numbers are multiplied.

$$
\begin{aligned}
(a + bi)(c + di) &= a(c + di) + bi(c + di) && \textit{Distributive}\\
&= ac + (ad)i + (bc)i + (bd)i^2 && \textit{Distributive}\\
&= ac + (ad)i + (bc)i + (bd)(-1) && i^2 = -1\\
&= ac - bd + (ad)i + (bc)i && \textit{Commutative}\\
&= (ac - bd) + (ad + bc)i && \textit{Distributive}
\end{aligned}
$$

Rather than trying to memorize this multiplication rule, just remember how the distributive property is used to multiply two complex numbers. The procedure is similar to multiplying two polynomials and combining like terms (as in the FOIL method).

EXAMPLE 2   Multiplying Complex Numbers

Find the products.

A. $(i)(-3i)$ **B.** $(2 - i)(4 + 3i)$ **C.** $(3 + 2i)(3 - 2i)$

SOLUTION

A. $(i)(-3i) = -3i^2 = -3(-1) = 3$

B. $(2 - i)(4 + 3i) = 8 + 6i - 4i - 3i^2$ *Binomial product*

$\qquad\qquad\qquad\quad = 8 + 6i - 4i - 3(-1)$ $i^2 = -1$

$\qquad\qquad\qquad\quad = 8 + 3 + 6i - 4i$ *Collect terms*

$\qquad\qquad\qquad\quad = 11 + 2i$ *Standard form*

C. $(3 + 2i)(3 - 2i) = 9 - 6i + 6i - 4i^2$

$\qquad\qquad\qquad\quad = 9 - 4(-1)$

$\qquad\qquad\qquad\quad = 9 + 4$

$\qquad\qquad\qquad\quad = 13$

Complex Conjugates and Division

Notice in Example 2 (C) that the product of two complex numbers can be a real number. This occurs with pairs of complex numbers of the form $a + bi$ and $a - bi$, called **complex conjugates.** In general, the product of two complex conjugates can be written as follows.

$$(a + bi)(a - bi) = a^2 - abi + abi - b^2i^2$$
$$= a^2 - b^2(-1)$$
$$= a^2 + b^2$$

Complex conjugates can be used to divide one complex number by another. That is, to find the quotient

$$\frac{a + bi}{c + di}, \qquad c \text{ and } d \text{ not both zero,}$$

multiply the numerator and denominator by the conjugate of the denominator to obtain

$$\frac{a + bi}{c + di} = \frac{a + bi}{c + di}\left(\frac{c - di}{c - di}\right)$$
$$= \frac{(ac + bd) + (bc - ad)i}{c^2 + d^2}.$$

This procedure is demonstrated in Examples 3 and 4.

EXAMPLE 3   Dividing Complex Numbers

Write the complex number $\dfrac{1}{1 + i}$ in standard form.

SOLUTION

$$\frac{1}{1 + i} = \frac{1}{1 + i}\left(\frac{1 - i}{1 - i}\right)$$
$$= \frac{1 - i}{1^2 - i^2}$$
$$= \frac{1 - i}{1 - (-1)}$$
$$= \frac{1 - i}{2}$$
$$= \frac{1}{2} - \frac{1}{2}i$$

EXAMPLE 4 Dividing Complex Numbers

Write the complex number $\dfrac{2 + 3i}{4 - 2i}$ in standard form.

SOLUTION

$$\frac{2 + 3i}{4 - 2i} = \frac{2 + 3i}{4 - 2i}\left(\frac{4 + 2i}{4 + 2i}\right)$$

$$= \frac{8 + 4i + 12i + 6i^2}{16 - 4i^2}$$

$$= \frac{8 - 6 + 16i}{16 + 4}$$

$$= \frac{1}{20}(2 + 16i)$$

$$= \frac{1}{10} + \frac{4}{5}i$$

Using the Quadratic Formula to solve a quadratic equation can produce a result such as $\sqrt{-3}$, which is not a real number. By factoring out $i = \sqrt{-1}$, you can write this number in standard form.

$$\sqrt{-3} = \sqrt{3(-1)} = \sqrt{3}\sqrt{-1} = \sqrt{3}\,i$$

The number $\sqrt{3}\,i$ is the principal square root of -3.

PRINCIPAL SQUARE ROOT OF A NEGATIVE NUMBER

If a is a positive number, then the **principal square root** of the negative number $-a$ is defined as

$$\sqrt{-a} = \sqrt{a}\,i.$$

Technology Note _____

If your graphing utility permits complex arithmetic, you can verify that $\sqrt{-5}\sqrt{-5} \neq \sqrt{(-5)(-5)}$. On the TI-85, you would have $\sqrt{(-5, 0)}\sqrt{(-5, 0)} = (-5, 0)$ and $\sqrt{(-5, 0)(-5, 0)} = (5, 0)$.

In this definition you are using the rule $\sqrt{ab} = \sqrt{a}\sqrt{b}$, for $a > 0$ and $b < 0$. This rule is not valid if *both* a and b are negative. For example,

$$\sqrt{-5}\sqrt{-5} = \left(\sqrt{5}\,i\right)\left(\sqrt{5}\,i\right) = 5i^2 = -5$$

whereas

$$\sqrt{(-5)(-5)} = \sqrt{25} = 5.$$

Consequently, $\sqrt{(-5)(-5)} \neq \sqrt{-5}\sqrt{-5}$.

REMARK When working with square roots of negative numbers, be sure to convert to standard form *before* multiplying.

EXAMPLE 5   **Writing Complex Numbers in Standard Form**

A. $\sqrt{-3}\sqrt{-12} = \sqrt{3}\,i\,\sqrt{12}\,i = \sqrt{36}\,i^2 = 6(-1) = -6$

B. $\sqrt{-48} - \sqrt{-27} = \sqrt{48}\,i - \sqrt{27}\,i = 4\sqrt{3}\,i - 3\sqrt{3}\,i = \sqrt{3}\,i$

C. $\left(-1 + \sqrt{-3}\,\right)^2 = \left(-1 + \sqrt{3}\,i\right)^2$
$$= (-1)^2 - 2\sqrt{3}\,i + \left(\sqrt{3}\,\right)^2(i^2)$$
$$= 1 - 2\sqrt{3}\,i + 3(-1)$$
$$= -2 - 2\sqrt{3}\,i$$

Applications

Just as every real number corresponds to a point on the real number line, every complex number corresponds to a point in the **complex plane,** as shown in Figure 2.15. In this figure, note that the vertical axis is called the **imaginary axis** and the horizontal axis is called the **real axis.**

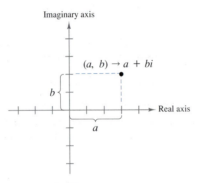

FIGURE 2.15

EXAMPLE 6   **Plotting Complex Numbers**

Plot the following complex numbers in the complex plane.

A. $2 + 3i$ **B.** $-1 + 2i$ **C.** 4

SOLUTION

A. To plot the complex number $2 + 3i$, move (from the origin) two units to the right on the real axis and then three units up, as shown in Figure 2.16. In other words, plotting the complex number $2 + 3i$ in the complex plane is comparable to plotting the point $(2, 3)$ in the Cartesian plane.

B. The complex number $-1 + 2i$ corresponds to the point $(-1, 2)$, as shown in Figure 2.16.

C. The complex number 4 corresponds to the point $(4, 0)$, as shown in Figure 2.16.

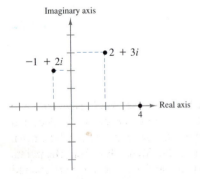

FIGURE 2.16

In the hands of a person who understands "fractal geometry," the complex plane can become an easel on which stunning pictures, called **fractals,** can be drawn. The most famous such picture is called the **Mandelbrot Set,** named after the Polish-born mathematician Benoit Mandelbrot. To draw the Mandelbrot Set, consider the sequence of numbers

$$c, c^2 + c, (c^2 + c)^2 + c, [(c^2 + c)^2 + c]^2 + c, \dots .$$

The behavior of this sequence depends on the value of the complex number c. For some values of c this sequence is **bounded,** and for other values it is **unbounded.** If the sequence is bounded, then the complex number c is in the Mandelbrot Set, and if the sequence is unbounded, then the complex number c is not in the Mandelbrot Set.

EXAMPLE 7   **Members of the Mandelbrot Set**

A. The complex number -2 is in the Mandelbrot Set because for $c = -2$, the corresponding Mandelbrot sequence is

$$-2, 2, 2, 2, 2, 2, \dots ,$$

which is bounded.

B. The complex number i is also in the Mandelbrot Set because for $c = i$, the corresponding Mandelbrot sequence is

$$i, -1 + i, -i, -1 + i, -i, -1 + i, \dots ,$$

which is bounded.

C. The complex number $1 + i$ is *not* in the Mandelbrot Set because for $c = 1 + i$, the corresponding Mandelbrot sequence is

$$1 + i, 1 + 3i, -7 + 7i, 1 - 97i, -9407 - 193i,$$
$$88454401 + 3631103i, \dots ,$$

which is unbounded.

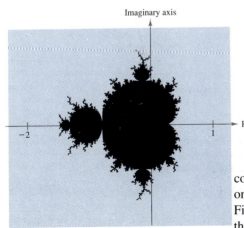

Imaginary axis

Real axis

-2 1

FIGURE 2.17

With this definition, a picture of the Mandelbrot Set would have only two colors—one color for points that are in the set (the sequence is bounded) and one color for points that are outside the set (the sequence is not bounded). Figure 2.17 shows a black and blue picture of the Mandelbrot Set. The points that are colored black are in the Mandelbrot Set, and the points that are colored blue are not.

To add more interest to the picture, computer scientists discovered that the points that are not in the Mandelbrot Set can be assigned a variety of colors, depending on how "quickly" their sequences diverge. Figure 2.18 shows three different appendages of the Mandelbrot Set. (The black portions of the picture represent points that are in the Mandelbrot Set.)

FIGURE 2.18

Figures 2.19, 2.20, and 2.21 show other types of fractals. From the pictures, you can see why fractals have fascinated people since their discovery (around 1980). The fractals shown were produced on a graphing calculator, using programs written by Chuck Biehl. Listings of the programs are in Appendix C.

The Sierpinski Gasket   The Fractal "Dragon"   A Fractal Fern

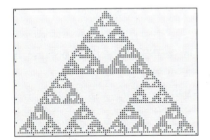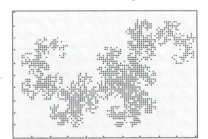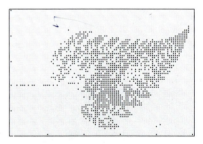

FIGURE 2.19   **FIGURE 2.20**   **FIGURE 2.21**

DISCUSSION PROBLEM

························

CUBE ROOTS OF UNITY

In the real number system, the equation $x^3 = 1$ has 1 as its only solution. However, in the complex number system, this equation has three solutions.

$$1, \quad \frac{-1 + \sqrt{3}\,i}{2}, \quad \text{and} \quad \frac{-1 - \sqrt{3}\,i}{2}$$

Try cubing each of these numbers to show that each number has the property that $x^3 = 1$.

WARM-UP

The following warm-up exercises involve skills that were covered in earlier sections. You will use these skills in the exercise set for this section.

In Exercises 1–8, simplify the expression.

1. $\sqrt{12}$

2. $\sqrt{500}$

3. $\sqrt{20} - \sqrt{5}$

4. $\sqrt{27} - \sqrt{243}$

5. $\sqrt{24}\sqrt{6}$

6. $2\sqrt{18}\sqrt{32}$

7. $\dfrac{1}{\sqrt{3}}$

8. $\dfrac{2}{\sqrt{2}}$

In Exercises 9 and 10, solve the quadratic equation.

9. $x^2 + x - 1 = 0$

10. $x^2 + 2x - 1 = 0$

SECTION 2.3 · EXERCISES

1. Write the first 16 positive integer powers of i (that is, i, i^2, i^3, ..., i^{16}), and express each as i, $-i$, 1, or -1.

2. Express each of the following powers of i as i, $-i$, 1, or -1.
(a) i^{40} (b) i^{25}
(c) i^{50} (d) i^{67}

In Exercises 3–6, find real numbers a and b so that the equation is true.

3. $a + bi = -10 + 6i$ **4.** $a + bi = 13 + 4i$

5. $(a - 1) + (b + 3)i = 5 + 8i$

6. $(a + 6) + 2bi = 6 - 5i$

In Exercises 7–18, write the complex number in standard form.

7. $4 + \sqrt{-9}$ **8.** $3 + \sqrt{-16}$

9. $2 - \sqrt{-27}$ **10.** $1 + \sqrt{-8}$

11. $\sqrt{-75}$ **12.** 45

13. $-6i + i^2$ **14.** $-4i^2 + 2i$

15. $-5i^5$ **16.** $(-i)^3$

17. 8   **18.** $\left(\sqrt{-4}\right)^2 - 5$

In Exercises 19–26, perform the indicated addition or subtraction and write the result in standard form.

19. $(5 + i) + (6 - 2i)$ **20.** $(13 - 2i) + (-5 + 6i)$

21. $(8 - i) - (4 - i)$ **22.** $(3 + 2i) - (6 + 13i)$

23. $\left(-2 + \sqrt{-8}\right) + \left(5 - \sqrt{-50}\right)$

24. $\left(8 + \sqrt{-18}\right) - \left(4 + 3\sqrt{2}\,i\right)$

25. $-\left(\frac{3}{2} + \frac{5}{2}i\right) + \left(\frac{5}{3} + \frac{11}{3}i\right)$

26. $(1.6 + 3.2i) + (-5.8 + 4.3i)$

In Exercises 27–30, write the conjugate of the complex number, and find the product of the number and its conjugate.

27. $5 + 3i$ **28.** $9 - 12i$

29. $-2 - \sqrt{5}\,i$ **30.** $20i$

In Exercises 31–54, perform the specified operation and write the result in standard form.

31. $\sqrt{-6} \cdot \sqrt{-2}$ **32.** $\sqrt{-5} \cdot \sqrt{-10}$

33. $\left(\sqrt{-10}\right)^2$ **34.** $\left(\sqrt{-75}\right)^2$

35. $(1 + i)(3 - 2i)$ **36.** $(6 - 2i)(2 - 3i)$

37. $6i(5 - 2i)$ **38.** $-8i(9 + 4i)$

39. $\left(\sqrt{14} + \sqrt{10}\,i\right)\left(\sqrt{14} - \sqrt{10}\,i\right)$

40. $\left(3 + \sqrt{-5}\right)\left(7 - \sqrt{-10}\right)$

41. $(4 + 5i)^2$ **42.** $(2 - 3i)^2$

43. $(2 + 3i)^2 + (2 - 3i)^2$ **44.** $(1 - 2i)^2 - (1 + 2i)^2$

45. $\dfrac{4}{4 - 5i}$ **46.** $\dfrac{3}{1 - i}$

47. $\dfrac{2 + i}{2 - i}$

48. $\dfrac{8 - 7i}{1 - 2i}$

49. $\dfrac{6 - 7i}{i}$

50. $\dfrac{8 + 20i}{2i}$

51. $\dfrac{5}{(1 + i)^3}$

52. $\dfrac{1}{i^3}$

53. $\dfrac{(21 - 7i)(4 + 3i)}{2 - 5i}$

54. $\dfrac{(2 - 3i)(5i)}{2 + 3i}$

In Exercises 55–60, plot the given complex number in the complex plane.

55. $-2 + i$

56. i

57. 3

58. $-2 - 3i$

59. $1 - 2i$

60. $-2i$

In Exercises 61–66, find the first six terms in the following sequence.

$$c, \quad c^2 + c, \quad (c^2 + c)^2 + c, \quad [(c^2 + c)^2 + c]^2 + c, \quad \ldots$$

From these terms, do you think the given complex number is in the Mandelbrot Set?

61. $c = 0$

62. $c = 2$

63. $c = \frac{1}{2}i$

64. $c = -i$

65. $c = 1$

66. $c = -1$

67. Cube the complex numbers 2, $-1 + \sqrt{3}\,i$, and $-1 - \sqrt{3}\,i$. What do you notice?

68. Raise the complex numbers 2, -2, $2i$, and $-2i$ to the fourth power. What do you notice?

69. Prove that the sum of a complex number $a + bi$ and its conjugate is a real number.

70. Prove that the difference of a complex number $a + bi$ and its conjugate is an imaginary number.

71. Prove that the product of a complex number $a + bi$ and its conjugate is a real number.

72. Prove that the conjugate of the product of two complex numbers $a_1 + b_1 i$ and $a_2 + b_2 i$ is the product of their conjugates.

2.4 SOLVING EQUATIONS ALGEBRAICALLY

Quadratic Equations / Polynomial Equations of Higher Degree / Equations Involving Radicals / Equations Involving Fractions or Absolute Values / Applications

Quadratic Equations

A **quadratic equation** in x is an equation that can be written in the standard form

$$ax^2 + bx + c = 0,$$

where a, b, and c are real numbers with $a \neq 0$. Another name for a quadratic equation in x is a **second-degree polynomial equation in x.** You should be familiar with the following four methods for solving quadratic equations.

Technology Note

Try programming the Quadratic Formula into a programmable calculator. For instance, the following program is for the TI-81.

```
:Disp "ENTER,A"
:Input A
:Disp "ENTER,B"
:Input B
:Disp "ENTER,C"
:Input C
:B² − 4AC→D
:If D<0
:Goto 1
:((−B+√D)/(2A))→S
:Disp S
:((−B − √D)/(2A))→S
:Disp S
:End
:Lbl 1
:Disp "NO REAL SOLUTION"
:End
```

To use this program, you must first write the equation in standard form. Then, enter the values of a, b, and c. After the final value has been entered, the program will display either two solutions *or* the words "No Real Solution."

SOLVING QUADRATIC EQUATIONS

Method	*Example*
Factoring If $ab = 0$, then $a = 0$ or $b = 0$.	$x^2 - x - 6 = 0$ $(x - 3)(x + 2) = 0$ $x = 3$ or -2
Square Root Principle If $u^2 = c$, where $c > 0$, then $u = \pm\sqrt{c}$.	$(x + 3)^2 = 16$ $x + 3 = \pm 4$ $x = 1$ or -7
Completing the Square If $x^2 + bx = c$, then $x^2 + bx + \left(\dfrac{b}{2}\right)^2 = c + \left(\dfrac{b}{2}\right)^2$ $\left(x + \dfrac{b}{2}\right)^2 = c + \dfrac{b^2}{4}.$	$x^2 - 6x = 5$ $x^2 - 6x + 3^2 = 5 + 3^2$ $(x - 3)^2 = 14$ $x - 3 = \pm\sqrt{14}$ $x = 3 \pm \sqrt{14}$
Quadratic Formula If $ax^2 + bx + c = 0$, then $x = \dfrac{-b \pm \sqrt{b^2 - 4ac}}{2a}.$	$2x^2 + 3x - 1 = 0$ $x = \dfrac{-3 \pm \sqrt{3^2 - 4(2)(-1)}}{2(2)}$ $= \dfrac{-3 \pm \sqrt{17}}{4}$

EXAMPLE 1   Solving Quadratic Equations by Factoring

Solve the quadratic equations.

A. $6x^2 = 3x$ **B.** $9x^2 - 6x + 1 = 0$

SOLUTION

A.

$6x^2 = 3x$	*Original equation*
$6x^2 - 3x = 0$	*Standard form*
$3x(2x - 1) = 0$	*Factored form*
$3x = 0 \rightarrow x = 0$	*Set 1st factor equal to 0*
$2x - 1 = 0 \rightarrow x = \dfrac{1}{2}$	*Set 2nd factor equal to 0*

B.

$9x^2 - 6x + 1 = 0$	*Standard form*
$(3x - 1)^2 = 0$	*Factored form*
$3x - 1 = 0$	*Set repeated factor equal to 0*
$x = \dfrac{1}{3}$	*Solution*

Throughout the text, when solving equations, be sure to check your solutions—either *algebraically,* by substituting in the original equation, or *graphically.* For instance, the graphs shown in Figure 2.22 visually reinforce the solutions obtained in Example 1.

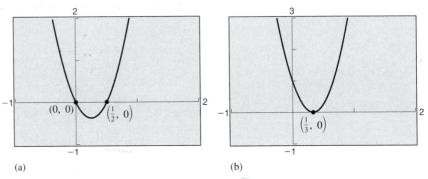

(a)   (b)

FIGURE 2.22

EXAMPLE 2   Extracting Square Roots

Solve the quadratic equations.

A. $4x^2 = 12$ **B.** $(x - 3)^2 = 7$

SOLUTION

A. $4x^2 = 12$ *Original equation*
$\quad\ x^2 = 3$ *Divide both sides by 4*
$\quad\quad x = \pm\sqrt{3}$ *Extract square roots*
B. $(x - 3)^2 = 7$ *Original equation*
$\quad x - 3 = \pm\sqrt{7}$ *Extract square roots*
$\quad\quad\ x = 3 \pm \sqrt{7}$ *Add 3 to both sides*

The graphs of $y = 4x^2 - 12$ and $y = (x - 3)^2 - 7$, shown in Figure 2.23, reinforce these solutions.

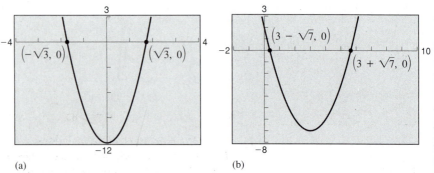

(a)   (b)

FIGURE 2.23

To complete the square when the leading coefficient is not 1, begin by dividing both sides of the equation by this coefficient.

EXAMPLE 3   Completing the Square: Leading Coefficient Is Not 1

$$3x^2 - 4x - 5 = 0 \qquad \text{\textit{Original equation}}$$

$$3x^2 - 4x = 5 \qquad \text{\textit{Add 5 to both sides}}$$

$$x^2 - \frac{4}{3}x = \frac{5}{3} \qquad \text{\textit{Divide both sides by 3}}$$

$$x^2 - \frac{4}{3}x + \left(\frac{2}{3}\right)^2 = \frac{5}{3} + \left(\frac{2}{3}\right)^2 \qquad \text{\textit{Add } } \left(\tfrac{2}{3}\right)^2 \text{ \textit{to both sides}}$$

$$\underbrace{\qquad\qquad}_{(\text{half})^2}$$

$$\left(x - \frac{2}{3}\right)^2 = \frac{19}{9}$$

$$x - \frac{2}{3} = \pm \frac{\sqrt{19}}{3} \qquad \text{\textit{Extract square roots}}$$

$$x = \frac{2}{3} \pm \frac{\sqrt{19}}{3} \qquad \text{\textit{Solutions}}$$

Using a calculator, you will find that the two solutions are approximately 2.11963 and -0.78630, which agree with the graphical solution shown in Figure 2.24.

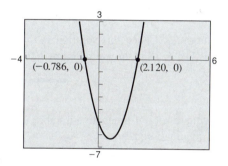

$(-0.786, 0)$ $(2.120, 0)$

FIGURE 2.24

EXAMPLE 4   Using the Quadratic Formula: Two Distinct Solutions

$$x^2 + 3x = 9 \qquad \text{\textit{Original equation}}$$

$$x^2 + 3x - 9 = 0$$

$$x = \frac{-b \pm \sqrt{b^2 - 4ac}}{2a} \qquad \text{\textit{Quadratic Formula}}$$

$$x = \frac{-3 \pm \sqrt{3^2 - 4(1)(-9)}}{2(1)} \qquad \text{\textit{Substitute}}$$

$$x = \frac{-3 \pm \sqrt{45}}{2}$$

$$x = \frac{-3 \pm 3\sqrt{5}}{2}$$

$$x \approx 1.85 \text{ or } -4.85 \qquad \text{\textit{Solutions}}$$

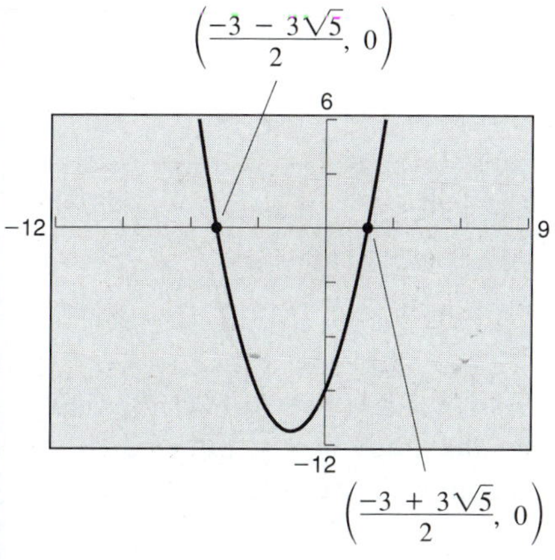

The equation has two solutions: $x \approx 1.85$ and $x \approx -4.85$. Check these solutions in the original equation. The graph of $y = x^2 + 3x - 9$, as shown in Figure 2.25, reinforces the solutions graphically.

$\left(\dfrac{-3 - 3\sqrt{5}}{2}, 0\right)$

$\left(\dfrac{-3 + 3\sqrt{5}}{2}, 0\right)$

FIGURE 2.25

EXAMPLE 5   Using the Quadratic Formula: One Repeated Solution

Use the Quadratic Formula to solve $8x^2 - 24x + 18 = 0$.

SOLUTION

This equation has a common factor of 2. To simplify things, first divide both sides of the equation by 2.

$$8x^2 - 24x + 18 = 0 \qquad \textit{Common factor of 2}$$
$$4x^2 - 12x + 9 = 0 \qquad \textit{Standard form}$$
$$x = \frac{-b \pm \sqrt{b^2 - 4ac}}{2a} \qquad \textit{Quadratic Formula}$$
$$x = \frac{-(-12) \pm \sqrt{(-12)^2 - 4(4)(9)}}{2(4)} \qquad \textit{Substitute}$$
$$x = \frac{12 \pm \sqrt{0}}{8} = \frac{3}{2} \qquad \textit{Repeated solution}$$

This quadratic equation has only one solution: $\frac{3}{2}$. Check this solution in the original equation. The graph of $y = 8x^2 - 24x + 18$ is shown in Figure 2.26.

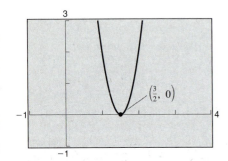

FIGURE 2.26

Example 6 shows how the principal square root of a negative number is used to represent complex solutions of a quadratic equation.

EXAMPLE 6   Complex Solutions of a Quadratic Equation

Solve the equation $3x^2 - 2x + 5 = 0$.

SOLUTION

By the Quadratic Formula, you can write the solutions as follows.

$$x = \frac{-(-2) \pm \sqrt{(-2)^2 - 4(3)(5)}}{2(3)}$$
$$= \frac{2 \pm \sqrt{-56}}{6}$$
$$= \frac{2 \pm 2\sqrt{14}\,i}{6}$$
$$= \frac{1}{3} \pm \frac{\sqrt{14}}{3}\,i$$

The equation has two solutions: $\frac{1}{3}\left(1 + \sqrt{14}\,i\right)$ and $\frac{1}{3}\left(1 - \sqrt{14}\,i\right)$. Because the original equation has no real solutions, the graph of $y = 3x^2 - 2x + 5$ should have no x-intercepts. This is confirmed in Figure 2.27.

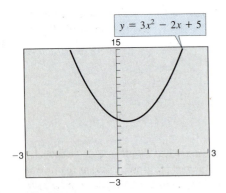

FIGURE 2.27

Polynomial Equations of Higher Degree

The methods used to solve quadratic equations can sometimes be extended to polynomial equations of higher degree, as shown in the next two examples.

EXAMPLE 7   Solving an Equation of Quadratic Type

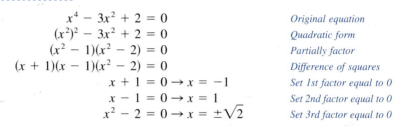

$$x^4 - 3x^2 + 2 = 0 \qquad \text{\textit{Original equation}}$$
$$(x^2)^2 - 3x^2 + 2 = 0 \qquad \text{\textit{Quadratic form}}$$
$$(x^2 - 1)(x^2 - 2) = 0 \qquad \text{\textit{Partially factor}}$$
$$(x + 1)(x - 1)(x^2 - 2) = 0 \qquad \text{\textit{Difference of squares}}$$
$$x + 1 = 0 \rightarrow x = -1 \qquad \text{\textit{Set 1st factor equal to 0}}$$
$$x - 1 = 0 \rightarrow x = 1 \qquad \text{\textit{Set 2nd factor equal to 0}}$$
$$x^2 - 2 = 0 \rightarrow x = \pm\sqrt{2} \qquad \text{\textit{Set 3rd factor equal to 0}}$$

The equation has four solutions: -1, 1, $\sqrt{2}$, and $-\sqrt{2}$. Check these solutions in the original equation. The graph of $y = x^4 - 3x^2 + 2$, shown in Figure 2.28, verifies the algebraic solution.

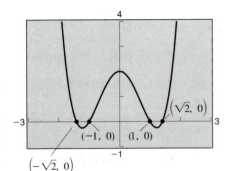

FIGURE 2.28

EXAMPLE 8   Solving a Polynomial Equation by Factoring

$$x^3 - 3x^2 - 3x + 9 = 0 \qquad \text{\textit{Original equation}}$$
$$x^2(x - 3) - 3(x - 3) = 0 \qquad \text{\textit{Group terms}}$$
$$(x - 3)(x^2 - 3) = 0 \qquad \text{\textit{Factor by grouping}}$$
$$x - 3 = 0 \rightarrow x = 3 \qquad \text{\textit{Set 1st factor equal to 0}}$$
$$x^2 - 3 = 0 \rightarrow x = \pm\sqrt{3} \qquad \text{\textit{Set 2nd factor equal to 0}}$$

The equation has three solutions: 3, $\sqrt{3}$, and $-\sqrt{3}$. Check these solutions in the original equation. The graph of $y = x^3 - 3x^2 - 3x + 9$, shown in Figure 2.29, verifies the algebraic solution.

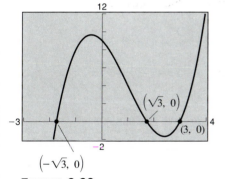

FIGURE 2.29

Technology Note

Some graphing utilities can solve polynomial equations directly. For instance, on the TI-85, you can use the POLY menu to verify that the solutions to Example 8 are 3, 1.732, and -1.732.

Equations Involving Radicals

An equation involving a radical expression can often be cleared of radicals by raising both sides of the equation to an appropriate power. When using this procedure, remember that it can introduce extraneous solutions—so be sure to check each solution in the original equation, or graphically.

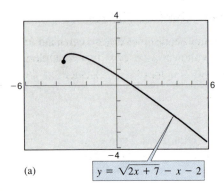

(a)

$$y = \sqrt{2x + 7} - x - 2$$

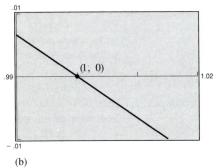

(b)

FIGURE 2.30

EXAMPLE 9   An Equation Involving a Radical

Solve $\sqrt{2x + 7} - x = 2$.

SOLUTION

The graph of $y = \sqrt{2x + 7} - x - 2$ is shown in Figure 2.30(a). Notice that the domain is $x \geq -\frac{7}{2}$, because the expression under the radical cannot be negative. Furthermore, there seems to be one solution near $x = 1$. Using the zoom and trace features shown in Figure 2.30(b), you can verify that $x = 1$ is the only solution. The equation can also be solved algebraically. To do this, first isolate the radical. Then, eliminate the square root by squaring both sides of the equation. You will obtain two solutions. Check both in the original equation.

CHECK

$$\sqrt{2(-3) + 7} + 3 = \sqrt{1} + 3 \neq 2 \qquad -3 \text{ does not check}$$
$$\sqrt{2(1) + 7} - 1 = \sqrt{9} - 1 = 2 \qquad 1 \text{ checks}$$

To solve equations with two or more radicals algebraically, it may be necessary to repeat the "isolate a radical and square both sides" routine discussed in Example 9. The benefits of a graphing utility are obvious in such cases.

EXAMPLE 10   An Equation Involving Two Radicals

$$\sqrt{2x + 6} - \sqrt{x + 4} = 1 \qquad \text{Original equation}$$
$$\sqrt{2x + 6} = 1 + \sqrt{x + 4} \qquad \text{Isolate radical}$$
$$2x + 6 = 1 + 2\sqrt{x + 4} + (x + 4) \qquad \text{Square both sides}$$
$$x + 1 = 2\sqrt{x + 4} \qquad \text{Isolate radical}$$
$$x^2 + 2x + 1 = 4(x + 4) \qquad \text{Square both sides}$$
$$x^2 - 2x - 15 = 0 \qquad \text{Standard form}$$
$$(x - 5)(x + 3) = 0 \qquad \text{Factor}$$
$$x - 5 = 0 \rightarrow x = 5 \qquad \text{Set 1st factor equal to 0}$$
$$x + 3 = 0 \rightarrow x = -3 \qquad \text{Set 2nd factor equal to 0}$$

From the graph of $y = \sqrt{2x + 6} - \sqrt{x - 4} - 1$ in Figure 2.31, you can conclude that $x = 5$ is the only solution.

FIGURE 2.31

Equations Involving Fractions or Absolute Values

To solve algebraically an equation involving fractions, multiply both sides of the equation by the least common denominator of each term in the equation. This procedure will "clear the equation of fractions." For instance, in the equation

$$\frac{2}{x^2 + 1} + \frac{1}{x} = \frac{2}{x}$$

you can multiply both sides of the equation by $x(x^2 + 1)$ to obtain

$$(2x) + (x^2 + 1) = 2(x^2 + 1).$$

Try solving this equation. You should obtain one solution: $x = 1$.

EXAMPLE 11 **An Equation Involving Fractions**

Solve $\dfrac{2}{x} = \dfrac{3}{x - 2} - 1$.

SOLUTION

For this equation, the least common denominator of the three terms is $x(x - 2)$, so you can begin by multiplying each term in the equation by this expression.

$$\frac{2}{x} = \frac{3}{x - 2} - 1$$

$$x(x - 2)\frac{2}{x} = x(x - 2)\frac{3}{x - 2} - x(x - 2)(1)$$

$$2(x - 2) = 3x - x(x - 2), \qquad x \neq 0, 2$$

$$2x - 4 = -x^2 + 5x$$

$$x^2 - 3x - 4 = 0$$

$$(x - 4)(x + 1) = 0$$

$$x - 4 = 0 \rightarrow x = 4$$

$$x + 1 = 0 \rightarrow x = -1$$

The equation has two solutions: 4 and -1. Check these solutions in the original equation. Use a graphing utility to graph

$$y = \frac{2}{x} - \frac{3}{x - 2} + 1$$

to verify these solutions.

REMARK   Graphs of functions involving variable denominators can be tricky because of the way graphing utilities skip over points where the denominator is zero. You will study graphs of this type of function in Sections 3.5 and 3.6.   ▬▬▬

EXAMPLE 12   An Equation Involving Absolute Value

Solve $|x^2 - 3x| = -4x + 6$.

SOLUTION

From the graph of $y = |x^2 - 3x| + 4x - 6$ in Figure 2.32, you can estimate the solutions to be -3 and 1. These can be verified by substitution into the equation. To solve *algebraically* an equation involving an absolute value, you must consider the fact that the expression inside the absolute value bars can be positive or negative. This consideration results in *two* separate equations, each of which must be solved.

First equation

$$x^2 - 3x = -4x + 6 \qquad \text{\textit{Use positive expression}}$$
$$x^2 + x - 6 = 0 \qquad \text{\textit{Standard form}}$$
$$(x + 3)(x - 2) = 0 \qquad \text{\textit{Factor}}$$
$$x + 3 = 0 \rightarrow x = -3 \qquad \text{\textit{Set 1st factor equal to 0}}$$
$$x - 2 = 0 \rightarrow x = 2 \qquad \text{\textit{Set 2nd factor equal to 0}}$$

Second equation

$$-(x^2 - 3x) = -4x + 6 \qquad \text{\textit{Use negative expression}}$$
$$x^2 - 7x + 6 = 0 \qquad \text{\textit{Standard form}}$$
$$(x - 1)(x - 6) = 0 \qquad \text{\textit{Factor}}$$
$$x - 1 = 0 \rightarrow x = 1 \qquad \text{\textit{Set 1st factor equal to 0}}$$
$$x - 6 = 0 \rightarrow x = 6 \qquad \text{\textit{Set 2nd factor equal to 0}}$$

CHECK

$$|(-3)^2 - 3(-3)| = -4(-3) + 6 \qquad \text{\textit{-3 checks}}$$
$$|2^2 - 3(2)| \neq -4(2) + 6 \qquad \text{\textit{2 does not check}}$$
$$|1^2 - 3(1)| = -4(1) + 6 \qquad \text{\textit{1 checks}}$$
$$|6^2 - 3(6)| \neq -4(6) + 6 \qquad \text{\textit{6 does not check}}$$

The equation has only two solutions: -3 and 1, just as you obtained by graphing.   ▬▬▬▬▬

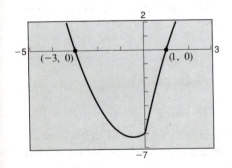

FIGURE 2.32

Applications

A common application of quadratic equations involves an object that is falling (or projected vertically into the air). The general equation that gives the height of such an object is a **position equation,** and (on the earth's surface) it has the form

$$s = -16t^2 + v_0 t + s_0.$$

In this equation, s is the height of the object (in feet), v_0 is the original velocity of the object (in feet per second), s_0 is the original height of the object (in feet), and t is the time (in seconds).

EXAMPLE 13   Falling Time

A construction worker on the 24th floor of a building project (see Figure 2.33) accidentally drops a wrench and immediately yells "Look out below!" Could a person at ground level hear this warning in time to get out of the way of the falling wrench?

SOLUTION

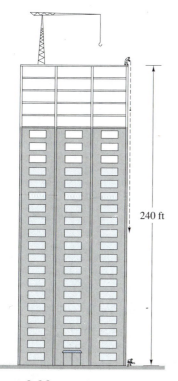

240 ft

Assume that each floor of the building is 10 feet high, so that the wrench is dropped from a height of 240 feet. Because sound travels at about 1100 feet per second, you can estimate that a person on ground level hears the warning within 1 second of the time the wrench is dropped. To set up a mathematical model for the height of the wrench, use the position equation $s = -16t^2 + v_0 t + s_0$. Because the object is dropped (rather than thrown), you can conclude that the initial velocity is $v_0 = 0$. Moreover, because the initial height is $s_0 = 240$ feet, you have the following model.

$$s = -16t^2 + 240, \qquad t \geq 0$$

Note that after falling for 1 second, the wrench is at a height of $-16(1^2) + 240 = 224$. After 2 seconds, the height of the wrench is $-16(2^2) + 240 = 176$. To find the number of seconds it takes the wrench to hit the ground, let the height s be zero and solve the resulting equation for t.

$s = -16t^2 + 240$	*Position equation*
$0 = -16t^2 + 240$	*Set height equal to 0*
$16t^2 = 240$	*Add $16t^2$ to both sides*
$t^2 = 15$	*Divide both sides by 16*
$t = \sqrt{15} \approx 3.87$	*Extract positive square root*

FIGURE 2.33

Thus, the wrench will take about 3.87 seconds to hit the ground. If the person hears the warning 1 second after the wrench is dropped, the person still has almost 3 seconds to get out of the way.

EXAMPLE 14   The Cost of a New Car

Between 1970 and 1990, the average cost of a new car increased according to the model

$$C = 30.5t^2 + 4192, \qquad 0 \le t \le 20,$$

as shown in Figure 2.34. In this model the cost is measured in dollars and the time t represents the year with $t = 0$ corresponding to 1970. If the average cost of a new car continued to increase according to this model, when would the average cost reach $20,000? (*Source:* Commerce Department, American Auto-datum)

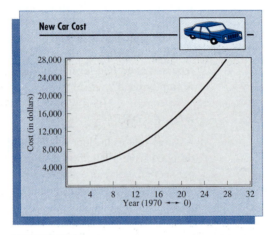

FIGURE 2.34

SOLUTION

To solve this problem, let the cost be $20,000 and solve the equation $20,000 = 30.5t^2 + 4192$ for t.

$$30.5t^2 + 4192 = 20,000 \qquad \textit{Set cost equal to 20,000}$$
$$30.5t^2 = 15,808 \qquad \textit{Subtract 4192 from both sides}$$
$$t^2 \approx 518.295 \qquad \textit{Divide both sides by 30.5}$$
$$t \approx \sqrt{518.295} \qquad \textit{Extract positive square root}$$

Thus, the solution is $t \approx 23$. Because $t = 0$ represents 1970, you can conclude that the average cost of a new car reached $20,000 in 1993. You could use a graphing utility to solve this problem. By finding the positive zero of $30.5t^2 + 4192 - 20,000 = 0$, you can see that $t \approx 23$.

EXAMPLE 15   Market Research

The marketing department at a publishing firm is asked to determine the price of a book. The department determines that the demand for the book depends upon the price of the book according to the formula

$$p = 40 - \sqrt{0.0001x + 1},$$

where p is the price per book in dollars and x is the number of books sold at the given price. For instance, in Figure 2.35 note that if the price is $39, then (according to the model) no one would be willing to buy the book. On the other hand, if the price were $17.60, then 5 million copies could be sold. If the publisher sets the price at $12.95, how many copies could be sold?

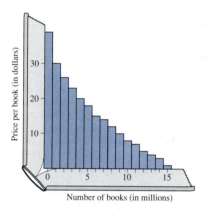

Price per book (in dollars)

Number of books (in millions)

FIGURE 2.35

SOLUTION

$$p = 40 - \sqrt{0.0001x + 1} \qquad \textit{Original model}$$
$$12.95 = 40 - \sqrt{0.0001x + 1} \qquad \textit{Set price at \$12.95}$$
$$\sqrt{0.0001x + 1} = 27.05 \qquad \textit{Isolate radical}$$
$$0.0001x + 1 = 731.7025 \qquad \textit{Square both sides}$$
$$0.0001x = 730.7025 \qquad \textit{Subtract 1 from both sides}$$
$$x = 7,307,025 \qquad \textit{Divide both sides by 0.0001}$$

With a price of $12.95, the publisher could expect to sell about 7.3 million copies.

DISCUSSION PROBLEM

MARKET ANALYSIS REVISITED

In Example 15, suppose the cost of producing the book were $450,000 plus $9.50 per book. Then the **cost equation** for the book would be

$$C = 9.5x + 450,000,$$

where C is measured in dollars and x represents the number of books produced. From Example 15, the total revenue would be

$$R = xp = x\left(40 - \sqrt{0.0001x + 1}\right).$$

How much profit (Profit = $R - C$) would the company make if the price of the book were set at $12.95?

WARM-UP

The following warm-up exercises involve skills that were covered in earlier sections. You will use these skills in the exercise set for this section.

In Exercises 1–4, simplify the expression.

1. $\sqrt{\frac{7}{50}}$

2. $\sqrt{32}$

3. $\sqrt{7^2 + 3 \cdot 7^2}$

4. $\sqrt{\frac{1}{4} + \frac{3}{8}}$

In Exercises 5–10, factor the algebraic expression.

5. $3x^2 + 7x$

6. $4x^2 - 25$

7. $16 - (x - 11)^2$

8. $x^2 + 7x - 18$

9. $10x^2 + 13x - 3$

10. $6x^2 - 73x + 12$

SECTION 2.4 · EXERCISES

In Exercises 1–12, solve the quadratic equation by factoring. (In Exercises 1–10, check your solution graphically.)

1. $6x^2 + 3x = 0$

2. $9x^2 - 1 = 0$

3. $x^2 - 2x - 8 = 0$

4. $x^2 - 10x + 9 = 0$

5. $x^2 + 10x + 25 = 0$

6. $16x^2 + 56x + 49 = 0$

7. $3 + 5x - 2x^2 = 0$

8. $2x^2 = 19x + 33$

9. $2x^2 + 8x - 6 = x^2 + 4x + 6$

10. $x(8 - x) = 12$

11. $x^2 + 2ax + a^2 = 0$

12. $(x + a)^2 - b^2 = 0$

In Exercises 13–20, solve the equation by extracting square roots. List both the exact answer *and* the decimal answer rounded to two decimal places.

13. $3x^2 = 36$

14. $9x^2 = 25$

15. $(x - 12)^2 = 18$

16. $(x + 13)^2 = 21$

17. $(x + 2)^2 = 12$

18. $(x - 5)^2 = 20$

19. $(x - 7)^2 = (x + 3)^2$

20. $(x + 5)^2 = (x + 4)^2$

In Exercises 21–26, solve the quadratic equation by completing the square.

21. $x^2 + 6x + 2 = 0$

22. $x^2 + 8x + 14 = 0$

23. $x(9x - 18) = -3$

24. $10x^2 - 5x + 1 = x^2 + 7x + 15$

25. $8 + 4x - x^2 = 0$

26. $4x^2 - 4x - 99 = 0$

In Exercises 27–48, use the Quadratic Formula to solve the equation.

27. $2x^2 + x - 1 = 0$

28. $2x^2 - x - 1 = 0$

29. $16x^2 + 8x - 3 = 0$

30. $25x^2 - 20x + 3 = 0$

31. $2 + 2x - x^2 = 0$

32. $x^2 - 10x + 22 = 0$

33. $x^2 + 14x + 44 = 0$

34. $6x = 4 - x^2$

35. $x^2 + 8x - 4 = 0$

36. $4x^2 - 4x - 4 = 0$

37. $12x - 9x^2 = -3$

38. $16x^2 + 22 = 40x$

39. $36x^2 + 24x - 7 = 0$

40. $3x + x^2 - 1 = 0$

41. $4x^2 + 4x = 7$

42. $16x^2 - 40x + 5 = 0$

43. $28x - 49x^2 = 4$

44. $9x^2 + 24x + 16 = 0$

45. $25h^2 + 80h + 61 = 0$

46. $8t = 5 + 2t^2$

47. $(y - 5)^2 = 2y$

48. $(z + 6)^2 = -2z$

In Exercises 49–80, find all real solutions of the equation. Check your answers in the original equation or use a graphic check.

49. $x^3 - 2x^2 - 3x = 0$

50. $2x^4 - 15x^3 + 18x^2 = 0$

51. $5x^3 + 30x^2 + 45x = 0$

52. $20y^3 - 125y = 0$

53. $x^3 - 3x^2 - x + 3 = 0$

54. $x^3 + 2x^2 + 3x + 6 = 0$

55. $x^4 - 10x^2 + 9 = 0$

56. $36t^4 + 29t^2 - 7 = 0$

57. $\dfrac{1}{t^2} + \dfrac{8}{t} + 15 = 0$

58. $6\left(\dfrac{s}{s+1}\right)^2 + 5\left(\dfrac{s}{s+1}\right) - 6 = 0$

59. $5 - 3x^{1/3} - 2x^{2/3} = 0$

60. $9t^{2/3} + 24t^{1/3} + 16 = 0$

61. $\sqrt{2x} - 10 = 0$

62. $\sqrt{5-x} - 3 = 0$

63. $\sqrt[3]{2x+5} + 3 = 0$

64. $\sqrt[3]{3x+1} - 5 = 0$

65. $\sqrt{x+1} - 3x = 1$

66. $\sqrt{x+5} = \sqrt{x-5}$

67. $\sqrt{x} + \sqrt{x-20} = 10$

68. $\sqrt{x} - \sqrt{x-5} = 1$

69. $\sqrt{x+5} + \sqrt{x-5} = 10$

70. $2\sqrt{x+1} - \sqrt{2x+3} = 1$

71. $\dfrac{20-x}{x} = x$

72. $\dfrac{4}{x} - \dfrac{5}{3} = \dfrac{x}{6}$

73. $\dfrac{1}{x} - \dfrac{1}{x+1} = 3$

74. $\dfrac{x}{x^2-4} + \dfrac{1}{x+2} = 3$

75. $\dfrac{4}{x+1} - \dfrac{3}{x+2} = 1$

76. $\dfrac{x+1}{3} - \dfrac{x+1}{x+2} = 0$

77. $|2x-1| = 5$

78. $|3x+2| = 7$

79. $|x| = x^2 + x - 3$

80. $|x^2 + 6x| = 3x + 18$

In Exercises 81–100, use a graphing utility to approximate the real solutions of the equation accurate to two decimal places.

81. $x^4 - x^3 + x - 1 = 0$

82. $x^4 + 2x^3 - 8x - 16 = 0$

83. $x^4 + 5x^2 - 36 = 0$

84. $x^4 - 4x^2 + 3 = 0$

85. $2x + 9\sqrt{x} - 5 = 0$

86. $6x - 7\sqrt{x} - 3 = 0$

87. $-\sqrt{26 - 11x} + 4 = x$

88. $x + \sqrt{31 - 9x} = 5$

89. $\sqrt{7x+36} - \sqrt{5x+16} = 2$

90. $3\sqrt{x} - \dfrac{4}{\sqrt{x}} = 4$

91. $x = \dfrac{3}{x} + \dfrac{1}{2}$

92. $4x + 1 = \dfrac{3}{x}$

93. $\dfrac{1}{x} = \dfrac{4}{x-1} + 1$

94. $x + \dfrac{9}{x+1} = 5$

95. $|x - 10| = x^2 - 10x$

96. $|x + 1| = x^2 - 5$

97. $3.2x^4 - 1.5x^2 - 2.1 = 0$

98. $7.08x^6 + 4.15x^3 - 9.6 = 0$

99. $1.8x - 6\sqrt{x} - 5.6 = 0$

100. $4x^{2/3} + 8x^{1/3} + 3.6 = 0$

In Exercises 101 and 102, use the position equation in Example 13 as the model for the problem.

101. *CN Tower*   The CN Tower in Toronto, Ontario is the world's tallest self-supporting structure. It is 1821 feet tall. An object is dropped from the top of the tower. How long will it take for it to hit the ground?

102. *Warfare*   A bomber flying at 32,000 feet over level terrain drops a 500-pound bomb.
   (a) How long will it take for the bomb to strike the ground?
   (b) If the plane is flying at 600 miles per hour, how far will the bomb travel horizontally during its descent?

103. *Dimensions of a Corral*   A rancher has 200 feet of fencing to enclose two adjacent rectangular corrals (see figure). Find the dimensions such that the enclosed area will be 1400 square feet.

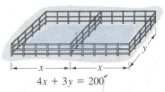

$4x + 3y = 200$

FIGURE FOR 103

104. *Dimensions of a Box*   An open box is to be made from a square piece of material by cutting 2-inch squares from each corner and turning up the sides (see figure). The volume of the finished box is to be 200 cubic inches. Find the size of the original piece of material.

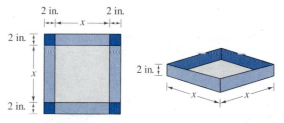

FIGURE FOR 104

105. *Oxygen Consumption*   The metabolic rate of ectothermic organisms increases with increasing temperature within a certain range. The figure shows experimental data for oxygen consumption (microliters per gram per hour) of a beetle for certain temperatures. These data can be approximated by the model

$$C = 0.45x^2 - 1.65x + 50.75, \qquad 10 \le x \le 25,$$

where x is the air temperature in degrees Celsius.

(a) Use a graphing utility to graph the consumption function over the specified domain.

(b) Move the cursor along the graph to approximate the air temperature if the oxygen consumption is 150 microliters per gram per hour.

(c) If the temperature is increased from 10 to 20 degrees, the oxygen consumption is increased by approximately what factor?

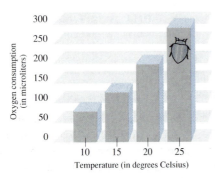

FIGURE FOR 105

106. *Sharing the Cost* Three students are planning to rent an apartment for a year and share equally in the monthly rent. By adding a fourth person to the group, each person could save $75 a month. How much is the monthly rent?

107. *Airspeed* An airline runs a commuter flight between two cities that are 720 miles apart. If the average speed of the planes is increased by 40 miles per hour, the travel time is decreased by 12 minutes. What airspeed is required to obtain this decrease in travel time?

108. *Average Speed* A family drove 1080 miles to their vacation lodge. Because of increased traffic density, their average speed on the return trip was decreased by 6 miles per hour and the trip took $2\frac{1}{2}$ hours longer. Determine their average speed on the way to the lodge.

109. *Saturated Steam* The temperature (in degrees Fahrenheit) of saturated steam increases as pressure increases. This relationship is approximated by the model

$$\text{Temperature} = 75.82 - 2.11x + 43.51\sqrt{x},$$
$$5 \le x \le 40,$$

where x is the absolute pressure in pounds per square inch.

(a) Use a graphing utility to graph the temperature function over the specified domain.

(b) The temperature of steam at sea level ($x = 14.696$) is $212°$. Evaluate the model at this pressure.

(c) Use the model to approximate the pressure if the temperature of the steam is $240°$.

110. *Airline Passengers* An airline offers daily flights between Chicago and Denver. The total monthly cost of these flights is given by

$$C = \sqrt{0.2x + 1},$$

where C is measured in millions of dollars and x is measured in thousands of passengers.

(a) Use a graphing utility to graph the cost function.

(b) Approximate the number of passengers that flew during a month when the total cost of the flights was 2.5 million dollars.

111. *Market Research* The demand equation for a product is

$$p = 40 - \sqrt{0.01x + 1},$$

where x is the number of units demanded per day and p is the price per unit. Find the demand if the price is set at $37.55.

112. *Market Research* The demand equation for a product is

$$p = 30 - \sqrt{0.0001x + 1},$$

where x is the number of units demanded per day and p is the price per unit. Find the demand if the price is set at $13.95.

2.5   SOLVING INEQUALITIES ALGEBRAICALLY AND GRAPHICALLY

Properties of Inequalities / Solving a Linear Inequality / Inequalities Involving Absolute Values / Polynomial Inequalities / Rational Inequalities / Applications

Properties of Inequalities

Simple inequalities were reviewed in Section 1 of the prerequisites chapter. There, inequality symbols $<$, $\leq$, $>$, and $\geq$ were used to compare two numbers and to denote subsets of real numbers. For instance, the simple inequality $x \geq 3$ denotes all real numbers x that are greater than or equal to 3. In this section you will study inequalities that contain more involved statements such as

$$5x - 7 < 3x + 9$$

and

$$-3 \leq 6x - 1 < 3.$$

As with an equation, you **solve an inequality** in the variable x by finding all values of x for which the inequality is true. These values are **solutions** and are said to **satisfy** the inequality. The set of all real numbers that are solutions of an inequality is the **solution set** of the inequality.

The set of all points on the real number line that represent the solution set is the **graph** of the inequality. Graphs of many types of inequalities consist of intervals on the real number line.

The procedures for solving linear inequalities in one variable are much like those for solving linear equations. To isolate the variable you can use **properties of inequalities.** These properties are similar to the properties of equality, but there are two important exceptions. When both sides of an inequality are multiplied or divided by a negative number, *the direction of the inequality symbol must be reversed.* Here is an example.

$-2 < 5$	*Original inequality*
$(-3)(-2) > (-3)(5)$	*Multiply both sides by -3*
$6 > -15$	*and reverse the inequality*

Two inequalities that have the same solution set are **equivalent.** The following list describes operations that can be used to create equivalent inequalities.

D I S C O V E R Y

Use a graphing utility to graph $f(x) = 5x - 7$ and $g(x) = 3x + 9$ on the same viewing rectangle. (Use $-1 \le x \le 15$ and $-5 \le y \le 50$.) For which values of x does the graph of f lie above the graph of g? Explain how the answer to this question can be used to solve the inequality in Example 1.

PROPERTIES OF INEQUALITIES

Let a, b, c, and d be real numbers.

Property	Example
1. Transitive Property $$a < b \text{ and } b < c \Rightarrow a < c$$	Because $-2 < 5$ and $5 < 7$, it follows that $-2 < 7$.
2. Addition of Inequalities $$a < b \text{ and } c < d \Rightarrow a + c < b + d$$	Because $2 < 4$ and $3 < 5$, it follows that $2 + 3 < 4 + 5$.
3. Addition of a Constant $$a < b \Rightarrow a + c < b + c$$	Because $-3 < 7$, it follows that $-3 + 2 < 7 + 2$.
4. Multiplying by a Constant (a) For $c > 0$, $$a < b \Rightarrow ac < bc.$$	Because $5 > 0$ and $3 < 9$, it follows that $3(5) < 9(5)$.
(b) For $c < 0$, $$a < b \Rightarrow ac > bc.$$	Because $-5 < 0$ and $3 < 9$, it follows that $3(-5) > 9(-5)$.

REMARK   Each of the above properties is true if the symbol $<$ is replaced by $\le$.

Solving a Linear Inequality

The simplest type of inequality to solve is a **linear inequality** in a single variable, such as $2x + 3 > 4$.

EXAMPLE 1   Solving a Linear Inequality

$5x - 7 > 3x + 9$	*Original inequality*
$5x > 3x + 16$	*Add 7 to both sides*
$5x - 3x > 16$	*Subtract 3x from both sides*
$2x > 16$	*Combine terms*
$x > 8$	*Divide both sides by 2*

Thus, the solution set consists of all real numbers that are greater than 8. The interval notation for this solution set is $(8, \infty)$. The number line graph of this solution set is shown in Figure 2.36.

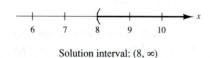

Solution interval: $(8, \infty)$

FIGURE 2.36

REMARK   The five inequalities forming the solution steps of Example 1 are all **equivalent** in the sense that each has the same solution set.

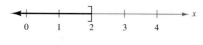

Solution interval: $(-\infty, 2]$

FIGURE 2.37

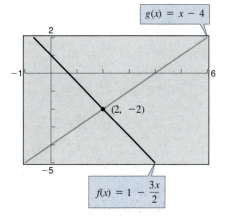

FIGURE 2.38

EXAMPLE 2   Solving an Inequality Algebraically and Graphically

$$1 - \frac{3x}{2} \geq x - 4 \qquad \text{\textit{Original inequality}}$$

$$2 - 3x \geq 2x - 8 \qquad \text{\textit{Multiply both sides by LCD}}$$

$$-3x \geq 2x - 10 \qquad \text{\textit{Subtract 2 from both sides}}$$

$$-5x \geq -10 \qquad \text{\textit{Subtract 2x from both sides}}$$

$$x \leq 2 \qquad \text{\textit{Divide both sides by} } -5 \text{ \textit{and reverse inequality}}$$

The solution set consists of all real numbers that are less than or equal to 2. The interval notation for this solution set is $(-\infty, 2]$. The number line graph of this solution set is shown in Figure 2.37. Graphically, you can solve the given inequality by sketching the graphs of the left and right sides

$$f(x) = 1 - \frac{3x}{2} \quad \text{and} \quad g(x) = x - 4.$$

From Figure 2.38 the graphs appear to intersect at the point $(2, -2)$. This is confirmed by the fact that $f(2) = -2 = g(2)$. Moreover, the graph of f lies above the graph of g to the left of their point of intersection, which implies that $f(x) \geq g(x)$ for all $x \leq 2$.

Sometimes it is convenient to write two inequalities as a **double inequality**. For instance, you can write the two inequalities $-4 \leq 5x - 2$ and $5x - 2 < 7$ more simply as $-4 \leq 5x - 2 < 7$. This form allows you to solve the two given inequalities together.

EXAMPLE 3   Solving a Double Inequality

$$-3 \leq 6x - 1 < 3 \qquad \text{\textit{Original inequality}}$$

$$-2 \leq 6x < 4 \qquad \text{\textit{Add 1 to all three parts}}$$

$$-\frac{1}{3} \leq x < \frac{2}{3} \qquad \text{\textit{Divide by 6 and reduce}}$$

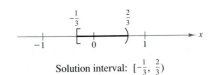

Solution interval: $[-\frac{1}{3}, \frac{2}{3})$

FIGURE 2.39

The solution set consists of all real numbers that are greater than or equal to $-\frac{1}{3}$ and less than $\frac{2}{3}$. The interval notation for this solution set is $[-\frac{1}{3}, \frac{2}{3})$. The number line graph of this solution set is shown in Figure 2.39.

To see how to solve inequalities involving absolute values, consider the following comparisons.

Equation or Inequality	Solution	Verbal Description	Graph
$\|x\| = 2$	$x = -2$ and $x = 2$	Values of x that lie 2 units from 0	
$\|x\| < 2$	$-2 < x < 2$	Values of x that lie *less than* 2 units from 0	
$\|x\| > 2$	$x < -2$ or $x > 2$	Values of x that lie *more than* 2 units from 0	

SOLVING AN ABSOLUTE VALUE INEQUALITY

Let x be a variable or an algebraic expression and let a be a real number such that $a \geq 0$.

1. The solutions of $|x| < a$ are all values of x that lie between $-a$ and a. That is,

$$|x| < a \quad \text{if and only if} \quad -a < x < a.$$

2. The solutions of $|x| > a$ are all values of x that are less than $-a$ or greater than a. That is,

$$|x| > a \quad \text{if and only if} \quad x < -a \quad \text{or} \quad x > a.$$

These two rules are also valid if $<$ is replaced by $\leq$ and $>$ is replaced by $\geq$.

EXAMPLE 4   Solving an Absolute Value Inequality

$$|x - 5| < 2 \qquad \textit{Original inequality}$$
$$-2 < x - 5 < 2 \qquad \textit{Equivalent inequalities}$$
$$-2 + 5 < x - 5 + 5 < 2 + 5 \qquad \textit{Add 5 to all three parts}$$
$$3 < x < 7 \qquad \textit{Solution set}$$

The solution set consists of all real numbers that are greater than 3 and less than 7. The interval notation for this solution set is $(3, 7)$. The graph of this solution set is shown in Figure 2.40.

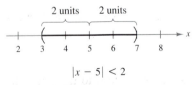

$|x - 5| < 2$

FIGURE 2.40

EXAMPLE 5   Solving An Absolute Value Inequality Graphically

Use a graphing utility to solve the following inequality.

$$\left| 2 - \frac{x}{3} \right| < 0.01$$

SOLUTION

You can solve the given absolute value inequality graphically by graphing the function

$$f(x) = \left| 2 - \frac{x}{3} \right| - 0.01$$

and determining the values of x for which the graph lies *below* the x-axis, as shown in Figure 2.41. Using the zoom and trace features, you can determine that the x-intercepts occur at $x \approx 5.97$ and $x \approx 6.03$. Because the graph of f lies below the x-axis between these two values, the solution interval is $(5.97, 6.03)$. Try solving this inequality algebraically to verify the solution interval.

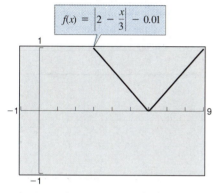

$$f(x) = \left| 2 - \frac{x}{3} \right| - 0.01$$

FIGURE 2.41

Polynomial Inequalities

To solve a polynomial inequality such as

$$x^2 - 2x - 3 < 0,$$

use the fact that a polynomial can change signs only at its zeros (the x-values that make the polynomial zero). Between two consecutive zeros a polynomial must be entirely positive or entirely negative. This means that when the real zeros of a polynomial are put in order, they divide the real line into intervals in which the polynomial has no sign changes. These zeros are the **critical numbers** of the inequality and the resulting intervals are the **test intervals** for the inequality. For example, the polynomial

$$x^2 - 2x - 3 = (x + 1)(x - 3)$$

has two zeros, $x = -1$ and $x = 3$, and these zeros divide the real line into three test intervals,

$$(-\infty, -1), \quad (-1, 3), \quad \text{and} \quad (3, \infty).$$

To solve the inequality $x^2 - 3x - 3 < 0$, you only need to test one value from each of these test intervals.

FINDING TEST INTERVALS FOR A POLYNOMIAL

To determine the intervals on which the values of a polynomial are entirely negative or entirely positive, use the following steps.

1. Find all real zeros of the polynomial, and arrange the zeros in increasing order (from smallest to largest). The zeros of a polynomial are its **critical numbers.**
2. Use the critical numbers of the polynomial to determine its **test intervals.**
3. Choose one representative x-value in each test interval and evaluate the polynomial at that value. If the value of the polynomial is negative, then the polynomial will have negative values for *every* x-value in the interval. If the value of the polynomial is positive, then the polynomial will have positive values for *every* x-value in the interval.

EXAMPLE 6   Investigating Polynomial Behavior

Determine the intervals on which $x^2 - x - 6$ is entirely negative and those on which it is entirely positive.

SOLUTION

By factoring the quadratic as $x^2 - x - 6 = (x + 2)(x - 3)$, you can see that the critical numbers occur at $x = -2$ and $x = 3$. Therefore, the test intervals for the quadratic are $(-\infty, -2)$, $(-2, 3)$, and $(3, \infty)$. In each test interval, choose a representative x-value and evaluate the polynomial, as shown in Table 2.1.

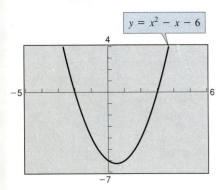

FIGURE 2.42

TABLE 2.1

Test interval	Representative x-value	Value of polynomial	Conclusion
$(-\infty, -2)$	$x = -3$	$(-3)^2 - (-3) - 6 = 6$	Polynomial is positive
$(-2, 3)$	$x = 0$	$(0)^2 - (0) - 6 = -6$	Polynomial is negative
$(3, \infty)$	$x = 4$	$(4)^2 - (4) - 6 = 6$	Polynomial is positive

The polynomial has positive values for every x in the intervals $(-\infty, -2)$ and $(3, \infty)$, and negative values for every x in the interval $(-2, 3)$. This result is shown graphically in Figure 2.42.

To determine the test intervals for a polynomial inequality, the inequality must first be written in standard form (with the polynomial on the left and zero on the right).

EXAMPLE 7   Solving a Polynomial Inequality

$$2x^2 + 5x > 12 \qquad \textit{Original inequality}$$
$$2x^2 + 5x - 12 > 0 \qquad \textit{Write in standard form}$$
$$(x + 4)(2x - 3) > 0 \qquad \textit{Factor}$$

Critical numbers: $x = -4, x = \frac{3}{2}$

Test intervals: $(-\infty, -4), (-4, \frac{3}{2}), (\frac{3}{2}, \infty)$

Test: Is $(x + 4)(2x - 3) > 0$?

After testing these intervals, you can see that the polynomial $2x^2 + 5x - 12$ is positive in the open intervals $(-\infty, -4)$ and $(\frac{3}{2}, \infty)$. Therefore, the solution set of the inequality is $(-\infty, -4) \cup (\frac{3}{2}, \infty)$. You could have solved this example with a graphing utility by graphing the polynomial function $y = 2x^2 + 5x - 12$, and noting where the graph is *above* the x-axis. You can see in Figure 2.43 that the polynomial lies above the x-axis for $x < -4$ or $x > \frac{3}{2}$.

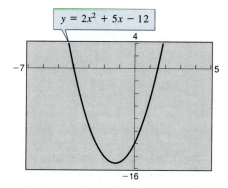

$y = 2x^2 + 5x - 12$

FIGURE 2.43

EXAMPLE 8   Unusual Solution Sets

A. The solution set of $x^2 + 2x + 4 > 0$ consists of the entire set of real numbers, $(-\infty, \infty)$. In other words, the quadratic $x^2 + 2x + 4$ is positive for every real value of x as indicated in Figure 2.44(a). (Note that this quadratic inequality has *no* critical numbers. In such a case, there is only one test interval—the entire real line.)

B. The solution set of $x^2 + 2x + 1 \le 0$ consists of the single real number, -1, because the graph touches the x-axis just at -1, as shown in Figure 2.44(b).

C. The solution set of $x^2 + 3x + 5 < 0$ is empty. In other words, the quadratic $x^2 + 3x + 5$ is not less than 0 for any value of x, as indicated in Figure 2.44(c).

D. The solution set of $x^2 - 4x + 4 > 0$ consists of all real numbers *except* the number 2. In interval notation, this solution set can be written as $(-\infty, 2) \cup (2, \infty)$. The graph of $x^2 - 4x + 4$ is above the x-axis except at $x = 2$, where it touches, as indicated in Figure 2.44(d).

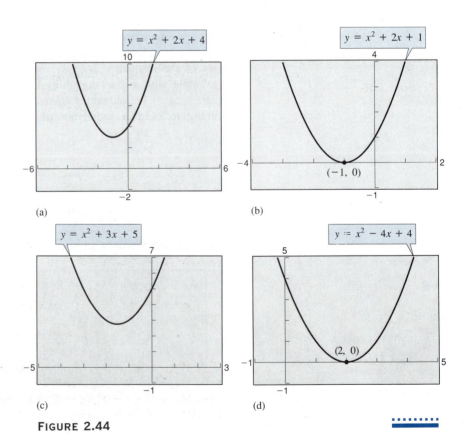

(a)

(b)

(c)

(d)

FIGURE 2.44

Rational Inequalities

The concepts of critical numbers and test intervals can be extended to inequalities involving rational expressions. To do this, use the fact that the value of a rational expression can change sign only at its *zeros* (the x-values for which its numerator is zero) and its *undefined values* (the x-values for which its denominator is zero). These two types of numbers make up the **critical numbers** of a rational inequality.

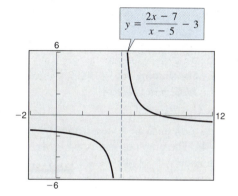

$$y = \frac{2x - 7}{x - 5} - 3$$

FIGURE 2.45

EXAMPLE 9   Solving a Rational Inequality

Solve $\dfrac{2x - 7}{x - 5} \le 3$.

SOLUTION

From Figure 2.45, you can see that the graph of

$$y = \frac{2x - 7}{x - 5} - 3$$

is negative on the intervals $(-\infty, 5)$ and $(8, \infty)$. To solve the inequality algebraically, use the following steps.

$$\frac{2x - 7}{x - 5} \le 3 \qquad \text{\textit{Original inequality}}$$

$$\frac{2x - 7}{x - 5} - 3 \le 0 \qquad \text{\textit{Standard form}}$$

$$\frac{2x - 7 - 3x + 15}{x - 5} \le 0 \qquad \text{\textit{Add fractions}}$$

$$\frac{-x + 8}{x - 5} \le 0 \qquad \text{\textit{Simplify}}$$

Now, in standard form you can see that the critical numbers are 5 and 8, and proceed as follows.

Critical numbers: $x = 5,\ x = 8$

Test intervals: $(-\infty, 5),\ (5, 8),\ (8, \infty)$

Test: Is $\dfrac{-x + 8}{x - 5} \le 0$?

By testing these intervals, you can determine that the rational expression $(-x + 8)/(x - 5)$ is negative in the open intervals $(-\infty, 5)$ and $(8, \infty)$. Moreover, because $(-x + 8)/(x - 5) = 0$ when $x = 8$, you can conclude that the solution set of the inequality is $(-\infty, 5) \cup [8, \infty)$.

Technology Note _____

To graph the function

$$y = \frac{2x - 7}{x - 5} - 3,$$

you may find it helpful to change your graphing utility from the standard "connected" mode to "dot" mode.

REMARK   It is incorrect to write the solution set $(-\infty, 5) \cup [8, \infty)$ as a single inequality, $8 \leq x < 5$. Do you see why?

Applications

EXAMPLE 10   Finding the Domain of an Expression

Find the domain of the expression $\sqrt{64 - 4x^2}$.

SOLUTION

Remember that the domain of an expression is the set of all x-values for which the expression is defined (has real values). Because $\sqrt{64 - 4x^2}$ is defined only if $64 - 4x^2$ is nonnegative, the domain is given by $64 - 4x^2 \geq 0$.

$$64 - 4x^2 \geq 0 \qquad \textit{Standard form}$$
$$16 - x^2 \geq 0 \qquad \textit{Divide both sides by 4}$$
$$(4 - x)(4 + x) \geq 0 \qquad \textit{Factor}$$

The inequality has two critical numbers: -4 and 4. A test shows that $64 - 4x^2 \geq 0$ in the *closed interval* $[-4, 4]$. The graph of $y = \sqrt{64 - 4x^2}$, shown in Figure 2.46, confirms that the domain is $[-4, 4]$.

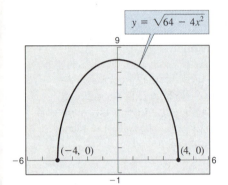

$y = \sqrt{64 - 4x^2}$

$(-4, 0)$ $(4, 0)$

FIGURE 2.46

EXAMPLE 11   The Height of a Projectile

A projectile is fired straight upward from ground level with an initial velocity of 384 feet per second. During what time period will its height exceed 2000 feet?

SOLUTION

The position of an object moving in a vertical path is given by $s = -16t^2 + v_0 t + s_0$, where s is the height in feet and t is the time in seconds. In this case, $s_0 = 0$ and $v_0 = 384$. Thus, you need to solve the inequality $-16t^2 + 384t > 2000$. Using a graphing utility, graph $s = -16t^2 + 384t$ and $s = 2000$, as shown in Figure 2.47. From the graph, you can determine that $-16t^2 + 384t > 2000$ for t between 7.64 and 16.36. You can verify this result algebraically as follows.

$$-16t^2 + 384t > 2000 \qquad \textit{Original inequality}$$
$$t^2 - 24t < -125 \qquad \textit{Divide by } -16 \textit{ and reverse inequality}$$
$$t^2 - 24t + 125 < 0 \qquad \textit{Standard form}$$

By the Quadratic Formula, the critical numbers are 7.64 and 16.36. A test will verify that the height of the projectile will exceed 2000 feet during the time interval 7.64 sec $< t < 16.36$ sec.

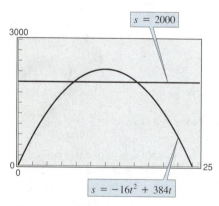

$s = 2000$

$s = -16t^2 + 384t$

FIGURE 2.47

DISCUSSION PROBLEM

A COMPUTER EXPERIMENT

If you have access to a computer that has the BASIC language, try running the following program.

```
10  DEF FNY(X)=X∧2−9
20  FOR I=1 TO 80
30  X=(I−38)/6
40  IF FNY(X)<0 THEN PRINT "−";
50  IF FNY(X)=0 THEN PRINT "0";
60  IF FNY(X)<0 THEN PRINT "+";
70  NEXT
80  PRINT: FOR I=−6 TO 6: PRINT "−|−−−−";: NEXT:PRINT "−|";
90  FOR I=−6 TO 6: PRINT USING "##";I;:PRINT " ";: NEXT:PRINT "7"
100 END
```

The printout from this program is shown below. Notice that the program evaluates the expression $x^2 - 9$ at several values between -6 and 7. If the value of $x^2 - 9$ is negative, a minus sign is printed above the number line; if the value of $x^2 - 9$ is zero, a zero is printed, and if the value of $x^2 - 9$ is positive, a plus sign is printed.

WARM-UP

The following warm-up exercises involves skills that were covered in earlier sections. You will use these skills in the exercise set for this section.

In Exercises 1–4, determine which of the two numbers is larger.

1. $-\frac{1}{2}, -7$ **2.** $-\frac{1}{3}, -\frac{1}{6}$

3. $-\pi, -3$ **4.** $-6, -\frac{13}{2}$

In Exercises 5–8, use inequality notation to describe the statement.

5. x is nonnegative.   **6.** z is strictly between -3 and 10.

7. P is no more than 2.   **8.** W is at least 200.

In Exercises 9 and 10, evaluate the expression for the given values of x.

9. $|x - 10|$, $x = 12, x = 3$ **10.** $|2x - 3|$, $x = \frac{3}{2}, x = 1$

SECTION 2.5 · EXERCISES

In Exercises 1–8, match the inequality with its graph. [The graphs are labeled (a), (b), (c), (d), (e), (f), (g), and (h).]

1. $x < 4$

2. $x \geq 6$

3. $-2 < x \leq 5$

4. $0 \leq x \leq \frac{7}{2}$

5. $|x| < 4$

6. $|x| > 3$

7. $|x - 5| > 2$

8. $|x + 6| < 3$

(a)
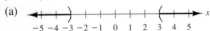

(b)

(c)

(d)
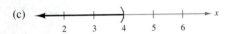

(e)

(f)

(g)
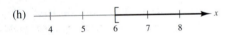

(h)

In Exercises 9–34, solve the inequality algebraically and sketch the solution on the real number line.

9. $4x < 12$

10. $2x > 3$

11. $-10x < 40$

12. $-6x > 15$

13. $x - 5 \geq 7$

14. $x + 7 \leq 12$

15. $4(x + 1) < 2x + 3$

16. $3x + 1 \geq 2$

17. $1 < 2x + 3 < 9$

18. $-8 \leq 1 - 3(x - 2) < 13$

19. $-4 < \dfrac{2x - 3}{3} < 4$

20. $0 \leq \dfrac{x + 3}{2} < 5$

21. $\dfrac{3}{4} > x + 1 > \dfrac{1}{4}$

22. $-1 < -\dfrac{x}{3} < 1$

23. $|x| < 5$

24. $|2x| < 6$

25. $\left|\dfrac{x}{2}\right| > 3$

26. $|5x| > 10$

27. $|x - 20| \leq 4$

28. $|x - 7| < 6$

29. $|x - 20| \geq 4$

30. $|x + 14| + 3 > 17$

31. $|9 - 2x| - 2 < -1$

32. $\left|1 - \dfrac{2x}{3}\right| < 1$

33. $|x - 5| < 0$

34. $|x - 5| \geq 0$

In Exercises 35–44, use a graphing utility to solve the inequality.

35. $4 - 2x < 3$

36. $2x + 7 < 3$

37. $\dfrac{-4x + 25}{5} \geq \dfrac{3x - 10}{10}$

38. $2x - 9 < \dfrac{-4x + 3}{3}$

39. $-\dfrac{5}{2} < 0.75x + 2 < 4$

40. $-1 \leq 13 - 2x \leq 9$

41. $\left|\dfrac{x - 3}{2}\right| \geq 5$

42. $|1 - 2x| < 5$

43. $2|x + 10| \geq 9$

44. $3|4 - 5x| \leq 9$

In Exercises 45–52, use absolute value notation to define each interval (or pair of intervals) on the real line.

45.

46.

47.

48.

49. All real numbers within 10 units of 12

50. All real numbers at least 5 units from 8

51. All real numbers whose distances from -3 are more than 5

52. All real numbers whose distances from -6 are no more than 7

In Exercises 53–76, solve the inequality algebraically or graphically.

53. $x^2 \leq 9$

54. $(x - 3)^2 \geq 1$

55. $x^2 + 4x + 4 \geq 9$

56. $x^2 - 6x + 9 < 16$

57. $x^2 + x < 6$

58. $x^2 + 2x > 3$

59. $3(x - 1)(x + 1) > 0$

60. $6(x + 2)(x - 1) < 0$

61. $x^2 + 2x - 3 < 0$

62. $x^2 - 4x - 1 > 0$

63. $4x^3 - 6x^2 < 0$

64. $4x^3 - 12x^2 > 0$

65. $x^3 - 4x \geq 0$

66. $2x^3 - x^4 \leq 0$

67. $\dfrac{1}{x} > x$

68. $\dfrac{1}{x} < 4$

69. $\dfrac{x + 6}{x + 1} < 2$

70. $\dfrac{x + 12}{x + 2} \geq 3$

71. $\dfrac{3x - 5}{x - 5} > 4$

72. $\dfrac{5 + 7x}{1 + 2x} < 4$

73. $\dfrac{4}{x + 5} > \dfrac{1}{2x + 3}$

74. $\dfrac{1}{x - 3} \leq \dfrac{9}{4x + 3}$

75. $|x^2 - 4x + 5| < 2x^{-1}$

76. $|x^3 + 6x - 1| \geq 2x + 4$

In Exercises 77–84, use a graphing utility to solve the inequality.

77. $(x - 1)^2(x + 2)^3 \geq 0$

78. $x^4(x - 3) \leq 0$

79. $-0.5x^2 + 12.5x + 1.6 > 0$

80. $1.2x^2 + 4.8x + 3.1 < 5.3$

81. $\dfrac{5}{x - 6} > \dfrac{3}{x + 2}$

82. $\dfrac{1}{x} \geq \dfrac{1}{x + 3}$

83. $\dfrac{3}{x - 1} - \dfrac{2}{x + 1} < 1$

84. $\dfrac{x^2 + x - 6}{x} \geq 0$

In Exercises 85–96, find the interval(s) on the real number line for which the radicand is nonnegative (greater than or equal to zero).

85. $\sqrt{3 - x}$

86. $\sqrt{x - 10}$

87. $\sqrt[4]{7 - 2x}$

88. $\sqrt[4]{6x + 15}$

89. $\sqrt[4]{4 - x^2}$

90. $\sqrt{x^2 - 4}$

91. $\sqrt{x^2 - 7x + 12}$

92. $\sqrt{144 - 9x^2}$

93. $\sqrt{12 - x - x^2}$

94. $\sqrt{x^2 + 4}$

95. $\sqrt{\dfrac{x}{x^2 - 9}}$

96. $\sqrt{\dfrac{x}{x + 4} - \dfrac{3x}{x - 1}}$

Break-Even Analysis   In Exercises 97 and 98, the functions giving the cost of producing x units of a product and the revenue for selling x units are given. In order to obtain a profit, the revenue must be *greater than* the cost. Use a graphing utility to graph the cost and revenue functions and determine the values of x for which the product will return a profit.

Cost Function	Revenue Function
97. $C = 95x + 750$	$R = 115.95x$
98. $C = 15.40x + 150,000$	$R = 24.55x$

99. *Daily Sales*   A doughnut shop at a shopping mall sells a dozen doughnuts for $2.95. Beyond the fixed costs (for rent, utilities, and insurance) of $150 per day, it costs $1.45 for enough materials (flour, sugar, and so on) and labor to produce a dozen doughnuts.

(a) Write the daily cost C as a function of the number of dozens x of doughnuts produced.

(b) Write the daily revenue R as a function of the number of dozens x of doughnuts sold.

(c) Write the daily profit as a function of x.

(d) Use a graphing utility to graph the profit function. If the daily profit varies between $50 and $200, between what levels (in dozens) do the daily sales vary?

(e) Rework parts (c) and (d) if on average there are 10 dozen unsold doughnuts at the end of each day.

100. *Annual Operating Cost*   A utility company has a fleet of vans. The annual operating cost per van is

$$C = 0.32m + 2300,$$

where m is the number of miles traveled by a van in a year. What number of miles will yield an annual operating cost that is less than $10,000?

101. *Cable Television Subscribers* The number of cable television subscribers in the United States between 1980 and 1989 is approximated by the model

Subscribers = $3.657t + 14.784$,

where the number of subscribers is given in millions and the time t represents the year, with $t = 0$ corresponding to 1980 (see figure). Assuming this model is correct, when will the number of subscribers *exceed* 55 million? (*Source:* Television and Cable Fact Book)

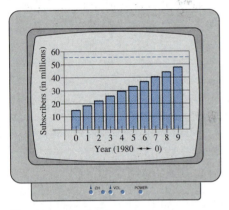

FIGURE FOR 101

102. *Teachers' Salaries* The average salary for elementary and secondary teachers in the United States from 1980 to 1989 is approximated by the model

Salary = $16,116 + 1496t$,

where the salary is given in dollars and the time t represents the year, with $t = 0$ corresponding to 1980 (see figure). Use a graphing utility to graph the salary function. Assuming the model is correct, move the cursor along the graph to determine when the average salary will *exceed* $35,000. (*Source:* National Education Association)

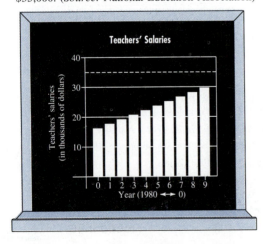

FIGURE FOR 102

103. *Heights* The heights h of two-thirds of the members of a certain population satisfy the inequality

$$\left| \frac{h - 68.5}{2.7} \right| \le 1,$$

where h is measured in inches. Determine the interval on the real line in which these heights lie.

104. *Relative Humidity* A certain electronic device is to be operated in an environment with relative humidity h in the interval defined by

$$|h - 50| \le 30.$$

What are the minimum and maximum relative humidities for the operation of this device?

105. *Height of a Projectile* A projectile is fired straight upward from ground level with an initial velocity of 160 feet per second.
  (a) At what instant will it be back at ground level?
  (b) During what time period will its height exceed 384 feet?

106. *Dimensions of a Field* A rectangular playing field with a perimeter of 100 meters is to have an area of at least 500 square meters. Within what bounds must the length of the rectangle lie?

107. *Resistors* When two resistors of resistance R_1 and R_2 are connected in parallel, the total resistance R satisfies the equation

$$\frac{1}{R} = \frac{1}{R_1} + \frac{1}{R_2}.$$

  (a) Solve the equation for R.
  (b) If $R_2 = 2$ ohms in a given parallel circuit, use the result of part (a) to write R as a function of R_1. Use a graphing utility to graph the function.
  (c) If R must be at least 1 ohm, find the value of R_1 that is a solution to the resulting inequality.

108. *Company Profits* The revenue and cost equations for a product are given by

$$R = x(50 - 0.0002x) \quad \text{and} \quad C = 12x + 150,000,$$

where R and C are measured in dollars and x represents the number of units sold.
  (a) Write the profit P as a function of x.
  (b) Use a graphing utility to graph the profit function.
  (c) How many units must be sold to obtain a profit of at least $1,650,000?

2.6   EXPLORING DATA: LINEAR MODELS AND SCATTER PLOTS

Scatter Plots   /   Fitting a Line to Data

Technology Note

Most graphing utilities have built-in statistical programs that can create scatter plots. Use your graphing utility to plot the points given in Table 2.2.

Scatter Plots

Many real-life situations involve finding relationships between two variables, such as the year and the number of people in the labor force. In a typical situation, data are collected and written as a set of ordered pairs. The graph of such a set is called a **scatter plot.**

EXAMPLE 1   Constructing a Scatter Plot

The data in Table 2.2 show the number of people P (in millions) in the United States who were part of the labor force from 1980 through 1990. In the table, t represents the year, with $t = 0$ corresponding to 1980. Sketch a scatter plot of the data. (*Source:* U.S. Bureau of Labor Statistics)

TABLE 2.2

t	0	1	2	3	4	5	6	7	8	9	10
P	109	110	112	113	115	117	120	122	123	126	126

SOLUTION

Begin by representing the data with a set of ordered pairs.

(0, 109), (1, 110), (2, 112), (3, 113), (4, 115), (5, 117),
(6, 120), (7, 122), (8, 123), (9, 126), (10, 126)

Then plot each point in a coordinate plane, as shown in Figure 2.48.

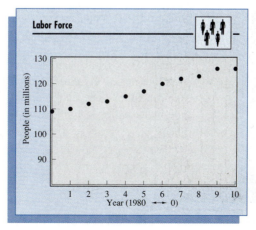

FIGURE 2.48

From the scatter plot in Figure 2.48, it appears that the points describe a relationship that is nearly linear. (The relationship is not *exactly* linear because the labor force did not increase by precisely the same amount each year.) A mathematical equation that approximates the relationship between t and P is called a *mathematical model.* When developing a mathematical model, you strive for two (often conflicting) goals—accuracy and simplicity. For the data above, a linear model of the form $P = at + b$ appears to be best. It is simple and relatively accurate.

Consider a collection of ordered pairs of the form (x, y). If y tends to increase as x increases, then the collection is said to have a **positive correlation.** If y tends to decrease as x increases, then the collection is said to have a **negative correlation.** Figure 2.49 shows three examples: one with a positive correlation, one with a negative correlation, and one with no (discernible) correlation.

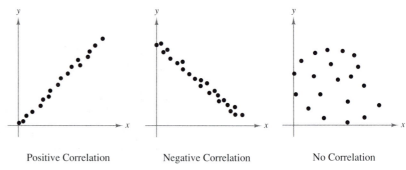

Positive Correlation   Negative Correlation   No Correlation

FIGURE 2.49

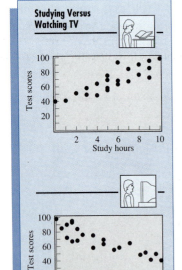

FIGURE 2.50

EXAMPLE 2   Interpreting Correlation

On a Friday, 22 students in a class were asked to keep track of the number of hours spent studying for a test on Monday and the number of hours spent watching television. The numbers are shown below. Construct a scatter plot for each set of data. Then, determine whether the points are positively correlated, are negatively correlated, or have no discernible correlation. What can you conclude? (The first coordinate is the number of hours and the second coordinate is the score obtained on Monday's test.)

Study hours: (0, 40), (1, 41), (2, 51), (3, 58), (3, 49), (4, 48), (4, 64),
(5, 55), (5, 69), (5, 58), (5, 75), (6, 68), (6, 63), (6, 93), (7, 84), (7, 67),
(8, 90), (8, 76), (9, 95), (9, 72), (9, 85), (10, 98)

TV hours: (0, 98), (1, 85), (2, 72), (2, 90), (3, 67), (3, 93), (3, 95),
(4, 68), (4, 84), (5, 76), (7, 75), (7, 58), (9, 63), (9, 69), (11, 55),
(12, 58), (14, 64), (16, 48), (17, 51), (18, 41), (19, 49), (20, 40)

SOLUTION

Scatter plots for the two sets of data are shown in Figure 2.50. The scatter plot relating study hours and test scores has a positive correlation. This means that the more a student studied, the higher his or her score tended to be. The scatter plot relating television hours and test scores has a negative correlation. This means that the more time a student spent watching television, the lower his or her score tended to be.

Fitting a Line to Data

Finding a linear model to represent the relationship described by a scatter plot is called **fitting a line to data.** You can do this graphically by simply sketching the line that appears to fit the points, finding two points on the line, and then finding the equation of the line that passes through the two points.

EXAMPLE 3   Fitting a Line to Data

Find a linear model that relates the year with the number of people in the United States labor force. (See Example 1.)

TABLE 2.3

t	0	1	2	3	4	5	6	7	8	9	10
P	109	110	112	113	115	117	120	122	123	126	126

SOLUTION

After plotting the data from Table 2.3, draw the line that you think best represents the data, as shown in Figure 2.51. Two points that lie on this line are (0, 109) and (9, 126). Using the point-slope form, you can find the equation of the line to be

$$P = \frac{17}{9}t + 109.$$ *Linear model*

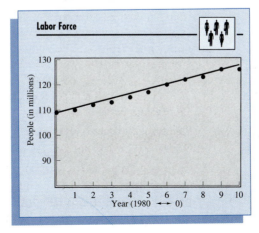

FIGURE 2.51

Once you have found a model, you can measure how well the model fits the data by comparing the actual values with the values given by the model, as shown in Table 2.4.

TABLE 2.4

t	0	1	2	3	4	5	6	7	8	9	10
Actual → P	109	110	112	113	115	117	120	122	123	126	126
Model → P	109	110.9	112.8	114.7	116.6	118.4	120.3	122.2	124.1	126	127.9

The sum of the squares of the differences between the actual values and the model's values is the **sum of the squared differences.** The model that has the least sum is called the **least squares regression line** for the data. For the model in Example 3, the sum of the squared differences is 13.81. The least squares regression line for the data is

$$P = 1.864t + 108.2. \qquad \textit{Best-fitting linear model}$$

Its sum of squared differences is 4.7.

LEAST SQUARES REGRESSION LINE

The least squares regression line, $y = ax + b$, for the points (x_1, y_1), (x_2, y_2), (x_3, y_3), . . . , (x_n, y_n) is given by

$$a = \frac{n \sum_{i=1}^{n} x_i y_i - \sum_{i=1}^{n} x_i \sum_{i=1}^{n} y_i}{n \sum_{i=1}^{n} x_i^2 - \left(\sum_{i=1}^{n} x_i\right)^2} \quad \text{and} \quad b = \frac{1}{n}\left(\sum_{i=1}^{n} y_i - a \sum_{i=1}^{n} x_i\right).$$

EXAMPLE 4 **Finding a Least Squares Regression Line**

Find the least squares regression line for the points $(-3, 0)$, $(-1, 1)$, $(0, 2)$, and $(2, 3)$.

SOLUTION

Begin by constructing a table like that shown in Table 2.5.

TABLE 2.5

x	y	xy	x^2
-3	0	0	9
-1	1	-1	1
0	2	0	0
2	3	6	4
$\displaystyle\sum_{i=1}^{n} x_i = -2$	$\displaystyle\sum_{i=1}^{n} y_i = 6$	$\displaystyle\sum_{i=1}^{n} x_i y_i = 5$	$\displaystyle\sum_{i=1}^{n} x_i^2 = 14$

Applying the formulas for the least squares regression line with $n = 4$ produces

$$a = \frac{n \displaystyle\sum_{i=1}^{n} x_i y_i - \displaystyle\sum_{i=1}^{n} x_i \displaystyle\sum_{i=1}^{n} y_i}{n \displaystyle\sum_{i=1}^{n} x_i^2 - \left(\displaystyle\sum_{i=1}^{n} x_i\right)^2} = \frac{4(5) - (-2)(6)}{4(14) - (-2)^2} = \frac{8}{13}$$

and

$$b = \frac{1}{n}\left(\sum_{i=1}^{n} y_i - a \sum_{i=1}^{n} x_i\right) = \frac{1}{4}\left[6 - \frac{8}{13}(-2)\right] = \frac{47}{26}.$$

Thus, the least squares regression line is $y = \frac{8}{13}x + \frac{47}{26}$, as shown in Figure 2.52.

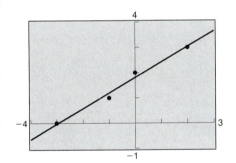

FIGURE 2.52

EXAMPLE 5   A Mathematical Model

The total amount of advertising expenses (in billions of dollars) in the United States from 1980 to 1989 is given in Table 2.6. (*Source:* McCann Erickson)

TABLE 2.6

Year	1980	1981	1982	1983	1984
Expenses	54.78	60.43	66.58	75.85	88.10

Year	1985	1986	1987	1988	1989
Expenses	94.75	102.14	109.79	118.32	125.55

The least squares regression line for the data is

$$y = 8.1377t + 53.029, \qquad 0 \le t \le 9,$$

where y represents the advertising expenses (in billions of dollars) and t represents the year, with $t = 0$ corresponding to 1980. Plot the actual data *and* the least squares regression line on the same graph. How closely does the line represent the data?

SOLUTION

The actual data are plotted in Figure 2.53, along with the least squares regression line. From the figure, it appears that the line is a "good fit" for the actual data. You can see how well the line fits by comparing the actual values of y with the values of y given by the equation of the line, as shown in Table 2.7.

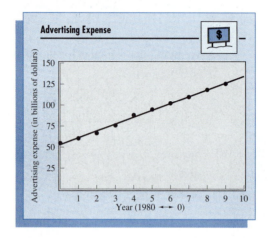

FIGURE 2.53

TABLE 2.7

t	0	1	2	3	4
Actual y	54.78	60.43	66.58	75.85	88.10
Model y	53.03	61.17	69.30	77.44	85.58

t	5	6	7	8	9
Actual y	94.75	102.14	109.79	118.32	125.55
Model y	93.72	101.86	109.99	118.13	126.27

Many calculators have "built-in" least squares regression programs. If your calculator has such a program, try using it to duplicate the results shown in the following example.

EXAMPLE 6   Finding a Least Squares Regression Line

The following ordered pairs (w, h) represent the shoe sizes w, and the heights h (in inches), of 25 men. Use a computer program or a statistical calculator to find the least squares regression line for these data.

(10.0, 70.0), (10.5, 71.0), (9.5, 70.0), (11.0, 72.0), (12.0, 74.0),
(8.5, 66.0), (9.0, 68.5), (13.0, 76.0), (10.5, 71.5), (10.5, 70.5),
(10.0, 72.0), (9.5, 70.0), (10.0, 71.0), (10.5, 69.5), (11.0, 71.5),
(12.0, 73.5), (12.5, 74.0), (11.0, 71.5), (9.0, 67.5), (10.0, 70.0),
(13.0, 73.5), (10.5, 72.5), (10.5, 71.0), (11.0, 73.0), (8.5, 68.0)

SOLUTION

A scatter plot for the data is shown in Figure 2.54. Note that the plot does not have 25 points because some of the ordered pairs graph as the same point. After entering the data into a statistical calculator, you can obtain

$a = 53.57$ and $b = 1.67$.

Thus, the least squares regression line for the data is

$h = 1.67w + 53.57$.

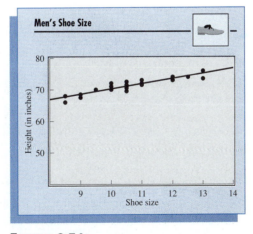

FIGURE 2.54

If you use a statistical calculator or computer program to duplicate the results of Example 6, you will notice that the program also outputs a value of $r \approx 0.918$. This number is called the **correlation coefficient** of the data.

Correlation coefficients vary between -1 and 1. Basically, the closer $|r|$ is to 1, the better the points can be described by a line. Three examples are shown in Figure 2.55.

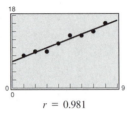

$r = 0.981$

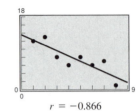

$r = -0.866$

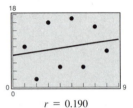

$r = 0.190$

FIGURE 2.55

DISCUSSION PROBLEM

·····················

AN EXPERIMENT

Go to the library, or some other reference source, and find some data that you think describe a linear relationship. Plot the points and find the equation of a line that approximately represents the points.

WARM-UP

··················

The following warm-up exercises involve skills that were covered in earlier sections. You will use these skills in the exercise set for this section.

In Exercises 1–4, sketch a graph of the equation.

1. $4x - 3y - 9 = 0$

2. $2x + 5y - 10 = 0$

3. $y = 25 - 2.25x$

4. $\dfrac{x}{2} + \dfrac{y}{4} = 1$

In Exercises 5–8, find an equation of the line through the two points.

5. $(0, 5)$, $(8, 0)$

6. $(0, -2)$, $(4, 6)$

7. $(1, -3)$, $(10, 6)$

8. $\left(\frac{3}{4}, 6\right)$, $\left(\frac{7}{2}, 3\right)$

In Exercises 9 and 10, find an equation of the line with the given slope and y-intercept.

9. Slope: $-\frac{4}{5}$; y-intercept: $(0, 8)$

10. Slope: $\frac{3}{2}$; y-intercept: $(0, 2)$

SECTION 2.6 · EXERCISES

Crop Yield   In Exercises 1–4, use the data in the table, where *x* is the number of units of fertilizer applied to sample plots and *y* is the yield (in bushels) of a crop.

x	0	1	2	3	4	5	6	7	8
y	58	60	59	61	63	66	65	67	70

1. Sketch a scatter plot of the data.
2. Determine whether the points are positively correlated, are negatively correlated, or have no discernible correlation.
3. Sketch a linear model that you think best represents the data. Find an equation of the line you sketched. Use the line to predict the yield if 10 units of fertilizer are used.
4. Can the model found in Exercise 3 be used to predict yield for arbitrarily large values of *x*? Explain.

Speed of Sound   In Exercises 5–8, use the data in the table, where *h* is altitude in thousands of feet and *v* is the speed of sound in feet per second.

h	0	5	10	15	20	25	30	35
v	1116	1097	1077	1057	1036	1015	995	973

5. Sketch a scatter plot of the data.
6. Determine whether the points are positively correlated, are negatively correlated, or have no discernible correlation.
7. Sketch a linear model that you think best represents the data. Find an equation of the line you sketched. Use the line to predict the speed of sound at an altitude of 27,000 feet.
8. The speed of sound at an altitude of 70,000 feet is approximately 971 feet per second. What does that suggest about the validity of using the model in Exercise 7 to extrapolate beyond the data given in the table?

In Exercises 9 and 10, (a) sketch a scatter plot of the points, (b) find an equation of the linear model you think best represents the data and find the sum of the squared differences, and (c) use the formulas of this section to find the least squares regression line and find the sum of the squared differences.

9. $(-1, 0)$, $(0, 1)$, $(1, 3)$, $(2, 3)$
10. $(0, 4)$, $(1, 3)$, $(2, 2)$, $(4, 1)$

In Exercises 11–14, (a) sketch a scatter plot of the points, (b) use the formulas of this section to find the least squares regression line, and (c) sketch the graph of the line.

11. $(-2, 0)$, $(-1, 1)$, $(0, 1)$, $(2, 2)$
12. $(-3, 1)$, $(-1, 2)$, $(0, 2)$, $(1, 3)$, $(3, 5)$
13. $(1, 5)$, $(2, 8)$, $(3, 13)$, $(4, 16)$, $(5, 22)$, $(6, 26)$
14. $(1, 10)$, $(2, 8)$, $(3, 8)$, $(4, 6)$, $(5, 5)$, $(6, 3)$

In Exercises 15–18, use a graphing utility to find the least squares regression line for the data. Sketch a scatter plot and the regression line.

15. $(0, 23)$, $(1, 20)$, $(2, 19)$, $(3, 17)$, $(4, 15)$, $(5, 11)$, $(6, 10)$
16. $(4, 52.8)$, $(5, 54.7)$, $(6, 55.7)$, $(7, 57.8)$, $(8, 60.2)$, $(9, 63.1)$, $(10, 66.5)$
17. $(-10, 5.1)$, $(-5, 9.8)$, $(0, 17.5)$, $(2, 25.4)$, $(4, 32.8)$, $(6, 38.7)$, $(8, 44.2)$, $(10, 50.5)$
18. $(-10, 213.5)$, $(-5, 174.9)$, $(0, 141.7)$, $(5, 119.7)$, $(8, 102.4)$, $(10, 87.6)$

19. *Advertising*   The management of a department store ran an experiment to determine if a relationship existed between sales *S* (in thousands of dollars) and the amount spent on advertising *x* (in thousands of dollars). The following data were collected.

x	1	2	3	4	5	6	7	8
S	405	423	455	466	492	510	525	559

(a) Use a graphing utility to find the least squares regression line. Use the equation to estimate sales if $4500 is spent on advertising.
(b) Make a scatter plot of the data and sketch the graph of the regression line.
(c) Use a computer or calculator to determine the correlation coefficient.

20. *School Enrollment*   The table gives the preprimary school enrollments y (in millions) for the years 1985 through 1991 where $t = 5$ corresponds to 1985. (*Source:* U.S. Bureau of the Census)

t	5	6	7	8	9	10	11
y	10.73	10.87	10.87	11.00	11.04	11.21	11.37

(a) Use a computer or calculator to find the least squares regression line. Use the equation to estimate enrollment in 1992.

(b) Make a scatter plot of the data and sketch the graph of the regression line.

(c) Use the computer or calculator to determine the correlation coefficient.

CHAPTER 2 · REVIEW EXERCISES

In Exercises 1 and 2, determine whether the equation is an identity or a conditional equation.

1. $6 - (x - 2)^2 = 2 + 4x - x^2$

2. $3(x - 2) + 2x = 2(x + 3)$

In Exercises 3 and 4, determine whether the values of x are solutions of the equation.

Equation	Values

3. $3x^2 + 7x + 5 = x^2 + 9$ (a) $x = 0$ (b) $x = -4$
    (c) $x = \frac{1}{2}$ (d) $x = -1$

4. $6 + \dfrac{3}{x - 4} = 5$ (a) $x = 4$ (b) $x = 0$
    (c) $x = -2$ (d) $x = 1$

In Exercises 5–28, solve the equation (if possible) and check your answer either algebraically or graphically.

5. $3x - 2(x + 5) = 10$

6. $4(x + 3) - 3 = 2(4 - 3x) - 4$

7. $3\left(1 - \dfrac{1}{5t}\right) = 0$ **8.** $\dfrac{1}{x - 2} = 3$

9. $6x^2 = 5x + 4$ **10.** $15 + x - 2x^2 = 0$

11. $(x + 4)^2 = 18$ **12.** $16x^2 = 25$

13. $x^2 - 12x + 30 = 0$ **14.** $5x^4 - 12x^3 = 0$

15. $4t^3 - 12t^2 + 8t = 0$ **16.** $2 - x^{-2} = 0$

17. $\dfrac{4}{(x - 4)^2} = 1$ **18.** $\dfrac{4}{x - 3} - \dfrac{4}{x} = 1$

19. $\sqrt{x + 4} = 3$ **20.** $\sqrt{x - 2} - 8 = 0$

21. $2\sqrt{x} - 5 = 0$

22. $\sqrt{2x + 3} + \sqrt{x - 2} = 2$

23. $(x - 1)^{2/3} - 25 = 0$

24. $(x + 2)^{3/4} = 27$

25. $(x + 4)^{1/2} + 5x(x + 4)^{3/2} = 0$

26. $8x^2(x^2 - 4)^{1/3} + (x^2 - 4)^{4/3} = 0$

27. $|x - 5| = 10$

28. $|x^2 - 6| = x$

In Exercises 29–36, use a graphing utility to solve the equation (if possible).

29. $x^2 + 6x - 3 = 0$

30. $12t^3 - 84t^2 + 120t = 0$

31. $5\sqrt{x} - \sqrt{x - 1} = 6$ **32.** $\sqrt{3x - 2} = 4 - x$

33. $\dfrac{1}{x} + \dfrac{1}{x + 1} = 2$ **34.** $\dfrac{1}{(t + 1)^2} = 1$

35. $|x^2 - 3| = 2x$ **36.** $|2x + 3| = 7$

In Exercises 37–40, solve the equation for the indicated variable.

37. Solve for r: $V = \frac{1}{3}\pi r^2 h$

38. Solve for X: $Z = \sqrt{R^2 - X^2}$

39. Solve for p: $L = \dfrac{k}{3\pi r^2 p}$

40. Solve for v: $E = 2kw\left(\dfrac{v}{2}\right)^2$

In Exercises 41–46, perform the indicated operations and write the result in standard form.

41. $-(6 - 2i) + (-8 + 3i)$

42. $\left(\dfrac{\sqrt{2}}{2} - \dfrac{\sqrt{2}}{2}i\right) - \left(\dfrac{\sqrt{2}}{2} + \dfrac{\sqrt{2}}{2}i\right)$

43. $5i(13 - 8i)$

44. $i(6 + i)(3 - 2i)$

45. $\dfrac{6 + i}{i}$

46. $\dfrac{3 + 2i}{5 + i}$

In Exercises 47–50, find a polynomial with integer coefficients that has the given zeros.

47. $-1, -1, \frac{1}{3}, -\frac{1}{2}$

48. $5, 1 - \sqrt{2}, 1 + \sqrt{2}$

49. $\frac{2}{3}, 4, \sqrt{3}i, -\sqrt{3}i$

50. $2, -3, 1 - 2i, 1 + 2i$

In Exercises 51–58, solve the inequality.

51. $\frac{1}{2}(3 - x) > \frac{1}{3}(2 - 3x)$

52. $x^2 - 2x \geq 3$

53. $\dfrac{x - 5}{3 - x} < 0$

54. $\dfrac{2}{x + 1} \leq \dfrac{3}{x - 1}$

55. $|x - 2| < 1$

56. $|x| \leq 4$

57. $|x - \frac{3}{2}| \geq \frac{3}{2}$

58. $|x + 3| > 4$

In Exercises 59–62, use a graphing utility to solve the inequality.

59. $\dfrac{x}{5} - 6 \leq -\dfrac{x}{2} + 6$

60. $2x^2 + x \geq 15$

61. $(x - 4)|x| > 0$

62. $|x(x - 6)| < 5$

In Exercises 63 and 64, find the domain of the expression by finding the interval(s) on the real number line for which the radicand is nonnegative.

63. $\sqrt{2x - 10}$

64. $\sqrt{x(x - 4)}$

65. *Monthly Profit* In October, a company's total profit was 12% more than it was in September. The total profit for the two months was $689,000. Find the profit for each month.

66. *Discount Rate* The price of a television set has been discounted $85. The sale price is $340. What percentage is the discount?

67. *Mixture Problem* A car radiator contains 10 quarts of a 30% antifreeze solution. How many quarts will have to be replaced with pure antifreeze if the resulting solution is to be 50% antifreeze?

68. *Starting Position* A fitness center has two running tracks around a rectangular playing floor. The tracks are 3 feet wide and form semicircles at the narrow end of the rectangular floor (see figure). Determine the distance between the starting positions if two runners must run the same distance to the finish line in one lap around the track.

69. *Depth of an Underwater Cable* The sonar of a ship locates a cable 2000 feet from the ship (see figure). The angle between the water level and the cable is 45 degrees. How deep is the cable?

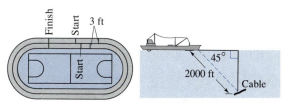

FIGURE FOR 68   **FIGURE FOR 69**

70. *Average Speed* You drove 56 miles on a service call for your company and it took 10 minutes longer than the return trip when you drove an average of 8 miles per hour faster. What was your average speed on the return trip?

71. *Cost Sharing* A group of farmers agree to share equally in the cost of a $48,000 piece of machinery. If they could find two more farmers to join the group, each person's share of the cost would decrease by $4000. How many farmers are presently in the group?

72. *Venture Capital* An individual is planning to start a small business that will require $90,000 before any income can be generated. Because it is difficult to borrow for new ventures, the individual wants a group of friends to divide the cost equally for a future share of the profit. Some are willing, but three more are needed so the price per person will be $2500 less. How many investors are needed?

73. *Market Research* The demand equation for a product is

$$p = 42 - \sqrt{0.001x + 2},$$

where x is the number of units demanded per day and p is the price per unit. Find the demand if the price is set at $29.95.

74. *Compound Interest* P dollars invested at interest rate r compounded annually for 5 years increases to an amount $A = P(1 + r)^5$. If an investment of $1000 is to increase to an amount greater than $1400 in 5 years, then the interest rate must be greater than what percentage?

75. *Break-Even Analysis*   The revenue for selling x units of a product is $R = 125.95x$. The cost of producing x units is $C = 92x + 1200$. In order to obtain a profit, the revenue must be greater than the cost. For what values of x will this product return a profit?

76. *Simply Supported Beam*   A simply supported beam of length 20 feet supports a uniformly distributed load of 1000 pounds per foot (see figure). The bending moment M in foot-pounds x feet from one end of the beam is given by

$$M = 500x(20 - x).$$

Load (1000 lb/ft)

x

|← 20 ft →|

FIGURE FOR 76

(a) Use a graphing utility to graph the function. Because of the physical constraints of the problem, give the domain of the function.

(b) Determine any points on the beam where the bending moment is zero.

(c) Use the graph to determine the point on the beam where the bending moment is greatest. What is the moment at that point?

(d) Determine the positions on the beam where the bending moment is less than 40,000 foot-pounds.

77. *Pendulum*   The period of a pendulum is

$$T = 2\pi \sqrt{\frac{L}{32}},$$

where T is the time in seconds and L is the length of the pendulum in feet. If the period is to be at least 2 seconds, determine the minimum length of the pendulum.

78. *Grade Point Average*   A study is being done to determine a relationship between the IQs of incoming college freshmen and their grade point averages after their freshman year. The data for a sample of 12 students are given in the table, where x is the IQ and y is the grade point average.

x	120	133	126	124	134	137
y	2.0	2.3	3.0	2.2	3.5	3.2

x	129	125	117	121	135	132
y	2.8	2.8	2.0	1.8	3.4	3.8

(a) Use a computer or calculator to find the least squares regression line for the data and the correlation coefficient.

(b) Plot the points and sketch a graph of the regression line.

(c) Use the line to predict the grade point average for an incoming freshman with an IQ of 128.

(d) It may appear to you that the line is not a particularly good fit to the data. What are some possible reasons for this?

CUMULATIVE TEST FOR PREREQUISITES, CHAPTERS 1 AND 2

Take this test as you would take a test in class. After you are done, check your work against the answers in the back of the book.

1. Evaluate: $(\frac{12}{5} \div 9) - (10 - \frac{1}{45})$

2. Simplify: $\dfrac{8x^2y^{-3}}{30x^{-1}y^2}$

3. Simplify: $\sqrt{24x^4y^3}$

4. Expand and simplify: $(x - 2)(x^2 + x - 3)$

5. Factor completely: $x - 5x^2 - 6x^3$

6. Subtract and simplify: $\dfrac{2}{s + 3} - \dfrac{1}{s + 1}$

7. Simplify: $\dfrac{\dfrac{2x + 1}{2\sqrt{x}} - \sqrt{x}}{2x + 1}$

8. Sketch a graph of the following equations without the aid of a graphing utility.

(a) $x - 3y + 12 = 0$ (b) $y = x^2 - 4x + 1$
(c) $y = \sqrt{4 - x}$ (d) $y = 3 - |x - 2|$

9. Find the constant C such that $(2, -3)$ is a solution point of the equation $2x - C(y + 1) = 10$.

10. Use the given point $(4, 9)$ on a line and the slope $m = -\frac{2}{3}$ of the line to find the coordinates of three additional points on the line. (The answers are not unique.)

11. Find an equation for the line passing through the points $(-\frac{1}{2}, 1)$ and $(3, 8)$.

12. Evaluate (if possible) the function

$$f(x) = \frac{x}{x - 2}$$

at the specified values of the independent variable.

(a) $f(6)$ (b) $f(2)$
(c) $f(2t)$ (d) $f(s + 2)$

13. Determine the domain and range of the function $g(x) = \dfrac{4}{\sqrt{3 - x}}$.

14. Express the area A of an equilateral triangle as a function of the length s of its sides.

15. Explain why the graph on the left does not represent y as a function of x.

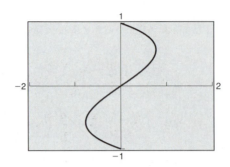

16. Describe how the graphs of each of the following functions would differ from the graph of $y = \sqrt[3]{x}$. It is not necessary to sketch the graphs.

(a) $f(x) = \sqrt[3]{\dfrac{x}{2}}$

(b) $r(x) = \dfrac{1}{2}\sqrt[3]{x}$

(c) $h(x) = \sqrt[3]{x} + 2$

(d) $g(x) = \sqrt[3]{x + 2}$

17. Find $f \circ g$ if $f(x) = \sqrt{x}$ and $g(x) = x^2 + 3$.

18. Determine whether the function $h(x) = 5x - 2$ is one-to-one. If it is, find its inverse.

19. Solve (if possible) the following equations.

(a) $2x - 3(x - 4) = 5$

(b) $\dfrac{2}{t - 3} + \dfrac{2}{t - 2} = \dfrac{10}{t^2 - 5t + 6}$

(c) $3y^2 + 6y + 2 = 0$

(d) $\sqrt{x + 10} = x - 2$

20. Use a graphing utility to approximate (accurate to one decimal place) any solutions of the equation $\sqrt{x + 4} + \sqrt{x} = 6$. [Remember to write the equation in the form $f(x) = 0$.]

21. Solve and sketch the solution on the real number line: $\left| \dfrac{x - 2}{3} \right| < 1$.

22. Solve and sketch the solution on the real number line: $x^2 - x < 6$.

23. An inheritance of \$12,000 is divided between two investments earning 7.5% and 9% simple interest. How much is in each investment if the total interest for 1 year is \$960?

24. Three gallons of a mixture is 60% water by volume. Determine the number of gallons of water that must be added to bring the mixture to 75% water.

25. A group of n people decide to buy a \$36,000 minibus for a charitable organization. Each person will pay an equal share of the cost. If three additional people were to join the group, the cost per person would decrease by \$1000. Find n.

26. Find the least squares regression line for the points $(0, 2)$, $(2, 4)$, $(4, 5)$.

POLYNOMIAL AND RATIONAL FUNCTIONS

3.1 QUADRATIC FUNCTIONS

The Graph of a Quadratic Function / The Standard Form of a Quadratic Function / Applications

The Graph of a Quadratic Function

Polynomial functions are the most commonly used functions in algebra. Let n be a nonnegative integer and let $a_n, a_{n-1}, \ldots, a_2, a_1, a_0$ be real numbers with $a_n \neq 0$. The function given by

$$f(x) = a_n x^n + a_{n-1} x^{n-1} + \cdots + a_2 x^2 + a_1 x + a_0$$

is called a **polynomial function of x with degree n.**

The function $f(x) = a$, where $a \neq 0$, has degree 0 and is a **constant function.** The polynomial function $f(x) = ax + b$, where $a \neq 0$, has degree 1 and is a **linear function.** In this section you will study second-degree polynomial functions, called **quadratic functions.**

DEFINITION OF A QUADRATIC FUNCTION

Let a, b, and c be real numbers with $a \neq 0$. The function

$$f(x) = ax^2 + bx + c$$

is a **quadratic function.** The graph of a quadratic function is called a **parabola.**

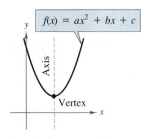

$a > 0$: Parabola opens upward.

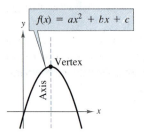

$a < 0$: Parabola opens downward.

FIGURE 3.1

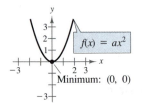

$a > 0$: Parabola opens upward.

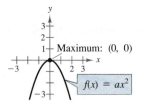

$a < 0$: Parabola opens downward.

FIGURE 3.2

All parabolas are symmetric with respect to a line called the **axis of symmetry,** or simply the **axis** of the parabola. The point where the axis intersects the parabola is the **vertex** of the parabola, as shown in Figure 3.1. If the leading coefficient is positive, then the graph of $f(x) = ax^2 + bx + c$ is a parabola that opens upward, and if the leading coefficient is negative, then the graph is a parabola that opens downward.

The simplest type of quadratic function is $f(x) = ax^2$. Its graph is a parabola whose vertex is $(0, 0)$. If $a > 0$, then the vertex is the *minimum* point on the graph, and if $a < 0$, then the vertex is the *maximum* point on the graph, as shown in Figure 3.2. When sketching the graph of $f(x) = ax^2$, it is helpful to use the graph of $y = x^2$ as a reference, as discussed in Section 1.5.

EXAMPLE 1   **Sketching the Graphs of Simple Quadratic Functions**

Describe how the graphs of the following functions are related to the graph of $y = x^2$.

A. $f(x) = \dfrac{1}{3}x^2$ **B.** $g(x) = 2x^2$

SOLUTION

A. Compared with $y = x^2$, each output of f "shrinks" by a factor of $\frac{1}{3}$. The result is a parabola that opens upward and is broader than the parabola represented by $y = x^2$, as shown in Figure 3.3.

B. Compared with $y = x^2$, each output of g "stretches" by a factor of 2, creating the more narrow parabola shown in Figure 3.4.

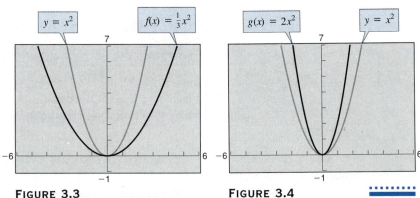

FIGURE 3.3   **FIGURE 3.4**

Recall from Section 1.5 that the graphs of $y = f(x \pm c)$, $y = f(x) \pm c$, and $y = -f(x)$ are rigid transformations of the graph of $y = f(x)$.

$y = f(x \pm c)$ *Horizontal shift*
$y = f(x) \pm c$ *Vertical shift*
$y = -f(x)$ *Reflection*

REMARK   In Example 1, note that the coefficient a determines how widely the parabola given by $f(x) = ax^2$ opens. If $|a|$ is small, the parabola opens more widely than if $|a|$ is large.

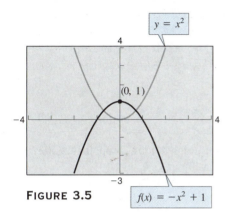

FIGURE 3.5

$y = x^2$

$(0, 1)$

$f(x) = -x^2 + 1$

EXAMPLE 2   Sketching a Parabola

Describe how the graphs of the following quadratic functions are related to the graph of $y = x^2$.

A. $f(x) = -x^2 + 1$ **B.** $g(x) = (x + 2)^2 - 3$

SOLUTION

A. With respect to the graph of $y = x^2$, the negative coefficient in $f(x) = -x^2 + 1$ reflects the graph *downward* and the positive constant term shifts the vertex *up one unit*. The graph of f is shown in Figure 3.5. Note that the axis of the parabola is the y-axis and the vertex is $(0, 1)$.

B. With respect to the graph of $y = x^2$, the graph of $g(x) = (x + 2)^2 - 3$ is obtained by a horizontal shift two units *to the left* and a vertical shift three units *down*, as shown in Figure 3.6. Note that the axis of the parabola is the vertical line $x = -2$ and the vertex is $(-2, -3)$.

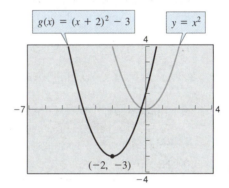

$g(x) = (x + 2)^2 - 3$ $y = x^2$

$(-2, -3)$

FIGURE 3.6

The Standard Form of a Quadratic Function

The equation in Example 2(B) is written in **standard form**

$$f(x) = a(x - h)^2 + k.$$

This form is especially convenient for sketching a parabola because it identifies the vertex of the parabola as (h, k).

STANDARD FORM OF A QUADRATIC FUNCTION

The quadratic function

$$f(x) = a(x - h)^2 + k, \qquad a \neq 0$$

is said to be in **standard form.** The graph of f is a parabola whose axis is the vertical line $x = h$ and whose vertex is the point (h, k). If $a > 0$, the parabola opens upward, and if $a < 0$, the parabola opens downward.

EXAMPLE 3   **Writing a Quadratic Function in Standard Form**

Describe the graph of $f(x) = 2x^2 + 8x + 7$.

SOLUTION

Write the quadratic function in standard form by completing the square. Notice that the first step is to factor out any coefficient of x^2 that is different from 1.

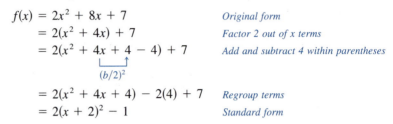

$$
\begin{aligned}
f(x) &= 2x^2 + 8x + 7 && \textit{Original form} \\
&= 2(x^2 + 4x) + 7 && \textit{Factor 2 out of x terms} \\
&= 2(x^2 + 4x + 4 - 4) + 7 && \textit{Add and subtract 4 within parentheses} \\
&\qquad \underset{(b/2)^2}{\underline{\quad\uparrow\quad}} \\
&= 2(x^2 + 4x + 4) - 2(4) + 7 && \textit{Regroup terms} \\
&= 2(x + 2)^2 - 1 && \textit{Standard form}
\end{aligned}
$$

From the standard form, you can see that the graph of f is a parabola that opens upward with vertex $(-2, -1)$, as shown in Figure 3.7.

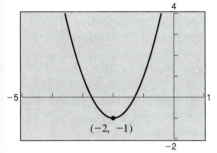

FIGURE 3.7

EXAMPLE 4   **Writing a Quadratic Function in Standard Form**

Describe the graph of $f(x) = -x^2 + 6x - 8$.

SOLUTION

$$
\begin{aligned}
f(x) &= -x^2 + 6x - 8 && \textit{Original form} \\
&= -(x^2 - 6x) - 8 && \textit{Factor } -1 \textit{ out of x terms} \\
&= -(x^2 - 6x + 9 - 9) - 8 && \textit{Add and subtract 9 within parentheses} \\
&\qquad \underset{b/2^2}{\underline{\quad\uparrow\quad}} \\
&= -(x^2 - 6x + 9) - (-9) - 8 && \textit{Regroup terms} \\
&= -(x - 3)^2 + 1 && \textit{Standard form } a < 0
\end{aligned}
$$

The graph of f is a parabola that opens downward with vertex at $(3, 1)$, as shown in Figure 3.8.

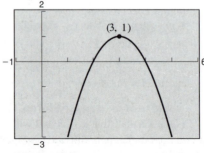

FIGURE 3.8

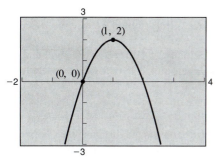

FIGURE 3.9

EXAMPLE 5   **Finding the Equation of a Parabola**

Find an equation for the parabola that has its vertex at $(1, 2)$ and that passes through the point $(0, 0)$, as shown in Figure 3.9.

SOLUTION

Because the parabola has a vertex at $(h, k) = (1, 2)$, the equation has the form

$$\overset{h}{\underset{\downarrow}{}}\quad\overset{k}{\underset{\downarrow}{}}$$
$$f(x) = a(x - 1)^2 + 2. \qquad \textit{Standard form}$$

Because the parabola passes through the point $(0, 0)$, it follows that $f(0) = 0$. Thus, you obtain

$$0 = a(0 - 1)^2 + 2,$$

which implies that $a = -2$. The equation is

$$f(x) = -2(x - 1)^2 + 2 = -2x^2 + 4x.$$

Try graphing $f(x) = -2x^2 + 4x$ with a graphing utility to confirm that its vertex is $(1, 2)$ and that it passes through the point $(0, 0)$.

To find the x-intercepts of the graph of $f(x) = ax^2 + bx + c$, solve the equation $ax^2 + bx + c = 0$. For example, from the equation

$$-2x^2 + 4x = -2x(x - 2) = 0$$

you can see that the parabola shown in Figure 3.9 has x-intercepts at $(0, 0)$ and $(2, 0)$. If $ax^2 + bx + c$ does not factor, you can use the Quadratic Formula or a graphing utility to find the x-intercepts. Remember, however, that a parabola may have no x-intercept.

Applications

Many applications involve finding the maximum or minimum value of a quadratic function. By writing the quadratic function $f(x) = ax^2 + bx + c$ in standard form,

$$f(x) = a\left(x + \frac{b}{2a}\right)^2 + \left(c - \frac{b^2}{4a}\right),$$

you can see that the vertex occurs as $x = -b/2a$, which implies the following.

1. If $a > 0$ then the quadratic function $f(x) = ax^2 + bx + c$ has a *minimum* that occurs at $x = -b/2a$.
2. If $a < 0$ then the quadratic function $f(x) = ax^2 + bx + c$ has a *maximum* that occurs at $x = -b/2a$.

In either case, you can find the minimum or maximum value of f by evaluating the function at $x = -b/2a$ or by graphing the function.

EXAMPLE 6   The Maximum Height of a Baseball

A baseball is hit 3 feet above ground at a velocity of 100 feet per second and at an angle of 45 degrees with respect to level ground. The path of the baseball is given by the function

$$f(x) = -0.0032x^2 + x + 3,$$

where $f(x)$ is the height of the baseball (in feet) and x is the distance from home plate (in feet). What is the maximum height reached by the baseball?

SOLUTION

For this quadratic function, you have

$$f(x) = ax^2 + bx + c = -0.0032x^2 + x + 3,$$

which implies that $a = -0.0032$ and $b = 1$. Because the function has a maximum when $x = -b/2a$, you can conclude that the baseball reaches its maximum height when x is

$$x = -\frac{b}{2a} = -\frac{1}{2(-0.0032)} = 156.25$$

from home plate. At this distance, the maximum height is

$$f(156.25) = -0.0032(156.25)^2 + 156.25 + 3 = 81.125.$$

The path of the baseball is shown in Figure 3.10. You can also solve this problem by using a graphing utility. By using the zoom and trace feature, you can determine that the maximum height on the graph occurs when $x \approx 156.25$. Note that you might have to change the y-scale in order to avoid a graph that is "too flat."

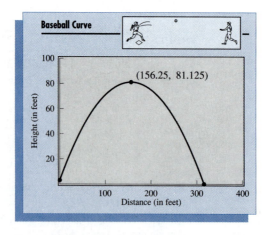

Baseball Curve

(156.25, 81.125)

Height (in feet)

Distance (in feet)

FIGURE 3.10

EXAMPLE 7   Charitable Contributions

According to a survey conducted in 1990 by *Independent Sector,* the percentage of their income that Americans give to charities is related to their household income. For families with an annual income of $100,000 or less, the percentage is approximately given by

$$P = 0.0014x^2 - 0.1529x + 5.855, \qquad 5 \le x \le 100,$$

where P is the percentage of annual income given and x is the annual income in thousands of dollars. According to this model, what income level corresponds to the least percentage of charitable contributions?

SOLUTION

There are two ways to answer this question. One is to use a graphing utility to graph the quadratic function, as shown in Figure 3.11. From this graph, it appears that the minimum percentage corresponds to an income level of about $55,000. By using the zoom feature, you can improve the approximation to $x \approx 54,600$. The other way to answer the question is to use the fact that the minimum point of the parabola occurs when $x = -b/2a$. For this function, you have $a = 0.0014$ and $b = -0.1529$. Thus,

$$x = -\frac{b}{2a} = -\frac{-0.1529}{2(0.0014)} \approx 54.6.$$

From this x-value, you can conclude that the minimum percentage corresponds to an income level of about $54,600.

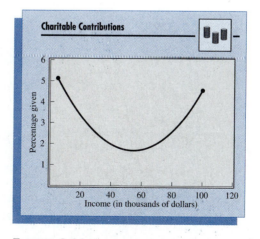

FIGURE 3.11

DISCUSSION PROBLEM

FINDING AN EQUATION FOR A CURVE

The parabola in the figure has an equation of the form $y = x^2 + bx + c$. Find the equation for this parabola, and write a short paragraph describing the method you used.

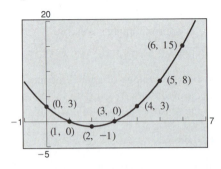

WARM-UP

The following warm-up exercises involve skills that were covered in earlier sections. You will use these skills in the exercise set for this section.

In Exercises 1–4, solve the quadratic equation by factoring.

1. $2x^2 + 11x - 6 = 0$ **2.** $5x^2 - 12x - 9 = 0$

3. $3 + x - 2x^2 = 0$ **4.** $x^2 + 20x + 100 = 0$

In Exercises 5–8, solve the quadratic equation by completing the square.

5. $x^2 - 6x + 4 = 0$ **6.** $x^2 + 4x + 1 = 0$

7. $2x^2 - 16x + 25 = 0$ **8.** $3x^2 + 30x + 74 = 0$

In Exercises 9 and 10, use the Quadratic Formula to solve the quadratic equation.

9. $x^2 + 3x + 1 = 0$ **10.** $x^2 + 3x - 3 = 0$

SECTION 3.1 · EXERCISES

In Exercises 1–6, match the quadratic function with its graph, and describe the viewing rectangle. [The graphs are labeled (a), (b), (c), (d), (e), and (f).]

1. $f(x) = (x - 3)^2$
2. $f(x) = (x + 5)^2$
3. $f(x) = x^2 - 4$
4. $f(x) = 5 - x^2$
5. $f(x) = 4 - (x - 1)^2$
6. $f(x) = (x + 2)^2 - 2$

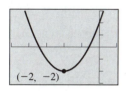

(a)

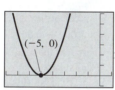

(b)

(c)

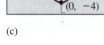

(d)

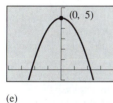

(e)

(f)

In Exercises 7–10, use a graphing utility to compare the graph of each function with the graph of $y = x^2$.

7. (a) $f(x) = \frac{1}{6}x^2$
   (b) $g(x) = -3x^2$
8. (a) $f(x) = -\frac{1}{2}x^2 + 4$
   (b) $g(x) = 2x^2 - 5$
9. (a) $f(x) = -(x + 3)^2 + 3$
   (b) $g(x) = \frac{4}{3}(x - 4)^2 - 3$
10. (a) $f(x) = 0.75(x - 4.5)^2 + 5.2$
    (b) $g(x) = -0.2(x + 1.4)^2 + 3.8$

In Exercises 11–16, find an equation for the given parabola.

11.

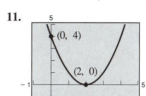

12.

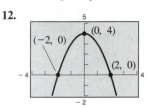

13.

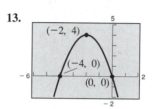

14.

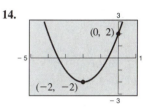

15.

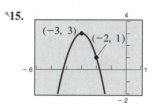

16.

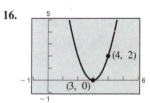

In Exercises 17–26, sketch the graph of the given quadratic function without the aid of a graphing utility. Identify the vertex and intercepts. Then use a graphing utility to check your result.

17. $f(x) = x^2 - 5$
18. $h(x) = 25 - x^2$
19. $f(x) = (x + 5)^2 - 6$
20. $f(x) = (x - 6)^2 + 3$
21. $h(x) = x^2 - 8x + 16$
22. $f(x) = x^2 + 3x + \frac{1}{4}$
23. $f(x) = -x^2 + 2x + 5$
24. $f(x) = -x^2 - 4x + 1$
25. $h(x) = 4x^2 - 4x + 21$
26. $f(x) = 2x^2 - x + 1$

In Exercises 27–30, use a graphing utility to graph the quadratic function. Identify the vertex and intercepts. Then check your result algebraically by completing the square.

27. $f(x) = -(x^2 + 2x - 3)$
28. $g(x) = x^2 + 8x + 11$
29. $f(x) = 2x^2 - 16x + 31$
30. $g(x) = \frac{1}{2}(x^2 + 4x - 2)$

In Exercises 31–34, find the quadratic function that has the indicated vertex and whose graph passes through the given point. Confirm your result with a graphing utility.

31. Vertex: $(3, 4)$;   point: $(1, 2)$

32. Vertex: $(2, 3)$;   point: $(0, 2)$

33. Vertex: $(5, 12)$;   point: $(7, 15)$

34. Vertex: $(-2, -2)$;   point: $(-1, 0)$

In Exercises 35–40, find two quadratic functions whose graphs have the given x-intercepts. (Find one function that has a graph that opens upward and another that has a graph that opens downward.)

35. $(-1, 0)$, $(3, 0)$

36. $(-\frac{5}{2}, 0)$, $(2, 0)$

37. $(0, 0)$, $(10, 0)$

38. $(4, 0)$, $(8, 0)$

39. $(-3, 0)$, $(-\frac{1}{2}, 0)$

40. $(-5, 0)$, $(5, 0)$

In Exercises 41 and 42, find two positive real numbers that satisfy the given requirements.

41. The sum is 110 and the product is a maximum.

42. The sum is S and the product is a maximum.

Maximum Area In Exercises 43 and 44, consider a rectangle of length x and perimeter P (see figure). (a) Express the area A as a function of x and determine the domain of the function. (b) Graph the area function. (c) Find the length and width of the rectangle of maximum area.

43. $P = 100$ feet

44. $P = 36$ meters

FIGURE FOR 43 AND 44

45. *Maximum Area* A rancher has 200 feet of fencing to enclose two adjacent rectangular corrals (see figure). What dimensions will produce a maximum enclosed area?

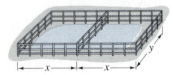

FIGURE FOR 45

46. *Maximum Area* An indoor physical fitness room consists of a rectangular region with a semicircle on each end (see figure). The perimeter of the room is to be a 200-meter running track. What dimensions will produce a maximum area of the rectangle?

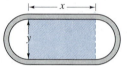

FIGURE FOR 46

47. *Maximum Revenue* Find the number of units that produce a maximum revenue,

$R = 900x - 0.1x^2$,

where R is the total revenue in dollars and x is the number of units sold.

48. *Maximum Revenue* Find the number of units that produce a maximum revenue,

$R = 100x - 0.0002x^2$,

where R is the total revenue in dollars and x is the number of units sold.

49. From 1950 to 1990, the average annual per capita consumption C of cigarettes by Americans (18 and older) can be modeled by

$C = 4024.5 + 51.4t - 3.1t^2$,

where t is the year, with $t = 0$ corresponding to 1960. The graph of the model is shown in the figure. (*Source:* U.S. Center for Disease Control)

(a) Beginning in 1966, all cigarette packages were required by law to carry a health warning. Do you think the warnings had any effect? Explain.

(b) In 1960, the U.S. had a population (18 and over) of 116,530,000. Of those, about 48,500,000 were smokers. What was the average annual cigarette consumption *per smoker* in 1960? What was the average daily cigarette consumption *per smoker*?

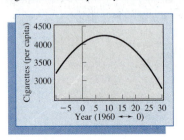

FIGURE FOR 49

50. *Maximum Profit* Let x be the amount (in hundreds of dollars) a company spends on advertising, and let P be the profit, where

$$P = 230 + 20x - 0.5x^2.$$

What expenditure for advertising gives the maximum profit?

51. *Trajectory of a Ball* The height y (in feet) of a ball thrown by a child is

$$y = -\tfrac{1}{12}x^2 + 2x + 4,$$

where x is the horizontal distance (in feet) from where the ball is thrown (see figure).
(a) Use a graphing utility to sketch the path of the ball.
(b) How high is the ball when it leaves the child's hand? (*Note:* Find y when $x = 0$.)
(c) How high is the ball when it reaches its maximum height?
(d) How far from the child does the ball strike the ground?

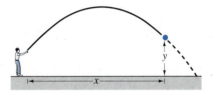

FIGURE FOR 51

52. *Maximum Height of a Dive* The path of a dive is given by

$$y = -\tfrac{4}{9}x^2 + \tfrac{24}{9}x + 10,$$

where y is the height in feet and x is the horizontal distance from the end of the diving board in feet (see figure). What is the maximum height of the dive?

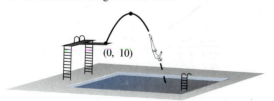

(0, 10)

FIGURE FOR 52

53. *Forestry* The number of board feet in a 16-foot log is approximated by the model

$$V = 0.77x^2 - 1.32x - 9.31, \qquad 0 \le x \le 40,$$

where V is the number of board feet and x is the diameter of the log at the small end in inches. (One board foot is a measure of volume equivalent to a board that is 12 inches wide, 12 inches long, and 1 inch thick.)
(a) Use a graphing utility to graph the function.
(b) Estimate the number of board feet in a 16-foot log with a diameter of 16 inches.
(c) Estimate the diameter of a 16-foot log that yielded 500 board feet when the lumber was sold.

54. *Automobile Aerodynamics* The amount of horsepower y required to overcome wind drag on a certain automobile is approximated by

$$y = 0.002x^2 + 0.005x - 0.0029, \qquad 0 \le x \le 100,$$

where x is the speed of the car in miles per hour.
(a) Use a graphing utility to graph the function.
(b) Estimate the maximum speed of the car if the power required to overcome wind drag is not to exceed 10 horsepower.

55. Complete the square for the quadratic function $f(x) = ax^2 + bx + c$ $(a \neq 0)$ and show that the vertex is at

$$\left(-\frac{b}{2a}, \ -\frac{b^2 - 4ac}{4a} \right).$$

56. Use Exercise 55 to verify the vertices found in Exercises 21 and 22.

57. Assume that the function $f(x) = ax^2 + bx + c$ $(a \neq 0)$ has two real zeros. Show that the x-coordinate of the vertex of the graph is the average of the zeros of f. (*Hint:* Use the Quadratic Formula.)

58. Use a graphing utility to demonstrate the result of Exercise 57 for each of the following equations.
(a) $f(x) = \tfrac{1}{2}(x - 3)^2 - 2$ (b) $f(x) = 6 - \tfrac{2}{3}(x + 1)^2$

3.2   POLYNOMIAL FUNCTIONS OF HIGHER DEGREE

**Graphs of Polynomial Functions / The Leading Coefficient Test /
Zeros of Polynomial Functions / The Intermediate Value Theorem**

Graphs of Polynomial Functions

At this point you should be able to sketch an accurate graph of polynomial functions of degrees 0, 1, and 2.

Function	Graph
$f(x) = a$	Horizontal line
$f(x) = ax + b$	Line of slope a
$f(x) = ax^2 + bx + c$	Parabola

The graphs of polynomial functions of degree greater than 2 are more difficult to sketch by hand. However, in this section you will learn how to recognize some of the basic features of the graphs of polynomial functions.

The graph of a polynomial function is **continuous.** Essentially, this means that the graph of a polynomial function has no breaks, as shown in Figure 3.12.

Another feature of the graph of a polynomial function is that it has only smooth, rounded turns, as shown in Figure 3.13(a); it cannot have a sharp, pointed turn such as the one shown in Figure 3.13(b). A specific example of a nonpolynomial function is $f(x) = |x|$, the graph of which has a sharp turn at the point (0, 0).

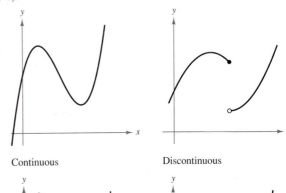

FIGURE 3.12   Continuous   Discontinuous

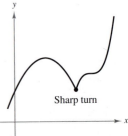

FIGURE 3.13

(a) Polynomial functions have smooth, rounded graphs.

(b) The graph of a polynomial function *cannot* have a sharp pointed turn.

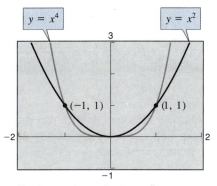

If n is even, the graph of $y = x^n$
touches axis at x-intercept.

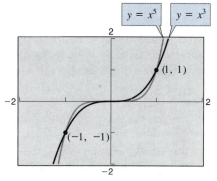

If n is odd, the graph of $y = x^n$
crosses axis at x-intercept.

FIGURE 3.14

The polynomial functions that have the simplest graphs are monomials of the form

$$f(x) = x^n,$$

where n is an integer greater than zero. From Figure 3.14, you can see that when n is *even* the graph is similar to the graph of $f(x) = x^2$, and when n is *odd* the graph is similar to the graph of $f(x) = x^3$. Moreover, the greater the value of n, the flatter the graph is on the interval $[-1, 1]$.

EXAMPLE 1   **Sketching Transformations of Monomial Functions**

Sketch the graphs of the following polynomial functions.

A. $f(x) = -x^5$ **B.** $g(x) = x^4 + 1$ **C.** $h(x) = (x + 1)^4$

SOLUTION

A. Because the degree of f is odd, the graph is similar to the graph of $y = x^3$. Moreover, the negative coefficient reflects the graph in the x-axis. Plotting the intercept $(0, 0)$ and the points $(1, -1)$ and $(-1, 1)$, you can obtain the graph shown in Figure 3.15.

B. In this case, the graph of g is an upward shift, by one unit, of the graph of $y = x^4$ (see Figure 3.14). Thus, you can obtain the graph shown in Figure 3.16.

C. The graph of h is a left shift, by one unit, of the graph of $y = x^4$ and it is shown in Figure 3.17.

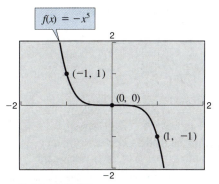

FIGURE 3.15

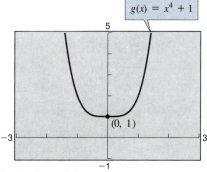

FIGURE 3.16

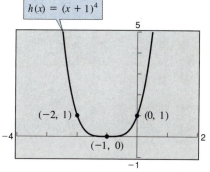

FIGURE 3.17

D I S C O V E R Y

Use a graphing utility to investigate the behavior of the graph of

$y = x^3 - 105x^2 + 21.$

First use a viewing rectangle in which $-2 \leq x \leq 2$ and $-10 \leq y \leq 30$. How complete a view of the graph does this viewing rectangle show? Does the graph move down as x increases indefinitely? Find a viewing rectangle that gives a good view of the basic characteristics of the graph.

The Leading Coefficient Test

In Example 1, note that the three graphs eventually rise or fall without bound as x moves to the right or left. Symbolically, this is written as

$$f(x) \longrightarrow \infty \quad \text{as} \quad x \longrightarrow \infty,$$

which means that $f(x)$ increases without bound as x moves to the right without bound. (The infinity symbol ∞ is used to indicate unboundedness.) Whether the graph of a polynomial eventually rises or falls can be determined by the function's degree (even or odd) and by its leading coefficient, as indicated by the **Leading Coefficient Test.**

LEADING COEFFICIENT TEST

As x increases or decreases without bound, the graph of the polynomial function $f(x) = a_n x^n + \cdots + a_1 x + a_0$ eventually rises or falls in the following manner. (*Note:* The dashed portions of the graphs indicate that the test determines *only* the right and left behavior of the graph.)

1. When n is *odd:*

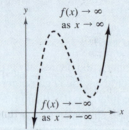

If the leading coefficient is positive ($a_n > 0$), then the graph falls to the left and rises to the right.

If the leading coefficient is negative ($a_n < 0$), then the graph rises to the left and falls to the right.

2. When n is *even:*

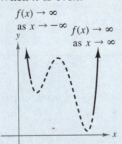
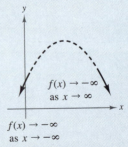

If the leading coefficient is positive ($a_n > 0$), then the graph rises to the left and right.

If the leading coefficient is negative ($a_n < 0$), then the graph falls to the left and right.

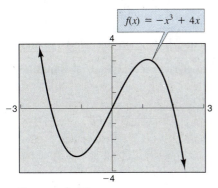

$f(x) = -x^3 + 4x$

FIGURE 3.18

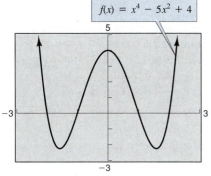

$f(x) = x^4 - 5x^2 + 4$

FIGURE 3.19

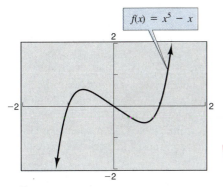

$f(x) = x^5 - x$

FIGURE 3.20

EXAMPLE 2   Applying the Leading Coefficient Test

Use the Leading Coefficient Test to determine the right and left behavior of the graphs of the following polynomial functions.

A. $f(x) = -x^3 + 4x$ **B.** $f(x) = x^4 - 5x^2 + 4$
C. $f(x) = x^5 - x$

SOLUTION

A. Because the degree is odd and the leading coefficient is negative, the graph rises to the left and falls to the right, as shown in Figure 3.18.
B. Because the degree is even and the leading coefficient is positive, the graph rises to the left and right, as shown in Figure 3.19.
C. Because the degree is odd and the leading coefficient is positive, the graph falls to the left and rises to the right, as shown in Figure 3.20.

Zeros of Polynomial Functions

It can be shown that for a polynomial function f of degree n, the following statements are true.

1. The function f has at most n real zeros. (This result is discussed in detail in Section 3.3.)
2. The graph of f has at most $n - 1$ relative extrema (local minimums or maximums).

Recall that a **zero** of a function f is a number x for which $f(x) = 0$. For instance, 0 and -3 are zeros of the function $f(x) = x^2 + 3x$ because $f(0) = 0^2 + 3(0) = 0$ and $f(-3) = (-3)^2 + 3(-3) = 0$.

Finding the zeros of polynomial functions is one of the most important problems in algebra. You have already seen that there is a strong interplay between graphical and algebraic approaches to this problem. Sometimes you can use information about the graph of a function to help find its zeros, and in other cases you can use information about the zeros of a function to help find a good viewing rectangle.

REAL ZEROS OF POLYNOMIAL FUNCTIONS

If f is a polynomial function and a is a real number, then the following statements are equivalent.

1. $x = a$ is a *zero* of the function f.
2. $x = a$ is a *solution* of the polynomial equation $f(x) = 0$.
3. $(x - a)$ is a *factor* of the polynomial $f(x)$.
4. $(a, 0)$ is an *x-intercept* of the graph of f.

Finding zeros of polynomial functions is closely related to factoring and finding x-intercepts, as demonstrated in Examples 3, 4, and 5.

EXAMPLE 3   Finding Zeros of a Third-Degree Polynomial Function

Find all real zeros of $f(x) = x^3 - x^2 - 2x$.

SOLUTION

$$
\begin{aligned}
f(x) &= x^3 - x^2 - 2x && \textit{Original function} \\
&= x(x^2 - x - 2) && \textit{Remove common monomial factor} \\
&= x(x - 2)(x + 1) && \textit{Factor completely}
\end{aligned}
$$

Thus, the real zeros are $x = 0$, $x = 2$, and $x = -1$, and the corresponding x-intercepts are $(0, 0)$, $(2, 0)$, and $(-1, 0)$, as shown in Figure 3.21. In the figure, note that the graph has two relative (local) extrema. This is consistent with the fact that a third-degree polynomial can have *at most* two relative extrema.

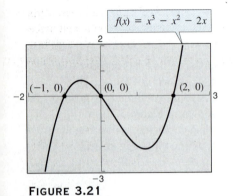

$f(x) = x^3 - x^2 - 2x$

FIGURE 3.21

EXAMPLE 4   Finding Zeros of a Fourth-Degree Polynomial Function

Find all real zeros of $f(x) = -2x^4 + 2x^2$.

SOLUTION

$$
\begin{aligned}
f(x) &= -2x^4 + 2x^2 && \textit{Original function} \\
&= -2x^2(x^2 - 1) && \textit{Remove common monomial factor} \\
&= -2x^2(x - 1)(x + 1) && \textit{Factor completely}
\end{aligned}
$$

Thus, the real zeros are $x = 0$, $x = 1$, and $x = -1$, and the corresponding x-intercepts are $(0, 0)$, $(1, 0)$, and $(-1, 0)$, as shown in Figure 3.22. Note in the figure that the graph has three relative extrema, which is consistent with the fact that a fourth-degree polynomial can have at most three relative extrema.

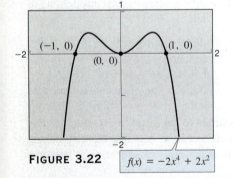

FIGURE 3.22 $f(x) = -2x^4 + 2x^2$

In Example 4, the real zero arising from $-2x^2 = 0$ is a **repeated zero**. In general, a factor of $(x - a)^k$ yields a repeated zero $x = a$ of **multiplicity** k. If k is odd, then the graph *crosses* the x-axis at $x = a$. If k is even, then the graph *touches* (but does not cross) the x-axis at $x = a$, as shown in Figure 3.22.

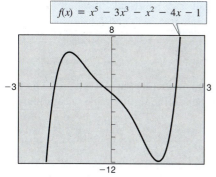

$f(x) = x^5 - 3x^3 - x^2 - 4x - 1$

FIGURE 3.23

EXAMPLE 5   Finding Zeros of a Fifth-Degree Polynomial Function

Find all zeros of $f(x) = x^5 - 3x^3 - x^2 - 4x - 1$.

SOLUTION

Use a graphing utility to obtain the graph shown in Figure 3.23. From the graph, you can see that there are three zeros. Using the zoom and trace features, you can determine that the zeros are approximately $x = -1.861$, $x = -0.254$, and $x = 2.115$. It should be noted that this fifth-degree polynomial factors as

$$f(x) = x^5 - 3x^3 - x^2 - 4x - 1$$
$$= (x^2 + 1)(x^3 - 4x - 1).$$

The three zeros obtained above are the zeros to the cubic on the right (the quadratic $x^2 + 1$ has no real zeros).

To find mathematical models, you may need to create a polynomial function that has a particular set of zeros.

EXAMPLE 6   Finding a Polynomial Function with Given Zeros

Find polynomial functions with the following zeros. (There are many correct solutions.)

A. $-2, -1, 1, 2$ **B.** $-\dfrac{1}{2}, 3, 3$

SOLUTION

A. For each of the given zeros, form a corresponding factor. For instance, the zero given by $x = -2$ corresponds to the factor $(x + 2)$. Thus, you can write the function as

$$f(x) = (x + 2)(x + 1)(x - 1)(x - 2)$$
$$= (x^2 - 4)(x^2 - 1)$$
$$= x^4 - 5x^2 + 4.$$

B. Note that the zero $x = -\frac{1}{2}$ corresponds to either $(x + \frac{1}{2})$ or $(2x + 1)$. To avoid fractions, choose the second factor and write

$$f(x) = (2x + 1)(x - 3)^2$$
$$= (2x + 1)(x^2 - 6x + 9)$$
$$= 2x^3 - 11x^2 + 12x + 9.$$

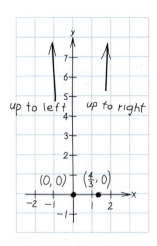

FIGURE 3.24

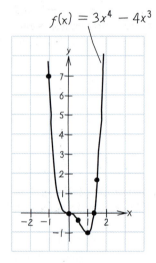

FIGURE 3.25

EXAMPLE 7   Sketching the Graph of a Polynomial Function

Sketch the graph of $f(x) = 3x^4 - 4x^3$ by hand.

SOLUTION

Because the leading coefficient is positive and the degree is even, you know that the graph eventually rises to the left and to the right, as shown in Figure 3.24. By factoring to obtain

$$f(x) = 3x^4 - 4x^3 = x^3(3x - 4),$$

you can see that the zeros of f are $x = 0$ and $x = \frac{4}{3}$ (both of odd multiplicity). Thus, the x-intercepts occur at $(0, 0)$ and $(\frac{4}{3}, 0)$. Finally, plot a few additional points as indicated in the accompanying table and obtain the graph shown in Figure 3.25.

x	-1	0.5	1	1.5
$f(x)$	7	-0.3125	-1	1.6875

The Intermediate Value Theorem

The **Intermediate Value Theorem** concerns the existence of real zeros of polynomial functions. The theorem states that if $(a, f(a))$ and $(b, f(b))$ are two points on the graph of a polynomial such that $f(a) \neq f(b)$, then for any number d between $f(a)$ and $f(b)$ there must be a number c between a and b such that $f(c) = d$. (See Figure 3.26.)

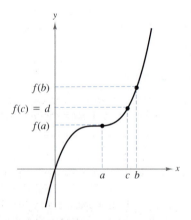

If d lies between $f(a)$ and $f(b)$, then there exists c between a and b such that $f(c) = d$.

FIGURE 3.26

INTERMEDIATE VALUE THEOREM

Let a and b be real numbers such that $a < b$. If f is a polynomial function such that $f(a) \neq f(b)$, then in the interval $[a, b]$, f takes on every value between $f(a)$ and $f(b)$.

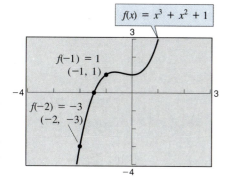

$f(x) = x^3 + x^2 + 1$

$f(-1) = 1$
$(-1, 1)$

$f(-2) = -3$
$(-2, -3)$

f has a zero between -2 and -1.

FIGURE 3.27

This theorem helps locate the real zeros of a polynomial function in the following way. If you can find a value $x = a$ where a polynomial function is positive, and another value $x = b$ where it is negative, then you can conclude that the function has at least one real zero between these two values. For example, the function $f(x) = x^3 + x^2 + 1$ is negative when $x = -2$ and positive when $x = -1$. Therefore, it follows from the Intermediate Value Theorem that f must have a real zero somewhere between -2 and -1, as shown in Figure 3.27.

EXAMPLE 8   Locating Zeros with the Intermediate Value Theorem

Use the Intermediate Value Theorem to find three intervals of length 1 in which the polynomial $f(x) = 12x^3 - 32x^2 + 3x + 5$ is guaranteed to have a zero.

SOLUTION

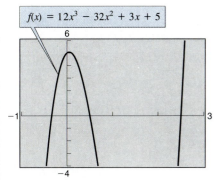

$f(x) = 12x^3 - 32x^2 + 3x + 5$

FIGURE 3.28

With a graphing utility, you can see that there are three zeros, as shown in Figure 3.28. The following table shows several function values.

x	-1	0	1	2	3
$f(x)$	-42	5	-12	-21	50

Because $f(-1)$ is negative and $f(0)$ is positive, you can conclude from the Intermediate Value Theorem that the function has a zero between -1 and 0, as indicated in the figure. Similarly, the function has a zero between 0 and 1, and another between 2 and 3. You could obtain more accurate approximations to these zeros using the zoom and trace features.

**DISCUSSION
PROBLEM**
....

**THE GRAPHS
OF CUBIC
POLYNOMIALS**

The graphs of cubic polynomials can be categorized according to the four basic shapes shown in the figure.

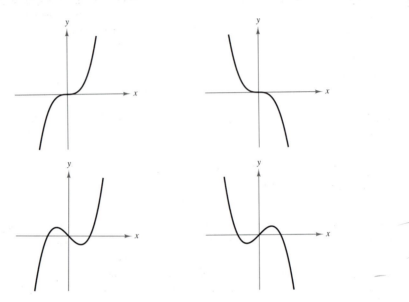

Match the graph of each of the following four functions with one of the basic shapes and write a short paragraph describing how you reached your conclusions.

(a) $f(x) = -x^3$ (b) $f(x) = -x^3 + x$

(c) $f(x) = x^3$ (d) $f(x) = x^3 - x$

WARM-UP
....

The following warm-up exercises involve skills that were covered in earlier sections. You will use these skills in the exercise set for this section.

In Exercises 1–6, factor the expression completely.

1. $12x^2 + 7x - 10$ **2.** $25x^3 - 60x^2 + 36x$ **3.** $12z^4 + 17z^3 + 5z^2$

4. $y^3 + 125$ **5.** $x^3 + 3x^2 - 4x - 12$ **6.** $x^3 + 2x^2 + 3x + 6$

In Exercises 7–10, find all real solutions to the equation.

7. $5x^2 + 8 = 0$ **8.** $x^2 - 6x + 4 = 0$

9. $4x^2 + 4x - 11 = 0$ **10.** $x^4 - 18x^2 + 81 = 0$

SECTION 3.2 · EXERCISES

1. Compare the graph of f with the graph of $y = x^3$.
  (a) $f(x) = (x - 2)^3$
  (b) $f(x) = x^3 - 2$
  (c) $f(x) = (x - 2)^3 - 2$
  (d) $f(x) = -\frac{1}{2}x^3$

2. Compare the graph of f with the graph of $y = x^4$.
  (a) $f(x) = (x + 3)^4$
  (b) $f(x) = x^4 - 3$
  (c) $f(x) = 4 - x^4$
  (d) $f(x) = \frac{1}{2}(x - 1)^4$

In Exercises 3–10, match the polynomial function with its graph and describe the viewing rectangle. [The graphs are labeled (a), (b), (c), (d), (e), (f), (g), and (h).]

3. $f(x) = -3x + 5$
4. $f(x) = x^2 - 2x$
5. $f(x) = -2x^2 - 8x - 9$
6. $f(x) = 3x^3 - 9x + 1$
7. $f(x) = -\frac{1}{3}x^3 + x - \frac{2}{3}$
8. $f(x) = -\frac{1}{4}x^4 + 2x^2$
9. $f(x) = 3x^4 + 4x^3$
10. $f(x) = x^5 - 5x^3 + 4x$

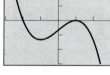

(a)

(b)

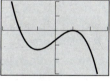

(c)

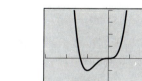

(d)

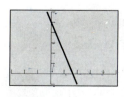

(e)

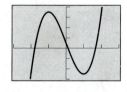

(f)

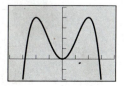

(g)

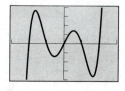

(h)

In Exercises 11–14, use a graphing utility to graph the functions f and g in the same viewing rectangle. Zoom out sufficiently far to show that the right-hand and left-hand behavior of f and g are identical.

11. $f(x) = x^3 - 9x + 1$ $g(x) = x^3$
12. $f(x) = -\frac{1}{3}(x^3 - 3x + 2)$ $g(x) = -\frac{1}{3}x^3$
13. $f(x) = -(x^4 - 4x^3 + 16x)$ $g(x) = -x^4$
14. $f(x) = 3x^4 - 6x^2$ $g(x) = 3x^4$

In Exercises 15–20, determine the right-hand and left-hand behavior of the graph of the polynomial function based on the Leading Coefficient Test. Use a graphing utility to verify your result.

15. $f(x) = 2x^2 - 3x + 1$
16. $g(x) = 5 - \frac{7}{2}x - 3x^2$
17. $h(t) = -\frac{2}{3}(t^2 - 5t + 3)$
18. $f(s) = -\frac{7}{8}(s^3 + 5s^2 - 7s + 1)$
19. $f(x) = 6 - 2x + 4x^2 - 5x^3$
20. $f(x) = \dfrac{3x^4 - 2x + 5}{4}$

In Exercises 21–36, find all the real zeros of the polynomial function. In each case, state whether you solved the problem algebraically or graphically and give reasons for your choice.

21. $f(x) = x^2 - 25$
22. $f(x) = 49 - x^2$
23. $h(t) = t^2 - 6t + 9$
24. $f(x) = x^2 + 10x + 25$
25. $f(x) = x^2 + x - 2$
26. $f(x) = \frac{1}{2}x^2 + \frac{5}{2}x - \frac{3}{2}$
27. $f(x) = 3x^2 - 12x + 3$
28. $g(x) = 5(x^2 - 2x - 1)$
29. $f(t) = t^3 - 4t^2 + 4t$
30. $f(x) = x^4 - x^3 - 20x^2$
31. $g(t) = \frac{1}{2}t^4 - \frac{1}{2}$
32. $f(x) = x^5 + x^3 - 6x$
33. $f(x) = 2x^4 - 2x^2 - 40$
34. $g(t) = t^5 - 6t^3 + 9t$
35. $f(x) = 5x^4 + 15x^2 + 10$
36. $f(x) = x^3 - 4x^2 - 25x + 100$

In Exercises 37–46, find a polynomial function that has the given zeros. (There are many correct answers.)

37. 0, 10

38. 0, −3

39. 2, −6

40. −4, 5

41. 0, −2, −3

42. 0, 2, 5

43. 4, −3, 3, 0

44. −2, −1, 0, 1, 2

45. $1 + \sqrt{3}, 1 - \sqrt{3}$

46. $2, 4 + \sqrt{5}, 4 - \sqrt{5}$

In Exercises 47–62, sketch the graph of the function. Describe a viewing rectangle that gives a good view of the basic characteristics of the graph.

47. $f(x) = -\frac{3}{2}$

48. $h(x) = \frac{1}{3}x - 3$

49. $f(t) = \frac{1}{4}(t^2 - 2t + 15)$

50. $g(x) = -x^2 + 10x - 16$

51. $f(x) = x^3 - 3x^2$

52. $f(x) = 1 - x^3$

53. $f(x) = x^3 - 4x$

54. $f(x) = \frac{1}{4}x^4 - 2x^2$

55. $g(t) = -\frac{1}{4}(t - 2)^2(t + 2)^2$

56. $f(x) = x^2(x - 4)$

57. $f(x) = \frac{1}{5}(x + 2)^2(x - 3)(2x - 9)$

58. $f(x) = \frac{1}{5}(x + 2)^2(3x - 5)^2$

59. $h(x) = \frac{1}{3}x^3(x - 4)^2$

60. $g(x) = \frac{1}{10}(x + 1)^2(x - 3)^3$

61. $f(x) = 1 - x^6$

62. $g(x) = 1 - (x + 1)^6$

In Exercises 63–66, use the Intermediate Value Theorem and a graphing utility to find the intervals of length 1 in which the polynomial function is guaranteed to have a zero. (See Example 8.)

63. $f(x) = x^3 - 3x^2 + 3$

64. $f(x) = 0.11x^3 - 2.07x^2 + 9.81x - 6.88$

65. $g(x) = 3x^4 + 4x^3 - 3$

66. $h(x) = x^4 - 10x^2 + 2$

67. *Volume of a Box* An open box is made from a 12-inch square piece of material by cutting equal squares from all corners and turning up the sides (see figure).
 (a) Verify that the volume of the box is $V(x) = 4x(6 - x)^2$.
 (b) Determine the domain of the function V.
 (c) Sketch a graph of the function and use the graph to estimate the value of x for which $V(x)$ is maximum.

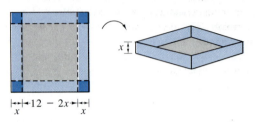

$\overset{x}{\underset{x}{\longmapsto}} \overset{}{12 - 2x} \overset{x}{\longmapsto}$

FIGURE FOR 67

68. *Volume of a Box* An open box with locking tabs is made from a 12-inch square piece of material. This is done by cutting equal squares from all corners and folding along the dashed lines shown in the figure.
 (a) Verify that the volume of the box is $V(x) = 8x(3 - x)(6 - x)$.
 (b) Determine the domain of the function V.
 (c) Sketch the graph of the function and use the graph to estimate the value of x for which $V(x)$ is maximum.

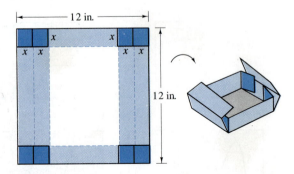

FIGURE FOR 68

69. *Advertising Expenses* The total revenue for a soft-drink company is related to its advertising expense by the function

$$R = \frac{1}{50,000}(-x^3 + 600x^2), \qquad 0 \le x \le 400,$$

where R is the total revenue in millions of dollars and x is the amount spent on advertising (in 10,000s of dollars). Use a graphing utility to obtain a graph of this function. Estimate the point on the graph where the function is increasing most rapidly. This point is called the **point of diminishing returns** because any expense above this amount will yield less return per dollar invested in advertising. (*Hint:* Use a viewing rectangle in which $-200 \le x \le 600$ and $0 \le y \le 650$.)

70. *Tree Growth* The growth of a red oak tree is approximated by the function

$$G = -0.003t^3 + 0.137t^2 + 0.458t - 0.839,$$
$$2 \le t \le 34,$$

where G is the height of the tree in feet and t is its age in years. Use a graphing utility to obtain a graph of this function. Estimate the age of the tree when it is growing most rapidly. This point is called the **point of diminishing returns** because the increase in yield will be less with each additional year. (*Hint:* Use a viewing rectangle in which $-10 \le x \le 45$ and $-5 \le y \le 60$.)

3.3   REAL ZEROS OF POLYNOMIAL FUNCTIONS

Long Division and Synthetic Division / The Remainder and Factor Theorems / Descartes's Rule of Signs / The Rational Zero Test / Bounds for Real Zeros of Polynomial Functions / Bisection Method

Long Division and Synthetic Division

Consider the graph of

$$f(x) = 6x^3 - 19x^2 + 16x - 4.$$

Notice that a zero of f occurs at $x = 2$, as shown in Figure 3.29. Because $x = 2$ is a zero of the polynomial function f, you know that $(x - 2)$ is a factor of $f(x)$. This means that there exists a second-degree polynomial $q(x)$ such that

$$f(x) = (x - 2) \cdot q(x).$$

To find $q(x)$, you can use **long division of polynomials.**

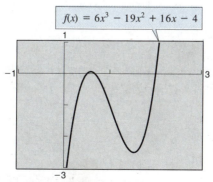

$f(x) = 6x^3 - 19x^2 + 16x - 4$

FIGURE 3.29

EXAMPLE 1   Long Division of Polynomials

Divide $f(x) = 6x^3 - 19x^2 + 16x - 4$ by $x - 2$, and use the result to factor $f(x)$ completely.

SOLUTION

Partial quotients

$$
\begin{array}{r}
6x^2 - \quad 7x + \quad 2 \\
x - 2 \overline{)6x^3 - \quad 19x^2 + 16x - 4} \\
\underline{6x^3 - \quad 12x^2} \\
-7x^2 + 16x \\
\underline{-7x^2 + 14x} \\
2x - 4 \\
\underline{2x - 4} \\
0
\end{array}
$$

Multiply: $6x^2(x - 2)$
Subtract
Multiply: $-7x(x - 2)$
Subtract
Multiply: $2(x - 2)$
Subtract

You can see that

$$6x^3 - 19x^2 + 16x - 4 = (x - 2)(6x^2 - 7x + 2)$$
$$= (x - 2)(2x - 1)(3x - 2).$$

Note that this factorization agrees with the graph of f (Figure 3.29) in that the three x-intercepts occur at $x = 2$, $x = \frac{1}{2}$, and $x = \frac{2}{3}$.

In Example 1, $x - 2$ is a factor of the polynomial $6x^3 - 19x^2 + 16x - 4$, and the long division process produces a remainder of zero. Often, long division will produce a nonzero remainder. For instance, if you divide $x^2 + 3x + 5$ by $x + 1$, you obtain the following.

$$
\begin{array}{r}
x \ + 2 \quad \longleftarrow \text{Quotient} \\
\text{Divisor} \longrightarrow x + 1 \overline{)x^2 + 3x + 5} \quad \longleftarrow \text{Dividend} \\
\underline{x^2 + \ x} \\
2x + 5 \\
\underline{2x + 2} \\
3 \quad \longleftarrow \text{Remainder}
\end{array}
$$

You can write this result as

$$
\overbrace{\frac{x^2 + 3x + 5}{\underbrace{x + 1}_{\text{Divisor}}}}^{\text{Dividend}} = x + 2 + \overbrace{\frac{3}{\underbrace{x + 1}_{\text{Divisor}}}}^{\text{Quotient} \ \text{Remainder}}
$$

or

$$x^2 + 3x + 5 = (x + 2)(x + 1) + 3.$$

This example illustrates the following **Division Algorithm.**

THE DIVISION ALGORITHM

If $f(x)$ and $d(x)$ are polynomials such that $d(x) \neq 0$, and the degree of $d(x)$ is less than or equal to the degree of $f(x)$, then there exist unique polynomials $q(x)$ and $r(x)$ such that

$$f(x) = d(x)q(x) + r(x)$$

Dividend   Divisor   Quotient   Remainder

where $r(x) = 0$ or the degree of $r(x)$ is less than the degree of $d(x)$. If the remainder $r(x)$ is zero, then $d(x)$ **divides evenly** into $f(x)$.

REMARK   The Division Algorithm can be written as

$$\frac{f(x)}{d(x)} = q(x) + \frac{r(x)}{d(x)}.$$

In the Division Algorithm the rational expression $f(x)/d(x)$ is **improper** because the degree of $f(x)$ is greater than or equal to the degree of $d(x)$. On the other hand, the rational expression $r(x)/d(x)$ is **proper** because the degree of $r(x)$ is less than the degree of $d(x)$.

    Synthetic division is a shortcut for long division by polynomials of the form $x - k$.

SYNTHETIC DIVISION (OF A CUBIC POLYNOMIAL)

To divide $ax^3 + bx^2 + cx + d$ by $x - k$, use the following pattern.

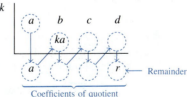

Coefficients of quotient

Vertical pattern: Add terms.

Diagonal pattern: Multiply by k.

REMARK   Synthetic division works *only* for divisors of the form $x - k$. You cannot use synthetic division to divide a polynomial by a quadratic such as $x^2 - 3$.

EXAMPLE 2   Using Synthetic Division

Use synthetic division to divide $x^4 - 10x^2 - 2x + 4$ by $x + 3$.

SOLUTION

You can set up the array as follows. (Note that a zero is included for each missing term in the dividend.)

Divisor: Dividend: $x^4 - 10x^2 + 4$
$x - (-3)$

$$
\begin{array}{r|rrrrr}
-3 & 1 & 0 & -10 & -2 & 4 \\
   &   & -3 & 9 & 3 & -3 \\
\hline
   & 1 & -3 & -1 & 1 & 1 \\
\end{array}
$$

← Remainder: 1

Quotient: $x^3 - 3x^2 - x + 1$

Thus, you have

$$\frac{x^4 - 10x^2 - 2x + 4}{x + 3} = x^3 - 3x^2 - x + 1 + \frac{1}{x + 3}.$$

The Remainder and Factor Theorems

Notice that in Example 2, if $f(x) = x^4 - 10x^2 - 2x + 4$, then

$$f(-3) = (-3)^4 - 10(-3)^2 - 2(-3) + 4 = 1,$$

which is the remainder $r = 1$. This observation leads to the following **Remainder Theorem.**

THE REMAINDER THEOREM

If a polynomial $f(x)$ is divided by $x - k$, then the remainder is $r = f(k)$.

EXAMPLE 3   Evaluating a Polynomial by the Remainder Theorem

Use the Remainder Theorem to evaluate the function at $x = -2$.

$$f(x) = 3x^3 + 8x^2 + 5x - 7$$

SOLUTION

Using synthetic division, you obtain the following.

$$
\begin{array}{r|rrrr}
-2 & 3 & 8 & 5 & -7 \\
   &   & -6 & -4 & -2 \\
\hline
   & 3 & 2 & 1 & -9 \\
\end{array}
$$

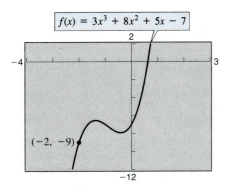

$f(x) = 3x^3 + 8x^2 + 5x - 7$

FIGURE 3.30

Because the remainder is $r = -9$, you can conclude that

$$f(-2) = -9.$$

This means that $(-2, -9)$ is a point on the graph of f, as shown in Figure 3.30. Check this by substituting $x = -2$ in the original function.

The Remainder Theorem can be used to prove the following important result.

FACTOR THEOREM

A polynomial $f(x)$ has a factor $(x - k)$ if and only if $f(k) = 0$.

EXAMPLE 4 Using Synthetic Division to Factor a Polynomial

Show that $(x - 2)$ and $(x + 3)$ are factors of the polynomial

$$f(x) = 2x^4 + 7x^3 - 4x^2 - 27x - 18.$$

Then find the remaining factors of $f(x)$.

SOLUTION

Using synthetic division by 2 and -3 successively produces the following.

$$
\begin{array}{r|rrrrr}
2 & 2 & 7 & -4 & -27 & -18 \\
  &   & 4 & 22 & 36 & 18 \\
\hline
  & 2 & 11 & 18 & 9 & 0
\end{array}
$$

$$
\begin{array}{r|rrrr}
-3 & 2 & 11 & 18 & 9 \\
   &   & -6 & -15 & -9 \\
\hline
   & 2 & 5 & 3 & 0
\end{array}
$$

Because the resulting quadratic $2x^2 + 5x + 3$ factors as

$$2x^2 + 5x + 3 = (2x + 3)(x + 1),$$

the complete factorization of $f(x)$ is

$$f(x) = (x - 2)(x + 3)(2x + 3)(x + 1).$$

This factorization implies that f has four real zeros: 2, -3, $-\frac{3}{2}$, and -1. This can be verified by a graph, as shown in Figure 3.31.

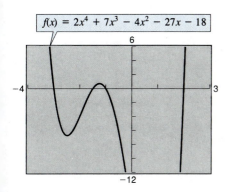

$f(x) = 2x^4 + 7x^3 - 4x^2 - 27x - 18$

FIGURE 3.31

In summary, the remainder r, obtained in the synthetic division of $f(x)$ by $x - k$, provides the following information.

1. The remainder r gives the value of f at $x = k$. That is, $r = f(k)$.
2. If $r = 0$, then $(x - k)$ is a factor of $f(x)$.
3. If $r = 0$, then $(k, 0)$ is an x-intercept of the graph of f.

Descartes's Rule of Signs

In Section 3.2, you learned that an nth degree polynomial function can have *at most n* real zeros. Of course, many nth degree polynomials do not have that many real zeros. For instance, $f(x) = x^2 + 1$ has no real zeros, and $f(x) = x^3 + 1$ has only one real zero. **Descartes's Rule of Signs** sheds more light on the number of real zeros that a polynomial can have.

DESCARTES'S RULE OF SIGNS

Let $f(x) = a_nx^n + a_{n-1}x^{n-1} + \cdots + a_2x^2 + a_1x + a_0$ be a polynomial with real coefficients and $a_0 \neq 0$.

1. The number of *positive real zeros* of f is either equal to the number of variations in sign of $f(x)$ or is less than that number by an even integer.
2. The number of *negative real zeros* of f is either equal to the number of variations in sign of $f(-x)$ or is less than that number by an even integer.

REMARK   When there is only one variation in sign, Descartes's Rule of Signs guarantees the existence of exactly one positive (or negative) real zero.

A **variation in sign** means that two consecutive coefficients have opposite signs.

EXAMPLE 5   Using Descartes's Rule of Signs

Apply Descartes's Rule of Signs to the polynomial function

$$f(x) = 3x^3 - 5x^2 + 6x - 4.$$

SOLUTION

Because $f(x)$ has three variations in sign, it follows that f can have either three or one positive real zeros. Moreover, because

$$f(-x) = 3(-x)^3 - 5(-x)^2 + 6(-x) - 4$$
$$= -3x^3 - 5x^2 - 6x - 4$$

has no variations in sign, it follows that f has no negative real zeros. From the graph of f in Figure 3.32, you can see that f has only one positive real zero.

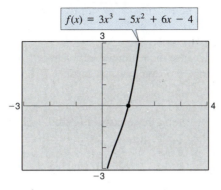

$$f(x) = 3x^3 - 5x^2 + 6x - 4$$

FIGURE 3.32

The Rational Zero Test

The **Rational Zero Test** relates the possible rational zeros of a polynomial (having integer coefficients) to the leading coefficient and to the constant term of the polynomial.

THE RATIONAL ZERO TEST

If the polynomial $f(x) = a_n x^n + a_{n-1} x^{n-1} + \cdots + a_2 x^2 + a_1 x + a_0$ has *integer* coefficients, then every rational zero of f has the form

$$\text{Rational zero} = \frac{p}{q},$$

where p and q have no common factors other than 1, p is a factor of the constant term a_0, and q is a factor of the leading coefficient a_n.

To use the Rational Zero Test, first list all rational numbers whose numerators are factors of the constant term and whose denominators are factors of the leading coefficient.

$$\text{Possible rational zeros} = \frac{\text{factors of constant term}}{\text{factors of leading coefficient}}$$

Having formed this list of *possible rational zeros,* use a trial and error method to determine which, if any, are actual zeros of the polynomial. Note that when the leading coefficient is 1, the possible rational zeros are simply the factors of the constant term.

EXAMPLE 6 **Rational Zero Test with Leading Coefficient of 1**

Find the rational zeros of $f(x) = x^3 + x + 1$.

SOLUTION

Because the leading coefficient is 1, the possible rational zeros are simply the factors of the constant term.

Possible rational zeros: ± 1

By testing these possible zeros, you can see that neither works.

$$f(1) = (1)^3 + 1 + 1 = 3$$
$$f(-1) = (-1)^3 + (-1) + 1 = -1$$

Thus, the polynomial has *no* rational zeros. Note from the graph of f in Figure 3.33 that f does have one real zero (between -1 and 0). However, by the Rational Zero Test, you know that this real zero is *not* a rational number.

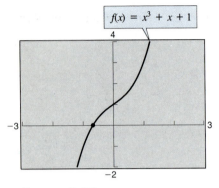

$f(x) = x^3 + x + 1$

FIGURE 3.33

If the leading coefficient of a polynomial is not 1, the list of possible rational zeros can increase dramatically. In such cases the search can be shortened in several ways. (1) A programmable calculator can be used to speed up the calculations. (2) A graphing utility can give a good estimate of the location of the zeros. (3) Synthetic division can be used to test the possible rational zeros.

Finding the first zero is often the most difficult part. After that, the search is simplified by working with the lower-degree polynomial obtained in synthetic division.

A graphing utility can help you determine which possible rational zeros to test, as demonstrated in Example 7.

EXAMPLE 7 **Using the Rational Zero Test**

Find all the real zeros of $f(x) = 10x^3 - 15x^2 - 16x + 12$.

SOLUTION

Because the leading coefficient is 10 and the constant term is 12, there is a long list of possible rational zeros.

Possible rational zeros

$$\frac{\text{Factors of 12}}{\text{Factors of 10}} = \frac{\pm 1, \pm 2, \pm 3, \pm 4, \pm 6, \pm 12}{\pm 1, \pm 2, \pm 5, \pm 10}$$

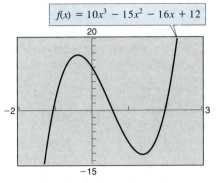

$f(x) = 10x^3 - 15x^2 - 16x + 12$

FIGURE 3.34

With so many possibilities (32, in fact), it is worth your time to use a graphing utility to focus on just a few. From Figure 3.34, it looks like three reasonable choices would be $x = -\frac{6}{5}$, $x = \frac{1}{2}$, and $x = 2$. Testing these by synthetic division shows that only $x = 2$ works. Thus, you have

$$f(x) = (x - 2)(10x^2 + 5x - 6).$$

Using the Quadratic Formula, you can find that the two additional zeros are irrational numbers.

$$x = \frac{-5 + \sqrt{265}}{20} \approx 0.5639$$

and

$$x = \frac{-5 - \sqrt{265}}{20} \approx -1.0639$$

Bounds for Real Zeros of Polynomial Functions

The final test for zeros of a polynomial function is related to the sign pattern in the last row of the synthetic division tableau. This test can give you an upper or a lower bound of the real zeros of f. A real number b is an **upper bound** for the real zeros of f if no zero is greater than b. Similarly, b is a **lower bound** if no real zero of f is less than b.

LOWER AND UPPER BOUND RULE

Let $f(x)$ be a polynomial with real coefficients and a positive leading coefficient. Suppose $f(x)$ is divided by $x - c$, using synthetic division.

1. If $c > 0$ and each number in the last row is either positive or zero, then c is an *upper bound* for the real zeros of f.
2. If $c < 0$ and the numbers in the last row are alternatively positive and negative (zero entries count as positive or negative), then c is a *lower bound* for the real zeros of f.

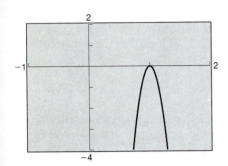

FIGURE 3.35

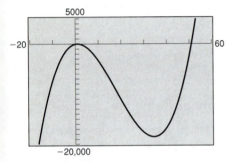

FIGURE 3.36

EXAMPLE 8   Finding the Real Zeros of a Polynomial

Find the real zeros of $f(x) = x^3 - 53x^2 + 103x - 51$.

SOLUTION

A preliminary graph (see Figure 3.35) indicates that there is a zero between 0 and 2. You can apply the lower and upper bound rule with $c = -1$ and $c = 2$, as follows.

$c = -1$:

$$
\begin{array}{r|rrrr}
-1 & 1 & -53 & 103 & -51 \\
   &   & -1 & 54 & -157 \\
\hline
   & 1 & -54 & 157 & -208
\end{array}
$$

Signs alternate

$c = 2$:

$$
\begin{array}{r|rrrr}
2 & 1 & -53 & 103 & -51 \\
  &   & 2 & -102 & 2 \\
\hline
  & 1 & -51 & 1 & -49
\end{array}
$$

Not all positive or zero

Because $c = -1 < 0$ and the numbers in the last row alternate in sign, -1 is a lower bound for the real zeros of f. That is, you can be sure that there are no zeros less than -1. For $c = 2$, because the numbers in the last row are not all positive, you cannot be sure that there are no zeros to the right of 2. In fact, if you zoom-out a couple of times, you can observe that there is a zero at $x = 51$ (see Figure 3.36). This example illustrates one of the limitations of a graphing utility: you may miss important features of a graph the first time you graph it. From Figure 3.35, you might have concluded that $x = 1$ was the only zero of f, but a more complete graph of this cubic brings to light its essential feature, as indicated in Figure 3.36.

Example 9 uses all three tests presented in this section to search for the real zeros of a polynomial function. In addition, the example shows how to handle rational coefficients by factoring out the reciprocal of their least common denominator.

D I S C O V E R Y

Graph $y = x^3 + 4.9x^2 - 126x + 382.5$ on the standard viewing rectangle. From this graph, what do you think the real zeros of y are? Now use the zoom feature to show that -15.5 and 5.1 are zeros of y. Discuss the "hidden behavior" of this graph.

EXAMPLE 9   A Polynomial Function with Rational Coefficients

Find the real zeros of $f(x) = x^3 - \dfrac{2}{3}x^2 + \dfrac{1}{2}x - \dfrac{1}{3}$.

SOLUTION

To find the rational zeros, rewrite $f(x)$ by factoring out the reciprocal of the least common denominator of the coefficients.

$$f(x) = \frac{6}{6}x^3 - \frac{4}{6}x^2 + \frac{3}{6}x - \frac{2}{6} = \frac{1}{6}(6x^3 - 4x^2 + 3x - 2)$$

Now, for the purpose of finding the zeros of f, you can drop the factor $\frac{1}{6}$. This is legitimate because the zeros of f are the same as the zeros of

$$g(x) = 6x^3 - 4x^2 + 3x - 2,$$

which has the following possible rational zeros.

$$\frac{\text{Factors of 2}}{\text{Factors of 6}} = \frac{\pm 1, \pm 2}{\pm 1, \pm 2, \pm 3, \pm 6,} = \pm 1, \pm \frac{1}{2}, \pm \frac{1}{3}, \pm \frac{1}{6}, \pm \frac{2}{3}, \pm 2$$

Because $g(x)$ has three variations in sign and $g(-x)$ has none, you can conclude by Descartes's Rule of Signs that there are three positive real zeros or one positive real zero, and no negative zeros. Trying $x = 1$, you obtain

$$
\begin{array}{r|rrr}
1 & 6 & -4 & 3 & -2 \\
  &   & 6 & 2 & 5 \\
\hline
  & 6 & 2 & 5 & 3 \leftarrow \text{All positive entries}
\end{array}
$$

Thus, $x = 1$ is not a zero, but because the last row has all positive entries, you know that $x = 1$ is an upper bound for the real zeros. Thus, you can restrict your search to zeros between 0 and 1. Choosing $x = \frac{2}{3}$, you obtain

$$
\begin{array}{r|rrr}
\frac{2}{3} & 6 & -4 & 3 & -2 \\
  &   & 4 & 0 & 2 \\
\hline
  & 6 & 0 & 3 & 0
\end{array}
$$

Thus, $f(x)$ factors as

$$f(x) = \frac{1}{6}\left(x - \frac{2}{3}\right)(6x^2 + 3) = \frac{1}{6}(3x - 2)(2x^2 + 1).$$

Because $2x^2 + 1$ has no real zeros, you can conclude that $x = \frac{2}{3}$ is the only real zero of f. A graph of f is shown in Figure 3.37. ■■■■■■■

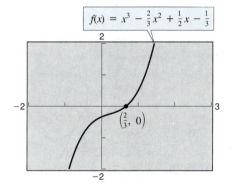

$$f(x) = x^3 - \frac{2}{3}x^2 + \frac{1}{2}x - \frac{1}{3}$$

$\left(\frac{2}{3}, 0\right)$

FIGURE 3.37

Here are two additional hints that should be helpful.

1. If the terms of $f(x)$ have a common monomial factor, it should be factored out before applying the tests in this section. For instance, by writing

$$
\begin{aligned}
f(x) &= x^4 - 5x^3 + 3x^2 + x \\
     &= x(x^3 - 5x^2 + 3x + 1),
\end{aligned}
$$

you can see that $x = 0$ is a zero of f and the remaining zeros can be obtained by analyzing the cubic factor.

2. If you are able to find all but two zeros of $f(x)$, you are home free because you can always use the Quadratic Formula on the remaining quadratic factor. For instance, if you succeed in writing

$$
\begin{aligned}
f(x) &= x^4 - 5x^3 + 3x^2 + x \\
     &= x(x - 1)(x^2 - 4x - 1),
\end{aligned}
$$

you can apply the Quadratic Formula to $x^2 - 4x - 1$ to find the two remaining zeros.

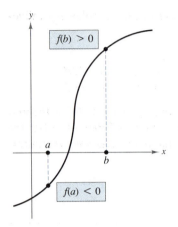

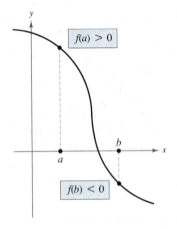

FIGURE 3.38

Bisection Method

One of the simplest methods for approximating the zeros of a polynomial is the **Bisection Method.** To apply this method, you must find two x-values—one at which the function is positive and one at which it is negative. Because the function is a polynomial, you know that it has at least one zero between the two x-values, as shown in Figure 3.38.

In Figure 3.38, you can see that a zero must occur somewhere in the interval (a, b). To apply the Bisection Method, cut this interval in half and consider the two intervals

$$\left(a, \frac{a + b}{2}\right) \quad \text{and} \quad \left(\frac{a + b}{2}, b\right).$$

Depending on the value of $f(x)$ at the midpoint

$$\frac{a + b}{2},$$

apply the Bisection Method again to one of these intervals. This process is illustrated in Example 10.

EXAMPLE 10   Using the Bisection Method

Use the Bisection Method to approximate the real zero of

$$f(x) = x^3 - x^2 + 1$$

to within 0.001 unit.

SOLUTION

Begin by graphing f, as shown in Figure 3.39. Note that

$$f(-0.8) = -0.152 < 0 \quad \text{and} \quad f(-0.7) = 0.167 > 0,$$

which implies that f has a zero between -0.8 and -0.7. Using the midpoint of this interval, you can approximate the zero to be

$$c = \frac{-0.8 + (-0.7)}{2} = -0.75.$$

The maximum error of this approximation is one-half the length of the interval. That is,

$$\text{Maximum error} \leq \frac{-0.7 - (-0.8)}{2} = 0.05.$$

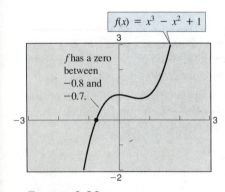

FIGURE 3.39

Next, calculate the value of $f(-0.75)$ and make one of the following deductions.

1. If $f(-0.75) = 0$, then -0.75 is a zero of f.
2. If $f(-0.75) > 0$, then a zero occurs between -0.8 and -0.75.
3. If $f(-0.75) < 0$, then a zero occurs between -0.75 and -0.7.

Because $f(-0.75) = 0.015626$ is positive, choose $(-0.8, -0.75)$ as the new interval. The results of several iterations are shown in Table 3.1. After the seventh iteration, the maximum error is less that 0.001, and you can approximate the zero of f to be

$$c = -0.75546875 \approx -0.755. \qquad \textit{To three-place accuracy}$$

TABLE 3.1

	a	c	b	$f(a)$	$f(c)$	$f(b)$	Maximum error
1	-0.8	-0.75	-0.7	-0.1520	0.0156	0.1670	0.05
2	-0.8	-0.775	-0.75	-0.1520	-0.0661	0.0156	0.025
3	-0.775	-0.7625	-0.75	-0.0661	-0.0247	0.0156	0.0125
4	-0.7625	-0.7563	-0.75	-0.0247	-0.0046	0.0156	0.0063
5	-0.7563	-0.7531	-0.75	-0.0046	0.0057	0.0156	0.0031
6	-0.7563	-0.7547	-0.7531	-0.0046	0.0006	0.0156	0.0016
7	-0.7563	-0.7555	-0.7547	-0.0046	-0.0020	0.0006	0.0008

By continuing the process in Table 3.1, you could approximate the zero to *any* desired accuracy, and we say that the sequence of successively better approximations *converges* to the zero of the function. The convergence of the Bisection Method is relatively slow and several iterations are usually necessary to obtain a very fine accuracy.

DISCUSSION PROBLEM

COMPARING REAL ZEROS AND RATIONAL ZEROS

Compare the *real* zeros of a polynomial function with the *rational* zeros of a polynomial function. Then answer the following questions.

(a) Is it possible for a polynomial function to have no rational zeros but to have real zeros? If so, give an example.

(b) If a polynomial function has three real zeros, and only one of them is a rational number, then must the other two zeros be irrational numbers?

(c) Consider a cubic polynomial function, $f(x) = ax^3 + bx^2 + cx + d$, where $a \neq 0$. Is it possible that f has no real zeros? If so, give an example. Is it possible that f has no rational zeros? If so, give an example.

WARM-UP

The following warm-up exercises involve skills that were covered in earlier sections. You will use these skills in the exercise set for this section.

In Exercises 1–4, write the expression in standard polynomial form.

1. $(x - 1)(x^2 + 2) + 5$ **2.** $(x^2 - 3)(2x + 4) + 8$

3. $(x^2 + 1)(x^2 - 2x + 3) - 10$ **4.** $(x + 6)(2x^3 - 3x) - 5$

In Exercises 5 and 6, factor the polynomial.

5. $x^2 - 4x + 3$ **6.** $4x^3 - 10x^2 + 6x$

In Exercises 7–10, find a polynomial function that has the given zeros.

7. 0, 3, 4   **8.** -6, 1

9. $-3, 1 + \sqrt{2}, 1 - \sqrt{2}$ **10.** $1, -2, 2 + \sqrt{3}, 2 - \sqrt{3}$

SECTION 3.3 · EXERCISES

In Exercises 1–14, divide by long division.

1. $\dfrac{2x^2 + 10x + 12}{x + 3}$

2. $\dfrac{5x^2 - 17x - 12}{x - 4}$

3. $\dfrac{4x^3 - 7x^2 - 11x + 5}{4x + 5}$

4. $\dfrac{6x^3 - 16x^2 + 17x - 6}{3x - 2}$

5. $\dfrac{x^4 + 5x^3 + 6x^2 - x - 2}{x + 2}$

6. $\dfrac{x^3 + 4x^2 - 3x - 12}{x^2 - 3}$

7. $\dfrac{7x + 3}{x + 2}$

8. $\dfrac{8x - 5}{2x + 1}$

9. $\dfrac{6x^3 + 10x^2 + x + 8}{2x^2 + 1}$

10. $\dfrac{x^3 - 9}{x^2 + 1}$

11. $\dfrac{x^4 + 3x^2 + 1}{x^2 - 2x + 3}$

12. $\dfrac{x^5 + 7}{x^3 - 1}$

13. $\dfrac{2x^3 - 4x^2 - 15x + 5}{(x - 1)^2}$

14. $\dfrac{x^4}{(x - 1)^3}$

In Exercises 15–32, divide by synthetic division.

15. $\dfrac{3x^3 - 17x^2 + 15x - 25}{x - 5}$

16. $\dfrac{5x^3 + 18x^2 + 7x - 6}{x + 3}$

17. $\dfrac{4x^3 - 9x + 8x^2 - 18}{x + 2}$

18. $\dfrac{9x^3 - 16x - 18x^2 + 32}{x - 2}$

19. $\dfrac{-x^3 + 75x - 250}{x + 10}$

20. $\dfrac{3x^3 - 16x^2 - 72}{x - 6}$

21. $\dfrac{5x^3 - 6x^2 + 8}{x - 4}$

22. $\dfrac{5x^3 + 6x + 8}{x + 2}$

23. $\dfrac{10x^4 - 50x^3 - 800}{x - 6}$

24. $\dfrac{x^5 - 13x^4 - 120x + 80}{x + 3}$

25. $\dfrac{x^3 + 512}{x + 8}$

26. $\dfrac{5x^3}{x + 3}$

27. $\dfrac{-3x^4}{x - 2}$

28. $\dfrac{-3x^4}{x + 2}$

29. $\dfrac{5 - 3x + 2x^2 - x^3}{x + 1}$ **30.** $\dfrac{180x - x^4}{x - 6}$

31. $\dfrac{4x^3 + 16x^2 - 23x - 15}{x + \frac{1}{2}}$ **32.** $\dfrac{3x^3 - 4x^2 + 5}{x - \frac{3}{2}}$

In Exercises 33–38, use synthetic division to show that x is a zero of the third-degree polynomial function, and use the result to factor the polynomial completely. List all the real zeros of the function, and use a graphing utility to verify your result.

Polynomial Function	Value of x
33. $f(x) = x^3 - 7x + 6$	$x = 2$
34. $g(x) = x^3 - 28x - 48$	$x = -4$
35. $g(x) = 2x^3 - 15x^2 + 27x - 10$	$x = \frac{1}{2}$
36. $h(x) = 48x^3 - 80x^2 + 41x - 6$	$x = \frac{2}{3}$
37. $h(x) = x^3 + 2x^2 - 3x - 6$	$x = \sqrt{3}$
38. $f(x) = x^3 + 2x^2 - 2x - 4$	$x = \sqrt{2}$

In Exercises 39–42, express the function in the form $f(x) = (x - k)q(x) + r$ for the given value of k, and demonstrate that $f(k) = r$.

Function	Value of k
39. $f(x) = x^3 - x^2 - 14x + 11$	$k = 4$
40. $f(x) = \frac{1}{3}(15x^4 + 10x^3 - 6x^2 + 17x + 14)$	$k = -\frac{2}{3}$
41. $f(x) = x^3 + 3x^2 - 2x - 14$	$k = \sqrt{2}$
42. $f(x) = 4x^3 - 6x^2 - 12x - 4$	$k = 1 - \sqrt{3}$

In Exercises 43–46, use synthetic division and the Remainder Theorem to find the function values.

43. $f(x) = 4x^3 - 13x + 10$
  (a) $f(1)$ (b) $f(-2)$
  (c) $f(\frac{1}{2})$ (d) $f(8)$
44. $g(x) = x^6 - 4x^4 + 3x^2 + 2$
  (a) $g(2)$ (b) $g(-4)$
  (c) $g(3)$ (d) $g(-1)$
45. $h(x) = 3x^3 + 5x^2 - 10x + 1$
  (a) $h(3)$ (b) $h(\frac{1}{3})$
  (c) $h(-2)$ (d) $h(-5)$
46. $f(x) = 0.4x^4 - 1.6x^3 + 0.7x^2 - 2$
  (a) $f(1)$ (b) $f(-2)$
  (c) $f(5)$ (d) $f(-10)$

In Exercises 47–56, use Descartes's Rule of Signs to determine the possible number of positive and negative zeros of the function.

47. $f(x) = x^3 + 3$
48. $g(x) = x^3 + 3x^2$
49. $h(x) = 3x^4 + 2x^2 + 1$
50. $h(x) = 2x^4 - 3x + 2$
51. $g(x) = 2x^3 - 3x^2 - 3$
52. $f(x) = 4x^3 - 3x^2 + 2x - 1$
53. $f(x) = -5x^3 + x^2 - x + 5$
54. $g(x) = 5x^5 + 10x$
55. $h(x) = 4x^2 - 8x + 3$
56. $f(x) = 3x^3 + 2x^2 + x + 3$

In Exercises 57–60, use the Rational Zero Test to list all possible rational zeros of f. Verify that the zeros of f on the graph are contained in the list.

57. $f(x) = x^3 + x^2 - 4x - 4$

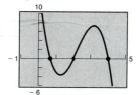

58. $f(x) = -3x^3 + 20x^2 - 36x + 16$

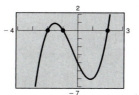

59. $f(x) = -4x^3 + 15x^2 - 8x - 3$

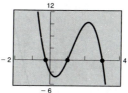

60. $f(x) = 4x^3 - 12x^2 - x + 15$

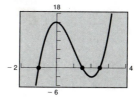

In Exercises 61–66, (a) list the possible rational zeros of f, (b) use a graphing utility to obtain the graph of f so that some of the possible zeros in part (a) can be disregarded, and then (c) determine all real zeros of f.

61. $f(x) = -2x^4 + 13x^3 - 21x^2 + 2x + 8$

62. $f(x) = 4x^4 - 17x^2 + 4$

63. $f(x) = 32x^3 - 52x^2 + 17x + 3$

64. $f(x) = 6x^3 - x^2 - 13x + 8$

65. $f(x) = 4x^3 + 7x^2 - 11x - 18$

66. $f(x) = 2x^3 + 5x^2 - 21x - 10$

In Exercises 67–70, use synthetic division to determine if each given x-value is an upper bound of the zeros of f, a lower bound of the zeros of f, or neither.

67. $f(x) = x^4 - 4x^3 + 15$
   (a) $x = 4$
   (b) $x = -1$
   (c) $x = 3$

68. $f(x) = 2x^3 - 3x^2 - 12x + 8$
   (a) $x = 2$
   (b) $x = 4$
   (c) $x = -1$

69. $f(x) = x^4 - 4x^3 + 16x - 16$
   (a) $x = -1$
   (b) $x = -3$
   (c) $x = 5$

70. $f(x) = 2x^4 - 8x + 3$
   (a) $x = 1$
   (b) $x = 3$
   (c) $x = -4$

In Exercises 71–86, find the real zeros of the function.

71. $f(x) = x^3 - 6x^2 + 11x - 6$

72. $f(x) = x^3 - 7x - 6$

73. $g(x) = x^3 - 4x^2 - x + 4$

74. $h(x) = x^3 - 9x^2 + 20x - 12$

75. $h(t) = t^3 + 12t^2 + 21t + 10$

76. $f(x) = x^3 + 6x^2 + 12x + 8$

77. $f(x) = x^3 - 4x^2 + 5x - 2$

78. $p(x) = x^3 - 9x^2 + 27x - 27$

79. $C(x) = 2x^3 + 3x^2 - 1$

80. $f(x) = 3x^3 - 19x^2 + 33x - 9$

81. $f(x) = 4x^3 - 3x - 1$

82. $f(z) = 12z^3 - 4z^2 - 27z + 9$

83. $f(y) = 4y^3 + 3y^2 + 8y + 6$

84. $g(x) = 3x^3 - 2x^2 + 15x - 10$

85. $f(x) = x^4 - 3x^2 + 2$

86. $P(t) = t^4 - 7t^2 + 12$

In Exercises 87–90, find the rational zeros of the polynomial function.

87. $P(x) = x^4 - \frac{25}{4}x^2 + 9$

88. $f(x) = x^3 - \frac{3}{2}x^2 - \frac{23}{2}x + 6$

89. $f(x) = x^3 - \frac{1}{4}x^2 - x + \frac{1}{4}$

90. $f(z) = z^3 + \frac{11}{6}z^2 - \frac{1}{2}z - \frac{1}{3}$

In Exercises 91–94, match the cubic equation with the numbers of rational and irrational zeros (a), (b), (c), or (d).

91. $f(x) = x^3 - 1$ **92.** $f(x) = x^3 - 2$

93. $f(x) = x^3 - x$ **94.** $f(x) = x^3 - 2x$

   (a) Rational zeros: 0   (b) Rational zeros: 3
       Irrational zeros: 1   Irrational zeros: 0
   (c) Rational zeros: 1   (d) Rational zeros: 1
       Irrational zeros: 2   Irrational zeros: 0

In Exercises 95–98, (a) list the possible rational zeros of f and show that none are zeros, (b) use a graphing utility to verify that there are real zeros of f, and (c) move the cursor along the graph to approximate the irrational zero(s) of the function accurate to two decimal places.

95. $f(x) = x^3 + x - 1$ **96.** $f(x) = x^3 - 3x - 1$

97. $f(x) = x^4 - x - 3$ **98.** $f(x) = x^5 + x - 1$

99. *Dimensions of a Box* An open box is made from a rectangular piece of material, 9 inches by 5 inches, by cutting equal squares from all corners and turning up the sides (see figure). Find the dimensions of the box, given that the volume is to be 18 cubic inches.

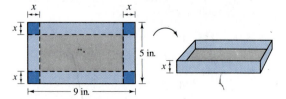

FIGURE FOR 99

100. *Dimensions of a Package*   A rectangular package to be sent by a postal service can have a maximum combined length and girth (perimeter of a cross section) of 108 inches (see figure). Find the dimensions of the package, given that the volume is 11,664 cubic inches.

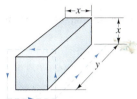

FIGURE FOR 100

101. *Advertising Costs*   A company that produces portable cassette players estimates that the profit for selling a particular model is given by

$$P = -76x^3 + 4830x^2 - 320,000, \qquad 0 \le x \le 60,$$

where P is the profit in dollars and x is the advertising expense (in 10,000s of dollars). According to this model, find the smaller of two advertising amounts that yield a profit of $2,500,000.

102. *Advertising Costs*   A company that manufactures bicycles estimates that the profit for selling a particular model is given by

$$P = -45x^3 + 2500x^2 - 275,000, \qquad 0 \le x \le 50,$$

where P is the profit in dollars and x is the advertising expense (in 10,000s of dollars). According to this model, find the smaller of two advertising amounts that yield a profit of $800,000.

103. *Transportation Cost*   The ordering and transportation cost C of the components used in manufacturing a certain product is given by

$$C = 100\left(\frac{200}{x^2} + \frac{x}{x + 30}\right), \qquad 1 \le x,$$

where C is measured in thousands of dollars and x is the order size in hundreds. In calculus, it can be shown that the cost is a minimum when $3x^3 - 40x^2 - 2400x - 36,000 = 0$. Approximate the optimal order size to the nearest hundred units.

104. *Imports into the United States*   The value of goods imported into the United States between 1984 and 1992 follows the model

$$I = -0.467t^3 + 11.457t^2 - 53.319t + 397.783,$$

where I is the annual value of goods imported (in billions of dollars) and t represents the year, with $t = 4$ corresponding to 1984 (see figure). According to this model, when did the annual value of goods imported reach 525 billion dollars? (*Source:* U.S. General Imports and Imports for Consumption)

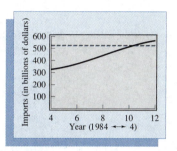

FIGURE FOR 104

In Exercises 105–108, use the bisection method to find the zero of the function (in the given interval) to the nearest hundredth.

105. $f(x) = 4x^3 + 14x - 8, \quad [0, 1]$

106. $f(x) = 4x^3 - 14x^2 - 2, \quad [3, 4]$

107. $f(x) = 7x^4 - 42x^3 + 43x^2 + 216x - 324, \quad [1, 2]$

108. $f(x) = 3x^4 - 12x^3 + 27x^2 + 4x - 4, \quad [0, 1]$

3.4   COMPLEX ZEROS AND THE FUNDAMENTAL THEOREM OF ALGEBRA

The Fundamental Theorem of Algebra / Conjugate Pairs / Factoring a Polynomial

The Fundamental Theorem of Algebra

You already know that an nth degree polynomial can have at most n real zeros. In this section you will learn that, in the complex number system, every nth degree polynomial function has *precisely* n zeros. This important result is derived from the **Fundamental Theorem of Algebra,** first proved by the famous German mathematician Carl Friedrich Gauss (1777–1855).

THE FUNDAMENTAL THEOREM OF ALGEBRA

If $f(x)$ is a polynomial of degree n, where $n > 0$, then f has at least one zero in the complex number system.

Using the Fundamental Theorem of Algebra and the equivalence of zeros and factors, you can prove the following theorem.

LINEAR FACTORIZATION THEOREM

If $f(x)$ is a polynomial of degree n, where $n > 0$, then f has precisely n linear factors,

$$f(x) = a(x - c_1)(x - c_2) \cdots (x - c_n),$$

where $c_1, c_2, \ldots, c_n$ are complex numbers and a is the leading coefficient of $f(x)$.

Note that neither the Fundamental Theorem of Algebra nor the Linear Factorization Theorem explains *how* to find the zeros or factors of a polynomial. Such theorems are called **existence theorems.** To find the zeros of a polynomial function, you must still rely on the techniques developed in the earlier parts of the text.

The Linear Factorization Theorem states that an nth degree polynomial has precisely n linear factors. Therefore, it follows that an nth degree polynomial *function* has precisely n zeros. Remember, however, that these n zeros can be real or complex, and that they may be repeated.

EXAMPLE 1   Zeros of Polynomial Functions

A. The first-degree polynomial function

$$f(x) = x - 2$$

has exactly *one* zero: $x = 2$.

B. Counting multiplicity, the second-degree polynomial function

$$f(x) = x^2 - 6x + 9 = (x - 3)(x - 3)$$

has exactly *two* zeros: $x = 3$ and $x = 3$.

C. The third-degree polynomial function

$$f(x) = x^3 + 4x = x(x - 2i)(x + 2i)$$

has exactly *three* zeros: $x = 0$, $x = 2i$, and $x = -2i$.

D. The fourth-degree polynomial function

$$f(x) = x^4 - 1 = (x - 1)(x + 1)(x - i)(x + i)$$

has exactly *four* zeros: $x = 1$, $x = -1$, $x = i$, and $x = -i$.

Example 2 shows how you can use the methods described in previous sections—Descartes's Rule of Signs, Rational Zero Test, synthetic division, and factoring—to find all the zeros of a polynomial function, including the complex zeros.

EXAMPLE 2   Finding the Zeros of a Polynomial Function

Write the polynomial function $f(x) = x^5 + x^3 + 2x^2 - 12x + 8$ as the product of linear factors and list all of its zeros.

SOLUTION

Descartes's Rule of Signs indicates two or no positive real zeros and one negative real zero. The graph of f shown in Figure 3.40 indicates that the negative real zero is near -2, and a positive real zero is perhaps $x = 1$. Choose these values from the possible rational zeros of ± 1, ± 2, ± 4, and ± 8. Synthetic division can be used to show that -2 is a zero and 1 is a repeated zero. Thus, you can write

$$f(x) = x^5 + x^3 + 2x^2 - 12x + 8$$
$$= (x - 1)(x - 1)(x + 2)(x^2 + 4)$$
$$= (x - 1)(x - 1)(x + 2)(x - 2i)(x + 2i),$$

which tells you that the five zeros of f are 1, 1, -2, $2i$, and $-2i$. Note from the graph of f shown in Figure 3.40 that the *real* zeros are the only ones that appear as x-intercepts.

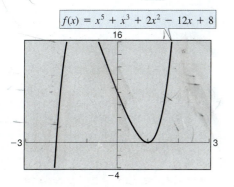

$$f(x) = x^5 + x^3 + 2x^2 - 12x + 8$$

FIGURE 3.40

Conjugate Pairs

In Example 2, note that the two complex zeros are **conjugates.** That is, they are of the form $a + bi$ and $a - bi$.

COMPLEX ZEROS OCCUR IN CONJUGATE PAIRS

Let $f(x)$ be a polynomial function that has real coefficients. If $a + bi$, where $b \neq 0$, is a zero of the function, then the conjugate $a - bi$ is also a zero of the function.

REMARK Be sure you see that this result is true only if the polynomial function has *real coefficients*. For instance, the result applies to the function $f(x) = x^2 + 1$, but not to the function $g(x) = x - i$.

EXAMPLE 3   Finding a Polynomial with Given Zeros

Find a *fourth-degree* polynomial function, with real coefficients, that has -1, -1, and $3i$ as zeros.

SOLUTION

Because $3i$ is a zero *and* the polynomial is stated to have real coefficients, the conjugate $-3i$ must also be a zero. Thus, from the Linear Factorization Theorem, $f(x)$ can be written as

$$f(x) = a(x + 1)(x + 1)(x - 3i)(x + 3i).$$

For simplicity, let $a = 1$, and obtain

$$f(x) = (x^2 + 2x + 1)(x^2 + 9) = x^4 + 2x^3 + 10x^2 + 18x + 9.$$

EXAMPLE 4   Finding a Polynomial with Given Zeros

Find a *cubic* polynomial function f, with real coefficients, that has 2 and $1 - i$ as zeros, and such that $f(1) = 3$.

SOLUTION

Because $1 - i$ is a zero of f, so is $1 + i$.

$$
\begin{aligned}
f(x) &= a(x - 2)[x - (1 - i)][x - (1 + i)] \\
&= a(x - 2)[x^2 - x(1 + i) - x(1 - i) + 1 - i^2] \\
&= a(x - 2)(x^2 - 2x + 2) \\
&= a(x^3 - 4x^2 + 6x - 4)
\end{aligned}
$$

To find the value of a, use the fact that $f(1) = 3$ and obtain $f(1) = a(1 - 4 + 6 - 4) = 3$. Thus, $a = -3$ and you can conclude that

$$f(x) = -3(x^3 - 4x^2 + 6x - 4) = -3x^3 + 12x^2 - 18x + 12.$$

Factoring a Polynomial

The Linear Factorization Theorem states that any nth degree polynomial can be written as the product of n linear factors,

$$f(x) = a(x - c_1)(x - c_2)(x - c_3) \cdots (x - c_n).$$

However, this result includes the possibility that some of the values of c_i are complex. The following result tells you that even if you do not want to get involved with "complex factors," you can still write $f(x)$ as the product of linear and/or quadratic factors.

FACTORS OF A POLYNOMIAL

Every polynomial of degree $n > 0$ with real coefficients can be written as the product of linear and quadratic factors with real coefficients, where the quadratic factors have no real zeros.

A quadratic factor with no real zeros is **irreducible over the reals.** Be sure you see that this is not the same as being *irreducible over the rationals.* For example, the quadratic

$$x^2 + 1 = (x - i)(x + i)$$

is irreducible over the reals (and therefore over the rationals). On the other hand, the quadratic

$$x^2 - 2 = (x - \sqrt{2})(x + \sqrt{2})$$

is irreducible over the rationals, but *reducible* over the reals.

EXAMPLE 5   Factoring a Polynomial

Write the polynomial $f(x) = x^4 - x^2 - 20$ in the following forms.

A. As the product of factors that are irreducible over the *rationals*
B. As the product of linear factors and quadratic factors that are irreducible over the *reals*
C. In completely factored form

SOLUTION

A. Begin by factoring the polynomial into the product of two quadratic polynomials.

$$x^4 - x^2 - 20 = (x^2 - 5)(x^2 + 4)$$

Both of these factors are irreducible over the rationals.

B. By factoring over the reals, you have

$$x^4 - x^2 - 20 = \left(x + \sqrt{5}\right)\left(x - \sqrt{5}\right)(x^2 + 4),$$

where the quadratic factor is irreducible over the reals.

C. In completely factored form, you have

$$x^4 - x^2 - 20 = \left(x + \sqrt{5}\right)\left(x - \sqrt{5}\right)(x - 2i)(x + 2i). \quad \text{\textbf{·········}}$$

EXAMPLE 6   **Finding the Zeros of a Polynomial Function**

Find all the zeros of $f(x) = x^4 - 3x^3 + 6x^2 + 2x - 60$ given that $1 + 3i$ is a zero of f.

SOLUTION

Because complex zeros occur in conjugate pairs, you know that $1 - 3i$ is also a zero of f. This means that both

$$[x - (1 + 3i)] \quad \text{and} \quad [x - (1 - 3i)]$$

are factors of $f(x)$. Multiplying these two factors produces

$$[x - (1 + 3i)][x - (1 - 3i)] = [(x - 1) - 3i][(x - 1) + 3i]$$
$$= (x - 1)^2 - 9i^2$$
$$= x^2 - 2x + 10.$$

Using long division, you can divide $x^2 - 2x + 10$ into $f(x)$ to obtain the following.

$$
\begin{array}{r}
x^2 - x - 6 \\
x^2 - 2x + 10 \overline{)\, x^4 - 3x^3 + 6x^2 + 2x - 60} \\
\underline{x^4 - 2x^3 + 10x^2} \\
-x^3 - 4x^2 + 2x \\
\underline{-x^3 + 2x^2 - 10x} \\
-6x^2 + 12x - 60 \\
\underline{-6x^2 + 12x - 60} \\
0
\end{array}
$$

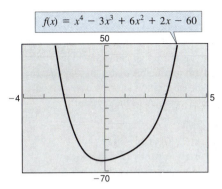

$$f(x) = x^4 - 3x^3 + 6x^2 + 2x - 60$$

FIGURE 3.41

Therefore, you have

$$f(x) = (x^2 - 2x + 10)(x^2 - x - 6)$$
$$= (x^2 - 2x + 10)(x - 3)(x + 2),$$

and you conclude that the zeros of f are $1 + 3i$, $1 - 3i$, 3, and -2. The graph in Figure 3.41 confirms these results. ....

Throughout this chapter, we have basically stated the results and examples in terms of *zeros of polynomial functions*. Be sure you see that the same results could have been stated in terms of *solutions of polynomial equations*. This is true because the zeros of the polynomial function

$$f(x) = a_n x^n + a_{n-1} x^{n-1} + \cdots + a_2 x^2 + a_1 x + a_0$$

are precisely the solutions of the polynomial equation

$$a_n x^n + a_{n-1} x^{n-1} + \cdots + a_2 x^2 + a_1 x + a_0 = 0.$$

DISCUSSION PROBLEM

FACTORING A POLYNOMIAL

Write a short paper summarizing the various techniques used to factor a polynomial. Include all of the techniques covered up to this point in the text. In what ways does a graphing utility help in factoring?

WARM-UP

The following warm-up exercises involve skills that were covered in earlier sections. You will use these skills in the exercise set for this section.

In Exercises 1–4, write each complex number in standard form and give its complex conjugate.

1. $4 - \sqrt{-29}$ **2.** $-5 - \sqrt{-144}$

3. $-1 + \sqrt{-32}$ **4.** $6 + \sqrt{-\frac{1}{4}}$

In Exercises 5–10, perform the indicated operations and write the result in standard form.

5. $(-3 + 6i) - (10 - 3i)$ **6.** $(12 - 4i) + 20i$

7. $(4 - 2i)(3 + 7i)$ **8.** $(2 - 5i)(2 + 5i)$

9. $\dfrac{1 + i}{1 - i}$ **10.** $(3 + 2i)^3$

SECTION 3.4 · EXERCISES

In Exercises 1–20, find all the zeros of the function and write the polynomial as a product of linear factors. When there is an extended list of possible rational zeros, use a graphing utility to graph the function in order to discard any rational number that is obviously not a zero of the function.

1. $f(x) = x^2 + 25$
2. $f(x) = x^2 - x + 56$
3. $h(x) = x^2 - 4x + 1$
4. $g(x) = x^2 + 10x + 23$
5. $f(x) = x^4 - 81$
6. $f(y) = y^4 - 625$
7. $f(z) = z^2 - 2z + 2$
8. $h(x) = x^3 - 3x^2 + 4x - 2$
9. $f(t) = t^3 - 3t^2 - 15t + 125$
10. $f(x) = x^3 + 11x^2 + 39x + 29$
11. $f(x) = 16x^3 - 20x^2 - 4x + 15$
12. $f(s) = 2s^3 - 5s^2 + 12s - 5$
13. $f(x) = 5x^3 - 9x^2 + 28x + 6$
14. $g(x) = 3x^3 - 4x^2 + 8x + 8$
15. $f(x) = x^4 + 10x^2 + 9$
16. $f(x) = x^4 + 29x^2 + 100$
17. $g(x) = x^4 - 4x^3 + 8x^2 - 16x + 16$
18. $h(x) = x^4 + 6x^3 + 10x^2 + 6x + 9$
19. $f(x) = 2x^4 + 5x^3 + 4x^2 + 5x + 2$
20. $g(x) = x^5 - 8x^4 + 28x^3 - 56x^2 + 64x - 32$

In Exercises 21–30, find a polynomial with integer coefficients that has the given zeros. (There are many correct answers.)

21. $1, 5i, -5i$
22. $4, 3i, -3i$
23. $2, 4 + i, 4 - i$
24. $6, -5 + 2i, -5 - 2i$
25. $i, -i, 6i, -6i$
26. $2, 2, 2, 4i, -4i$
27. $-5, -5, 1 + \sqrt{3}\,i$
28. $\frac{2}{3}, -1, 3 + \sqrt{2}\,i$
29. $\frac{3}{4}, -2, -\frac{1}{2} + i$
30. $0, 0, 4, 1 + i$

In Exercises 31–34, write the polynomial function (a) as the product of factors that are irreducible over the *rationals,* (b) as the product of linear and quadratic factors that are irreducible over the *reals,* and (c) in completely factored form.

31. $f(x) = x^4 + 6x^2 - 27$
32. $f(x) = x^4 - 2x^3 - 3x^2 + 12x - 18$
    (*Hint:* One factor is $x^2 - 6$.)
33. $f(x) = x^4 - 4x^3 + 5x^2 - 2x - 6$
    (*Hint:* One factor is $x^2 - 2x - 2$.)
34. $f(x) = x^4 - 3x^3 - x^2 - 12x - 20$
    (*Hint:* One factor is $x^2 + 4$.)

In Exercises 35–44, use the given zero of f to find all the zeros of f.

Function	Zero of the Function
35. $f(x) = 2x^3 + 3x^2 + 50x + 75$	$r = 5i$
36. $f(x) = x^3 + x^2 + 9x + 9$	$r = 3i$
37. $f(x) = 2x^4 - x^3 + 7x^2 - 4x - 4$	$r = 2i$
38. $g(x) = x^3 - 7x^2 - x + 87$	$r = 5 + 2i$
39. $g(x) = 4x^3 + 23x^2 + 34x - 10$	$r = -3 + i$
40. $h(x) = 3x^3 - 4x^2 + 8x + 8$	$r = 1 - \sqrt{3}\,i$
41. $f(x) = x^4 + 2x^3 - 9x^2 - 20x + 44$	$r = -3 + \sqrt{2}\,i$
42. $f(x) = x^3 + 4x^2 + 14x + 20$	$r = -1 - 3i$
43. $h(x) = 8x^3 - 14x^2 + 18x - 9$	$r = \dfrac{1 - \sqrt{5}\,i}{2}$
44. $f(x) = 25x^3 - 55x^2 - 54x - 18$	$r = \dfrac{-2 + \sqrt{2}\,i}{5}$

45. *Maximum Height* A baseball is thrown upward from ground level with an initial velocity of 48 feet per second and its height h in feet is given by

$$h = -16t^2 + 48t, \qquad 0 \le t \le 3,$$

where t is the time in seconds. Suppose you are told the ball reaches a height of 64 feet. Is this possible? Explain.

46. *Profit*   The demand equation for a certain product is given by $p = 140 - 0.0001x$, where p is the unit price (in dollars) of the product and x is the number of units produced and sold. The cost equation for the product is $C = 80x + 150,000$, where C is the total cost (in dollars) and x is the number of units produced. The total profit obtained by producing and selling x units is given by

$$P = R - C = xp - C.$$

Suppose you are working in the marketing department of the company that produces this product, and you are asked to determine a price p that would yield a profit of 9 million dollars. Is this possible? Explain.

47. Find a quadratic function f (with integer coefficients) that has $\pm \sqrt{b}\, i$ as zeros. Assume that b is a positive integer.

48. Find a quadratic function f (with integer coefficients) that has $a \pm bi$ as zeros. Assume that b is a positive integer.

3.5   RATIONAL FUNCTIONS AND ASYMPTOTES

Introduction to Rational Functions   /   Horizontal and Vertical Asymptotes   /   Applications

Technology Note

The standard mode on most graphing utilities is the "connected mode" in which the plotted points are connected by line segments. When graphing rational functions, it may be better to use the "dot" or "plot" mode. Try graphing $y = 1/(x - 3)$ in both modes to see which you prefer.

It is also possible to choose a "friendly window" in connected mode and achieve a good graph of a rational function. For example, on the TI-81, try graphing the function $y = 1/(x - 2)$ using the window $-4.5 \le x \le 5$, $-5 \le y \le 5$. Trace along the curve near $x = 2$. What happens when x is precisely 2? Similarly, try the viewing rectangle, $-6.3 \le x \le 6.3$, $-5 \le y \le 5$, on the TI-85.

Introduction to Rational Functions

A **rational function** is one that can be written in the form

$$f(x) = \frac{p(x)}{q(x)},$$

where $p(x)$ and $q(x)$ are polynomials and $q(x)$ is not the zero polynomial. In this section we assume $p(x)$ and $q(x)$ have no common factors. Some examples of rational functions are

$$f(x) = \frac{1}{x + 2}, \quad g(x) = \frac{x - 1}{(x + 1)(x - 2)}, \quad \text{and} \quad h(x) = \frac{x}{x^2 + 1}.$$

The domain of f excludes $x = -2$, and the domain of g excludes $x = 2$ and $x = -1$. The domain of h is all real numbers because there are no real values of x that make the denominator $x^2 + 1$ equal to zero.

In general, the *domain* of a rational function of x includes all real numbers except x-values that make the denominator zero.

EXAMPLE 1   Finding the Domain of a Rational Function

Find the domain of $f(x) = 1/x$. Discuss the behavior of f near any excluded x-values.

SOLUTION

Because the denominator is zero when $x = 0$, the domain of f is all real numbers except $x = 0$. To determine the behavior of f near this excluded value, evaluate $f(x)$ to the left and right of $x = 0$, as indicated in the tables.

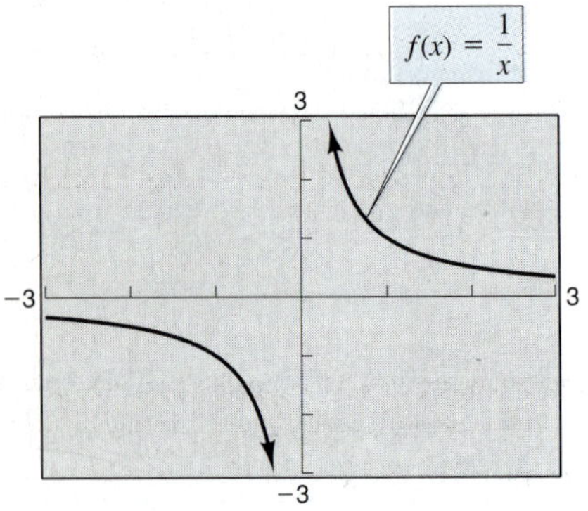

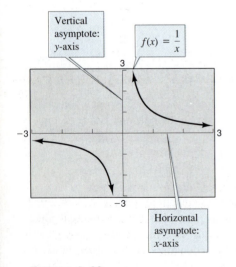

$f(x) = \dfrac{1}{x}$

FIGURE 3.42

x Approaches 0 from the Left		x Approaches 0 from the Right	
x	$f(x)$	x	$f(x)$
-1	-1	1	1
-0.5	-2	0.5	2
-0.1	-10	0.1	10
-0.01	-100	0.01	100
-0.001	-1000	0.001	1000

Note that as x approaches 0 *from the left*, $f(x)$ appears to decrease without bound, whereas as x approaches 0 *from the right*, $f(x)$ appears to increase without bound. This behavior is confirmed by the graph of f that is shown in Figure 3.42.

Horizontal and Vertical Asymptotes

In Example 1, the behavior of $f(x) = 1/x$ near $x = 0$ is denoted as follows.

$$f(x) \to -\infty \text{ as } x \to 0^- \qquad f(x) \to \infty \text{ as } x \to 0^+$$

$f(x)$ decreases without bound as x approaches 0 from the left. $f(x)$ increases without bound as x approaches 0 from the right.

The line $x = 0$ is a **vertical asymptote** of the graph of f, as shown in Figure 3.43. The graph of f also has a **horizontal asymptote**—the line $y = 0$. This means that the values of $f(x) = 1/x$ approach zero as x increases or decreases without bound.

$$f(x) \to 0 \text{ as } x \to -\infty \qquad f(x) \to 0 \text{ as } x \to \infty$$

$f(x)$ approaches 0 as x decreases without bound. $f(x)$ approaches 0 as x increases without bound.

Vertical asymptote: y-axis

$f(x) = \dfrac{1}{x}$

Horizontal asymptote: x-axis

FIGURE 3.43

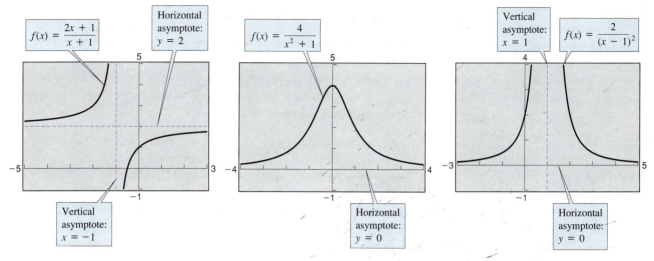

FIGURE 3.44

In general, the line $x = a$ is a **vertical asymptote** of the graph of f if $f(x) \to \infty$ or $f(x) \to -\infty$ as x approaches a either from the right or from the left. The line $y = b$ is a **horizontal asymptote** of the graph of f if $f(x)$ approaches b as $x \to \infty$ or $x \to -\infty$.

Though the graph of a rational function never intersects its vertical asymptote(s), it may intersect its horizontal asymptote. Eventually however, the distance between the horizontal asymptote and the points on the graph must approach zero (as $x \to \infty$ or $x \to -\infty$). Figure 3.44 shows the horizontal and vertical asymptotes of the graphs of three rational functions.

A rational function can have several vertical asymptotes, but it can have at most one horizontal asymptote.

ASYMPTOTES OF A RATIONAL FUNCTION

Let f be the rational function given by

$$f(x) = \frac{p(x)}{q(x)} = \frac{a_n x^n + a_{n-1} x^{n-1} + \cdots + a_1 x + a_0}{b_m x^m + b_{m-1} x^{m-1} + \cdots + b_1 x + b_0},$$

where $p(x)$ and $q(x)$ have no common factors.

1. The graph of f has vertical asymptotes at the zeros of $q(x)$.
2. The graph of f has at most one horizontal asymptote, as follows.
   (a) If $n < m$, then the x-axis ($y = 0$) is a horizontal asymptote.
   (b) If $n = m$, then the line $y = a_n / b_m$ is a horizontal asymptote.
   (c) If $n > m$, then the graph of f has no horizontal asymptote.

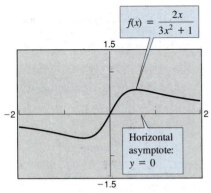

Vertical asymptote: None
Horizontal asymptote: $y = 0$

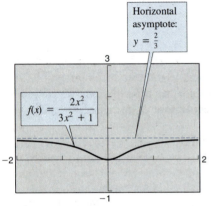

Vertical asymptote: None
Horizontal asymptote: $y = \frac{2}{3}$

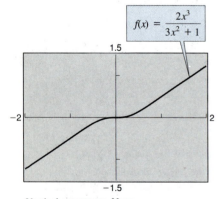

Vertical asymptote: None
Horizontal asymptote: None

FIGURE 3.45

You can apply this theorem by comparing the degrees of the numerator and denominator, as illustrated in Figure 3.45.

EXAMPLE 2   A Calculator Experiment

Find the horizontal asymptote of the graph of

$$f(x) = \frac{3x^3 + 7x^2 + 2}{-4x^3 + 5}.$$

SOLUTION

Because the degree of the numerator and denominator are the same, the horizontal asymptote is given by the ratio of the leading coefficients,

$$y = \frac{\text{leading coefficient of numerator}}{\text{leading coefficient of denominator}} = \frac{3}{-4} = -\frac{3}{4}.$$

The following table shows how the values of $f(x)$ become closer and closer to $-\frac{3}{4}$ as x becomes increasingly large or small.

x Decreases Without Bound		**x Increases Without Bound**	
x	$f(x)$	x	$f(x)$
-1	0.6667	1	12
-10	-0.5738	10	-0.9267
-100	-0.7325	100	-0.7675
-1000	-0.7482	1000	-0.7518
$-10,000$	-0.7498	$10,000$	-0.7502

Graphs of rational functions that are produced by a graphing utility can be misleading. Try using a graphing utility to confirm the graph shown in Figure 3.46.

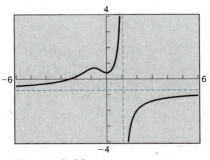

FIGURE 3.46

A function that is not rational can have two horizontal asymptotes—one to the left and one to the right.

EXAMPLE 3   A (Nonrational) Graph with Two Horizontal Asymptotes

Use a graphing utility to find the horizontal asymptotes of

$$f(x) = \frac{x + 10}{|x| + 2}.$$

SOLUTION

The graph of f is shown in Figure 3.47. Note that the graph appears to have two different horizontal asymptotes. You can confirm this by rewriting the function as follows.

$$f(x) = \begin{cases} \dfrac{x + 10}{-x + 2}, & x \leq 0 \\[2mm] \dfrac{x + 10}{x + 2}, & x > 0 \end{cases}$$

In this form, you can see that the line $y = -1$ is a horizontal asymptote to the left and the line $y = 1$ is a horizontal asymptote to the right.

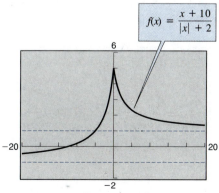

$$f(x) = \frac{x + 10}{|x| + 2}$$

FIGURE 3.47

Applications

EXAMPLE 4   Cost-Benefit Model for Smokestack Emission

A utility company burns coal to generate electricity. The cost C of removing $p\%$ of the smokestack pollutants is given by

$$C = \frac{80,000p}{100 - p}.$$

Sketch the graph of this function. The state legislature is considering a law that will require utility companies to remove 90% of the pollutants from their smokestack emissions. The current law requires 85% removal. How much additional expense is the new law imposing on the utility company?

SOLUTION

The graph of this function is shown in Figure 3.48. Note that the graph has a vertical asymptote at $p = 100$. Because the current law requires 85% removal, the current cost to the utility company is

$$C = \frac{80,000(85)}{100 - 85} \approx \$453,333.$$

At 90% removal, the cost to the utility company will be

$$C = \frac{80,000(90)}{100 - 90} = \$720,000.$$

Therefore, the new law requires the utility company to spend an additional

$$720,000 - 453,333 = \$266,667.$$

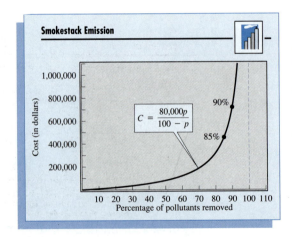

The cost of removing a certain percentage of the pollutants from the stack emission is typically not a linear function. As the percentage of removed pollutants approaches 100%, the cost tends to become prohibitive.

FIGURE 3.48

EXAMPLE 5 Average Cost of Producing a Product

A business has a cost function of $C = 0.5x + 5000$, where C is measured in dollars and x is the number of units produced. The *average cost per unit* is given by

$$\overline{C} = \frac{C}{x} = \frac{0.5x + 5000}{x}.$$

Find the average cost per unit when $x = 1000$, 10,000, and 100,000. What is the horizontal asymptote for this function, and what does it represent?

SOLUTION

When $x = 1000$, $\overline{C} = \dfrac{0.5(1000) + 5000}{1000} = \$5.50$.

When $x = 10{,}000$, $\overline{C} = \dfrac{0.5(10{,}000) + 5000}{10{,}000} = \$1.00$.

When $x = 100{,}000$, $\overline{C} = \dfrac{0.5(100{,}000) + 5000}{100{,}000} = \$0.55$.

As shown in Figure 3.49, the horizontal asymptote is given by the line $C = 0.50$. This line represents the least possible unit cost for the product. Note that this example points out one of the major problems of a small business—that it is difficult to have competitively low prices when the production level is low.

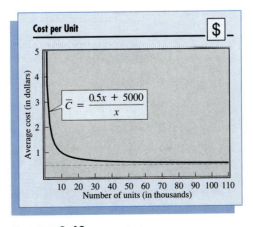

FIGURE 3.49

DISCUSSION PROBLEM

COMMON FACTORS IN THE NUMERATOR AND DENOMINATOR

In the guidelines for graphing a rational function, we noted that the rational function should have no factor that is common to its numerator and denominator. To see why we required this, consider the function given by

$$f(x) = \frac{x(x-1)}{x},$$

which has a common factor of x in the numerator and denominator. Graph this function. Does it have a vertical asymptote at $x = 0$? Does your graphing utility indicate the domain of the function f?

The following warm-up exercises involve skills that were covered in earlier sections. You will use these skills in the exercise set for this section.

In Exercises 1–4, factor the polynomial.

1. $x^2 - 3x - 10$ **2.** $x^2 - 7x + 10$

3. $x^3 + 4x^2 + 3x$ **4.** $x^3 - 4x^2 - 2x + 8$

In Exercises 5–8, graph the equation.

5. $y = 2$ **6.** $x = -1$

7. $y = x - 2$ **8.** $y = -x + 1$

In Exercises 9 and 10, use long division to write the rational expression as the sum of a polynomial and a rational expression.

9. $\dfrac{x^2 + 5x + 6}{x - 4}$ **10.** $\dfrac{x^2 + 5x + 6}{x + 4}$.

SECTION 3.5 · EXERCISES

In Exercises 1–6, (a) complete each table, (b) determine the vertical and horizontal asymptotes of the graph, and (c) find the domain of the function.

x	f(x)
1	
1.5	
1.9	
1.99	
1.999	

x	f(x)
3	
2.5	
2.1	
2.01	
2.001	

x	f(x)
3	
5	
10	
100	
1000	

1. $f(x) = \dfrac{1}{x - 2}$

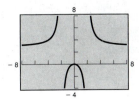

2. $f(x) = \dfrac{5x}{x - 2}$

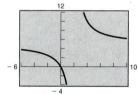

3. $f(x) = \dfrac{3x}{|x - 2|}$

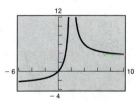

4. $f(x) = \dfrac{3}{|x - 2|}$

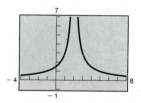

5. $f(x) = \dfrac{3x^2}{x^2 - 4}$

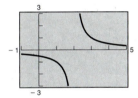

6. $f(x) = \dfrac{4x}{x^2 - 4}$

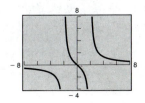

In Exercises 7–14, find the domain of the function and identify any horizontal and vertical asymptotes.

7. $f(x) = \dfrac{1}{x^2}$

8. $f(x) = \dfrac{4}{(x - 2)^3}$

9. $f(x) = \dfrac{2 + x}{2 - x}$

10. $f(x) = \dfrac{1 - 5x}{1 + 2x}$

11. $f(x) = \dfrac{x^3}{x^2 - 1}$

12. $f(x) = \dfrac{2x^2}{x + 1}$

13. $f(x) = \dfrac{3x^2 + 1}{x^2 + 9}$

14. $f(x) = \dfrac{3x^2 + x - 5}{x^2 + 1}$

In Exercises 15–20, match the rational function with its graph, and describe the given viewing rectangle. [The graphs are labeled (a), (b), (c), (d), (e), and (f).]

15. $f(x) = \dfrac{2}{x + 1}$

16. $f(x) = \dfrac{1}{x - 4}$

17. $f(x) = \dfrac{x + 1}{x}$

18. $f(x) = \dfrac{1 - 2x}{x}$

19. $f(x) = \dfrac{x - 2}{x - 1}$

20. $f(x) = -\dfrac{x + 2}{x + 1}$

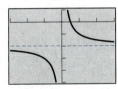

(a)

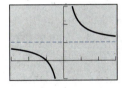

(b)

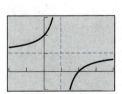

(c)

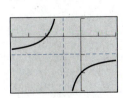

(d)

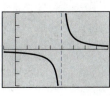

(e)

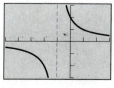

(f)

In Exercises 21–24, (a) determine the domain of f and g, (b) find any vertical asymptotes of f, (c) use a graphing utility to obtain the graphs of f and g in the same viewing rectangle, and (d) explain why the graphing utility does or does not show the difference in the domains of f and g.

21. $f(x) = \dfrac{x^2 - 4}{x + 2}$

$g(x) = x - 2$

22. $f(x) = \dfrac{x^2(x - 3)}{x^2 - 3x}$

$g(x) = x$

23. $f(x) = \dfrac{x - 3}{x^2 - 3x}$

$g(x) = \dfrac{1}{x}$

24. $f(x) = \dfrac{2x - 8}{x^2 - 9x + 20}$

$g(x) = \dfrac{2}{x - 5}$

25. *Cost of Clean Water* The cost in millions of dollars for removing $p\%$ of the industrial and municipal pollutants discharged into a river is given by

$$C = \frac{255p}{100 - p}, \qquad 0 \le p < 100.$$

(a) Find the cost of removing 10% of the pollutants.
(b) Find the cost of removing 40%.
(c) Find the cost of removing 75%.
(d) According to this model, would it be possible to remove 100% of the pollutants?

26. *Recycling Costs* In a pilot project, a rural township was given recycling bins for separating and storing recyclable products. The cost in dollars for giving bins to $p\%$ of the population is given by

$$C = \frac{25{,}000p}{100 - p}, \qquad 0 \le p < 100.$$

(a) Find the cost of giving bins to 15% of the population.
(b) Find the cost of giving bins to 50% of the population.
(c) Find the cost of giving bins to 90% of the population.
(d) According to this model, would it be possible to give bins to 100% of the residents?

27. *Population of Deer* The game commission introduces 50 deer into newly acquired state game lands. The population of the herd is given by

$$N = \frac{10(5 + 3t)}{1 + 0.04t}, \qquad 0 \le t,$$

where t is the time in years (see figure).
(a) Find the population when t is 5, 10, and 25.
(b) What is the limiting size of the herd as time increases?

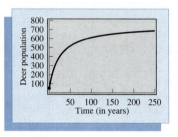

FIGURE FOR 27

28. *Food Consumption* A biology class performs an experiment comparing the quantity of food consumed by a certain kind of moth with the quantity supplied (see figure). The model for their experimental data is given by

$$y = \frac{1.568x - 0.001}{6.360x + 1},$$

where x is the quantity (in milligrams) of food supplied and y is the quantity (in milligrams) eaten. At what level of consumption will the moth become filled?

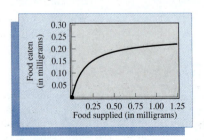

FIGURE FOR 28

29. *Human Memory Model* Psychologists have developed mathematical models to predict performance as a function of the number of trials n for a certain task. Consider the learning curve given by

$$P = \frac{0.5 + 0.9(n - 1)}{1 + 0.9(n - 1)}, \qquad 0 < n,$$

where P is the percentage of correct responses after n trials.

(a) Complete the following table for this model.

n	1	2	3	4	5	6	7	8	9	10
P										

(b) According to this model, what is the limiting percentage of correct responses as n increases?

30. As the magnitude of x increases, what value does the function f approach in each of the following? Is $f(x)$ greater than or less than this functional value when (i) x is positive and large in magnitude, and (ii) x is negative and large in magnitude?

(a) $f(x) = 4 - \dfrac{1}{x}$ (b) $f(x) = 2 + \dfrac{1}{x - 3}$

(c) $f(x) = \dfrac{2x - 1}{x - 3}$ (d) $f(x) = \dfrac{2x - 1}{x^2 + 1}$

3.6   GRAPHS OF RATIONAL FUNCTIONS

The Graph of a Rational Function / Slant Asymptotes / Application

The Graph of a Rational Function

In this section you will study techniques for analyzing the graph of a rational function.

REMARK Testing for symmetry can be useful, especially for simple rational functions. For example, the graph of $f(x) = 1/x$ is symmetrical with respect to the origin, and the graph of $g(x) = 1/x^2$ is symmetrical with respect to the y-axis.

ANALYZING THE GRAPHS OF RATIONAL FUNCTIONS

Let $f(x) = p(x)/q(x)$, when $p(x)$ and $q(x)$ are polynomials with no common factors.

1. The y-intercept (if any) is the value $f(0)$.
2. The x-intercepts (if any) are the zeros of the numerator—that is, the solutions to the equation $p(x) = 0$.
3. The vertical asymptotes (if any) are the zeros of the denominator—that is, the solutions to the equation $q(x) = 0$.
4. The horizontal asymptote (if any) is the value that $f(x)$ approaches as x increases or decreases without bound.
5. Determining the behavior of the graph *between* and *beyond* each x-intercept and vertical asymptote is crucial for describing the complete graph of a rational function.

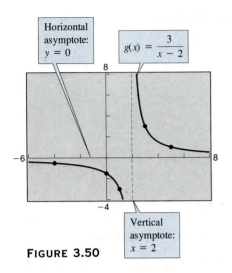

FIGURE 3.50

REMARK The graph of g is a vertical stretch and a right shift of the graph of $f(x) = 1/x$ because

$$g(x) = \frac{3}{x-2} = 3\left(\frac{1}{x-2}\right)$$

$$= 3f(x-2).$$

EXAMPLE 1   Graphing a Rational Function

Graph the function

$$g(x) = \frac{3}{x-2}.$$

SOLUTION

Begin by noting that the numerator and denominator have no common factors.

y-intercept: $\left(0, -\frac{3}{2}\right)$, from $g(0) = -\frac{3}{2}$

x-intercept: None, because $3 \neq 0$

Vertical asymptote: $x = 2$, zero of denominator

Horizontal asymptote: $y = 0$, degree of $p(x) <$ degree of $q(x)$

Additional points:

x	-4	1	3	5
$g(x)$	-0.5	-3	3	1

By plotting the intercepts, asymptotes, and a few additional points, we obtain the graph shown in Figure 3.50. Use a graphing utility to confirm these results.

EXAMPLE 2   Graphing a Rational Function

Graph the function

$$f(x) = \frac{2x-1}{x}.$$

SOLUTION

Begin by noting that the numerator and denominator have no common factors.

y-intercept: None, because $x = 0$ is not in the domain

x-intercept: $\left(\frac{1}{2}, 0\right)$, from $2x - 1 = 0$

Vertical asymptote: $x = 0$, zero of denominator

Horizontal asymptote: $y = 2$, degree of $p(x) =$ degree of $q(x)$

Additional points:

x	-4	-1	$\frac{1}{4}$	4
$f(x)$	2.25	3	-2	1.75

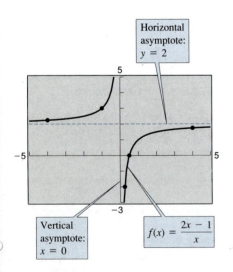

FIGURE 3.51

By plotting the intercepts, asymptotes, and a few additional points, we obtain the graph shown in Figure 3.51. Use a graphing utility to confirm these results.

EXAMPLE 3   Graphing a Rational Function

Graph the function

$$f(x) = \frac{x}{x^2 - x - 2}.$$

SOLUTION

Factoring the denominator produces

$$f(x) = \frac{x}{x^2 - x - 2} = \frac{x}{(x + 1)(x - 2)}.$$

Thus, the numerator and denominator have no common factors.

y-intercept: (0, 0),   because $f(0) = 0$
x-intercept: (0, 0)
Vertical asymptotes: $x = -1$, $x = 2$,   zeros of denominator
Horizontal asymptote: $y = 0$,   degree of $p(x)$ < degree of $q(x)$

Additional points:

x	-3	-0.5	1	3
$f(x)$	-0.3	0.4	-0.5	0.75

By plotting the intercept, asymptotes, and a few additional points, we obtain the graph shown in Figure 3.52. Use a graphing utility to confirm these results.

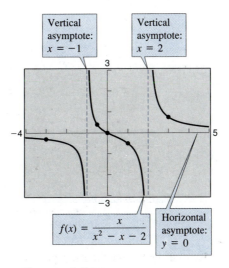

Vertical asymptote: $x = -1$

Vertical asymptote: $x = 2$

$$f(x) = \frac{x}{x^2 - x - 2}$$

Horizontal asymptote: $y = 0$

FIGURE 3.52

EXAMPLE 4   **Graphing a Rational Function**

Graph the function

$$f(x) = \frac{2(x^2 - 9)}{x^2 - 4}.$$

SOLUTION

Factoring the numerator and denominator produces

$$f(x) = \frac{2(x^2 - 9)}{x^2 - 4} = \frac{2(x - 3)(x + 3)}{(x - 2)(x + 2)}.$$

Thus, the numerator and denominator have no common factors.

$$y\text{-intercept:} \quad \left(0, \frac{9}{2}\right), \quad \text{from } f(0) = \frac{9}{2}$$

$$x\text{-intercepts:} \quad (-3, 0) \text{ and } (3, 0)$$

Vertical asymptotes: $x = -2$, $x = 2$,   zeros of denominator

Horizontal asymptote: $y = 2$,   degree of $p(x)$ = degree of $q(x)$

Symmetry: With respect to y-axis, because $f(-x) = f(x)$

Additional points:	x	0.5	2.5	6
	$f(x)$	4.67	-2.44	1.69

By plotting the intercepts, asymptotes, and a few additional points, we obtain the graph shown in Figure 3.53. Use a graphing utility to confirm these results.

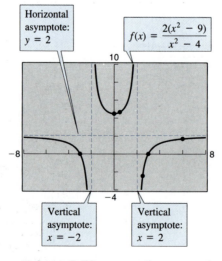

Horizontal asymptote: $y = 2$

$f(x) = \dfrac{2(x^2 - 9)}{x^2 - 4}$

Vertical asymptote: $x = -2$

Vertical asymptote: $x = 2$

FIGURE 3.53

Slant Asymptotes

If the degree of the numerator of a rational function is exactly *one more* than
the degree of its denominator, then the graph of the function has a **slant
asymptote.** For example, the graphs of

$$f(x) = \frac{x^2 - x}{x + 1} \quad \text{and} \quad g(x) = \frac{-x^3 + x^2 + 4}{x^2}$$

have slant asymptotes, as shown in Figure 3.54. You can use long division to
find the equation of a slant asymptote. For instance, dividing $x + 1$ into
$x^2 - x$ produces

$$f(x) = \frac{x^2 - x}{x + 1} = \underbrace{x - 2}_{\substack{\text{Slant asymptote} \\ (y = x - 2)}} + \frac{2}{x + 1}.$$

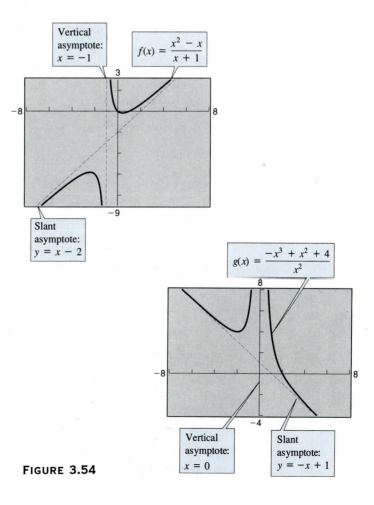

FIGURE 3.54

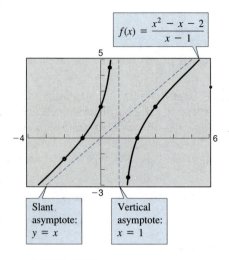

$$f(x) = \frac{x^2 - x - 2}{x - 1}$$

Slant asymptote: $y = x$

Vertical asymptote: $x = 1$

FIGURE 3.55

EXAMPLE 5   A Rational Function with a Slant Asymptote

Graph the function

$$f(x) = \frac{x^2 - x - 2}{x - 1}.$$

SOLUTION

As a preliminary step, write $f(x)$ in two different ways. Factoring the numerator

$$f(x) = \frac{x^2 - x - 2}{x - 1} = \frac{(x - 2)(x + 1)}{x - 1}$$

allows you to recognize the x-intercepts, and long division

$$f(x) = \frac{x^2 - x - 2}{x - 1} = x - \frac{2}{x - 1}$$

allows you to recognize that the line $y = x$ is a slant asymptote of the graph.

y-intercept:	$(0, 2)$, because $f(0) = 2$
x-intercepts:	$(-1, 0)$ and $(2, 0)$
Vertical asymptote:	$x = 1$
Slant asymptote:	$y = x$

Additional points:

x	-2	0.5	1.5	3
$f(x)$	-1.33	4.5	-2.5	2

The graph of f is shown in Figure 3.55. Use a graphing utility to graph f and its slant asymptote $y = x$ on the same screen and confirm the above results.

Note that it is possible for the graph of a rational function to cross its horizontal asymptote or its slant asymptote, as shown in Figure 3.56.

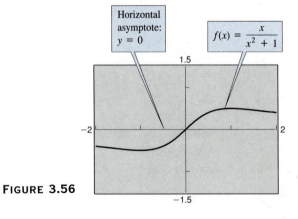

Horizontal asymptote: $y = 0$

$$f(x) = \frac{x}{x^2 + 1}$$

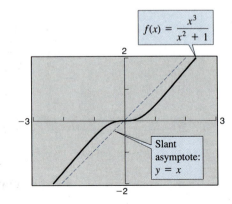

$$f(x) = \frac{x^3}{x^2 + 1}$$

Slant asymptote: $y = x$

FIGURE 3.56

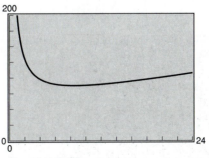

FIGURE 3.57

Application

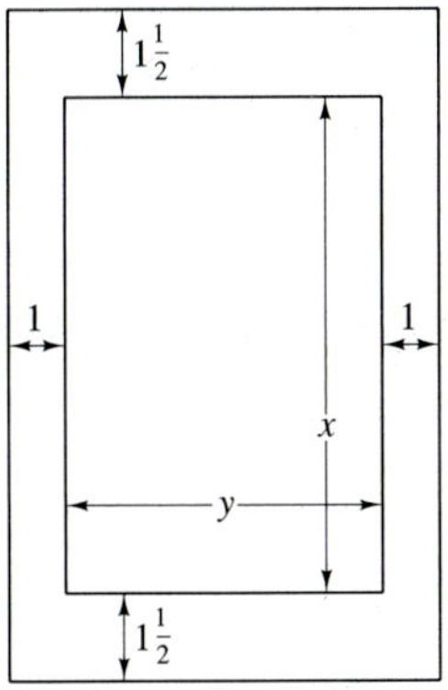

EXAMPLE 6   Finding the Minimum Area

A rectangular page is to contain 48 square inches of print. The margins at the top and bottom of the page are each $1\frac{1}{2}$ inches. The margins on each side are 1 inch. What should the dimensions of the page be so that the least amount of paper is used?

SOLUTION

Let A be the area to be minimized. From Figure 3.57, you can write

$$A = (x + 3)(y + 2).$$

The printed area inside the margins is given by

$$48 = xy \quad \text{or} \quad y = \frac{48}{x}.$$

To find the minimum area, rewrite the equation for A in terms of just one variable by substituting $y = 48/x$.

$$A = (x + 3)\left(\frac{48}{x} + 2\right) = \frac{(x + 3)(48 + 2x)}{x}, \quad x > 0$$

The graph of this rational function is shown in Figure 3.58. Because x represents the height of the printed area, you need consider only the portion of the graph for which x is positive. Using the zoom and trace features of a graphing utility, you can approximate the minimum value of A to occur when $x \approx 8.5$ inches. The corresponding value of y is $y \approx 48/8.5 \approx 5.6$ inches. Thus, the dimensions of the page should be

$$x + 3 \approx 11.5 \text{ inches} \quad \text{by} \quad y + 2 \approx 7.6 \text{ inches.}$$

If you go on to take a course in calculus, you will learn a technique for finding the exact value of x that produces a minimum area. In this case, that value is $x = 6\sqrt{2} \approx 8.485$.

200

0
0   24

FIGURE 3.58

DISCUSSION PROBLEM

········

A PARABOLIC ASYMPTOTE

The technique used to find slant asymptotes works for rational functions in which the degree of the numerator is one more than the degree of the denominator. If the degree of the numerator is two more than the degree of the denominator, then the graph of the rational function has a *parabolic asymptote*. For instance, consider the rational function given by

$$f(x) = \frac{x^3 + 1}{x} = x^2 + \frac{1}{x}.$$

The graphs of f and $y = x^2$ are both shown in the figure. Note that as x approaches positive or negative infinity, the two graphs nearly coincide. Write a short paragraph describing how you can use a parabolic asymptote as an aid to analyzing the graph of a rational function.

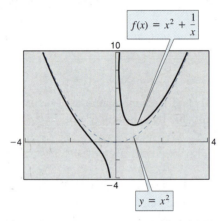

$$f(x) = x^2 + \frac{1}{x}$$

$$y = x^2$$

WARM-UP

·················

The following warm-up exercises involve skills that were covered in earlier sections. You will use these skills in the exercise set for this section.

In Exercises 1 and 2, find any intercepts of the graph of the equation.

1. $2x - xy - y + 6 = 0$ **2.** $4x - 5y + 12 = 0$

In Exercises 3 and 4, determine whether the function is even, odd, or neither.

3. $g(x) = x(x^2 - 5)$ **4.** $h(x) = \frac{1}{8}x^4 - 3x^2 - 4$

In Exercises 5–8, find the domain of the function and identify any horizontal or vertical asymptotes.

5. $f(x) = \dfrac{6}{x - 8}$ **6.** $f(x) = \dfrac{3x - 1}{4x + 1}$ **7.** $h(x) = \dfrac{2x^2 + 5}{x^2 - 9}$ **8.** $g(x) = 4 - \dfrac{1}{x}$

In Exercises 9 and 10, perform the specified division.

9. $\dfrac{4x^2 + 5x + 8}{2x - 1}$ **10.** $\dfrac{x^3 + 1}{x^2}$

SECTION 3.6 · EXERCISES

In Exercises 1–4, use the graph of $f(x) = 1/x$ to sketch the graph of g.

1. $g(x) = \dfrac{1}{x} + 1$

2. $g(x) = \dfrac{1}{x - 1}$

3. $g(x) = -\dfrac{1}{x}$

4. $g(x) = \dfrac{1}{x + 2}$

In Exercises 5–8, use the graph of $f(x) = 4/x^2$ to sketch the graph of g.

5. $g(x) = \dfrac{4}{x^2} - 2$

6. $g(x) = -\dfrac{4}{x^2}$

7. $g(x) = \dfrac{4}{(x - 2)^2}$

8. $g(x) = \dfrac{1}{x^2}$

In Exercises 9–12, use the graph of $f(x) = 8/x^3$ to sketch the graph of g.

9. $g(x) = \dfrac{8}{(x + 2)^3}$

10. $g(x) = \dfrac{8}{x^3} + 1$

11. $g(x) = -\dfrac{8}{x^3}$

12. $g(x) = \dfrac{1}{x^3}$

In Exercises 13–28, sketch the graph of the rational function without the aid of a graphing utility. As sketching aids, check for intercepts, symmetry, vertical asymptotes, and horizontal asymptotes. Use a graphing utility to verify your graph.

13. $f(x) = \dfrac{1}{x + 2}$

14. $f(x) = \dfrac{1}{x - 3}$

15. $h(x) = \dfrac{-1}{x + 2}$

16. $g(x) = \dfrac{1}{3 - x}$

17. $C(x) = \dfrac{5 + 2x}{1 + x}$

18. $P(x) = \dfrac{1 - 3x}{1 - x}$

19. $g(x) = \dfrac{1}{x + 2} + 2$

20. $f(t) = \dfrac{1 - 2t}{t}$

21. $f(x) = \dfrac{x^2}{x^2 + 9}$

22. $f(x) = 2 - \dfrac{3}{x^2}$

23. $h(x) = \dfrac{x^2}{x^2 - 9}$

24. $g(x) = \dfrac{x}{x^2 - 9}$

25. $g(x) = -\dfrac{1}{(x - 2)^2} + 3$

26. $f(x) = -\dfrac{1}{(x - 2)^2}$

27. $f(x) = \dfrac{3x}{x^2 - x - 2}$

28. $f(x) = \dfrac{2x}{x^2 + x - 2}$

In Exercises 29–36, use a graphing utility to graph the function, and give its domain and range.

29. $f(x) = \dfrac{2 + x}{1 - x}$

30. $f(x) = \dfrac{3 - x}{2 - x}$

31. $f(t) = \dfrac{3t + 1}{t}$

32. $h(x) = \dfrac{1}{x - 3} + 1$

33. $h(t) = \dfrac{4}{t^2 + 1}$

34. $g(x) = -\dfrac{x}{(x - 2)^2}$

35. $f(x) = \dfrac{20x}{x^2 + 1} - \dfrac{1}{x}$

36. $f(x) = 5\left(\dfrac{1}{x - 4} - \dfrac{1}{x + 2}\right)$

In Exercises 37–46, sketch the graph of the rational function. As sketching aids, check for intercepts, symmetry, vertical asymptotes, and slant asymptotes. Use a graphing utility to verify your graph.

37. $f(x) = \dfrac{2x^2 + 1}{x}$

38. $f(x) = \dfrac{1 - x^2}{x}$

39. $g(x) = \dfrac{x^2 + 1}{x}$

40. $h(x) = \dfrac{x^2}{x - 1}$

41. $f(x) = \dfrac{x^3}{x^2 - 1}$

42. $g(x) = \dfrac{x^3}{2x^2 - 8}$

43. $f(x) = \dfrac{x^2 - x + 1}{x - 1}$

44. $f(x) = \dfrac{2x^2 - 5x + 5}{x - 2}$

45. $f(x) = \dfrac{x^2 + 5x + 8}{x + 3}$

46. $f(x) = \dfrac{2x^2 + x}{x + 1}$

In Exercises 47 and 48, use a graphing utility to graph the function and note that a graph may have two horizontal asymptotes.

47. $h(x) = \dfrac{6x}{\sqrt{x^2 + 1}}$

48. $g(x) = \dfrac{4|x - 2|}{x + 1}$

In Exercises 49 and 50, use a graphing utility to graph the function and note that a graph may cross its horizontal asymptote.

49. $f(x) = \dfrac{4(x - 1)^2}{x^2 - 4x + 5}$

50. $g(x) = \dfrac{3x^4 - 5x + 3}{x^4 + 1}$

In Exercises 51 and 52, use a graphing utility to graph the function. Explain why there is no vertical asymptote when a superficial examination of the function may indicate that there should be one.

51. $h(x) = \dfrac{6 - 2x}{3 - x}$

52. $g(x) = \dfrac{x^2 + x - 2}{x - 1}$

In Exercises 53 and 54, use a graphing utility to graph the function and locate any relative maximum or minimum points on the graph.

53. $f(x) = \dfrac{3(x + 1)}{x^2 + x + 1}$

54. $C(x) = x + \dfrac{16}{x}$

55. *Concentration of a Mixture*   A 250-gallon storage tank contains 10 gallons of a 25% brine solution. A person adds x gallons of a 75% brine solution to the tank.

(a) Show that the concentration C of the final mixture is given by
$$C = \frac{3x + 10}{4(x + 10)}.$$

(b) Determine the domain of the function based on the physical constraints of the problem.

(c) Use a graphing utility to graph the function. As the tank is filled, what happens to the rate at which the concentration of brine is increasing? What does the concentration of brine appear to approach?

56. *Dimensions of a Rectangle*   A rectangular region of length x and width y has an area of 500 square meters.

(a) Express the width y as a function of x.

(b) Determine the domain of the function based on the physical constraints of the problem.

(c) Sketch a graph of the function and determine the width of the rectangle if $x = 30$ meters.

57. *Page Design*   A page that is x inches wide and y inches high contains 30 square inches of print. The margins at the top and bottom are 2 inches, and the margins on each side are 1 inch (see figure).

(a) Show that the total area A of the page is given by
$$A = \frac{2x(2x + 11)}{x - 2}.$$

(b) Determine the domain of the function based on the physical constraints of the problem.

(c) Use a graphing utility to graph the area function and approximate the page size so the least amount of paper will be used.

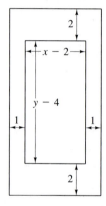

FIGURE FOR 57

58. *Minimum Area*   A right triangle is formed in the first quadrant by the x-axis, the y-axis, and a line segment through the point (2, 3) (see figure).

(a) Show that an equation of the line segment is
$$y = \frac{3(x - a)}{2 - a}, \qquad 0 \le x \le a.$$

(b) Show that the area of the triangle is
$$A = \frac{-3a^2}{2(2 - a)}.$$

(c) Use a graphing utility to graph the area function, and, from the graph, estimate the value of a that yields a minimum area.

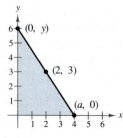

FIGURE FOR 58

CHAPTER 3 · REVIEW EXERCISES

In Exercises 1–4, sketch the graph of the quadratic function. Identify the vertex and the intercepts.

1. $f(x) = (x + \frac{3}{2})^2 + 1$ **2.** $f(x) = (x - 4)^2 - 4$

3. $f(x) = \frac{1}{3}(x^2 + 5x - 4)$ **4.** $f(x) = 3x^2 - 12x + 11$

In Exercises 5 and 6, find the quadratic function that has the indicated vertex and whose graph passes through the given point.

5. Vertex: $(1, -4)$;   point: $(2, -3)$

6. Vertex: $(2, 3)$;   point; $(-1, 6)$

In Exercises 7–14, find the maximum or minimum value of the quadratic function.

7. $g(x) = x^2 - 2x$

8. $f(x) = x^2 + 8x + 10$

9. $f(x) = 6x - x^2$

10. $h(x) = 3 + 4x - x^2$

11. $f(t) = -2t^2 + 4t + 1$

12. $h(x) = 4x^2 + 4x + 13$

13. $h(x) = x^2 + 5x - 4$

14. $f(x) = 4x^2 + 4x + 5$

15. *Maximum Area*   A rectangle is inscribed in the region bounded by the x-axis, the y-axis, and the graph of $x + 2y - 6 = 0$ (see figure). Find the coordinates (x, y) that yield a maximum area for the rectangle.

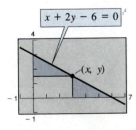

FIGURE FOR 15

16. *Maximum Area*   The perimeter of a rectangle is 200 feet. Let x represent the width of the rectangle and write a quadratic function that expresses the area of the rectangle in terms of x. Of all possible rectangles with a perimeter of 200 feet, find the dimensions of the one that has the greatest area.

17. *Maximum Revenue*   Find the number of units x that produces a maximum revenue R, where

$R = 900x - 0.1x^2$.

18. *Maximum Profit*   Let x be the amount (in hundreds of dollars) that a company spends on advertising, and let P be the profit, where

$P = 230 + 20x - \frac{1}{2}x^2$.

What amount of advertising will yield a maximum profit?

19. *Maximum Profit*   A real estate office handles 50 apartment units. When the rent is $540 per month, all units are occupied. However, for each $30 increase in rent, one unit becomes vacant. Each occupied unit requires an average of $18 per month for service and repairs. What rent should be charged to realize the most profit?

20. *Minimum Cost*   A manufacturer has daily production costs of

$C = 20,000 - 120x + 0.055x^2$,

where C is the total cost in dollars and x is the number of units produced. How many units should be produced each day to yield a minimum cost?

In Exercises 21–24, determine the right-hand and left-hand behavior of the graph of the polynomial function. Use a graphing utility to verify your answer.

21. $f(x) = -x^2 + 6x + 9$ **22.** $f(x) = \frac{1}{2}x^3 + 2x$

23. $g(x) = \frac{3}{4}(x^4 + 3x^2 + 2)$

24. $h(x) = -x^5 - 7x^2 + 10x$

In Exercises 25–32, sketch a graph of the function.

25. $f(x) = -(x - 2)^3$ **26.** $f(x) = (x + 1)^3$

27. $g(x) = x^4 - x^3 - 2x^2$ **28.** $h(x) = -2x^3 - x^2 + x$

29. $f(t) = t^3 - 3t$ **30.** $f(x) = -x^3 + 3x - 2$

31. $f(x) = x(x + 3)^2$ **32.** $f(t) = t^4 - 4t^2$

In Exercises 33–38, divide by long division.

33. $\dfrac{24x^2 - x - 8}{3x - 2}$ **34.** $\dfrac{5x^3 - 13x^2 - x + 2}{x^2 - 3x + 1}$

35. $\dfrac{x^4 - 3x^2 + 2}{x^2 - 1}$ **36.** $\dfrac{3x^4}{x^2 - 1}$

37. $\dfrac{x^4 - 3x^3 + 4x^2 - 6x + 3}{x^2 + 2}$

38. $\dfrac{6x^4 + 10x^3 + 13x^2 - 5x + 2}{2x^2 - 1}$

In Exercises 39–42, use synthetic division to perform the division.

39. $\dfrac{0.25x^4 - 4x^3}{x - 2}$

40. $\dfrac{2x^3 + 2x^2 - x + 2}{x - \left(\frac{1}{2}\right)}$

41. $\dfrac{2x^3 - 5x^2 + 12x - 5}{x - (1 + 2i)}$

42. $\dfrac{9x^3 - 15x^2 - 11x - 5}{x - \left[\left(\frac{1}{3}\right) + \left(\frac{2}{3}\right)i\right]}$

In Exercises 43–46, use synthetic division to determine whether the values of x are zeros of the function.

43. $f(x) = 2x^3 + 3x^2 - 20x - 21$
    (a) $x = 4$ (b) $x = -1$
    (c) $x = -\frac{7}{2}$ (d) $x = 0$

44. $f(x) = 20x^4 + 9x^3 - 14x^2 - 3x$
    (a) $x = -1$ (b) $x = \frac{3}{4}$
    (c) $x = 0$ (d) $x = 1$

45. $f(x) = 2x^3 + 7x^2 - 18x - 30$
    (a) $x = 1$ (b) $x = \frac{5}{2}$
    (c) $x = -3 + \sqrt{3}$ (d) $x = 0$

46. $f(x) = 3x^3 - 26x^2 + 364x - 232$
    (a) $x = 4 - 10i$ (b) $x = 4$
    (c) $x = \frac{2}{3}$ (d) $x = -1$

In Exercises 47–50, use synthetic division and the Remainder Theorem to find the specified value of the function.

47. $g(x) = 2x^4 - 17x^3 + 58x^2 - 77x + 26$
    (a) $g(-2)$ (b) $g(\frac{1}{2})$

48. $h(x) = 5x^5 - 2x^4 - 45x + 18$
    (a) $h(2)$ (b) $h\left(\sqrt{3}\right)$

49. $f(x) = x^4 + 10x^3 - 24x^2 + 20x + 44$
    (a) $f(-3)$ (b) $f\left(\sqrt{2}\,i\right)$

50. $g(t) = 2t^5 - 5t^4 - 8t + 20$
    (a) $g(-4)$ (b) $g\left(\sqrt{2}\right)$

In Exercises 51 and 52, use Descartes's Rule of Signs to determine the possible number of positive and negative zeros of the function.

51. $g(x) = 5x^3 + 3x^2 - 6x + 9$

52. $h(x) = -2x^5 + 4x^3 - 2x^2 + 5$

In Exercises 53 and 54, use the Rational Zero Test to list all possible rational zeros of f.

53. $f(x) = -4x^3 + 8x^2 - 3x + 15$

54. $f(x) = 3x^4 + 4x^3 - 5x^2 - 8$

In Exercises 55–60, find all the zeros of the function. When there is an extended list of possible rational zeros, use a graphing utility to graph the function in order to discard any rational number that is obviously not a zero of the function.

55. $f(x) = 4x^3 - 11x^2 + 10x - 3$

56. $f(x) = 10x^3 + 21x^2 - x - 6$

57. $f(x) = 6x^3 - 5x^2 + 24x - 20$

58. $f(x) = x^3 - 1.3x^2 - 1.7x + 0.6$

59. $f(x) = 6x^4 - 25x^3 + 14x^2 + 27x - 18$

60. $f(x) = 5x^4 + 126x^2 + 25$

In Exercises 61–64, (a) list the possible rational zeros of f and show that none are zeros, (b) use a graphing utility to verify that there are real zeros of f, and (c) move the cursor along the graph to approximate the irrational zeros(s) of the function accurate to two decimal places.

61. $f(x) = x^4 + 2x - 1$ **62.** $g(x) = x^3 - 3x^2 + 3x + 2$

63. $h(x) = x^3 - 6x^2 + 12x - 10$

64. $f(x) = x^5 + 2x^3 - 3x - 20$

65. *Volume Marker* A spherical tank of radius 50 feet (see figure) will be two-thirds full when the depth of the fluid is $x + 50$ feet, where
$$3x^3 - 22,500x + 250,000 = 0.$$
Use a graphing utility to approximate x to within 0.01 unit.

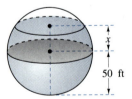

50 ft

FIGURE FOR 65

66. *Age of the Groom* The average age of the groom in a wedding for the given age of the bride can be approximated by the model
$$y = -0.00428x^2 + 1.442x - 3.136, \qquad 20 \le x \le 55,$$
where y is the age of the groom and x is the age of the bride. For what age of the bride is the average age of the groom 30? (*Source: U.S. National Center for Health Statistics*)

In Exercises 67–70, find the domain of the function and identify any horizontal or vertical asymptotes.

67. $f(x) = \dfrac{4}{x + 3}$

68. $f(x) = \dfrac{2x^2 + 5x - 3}{x^2 + 2}$

69. $g(x) = \dfrac{x^2}{x^2 - 4}$

70. $g(x) = \dfrac{1}{(x - 3)^2}$

In Exercises 71–80, sketch the graph of the rational function. As sketching aids, check for intercepts, symmetry, vertical asymptotes, horizontal asymptotes, and slant asymptotes.

71. $f(x) = \dfrac{-5}{x^2}$

72. $f(x) = \dfrac{4}{x}$

73. $g(x) = \dfrac{2 + x}{1 - x}$

74. $g(x) = \dfrac{-2}{(x + 3)^2}$

75. $P(x) = \dfrac{x^2}{x^2 + 1}$

76. $f(x) = \dfrac{2x}{x^2 + 4}$

77. $y = \dfrac{x}{x^2 - 1}$

78. $h(x) = \dfrac{4}{(x - 1)^2}$

79. $f(x) = \dfrac{2x^3}{x^2 + 1}$

80. $y = \dfrac{2x^2}{x^2 - 4}$

In Exercises 81–84, use a graphing utility to graph the function.

81. $s(x) = \dfrac{8x^2}{x^2 + 4}$

82. $y = \dfrac{5x}{x^2 - 4}$

83. $g(x) = \dfrac{x^2 + 1}{x + 1}$

84. $y = \dfrac{1}{x + 3} + 2$

85. *Average Cost*   A business has a cost of $C = 0.5x + 500$ for producing x units. The average cost per unit is

$$\overline{C} = \frac{C}{x} = \frac{0.5x + 500}{x}, \qquad 0 < x.$$

Determine the average cost per unit as x increases without bound. (Find the horizontal asymptote.)

86. *Average Cost*   The cost of producing x units of a product is C, and therefore the average cost per unit is

$$\overline{C} = \frac{C}{x} = \frac{100{,}000 + 0.9x}{x}, \qquad 0 < x.$$

Use a graphing utility to graph the average cost function and find the average cost of producing $x = 1000$, $x = 10{,}000$, and $x = 100{,}000$ units.

87. *Seizure of Illegal Drugs*   The cost, in millions of dollars, for the federal government to seize $p\%$ of an illegal drug as it enters the country is given by

$$C = \frac{528p}{100 - p}, \qquad 0 \le p < 100.$$

(a) Find the cost of seizing 25% of an illegal drug.
(b) Find the cost of seizing 50%.
(c) Find the cost of seizing 75%.
(d) According to this model, would it be possible to seize 100% of the drug?

88. *Population of Fish*   The Parks and Wildlife Commission introduces 80,000 fish into a large man-made lake. The population of the fish, in thousands, is given by

$$N = \frac{20(4 + 3t)}{1 + 0.05t}, \qquad 0 \le t,$$

where t is the time in years.
(a) Find the population when t is 5, 10, and 25.
(b) What is the limiting number of fish in the lake as time increases?

89. *Capillary Attraction*   The rise of distilled water in tubes x inches in diameter is approximated by the model

$$y = \left(\frac{0.80 - 0.54x}{1 + 2.72x}\right)^2, \qquad 0 < x,$$

where y is measured in inches. Use a graphing utility to graph the model and approximate the diameter of the tube that will cause the water to rise 0.1 inch.

90. *Photosynthesis*   The amount y of CO_2 uptake in milligrams per square decimeter per hour at optimal temperatures and with the natural supply of CO_2 is approximated by the model

$$y = \frac{18.47x - 2.96}{0.23x + 1}, \qquad 0 < x,$$

where x is the light intensity in watts per square meter. Use a graphing utility to graph the function and determine the limiting amount of CO_2 uptake.

SOLUTION

Table 4.2 lists some values for each function, and Figure 4.2 shows their graphs. Note that both graphs are decreasing. Moreover, the graph of

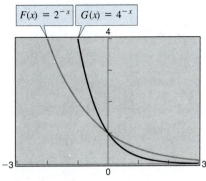

TABLE 4.2

x	-3	-2	-1	0	1	2
A. $F(x) = 2^{-x}$	8	4	2	1	$\frac{1}{2}$	$\frac{1}{4}$
B. $G(x) = 4^{-x}$	64	16	4	1	$\frac{1}{4}$	$\frac{1}{16}$

FIGURE 4.2

D I S C O V E R Y

Use a graphing utility to graph $y = a^x$ with $a = 2, 3,$ and 5 on the same viewing rectangle. (Use a viewing rectangle in which $-2 \le x \le 1$ and $0 \le y \le 2$.) How do the graphs compare with each other? Which graph is on the top in the interval $(-\infty, 0)$? Which is on the bottom? Which graph is on the top in the interval $(0, \infty)$? Which is on the bottom?

Repeat this experiment with the graphs of $y = a^{-x}$ with $a = 2, 3,$ and 5.

The graphs in Figures 4.1 and 4.2 are typical of the exponential functions a^x and a^{-x}. They have $(0, 1)$ as their y-intercept, they have the x-axis as a horizontal asymptote, and they are continuous. The basic characteristics of these exponential functions are summarized in Figure 4.3. Try using a graphing utility to verify the characteristics listed in this summary.

Graph of $y = a^x$
- Domain: $(-\infty, \infty)$
- Range: $(0, \infty)$
- Intercept: $(0, 1)$
- Increasing
- x-axis is a horizontal asymptote
  $(a^x \to 0$ as $x \to -\infty)$
- Continuous

Graph of $y = a^{-x}$
- Domain: $(-\infty, \infty)$
- Range: $(0, \infty)$
- Intercept: $(0, 1)$
- Decreasing
- x-axis is a horizontal asymptote
  $(a^{-x} \to 0$ as $x \to \infty)$
- Continuous
- Reflection of graph of
  $y = a^x$ about y-axis

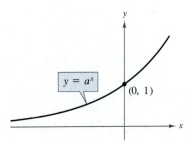

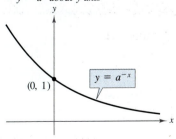

Characteristics of the Exponential Functions a^x and $a^{-x} (a > 1)$

FIGURE 4.3

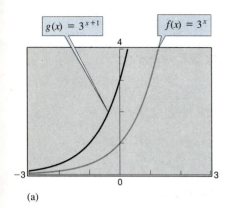

(a)

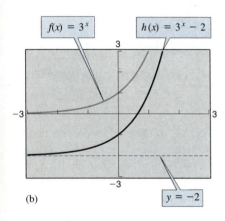

(b)

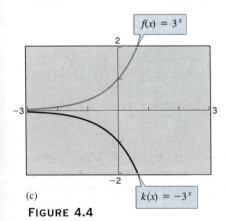

(c)

FIGURE 4.4

EXAMPLE 4 Graphing Exponential Functions

Compare the graph of each of the following with the graph of $f(x) = 3^x$. Identify the domain and range of each function.

A. $g(x) = 3^{x+1}$
B. $h(x) = 3^x - 2$
C. $k(x) = -3^x$

SOLUTION

A. Because $g(x) = 3^{x+1} = f(x + 1)$, the graph of g can be obtained by shifting the graph of f one unit to the left, as shown in Figure 4.4(a). The domain is $(-\infty, \infty)$ and the range is $(0, \infty)$.

B. Because $h(x) = 3^x - 2 = f(x) - 2$, the graph of h can be obtained by shifting the graph of f down two units, as shown in Figure 4.4(b). The domain is $(-\infty, \infty)$ and the range is $(-2, \infty)$.

C. Because $k(x) = -3^x = -f(x)$, the graph of k can be obtained by reflecting the graph of f in the x-axis, as shown in Figure 4.4(c). The domain is $(-\infty, \infty)$ and the range is $(-\infty, 0)$.

The Natural Base e

For many applications, the convenient choice for a base is the irrational number

$$e \approx 2.71828 \ldots ,$$

called the **natural base.** The function $f(x) = e^x$ is the **natural exponential function.** Its graph is shown in Figure 4.5. Be sure you see that for the exponential function $f(x) = e^x$, e is the constant 2.71828 . . . , whereas x is the variable.

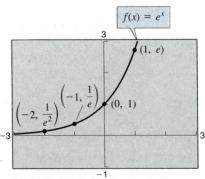

FIGURE 4.5

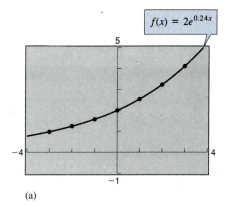

(a)

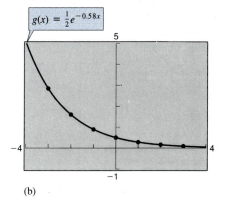

(b)

FIGURE 4.6

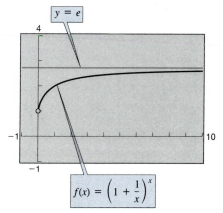

FIGURE 4.7

EXAMPLE 5   Evaluating the Natural Exponential Function

Number	Display	Rounded to 3 Decimal Places
A. e^2	7.389056099	7.389
B. e^{-1}	0.3678794412	0.368
C. $e^{0.48}$	1.616074402	1.616

EXAMPLE 6   Graphing a Natural Exponential Function

Graph the following natural exponential functions.

A. $f(x) = 2e^{0.24x}$ **B.** $g(x) = \dfrac{1}{2}e^{-0.58x}$

SOLUTION

In Figure 4.6, note that each graph has the x-axis as a horizontal asymptote.

EXAMPLE 7   Approximation of the Number e

Evaluate the expression

$$\left(1 + \frac{1}{x}\right)^x$$

for several large values of x to see that the values approach $e \approx 2.71828$ as x increases without bound.

SOLUTION

Using a calculator, you can complete Table 4.3. From this table, it seems reasonable to conclude that

$$\left(1 + \frac{1}{x}\right)^x \rightarrow e \quad \text{as} \quad x \rightarrow \infty.$$

You can further confirm this conclusion by using a graphing utility to graph $f(x) = [1 + (1/x)]^x$ and $y = e$ on the same display, as shown in Figure 4.7. Note that as x increases, the graph of f gets closer and closer to the line given by $y = e$.

TABLE 4.3

x	10	100	1000	10,000	100,000	1,000,000
$\left(1 + \dfrac{1}{x}\right)^x$	2.59374	2.70481	2.71692	2.71815	2.71827	2.71828

Compound Interest

One of the most familiar examples of exponential growth is that of an investment earning *continuously compounded interest*. Suppose a principal P is invested at an annual percentage rate r, compounded once a year. If the interest is added to the principal at the end of the year, then the balance is

$$P_1 = P + Pr = P(1 + r).$$

This pattern of multiplying the previous principal by $1 + r$ is then repeated each successive year, as shown in Table 4.4.

TABLE 4.4

Time in years	Balance after each compounding
0	$P = P$
1	$P_1 = P(1 + r)$
2	$P_2 = P_1(1 + r) = P(1 + r)(1 + r) = P(1 + r)^2$
3	$P_3 = P_2(1 + r) = P(1 + r)^2(1 + r) = P(1 + r)^3$
.	.
.	.
.	.
n	$P_n = P(1 + r)^n$

To accommodate more frequent (quarterly, monthly, or daily) compounding of interest, let n be the number of compoundings per year and let t be the number of years. Then the rate per compounding is r/n, and the account balance after t years is

$$A = P\left(1 + \frac{r}{n}\right)^{nt}.$$ *Amount with n compoundings per year*

If you let the number of compoundings, n, increase without bound, you approach **continuous compounding.** In the formula for n compoundings per year, let $m = n/r$. This produces

$$A = P\left(1 + \frac{r}{n}\right)^{nt} = P\left(1 + \frac{1}{m}\right)^{mrt} = P\left[\left(1 + \frac{1}{m}\right)^{m}\right]^{rt}.$$

As m increases without bound, you know from Example 7 that $[1 + (1/m)]^m$ approaches e. Hence, for continuous compounding, it follows that

$$P\left[\left(1 + \frac{1}{m}\right)^{m}\right]^{rt} \rightarrow P[e]^{rt},$$

and you can write $A = Pe^{rt}$. This result is part of the reason that e is the "natural" choice for a base of an exponential function.

FORMULAS FOR COMPOUND INTEREST

After t years, the balance A in an account with principal P and annual percentage rate r (expressed as a decimal) is given by the following formulas.

1. For n compoundings per year: $A = P\left(1 + \dfrac{r}{n}\right)^{nt}$

2. For continuous compounding: $A = Pe^{rt}$

EXAMPLE 8   **Finding the Balance for Compound Interest**

A sum of $9000 is invested at an annual percentage rate of 8.5%, compounded annually. Find the balance in the account after 3 years.

SOLUTION

In this case, $P = 9000$, $r = 8.5\% = 0.085$, $n = 1$, and $t = 3$. Using the formula

$$A = P\left(1 + \frac{r}{n}\right)^{nt},$$

you have

$$A = 9000(1 + 0.085)^3 = 9000(1.085)^3 \approx \$11{,}495.60.$$

The graph of the balance in the account after t years is shown in Figure 4.8.

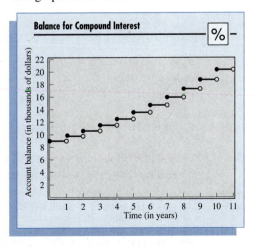

FIGURE 4.8

EXAMPLE 9 **Compounding n Times and Compounding Continuously**

A total of $12,000 is invested at an annual percentage rate of 9%. Find the balance after 5 years if it is compounded

A. quarterly.
B. continuously.

SOLUTION

A. For quarterly compoundings, you have $n = 4$. Thus, in 5 years at 9%, the balance is

$$A = P\left(1 + \frac{r}{n}\right)^{nt}$$

$$= 12{,}000\left(1 + \frac{0.09}{4}\right)^{4(5)}$$

$$= \$18{,}726.11.$$

B. Compounding continuously, the balance is

$$A = Pe^{rt} = 12{,}000e^{0.09(5)} = \$18{,}819.75.$$

Note that continuous compounding yields

$$\$18{,}819.75 - \$18{,}726.11 = \$93.64$$

more than quarterly compounding.

Other Applications

EXAMPLE 10 **An Application Involving Radioactive Decay**

Let y represent the mass of a particular radioactive element whose half-life is 25 years. After t years, the mass (in grams) is given by

$$y = 10\left(\frac{1}{2}\right)^{t/25}.$$

A. What is the initial mass (when $t = 0$)?
B. How much of the initial mass is present after 80 years?

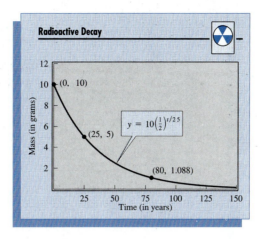

FIGURE 4.9

*Technology Note*_____

When will the mass be four grams in Example 10? You can easily determine this by graphing

$y_1 = 10(.5)^{t/25}$

$y_2 = 4$

on the same viewing rectangle. You should be able to verify that the point of intersection is approximately $t = 33$.

SOLUTION

A. When $t = 0$, the mass is

$$y = 10\left(\frac{1}{2}\right)^0 = 10(1) = 10 \text{ grams.}$$

B. When $t = 80$, the mass is

$$y = 10\left(\frac{1}{2}\right)^{80/25} = 10(0.5)^{3.2} \approx 1.088 \text{ grams.}$$

The graph of this function is shown in Figure 4.9.

EXAMPLE 11   **Population Growth**

The approximate number of fruit flies in an experimental population after t hours is given by

$$Q(t) = 20e^{0.03t}, \qquad t \geq 0.$$

A. Find the initial number of fruit flies in the population.
B. How large is the population of fruit flies after 72 hours?
C. Sketch the graph of Q.

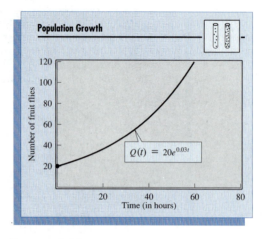

FIGURE 4.10

SOLUTION

A. To find the initial population, evaluate $Q(t)$ at $t = 0$.

$$Q(0) = 20e^{0.03(0)} = 20e^0 = 20(1) = 20 \text{ flies}$$

B. After 72 hours, the population size is

$$Q(72) = 20e^{(0.03)(72)} = 20e^{2.16} \approx 173 \text{ flies.}$$

C. The graph of Q is shown in Figure 4.10.

DISCUSSION PROBLEM

EXPONENTIAL GROWTH

Consider the following sequences of numbers.

  Sequence 1: 2, 4, 6, 8, 10, 12, . . . , $2n$

  Sequence 2: 2, 4, 8, 16, 32, 64, . . . , 2^n

The first sequence, $f(n) = 2n$, is an example of **linear growth.** The second sequence, $f(n) = 2^n$, is an example of **exponential growth.** Which of the following sequences represents linear growth and which represents exponential growth? Can you find a linear function and an exponential function that represent the two sequences?

(a) 3, 6, 9, 12, 15, . . .
(b) 3, 9, 27, 81, 243, . . .

Later you will see that sequences that represent linear growth are *arithmetic sequences* and sequences that represent exponential growth are *geometric sequences.*

WARM-UP

The following warm-up exercises involve skills that were covered in earlier sections. You will use these skills in the exercise set for this section.

In Exercises 1–10, use the properties of exponents to simplify the expression.

1. $5^{2x}(5^{-x})$

2. $3^{-x}(3^{3x})$

3. $\dfrac{4^{5x}}{4^{2x}}$

4. $\dfrac{10^{2x}}{10^{x}}$

5. $(4^x)^2$

6. $(4^{2x})^5$

7. $\left(\dfrac{2^x}{3^x}\right)^{-1}$

8. $(4^{6x})^{1/2}$

9. $(2^{3x})^{-1/3}$

10. $(16^x)^{1/4}$

SECTION 4.1 · EXERCISES

In Exercises 1–10, use a calculator to evaluate the expression. Round your result to three decimal places.

1. $(3.4)^{5.6}$

2. $5000(2^{-1.5})$

3. $1000(1.06)^{-5}$

4. $(1.005)^{400}$

5. $5^{-\pi}$

6. $\sqrt[3]{4395}$

7. $100^{\sqrt{2}}$

8. $e^{1/2}$

9. $e^{-3/4}$

10. $e^{3.2}$

In Exercises 11–18, match the exponential function with its graph, and describe the given viewing rectangle. [The graphs are labeled (a), (b), (c), (d), (e), (f), (g), and (h).]

11. $f(x) = 3^x$

12. $f(x) = -3^x$

13. $f(x) = 3^{-x}$

14. $f(x) = -3^{-x}$

15. $f(x) = 3^x - 4$

16. $f(x) = 3^x + 1$

17. $f(x) = -3^{x-2}$

18. $f(x) = 3^{x-2}$

(a)

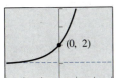

(b)

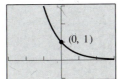

(c)

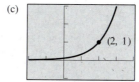

(d)

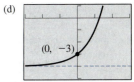

(e)

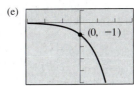

(f)

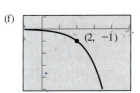

(g)

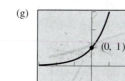

(h)

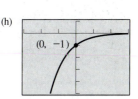

19. Use a graphing utility to graph $y = 3^x$ and $y = 4^x$ in the same viewing rectangle. Use the graphs to solve the inequality

$$4^x < 3^x.$$

20. Use a graphing utility to graph $y = \left(\frac{1}{2}\right)^x$ and $y = \left(\frac{1}{4}\right)^x$ in the same viewing rectangle. Use the graphs to solve the inequality

$$\left(\tfrac{1}{4}\right)^x < \left(\tfrac{1}{2}\right)^x.$$

In Exercises 21–30, sketch the graph of the exponential function *by hand*. Then use a graphing utility to confirm your graph.

21. $g(x) = 5^x$

22. $f(x) = (\frac{3}{2})^x$

23. $f(x) = (\frac{1}{5})^x = 5^{-x}$

24. $h(x) = (\frac{3}{2})^{-x}$

25. $h(x) = 5^{x-2}$

26. $g(x) = (\frac{3}{2})^{x+2}$

27. $g(x) = 5^{-x} - 3$

28. $f(x) = (\frac{3}{2})^{-x} + 2$

29. $y = 3^{x-2} + 1$

30. $y = 4^{x+1} - 2$

In Exercises 31–36, use a graphing utility to graph the exponential function.

31. $y = 1.08^{-5x}$

32. $y = 1.08^{5x}$

33. $s(t) = 2e^{0.12t}$

34. $s(t) = 3e^{-0.2t}$

35. $g(x) = 1 + e^{-x}$

36. $h(x) = e^{x-2}$

37. Graph the following functions in the same viewing rectangle of a graphing utility.
(a) $f(x) = 3^x$
(b) $g(x) = f(x - 2) = 3^{x-2}$
(c) $h(x) = -\frac{1}{2}f(x) = -\frac{1}{2}3^x$
(d) $q(x) = f(-x) + 3 = 3^{-x} + 3$

38. Use a graphing utility to graph each of the following functions. Use the graphs to determine any asymptotes of the functions.
(a) $f(x) = \dfrac{8}{1 + e^{-0.5x}}$ (b) $g(x) = \dfrac{8}{1 + e^{-0.5/x}}$

39. Use a graphing utility to graph each of the following functions. Use the graphs to determine where each function is increasing and decreasing, and approximate any relative maximum or minimum values of each function.
(a) $f(x) = x^2 e^{-x}$ (b) $g(x) = x2^{3-x}$

40. Use a graphing utility to demonstrate that
$$\left(1 + \frac{0.5}{x}\right)^x \to e^{0.5}$$
as x increases without bound.

In Exercises 41–44, complete the table to determine the balance A for P dollars invested at rate r for t years and compounded n times per year.

n	1	2	4	12	365	Continuous compounding
A						

41. $P = \$2500$, $r = 12\%$, $t = 10$ years

42. $P = \$1000$, $r = 10\%$, $t = 10$ years

43. $P = \$2500$, $r = 12\%$, $t = 20$ years

44. $P = \$1000$, $r = 10\%$, $t = 40$ years

In Exercises 45–48, complete the table to determine the amount of money P that should be invested at rate r to produce a final balance of $\$100,000$ in t years.

t	1	10	20	30	40	50
P						

45. $r = 9\%$, compounded continuously

46. $r = 12\%$, compounded continuously

47. $r = 10\%$, compounded monthly

48. $r = 7\%$, compounded daily

49. *Trust Fund*   On the day of your grandchild's birth, you deposited $\$25,000$ in a trust fund that pays 8.75% interest, compounded continuously. Determine the balance in this account on your grandchild's 25th birthday.

50. *Trust Fund*   Suppose you deposit $\$5000$ in a trust fund that pays 7.5% interest, compounded continuously. In the trust fund, you specify that the balance will be given to the college from which you graduated after the money has earned interest for 50 years. How much will your college receive after 50 years?

51. *Demand Function*   The demand equation for a certain product is
$$p = 500 - 0.5e^{0.004x}.$$
Find the price p for a demand of (a) $x = 1000$ units and (b) $x = 1500$ units. Describe the graph of p. What is the domain? What is the range?

52. *Demand Function*   The demand equation for a certain product is
$$p = 5000\left(1 - \frac{4}{4 + e^{-0.002x}}\right).$$
Find the price p for a demand of (a) $x = 100$ units and (b) $x = 500$ units. Describe the graph of p. What is the domain? What is the range?

53. *Bacteria Growth*   A certain type of bacteria increases according to the model
$$P(t) = 100e^{0.2197t},$$
where t is the time in hours. Find (a) $P(0)$, (b) $P(5)$, and (c) $P(10)$.

54. *Population Growth*   The population of a town increases according to the model
$$P(t) = 2500e^{0.0293t},$$
where t is the time in years, with $t = 0$ corresponding to 1990. Use the model to approximate the population in (a) 1995, (b) 2000, and (c) 2010.

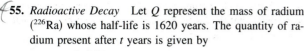

55. *Radioactive Decay*   Let Q represent the mass of radium (^{226}Ra) whose half-life is 1620 years. The quantity of radium present after t years is given by

$$Q = 25 \left(\tfrac{1}{2}\right)^{t/1620}.$$

(a) Determine the initial quantity (when $t = 0$).

(b) Determine the quantity present after 1000 years.

(c) Use a graphing utility to graph this function over the interval $t = 0$ to $t = 5000$.

56. *Radioactive Decay*   Let Q represent the mass of carbon-14 (^{14}C) whose half-life is 5730 years. The quantity of carbon-14 present after t years is given by

$$Q = 10\left(\tfrac{1}{2}\right)^{t/5730}.$$

(a) Determine the initial quantity (when $t = 0$).

(b) Determine the quantity present after 2000 years.

(c) Sketch the graph of this function over the interval $t = 0$ to $t = 10,000$.

57. *Forest Defoliation*   To estimate the amount of defoliation caused by the gypsy moth during a given year, a forester counts the number of egg masses on $\frac{1}{40}$ of an acre the preceding fall. The percentage of defoliation y is approximated by

$$y = \frac{300}{3 + 17e^{-1.57x}},$$

where x is the number of egg masses in thousands.

(a) Use a graphing utility to graph the function.

(b) Estimate the percentage of defoliation if 2000 egg masses are counted.

(c) Estimate the number of egg masses that existed if you observe that approximately $\frac{2}{3}$ of a forest is defoliated. (*Source:* Department of Environmental Resources)

58. *Inflation*   If the annual rate of inflation averages 5% over the next 10 years, then the approximate cost C of goods or services during any year in that decade will be given by

$$C(t) = P(1.05)^t,$$

where t is the time in years and P is the present cost. If the price of an oil change for your car is presently $19.95, estimate the price 10 years from now.

59. *Depreciation*   After t years, the value of a car that cost $20,000 is given by

$$V(t) = 20,000 \left(\tfrac{3}{4}\right)^t.$$

Graph the function and determine the value of the car 2 years after it was purchased.

60. Given the exponential function $f(x) = a^x$, show that

(a) $f(u + v) = f(u) \cdot f(v)$ (b) $f(2x) = [f(x)]^2.$

4.2   LOGARITHMIC FUNCTIONS AND THEIR GRAPHS

**Logarithmic Functions / Graphs of Logarithmic Functions /
The Natural Logarithmic Function / Application**

Logarithmic Functions

In Section 1.7, you learned that if a function has the property that no horizontal line intersects its graph more than once, then the function must have an inverse. By looking back at the graphs of the exponential functions introduced in Section 4.1, you will see that every function of the form $f(x) = a^x$ passes the "horizontal line test," and therefore must have an inverse. This inverse function is the **logarithmic function with base** a.

REMARK   The equations $y = \log_a x$ and $a^y = x$ are equivalent. The first equation is in logarithmic form and the second is in exponential form.

DEFINITION OF LOGARITHMIC FUNCTION

For $x > 0$ and $0 < a \neq 1$,

$$y = \log_a x \quad \text{if and only if} \quad a^y = x.$$

The function $f(x) = \log_a x$ is the **logarithmic function with base a.**

When evaluating logarithms, remember that *a logarithm is an exponent.* This means that $\log_a x$ is the exponent to which a must be raised to obtain x. For instance, $\log_2 8 = 3$ because 2 must be raised to the third power to obtain 8.

EXAMPLE 1   Evaluating Logarithms

A. $\log_2 32 = 5$ because $2^5 = 32$.

B. $\log_3 27 = 3$ because $3^3 = 27$.

C. $\log_4 2 = \dfrac{1}{2}$ because $4^{1/2} = \sqrt{4} = 2$.

D. $\log_{10} \dfrac{1}{100} = -2$ because $10^{-2} = \dfrac{1}{10^2} = \dfrac{1}{100}$.

E. $\log_3 1 = 0$ because $3^0 = 1$.

F. $\log_2 2 = 1$ because $2^1 = 2$.

The logarithmic function with base 10 is the **common logarithmic function.** On most calculators, this function is denoted by *log*. You can tell whether this key denotes base 10 by evaluating log 10. The display should be 1.

EXAMPLE 2   Evaluating Logarithms on a Calculator

Number	Display	Rounded to 3 Decimal Places
A. $\log_{10} 54$	1.73239376	1.732
B. $2 \log_{10} 2.5$	0.7958800173	0.796
C. $\log_{10}(-2)$	ERROR	

Note that most calculators display an error message when you try to evaluate $\log_{10}(-2)$. The reason for this is that the domain of every logarithmic function is the set of *positive real numbers.*

PROPERTIES OF LOGARITHMS

1. $\log_a 1 = 0$ because 0 is the power to which a must be raised to obtain 1.
2. $\log_a a = 1$ because 1 is the power to which a must be raised to obtain a.
3. $\log_a a^x = x$ because x is the power to which a must be raised to obtain a^x.

D I S C O V E R Y

Use a graphing utility to graph $y = \log(x)$ on the viewing rectangle $-2 \le x \le 10$, $-8 \le y \le 4$. What are the domain and range of this function? On what interval does y increase? Does it have a maximum or minimum? What are the intercepts and asymptotes?

Graphs of Logarithmic Functions

To sketch the graph of $y = \log_a x$, you can use the fact that the graphs of inverse functions are reflections of each other in the line $y = x$.

EXAMPLE 3   Graphs of Exponential and Logarithmic Functions

On the same coordinate plane, sketch graphs of the following functions by hand.

A. $f(x) = 2^x$
B. $g(x) = \log_2 x$

SOLUTION

A. For $f(x) = 2^x$, construct a table of values.

x	-2	-1	0	1	2	3
$f(x) = 2^x$	$\frac{1}{4}$	$\frac{1}{2}$	1	2	4	8

By plotting these points and connecting them with a smooth curve, you obtain the graph shown in Figure 4.11.

B. Because $g(x) = \log_2 x$ is the inverse of $f(x) = 2^x$, the graph of g is obtained by reflecting the graph of f in the line $y = x$, as shown in Figure 4.11. This reflection is formed by interchanging the x- and y-coordinates of the points in the above table of values.

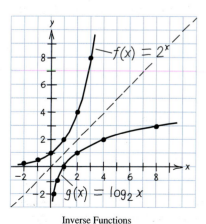

Inverse Functions

FIGURE 4.11

EXAMPLE 4   Sketching the Graph of a Logarithmic Function

Sketch the graph of $f(x) = \log_{10} x$ by hand. Then verify your result with a graphing utility.

SOLUTION

Begin by making a table of values. Note that some of the values can be obtained without a calculator, whereas others require a calculator. Plot the corresponding points and sketch the graph in Figure 4.12.

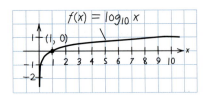

FIGURE 4.12

x	Without a calculator				With a calculator		
x	$\frac{1}{100}$	$\frac{1}{10}$	1	10	2	5	8
$\log_{10} x$	-2	-1	0	1	0.301	0.699	0.903

Compare this graph with that obtained with a graphing utility, and note that the domain and range are $(0, \infty)$ and $(-\infty, \infty)$, respectively.

The graph in Figure 4.12 is typical of functions of the form $f(x) = \log_a x$, where $a > 1$. They have one x-intercept, $(1, 0)$, and one vertical asymptote, $x = 0$. Notice how slowly the graph rises for $x > 1$. In Figure 4.12 you would need to move out to $x = 1000$ before the graph would rise to $y = 3$. The basic characteristics of logarithmic graphs are summarized in Figure 4.13.

In Example 5, the graph of $\log_a x$ is used to sketch the graphs of functions of the form $y = b \pm \log_a(x + c)$. The function $f(x) = \log_a(bx + c)$ has a domain that consists of all x such that $bx + c > 0$. The vertical asymptote occurs when $bx + c = 0$, and the x-intercept occurs when $bx + c = 1$.

REMARK   In Figure 4.13, note that the vertical asymptote occurs at $x = 0$, where $\log_a x$ os *undefined*. As x gets close to 0, the value of $\log_a$ approaches negative infinity.

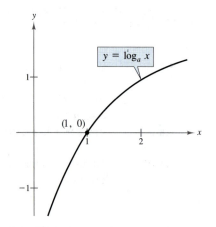

Graph of $y = \log_a x$, $a > 1$
- Domain: $(0, \infty)$
- Range: $(-\infty, \infty)$
- Intercept: $(1, 0)$
- Increasing
- y-axis is a vertical asymptote
  ($\log_a x \to -\infty$ as $x \to 0^+$)
- Continuous
- Reflection of graph of $y = a^x$ about the line $y = x$

FIGURE 4.13

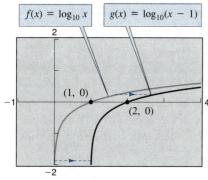

(a) Right shift of 1 unit
Vertical asymptote is $x = 1$.

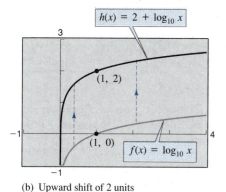

(b) Upward shift of 2 units
Vertical asymptote remains $x = 0$.

FIGURE 4.14

EXAMPLE 5   **Sketching the Graphs of Logarithmic Functions**

Compare the graphs of the following functions with the graph of $f(x) = \log_{10} x$.

A. $g(x) = \log_{10}(x - 1)$ **B.** $h(x) = 2 + \log_{10} x$

SOLUTION

The graph of each of these functions is similar to the graph of $f(x) = \log_{10} x$, as shown in Figure 4.14.

A. Because $g(x) = \log_{10}(x - 1) = f(x - 1)$, the graph of g can be obtained by shifting the graph of f one unit to the right.
B. Because $h(x) = 2 + \log_{10} x = 2 + f(x)$, the graph of h can be obtained by shifting the graph of f two units up.

The Natural Logarithmic Function

As with exponential functions, the most widely used base for logarithmic functions is the number e. The logarithmic function with base e is called the **natural logarithmic function.** It is denoted by the symbol $\ln x$, read as "el en of x."

THE NATURAL LOGARITHMIC FUNCTION

The function defined by

$$f(x) = \log_e x = \ln x, \qquad x > 0$$

is called the **natural logarithmic function.**

The three properties of logarithms listed earlier in this section are also valid for natural logarithms.

PROPERTIES OF NATURAL LOGARITHMS

1. $\ln 1 = 0$ because 0 is the power to which e must be raised to obtain 1.
2. $\ln e = 1$ because 1 is the power to which e must be raised to obtain e.
3. $\ln e^x = x$ because x is the power to which e must be raised to obtain e^x.

FIGURE 4.15

The graph of the natural logarithmic function is shown in Figure 4.15.

EXAMPLE 6   **Evaluating the Natural Logarithmic Function**

A. $\ln \dfrac{1}{e} = \ln e^{-1} = -1$ *Property 3*

B. $\ln e^2 = 2$ *Property 3*

On most calculators, the natural logarithm is denoted by *ln.* Try using your calculator to evaluate the natural logarithmic expressions shown in the next example.

EXAMPLE 7   **Evaluating the Natural Logarithmic Function**

Number	*Display*	*Rounded to 3 Decimal Places*
A. $\ln 2$	0.6931471806	0.693
B. $\ln 0.3$	−1.203972804	−1.204
C. $\ln e$	1	1.000
D. $\ln(-1)$	ERROR	

Be sure you see that $\ln(-1)$ gives an error. This occurs because the domain of $\ln x$ is the set of positive real numbers. (See Figure 4.15.) Hence, $\ln(-1)$ is undefined.

EXAMPLE 8   **Finding the Domains of Logarithmic Functions**

Find the domains and ranges of the following functions.

A. $f(x) = \ln(x - 2)$
B. $g(x) = \ln(2 - x)$
C. $h(x) = \ln x^2$

Then use a graphing utility to graph each function.

SOLUTION

A. Because $\ln(x - 2)$ is defined only if $x - 2 > 0$, it follows that the domain of f is $(2, \infty)$.
B. Because $\ln(2 - x)$ is defined only if $2 - x > 0$, it follows that the domain of g is $(-\infty, 2)$. The graph of g is shown in Figure 4.16(b).
C. Because $\ln x^2$ is defined only if $x^2 > 0$, it follows that the domain of h is all real numbers except $x = 0$. Note that $\ln x^2$ means $\ln(x^2)$.

REMARK   Note how the graphing utility confirms that the domains are correct.

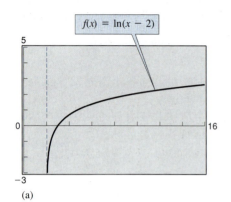

(a)

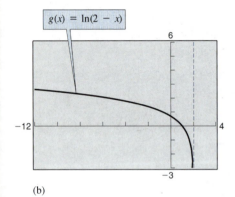

(b)

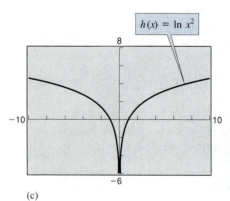

(c)

FIGURE 4.16

Application

EXAMPLE 9   Human Memory Model

Students participating in a psychological experiment attended several lectures on a subject. Every month for a year after that, the students were tested to see how much of the material they remembered. The average scores for the group were given by the *human memory model*

$$f(t) = 75 - 6 \ln(t + 1), \qquad 0 \le t \le 12,$$

where t is the time in months.

A. What was the average score on the original ($t = 0$) exam?
B. What was the average score at the end of $t = 2$ months?
C. What was the average score at the end of $t = 6$ months?
D. Choose an appropriate viewing rectangle and sketch the graph of f.

SOLUTION

A. The original average score was

$$f(0) = 75 - 6 \ln(0 + 1) = 75 - 6(0) = 75.$$

B. After 2 months, the average score was

$$f(2) = 75 - 6 \ln 3 \approx 75 - 6(1.0986) \approx 68.4.$$

C. After 6 months, the average score was

$$f(6) = 75 - 6 \ln 7 \approx 75 - 6(1.9459) \approx 63.3.$$

D. The graph of f is shown in Figure 4.17.

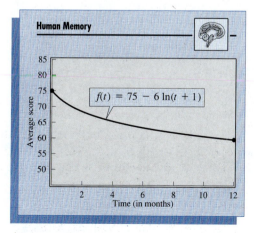

FIGURE 4.17

DISCUSSION PROBLEM

· · · · · · · · · · · · · · · · · · ·

THE GRAPH OF A LOGARITHMIC FUNCTION

Because the range of the logarithmic function $f(x) = \log_a x$ is $(-\infty, \infty)$, you can make the value of $\log_a x$ as large as you want. Can you find values of x that satisfy the following equations?

(a) $\log_{10} x = 10$ (b) $\log_{10} x = 1000$ (c) $\log_{10} x = 10,000,000$

WARM-UP

· · · · · · · · · · · · · · · · · · ·

The following warm-up exercises involve skills that were covered in earlier sections. You will use these skills in the exercise set for this section.

In Exercises 1–4, solve for x.

1. $2^x = 8$ **2.** $4^x = 1$

3. $10^x = 0.1$ **4.** $e^x = e$

In Exercises 5 and 6, evaluate the expression. (Round your result to three decimal places.)

5. e^2 **6.** e^{-1}

In Exercises 7–10, describe how the graph of g is related to the graph of f.

7. $g(x) = f(x + 2)$ **8.** $g(x) = -f(x)$

9. $g(x) = -1 + f(x)$ **10.** $g(x) = f(-x)$

SECTION 4.2 · EXERCISES

In Exercises 1–12, evaluate the expression without using a calculator.

1. $\log_2 16$

2. $\log_2 \frac{1}{8}$

3. $\log_{16} 4$

4. $\log_{27} 9$

5. $\log_7 1$

6. $\log_{10} 1000$

7. $\log_{10} 0.01$

8. $\log_{10} 10$

9. $\ln e^3$

10. $\ln 1$

11. $\log_a a^2$

12. $\log_a \frac{1}{a}$

In Exercises 13–20, use the definition of a logarithm to rewrite the exponential equation as a logarithmic equation. For instance, the logarithmic form of $2^3 = 8$ is $\log_2 8 = 3$.

13. $5^3 = 125$ **14.** $8^2 = 64$

15. $81^{1/4} = 3$ **16.** $9^{3/2} = 27$

17. $6^{-2} = \frac{1}{36}$ **18.** $10^{-3} = 0.001$

19. $e^3 = 20.0855 \ldots$ **20.** $e^0 = 1$

In Exercises 21–26, use a calculator to evaluate the logarithm. (Round your result to three decimal places.)

21. $\log_{10} 345$ **22.** $\log_{10}(\frac{4}{5})$

23. $\log_{10} 0.48$ **24.** $\log_{10} 12.5$

25. $\ln 18.42$ **26.** $\ln(\sqrt{5} - 2)$

In Exercises 27–30, graph f and g on the same coordinate plane to demonstrate that one is the inverse of the other.

27. $f(x) = 3^x$, $g(x) = \log_3 x$

28. $f(x) = 5^x$, $g(x) = \log_5 x$

29. $f(x) = e^x$, $g(x) = \ln x$

30. $f(x) = 10^x$, $g(x) = \log_{10} x$

In Exercises 31–36, use the graph of $y = \ln x$ to match the function with its graph, and describe the given viewing rectangle. [The graphs are labeled, (a), (b), (c), (d), (e), and (f).]

31. $f(x) = \ln x + 2$ **32.** $f(x) = -\ln x$

33. $f(x) = -\ln(x + 2)$ **34.** $f(x) = \ln(x - 1)$

35. $f(x) = \ln(1 - x)$ **36.** $f(x) = -\ln(-x)$

(a)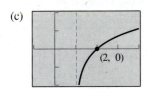

(b)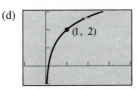

(c)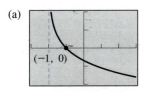

(d)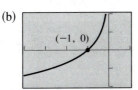

(e)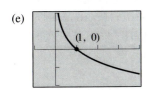

(f)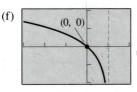

In Exercises 37–42, find the domain, vertical asymptote, and x-intercept of the logarithmic function, and sketch its graph.

37. $f(x) = \log_4 x$ **38.** $f(x) = -\log_6(x + 2)$

39. $y = -\log_3(x + 2)$ **40.** $y = \log_{10}\left(\frac{x}{5}\right)$

41. $f(x) = \ln(x - 2)$ **42.** $g(x) = \ln(-x)$

In Exercises 43 and 44, use a graphing utility to graph the function. Use the graph to determine the intervals in which the function is increasing and decreasing and approximate any relative maximum or minimum values of the function.

43. $f(x) = \dfrac{x}{2} - \ln\dfrac{x}{4}$ **44.** $g(x) = \dfrac{12 \ln x}{x}$

45. Use a graphing utility to graph f and g on the same screen. Then determine which is increasing at the greater rate for "large" values of x. What can you conclude about the rate of growth of the natural logarithmic function?
  (a) $f(x) = \ln x$, $g(x) = \sqrt{x}$
  (b) $f(x) = \ln x$, $g(x) = \sqrt[4]{x}$

46. The table of values was obtained by evaluating a function. Determine which of the statements may be true and which must be false.

x	1	2	8
y	0	1	3

  (a) y is an exponential function of x.
  (b) y is a logarithmic function of x.
  (c) x is an exponential function of y.
  (d) y is a linear function of x.

47. *Human Memory Model*   Students in a mathematics class were given an exam and then tested monthly with an equivalent exam. The average score for the class was given by the human memory model

$$f(t) = 80 - 17 \log_{10}(t + 1), \qquad 0 \le t \le 12,$$

where t is the time in months.
  (a) What was the average score on the original exam ($t = 0$)?
  (b) What was the average score after 4 months?
  (c) What was the average score after 10 months?

48. *Population Growth* The population of a town will double in

$$t = \frac{10 \ln 2}{\ln 67 - \ln 50} \text{ years.}$$

Find t.

49. *World Population Growth* The time in years required for the world population to double if it is increasing at a continuous rate of r is given by

$$t = \frac{\ln 2}{r}.$$

Complete the table.

r	0.005	0.010	0.015	0.020	0.025	0.030
t						

50. *Investment Time* A principal P, invested at $9\frac{1}{2}\%$ and compounded continuously, increases to an amount K times the original principal after t years, where t is given by

$$t = \frac{\ln K}{0.095}.$$

(a) Complete the table.

K	1	2	4	6	8	10	12
t							

(b) Use a graphing utility to graph this function.

Ventilation Rates In Exercises 51 and 52, use the model

$$y = 80.4 - 11 \ln x, \qquad 100 \le x \le 1500,$$

which approximates the minimum required ventilation rate in terms of the air space per child in a public school classroom. In the model, x is the air space per child in cubic feet and y is the ventilation rate in cubic feet per minute.

51. Use a graphing utility to graph the function and approximate the required ventilation rate if there are 300 cubic feet of air space per child.

52. A classroom is designed for 30 students. The air-conditioning system in the room has the capacity of moving 450 cubic feet of air per minute.
(a) Determine the ventilation rate per child assuming the room is filled to capacity.
(b) Use the graph from Exercise 51 to estimate the air space required per child.
(c) Determine the minimum number of square feet of floor space required for the room if the ceiling height is 30 feet.

Monthly Payment In Exercises 53–56, use the model

$$t = \frac{5.315}{-6.7968 + \ln x}, \qquad 1000 < x,$$

which approximates the length of a home mortgage of $120,000 at 10% in terms of the monthly payment. In the model, t is the length of the mortgage in years and x is the monthly payment in dollars (see figure).

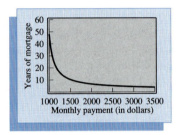

FIGURE FOR 53–56

53. Use the model to approximate the length of a home mortgage (for $120,000 at 10%) that has a monthly payment of $1167.41.

54. Use the model to approximate the length of a home mortgage (for $120,000 at 10%) that has a monthly payment of $1068.45.

55. Approximate the total amount paid over the term of a mortgage with a monthly payment of $1167.41.

56. Approximate the total amount paid over the term of a mortgage with a monthly payment of $1068.45.

57. *Work* The work (in foot-pounds) done in compressing an initial volume of 9 cubic feet at a pressure of 15 pounds per square inch to a volume of 3 cubic feet is

$$W = 19,440(\ln 9 - \ln 3).$$

Find W.

58. *Sound Intensity*   The relationship between the number of decibels β and the intensity of a sound I in watts per square meter is given by

$$\beta = 10 \log_{10}\left(\frac{I}{10^{-16}}\right).$$

Determine the number of decibels of a sound with an intensity of 10^{-4} watts per square meter.

59. (a) Use a calculator to complete the table for the function

$$f(x) = \frac{\ln x}{x}.$$

x	1	5	10	10^2	10^4	10^6
$f(x)$						

(b) Use the table in part (a) to determine what $f(x)$ approaches as x increases without bound.

(c) Use a graphing utility to confirm the result of part (b).

60. Answer the following for the function $f(x) = \log_{10} x$. Do not use a calculator.

(a) What is the domain of f?

(b) Find f^{-1}.

(c) If x is a real number between 1000 and 10,000, determine the interval in which $f(x)$ will be found.

(d) Determine the interval in which x will be found if $f(x)$ is negative.

(e) If $f(x)$ is increased by one unit, then x must have been increased by what factor?

(f) If $f(x_1) = 3n$ and $f(x_2) = n$, find the ratio of x_1 to x_2.

61. Use a graphing utility to determine whether $y = 2 \ln x$ and $y = \ln x^2$ are identical. Explain your reasoning.

4.3   PROPERTIES OF LOGARITHMS
···

**Change of Base / Properties of Logarithms /
Rewriting Logarithmic Expressions**

Change of Base

Most calculators have only two types of "log keys," one for common logarithms (base 10) and one for natural logarithms (base e). Although common logs and natural logs are the most frequently used, you may occasionally need to evaluate logarithms to other bases. To do this, the following *change of base formula* is useful. (This formula is derived in Example 9 in Section 4.4.)

CHANGE OF BASE FORMULA

Let a, b, and x be positive real numbers such that $a \neq 1$ and $b \neq 1$. Then $\log_a x$ is given by

$$\log_a x = \frac{\log_b x}{\log_b a}.$$

REMARK   One way to look at the change of base formula is that logarithms to base a are simply *constant multiples* of logarithms to base b. The constant multiplier is $1/(\log_b a)$.

EXAMPLE 1 Changing Bases

Use *common logarithms* to evaluate the following.

A. $\log_4 30$ **B.** $\log_2 14$

SOLUTION

A. Using the change of base formula with $a = 4$, $b = 10$, and $x = 30$, convert to common logarithms and obtain

$$\log_4 30 = \frac{\log_{10} 30}{\log_{10} 4} \approx \frac{1.47712}{0.60206} \approx 2.4534.$$

B. Using the change of base formula with $a = 2$, $b = 10$, and $x = 14$, convert to common logarithms and obtain

$$\log_2 14 = \frac{\log_{10} 14}{\log_{10} 2} \approx \frac{1.14613}{0.30103} \approx 3.8074.$$

EXAMPLE 2 Changing Bases

Use *natural logarithms* to evaluate the following.

A. $\log_4 30$ **B.** $\log_2 14$

SOLUTION

A. Using the change of base formula with $a = 4$, $b = e$, and $x = 30$, convert to natural logarithms and obtain

$$\log_4 30 = \frac{\ln 30}{\ln 4} \approx \frac{3.40120}{1.38629} \approx 2.4534.$$

B. Using the change of base formula with $a = 2$, $b = e$, and $x = 14$, convert to natural logarithms and obtain

$$\log_2 14 = \frac{\ln 14}{\ln 2} \approx \frac{2.63906}{0.693147} \approx 3.8074.$$

Note that the results agree with those obtained in Example 1, using common logarithms.

Properties of Logarithms

You know from the previous section that the logarithmic function with base a is the *inverse* of the exponential function with base a. Thus, it makes sense that the properties of exponents should have corresponding properties involving logarithms. For instance, the exponential property $a^0 = 1$ corresponds to the logarithmic property $\log_a 1 = 0$.

In this section you will learn how to use the logarithmic properties that correspond to the following three exponential properties.

1. $a^n a^m = a^{n+m}$

2. $\dfrac{a^n}{a^m} = a^{n-m}$

3. $(a^n)^m = a^{nm}$

PROPERTIES OF LOGARITHMS

Let a be a positive number such that $a \ne 1$, and let n be a real number. If u and v are positive real numbers, then the following properties are true.

Base a Logarithm	*Natural Logarithm*
1. $\log_a(uv) = \log_a u + \log_a v$	1. $\ln(uv) = \ln u + \ln v$
2. $\log_a \dfrac{u}{v} = \log_a u - \log_a v$	2. $\ln \dfrac{u}{v} = \ln u - \ln v$
3. $\log_a u^n = n \log_a u$	3. $\ln u^n = n \ln u$

PROOF

We give a proof of Property 1 and leave the other two proofs for you. To prove Property 1, let

$$x = \log_a u \quad \text{and} \quad y = \log_a v.$$

The corresponding exponential forms of these two equations are

$$a^x = u \quad \text{and} \quad a^y = v.$$

Multiplying u and v produces $uv = a^x a^y = a^{x+y}$. The corresponding logarithmic form of $uv = a^{x+y}$ is

$$\log_a(uv) = x + y.$$

Hence, $\log_a(uv) = \log_a u + \log_a v$.

REMARK   There is no general property that can be used to rewrite $\log_a(u \pm v)$. Specifically,

$$\log_a(x + y) \quad \text{DOES NOT EQUAL} \quad \log_a x + \log_a y.$$

EXAMPLE 3   Using Properties of Logarithms

Given $\ln 2 \approx 0.693$, $\ln 3 \approx 1.099$, and $\ln 7 \approx 1.946$, use the properties of logarithms to approximate the following. Then use a calculator to verify the result.

A. $\ln 6$

B. $\ln \dfrac{7}{27}$

SOLUTION

A. $\ln 6 = \ln(2 \cdot 3)$

$\qquad = \ln 2 + \ln 3 \qquad\qquad$ *Property 1*

$\qquad \approx 0.693 + 1.099$

$\qquad = 1.792$

B. $\ln \dfrac{7}{27} = \ln 7 - \ln 27 \qquad\qquad$ *Property 2*

$\qquad = \ln 7 - \ln 3^3$

$\qquad = \ln 7 - 3 \ln 3 \qquad\qquad$ *Property 3*

$\qquad \approx 1.946 - 3(1.099)$

$\qquad = -1.351$

EXAMPLE 4   Using Properties of Logarithms

Use the properties of logarithms to verify that

$$-\ln \frac{1}{2} = \ln 2.$$

SOLUTION

$$-\ln \frac{1}{2} = -\ln(2^{-1}) = -(-1) \ln 2 = \ln 2$$

Try verifying this result on your calculator.

Rewriting Logarithmic Expressions

The properties of logarithms are useful for rewriting logarithmic expressions in forms that simplify the operations of algebra. This is true because they convert complicated products, quotients, and exponential forms into simpler sums, differences, and products, respectively.

EXAMPLE 5 Rewriting the Logarithm of a Product

Use the properties of logarithms to rewrite

$$\log_{10} 5x^3 y$$

as the sum of logarithms.

SOLUTION

$$
\begin{aligned}
\log_{10} 5x^3 y &= \log_{10} 5 + \log_{10} x^3 y && \textit{Property 1}\\
&= \log_{10} 5 + \log_{10} x^3 + \log_{10} y && \textit{Property 1}\\
&= \log_{10} 5 + 3 \log_{10} x + \log_{10} y && \textit{Property 3}
\end{aligned}
$$

EXAMPLE 6 Rewriting the Logarithm of a Quotient

Use the properties of logarithms to rewrite

$$\ln \frac{\sqrt{3x - 5}}{7}$$

as the sum and/or difference of logarithms.

SOLUTION

$$
\begin{aligned}
\ln \frac{\sqrt{3x - 5}}{7} &= \ln(3x - 5)^{1/2} - \ln 7 && \textit{Property 2}\\
\\
&= \frac{1}{2} \ln(3x - 5) - \ln 7 && \textit{Property 3}
\end{aligned}
$$

Technology Note

When you rewrite a logarithmic expression, be careful to check that the domains are the same. For example, the domain of $\ln x^4$ is all $x \neq 0$, whereas the domain of $4 \ln x$ is $x > 0$. Verify this on your graphing utility by graphing the two functions $y_1 = \ln x^4$ and $y_2 = 4 \ln x$.

Examples 5 and 6 use the properties of logarithms to *expand* logarithmic expressions. Examples 7 and 8 reverse the procedure by using properties of logarithms to *condense* logarithmic expressions.

EXAMPLE 7 Condensing a Logarithmic Expression

Rewrite the following expression as the logarithm of a single quantity.

$$\frac{1}{2} \log_{10} x - 3 \log_{10}(x + 1)$$

SOLUTION

$$
\begin{aligned}
\frac{1}{2} \log_{10} x - 3 \log_{10}(x + 1) &= \log_{10} x^{1/2} - \log_{10}(x + 1)^3\\
\\
&= \log_{10} \frac{\sqrt{x}}{(x + 1)^3}
\end{aligned}
$$

EXAMPLE 8   Condensing a Logarithmic Expression

Rewrite the following expression as the logarithm of a single quantity.

$$2 \ln(x + 2) - \ln x$$

SOLUTION

$$2 \ln(x + 2) - \ln x = \ln(x + 2)^2 - \ln x$$

$$= \ln \frac{(x + 2)^2}{x}$$

When expanding or condensing logarithmic expressions, you should compare the domain of the original expression with the domain of the expanded or condensed expression. For instance, the domain of $\ln x^2$ is all nonzero real numbers, whereas the domain of $2 \ln x$ is all positive real numbers.

DISCUSSION PROBLEM

DEMONSTRATING PROPERTIES OF LOGARITHMS

Use a calculator to demonstrate that

$$\frac{\ln x}{\ln y} \neq \ln \frac{x}{y} = \ln x - \ln y$$

by completing the table.

x	y	$\dfrac{\ln x}{\ln y}$	$\ln \dfrac{x}{y}$	$\ln x - \ln y$
1	2			
3	4			
10	5			
4	0.5			

WARM-UP
....

The following warm-up exercises involve skills that were covered in earlier sections. You will use these skills in the exercise set for this section.

In Exercises 1–4, evaluate the expression without using a calculator.

1. $\log_7 49$ **2.** $\log_2(\frac{1}{32})$ **3.** $\ln \dfrac{1}{e^2}$ **4.** $\log_{10} 0.001$

In Exercises 5–8, simplify the expression.

5. $e^2 e^3$ **6.** $\dfrac{e^2}{e^3}$ **7.** $(e^2)^3$ **8.** $(e^2)^0$

In Exercises 9 and 10, rewrite the expression in exponential form.

9. $\dfrac{1}{x^2}$ **10.** $\sqrt{x}$

SECTION 4.3 · EXERCISES
....

In Exercises 1–4, use the change of base formula to write the logarithm as a quotient of common logarithms. For instance, $\log_2 3 = (\log_{10} 3)/(\log_{10} 2)$.

1. $\log_3 5$ **2.** $\log_4 10$ **- 3.** $\log_2 x$ **4.** $\ln 5$

In Exercises 5–8, use the change of base formula to write the logarithm as a quotient of natural logarithms. For instance, $\log_2 3 = (\ln 3)/(\ln 2)$.

5. $\log_3 5$ **6.** $\log_4 10$ **7.** $\log_2 x$ **8.** $\log_{10} 5$

In Exercises 9–16, evaluate the logarithm using the change of base formula. Do the problem twice, once with common logarithms and once with natural logarithms. Round your result to three decimal places.

9. $\log_3 7$ **10.** $\log_7 4$

11. $\log_{1/2} 4$ **12.** $\log_4 0.55$

13. $\log_9 0.4$ **14.** $\log_{20} 125$

15. $\log_{15} 1250$ **16.** $\log_{1/3} 0.015$

In Exercises 17–36, use the properties of logarithms to write the expression as a sum, difference, and/or constant multiple of logarithms.

17. $\log_{10} 5x$ **18.** $\log_{10} 10z$

19. $\log_{10} \dfrac{5}{x}$ **20.** $\log_{10} \dfrac{y}{2}$

21. $\log_8 x^4$ **22.** $\log_6 z^{-3}$

23. $\ln \sqrt{z}$ **24.** $\ln \sqrt[3]{t}$

25. $\ln xyz$ **26.** $\ln \dfrac{xy}{z}$

27. $\ln \sqrt{a-1}$ **28.** $\ln\left(\dfrac{x^2-1}{x^3}\right)$

29. $\ln z(z-1)^2$ **30.** $\ln \sqrt{\dfrac{x^2}{y^3}}$

31. $\ln \sqrt[3]{\dfrac{x}{y}}$ **32.** $\ln \dfrac{x}{\sqrt{x^2+1}}$

33. $\ln \dfrac{x^4 \sqrt{y}}{z^5}$ **34.** $\ln \sqrt{x^2(x+2)}$

35. $\log_b \dfrac{x^2}{y^2 z^3}$ **36.** $\log_b \dfrac{\sqrt{xy^4}}{z^4}$

In Exercises 37–56, write the expression as the logarithm of a single quantity.

37. $\ln x + \ln 2$

38. $\ln y + \ln z$

39. $\log_4 z - \log_4 y$

40. $\log_5 8 - \log_5 t$

41. $2 \log_2(x + 4)$

42. $-4 \log_6 2x$

43. $\frac{1}{3} \log_3 5x$

44. $\frac{3}{2} \log_7(z - 2)$

45. $\ln x - 3 \ln(x + 1)$

46. $2 \ln 8 + 5 \ln z$

47. $\ln(x - 2) - \ln(x + 2)$

48. $3 \ln x + 2 \ln y - 4 \ln z$

49. $\ln x - 2[\ln(x + 2) + \ln(x - 2)]$

50. $4[\ln z + \ln(z + 5)] - 2 \ln(z - 5)$

51. $\frac{1}{3}[2 \ln(x + 3) + \ln x - \ln(x^2 - 1)]$

52. $2[\ln x - \ln(x + 1) - \ln(x - 1)]$

53. $\frac{1}{3}[\ln y + 2 \ln(y + 4)] - \ln(y - 1)$

54. $\frac{1}{2}[\ln(x + 1) + 2 \ln(x - 1)] + 3 \ln x$

55. $2 \ln 3 - \frac{1}{2} \ln(x^2 + 1)$

56. $\frac{3}{2} \ln 5t^6 - \frac{3}{4} \ln t^4$

In Exercises 57–66, approximate the logarithm using the properties of logarithms, given $\log_b 2 \approx 0.3562$, $\log_b 3 \approx 0.5646$, and $\log_b 5 \approx 0.8271$.

57. $\log_b 6$

58. $\log_b\left(\frac{5}{3}\right)$

59. $\log_b 40 \quad (40 = 2^3 \cdot 5)$ **60.** $\log_b 18$

61. $\log_b \dfrac{\sqrt{2}}{2}$

62. $\log_b \sqrt[3]{75}$

63. $\log_b \sqrt{5b}$

64. $\log_b(3b^2)$

65. $\log_b \dfrac{(4.5)^3}{\sqrt{3}}$

66. $\log_b 1$

In Exercises 67–72, find the exact value of the logarithm.

67. $\log_3 9$

68. $\log_6 \sqrt[3]{6}$

69. $\log_4 16^{1.2}$

70. $\log_5\left(\frac{1}{125}\right)$

71. $\ln e^{4.5}$

72. $\ln \sqrt[4]{e^3}$

In Exercises 73–80, use the properties of logarithms to simplify the logarithmic expression.

73. $\log_4 8$

74. $\log_5\left(\frac{1}{15}\right)$

75. $\log_7 \sqrt{70}$

76. $\log_2(4^2 \cdot 3^4)$

77. $\log_5\left(\frac{1}{250}\right)$

78. $\log_{10}\left(\frac{9}{300}\right)$

79. $\ln(5e^6)$

80. $\ln \dfrac{6}{e^2}$

81. *Sound Intensity*   The relationship between the number of decibels β and the intensity of a sound I in watts per square meter is given by

$$\beta = 10 \log_{10}\left(\frac{I}{10^{-16}}\right).$$

Use properties of logarithms to write the formula in simpler form, and determine the number of decibels of a sound with an intensity of 10^{-10} watts per square meter.

82. Approximate the natural logarithms of as many integers as possible between 1 and 20 given that $\ln 2 \approx 0.6931$, $\ln 3 \approx 1.0986$, and $\ln 5 \approx 1.6094$.

83. Use a graphing utility to graph

$$f(x) = \ln \frac{x}{2}, \qquad g(x) = \frac{\ln x}{\ln 2}, \qquad h(x) = \ln x - \ln 2$$

in the same viewing rectangle. Which two functions have identical graphs?

84. Prove that $\log_b \dfrac{u}{v} = \log_b u - \log_b v$.

85. Prove that $\log_b u^n = n \log_b u$.

4.4 SOLVING EXPONENTIAL AND LOGARITHMIC EQUATIONS

Introduction / Solving Exponential Equations / Solving Logarithmic Equations / Approximating Solutions / Application

Introduction

So far in this chapter, you have studied the definitions, graphs, and properties of exponential and logarithmic functions. In this section, you will study procedures for *solving equations* involving these exponential and logarithmic functions. As a simple example, consider the exponential equation $2^x = 32$. You can solve this equation by rewriting it as $2^x = 2^5$, which implies that $x = 5$. Although this method works in some cases, it does not work for an equation as simple as $e^x = 7$. To solve for x in this case, you can take the natural logarithm of both sides to obtain

$$e^x = 7 \qquad \text{\textit{Original equation}}$$
$$\ln e^x = \ln 7 \qquad \text{\textit{Take ln of both sides}}$$
$$x = \ln 7. \qquad \text{\textit{Solution}}$$

Technology Note

When solving exponential and logarithmic equations you need to be aware that you can introduce "extraneous solutions" during the solution process. Show how your graphing utility can help you eliminate the extraneous solution in the following example.

$2 \ln x = \ln 4$
$\ln x^2 = \ln 4$
$x^2 = 4$
$x = \pm 2$
$x = 2$ ($x = -2$ is extraneous)

GUIDELINES FOR SOLVING EXPONENTIAL AND LOGARITHMIC EQUATIONS

1. *To solve an exponential equation,* first isolate the exponential expression, then take the logarithm of both sides and solve for the variable.
2. *To solve a logarithmic equation,* rewrite the equation in exponential form and solve for the variable.

Note that these two guidelines are based on the **inverse properties** of exponential and logarithmic functions.

Base a	Base e
1. $\log_a a^x = x$	$\ln e^x = x$
2. $a^{\log_a x} = x$	$e^{\ln x} = x$

Solving Exponential Equations

EXAMPLE 1   Solving an Exponential Equation

$$e^x = 72 \qquad \text{\textit{Original equation}}$$
$$\ln e^x = \ln 72 \qquad \text{\textit{Take ln of both sides}}$$
$$x = \ln 72 \qquad \text{\textit{Inverse property of logs and exponents}}$$
$$x \approx 4.277$$

The solution is $x = \ln 72$. Check this solution in the original equation.

EXAMPLE 2   Solving an Exponential Equation

Solve $e^x + 5 = 60$.

SOLUTION

$e^x + 5 = 60$	*Original equation*
$e^x = 55$	*Subtract 5 from both sides*
$\ln e^x = \ln 55$	*Take ln of both sides*
$x = \ln 55$	*Inverse property of logs and exponents*
$x \approx 4.007$	

The solution is $x = \ln 55$. Check this solution in the original equation. The graph of $y = e^x - 55$ is shown in Figure 4.18. Use the zoom feature of your graphing utility to find the x-intercept and thus confirm the solution.

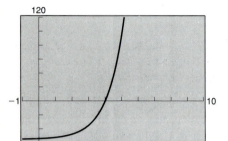

FIGURE 4.18

EXAMPLE 3   Solving an Exponential Equation

$4e^{2x} = 5$	*Original equation*
$e^{2x} = \dfrac{5}{4}$	*Divide both sides by 4*
$\ln e^{2x} = \ln \dfrac{5}{4}$	*Take ln of both sides*
$2x = \ln \dfrac{5}{4}$	*Inverse property of logs and exponents*
$x = \dfrac{1}{2} \ln \dfrac{5}{4}$	*Divide both sides by 2*
$x \approx 0.112$	

The solution is $x = \frac{1}{2} \ln \frac{5}{4}$. Check this solution in the original equation. The graph of $y = 4e^{2x} - 5$, shown in Figure 4.19, helps to confirm this solution.

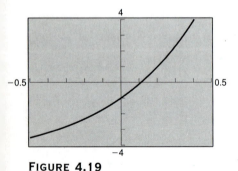

FIGURE 4.19

When an equation involves two or more exponential expressions, you can still use a procedure similar to that demonstrated in the first three examples. However, the algebra is a bit more complicated and a graphical approach is often easier. Study the next example carefully.

EXAMPLE 4   Solving an Exponential Equation

Solve for x in the equation $e^{2x} - 3e^x + 2 = 0$.

SOLUTION

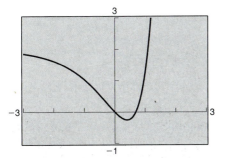

FIGURE 4.20

The graph of the function $f(x) = e^{2x} - 3e^x + 2$ in Figure 4.20 indicates that there are two solutions: one near $x = 0$ and one near $x = 0.7$. You can verify that $x = 0$ is indeed a solution by noting that $f(0) = e^{2(0)} - 3e^0 + 2 = 1 - 3 + 2 = 0$. Similarly, you can use the zoom and trace features to determine that $x \approx 0.693$. You can also solve this problem algebraically.

$e^{2x} - 3e^x + 2 = 0$	*Original equation*
$(e^x)^2 - 3e^x + 2 = 0$	*Quadratic form*
$(e^x - 2)(e^x - 1) = 0$	*Factor*
$e^x - 2 = 0 \qquad e^x - 1 = 0$	*Set factors to zero*
$e^x = 2 \qquad\qquad e^x = 1$	
$x = \ln 2 \qquad\quad x = 0$	*Solutions*

The equation has two solutions: $x = \ln 2 \approx 0.693$ and $x = 0$, which confirms the graphical analysis.

Examples 1 through 4 all deal with exponential equations in which the base is e. The same approach can be used to solve exponential equations involving other bases, as shown in Example 5.

EXAMPLE 5   A Base Other Than e

Solve $2^x = 10$.

SOLUTION

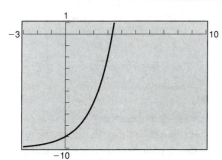

FIGURE 4.21

$2^x = 10$	*Original equation*
$\ln 2^x = \ln 10$	*Take log of both sides*
$x \ln 2 = \ln 10$	*Property of logarithms*
$x = \dfrac{\ln 10}{\ln 2}$	*Divide both sides by ln 2*

The equation has one solution: $x = \ln 10 / \ln 2 \approx 3.32$. Check this solution in the original equation. The graph of $y = 2^x - 10$, shown in Figure 4.21, helps to confirm this solution. (*Note:* Using the change of base formula, the solution could be written as $x = \log_2 10$.)

Solving Logarithmic Equations

To solve a logarithmic equation such as $\ln x = 3$, you can write the equation in exponential form as follows.

$$\ln x = 3 \qquad\qquad \textit{Logarithmic form}$$

$$e^{\ln x} = e^3 \qquad\qquad \textit{Exponentiate both sides}$$

$$x = e^3 \qquad\qquad \textit{Exponential form}$$

This procedure is called *exponentiating* both sides of an equation. It is applied after isolating the logarithmic expression.

EXAMPLE 6   Solving a Logarithmic Equation

Solve $5 + 2 \ln x = 4$.

SOLUTION

$$5 + 2 \ln x = 4 \qquad\qquad \textit{Original equation}$$

$$2 \ln x = -1 \qquad\qquad \textit{Subtract 5 from both sides}$$

$$\ln x = -\frac{1}{2} \qquad\qquad \textit{Divide both sides by 2}$$

$$e^{\ln x} = e^{-1/2} \qquad\qquad \textit{Exponentiate both sides}$$

$$x = e^{-1/2} \qquad\qquad \textit{Inverse property of exponents and logs}$$

$$x \approx 0.607$$

The equation has one solution: $x = e^{-1/2}$. Check this solution in the original equation. The graph of $y = 1 + 2 \ln x$, shown in Figure 4.22, helps to confirm this solution.

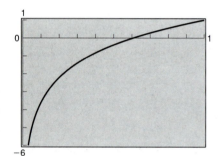

FIGURE 4.22

EXAMPLE 7   Solving a Logarithmic Equation

Solve $2 \ln 3x = 4$.

SOLUTION

$$2 \ln 3x = 4 \qquad\qquad \textit{Original equation}$$

$$\ln 3x = 2 \qquad\qquad \textit{Divide both sides by 2}$$

$$e^{\ln 3x} = e^2 \qquad\qquad \textit{Exponentiate both sides}$$

$$3x = e^2 \qquad\qquad \textit{Inverse property of exponents and logs}$$

$$x = \frac{1}{3}e^2 \qquad\qquad \textit{Divide both sides by 3}$$

$$x \approx 2.463$$

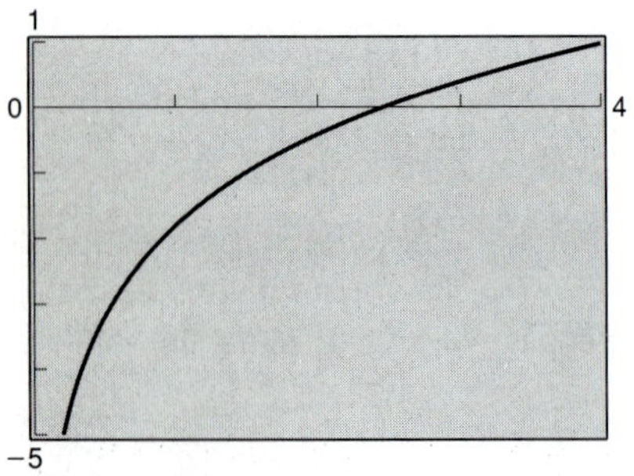

The equation has one solution: $x = \frac{1}{3}e^2$. Check this solution in the original equation. The graph of $y = -4 + 2 \ln 3x$, shown in Figure 4.23, helps to confirm this solution.

FIGURE 4.23

Complicated equations involving logarithmic expressions can be solved using a graphing utility, as demonstrated in Example 8.

EXAMPLE 8 **Solving a Logarithmic Equation**

Solve for x in the equation $\ln(x - 2) + \ln(2x - 3) = 2 \ln x$.

SOLUTION

The graph of the function $f(x) = \ln(x - 2) + \ln(2x - 3) - 2 \ln x$, shown in Figure 4.24, indicates that the only zero is $x = 6$. This can be verified by substituting $x = 6$ into the original equation. You could also solve this equation algebraically. Notice that in this case the technique produces an extraneous solution.

$$\ln(x - 2) + \ln(2x - 3) = 2 \ln x \qquad \textit{Original equation}$$
$$\ln(x - 2)(2x - 3) = \ln x^2 \qquad \textit{Properties of logarithms}$$
$$\ln(2x^2 - 7x + 6) = \ln x^2$$
$$e^{\ln(2x^2 - 7x + 6)} = e^{\ln x^2} \qquad \textit{Exponentiate both sides}$$
$$2x^2 - 7x + 6 = x^2 \qquad \textit{Inverse property of exponents and logs}$$
$$x^2 - 7x + 6 = 0 \qquad \textit{Quadratic form}$$
$$(x - 6)(x - 1) = 0 \qquad \textit{Factor}$$
$$x - 6 = 0 \longrightarrow x = 6 \qquad \textit{Set 1st factor equal to 0}$$
$$x - 1 = 0 \longrightarrow x = 1 \qquad \textit{Set 2nd factor equal to 0}$$

Finally, by checking these two "solutions" in the original equation, you can conclude that $x = 1$ is not valid. Can you see why? Thus, the only solution is $x = 6$.

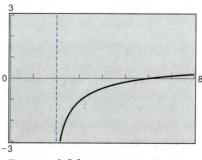

FIGURE 4.24

EXAMPLE 9   **The Change of Base Formula**

Prove the change of base formula given in Section 4.3.

$$\log_a x = \frac{\log_b x}{\log_b a}$$

SOLUTION

Begin by letting $y = \log_a x$ and writing the equivalent exponential form $a^y = x$. Now, taking the logarithm *with base b* of both sides produces the following.

$$\log_b a^y = \log_b x$$
$$y \log_b a = \log_b x$$
$$y = \frac{\log_b x}{\log_b a}$$
$$\log_a x = \frac{\log_b x}{\log_b a}$$

When solving exponential or logarithmic equations, the following properties are useful.

1. $x = y$ if and only if $\log_a x = \log_a y$.
2. $x = y$ if and only if $a^x = a^y, a > 0, a \neq 1$.

Can you see where these properties were used in the examples in this section?

Approximating Solutions

Equations that involve combinations of algebraic functions, exponential functions, and/or logarithmic functions can be very difficult to solve by algebraic procedures. Here again you can take advantage of a graphing utility.

EXAMPLE 10   **Approximating the Solution of an Equation**

Approximate the solutions of $\ln x = x^2 - 2$.

SOLUTION

To begin, use a graphing utility to graph

$$y = -x^2 + 2 + \ln x,$$

as shown in Figure 4.25. From this graph, you can see that the equation has two solutions. Next, using the zoom and trace features, you can approximate the two solutions to be $x \approx 0.138$ and $x \approx 1.564$.

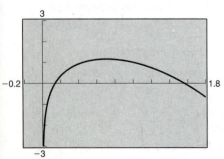

FIGURE 4.25

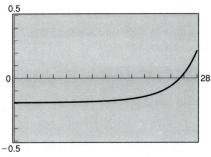

FIGURE 4.26

Application

EXAMPLE 11   Waste Processed for Energy Recovery

From 1960 to 1986, the amount of municipal waste processed for energy recovery in the United States can be approximated by the equation

$$y = 0.00643e^{0.00533t^2},$$

where y is the amount of waste (in pounds per person) that was processed for energy recovery and t represents the year, with $t = 0$ corresponding to 1960. According to this model, during which year did the amount of waste reach 0.2 pound? (*Source:* Franklin Associates *Characterization of Municipal Solid Waste in U.S.*)

SOLUTION

To solve for t in the equation $0.00643e^{0.00533t^2} = 0.2$, use a graphing utility to graph

$$y = 0.00643e^{0.00533t^2} - 0.2$$

on the domain $0 \leq t \leq 28$. From the graph (Figure 4.26), you can see that there is a zero between $t = 25$ and $t = 26$. Using the zoom and trace features, you can determine that $t \approx 25.4$ years. Because $t = 0$ represents 1960, it follows that the amount of waste would have reached 0.2 pound per person in 1985.

DISCOVERY

Use a graphing utility to graph $y = \ln x$ and $y = x^2 - 2$ on the same viewing rectangle. (Use a viewing rectangle in which $0 \leq x \leq 3$ and $-5 \leq y \leq 5$.) Then zoom in to approximate the two points of intersection. Compare your results with those given in Example 10. Which solution technique do you prefer? Explain your reasons.

DISCUSSION PROBLEM

VERIFYING INVERSE RELATIONSHIPS

Use a graphing utility to verify the following inverse relationships between logarithmic and exponential functions.

Base 10	*Base e*
1. $\log_{10} 10^x = x$	$\ln e^x = x$
2. $10^{\log_{10} x} = x$	$e^{\ln x} = x$

You can do this by sketching the graph of each relationship. For instance, to verify that $\ln e^x = x$, try sketching the graphs of $y = \ln e^x$ and $y = x$. The two graphs should be identical. Be sure to determine the domain of each function.

WARM-UP

The following warm-up exercises involve skills that were covered in earlier sections. You will use these skills in the exercise set for this section.

In Exercises 1–6, solve for x.

1. $x \ln 2 = \ln 3$ **2.** $(x - 1)\ln 4 = 2$ **3.** $2xe^2 = e^3$
4. $4xe^{-1} = 8$ **5.** $x^2 - 4x + 5 = 0$ **6.** $2x^2 - 3x + 1 = 0$

In Exercises 7–10, simplify the expression.

7. $\log_{10} 100^x$ **8.** $\log_4 64^x$ **9.** $\ln e^{2x}$ **10.** $\ln e^{-x^2}$

SECTION 4.4 · EXERCISES

In Exercises 1–10, solve for x.

1. $4^x = 16$ **2.** $3^x = 243$
3. $7^x = \frac{1}{49}$ **4.** $8^x = 4$
5. $(\frac{3}{4})^x = \frac{27}{64}$ **6.** $3^{x-1} = 27$
7. $\log_4 x = 3$ **8.** $\log_5 5x = 2$
9. $\log_{10} x = -1$ **10.** $\ln(2x - 1) = 0$

In Exercises 11–16, apply the inverse properties of $\ln x$ and e^x to simplify the expression.

11. $\ln e^{x^2}$ **12.** $\ln e^{2x-1}$
13. $e^{\ln(5x+2)}$ **14.** $-1 + \ln e^{2x}$
15. $e^{\ln x^2}$ **16.** $-8 + e^{\ln x^3}$

In Exercises 17–36, solve the exponential equation algebraically. (Round your result to three decimal places.)

17. $e^x = 10$ **18.** $4e^x = 91$
19. $7 - 2e^x = 5$ **20.** $-14 + 3e^x = 11$
21. $e^{3x} = 12$ **22.** $e^{2x} = 50$
23. $500e^{-x} = 300$ **24.** $1000e^{-4x} = 75$
25. $e^{2x} - 4e^x - 5 = 0$ **26.** $e^{2x} - 5e^x + 6 = 0$
27. $20(100 - e^{x/2}) = 500$ **28.** $\dfrac{400}{1 + e^{-x}} = 200$
29. $10^x = 42$ **30.** $10^x = 570$
31. $3^{2x} = 80$ **32.** $6^{5x} = 3000$
33. $5^{-t/2} = 0.20$ **34.** $4^{-3t} = 0.10$
35. $\left(1 + \dfrac{0.10}{12}\right)^{12t} = 2$ **36.** $2^{3-x} = 565$

In Exercises 37–44, use a graphing utility to solve the exponential equation. (Round your result to three decimal places.)

37. $3e^{3x/2} = 962$ **38.** $6e^{1-x} = 25$
39. $e^{0.09t} = 3$ **40.** $e^{0.125t} = 8$
41. $8(10^{3x}) = 12$ **42.** $3(5^{x-1}) = 21$
43. $\left(1 + \dfrac{0.065}{365}\right)^{365t} = 4$ **44.** $\dfrac{3000}{2 + e^{2x}} = 2$

In Exercises 45–60, solve the logarithmic equation algebraically. (Round your result to three decimal places.)

45. $\ln x = -3$ **46.** $\ln x = 2$
47. $\ln 2x = 2.4$ **48.** $3 \ln 5x = 10$
49. $\ln \sqrt{x + 2} = 1$ **50.** $\ln(x + 1)^2 = 2$
51. $\ln x + \ln(x - 2) = 1$ **52.** $\ln x + \ln(x + 3) = 1$
53. $\log_{10}(z - 3) = 2$ **54.** $\log_{10} x^2 = 6$
55. $\log_{10}(x + 4) - \log_{10} x = \log_{10}(x + 2)$
56. $\log_4 x - \log_4(x - 1) = \frac{1}{2}$
57. $\log_3 x + \log_3(x^2 - 8) = \log_3 8x$
58. $\log_2 x + \log_2(x + 2) = \log_2(x + 6)$
59. $\ln(x + 5) = \ln(x - 1) - \ln(x + 1)$
60. $\ln(x + 1) - \ln(x - 2) = \ln x^2$

In Exercises 61–64, use a graphing utility to solve the logarithmic equation. (Round your result to three decimal places.)

61. $2 \ln x = 7$ **62.** $\ln 4x = 1$
63. $\ln x + \ln(x^2 + 1) = 8$
64. $\log_{10} 8x - \log_{10}(1 + \sqrt{x}) = 2$

Compound Interest   In Exercises 65 and 66, find the time required for a $1000 investment to double at interest rate r, compounded continuously.

65. $r = 0.085$ **66.** $r = 0.12$

Compound Interest   In Exercises 67 and 68, find the time required for a $1000 investment to triple at interest rate r, compounded continuously.

67. $r = 0.085$ **68.** $r = 0.12$

69. *Demand Function*   The demand equation for a product is

$$p = 500 - 0.5(e^{0.004x}).$$

Find the demand x for a price of (a) $p = \$350$ and (b) $p = \$300$.

70. *Demand Function*   The demand equation for a product is

$$p = 5000\left(1 - \frac{4}{4 + e^{-0.002x}}\right).$$

Find the demand x for a price of (a) $p = \$600$ and (b) $p = \$400$.

71. *Forest Yield*   The yield V (in millions of cubic feet per acre) for a forest at age t years is given by

$$V = 6.7e^{-48.1/t}.$$

(a) Use a graphing utility to graph the function.
(b) Determine the horizontal asymptote of the function. Interpret its meaning in the context of the problem.
(c) Find the time necessary to have a yield of 1.3 million cubic feet.

72. *Trees per Acre*   The number of trees per acre N of a certain species is approximated by the model

$$N = 68 \cdot 10^{-0.04x}, \qquad 5 \le x \le 40,$$

where x is the average diameter of the trees 3 feet above the ground. Use the model to approximate the average diameter of the trees in a test plot when $N = 21$.

73. *Average Heights*   The percentage of American males between the ages of 18 and 24 who are no more than x inches tall is given by

$$m(x) = \frac{100}{1 + e^{-0.6114(x - 69.71)}}, \qquad 60 \le x \le 80,$$

where m is the percentage and x is the height in inches. (*Source:* U.S. National Center for Health Statistics) The function giving the percentages f for females for the same ages is given by

$$f(x) = \frac{100}{1 + e^{-0.66607(x - 64.51)}}, \qquad 55 \le x \le 75.$$

(a) Use a graphing utility to graph each function in the same viewing rectangle.
(b) Determine the horizontal asymptotes of the functions.
(c) What is the median height of each sex?

74. *Human Memory Model*   In a group project in learning theory, a mathematical model for the proportion P of correct responses after n trials was found to be

$$P = \frac{0.83}{1 + e^{-0.2n}}.$$

(a) Use a graphing utility to graph the function.
(b) Determine the horizontal asymptotes of the function. Interpret the meaning of the upper asymptote in the context of this problem.
(c) After how many trials will 60% of the responses be correct?

4.5 APPLICATIONS OF EXPONENTIAL AND LOGARITHMIC FUNCTIONS

Compound Interest / Growth and Decay / Logistics Growth Models / Logarithmic Models

Compound Interest

In this section, you will study four basic types of applications: (1) compound interest, (2) growth and decay, (3) logistics models, and (4) intensity models. The problems presented in this section involve the full range of solution techniques studied in this chapter.

EXAMPLE 1 Doubling Time for an Investment

An investment is made in a trust fund at an annual percentage rate of 9.5%, compounded quarterly. How long will it take for the investment to double in value?

SOLUTION

For quarterly compounding, use the formula

$$A = P\left(1 + \frac{r}{4}\right)^{4t}.$$

Using $r = 0.095$, the time required for the investment to double is given by solving for t in the equation $2P = A$.

$$2P = P\left(1 + \frac{0.095}{4}\right)^{4t} \qquad 2P = A$$

$$2 = (1.02375)^{4t} \qquad \textit{Divide both sides by P}$$

$$\ln 2 = \ln(1.02375)^{4t} \qquad \textit{Take ln of both sides}$$

$$\ln 2 = 4t \ln(1.02375)$$

$$t = \frac{\ln 2}{4 \ln(1.02375)} \approx 7.4$$

Therefore, it will take approximately 7.4 years for the investment to double in value with quarterly compounding.

Try reworking Example 1 using continuous compounding. To do this you will need to solve the equation

$$2P = Pe^{0.095t}.$$

The solution is $t \approx 7.3$ years, which makes sense because the principal should double more quickly with continuous compounding than with quarterly compounding.

DISCOVERY

A person deposits $1000 in an account that pays 9.5% per year, compounded quarterly. The balance after t years is

$$A = 1000\left(1 + \frac{0.095}{4}\right)^{4t}.$$

Another person deposits $1000 in an account that pays 9.5% per year, compounded continuously. The balance after t years is $A = 1000e^{0.095t}$. Use a graphing utility to graph both equations on the same viewing rectangle. (Use a viewing rectangle in which $0 \le x \le 8$ and $1000 \le y \le 2200$.) What do you conclude? Which of the two accounts will reach a balance of $2000 first? Justify your answer by zooming in near $x = 7$.

EXAMPLE 2 Finding an Annual Percentage Rate

An investment of $10,000 is compounded continuously. What annual percentage rate will produce a balance of $25,000 in 10 years?

SOLUTION

Use the formula $A = Pe^{rt}$ with $P = 10,000$, $A = 25,000$, and $t = 10$, and solve the equation for r.

$$10,000e^{10r} = 25,000$$
$$e^{10r} = 2.5$$
$$10r = \ln 2.5$$
$$r = \frac{1}{10}\ln 2.5 \approx 0.0916$$

Thus, the annual percentage rate must be approximately 9.16%. You can verify this result using the zoom and trace features of a graphing utility. By graphing $y = 10,000e^{10x}$ and $y = 25,000$ in the same viewing rectangle with an x scale of 0.1, you can confirm that $r \approx 0.0916$, as shown in Figure 4.27.

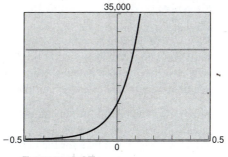

FIGURE 4.27

EXAMPLE 3 The Effective Yield for an Investment

A deposit is compounded continuously at an annual percentage rate of 7.5%. Find the **effective yield.** That is, find the simple interest rate that would yield the same balance at the end of 1 year.

SOLUTION

Using the formula $A = Pe^{rt}$ with $r = 0.075$ and $t = 1$, the balance at the end of 1 year is

$$A = Pe^{0.075(1)}$$
$$\approx P(1.0779)$$
$$= P(1 + 0.0779). \qquad \qquad A = P(1 + r)$$

Because the formula for simple interest after 1 year is $A = P(1 + r)$, it follows that the effective yield is approximately 7.79%.

Growth and Decay

The balance in an account earning *continuously compounded* interest is one example of a quantity that increases over time according to the **exponential growth model**

$$Q(t) = Ce^{kt}.$$

In general, $Q(t)$ is the size of the "population" at any time t, C is the original population (when $t = 0$), and k is a constant determined by the rate of growth. If $k > 0$, the population *grows* (increases) over time, and if $k < 0$ it *decays* (decreases) over time. Example 11 in Section 4.1 is an example of population growth. It may help to remember this growth model as $(Then) = (Now)\,(e^{kt})$.

EXAMPLE 4   Exponential Decay

Radioactive iodine is a by-product of some types of nuclear reactors. Its **half-life** is 60 days. That is, after 60 days, a given amount of radioactive iodine will have decayed to half the original amount. Suppose a contained nuclear accident occurs and gives off an initial amount C of radioactive iodine.

A. Write an equation for the amount of radioactive iodine present at any time t following the accident.

B. How long will it take for the radioactive iodine to decay to a level of 20% of the original amount?

SOLUTION

A. We first need to find the rate k in the exponential model $Q(t) = Ce^{kt}$. Knowing that half the original amount remains after $t = 60$ days, you obtain

$$Q(60) = Ce^{k(60)} = \frac{1}{2}C$$

$$e^{60k} = \frac{1}{2}$$

$$60k = -\ln 2$$

$$k = \frac{-\ln 2}{60} \approx -0.0116.$$

Thus, the exponential model is

$$Q(t) = Ce^{-0.0116t}.$$

B. The time required to decay to 20% of the original amount is given by

$$Q(t) = Ce^{-0.0116t} = (0.2)C$$
$$e^{-0.0116t} = 0.2$$
$$-0.0116t = \ln 0.2$$
$$t = \frac{\ln 0.2}{-0.0116} \approx 139 \text{ days.}$$

In living organic material, the ratio of radioactive carbon isotopes (carbon-14) to the number of nonradioactive carbon isotopes (carbon-12) is about 1 to 10^{12}. When organic material dies, its carbon-12 content remains fixed, whereas its radioactive carbon-14 begins to decay with a half-life of about 5700 years. To estimate the age of dead organic material, scientists use the following formula, which denotes the ratio of carbon-14 to carbon-12 present at any time t (in years).

$$R = \frac{1}{10^{12}} e^{-t/8223}$$

The graph of R is shown in Figure 4.28. Note that R decreases as the time t increases.

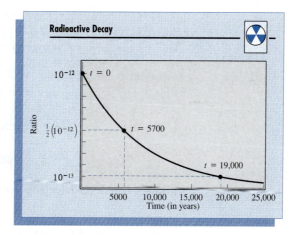

FIGURE 4.28

EXAMPLE 5   **Carbon Dating**

Suppose the carbon-14/carbon-12 ratio of a newly discovered fossil is

$$R = \frac{1}{10^{13}}.$$

Estimate the age of the fossil.

SOLUTION

In the carbon dating model, substitute the given value of R to obtain the following.

$$\frac{1}{10^{12}}e^{-t/8223} = R \qquad \text{Original model}$$

$$\frac{e^{-t/8223}}{10^{12}} = \frac{1}{10^{13}} \qquad \text{Let R equal } 1/10^{13}$$

$$e^{-t/8223} = \frac{1}{10} \qquad \text{Multiply both sides by } 10^{12}$$

$$\ln e^{-t/8223} = \ln \frac{1}{10} \qquad \text{Take log of both sides}$$

$$-\frac{t}{8223} \approx -2.3026 \qquad \text{Inverse property of logs and exponents}$$

$$t \approx 18,934 \qquad \text{Multiply both sides by } 8223$$

Thus, to the nearest thousand years, you can estimate the age of the fossil to be 19,000 years. You could obtain this solution graphically by graphing the equation $y = e^{-t/8223} - \frac{1}{10}$, as indicated in Figure 4.29. ▪▪▪▪▪▪▪▪▪

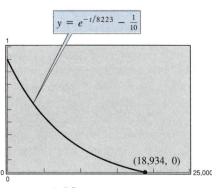

$$y = e^{-t/8223} - \frac{1}{10}$$

(18,934, 0)

25,000

FIGURE 4.29

REMARK   The carbon dating model in Example 5 assumes that the carbon-14/carbon-12 ratio was one part in 10,000,000,000,000. Suppose an error in measurement occurred and the actual ratio was only one part in 8,000,000,000,000. The fossil age corresponding to the actual ratio would then be approximately 17,000 years. Try checking this result. ▪▪▪▪▪

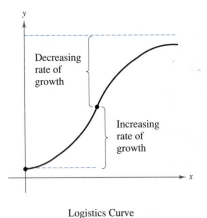

Decreasing rate of growth

Increasing rate of growth

Logistics Curve

FIGURE 4.30

Logistics Growth Models

Some populations initially have rapid growth, followed by a declining rate of growth, as indicated by the graph in Figure 4.30. One model for describing this type of growth pattern is the **logistics curve** given by the function

$$y = \frac{a}{1 + be^{-(t-c)/d}},$$

where y is the population size and t is the time. An example would be a bacteria culture allowed to grow initially under ideal conditions, followed by less favorable conditions that inhibit growth, such as overcrowding. A logistics growth curve is also called a **sigmoidal curve**.

EXAMPLE 6   Spread of a Virus

On a college campus of 5000 students, one student returned from vacation with a contagious flu virus. The spread of the virus through the student body is given by

$$y = \frac{5000}{1 + 4999e^{-0.8t}},$$

where y is the total number infected after t days. The college will cancel classes when 40% or more of the students are ill.

A. How many are infected after 5 days?

B. After how many days will the college cancel classes?

SOLUTION

A. After 5 days, the number of students infected is

$$y = \frac{5000}{1 + 4999e^{-0.8(5)}} = \frac{5000}{1 + 4999e^{-4}} \approx 54.$$

B. In this case, the number of students infected is $(0.40)(5000) = 2000$. Therefore, you can solve for t in the following equation.

$$2000 = \frac{5000}{1 + 4999e^{-0.8t}}$$

Using a graphing utility, graph $y = 2000$ and $y = 5000/(1 + 4999e^{-0.8t})$ and find the point of intersection. In Figure 4.31, you can see that the point of intersection occurs near $t \approx 10.1$. Hence, after 10 days, at least 40% of the students will be infected, and the college will cancel classes.

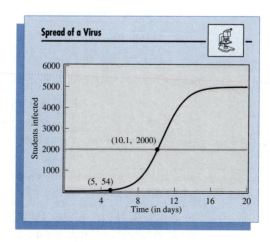

Spread of a Virus

FIGURE 4.31

Logarithmic Models

Sound and shock waves can be measured by the **intensity model**

$$S = K \log_{10} \frac{I}{I_0},$$

where I is the intensity of the stimulus wave, I_0 is the **threshold intensity** (the smallest value of I that can be detected by the listening device), and K determines the units in which S is measured. Sound heard by the human ear is measured in decibels. One **decibel** is considered to be the smallest detectable difference in the loudness of two sounds.

EXAMPLE 7   **Magnitude of Earthquakes**

On the Richter Scale, the magnitude R of an earthquake of intensity I is given by

$$R = \log_{10} \frac{I}{I_0},$$

where $I_0 = 1$ is the minimum intensity used for comparison. Find the intensity per unit of area for the following earthquakes. (Intensity is a measure of the wave energy of an earthquake.)

A. San Francisco in 1906, $R = 8.6$
B. Mexico City in 1978, $R = 7.85$
C. San Francisco Bay Area in 1989, $R = 7.1$

SOLUTION

A. Because $I_0 = 1$ and $R = 8.6$, you have

$$8.6 = \log_{10} I$$
$$I = 10^{8.6} \approx 398,107,171.$$

B. For Mexico City, you have $7.85 = \log_{10} I$, and

$$I = 10^{7.85} \approx 70,795,000.$$

C. For $R = 7.1$, you have $7.1 = \log_{10} I$, and

$$I = 10^{7.1} \approx 12,589,254.$$

Note that an increase of 1.5 units on the Richter Scale (from 7.1 to 8.6) represents an intensity change by a factor of

$$\frac{398,107,171}{12,589,254} \approx 31.6.$$

In other words, the "great San Francisco earthquake" in 1906 had a magnitude that was about 32 times greater than the one in 1989.

DISCUSSION PROBLEM

••••••••••••••••••••••

COMPARING POPULATION MODELS

The population (in millions) of the United States from 1800 to 1990 is given in the accompanying table.

t	0	1	2	3	4	5	6	7	8	9
Year	1800	1810	1820	1830	1840	1850	1860	1870	1880	1890
Population	5.31	7.23	9.64	12.87	17.07	23.19	31.44	39.82	50.16	62.95

t	10	11	12	13	14	15	16	17	18	19
Year	1900	1910	1920	1930	1940	1950	1960	1970	1980	1990
Population	75.99	91.97	105.71	122.78	131.67	151.33	179.32	203.30	226.55	250.00

Using a statistical procedure called *least squares regression analysis,* we found the best quadratic and exponential models for this data. Which of the following two equations is a better model for the population of the United States between 1800 and 1990? Write a short paragraph describing the method you used to reach your conclusion.

Quadratic Model	*Exponential Model*

$$P = 0.662t^2 + 0.211t + 6.165 \qquad P = 7.7899e^{0.2013t}$$

WARM-UP

•••••••••••••••••••

The following warm-up exercises involve skills that were covered in earlier sections. You will use these skills in the exercise set for this section.

In Exercises 1–6, sketch the graph of the equation.

1. $y = 2^{0.25x}$
2. $y = 2^{-0.25x}$
3. $y = 4 \log_2 x$
4. $y = \ln(x - 3)$
5. $y = e^{-x^2/5}$
6. $y = \dfrac{2}{1 + e^{-x}}$

In Exercises 7–10, solve the given equation for *x*. (Round your result to three decimal places.)

7. $3e^{2x} = 7$
8. $2e^{-0.2x} = 0.002$
9. $4 \ln 5x = 14$
10. $6 \ln 2x = 12$

SECTION 4.5 · EXERCISES

Compound Interest In Exercises 1–10, complete the table for a savings account in which interest is compounded continuously.

	Initial Investment	Annual % Rate	Effective Yield	Time to Double	Amount After 10 Years
1.	$1000	12%			
2.	$20,000	$10\frac{1}{2}$%			
3.	$750			$7\frac{3}{4}$ yr	
4.	$10,000			5 yr	
5.	$500				$1292.85
6.	$2000		4.5%		
7.		11%			$19,205.00
8.		8%			$20,000.00
9.	$5000		8.33%		
10.	$250		12.19%		

Compound Interest In Exercises 11 and 12, determine the principal P that must be invested at rate r, compounded monthly, so that $500,000 will be available for retirement in t years.

11. $r = 7\frac{1}{2}$%, $t = 20$ **12.** $r = 12$%, $t = 40$

Compound Interest In Exercises 13 and 14, determine the time necessary for $1000 to double if it is invested at interest rate r compounded (a) annually, (b) monthly, (c) daily, and (d) continuously.

13. $r = 11$%   **14.** $r = 10\frac{1}{2}$%

15. *Compound Interest* Complete the table for the time t necessary for P dollars to triple if interest is compounded continuously at rate r.

r	2%	4%	6%	8%	10%	12%
t						

16. *Compound Interest* Complete the table for the time t necessary for P dollars to triple if interest is compounded annually at rate r.

r	2%	4%	6%	8%	10%	12%
t						

17. *Comparing Investments* If $1 is invested in an account over a 10-year period, the amount in the account is given by

$$A = 1 + 0.075t \quad \text{or} \quad A = e^{0.07t}$$

depending on whether it is simple interest at $7\frac{1}{2}$% or continuous compound interest at 7%. Use a graphing utility to graph each function in the same viewing rectangle, and determine which grows at the higher rate.

18. *Comparing Investments* If $1 is invested in an account over a 10-year period, the amount in the account is given by

$$A = 1 + 0.06t \quad \text{or} \quad A = \left(1 + \frac{0.055}{365}\right)^{365t}$$

depending on whether it is simple interest at 6% or compound interest at $5\frac{1}{2}$% compounded daily. Use a graphing utility to graph each function in the same viewing rectangle, and determine which grows at the higher rate.

Radioactive Decay In Exercises 19–24, complete the table for the given radioactive isotope.

	Isotope	Half-life (Years)	Initial Quantity	Amount After 1000 Years	Amount After 10,000 Years
19.	^{226}Ra	1620	10 g		
20.	^{226}Ra	1620		1.5 g	
21.	^{14}C	5730			2 g
22.	^{14}C	5730	3 g		
23.	^{230}Pu	24,360		2.1 g	
24.	^{230}Pu	24,360			0.4 g

In Exercises 25–28, find the constant k such that the exponential function $y = Ce^{kt}$ passes through the given points on the graph.

25.
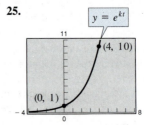

26.
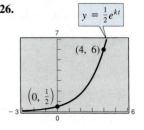

27.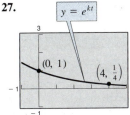

$y = e^{kt}$

(0, 1) $\left(4, \frac{1}{4}\right)$

28.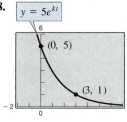

$y = 5e^{kt}$

(0, 5)   (3, 1)

29. Population   The population P of a city is given by
$$P = 105,300e^{0.015t},$$
where t is the time in years, with $t = 0$ corresponding to 1990. What was the population in 1990? According to this model, in what year will the city have a population of 150,000?

30. Population   The population P of a city is given by
$$P = 240,360e^{0.012t},$$
where t is the time in years, with $t = 0$ corresponding to 1990. What was the population in 1990? According to this model, in what year will the city have a population of 250,000?

31. Population   The population P of a city is given by
$$P = 2500e^{kt},$$
where t is the time in years, with $t = 0$ corresponding to the year 1990. In 1945, the population was 1350. Find the value of k and use this result to predict the population in the year 2010.

32. Population   The population P of a city is given by
$$P = 140,500e^{kt},$$
where t is the time in years, with $t = 0$ corresponding to the year 1990. In 1960, the population was 100,250. Find the value of k and use this result to predict the population in the year 2000.

33. Population   The population of Dhaka, Bangladesh was 4.22 million in 1990 and its projected population for the year 2000 is 6.49 million. Find the exponential growth model $y = Ce^{kt}$ for the population growth of Dhaka by letting $t = 0$ correspond to 1990. Use the model to predict the population of the city in 2010. (*Source:* United Nations)

34. Population   The population of Houston, Texas was 2.30 million in 1990 and its projected population for the year 2000 is 2.65 million. Find the exponential growth model $y = Ce^{kt}$ for the population growth of Houston by letting $t = 0$ correspond to 1990. Use the model to predict the population of the city in 2010. (*Source:* U. S. Bureau of the Census)

35. Bacteria Growth   The number of bacteria N in a culture is given by the model
$$N = 100e^{kt},$$
where t is the time in hours, with $t = 0$ corresponding to the time when $N = 100$. If $N = 300$ when $t = 5$, estimate the time required for the population to double in size.

36. Bacteria Growth   The number of bacteria N in a culture is given by the model
$$N = 250e^{kt},$$
where t is the time in hours, with $t = 0$ corresponding to the time when $N = 250$. If $N = 280$ when $t = 10$, estimate the time required for the population to double in size.

37. Radioactive Decay   The half-life of radioactive radium (^{226}Ra) is 1620 years. What percentage of a present amount of radioactive radium will remain after 100 years?

38. Radioactive Decay ^{14}C dating assumes that the carbon dioxide on earth today has the same radioactive content as it did centuries ago. If this is true, then the amount of ^{14}C absorbed by a tree that grew several centuries ago should be the same as the amount of ^{14}C absorbed by a tree growing today. A piece of ancient charcoal contains only 15% as much of the radioactive carbon as a piece of modern charcoal. How long ago was the tree burned to make the ancient charcoal if the half-life of ^{14}C is 5730 years?

39. Depreciation   A certain car that cost $22,000 new has a depreciated value of $16,500 after 1 year. Find the value of the car when it is 3 years old by using the exponential model $y = Ce^{kt}$.

40. Depreciation   A computer that cost $4600 new has a depreciated value of $3000 after 2 years. Find the value of the computer after 3 years by using the exponential model $y = Ce^{kt}$.

41. Sales   The sales S (in thousands of units) of a new product after it has been on the market t years are given by
$$S(t) = 100(1 - e^{kt}).$$
(a) Find S as a function of t if 15,000 units have been sold after 1 year.
(b) How many units will be sold after 5 years?

42. Learning Curve   The management at a factory has found that the maximum number of units a worker can produce in a day is 30. The learning curve for the number of units N produced per day after a new employee has worked t days is given by $N = 30(1 - e^{kt})$. After 20 days on the job, a particular worker produced 19 units.
(a) Find the learning curve for this worker (that is, find the value of k).
(b) How many days should pass before this worker is producing 25 units per day?

43. *Women's Heights*   The distribution of the heights of American women between the ages of 25 and 34 can be approximated by the function

$$p = 0.166e^{-(x-64.5)^2/11.5},$$

where x is the height in inches. Use a graphing utility to graph this function. What is the average height of women in this age bracket? (*Source:* U.S. National Center for Health Statistics)

44. *Men's Heights*   The distribution of the heights of American men between the ages of 25 and 34 can be approximated by the function

$$p = 0.193e^{-(x-70)^2/14.32},$$

where x is the height in inches. Use a graphing utility to graph this function. What is the average height of men in this age bracket? (*Source:* U.S. National Center for Health Statistics)

45. *Stocking a Lake with Fish*   A certain lake was stocked with 500 fish and the fish population increased according to the logistics curve

$$p(t) = \frac{10,000}{1 + 19e^{-t/5}},$$

where t is the time in months.

(a) Use a graphing utility to graph the function. Determine the larger of the two horizontal asymptotes and interpret its meaning in the context of the problem.

(b) Estimate the fish population after 5 months.

(c) After how many months will the fish population be 2000?

46. *Endangered Species*   A conservation organization releases 100 animals of an endangered species into a game preserve. The organization believes that the preserve has a carrying capacity of 1000 animals and that the growth of the herd will be modeled by the logistics curve

$$p(t) = \frac{1000}{1 + 9e^{-0.1656t}},$$

where t is the time in months.

(a) Use a graphing utility to graph the function. Determine the larger of the two horizontal asymptotes and interpret its meaning in the context of the problem.

(b) Estimate the population after 5 months.

(c) After how many months will the population be 500?

47. *Sales and Advertising*   The sales S (in thousands of units) of a product after x hundred dollars is spent on advertising is given by

$$S = 10(1 - e^{kx}).$$

(a) Find S as a function of x if 2500 units are sold when $500 is spent on advertising.

(b) Estimate the number of units that will be sold if advertising expenditures are raised to $700.

48. *Sales and Advertising*   After discontinuing all advertising for a certain product in 1988, the manufacturer noted that sales began to drop according to the model

$$S = \frac{500,000}{1 + 0.6e^{kt}},$$

where S represents the number of units sold and t represents the year, with $t = 0$ corresponding to 1988.

(a) Find k if the company sold 300,000 units in 1990.

(b) According to this model, what will sales be in 1993?

Earthquake Magnitudes   In Exercises 49 and 50, use the Richter Scale (see Example 7) for measuring the magnitude of earthquakes.

49. Find the magnitude R of an earthquake of intensity I (let $I_0 = 1$).

(a) $I = 80,500,000$

(b) $I = 48,275,000$

50. Find the intensity I of an earthquake measuring R on the Richter Scale (let $I_0 = 1$).

(a) Mexico City in 1985, $R = 8.1$

(b) Los Angeles in 1971, $R = 6.7$

Intensity of Sound   In Exercises 51–54, use the following information to determine the level of sound (in decibels) for the given sound intensity. The level of sound β, in decibels, with an intensity of I is given by

$$\beta(I) = 10 \log_{10} \frac{I}{I_0},$$

where I_0 is an intensity of 10^{-16} watts per square centimeter, corresponding roughly to the faintest sound that can be heard by the human ear.

51. (a) $I = 10^{-14}$ watts per square centimeter (faint whisper)

(b) $I = 10^{-9}$ watts per square centimeter (busy street corner)

(c) $I = 10^{-6.5}$ watts per square centimeter (air hammer)

(d) $I = 10^{-4}$ watts per square centimeter (threshold of pain)

52. (a) $I = 10^{-13}$ watts per square centimeter (whisper)

(b) $I = 10^{-7.5}$ watts per square centimeter (DC-8 4 miles from takeoff)

(c) $I = 10^{-7}$ watts per square centimeter (diesel truck at 25 feet)

(d) $I = 10^{-4.5}$ watts per square centimeter (auto horn at 3 feet)

53. *Noise Level*   Due to the installation of noise suppression materials, the noise level in an auditorium was reduced from 93 to 80 decibels. Find the percentage decrease in the intensity level of the noise because of the installation of these materials.

54. *Noise Level*   Due to the installation of a muffler, the noise level in an engine was reduced from 88 to 72 decibels. Find the percentage decrease in the intensity level of the noise because of the installation of the muffler.

Acidity   In Exercises 55–60, use the acidity model given by

$$pH = -\log_{10}[H^+],$$

where acidity (pH) is a measure of the hydrogen ion concentration $[H^+]$ (measured in moles of hydrogen per liter) of a solution.

55. Find the pH if $[H^+] = 2.3 \times 10^{-5}$.

56. Find the pH if $[H^+] = 11.3 \times 10^{-6}$.

57. Compute $[H^+]$ for a solution in which pH $= 5.8$.

58. Compute $[H^+]$ for a solution in which pH $= 3.2$.

59. A certain fruit has a pH of 2.5 and an antacid tablet has a pH of 9.5. The hydrogen ion concentration of the fruit is how many times the concentration of the tablet?

60. If the pH of a solution is decreased by one unit, the hydrogen ion concentration is increased by what factor?

61. *Home Mortgage*   An $80,000 home mortgage for 35 years at $9\frac{1}{2}\%$ has a monthly payment of $657.28. Part of the monthly payment goes for the interest charge on the unpaid balance and the remainder of the payment is used to reduce the principal. The amount that goes for interest is given by

$$u = M - \left(M - \frac{Pr}{12}\right)\left(1 + \frac{r}{12}\right)^{12t},$$

and the amount that goes toward reduction of the principal is given by

$$v = \left(M - \frac{Pr}{12}\right)\left(1 + \frac{r}{12}\right)^{12t}.$$

In these formulas, P is the size of the mortgage, r is the interest rate, M is the monthly payment, and t is the time in years.

(a) Use a graphing utility to graph each function on the same viewing rectangle. (The viewing rectangle should show all 35 years of mortgage payments.)

(b) In the early years of the mortgage, the larger part of the monthly payment goes for what purpose? Approximate the time when the monthly payment is evenly divided between interest and principal reduction.

62. *Home Mortgage*   The total interest paid on a home mortgage of P dollars, at interest rate r for t years, is given by

$$u = P\left[\frac{rt}{1 - \left(\dfrac{1}{1 + r/12}\right)^{12t}} - 1\right].$$

Consider an $80,000 home mortgage at $9\frac{1}{2}\%$.

(a) Use a graphing utility to graph the total interest function.

(b) Approximate the length of the mortgage when the total interest paid is the same as the size of the mortgage. Is it possible that some people are paying twice as much in interest charges as the size of the mortgage?

63. *Estimating the Time of Death*   At 8:30 A.M., a coroner was called to the home of a person who had died during the night. In order to estimate the time of death, the coroner took the person's temperature twice. At 9:00 A.M., the temperature was $85.7°$, and at 9:30 A.M., the temperature was $82.8°$. From these two temperatures, the coroner was able to determine that

$$t = -2.5 \ln \frac{T - 70}{98.6 - 70},$$

where t is the time in hours that has elapsed since the person died and T is the temperature (in degrees Fahrenheit) of the person's body at 9:00 A.M. (The person had a normal body temperature of $98.6°$ at death, and the room temperature was a constant $70°$. This formula is derived from a general cooling principle called Newton's Law of Cooling.) Use this formula to estimate the time of death of the person.

64. *Population Growth*   In Exercises 33 and 34, you can see that the populations of Dhaka and Houston are growing at different rates. What constant in the equation $y = Ce^{kt}$ is affected by these different growth rates? Discuss the relationship between the different growth rates and the magnitude of the constant.

4.6 EXPLORING DATA: NONLINEAR MODELS

Classifying Scatter Plots / Fitting Nonlinear Models to Data / Applications

Classifying Scatter Plots

In Section 2.6, you saw how to fit linear models to data. In real life, many relationships between two variables are nonlinear. A scatter plot can be used to give you an idea of which type of model can be used to best fit a set of data.

EXAMPLE 1 Classifying Scatter Plots

Sketch a scatter plot for each set of data. Then decide whether the data could be best modeled by an exponential model, $y = ae^{bx}$, or a logarithmic model, $y = a + b \ln x$.

A. (0.9, 1.9), (1.3, 2.4), (1.3, 2.2), (1.4, 2.4), (1.6, 2.7), (1.8, 3.0), (2.1, 3.4), (2.1, 3.3), (2.5, 4.2), (2.9, 5.1), (3.2, 6.0), (3.3, 6.2), (3.6, 7.3), (4.0, 8.8), (4.2, 9.8), (4.3, 10.2)

B. (0.9, 3.2), (1.3, 4.0), (1.3, 3.8), (1.4, 4.2), (1.6, 4.5), (1.8, 4.8), (2.1, 5.1), (2.1, 5.0), (2.5, 5.5), (2.9, 5.8), (3.2, 6.0), (3.3, 6.2), (3.6, 6.2), (4.0, 6.6), (4.2, 6.7), (4.3, 6.8)

SOLUTION

Begin by entering the data into a graphing utility. You should obtain the scatter plots shown in Figure 4.32.

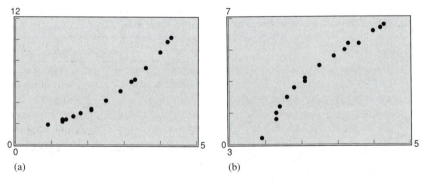

(a)   (b)

FIGURE 4.32

From the scatter plots, it appears that the data in part (a) can be modeled by an exponential function and the data in part (b) can be modeled by a logarithmic function.

Fitting Nonlinear Models to Data

Once you have used a scatter plot to determine the type of model to be fit to a set of data, there are several ways that you can actually find the model. Each method is best used with a computer or calculator, rather than with hand calculations.

The classic method is to transform one or both of the coordinates of the data points, so that the resulting points can be fit with a linear model. For instance, you can fit an *exponential* model to a set of points of the form (x, y) by fitting a *linear* model to points of the form $(x, \ln y)$.

EXAMPLE 2   **Fitting an Exponential Model to Data**

Find an exponential model for the points given in Example 1(A).

(0.9, 1.9), (1.3, 2.4), (1.3, 2.2), (1.4, 2.4), (1.6, 2.7), (1.8, 3.0),
(2.1, 3.4), (2.1, 3.3), (2.5, 4.2), (2.9, 5.1), (3.2, 6.0), (3.3, 6.2),
(3.6, 7.3), (4.0, 8.8), (4.2, 9.8), (4.3, 10.2)

SOLUTION

Begin by transforming the points by taking the natural logarithm of each y-coordinate, as shown in Table 4.5.

TABLE 4.5

x	0.9	1.3	1.3	1.4	1.6	1.8	2.1	2.1
y	1.9	2.4	2.2	2.4	2.7	3.0	3.4	3.3
$\ln y$	0.642	0.875	0.788	0.875	0.993	1.099	1.224	1.194

x	2.5	2.9	3.2	3.3	3.6	4.0	4.2	4.3
y	4.2	5.1	6.0	6.2	7.3	8.8	9.8	10.2
$\ln y$	1.435	1.629	1.792	1.825	1.988	2.175	2.282	2.322

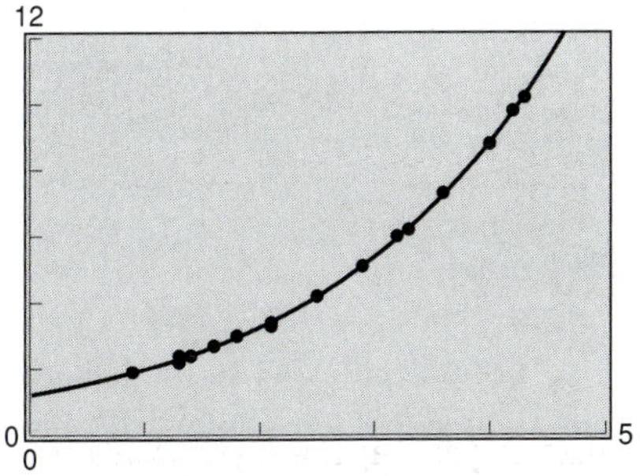

Next, use a graphing utility to sketch a scatter plot of the points $(x, \ln y)$, as shown in Figure 4.33. Notice that the points appear to fit a linear model. By applying the least squares regression formulas to the transformed points, you obtain

$$\ln y = 0.1847 + 0.4984x \qquad \text{\textit{Least squares regression "line"}}$$
$$y = e^{0.1847+0.4984x} \qquad \text{\textit{Exponentiate both sides}}$$
$$y = 1.203e^{0.4984x} \qquad \text{\textit{Exponential model}}$$

Figure 4.34 shows the graph of this model with the original data points.

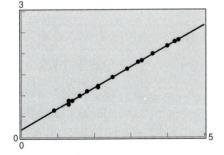

FIGURE 4.33

FIGURE 4.34

You can fit a *logarithmic* model to a set of points of the form (x, y) by fitting a *linear* model to points of the form $(\ln x, y)$, as illustrated in Example 3.

EXAMPLE 3   Fitting a Logarithmic Model to Data

Find a logarithmic model for the points given in Example 1(B).

(0.9, 3.2), (1.3, 4.0), (1.3, 3.8), (1.4, 4.2), (1.6, 4.5), (1.8, 4.8), (2.1, 5.1), (2.1, 5.0), (2.5, 5.5), (2.9, 5.8), (3.2, 6.0), (3.3, 6.2), (3.6, 6.2), (4.0, 6.6), (4.2, 6.7), (4.3, 6.8)

SOLUTION

Begin by transforming the points by taking the natural logarithm of each x-coordinate, as shown in Table 4.6.

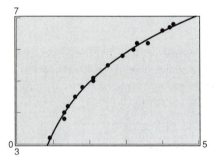

FIGURE 4.35

TABLE 4.6

x	0.9	1.3	1.3	1.4	1.6	1.8	2.1	2.1
$\ln x$	-0.105	0.262	0.262	0.336	0.470	0.588	0.742	0.742
y	3.2	4.0	3.8	4.2	4.5	4.8	5.1	5.0

x	2.5	2.9	3.2	3.3	3.6	4.0	4.2	4.3
$\ln x$	0.916	1.065	1.163	1.194	1.281	1.386	1.435	1.459
y	5.5	5.8	6.0	6.2	6.2	6.6	6.7	6.8

Next, use a graphing utility to sketch a scatter plot of the points $(\ln x, y)$, as shown in Figure 4.35. Notice that the points appear to fit a linear model. By applying the least squares regression formulas to the transformed points, you obtain

$$y = 3.377 + 2.301 \ln x. \qquad \textit{Logarithmic model}$$

Figure 4.36 shows the graph of this model with the original data points.

FIGURE 4.36

········

Applications

EXAMPLE 4   Finding an Exponential Model

The total amount A (in billions of dollars) spent on health care in the United States from 1970 through 1989 is shown in Table 4.7. Find a model for the data. Then use the model to predict the amount spent in 1998. In the table, t represents the year, with $t = 0$ corresponding to 1970. (*Source:* U.S. Health Care Financing Administration)

TABLE 4.7

t	0	1	2	3	4	5	6	7	8	9
A	74.4	82.3	92.3	102.5	116.1	132.9	152.2	172.0	193.4	216.6

t	10	11	12	13	14	15	16	17	18	19
A	249.1	288.6	323.8	356.1	387.0	420.1	452.3	492.5	544.0	604.1

SOLUTION

Begin by entering the data into a computer or graphing calculator. Then, use the computer to sketch a scatter plot of the data, as shown in Figure 4.37. From the scatter plot, it appears that an exponential model is a good fit. After running the exponential regression program, you should obtain

$$A = 76.26(1.12)^t \quad \text{or} \quad A = 76.25e^{0.1133t}.$$

(The correlation coefficient is $r = 0.997$, which implies that the model is a good fit to the data.) From the model, you can see that the amount spent on health care from 1970 through 1989 had an average annual increase of 12%. From this model, you can predict the 1998 amount to be

$$A = 76.26(1.12)^{28} \approx 1821.4 \text{ billion dollars,}$$

which is over three times the amount spent in 1989.

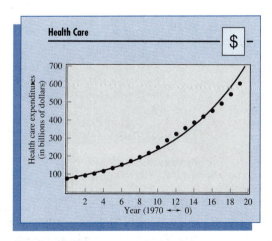

FIGURE 4.37

There are many other types of nonlinear models. One type is called a power model,

$$y = ax^b.$$ *Power model*

To fit data to this type of model, you can use a built-in program on a computer or calculator. Or you can transform the points (x, y) to the form $(\ln x, \ln y)$ and fit a linear model to the transformed points. This technique is illustrated in Example 5.

EXAMPLE 5   Using Logarithms to Find a Power Model

Table 4.8 gives the mean distance x and the period y of the six planets that are closest to the Sun. In the table the mean distance is given in terms of astronomical units (where the earth's mean distance is defined to be 1.0), and the period is given in years. Find a model for these data.

TABLE 4.8

Planet	Mercury	Venus	Earth	Mars	Jupiter	Saturn
Period, y	0.241	0.615	1.0	1.881	11.861	29.457
Distance, x	0.387	0.723	1.0	1.523	5.203	9.541

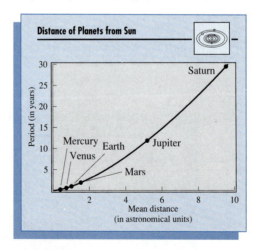

FIGURE 4.38

SOLUTION

Figure 4.38 shows a scatter plot for the points in the table. From the plot, it is not clear what type of model would best fit the data. (An exponential model has a correlation coefficient of only 0.895. With scientific data, one would

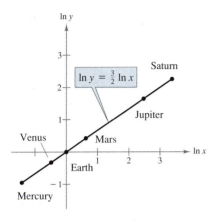

FIGURE 4.39

expect a better correlation.) To help classify the type of scatter plot, you ⌐ take the logarithm of each coordinate so that you obtain points of the form $(\ln x, \ln y)$, as shown in Table 4.9.

TABLE 4.9

Planet	Mercury	Venus	Earth	Mars	Jupiter	Saturn
$\ln y$	-1.423	-0.486	0	0.632	2.473	3.383
$\ln x$	-0.949	-0.324	0.0	0.421	1.649	2.256

Figure 4.39 shows a scatter plot for the transformed points. Note that the points fit a linear model. The least squares regression line for the transformed points is

$$\ln y = \frac{3}{2}\ln x \quad \text{or} \quad y = x^{3/2}.$$

DISCUSSION PROBLEM

RESEARCH PROJECT

Use your school's library or some other reference source to find data that can be modeled with a *nonlinear* model. Find the model that you think best fits the data and explain how you could use the model to answer questions about the data.

WARM-UP

The following warm-up exercises involve skills that were covered in earlier sections. You will use these skills in the exercise set for this section.

In Exercises 1–6, sketch a graph of the equation.

1. $y = 10e^{-0.2x}$ **2.** $y = 2e^{0.2x}$

3. $y = 3 + 0.4\ln x$ **4.** $y = 15 - 2\ln x$

5. $y = 600(1.08)^t$ **6.** $y = 1000(0.8)^t$

In Exercises 7–10, find an equation of the regression line through the points.

7. $(-2, 4)$, $(-1, 3)$, $(0, 3)$, $(1, 2)$, $(2, 1)$

8. $(-6, 0)$, $(-3, 4)$, $(3, 6)$, $(6, 9)$

9. $(0, 10)$, $(1, 14)$, $(2, 17)$, $(3, 22)$

10. $(0, 8)$, $(2, 7)$, $(3, 5)$, $(5, 4)$, $(8, 0)$

SECTION 4.6 · EXERCISES

In Exercises 1–6, sketch a scatter plot for the data set. Decide whether the data could be best modeled by a linear model, $y = mx + b$, an exponential model, $y = ae^{bx}$, or a logarithmic model, $y = a + b \ln x$.

1. (1, 2.0), (1.5, 3.5), (2, 4.0), (4, 5.8), (6, 7.0), (8, 7.8)
2. (1, 5.8), (1.5, 6.0), (2, 6.5), (4, 7.6), (6, 8.9), (8, 10.0)
3. (1, 4.4), (1.5, 4.7), (2, 5.5), (4, 9.9), (6, 18.1), (8, 33.0)
4. (1, 11.0), (1.5, 9.6), (2, 8.2), (4, 4.5), (6, 2.5), (8, 1.4)
5. (1, 7.5), (1.5, 7.0), (2, 6.8), (4, 5.0), (6, 3.5), (8, 2.0)
6. (1, 5.0), (1.5, 6.0), (2, 6.4), (4, 7.8), (6, 8.6), (8, 9.0)

In Exercises 7–10, use a graphing utility to find an exponential model, $y = ae^{bx}$, through the points. Sketch a scatter plot and the exponential model.

7. (0, 4), (1, 5), (2, 6), (3, 8), (4, 12)
8. (0, 6), (2, 8.9), (4, 20.0), (6, 34.3), (8, 61.1), (10, 120.5)
9. (0, 10.0), (1, 6.1), (2, 4.2), (3, 3.8), (4, 3.6)
10. (−3, 120.2), (0, 80.5), (3, 64.8), (6, 58.2), (10, 55.0)

In Exercises 11–14, use a graphing utility to find a logarithmic model, $y = a + b \ln x$, through the points. Sketch a scatter plot and the exponential model.

11. (1, 2), (2, 3), (3, 3.5), (4, 4), (5, 4.1), (6, 4.2), (7, 4.5)
12. (1, 8.5), (2, 11.4), (4, 12.8), (6, 13.6), (8, 14.2), (10, 14.6)
13. (1, 10), (2, 6), (3, 6), (4, 5), (5, 3), (6, 2)
14. (3, 14.6), (6, 11.0), (9, 9.0), (12, 7.6), (15, 6.5)

In Exercises 15–18, use a graphing utility to find a power model, $y = ax^b$, through the points. Sketch a scatter plot and the power model.

15. (1, 2.0), (2, 3.4), (5, 6.7), (6, 7.3), (10, 12.0)
16. (0.5, 1.0), (2, 12.5), (4, 33.2), (6, 65.7), (8, 98.5), (10, 150.0)
17. (1, 10.0), (2, 4.0), (3, 0.7), (4, 0.1)
18. (2, 450), (4, 385), (6, 345), (8, 332), (10, 312), (12, 300)

In Exercises 19 and 20, use a computer or calculator to (a) fit an exponential function $y = ab^x$ to the data, and (b) plot the data and the exponential function.

19. *Breaking Strength* The breaking strength y (in tons) of steel cable of diameter d (in inches) is given in the accompanying table.

d	0.50	0.75	1.00	1.25	1.50	1.75
y	9.85	21.8	38.3	59.2	84.4	114.0

20. *World Population* The world population y (in billions) for the 10 years from 1982 through 1991 is given in the accompanying table where $x = 2$ corresponds to 1982.

x	2	3	4	5	6
y	4.60	4.68	4.77	4.85	4.94

x	7	8	9	10	11
y	5.02	5.11	5.20	5.33	5.42

In Exercises 21 and 22, use a computer or calculator to (a) fit a logarithmic function $y = a + b \ln t$ to the data, and (b) plot the data and the logarithmic function.

21. *Water Pollution* The amount y (in billions of dollars) spent on water pollution abatement in constant 1982 dollars in the United States from 1981 through 1989 is given in the following table. In the table, $t = 1$ represents the year 1981. (*Source:* U.S. Bureau of Economic Analysis)

t	1	2	3	4	5
y	22.0	21.2	21.5	23.3	24.7

t	6	7	8	9
y	26.4	28.0	27.2	28.8

22. *Rail Travel Receipts*   The receipts *y* (in millions of dollars) of the railroad industry in the United States from 1981 through 1990 is given in the following table. In the table, *t* = 1 represents the year 1981. (*Source:* U.S. Travel Data Center)

t	1	2	3	4	5
y	429	436	481	524	563

t	6	7	8	9	10
y	592	639	738	842	888

23. *Lumber*   The accompanying table gives the total domestic production *x*, and the domestic consumption *y*, of lumber in the United States from 1983 through 1990. The measurements are in millions of board feet. (*Source:* Current Industrial Reports) Find a power model for the relationship between production and consumption.

x	34.6	37.1	36.4	42.0
y	48.7	52.7	53.5	57.4

x	44.9	44.6	43.6	43.9
y	61.8	58.8	58.8	54.5

CHAPTER 4 · REVIEW EXERCISES

In Exercises 1–6, match the function with its graph and describe the given viewing rectangle. [The graphs are labeled (a), (b), (c), (d), (e), and (f).]

1. $f(x) = 2^x$

2. $f(x) = 2^{-x}$

3. $f(x) = -2^x$

4. $f(x) = 2^x + 1$

5. $f(x) = \log_2 x$

6. $f(x) = \log_2(x - 1)$

(a)

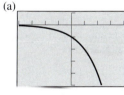

(b)

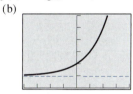

(c)

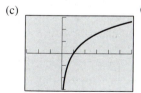

(d)

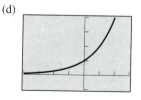

(e)

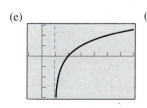

(f)

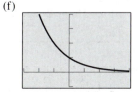

In Exercises 7–14, sketch the graph of the function.

7. $f(x) = 6^x$

8. $f(x) = 0.3^x$

9. $g(x) = 6^{-x}$

10. $g(x) = 0.3^{-x}$

11. $h(x) = e^{-x/2}$

12. $h(x) = 2 - e^{-x/2}$

13. $f(x) = e^{x+2}$

14. $s(t) = 4e^{-2/t}, \ t > 0$

In Exercises 15 and 16, complete the table to determine the balance *A* for *P* dollars invested at rate *r* for *t* years and compounded *n* times per year.

n	1	2	4	12	365	Continuous compounding
A						

15. $P = \$3500, \ r = 10.5\%, \ t = 10$ years

16. $P = \$2000, \ r = 12\%, \ t = 30$ years

In Exercises 17 and 18, complete the following table to determine the amount of money P that should be invested at rate r to produce a final balance of $200,000 in t years.

t	1	10	20	30	40	50
P						

17. $r = 8\%$, compounded continuously
18. $r = 10\%$, compounded monthly

19. *Trust Fund*   On the day your child was born you deposited $50,000 in a trust fund that pays 8.75% interest, compounded continuously. Determine the balance in the account at the time of your child's 35th birthday.

20. *Depreciation*   After t years, the value of a car that cost $14,000 is given by

$$V(t) = 14,000(\tfrac{3}{4})^t.$$

Use a graphing utility to graph the function and determine the value of the car 2 years after it was purchased.

21. *Drug Decomposition*   A solution of a certain drug contained 500 units per milliliter when prepared. It was analyzed after 40 days and found to contain 300 units per milliliter. Assuming that the rate of decomposition is proportional to the amount present, the equation giving the amount A after t days is

$$A = 500e^{-0.013t}.$$

Use this model to find A when $t = 60$. Solve this problem algebraically and graphically. Which method do you prefer?

22. *Waiting Times*   The average time between incoming calls at a switchboard is 3 minutes. The probability of waiting less than t minutes until the next incoming call is approximated by the model

$$P(t) = 1 - e^{-t/3}.$$

If a call has just come in, find the probability that the next call will be within
(a) $\frac{1}{2}$ minute.   (b) 2 minutes.   (c) 5 minutes.
Solve the problem algebraically and graphically. Which method do you prefer?

23. *Fuel Efficiency*   A certain automobile gets 28 miles per gallon of gasoline for speeds up to 50 miles per hour. Over 50 miles per hour, the number of miles per gallon drops at the rate of 12% for each 10 miles per hour. If s is the speed and y is the number of miles per gallon, then

$$y = 28e^{0.6-0.012s}, \quad s \geq 50.$$

Use this function to complete the table. Write a short paragraph describing the results.

Speed	50	55	60	65	70
Miles per gallon					

24. *Inflation*   If the annual rate of inflation averages 4.5% over the next 10 years, then the approximate cost C of goods or services during any year in that decade will be given by

$$C(t) = P(1.05)^t,$$

where t is the time in years and P is the present cost. If the price of a tire for your car is presently $69.95, estimate the price 10 years from now.

In Exercises 25–30, sketch the graph of the function.

25. $g(x) = \log_3 x$ **26.** $g(x) = \log_5 x$
27. $f(x) = \ln x + 3$ **28.** $f(x) = \ln(x - 3)$
29. $h(x) = \ln(e^{x-1})$ **30.** $f(x) = \tfrac{1}{4} \ln x$

In Exercises 31 and 32, use the definition of a logarithm to write the given equation in logarithmic form.

31. $4^3 = 64$ **32.** $25^{3/2} = 125$

In Exercises 33–40, evaluate the expression.

33. $\log_{10} 1000$

34. $\log_9 3$

35. $\log_3 \frac{1}{9}$

36. $\log_4 \frac{1}{16}$

37. $\ln e^7$

38. $\log_a \dfrac{1}{a}$

39. $\ln 1$

40. $\ln e^{-3}$

In Exercises 41–44, evaluate the logarithm using the change of base formula. Do each problem twice, once with common logarithms and once with natural logarithms. (Round your result to three decimal places.)

41. $\log_4 9$

42. $\log_{1/2} 5$

43. $\log_{12} 200$

44. $\log_3 0.28$

In Exercises 45–50, use the properties of logarithms to write the expression as a sum, difference, and/or multiple of logarithms.

45. $\log_5 5x^2$

46. $\log_7 \dfrac{\sqrt{x}}{4}$

47. $\log_{10} \dfrac{5\sqrt{y}}{x^2}$

48. $\ln \left| \dfrac{x-1}{x+1} \right|$

49. $\ln[(x^2 + 1)(x - 1)]$

50. $\ln \sqrt[5]{\dfrac{4x^2 - 1}{4x^2 + 1}}$

In Exercises 51–56, write the expression as the logarithm of a single quantity.

51. $\log_2 5 + \log_2 x$

52. $\log_6 y - 2 \log_6 z$

53. $\frac{1}{2} \ln |2x - 1| - 2 \ln |x + 1|$

54. $5 \ln |x - 2| - \ln |x + 2| - 3 \ln |x|$

55. $\ln 3 + \frac{1}{3} \ln(4 - x^2) - \ln x$

56. $3[\ln x - 2 \ln(x^2 + 1)] + 2 \ln 5$

In Exercises 57–60, determine whether the statement or equation is true or false. Explain how to use a graphing utility to support your conclusion.

57. The domain of $f(x) = \ln x$ is the set of all real numbers.

58. $\log_b b^{2x} = 2x$

59. $\ln(x + 2) = \ln x + \ln 2$

60. $e^{x-1} = \dfrac{e^x}{e}$

In Exercises 61–64, approximate the logarithm using the properties of logarithms given $\log_b 2 \approx 0.3562$, $\log_b 3 \approx 0.5646$, and $\log_b 5 \approx 0.8271$.

61. $\log_b 25$

62. $\log_b\left(\frac{25}{9}\right)$

63. $\log_b \sqrt{3}$

64. $\log_b 30$

65. *Snow Removal*   The number of miles s of roads cleared of snow is approximated by the model

$$s = 25 - \dfrac{13 \ln \dfrac{h}{12}}{\ln 3}, \quad 2 \le h \le 15,$$

where h is the depth of the snow in inches. Use this model to find s when $h = 10$ inches.

66. *Climb Rate*   The time t, in minutes, required for a small plane to climb to an altitude of h feet is given by

$$t = 50 \log_{10} \dfrac{18,000}{18,000 - h},$$

where 18,000 feet is the plane's absolute ceiling. Find the time for the plane to climb to an altitude of 4000 feet.

In Exercises 67–72, solve the exponential equation. (Round your result to three decimal places.)

67. $e^x = 12$

68. $e^{3x} = 25$

69. $3e^{-5x} = 132$

70. $14e^{3x+2} = 560$

71. $e^{2x} - 7e^x + 10 = 0$

72. $e^{2x} - 6x + 8 = 0$

In Exercises 73–76, solve the logarithmic equation. (Round your result to three decimal places.)

73. $\ln 3x = 8.2$ **74.** $2 \ln 4x = 15$

75. $\ln x - \ln 3 = 2$ **76.** $\ln \sqrt{x + 1} = 2$

In Exercises 77–80, find the exponential function $y = Ce^{kt}$ that passes through the two points.

77. $(0, 2), (4, 3)$ **78.** $(0, \frac{1}{2}), (5, 5)$

79. $(0, 4), (5, \frac{1}{2})$ **80.** $(0, 2), (5, 1)$

81. *Demand Function* The demand equation for a certain product is given by

$$p = 500 - 0.5e^{0.004x}.$$

Find the demand x for a price of (a) $p = \$450$ and (b) $p = \$400$.

82. *Typing Speed* In a typing class, the average number of words per minute typed after t weeks of lessons was found to be

$$N = \frac{157}{1 + 5.4e^{-0.12t}}.$$

Find the time necessary to type (a) 50 words per minute and (b) 75 words per minute.

83. *Compound Interest* A deposit of $750 is made in a savings account for which the interest is compounded continuously. The balance will double in $7\frac{3}{4}$ years.
(a) What is the annual percentage rate for this account?
(b) Find the balance in the account after 10 years.
(c) Find the effective yield.

84. *Compound Interest* A deposit of $10,000 is made in a savings account for which the interest is compounded continuously. The balance will double in 5 years.
(a) What is the annual percentage rate for this account?
(b) Find the balance after 1 year.
(c) Find the effective yield.

85. *Sound Intensity* The relationship between the number of decibels β and the intensity of a sound I in watts per square centimeter is given by

$$\beta = 10 \log_{10}\left(\frac{I}{10^{-16}}\right).$$

Determine the intensity of a sound in watts per square centimeter if the decibel level is 125.

86. *Earthquake Magnitudes* On the Richter Scale, the magnitude R of an earthquake of intensity I is given by

$$R = \log_{10}\frac{I}{I_0},$$

where $I_0 = 1$ is the minimum intensity used for comparison. Find the intensity per unit of area for the following values of R.
(a) $R = 8.4$ (b) $R = 6.85$ (c) $R = 9.1$

87. Use a graphing utility to find an exponential model $y = ae^{bx}$ through the points $(0, 250)$, $(4, 135)$, $(6, 92)$, $(10, 67)$. Sketch a scatter plot and the exponential model.

88. The data in the accompanying table give the yield y (in milligrams) of a chemical reaction after t minutes.

t	1	2	3	4
y	1.5	7.4	10.2	13.4

t	5	6	7	8
y	15.8	16.3	18.2	18.3

(a) Use a graphing utility to find the least squares regression line for the data.
(b) Use a graphing utility to fit the logarithmic equation $y = a + b \ln x$ to the data.
(c) Create a scatter plot of the data and sketch the graphs of the equations found in parts (a) and (b). Which is a better model of the data?

TRIGONOMETRIC FUNCTIONS

5.1 RADIAN AND DEGREE MEASURE

Angles / Radian Measure / Degree Measure / Applications

Angles

As derived from the Greek language, the word **trigonometry** means "measurement of triangles." Initially, trigonometry dealt with relationships among the sides and angles of triangles. As such, it was used in the development of astronomy, navigation, and surveying.

With the advent of calculus in the 17th century and a resulting expansion of knowledge in the physical sciences, a different perspective arose—one that viewed the classic trigonometric relationships as *functions* with the set of real numbers as their domains. Consequently, the applications of trigonometry expanded to include a vast number of physical phenomena involving rotations or vibrations. These include sound waves, light rays, planetary orbits, vibrating strings, pendulums, and orbits of atomic particles. Our approach to trigonometry incorporates *both* perspectives, starting with angles and their measure.

An **angle** is determined by rotating a ray (half-line) about its endpoint. The starting position of the ray is called the **initial side** of the angle, and the position after rotation is called the **terminal side,** as shown in Figure 5.1. The endpoint of the ray is called the **vertex** of the angle.

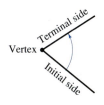

FIGURE 5.1

In a coordinate system, an angle is in **standard position** if its vertex is the origin and its initial side coincides with the positive x-axis, as shown in Figure 5.2. **Positive angles** are generated by counterclockwise rotation, and **negative angles** by clockwise rotation, as shown in Figures 5.3 and 5.4. To label angles in trigonometry, we use the Greek letters α (alpha), β (beta), and θ (theta), as well as uppercase letters A, B, and C. In Figure 5.4, note that the angles α and β have the same initial and terminal sides. Such angles are **coterminal.**

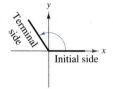

Standard Position of an Angle

FIGURE 5.2

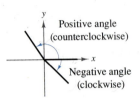

Positive Angle

FIGURE 5.3

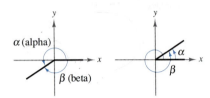

Coterminal Angles

FIGURE 5.4

Radian Measure

The **measure of an angle** is determined by the amount of rotation from the initial to the terminal side. One way to measure angles is in radians. This type of measure is needed in calculus. To define a radian we use a **central angle** of a circle, one whose vertex is the center of the circle, as shown in Figure 5.5.

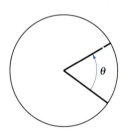

Central Angle θ

FIGURE 5.5

DEFINITION OF A RADIAN

One **radian** is the measure of a central angle θ that subtends (intercepts) an arc s equal in length to the radius r of the circle (see figure).

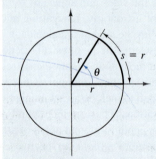

arc length = radius when $\theta = 1$ radian

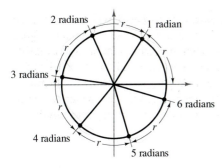

FIGURE 5.6

Because the circumference of a circle is $2\pi r$, it follows that a central angle of one full revolution (counterclockwise) corresponds to an arc length of $s = 2\pi r$. Moreover, because each radian intercepts an arc of length r, we conclude that one full revolution corresponds to an angle of

$$\frac{2\pi r}{r} = 2\pi \text{ radians.}$$

Note that since $2\pi \approx 6.28$, there are a little more than six radius lengths in a full circle, as shown in Figure 5.6.

In general, the radian measure of a central angle θ is obtained by dividing the arc length s by r. That is,

$$\frac{s}{r} = \theta, \qquad\qquad \textit{Radian measure}$$

where θ *is measured in radians.* Because the units of measure for s and r are the same, this ratio is unitless—it is simply a real number. The radian measures of several common angles are shown in Figure 5.7.

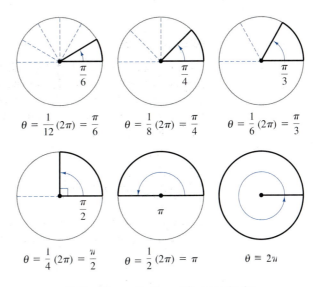

Radian Measure for Several Common Angles

FIGURE 5.7

Recall that the four quadrants in a coordinate system are numbered counterclockwise as I, II, III, and IV. Figure 5.8 shows which angles between 0 and 2π lie in each of the four quadrants.

You can find an angle that is coterminal to a given angle θ by adding or subtracting 2π (one revolution), as demonstrated in Example 1. (Note that a given angle has many coterminal angles. For instance, $\theta = \pi/6$ is coterminal with both $13\pi/6$ and $-11\pi/6$.)

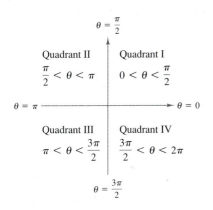

FIGURE 5.8

EXAMPLE 1 Sketching and Finding Coterminal Angles

A. To find an angle that is coterminal to the positive angle $\theta = 13\pi/6$, you can subtract 2π to obtain

$$\frac{13\pi}{6} - 2\pi = \frac{\pi}{6}.$$

Thus, the terminal side of θ lies in Quadrant I. Its sketch is shown in Figure 5.9(a).

B. To find an angle that is coterminal to the positive angle $\theta = 3\pi/4$, you can subtract 2π to obtain

$$\frac{3\pi}{4} - 2\pi = -\frac{5\pi}{4},$$

as shown in Figure 5.9(b).

C. To find an angle that is coterminal to the negative angle $\theta = -2\pi/3$, you can add 2π to obtain

$$\theta = -\frac{2\pi}{3} + 2\pi = \frac{4\pi}{3},$$

as shown in Figure 5.9(c).

REMARK The phrase "the terminal side of θ lies in a quadrant" can be abbreviated by simply saying that "θ lies in a quadrant." The terminal sides of the quadrant angles 0, $\pi/2$, π, and $3\pi/2$ do not lie within any of the four quadrants.

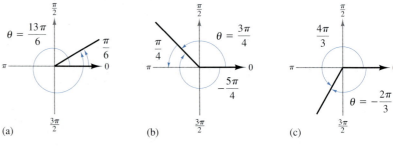

(a) (b) (c)

FIGURE 5.9

Figure 5.10 shows several common angles with their radian measures. Note that we classify angles between 0 and $\pi/2$ radians as **acute** and angles between $\pi/2$ and π as **obtuse**.

Acute angle:
between 0 and $\dfrac{\pi}{2}$

Right angle:
quarter revolution

Obtuse angle:
between $\dfrac{\pi}{2}$ and π

Straight angle:
half revolution

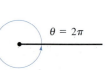

Full revolution

FIGURE 5.10

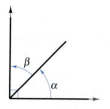

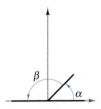

(a) Complementary angles

(b) Supplementary angles

FIGURE 5.11

Two *positive* angles α and β are said to be **complementary** (or complements of each other) if their sum is $\pi/2$, as shown in Figure 5.11(a). For example, $\pi/6$ and $\pi/3$ are complementary angles because

$$\frac{\pi}{6} + \frac{\pi}{3} = \frac{\pi}{2}.$$

Two positive angles are **supplementary** (or supplements of each other) if their sum is π, as shown in Figure 5.11(b). For example, $2\pi/3$ and $\pi/3$ are supplementary angles because

$$\frac{2\pi}{3} + \frac{\pi}{3} = \pi.$$

EXAMPLE 2   **Complementary, Supplementary, and Coterminal Angles**

A. The complement of $\theta = \pi/12$ is

$$\frac{\pi}{2} - \frac{\pi}{12} = \frac{6\pi}{12} - \frac{\pi}{12} = \frac{5\pi}{12},$$

as shown in Figure 5.12(a).

B. The supplement of $\theta = 5\pi/6$ is

$$\pi - \frac{5\pi}{6} = \frac{6\pi}{6} - \frac{5\pi}{6} = \frac{\pi}{6},$$

as shown in Figure 5.12(b).

C. In radian measure, a coterminal angle is found by adding or subtracting 2π. For $\theta = 17\pi/6$, you subtract 2π to obtain

$$\frac{17\pi}{6} - 2\pi = \frac{17\pi}{6} - \frac{12\pi}{6} = \frac{5\pi}{6},$$

as shown in Figure 5.12(c). Thus, $17\pi/6$ and $5\pi/6$ are coterminal.

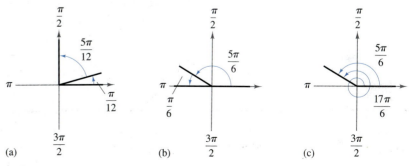

(a)   (b)   (c)

FIGURE 5.12

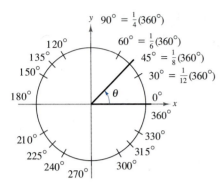

Degree Measure of an Angle

FIGURE 5.13

Degree Measure

A second way to measure angles is in terms of degrees. A measure of **one degree (1°)** is equivalent to $1/360$ of a complete revolution about the vertex. To measure angles in degrees, it is convenient to mark degrees on the circumference of a circle as shown in Figure 5.13. Thus, a full revolution (counterclockwise) corresponds to 360°, a half revolution to 180°, and a quarter revolution to 90°.

Because 2π radians is the measure of an angle of one complete revolution, degrees and radians are related by the equations

$$360° = 2\pi \text{ rad} \quad \text{and} \quad 180° = \pi \text{ rad}.$$

From the latter equation, we obtain

$$1° = \frac{\pi}{180} \text{ rad} \quad \text{and} \quad 1 \text{ rad} = \left(\frac{180}{\pi}\right)°,$$

which lead to the following conversion rules.

REMARK Note that when no units of angle measure are specified, *radian measure is implied.* For instance, if we write $\theta = \pi$ or $\theta = 2$, we mean $\theta = \pi$ radians or $\theta = 2$ radians.

CONVERSIONS: DEGREES ↔ RADIANS

1. To convert degrees to radians, multiply degrees by $\dfrac{\pi \text{ rad}}{180°}$.

2. To convert radians to degrees, multiply radians by $\dfrac{180°}{\pi \text{ rad}}$.

To apply these two conversion rules, use the basic relationship π rad $=$ 180°.

EXAMPLE 3   Converting from Degrees to Radians

A. $135° = (135 \text{ deg})\left(\dfrac{\pi \text{ rad}}{180 \text{ deg}}\right)$ *Multiply by $\pi/180$*

$\qquad = \dfrac{3\pi}{4} \text{ rad}$

B. $540° = (540 \text{ deg})\left(\dfrac{\pi \text{ rad}}{180 \text{ deg}}\right)$ *Multiply by $\pi/180$*

$\qquad = 3\pi \text{ rad}$

C. $-270° = (-270 \text{ deg})\left(\dfrac{\pi \text{ rad}}{180 \text{ deg}}\right)$ *Multiply by $\pi/180$*

$\qquad = -\dfrac{3\pi}{2} \text{ rad}$

Technology Note

Calculators and graphing utilities have both degree and radian modes. When you are using a calculator or graphing utility, be sure you use the correct mode. Most graphing utilities allow you to convert directly from degrees to radians and from radians to degrees. Try using a graphing utility to convert $135°$ to radians and $-\pi/2$ radians to degrees.

EXAMPLE 4   Converting from Radians to Degrees

A. $-\dfrac{\pi}{2}$ rad $= \left(-\dfrac{\pi}{2}\text{ rad}\right)\left(\dfrac{180\text{ deg}}{\pi\text{ rad}}\right)$ *Multiply by $180/\pi$*

$= -90°$

B. $\dfrac{9\pi}{2}$ rad $= \left(\dfrac{9\pi}{2}\text{ rad}\right)\left(\dfrac{180\text{ deg}}{\pi\text{ rad}}\right)$ *Multiply by $180/\pi$*

$= 810°$

C. 2 rad $= (2\text{ rad})\left(\dfrac{180\text{ deg}}{\pi\text{ rad}}\right) = \dfrac{360}{\pi}$ deg   *Multiply by $180/\pi$*

$\approx 114.59°$

With calculators it is convenient to use *decimal* degrees to denote fractional parts of degrees. Historically, however, fractional parts of degrees were expressed in *minutes* and *seconds*, using the prime (′) and double prime (″) notations, respectively. That is,

$$1' = \text{one minute} = \frac{1}{60}(1°)$$

$$1'' = \text{one second} = \frac{1}{60}(1') = \frac{1}{3600}(1°).$$

Consequently, an angle of 64 degrees, 32 minutes, and 47 seconds is represented by $\theta = 64°\ 32'\ 47''$.

Many calculators have special keys for converting an angle in degrees, minutes, and seconds ($D°\ M'\ S''$) into decimal degree form, and conversely. If your calculator does not have these special keys, you can use the techniques demonstrated in the next example to make the conversions.

EXAMPLE 5   Converting an Angle from $D°\ M'\ S''$ to Decimal Form

Convert $152°\ 15'\ 29''$ to decimal degree form.

SOLUTION

$$152°\ 15'\ 29'' = 152° + \left(\frac{15}{60}\right)° + \left(\frac{29}{3600}\right)°$$

$$\approx 152° + 0.25° + 0.00806°$$

$$= 152.25806°.$$

Applications

The *radian measure* formula, $\theta = s/r$, can be used to measure arc length along a circle. Specifically, for a circle of radius r, a central angle θ subtends an arc of length s given by

$$s = r\theta, \qquad \textit{Length of circular arc}$$

where θ is measured in radians.

EXAMPLE 6   Finding Arc Length

A circle has a radius of 4 inches. Find the length of the arc cut off (subtended) by a central angle of 240°, as shown in Figure 5.14.

SOLUTION

To use the formula $s = r\theta$, we must first convert 240° to radian measure.

$$240° = (240 \text{ deg})\left(\frac{\pi \text{ rad}}{180 \text{ deg}}\right) = \frac{4\pi}{3} \text{ rad}$$

Then, using a radius of $r = 4$ inches, we find the arc length to be

$$s = r\theta$$
$$= 4\left(\frac{4\pi}{3}\right)$$
$$= \frac{16\pi}{3}$$
$$\approx 16.76 \text{ inches.}$$

Note that the units for $r\theta$ are determined by the units for r because θ has no units.

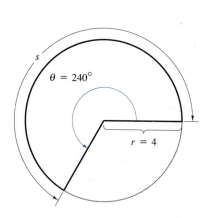

FIGURE 5.14

The formula for the length of a circular arc can be used to analyze the motion of a particle moving at a *constant speed* along a circular path. Assume the particle is moving at a constant speed along a circular path (of radius r). If s is the length of the arc traveled in time t, then we say that the **speed** of the particle is

$$\text{Speed} = \frac{\text{distance}}{\text{time}} = \frac{s}{t}.$$

Moreover, if θ is the angle (in radian measure) corresponding to the arc length s, then the **angular speed** of the particle is

$$\text{Angular speed} = \frac{\theta}{t}.$$

EXAMPLE 7   Finding the Speed of an Object

The second hand on a clock is 4 inches long, as shown in Figure 5.15. Find the speed of the tip of this second hand.

SOLUTION

The time required for the second hand to make one full revolution is

$$t = 60 \text{ seconds} = 1 \text{ minute}.$$

The distance traveled by the tip of the second hand in one revolution is

$$s = 2\pi(\text{radius}) = 2\pi(4) = 8\pi \text{ inches}.$$

Therefore, the speed of the tip of the second hand is

$$\text{Speed} = \frac{s}{t} = \frac{8\pi \text{ inches}}{60 \text{ seconds}} \approx 0.419 \text{ in./sec.}$$

FIGURE 5.15

EXAMPLE 8   Finding Angular Speed and Linear Speed

A lawn roller that is 30 inches in diameter makes one revolution every $\frac{5}{6}$ second, as shown in Figure 5.16.

A. Find the angular speed of the roller in radians per second.
B. How fast is the roller moving across the lawn?

SOLUTION

A. Because there are 2π radians in one revolution, it follows that the angular speed is

$$\text{Angular speed} = \frac{\theta}{t}$$

$$= \frac{2\pi \text{ rad}}{\frac{5}{6} \text{ sec}}$$

$$= 2.4\pi \text{ rad/sec.}$$

B. Because the diameter is 30 inches, $r = 15$ and $s = 2\pi r = 30\pi$ inches. Thus,

$$\text{Speed} = \frac{s}{t}$$

$$= \frac{30\pi \text{ in.}}{\frac{5}{6} \text{ sec}}$$

$$= 36\pi \text{ in./sec}$$

$$\approx 113.1 \text{ in./sec.}$$

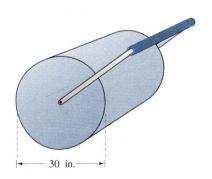

— 30 in. —

FIGURE 5.16

DISCUSSION PROBLEM

AN ANGLE-DRAWING PROGRAM

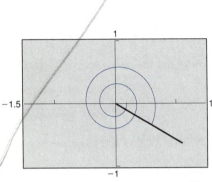

If you have a programmable calculator, try entering an "angle-drawing program." Then use the program to draw several different angles. The following program is for the TI-81.

```
:Disp "ENTER,MODE"            :0→Tmin
:Disp "0,RADIAN"              :abs T→Tmax
:Disp "1,DEGREE"             :.15→Tstep
:Input M                      :cos T→A
:Disp "ENTER,ANGLE"           :sin T→B
:Input T                      :Param
:If M=1                       :1→S
:πT/180→T                     :If T<0
:Rad                          :-1→S
:ClrDraw                      :"(.25+.04T)cos T"→X₁ₜ
:All-Off                      :"S(.25+.04T)sin T"→Y₁ₜ
:-1.5→Xmin                    :DispGraph
:1.5→Xmax                     :Line(0,0,A,B)
:1→Xscl                       :Pause
:-1→Ymin                      :Function
:1→Ymax                       :End
:1→Yscl
```

To run this program, enter 0 for radian mode or 1 for degree mode. Then enter any angle (the angle can be negative or larger than 360°). After the angle is displayed, press ENTER to clear the display. The figure shows an angle of −750°, as drawn by this program.

WARM-UP

The following warm-up exercises involve skills that were covered in earlier sections. You will use these skills in the exercise set for this section.

In Exercises 1–10, solve for x.

1. $x + 135 = 180$

2. $790 = 720 + x$

3. $\pi = \dfrac{5\pi}{6} + x$

4. $2\pi - x = \dfrac{5\pi}{3}$

5. $\dfrac{45}{180} = \dfrac{x}{\pi}$

6. $\dfrac{240}{180} = \dfrac{x}{\pi}$

7. $\dfrac{\pi}{180} = \dfrac{x}{20}$

8. $\dfrac{180}{\pi} = \dfrac{330}{x}$

9. $\dfrac{x}{60} = \dfrac{3}{4}$

10. $\dfrac{x}{3600} = 0.0125$

SECTION 5.1 · EXERCISES

In Exercises 1–4, determine the quadrant in which the terminal side of the angle lies. (The angle is given in radians.)

1. (a) $\dfrac{\pi}{5}$ (b) $\dfrac{7\pi}{5}$

2. (a) $-\dfrac{\pi}{12}$ (b) $-\dfrac{11\pi}{9}$

3. (a) -1 (b) -2

4. (a) 5.63 (b) -2.25

In Exercises 5 and 6, determine the quadrant in which the terminal side of the angle lies.

5. (a) $130°$ (b) $285°$

6. (a) $-260°$ (b) $-3.4°$

In Exercises 7–10, sketch the angle in standard position.

7. (a) $\dfrac{5\pi}{4}$

     (b) $\dfrac{2\pi}{3}$

8. (a) $-\dfrac{7\pi}{4}$

     (b) $-\dfrac{5\pi}{2}$

9. (a) $30°$

     (b) $150°$

10. (a) $405°$

     (b) $-480°$

In Exercises 11 and 12, determine two coterminal angles (one positive and one negative) for the angle. Give your results in radians.

11. (a)
$\theta = \dfrac{\pi}{9}$
     (b)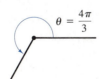
$\theta = \dfrac{4\pi}{3}$

12. (a)
$\theta = -\dfrac{9\pi}{4}$
     (b)
$\theta = -\dfrac{2\pi}{15}$

In Exercises 13–16, determine two coterminal angles (one positive and one negative) for the angle. Give your results in degrees.

13. (a)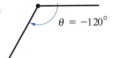
$\theta = 36°$
     (b)
$\theta = -45°$

14. (a) $\theta = -120°$ (b) $\theta = 390°$

15. (a) $\theta = 300°$ (b) $\theta = 740°$

16. (a) $\theta = -420°$ (b) $\theta = 230°$

In Exercises 17–20, find (if possible) the positive angle complement and the positive angle supplement of the angle.

17. (a) $\dfrac{\pi}{3}$ (b) $\dfrac{3\pi}{4}$

18. (a) 1 (b) 2

19. (a) $18°$ (b) $115°$

20. (a) $79°$ (b) $150°$

In Exercises 21–24, express the angle in degree measure. (Do not use a calculator.)

21. (a) $\dfrac{3\pi}{2}$ (b) $\dfrac{7\pi}{6}$

22. (a) $-\dfrac{7\pi}{12}$ (b) $\dfrac{\pi}{9}$

23. (a) $\dfrac{7\pi}{3}$ (b) $-\dfrac{11\pi}{30}$

24. (a) $\dfrac{11\pi}{6}$ (b) $\dfrac{34\pi}{15}$

In Exercises 25–28, express the angle in radian measure as a multiple of π. (Do not use a calculator.)

25. (a) $30°$ (b) $150°$
26. (a) $315°$ (b) $120°$
27. (a) $-20°$ (b) $-240°$
28. (a) $-270°$ (b) $144°$

In Exercises 29–32, convert the angle from degrees to radian measure. Express your result to three decimal places.

29. (a) $115°$ (b) $87.4°$
30. (a) $-216.35°$ (b) $-48.27°$
31. (a) $532°$ (b) $0.54°$
32. (a) $-0.83°$ (b) $345°$

In Exercises 33–36, convert the angle from radian to degree measure. Express your result to three decimal places.

33. (a) $\dfrac{\pi}{7}$ (b) $\dfrac{5\pi}{11}$

34. (a) $\dfrac{15\pi}{8}$ (b) 6.5π

35. (a) -4.2π (b) 4.8
36. (a) -2 (b) -0.57

In Exercises 37 and 38, convert the angle measurement to decimal form.

37. (a) $245°\ 10'$ (b) $2°\ 12'$
38. (a) $-135°\ 36''$ (b) $-408°\ 16'\ 25''$

In Exercises 39–42, convert the angle measurement to $D°\ M'\ S''$ form.

39. (a) $240.6°$ (b) $-145.8°$
40. (a) $-345.12°$ (b) 0.45
41. (a) 2.5 (b) -3.58
42. (a) -0.355 (b) 0.7865

In Exercises 43–46, find the radian measure of the central angle of a circle of radius r that intercepts an arc of length s.

Radius	Arc Length
43. 15 inches	4 inches
44. 16 feet	10 feet
45. 14.5 centimeters	25 centimeters
46. 80 kilometers	160 kilometers

In Exercises 47–50, on the circle of radius r find the length of the arc intercepted by the central angle θ.

Radius	Central Angle
47. 15 inches	$180°$
48. 9 feet	$60°$
49. 6 meters	2 radians
50. 40 centimeters	$\dfrac{3\pi}{4}$ radians

Distance Between Cities In Exercises 51–54, find the distance between the two cities. Assume that the earth is a sphere of radius 4000 miles and that the cities are on the same meridian (one city is due north of the other).

City	Latitude
51. Dallas	$32°\ 47'\ 9''$ N
Omaha	$41°\ 15'\ 42''$ N
52. San Francisco	$37°\ 46'\ 39''$ N
Seattle	$47°\ 36'\ 32''$ N
53. Miami	$25°\ 46'\ 37''$ N
Erie	$42°\ 7'\ 15''$ N
54. Johannesburg, South Africa	$26°\ 10'$ S
Jerusalem, Israel	$31°\ 47'$ N

55. *Difference in Latitudes* Assuming that the earth is a sphere of radius 4000 miles, what is the difference in latitude of two cities, one of which is 325 miles due north of the other?

56. *Difference in Latitudes* Assuming that the earth is a sphere of radius 4000 miles, what is the difference in latitude of two cities, one of which is 500 miles due north of the other?

57. *Instrumentation*   The pointer on a voltmeter is 2 inches long (see figure). Find the angle through which the pointer rotates when it moves $\frac{1}{2}$ inch on the scale.

FIGURE FOR 57

58. *Electric Hoist*   An electric hoist is used to lift a piece of equipment (see figure). The diameter of the drum on the hoist is 8 inches and the equipment must be raised 1 foot. Find the number of degrees through which the drum must rotate.

FIGURE FOR 58

59. *Angular Speed*   A car is moving at the rate of 50 miles per hour, and the diameter of each wheel is 2.5 feet. (a) Find the number of revolutions per minute of the rotating wheels. (b) Find the angular speed of the wheels in radians per minute.

60. *Angular Speed*   A truck is moving at the rate of 50 miles per hour, and the diameter of each wheel is 3 feet. (a) Find the number of revolutions per minute the wheels are rotating. (b) Find the angular speed of the wheels in radians per minute.

61. *Angular Speed*   A 2-inch-diameter pulley on an electric motor that runs at 1700 revolutions per minute is connected by a belt to a 4-inch-diameter pulley on a saw arbor. (a) Find the angular speed (in radians per minute) of each pulley. (b) Find the rotational speed (in rpm) of the saw.

62. *Angular Speed*   How long will it take a pulley rotating at 12 radians per second to make 100 revolutions?

63. *Circular Saw Speed*   The circular blade on a saw has a diameter of 7.5 inches and the blade rotates at 2400 revolutions per minute (see figure). (a) Find the angular speed in radians per second. (b) Find the speed of the saw teeth (in feet per second) as they contact the wood being cut.

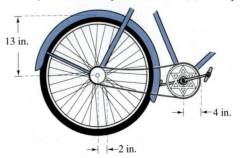

FIGURE FOR 63

64. *Speed of a Bicycle*   The radii of the sprocket assemblies and the wheel of the bicycle in the figure are 4 inches, 2 inches, and 13 inches, respectively. If the cyclist is pedaling at the rate of 1 revolution per second, find the speed of the bicycle in (a) feet per second and (b) miles per hour.

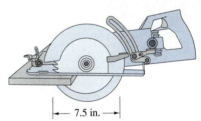

FIGURE FOR 64

5.2   THE TRIGONOMETRIC FUNCTIONS AND THE UNIT CIRCLE

The Unit Circle   /   The Trigonometric Functions   /   Domain and Period of Sine and Cosine   /   Evaluating Trig Functions with a Calculator

The Unit Circle

The two historical perspectives of trigonometry incorporate different methods for introducing the trigonometric functions. Our first introduction to these functions is based on the unit circle. Consider the **unit circle** given by

$$x^2 + y^2 = 1, \qquad \textit{Unit circle}$$

as shown in Figure 5.17. Imagine that the real number line is wrapped around this circle, with positive numbers corresponding to a counterclockwise wrapping and negative numbers corresponding to a clockwise wrapping, as shown in Figure 5.18.

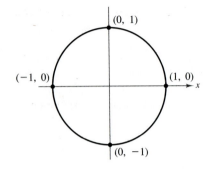

Unit Circle: $x^2 + y^2 = 1$

FIGURE 5.17

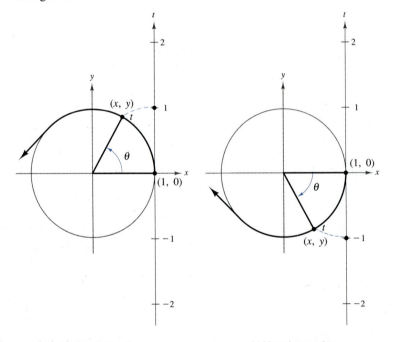

(a) Positive numbers   (b) Negative numbers

FIGURE 5.18

As the real number line is wrapped around the unit circle, each real number t will correspond with a point (x, y) on the circle. For example, the real number 0 corresponds to the point $(1, 0)$. Moreover, because the unit circle has a circumference of 2π, the real number 2π will also correspond to the point $(1, 0)$.

The Trigonometric Functions

From the preceding discussion, it follows that the coordinates x and y are two functions of the real variable t. These coordinates are used to define the six trigonometric functions of t.

sine	cosecant
cosine	secant
tangent	cotangent

These six functions are normally abbreviated as sin, csc, cos, sec, tan, and cot, respectively.

DEFINITION OF TRIGONOMETRIC FUNCTIONS

Let t be a real number at (x, y) the point on the unit circle corresponding to t.

$$\sin t = y \qquad\qquad \csc t = \frac{1}{y}, \quad y \neq 0$$

$$\cos t = x \qquad\qquad \sec t = \frac{1}{x}, \quad x \neq 0$$

$$\tan t = \frac{y}{x}, \quad x \neq 0 \qquad\qquad \cot t = \frac{x}{y}, \quad y \neq 0$$

REMARK   As an aid to memorizing these definitions, note that the functions in the second column are the *reciprocals* of the corresponding functions in the first column.

In the definition of the trigonometric functions note that the tangent or secant is not defined when $x = 0$. For instance, because $t = \pi/2$ corresponds to $(x, y) = (0, 1)$, it follows that $\tan(\pi/2)$ and $\sec(\pi/2)$ are *undefined*. Similarly, the cotangent or cosecant is not defined when $y = 0$. For instance, because $t = 0$ corresponds to $(x, y) = (1, 0)$, cot 0 and csc 0 are *undefined*.

In Figure 5.19, the unit circle has been divided into eight equal arcs, corresponding to t-values of

$$0, \frac{\pi}{4}, \frac{\pi}{2}, \frac{3\pi}{4}, \pi, \frac{5\pi}{4}, \frac{3\pi}{2}, \frac{7\pi}{4}, \text{ and } 2\pi.$$

Similarly, in Figure 5.20, the unit circle has been divided into 12 equal arcs, corresponding to t-values of

$$0, \frac{\pi}{6}, \frac{\pi}{3}, \frac{\pi}{2}, \frac{2\pi}{3}, \frac{5\pi}{6}, \pi, \frac{7\pi}{6}, \frac{4\pi}{3}, \frac{3\pi}{2}, \frac{5\pi}{3}, \frac{11\pi}{6}, \text{ and } 2\pi.$$

Using the (x, y) coordinates in Figures 5.19 and 5.20, you can easily evaluate the trigonometric functions for common t-values. This procedure is demonstrated in Examples 1, 2, and 3.

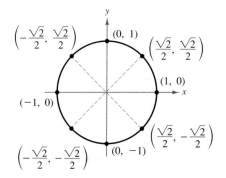

Unit Circle Divided into 8 Equal Arcs

FIGURE 5.19

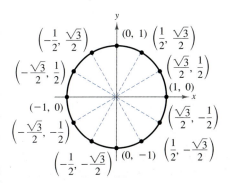

Unit Circle Divided into 12 Equal Arcs

FIGURE 5.20

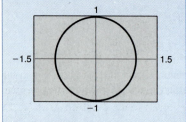
EXAMPLE 1   **Evaluating Trigonometric Functions of Real Numbers**

Evaluate the six trigonometric functions at the following real numbers.

A. $t = \dfrac{\pi}{6}$ **B.** $t = \dfrac{5\pi}{4}$

SOLUTION

A. Because $t = \pi/6$ corresponds to the first-quadrant point $(x, y) = \left(\sqrt{3}/2,\ 1/2\right)$, you can write the following.

$$\sin \frac{\pi}{6} = y = \frac{1}{2} \qquad\qquad \csc \frac{\pi}{6} = 2$$

$$\cos \frac{\pi}{6} = x = \frac{\sqrt{3}}{2} \qquad\qquad \sec \frac{\pi}{6} = \frac{2}{\sqrt{3}} = \frac{2\sqrt{3}}{3}$$

$$\tan \frac{\pi}{6} = \frac{y}{x} = \frac{1/2}{\sqrt{3}/2} = \frac{1}{\sqrt{3}} \qquad\qquad \cot \frac{\pi}{6} = \sqrt{3}$$

B. Because $t = 5\pi/4$ corresponds to the third-quadrant point $(x, y) = \left(-\sqrt{2}/2,\ -\sqrt{2}/2\right)$, you can write the following.

$$\sin \frac{5\pi}{4} = y = -\frac{\sqrt{2}}{2} \qquad\qquad \csc \frac{5\pi}{4} = -\frac{2}{\sqrt{2}} = -\sqrt{2}$$

$$\cos \frac{5\pi}{4} = x = -\frac{\sqrt{2}}{2} \qquad\qquad \sec \frac{5\pi}{4} = -\frac{2}{\sqrt{2}} = -\sqrt{2}$$

$$\tan \frac{5\pi}{4} = \frac{y}{x} = \frac{-\sqrt{2}/2}{-\sqrt{2}/2} = 1 \qquad\qquad \cot \frac{5\pi}{4} = 1$$

EXAMPLE 2   **Evaluating Trigonometric Functions of Real Numbers**

Evaluate the six trigonometric functions at the following real numbers.

A. $t = 0$ **B.** $t = \pi$

SOLUTION

A. $t = 0$ corresponds to the point $(x, y) = (1, 0)$ on the unit circle.

$$\sin 0 = y = 0 \qquad \csc 0 \text{ is undefined}$$

$$\cos 0 = x = 1 \qquad \sec 0 = 1$$

$$\tan 0 = \frac{y}{x} = 0 \qquad \cot 0 \text{ is undefined}$$

B. $t = \pi$ corresponds to the point $(x, y) = (-1, 0)$ on the unit circle.

$$\sin \pi = y = 0 \qquad\qquad \csc \pi \text{ is undefined}$$

$$\cos \pi = x = -1 \qquad\qquad \sec \pi = -1$$

$$\tan \pi = \frac{y}{x} = \frac{0}{-1} = 0 \qquad\qquad \cot \pi \text{ is undefined}$$

EXAMPLE 3 **Evaluating Trigonometric Functions of Real Numbers**

Evaluate the six trigonometric functions at the following real numbers.

A. $t = -\dfrac{\pi}{3}$ **B.** $t = \dfrac{5\pi}{2}$

SOLUTION

A. Moving *clockwise* around the unit circle, you can see that $t = -\pi/3$ corresponds to the point $(x, y) = \left(1/2, -\sqrt{3}/2\right)$.

$$\sin\left(-\frac{\pi}{3}\right) = 4 = -\frac{\sqrt{3}}{2} \qquad \csc\left(-\frac{\pi}{3}\right) = -\frac{2}{\sqrt{3}}$$

$$\cos\left(-\frac{\pi}{3}\right) = x = \frac{1}{2} \qquad \sec\left(-\frac{\pi}{3}\right) = 2$$

$$\tan\left(-\frac{\pi}{3}\right) = \frac{4}{x} = -\sqrt{3} \qquad \cot\left(-\frac{\pi}{3}\right) = \frac{x}{y} = -\frac{1}{\sqrt{3}}$$

B. Moving *counterclockwise* around the unit circle one and a quarter revolutions, you can see that $t = 5\pi/2$ corresponds to the point $(x, y) = (0, 1)$.

$$\sin\frac{5\pi}{2} = y = 1 \qquad\qquad \csc\frac{5\pi}{2} = 1$$

$$\cos\frac{5\pi}{2} = x = 0 \qquad\qquad \sec\frac{5\pi}{2} \text{ is undefined}$$

$$\tan\frac{5\pi}{2} = \frac{y}{x} \text{ is undefined} \qquad \cot\frac{5\pi}{2} = \frac{x}{y} = 0$$

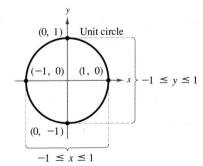

(0, 1) Unit circle

(−1, 0) (1, 0)

$-1 \le y \le 1$

(0, −1)

$-1 \le x \le 1$

FIGURE 5.21

Domain and Period of Sine and Cosine

The *domain* of the sine and cosine functions is the set of all real numbers. To determine the *range* of these two functions, consider the unit circle shown in Figure 5.21. Because $r = 1$, it follows that $\sin t = y$ and $\cos t = x$. Moreover, because (x, y) is on the unit circle, you know that $-1 \le y \le 1$ and $-1 \le x \le 1$, and it follows that the values of the sine and cosine also range between -1 and 1. That is,

$$-1 \le y \le 1 \qquad\qquad -1 \le x \le 1$$
$$-1 \le \sin t \le 1 \quad \text{and} \quad -1 \le \cos t \le 1.$$

Suppose you add 2π to each value of t in the interval $[0, 2\pi]$, thus completing a second revolution around the unit circle, as shown in Figure 5.22. The values of $\sin(t + 2\pi)$ and $\cos(t + 2\pi)$ correspond to those of $\sin t$ and $\cos t$. Similar results can be obtained for repeated revolutions (positive or negative) on the unit circle. This leads to the general result

$$\sin(t + 2\pi n) = \sin t \quad \text{and} \quad \cos(t + 2\pi n) = \cos t$$

for any integer n and real number t. Functions that behave in such a repetitive (or cyclic) manner are called **periodic.**

REMARK In Figure 5.22, *positive* multiples of 2π were added to the *t*-values. You could just as well have added *negative* multiples. For instance, $\pi/4 - 2\pi$ and $\pi/4 - 4\pi$ are also coterminal to $\pi/4$.

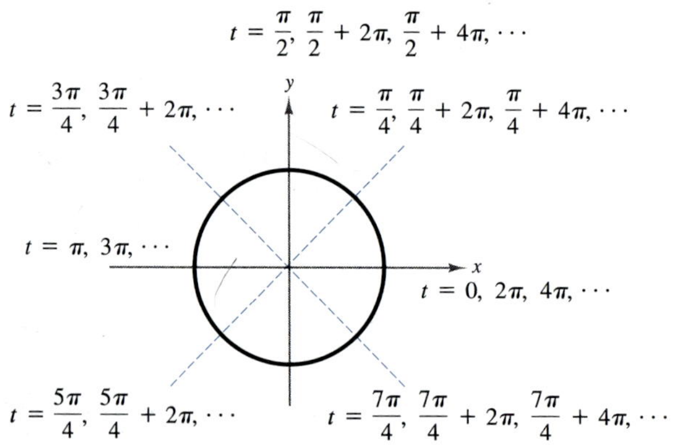

Repeated Revolutions on the Unit Circle

FIGURE 5.22

D I S C O V E R Y

As the real number line is wrapped around the unit circle, each real number t will correspond with a point $(x, y) = (\cos t, \sin t)$ on the circle. You can visualize this graphically by setting your graphing utility to *simultaneous* mode. Using radian and parametric modes as well, let

$$X_{1t} = \cos T$$
$$Y_{1t} = \sin T$$
$$X_{2t} = T$$
$$Y_{2t} = \sin T.$$

Select the following ranges: *Tmin* = 0, *Tmax* = 6.3, *Tstep* = .1, *Xmin* = -2, *Xmax* = 7, *Ymin* = -3, *Ymax* = 3. Notice how the graphing utility traces out the unit circle and the sine function simultaneously. Try changing Y_{2t} to cos T or to tan T.

DEFINITION OF A PERIODIC FUNCTION

A function f is **periodic** if there exists a positive real number c such that

$$f(t + c) = f(t)$$

for all t in the domain of f. The least number c for which f is periodic is called the **period** of f.

From this definition it follows that the sine and cosine functions are periodic and have a period of 2π. The other four trigonometric functions are also periodic, and we will say more about this in Section 5.6.

EXAMPLE 4   Using the Period to Evaluate Sine and Cosine

A. Because $\dfrac{13\pi}{6} = 2\pi + \dfrac{\pi}{6}$, you have

$$\sin \frac{13\pi}{6} = \sin\left(2\pi + \frac{\pi}{6}\right) = \sin \frac{\pi}{6} = \frac{1}{2}.$$

B. Because $-\dfrac{7\pi}{2} = -4\pi + \dfrac{\pi}{2}$, you have

$$\cos\left(-\frac{7\pi}{2}\right) = \cos\left(-4\pi + \frac{\pi}{2}\right) = \cos \frac{\pi}{2} = 0.$$

Recall from Section 1.4 that a function f is *even* if

$$f(-t) = f(t) \qquad \text{\textit{Even function}}$$

and is *odd* if

$$f(-t) = -f(t). \qquad \text{\textit{Odd function}}$$

Of the six trigonometric functions, two are even and four are odd, as stated in the following theorem.

EVEN AND ODD TRIGONOMETRIC FUNCTIONS

The cosine and secant functions are *even*.

$$\cos(-t) = \cos t \qquad \sec(-t) = \sec t$$

The sine, cosecant, tangent, and cotangent functions are *odd*.

$$\sin(-t) = -\sin t \qquad \csc(-t) = -\csc t$$
$$\tan(-t) = -\tan t \qquad \cot(-t) = -\cot t$$

Evaluating Trig Functions with a Calculator

From the arc length formula $s = r\theta$, with $r = 1$, you can see that each real number t measures a central angle (in radians). That is, $t = r\theta = 1(\theta) = \theta$ radians. Thus, when you are evaluating trigonometric functions, *it doesn't make any difference whether you consider t to be a real number or an angle given in radians.*

A scientific calculator can be used to obtain decimal approximations of the values of the trigonometric functions. Before doing this, however, you must be sure that the calculator is set to the correct mode: degrees or radians. Here are two examples.

DISCOVERY

Set your graphing utility in degree mode and enter tan 90. What happens? Why? Now set your graphing utility to radian mode and enter tan(3.14159/2). Explain the calculator's answer.

Degree mode: cos 28   *Display 0.8829475929*

Radian mode: tan $\dfrac{\pi}{12}$ *Display 0.2679491924*

Most calculators do not have keys for the cosecant, secant, and cotangent functions. To evaluate these functions, use the reciprocal key with the functions sine, cosine, and tangent. Here is an example.

$$\csc \frac{\pi}{8} = \frac{1}{\sin (\pi/8)} \approx 2.613.$$

EXAMPLE 5   **Using a Calculator to Evaluate Trigonometric Functions**

Use a calculator to evaluate the following. (Round to three decimal places.)

A. $\sin(-76.4)°$ **B.** cot 1.5   **C.** $\sec(5° \ 40' \ 12'')$

SOLUTION

Function	*Mode*	*Display*	*Rounded to 3 Decimal Places*
A. $\sin(-76.4)°$	Degree	−0.9719610006	−0.972
B. cot 1.5	Radian	0.0709148443	0.071
C. sec 5° 40′ 12″	Degree	1.004916618	1.005

Note that 5° 40′ 12″ = 5.67°.

DISCUSSION PROBLEM

YOU BE THE INSTRUCTOR

Suppose you are tutoring a student who is just learning trigonometry. Your student was asked to evaluate the cosine of 2 radians and, using a calculator, obtained the following.

Keystrokes	*Display*
COS 2	0.999390827

You know that 2 radians lies in the second quadrant. You also know that this implies that the cosine of 2 radians should be negative. What did your student do wrong?

WARM-UP

The following warm-up exercises involve skills that were covered in earlier sections. You will use these skills in the exercise set for this section.

In Exercises 1 and 2, simplify the expression.

1. $\dfrac{\frac{1}{2}}{\frac{-\sqrt{3}}{2}}$ 2. $\dfrac{\frac{\sqrt{2}}{2}}{\frac{-\sqrt{2}}{2}}$

In Exercises 3 and 4, find an angle θ in the interval $[0, 2\pi]$ that is coterminal with the given angle.

3. $\dfrac{8\pi}{3}$ 4. $-\dfrac{\pi}{4}$

In Exercises 5 and 6, convert the angle to radian measure.

5. $30°$ 6. $135°$

In Exercises 7 and 8, convert the angle to degree measure.

7. $\dfrac{\pi}{3}$ radians   8. $-\dfrac{3\pi}{2}$ radians

9. Determine the circumference of a circle with radius 1.
10. Determine the arc length of a semicircle with radius 1.

SECTION 5.2 · EXERCISES

In Exercises 1–8, find the point (x, y) on the unit circle that corresponds to the real number t (see Figures 5.19 and 5.20).

1. $t = \dfrac{\pi}{4}$ 2. $t = \dfrac{\pi}{3}$

3. $t = \dfrac{5\pi}{6}$ 4. $t = \dfrac{5\pi}{4}$

5. $t = \dfrac{4\pi}{3}$ 6. $t = \dfrac{11\pi}{6}$

7. $t = \dfrac{3\pi}{2}$ 8. $t = \pi$

In Exercises 9–16, evaluate the sine, cosine, and tangent of the real number.

9. $t = \dfrac{\pi}{4}$ 10. $t = -\dfrac{\pi}{4}$

11. $t = -\dfrac{5\pi}{4}$ 12. $t = -\dfrac{5\pi}{6}$

13. $t = \dfrac{11\pi}{6}$ 14. $t = \dfrac{2\pi}{3}$

15. $t = \dfrac{4\pi}{3}$ 16. $t = \dfrac{7\pi}{4}$

In Exercises 17–22, evaluate (if possible) the six trigonometric functions of the real number.

17. $t = \dfrac{3\pi}{4}$

18. $t = -\dfrac{2\pi}{3}$

19. $t = \dfrac{\pi}{2}$

20. $t = \dfrac{3\pi}{2}$

21. $t = -\dfrac{4\pi}{3}$

22. $t = -\dfrac{11\pi}{6}$

In Exercises 23–30, evaluate the trigonometric function using its period as an aid.

23. $\sin 3\pi$

24. $\cos 3\pi$

25. $\cos \dfrac{8\pi}{3}$

26. $\sin \dfrac{9\pi}{4}$

27. $\cos \dfrac{19\pi}{6}$

28. $\sin\left(-\dfrac{13\pi}{6}\right)$

29. $\sin\left(-\dfrac{9\pi}{4}\right)$

30. $\cos\left(-\dfrac{8\pi}{4}\right)$

In Exercises 31–36, use the value of the trigonometric function to evaluate the indicated functions.

31. $\sin t = \tfrac{1}{3}$
   (a) $\sin(-t)$ (b) $\csc(-t)$

32. $\sin(-t) = \tfrac{2}{5}$
   (a) $\sin t$ (b) $\csc t$

33. $\cos(-t) = -\tfrac{7}{8}$
   (a) $\cos t$ (b) $\sec(-t)$

34. $\cos t = -\tfrac{3}{4}$
   (a) $\cos(-t)$ (b) $\sec(-t)$

35. $\sin t = \tfrac{4}{5}$
   (a) $\sin(\pi - t)$ (b) $\sin(t + \pi)$

36. $\cos t = \tfrac{4}{5}$
   (a) $\cos(\pi - t)$ (b) $\cos(t + \pi)$

In Exercises 37–44, use a calculator to evaluate the trigonometric function. (Set your calculator in the correct mode and round your answer to four decimal places.)

37. $\sin \dfrac{\pi}{4}$

38. $\tan \pi$

39. $\cos 34.2°$

40. $\cot 1$

41. $\tan 110.5°$

42. $\sec 54.9°$

43. $\csc 0.8$

44. $\sin(-0.9)$

In Exercises 45 and 46, use the accompanying figure and a straightedge to approximate the values of the trigonometric functions. In Exercises 47 and 48, approximate the solutions of the equations. Use $0 \le t \le 2\pi$.

45. (a) $\sin 5$ (b) $\cos 2$

46. (a) $\sin 0.75$ (b) $\cos 2.5$

47. (a) $\sin t = 0.25$ (b) $\cos t = -0.25$

48. (a) $\sin t = -0.75$ (b) $\cos t = 0.75$

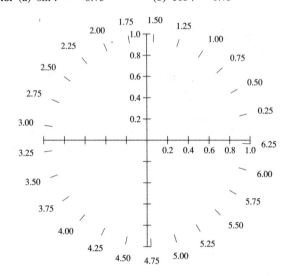

FIGURE FOR 45–48

49. *Harmonic Motion* The displacement from equilibrium of an oscillating weight suspended by a spring is

$$y(t) = \tfrac{1}{4}\cos 6t,$$

where y is the displacement in feet and t is the time in seconds. Find the displacement when (a) $t = 0$, (b) $t = \tfrac{1}{4}$, and (c) $t = \tfrac{1}{2}$.

50. *Harmonic Motion* The displacement from equilibrium of an oscillating weight suspended by a spring and subject to the damping effect of friction is

$$y(t) = \tfrac{1}{4}e^{-t}\cos 6t,$$

where y is the displacement in feet and t is the time in seconds. Find the displacement when (a) $t = 0$, (b) $t = \tfrac{1}{4}$, and (c) $t = \tfrac{1}{2}$.

51. *Electric Circuits*   The initial current and charge in the electrical circuit shown in the accompanying figure is zero. The current when 100 volts is applied to the circuit is given by

$$I = 5e^{-2t} \sin t$$

if the resistance, inductance, and capacitance are 80 ohms, 20 henrys, and 0.01 farads, respectively. Approximate the current $t = 0.7$ seconds after the voltage is applied.

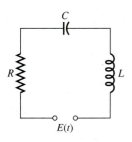

FIGURE FOR 51

52. Use the unit circle to verify that the cosine and secant functions are even and the sine, cosecant, tangent, and cotangent functions are odd.

5.3   TRIGONOMETRIC FUNCTIONS AND RIGHT TRIANGLES

Trigonometric Functions of an Acute Angle   /   Trigonometric Identities   /   Applications Involving Right Triangles

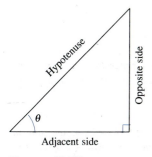

FIGURE 5.23

Trigonometric Functions of an Acute Angle

Our second look at the trigonometric functions is from a *right triangle* perspective. In Figure 5.23, the three sides of the triangle are the **hypotenuse,** the **opposite side** (the side opposite the angle θ), and the **adjacent side** (the side adjacent to the angle θ). Using the lengths of these three sides, you can form six ratios that define the six trigonometric functions of the acute angle θ.

In the following definition it is important to see that $0° < \theta < 90°$ and that for such angles the value of each of the six trigonometric functions is *positive.*

RIGHT TRIANGLE DEFINITION OF TRIGONOMETRIC FUNCTIONS

Let θ be an *acute* angle of a right triangle. Then the six trigonometric functions *of the angle* θ are defined as follows.

$$\sin \theta = \frac{\text{opp}}{\text{hyp}} \qquad \csc \theta = \frac{\text{hyp}}{\text{opp}}$$

$$\cos \theta = \frac{\text{adj}}{\text{hyp}} \qquad \sec \theta = \frac{\text{hyp}}{\text{adj}}$$

$$\tan \theta = \frac{\text{opp}}{\text{adj}} \qquad \cot \theta = \frac{\text{adj}}{\text{opp}}$$

The abbreviations opp, adj, and hyp represent the lengths of the three sides of a right triangle.

opp = the length of the side *opposite* θ
adj = the length of the side *adjacent* to θ
hyp = the length of the *hypotenuse*

EXAMPLE 1   Evaluating Trigonometric Functions

Find the values of the six trigonometric functions of θ in the right triangle shown in Figure 5.24.

SOLUTION

By the Pythagorean Theorem, $(\text{hyp})^2 = (\text{opp})^2 + (\text{adj})^2$, it follows that

$$\text{hyp} = \sqrt{3^2 + 4^2} = \sqrt{25} = 5.$$

Thus, adj = 3, opp = 4, and hyp = 5.

$$\sin \theta = \frac{\text{opp}}{\text{hyp}} = \frac{4}{5} \qquad \csc \theta = \frac{\text{hyp}}{\text{opp}} = \frac{5}{4}$$

$$\cos \theta = \frac{\text{adj}}{\text{hyp}} = \frac{3}{5} \qquad \sec \theta = \frac{\text{hyp}}{\text{adj}} = \frac{5}{3}$$

$$\tan \theta = \frac{\text{opp}}{\text{adj}} = \frac{4}{3} \qquad \cot \theta = \frac{\text{adj}}{\text{opp}} = \frac{3}{4}$$

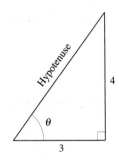

FIGURE 5.24

In Example 1, you were given the lengths of the sides of the right triangle, but not the angle θ. A much more common problem in trigonometry is to be asked to find the trigonometric functions for a *given* acute angle θ. To do this, construct a right triangle having θ as one of its angles.

EXAMPLE 2   Evaluating Trigonometric Functions of 45°

Find the values of sin 45°, cos 45°, and tan 45°.

SOLUTION

Construct a right triangle having 45° as one of its acute angles, as shown in Figure 5.25. Choose the length of the adjacent side to be 1. From geometry, you know that the other acute angle is also 45°. Hence, the triangle is isosceles and the length of the opposite side is also 1. Using the Pythagorean Theorem, you can find the length of the hypotenuse to be

$$\text{hyp} = \sqrt{1^2 + 1^2} = \sqrt{2}.$$

Finally, you can write the following.

$$\sin 45° = \frac{\text{opp}}{\text{hyp}} = \frac{1}{\sqrt{2}} = \frac{\sqrt{2}}{2}$$

$$\cos 45° = \frac{\text{adj}}{\text{hyp}} = \frac{1}{\sqrt{2}} = \frac{\sqrt{2}}{2}$$

$$\tan 45° = \frac{\text{opp}}{\text{adj}} = \frac{1}{1} = 1$$

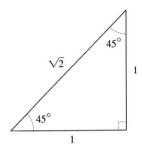

FIGURE 5.25

EXAMPLE 3   Evaluating Trigonometric Functions of 30° and 60°

Use the equilateral triangle shown in Figure 5.26 to find the values of sin 60°, cos 60°, sin 30°, and cos 30°.

SOLUTION

Try using the Pythagorean Theorem to verify the lengths of the sides given in Figure 5.26. For $\theta = 60°$, you have adj = 1, opp = $\sqrt{3}$, and hyp = 2, which implies that

$$\sin 60° = \frac{\text{opp}}{\text{hyp}} = \frac{\sqrt{3}}{2} \quad \text{and} \quad \cos 60° = \frac{\text{adj}}{\text{hyp}} = \frac{1}{2}.$$

For $\theta = 30°$, you have adj = $\sqrt{3}$, opp = 1, and hyp = 2, which implies that

$$\sin 30° = \frac{\text{opp}}{\text{hyp}} = \frac{1}{2} \quad \text{and} \quad \cos 30° = \frac{\text{adj}}{\text{hyp}} = \frac{\sqrt{3}}{2}.$$

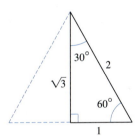

FIGURE 5.26

DISCOVERY

Set your graphing utility in degree mode and choose an angle x. Now evaluate $\cos(x)$ and $\sin(90 - x)$. What do you observe? Repeat this experiment with $\sin(x)$ and $\cos(90 - x)$.

Because the angles 30°, 45°, and 60° ($\pi/6$, $\pi/4$, and $\pi/3$) occur frequently in trigonometry, we suggest that you learn to construct the triangles shown in Figures 5.25 and 5.26.

SINE, COSINE, AND TANGENT OF SPECIAL ANGLES

$$\sin 30° = \sin \frac{\pi}{6} = \frac{1}{2}, \qquad \cos 30° = \cos \frac{\pi}{6} = \frac{\sqrt{3}}{2}, \qquad \tan 30° = \tan \frac{\pi}{6} = \frac{\sqrt{3}}{3}$$

$$\sin 45° = \sin \frac{\pi}{4} = \frac{\sqrt{2}}{2}, \qquad \cos 45° = \cos \frac{\pi}{4} = \frac{\sqrt{2}}{2}, \qquad \tan 45° = \tan \frac{\pi}{4} = 1$$

$$\sin 60° = \sin \frac{\pi}{3} = \frac{\sqrt{3}}{2}, \qquad \cos 60° = \cos \frac{\pi}{3} = \frac{1}{2}, \qquad \tan 60° = \tan \frac{\pi}{3} = \sqrt{3}$$

Trigonometric Identities

In the preceding list, note that $\sin 30° = \frac{1}{2} = \cos 60°$. This occurs because $30°$ and $60°$ are complementary angles. In general, it can be shown that *cofunctions of complementary angles are equal.* That is, if θ is an acute angle, then the following relationships are true.

$$\sin(90° - \theta) = \cos \theta \qquad \cos(90° - \theta) = \sin \theta$$
$$\tan(90° - \theta) = \cot \theta \qquad \cot(90° - \theta) = \tan \theta$$
$$\sec(90° - \theta) = \csc \theta \qquad \csc(90° - \theta) = \sec \theta$$

For instance, because $10°$ and $80°$ are complementary angles, it follows that $\sin 10° = \cos 80°$ and $\tan 10° = \cot 80°$.

FUNDAMENTAL TRIGONOMETRIC IDENTITIES

Reciprocal Identities

$$\sin \theta = \frac{1}{\csc \theta} \qquad \sec \theta = \frac{1}{\cos \theta} \qquad \tan \theta = \frac{1}{\cot \theta}$$

$$\csc \theta = \frac{1}{\sin \theta} \qquad \cos \theta = \frac{1}{\sec \theta} \qquad \cot \theta = \frac{1}{\tan \theta}$$

Quotient Identities

$$\tan \theta = \frac{\sin \theta}{\cos \theta} \qquad \cot \theta = \frac{\cos \theta}{\sin \theta}$$

Pythagorean Identities

$$\sin^2 \theta + \cos^2 \theta = 1 \qquad 1 + \tan^2 \theta = \sec^2 \theta \qquad 1 + \cot^2 \theta = \csc^2 \theta$$

Technology Note

Use your calculator to confirm several of the trigonometric identities at the right for various values of θ. For instance, calculate $(\sin 0.5)^2 + (\cos 0.5)^2$ and observe that the value is 1.

REMARK We use $\sin^2 \theta$ to represent $(\sin \theta)^2$, $\cos^2 \theta$ to represent $(\cos \theta)^2$, and so on.

EXAMPLE 4   Applying Trigonometric Identities

Let θ be the acute angle such that $\sin \theta = 0.6$. Use trigonometric identities to find the values of the following.

A. $\cos \theta$ **B.** $\tan \theta$

SOLUTION

A.
$$\sin^2 \theta + \cos^2 \theta = 1 \qquad \textit{Pythagorean Identity}$$
$$(0.6)^2 + \cos^2 \theta = 1$$
$$\cos^2 \theta = 1 - (0.6)^2 = 0.64$$
$$\cos \theta = \sqrt{0.64} = 0.8.$$

B. Now, knowing the values of $\sin \theta$ and $\cos \theta$, you can find the value of $\tan \theta$.

$$\tan \theta = \frac{\sin \theta}{\cos \theta} = \frac{0.6}{0.8} = 0.75.$$

Try using the definitions of $\cos \theta$ and $\tan \theta$, and the triangle shown in Figure 5.27, to check these results.

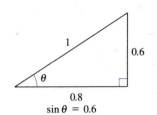

$$\sin \theta = 0.6$$

FIGURE 5.27

EXAMPLE 5   Applying Trigonometric Identities

Let θ be an acute angle such that $\tan \theta = 3$. Use trigonometric identities to find the values of the following.

A. $\cot \theta$ **B.** $\sec \theta$

SOLUTION

A.
$$\cot \theta = \frac{1}{\tan \theta} \qquad \textit{Reciprocal Identity}$$
$$= \frac{1}{3}$$

B.
$$\sec^2 \theta = \tan^2 \theta + 1 \qquad \textit{Pythagorean Identity}$$
$$\sec^2 \theta = 3^3 + 1$$
$$\sec^2 \theta = 10$$
$$\sec \theta = \sqrt{10}$$

Try using the definitions of $\cot \theta$ and $\sec \theta$, and the triangle shown in Figure 5.28, to check these results.

$$\tan \theta = 3$$

FIGURE 5.28

Applications Involving Right Triangles

Many applications of trigonometry involve a process called **solving right triangles.** In this type of application, you are usually given two sides of a right triangle and asked to find one of its acute angles, *or* you are given one side and one of the acute angles and asked to find one of the other sides.

EXAMPLE 6 Using Trigonometry to Solve a Right Triangle

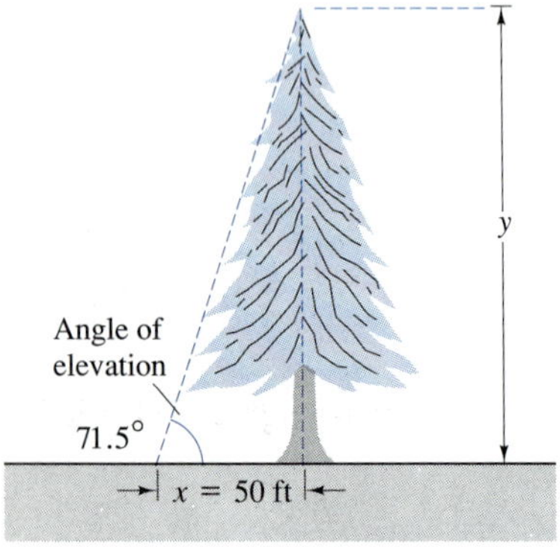

A surveyor is standing 50 feet from the base of a large tree, as shown in Figure 5.29. The surveyor measures the angle of elevation to the top of the tree as 71.5°. How tall is the tree?

SOLUTION

From Figure 5.29, you can see that

$$\tan 71.5° = \frac{\text{opp}}{\text{adj}} = \frac{y}{x},$$

where $x = 50$ and y is the height of the tree. Thus, you can determine the height of the tree to be

$$y = x \tan 71.5° \approx 50(2.98868) \approx 149.4 \text{ feet.}$$

Angle of elevation

71.5°

$x = 50$ ft

y

FIGURE 5.29

EXAMPLE 7 Using Trigonometry to Solve a Right Triangle

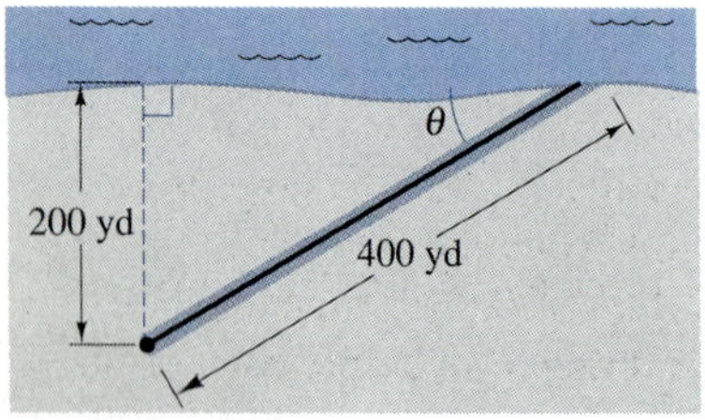

A person is standing 200 yards from a river. Rather than walking directly to the river, the person walks 400 yards along a straight path to the river's edge. Find the acute angle θ between this path and the river's edge, as indicated in Figure 5.30.

SOLUTION

From Figure 5.30, you can see that the sine of the angle θ is

$$\sin \theta = \frac{\text{opp}}{\text{hyp}} = \frac{200}{400} = \frac{1}{2}.$$

Now, you can recognize that $\theta = 30°$.

θ

200 yd

400 yd

FIGURE 5.30

In Example 7, you were able to recognize that the acute angle that satisfies the equation $\sin \theta = \frac{1}{2}$ is $\theta = 30°$. Suppose, however, that you were given the equation $\sin \theta = 0.6$ and asked to find the acute angle θ. Because

$$\sin 30° = \frac{1}{2} = 0.5000 \quad \text{and} \quad \sin 45° = \frac{1}{\sqrt{2}} \approx 0.7071,$$

you can figure that θ lies somewhere between $30°$ and $45°$. A more precise value of θ can be found using the inverse key on a calculator. (Consult your calculator manual to see how this key works on your own calculator.) For most calculators, one of the following keystroke sequences will work.

Degree mode: .6 INV sin   *Display 36.86989765*

Degree mode: 2nd sin .6 ENTER   *Display 36.86989765*

Thus, you can conclude that if $\sin \theta = 0.6$, then $\theta \approx 36.87°$.

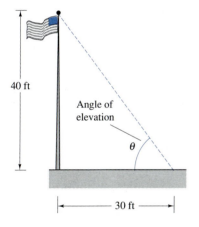

40 ft

Angle of
elevation

θ

30 ft

FIGURE 5.31

EXAMPLE 8   **Using Trigonometry to Solve a Right Triangle**

A 40-foot flagpole casts a 30-foot shadow, as shown in Figure 5.31. Find θ, the angle of elevation of the sun.

SOLUTION

Figure 5.31 shows that the *opposite* and *adjacent* sides are known. Thus,

$$\tan \theta = \frac{\text{opp}}{\text{adj}} = \frac{40}{30}.$$

With a calculator in degree mode you can obtain $\theta \approx 53.13°$.

**DISCUSSION
PROBLEM**

**COMPARING
DEFINITIONS
OF TRIGONO-
METRIC FUNC-
TIONS**

In Section 5.2 and in this section, we presented two different definitions of trigonometric functions. One was the "unit circle" definition and the other was the "right triangle" definition. Write a short paper that compares the two definitions. Then use both definitions to find the values of the six trigonometric functions at $\theta = 30°$. For this value of θ, which definition do you prefer? For $\theta = 3\pi$, which do you prefer and why?

The following warm-up exercises involve skills that were covered in earlier sections. You will use these skills in the exercise set for this section.

In Exercises 1–4, find the distance between the points.

1. (3, 8), (1, 4)   **2.** (5, 2), (2, −7)

3. (−4, 0), (2, 8)   **4.** (−3, −3), (0, 0)

In Exercises 5–10, perform the indicated operation(s). (Round your result to two decimal places.)

5. 0.300 × 4.125   **6.** 7.30 × 43.50

7. $\dfrac{151.5}{2.40}$ **8.** $\dfrac{3740}{28.0}$

9. $\dfrac{19,500}{0.007}$ **10.** $\dfrac{(10.5)(3401)}{1240}$

SECTION 5.3 · EXERCISES

In Exercises 1–8, find the exact values of the six trigonometric functions of the angle θ. (Use the Pythagorean Theorem to find the third side of the triangle.)

1.
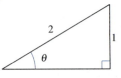

2.
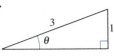

5.
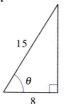

6.
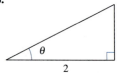

3.
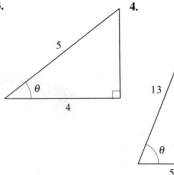

4.

7.
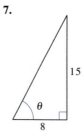

8.
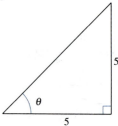

In Exercises 9–16, sketch a right triangle corresponding to the trigonometric function of the acute angle θ. Use the Pythagorean Theorem to determine the third side and then find the other five trigonometric functions of θ.

9. $\sin \theta = \frac{2}{3}$ **10.** $\cot \theta = 5$

11. $\sec \theta = 2$ **12.** $\cos \theta = \frac{5}{7}$

13. $\tan \theta = 3$ **14.** $\csc \theta = \frac{17}{4}$

15. $\cot \theta = \frac{3}{2}$ **16.** $\sin \theta = \frac{3}{8}$

In Exercises 17–20, use the given values and the trigonometric identities to evaluate the trigonometric functions.

17. $\sin 60° = \dfrac{\sqrt{3}}{2}$, $\cos 60° = \dfrac{1}{2}$

   (a) $\tan 60°$ (b) $\sin 30°$
   (c) $\cos 30°$ (d) $\cot 60°$

18. $\sin 30° = \dfrac{1}{2}$, $\tan 30° = \dfrac{\sqrt{3}}{3}$

   (a) $\csc 30°$ (b) $\cot 60°$
   (c) $\cos 30°$ (d) $\cot 30°$

19. $\csc \theta = 3$, $\sec \theta = \dfrac{3\sqrt{2}}{4}$

   (a) $\sin \theta$ (b) $\cos \theta$
   (c) $\tan \theta$ (d) $\sec(90° - \theta)$

20. $\sec \theta = 5$, $\tan \theta = 2\sqrt{6}$

   (a) $\cos \theta$ (b) $\cot \theta$
   (c) $\cot(90° - \theta)$ (d) $\sin \theta$

In Exercises 21–24, evaluate the trigonometric functions.

21. (a) $\cos 60°$ (b) $\tan \dfrac{\pi}{4}$

22. (a) $\csc 30°$ (b) $\sin \dfrac{\pi}{4}$

23. (a) $\cot 45°$ (b) $\cos 45°$

24. (a) $\sin \dfrac{\pi}{3}$ (b) $\csc 45°$

In Exercises 25–34, use a calculator to evaluate each function. Round your result to four decimal places. (Be sure the calculator is in the correct mode.)

25. (a) $\sin 10°$ (b) $\cos 80°$

26. (a) $\tan 23.5°$ (b) $\cot 66.5°$

27. (a) $\sin 16.35°$ (b) $\csc 16.35°$

28. (a) $\cos 16° \, 18'$ (b) $\sin 73° \, 56'$

29. (a) $\sec 42° \, 12'$ (b) $\csc 48° \, 7'$

30. (a) $\cos 4° \, 50' \, 15''$ (b) $\sec 4° \, 50' \, 15''$

31. (a) $\cot \dfrac{\pi}{16}$ (b) $\tan \dfrac{\pi}{16}$

32. (a) $\sec 0.75$ (b) $\cos 0.75$

33. (a) $\csc 1$ (b) $\sec\left(\dfrac{\pi}{2} - 1\right)$

34. (a) $\tan \dfrac{1}{2}$ (b) $\cot\left(\dfrac{\pi}{2} - \dfrac{1}{2}\right)$

In Exercises 35–40, find the value of θ in degrees $(0° < \theta < 90°)$ and radians $(0 < \theta < \pi/2)$ without using a calculator.

35. (a) $\sin \theta = \dfrac{1}{2}$ **36.** (a) $\cos \theta = \dfrac{\sqrt{2}}{2}$

   (b) $\csc \theta = 2$ (b) $\tan \theta = 1$

37. (a) $\sec \theta = 2$ **38.** (a) $\tan \theta = \sqrt{3}$

   (b) $\cot \theta = 1$ (b) $\cos \theta = \frac{1}{2}$

39. (a) $\csc \theta = \dfrac{2\sqrt{3}}{3}$ **40.** (a) $\cot \theta = \dfrac{\sqrt{3}}{3}$

   (b) $\sin \theta = \dfrac{\sqrt{2}}{2}$ (b) $\sec \theta = \sqrt{2}$

In Exercises 41–44, find the value of θ in degrees $(0° < \theta < 90°)$ and radians $(0 < \theta < \pi/2)$ by using the inverse key on a calculator.

41. (a) $\sin \theta = 0.8191$ **42.** (a) $\cos \theta = 0.9848$
   (b) $\cos \theta = 0.0175$ (b) $\cos \theta = 0.8746$

43. (a) $\tan \theta = 1.1920$ **44.** (a) $\sin \theta = 0.3746$
   (b) $\tan \theta = 0.4663$ (b) $\cos \theta = 0.3746$

45. Solve for *y*.

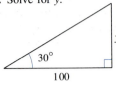

46. Solve for *x*.

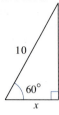

47. Solve for *x*.

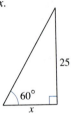

48. Solve for *r*.

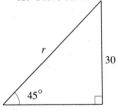

49. Solve for *r*.

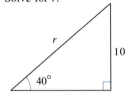

50. Solve for *x*.

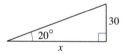

51. Solve for *y*.

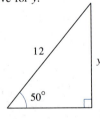

52. Solve for *r*.

53. *Height* A 6-foot person standing 12 feet from a streetlight casts an 8-foot shadow (see figure). What is the height of the streetlight?

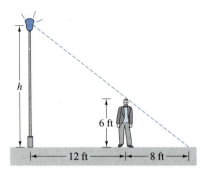

FIGURE FOR 53

54. *Height* A 6-foot man walked from the base of a broadcasting tower directly toward the tip of the shadow cast by the tower. When he was 132 feet from the tower, his shadow started to appear beyond the tower's shadow. What is the height of the tower if the man's shadow is 3 feet long?

55. *Length* A 20-feet ladder leaning against the side of a house makes a 75° angle with the ground (see figure). How far up the side of the house does the ladder reach?

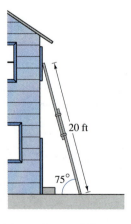

FIGURE FOR 55

56. *Width of a River*   A biologist wants to know the width w of a river in order to properly set instruments for studying the pollutants in the water. From point A, the biologist walks downstream 100 feet and sights to point C. From this sighting, it is determined that $\theta = 50°$ (see figure). How wide is the river?

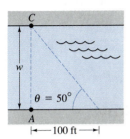

FIGURE FOR 56

57. *Distance*   From a 150-foot observation tower on the coast, a Coast Guard officer sights a boat in difficulty. The angle of depression of the boat is $4°$ (see figure). How far is the boat from the shoreline?

FIGURE FOR 57

58. *Angle of Elevation*   A ramp $17\frac{1}{2}$ feet in length rises to a loading platform that is $3\frac{1}{3}$ feet off the ground (see figure). Find the angle θ that the ramp makes with the ground.

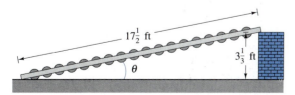

FIGURE FOR 58

59. *Machine Shop Calculations*   A steel plate has the form of a quarter circle with a radius of 24 inches. Two $\frac{3}{8}$-inch holes are to be drilled in the plate positioned as shown in the figure. Find the coordinates of the center of each hole.

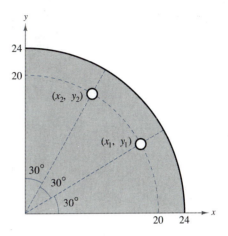

FIGURE FOR 59

60. *Machine Shop Calculations*   A tapered shaft has a diameter of 2 inches at the small end and is 6 inches long (see figure). If the taper is $3°$, find the diameter d of the large end of the shaft.

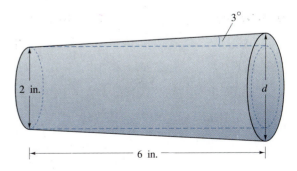

FIGURE FOR 60

61. *Trigonometric Functions by Actual Measurement*   Use a compass to sketch a quarter of a circle of radius 10 centimeters. Using a protractor, construct an angle of 25° in standard position (see figure). Drop a perpendicular line from the point of intersection of the terminal side of the angle and the arc of the circle. By actual measurement, calculate the coordinates (x, y) of the point of intersection and use these measurements to approximate the six trigonometric functions of a 25° angle.

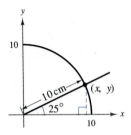

FIGURE FOR 61

62. *Trigonometric Functions by Actual Measurement*   Repeat Exercise 61 using an angle of 75°.

In Exercises 63–68, determine whether the statement is true or false, and give a reason for your answer.

63. $\sin 60° \csc 60° = 1$

64. $\sec 30° = \csc 60°$

65. $\sin 45° + \cos 45° = 1$

66. $\cot^2 10° - \csc^2 10° = -1$

67. $\dfrac{\sin 60°}{\sin 30°} = \sin 2°$

68. $\tan[(0.8)^2] = \tan^2(0.8)$

69. A 30-foot ladder leaning against the side of a house is 4 feet from the house at the base (see figure).
 (a) How far up the side of the house does the ladder reach? Express your answer accurate to two decimal digits.
 (b) Use the fact that $\cos \theta = 4/30$ to find the angle that the ladder makes with the ground. Express your answer in degrees accurate to two decimal digits.
 (c) Use the tangent of the angle found in part (b) to answer part (a) again.
 (d) Why do your answers to parts (a) and (c) differ slightly?

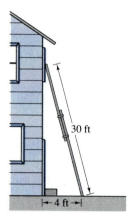

30 ft

4 ft

FIGURE FOR 69

5.4   TRIGONOMETRIC FUNCTIONS OF ANY ANGLE
Trigonometric Functions of Any Angle   /   Reference Angles

Trigonometric Functions of Any Angle

In Section 5.3, you learned to evaluate trigonometric functions of an acute angle. In this section you will learn to evaluate trigonometric functions of any angle.

DEFINITION OF TRIGONOMETRIC FUNCTIONS OF ANY ANGLE

Let θ be an angle in standard position with (x, y) any point (except the origin) on the terminal side of θ and $r = \sqrt{x^2 + y^2}$ (see figure).

$$\sin \theta = \frac{y}{r} \qquad\qquad \csc \theta = \frac{r}{y}, \quad y \neq 0$$

$$\cos \theta = \frac{x}{r} \qquad\qquad \sec \theta = \frac{r}{x}, \quad x \neq 0$$

$$\tan \theta = \frac{y}{x}, \quad x \neq 0 \qquad \cot \theta = \frac{x}{y}, \quad y \neq 0$$

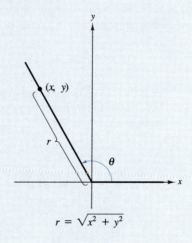

$$r = \sqrt{x^2 + y^2}$$

REMARK   Because $r = \sqrt{x^2 + y^2}$ *cannot* be zero, it follows that the sine and cosine functions are defined for any real value of θ. However, if $x = 0$, the tangent and secant of θ are undefined. For example, the tangent of $90°$ is undefined. Try calculating the tangent of $90°$ with your calculator. Similarly, if $y = 0$, the cotangent and cosecant of θ are undefined.

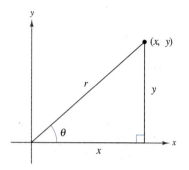

FIGURE 5.32

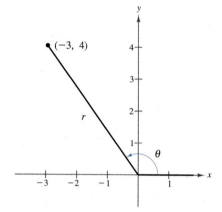

FIGURE 5.33

If θ is an *acute* angle, then these definitions coincide with those given in the previous section. To see this, note in Figure 5.32 that for an acute angle θ, $x = $ adj, $y = $ opp, and $r = $ hyp.

EXAMPLE 1   Evaluating Trigonometric Functions

Let $(-3, 4)$ be a point on the terminal side of θ. Find the sine, cosine, and tangent of θ.

SOLUTION

Referring to Figure 5.33, you can see that $x = -3$, $y = 4$, and

$$r = \sqrt{x^2 + y^2} = \sqrt{(-3)^2 + 4^2} = \sqrt{25} = 5.$$

Thus, you can write the following.

$$\sin \theta = \frac{y}{r} = \frac{4}{5}$$

$$\cos \theta = \frac{x}{r} = \frac{-3}{5} = -\frac{3}{5}$$

$$\tan \theta = \frac{y}{x} = \frac{4}{-3} = -\frac{4}{3}$$

The *signs* of trigonometric function values in the four quadrants can be determined easily from the definitions of the functions. For instance, because

$$\cos \theta = \frac{x}{r},$$

it follows that $\cos \theta$ is positive wherever $x > 0$, which is in Quadrants I and IV. (Remember, r is always positive.) In a similar manner you can verify the results shown in Figure 5.34.

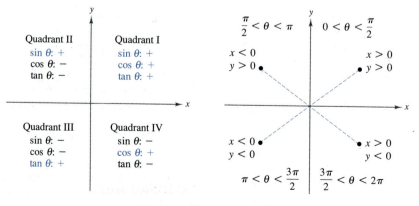

Signs of Trigonometric Functions

FIGURE 5.34

EXAMPLE 2   **Evaluating Trigonometric Functions**

Given $\tan \theta = -\frac{5}{4}$ and $\cos \theta > 0$, find $\sin \theta$ and $\sec \theta$.

SOLUTION

Note that θ lies in Quadrant IV because that is the only quadrant in which the tangent is negative and the cosine is positive. Moreover, using

$$\tan \theta = \frac{y}{x} = -\frac{5}{4}$$

and the fact that y is negative in Quadrant IV, you can let $y = -5$ and $x = 4$. Hence, $r = \sqrt{16 + 25} = \sqrt{41}$ and you have

$$\sin \theta = \frac{y}{r} = \frac{-5}{\sqrt{41}} \approx -0.7809 \quad \text{and} \quad \sec \theta = \frac{r}{x} = \frac{\sqrt{41}}{4} \approx 1.6008.$$

EXAMPLE 3   **Trigonometric Functions of Quadrant Angles**

Evaluate the sine function at the four quadrant angles 0, $\pi/2$, π, and $3\pi/2$.

SOLUTION

To begin, choose a point on the terminal side of each angle, as shown in Figure 5.35. For each of the four given points, $r = 1$, and you have

$$\sin 0 = \frac{y}{r} = \frac{0}{1} = 0 \qquad\qquad (x, y) = (1, 0)$$

$$\sin \frac{\pi}{2} = \frac{y}{r} = \frac{1}{1} = 1 \qquad\qquad (x, y) = (0, 1)$$

$$\sin \pi = \frac{y}{r} = \frac{0}{1} = 0 \qquad\qquad (x, y) = (-1, 0)$$

$$\sin \frac{3\pi}{2} = \frac{y}{r} = \frac{-1}{1} = -1. \qquad\qquad (x, y) = (0, -1)$$

Trying using Figure 5.35 to evaluate some of the other trigonometric functions at the four quadrant angles and check them on your calculator.

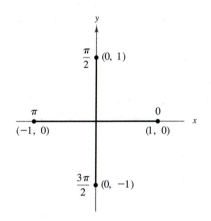

FIGURE 5.35

Reference Angles

The values of the trigonometric functions of angles greater than $90°$ (or less than $0°$) can be determined from their values at corresponding acute angles called **reference angles**.

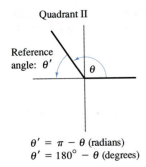

Quadrant II

Reference angle: θ'

$\theta' = \pi - \theta$ (radians)
$\theta' = 180° - \theta$ (degrees)

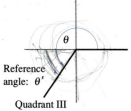

Reference angle: θ'

Quadrant III

$\theta' = \theta - \pi$ (radians)
$\theta' = \theta - 180°$ (degrees)

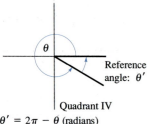

Reference angle: θ'

Quadrant IV

$\theta' = 2\pi - \theta$ (radians)
$\theta' = 360° - \theta$ (degrees)

FIGURE 5.36

DEFINITION OF REFERENCE ANGLES

Let θ be an angle in standard position. Its **reference angle** is the acute angle θ' formed by the terminal side of θ and the horizontal axis.

Figure 5.36 shows the reference angles for θ in Quadrants II, III, and IV.

EXAMPLE 4   Finding Reference Angles

Find the reference angle θ' for each of the following.

A. $\theta = 300°$ **B.** $\theta = 2.3$ **C.** $\theta = -135°$

SOLUTION

A. Because $\theta = 300°$ lies in Quadrant IV, the angle it makes with the x-axis is

$$\theta' = 360° - 300° = 60°. \qquad \textit{Degrees}$$

B. Because $\theta = 2.3$ lies between $\pi/2 \approx 1.5708$ and $\pi \approx 3.1416$, it follows that θ is in Quadrant II and its reference angle is

$$\theta' = \pi - 2.3 \approx 0.8416. \qquad \textit{Radians}$$

C. Because $\theta = -135°$ is coterminal with $225°$, it lies in Quadrant III. Hence, the reference angle is

$$\theta' = 225° - 180° = 45°. \qquad \textit{Degrees}$$

Figure 5.37 shows each angle θ and its reference angle θ'.

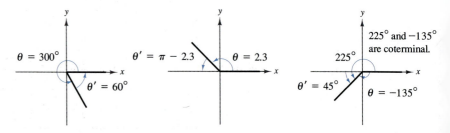

(a) θ in Quadrant IV   (b) θ in Quadrant II   (c) θ in Quadrant III

FIGURE 5.37

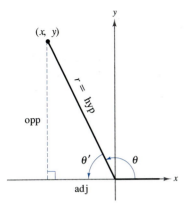

$$opp = |y|, \ adj = |x|$$

FIGURE 5.38

To see how a reference angle is used to evaluate a trigonometric function, consider the point (x, y) on the terminal side of θ, as shown in Figure 5.38. By definition, you know that

$$\sin \theta = \frac{y}{r} \quad \text{and} \quad \tan \theta = \frac{y}{x}.$$

For the right triangle with acute angle θ' and sides of lengths $|x|$ and $|y|$, you have

$$\sin \theta' = \frac{\text{opp}}{\text{hyp}} = \frac{|y|}{r} \quad \text{and} \quad \tan \theta' = \frac{\text{opp}}{\text{adj}} = \frac{|y|}{|x|}.$$

Thus, it follows that $\sin \theta$ and $\sin \theta'$ are equal, *except possibly in sign*. The same is true for $\tan \theta$ and $\tan \theta'$ *and* for the other four trigonometric functions. In all cases, the sign of the function value can be determined by the quadrant in which θ lies.

EVALUATING TRIGONOMETRIC FUNCTIONS OF ANY ANGLE

To find the value of a trigonometric function of any angle θ,

1. Determine the function value for the associated reference angle θ'.
2. Depending on the quadrant in which θ lies, prefix the appropriate sign to the function value.

By using reference angles and the special angles discussed in the previous section, you can greatly extend your scope of *exact* trigonometric values. Table 5.1 lists the function values for selected reference angles. For instance, knowing the function values of 30° means that you know the function values of all angles for which 30° is a reference angle.

TABLE 5.1

θ (degrees)	0°	30°	45°	60°	90°	180°	270°
θ (radians)	0	$\dfrac{\pi}{6}$	$\dfrac{\pi}{4}$	$\dfrac{\pi}{3}$	$\dfrac{\pi}{2}$	π	$\dfrac{3\pi}{2}$
$\sin \theta$	0	$\dfrac{1}{2}$	$\dfrac{\sqrt{2}}{2}$	$\dfrac{\sqrt{3}}{2}$	1	0	-1
$\cos \theta$	1	$\dfrac{\sqrt{3}}{2}$	$\dfrac{\sqrt{2}}{2}$	$\dfrac{1}{2}$	0	-1	0
$\tan \theta$	0	$\dfrac{\sqrt{3}}{3}$	1	$\sqrt{3}$	undef.	0	undef.

EXAMPLE 5 Trigonometric Functions of Nonacute Angles

Evaluate the following.

A. $\cos\dfrac{4\pi}{3}$ **B.** $\tan(-210°)$ **C.** $\csc\dfrac{11\pi}{4}$

SOLUTION

A. Because $\theta = 4\pi/3$ lies in Quadrant III, the reference angle is $\theta' = (4\pi/3) - \pi = \pi/3$, as shown in Figure 5.39(a). Moreover, the cosine is negative in Quadrant III, so that

$$\cos\frac{4\pi}{3} = (-)\cos\frac{\pi}{3} = -\frac{1}{2}. \qquad \textit{Reference angle, } \pi/3$$

B. Because $-210° + 360° = 150°$, it follows that $-210°$ is coterminal with the second-quadrant angle $150°$. Therefore, the reference angle is $\theta = 180° - 150° = 30°$, as shown in Figure 5.39(b). Finally, because the tangent is negative in Quadrant II, you have

$$\tan(-210°) = (-)\tan 30° = -\frac{\sqrt{3}}{3}. \quad \textit{Reference angle, } 30°$$

C. Because $(11\pi/4) - 2\pi = 3\pi/4$, it follows that $11\pi/4$ is coterminal with the second-quadrant angle $3\pi/4$. Therefore, the reference angle is $\theta' = \pi - (3\pi/4) = \pi/4$, as shown in Figure 5.39(c). Because the cosecant is positive in Quadrant II, you have

$$\csc\frac{11\pi}{4} = (+)\csc\frac{\pi}{4} = \frac{1}{\sin \pi/4} = \sqrt{2}. \quad \textit{Reference angle, } \pi/4$$

Check the above values with your calculator.

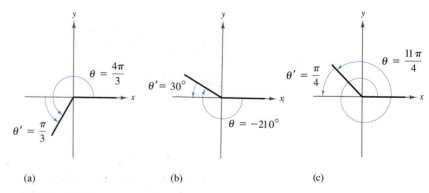

(a)   (b)   (c)

FIGURE 5.39

The fundamental trigonometric identities listed in the previous section (for an acute angle θ) are also valid when θ is any angle.

EXAMPLE 6   **Using Identities to Evaluate Trigonometric Functions**

Let θ be an angle in Quadrant II such that $\sin \theta = \frac{1}{3}$. Find the following.

A. $\cos \theta$ **B.** $\tan \theta$

SOLUTION

A. Because $\sin \theta = \frac{1}{3}$, use the Pythagorean Identity $\sin^2 \theta + \cos^2 \theta = 1$ to obtain

$$\left(\frac{1}{3}\right)^2 + \cos^2 \theta = 1$$

$$\cos^2 \theta = 1 - \frac{1}{9} = \frac{8}{9}.$$

Because $\cos \theta < 0$ in Quadrant II, use the negative root

$$\cos \theta = -\frac{\sqrt{8}}{\sqrt{9}} = -\frac{2\sqrt{2}}{3}.$$

B. Using the result from part A and the trigonometric identity $\tan \theta = \sin \theta / \cos \theta$, you obtain

$$\tan \theta = \frac{\frac{1}{3}}{-\frac{2\sqrt{2}}{3}} = -\frac{1}{2\sqrt{2}} = -\frac{\sqrt{2}}{4}.$$

Scientific calculators can be used to approximate the values of trigonometric functions of any angle, as demonstrated in Example 7.

EXAMPLE 7   **Evaluating Trigonometric Functions with a Calculator**

Use a calculator to approximate the following values. (Round your answers to three decimal places.)

A. $\cot 410°$ **B.** $\sin(-7)$ **C.** $\tan \dfrac{14\pi}{5}$

SOLUTION

	Function	Mode	Display	Rounded to 3 Decimal Places
A.	$\cot 410°$	Degree	0.8390996312	0.839
B.	$\sin(-7)$	Radian	−0.6569865987	−0.657
C.	$\tan \dfrac{14\pi}{5}$	Radian	−0.726542528	−0.727

At this point, you have completed your introduction to basic trigonometry. You have measured angles in both degrees and radians. You have studied the definitions of the six trigonometric functions from a right triangle perspective and as functions of real numbers. In your remaining study of trigonometry, you will continue to rely on both perspectives.

For your convenience we have included on the endpapers of this text a summary of basic trigonometry.

DISCUSSION PROBLEM
....

MEMORIZATION AIDS

There are many different techniques that people use to memorize (or reconstruct) trigonometric formulas. Here is one that we like to use for the sines and cosines of common angles.

θ	$0°$	$30°$	$45°$	$60°$	$90°$
$\sin \theta$	$\dfrac{\sqrt{0}}{2}$	$\dfrac{\sqrt{1}}{2}$	$\dfrac{\sqrt{2}}{2}$	$\dfrac{\sqrt{3}}{2}$	$\dfrac{\sqrt{4}}{2}$
$\cos \theta$	$\dfrac{\sqrt{4}}{2}$	$\dfrac{\sqrt{3}}{2}$	$\dfrac{\sqrt{2}}{2}$	$\dfrac{\sqrt{1}}{2}$	$\dfrac{\sqrt{0}}{2}$

Write a paragraph describing the pattern indicated by this table. Discuss other memory aids for trigonometric formulas.

WARM-UP
....

The following warm-up exercises involve skills that were covered in earlier sections. You will use these skills in the exercise set for this section.

In Exercises 1–6, evaluate the trigonometric function from memory.

1. $\sin 30°$ **2.** $\tan 45°$

3. $\cos \dfrac{\pi}{4}$ **4.** $\cot \dfrac{\pi}{3}$

5. $\sec \dfrac{\pi}{6}$ **6.** $\csc \dfrac{\pi}{4}$

In Exercises 7–10, use the given trigonometric function of an acute angle θ to find the values of the remaining trigonometric functions.

7. $\tan \theta = \frac{3}{2}$ **8.** $\cos \theta = \frac{2}{3}$

9. $\sin \theta = \frac{1}{5}$ **10.** $\sec \theta = 3$

SECTION 5.4 · EXERCISES

In Exercises 1–4, determine the exact values of the six trigono-metric functions of the angle θ.

1. (a)   (b)

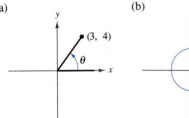

2. (a)   (b)

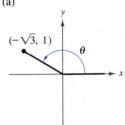 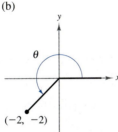

3. (a)   (b)

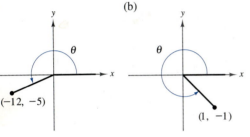

4. (a)   (b)

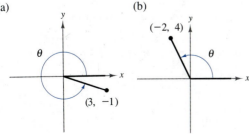

In Exercises 5–8, the point is on the terminal side of an angle in standard position. Determine the exact values of the six trigonometric functions of the angle.

5. (a) $(7, 24)$ (b) $(7, -24)$
6. (a) $(8, 15)$ (b) $(-9, -40)$
7. (a) $(-4, 10)$ (b) $(3, -5)$
8. (a) $(-5, -2)$ (b) $\left(-\frac{3}{2}, 3\right)$

$R = \sqrt{x^2 + y^2}$

In Exercises 9–12, use the two similar triangles in the figure to find (a) the unknown sides of the triangles and (b) the six trigonometric functions of the angles α_1 and α_2.

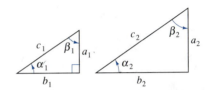

FIGURE FOR 9–12

9. $a_1 = 3$, $b_1 = 4$, $a_2 = 9$
10. $b_1 = 12$, $c_1 = 13$, $c_2 = 26$
11. $a_1 = 1$, $c_1 = 2$, $b_2 = 5$
12. $b_1 = 4$, $a_2 = 4$, $b_2 = 10$

In Exercises 13–16, determine the quadrant in which θ lies.

13. (a) $\sin \theta < 0$ and $\cos \theta < 0$
(b) $\sin \theta > 0$ and $\cos \theta < 0$
14. (a) $\sin \theta > 0$ and $\cos \theta > 0$
(b) $\sin \theta < 0$ and $\cos \theta > 0$
15. (a) $\sin \theta > 0$ and $\tan \theta < 0$
(b) $\cos \theta > 0$ and $\tan \theta < 0$
16. (a) $\sec \theta > 0$ and $\cot \theta < 0$
(b) $\csc \theta < 0$ and $\tan \theta > 0$

In Exercises 17–26, find the values (if possible) of the six trigonometric functions of θ using the given functional value and constraint.

Functional Value	Constraint
17. $\sin \theta = \frac{3}{5}$	θ lies in Quadrant II
18. $\cos \theta = -\frac{4}{5}$	θ lies in Quadrant III
19. $\tan \theta = -\frac{15}{8}$	$\sin \theta < 0$
20. $\cos \theta = \frac{8}{17}$	$\tan \theta < 0$
21. $\sec \theta = -2$	$\sin \theta > 0$
22. $\cot \theta$ is undefined	$\frac{\pi}{2} \leq \theta \leq \frac{3\pi}{2}$
23. $\sin \theta = 0$	$\sec \theta = -1$
24. $\tan \theta$ is undefined	$\pi \leq \theta \leq 2\pi$

25. The terminal side of θ is in Quadrant III and lies on the line $y = 2x$.

26. The terminal side of θ is in Quadrant IV and lies on the line $4x + 3y = 0$.

In Exercises 27–34, find the reference angle θ', and sketch θ and θ' in standard position.

27. (a) $\theta = 203°$ (b) $\theta = 127°$
28. (a) $\theta = 309°$ (b) $\theta = 226°$
29. (a) $\theta = -245°$ (b) $\theta = -72°$
30. (a) $\theta = -145°$ (b) $\theta = -239°$
31. (a) $\theta = \frac{2\pi}{3}$ (b) $\theta = \frac{7\pi}{6}$
32. (a) $\theta = \frac{7\pi}{4}$ (b) $\theta = \frac{8\pi}{9}$
33. (a) $\theta = 3.5$ (b) $\theta = 5.8$
34. (a) $\theta = \frac{11\pi}{3}$ (b) $\theta = -\frac{7\pi}{10}$

In Exercises 35–44, evaluate the sine, cosine, and tangent of each angle without using a calculator.

35. (a) $225°$ (b) $-225°$
36. (a) $300°$ (b) $330°$
37. (a) $750°$ (b) $510°$
38. (a) $-405°$ (b) $-120°$
39. (a) $\frac{4\pi}{3}$ (b) $\frac{2\pi}{3}$

40. (a) $\frac{\pi}{4}$ (b) $\frac{5\pi}{4}$
41. (a) $-\frac{\pi}{6}$ (b) $\frac{5\pi}{6}$
42. (a) $-\frac{\pi}{2}$ (b) $\frac{\pi}{2}$
43. (a) $\frac{11\pi}{4}$ (b) $-\frac{13\pi}{6}$
44. (a) $\frac{10\pi}{3}$ (b) $\frac{17\pi}{3}$

In Exercises 45–52, use a calculator to evaluate the trigonometric functions to four decimal places. (Be sure the calculator is set in the correct mode.)

45. (a) $\sin 10°$ (b) $\csc 10°$
46. (a) $\sec 225°$ (b) $\sec 135°$
47. (a) $\cos(-110°)$ (b) $\cos 250°$
48. (a) $\csc 330°$ (b) $\csc 150°$
49. (a) $\tan 240°$ (b) $\cot 210°$
50. (a) $\cot 1.35$ (b) $\tan 1.35$
51. (a) $\tan \frac{\pi}{9}$ (b) $\tan \frac{10\pi}{9}$
52. (a) $\sin(-0.65)$ (b) $\sin 5.63$

In Exercises 53–58, find two values of θ that satisfy the equation. Give your answers in degrees ($0° \leq \theta < 360°$) and radians ($0 \leq \theta < 2\pi$). Do not use a calculator.

53. (a) $\sin \theta = \frac{1}{2}$ (b) $\sin \theta = -\frac{1}{2}$
54. (a) $\cos \theta = \frac{\sqrt{2}}{2}$ (b) $\cos \theta = -\frac{\sqrt{2}}{2}$
55. (a) $\csc \theta = \frac{2\sqrt{3}}{3}$ (b) $\cot \theta = -1$
56. (a) $\sec \theta = 2$ (b) $\sec \theta = -2$
57. (a) $\tan \theta = 1$ (b) $\cot \theta = -\sqrt{3}$
58. (a) $\sin \theta = \frac{\sqrt{3}}{2}$ (b) $\sin \theta = -\frac{\sqrt{3}}{2}$

In Exercises 59 and 60, use a calculator to approximate two values of $\theta(0° \leq \theta < 360°)$ that satisfy the equation. Round to two decimal places.

59. (a) $\sin \theta = 0.8191$ (b) $\sin \theta = -0.2589$
60. (a) $\cos \theta = 0.8746$ (b) $\cos \theta = -0.2419$

In Exercises 61–64, use a calculator to approximate two values of $\theta(0 \leq \theta < 2\pi)$ that satisfy the equation. Round to three decimal places.

61. (a) $\cos \theta = 0.9848$ (b) $\cos \theta = -0.5890$
62. (a) $\sin \theta = 0.0175$ (b) $\sin \theta = -0.6691$
63. (a) $\tan \theta = 1.192$ (b) $\tan \theta = -8.144$
64. (a) $\cot \theta = 5.671$ (b) $\cot \theta = -1.280$

In Exercises 65–68, use the value of the given trigonometric function and trigonometric identities to find the required trigonometric function of the angle θ in the specified quadrant.

Given Function	Quadrant	Find
65. $\sin \theta = -\frac{3}{5}$	IV	$\cos \theta$
66. $\tan \theta = \frac{3}{2}$	III	$\sec \theta$
67. $\csc \theta = -2$	IV	$\cot \theta$
68. $\sec \theta = -\frac{9}{4}$	III	$\tan \theta$

In Exercises 69 and 70, evaluate the expression without using a calculator.

69. $\sin^2 2 + \cos^2 2$ **70.** $\tan^2 20° - \sec^2 20°$

71. *Average Temperature*   The average daily temperature T (in degrees Fahrenheit) for a city is

$$T = 45 - 23 \cos\left[\frac{2\pi}{365}(t - 32)\right],$$

where t is the time in days with $t = 1$ corresponding to January 1. Find the average daily temperature on the following days.
(a) January 1   (b) July 4 ($t = 185$)
(c) October 18 ($t = 291$)

72. *Sales*   A company that produces a seasonal product forecasts monthly sales over the next two years to be

$$S = 23.1 + 0.442t + 4.3 \sin \frac{\pi t}{6},$$

where S is measured in thousands of units and t is the time in months, with $t = 1$ representing January 1991. Predict sales for the following months.
(a) February 1991   (b) February 1992
(c) September 1991   (d) September 1992

73. *Distance*   An airplane flying at an altitude of 5 miles is on a flight path that passes directly over an observer (see figure). If θ is the angle of elevation from the observer to the plane, find the distance from the observer to the plane when (a) $\theta = 30°$, (b) $\theta = 75°$, and (c) $\theta = 90°$.

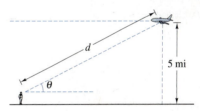

FIGURE FOR 73

74. Consider an angle in standard position with $r = 10$ cm, as shown in the figure. Write a short paragraph describing the changes in the magnitudes of x, y, $\sin \theta$, $\cos \theta$, and $\tan \theta$, as θ increases continuously from $0°$ to $90°$.

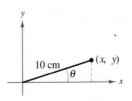

FIGURE FOR 74

5.5   GRAPHS OF SINE AND COSINE FUNCTIONS

Basic Sine and Cosine Curves / Key Points on Basic Sine and Cosine Curves / Amplitude and Period of Sine and Cosine Curves / Translations of Sine and Cosine Curves

Basic Sine and Cosine Curves

In this section you will study techniques for sketching the graphs of the sine and cosine functions. The graph of the sine function is called a **sine curve**. In Figure 5.40, the solid portion of the graph represents one period of the function and is called **one cycle** of the sine curve. The gray portion of the graph indicates that the basic sine wave repeats indefinitely to the right and left. The graph of the cosine function is shown in Figure 5.41. To produce these graphs with a graphing utility, make sure you have set the mode to *radians*.

Recall from Section 5.2 that the domain of the sine and cosine function is the set of all real numbers. Moreover, the range of each function is the interval $[-1, 1]$, and each function has a period of 2π. Do you see how this information is consistent with the basic graphs given in Figures 5.40 and 5.41?

Note from Figures 5.40 and 5.41 that the sine graph is symmetric with respect to the *origin*, whereas the cosine graph is symmetric with respect to the y-axis. These properties of symmetry follow from the fact that the sine function is odd whereas the cosine function is even.

x	0	$\frac{\pi}{6}$	$\frac{\pi}{3}$	$\frac{\pi}{2}$	$\frac{3\pi}{4}$	π	$\frac{3\pi}{2}$	2π
$\sin x$	0	$\frac{1}{2}$	$\frac{\sqrt{3}}{2}$	1	$\frac{\sqrt{2}}{2}$	0	-1	0

FIGURE 5.40

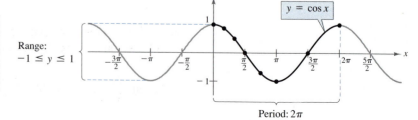

Range: $-1 \le y \le 1$

$y = \sin x$

Period: 2π

x	0	$\frac{\pi}{6}$	$\frac{\pi}{3}$	$\frac{\pi}{2}$	$\frac{3\pi}{4}$	π	$\frac{3\pi}{2}$	2π
$\cos x$	1	$\frac{\sqrt{3}}{2}$	$\frac{1}{2}$	0	$-\frac{\sqrt{2}}{2}$	-1	0	1

FIGURE 5.41

Range: $-1 \le y \le 1$

$y = \cos x$

Period: 2π

Key Points on Basic Sine and Cosine Curves

To construct the graphs of the basic sine and cosine functions *by hand,* it helps to note five **key points** in one period of each graph: the intercepts, maximum points, and minimum points. For the sine function, the key points are

Intercept Maximum Intercept Minimum Intercept

$$(0, 0), \quad \left(\frac{\pi}{2}, 1\right), \quad (\pi, 0), \quad \left(\frac{3\pi}{2}, -1\right), \quad (2\pi, 0).$$

For the cosine function, the key points are

Maximum Intercept Minimum Intercept Maximum

$$(0, 1), \quad \left(\frac{\pi}{2}, 0\right), \quad (\pi, -1), \quad \left(\frac{3\pi}{2}, 0\right), \quad (2\pi, 1).$$

Note how the x-coordinates of these points divide the period of $\sin x$ and $\cos x$ into *four* equal parts, as indicated in Figure 5.42.

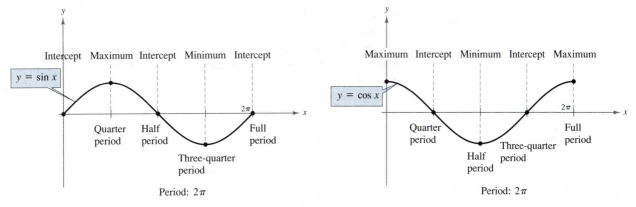

FIGURE 5.42

Technology Note _____

When using a graphing utility to graph trigonometric functions, pay special attention to the viewing rectangle you use. For instance, try graphing $y = [\sin(30x)]/30$ on the standard viewing rectangle. What do you observe? Use the zoom feature to find a viewing rectangle that displays a good view of the graph.

EXAMPLE 1 Using Key Points to Sketch a Sine Curve

Sketch the graph of $y = 2 \sin x$ on the interval $[-\pi, 4\pi]$.

SOLUTION

The y-values for the key points of $y = 2 \sin x$ have twice the magnitude of those of $y = \sin x$. Thus, the key points for $y = 2 \sin x$ are

$$(0, 0), \quad \left(\frac{\pi}{2}, 2\right), \quad (\pi, 0), \quad \left(\frac{3\pi}{2}, -2\right), \quad \text{and} \quad (2\pi, 0).$$

By connecting these key points with a smooth curve and extending the curve in both directions over the interval $[-\pi, 4\pi]$, you obtain the graph shown in Figure 5.43. Try using a graphing utility to check this result.

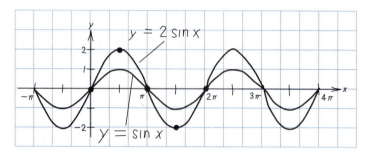

FIGURE 5.43

Amplitude and Period of Sine and Cosine Curves

In the rest of this section, you will study the graphic effect of each of the constants a, b, c, and d in equations of the forms

$$y = d + a \sin(bx - c) \quad \text{and} \quad y = d + a \cos(bx - c).$$

A quick review of the transformations studied in Section 1.5 should help in this investigation.

The constant factor a in $y = a \sin x$ acts as a *vertical stretch* or *vertical shrink* of the basic sine curve, as shown in Example 1. (If $|a| > 1$, the basic sine curve is stretched, and if $|a| < 1$, the basic sine curve is shrunk.) The result is that the graph of $y = a \sin x$ ranges between $-a$ and a instead of between -1 and 1. The absolute value of a is the **amplitude** of the function $y = a \sin x$.

DEFINITION OF AMPLITUDE OF SINE AND COSINE CURVES

The **amplitude** of $y = a \sin x$ and $y = a \cos x$ is the largest value of y and is given by

Amplitude $= |a|$.

EXAMPLE 2   Vertical Shrinking and Stretching

Sketch the graphs of $y = \frac{1}{2} \cos x$ and $y = 3 \cos x$.

SOLUTION

Because the amplitude of $y = \frac{1}{2} \cos x$ is $\frac{1}{2}$, the maximum value is $\frac{1}{2}$ and the minimum value is $-\frac{1}{2}$. For one cycle, $0 \leq x \leq 2\pi$, the key points are

$$\left(0, \frac{1}{2}\right), \quad \left(\frac{\pi}{2}, 0\right), \quad \left(\pi, -\frac{1}{2}\right), \quad \left(\frac{3\pi}{2}, 0\right), \quad \text{and} \quad \left(2\pi, \frac{1}{2}\right).$$

The amplitude of $y = 3 \cos x$ is 3, and the key points are

$$(0, 3), \quad \left(\frac{\pi}{2}, 0\right), \quad (\pi, -3), \quad \left(\frac{3\pi}{2}, 0\right), \quad \text{and} \quad (2\pi, 3).$$

The graphs of these two functions are shown in Figure 5.44. Try using a graphing utility to check this result.

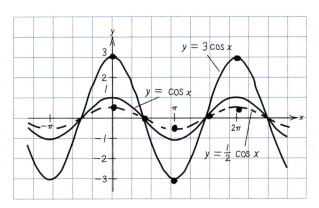

Amplitude Determines Vertical Stretch or Shrink

FIGURE 5.44

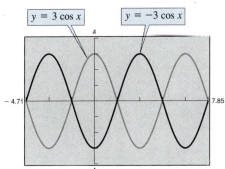

Reflection in the *x*-Axis

FIGURE 5.45

You know from Section 1.5 that the graph of $y = -f(x)$ is a **reflection** (in the *x*-axis) of the graph of $y = f(x)$. For instance, the graph of $y = -3 \cos x$ is a reflection of the graph of $y = 3 \cos x$, as shown in Figure 5.45.

Because $y = a \sin x$ completes one cycle from $x = 0$ to $x = 2\pi$, it follows that $y = a \sin bx$ completes one cycle from $bx = 0$ to $bx = 2\pi$. This implies that $y = a \sin bx$ completes one cycle from $x = 0$ to $x = 2\pi/b$.

Use a graphing utility to draw the graph of $y = \sin bx$, where $b = 0.5$, 1, and 2. (Use a viewing rectangle in which $0 \leq X \leq 6.3$ and $-2 \leq Y \leq 2$.) How does the value of b affect the graph?

PERIOD OF SINE AND COSINE FUNCTIONS

Let b be a positive real number. The **period** of $y = a \sin bx$ and $y = a \cos bx$ is $2\pi/b$.

Note that if $0 < b < 1$, the period of $y = a \sin bx$ is greater than 2π and represents a *horizontal stretching* of the graph of $y = a \sin x$. Similarly, if $b > 1$, the period of $y = a \sin bx$ is less than 2π and represents a *horizontal shrinking* of the graph of $y = a \sin x$.

If b is negative, use the identities $\sin(-x) = -\sin x$ and $\cos(-x) = \cos x$ to rewrite the function.

EXAMPLE 3   **Horizontal Stretching and Shrinking**

Sketch the graph of

$$y = \sin \frac{x}{2}.$$

SOLUTION

The amplitude is 1. Moreover, because $b = \frac{1}{2}$, the period is $2\pi/(\frac{1}{2}) = 4\pi$. By dividing the period-interval $[0, 4\pi]$ into four equal parts with the values $\pi, 2\pi$, and 3π, you obtain the following key points on the graph.

$$(0, 0), \quad (\pi, 1), \quad (2\pi, 0), \quad (3\pi, -1), \quad \text{and} \quad (4\pi, 0)$$

The graph is shown in Figure 5.46. Use a graphing utility to check this result.

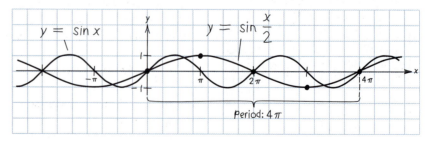

FIGURE 5.46

EXAMPLE 4   Horizontal Stretching and Shrinking

Sketch the graph of $y = \sin 3x$.

SOLUTION

The amplitude is 1, and because $b = 3$, the period is

$$\frac{2\pi}{b} = \frac{2\pi}{3}.$$

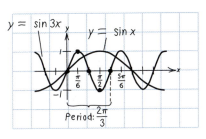

$y = \sin 3x$

$y = \sin x$

Period: $\frac{2\pi}{3}$

FIGURE 5.47

Dividing the period-interval $[0, 2\pi/3]$ into four equal parts, you obtain the following key points on the graph.

$$(0, 0), \quad \left(\frac{\pi}{6}, 1\right), \quad \left(\frac{\pi}{3}, 0\right), \quad \left(\frac{\pi}{2}, -1\right), \quad \text{and} \quad \left(\frac{2\pi}{3}, 0\right)$$

The graph is shown in Figure 5.47. Use a graphing utility to check this result.

EXAMPLE 5   Using a Graphing Utility

Use a graphing utility to graph $y = 2 \cos 0.4x$ and $y = 2 \cos 4x$ on the same screen. Determine the period of each function from its graph.

SOLUTION

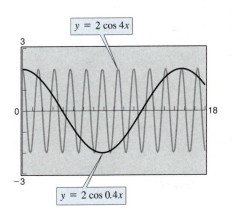

$y = 2 \cos 4x$

$y = 2 \cos 0.4x$

FIGURE 5.48

The standard viewing rectangle is not appropriate for graphing most sine and cosine functions. You should choose a viewing rectangle that accommodates the amplitudes of the functions as well as any stretching or shrinking that may occur. For these graphs, a y-scale ranging between -3 and 3 will accommodate the amplitudes of 2. Figure 5.48 shows the two graphs and an appropriate viewing rectangle. From the graph, you can estimate the period of $y = 2 \cos 0.4x$ to be about 16 and the period of $y = 2 \cos 4x$ to be about 1.5. Algebraically, the period of $y = 2 \cos 0.4x$ is

$$\frac{2\pi}{0.4} = 5\pi \approx 15.71,$$

and the period of $y = 2 \cos 4x$ is

$$\frac{2\pi}{4} = \frac{\pi}{2} \approx 1.57.$$

Translations of Sine and Cosine Curves

The constant c in the general equations

$$y = a \sin(bx - c) \quad \text{and} \quad y = a \cos(bx - c)$$

creates a horizontal shift of the basic sine and cosine curves. The graph of $y = a \sin(bx - c)$ completes one cycle from $bx - c = 0$ to $bx - c = 2\pi$. By solving for x in the inequality $0 \le bx - c \le 2\pi$, you can find the interval for one cycle to be

Left endpoint   Right endpoint

$$\frac{c}{b} \le x \le \frac{c}{b} + \frac{2\pi}{b}.$$

Period

This implies that the period of $y = a \sin(bx - c)$ is $2\pi/b$, and the graph of $y = a \sin bx$ is shifted by an amount c/b. The number c/b is the **phase shift.**

D I S C O V E R Y

Use a graphing utility to draw the graph of $y = \sin(x + c)$, where $c = -\pi/4$, 0, and $\pi/4$. (Use a viewing rectangle in which $0 \le X \le 6.3$ and $-2 \le Y \le 2$.) How does the value of c affect the graph?

D I S C O V E R Y

Use your graphing utility to compare the graphs of $y_1 = \cos x$ and $y_2 = \cos(-x)$. Discuss the symmetry of the cosine function. Is $\cos x$ an even function, odd function, or neither? Repeat the experiment for $y_1 = \sin x$ and $y_2 = \sin(-x)$.

GRAPHS OF THE SINE AND COSINE FUNCTIONS

The graphs of $y = a \sin(bx - c)$ and $y = a \cos(bx - c)$ have the following characteristics. (Assume $b > 0$.)

Amplitude $= |a|$

Period $= 2\pi/b$

The left and right endpoints corresponding to a one-cycle interval of the graphs can be determined by solving the equations $bx - c = 0$ and $bx - c = 2\pi$.

EXAMPLE 6 Horizontal Shift

Describe the graph of

$$y = \frac{1}{2} \sin\left(x - \frac{\pi}{3}\right).$$

SOLUTION

Because $a = \frac{1}{2}$ and $b = 1$, the amplitude is $\frac{1}{2}$ and the period is 2π. By solving the inequality

$$0 \le x - \frac{\pi}{3} \le 2\pi$$

$$\frac{\pi}{3} \le x \qquad \le \frac{7\pi}{3},$$

you can see that the interval $[\pi/3, 7\pi/3]$ corresponds to one cycle of the graph. Dividing this interval into four equal parts produces the following key points.

$$\left(\frac{\pi}{3}, 0\right), \quad \left(\frac{5\pi}{6}, \frac{1}{2}\right), \quad \left(\frac{4\pi}{3}, 0\right), \quad \left(\frac{11\pi}{6}, -\frac{1}{2}\right), \quad \text{and} \quad \left(\frac{7\pi}{3}, 0\right)$$

The graph is shown in Figure 5.49.

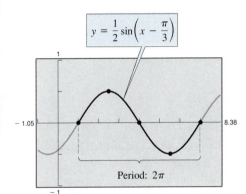

$$y = \frac{1}{2} \sin\left(x - \frac{\pi}{3}\right)$$

Period: 2π

FIGURE 5.49

The final type of transformation is the *vertical shift* caused by the constant d in the equations

$$y = d + a \sin(bx - c) \quad \text{and} \quad y = d + a \cos(bx - c).$$

The shift is d units upward for $d > 0$ and downward for $d < 0$. In other words, the graph oscillates about the horizontal line $y = d$ instead of the x-axis.

EXAMPLE 7 Vertical Shift

Describe the graph of $y = 2 + 3 \sin 2x$.

SOLUTION

The amplitude is 3 and the period is π. The key points over the interval $[0, \pi]$ are

$$(0, 2), \quad \left(\frac{\pi}{4}, 5\right), \quad \left(\frac{\pi}{2}, 2\right), \quad \left(\frac{3\pi}{4}, -1\right), \quad \text{and} \quad (\pi, 2).$$

The graph is shown in Figure 5.50.

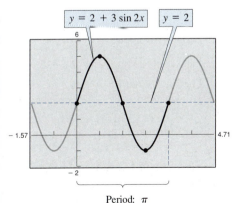

$y = 2 + 3 \sin 2x$ $y = 2$

Period: π

FIGURE 5.50

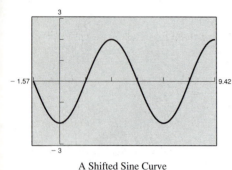

A Shifted Sine Curve

FIGURE 5.51

EXAMPLE 8   Finding an Equation for a Given Graph

Find the amplitude, period, and phase shift for the sine function whose graph is shown in Figure 5.51. Write an equation for this graph.

SOLUTION

The amplitude for this sine curve is 2. The period is 2π, and there is a right phase shift of $\pi/2$. Thus, you can write the following equation.

$$y = 2 \sin\left(x - \frac{\pi}{2}\right)$$

Try finding a cosine function with the same graph.

DISCUSSION PROBLEM

A SINE SHOW

If you have a programmable calculator, try running the following "sine-show program." This program simultaneously draws the unit circle and the corresponding points on the sine curve. After the circle and sine curve are drawn, you can connect points on the unit circle with their corresponding points on the sine curve by repeatedly pressing ENTER. After the program is complete, the screen should look like that shown in the figure. (This program is for the TI-81. Program steps for other calculators are given in Appendix B.)

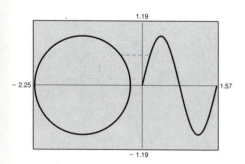

```
:Rad                        :Line(-1.25,1.19,-1.25,1.19)
:ClrDraw                    :DispGraph
:All-Off                    :0→N
:Param                      :Lbl 1
:Simul                      :N+1→N
:-2.25→Xmin                 :Nπ/6.5→T
:π/2→Xmax                   :-1.25+cos T→A
:3→Xscl                     :sin T→B
:-1.19→Ymin                 :T/4→C
:1.19→Ymax                  :Line(A,B,C,B)
:1→Yscl                     :Pause
:0→Tmin                     :If N=12
:6.3→Tmax                   :Goto 2
:.15→Tstep                  :Goto 1
:"-1.25+cos T"→X₁T          :Lbl 2
:"sin T"→Y₁T                :Function
:"T/4"→X₂T                  :Sequence
:"sin T"→Y₂T                :End
```

WARM-UP

The following warm-up exercises involve skills that were covered in earlier sections. You will use these skills in the exercise set for this section.

In Exercises 1 and 2, simplify the expression.

1. $\dfrac{2\pi}{\frac{1}{3}}$ **2.** $\dfrac{2\pi}{4\pi}$

In Exercises 3–6, solve for x.

3. $2x - \dfrac{\pi}{3} = 0$ **4.** $2x - \dfrac{\pi}{3} = 2\pi$

5. $3\pi x + 6\pi = 0$ **6.** $3\pi x + 6\pi = 2\pi$

In Exercises 7–10, evaluate the trigonometric function without using a calculator.

7. $\sin \dfrac{\pi}{2}$ **8.** $\sin \pi$ **9.** $\cos 0$ **10.** $\cos \dfrac{\pi}{2}$

SECTION 5.5 · EXERCISES

In Exercises 1–10, determine the period and amplitude of the function. Then describe the viewing rectangle for Exercises 1–6.

5. $y = \frac{1}{2} \sin \pi x$ **6.** $y = \dfrac{5}{2} \cos \dfrac{\pi x}{2}$

1. $y = 2 \sin 2x$ **2.** $y = 3 \cos 3x$

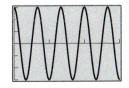   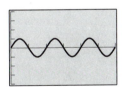

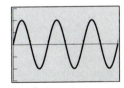

7. $y = -3 \sin 10x$

3. $y = \dfrac{3}{2} \cos \dfrac{x}{2}$ **4.** $y = -2 \sin \dfrac{x}{3}$

8. $u = -\dfrac{5}{2} \cos \dfrac{x}{4}$

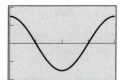   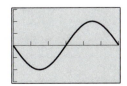

9. $y = 3 \sin 4\pi x$

10. $y = \dfrac{2}{3} \cos \dfrac{\pi x}{10}$

In Exercises 11–16, describe the relationship between the graphs of f and g.

11. $f(x) = \sin x$
  $g(x) = \sin(x - \pi)$

12. $f(x) = \cos x$
  $g(x) = \cos(x + \pi)$

13. $f(x) = \cos 2x$
  $g(x) = -\cos 2x$

14. $f(x) = \sin 3x$
  $g(x) = \sin(-3x)$

15. $f(x) = \sin x$
  $g(x) = 2 + \sin x$

16. $f(x) = \cos 4x$
  $g(x) = -2 + \cos 4x$

In Exercises 17–24, sketch the graphs of the two functions on the same coordinate plane. (Include two full periods, and use a graphing utility to verify your result.)

17. $f(x) = -2 \sin x$
  $g(x) = 4 \sin x$

18. $f(x) = \sin x$
  $g(x) = \sin \dfrac{x}{3}$

19. $f(x) = \cos x$
  $g(x) = 1 + \cos x$

20. $f(x) = 2 \cos 2x$
  $g(x) = -\cos 4x$

21. $f(x) = -\dfrac{1}{2} \sin \dfrac{x}{2}$
  $g(x) = 3 - \dfrac{1}{2} \sin \dfrac{x}{2}$

22. $f(x) = 4 \sin \pi x$
  $g(x) = 4 \sin \pi x - 3$

23. $f(x) = 2 \cos x$
  $g(x) = 2 \cos(x + \pi)$

24. $f(x) = -\cos x$
  $g(x) = -\cos(x - \pi)$

In Exercises 25–28, sketch the graphs of f and g on the same coordinate axes and show that $f(x) = g(x)$ for all x. (Include two full periods, and use a graphing utility to verify your result.)

25. $f(x) = \sin x$
  $g(x) = \cos\left(x - \dfrac{\pi}{2}\right)$

26. $f(x) = \sin x$
  $g(x) = -\cos\left(x + \dfrac{\pi}{2}\right)$

27. $f(x) = \cos x$
  $g(x) = -\sin\left(x - \dfrac{\pi}{2}\right)$

28. $f(x) = \cos x$
  $g(x) = -\cos(x - \pi)$

In Exercises 29–42, sketch the graph of the function. (Include two full periods, and use a graphing utility to verify your result.)

29. $y = -2 \sin 6x$

30. $y = -3 \cos 4x$

31. $y = \cos 2\pi x$

32. $y = \dfrac{3}{2} \sin \dfrac{\pi x}{4}$

33. $y = -\sin \dfrac{2\pi x}{3}$

34. $y = 10 \cos \dfrac{\pi x}{6}$

35. $y = 2 - \sin \dfrac{2\pi x}{3}$

36. $y = 2 \cos x - 3$

37. $y = \sin\left(x - \dfrac{\pi}{4}\right)$

38. $y = \dfrac{1}{2} \sin(x - \pi)$

39. $y = 3 \cos(x + \pi)$

40. $y = 4 \cos\left(x + \dfrac{\pi}{4}\right)$

41. $y = \dfrac{1}{10} \cos 60\pi x$

42. $y = -3 + 5 \cos \dfrac{\pi t}{12}$

In Exercises 43–54, use a graphing utility to graph the function. (Include two full periods.)

43. $y = 3 \cos(x + \pi) - 3$

44. $y = 4 \cos\left(x + \dfrac{\pi}{4}\right) + 4$

45. $y = \dfrac{2}{3} \cos\left(\dfrac{x}{2} - \dfrac{\pi}{4}\right)$

46. $y = -3 \cos(6x + \pi)$

47. $y = -2 \sin(4x + \pi)$

48. $y = -4 \sin\left(\dfrac{2}{3}x - \dfrac{\pi}{3}\right)$

49. $y = \cos\left(2\pi x - \dfrac{\pi}{2}\right) + 1$

50. $y = 3 \cos\left(\dfrac{\pi x}{2} + \dfrac{\pi}{2}\right) - 2$

51. $y = -0.1 \sin\left(\dfrac{\pi x}{10} + \pi\right)$

52. $y = 5 \sin(\pi - 2x) + 10$

53. $y = 5 \cos(\pi - 2x) + 2$

54. $y = \dfrac{1}{100} \sin 120\pi t$

In Exercises 55–58, use the graph of the trigonometric function to find all real numbers x in the interval $[-2\pi, 2\pi]$ that give the specified function value.

Function	Function Value
55. $\sin x$	$-\dfrac{1}{2}$
56. $\cos x$	-1
57. $\cos x$	$\dfrac{\sqrt{2}}{2}$
58. $\sin x$	$\dfrac{\sqrt{3}}{2}$

In Exercises 59 and 60, find a and d for the function $f(x) = a \cos x + d$ so that the graph of f matches the figure.

59.

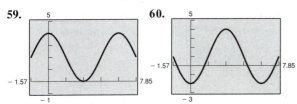

60.

In Exercises 61–64, find a, b, and c so that the graph of the function matches the graph in the figure.

61. $y = a \sin(bx - c)$

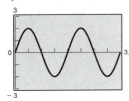

62. $y = a \sin(bx - c)$

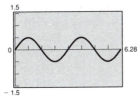

63. $y = a \cos(bx - c)$

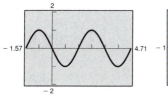

64. $y = a \sin(bx - c)$

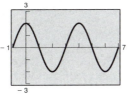

65. Use a graphing utility to graph the functions $f(x) = 2e^x$ and $g(x) = 5 \cos x$. Approximate any points of intersection of the graphs in the interval $[-\pi, \pi]$.

66. Use a graphing utility to determine the smallest *integer* value of a such that the graphs of $f(x) = 2 \ln x$ and $g(x) = a \cos x$ intersect more than once.

67. *Respiratory Cycle*   For a person at rest, the velocity v (in liters per second) of air flow during a respiratory cycle is

$$v = 0.85 \sin \frac{\pi t}{3},$$

where t is the time in seconds. (Inhalation occurs when $v > 0$, and exhalation occurs when $v < 0$.)
(a) Find the time for one full respiratory cycle.
(b) Find the number of cycles per minute.
(c) Use a graphing utility to graph the velocity function.

68. *Respiratory Cycle*   After exercising for a few minutes, a person has a respiratory cycle for which the velocity of air flow is approximated by

$$v = 1.75 \sin \frac{\pi t}{2}.$$

Use this model to repeat Exercise 67.

69. *Piano Tuning*   When tuning a piano, a technician strikes a tuning fork for the A above middle C and sets up wave motion that can be approximated by

$$y = 0.001 \sin 880\pi t,$$

where t is the time in seconds.
(a) What is the period p of this function?
(b) The frequency f is given by $f = 1/p$. What is the frequency of this note?
(c) Use a graphing utility to graph this function.

70. *Blood Pressure*   The function

$$P = 100 - 20 \cos \frac{5\pi t}{3}$$

approximates the blood pressure P in millimeters of mercury at time t in seconds for a person at rest.
(a) Find the period of the function.
(b) Find the number of heartbeats per minute.
(c) Use a graphing utility to graph the pressure function.

Sales   In Exercises 71 and 72, use a graphing utility to graph the sales function over 1 year where S is sales in thousands of units and t is the time in months, with $t = 1$ corresponding to January. Use the graph to determine the month of maximum sales and the month of minimum sales.

71. $S = 22.3 - 3.4 \cos \dfrac{\pi t}{6}$

72. $S = 74.50 + 43.75 \sin \dfrac{\pi t}{6}$

In Exercises 73–76, describe the relationship between the graph of the functions f and g.

73.

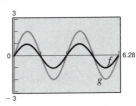

74.

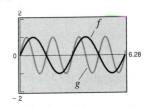

75.

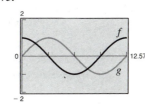

76.

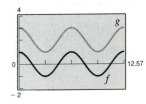

5.6   OTHER TRIGONOMETRIC GRAPHS

Graphs of Tangent and Cotangent Functions / Graphs of the Reciprocal Functions / Graphs of Combinations of Trigonometric Functions / Combinations of Algebraic and Trigonometric Functions / Damped Trigonometric Graphs

Graphs of Tangent and Cotangent Functions

Recall from Section 5.2 that the tangent function is odd. That is, $\tan(-x) = -\tan x$. Consequently, the graph of $y = \tan x$ is symmetric with respect to the origin. From the identity $\tan x = \sin x/\cos x$, you know that the tangent is undefined when $\cos x = 0$. Two such values are $x = \pm\pi/2 \approx \pm 1.5708$.

x	$-\dfrac{\pi}{2}$	-1.57	-1.5	-1	0	1	1.5	1.57	$\dfrac{\pi}{2}$
$\tan x$	undef.	-1255.8	-14.1	-1.56	0	1.56	14.1	1255.8	undef.

tan x approaches $-\infty$ as x approaches $-\pi/2$ from the right

tan x approaches ∞ as x approaches $\pi/2$ from the left

As indicated in the table, $\tan x$ increases without bound as x approaches $\pi/2$ from the left, and decreases without bound as x approaches $-\pi/2$ from the right. Thus, the graph of $y = \tan x$ has *vertical asymptotes* as $x = \pi/2$ and $-\pi/2$, as shown in Figure 5.52. Moreover, because the period of the tangent function is π, vertical asymptotes also occur when $x = \pi/2 \pm n\pi$. The domain of the tangent function is the set of all real numbers other than $x = \pi/2 \pm n\pi$, and the range is the set of all real numbers.

Sketching the graph of a function with the form $y = a\tan(bx - c)$ is similar to sketching the graph of $y = a\sin(bx - c)$ in that you locate key points which identify the intercepts and asymptotes. Two consecutive asymptotes can be found by solving the equations

$$bx - c = -\frac{\pi}{2} \quad \text{and} \quad bx - c = \frac{\pi}{2}.$$

The midpoint between two consecutive asymptotes is an x-intercept of the graph. After plotting the asymptotes and the x-intercept, plot a few additional points between the two asymptotes and sketch one cycle. Finally, sketch one or two additional cycles to the left and right.

Period: π
Domain: all $x \neq \dfrac{\pi}{2} \pm n\pi$

Range: $(-\infty, \infty)$
Vertical asymptotes: $x = \dfrac{\pi}{2} \pm n\pi$

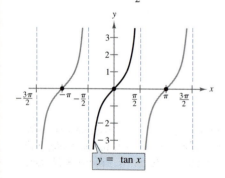

$y = \tan x$

FIGURE 5.52

REMARK   The period of the function $y = a\tan(bx - c)$ is the distance between two consecutive asymptotes. The amplitude of a tangent function is not defined.

EXAMPLE 1   Sketching the Graph of a Tangent Function

Sketch the graph of $y = \tan \dfrac{x}{2}$.

SOLUTION

By solving the inequality

$$-\frac{\pi}{2} < \frac{x}{2} < \frac{\pi}{2}$$

$$-\pi < x < \pi$$

you can see that two consecutive asymptotes occur at $x = -\pi$ and $x = \pi$. Between these two asymptotes, plot a few points including the x-intercept, as shown in the table. Three cycles of the graph are shown in Figure 5.53. Use a graphing utility to confirm this result.

x	$-\dfrac{\pi}{2}$	0	$\dfrac{\pi}{2}$
$\tan \dfrac{x}{2}$	-1	0	1

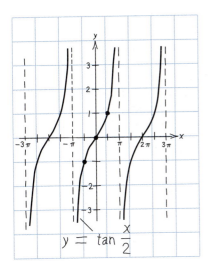

FIGURE 5.53

EXAMPLE 2   Sketching the Graph of a Tangent Function

Sketch the graph of $y = -3 \tan 2x$.

SOLUTION

By solving the inequality

$$-\frac{\pi}{2} < 2x < \frac{\pi}{2}$$

$$-\frac{\pi}{4} < x < \frac{\pi}{4}$$

you can see that two consecutive asymptotes occur at $x = -\pi/4$ and $x = \pi/4$. Between these two asymptotes, plot a few points as shown in the table, and complete one cycle. Four cycles of the graph are shown in Figure 5.54. Use a graphing utility to confirm this result.

x	$-\dfrac{\pi}{8}$	0	$\dfrac{\pi}{8}$
$-3 \tan 2x$	3	0	-3

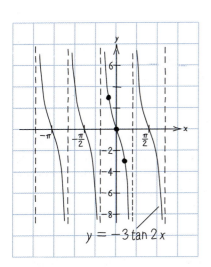

FIGURE 5.54

Period: π
Domain: all $x \neq n\pi$
Range: $(-\infty, \infty)$
Vertical asymptotes: $x = n\pi$

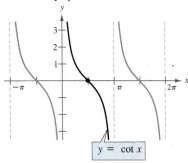

FIGURE 5.55

Technology Note _____

You can use the tangent function on your graphing utility to obtain the graph of the cotangent function. For example, to graph the function $y = 2 \cot (x/3)$ from Example 3, let

$y_1 = 2/\tan(x/3)$.

If you select the viewing rectangle $-9 \leq x \leq 18$ and $-6 \leq y \leq 6$, you should obtain a graph similar to that of Figure 5.56.

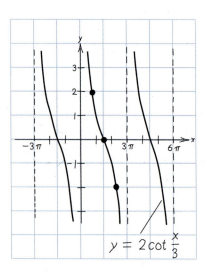

FIGURE 5.56

By comparing the graphs in Examples 1 and 2, you can see that the graph of

$$y = a \tan(bx + c)$$

is increasing between consecutive vertical asymptotes if $a > 0$ and decreasing between consecutive vertical asymptotes if $a < 0$. In other words, the graph for $a < 0$ is a reflection of the graph for $a > 0$.

The graph of the cotangent function is similar to the graph of the tangent function. It also has a period of π. However, from the identity

$$y = \cot x = \frac{\cos x}{\sin x}$$

you can see that the cotangent function has vertical asymptotes at $x = n\pi$ (because $\sin x$ is zero at these x-values). The graph of the cotangent function is shown in Figure 5.55.

EXAMPLE 3   Sketching the Graph of a Cotangent Function
················

Sketch the graph of $y = 2 \cot \dfrac{x}{3}$.

SOLUTION

To locate two consecutive vertical asymptotes of the graph, you can solve the following inequality.

$$0 < \frac{x}{3} < \pi$$

$$0 < x < 3\pi$$

Then, between these two asymptotes, plot the points shown in the following table, and complete one cycle of the graph. (Note that the period is 3π, the distance between consecutive asymptotes.) Three cycles of the graph are shown in Figure 5.56.

x	$\dfrac{3\pi}{4}$	$\dfrac{3\pi}{2}$	$\dfrac{9\pi}{4}$
$2 \cot \dfrac{x}{3}$	2	0	-2

········

Graphs of the Reciprocal Functions

The graphs of the two remaining trigonometric functions can be obtained from the graphs of the sine and cosine functions using the reciprocal identities

$$\csc x = \frac{1}{\sin x} \quad \text{and} \quad \sec x = \frac{1}{\cos x}.$$

For instance, at a given value for x, the y-coordinate for $\sec x$ is the reciprocal of the y-coordinate for $\cos x$. Of course, when $\cos x = 0$, the reciprocal does not exist. Near such values for x, the behavior of the secant function is similar to that of the tangent function. In other words, the graphs of $\tan x = (\sin x)/(\cos x)$ and $\sec x = 1/(\cos x)$ have vertical asymptotes at $x = (\pi/2) + n\pi$, because the cosine is zero at these x-values. Similarly, $\cot x = (\cos x)/(\sin x)$ and $\csc x = 1/(\sin x)$ have vertical asymptotes where $\sin x = 0$, that is, at $x = n\pi$.

To sketch the graph of a secant or cosecant function, we suggest that you first make a sketch of its reciprocal function. For instance, to sketch the graph of $y = \csc x$, first sketch the graph of $y = \sin x$. Then take reciprocals of the y-coordinates to obtain points on the graph of $y = \csc x$. You can use this procedure to obtain the graphs shown in Figure 5.57.

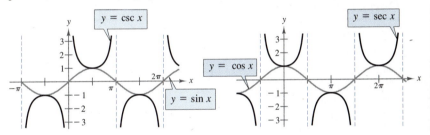

Period: 2π
Domain: all $x \neq n\pi$
Range: all y not in $(-1, 1)$
Vertical asymptotes: $x = n\pi$
Symmetry: origin

Period: 2π
Domain: all $x \neq \dfrac{\pi}{2} + n\pi$
Range: all y not in $(-1, 1)$
Vertical asymptotes: $x = \dfrac{\pi}{2} + n\pi$
Symmetry: y-axis

FIGURE 5.57

In comparing the graphs of the secant and cosecant functions with those of the sine and cosine functions, note that the "hills" and "valleys" are interchanged. For example, a hill (or maximum point) on the sine curve corresponds to a valley (a local minimum) on the cosecant curve. Similarly, a valley (or minimum point) on the sine curve corresponds to a hill (a local maximum) on the cosecant curve, as shown in Figure 5.58.

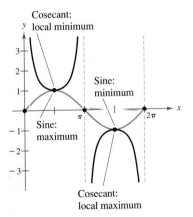

FIGURE 5.58

EXAMPLE 4   Graphing a Cosecant Function

Use a graphing utility to graph

$$y = 2 \sin\left(x + \frac{\pi}{4}\right) \quad \text{and} \quad y = 2 \csc\left(x + \frac{\pi}{4}\right).$$

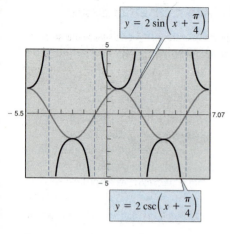

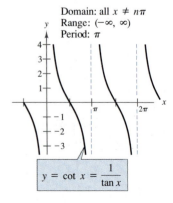

FIGURE 5.59

SOLUTION

The two graphs are shown in Figure 5.59. Note how the "hills" and "valleys" of each graph are related. For the function $y = 2 \sin[x + (\pi/4)]$, the amplitude is 2 and the period is 2π. One cycle of the sine function corresponds to the interval from $x = -\pi/4$ to $x = 7\pi/4$. Because the sine function is zero at the endpoints of this interval, the corresponding cosecant function

$$y = 2 \csc\left(x + \frac{\pi}{4}\right) = 2\left[\frac{1}{\sin\left(x + \frac{\pi}{4}\right)}\right]$$

has vertical asymptotes at $x = -\pi/4$, $x = 3\pi/4$, and $7\pi/4$.

In Figure 5.60, we summarize the graphs, domains, ranges, and periods of the six basic trigonometric functions.

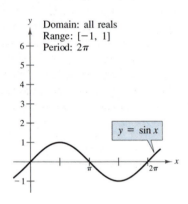

Domain: all reals
Range: $[-1, 1]$
Period: 2π

$y = \sin x$

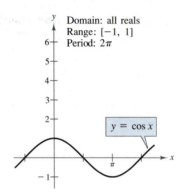

Domain: all reals
Range: $[-1, 1]$
Period: 2π

$y = \cos x$

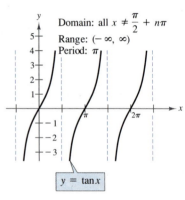

Domain: all $x \neq \dfrac{\pi}{2} + n\pi$
Range: $(-\infty, \infty)$
Period: π

$y = \tan x$

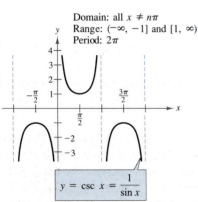

Domain: all $x \neq n\pi$
Range: $(-\infty, -1]$ and $[1, \infty)$
Period: 2π

$y = \csc x = \dfrac{1}{\sin x}$

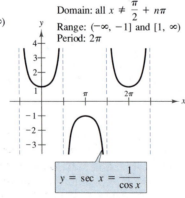

Domain: all $x \neq \dfrac{\pi}{2} + n\pi$
Range: $(-\infty, -1]$ and $[1, \infty)$
Period: 2π

$y = \sec x = \dfrac{1}{\cos x}$

Domain: all $x \neq n\pi$
Range: $(-\infty, \infty)$
Period: π

$y = \cot x = \dfrac{1}{\tan x}$

Graphs of the Six Trigonometric Functions

FIGURE 5.60

Graphs of Combinations of Trigonometric Functions

Sums, differences, products, and quotients of periodic functions are also periodic. The period of the combined function is the least common multiple of the periods of the component functions. This period is important for determining an appropriate viewing rectangle for a graphing utility.

EXAMPLE 5   Finding the Period and Relative Extrema of a Function

Graph $y = \sin x - \cos 2x$. Find the period and the relative minimums and maximums of this function.

SOLUTION

The period of $\sin x$ is 2π and the period of $\cos 2x$ is π. Thus, the period of the given function is 2π, because 2π is the least common multiple of 2π and π. This conclusion is further reinforced by graphing the function, as shown in Figure 5.61. From the graph, it appears that the function has two relative maximums and two relative minimums in each complete cycle. For instance, between 0 and 2π, the graph appears to have relative maximums when $x = \pi/2 \approx 1.57$ and when $x = 3\pi/2 \approx 4.71$. Using the zoom and trace features of the graphing utility, you can find that the relative minimums occur when $x \approx 3.40$ and when $x \approx 6.03$.

Relative maximums:	(1.57, 2.00)	(4.71, 0.00)
Relative minimums:	(3.40, −1.13)	(6.03, −1.13)

$y = \sin x - \cos 2x$

FIGURE 5.61

EXAMPLE 6   Finding the Period and Range of a Function

Graph $y = 2 \sin 6x + \sin 4x$. Find the period and range of this function.

SOLUTION

The period of $2 \sin 6x$ is $\pi/3$ and the period of $\sin 4x$ is $\pi/2$. Thus, the period of the given function is π, because π is the least common multiple of $\pi/3$ and $\pi/2$. This conclusion is further reinforced by graphing the function, as shown in Figure 5.62. In the interval from 0 to π, the maximum y-value occurs when $x \approx 0.29$ and $y \approx 2.89$. The minimum value in this interval occurs when $x \approx 2.86$ and $y \approx -2.89$. Thus, the range of the function is approximated by

$$-2.89 \le y \le 2.89. \qquad \textit{Range}$$

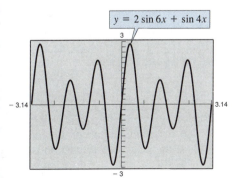

$y = 2 \sin 6x + \sin 4x$

FIGURE 5.62

Combinations of Algebraic and Trigonometric Functions

Functions that are combinations of algebraic and trigonometric functions are not, in general, periodic.

EXAMPLE 7 The Graph of a Nonperiodic Function

Graph $y = x + \cos x$. Find the domain and range of the function. Approximate any zeros of the graph.

SOLUTION

The graph of the function is shown in Figure 5.63. Notice that even though the function is not periodic, it does have a pattern that repeats. Notice that the graph of $y = x + \cos x$ oscillates about the line $y = x$. Both the domain and range of the function is the set of all real numbers. Using the zoom feature, you can find that the zero of $y = x + \cos x$ is approximately $x = -0.739$.

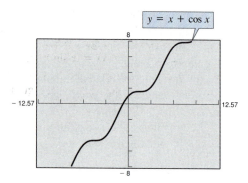

FIGURE 5.63

EXAMPLE 8 A Function Involving Absolute Value

Graph $y = |\sin x|$, and find the domain and range of the function.

SOLUTION

The domain of the function is the set of all real numbers. The graph of the function is shown in Figure 5.64. Notice that the minimum value of the function is 0 and the maximum value is 1. Thus, the range of the function is given by $0 \le y \le 1$.

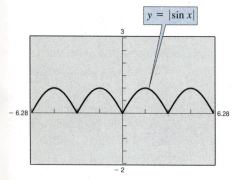

FIGURE 5.64

Damped Trigonometric Graphs

A *product* of two functions can be graphed using properties of the individual functions involved. For instance, consider the function

$$f(x) = x \sin x$$

as the product of the functions $y = x$ and $y = \sin x$. Using properties of absolute value and the fact that $|\sin x| \le 1$, you obtain $0 \le |x||\sin x| \le |x|$. Consequently,

$$-|x| \le x \sin x \le |x|,$$

which means that the graph of $f(x) = x \sin x$ lies between the lines $y = -x$ and $y = x$. Furthermore, since

$$f(x) = x \sin x = \pm x \quad \text{at} \quad x = \frac{\pi}{2} + n\pi$$

$$f(x) = x \sin x = 0 \quad \text{at} \quad x = n\pi,$$

the graph of f touches the line $y = -x$ or the line $y = x$ at $x = (\pi/2) + n\pi$ and has x-intercepts at $x = n\pi$. The graph of f, together with $y = x$ and $y = -x$, is shown in Figure 5.65.

In the function $f(x) = x \sin x$, the factor x is called the **damping factor.** By changing the damping factor, you can change the graph significantly. For example, look in Figure 5.66 at the graphs of

$$y = \frac{1}{x} \sin x$$

and

$$y = e^{-x} \sin 3x.$$

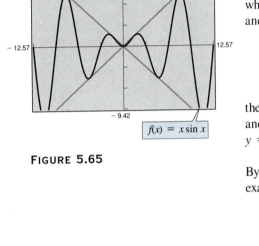

FIGURE 5.65

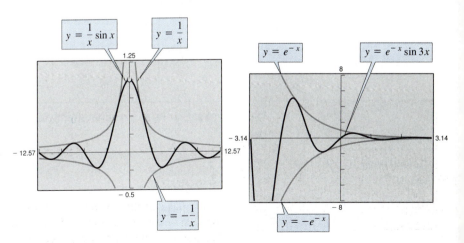

FIGURE 5.66

EXAMPLE 9 **Damped Cosine Wave**

Graph $f(x) = 2^{-x/2} \cos x$.

SOLUTION

A graph of $f(x) = 2^{-x/2} \cos x$ is shown in Figure 5.67. To analyze this function further, consider $f(x)$ as the product of the two functions

$$y = 2^{-x/2} \quad \text{and} \quad y = \cos x,$$

each of which has the set of real numbers as its domain. For any real number x, you know that $2^{-x/2} \geq 0$ and $|\cos x| \leq 1$. Therefore, $|2^{-x/2}||\cos x| \leq 2^{-x/2}$, which means that

$$-2^{-x/2} \leq 2^{-x/2} \cos x \leq 2^{-x/2}.$$

Furthermore, since

$$f(x) = 2^{-x/2} \cos x = \pm 2^{-x/2} \quad \text{at} \quad x = n\pi$$

and

$$f(x) = 2^{-x/2} \cos x = 0 \quad \text{at} \quad x = \frac{\pi}{2} + n\pi,$$

the graph of f touches the curve $y = -2^{-x/2}$ or the curve $y = 2^{-x/2}$ at $x = n\pi$ and has intercepts at $x = (\pi/2) + n\pi$.

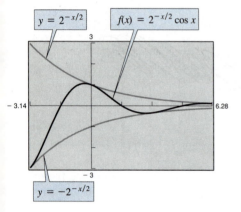

$y = 2^{-x/2}$ $f(x) = 2^{-x/2} \cos x$

$y = -2^{-x/2}$

FIGURE 5.67

DISCUSSION PROBLEM

GRAPHING UTILITIES

Use a graphing utility to graph the following functions for varying values of a, b, c, and d.

$$y = d + a \sin(bx + c) \qquad y = d + a \cos(bx + c)$$
$$y = d + a \tan(bx + c) \qquad y = d + a \cot(bx + c)$$
$$y = d + a \sec(bx + c) \qquad y = d + a \csc(bx + c)$$

In a paper or a discussion, summarize the effects of the constants a, b, c, and d in these graphs.

WARM-UP

The following warm-up exercises involve skills that were covered in earlier sections. You will use these skills in the exercise set for this section.

In Exercises 1–4, find the x-values in the interval $[0, 2\pi]$ for which $f(x)$ is -1, 0, or 1.

1. $f(x) = \sin x$ **2.** $f(x) = \cos x$ **3.** $f(x) = \sin 2x$ **4.** $f(x) = \cos \dfrac{x}{2}$

In Exercises 5–8, graph the function.

5. $y = |x|$ **6.** $y = e^{-x}$ **7.** $y = \sin \pi x$ **8.** $y = \cos 2x$

In Exercises 9 and 10, evaluate $f(x)$ when $x = 0, \pi/6, \pi/4, \pi/3,$ and $\pi/2$.

9. $f(x) = x \cos x$ **10.** $f(x) = x + \sin x$

SECTION 5.6 · EXERCISES

In Exercises 1–8, match the trigonometric function with its graph and describe the viewing rectangle. [The graphs are labeled (a), (b), (c), (d), (e), (f), (g), and (h).]

1. $y = \sec 2x$

2. $y = \tan 3x$

3. $y = \tan \dfrac{x}{2}$

4. $y = 2 \csc \dfrac{x}{2}$

5. $y = \cot \pi x$

6. $y = \frac{1}{2} \sec \pi x$

7. $y = -\sec x$

8. $y = -2 \csc 2\pi x$

(c)
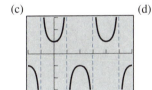

(d)

(e)
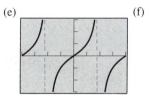

(f)
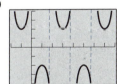

(a)   (b)
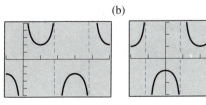

(g)   (h)
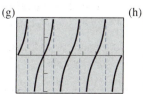
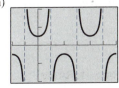

In Exercises 9–30, sketch the graph of the function. (Include two full periods, and use a graphing utility to verify your result.)

9. $y = \frac{1}{3} \tan x$

10. $y = \frac{1}{4} \tan x$

11. $y = \tan 2x$

12. $y = -3 \tan \pi x$

13. $y = -\frac{1}{2} \sec x$

14. $y = \frac{1}{4} \sec x$

15. $y = -\sec \pi x$

16. $y = 2 \sec 4x$

17. $y = \sec \pi x - 1$

18. $y = -2 \sec 4x + 2$

19. $y = \csc \dfrac{x}{2}$

20. $y = \csc \dfrac{x}{3}$

21. $y = \cot \dfrac{x}{2}$

22. $y = 3 \cot \dfrac{\pi x}{2}$

23. $y = \frac{1}{2} \sec 2x$

24. $y = -\frac{1}{2} \tan x$

25. $y = \tan \dfrac{\pi x}{4}$

26. $y = \sec(x + \pi)$

27. $y = \csc(\pi - x)$

28. $y = \sec(\pi - x)$

29. $y = \dfrac{1}{4} \csc\left(x + \dfrac{\pi}{4}\right)$

30. $y = 2 \cot\left(x + \dfrac{\pi}{2}\right)$

In Exercises 31–40, use a graphing utility to graph the function. (Include two full periods.)

31. $y = \tan \dfrac{x}{3}$

32. $y = -\tan 2x$

33. $y = -2 \sec 4x$

34. $y = \sec \pi x$

35. $y = \tan\left(x - \dfrac{\pi}{4}\right)$

36. $y = -\csc(4x - \pi)$

37. $y = \dfrac{1}{4} \cot\left(x - \dfrac{\pi}{2}\right)$

38. $y = \dfrac{1}{3} \sec\left(\dfrac{\pi x}{2} + \dfrac{\pi}{2}\right)$

39. $y = 2 \sec(2x - \pi)$

40. $y = 0.1 \tan\left(\dfrac{\pi x}{4} + \dfrac{\pi}{4}\right)$

In Exercises 41–44, use the graph of the trigonometric function to find all real numbers x in the interval $[-2\pi, 2\pi]$ that give the specified function value.

Function	Function Value
41. $\tan x$	1
42. $\cot x$	$-\sqrt{3}$
43. $\sec x$	-2
44. $\csc x$	$\sqrt{2}$

In Exercises 45 and 46, use the graph of the function to determine whether the function is even, odd, or neither.

45. $f(x) = \sec x$

46. $f(x) = \tan x$

47. Consider the functions $f(x) = 2 \sin x$ and $g(x) = \frac{1}{2} \csc x$ over the interval $(0, \pi)$.

(a) Use a graphing utility to graph f and g in the same viewing rectangle.

(b) Approximate the interval where $f > g$.

(c) Describe the behavior of each of the functions as x approaches π. How is the behavior of g related to the behavior of f as x approaches π?

48. Consider the functions $f(x) = \tan(\pi x/2)$ and $g(x) = \frac{1}{2} \sec(\pi x/2)$ over the interval $(-1, 1)$.

(a) Use a graphing utility to graph f and g in the same viewing rectangle.

(b) Approximate the interval where $f < g$.

In Exercises 49–54, sketch the graph of the function. (Use a graphing utility to verify your result.)

49. $y = 2 - 2 \sin \dfrac{x}{2}$

50. $y = -3 + \cos x$

51. $y = 4 - 2 \cos \pi x$

52. $y = 5 - \frac{1}{2} \sin 2\pi x$

53. $y = 1 + \csc x$

54. $y = 2 + \tan \pi x$

In Exercises 55–60, use a graphing utility to graph the function. Determine the period of the function and approximate any relative minimums and maximums of the function through one period.

55. $y = \sin x + \cos x$

56. $y = \cos x + \cos 2x$

57. $f(x) = 2 \sin x + \sin 2x$

58. $f(x) = 2 \sin x + \cos 2x$

59. $g(x) = \cos x - \cos \dfrac{x}{2}$

60. $g(x) = \sin x - \dfrac{1}{2} \sin \dfrac{x}{2}$

In Exercises 61–64, use a graphing utility to graph the function through three periods.

61. $h(x) = \sin x + \frac{1}{3} \sin 5x$

62. $h(x) = \cos x - \frac{1}{4} \cos 2x$

63. $y = -3 + \cos x + 2 \sin 2x$

64. $y = \sin \pi x + \sin \dfrac{\pi x}{2}$

In Exercises 65–72, use a graphing utility to graph the given function and the algebraic component of the function in the same viewing rectangle. Notice that the graph of the function oscillates about the graph of the algebraic component.

65. $y = x + \sin x$ **66.** $y = x + \cos x$

67. $f(x) = \frac{1}{2}x - 2 \cos x$ **68.** $f(x) = 2x - \sin x$

69. $g(t) = -t + \sin \frac{\pi t}{2}$ **70.** $h(s) = -\frac{s}{2} - \frac{1}{2} \sin \frac{\pi s}{4}$

71. $y = \frac{x^2}{8} + \sin \frac{\pi x}{2}$ **72.** $y = 4 - \frac{x^2}{16} + 4 \cos \pi x$

In Exercises 73–76, use a graphing utility to graph the function and the damping factor of the function in the same viewing rectangle. Describe the behavior of the function as x increases without bound.

73. $f(x) = 2^{-x/4} \cos \pi x$ **74.** $f(t) = e^{-t} \cos t$

75. $g(x) = e^{-x^2/2} \sin x$ **76.** $h(t) = 2^{-t^2/4} \sin t$

In Exercises 77–80, use a graphing utility to graph the function and the equations $y = x$ and $y = -x$ in the same viewing rectangle. Describe the behavior of the given function as x approaches zero.

77. $f(x) = x \cos x$ **78.** $f(x) = |x \sin x|$

79. $g(x) = |x| \sin x$ **80.** $g(x) = |x| \cos x$

In Exercises 81–86, use a graphing utility to graph the function. Describe the behavior of the function as x approaches zero.

81. $y = \frac{6}{x} + \cos x,\quad x > 0$ **82.** $y = \frac{4}{x} + \sin 2x,\quad x > 0$

83. $g(x) = \frac{\sin x}{x}$ **84.** $f(x) = \frac{1 - \cos x}{x}$

85. $f(x) = \sin \frac{1}{x}$ **86.** $h(x) = x \sin \frac{1}{x}$

In Exercises 87–90, use a graphing utility to graph the functions f and g in the same viewing rectangle. Use the graphs to determine the relationship between the functions.

87. $f(x) = \sin x + \cos\left(x + \frac{\pi}{2}\right),\quad g(x) = 0$

88. $f(x) = \sin x - \cos\left(x + \frac{\pi}{2}\right),\quad g(x) = 2 \sin x$

89. $f(x) = \sin^2 x,\quad g(x) = \frac{1}{2}(1 - \cos 2x)$

90. $f(x) = \cos^2 \frac{\pi x}{2},\quad g(x) = \frac{1}{2}(1 + \cos \pi x)$

In Exercises 91–94, use a graphing utility to graph the function over the specified interval.

Function	Interval
91. $f(t) = t^2 \sin t$	$[0, 2\pi]$
92. $f(x) = \sqrt{2x} \sin x$	$[0, 4\pi]$
93. $f(x) = \sin x - \frac{1}{3} \sin 3x + \frac{1}{5} \sin 5x$	$[0, \pi]$
94. $f(x) = \frac{1}{2} - \frac{4}{\pi^2}\left(\cos \pi x + \frac{1}{9} \cos 3\pi x\right)$	$[0, 2]$

95. *Distance* A plane flying at an altitude of 6 miles over level ground will pass directly over a radar antenna (see figure). Let d be the ground distance from the antenna to the point directly under the plane and let x be the angle of elevation to the plane from the antenna. Write d as a function of x, and graph the function over the interval $0 < x < \pi$.

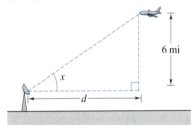

FIGURE FOR 95

96. *Television Coverage* A television camera is on a reviewing platform 100 feet from the street on which a parade will be passing from left to right (see figure). Express the distance d from the camera to a particular unit in the parade as a function of the angle x, and graph the function over the interval $\pi/2 < x < \pi/2$. (Consider x as negative when a unit in the parade approaches from the left.)

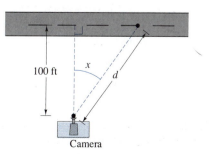

Camera

FIGURE FOR 96

97. *Sales* The projected monthly sales S (in thousands of units) of a seasonal product is modeled by

$$S = 74 + 3t + 40 \sin \frac{\pi t}{6},$$

where t is the time in months, with $t = 1$ corresponding to January. Graph this sales function over 1 year.

98. *Sales* The projected monthly sales S (in thousands of units) of a seasonal product is modeled by

$$S = 25 + 2t + 20 \sin \frac{\pi t}{6},$$

where t is the time in months, with $t = 1$ corresponding to January. Graph this sales function over 1 year.

99. *Predator-Prey Problem* Suppose the population of a certain predator at time t (in months) in a given region is estimated to be

$$P = 10{,}000 + 3000 \sin \frac{2\pi t}{24},$$

and the population of its primary food source (its prey) is estimated to be

$$p = 15{,}000 + 5000 \cos \frac{2\pi t}{24}.$$

Graph both of these functions in the same viewing rectangle, and explain the oscillations in the size of each population.

100. *Normal Temperatures* The normal monthly high temperature for Erie, Pennsylvania, is approximated by

$$H(t) = 54.33 - 20.38 \cos \frac{\pi t}{6} - 15.69 \sin \frac{\pi t}{6},$$

and the normal monthly low temperature is approximated by

$$L(t) = 39.36 - 15.70 \cos \frac{\pi t}{6} - 14.16 \sin \frac{\pi t}{6},$$

where t is the time in months, with $t = 1$ corresponding to January. (*Source:* NOAA) Use a graphing utility to graph the functions over a period of 1 year, and use the graphs to answer the following questions.

(a) During what part of the year is the difference between the normal high and low temperatures greatest? When is it smallest?

(b) The Sun is farthest north in the sky around June 21, but the graph shows the warmest temperatures at a later date. Approximate the lag time of the temperatures relative to the position of the Sun.

101. *Harmonic Motion* An object weighing W pounds is suspended from the ceiling by a steel spring (see figure). The weight is pulled downward (positive direction) from its equilibrium position and released. The resulting motion of the weight is described by the function

$$y = \tfrac{1}{2} e^{-t/4} \cos 4t, \qquad t > 0,$$

where y is the distance in feet and t is the time in seconds. Graph the function.

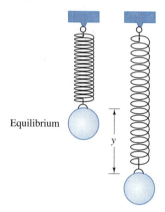

FIGURE FOR 101

5.7   INVERSE TRIGONOMETRIC FUNCTIONS

Inverse Sine Function / Other Inverse Trigonometric Functions /
Compositions of Trigonometric and Inverse Trigonometric Functions

Inverse Sine Function

Recall that for a function to have an inverse, it must be one-to-one. From
Figure 5.68 it is clear that $y = \sin x$ is not one-to-one because different values
of x yield the same y-value. If, however, you restrict the domain to the interval
$-\pi/2 \leq x \leq \pi/2$ (corresponding to the darker portion of the graph in Figure
5.68), the following properties hold.

1. On the interval $[-\pi/2, \pi/2]$, the function $y = \sin x$ is increasing.
2. On the interval $[-\pi/2, \pi/2]$, $y = \sin x$ takes on its full range of values,
   $-1 \leq \sin x \leq 1$.
3. On the interval $[-\pi/2, \pi/2]$, $y = \sin x$ is a one-to-one function.

Thus, on the restricted domain $-\pi/2 \leq x \leq \pi/2$, $y = \sin x$ has a unique
inverse called the **inverse sine function.** It is denoted by

$$y = \arcsin x \quad \text{or} \quad y = \sin^{-1} x.$$

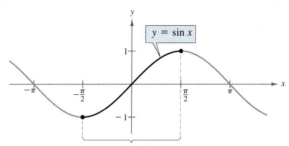

Sin x is one-to-one on this interval.

FIGURE 5.68

The notation $\sin^{-1} x$ is consistent with the inverse function notation $f^{-1}(x)$,
which is used in Section 1.7. The arcsin x notation (read as "the arcsine of x")
comes from the association of a central angle with its subtended *arc length* on
a unit circle. Thus, arcsin x means the angle (or arc) whose sine is x. Both
notations, arcsin x and $\sin^{-1} x$, are commonly used in mathematics, so remember that $\sin^{-1} x$ denotes the *inverse* sine function rather than $1/\sin x$.

The values of arcsin x lie in the interval $-\pi/2 \leq \arcsin x \leq \pi/2$. The
graph of $y = \arcsin x$ is shown in Example 2. Try producing this graph with
a graphing utility. Note that the domain of $y = \arcsin x$ is $-1 \leq x \leq 1$.

REMARK When evaluating the inverse sine function, it helps to remember the phrase "the arcsine of x is the
angle (or number) whose sine is x."

DEFINITION OF INVERSE SINE FUNCTION

The **inverse sine function** is defined by

$$y = \arcsin x \quad \text{if and only if} \quad \sin y = x,$$

where $-1 \le x \le 1$ and $-\pi/2 \le y \le \pi/2$. The domain of $y = \arcsin x$ is $[-1, 1]$, and the range is $[-\pi/2, \pi/2]$.

As with trigonometric functions, much of the work with inverse trigonometric functions can be done by *exact* calculations rather than by calculator approximations. Exact calculations help to increase your understanding of the inverse functions by relating them to the triangle definitions of the trigonometric functions.

EXAMPLE 1   Evaluating the Inverse Sine Function

Find the values of the following (if possible).

REMARK Try verifying the answers
in Example 1 with your calculator.
What happens when you try to calculate $\sin^{-1} 2$?

A. $\arcsin \left(-\dfrac{1}{2} \right)$ **B.** $\sin^{-1} \dfrac{\sqrt{3}}{2}$ **C.** $\sin^{-1} 2$

SOLUTION

A. By definition, $y = \arcsin(-\frac{1}{2})$ implies that

$$\sin y = -\frac{1}{2}, \quad \text{for} \quad -\frac{\pi}{2} \le y \le \frac{\pi}{2}.$$

Because $\sin(-\pi/6) = -\frac{1}{2}$, you can conclude that $y = -\pi/6$ and

$$\arcsin \left(-\frac{1}{2} \right) = -\frac{\pi}{6}.$$

B. By definition, $y = \sin^{-1}\left(\sqrt{3}/2 \right)$ implies that

$$\sin y = \frac{\sqrt{3}}{2}, \quad \text{for} \quad -\frac{\pi}{2} \le y \le \frac{\pi}{2}.$$

Because $\sin(\pi/3) = \sqrt{3}/2$, you can conclude that $y = \pi/3$ and

$$\sin^{-1} \frac{\sqrt{3}}{2} = \frac{\pi}{3}.$$

C. It is not possible to evaluate $y = \sin^{-1} x$ when $x = 2$, because there is no angle whose sine is 2. Remember that the domain of the inverse sine function is $[-1, 1]$.

From Section 1.7, you know that graphs of inverse functions are reflections of each other in the line $y = x$.

EXAMPLE 2   Sketching the Graph of the Arcsine Function

Sketch a graph of $y = \arcsin x$ *by hand.*

SOLUTION

By definition, the equations

$$y = \arcsin x \quad \text{and} \quad \sin y = x$$

are equivalent for $-\pi/2 \le y \le \pi/2$. Hence, their graphs are the same. By assigning values to y in the second equation, you can construct the following table of values.

y	$-\dfrac{\pi}{2}$	$-\dfrac{\pi}{4}$	$-\dfrac{\pi}{6}$	0	$\dfrac{\pi}{6}$	$\dfrac{\pi}{4}$	$\dfrac{\pi}{2}$
$x = \sin y$	-1	$-\dfrac{\sqrt{2}}{2}$	$-\dfrac{1}{2}$	0	$\dfrac{1}{2}$	$\dfrac{\sqrt{2}}{2}$	1

The resulting graph of $y = \arcsin x$ is shown in Figure 5.69. Note that it is the reflection (in the line $y = x$) of the darker part of Figure 5.68. Be sure you see that Figure 5.69 shows the *entire* graph of the inverse sine function. Remember that the range of $y = \arcsin x$ is the closed interval $[-\pi/2, \pi/2]$. You can verify this graph with a graphing utility for the function $y = \arcsin x$.

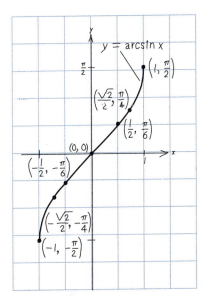

FIGURE 5.69

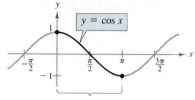

Cos x is one-to-one on this interval.

FIGURE 5.70

Other Inverse Trigonometric Functions

The cosine function is decreasing on the interval $0 \leq x \leq \pi$, as shown in Figure 5.70. Consequently, on this interval the cosine has an inverse function, which is called the **inverse cosine function** and is denoted by

$$y = \arccos x \quad \text{or} \quad y = \cos^{-1} x.$$

Similarly, you can define an **inverse tangent function** by restricting the domain of $y = \tan x$ to the interval $(-\pi/2, \pi/2)$. The following list summarizes the definitions of the three most common inverse trigonometric functions. The remaining three are discussed in the exercise set.

DEFINITION OF THE INVERSE TRIGONOMETRIC FUNCTIONS

Function	Domain	Range
$y = \arcsin x$ if and only if $\sin y = x$	$-1 \leq x \leq 1$	$-\dfrac{\pi}{2} \leq y \leq \dfrac{\pi}{2}$
$y = \arccos x$ if and only if $\cos y = x$	$-1 \leq x \leq 1$	$0 \leq y \leq \pi$
$y = \arctan x$ if and only if $\tan y = x$	$-\infty < x < \infty$	$-\dfrac{\pi}{2} < y < \dfrac{\pi}{2}$

The graphs of these three inverse trigonometric functions are shown in Figure 5.71.

Domain: $[-1, 1]$
Range: $\left[-\dfrac{\pi}{2}, \dfrac{\pi}{2}\right]$

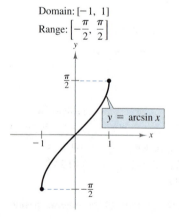

Domain: $[-1, 1]$
Range: $[0, \pi]$

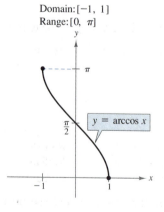

Domain: $(-\infty, \infty)$
Range: $\left(-\dfrac{\pi}{2}, \dfrac{\pi}{2}\right)$

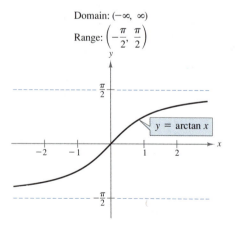

FIGURE 5.71

EXAMPLE 3   **Evaluating Inverse Trigonometric Functions**

Find the exact values of the following.

A. $\arccos \dfrac{\sqrt{2}}{2}$ **B.** $\arccos(-1)$ **C.** $\arctan 0$

SOLUTION

A. Because $\cos(\pi/4) = \sqrt{2}/2$ and $\pi/4$ lies in $[0, \pi]$, it follows that

$$\arccos \frac{\sqrt{2}}{2} = \frac{\pi}{4}.$$

B. Because $\cos \pi = -1$ and π lies in $[0, \pi]$, it follows that

$$\arccos(-1) = \pi.$$

C. Because $\tan 0 = 0$ and 0 lies in $(-\pi/2, \pi/2)$, it follows that

$$\arctan 0 = 0.$$

In Example 3, you were able to find the *exact* values of the given inverse trigonometric functions without a calculator. In the next example, a calculator is necessary to approximate the function values.

REMARK In Example 4, had you set the calculator to the degree mode, the display would have been in degrees rather than radians. This convention is peculiar to calculators. By definition, the values of inverse trigonometric functions are always *in radians*.

Technology Note _____

Most graphing utilities do not have keys for evaluating the inverse cotangent function, inverse secant function, or the inverse cosecant function. Is it still possible to calculate arcsec 3.4? If you let $x = \text{arcsec } 3.4$, then $\sec x = 3.4$ and $\cos x = 1/\sec x = 1/3.4$. Hence, using the inverse cosine function key, $x \approx 1.272$.

EXAMPLE 4   **Using a Calculator to Evaluate Inverse Trigonometric Functions**

Use a calculator to approximate the values (if possible).

A. $\arctan(-8.45)$ **B.** $\arcsin 0.2447$ **C.** $\arccos 2$

SOLUTION

Function	Mode	Display	Rounded to 3 Decimal Places
A. $\arctan(-8.45)$	Radian	-1.453001005	-1.453
B. $\arcsin 0.2447$	Radian	0.2472102741	0.247
C. $\arccos 2$	Radian	ERROR	

Note that the *error* in part (C) occurs because the domain of the inverse cosine function is $[-1, 1]$.

Compositions of Trigonometric and Inverse Trigonometric Functions

Recall from Section 1.7 that inverse functions possess the properties

$$f(f^{-1}(x)) = x \quad \text{and} \quad f^{-1}(f(x)) = x.$$

The inverse trigonometric versions of these properties are as follows.

REMARK Keep in mind that these inverse properties do not apply for arbitrary values of x and y. For instance,

$$\arcsin\left(\sin\frac{3\pi}{2}\right) = \arcsin(-1)$$

$$= -\frac{\pi}{2} \neq \frac{3\pi}{2}.$$

In other words, the property $\arcsin(\sin y) = y$ is not valid for values of y outside the interval $[-\pi/2, \pi/2]$.

INVERSE PROPERTIES

If $-1 \leq x \leq 1$ and $-\pi/2 \leq y \leq \pi/2$, then

$$\sin(\arcsin x) = x \quad \text{and} \quad \arcsin(\sin y) = y.$$

If $-1 \leq x \leq 1$ and $0 \leq y \leq \pi$, then

$$\cos(\arccos x) = x \quad \text{and} \quad \arccos(\cos y) = y.$$

If $-\pi/2 < y < \pi/2$, then

$$\tan(\arctan x) = x \quad \text{and} \quad \arctan(\tan y) = y.$$

DISCOVERY

(a) Use a graphing utility to graph $y = \arcsin(\sin x)$. What are the domain and range of the function? Explain why $\arcsin(\sin 4)$ does not equal 4.

(b) Use a graphing utility to graph $y = \sin(\arcsin x)$. What are the domain and range of the function? Explain why $\sin(\arcsin 4)$ is not defined.

EXAMPLE 5 **Using Inverse Properties**

If possible, find the exact values.

A. $\tan[\arctan(-5)]$ **B.** $\arcsin\left(\sin\frac{5\pi}{3}\right)$ **C.** $\cos(\cos^{-1}\pi)$

SOLUTION

A. Because -5 lies in the domain of arctan x, the inverse property applies, and you have $\tan[\arctan(-5)] = -5$.

B. In this case, $5\pi/3$ does not lie within the range of the arcsine function, $-\pi/2 \leq x \leq \pi/2$. However, $5\pi/3$ is coterminal with

$$\frac{5\pi}{3} - 2\pi = -\frac{\pi}{3},$$

which does lie in the range of the arcsine function, and you have

$$\arcsin\left(\sin\frac{5\pi}{3}\right) = \arcsin\left[\sin\left(-\frac{\pi}{3}\right)\right] = -\frac{\pi}{3}.$$

C. The expression $\cos(\cos^{-1}\pi)$ is not defined, because $\cos^{-1}\pi$ is not defined. Remember that the domain of the inverse cosine function is $[-1, 1]$.

REMARK Try verifying these results with your calculator.

Example 6 shows how to use right triangles to find exact values of functions of inverse functions. Example 7 shows how to use triangles to convert a trigonometric expression into an algebraic one. This conversion technique is used frequently in calculus.

EXAMPLE 6   **Evaluating Functions of Inverse Trigonometric Functions**

Find the exact values of the following.

A. $\tan\left(\arccos\dfrac{2}{3}\right)$ **B.** $\cos\left[\arcsin\left(-\dfrac{3}{5}\right)\right]$

SOLUTION

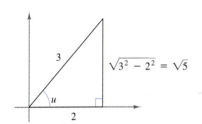

FIGURE 5.72

A. If you let $u = \arccos\frac{2}{3}$, then $\cos u = \frac{2}{3}$. Because $\cos u$ is positive, u is a *first*-quadrant angle. You can sketch and label angle u as shown in Figure 5.72. Consequently,

$$\tan\left(\arccos\frac{2}{3}\right) = \tan u = \frac{\text{opp}}{\text{adj}} = \frac{\sqrt{5}}{2}.$$

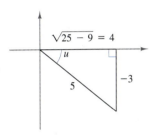

FIGURE 5.73

B. If you let $u = \arcsin(-\frac{3}{5})$, then $\sin u = -\frac{3}{5}$. Because $\sin u$ is negative, u is a *fourth*-quadrant angle. You can sketch and label angle u as shown in Figure 5.73. Consequently,

$$\cos\left[\arcsin\left(-\frac{3}{5}\right)\right] = \cos u = \frac{\text{adj}}{\text{hyp}} = \frac{4}{5}.$$

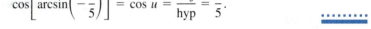

EXAMPLE 7   **Some Problems from Calculus**

Write each of the following as an algebraic expression in x.

A. $\sin(\arccos 3x), \quad 0 \leq x \leq \dfrac{1}{3}$ **B.** $\cot(\arccos 3x), \quad 0 \leq x < \dfrac{1}{3}$

SOLUTION

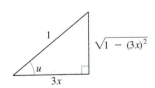

FIGURE 5.74

If you let $u = \arccos 3x$, then $\cos u = 3x$. Because

$$\cos u = \frac{3x}{1} = \frac{\text{adj}}{\text{hyp}},$$

you can sketch a right triangle with acute angle u, as shown in Figure 5.74. From this triangle, you can convert each expression to algebraic form.

A. $\sin(\arccos 3x) = \sin u = \dfrac{\text{opp}}{\text{hyp}} = \sqrt{1 - 9x^2}, \quad 0 \leq x \leq \dfrac{1}{3}$

B. $\cot(\arccos 3x) = \cot u = \dfrac{\text{adj}}{\text{opp}} = \dfrac{3x}{\sqrt{1 - 9x^2}}, \quad 0 \leq x < \dfrac{1}{3}$

REMARK   In Example 7, a similar argument can be made for x-values lying in the interval $[-1/3, 0]$. Why do we restrict $x < \frac{1}{3}$ in part (B)?

DISCUSSION PROBLEM

INVERSE FUNCTIONS

You have studied inverses for several types of functions. Match each of the functions in the left column with its inverse function in the right column.

1. $f(x) = x$
2. $f(x) = x^2, \quad 0 \le x$
3. $f(x) = x^3$
4. $f(x) = e^x$
5. $f(x) = \ln x$
6. $f(x) = \sin x, \quad -\dfrac{\pi}{2} \le x \le \dfrac{\pi}{2}$
7. $f(x) = \cos x, \quad 0 \le x \le \pi$
8. $f(x) = \tan x, \quad -\dfrac{\pi}{2} < x < \dfrac{\pi}{2}$

(a) $f^{-1}(x) = \arcsin x$
(b) $f^{-1}(x) = \ln x$
(c) $f^{-1}(x) = \sqrt{x}$
(d) $f^{-1}(x) = \arctan x$
(e) $f^{-1}(x) = \arccos x$
(f) $f^{-1}(x) = \sqrt[3]{x}$
(g) $f^{-1}(x) = e^x$
(h) $f^{-1}(x) = x$

Provide reasons for your answers.

WARM-UP

The following warm-up exercises involve skills that were covered in earlier sections. You will use these skills in the exercise set for this section.

In Exercises 1–4, evaluate the trigonometric function without using a calculator.

1. $\sin\left(-\dfrac{\pi}{2}\right)$ 2. $\cos \pi$

3. $\tan\left(-\dfrac{\pi}{4}\right)$ 4. $\sin \dfrac{\pi}{4}$

In Exercises 5 and 6, find a real number x in the interval $[-\pi/2, \pi/2]$ that has the same sine value as the given value.

5. $\sin 2\pi$ 6. $\sin \dfrac{5\pi}{6}$

In Exercises 7 and 8, find a real number x in the interval $[0, \pi]$ that has the same cosine value as the given value.

7. $\cos 3\pi$ 8. $\cos\left(-\dfrac{\pi}{4}\right)$

In Exercises 9 and 10, find a real number x in the interval $(-\pi/2, \pi/2)$ that has the same tangent value as the given value.

9. $\tan 4\pi$ 10. $\tan \dfrac{3\pi}{4}$

SECTION 5.7 · EXERCISES

In Exercises 1–16, evaluate the expression without using a calculator.

1. $\arcsin \frac{1}{2}$ **2.** $\arcsin 0$

3. $\arccos \frac{1}{2}$ **4.** $\arccos 0$

5. $\arctan \dfrac{\sqrt{3}}{3}$ **6.** $\arctan(-1)$

7. $\arccos\left(-\dfrac{\sqrt{3}}{2}\right)$ **8.** $\arcsin\left(-\dfrac{\sqrt{2}}{2}\right)$

9. $\arctan\left(-\sqrt{3}\right)$ **10.** $\arctan\left(\sqrt{3}\right)$

11. $\arccos\left(-\dfrac{1}{2}\right)$ **12.** $\arcsin \dfrac{\sqrt{2}}{2}$

13. $\arcsin \dfrac{\sqrt{3}}{2}$ **14.** $\arctan\left(-\dfrac{\sqrt{3}}{3}\right)$

15. $\arctan 0$ **16.** $\arccos 1$

In Exercises 17–28, use a calculator to approximate the given value. (Round your result to two decimal places.)

17. $\arccos 0.28$ **18.** $\arcsin 0.45$

19. $\arcsin(-0.75)$ **20.** $\arccos(-0.8)$

21. $\arctan(-2)$ **22.** $\arctan 15$

23. $\arcsin 0.31$ **24.** $\arccos 0.26$

25. $\arccos(-0.41)$ **26.** $\arcsin(-0.125)$

27. $\arctan 0.92$ **28.** $\arctan 2.8$

In Exercises 29 and 30, use a graphing utility to graph f, g, and $y = x$ in the same viewing rectangle to verify geometrically that g is the inverse of f. (Be sure to properly restrict the domain of f.)

29. $f(x) = \tan x$, $g(x) = \arctan x$

30. $f(x) = \sin x$, $g(x) = \arcsin x$

In Exercises 31–36, use the properties of inverse trigonometric functions to evaluate the expression.

31. $\sin(\arcsin 0.3)$ **32.** $\tan(\arctan 25)$

33. $\cos[\arccos(-0.1)]$ **34.** $\sin[\arcsin(-0.2)]$

35. $\arcsin(\sin 3\pi)$ **36.** $\arccos\left(\cos \dfrac{7\pi}{2}\right)$

In Exercises 37–44, find the exact value of the expression without using a calculator. (*Hint:* Make a sketch of a right triangle, as illustrated in Example 6.)

37. $\sin(\arctan \frac{3}{4})$ **38.** $\sec(\arcsin \frac{4}{5})$

39. $\cos(\arctan 2)$ **40.** $\sin\left(\arccos \dfrac{\sqrt{5}}{5}\right)$

41. $\cos(\arcsin \frac{5}{13})$ **42.** $\csc[\arctan(-\frac{5}{12})]$

43. $\sec[\arctan(-\frac{3}{5})]$ **44.** $\tan[\arcsin(-\frac{3}{4})]$

In Exercises 45 and 46, use a graphing utility to graph f and g in the same viewing rectangle to verify that the two are equal. Identify any asymptotes of the graphs.

45. $f(x) = \sin(\arctan 2x)$, $g(x) = \dfrac{2x}{\sqrt{1 + 4x^2}}$

46. $f(x) = \tan\left(\arccos \dfrac{x}{2}\right)$, $g(x) = \dfrac{\sqrt{4 - x^2}}{x}$

In Exercises 47–56, write an algebraic expression that is equivalent to the given expression. (*Hint*: Sketch a right triangle, as demonstrated in Example 7.)

47. $\cot(\arctan x)$ **48.** $\sin(\arctan x)$

49. $\cos(\arcsin 2x)$ **50.** $\sec(\arctan 3x)$

51. $\sin(\arccos x)$ **52.** $\cot\left(\arctan \dfrac{1}{x}\right)$

53. $\tan\left(\arccos \dfrac{x}{3}\right)$ **54.** $\sec[\arcsin(x - 1)]$

55. $\csc\left(\arccos \dfrac{x}{\sqrt{2}}\right)$ **56.** $\cos\left(\arcsin \dfrac{x - h}{r}\right)$

In Exercises 57–60, fill in the blanks.

57. $\arctan \dfrac{9}{x} = \arcsin\left(\boxed{}\right)$, $x \neq 0$

58. $\arcsin \dfrac{\sqrt{36 - x^2}}{6} = \arccos\left(\boxed{}\right)$, $0 \le x \le 6$

59. $\arccos \dfrac{3}{\sqrt{x^2 - 2x + 10}} = \arcsin\left(\boxed{}\right)$

60. $\arccos \dfrac{x - 2}{2} = \arctan\left(\boxed{}\right)$, $|x - 2| \le 2$

In Exercises 61–64, graph the function.

61. $f(x) = \arcsin(x - 1)$

62. $f(x) = \dfrac{\pi}{2} + \arctan x$

63. $f(x) = \arctan 2x$

64. $f(x) = \arccos \dfrac{x}{4}$

In Exercises 65 and 66, write the given function in terms of the sine function by using the identity

$$A \cos \omega t + B \sin \omega t = \sqrt{A^2 + B^2} \sin\left(\omega t + \arctan \dfrac{B}{A}\right).$$

Verify your result by using a graphing utility to graph both forms of the function.

65. $f(t) = 3 \cos 2t + 3 \sin 2t$

66. $f(t) = 4 \cos \pi t + 3 \sin \pi t$

67. *Photography*   A photographer is taking a picture of a 4-foot-square painting hung in an art gallery. The camera lens is 1 foot below the lower edge of the painting (see figure). The angle β subtended by the camera lens x feet from the painting is given by

$$\beta = \arctan \dfrac{4x}{x^2 + 5}, \qquad x > 0.$$

(a) Use a graphing utility to graph β as a function of x.

(b) Move the cursor along the graph to approximate the distance from the picture when β is maximum.

(c) Identify any asymptote of the graph and discuss its meaning in the context of the problem.

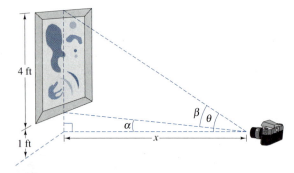

FIGURE FOR 67

68. *Photography*   A television camera at ground level is filming the lift-off of the space shuttle at a point 2000 feet from the launch pad (see figure). If θ is the angle of elevation to the shuttle and s is the height of the shuttle in feet, write θ as a function of s. Find θ when (a) $s = 1000$ and (b) $s = 4000$.

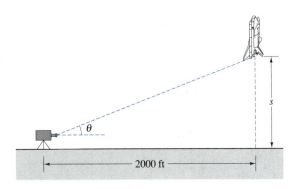

FIGURE FOR 68

69. *Docking a Boat*   A boat is pulled in by means of a winch located on a dock 12 feet above the deck of the boat (see figure). If θ is the angle of elevation from the boat to the winch and s is the length of the rope from the winch to the boat, write θ as a function of s. Find θ when (a) $s = 48$ feet and (b) $s = 24$ feet.

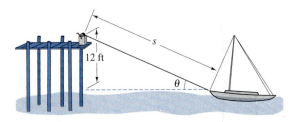

FIGURE FOR 69

70. *Area*   In calculus, it is shown that the area of the region bounded by the graphs of $y = 0$, $y = 1/(x^2 + 1)$, $x = a$, and $x = b$ is given by

Area $= \arctan b - \arctan a$

(see figure). Find the area for the following values of a and b.

(a) $a = 0$, $b = 1$ (b) $a = -1$, $b = 1$

(c) $a = 0$, $b = 3$ (d) $a = -1$, $b = 3$

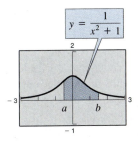

FIGURE FOR 70

71. Define the inverse cotangent function by restricting the domain of the cotangent to the interval $(0, \pi)$, and sketch its graph.

72. Define the inverse secant function by restricting the domain of the secant to the intervals $[0, \pi/2)$ and $(\pi/2, \pi]$, and sketch its graph.

73. Define the inverse cosecant function by restricting the domain of the cosecant to the intervals $[-\pi/2, 0)$ and $(0, \pi/2]$, and sketch its graph.

74. Use the results of Exercises 71–73 to evaluate the following without using a calculator.

(a) $\operatorname{arcsec} \sqrt{2}$ (b) $\operatorname{arcsec} 1$

(c) $\operatorname{arccot} -\sqrt{3}$ (d) $\operatorname{arcsec} 2$

In Exercises 75–79, verify the identity with a calculator, then prove the identity.

75. $\arcsin(-x) = -\arcsin x$

76. $\arctan(-x) = -\arctan x$

77. $\arccos(-x) = \pi - \arccos x$

78. $\arctan x + \arctan \dfrac{1}{x} = \dfrac{\pi}{2}, \quad x > 0$

79. $\arcsin x + \arccos x = \dfrac{\pi}{2}$

80. The Chebyshev polynomial of degree n is defined by the formula $T_n(x) = \cos(n \arccos x)$ for $-1 \le x \le 1$ and $n = 1, 2, 3, \ldots$.

(a) Show that $T_0(x) = 1$.

(b) Show that $T_1(x) = x$.

(c) Find the quadratic polynomial $T_2(x)$.

(d) Graph $T_3(x)$ and $4x^3 - 3x$ on the same viewing rectangle. What do you observe?

5.8   APPLICATIONS OF TRIGONOMETRY
. . . .

Applications Involving Right Triangles   /   **Trigonometry and Bearings**   /
Harmonic Motion

Applications Involving Right Triangles

In keeping with our twofold perspective of trigonometry, this section includes
both right triangle applications and applications that emphasize the periodic
nature of the trigonometric functions.

In this section, the three angles of a right triangle are denoted by the letters
A, B, and C (where C is the right angle), and the lengths of the sides opposite
these angles are denoted by the letters a, b, and c (where c is the hypotenuse).

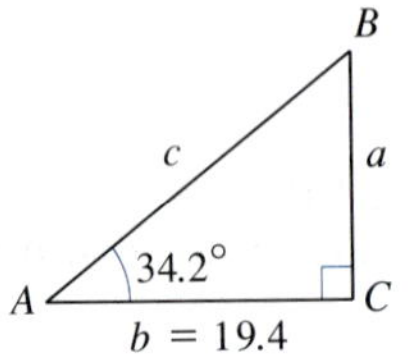

EXAMPLE 1   **Solving a Right Triangle, Given One Acute Angle and**
. . . .   **One Side**

Solve the right triangle having $A = 34.2°$ and $b = 19.4$, as shown in Figure
5.75.

SOLUTION

Because $C = 90°$, it follows that $A + B = 90°$ and

$$B = 90° - 34.2° = 55.8°.$$

To solve for a, use the fact that

$$\tan A = \frac{\text{opp}}{\text{adj}} = \frac{a}{b} \longrightarrow a = b \tan A.$$

Thus,

$$a = 19.4 \tan 34.2° \approx 13.18.$$

Similarly, to solve for c, use the fact that

$$\cos A = \frac{\text{adj}}{\text{hyp}} = \frac{b}{c} \longrightarrow c = \frac{b}{\cos A}.$$

Thus,

$$c = \frac{19.4}{\cos 34.2°} \approx 23.46.$$

FIGURE 5.75

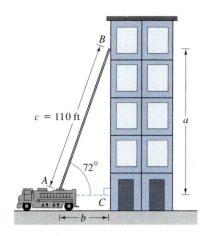

FIGURE 5.76

EXAMPLE 2   Finding a Side of a Right Triangle

A safety regulation states that the maximum angle of elevation for a rescue ladder is 72°. If a fire department's longest ladder is 110 feet, what is the maximum safe rescue height?

SOLUTION

A sketch is shown in Figure 5.76. From the equation

$$\sin A = \frac{a}{c},$$

it follows that

$$a = c \sin A = 110(\sin 72°) \approx 104.6 \text{ feet.}$$

EXAMPLE 3   Finding a Side of a Right Triangle

At point 200 feet from the base of a building, the angle of elevation to the *bottom* of a smokestack is 35°, whereas the angle of elevation to the *top* is 53°, as shown in Figure 5.77. Find the height s of the smokestack alone.

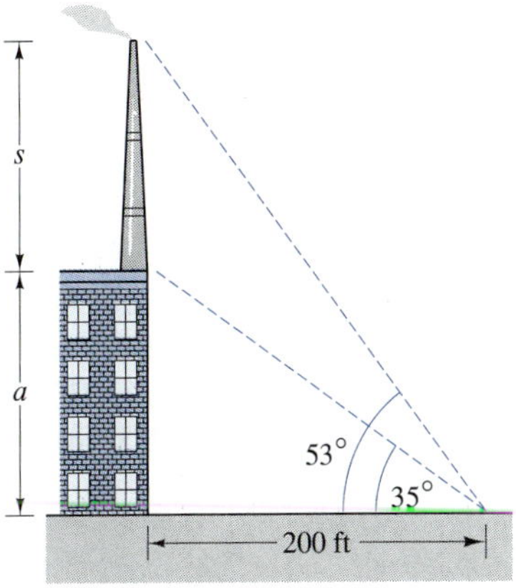

SOLUTION

Note from Figure 5.77 that this problem involves two right triangles. In the smaller right triangle, use the fact that $\tan 35° = a/200$ to conclude that the height of the building is

$$a = 200 \tan 35°.$$

Now, from the larger right triangle, use the equation

$$\tan 53° = \frac{a + s}{200}$$

to conclude that $a + s = 200 \tan 53°$. Hence, the height of the smokestack is

$$s = 200 \tan 53° - a = 200 \tan 53° - 200 \tan 35° \approx 125.4 \text{ feet.}$$

FIGURE 5.77

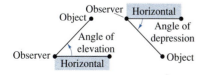

FIGURE 5.78

Examples 2 and 3 used the term **angle of elevation** to represent the angle from the horizontal upward to an object. For objects that lie below the horizontal, it is common to use the term **angle of depression,** as shown in Figure 5.78.

In Examples 1 through 3, you found the lengths of the sides of a right triangle, given an acute angle and the length of one of the sides. You can also find the angles of a right triangle given only the lengths of two sides, as demonstrated in Example 4.

EXAMPLE 4   Finding an Acute Angle of a Right Triangle

A swimming pool is 20 meters long and 12 meters wide. The bottom of the pool is slanted so that the water depth is 1.3 meters at the shallow end and 4 meters at the deep end, as shown in Figure 5.79. Find the angle of depression of the bottom of the pool.

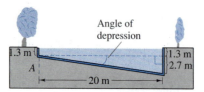

FIGURE 5.79

REMARK Note that the width of the pool, 12 meters, is irrelevant to the problem.

SOLUTION

Using the tangent function, you can see that

$$\tan A = \frac{\text{opp}}{\text{adj}} = \frac{2.7}{20} = 0.135.$$

Thus, the angle of depression is given by

$$A = \arctan 0.135 \approx 0.13419 \text{ radians} \approx 7.69°.$$

Trigonometry and Bearings

In surveying and navigation, directions are generally given in terms of **bearings.** A bearing measures the acute angle a path or line of sight makes with a fixed north-south line. For instance, in Figure 5.80(a), the bearing is S 35° E, meaning 35 *degrees east of south.* Similarly, the bearings in Figure 5.80(b) and (c) are N 80° W and N 45° E, respectively.

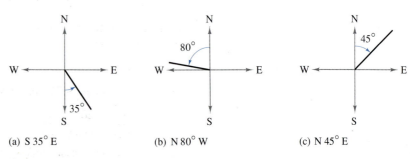

(a) S 35° E   (b) N 80° W   (c) N 45° E

FIGURE 5.80

EXAMPLE 5   Finding Directions in Terms of Bearings

A ship leaves port at noon and heads due west at 20 knots (nautical miles per hour). At 2 P.M., to avoid a storm, the ship changes course to N 54° W, as shown in Figure 5.81. Find the ship's bearing and distance from the port of departure at 3 P.M.

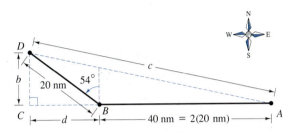

FIGURE 5.81

SOLUTION

Because the ship travels west at 20 knots for two hours, the length of AB is 40. Similarly, BD is 20. In triangle BCD, you have $B = 90° - 54° = 36°$. The two sides of this triangle are determined as follows.

$$\sin B = \frac{b}{20} \qquad \text{and} \quad \cos B = \frac{d}{20}$$

$$b = 20 \sin 36° \qquad\qquad d = 20 \cos 36°$$

Now, in triangle ACD, you can determine angle A as follows.

$$\tan A = \frac{b}{d + 40} = \frac{20 \sin 36°}{20 \cos 36° + 40} \approx 0.2092494$$

$$A \approx \arctan 0.2092494 \approx 0.2062732 \text{ radians} \approx 11.82°$$

The angle with the north-south line is $90° - 11.82° = 78.18°$. Therefore, the bearing of the ship is

    N 78.18° W.   *Bearing*

Finally, from triangle ACD, you have $\sin A = b/c$, which yields

$$c = \frac{b}{\sin A} = \frac{20 \sin 36°}{\sin(11.82)}$$

$$c \approx 57.4 \text{ nautical miles.} \qquad \textit{Distance from port}$$

Harmonic Motion

The periodic nature of the trigonometric functions is useful for describing the motion of a point on an object that vibrates, oscillates, rotates, or is moved by wave motion.

For example, consider a ball that is bobbing up and down on the end of a spring, as shown in Figure 5.82. Suppose that 10 centimeters is the maximum distance the ball moves vertically upward or downward from its equilibrium (at rest) position. Suppose further that the time it takes for the ball to move from its maximum displacement above zero to its maximum displacement below zero and back again is $t = 4$ seconds. Assuming the ideal conditions of perfect elasticity and no friction or air resistance, the ball would continue to move up and down in a uniform and regular manner.

From this spring you can conclude that the period (time for one complete cycle) of the motion is

Period = 4 seconds

and that its amplitude (maximum displacement from equilibrium) is

Amplitude = 10 centimeters.

Motion of this nature can be described by a sine or cosine function, and is called **simple harmonic motion.**

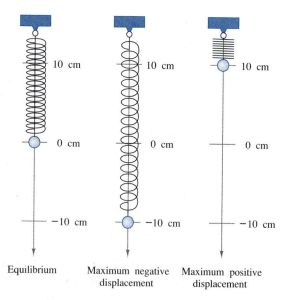

Equilibrium   Maximum negative   Maximum positive
                      displacement   displacement

Simple Harmonic Motion

FIGURE 5.82

DEFINITION OF SIMPLE HARMONIC MOTION

A point that moves on a coordinate line is in **simple harmonic motion** if its distance d from the origin at time t is given by either

$$d = a \sin \omega t \quad \text{or} \quad d = a \cos \omega t,$$

where a and ω are real numbers such that $\omega > 0$. The motion has **amplitude** $|a|$, **period** $2\pi/\omega$, and **frequency** $\omega/2\pi$.

EXAMPLE 6   Simple Harmonic Motion

Write the equation for the simple harmonic motion of the ball described in Figure 5.82, where the period is 4 seconds. What is the frequency of this motion?

SOLUTION

Because the spring is at equilibrium ($d = 0$) when $t = 0$, use the equation

$$d = a \sin \omega t.$$

Moreover, because the maximum displacement from zero is 10 and the period is 4, you have

Amplitude $= |a| = 10$

$$\text{Period} = \frac{2\pi}{\omega} = 4 \quad \longrightarrow \quad \omega = \frac{\pi}{2}.$$

Consequently, the equation of motion is

$$d = 10 \sin \frac{\pi}{2} t.$$

Note that the choice of $a = 10$ or $a = -10$ depends on whether the ball initially moves up or down. The frequency is given by

$$\text{Frequency} = \frac{\omega}{2\pi} = \frac{\pi/2}{2\pi} = \frac{1}{4} \text{ cycle per second.}$$

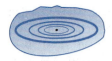

FIGURE 5.83

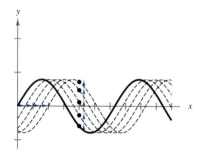

A fishing bob moves in a vertical
direction as waves move to the right.

FIGURE 5.84

One illustration of the relation between sine waves and harmonic motion is seen in the wave motion resulting from dropping a stone into a calm pool of water. The waves move outward in roughly the shape of sine (or cosine) waves, as shown in Figure 5.83. As an example, suppose you are fishing and your fishing bob is attached so that it does not move horizontally. As the waves move outward from the dropped stone, your fishing bob will move up and down in simple harmonic motion, as shown in Figure 5.84.

EXAMPLE 7   **Simple Harmonic Motion**

Given the equation for simple harmonic motion,

$$d = 6 \cos \frac{3\pi}{4} t,$$

find the following.

A. The maximum displacement
B. The frequency
C. The value of d when $t = 4$
D. The least positive value of t for which $d = 0$

SOLUTION

The given equation has the form $d = a \cos \omega t$, with $a = 6$ and $\omega = 3\pi/4$.

A. The maximum displacement (from the point of equilibrium) is given by the amplitude. Thus, the maximum displacement is 6.

B. Frequency $= \dfrac{\omega}{2\pi} = \dfrac{3\pi/4}{2\pi} = \dfrac{3}{8}$ cycle per unit of time

C. $d = 6 \cos\left[\dfrac{3\pi}{4}(4)\right] = 6 \cos 3\pi = 6(-1) = -6$

D. To find the least positive value of t for which $d = 0$, solve the equation

$$d = 6 \cos \frac{3\pi}{4} t = 0$$

to obtain

$$\frac{3\pi}{4} t = \frac{\pi}{2}, \frac{3\pi}{2}, \frac{5\pi}{2}, \ldots$$

$$t = \frac{2}{3}, 2, \frac{10}{3}, \ldots.$$

Thus, the least positive value of t is $t = \frac{2}{3}$.

Technology Note

You can use your graphing utility to solve part D as follows. Graph the function $y_1 = 6 \cos(\frac{3\pi}{4})$ on the viewing rectangle $0 \le x \le 6$, $-7 \le y \le 7$, and observe that the first x-intercept is approximately $x = 0.7$. Using the zoom and trace features, you will see that $x \approx 0.667$.

Many other physical phenomena can be characterized by wave motion. These include electromagnetic waves such as radio waves, television waves, and microwaves. Radio waves transmit sound in two different ways. For an AM station, the *amplitude* of the wave is modified to carry sound (AM stands for **amplitude modulation**). See Figure 5.85(a). An FM radio signal has its *frequency* modified in order to carry sound, hence the term **frequency modulation.** See Figure 5.85(b).

(a) AM: Amplitude modulation

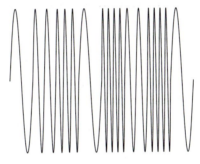

(b) FM: Frequency modulation

FIGURE 5.85

Radio Waves

Suppose you are teaching a class in trigonometry. Write two "right triangle problems" that you think would be reasonable to ask your students to solve. (Assume that your students have 5 minutes to solve each problem.)

WARM-UP

The following warm-up exercises involve skills that were covered in earlier sections. You will use these skills in the exercise set for this section.

In Exercises 1–4, evaluate the expression and round to two decimal places.

1. $20 \sin 25°$

2. $42 \tan 62°$

3. $\arcsin 0.8723$

4. $\arctan 2.8703$

In Exercises 5 and 6, solve for x and round to two decimal places.

5. $\cos 22° = \dfrac{x + 13 \sin 22°}{13 \sin 54°}$

6. $\tan 36° = \dfrac{x + 85 \tan 18°}{85}$

In Exercises 7–10, find the amplitude and period of the function.

7. $f(x) = -4 \sin 2x$

8. $f(x) = \frac{1}{2} \sin \pi x$

9. $g(x) = 3 \cos 3\pi x$

10. $g(x) = 0.2 \cot \dfrac{x}{4}$

SECTION 5.8 · EXERCISES

In Exercises 1–10, solve the right triangle shown in the figure. (Round your result to two decimal places.)

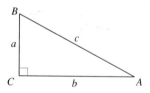

FIGURE FOR 1-10

1. $A = 20°$, $b = 10$

2. $B = 54°$, $c = 15$

3. $B = 71°$, $b = 24$

4. $A = 8.4°$, $a = 40.5$

5. $A = 12° 15'$, $c = 430.5$

6. $B = 65° 12'$, $a = 14.2$

7. $a = 6$, $b = 10$

8. $a = 25$, $c = 35$

9. $b = 16$, $c = 52$

10. $b = 1.32$, $c = 9.45$

In Exercises 11 and 12, find the altitude of the isosceles triangle shown in the figure. (Round your result to two decimal places.)

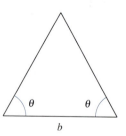

FIGURE FOR 11 AND 12

11. $\theta = 52°$, $b = 4$ inches

12. $\theta = 18°$, $b = 10$ meters

13. *Length of a Shadow*   If the sun is 30° above the horizon, find the length of a shadow cast by a silo that is 70 feet high (see figure).

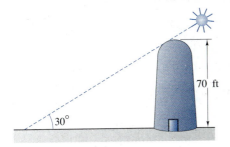

FIGURE FOR 13

14. *Length of a Shadow*   The sun is 20° above the horizon. Find the length of a shadow cast by a building that is 600 feet high (see figure).

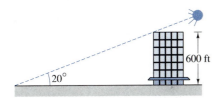

FIGURE FOR 14

15. *Height*   A ladder of length 16 feet leans against the side of a house (see figure). Find the height h of the top of the ladder if the angle of elevation of the ladder is 74°.

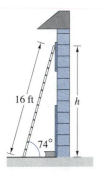

FIGURE FOR 15

16. *Height*   The length of a shadow of a tree is 125 feet when the angle of elevation of the sun is 33° (see figure). Approximate the height h of the tree.

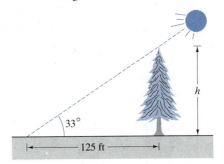

FIGURE FOR 16

17. *Angle of Elevation*   An amateur radio operator erects a 75-foot vertical tower for his antenna (see figure). Find the angle of elevation to the top of the tower at a point on level ground 50 feet from the base.

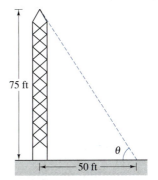

FIGURE FOR 17

18. *Angle of Elevation*   The height of an outdoor basketball backboard is $12\frac{1}{2}$ feet, and the backboard casts a shadow $17\frac{1}{3}$ feet long (see figure). Find the angle of elevation of the sun.

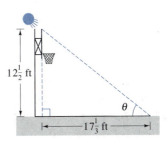

FIGURE FOR 18

19. *Angle of Depression* A spacecraft is traveling in a circular orbit 100 miles above the surface of the earth (see figure). Find the angle of depression from the spacecraft to the horizon. Assume that the radius of the earth is 4000 miles.

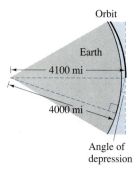

Orbit

Earth

4100 mi

4000 mi

Angle of depression

FIGURE FOR 19

20. *Angle of Depression* Find the angle of depression from the top of a lighthouse 250 feet above water level to the water line of a ship 2 miles offshore.

21. *Airplane Ascent* When an airplane leaves the runway (see figure), its angle of climb is 18° and its speed is 275 feet per second. Find the altitude of the plane after 1 minute.

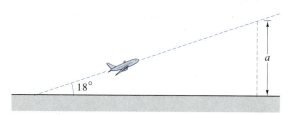

18°

a

FIGURE FOR 21

22. *Mountain Descent* A sign on the roadway at the top of a mountain indicates that for the next 4 miles the grade is 12.5° (see figure). Find the change in elevation for a car descending the mountain.

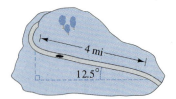

4 mi

12.5°

FIGURE FOR 22

23. *Height* From a point 50 feet in front of a church, the angles of elevation to the base of the steeple and to the top of the steeple are 35° and 47° 40′, respectively (see figure). Find the height of the steeple.

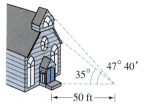

47° 40′

35°

50 ft

FIGURE FOR 23

24. *Height* From a point 100 feet in front of a public library, the angles of elevation to the base of the flagpole and to the top of the pole are 28° and 39° 45′, respectively. The flagpole is mounted on the front of the library's roof (see figure). Find the height of the pole.

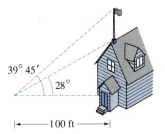

39° 45′

28°

100 ft

FIGURE FOR 24

25. *Navigation* An airplane flying at 550 miles per hour has a bearing of N 52° E. After flying for 1.5 hours, how far north and how far east has the plane traveled from its point of departure?

26. *Navigation* A ship leaves port at noon and has a bearing of S 27° W. If the ship is sailing at 20 knots, how many nautical miles south and how many nautical miles west has the ship traveled by 6:00 P.M.?

27. *Navigation* A ship is 45 miles east and 30 miles south of port. If the captain wants to sail directly to port, what bearing should be taken?

28. *Navigation* A plane is 120 miles north and 85 miles east of an airport. If the pilot wants to fly directly to the airport, what bearing should be taken?

29. *Surveying*   A surveyor wishes to find the distance across a swamp (see figure). The bearing from A to B is N 32° W. The surveyor walks 50 yards from A, and at point C the bearing to B is N 68° W. (a) Find the bearing from A to C. (b) Find the distance from A to B.

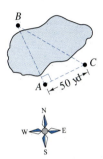

FIGURE FOR 29

30. *Location of a Fire*   Two fire towers are 20 miles apart, tower A being due west of tower B. A fire is spotted from the towers, and the bearings from A and B are N 76° E and N 56° W (see figure). Find the distance d of the fire from the line segment AB. [*Hint:* Use the fact that $d = 20/(\cot 14° + \cot 34°).$]

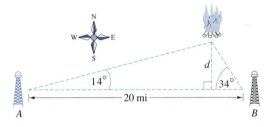

FIGURE FOR 30

31. *Distance Between Ships*   An observer in a lighthouse 300 feet above sea level spots two ships directly offshore. The angles of depression to the ships are 4° and 6.5° (see figure). How far apart are the ships?

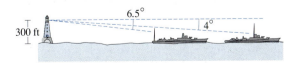

FIGURE FOR 31

32. *Distance Between Towns*   A passenger in an airplane flying at 30,000 feet sees two towns directly to the left of the airplane. The angles of depression to the towns are 28° and 55° (see figure). How far apart are the towns?

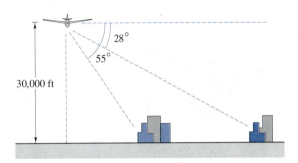

FIGURE FOR 32

33. *Altitude of a Plane*   A plane is observed approaching your home, and you assume it is traveling at 550 miles per hour. If the angle of elevation of the plane is 16° at one time, and 1 minute later the angle is 57°, approximate the altitude of the plane.

34. *Height of a Mountain*   In traveling across flat land, you notice a mountain directly in front of you. The angle of elevation (to the peak) is 3.5°. After you drive 13 miles closer to the mountain, the angle of elevation is 9°. Approximate the height of the mountain.

35. *Length*   A regular pentagon is inscribed in a circle of radius 25 inches. Find the length of the sides of the pentagon.

36. *Length*   A regular hexagon is inscribed in a circle of radius 25 inches. Find the length of the sides of the hexagon.

37. *Wrench Size*   Use the figure to find the distance y across the flat sides of the hexagonal nut as a function of r.

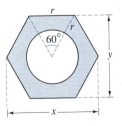

FIGURE FOR 37

38. *Bolt Circle*   The figure shows a circular sheet 25 cm in diameter containing 12 equally spaced bolt holes. Determine the straight-line distance between the centers of adjacent bolt holes.

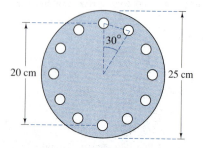

FIGURE FOR 38

Trusses   In Exercises 39 and 40, find the lengths of all the pieces of the truss.

39.

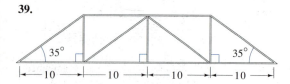

40.

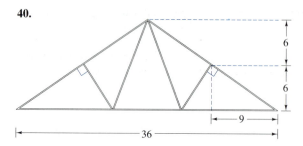

Harmonic Motion   In Exercises 41–44, for the simple harmonic motion described by the given trigonometric function, find (a) the maximum displacement, (b) the frequency, and (c) the least possible value of t for which $d = 0$.

41. $d = 4 \cos 8\pi t$ **42.** $d = \frac{1}{2} \cos 20\pi t$

43. $d = \frac{1}{16} \sin 120\pi t$ **44.** $d = \frac{1}{64} \sin 792\pi t$

45. *Tuning Fork*   A point on the end of a tuning fork moves in simple harmonic motion described by $d = a \sin \omega t$. Find ω given that the tuning fork for middle C has a frequency of 264 vibrations per second.

46. *Wave Motion*   A buoy oscillates in simple harmonic motion as waves go past. At a given time it is noted that the buoy moves a total of 3.5 feet from its low point to its high point, and that it returns to its high point every 10 seconds. Write an equation that describes the motion of the buoy if, at $t = 0$, it is at its high point.

CHAPTER 5 · REVIEW EXERCISES

In Exercises 1–4, sketch the angle in standard position, and list one positive and one negative coterminal angle.

1. $\dfrac{11\pi}{4}$

2. $\dfrac{2\pi}{9}$

3. $-110°$

4. $-405°$

In Exercises 5–8, convert the angle measurement to decimal form. (Round your result to two decimal places.)

5. $135°\ 16'\ 45''$

6. $-234°\ 50''$

7. $5°\ 22'\ 53''$

8. $280°\ 8'\ 50''$

In Exercises 9–12, convert the angle measurement to $D°\ M'\ S''$ form.

9. $135.27°$

10. $25.1°$

11. $-85.15°$

12. $-327.85°$

In Exercises 13–16, convert the angle measurement from radians to degrees. (Round your result to two decimal places.)

13. $\dfrac{5\pi}{7}$

14. $-\dfrac{3\pi}{5}$

15. -3.5

16. 1.75

In Exercises 17–20, convert the angle measurement from degrees to radians. (Round your result to four decimal places.)

17. $480°$

18. $-16.5°$

19. $-33°\ 45'$

20. $84°\ 15'$

In Exercises 21–24, find the reference angle for the given angle.

21. $252°$

22. $640°$

23. $-\dfrac{6\pi}{5}$

24. $\dfrac{17\pi}{3}$

In Exercises 25–30, find the six trigonometric functions of the angle θ (in standard position) whose terminal side passes through the point.

25. $(12, 16)$

26. $(x, 4x)$

27. $(-7, 2)$

28. $(4, -8)$

29. $(-4, -6)$

30. $\left(\tfrac{2}{3}, \tfrac{5}{2}\right)$

In Exercises 31–34, find the remaining five trigonometric functions of θ satisfying the given conditions. (*Hint:* Sketch a right triangle.)

31. $\sec\theta = \tfrac{6}{5}$, $\tan\theta < 0$

32. $\tan\theta = -\tfrac{12}{5}$, $\sin\theta > 0$

33. $\sin\theta = \tfrac{3}{8}$, $\cos\theta < 0$

34. $\cos\theta = -\tfrac{2}{5}$, $\sin\theta > 0$

In Exercises 35–40, evaluate the trigonometric function without using a calculator.

35. $\tan\dfrac{\pi}{3}$

36. $\sec\dfrac{\pi}{4}$

37. $\sin\dfrac{5\pi}{3}$

38. $\cot\left(\dfrac{5\pi}{6}\right)$

39. $\cos 495°$

40. $\csc 270°$

In Exercises 41–44, use a calculator to evaluate the trigonometric function. (Round your result to two decimal places.)

41. $\tan 33°$

42. $\csc 105°$

43. $\sec\dfrac{12\pi}{5}$

44. $\sin\left(-\dfrac{\pi}{9}\right)$

In Exercises 45–48, find two values of θ in degrees $(0° \le \theta < 360°)$ and in radians $(0 \le \theta < 2\pi)$ without using a calculator.

45. $\cos\theta = -\dfrac{\sqrt{2}}{2}$

46. $\sec\theta$ is undefined

47. $\csc\theta = -2$

48. $\tan\theta = \dfrac{\sqrt{3}}{3}$

In Exercises 49–52, find two values of θ in degrees $(0° \le \theta < 360°)$ and in radians $(0 \le \theta < 2\pi)$ by using a calculator.

49. $\sin\theta = 0.8387$

50. $\cot\theta = -1.5399$

51. $\sec\theta = -1.0353$

52. $\csc\theta = 11.4737$

In Exercises 53–56, write an algebraic expression for the given expression.

53. $\sec\left[\arcsin(x-1)\right]$ **54.** $\tan\left(\arccos\dfrac{x}{2}\right)$

55. $\sin\left(\arccos\dfrac{x^2}{4-x^2}\right)$ **56.** $\csc(\arcsin 10x)$

In Exercises 57–70, sketch a graph of the function through two full periods. (Use a graphing utility to verify your result.)

57. $y = 3\cos 2\pi x$ **58.** $y = -2\sin \pi x$

59. $f(x) = 5\sin\dfrac{2x}{5}$ **60.** $f(x) = 8\cos\left(-\dfrac{x}{4}\right)$

61. $f(x) = -\dfrac{1}{4}\cos\dfrac{\pi x}{4}$ **62.** $f(x) = -\tan\dfrac{\pi x}{4}$

63. $g(t) = \dfrac{5}{2}\sin(t-\pi)$ **64.** $g(t) = 3\cos(t+\pi)$

65. $h(t) = \tan\left(t-\dfrac{\pi}{4}\right)$ **66.** $h(t) = \sec\left(t-\dfrac{\pi}{4}\right)$

67. $f(t) = \csc\left(3t-\dfrac{\pi}{2}\right)$ **68.** $f(t) = 3\csc\left(2t+\dfrac{\pi}{4}\right)$

69. $f(\theta) = \cot\dfrac{\pi\theta}{8}$

70. $E(t) = 110\cos\left(120\pi t - \dfrac{\pi}{3}\right)$

In Exercises 71–80, use a graphing utility to graph the function.

71. $f(x) = \dfrac{x}{4} - \sin x$ **72.** $g(x) = 3\left(\sin\dfrac{\pi x}{3} + 1\right)$

73. $y = \dfrac{x}{3} + \cos \pi x$ **74.** $y = 4 - \dfrac{x}{4} + \cos \pi x$

75. $h(\theta) = \theta\sin\pi\theta$ **76.** $f(t) = 2.5e^{-t/4}\sin 2\pi t$

77. $y = \arcsin\dfrac{x}{2}$ **78.** $y = 2\arccos x$

79. $f(x) = \dfrac{\pi}{2} + \arctan x$ **80.** $f(x) = \arccos(x-\pi)$

In Exercises 81–84, use a graphing utility to graph the function. Use the graph to determine if the function is periodic. If the function is periodic, find any relative maximum or minimum points through one period.

81. $f(x) = e^{\sin x}$ **82.** $g(x) = \sin e^x$

83. $g(x) = 2\sin x\cos^2 x$ **84.** $h(x) = 4\sin^2 x\cos^2 x$

85. *Altitude of a Triangle*   Find the altitude of the triangle in the figure.

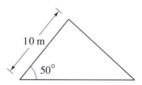

FIGURE FOR 85

86. *Angle of Elevation*   The height of a radio transmission tower is 225 feet, and it casts a shadow of length 105 feet (see figure). Find the angle of elevation of the Sun.

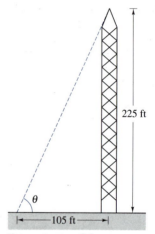

FIGURE FOR 86

87. *Shuttle Height*   An observer 2.5 miles from the launch pad of a space shuttle measures the angle of elevation to the base of the vehicle to be 28° soon after lift-off (see figure). How high is the shuttle at that instant? (Assume the shuttle is still moving vertically.)

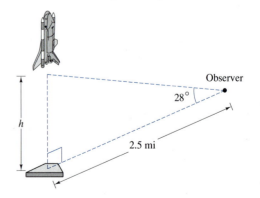

FIGURE FOR 87

88. *Distance*   From city A to city B, a plane flies 650 miles at a bearing of N 48° E. From city B to city C, the plane flies 810 miles at a bearing of S 65° E. Find the distance from A to C and the bearing from A to C.

89. *Railroad Grade*   A train travels 2.5 miles on a straight track with a grade of 1° 10′ (see figure). What is the vertical rise of the train in that distance?

1° 10′

2.5 mi

FIGURE FOR 89

90. *Distance Between Towns*   A passenger in an airplane flying at 35,000 feet sees two towns directly to the left of the airplane. The angles of depression to the towns are 32° and 76° (see figure). How far apart are the towns?

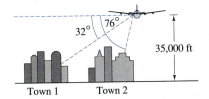

32°   76°

35,000 ft

Town 1   Town 2

FIGURE FOR 90

91. Using calculus, it can be shown that the sine and cosine functions can be approximated by the polynomials

$$\sin x \approx x - \frac{x^3}{3!} + \frac{x^5}{5!} - \frac{x^7}{7!}$$

and

$$\cos x \approx 1 - \frac{x^2}{2!} + \frac{x^4}{4!} - \frac{x^6}{6!},$$

where x is in radians.

(a) Use a graphing utility to graph the sine function and its polynomial approximation in the same viewing rectangle.

(b) Use a graphing utility to graph the cosine function and its polynomial approximation in the same viewing rectangle.

(c) Study the patterns in the polynomial approximations of the sine and cosine functions and guess the next term in each. Then repeat parts (a) and (b). Do you think your guesses were correct? How did the accuracy of the approximations change when additional terms were added?

CUMULATIVE TEST FOR CHAPTERS 3–5

Take this test as you would take a test in class. After you are done, check your work against the answers in the back of the book.

1. Use a graphing utility to graph the quadratic function
$$f(x) = \tfrac{1}{4}(4x^2 - 12x + 17).$$
Find the coordinates of the vertex of the parabola.

2. Find a quadratic function whose graph is a parabola with vertex at $(0, 6)$ and passes through the point $(2, 5)$.

3. Describe the right-hand and left-hand behavior of the polynomial function
$$f(x) = -\tfrac{2}{3}x^3 + 3x^2 - 2x + 1.$$

4. Find a polynomial function with integer coefficients whose zeros are $-4, \tfrac{1}{2}$, and 2.

5. Sketch a graph of the function $f(t) = \tfrac{1}{4}t(t - 2)^2$ without the aid of a graphing utility.

6. Perform the division: $\dfrac{6x^3 - 4x^2}{2x^2 + 1}$

7. Use synthetic division to perform the division: $\dfrac{3x^3 - 5x + 4}{x - 2}$

8. Find the rational zeros of the function
$$f(x) = 6x^3 - 25x^2 - 8x + 48.$$
(*Hint:* Use a graphing utility to eliminate some of the possible rational zeros.)

9. Use a graphing utility to approximate (accurate to one decimal place) the zero of the function $g(t) = t^3 - 5t - 2$ in the interval $[2, 3]$.

10. Sketch a graph of each of the following.
(a) $g(s) = \dfrac{2s}{s - 3}$ (b) $g(s) = \dfrac{2s^2}{s - 3}$

11. Sketch a graph of each of the following.
(a) $f(x) = 6(2^{-x})$ (b) $g(x) = \log_3 x$

12. Evaluate without the aid of a calculator: $\log_5 125$

13. Use the properties of a logarithm to write the expression
$$2 \ln x - \tfrac{1}{2} \ln(x + 5)$$
as the logarithm of a single quantity.

14. Solve each of the following, giving your answers accurate to two decimal places.
(a) $6e^{2x} = 72$ (b) $\log_2 x + \log_2 5 = 6$

15. On the day a grandchild is born, a grandparent deposits $2500 into a fund earning 7.5%, compounded continuously. Determine the balance in the account at the time of the grandchild's 25th birthday.

16. Express the angle $4\pi/9$ in degree measure and sketch the angle in standard position.

17. Express the angle $-120°$ in radian measure as a multiple of π and sketch the angle in standard position.

18. The terminal side of an angle θ in standard position passes through the point $(12, 5)$. Evaluate the six trigonometric functions of the angle.

19. If $\cos t = -\frac{2}{3}$ where $\pi/2 < t < \pi$, find $\sin t$ and $\tan t$.

20. Use a calculator to approximate $\sin(-1.25)$. Round your answer to four decimal places.

21. Use a calculator to approximate two values of $\theta (0° \leq \theta < 360°)$ such that $\sec \theta = 2.125$. Round your answers to two decimal places.

22. Sketch the graph of each of the following functions through two periods.

(a) $y = -3 \sin 2x$ (b) $f(x) = 2 \cos\left(x - \dfrac{\pi}{2}\right)$

(c) $g(x) = \tan\left(\dfrac{\pi x}{2}\right)$ (d) $h(t) = \sec t$

23. Write a sentence describing the relationship between the graphs of the functions $f(x) = \sin x$ and g.

(a) $g(x) = 10 + \sin x$ (b) $g(x) = \sin \dfrac{\pi x}{2}$

(c) $g(x) = \sin\left(x + \dfrac{\pi}{4}\right)$ (d) $g(x) = -\sin x$

24. Find a, b, and c so that the graph of $f(x) = a \sin(bx + c)$ matches the graph in the figure.

25. Consider the function $f(x) = \sin 3x - 2 \cos x$.
(a) Use a graphing utility to graph the function.
(b) Determine the period of the function.
(c) Approximate (accurate to one decimal place) the zero of the function in the interval $[0, 3]$.
(d) Approximate (accurate to one decimal place) the maximum value of the function in the interval $[0, 3]$.

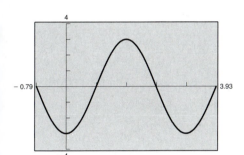

FIGURE FOR 24

26. Consider the function

$$f(t) = 2^{-t/2} \cos\left(\dfrac{\pi t}{2}\right)$$

where t is the time in seconds.
(a) Use a graphing utility to graph the function.
(b) Is the function periodic?
(c) Beyond what time t is the maximum value of the function less than 0.3?

27. Evaluate the expression *without* the aid of a calculator.

(a) $\arcsin(\frac{1}{2})$ (b) $\arctan \sqrt{3}$

28. Write an algebraic expression that is equivalent to $\sin(\arccos 2x)$.

29. From a point on the ground 600 feet from the foot of a cliff, the angle of elevation of the top of the cliff is $32° 30'$. How high is the cliff?

ADDITIONAL TOPICS IN TRIGONOMETRY

6.1 APPLICATIONS OF FUNDAMENTAL IDENTITIES

Introduction / Some Uses of the Fundamental Identities

Introduction

In Chapter 5 you studied the basic definitions, properties, graphs, and applications of the individual trigonometric functions. In this chaper, you will learn how to use the fundamental identities to perform the following.

1. Evaluate trigonometric functions.
2. Simplify trigonometric expressions.
3. Develop additional trigonometric identities.
4. Solve trigonometric equations.

You will also make use of many algebraic skills, such as finding special products, factoring, performing operations with fractional expressions, rationalizing denominators, and solving equations.

As you study this chapter, remember that the best problem solvers are those that can solve problems by a variety of means. For instance, in this chapter many of the problems can be solved graphically, analytically, and numerically (using a table). Even after you have solved a problem one way, we encourage you to check your solution by solving the problem another way.

For convenience, we have summarized the fundamental trigonometric identities on the next page. Because these identities are used so frequently, it would be a good idea to commit them to memory.

FUNDAMENTAL TRIGONOMETRIC IDENTITIES

Reciprocal Identities

$$\sin u = \frac{1}{\csc u} \qquad \sec u = \frac{1}{\cos u} \qquad \tan u = \frac{1}{\cot u}$$

$$\csc u = \frac{1}{\sin u} \qquad \cos u = \frac{1}{\sec u} \qquad \cot u = \frac{1}{\tan u}$$

Quotient Identities

$$\tan u = \frac{\sin u}{\cos u} \qquad \cot u = \frac{\cos u}{\sin u}$$

Pythagorean Identities

$$\sin^2 u + \cos^2 u = 1 \qquad 1 + \tan^2 u = \sec^2 u \qquad 1 + \cot^2 u = \csc^2 u$$

Cofunction Identities

$$\sin\left(\frac{\pi}{2} - u\right) = \cos u \quad \sec\left(\frac{\pi}{2} - u\right) = \csc u \quad \tan\left(\frac{\pi}{2} - u\right) = \cot u$$

$$\cos\left(\frac{\pi}{2} - u\right) = \sin u \quad \csc\left(\frac{\pi}{2} - u\right) = \sec u \quad \cot\left(\frac{\pi}{2} - u\right) = \tan u$$

Even-Odd Identities

$$\sin(-u) = -\sin u \qquad \sec(-u) = \sec u \qquad \tan(-u) = -\tan u$$

$$\csc(-u) = -\csc u \qquad \cos(-u) = \cos u \qquad \cot(-u) = -\cot u$$

REMARK The Pythagorean identities are sometimes used in radical forms such as

$$\sin u = \pm\sqrt{1 - \cos^2 u} \quad \text{or}$$

$$\tan u = \pm\sqrt{\sec^2 u - 1},$$

where the sign depends on the choice of u.

Some Uses of the Fundamental Identities

One common use of trigonometric identities is to use given values of one or more trigonometric functions to find the values of the other trigonometric functions.

EXAMPLE 1   Using Identities to Evaluate a Function

Use the conditions $\sec u = -\frac{3}{2}$ and $\tan u > 0$ to find the values of all six trigonometric functions.

SOLUTION

Using a reciprocal identity, you can write

$$\cos u = \frac{1}{\sec u} = \frac{1}{-\dfrac{3}{2}} = -\frac{2}{3}$$

By a Pythagorean identity, you obtain

$$\sin^2 u = 1 - \cos^2 u = 1 - \left(-\frac{2}{3}\right)^2 = 1 - \frac{4}{9} = \frac{5}{9}.$$

Because $\sec u < 0$ and $\tan u > 0$, it follows that u lies in Quadrant III. Moreover, because $\sin u$ is negative when u is in Quadrant III, choose the negative root to obtain

$$\sin u = -\frac{\sqrt{5}}{3}.$$

Now, knowing the values of the sine and cosine, you can find the values of all six trigonometric functions.

$$\sin u = -\frac{\sqrt{5}}{3} \qquad\qquad \csc u = \frac{1}{\sin u} = -\frac{3}{\sqrt{5}}$$

$$\cos u = -\frac{2}{3} \qquad\qquad \sec u = -\frac{3}{2}$$

$$\tan u = \frac{\sin u}{\cos u} = \frac{-\dfrac{\sqrt{5}}{3}}{-\dfrac{2}{3}} = \frac{\sqrt{5}}{2}$$

Compare this approach with the triangle approach in Example 2, Section 5.3.

The next four examples use algebra and the fundamental identities to simplify, factor, and/or combine trigonometric *expressions* such as

$$\cot x - \cos x \sin x, \quad \frac{1 - \sin x}{\cos^2 x}, \quad \text{and} \quad \frac{\tan x}{\sec x - 1}.$$

EXAMPLE 2   Simplifying a Trigonometric Expression

Simplify the expression $\sin x \cos^2 x - \sin x$.

SOLUTION

First factor out a common monomial factor and then use a fundamental identity.

$$\sin x \cos^2 x - \sin x = \sin x(\cos^2 x - 1) \qquad \textit{Monomial factor}$$
$$= -\sin x(1 - \cos^2 x)$$
$$= -\sin x(\sin^2 x) \qquad \textit{Pythagorean Identity}$$
$$= -\sin^3 x \qquad \textit{Multiply}$$

You can graphically confirm this result with a graphing utility. Graph $y = \sin x \cos^2 x - \sin x$ and $y = -\sin^3 x$ on the same display screen, and notice that the two graphs coincide, as shown in Figure 6.1.

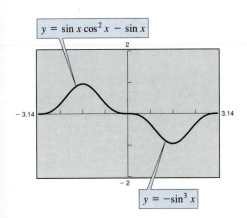

$y = \sin x \cos^2 x - \sin x$

$y = -\sin^3 x$

FIGURE 6.1

EXAMPLE 3   **Factoring Trigonometric Expressions**

Factor the following.

A. $\sec^2 \theta - 1$ **B.** $4 \tan^2 \theta + \tan \theta - 3$

SOLUTION

A. Using the difference of two squares pattern, you can factor the expression as

$$\sec^2 \theta - 1 = (\sec \theta - 1)(\sec \theta + 1).$$

B. This expression has the polynomial form, $ax^2 + bx + c$, where $x = \tan \theta$, and it factors as

$$4 \tan^2 \theta + \tan \theta - 3 = (4 \tan \theta - 3)(\tan \theta + 1).$$

On occasion, factoring or simplifying can best be done by first rewriting the expression in terms of just *one* trigonometric function or in terms of *sine and cosine alone.*

EXAMPLE 4   **Factoring a Trigonometric Expression**

Factor the expression $\csc^2 x - \cot x - 3$.

SOLUTION

As given, this expression cannot be factored. If, however, you use the identity $\csc^2 x = 1 + \cot^2 x$ to rewrite the expression in terms of the cotangent alone, you can factor to obtain

$$
\begin{aligned}
\csc^2 x - \cot x - 3 &= (1 + \cot^2 x) - \cot x - 3 && \textit{Pythagorean Identity} \\
&= \cot^2 x - \cot x - 2 && \textit{Combine terms} \\
&= (\cot x - 2)(\cot x + 1). && \textit{Factor}
\end{aligned}
$$

EXAMPLE 5   **Combining Fractional Expressions**

Perform the indicated addition and simplify the result.

$$\frac{\sin \theta}{1 + \cos \theta} + \frac{\cos \theta}{\sin \theta}$$

SOLUTION

$$\frac{\sin \theta}{1 + \cos \theta} + \frac{\cos \theta}{\sin \theta} = \frac{(\sin \theta)(\sin \theta) + (\cos \theta)(1 + \cos \theta)}{(1 + \cos \theta)(\sin \theta)}$$

$$= \frac{\sin^2 \theta + \cos^2 \theta + \cos \theta}{(1 + \cos \theta)\sin \theta} \qquad \textit{Multiply}$$

$$= \frac{1 + \cos \theta}{(1 + \cos \theta)\sin \theta} \qquad \textit{Pythagorean Identity}$$

$$= \frac{1}{\sin \theta} \qquad \textit{Cancel common factor}$$

$$= \csc \theta \qquad \textit{Reciprocal Identity}$$

EXAMPLE 6   **Verifying Trigonometric Identities Graphically**

Use a graphing utility to determine which of the following is an identity.

A. $\cos 3x \overset{?}{=} 4 \cos^3 x - 3 \cos x$ **B.** $\cos 3x \overset{?}{=} \sin\left(3x - \frac{\pi}{2}\right)$

SOLUTION

A. Using a graphing utility, you can see that the graphs of $y = \cos 3x$ and $y = 4 \cos^3 x - 3 \cos x$ appear to coincide, as shown in Figure 6.2(a). Therefore, this appears to be an identity.

B. From the graphs shown in Figure 6.2(b), you can see that this is not an identity.

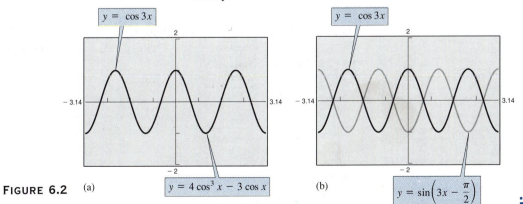

FIGURE 6.2   (a)   (b)

The last two examples of this section involve techniques for rewriting expressions into forms that are useful in calculus.

EXAMPLE 7   **Rewriting a Trigonometric Expression**

Rewrite

$$\frac{1}{1 + \sin x}$$

so that it is *not* in fractional form.

SOLUTION

$$\frac{1}{1 + \sin x} = \frac{1}{1 + \sin x} \cdot \frac{1 - \sin x}{1 - \sin x}$$ *Multiply numerator and denominator by* $(1 - \sin x)$

$$= \frac{1 - \sin x}{1 - \sin^2 x}$$ *Multiply*

$$= \frac{1 - \sin x}{\cos^2 x}$$ *Pythagorean Identity*

$$= \frac{1}{\cos^2 x} - \frac{\sin x}{\cos^2 x}$$ *Separate fractions*

$$= \frac{1}{\cos^2 x} - \frac{\sin x}{\cos x} \cdot \frac{1}{\cos x}$$

$$= \sec^2 x - \tan x \sec x$$ *Identities*

EXAMPLE 8   **Trigonometric Substitution**

Use the substitution $x = 2 \tan \theta$ to express $\sqrt{4 + x^2}$ as a trigonometric function of θ where $0 < \theta < \pi/2$.

SOLUTION

Letting $x = 2 \tan \theta$ produces the following.

$$\sqrt{4 + x^2} = \sqrt{4 + (2 \tan \theta)^2}$$

$$= \sqrt{4(1 + \tan^2 \theta)}$$

$$= \sqrt{4 \sec^2 \theta}$$ *Pythagorean Identity*

$$= 2 \sec \theta$$ $\sec \theta > 0$ *for* $0 < \theta < \dfrac{\pi}{2}$

Figure 6.3 shows a right triangle illustration of the trigonometric substitution in Example 8. For $0 < \theta < \pi/2$, you obtain opp $= x$, adj $= 2$, and hyp $= \sqrt{4 + x^2}$. Thus, you can write

$$\sec \theta = \frac{\sqrt{4 + x^2}}{2} \quad \text{or} \quad \sqrt{4 + x^2} = 2 \sec \theta.$$

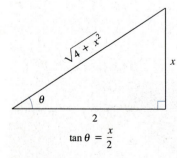

$\tan \theta = \dfrac{x}{2}$

FIGURE 6.3

DISCUSSION PROBLEM

....

REMEMBERING TRIGONOMETRIC IDENTITIES

Most people find the Pythagorean Identity involving sine and cosine to be fairly easy to remember: $\sin^2 u + \cos^2 u = 1$. The one involving tangent and secant, however, tends to give some people trouble. They can't remember if the identity is

$$1 + \tan^2 u = \sec^2 u$$

or

$$1 + \sec^2 u = \tan^2 u.$$

Which of these two is the correct Pythagorean Identity involving tangent and secant? Write a short paragraph describing how a person can remember (or derive) this identity.

WARM-UP

....

The following warm-up exercises involve skills that were covered in earlier sections. You will use these skills in the exercise set for this section.

In Exercises 1 and 2, use a right triangle to evaluate the other five trigonometric functions of the acute angle θ.

1. $\tan \theta = \frac{3}{2}$ **2.** $\sec \theta = 3$

In Exercises 3 and 4, determine the exact value of the six trigonometric functions of θ. Assume the given point is on the terminal side of an angle θ in standard position.

3. $(7, -3)$ **4.** $(-10, 5)$

In Exercises 5–8, simplify the expression.

5. $\sqrt{1 - \left(\dfrac{\sqrt{3}}{2}\right)^2}$ **6.** $\sqrt{\left(\dfrac{3}{4}\right)^2 + 1}$

7. $\sqrt{1 + \left(\dfrac{3}{8}\right)^2}$ **8.** $\sqrt{1 - \left(\dfrac{\sqrt{5}}{3}\right)^2}$

In Exercises 9 and 10, perform the indicated operations and simplify the result.

9. $\dfrac{4}{1 + x} + \dfrac{x}{4}$ **10.** $\dfrac{3}{1 - x} - \dfrac{5}{1 + x}$

SECTION 6.1 · EXERCISES

In Exercises 1–10, use the fundamental identities to evaluate (if possible) the other trigonometric functions.

1. $\sin x = \dfrac{1}{2}, \quad \cos x = \dfrac{\sqrt{3}}{2}$

2. $\tan x = \dfrac{\sqrt{3}}{3}, \quad \cos x = -\dfrac{\sqrt{3}}{2}$

3. $\sec \theta = \sqrt{2}, \quad \sin \theta = -\dfrac{\sqrt{2}}{2}$

4. $\csc \theta = \dfrac{5}{3}, \quad \tan \theta = \dfrac{3}{4}$

5. $\sin(-x) = -\dfrac{2}{3}, \quad \tan x = -\dfrac{2\sqrt{5}}{5}$

6. $\cos\left(\dfrac{\pi}{2} - x\right) = \dfrac{3}{5}, \quad \cos x = \dfrac{4}{5}$

7. $\tan \theta = 2, \quad \sin \theta < 0$

8. $\sec \theta = -3, \quad \tan \theta < 0$

9. $\sin \theta = -1, \quad \cot \theta = 0$

10. $\tan \theta$ is undefined, $\quad \sin \theta > 0$

In Exercises 11–16, match the trigonometric expression with one of the following.

(a) -1 (b) $\cos x$ (c) $\cot x$
(d) 1 (e) $-\tan x$ (f) $\sin x$

11. $\sec x \cos x$

12. $\dfrac{\sin(-x)}{\cos(-x)}$

13. $\tan^2 x - \sec^2 x$

14. $\dfrac{1 - \cos^2 x}{\sin x}$

15. $\cot x \sin x$

16. $\dfrac{\sin\left(\dfrac{\pi}{2} - x\right)}{\cos\left(\dfrac{\pi}{2} - x\right)}$

In Exercises 17–22, match the trigonometric expression with one of the following.

(a) $\csc x$ (b) $\tan x$ (c) $\sin^2 x$
(d) $\sin x \tan x$ (e) $\sec^2 x$ (f) $\sec^2 x + \tan^2 x$

17. $\sin x \sec x$

18. $\cos^2 x(\sec^2 x - 1)$

19. $\dfrac{\sec^2 x - 1}{\sin^2 x}$

20. $\cot x \sec x$

21. $\sec^4 x - \tan^4 x$

22. $\dfrac{\cos^2\left(\dfrac{\pi}{2} - x\right)}{\cos x}$

In Exercises 23–36, use fundamental identities to simplify the expression. Use a graphing utility to verify your result. (See Example 2.)

23. $\tan \phi \csc \phi$

24. $\sin \phi(\csc \phi - \sin \phi)$

25. $\cos \beta \tan \beta$

26. $\sec \alpha \dfrac{\sin \alpha}{\tan \alpha}$

27. $\dfrac{\cot x}{\csc x}$

28. $\dfrac{\csc \theta}{\sec \theta}$

29. $\sec^2 x(1 - \sin^2 x)$

30. $\dfrac{1}{\tan^2 x + 1}$

31. $\dfrac{\sin(-x)}{\cos x}$

32. $\dfrac{\tan^2 \theta}{\sec^2 \theta}$

33. $\cos\left(\dfrac{\pi}{2} - x\right)\sec x$

34. $\cot\left(\dfrac{\pi}{2} - x\right)\cos x$

35. $\dfrac{\cos^2 y}{1 - \sin y}$

36. $\cos t(1 + \tan^2 t)$

In Exercises 37–44, factor the expression and use fundamental identities to simplify. Use a graphing utility to verify your result.

37. $\tan^2 x - \tan^2 x \sin^2 x$

38. $\sec^2 x \tan^2 x + \sec^2 x$

39. $\sin^2 x \sec^2 x - \sin^2 x$

40. $\dfrac{\sec^2 x - 1}{\sec x - 1}$

41. $\tan^4 x + 2 \tan^2 x + 1$

42. $1 - 2 \cos^2 x + \cos^4 x$

43. $\sin^4 x - \cos^4 x$

44. $\csc^3 x - \csc^2 x - \csc x + 1$

In Exercises 45–48, perform the multiplication and use fundamental identities to simplify.

45. $(\sin x + \cos x)^2$

46. $(\cot x + \csc x)(\cot x - \csc x)$

47. $(\sec x + 1)(\sec x - 1)$

48. $(3 - 3 \sin x)(3 + 3 \sin x)$

In Exercises 49–52, perform the addition or subtraction and use fundamental identities to simplify.

49. $\dfrac{1}{1 + \cos x} + \dfrac{1}{1 - \cos x}$

50. $\dfrac{1}{\sec x + 1} - \dfrac{1}{\sec x - 1}$

51. $\dfrac{\cos x}{1 + \sin x} + \dfrac{1 + \sin x}{\cos x}$

52. $\tan x - \dfrac{\sec^2 x}{\tan x}$

In Exercises 53–56, rewrite the expression so that it is *not* in fractional form.

53. $\dfrac{\sin^2 y}{1 - \cos y}$

54. $\dfrac{5}{\tan x + \sec x}$

55. $\dfrac{3}{\sec x - \tan x}$

56. $\dfrac{\tan^2 x}{\csc x + 1}$

In Exercises 57–60, use a graphing utility to determine whether or not the equation is an identity. (See Example 6.)

57. $\csc x = \cot \dfrac{x}{2} - \cot x$

58. $\sec^2 x \csc x = \sec^2 x + \csc x$

59. $\sin 2x = 2 \sin x$

60. $\sin x \tan x = \csc x - \cos x$

61. Use a graphing utility to determine the values of x in the interval $[0, 2\pi]$ for which (a) $\cos x = \sqrt{1 - \sin^2 x}$, and (b) $\cos x = -\sqrt{1 - \sin^2 x}$.

62. Use a graphing utility to determine which of the six trigonometric functions is equal to the expression

$$\sin x \tan x + \cos x.$$

In Exercises 63–72, use the specified trigonometric substitution to write the algebraic expression as a trigonometric function of θ, where $0 < \theta < \pi/2$.

63. $\sqrt{25 - x^2}, \quad x = 5 \sin \theta$

64. $\sqrt{16 - 4x^2}, \quad x = 2 \sin \theta$

65. $\sqrt{x^2 - 9}, \quad x = 3 \sec \theta$

66. $\sqrt{x^2 - 4}, \quad x = 2 \sec \theta$

67. $\sqrt{x^2 + 25}, \quad x = 5 \tan \theta$

68. $\sqrt{x^2 + 100}, \quad x = 10 \tan \theta$

69. $\sqrt{1 - (x - 1)^2}, \quad x - 1 = \sin \theta$

70. $\sqrt{1 - e^{2x}}, \quad e^x = \sin \theta$

71. $\sqrt{(9 + x^2)^3}, \quad x = 3 \tan \theta$

72. $\sqrt{(x^2 - 16)^3}, \quad x = 4 \sec \theta$

In Exercises 73 and 74, determine the values of $\theta, 0 \le \theta < 2\pi$, for which the equation is true.

73. $\sec \theta = \sqrt{1 + \tan^2 \theta}$

74. $\sin \theta = -\sqrt{1 - \cos^2 \theta}$

In Exercises 75 and 76, rewrite the expression as a single logarithm and simplify the result.

75. $\ln |\cos \theta| - \ln |\sin \theta|$

76. $\ln |\cot t| + \ln(1 + \tan^2 t)$

In Exercises 77–80, determine whether or not the equation is an identity, and give a reason for your answer.

77. $\dfrac{\sin k\theta}{\cos k\theta} = \tan \theta, \quad k$ is constant

78. $\dfrac{1}{5 \cos \theta} = 5 \sec \theta$

79. $\sin \theta \csc \theta = 1$

80. $\sin \theta \csc \phi = 1$

In Exercises 81–84, use a calculator to demonstrate the identity for the given values of θ.

81. $\csc^2 \theta - \cot^2 \theta = 1$

(a) $\theta = 132°$ (b) $\theta = \dfrac{2\pi}{7}$

82. $\tan^2 \theta + 1 = \sec^2 \theta$

(a) $\theta = 346°$ (b) $\theta = 3.1$

83. $\cos\left(\dfrac{\pi}{2} - \theta\right) = \sin \theta$

(a) $\theta = 80°$ (b) $\theta = 0.8$

84. $\sin(-\theta) = -\sin \theta$

(a) $\theta = 250°$ (b) $\theta = \frac{1}{2}$

85. Express each of the other trigonometric functions of θ in terms of $\sin \theta$.

86. Express each of the other trigonometric functions of θ in terms of $\cos \theta$.

6.2 VERIFYING TRIGONOMETRIC IDENTITIES
Introduction / Verifying Trigonometric Identities

Introduction

In the previous section, you learned how to rewrite trigonometric expressions in equivalent forms. In this section, you will learn to prove or verify trigonometric identities. (In the next section, you will learn to solve trigonometric equations.) The key to verifying identities and solving equations is the ability to use the fundamental identities and the rules of algebra to rewrite trigonometric expressions.

Before going on, let's review some distinctions among expressions, equations, and identities. An *expression* has no equal sign. It is merely a combination of terms. When simplifying expressions, you use an equal sign only to indicate the equivalence of the original expression and the new form. An *equation* is a statement containing an equal sign that is true for a specific set of values. In this sense, it is really a *conditional* equation. For example, the equation

$$\sin x = -1$$

is true only for $x = (3\pi/2) \pm 2n\pi$. Hence, it is a conditional equation. On the other hand, an equation that is true for all real values in the domain of the variable is an *identity*. For example, the familiar equation

$$\sin^2 x = 1 - \cos^2 x$$

is true for all real numbers x. Hence, it is an identity.

Although there are similarities, proving that a trigonometric equation is an identity is quite different from solving an equation. There is no well-defined set of rules to follow in verifying trigonometric identities, and the process is best learned by practice. However, the following guidelines should be helpful.

GUIDELINES FOR VERIFYING TRIGONOMETRIC IDENTITIES

1. Work with one side of the equation at a time. It is often better to work with the more complicated side first.
2. Look for opportunities to factor an expression, add fractions, square a binomial, or create a monomial denominator.
3. Look for opportunities to use the fundamental identities. Note which functions are in the final expression you want. Sines and cosines pair up well, as do secants and tangents, and cosecants and cotangents.
4. If the preceding guidelines do not help, try converting all terms to sines and cosines.
5. Do not just sit and stare at the problem. Try something! Even paths that lead to dead ends can give you insights.

Verifying Trigonometric Identities

EXAMPLE 1   Verifying a Trigonometric Identity

Verify the identity

$$\frac{\sec^2 \theta - 1}{\sec^2 \theta} = \sin^2 \theta.$$

SOLUTION

Start with the left side, because it is more complicated.

$$\frac{\sec^2 \theta - 1}{\sec^2 \theta} = \frac{(\tan^2 \theta + 1) - 1}{\sec^2 \theta} \qquad \textit{Pythagorean Identity}$$

$$= \frac{\tan^2 \theta}{\sec^2 \theta} \qquad \textit{Simplify}$$

$$= \tan^2 \theta(\cos^2 \theta) \qquad \textit{Reciprocal Identity}$$

$$= \frac{\sin^2 \theta}{\cos^2 \theta}(\cos^2 \theta) \qquad \textit{Tangent Identity}$$

$$= \sin^2 \theta \qquad \textit{Reduce}$$

ALTERNATIVE SOLUTION

Sometimes it is helpful to separate a fraction into two parts.

$$\frac{\sec^2 \theta - 1}{\sec^2 \theta} = \frac{\sec^2 \theta}{\sec^2 \theta} - \frac{1}{\sec^2 \theta} \qquad \textit{Separate fractions}$$

$$= 1 - \cos^2 \theta \qquad \textit{Reciprocal Identity}$$

$$= \sin^2 \theta \qquad \textit{Pythagorean Identity}$$

Technology Note

A graphing utility can be used to confirm trigonometric identities graphically. For instance, in Example 1 you can graph

$$y = \frac{\sec^2 \theta - 1}{\sec^2 \theta} \text{ and } y = \sin^2 \theta$$

on the same viewing rectangle and observe that the graphs are the same. You can also use this technique to confirm intermediate steps in your verification process.

As you can see in Example 1, there can be more than one way to verify an identity. Your method may differ from that used by your instructor or fellow students. Here is a good chance to be creative and establish your own style, but try to be as efficient as possible.

EXAMPLE 2   Combining Fractions Before Using Identities

Verify the identity

$$\frac{1}{1 - \sin \alpha} + \frac{1}{1 + \sin \alpha} = 2 \sec^2 \alpha.$$

SOLUTION

$$\frac{1}{1 - \sin\alpha} + \frac{1}{1 + \sin\alpha} = \frac{1 + \sin\alpha + 1 - \sin\alpha}{(1 - \sin\alpha)(1 + \sin\alpha)} \qquad \textit{Add fractions}$$

$$= \frac{2}{1 - \sin^2\alpha} \qquad \textit{Simplify}$$

$$= \frac{2}{\cos^2\alpha} \qquad \textit{Pythagorean Identity}$$

$$= 2\sec^2\alpha \qquad \textit{Reciprocal Identity}$$

Technology Note

One way to verify the result of Example 3 is to graph

$$y_1 = ((\tan x)^2 + 1)((\cos x)^2 - 1)$$

$$y_2 = -(\tan x)^2$$

on the "trig" viewing rectangle, $-2\pi \le x \le 2\pi$, $-3 \le y \le 3$. Now graph $y_3 = y_1 - y_2 + 1$. What do you observe? Why is this better than simply graphing $y_3 = y_1 - y_2$?

EXAMPLE 3   **Verifying a Trigonometric Identity**

Verify the identity $(\tan^2 x + 1)(\cos^2 x - 1) = -\tan^2 x$.

SOLUTION

By applying identities before multiplying, you obtain the following.

$$(\tan^2 x + 1)(\cos^2 x - 1) = (\sec^2 x)(-\sin^2 x) \qquad \textit{Pythagorean identities}$$

$$= -\frac{\sin^2 x}{\cos^2 x} \qquad \textit{Reciprocal Identity}$$

$$= -\left(\frac{\sin x}{\cos x}\right)^2 \qquad \textit{Rule of exponents}$$

$$= -\tan^2 x \qquad \textit{Tangent Identity}$$

EXAMPLE 4   **Converting to Sines and Cosines**

Verify the identity $\tan x + \cot x = \sec x \csc x$.

SOLUTION

In this case, there appear to be no fractions to add, no products to find, and no opportunity to use one of the Pythagorean identities. Notice, however, what happens when you convert the left side to sines and cosines.

$$\tan x + \cot x = \frac{\sin x}{\cos x} + \frac{\cos x}{\sin x} \qquad \textit{Identities}$$

$$= \frac{\sin^2 x + \cos^2 x}{\cos x \sin x} \qquad \textit{Add fractions}$$

$$= \frac{1}{\cos x \sin x} \qquad \sin^2 x + \cos^2 x = 1$$

$$= \frac{1}{\cos x} \cdot \frac{1}{\sin x} \qquad \textit{Product of fractions}$$

$$= \sec x \csc x \qquad \textit{Reciprocal identities}$$

EXAMPLE 5   **Verifying a Trigonometric Identity**

Verify the identity $\sec y + \tan y = \dfrac{\cos y}{1 - \sin y}$.

SOLUTION

Work with the *right* side. Note that you can create a monomial denominator by multiplying the numerator and denominator by $(1 + \sin y)$.

$$\frac{\cos y}{1 - \sin y} = \frac{\cos y}{1 - \sin y}\left(\frac{1 + \sin y}{1 + \sin y}\right) \qquad \textit{Multiply numerator and denominator by } (1 + \sin y)$$

$$= \frac{\cos y + \cos y \sin y}{1 - \sin^2 y}$$

$$= \frac{\cos y + \cos y \sin y}{\cos^2 y} \qquad \textit{Pythagorean Identity}$$

$$= \frac{\cos y}{\cos^2 y} + \frac{\cos y \sin y}{\cos^2 y} \qquad \textit{Separate fractions}$$

$$= \frac{1}{\cos y} + \frac{\sin y}{\cos y} \qquad \textit{Reduce}$$

$$= \sec y + \tan y \qquad \textit{Identities}$$

So far in this section, you have been verifying trigonometric identities by working with one side of the equation and converting to the form given on the other side. On occasion, it is practical to work with each side *separately,* to obtain one common form equivalent to both sides.

EXAMPLE 6   **Working with Each Side Separately**

Verify the identity

$$\frac{\cot^2 \theta}{1 + \csc \theta} = \frac{1 - \sin \theta}{\sin \theta}.$$

SOLUTION

Working with the left side, you have

$$\frac{\cot^2 \theta}{1 + \csc \theta} = \frac{\csc^2 \theta - 1}{1 + \csc \theta} \qquad \cot^2 \theta = \csc^2 \theta - 1$$

$$= \frac{(\csc \theta - 1)(\csc \theta + 1)}{1 + \csc \theta} \qquad \textit{Factor}$$

$$= \csc \theta - 1. \qquad \textit{Reduce}$$

Now, simplifying the right side, you have

$$\frac{1 - \sin \theta}{\sin \theta} = \frac{1}{\sin \theta} - \frac{\sin \theta}{\sin \theta} = \csc \theta - 1.$$

The identity is verified, because both sides are equal to $\csc \theta - 1$.

In Example 7, powers of trigonometric functions are rewritten as more complicated sums or products of trigonometric functions. This is a common procedure in calculus.

EXAMPLE 7 Two Examples from Calculus

Verify the identities.

A. $\tan^4 x = \tan^2 x \sec^2 x - \tan^2 x$
B. $\sin^3 x \cos^4 x = (\cos^4 x - \cos^6 x)\sin x$

SOLUTION

Note the use of the Pythagorean identities in the verifications.

A. $\tan^4 x = (\tan^2 x)(\tan^2 x)$	*Separate factors*
$= \tan^2 x(\sec^2 x - 1)$	*Pythagorean Identity*
$= \tan^2 x \sec^2 x - \tan^2 x$	*Multiply*
B. $\sin^3 x \cos^4 x = \sin^2 x \cos^4 x \sin x$	*Separate factors*
$= (1 - \cos^2 x)\cos^4 x \sin x$	*Pythagorean Identity*
$= (\cos^4 x - \cos^6 x)\sin x$	*Multiply*

DISCUSSION PROBLEM

YOU BE THE INSTRUCTOR

Suppose you are tutoring a student in trigonometry. After working several homework problems, your student becomes discouraged because of an inability to get answers that agree with those in the back of the textbook. Which of the following answers has your student actually gotten right? Which are wrong? Why?

Student's Answers	*Text's Answers*
(a) $\dfrac{1}{2 \csc x}$	$\dfrac{1}{2} \sin x$
(b) $(1 - \cos x)(1 + \cos x)$	$\sin^2 x$
(c) $\dfrac{\sin^2 x}{1 - \cos x}$	$1 + \cos x$
(d) $\cot^2 x + 2$	$1 + \csc^2 x$
(e) $(\sec x - \tan x)(\sec x + \tan x)$	1

WARM-UP

The following warm-up exercises involve skills that were covered in earlier sections. You will use these skills in the exercise set for this section.

In Exercises 1–6, factor each expression and, if possible, simplify the results.

1. (a) $x^2 - x^2 y^2$
  (b) $\sin^2 x - \sin^2 x \cos^2 x$

2. (a) $x^2 + x^2 y^2$
  (b) $\cos^2 x + \cos^2 x \tan^2 x$

3. (a) $x^4 - 1$
  (b) $\tan^4 x - 1$

4. (a) $z^3 + 1$
  (b) $\tan^3 x + 1$

5. (a) $x^3 - x^2 + x - 1$
  (b) $\cot^3 x - \cot^2 x + \cot x - 1$

6. (a) $x^4 - 2x^2 + 1$
  (b) $\sin^4 x - 2 \sin^2 x + 1$

In Exercises 7–10, perform the additions or subtractions and, if possible, simplify the results.

7. (a) $\dfrac{y^2}{x} - x$

  (b) $\dfrac{\csc^2 x}{\cot x} - \cot x$

8. (a) $1 - \dfrac{1}{x^2}$

  (b) $1 - \dfrac{1}{\sec^2 x}$

9. (a) $\dfrac{y}{1 + z} + \dfrac{1 + z}{y}$

  (b) $\dfrac{\sin x}{1 + \cos x} + \dfrac{1 + \cos x}{\sin x}$

10. (a) $\dfrac{y}{z} - \dfrac{z}{1 + y}$

  (b) $\dfrac{\tan x}{\sec x} - \dfrac{\sec x}{1 + \tan x}$

SECTION 6.2 · EXERCISES

In Exercises 1–50, verify the identity, and confirm it with a graphing utility.

1. $\sin t \csc t = 1$

2. $\tan y \cot y = 1$

3. $(1 + \sin \alpha)(1 - \sin \alpha) = \cos^2 \alpha$

4. $\cot^2 y (\sec^2 y - 1) = 1$

5. $\cos^2 \beta - \sin^2 \beta = 1 - 2 \sin^2 \beta$

6. $\cos^2 \beta - \sin^2 \beta = 2 \cos^2 \beta - 1$

7. $\tan^2 \theta + 4 = \sec^2 \theta + 3$

8. $2 - \sec^2 z = 1 - \tan^2 z$

9. $\sin^2 \alpha - \sin^4 \alpha = \cos^2 \alpha - \cos^4 \alpha$

10. $\cos x + \sin x \tan x = \sec x$

11. $\dfrac{\sec^2 x}{\tan x} = \sec x \csc x$

12. $\dfrac{\cot^3 t}{\csc t} = \cos t (\csc^2 t - 1)$

13. $\dfrac{\cot^2 t}{\csc t} = \csc t - \sin t$

14. $\dfrac{1}{\sin x} - \sin x = \dfrac{\cos^2 x}{\sin x}$

15. $\sin^{1/2} x \cos x - \sin^{5/2} x \cos x = \cos^3 x \sqrt{\sin x}$

16. $\sec^6 x (\sec x \tan x) - \sec^4 x (\sec x \tan x) = \sec^5 x \tan^3 x$

17. $\dfrac{1}{\sec x \tan x} = \csc x - \sin x$

18. $\dfrac{\sec \theta - 1}{1 - \cos \theta} = \sec \theta$

19. $\csc x - \sin x = \cos x \cot x$

20. $\dfrac{\sec x + \tan x}{\sec x - \tan x} = (\sec x + \tan x)^2$

21. $\dfrac{1}{\tan x} + \dfrac{1}{\cot x} = \tan x + \cot x$

22. $\dfrac{1}{\sin x} - \dfrac{1}{\csc x} = \csc x - \sin x$

23. $\dfrac{\cos \theta \cot \theta}{1 - \sin \theta} - 1 = \csc \theta$

24. $\cos x - \dfrac{\cos x}{1 - \tan x} = \dfrac{\sin x \cos x}{\sin x - \cos x}$

25. $2 \sec^2 x - 2 \sec^2 x \sin^2 x - \sin^2 x - \cos^2 x = 1$

26. $\csc x(\csc x - \sin x) + \dfrac{\sin x - \cos x}{\sin x} + \cot x = \csc^2 x$

27. $2 + \cos^2 x - 3 \cos^4 x = \sin^2 x(2 + 3 \cos^2 x)$

28. $4 \tan^4 x + \tan^2 x - 3 = \sec^2 x(4 \tan^2 x - 3)$

29. $\sec^4 \theta - \tan^4 \theta = 1 + 2 \tan^2 \theta$

30. $\csc^4 \theta - \cot^4 \theta = 2 \csc^2 \theta - 1$

31. $\dfrac{\sin \beta}{1 - \cos \beta} = \dfrac{1 + \cos \beta}{\sin \beta}$

32. $\dfrac{\cot \alpha}{\csc \alpha - 1} = \dfrac{\csc \alpha + 1}{\cot \alpha}$

33. $\cos\left(\dfrac{\pi}{2} - x\right)\csc x = 1$

34. $\dfrac{\cos\left(\dfrac{\pi}{2} - x\right)}{\sin\left(\dfrac{\pi}{2} - x\right)}$

35. $\dfrac{\csc(-x)}{\sec(-x)} = -\cot x$

36. $(1 + \sin y)[1 + \sin(-y)] = \cos^2 y$

37. $\dfrac{\cos(-\theta)}{1 + \sin(-\theta)} = \sec \theta + \tan \theta$

38. $\dfrac{1 + \sec(-\theta)}{\sin(-\theta) + \tan(-\theta)} = -\csc \theta$

39. $\sin^2 x + \sin^2\left(\dfrac{\pi}{2} - x\right) = 1$

40. $\sec^2 y - \cot^2\left(\dfrac{\pi}{2} - y\right) = 1$

41. $\sqrt{\dfrac{1 + \sin \theta}{1 - \sin \theta}} = \dfrac{1 + \sin \theta}{|\cos \theta|}$

42. $\sqrt{\dfrac{1 - \cos \theta}{1 + \cos \theta}} = \dfrac{1 - \cos \theta}{|\sin \theta|}$

43. $\dfrac{\sin x \cos y + \cos x \sin y}{\cos x \cos y - \sin x \sin y} = \dfrac{\tan x + \tan y}{1 - \tan x \tan y}$

44. $\dfrac{\tan x + \tan y}{1 - \tan x \tan y} = \dfrac{\cot x + \cot y}{\cot x \cot y - 1}$

45. $\dfrac{\tan x + \cot y}{\tan x \cot y} = \tan y + \cot x$

46. $\dfrac{\cos x - \cos y}{\sin x + \sin y} + \dfrac{\sin x - \sin y}{\cos x + \cos y} = 0$

47. $\ln|\tan \theta| = \ln|\sin \theta| - \ln|\cos \theta|$

48. $\ln|\sec \theta| = -\ln|\cos \theta|$

49. $-\ln(1 + \cos \theta) = \ln(1 - \cos \theta) - 2 \ln|\sin \theta|$

50. $-\ln|\sec \theta + \tan \theta| = \ln|\sec \theta - \tan \theta|$

In Exercises 51–54, explain why the equation is *not* an identity, and find one value of the variable for which the equation is not true.

51. $\sin \theta = \sqrt{1 - \cos^2 \theta}$

52. $\tan \theta = \sqrt{\sec^2 \theta - 1}$

53. $\sqrt{\tan^2 x} = \tan x$

54. $\sqrt{\sin^2 x + \cos^2 x} = \sin x + \cos x$

55. *Friction*   The force acting on a stationary object weighing W units on an inclined plane positioned at an angle of θ with the horizontal is modeled by

$$\mu W \cos \theta = W \sin \theta,$$

where μ is the coefficient of friction (see figure). Solve the equation for μ and simplify the result.

FIGURE FOR 55

56. *Rate of Change*   The rate of change of the function $f(x) = \sin x + \csc x$ with respect to change in the variable x is given by the expression $\cos x - \csc x \cot x$. Show that the expression for the rate of change can also be given by $-\cos x \cot^2 x$.

6.3 SOLVING TRIGONOMETRIC EQUATIONS

Introduction / Equations of Quadratic Type / Functions Involving Multiple Angles / Using Inverse Functions and a Calculator

Introduction

In this section, you will switch from *verifying* trigonometric identities to *solving* trigonometric equations. To see the difference, consider the two equations

$$\sin^2 x + \cos^2 x = 1$$

and

$$\sin x = 1.$$

The first equation is an identity because it is true for *all* real values of x. The second equation, however, is true only for *some* values of x. When you find these values, you are solving the equation.

EXAMPLE 1   Solving a Trigonometric Equation

Solve $2 \sin x - 1 = 0$.

SOLUTION

$2 \sin x - 1 = 0$	*Original equation*
$2 \sin x = 1$	*Add 1 to both sides*
$\sin x = \dfrac{1}{2}$	*Divide both sides by 2*

To solve for x, note that the equation $\sin x = \frac{1}{2}$ has solutions $x = \pi/6$ and $x = 5\pi/6$ in the interval $[0, 2\pi)$. Moreover, because $\sin x$ has a period of 2π, there are infinitely many other solutions, which can be written as

$$x = \frac{\pi}{6} + 2n\pi \quad \text{and} \quad x = \frac{5\pi}{6} + 2n\pi, \quad \textit{General solution}$$

where n is an integer, as shown in Figure 6.4. This is called the **general form** of the solution.

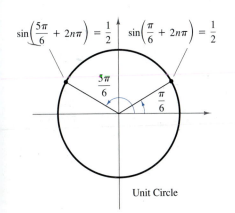

$$\sin\left(\frac{5\pi}{6} + 2n\pi\right) = \frac{1}{2} \qquad \sin\left(\frac{\pi}{6} + 2n\pi\right) = \frac{1}{2}$$

Unit Circle

FIGURE 6.4

Another way to see that the equation $\sin x = \frac{1}{2}$ has infinitely many solutions is indicated in Figure 6.5, which shows the graphs of $y_1 = \sin x$ and $y_2 = \frac{1}{2}$. For $0 \le x < 2\pi$, the solutions are $x = \pi/6$ and $x = 5\pi/6$. Any angles that are coterminal with $\pi/6$ or $5\pi/6$ will also be solutions of the equation.

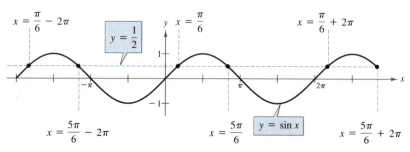

FIGURE 6.5

EXAMPLE 2   Collecting Like Terms

Solve $\sin x + \sqrt{2} = -\sin x$.

SOLUTION

$$\sin x + \sqrt{2} = -\sin x \qquad \textit{Original equation}$$

$$\sin x + \sin x = -\sqrt{2} \qquad \textit{Add } \sin x \textit{ to both sides}$$
$$\textit{and subtract } \sqrt{2} \textit{ from both sides}$$

$$2 \sin x = -\sqrt{2} \qquad \textit{Collect like terms}$$

$$\sin x = -\frac{\sqrt{2}}{2} \qquad \textit{Divide both sides by 2}$$

Because $\sin x$ has a period of 2π, you can begin by finding all solutions in the interval $[0, 2\pi)$. These are $x = 5\pi/4$ and $x = 7\pi/4$. Next, add $2n\pi$ to each of these solutions to obtain the general form

$$x = \frac{5\pi}{4} + 2n\pi \quad \text{and} \quad x = \frac{7\pi}{4} + 2n\pi, \qquad \textit{General solution}$$

where n is an integer. The graph of $y = 2 \sin x + \sqrt{2}$, shown in Figure 6.6, confirms this result.

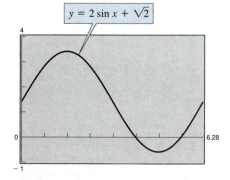

FIGURE 6.6

EXAMPLE 3   Extracting Square Roots

Solve $3 \tan^2 x - 1 = 0$.

SOLUTION

$$3 \tan^2 x - 1 = 0 \qquad \textit{Original equation}$$

$$3 \tan^2 x = 1 \qquad \textit{Add 1 to both sides}$$

$$\tan^2 x = \frac{1}{3} \qquad \textit{Divide both sides by 3}$$

$$\tan x = \pm \frac{1}{\sqrt{3}} \qquad \textit{Extract square roots}$$

Because tan x has a period of π, you can begin by finding all solutions in the interval $[0, \pi)$. These are $x = \pi/6 \approx 0.52$ and $x = 5\pi/6 \approx 2.62$. Next, add $n\pi$ to each to obtain the general form

$$x = \frac{\pi}{6} + n\pi \quad \text{and} \quad x = \frac{5\pi}{6} + n\pi, \qquad \textit{General solution}$$

where n is an integer. The graph of $y = 3 \tan^2 x - 1$, shown in Figure 6.7, confirms these results.

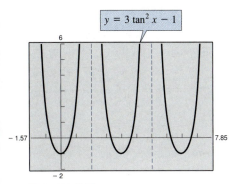

FIGURE 6.7

When two or more functions occur in the same equation, collect all terms on one side and try to separate the functions by factoring.

EXAMPLE 4   Factoring

Solve $\cot x \cos^2 x = 2 \cot x$.

SOLUTION

$$\cot x \cos^2 x = 2 \cot x \qquad \textit{Original equation}$$

$$\cot x \cos^2 x - 2 \cot x = 0 \qquad \textit{Subtract 2 cot x from both sides}$$

$$\cot x (\cos^2 x - 2) = 0 \qquad \textit{Factor left side}$$

By setting each of these factors equal to zero, you obtain the following.

$$\cot x = 0 \qquad \text{and} \qquad \cos^2 x - 2 = 0$$
$$\qquad\qquad\qquad\qquad\qquad \cos^2 x = 2$$
$$x = \frac{\pi}{2} \qquad\qquad\qquad \cos x = \pm\sqrt{2}$$

No solution is obtained from $\cos x = \pm\sqrt{2}$, because $\pm\sqrt{2}$ are outside the range of the cosine function. Therefore, the general form of the solution is obtained by adding multiples of π to $x = \pi/2$, to obtain

$$x = \frac{\pi}{2} + n\pi, \qquad \textit{General solution}$$

where n is an integer. The graph of $y = \cot x \cos^2 x - 2 \cot x$, shown in Figure 6.8, confirms these results.

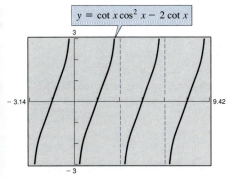

FIGURE 6.8

Equations of Quadratic Type

Many trigonometric equations are of quadratic type. Here are two examples.

Quadratic in sin x

$$2 \sin^2 x - \sin x - 1 = 0$$
$$2(\sin x)^2 - (\sin x) - 1 = 0$$

Quadratic in sec x

$$\sec^2 x - 3 \sec x - 2 = 0$$
$$(\sec x)^2 - 3(\sec x) - 2 = 0$$

To solve equations of this type, factor the quadratic or, if this is not possible, use the Quadratic Formula.

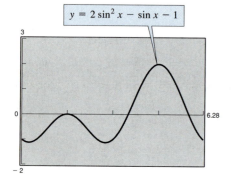

$y = 2 \sin^2 x - \sin x - 1$

FIGURE 6.9

EXAMPLE 5 Factoring an Equation of Quadratic Type

Solve $2 \sin^2 x - \sin x - 1 = 0$.

SOLUTION

The graph of $y = 2 \sin^2 x - \sin x - 1$, shown in Figure 6.9, indicates that there are three solutions in the interval $[0, 2\pi)$. Treating the equation as a quadratic in sin x and factoring, you can obtain the following.

$$2 \sin^2 x - \sin x - 1 = 0 \qquad \textit{Original equation}$$
$$(2 \sin x + 1)(\sin x - 1) = 0 \qquad \textit{Factor}$$

Setting each factor equal to zero produces the following solutions.

$$2 \sin x + 1 = 0 \qquad \text{and} \qquad \sin x - 1 = 0$$

$$\sin x = -\frac{1}{2} \qquad\qquad \sin x = 1$$

$$x = \frac{7\pi}{6}, \frac{11\pi}{6} \qquad\qquad x = \frac{\pi}{2}$$

The general solution is

$$x = \frac{7\pi}{6} + 2n\pi, \qquad x = \frac{11\pi}{6} + 2n\pi, \qquad x = \frac{\pi}{2} + 2n\pi,$$

where n is an integer.

EXAMPLE 6 Writing in Terms of a Single Trigonometric Function

Solve $2 \sin^2 x + 3 \cos x - 3 = 0$.

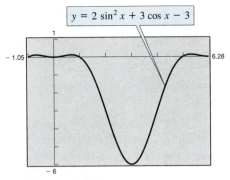

$y = 2 \sin^2 x + 3 \cos x - 3$

FIGURE 6.10

SOLUTION

The graph of $y = 2 \sin^2 x + 3 \cos x - 3$, shown in Figure 6.10, indicates that there are three solutions in the interval $[0, 2\pi)$. Proceeding algebraically, you can write the following.

$$2 \sin^2 x + 3 \cos x - 3 = 0 \qquad \textit{Original equation}$$
$$2(1 - \cos^2 x) + 3 \cos x - 3 = 0 \qquad \textit{Pythagorean Identity}$$
$$2 \cos^2 x - 3 \cos x + 1 = 0 \qquad \textit{Multiply both sides by } -1$$
$$(2 \cos x - 1)(\cos x - 1) = 0 \qquad \textit{Factor}$$

By setting each factor equal to zero, you can find the solutions in the interval $[0, 2\pi)$ to be $x = 0$, $x = \pi/3$, and $x = 5\pi/3$. The general solution is therefore

$$x = 2n\pi, \qquad x = \frac{\pi}{3} + 2n\pi, \qquad x = \frac{5\pi}{3} + 2n\pi, \qquad \textit{General solution}$$

where n is an integer.

Squaring both sides of an equation can introduce extraneous solutions, as indicated in Example 7.

EXAMPLE 7   Squaring and Converting to Quadratic Type

$$\cos x + 1 = \sin x \qquad \textit{Original equation}$$
$$\cos^2 x + 2 \cos x + 1 = \sin^2 x \qquad \textit{Square both sides}$$
$$\cos^2 x + 2 \cos x + 1 = 1 - \cos^2 x \qquad \textit{Identity}$$
$$2 \cos^2 x + 2 \cos x = 0 \qquad \textit{Collect terms}$$
$$2 \cos x(\cos x + 1) = 0 \qquad \textit{Factor}$$

Setting each factor equal to zero produces the following.

$$2 \cos x = 0 \qquad \text{and} \qquad \cos x + 1 = 0$$
$$\cos x = 0 \qquad\qquad\qquad \cos x = -1$$
$$x = \frac{\pi}{2}, \frac{3\pi}{2} \qquad\qquad\qquad x = \pi$$

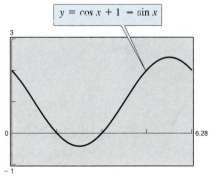

$y = \cos x + 1 - \sin x$

FIGURE 6.11

Because you squared both sides of the original equation, you must check for extraneous solutions. Of the three possible solutions, $x = 3\pi/2$ turns out to be extraneous. (Check this.) Thus, in the interval $[0, 2\pi)$, the only two solutions are $x = \pi/2$ and $x = \pi$. You can confirm this from the graph of $y = \cos x + 1 - \sin x$, shown in Figure 6.11. The general solution is $x = \pi/2 + 2n\pi$ and $x = \pi + 2n\pi$, where n is an integer.

Functions Involving Multiple Angles

EXAMPLE 8   Functions of Multiple Angles

Solve $2 \cos 3t - 1 = 0$.

SOLUTION

The graph of $y = 2 \cos 3t - 1$, shown in Figure 6.12, indicates that there are six solutions in the interval $[0, 2\pi)$. Proceeding algebraically, you can write the following.

$2 \cos 3t - 1 = 0$	*Original equation*
$2 \cos 3t = 1$	*Add 1 to both sides*
$\cos 3t = \dfrac{1}{2}$	*Divide both sides by 2*

In the interval $[0, 2\pi)$, you know that $3t = \pi/3$ and $3t = 5\pi/3$, so that

$$3t = \frac{\pi}{3} + 2n\pi \quad \text{and} \quad 3t = \frac{5\pi}{3} + 2n\pi$$

$$t = \frac{\pi}{9} + \frac{2n\pi}{3} \quad \text{and} \quad t = \frac{5\pi}{9} + \frac{2n\pi}{3}, \quad \textit{General solution}$$

where n is an integer. The six solutions in the interval $[0, 2\pi)$ are as follows.

$$\frac{\pi}{9}, \quad \frac{7\pi}{9}, \quad \frac{13\pi}{9}, \quad \frac{5\pi}{9}, \quad \frac{11\pi}{9}, \quad \frac{17\pi}{9}$$

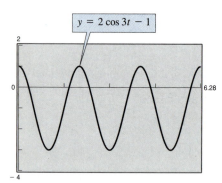

FIGURE 6.12

EXAMPLE 9   Functions of Multiple Angles

Solve $3 \tan(x/2) + 3 = 0$.

SOLUTION

The graph of $y = 3 \tan(x/2) + 3$, shown in Figure 6.13, indicates that there is only one solution in the interval $[0, 2\pi)$. Using the zoom and trace features, you can determine that the solution is $x \approx 4.712$. Algebraically, you can solve the equation as follows.

$3 \tan \dfrac{x}{2} + 3 = 0$	*Original equation*
$3 \tan \dfrac{x}{2} = -3$	*Subtract 3 from both sides*
$\tan \dfrac{x}{2} = -1$	*Divide both sides by 3*

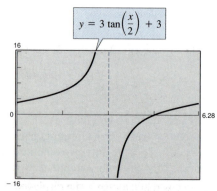

FIGURE 6.13

In the interval $[0, \pi)$, you know that $x/2 = 3\pi/4$, so that

$$\frac{x}{2} = \frac{3\pi}{4} + n\pi$$

$$x = \frac{3\pi}{2} + 2n\pi, \qquad\qquad \textit{General solution}$$

where n is an integer.

Using Inverse Functions and a Calculator

EXAMPLE 10   Using Inverse Functions

Solve $\sec^2 x - 2 \tan x = 4$.

SOLUTION

$$\sec^2 x - 2 \tan x = 4 \qquad \textit{Original equation}$$
$$1 + \tan^2 x - 2 \tan x - 4 = 0 \qquad \textit{Pythagorean Identity}$$
$$\tan^2 x - 2 \tan x - 3 = 0 \qquad \textit{Combine like terms}$$
$$(\tan x - 3)(\tan x + 1) = 0 \qquad \textit{Factor}$$

Setting each factor equal to zero produces two solutions in the interval $(-\pi/2, \pi/2)$.

$$\tan x = 3, \qquad\qquad \tan x = -1$$

$$x = \arctan 3 \qquad\qquad x = -\frac{\pi}{4}$$

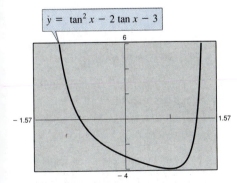

$\dot{y} = \tan^2 x - 2 \tan x - 3$

FIGURE 6.14

Adding multiples of π (the period of the tangent) produces the general solution

$$x = \arctan 3 + n\pi \quad \text{and} \quad x = -\frac{\pi}{4} + n\pi, \qquad \textit{General solution}$$

where n is an integer. Note that the graph of $y = \tan^2 x - 2 \tan x - 3$, shown in Figure 6.14, has the two x-intercepts

$$x = \arctan 3 \approx 1.2490 \quad \text{and} \quad x = -\frac{\pi}{4} \approx -0.7854$$

in the interval $(-\pi/2, \pi/2)$.

When a calculator is used for arcsin x, arccos x, and arctan x, the displayed solution may need to be adjusted to obtain solutions in the desired interval.

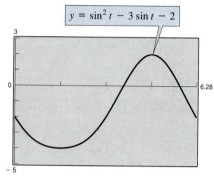

FIGURE 6.15

EXAMPLE 11   Using the Quadratic Formula

Solve $\sin^2 t - 3 \sin t - 2 = 0$ in the interval $[0, 2\pi)$.

SOLUTION

The graph of $y = \sin^2 t - 3 \sin t - 2$, shown in Figure 6.15, indicates that there are two solutions in the interval $[0, 2\pi)$. To find these solutions algebraically, you can use the Quadratic Formula as follows.

$$\sin^2 t - 3 \sin t - 2 = 0 \qquad \textit{Original equation}$$

$$\sin t = \frac{-(-3) \pm \sqrt{(-3)^2 - 4(1)(-2)}}{2(1)} \qquad \textit{Quadratic Formula}$$

$$= \frac{3 \pm \sqrt{17}}{2} \qquad \textit{Simplify}$$

$$\approx 3.561553 \quad \text{or} \quad -0.5615528$$

Because the range of the sine function is $[-1, 1]$, the equation $\sin t = 3.561553$ has no solution. To solve the equation $\sin t = -0.5615528$, use a calculator and the inverse sine function to obtain

$$t \approx \arcsin(-0.5615528) \approx -0.5962612.$$

Note that this solution is not in the interval $[0, 2\pi)$. In this interval, the two solutions are $t \approx \pi + 0.60 \approx 3.74$ and $t \approx 2\pi - 0.60 \approx 5.68$. ·········

For the next example, there is no analytic way to solve the equation: we rely solely on a graphing utility to approximate the solutions.

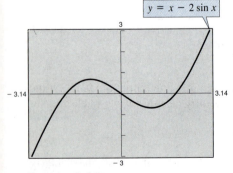

FIGURE 6.16

EXAMPLE 12   Approximating Solutions

Approximate the solutions of $x = 2 \sin x$.

SOLUTION

The graph of $y = x - 2 \sin x$ is shown in Figure 6.16. From the graph, you can see that one solution is $x = 0$. Using the zoom and trace features of a graphing utility, you can approximate the positive solution to be $x \approx 1.8955$. Finally, by symmetry, you can approximate the negative solution to be $x \approx -1.8955$. ·········

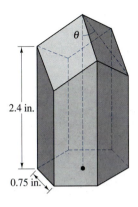

FIGURE 6.17

EXAMPLE 13   **Surface Area of a Honeycomb**

The surface area of a honeycomb is given by the equation

$$S = 6hs + \frac{3}{2}s^2\left(\frac{\sqrt{3} - \cos\theta}{\sin\theta}\right), \qquad 0 \le \theta \le 90°,$$

where $h = 2.4$ inches, $s = 0.75$ inches, and θ is the angle indicated in Figure 6.17.

A. What value of θ gives a surface area of 12 square inches?
B. What value of θ gives the minimum surface area?

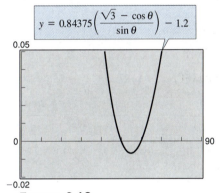

FIGURE 6.18

SOLUTION

A. Let $h = 2.4$, $s = 0.75$, and $S = 12$.

$$10.8 + 0.84375\left(\frac{\sqrt{3} - \cos\theta}{\sin\theta}\right) = 12$$

$$0.84375\left(\frac{\sqrt{3} - \cos\theta}{\sin\theta}\right) - 1.2 = 0$$

The graph of

$$y = 0.84375\left(\frac{\sqrt{3} - \cos\theta}{\sin\theta}\right) - 1.2$$

is shown in Figure 6.18. Using the zoom and trace features, you can determine that $\theta \approx 59.9°$ and $\approx 49.9°$.

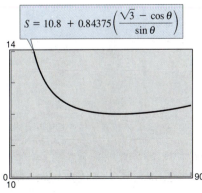

FIGURE 6.19

B. Graph the function

$$S = 10.8 + 0.84375\left(\frac{\sqrt{3} - \cos\theta}{\sin\theta}\right),$$

as shown in Figure 6.19. You can zoom in on the minimum point on the graph, which occurs at $\theta \approx 54.7°$. By using calculus, it can be shown that

$$\theta = \arccos\left(\frac{1}{\sqrt{3}}\right) \approx 54.7356$$

is the exact minimum value.

DISCUSSION PROBLEM

....

EQUATIONS WITH NO SOLUTIONS

One of the following three equations has solutions and the other two don't. Which two equations do not have solutions?

(a) $\sin^2 x - 5 \sin x + 6 = 0$
(b) $\sin^2 x - 4 \sin x + 6 = 0$
(c) $\sin^2 x - 5 \sin x - 6 = 0$

Can you find conditions involving the constants b and c that will guarantee that the equation $\sin^2 x + b \sin x + c = 0$ has at least one solution?

WARM-UP

....

The following warm-up exercises involve skills that were covered in earlier sections. You will use these skills in the exercise set for this section.

In Exercises 1–6, find the values of θ in the interval $0 \le \theta < 2\pi$ that satisfy the equation.

1. $\cos \theta = -\dfrac{1}{2}$ **2.** $\sin \theta = \dfrac{\sqrt{3}}{2}$ **3.** $\cos \theta = \dfrac{\sqrt{2}}{2}$

4. $\sin \theta = -\dfrac{\sqrt{2}}{2}$ **5.** $\tan \theta = \sqrt{3}$ **6.** $\tan \theta = -1$

In Exercises 7–10, solve for x.

7. $\dfrac{x}{3} + \dfrac{x}{5} = 1$ **8.** $2x(x + 3) - 5(x + 3) = 0$

9. $2x^2 - 4x - 5 = 0$ **10.** $\dfrac{1}{x} = \dfrac{x}{2x + 3}$

SECTION 6.3 · EXERCISES

In Exercises 1–6, verify that the given values of x are solutions of the equation.

1. $2 \cos x - 1 = 0$

(a) $x = \dfrac{\pi}{3}$ (b) $x = \dfrac{5\pi}{3}$

2. $\csc x - 2 = 0$

(a) $x = \dfrac{\pi}{6}$ (b) $x = \dfrac{5\pi}{6}$

3. $3 \tan^2 2x - 1 = 0$

(a) $x = \dfrac{\pi}{12}$ (b) $x = \dfrac{5\pi}{12}$

4. $2 \cos^2 4x - 1 = 0$

(a) $x = \dfrac{\pi}{16}$ (b) $x = \dfrac{3\pi}{16}$

5. $2 \sin^2 x - \sin x - 1 = 0$

(a) $x = \dfrac{\pi}{2}$ (b) $x = \dfrac{7\pi}{6}$

6. $\sec^4 x - 4 \sec^2 x = 0$

(a) $x = \dfrac{2\pi}{3}$ (b) $x = \dfrac{5\pi}{3}$

In Exercises 7–20, solve the equation. Verify your result with a graphing utility.

7. $2 \cos x + 1 = 0$ **8.** $2 \sin x - 1 = 0$

9. $\sqrt{3} \csc x - 2 = 0$ **10.** $\tan x + 1 = 0$

11. $2 \sin^2 x = 1$ **12.** $\tan^2 x = 3$

13. $3 \sec^2 x - 4 = 0$ **14.** $\csc^2 x - 2 = 0$

15. $\tan x(\tan x - 1) = 0$ **16.** $\cos x(2 \cos x + 1) = 0$

17. $\sin x(\sin x + 1) = 0$ **18.** $4 \sin^2 x - 3 = 0$

19. $\sin^2 x = 3 \cos^2 x$

20. $(3 \tan^2 x - 1)(\tan^2 x - 3) = 0$

In Exercises 21–32, find all solutions of the equation in the interval $[0, 2\pi)$. Verify your results with a graphing utility.

21. $\sec x \csc x - 2 \csc x = 0$

22. $\sec^2 x - \sec x - 2 = 0$

23. $2 \sin^2 x + 3 \sin x + 1 = 0$

24. $3 \tan^3 x - \tan x = 0$

25. $2 \sec^2 x + \tan^2 x - 3 = 0$

26. $2 \sin^2 x = 2 + \cos x$

27. $2 \sin x + \csc x = 0$ **28.** $\csc x + \cot x = 1$

29. $\sin 2x = -\dfrac{\sqrt{3}}{2}$ **30.** $\tan 3x = 1$

31. $\cos \dfrac{x}{2} = \dfrac{\sqrt{2}}{2}$ **32.** $\sec 4x = 2$

In Exercises 33–36, solve the algebraic and trigonometric equations. Restrict the solutions of the trigonometric equations to the interval $[0, 2\pi)$.

33. $6y^2 - 13y + 6 = 0$
$6 \cos^2 x - 13 \cos x + 6 = 0$

34. $y^2 + y - 20 = 0$
$\sin^2 x + \sin x - 20 = 0$

35. $y^2 - 8y + 13 = 0$
$\tan^2 x - 8 \tan x + 13 = 0$

36. $2y^2 + 6y - 1 = 0$
$2 \cos^2 x + 6 \cos x - 1 = 0$

In Exercises 37–50, use a graphing utility to approximate all solutions of the equation in the interval $[0, 2\pi)$.

37. $2 \cos x - \sin x = 0$

38. $4 \sin^3 x - 2 \sin x - 1 = 0$

39. $\dfrac{1 + \sin x}{\cos x} + \dfrac{\cos x}{1 + \sin x} = 4$ **40.** $\dfrac{\cos x \cot x}{1 - \sin x} = 3$

41. $2 \sin x - x = 0$ **42.** $x \cos x - 1 = 0$

43. $\sec^2 x + 0.5 \tan x - 1 = 0$

44. $\csc^2 x + 0.5 \cot x - 5 = 0$

45. $2 \tan^2 x + 7 \tan x - 15 = 0$

46. $12 \cos^2 x + 5 \cos x - 3 = 0$

47. $12 \sin^2 x - 13 \sin x + 3 = 0$

48. $3 \tan^2 x + 4 \tan x - 4 = 0$

49. $\sin^2 x + 2 \sin x - 1 = 0$

50. $4 \cos^2 x - 4 \cos x - 1 = 0$

Extrema of a Function   In Exercises 51 and 52, (a) use a graphing utility to graph the function f and approximate the maximum and minimum points on the graph in the interval $[0, 2\pi]$. (b) Solve the given trigonometric equation and verify that its solutions are the x-coordinates of the maximum and minimum points of f.

Function	*Trigonometric Equation*
51. $f(x) = \sin x + \cos x$	$\cos x - \sin x = 0$
52. $f(x) = 2 \sin x + \cos 2x$	$2 \cos x - 4 \sin x \cos x = 0$

53. *Picard Iteration*   Use the following procedure to solve the equation $\cos x = x$. Begin by calculating $\cos 0.5$. Then take the cosine of your result. Repeat this procedure over and over until you observe that the result does not change. Describe another way to solve the same equation.

54. *How Many Solutions?*   Use a graphing utility to graph $y = \sin(1/x)$.

(a) How many solutions does the equation

$$\sin \frac{1}{x} = 0$$

have in the interval $[-1, 1]$? Describe one or more viewing rectangles that can be used to help support your conclusion.

(b) Does the equation $\sin(1/x) = 0$ have a greatest solution? If so, approximate the solution. If not, explain.

55. *Maximum Area*   The area of a rectangle inscribed in one arch of the graph of $y = \cos x$, as shown in the figure, is given by

$$A = 2x \cos x, \qquad -\frac{\pi}{2} < x < \frac{\pi}{2}.$$

Use a graphing utility to graph the area function, and approximate the area of the largest inscribed rectangle.

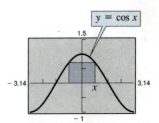

FIGURE FOR 55

56. *Sales* The monthly sales (in thousands of units) of a seasonal product are approximated by

$$S = 74.50 + 43.75 \sin \frac{\pi t}{6},$$

where t is the time in months, with $t = 1$ corresponding to January (see figure). Determine the months when sales exceed 100,000 units.

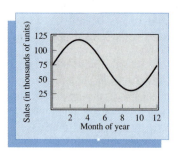

FIGURE FOR 56

57. *Harmonic Motion* A weight is oscillating on the end of a spring. The position of the weight relative to the point of equilibrium is given by

$$y = \tfrac{1}{4}(\cos 8t - 3 \sin 8t),$$

where y is the displacement in feet and t is the time in seconds (see figure). Find the times when the weight is at the point of equilibrium ($y = 0$) for $0 \leq t \leq 1$.

58. *Damped Harmonic Motion* The displacement from equilibrium of a weight oscillating on the end of a spring is given by

$$y = 1.56e^{-0.22t} \cos 4.9t,$$

where y is the displacement in feet and t is the time in seconds (see figure). Use a graphing utility to graph the displacement function for $0 \leq t \leq 10$. Find the time beyond which the displacement does not exceed 1 inch from equilibrium.

59. *Projectile Motion* A batted baseball leaves the bat at an angle of θ with the horizontal and an initial velocity of $v_0 = 100$ feet per second. The ball is caught by an outfielder 300 feet from home plate (see figure). Find θ if the range r of a projectile is given by

$$r = \tfrac{1}{32}v_0^2 \sin 2\theta.$$

60. *Projectile Motion* A marksman intends to hit a target at a distance of 1000 yards with a gun that has a muzzle velocity of 1200 feet per second (see figure). Neglecting air resistance, determine the minimum elevation of the gun if the range is given by

$$r = \tfrac{1}{32}v_0^2 \sin 2\theta.$$

A number c is a **fixed point** of a function f if $f(c) = c$. For example, 2 is a fixed point of $f(x) = x^3 - x - 4$. Finding a fixed point of f is equivalent to finding a solution to the equation $f(x) = x$, or $f(x) - x = 0$.

In Exercises 61–64, find the smallest positive fixed point of the following functions.

61. $f(x) = \tan \dfrac{\pi x}{4}$

62. $f(x) = e^{-x}$

63. $f(x) = \dfrac{4x - x^2 + 2}{4}$

64. $f(x) = \dfrac{5x + 2 - x^3}{5}$

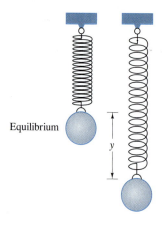

FIGURE FOR 57 AND 58

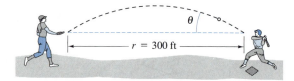

FIGURE FOR 59

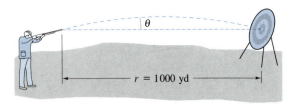

FIGURE FOR 60

6.4   SUM AND DIFFERENCE FORMULAS

Introduction / Using Sum and Difference Formulas

Introduction

In this and the following section, you will study several trigonometric identities that are important in scientific applications. We begin with six sum and difference formulas that express trigonometric functions of $(u \pm v)$ as functions of u and v alone.

SUM AND DIFFERENCE FORMULAS

$\sin(u + v) = \sin u \cos v + \cos u \sin v$ $\qquad$ $\sin(u - v) = \sin u \cos v - \cos u \sin v$

$\cos(u + v) = \cos u \cos v - \sin u \sin v$ $\qquad$ $\cos(u - v) = \cos u \cos v + \sin u \sin v$

$$\tan(u + v) = \frac{\tan u + \tan v}{1 - \tan u \tan v} \qquad\qquad \tan(u - v) = \frac{\tan u - \tan v}{1 + \tan u \tan v}$$

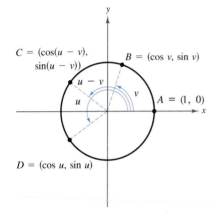

$C = (\cos(u - v), \sin(u - v))$

$B = (\cos v, \sin v)$

$A = (1, 0)$

$D = (\cos u, \sin u)$

FIGURE 6.20

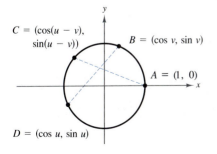

$C = (\cos(u - v), \sin(u - v))$

$B = (\cos v, \sin v)$

$A = (1, 0)$

$D = (\cos u, \sin u)$

FIGURE 6.21

PROOF

We prove only the formulas for $\cos(u \pm v)$. In Figure 6.20, let A be the point $(1, 0)$ and then use u and v to locate the points $B = (\cos v, \sin v)$, $C = (\cos(u - v), \sin(u - v))$, and $D = (\cos u, \sin u)$ on the unit circle. For convenience, assume that $0 < v < u < 2\pi$. From Figure 6.21, note that arcs AC and BD have the same length. Hence, *line segments AC and BD* are also equal in length. Let the length of $AC = d_1$, and the length of $BD = d_2$. Then $d_1 = d_2$ and $d_1{}^2 = d_2{}^2$. By the Distance Formula, you have

$$\begin{aligned}
d_1{}^2 &= [1 - \cos(u - v)]^2 + [0 - \sin(u - v)]^2 \\
&= 1 - 2\cos(u - v) + \cos^2(u - v) + \sin^2(u - v) \\
&= 2 - 2\cos(u - v)
\end{aligned}$$

and

$$\begin{aligned}
d_2{}^2 &= (\cos u - \cos v)^2 + (\sin u - \sin v)^2 \\
&= \cos^2 u - 2\cos u \cos v + \cos^2 v + \sin^2 u - 2\sin u \sin v + \sin^2 v \\
&= (\sin^2 u + \cos^2 u) + (\sin^2 v + \cos^2 v) - 2\cos u \cos v - 2\sin u \sin v \\
&= 2 - 2\cos u \cos v - 2\sin u \sin v.
\end{aligned}$$

Equating $d_1{}^2$ and $d_2{}^2$ yields

$$\begin{aligned}
2 - 2\cos(u - v) &= 2 - 2\cos u \cos v - 2\sin u \sin v \\
-2\,\cos(u - v) &= -2(\cos u \cos v + \sin u \sin v) \\
\cos(u - v) &= \cos u \cos v + \sin u \sin v.
\end{aligned}$$

REMARK Note that $\sin(u + v) \neq \sin u + \sin v$. Similar statements can be made for $\cos(u + v)$ and $\tan(u + v)$.

The formula for $\cos(u + v)$ can be established by considering $u + v = u - (-v)$ and using the formula just derived to obtain

$$\cos(u + v) = \cos[u - (-v)]$$
$$= \cos u \cos(-v) + \sin u \sin(-v) = \cos u \cos v - \sin u \sin v.$$

▪▪▪▪▪▪▪▪▪

DISCOVERY

(a) Use a graphing utility to graph $y = \cos(x + 2)$ and $y = \cos x + \cos 2$ in the same viewing rectangle. What can you conclude from the graphs? Is it true that $\cos(x + 2) = \cos x + \cos 2$?

(b) Use a graphing utility to graph $y = \sin(x + 4)$ and $y = \sin x + \sin 4$ in the same viewing rectangle. What can you conclude from the graphs? Is it true that $\sin(x + 4) = \sin x + \sin 4$?

Using Sum and Difference Formulas

EXAMPLE 1   Evaluating a Trigonometric Function

Find the exact value of $\cos 75°$.

SOLUTION

To find the *exact* value of $\cos 75°$, use the fact that $75° = 30° + 45°$. Consequently, the formula for $\cos(u + v)$ yields

$$\cos 75° = \cos(30° + 45°)$$
$$= \cos 30° \cos 45° - \sin 30° \sin 45°$$
$$= \frac{\sqrt{3}}{2}\left(\frac{\sqrt{2}}{2}\right) - \frac{1}{2}\left(\frac{\sqrt{2}}{2}\right) = \frac{\sqrt{6} - \sqrt{2}}{4}.$$

Check this with a calculator. You will find that $\cos 75° \approx 0.259 \approx (\sqrt{6} - \sqrt{2})/4$.

▪▪▪▪▪▪▪▪▪

EXAMPLE 2   Evaluating a Trigonometric Function

Find the exact value of

$$\cos \frac{\pi}{12}.$$

SOLUTION

Using the fact that

$$\frac{\pi}{12} = \frac{\pi}{3} - \frac{\pi}{4},$$

together with the formula for $\cos(u - v)$, you obtain

$$\cos \frac{\pi}{12} = \cos\left(\frac{\pi}{3} - \frac{\pi}{4}\right)$$
$$= \cos \frac{\pi}{3} \cos \frac{\pi}{4} + \sin \frac{\pi}{3} \sin \frac{\pi}{4}$$
$$= \frac{1}{2}\left(\frac{\sqrt{2}}{2}\right) + \frac{\sqrt{3}}{2}\left(\frac{\sqrt{2}}{2}\right) = \frac{\sqrt{2} + \sqrt{6}}{4}.$$

▪▪▪▪▪▪▪▪▪

EXAMPLE 3   Evaluating a Trigonometric Expression

Find the exact value of $\sin 42° \cos 12° - \cos 42° \sin 12°$.

SOLUTION

Recognizing that this expression fits the formula for $\sin(u - v)$, you can write

$$\sin 42° \cos 12° - \cos 42° \sin 12° = \sin(42° - 12°) = \sin 30° = \frac{1}{2}.$$

EXAMPLE 4   Evaluating a Trigonometric Expression

Find the exact value of

$$\frac{\tan 80° + \tan 55°}{1 - \tan 80° \tan 55°}.$$

SOLUTION

From the formula for $\tan(u + v)$, you have

$$\frac{\tan 80° + \tan 55°}{1 - \tan 80° \tan 55°} = \tan(80° + 55°) = \tan 135° = -\tan 45° = -1.$$

EXAMPLE 5   An Application of a Difference Formula

Find $\cos(u - v)$ given that

$$\cos u = -\frac{15}{17}, \quad \pi < u < \frac{3\pi}{2}, \quad \text{and} \quad \sin v = \frac{4}{5}, \quad 0 < v < \frac{\pi}{2}.$$

SOLUTION

Using the given values for $\cos u$ and $\sin v$, you can sketch angles u and v, as shown in Figure 6.22. This implies that

$$\cos v = \frac{3}{5} \quad \text{and} \quad \sin u = -\frac{8}{17}.$$

Therefore,

$$\cos(u - v) = \cos u \cos v + \sin u \sin v$$
$$= \left(-\frac{15}{17}\right)\left(\frac{3}{5}\right) + \left(-\frac{8}{17}\right)\left(\frac{4}{5}\right) = -\frac{77}{85}.$$

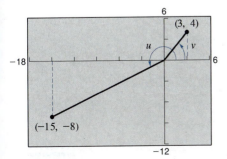

FIGURE 6.22

EXAMPLE 6 Proving a Cofunction Identity

Use the formula for $\cos(u - v)$ to prove the cofunction identity

$$\cos\left(\frac{\pi}{2} - x\right) = \sin x.$$

SOLUTION

Using the difference formula $\cos(u - v) = \cos u \cos v + \sin u \sin v$ produces

$$\begin{aligned}
\cos\left(\frac{\pi}{2} - x\right) &= \cos\frac{\pi}{2}\cos x + \sin\frac{\pi}{2}\sin x \\
&= (0)\cos x + (1)\sin x \\
&= \sin x.
\end{aligned}$$

You should verify this identity by graphing

$$y = \cos\left(\frac{\pi}{2} - x\right)$$

and $y = \sin x$ on the same screen.

Sum and difference formulas can be used to derive **reduction formulas** involving expressions such as

$$\sin\left(\theta + \frac{n\pi}{2}\right) \quad \text{and} \quad \cos\left(\theta + \frac{n\pi}{2}\right),$$

where n is an integer.

EXAMPLE 7 Deriving Reduction Formulas

Simplify

$$\cos\left(\theta - \frac{3\pi}{2}\right).$$

SOLUTION

Using the formula $\cos(u - v) = \cos u \cos v + \sin u \sin v$ produces

$$\begin{aligned}
\cos\left(\theta - \frac{3\pi}{2}\right) &= \cos\theta\cos\frac{3\pi}{2} + \sin\theta\sin\frac{3\pi}{2} \\
&= (\cos\theta)(0) + (\sin\theta)(-1) \\
&= -\sin\theta.
\end{aligned}$$

EXAMPLE 8   Solving a Trigonometric Equation

Solve

$$\sin\left(x + \frac{\pi}{4}\right) + \sin\left(x - \frac{\pi}{4}\right) = -1$$

in the interval $[0, 2\pi)$.

SOLUTION

Using sum and difference formulas, you can rewrite the given equation as

$$\sin x \cos \frac{\pi}{4} + \cos x \sin \frac{\pi}{4} + \sin x \cos \frac{\pi}{4} - \cos x \sin \frac{\pi}{4} = -1$$

$$2 \sin x \cos \frac{\pi}{4} = -1$$

$$2(\sin x)\left(\frac{\sqrt{2}}{2}\right) = -1$$

$$\sin x = -\frac{1}{\sqrt{2}}$$

$$\sin x = -\frac{\sqrt{2}}{2}.$$

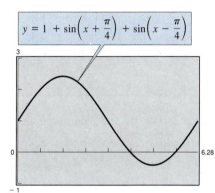

$$y = 1 + \sin\left(x + \frac{\pi}{4}\right) + \sin\left(x - \frac{\pi}{4}\right)$$

FIGURE 6.23

Therefore, the only solutions in the interval $[0, 2\pi)$ are $x = 5\pi/4$ and $x = 7\pi/4$. The graph of

$$y = 1 + \sin\left(x + \frac{\pi}{4}\right) + \sin\left(x - \frac{\pi}{4}\right),$$

shown in Figure 6.23, confirms these two solutions.

Example 9 shows how a sum formula can be used to rewrite a trigonometric expression in a form that is useful in calculus.

EXAMPLE 9   An Application from Calculus

Verify that

$$\frac{\sin(x + h) - \sin x}{h} = (\cos x)\left(\frac{\sin h}{h}\right) - (\sin x)\left(\frac{1 - \cos h}{h}\right),$$

where $h \neq 0$.

SOLUTION

$$\frac{\sin(x + h) - \sin x}{h} = \frac{\sin x \cos h + \cos x \sin h - \sin x}{h}$$

$$= \frac{\cos x \sin h - \sin x(1 - \cos h)}{h}$$

$$= (\cos x)\left(\frac{\sin h}{h}\right) - (\sin x)\left(\frac{1 - \cos h}{h}\right)$$

DISCUSSION PROBLEM

VERIFYING A SUM FORMULA

At the beginning of this section, we listed a proof of the formula for $\cos(u - v)$. Show how you can use this formula together with the identity

$$\sin x = \cos\left(\frac{\pi}{2} - x\right)$$

to prove the formula for $\sin(u + v)$. *Hint:* Start by writing

$$\sin(u + v) = \cos\left[\frac{\pi}{2} - (u + v)\right]$$

WARM-UP

The following warm-up exercises involve skills that were covered in earlier sections. You will use these skills in the exercise set for this section.

In Exercises 1–4, use the given information to find $\sin \theta$.

1. $\tan \theta = \dfrac{1}{3}$, $\quad \theta$ in Quadrant I

2. $\cot \theta = \dfrac{3}{5}$, $\quad \theta$ in Quadrant III

3. $\cos \theta = \dfrac{3}{4}$, $\quad \theta$ in Quadrant IV

4. $\sec \theta = -3$, $\quad \theta$ in Quadrant II

In Exercises 5 and 6, find all solutions in the interval $[0, 2\pi)$.

5. $\sin x = \dfrac{\sqrt{2}}{2}$

6. $\cos x = 0$

In Exercises 7–10, simplify the expression.

7. $\tan x \sec^2 x - \tan x$

8. $\dfrac{\cos x \csc x}{\tan x}$

9. $\dfrac{\cos x}{1 - \sin x} - \tan x$

10. $\dfrac{\cos^4 x - \sin^4 x}{\cos^2 x}$

SECTION 6.4 · EXERCISES

In Exercises 1–10, use the sum and difference identities to find the exact values of the sine, cosine, and tangent of the given angle.

1. $75° = 30° + 45°$

2. $15° = 45° - 30°$

3. $105° = 60° + 45°$

4. $165° = 135° + 30°$

5. $195° = 225° - 30°$

6. $255° = 300° - 45°$

7. $\dfrac{11\pi}{12} = \dfrac{3\pi}{4} + \dfrac{\pi}{6}$

8. $\dfrac{7\pi}{12} = \dfrac{\pi}{3} + \dfrac{\pi}{4}$

9. $\dfrac{17\pi}{12} = \dfrac{9\pi}{4} - \dfrac{5\pi}{6}$

10. $-\dfrac{\pi}{12} = \dfrac{\pi}{6} - \dfrac{\pi}{4}$

In Exercises 11–20, use the sum and difference identities to write the expression as the sine, cosine, or tangent of an angle.

11. $\cos 25° \cos 15° - \sin 25° \sin 15°$

12. $\sin 140° \cos 50° + \cos 140° \sin 50°$

13. $\sin 230° \cos 30° - \cos 230° \sin 30°$

14. $\cos 20° \cos 30° + \sin 20° \sin 30°$

15. $\dfrac{\tan 325° - \tan 86°}{1 + \tan 325° \tan 86°}$

16. $\dfrac{\tan 140° - \tan 60°}{1 + \tan 140° \tan 60°}$

17. $\sin 3 \cos 1.2 - \cos 3 \sin 1.2$

18. $\cos \dfrac{\pi}{7} \cos \dfrac{\pi}{5} - \sin \dfrac{\pi}{7} \sin \dfrac{\pi}{5}$

19. $\dfrac{\tan 2x + \tan x}{1 - \tan 2x \tan x}$

20. $\cos 3x \cos 2y + \sin 3x \sin 2y$

In Exercises 21–24, find the exact value of the trigonometric function given that

$$\sin u = \frac{5}{13}, \quad 0 < u < \frac{\pi}{2} \quad \text{and}$$

$$\cos v = -\frac{3}{5}, \quad \frac{\pi}{2} < v < \pi.$$

21. $\sin(u + v)$

22. $\cos(v - u)$

23. $\cos(u + v)$

24. $\sin(u - v)$

In Exercises 25–28, find the exact value of the trigonometric function given that

$$\sin u = \frac{7}{25}, \quad \frac{\pi}{2} < u < \pi \quad \text{and}$$

$$\cos v = \frac{4}{5}, \quad \frac{3\pi}{2} < v < 2\pi.$$

25. $\cos(u + v)$

26. $\sin(u + v)$

27. $\sin(v - u)$

28. $\cos(u - v)$

In Exercises 29–42, verify the identity.

29. $\sin\left(\dfrac{\pi}{2} + x\right) = \cos x$

30. $\sin(3\pi - x) = \sin x$

31. $\sin\left(\dfrac{\pi}{6} + x\right) = \dfrac{1}{2}(\cos x + \sqrt{3} \sin x)$

32. $\cos\left(\dfrac{5\pi}{4} - x\right) = -\dfrac{\sqrt{2}}{2}(\cos x + \sin x)$

33. $\cos(\pi - \theta) + \sin\left(\dfrac{\pi}{2} + \theta\right) = 0$

34. $\tan\left(\dfrac{\pi}{4} - \theta\right) = \dfrac{1 - \tan \theta}{1 + \tan \theta}$

35. $\cos(x + y) \cos(x - y) = \cos^2 x - \sin^2 y$

36. $\sin(x + y) \sin(x - y) = \sin^2 x - \sin^2 y$

37. $\sin(x + y) + \sin(x - y) = 2 \sin x \cos y$

38. $\cos(x + y) + \cos(x - y) = 2 \cos x \cos y$

39. $\cos(n\pi + \theta) = (-1)^n \cos \theta, \quad n$ is an integer

40. $\sin(n\pi + \theta) = (-1)^n \sin \theta, \quad n$ is an integer

41. $a \sin B\theta + b \cos B\theta = \sqrt{a^2 + b^2} \sin(B\theta + C)$, where $C = \arctan b/a, a > 0$

42. $a \sin B\theta + b \cos B\theta = \sqrt{a^2 + b^2} \cos(B\theta - C)$, where $C = \arctan a/b, b > 0$

In Exercises 43–46, use a graphing utility to verify the identity. (Graph both members of the equation and show that the graphs coincide.)

43. $\cos\left(\dfrac{3\pi}{2} - x\right) = -\sin x$

44. $\cos(\pi + x) = -\cos x$

45. $\sin\left(\dfrac{3\pi}{2} + \theta\right) + \sin(\pi - \theta) = \sin \theta - \cos \theta$

46. $\tan(\pi + \theta) = \tan \theta$

In Exercises 47–50, use the formulas given in Exercises 41 and 42 to write the trigonometric expression in the following forms. Use a graphing utility to verify your results.

(a) $\sqrt{a^2 + b^2}\, \sin(B\theta + C)$

(b) $\sqrt{a^2 + b^2}\, \cos(B\theta - C)$

47. $\sin \theta + \cos \theta$ 　　　　**48.** $3 \sin 2\theta + 4 \cos 2\theta$

49. $12 \sin 3\theta + 5 \cos 3\theta$ 　　**50.** $\sin 2\theta - \cos 2\theta$

In Exercises 51 and 52, use the formulas given in Exercises 41 and 42 to write the trigonometric expression in the form $a \sin B\theta + b \cos B\theta$. Use a graphing utility to verify your result.

51. $2 \sin\left(\theta + \dfrac{\pi}{4}\right)$ 　　**52.** $5 \cos\left(\theta + \dfrac{3\pi}{4}\right)$

In Exercises 53 and 54, write the trigonometric expression as an algebraic expression.

53. $\sin(\arcsin x + \arccos x)$ 　　**54.** $\sin(\arctan 2x - \arccos x)$

In Exercises 55–58, find all solutions of the equation in the interval $[0, 2\pi)$.

55. $\sin\left(x + \dfrac{\pi}{3}\right) + \sin\left(x - \dfrac{\pi}{3}\right) = 1$

56. $\sin\left(x + \dfrac{\pi}{6}\right) - \sin\left(x - \dfrac{\pi}{6}\right) = \dfrac{1}{2}$

57. $\cos\left(x + \dfrac{\pi}{4}\right) + \cos\left(x - \dfrac{\pi}{4}\right) = 1$

58. $\tan(x + \pi) - \cos\left(x + \dfrac{\pi}{2}\right) = 0$

In Exercises 59 and 60, use a graphing utility to approximate all solutions of the equation in the interval $[0, 2\pi)$.

59. $\cos\left(x + \dfrac{\pi}{4}\right) - \cos\left(x - \dfrac{\pi}{4}\right) = 1$

60. $\tan(x + \pi) + 2 \sin(x + \pi) = 0$

61. *Standing Waves* The equation of a standing wave is obtained by adding the displacements of two waves traveling in opposite directions (see figure). Assume that each of the waves has amplitude A, period T, and wavelength λ. If the models for these waves are

$$y_1 = A \cos\left[2\pi\left(\dfrac{t}{T} - \dfrac{x}{\lambda}\right)\right] \quad \text{and} \quad y_2 = A \cos\left[2\pi\left(\dfrac{t}{T} + \dfrac{x}{\lambda}\right)\right],$$

show that

$$y_1 + y_2 = 2A \cos \dfrac{2\pi t}{T} \cos \dfrac{2\pi x}{\lambda}.$$

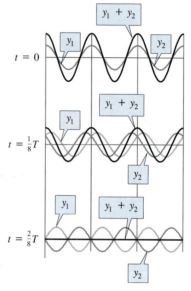

FIGURE FOR 61

62. *Harmonic Motion* A weight is attached to a spring suspended vertically from a ceiling. When a driving force is applied to the system, the weight moves vertically from its equilibrium position, and this motion is described by the model

$$y = \tfrac{1}{3} \sin 2t + \tfrac{1}{4} \cos 2t,$$

where y is the distance from equilibrium measured in feet and t is the time in seconds.

(a) Write the model in the form

$$y = \sqrt{a^2 + b^2}\, \sin(Bt + C).$$

(See Exercise 41.)

(b) Use a graphing utility to graph the model.

(c) Find the amplitude of the oscillations of the weight.

(d) Find the frequency of the oscillations of the weight.

63. Verify the following identity used in calculus.

$$\frac{\cos(x + h) - \cos x}{h} = \cos x\left(\frac{\cos h - 1}{h}\right) - \sin x\left(\frac{\sin h}{h}\right)$$

64. Use the sum formulas for the sine and cosine to derive the formula

$$\tan(u + v) = \frac{\tan u + \tan v}{1 - \tan u \tan v}.$$

6.5   MULTIPLE-ANGLE AND PRODUCT-TO-SUM FORMULAS

Multiple-Angle Formulas / Power-Reducing Formulas / Half-Angle Formulas / Product-to-Sum Formulas

Multiple-Angle Formulas

In this section, you will study four other categories of trigonometric identities. The first involves functions of multiple angles such as $\sin ku$ and $\cos ku$. The second involves squares of trigonometric functions such as $\sin^2 u$. The third involves functions of half-angles such as $\sin u/2$, and the fourth involves products of trigonometric functions such as $\sin u \cos v$.

> **DISCOVERY**
>
> (a) Use a graphing utility to graph $y = \cos(2x)$ and $y = 2 \cos x$ in the same viewing rectangle. What can you conclude from the graphs? Is it true that $\cos(2x) = 2 \cos x$?
>
> (b) Use a graphing utility to graph $y = \sin(4x)$ and $y = 4 \sin x$ in the same viewing rectangle. What can you conclude from the graphs? Is it true that $\sin(4x) = 4 \sin x$?

DOUBLE-ANGLE FORMULAS

$$\sin 2u = 2 \sin u \cos u$$
$$\cos 2u = \cos^2 u - \sin^2 u = 2 \cos^2 u - 1 = 1 - 2 \sin^2 u$$
$$\tan 2u = \frac{2 \tan u}{1 - \tan^2 u}$$

PROOF

To prove the first formula, let $v = u$ in the formula for $\sin(u + v)$, and obtain

$$\sin 2u = \sin(u + u)$$
$$= \sin u \cos u + \cos u \sin u$$
$$= 2 \sin u \cos u.$$

The other double-angle formulas can be proved in a similar way.

REMARK   Note that $\sin 2u \neq 2 \sin u$. Similar statements can be made for $\cos 2u$ and $\tan 2u$.

EXAMPLE 1   Solving a Trigonometric Equation

Solve $2 \cos x + \sin 2x = 0$.

SOLUTION

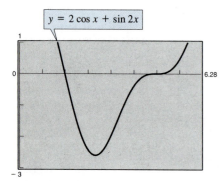
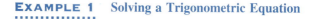

FIGURE 6.24

The graph of $y = 2 \cos x + \sin 2x$, shown in Figure 6.24, suggests that there are two solutions in the interval $[0, 2\pi)$. To find these solutions analytically, begin by rewriting the equation so that it involves functions of x (rather than $2x$). Then factor and solve as usual.

$$2 \cos x + \sin 2x = 0 \qquad \text{\textit{Original equation}}$$
$$2 \cos x + 2 \sin x \cos x = 0 \qquad \text{\textit{Double-angle formula}}$$
$$2 \cos x(1 + \sin x) = 0 \qquad \text{\textit{Factor}}$$

$$\cos x = 0, \qquad 1 + \sin x = 0 \qquad \text{\textit{Set factors to zero}}$$

$$x = \frac{\pi}{2}, \frac{3\pi}{2} \qquad\qquad x = \frac{3\pi}{2} \qquad \text{\textit{Solutions in } } [0, 2\pi)$$

Therefore, the general solution is

$$x = \frac{\pi}{2} + 2n\pi \quad \text{and} \quad x = \frac{3\pi}{2} + 2n\pi,$$

where n is an integer.

EXAMPLE 2   Locating Relative Minimums and Maximums

Find the relative minimums and relative maximums of $y = 4 \cos^2 x - 2$ in the interval $[0, \pi]$.

SOLUTION

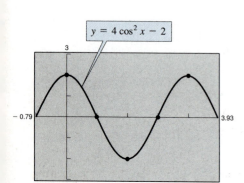

FIGURE 6.25

Using a double-angle identity, you can rewrite the given function as

$$y = 4 \cos^2 x - 2 = 2(2 \cos^2 x - 1) = 2 \cos 2x.$$

Using the techniques discussed in Section 5.5, you can recognize that the graph of this function has an amplitude of 2 and a period of π, as shown in Figure 6.25. The key points in the interval $[0, \pi]$ are as follows.

Maximum	Intercept	Minimum	Intercept	Maximum
$(0, 2)$	$\left(\dfrac{\pi}{4}, 0\right)$	$\left(\dfrac{\pi}{2}, -2\right)$	$\left(\dfrac{3\pi}{4}, 0\right)$	$(\pi, 2)$

EXAMPLE 3   **Evaluating Functions Involving Double Angles**

Use the fact that

$$\cos \theta = \frac{5}{13}, \qquad \frac{3\pi}{2} < \theta < 2\pi$$

to find $\sin 2\theta$, $\cos 2\theta$, and $\tan 2\theta$.

SOLUTION

In Figure 6.26, you can see that $\sin \theta = y/r = -12/13$. Consequently, you can write the following

$$\sin 2\theta = 2 \sin \theta \cos \theta = 2\left(\frac{-12}{13}\right)\left(\frac{5}{13}\right) = -\frac{120}{169}$$

$$\cos 2\theta = 2 \cos^2 \theta - 1 = 2\left(\frac{25}{169}\right) - 1 = -\frac{119}{169}$$

$$\tan 2\theta = \frac{\sin 2\theta}{\cos 2\theta} = \frac{120}{119}$$

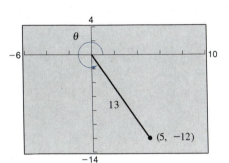

FIGURE 6.26

The double-angle formulas are not restricted to angles 2θ and θ. Other *double* combinations such as 4θ and 2θ are also valid. Here are two examples.

$$\sin 4\theta = 2 \sin 2\theta \cos 2\theta \quad \text{and} \quad \cos 6\theta = \cos^2 3\theta - \sin^2 3\theta$$

EXAMPLE 4   **Deriving a Triple-Angle Formula**

Express $\sin 3x$ in terms of $\sin x$.

SOLUTION

Consider that $3x = 2x + x$ produces

$$\sin 3x = \sin(2x + x)$$
$$= \sin 2x \cos x + \cos 2x \sin x$$
$$= 2 \sin x \cos x \, \text{co}$$
$$= 2 \sin x \cos^2 x$$
$$= 2 \sin$$
$$= 2 \sin$$
$$= 3 \sin$$

Try graphing y
rectangle. Thei

Power-Reducing Formulas

The double-angle formulas can be used to obtain the following **power-reducing formulas.**

POWER-REDUCING FORMULAS

$$\sin^2 u = \frac{1 - \cos 2u}{2} \qquad \cos^2 u = \frac{1 + \cos 2u}{2}$$

$$\tan^2 u = \frac{1 - \cos 2u}{1 + \cos 2u}$$

PROOF

The first two formulas can be verified by solving for $\sin^2 u$ and $\cos^2 u$, respectively, in the double-angle formulas

$$\cos 2u = 1 - 2 \sin^2 u \quad \text{and} \quad \cos 2u = 2 \cos^2 u - 1.$$

The third formula can be verified using the fact that

$$\tan^2 u = \frac{\sin^2 u}{\cos^2 u}.$$

Example 5 shows a typical power reduction that is used in calculus.

EXAMPLE 5   Reducing the Power of a Trigonometric Function

Rewrite $\sin^4 x$ as a sum involving first powers of the cosines of multiple angles.

SOLUTION

Note the repeated use of power-reducing formulas in the following procedure.

$$\sin^4 x = (\sin^2 x)^2 = \left(\frac{1 - \cos 2x}{2}\right)^2$$

$$= \frac{1}{4}(1 - 2\cos 2x + \cos^2 2x)$$

$$= \frac{1}{4}\left(1 - 2\cos 2x + \frac{1 + \cos 4x}{2}\right)$$

$$= \frac{1}{4} - \frac{1}{2}\cos 2x + \frac{1}{8} + \frac{1}{8}\cos 4x$$

$$= \frac{3}{8} - \frac{1}{2}\cos 2x + \frac{1}{8}\cos 4x$$

$$= \frac{1}{8}(3 - 4\cos 2x + \cos 4x)$$

Half-Angle Formulas

You can derive some useful alternative forms of the power-reducing formulas by replacing u with $u/2$. The results are **half-angle formulas.**

HALF-ANGLE FORMULAS

$$\sin \frac{u}{2} = \pm \sqrt{\frac{1 - \cos u}{2}}$$

$$\cos \frac{u}{2} = \pm \sqrt{\frac{1 + \cos u}{2}}$$

$$\tan \frac{u}{2} = \frac{1 - \cos u}{\sin u} = \frac{\sin u}{1 + \cos u}$$

The signs of $\sin u/2$ and $\cos u/2$ depend on the quadrant in which $u/2$ lies.

EXAMPLE 6   **Using a Half-Angle Formula**

Find the exact value of $\sin 105°$.

SOLUTION

Begin by noting that $105°$ is half of $210°$. Then, using the half-angle formula for $\sin(u/2)$ and the fact that $105°$ lies in Quadrant II, you have

$$\sin 105° = \sqrt{\frac{1 - \cos 210°}{2}}$$

$$= \sqrt{\frac{1 - (-\cos 30°)}{2}}$$

$$= \sqrt{\frac{1 + (\sqrt{3}/2)}{2}}$$

$$= \frac{\sqrt{2 + \sqrt{3}}}{2}.$$

Choose the positive square root because $\sin \theta$ is positive in Quad
your calculator to evaluate $\sin 105°$ and $\sqrt{2 + \sqrt{3}}/2$ to see th
are approximately 0.9659258.

EXAMPLE 7   Solving a Trigonometric Equation

Solve

$$2 - \sin^2 x = 2 \cos^2 \frac{x}{2}.$$

SOLUTION

The graph of $y = 2 - \sin^2 x - 2 \cos^2(x/2)$, shown in Figure 6.27, suggests that there are three solutions in the interval $[0, 2\pi)$. To find these analytically, solve the given equation as follows.

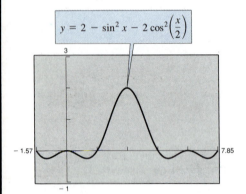

$y = 2 - \sin^2 x - 2 \cos^2\left(\dfrac{x}{2}\right)$

FIGURE 6.27

$2 - \sin^2 x = 2 \cos^2 \dfrac{x}{2}$		*Original equation*
$2 - \sin^2 x = 2\left(\dfrac{1 + \cos x}{2}\right)$		*Half-angle formula*
$2 - \sin^2 x = 1 + \cos x$		*Simplify*
$2 - (1 - \cos^2 x) = 1 + \cos x$		*Pythagorean Identity*
$\cos^2 x - \cos x = 0$		*Simplify*
$\cos x(\cos x - 1) = 0$		*Factor*

By setting the factors $\cos x$ and $(\cos x - 1)$ equal to zero, you can find that the solutions in the interval $[0, 2\pi)$ are $x = \pi/2 \approx 1.57$, $x = 3\pi/2 \approx 4.71$, and $x = 0$. Therefore, the general solution is

$$x = 2n\pi, \quad x = \frac{\pi}{2} + 2n\pi, \quad \text{and} \quad x = \frac{3\pi}{2} + 2n\pi,$$

where n is an integer.

Product-to-Sum Formulas

Each of the following **product-to-sum formulas** is easily verified using the sum and difference formulas discussed in the preceding section.

PRODUCT-TO-SUM FORMULAS

$$\sin u \sin v = \frac{1}{2}[\cos(u - v) - \cos(u + v)]$$

$$\cos u \cos v = \frac{1}{2}[\cos(u - v) + \cos(u + v)]$$

$$\sin u \cos v = \frac{1}{2}[\sin(u + v) + \sin(u - v)]$$

$$\cos u \sin v = \frac{1}{2}[\sin(u + v) - \sin(u - v)]$$

EXAMPLE 8   Writing Products as Sums

Rewrite $\cos 5x \sin 4x$ as a sum or difference.

SOLUTION

$$\cos 5x \sin 4x = \frac{1}{2}[\sin(5x + 4x) - \sin(5x - 4x)] = \frac{1}{2} \sin 9x - \frac{1}{2} \sin x$$

Occasionally, it is useful to reverse the procedure and write a sum of trigonometric functions as a product. This can be accomplished with the following **sum-to-product formulas.**

SUM-TO-PRODUCT FORMULAS

$$\sin x + \sin y = 2 \sin\left(\frac{x + y}{2}\right) \cos\left(\frac{x - y}{2}\right)$$

$$\sin x - \sin y = 2 \cos\left(\frac{x + y}{2}\right) \sin\left(\frac{x - y}{2}\right)$$

$$\cos x + \cos y = 2 \cos\left(\frac{x + y}{2}\right) \cos\left(\frac{x - y}{2}\right)$$

$$\cos x - \cos y = -2 \sin\left(\frac{x + y}{2}\right) \sin\left(\frac{x - y}{2}\right)$$

EXAMPLE 9   Using a Sum-to-Product Formula

Find the exact value of $\cos 195° + \cos 105°$.

SOLUTION

Using the appropriate sum-to-product formula produces

$$\cos 195° + \cos 105° = 2 \cos\left(\frac{195° + 105°}{2}\right) \cos\left(\frac{195° - 105°}{2}\right)$$

$$= 2 \cos 150° \cos 45°$$

$$= 2\left(-\frac{\sqrt{3}}{2}\right)\left(\frac{\sqrt{2}}{2}\right)$$

$$= -\frac{\sqrt{6}}{2}.$$

EXAMPLE 10   Solving a Trigonometric Equation

Solve $\sin 5x + \sin 3x = 0$.

SOLUTION

The graph of $y = \sin 5x + \sin 3x$, shown in Figure 6.28, suggests that there are eight solutions in the interval $[0, 2\pi)$. To find these analytically, proceed as follows.

$$\sin 5x + \sin 3x = 0 \qquad \textit{Original equation}$$

$$2 \sin\left(\frac{5x + 3x}{2}\right) \cos\left(\frac{5x - 3x}{2}\right) = 0 \qquad \textit{Sum-to-product formula}$$

$$2 \sin 4x \cos x = 0 \qquad \textit{Simplify}$$

By setting the factor $\sin 4x$ equal to zero, you find that the solutions in the interval $[0, 2\pi)$ are

$$x = 0, \quad \frac{\pi}{4}, \quad \frac{\pi}{2}, \quad \frac{3\pi}{4}, \quad \pi, \quad \frac{5\pi}{4}, \quad \frac{3\pi}{2}, \quad \frac{7\pi}{4}.$$

Moreover, the equation $\cos x = 0$ yields no additional solutions, and you can conclude the solutions are of the form

$$x = \frac{n\pi}{4},$$

where n is an integer.

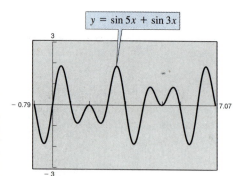

$y = \sin 5x + \sin 3x$

FIGURE 6.28

EXAMPLE 11   Verifying a Trigonometric Identity

Verify the identity

$$\frac{\sin t + \sin 3t}{\cos t + \cos 3t} = \tan 2t.$$

SOLUTION

Using appropriate sum-to-product formulas, you obtain

$$\frac{\sin t + \sin 3t}{\cos t + \cos 3t} = \frac{2 \sin 2t \cos(-t)}{2 \cos 2t \cos(-t)} = \frac{\sin 2t}{\cos 2t} = \tan 2t.$$

DISCUSSION PROBLEM

· · · · · · · · · · · · · · · ·

DERIVING A TRIPLE-ANGLE FORMULA

In Example 4, you saw how to derive a formula for sin 3x. Show how you can derive a similar formula for cos 3x. That is, find a formula that expresses cos 3x in terms of cos x, and then verify your formula graphically.

WARM-UP

· · · · · · · · · · · · · · · ·

The following warm-up exercises involve skills that were covered in earlier sections. You will use these skills in the exercise set for this section.

In Exercises 1 and 2, factor the trigonometric expression.

1. $2 \sin x + \sin x \cos x$
2. $\cos^2 x - \cos x - 2$

In Exercises 3–6, find all solutions of the equation in the interval $[0, 2\pi)$.

3. $\sin 2x = 0$
4. $\cos 2x = 0$

5. $\cos \dfrac{x}{2} = 0$
6. $\sin \dfrac{x}{2} = 0$

In Exercises 7–10, simplify the expression.

7. $\dfrac{1 - \cos \dfrac{\pi}{4}}{2}$
8. $\dfrac{1 + \cos \dfrac{\pi}{3}}{2}$

9. $\dfrac{2 \sin 3x \cos x}{2 \cos 3x \cos x}$
10. $(1 - 2 \sin^2 x) \cos x - 2 \sin^2 x \cos x$

SECTION 6.5 · EXERCISES

In Exercises 1–10, use a graphing utility to graph the function and approximate its zeros in the interval $[0, 2\pi)$. If possible, find the exact values of the zeros algebraically.

1. $f(x) = \sin 2x - \sin x$
2. $f(x) = \sin 2x + \cos x$
3. $g(x) = 4 \sin x \cos x - 1$
4. $g(x) = \sin 2x \sin x - \cos x$
5. $h(x) = \cos 2x - \cos x$
6. $h(x) = \cos 2x + \sin x$
7. $y = \tan 2x - \cot x$
8. $y = \tan 2x - 2 \cos x$
9. $h(t) = \sin 4t + 2 \sin 2t$
10. $f(s) = (\sin 2s + \cos 2s)^2 - 1$

In Exercises 11–14, use a double-angle identity to rewrite the function. Use a graphing utility to graph both versions of the function and note that the graphs coincide. Identify the relative minimums and relative maximums over the interval $[0, 2\pi)$.

11. $f(x) = 6 \sin x \cos x$
12. $g(x) = 4 \sin x \cos x + 2$
13. $g(x) = 4 - 8 \sin^2 x$
14. $f(x) = (\cos x + \sin x)(\cos x - \sin x)$

In Exercises 15–20, find the exact values of sin 2u, cos 2u, and tan 2u by using the double-angle formulas.

15. $\sin u = \dfrac{3}{5}, \quad 0 < u < \dfrac{\pi}{2}$

16. $\cos u = -\dfrac{2}{3}, \quad \dfrac{\pi}{2} < u < \pi$

17. $\tan u = \dfrac{1}{2}, \quad \pi < u < \dfrac{3\pi}{2}$

18. $\cot u = -4, \quad \dfrac{3\pi}{2} < u < 2\pi$

19. $\sec u = -\dfrac{5}{2}, \quad \dfrac{\pi}{2} < u < \pi$

20. $\csc u = 3, \quad \dfrac{\pi}{2} < u < \pi$

In Exercises 21–26, use the power-reducing formulas to write the expression in terms of the first power of the cosine.

21. $\cos^4 x$ **22.** $\sin^4 x$

23. $\sin^2 x \cos^2 x$ **24.** $\cos^6 x$

25. $\sin^2 x \cos^4 x$ **26.** $\sin^4 x \cos^2 x$

In Exercises 27–32, use the half-angle formulas to determine the exact values of the sine, cosine, and tangent of the given angle.

27. $105°$ **28.** $165°$

29. $112° \, 30'$ **30.** $67° \, 30'$

31. $\dfrac{\pi}{8}$ **32.** $\dfrac{\pi}{12}$

In Exercises 33–38, find the exact values of $\sin(u/2)$, $\cos(u/2)$, and $\tan(u/2)$ by using the half-angle formulas.

33. $\sin u = \dfrac{5}{13}, \quad \dfrac{\pi}{2} < u < \pi$

34. $\cos u = \dfrac{3}{5}, \quad 0 < u < \dfrac{\pi}{2}$

35. $\tan u = -\dfrac{5}{8}, \quad \dfrac{3\pi}{2} < u < 2\pi$

36. $\cot u = 3, \quad \pi < u < \dfrac{3\pi}{2}$

37. $\csc u = -\dfrac{5}{3}, \quad \pi < u < \dfrac{3\pi}{2}$

38. $\sec u = -\dfrac{7}{2}, \quad \dfrac{\pi}{2} < u < \pi$

In Exercises 39–42, use the half-angle formulas to simplify the expression.

39. $\sqrt{\dfrac{1 - \cos 6x}{2}}$ **40.** $\sqrt{\dfrac{1 + \cos 4x}{2}}$

41. $-\sqrt{\dfrac{1 - \cos 8x}{1 + \cos 8x}}$ **42.** $-\sqrt{\dfrac{1 - \cos(x - 1)}{2}}$

In Exercises 43–46, use a graphing utility to graph the function and approximate its zeros in the interval $[0, 2\pi)$. If possible, find the exact values of the zeros algebraically.

43. $f(x) = \sin \dfrac{x}{2} + \cos x$

44. $h(x) = \sin \dfrac{x}{2} + \cos x - 1$

45. $h(x) = \cos \dfrac{x}{2} - \sin x$ **46.** $g(x) = \tan \dfrac{x}{2} - \sin x$

In Exercises 47–56, use the product-to-sum formulas to write the product as a sum.

47. $6 \sin \dfrac{\pi}{4} \cos \dfrac{\pi}{4}$ **48.** $4 \sin \dfrac{\pi}{3} \cos \dfrac{5\pi}{6}$

49. $\sin 5\theta \cos 3\theta$ **50.** $3 \sin 2\alpha \sin 3\alpha$

51. $5 \cos(-5\beta) \cos 3\beta$ **52.** $\cos 2\theta \cos 4\theta$

53. $\sin(x + y) \sin(x - y)$ **54.** $\sin(x + y) \cos(x - y)$

55. $\sin(\theta + \pi) \cos(\theta - \pi)$ **56.** $10 \cos 75° \cos 15°$

In Exercises 57–66, use the sum-to-product formulas to write the sum or difference as a product.

57. $\sin 60° + \sin 30°$ **58.** $\cos 120° + \cos 30°$

59. $\cos \dfrac{3\pi}{4} - \cos \dfrac{\pi}{4}$ **60.** $\sin 5\theta - \sin 3\theta$

61. $\cos 6x + \cos 2x$ **62.** $\sin x + \sin 5x$

63. $\sin(\alpha + \beta) - \sin(\alpha - \beta)$

64. $\cos\left(\theta + \dfrac{\pi}{2}\right) - \cos\left(\theta - \dfrac{\pi}{2}\right)$

65. $\cos(\phi + 2\pi) + \cos \phi$

66. $\sin\left(x + \dfrac{\pi}{2}\right) + \sin\left(x - \dfrac{\pi}{2}\right)$

In Exercises 67–70, use a graphing utility to graph the function and approximate its zeros in the interval $[0, 2\pi)$. If possible, find the exact values of the zeros algebraically.

67. $g(x) = \sin 6x + \sin 2x$

68. $h(x) = \cos 2x - \cos 6x$

69. $f(x) = \dfrac{\cos 2x}{\sin 3x - \sin x} - 1$

70. $f(x) = \sin^2 3x - \sin^2 x$

In Exercises 71–88, verify the identity. Use a graphing utility to confirm the identity.

71. $\csc 2\theta = \dfrac{\csc \theta}{2 \cos \theta}$ **72.** $\sec 2\theta = \dfrac{\sec^2 \theta}{2 - \sec^2 \theta}$

73. $\cos^2 2\alpha - \sin^2 2\alpha = \cos 4\alpha$

74. $\cos^4 x - \sin^4 x = \cos 2x$

75. $(\sin x + \cos x)^2 = 1 + \sin 2x$

76. $\sin \dfrac{\alpha}{3} \cos \dfrac{\alpha}{3} = \dfrac{1}{2} \sin \dfrac{2\alpha}{3}$

77. $\cos 3\beta = \cos^3 \beta - 3 \sin^2 \beta \cos \beta$

78. $\sin 4\beta = 4 \sin \beta \cos \beta(1 - 2 \sin^2 \beta)$

79. $1 + \cos 10y = 2 \cos^2 5y$

80. $\dfrac{\cos 3\beta}{\cos \beta} = 1 - 4 \sin^2 \beta$

81. $\sec \dfrac{u}{2} = \pm \sqrt{\dfrac{2 \tan u}{\tan u + \sin u}}$

82. $\tan \dfrac{u}{2} = \csc u - \cot u$

83. $\dfrac{\cos 4x + \cos 2x}{\sin 4x + \sin 2x} = \cot 3x$

84. $\dfrac{\cos 3x - \cos x}{\sin 3x - \sin x} = -\tan 2x$

85. $\dfrac{\cos 4x - \cos 2x}{2 \sin 3x} = -\sin x$

86. $\dfrac{\sin x \pm \sin y}{\cos x + \cos y} = \tan \dfrac{x \pm y}{2}$

87. $\dfrac{\cos t + \cos 3t}{\sin 3t - \sin t} = \cot t$

88. $\sin\left(\dfrac{\pi}{6} + x\right) + \sin\left(\dfrac{\pi}{6} - x\right) = \cos x$

In Exercises 89 and 90, sketch the graph of the function by using the power-reducing formulas.

89. $f(x) = \sin^2 x$ **90.** $f(x) = \cos^2 x$

In Exercises 91 and 92, write the trigonometric expression as an algebraic expression.

91. $\sin(2 \arcsin x)$ **92.** $\cos(2 \arccos x)$

In Exercises 93 and 94, prove the product-to-sum formula.

93. $\cos u \sin v = \frac{1}{2}[\sin(u + v) - \sin(u - v)]$

94. $\cos u \cos v = \frac{1}{2}[\cos(u - v) + \cos(u + v)]$

95. *Area*   The lengths of the two equal sides of an isosceles triangle are 10 feet (see figure). The angle between the two sides is θ.
(a) Express the area of the triangle as a function of $\theta/2$.
(b) Express the area of the triangle as a function of θ and determine the value of θ so the area is maximum.

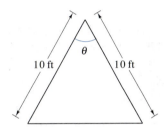

FIGURE FOR 95

96. *Projectile Motion*   The range of a projectile fired at an angle θ with the horizontal and with an initial velocity of v_0 feet per second is given by

$$r = \tfrac{1}{32} v_0^2 \sin 2\theta,$$

where r is measured in feet. Determine the expression for the range in terms of θ.

6.6   LAW OF SINES

Law of Sines   /   The Ambiguous Case (SSA)   /   Application   /
The Area of an Oblique Triangle

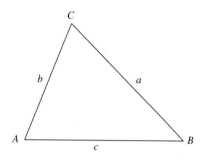

FIGURE 6.29

Law of Sines

In Chapter 5 you studied techniques for solving right triangles. In this section and the next, you will solve **oblique triangles**—triangles that have no right angles. As standard notation, we label the angles of a triangle as A, B, and C, and their opposite sides as a, b, and c, as shown in Figure 6.29.

To solve an oblique triangle, you need to know the measure of at least one side and any two other parts of the triangle—either two sides, two angles, or one angle and one side. This breaks down into four cases.

1. Two angles and any side (AAS or ASA).
2. Two sides and an angle opposite one of them (SSA).
3. Three sides (SSS).
4. Two sides and their included angle (SAS).

The first two cases can be solved using the **Law of Sines,** whereas the last two cases require the **Law of Cosines** (to be discussed in Section 6.7).

REMARK   The Law of Sines can also be written in the reciprocal form
$$\frac{\sin A}{a} = \frac{\sin B}{b} = \frac{\sin C}{c}.$$

LAW OF SINES

If ABC is a triangle with sides a, b, and c, then
$$\frac{a}{\sin A} = \frac{b}{\sin B} = \frac{c}{\sin C}.$$

A is acute.   A is obtuse.

Oblique Triangles

When using a calculator with the Law of Sines, remember to store all intermediate calculations. By not rounding until the final result, you minimize the round-off error.

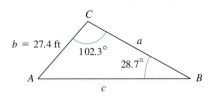

FIGURE 6.30

EXAMPLE 1   Given Two Angles and One Side—AAS

Given a triangle with $C = 102.3°$, $B = 28.7°$, and $b = 27.4$ feet, as shown in Figure 6.30, find the remaining angle and sides.

SOLUTION

The third angle of the triangle is

$$A = 180° - B - C = 180° - 28.7° - 102.3° = 49.0°.$$

By the Law of Sines, you have

$$\frac{a}{\sin 49°} = \frac{b}{\sin 28.7°} = \frac{c}{\sin 102.3°}.$$

Because $b = 27.4$, you obtain

$$a = \frac{27.4}{\sin 28.7°}(\sin 49°) \approx 43.06 \text{ feet}$$

and

$$c = \frac{27.4}{\sin 28.7°}(\sin 102.3°) \approx 55.75 \text{ feet}.$$

Note that the ratio $27.4/\sin 28.7°$ occurs in both solutions, and you can save time by storing this result for repeated use.

When solving triangles, a careful sketch is useful as a quick test for the feasibility of an answer. Remember that the longest side lies opposite the largest angle, and the shortest side lies opposite the smallest angle of a triangle.

EXAMPLE 2   Given Two Angles and One Side—ASA

A pole tilts *toward* the sun at an 8° angle from vertical, and it casts a 22-foot shadow. The angle of elevation from the tip of the shadow to the top of the pole is 43°. How tall is the pole?

SOLUTION

In Figure 6.31, note that $A = 43°$, and $B = 90° + 8° = 98°$. Thus, the third angle is

$$C = 180° - A - B = 180° - 43° - 98° = 39°.$$

By the Law of Sines, you have

$$\frac{a}{\sin 43°} = \frac{c}{\sin 39°}.$$

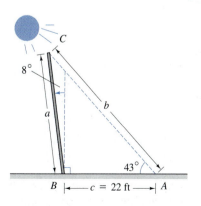

FIGURE 6.31

REMARK   For practice, try reworking Example 2 for a pole that tilts *away* from the sun under the same conditions.

Because $c = 22$ feet, the length of the pole is

$$a = \frac{22}{\sin 39°}(\sin 43°) \approx 23.84 \text{ feet.}$$

The Ambiguous Case (SSA)

In Examples 1 and 2, you saw that two angles (whose sum is less than 180°) and one side determine a unique triangle. However, if two sides and one opposite angle are given, three possible situations can occur: (1) no such triangle exists, (2) one such triangle exists, or (3) two distinct triangles may satisfy the conditions. The possibilities in this *ambiguous* case (SSA) are summarized in the following table.

The Ambiguous Case (SSA) (given: a, b, and A)

	A Is Acute				A Is Obtuse	
Sketch ($h = b \sin A$)						
Necessary Condition	$a < h$	$a = h$	$a > b$	$h < a < b$	$a \leq b$	$a > b$
Triangles Possible	None	One	One	Two	None	One

To determine which of the possibilities holds for a given pair of sides and opposite angle, it helps to make a sketch.

EXAMPLE 3   Single-Solution Case—SSA

Given a triangle with $a = 22$ inches, $b = 12$ inches, and $A = 42°$, find the remaining side and angles.

SOLUTION

Because A is acute and $a > b$, you know that B is also acute and there is only one triangle that satisfies the given conditions, as shown in Figure 6.32. Thus, by the Law of Sines, you have

$$\frac{22}{\sin 42°} = \frac{12}{\sin B},$$

which implies that

$$\sin B = 12\left(\frac{\sin 42°}{22}\right) \approx 0.3649803$$

$$B \approx 21.41°.$$

Now you can determine that $C \approx 180° - 42° - 21.41° = 116.59°$, and the remaining side is given by

$$\frac{c}{\sin 116.59°} = \frac{22}{\sin 42°}$$

$$c = \sin 116.59°\left(\frac{22}{\sin 42°}\right) \approx 29.40 \text{ inches.}$$

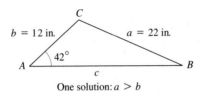

One solution: $a > b$

FIGURE 6.32

EXAMPLE 4   No-Solution Case—SSA

Show that there is no triangle that satisfies either of the following conditions.

A. $a = 15, \quad b = 25, \quad A = 85°$
B. $a = 15.2, \quad b = 20, \quad A = 110°$

SOLUTION

A. Begin by making the sketch shown in Figure 6.33. From this figure it appears that no triangle is formed. You can verify this using the Law of Sines.

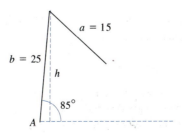

No solution: $a < h$

FIGURE 6.33

$$\frac{a}{\sin A} = \frac{b}{\sin B}$$

$$\frac{15}{\sin 85°} = \frac{25}{\sin B}$$

$$\sin B = 25\left(\frac{\sin 85°}{15}\right) \approx 1.660 > 1$$

This contradicts the fact that $|\sin B| \leq 1$. Hence, no triangle can be formed having sides $a = 15$ and $b = 25$ and an angle of $A = 85°$.

B. Because A is obtuse and $a = 15.2$ is less than $b = 20$, you can conclude that there is *no solution,* as shown in Figure 6.34. Try using the Law of Sines to verify this.

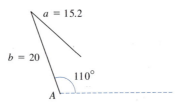

No solution: $a < b$ and $A > 90°$

FIGURE 6.34

EXAMPLE 5   Two-Solution Case—SSA

Find two triangles for which $a = 12$ meters, $b = 31$ meters, and $A = 20.5°$.

SOLUTION

To begin, note that

$$h = b \sin A = 31(\sin 20.5°) \approx 10.86 \text{ meters.}$$

Hence, $h < a < b$, and you can conclude that there are two possible triangles. By the Law of Sines, you obtain

$$\frac{a}{\sin A} = \frac{b}{\sin B},$$

which implies that

$$\sin B = b\left(\frac{\sin A}{a}\right) = 31\left(\frac{\sin 20.5°}{12}\right) \approx 0.9047 \text{ meters.}$$

There are two angles $B_1 \approx 64.8°$ and $B_2 \approx 115.2°$ between $0°$ and $180°$ whose sine is 0.9047. For $B_1 \approx 64.8°$, you obtain

$$C \approx 180° - 20.5° - 64.8° = 94.7°$$

$$c = \frac{a}{\sin A}(\sin C) = \frac{12}{\sin 20.5°}(\sin 94.7°) \approx 34.15.$$

For $B_2 \approx 115.2°$, you obtain

$$C \approx 180° - 20.5° - 115.2° = 44.3°$$

$$c = \frac{a}{\sin A}(\sin C) = \frac{12}{\sin 20.5°}(\sin 44.3°) \approx 23.93 \text{ meters.}$$

The resulting triangles are shown in Figures 6.35 and 6.36.

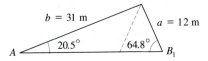

FIGURE 6.35

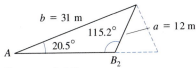

FIGURE 6.36

Application

EXAMPLE 6 **An Application of the Law of Sines**

The course for a boat race starts at point A and proceeds in the direction S 52° W to point B, then in the direction S 40° E to point C, and finally back to A, as shown in Figure 6.37. The point C lies 8 kilometers directly south of point A. Approximate the total distance of the race course.

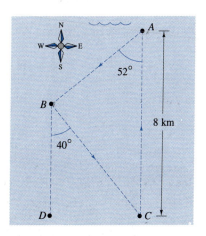

FIGURE 6.37

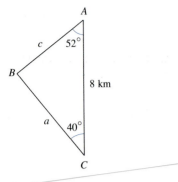

FIGURE 6.38

SOLUTION

Because lines BD and AC are parallel, it follows that $\angle BCA = \angle DBC$. Consequently, triangle ABC has the measures shown in Figure 6.38. For angle B, you have

$$B = 180° - 52° - 40° = 88°.$$

Using the Law of Sines

$$\frac{a}{\sin 52°} = \frac{b}{\sin 88°} = \frac{c}{\sin 40°},$$

you can let $b = 8$ and obtain the following.

$$a = \frac{8}{\sin 88°}(\sin 52°) \approx 6.308$$

$$c = \frac{8}{\sin 88°}(\sin 40°) \approx 5.145$$

Finally, the total length of the course is approximately

Length $\approx 8 + 6.308 + 5.145 = 19.453$ kilometers.

The Area of an Oblique Triangle

The procedure used to prove the Law of Sines leads to a simple formula for the area of an oblique triangle. Referring to Figure 6.39, note that each triangle has a height of

$$h = b \sin A.$$

Consequently, the area of each triangle is given by

$$\text{Area} = \frac{1}{2}(\text{base})(\text{height}) = \frac{1}{2}(c)(b \sin A) = \frac{1}{2}bc \sin A.$$

By similar arguments, you can develop the formula

$$\text{Area} = \frac{1}{2}ab \sin C = \frac{1}{2}ac \sin B.$$

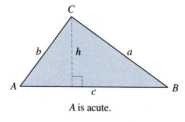

A is acute.   *A is obtuse.*

Oblique Triangles

FIGURE 6.39

AREA OF AN OBLIQUE TRIANGLE

The area of any triangle is given by one-half the product of the lengths of two sides times the sine of their included angle. That is,

$$\text{Area} = \frac{1}{2}bc \sin A = \frac{1}{2}ab \sin C = \frac{1}{2}ac \sin B.$$

EXAMPLE 7   Finding the Area of an Oblique Triangle

Find the area of a triangular lot having two sides of lengths 90 meters and 52 meters and an included angle of 102°.

SOLUTION

Consider $a = 90$ m, $b = 52$ m, and angle $C = 102°$, as shown in Figure 6.40. Then the area of the triangle is

$$\text{Area} = \frac{1}{2}ab \sin C = \frac{1}{2}(90)(52)(\sin 102°) \approx 2289 \text{ square meters.}$$

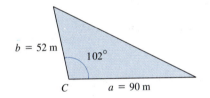

FIGURE 6.40

DISCUSSION PROBLEM

SOLVING RIGHT TRIANGLES

In this section, you have been using the Law of Sines to solve *oblique* triangles. Can the Law of Sines also be used to solve a right triangle? If so, write a short paragraph explaining how to use the Law of Sines to solve the following two triangles. Is there an easier way to solve these triangles?

(a) (AAS)

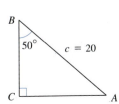

(b) (ASA)

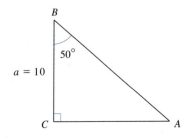

WARM-UP

The following warm-up exercises involve skills that were covered in earlier sections. You will use these skills in the exercise set for this section.

In Exercises 1–6, solve the indicated *right* triangle. (In the right triangle, *a* and *b* are the lengths of the sides and *c* is the length of the hypotenuse.)

1. $a = 3$, $c = 6$ **2.** $a = 5$, $b = 5$ **3.** $b = 15$, $c = 17$

4. $A = 42°$, $a = 7.5$ **5.** $B = 10°$, $b = 4$ **6.** $B = 72° \, 15'$, $c = 150$

In Exercises 7 and 8, find the altitude of the triangle.

7.

8.

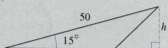

In Exercises 9 and 10, solve the equation for *x*.

9. $\dfrac{2}{\sin 30°} = \dfrac{9}{x}$ **10.** $\dfrac{100}{\sin 72°} = \dfrac{x}{\sin 60°}$

SECTION 6.6 · EXERCISES

In Exercises 1–16, use the given information to find the remaining sides and angles of the triangle labeled as shown in Figure 6.29.

1.

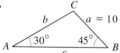

4.

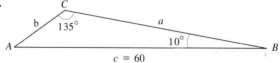

5. $A = 36°$, $a = 8$, $b = 5$

2.

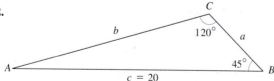

6. $A = 60°$, $a = 9$, $c = 10$

7. $A = 150°$, $C = 20°$, $a = 200$

8. $A = 24.3°$, $C = 54.6°$, $c = 2.68$

9. $A = 83° \, 20'$, $C = 54.6°$, $c = 18.1$

10. $A = 5° \, 40'$, $B = 8° \, 15'$, $b = 4.8$

11. $B = 15° \, 30'$, $a = 4.5$, $b = 6.8$

12. $C = 85° \, 20'$, $a = 35$, $c = 50$

13. $C = 145°$, $b = 4$, $c = 14$

14. $A = 100°$, $a = 125$, $c = 10$

3.

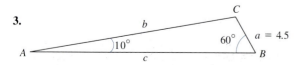

15. $A = 110° \, 15'$, $a = 48$, $b = 16$

16. $B = 2° \, 45'$, $b = 6.2$, $c = 5.8$

In Exercises 17–22, use the given information to find (if possible) the remaining sides and angles of the triangle. If two solutions exist, find both.

17. $A = 58°$, $a = 4.5$, $b = 12.8$
18. $A = 58°$, $a = 11.4$, $b = 12.8$
19. $A = 58°$, $a = 4.5$, $b = 5$
20. $A = 58°$, $a = 42.4$, $b = 50$
21. $A = 110°$, $a = 125$, $b = 200$
22. $A = 110°$, $a = 125$, $b = 100$

In Exercises 23 and 24, find a value for b such that the triangle has (a) one solution, (b) two solutions, and (c) no solution.

23. $A = 36°$, $a = 5$ **24.** $A = 60°$, $a = 10$

In Exercises 25–30, find the area of the triangle having the indicated sides and angles.

25. $C = 120°$, $a = 4$, $b = 6$
26. $B = 72° 30'$, $a = 105$, $c = 64$
27. $A = 43° 45'$, $b = 57$, $c = 85$
28. $A = 5° 15'$, $b = 4.5$, $c = 22$
29. $B = 130°$, $a = 62$, $c = 20$
30. $C = 84° 30'$, $a = 16$, $b = 20$

31. *Streetlight Design* Find the length d of the brace required to support the streetlight in the figure.

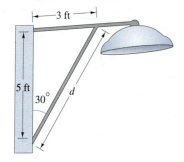

FIGURE FOR 31

32. *Height* Because of the prevailing winds, a tree grew so that it was leaning 6° from the vertical. At a point 100 feet away from the tree, the angle of elevation to the top of the tree is 22° 50′ (see figure). Find the height h of the tree.

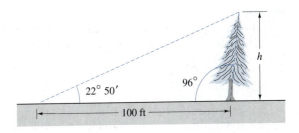

FIGURE FOR 32

33. *Bridge Design* A bridge is to be built across a small lake from B to C (see figure). The bearing from B to C is S 41° W. From a point A, 100 yards from B, the bearings to B and C are S 74° E and S 28° E, respectively. Find the distance from B to C.

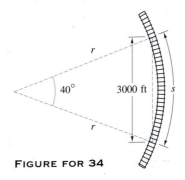

FIGURE FOR 33

34. *Railroad Track Design* The circular arc of a railroad curve has a chord of length 3000 feet, and a central angle of 40° (see figure). Find (a) the radius r of the circular arc and (b) the length s of the circular arc.

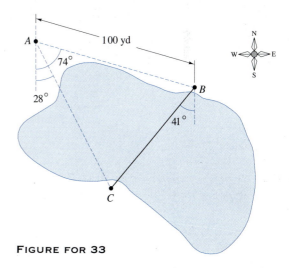

FIGURE FOR 34

35. *Altitude*   The angles of elevation to an airplane from two points *A* and *B* on level ground are 51° and 68°, respectively. The points *A* and *B* are 6 miles apart, and the airplane is between these positions in the same vertical plane. Find the altitude of the airplane.

36. *Altitude*   The angles of elevation to an airplane from two points *A* and *B* on level ground are 51° and 68°, respectively. The points *A* and *B* are 2.5 miles apart, and the airplane is east of both points in the same vertical plane. Find the altitude of the plane.

37. *Locating a Fire*   Two fire towers *A* and *B* are 18.5 miles apart. The bearing from *A* to *B* is N 65° E. A fire is spotted by the ranger in each tower, and its bearings from *A* and *B* are N 28° E and N 16.5° W, respectively (see figure). Find the distance of the fire from each tower.

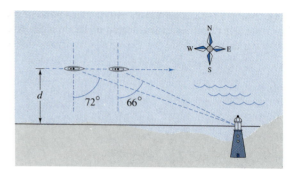

FIGURE FOR 38

39. *Distance*   A family is traveling due west on a road that passes a famous landmark. At a given time, the bearing to the landmark is N 62° W, and after the family travels 5 miles farther the bearing is N 38° W. What is the closest the family will come to the landmark while on the road?

40. *Engine Design*   The connecting rod in a certain engine is 6 inches long, and the radius of the crankshaft is $1\frac{1}{2}$ inches (see figure). The spark plug fires at 5° before top dead center. How far is the piston from the top of its stroke at this time?

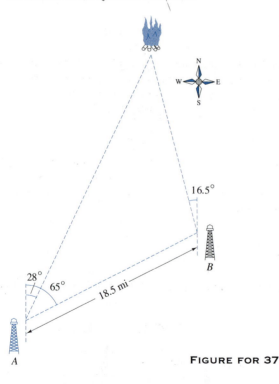

FIGURE FOR 37

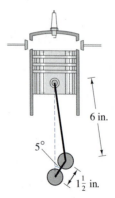

FIGURE FOR 40

38. *Distance*   A boat is sailing due east parallel to the shoreline at a speed of 10 miles per hour. At a given time, the bearing to the lighthouse is S 72° E, and 15 minutes later the bearing is S 66° E (see figure). Find the distance from the boat to the shoreline if the lighthouse is at the shoreline.

41. *Verification of Testimony*   The following information about a triangular parcel of land is given at a zoning board meeting: "One side is 450 feet long, and another is 120 feet long. The angle opposite the shorter side is 30°." Could this information be correct?

42. *Distance*   The angles of elevation to an airplane, θ and ϕ, are being continuously monitored at two observation points A and B, which are 2 miles apart (see figure). Write an equation giving the distance d between the plane and point B in terms of θ and ϕ.

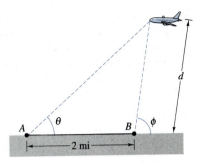

FIGURE FOR 42

6.7   LAW OF COSINES

Law of Cosines / Heron's Formula

Law of Cosines

Two cases remain in the list of conditions needed to solve an oblique triangle—SSS and SAS. The Law of Sines does not work in either of these cases. To see why, consider the three ratios given in the Law of Sines.

$$\frac{a}{\sin A} = \frac{b}{\sin B} = \frac{c}{\sin C}$$

To use the Law of Sines, you must know at least one side and its opposite angle. If you are given three sides (SSS), or two sides and their included angle (SAS), none of the above ratios would be complete. In such cases you must rely on the **Law of Cosines.**

LAW OF COSINES

If ABC is a triangle with sides a, b, and c, then the following equations are valid.

Standard Form	*Alternative Form*
$a^2 = b^2 + c^2 - 2bc \cos A$	$\cos A = \dfrac{b^2 + c^2 - a^2}{2bc}$
$b^2 = a^2 + c^2 - 2ac \cos B$	$\cos B = \dfrac{a^2 + c^2 - b^2}{2ac}$
$c^2 = a^2 + b^2 - 2ab \cos C$	$\cos C = \dfrac{a^2 + b^2 - c^2}{2ab}$

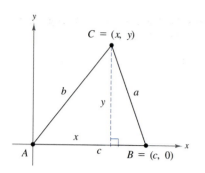

FIGURE 6.41

PROOF

We prove only the first equation for a triangle that has three acute angles, as shown in Figure 6.41. In the figure, note that vertex B has coordinates $(c, 0)$. Furthermore, C has coordinates (x, y), where $x = b \cos A$ and $y = b \sin A$. Because a is the distance from C to B, you have

$$a = \sqrt{(x - c)^2 + (y - 0)^2}$$
$$a^2 = (b \cos A - c)^2 + (b \sin A)^2$$
$$= b^2 \cos^2 A - 2bc \cos A + c^2 + b^2 \sin^2 A$$
$$= b^2(\sin^2 A + \cos^2 A) + c^2 - 2bc \cos A.$$

Using the identity $\sin^2 A + \cos^2 A = 1$ produces

$$a^2 = b^2 + c^2 - 2bc \cos A.$$

Similar arguments can be used to establish the other two equations.

Note that if $A = 90°$ in Figure 6.41, then $\cos A = 0$, and the first form of the Law of Cosines becomes the Pythagorean Theorem.

$$a^2 = b^2 + c^2$$

Thus, the Pythagorean Theorem is actually just a special case of the more general Law of Cosines.

EXAMPLE 1   Given Three Sides of a Triangle—SSS

Find the three angles of the triangle whose sides have lengths $a = 8$ feet, $b = 19$ feet, and $c = 14$ feet.

SOLUTION

It is a good idea first to find the angle opposite the longest side—side b in this case (see Figure 6.42). Using the Law of Cosines produces

$$\cos B = \frac{a^2 + c^2 - b^2}{2ac} = \frac{8^2 + 14^2 - 19^2}{2(8)(14)} \approx -0.45089.$$

Because $\cos B$ is negative, you know B is an *obtuse* angle given by $B \approx 116.80°$. At this point you could use the Law of Cosines to find $\cos A$ and $\cos C$. However, knowing that $B \approx 116.80°$, it is simpler to use the Law of Sines to obtain the following.

$$\frac{b}{\sin B} = \frac{a}{\sin A}$$

$$\sin A = a\left(\frac{\sin B}{b}\right) \approx 8\left(\frac{\sin 116.80°}{19}\right) \approx 0.37582$$

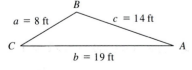

FIGURE 6.42

Because B is obtuse, you know that A must be acute, because a triangle can have, at most, one obtuse angle. Thus, $A \approx 22.08°$ and

$$C \approx 180° - 22.08° - 116.80° = 41.12°.$$

Do you see why it was wise to find the largest angle *first* in Example 1? Knowing the cosine of an angle, you can determine whether the angle is acute or obtuse. That is,

$\cos \theta > 0$ for $0° < \theta < 90°$ *Acute*

$\cos \theta < 0$ for $90° < \theta < 180°$.   *Obtuse*

So, in Example 1, once you found that B was obtuse, you could determine that A and C were both acute. If the largest angle is acute, then the remaining two angles will be acute also.

EXAMPLE 2   Given Two Sides and the Included Angle—SAS

The pitcher's mound on a softball field is 46 feet from home plate, and the distance between the bases is 60 feet, as shown in Figure 6.43. How far is the pitcher's mound from first base? (Note that the pitcher's mound is *not* halfway between home plate and second base.)

SOLUTION

From triangle HPF, you can see that $H = 45°$ (line HP bisects the right angle at H), $f = 46$, and $p = 60$. Using the Law of Cosines for this SAS case produces

$$\begin{aligned} h^2 &= f^2 + p^2 - 2fp \cos H \\ &= 46^2 + 60^2 - 2(46)(60) \cos 45° \\ &\approx 1812.8. \end{aligned}$$

Therefore, the approximate distance from the pitcher's mound to first base is

$$h \approx \sqrt{1812.8} \approx 42.58 \text{ feet.}$$

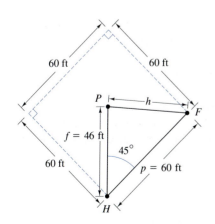

FIGURE 6.43

EXAMPLE 3   **Given Two Sides and the Included Angle—SAS**

A ship travels 60 miles due east, then adjusts its course 15° northward, as shown in Figure 6.44. After traveling 80 miles in that direction, how far is the ship from its point of departure?

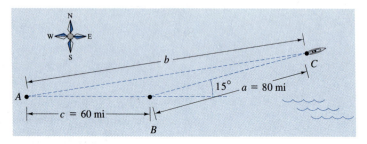

FIGURE 6.44

SOLUTION

You have $c = 60$, $B = 180° - 15° = 165°$, and $a = 80$. Consequently, by the Law of Cosines, you obtain

$$b^2 = a^2 + c^2 - 2ac \cos B$$
$$= 80^2 + 60^2 - 2(80)(60) \cos 165° \approx 19{,}273.$$

Therefore, the distance b is

$$b \approx \sqrt{19{,}273} \approx 138.8 \text{ miles.}$$

Heron's Formula

The following formula for the area of a triangle is credited to the Greek mathematician Heron (c. 100 B.C.).

HERON'S AREA FORMULA

Given any triangle with sides of lengths a, b, and c, the area of the triangle is

$$\text{Area} = \sqrt{s(s - a)(s - b)(s - c)},$$

where $s = (a + b + c)/2$ is one-half the perimeter of the triangle.

EXAMPLE 4   Using Heron's Area Formula

Find the area of the triangular region having sides of lengths $a = 47$ yards, $b = 58$ yards, and $c = 78.6$ yards.

SOLUTION

Because

$$s = \frac{1}{2}(a + b + c) = \frac{183.6}{2} = 91.8,$$

Heron's Formula yields

$$\begin{aligned}
\text{Area} &= \sqrt{s(s - a)(s - b)(s - c)} \\
&= \sqrt{91.8(44.8)(33.8)(13.2)} \\
&\approx 1354.58 \text{ square yards.}
\end{aligned}$$

DISCUSSION PROBLEM

THE AREA OF A TRIANGLE

We have now discussed three different formulas for the area of a triangle.

Standard formula: Area $= \frac{1}{2}bh$

Oblique triangle: Area $= \frac{1}{2}bc \sin A = \frac{1}{2}ab \sin C = \frac{1}{2}ac \sin B$

Heron's formula: Area $= \sqrt{s(s - a)(s - b)(s - c)}$, $s = \frac{1}{2}(a + b + c)$

Use the most appropriate formula to find the area of each of the following triangles. Show your work and give your reason for choosing each formula.

(a)

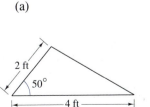

(b)

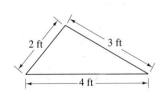

(c)

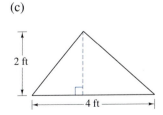

WARM-UP

The following warm-up exercises involve skills that were covered in earlier sections. You will use these skills in the exercise set for this section.

In Exercises 1 and 2, simplify the expression.

1. $\sqrt{(7-3)^2 + [1-(-5)]^2}$

2. $\sqrt{[-2-(-5)]^2 + (12-6)^2}$

In Exercises 3 and 4, find the distance between the two points.

3. $(4, -2)$, $(8, 10)$

4. $(1, 3)$, $(7, 12)$

In Exercises 5 and 6, find the area of the triangle.

5.

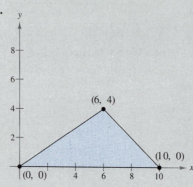

6.

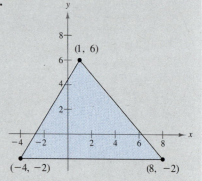

In Exercises 7–10, find (if possible) the remaining sides and angles of the triangle.

7. $A = 10°$, $C = 100°$, $b = 25$

8. $A = 20°$, $C = 90°$, $c = 100$

9. $B = 30°$, $b = 6.5$, $c = 15$

10. $A = 30°$, $b = 6.5$, $a = 10$

SECTION 6.7 · EXERCISES

In Exercises 1–14, use the Law of Cosines to solve the triangle.

1.

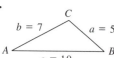

2.

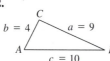

3.

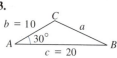

4.

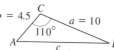

5. $a = 9$, $b = 12$, $c = 15$

6. $a = 55$, $b = 25$, $c = 72$

7. $a = 75.4$, $b = 52$, $c = 52$

8. $a = 1.42$, $b = 0.75$, $c = 1.25$

9. $A = 120°$, $b = 3$, $c = 10$

10. $A = 55°$, $b = 3$, $c = 10$

11. $B = 8° \, 45'$, $a = 25$, $c = 15$

12. $B = 75° \, 20'$, $a = 6.2$, $c = 9.5$

13. $C = 125° \, 40'$, $a = 32$, $b = 32$

14. $C = 15°$, $a = 6.25$, $b = 2.15$

In Exercises 15–20, complete the table by solving the parallelogram shown in the figure. (The lengths of the diagonals are c and d.)

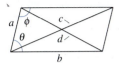

FIGURE FOR 15–20

	a	b	c	d	θ	ϕ
15.	4	6			30°	
16.	25	35				120°
17.	10	14	20			
18.	40	60		80		
19.	10		18	12		
20.		25	50	35		

In Exercises 21–26, use Heron's Formula to find the area of the triangle.

21. $a = 5$, $b = 7$, $c = 10$
22. $a = 2.5$, $b = 10.2$, $c = 9$
23. $a = 12$, $b = 15$, $c = 9$
24. $a = 75.4$, $b = 52$, $c = 52$
25. $a = 20$, $b = 20$, $c = 10$
26. $a = 4.25$, $b = 1.55$, $c = 3.00$

27. *Area*   The lengths of the sides of a triangular parcel of land are approximately 400 feet, 500 feet, and 700 feet. Approximate the area of the parcel.

28. *Area*   The lengths of two adjacent sides of a parallelogram are 4 yards and 6 yards. Find the area of the parallelogram if the angle between the two sides is 30°.

29. *Navigation*   A boat race is run along a triangular course marked by buoys A, B, and C. The race starts with the boats headed west. The other two sides of the course lie to the north of the first side, and their lengths are 3500 feet and 6500 feet (see figure). Find the bearings for the last two legs of the race.

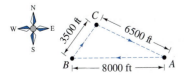

FIGURE FOR 29

30. *Navigation*   A plane flies 675 miles from A to B with a bearing of N 75° E. Then it flies 540 miles from B to C with a bearing of N 32° E (see figure). Find the straight-line distance and bearing from C to A.

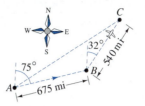

FIGURE FOR 30

31. *Distance*   Two ships leave a port at 9 A.M. One travels at a bearing of N 53° W at 12 miles per hour and the other at a bearing of S 67° W at 16 miles per hour. Approximately how far apart are they at noon that day?

32. *Distance*   A 100-foot vertical tower is to be erected on the side of a hill that makes an 8° angle with the horizontal (see figure). Find the length of each of the two guy wires that will be anchored 75 feet uphill and downhill from the base of the tower.

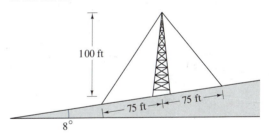

FIGURE FOR 32

33. *Surveying*   To approximate the length of a marsh, a surveyor walks 950 feet from point A to point B, then turns 80° and walks 800 feet to point C (see figure). Approximate the length $\overline{AC}$ of the marsh.

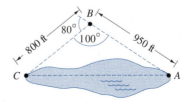

FIGURE FOR 33

34. *Surveying*   A triangular parcel of land has 375 feet of frontage, and the other boundaries have lengths of 250 feet and 300 feet. What angles does the frontage make with the two other boundaries?

35. *Streetlight Design*   Determine the angle θ in the design of the streetlight as shown in the figure.

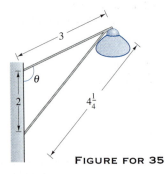

FIGURE FOR 35

36. *Aircraft Tracking*   In order to determine the distance between two aircraft, a tracking station continuously determines the distance to each aircraft and the angle α between them. Determine the distance a between the planes when $\alpha = 42°$, $b = 35$ miles, and $c = 20$ miles.

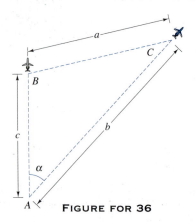

FIGURE FOR 36

37. *Engineering*   If Q is the midpoint of the line segment $\overline{PR}$, find the lengths of the line segments $\overline{PQ}$, $\overline{QS}$, and $\overline{RS}$ on the truss rafter shown in the figure.

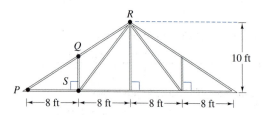

FIGURE FOR 37

38. *Paper Manufacturing*   In a certain process with continuous paper, the paper passes across three rollers of radii 3 inches, 4 inches, and 6 inches (see figure). The centers of the 3-inch and 6-inch rollers are d inches apart, and the length of the arc in contact with the paper on the 4-inch roller is s inches. Complete the following table.

d (inches)	9	10	12	13	14	15	16
θ (degrees)							
s (inches)							

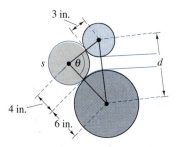

FIGURE FOR 38

39. *Navigation*   On a certain map, Orlando is 7 inches due south of Niagara Falls, Denver is 10.75 inches from Orlando, and Denver is 9.25 inches from Niagara Falls (see figure).
(a) Find the bearing of Denver from Orlando.
(b) Find the bearing of Denver from Niagara Falls.

FIGURE FOR 39

40. *Navigation*   On a certain map, Minneapolis is 6.5 inches due west of Albany, Phoenix is 8.5 inches from Minneapolis, and Phoenix is 14.5 inches from Albany (see figure).
(a) Find the bearing of Minneapolis from Phoenix.
(b) Find the bearing of Albany from Phoenix.

FIGURE FOR 40

41. *Baseball*   In a (square) baseball diamond with 90-foot sides, the pitcher's mound is 60 feet from home plate.
(a) How far is it from the pitcher's mound to third base?
(b) When a runner is halfway from second to third, how far is the runner from the pitcher's mound?

42. *Baseball*   The baseball player in center field is playing approximately 330 feet from the television camera that is behind home plate. A batter hits a fly ball that goes to the wall 420 feet from the camera (see figure). Approximate the number of feet that the center fielder had to run to make the catch if the camera turned 9° in following the play.

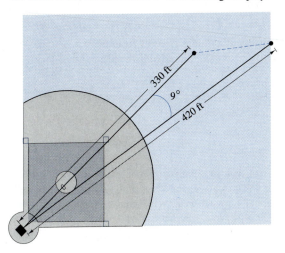

FIGURE FOR 42

43. *Awning Design*   A retractable awning lowers at an angle of 50° from the top of a patio door that is 7 feet high (see figure). Find the length x of the awning if no direct sunlight is to enter the door when the angle of elevation of the sun is greater than 65°.

44. *Circumscribed and Inscribed Circles*   Let R and r be the radii of the circumscribed and inscribed circles of a triangle ABC, respectively, and let $s = (a + b + c)/2$ (see figure). Prove the following.

(a) $2R = \dfrac{a}{\sin A} = \dfrac{b}{\sin B} = \dfrac{c}{\sin C}$

(b) $r = \sqrt{\dfrac{(s - a)(s - b)(s - c)}{s}}$

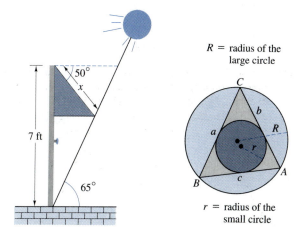

FIGURE FOR 43   **FIGURE FOR 44**

Circumscribed and Inscribed Circles   In Exercises 45 and 46, use the results of Exercise 44.

45. Given the triangle with $a = 25$, $b = 55$, and $c = 72$, find the areas of (a) the triangle, (b) the circumscribed circle, and (c) the inscribed circle.

46. Find the length of the largest circular track that can be built on a triangular piece of property whose sides measure 200 feet, 250 feet, and 325 feet.

47. Use the Law of Cosines to prove that
$$\frac{1}{2}bc(1 + \cos A) = \frac{a + b + c}{2} \cdot \frac{-a + b + c}{2}.$$

48. Use the Law of Cosines to prove that
$$\frac{1}{2}bc(1 - \cos A) = \frac{a - b + c}{2} \cdot \frac{a + b - c}{2}.$$

6.8   DeMoivre's Theorem and Nth Roots

Trigonometric Form of a Complex Number   /   **Multiplication and Division of Complex Numbers**   /   **Powers of Complex Numbers**   /   **Roots of Complex Numbers**

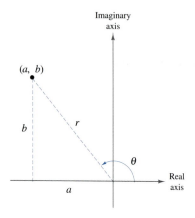

Complex Number: $a + bi$

FIGURE 6.45

Trigonometric Form of a Complex Number

In Section 2.3 you learned how to add, subtract, multiply, and divide complex numbers. To work effectively with *powers* and *roots* of complex numbers, it is helpful to write complex numbers in **trigonometric form.** In Figure 6.45, consider the nonzero complex number $a + bi$. By letting θ be the angle from the positive x-axis (measured counterclockwise) to the line segment connecting the origin and the point (a, b), you can write

$$a = r \cos \theta \quad \text{and} \quad b = r \sin \theta,$$

where $r = \sqrt{a^2 + b^2}$. Consequently, you have

$$a + bi = (r \cos \theta) + (r \sin \theta)i,$$

from which you obtain the following **trigonometric form of a complex number.**

TRIGONOMETRIC FORM OF A COMPLEX NUMBER

Let $z = a + bi$ be a complex number. The **trigonometric form** of z is

$$z = r(\cos \theta + i \sin \theta),$$

where $a = r \cos \theta$, $b = r \sin \theta$, $r = \sqrt{a^2 + b^2}$, and $\tan \theta = b/a$. The number r is the **modulus** of z, and θ is an **argument** of z. The modulus of a complex number is also called its **absolute value** and is denoted by

$$|z| = \sqrt{a^2 + b^2}.$$

REMARK   The trigonometric form of a complex number is also called the **polar form.** Because there are infinitely many choices for θ, the trigonometric form of a complex number is not unique. Normally, we use θ values in the interval $0 \le \theta < 2\pi$, although on occasion we may use $\theta < 0$. Note also that if $a = 0$, then $\theta = \pi/2$ even though $\tan \theta$ is not defined.

EXAMPLE 1   Writing Complex Numbers in Trigonometric Form

Write the following complex numbers in trigonometric form.

A. $z = -2 - 2\sqrt{3}\,i$ **B.** $z = 6 + 2i$

Use radian measure for part (a) and degree measure for part (b).

SOLUTION

The graphs of the complex numbers are shown in Figure 6.46.

A. The absolute value of z is

$$r = \left| -2 - 2\sqrt{3}\,i \right|$$
$$= \sqrt{(-2)^2 + \left(-2\sqrt{3}\right)^2}$$
$$= \sqrt{16}$$
$$= 4,$$

and the angle θ is given by

$$\tan\theta = \frac{b}{a} = \frac{-2\sqrt{3}}{-2} = \sqrt{3}.$$

Because $\tan \pi/3 = \sqrt{3}$ and $z = -2 - 2\sqrt{3}\,i$ lies in Quadrant III, choose θ to be $\theta = \pi + \pi/3 = 4\pi/3$. Thus, the trigonometric form is

$$z = r(\cos\theta + i\sin\theta)$$
$$= 4\left(\cos\frac{4\pi}{3} + i\sin\frac{4\pi}{3}\right).$$

B. Here you have $r = |6 + 2i| = 2\sqrt{10}$ with θ given by

$$\tan\theta = \frac{2}{6} = \frac{1}{3} \qquad\qquad \text{\textit{θ in Quadrant I}}$$

$$\theta = \arctan\frac{1}{3} \approx 18.4°.$$

Therefore the trigonometric form of z is

$$z = r(\cos\theta + i\sin\theta)$$
$$= 2\sqrt{10}\left[\cos\left(\arctan\frac{1}{3}\right) + i\sin\left(\arctan\frac{1}{3}\right)\right]$$
$$\approx 2\sqrt{10}(\cos 18.4° + i\sin 18.4°).$$

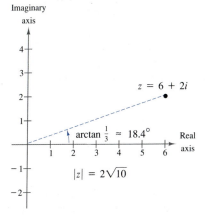

(a)

(b)

FIGURE 6.46

Technology Note _____

You can use your graphing utility to convert complex numbers to trigonometric form. For example, to convert $z = -2 - 2\sqrt{3}\,i$ using the TI-81, go to the $R \triangleright P($ key in the MATH menu. Upon entering $\left(-2, -2\sqrt{3}\,\right)$ as an ordered pair separated by a comma, and pressing ENTER, the calculator returns 4 as the absolute value of $-2 - 2\sqrt{3}\,i$. Pressing θ (above the 3 key) returns -2.094395102, which is $4\pi/3 - 2\pi$. Note that the calculator automatically stores the absolute value (4) in the R register, and the angle (-2.094395102) in the θ register. If you change the MODE to degrees, you can verify part (B) of Example 1.

EXAMPLE 2 **Writing a Complex Number in Standard Form**

Write the following complex number in standard form $a + bi$.

$$z = \sqrt{8}\left[\cos\left(-\frac{\pi}{3}\right) + i\,\sin\left(-\frac{\pi}{3}\right)\right]$$

SOLUTION

Because $\cos(-\pi/3) = \frac{1}{2}$ and $\sin(-\pi/3) = -\sqrt{3}/2$, you can write

$$z = \sqrt{8}\left[\cos\left(-\frac{\pi}{3}\right) + i\,\sin\left(-\frac{\pi}{3}\right)\right]$$

$$= \sqrt{8}\left(\frac{1}{2} - \frac{\sqrt{3}}{2}i\right)$$

$$= 2\sqrt{2}\left(\frac{1}{2} - \frac{\sqrt{3}}{2}i\right)$$

$$= \sqrt{2} - \sqrt{6}\,i.$$

Technology Note _____

You can also use your graphing utility to convert from trigonometric form to standard form. On the TI-81, use the $P \triangleright R($ key in the MATH menu. The ordered pair $\left(\sqrt{8}, -\pi/3\right)$ is converted to $x = \sqrt{2}$ and $y = -\sqrt{6}$.

Multiplication and Division of Complex Numbers

The trigonometric form adapts nicely to multiplication and division of complex numbers. Suppose you are given two complex numbers

$$z_1 = r_1(\cos\theta_1 + i\,\sin\theta_1) \quad \text{and} \quad z_2 = r_2(\cos\theta_2 + i\,\sin\theta_2).$$

The product of z_1 and z_2 is

$$z_1 z_2 = r_1 r_2(\cos\theta_1 + i\,\sin\theta_1)(\cos\theta_2 + i\,\sin\theta_2)$$

$$= r_1 r_2[(\cos\theta_1\cos\theta_2 - \sin\theta_1\sin\theta_2)$$

$$+ i(\sin\theta_1\cos\theta_2 + \cos\theta_1\sin\theta_2)].$$

Using the sum and difference formulas for cosine and sine, you can rewrite this equation as

$$z_1 z_2 = r_1 r_2[\cos(\theta_1 + \theta_2) + i\,\sin(\theta_1 + \theta_2)].$$

This establishes the first part of the following rule.

PRODUCT AND QUOTIENT OF TWO COMPLEX NUMBERS

Let $z_1 = r_1(\cos\theta_1 + i\sin\theta_1)$ and $z_2 = r_2(\cos\theta_2 + i\sin\theta_2)$ be complex numbers.

$$z_1 z_2 = r_1 r_2[\cos(\theta_1 + \theta_2) + i\sin(\theta_1 + \theta_2)] \qquad \textit{Product}$$

$$\frac{z_1}{z_2} = \frac{r_1}{r_2}[\cos(\theta_1 - \theta_2) + i\sin(\theta_1 - \theta_2)], \qquad z_2 \neq 0 \qquad \textit{Quotient}$$

Note that this rule says that to multiply two complex numbers, you multiply moduli and add arguments, whereas to divide two complex numbers you divide moduli and subtract arguments.

EXAMPLE 3 Multiplying Complex Numbers in Trigonometric Form

Find the product of the following complex numbers.

$$z_1 = 2\left(\cos\frac{2\pi}{3} + i\sin\frac{2\pi}{3}\right) \qquad z_2 = 8\left(\cos\frac{11\pi}{6} + i\sin\frac{11\pi}{6}\right)$$

Express your answer in standard form.

SOLUTION

$$z_1 z_2 = 2\left(\cos\frac{2\pi}{3} + i\sin\frac{2\pi}{3}\right) \cdot 8\left(\cos\frac{11\pi}{6} + i\sin\frac{11\pi}{6}\right)$$

$$= 16\left[\cos\left(\frac{2\pi}{3} + \frac{11\pi}{6}\right) + i\sin\left(\frac{2\pi}{3} + \frac{11\pi}{6}\right)\right]$$

$$= 16\left(\cos\frac{5\pi}{2} + i\sin\frac{5\pi}{2}\right)$$

$$= 16\left(\cos\frac{\pi}{2} + i\sin\frac{\pi}{2}\right)$$

$$= 16[0 + i(1)] = 16i$$

Check this result by first converting to the standard forms $z_1 = -1 + \sqrt{3}i$ and $z_2 = 4\sqrt{3} - 4i$ and then multiplying algebraically, as in Section 2.3.

EXAMPLE 4   Dividing Complex Numbers in Trigonometric Form

Find z_1/z_2 for the following two complex numbers.

$$z_1 = 24(\cos 300° + i \sin 300°) \qquad z_2 = 8(\cos 75° + i \sin 75°)$$

Express your answer in standard form.

SOLUTION

$$\frac{z_1}{z_2} = \frac{24(\cos 300° + i \sin 300°)}{8(\cos 75° + i \sin 75°)}$$

$$= \frac{24}{8}[\cos(300° - 75°) + i \sin(300° - 75°)]$$

$$= 3(\cos 225° + i \sin 225°)$$

$$= 3\left[\left(-\frac{\sqrt{2}}{2}\right) + i\left(-\frac{\sqrt{2}}{2}\right)\right]$$

$$= -\frac{3\sqrt{2}}{2} - \frac{3\sqrt{2}}{2}i$$

Powers of Complex Numbers

Consider the complex number (in trigonometric form) $z = r(\cos \theta + i \sin \theta)$. Repeated use of the multiplication rule from the previous section yields

$$z = r(\cos \theta + i \sin \theta)$$

$$z^2 = r(\cos \theta + i \sin \theta)r(\cos \theta + i \sin \theta) = r^2(\cos 2\theta + i \sin 2\theta)$$

$$z^3 = z^2(z) = r^2(\cos 2\theta + i \sin 2\theta)r(\cos \theta + i \sin \theta)$$

$$= r^3(\cos 3\theta + i \sin 3\theta).$$

Similarly,

$$z^4 = r^4(\cos 4\theta + i \sin 4\theta)$$

$$z^5 = r^5(\cos 5\theta + i \sin 5\theta)$$

$$\vdots$$

This pattern leads to the following important theorem, which is named after the French mathematician Abraham DeMoivre (1667–1754).

*Technology Note*_____

DeMoivre's Theorem is easily implemented on your graphing utility. For instance, to verify Example 5 on the TI-81, go to the $R \triangleright P($ key in the MATH menu. Entering $(-1, \sqrt{3})$ as an ordered pair separated by a comma, and pressing ENTER converts the complex number $-1 + \sqrt{3}i$ to trigonometric form. Then go to the $P \triangleright R($ key and enter $(R \wedge 12, 12\theta)$. Compare the result to the answer found in Example 5. Why does the Y register contain 0?

DEMOIVRE'S THEOREM

If $z = r(\cos \theta + i \sin \theta)$ is a complex number and n is a positive integer, then

$$z^n = [r(\cos \theta + i \sin \theta)]^n = r^n(\cos n\theta + i \sin n\theta).$$

EXAMPLE 5   **Finding Powers of a Complex Number**

Use DeMoivre's Theorem to find $\left(-1 + \sqrt{3}\,i\right)^{12}$.

SOLUTION

First convert to trigonometric form.

$$-1 + \sqrt{3}\,i = 2\left(\cos\frac{2\pi}{3} + i\,\sin\frac{2\pi}{3}\right)$$

Then, by DeMoivre's Theorem, you have

$$\left(-1 + \sqrt{3i}\right) = \left[2\left(\cos\frac{2\pi}{3} + i\,\sin\frac{2\pi}{3}\right)\right]^{12}$$

$$= 2^{12}\left[\cos\left(12\cdot\frac{2\pi}{3}\right) + i\,\sin\left(12\cdot\frac{2\pi}{3}\right)\right]$$

$$= 4096(\cos 8\pi + i\,\sin 8\pi)$$

$$= 4096(1 + 0)$$

$$= 4096.$$

Are you surprised to see a real number as the answer?

Roots of Complex Numbers

Recall that a consequence of the Fundamental Theorem of Algebra is that a polynomial equation of degree n has n solutions in the complex number system. Hence, an equation such as $x^6 = 1$ has six solutions, which can be found by factoring and using the Quadratic Formula.

$$x^6 - 1 = (x^3 - 1)(x^3 + 1)$$
$$= (x - 1)(x^2 + x + 1)(x + 1)(x^2 - x + 1) = 0$$

Consequently, the solutions are

$$x = \pm 1, \quad x = \frac{-1 \pm \sqrt{3}\,i}{2}, \quad \text{and} \quad x = \frac{1 \pm \sqrt{3}\,i}{2}.$$

Each of these numbers is a *sixth root of* 1.

DEFINITION OF NTH ROOT OF A COMPLEX NUMBER

The complex number $u = a + bi$ is an **nth root** of the complex number z if

$$z = u^n = (a + bi)^n.$$

If $z = 1$, then u is called an **nth root of unity.**

To find a formula for an nth root of a complex number, let u be an nth root of z, where

$$u = s(\cos \beta + i \sin \beta) \quad \text{and} \quad z = r(\cos \theta + i \sin \theta).$$

By DeMoivre's Theorem and the fact that $u^n = z$, you have

$$s^n(\cos n\beta + i \sin n\beta) = r(\cos \theta + i \sin \theta).$$

Now, taking the absolute value of both sides of this equation, it follows that $s^n = r$. Substituting back into the previous equation and dividing by r produces

$$\cos n\beta + i \sin n\beta = \cos \theta + i \sin \theta.$$

Thus, it follows that

$$\cos n\beta = \cos \theta \quad \text{and} \quad \sin n\beta = \sin \theta.$$

Because both sine and cosine have a period of 2π, these last two equations have solutions if and only if the angles differ by a multiple of 2π. Consequently, there must exist an integer k such that

$$n\beta = \theta + 2\pi k$$

$$\beta = \frac{\theta + 2\pi k}{n}.$$

By substituting this value for β into the trigonometric form of u, you obtain the result stated in the following theorem.

NTH ROOTS OF A COMPLEX NUMBER

For a positive integer n, the complex number $z = r(\cos \theta + i \sin \theta)$ has exactly n distinct nth roots given by

$$\sqrt[n]{r}\left(\cos \frac{\theta + 2\pi k}{n} + i \sin \frac{\theta + 2\pi k}{n}\right),$$

where $k = 0, 1, 2, \ldots, n - 1$.

REMARK Note that when k exceeds $n - 1$, the roots begin to repeat. For instance, if $k = n$, the angle $(\theta + 2\pi n)/n$ is coterminal with θ/n, which is also obtained when $k = 0$.

This formula for the nth roots of a complex number z has a nice geometrical interpretation, as shown in Figure 6.47. Note that because the nth roots of z all have the same magnitude $\sqrt[n]{r}$, they all lie on a circle of radius $\sqrt[n]{r}$ with center at the origin. Furthermore, the n roots are equally spaced along the circle, because successive nth roots have arguments that differ by $2\pi/n$.

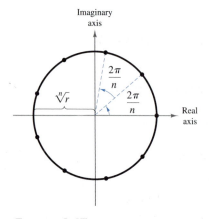

FIGURE 6.47

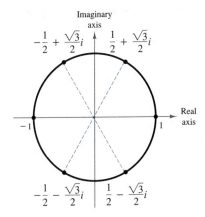

Imaginary
axis

$-\dfrac{1}{2} + \dfrac{\sqrt{3}}{2}i$ $\dfrac{1}{2} + \dfrac{\sqrt{3}}{2}i$

Real
axis

-1 1

$-\dfrac{1}{2} - \dfrac{\sqrt{3}}{2}i$ $\dfrac{1}{2} - \dfrac{\sqrt{3}}{2}i$

FIGURE 6.48

REMARK   In Figure 6.48, notice that the roots obtained in Example 6 all have a magnitude of 1 and are equally spaced around this unit circle. Also, notice that the complex roots occur in conjugate pairs, as previously discussed in Section 2.3.

EXAMPLE 6   Finding *n*th Roots of a Real Number

Find all the sixth roots of 1.

SOLUTION

First, write 1 in the trigonometric form $1 = 1(\cos 0 + i \sin 0)$. Then, by the *n*th root formula, with $n = 6$ and $r = 1$, the roots have the form

$$\sqrt[6]{1}\left(\cos \frac{0 + 2\pi k}{6} + i \sin \frac{0 + 2\pi k}{6}\right),$$

or simply $\cos(\pi k/3) + i \sin(\pi k/3)$. Thus, for $k = 0, 1, 2, 3, 4, 5$, the sixth roots are as follows. (See Figure 6.48.)

$$\cos 0 + i \sin 0 = 1$$

$$\cos \frac{\pi}{3} + i \sin \frac{\pi}{3} = \frac{1}{2} + \frac{\sqrt{3}}{2}i$$

$$\cos \frac{2\pi}{3} + i \sin \frac{2\pi}{3} = -\frac{1}{2} + \frac{\sqrt{3}}{2}i$$

$$\cos \pi + i \sin \pi = -1$$

$$\cos \frac{4\pi}{3} + i \sin \frac{4\pi}{3} = -\frac{1}{2} - \frac{\sqrt{3}}{2}i$$

$$\cos \frac{5\pi}{3} + i \sin \frac{5\pi}{3} = \frac{1}{2} - \frac{\sqrt{3}}{2}i$$

EXAMPLE 7   Finding the *n*th Roots of a Complex Number

Find the three cube roots of $z = -2 + 2i$.

SOLUTION

Because z lies in Quadrant II, the trigonometric form for z is

$$z = -2 + 2i = \sqrt{8}(\cos 135° + i \sin 135°).$$

By our formula for *n*th roots, the cube roots have the form

$$\sqrt[3]{8}\left(\cos \frac{135° + 360°k}{3} + i \sin \frac{135° + 360°k}{3}\right).$$

Finally, for $k = 0, 1, 2$, you obtain the roots

$$\sqrt{2}(\cos 45° + i \sin 45°) = 1 + i$$

$$\sqrt{2}(\cos 165° + i \sin 165°) \approx -1.3660 + 0.3660i$$

$$\sqrt{2}(\cos 285° + i \sin 285°) \approx 0.3660 - 1.3660i.$$

REMARK   In Example 7, note that the roots *do not* occur in conjugate pairs. Do you see why?

EXAMPLE 8   Finding the Roots of a Polynomial Equation

Solve $x^4 + 16 = 0$.

SOLUTION

The given equation can be written as

$$x^4 = -16 \quad \text{or} \quad 4 = 16(\cos \pi + i \sin \pi),$$

which means that you can solve the equation by finding the four fourth roots of -16. Each of these roots has the form

$$\sqrt[4]{16}\left(\cos \frac{\pi + 2\pi k}{4} + i \sin \frac{\pi + 2\pi k}{4}\right).$$

Finally, using $k = 0, 1, 2, 3$, you obtain the roots

$$2\left(\cos \frac{\pi}{4} + i \sin \frac{\pi}{4}\right) = 2\left(\frac{\sqrt{2}}{2} + \frac{\sqrt{2}}{2}i\right) = \sqrt{2} + \sqrt{2}\,i$$

$$2\left(\cos \frac{3\pi}{4} + i \sin \frac{3\pi}{4}\right) = 2\left(-\frac{\sqrt{2}}{2} + \frac{\sqrt{2}}{2}i\right) = -\sqrt{2} + \sqrt{2}\,i$$

$$2\left(\cos \frac{5\pi}{4} + i \sin \frac{5\pi}{4}\right) = 2\left(-\frac{\sqrt{2}}{2} - \frac{\sqrt{2}}{2}i\right) = -\sqrt{2} - \sqrt{2}\,i$$

$$2\left(\cos \frac{7\pi}{4} + i \sin \frac{7\pi}{4}\right) = 2\left(\frac{\sqrt{2}}{2} - \frac{\sqrt{2}}{2}i\right) = \sqrt{2} - \sqrt{2}\,i.$$

DISCUSSION PROBLEM

A FAMOUS MATHEMATICAL FORMULA

In this section you studied DeMoivre's Theorem, which gives a formula for raising a complex number to a positive integer power. Another famous formula that involves complex numbers and powers is called Euler's Formula, after the German mathematician Leonhard Euler (1707–1783). This formula states that

$$e^{a+bi} = e^a(\cos b + i \sin b).$$

While the interpretation of this formula is beyond the scope of this text, we decided to include it because it gives rise to one of the most wonderful equations in mathematics,

$$e^{\pi i} + 1 = 0.$$

The elegant equation relates the five most famous numbers in mathematics, 0, 1, π, e, and i, in a single equation. Show how Euler's Formula can be used to derive the equation.

WARM-UP

The following warm-up exercises involve skills that were covered in earlier sections. You will use these skills in the exercise set for this section.

In Exercises 1 and 2, simplify the expression.

1. $\sqrt[3]{54}$ **2.** $\sqrt[4]{16 + 48}$

In Exercises 3–6, write the complex number in trigonometric form.

3. $-5 + 5i$ **4.** $-3i$ **5.** -12 **6.** 12

In Exercises 7–10, perform the indicated operation. Leave the result in trigonometric form.

7. $\left(\cos\dfrac{\pi}{4} + i\sin\dfrac{\pi}{4}\right)\left(\cos\dfrac{\pi}{2} + i\sin\dfrac{\pi}{2}\right)$

8. $\left(\cos\dfrac{\pi}{12} + i\sin\dfrac{\pi}{12}\right)\left(\cos\dfrac{5\pi}{6} + i\sin\dfrac{5\pi}{6}\right)$

9. $\dfrac{6[\cos(2\pi/3) + i\sin(2\pi/3)]}{3[\cos(\pi/6) + i\sin(\pi/6)]}$

10. $\dfrac{2(\cos 55° + i\sin 55°)}{3(\cos 10° + i\sin 10°)}$

SECTION 6.8 · EXERCISES

In Exercises 1–6, represent the complex number graphically and find its absolute value.

1. $-5i$ **2.** -5

3. $-4 + 4i$ **4.** $5 - 12i$

5. $6 - 7i$ **6.** $-8 + 3i$

In Exercises 7–10, express the complex number in trigonometric form.

7.

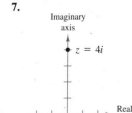

8.

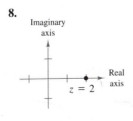

9.

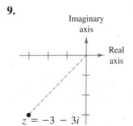

10.

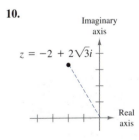

In Exercises 11–26, represent the complex number graphically, and find the trigonometric form of the number.

11. $3 - 3i$ **12.** $-2 - 2i$

13. $\sqrt{3} + i$ **14.** $-1 + \sqrt{3}i$

15. $-2(1 + \sqrt{3}i)$ **16.** $\frac{5}{2}(\sqrt{3} - i)$

17. $6i$ **18.** 4

19. $-7 + 4i$ **20.** $3 - i$

21. 7 **22.** $-2i$

23. $1 + 6i$ **24.** $2\sqrt{2} - i$

25. $-3 - i$ **26.** $1 + 3i$

In Exercises 27–36, represent the complex number graphically, and find the standard form of the number.

27. $2(\cos 150° + i \sin 150°)$

28. $5(\cos 135° + i \sin 135°)$

29. $\frac{3}{2}(\cos 300° + i \sin 300°)$

30. $\frac{3}{4}(\cos 315° + i \sin 315°)$

31. $3.75\left(\cos \dfrac{3\pi}{4} + i \sin \dfrac{3\pi}{4}\right)$

32. $8\left(\cos \dfrac{\pi}{12} + i \sin \dfrac{\pi}{12}\right)$

33. $4\left(\cos \dfrac{3\pi}{2} + i \sin \dfrac{3\pi}{2}\right)$

34. $7(\cos 0° + i \sin 0°)$

35. $3[\cos(18° \ 45') + i \sin(18° \ 45')]$

36. $6[\cos(230° \ 30') + i \sin(230° \ 30')]$

In Exercises 37–48, perform the indicated operation and leave the result in trigonometric form.

37. $\left[3\left(\cos \dfrac{\pi}{3} + i \sin \dfrac{\pi}{3}\right)\right]\left[4\left(\cos \dfrac{\pi}{6} + i \sin \dfrac{\pi}{6}\right)\right]$

38. $\left[\dfrac{3}{2}\left(\cos \dfrac{\pi}{2} + i \sin \dfrac{\pi}{2}\right)\right]\left[6\left(\cos \dfrac{\pi}{4} + i \sin \dfrac{\pi}{4}\right)\right]$

39. $\left[\frac{5}{3}(\cos 140° + i \sin 140°)\right]\left[\frac{2}{3}(\cos 60° + i \sin 60°)\right]$

40. $[0.5(\cos 100° + i \sin 100°)]$
$[0.8(\cos 300° + i \sin 300°)]$

41. $[0.45(\cos 310° + i \sin 310°)]$
$[0.60(\cos 200° + i \sin 200°)]$

42. $(\cos 5° + i \sin 5°)(\cos 20° + i \sin 20°)$

43. $\dfrac{2(\cos 120° + i \sin 120°)}{4(\cos 40° + i \sin 40°)}$

44. $\dfrac{\cos 40° + i \sin 40°}{\cos 10° + i \sin 10°}$

45. $\dfrac{\cos \dfrac{5\pi}{3} + i \sin \dfrac{5\pi}{3}}{\cos \pi + i \sin \pi}$

46. $\dfrac{5(\cos 4.3 + i \sin 4.3)}{4(\cos 2.1 + i \sin 2.1)}$

47. $\dfrac{12(\cos 52° + i \sin 52°)}{3(\cos 110° + i \sin 110°)}$

48. $\dfrac{9(\cos 20° + i \sin 20°)}{5(\cos 75° + i \sin 75°)}$

In Exercises 49–54, (a) give the trigonometric form of the complex number, (b) perform the indicated operation using the trigonometric form, and (c) perform the indicated operation using the standard form and check your result with the answer to part (b).

49. $(2 + 2i)(1 - i)$

50. $(\sqrt{3} + i)(1 + i)$

51. $-2i(1 + i)$

52. $\dfrac{3 + 4i}{1 + \sqrt{3}\,i}$

53. $\dfrac{5}{2 + 3i}$

54. $\dfrac{4i}{-4 + 2i}$

55. Given two complex numbers $z_1 = r_1(\cos \theta_1 + i \sin \theta_1)$ and $z_2 = r_2(\cos \theta_2 + i \sin \theta_2)$, $z_2 \neq 0$, prove that

$$\frac{z_1}{z_2} = \frac{r_1}{r_2}[\cos(\theta_1 - \theta_2) + i \sin(\theta_1 - \theta_2)].$$

56. Show that the complex conjugate of $z = r(\cos \theta + i \sin \theta)$ is $\bar{z} = r[\cos(-\theta) + i \sin(-\theta)]$.

57. Use the trigonometric form of z and $\bar{z}$ in Exercise 56 to find (a) $z\bar{z}$ and (b) $z/\bar{z}$, $z \neq 0$.

58. Show that the negative of $z = r(\cos \theta + i \sin \theta)$ is $-z = r[\cos(\theta + \pi) + i \sin(\theta + \pi)]$.

In Exercises 59 and 60, sketch the graph of all complex numbers z satisfying the given condition.

59. $|z| = 2$

60. $\theta = \pi/6$

In Exercises 61–72, use DeMoivre's Theorem to find the indicated power of the complex number. Express the result in standard form.

61. $(1 + i)^5$

62. $(2 + 2i)^6$

63. $(-1 + i)^{10}$

64. $(1 - i)^{12}$

65. $2(\sqrt{3} + i)^7$

66. $4(1 - \sqrt{3}\,i)^3$

67. $[5(\cos 20° + i \sin 20°)]^3$

68. $[3(\cos 150° + i \sin 150°)]^4$

69. $\left(\cos \dfrac{5\pi}{4} + i \sin \dfrac{5\pi}{4}\right)^{10}$

70. $\left[2\left(\cos \dfrac{\pi}{2} + i \sin \dfrac{\pi}{2}\right)\right]^8$

71. $[5(\cos 3.2 + i \sin 3.2)]^4$

72. $(\cos 0 + i \sin 0)^{20}$

In Exercises 73–84, (a) use DeMoivre's Theorem to find the indicated roots of the complex number, (b) represent each of the roots graphically, and (c) express each of the roots in standard form.

73. Square roots of: $9(\cos 120° + i \sin 120°)$

74. Square roots of: $16(\cos 60° + i \sin 60°)$

75. Fourth roots of: $16\left(\cos \dfrac{4\pi}{3} + i \sin \dfrac{4\pi}{3}\right)$

76. Fifth roots of: $32\left(\cos \dfrac{5\pi}{6} + i \sin \dfrac{5\pi}{6}\right)$

77. Square roots of: $-25i$

78. Fourth roots of: $625i$

79. Cube roots of: $-\dfrac{125}{2}\left(1 + \sqrt{3}\,i\right)$

80. Cube roots of: $-8\sqrt{2}(1 - i)$

81. Cube roots of: 8

82. Fourth roots of: i

83. Fifth roots of: 1

84. Cube roots of: 1000

In Exercises 85–92, find all the solutions of the equation and represent the solutions on a unit circle.

85. $x^4 - i = 0$

86. $x^3 + 1 = 0$

87. $x^5 + 243 = 0$

88. $x^4 - 81 = 0$

89. $x^3 + 64i = 0$

90. $x^6 - 64i = 0$

91. $x^3 - (1 - i) = 0$

92. $x^4 + (1 + i) = 0$

CHAPTER 6 · REVIEW EXERCISES

In Exercises 1–10, simplify the trigonometric expression.

1. $\dfrac{1}{\cot^2 x + 1}$

2. $\dfrac{\sin 2\alpha}{\cos^2 \alpha - \sin^2 \alpha}$

3. $\dfrac{\sin^2 \alpha - \cos^2 \alpha}{\sin^2 \alpha - \sin \alpha \cos \alpha}$

4. $\dfrac{\sin^3 \beta + \cos^3 \beta}{\sin \beta + \cos \beta}$

5. $\cos^2 \beta + \cos^2 \beta \tan^2 \beta$

6. $\dfrac{\sin \theta}{1 + \cos \theta} + \dfrac{1 + \cos \theta}{\sin \theta}$

7. $\tan^2 \theta(\csc^2 \theta - 1)$

8. $\dfrac{2 \tan(x + 1)}{1 - \tan^2(x + 1)}$

9. $1 - 4 \sin^2 x \cos^2 x$

10. $\sqrt{\dfrac{1 - \cos^2 x}{1 + \cos x}}$

In Exercises 11–30, verify the identity.

11. $\tan x(1 - \sin^2 x) = \frac{1}{2} \sin 2x$

12. $\cos x(\tan^2 x + 1) = \sec x$

13. $\sec^2 x \cot x - \cot x = \tan x$

14. $\sin^3 \theta + \sin \theta \cos^2 \theta = \sin \theta$

15. $\sin^5 x \cos^2 x = (\cos^2 x - 2 \cos^4 x + \cos^6 x) \sin x$

16. $\cos^3 x \sin^2 x = (\sin^2 x - \sin^4 x) \cos x$

17. $\sin 3\theta \sin \theta = \frac{1}{2}(\cos 2\theta - \cos 4\theta)$

18. $\sin 3x \cos 2x = \frac{1}{2}(\sin 5x + \sin x)$

19. $\sqrt{\dfrac{1 - \sin \theta}{1 + \sin \theta}} = \dfrac{1 - \sin \theta}{|\cos \theta|}$

20. $\sqrt{1 - \cos x} = \dfrac{|\sin x|}{\sqrt{1 + \cos x}}$

21. $\cos 3x = 4 \cos^3 x - 3 \cos x$

22. $\cos\left(x + \dfrac{\pi}{2}\right) = -\sin x$

23. $\cot\left(\dfrac{\pi}{2} - x\right) = \tan x$

24. $\sin(\pi - x) = \sin x$

25. $\dfrac{\sec x - 1}{\tan x} = \tan \dfrac{x}{2}$

26. $\dfrac{2 \cos 3x}{\sin 4x - \sin 2x} = \csc x$

27. $\dfrac{\cos 3x - \cos x}{\sin 3x - \sin x} = -\tan 2x$

28. $1 - \cos 2x = 2 \sin^2 x$

29. $2 \sin y \cos y \sec 2y = \tan 2y$

30. $\dfrac{\sin(\alpha + \beta)}{\cos \alpha \cos \beta} = \tan \alpha + \tan \beta$

In Exercises 31–36, use a graphing utility to verify the identity. (Graph each member of the equation, and note that the graphs coincide.)

31. $\sin\left(x - \dfrac{3\pi}{2}\right) = \cos x$

32. $\sin 4x = 8 \cos^3 x \sin x - 4 \cos x \sin x$

33. $\tan^2 x = \dfrac{1 - \cos 2x}{1 + \cos 2x}$

34. $\cos^2 5x - \cos^2 x = -\sin 4x \sin 6x$

35. $1 + \cos 2x + \cos 4x + \cos 6x = 4 \cos x \cos 2x$
$\cos 3x$

36. $\sin 2x + \sin 4x - \sin 6x = 4 \sin x \sin 2x \sin 3x$

In Exercises 37–40, find the exact value of the trigonometric function by using the sum, difference, or half-angle formulas.

37. $\sin \dfrac{5\pi}{12} = \sin\left(\dfrac{2\pi}{3} - \dfrac{\pi}{4}\right)$

38. $\cos 285° = \cos(225° + 60°)$

39. $\cos(157° \; 30') = \cos \dfrac{315°}{2}$

40. $\sin \dfrac{3\pi}{8} = \sin\left[\dfrac{1}{2}\left(\dfrac{3\pi}{4}\right)\right]$

In Exercises 41–46, find the exact value of the trigonometric function given that $\sin u = \frac{3}{4}$, $\cos v = -\frac{5}{13}$, and u and v are in Quadrant II.

41. $\sin(u + v)$　　　　**42.** $\tan(u + v)$

43. $\cos(u - v)$　　　　**44.** $\sin 2v$

45. $\cos \dfrac{u}{2}$　　　　**46.** $\tan 2v$

In Exercises 47–50, determine if the statement is true or false. If it is false, make the necessary correction.

47. If $\dfrac{\pi}{2} < \theta < \pi$, then $\cos \dfrac{\theta}{2} < 0$.

48. $\sin(x + y) = \sin x + \sin y$

49. $4 \sin(-x) \cos(-x) = -2 \sin 2x$

50. $4 \sin 45° \cos 15° = 1 + \sqrt{3}$

In Exercises 51–60, use a graphing utility to obtain a graph of the function and approximate its zeros in the interval $[0, 2\pi)$. If possible, find the exact values of the zeros algebraically.

51. $f(x) = \sin x - \tan x$

52. $f(x) = \csc x - 2 \cot x$

53. $g(x) = \sin 2x + \sqrt{2} \sin x$

54. $g(x) = \cos 4x - 7 \cos 2x - 8$

55. $h(x) = \cos^2 x + \sin x - 1$

56. $h(x) = \sin 4x - \sin 2x$

57. $y = \dfrac{1 + \sin x}{\cos x} + \dfrac{\cos x}{1 + \sin x} - 4$

58. $y = \cos x - \cos \dfrac{x}{2}$

59. $g(t) = \tan^3 t - \tan^2 t + 3 \tan t - 3$

60. $h(s) = \sin s + \sin 3s + \sin 5s$

In Exercises 61 and 62, write the trigonometric expression as a product.

61. $\cos 3\theta + \cos 2\theta$　　　**62.** $\sin\left(x + \dfrac{\pi}{4}\right) - \sin\left(x - \dfrac{\pi}{4}\right)$

In Exercises 63 and 64, write the trigonometric expression as a sum or difference.

63. $\sin 3\alpha \sin 2\alpha$　　　**64.** $\cos \dfrac{x}{2} \cos \dfrac{x}{4}$

In Exercises 65 and 66, write the trigonometric expression as an algebraic expression.

65. $\cos(2 \arccos 2x)$　　　**66.** $\sin(2 \arctan x)$

67. *Rate of Change*　The rate of change of the function $f(x) = 2\sqrt{\sin x}$ with respect to change in the variable x is given by the expression $\cos x / \sqrt{\sin x}$. Show that the expression for the rate of change can also be given by $\cot x \sqrt{\sin x}$.

68. *Projectile Motion*　A baseball leaves the hand of the first baseman at an angle of θ with the horizontal and an initial velocity of $v_0 = 80$ feet per second. The ball is caught by the second baseman 100 feet away. Find θ if the range r of a projectile is given by
$$r = \tfrac{1}{32} v_0^2 \sin 2\theta.$$

69. *Harmonic Motion*   A weight is attached to a spring suspended vertically from a ceiling. When a driving force is applied to the system, the weight moves vertically from its equilibrium position. This motion is described by the model.

$$y = 1.5 \sin 8t - 0.5 \cos 8t,$$

where y is the distance from equilibrium measured in feet and t is the time in seconds.

(a) Write the model in the form

$$y = \sqrt{a^2 + b^2} \sin(Bt + C).$$

(b) Use a graphing utility to obtain a graph of the model.

(c) Find the amplitude of the oscillations of the weight.

(d) Find the frequency of the oscillations of the weight.

70. *Volume*   A trough for feeding cattle is 16 feet long, and its cross sections are isosceles triangles with the two equal sides being 18 inches in length (see figure). The angle between the two equal sides is θ.

(a) Express the volume of the trough as a function of $\theta/2$.

(b) Express the volume of the trough as a function of θ and determine the value of θ so the volume is maximum.

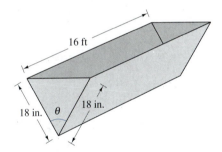

FIGURE FOR 70

In Exercises 71–86, solve the triangle (if possible). If two solutions exist, list both.

71. $a = 5, b = 8, c = 10$

72. $a = 6, b = 9, C = 45°$

73. $A = 12°, B = 58°, a = 5$

74. $B = 110°, C = 30°, c = 10.5$

75. $B = 110°, a = 4, c = 4$

76. $a = 80, b = 60, c = 100$

77. $A = 75°, a = 2.5, b = 16.5$

78. $A = 130°, a = 50, b = 30$

79. $B = 115°, a = 7, b = 14.5$

80. $C = 50°, a = 25, c = 22$

81. $A = 15°, a = 5, b = 10$

82. $B = 150°, a = 64, b = 10$

83. $B = 150°, a = 10, c = 20$

84. $a = 2.5, b = 15.0, c = 4.5$

85. $B = 25°, a = 6.2, b = 4$

86. $B = 90°, a = 5, c = 12$

In Exercises 87–90, find the area of the triangle.

87. $a = 4, b = 5, c = 7$ **88.** $a = 15, b = 8, c = 10$

89. $A = 27°, b = 5, c = 8$ **90.** $B = 80°, a = 4, c = 8$

91. *Height of a Tree*   Find the height of a tree that stands on a hillside of slope 32° (from the horizontal) if from a point 75 feet down the hill from the tree the angle of elevation to the top of the tree is 48° (see figure).

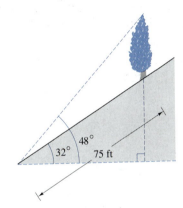

FIGURE FOR 91

92. *Surveying*   To approximate the length of a marsh, a surveyor walks 450 meters from point A to point B. Then the surveyor turns 65° and walks 325 meters to point C. Approximate the length AC of the marsh (see figure).

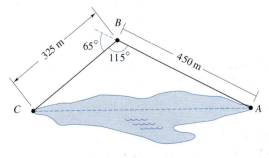

FIGURE FOR 92

93. *Height* From a certain distance, the angle of elevation to the top of a building is $17°$. At a point 50 meters closer to the building, the angle of elevation is $31°$. Approximate the height of the building.

94. *River Width* Determine the width of a river that flows due east, if a tree on the opposite bank has a bearing of N $22°$ $30'$ E and, after walking 400 feet downstream, a surveyor finds the tree has a bearing of N $15°$ W.

95. *Navigation* Two planes leave an airport at approximately the same time. One is flying at 425 miles per hour at a bearing of N $5°$ W, and the other is flying at 530 miles per hour at a bearing of N $67°$ E (see figure). How far apart are the planes after flying for 2 hours?

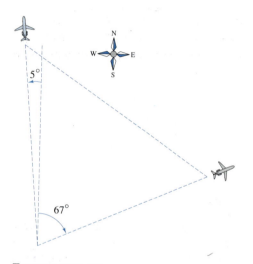

FIGURE FOR 95

96. *Geometry* The lengths of the diagonals of a parallelogram are 10 feet and 16 feet. Find the lengths of the sides of the parallelogram if the diagonals intersect at an angle of $28°$.

In Exercises 97–100, find the trigonometric form of the complex number.

97. $5 - 5i$

98. $-3\sqrt{3} + 3i$

99. $5 + 12i$

100. -7

In Exercises 101–104, write the complex number in standard form.

101. $100(\cos 240° + i \sin 240°)$

102. $24(\cos 330° + i \sin 330°)$

103. $13(\cos 0 + i \sin 0)$

104. $8\left(\cos \dfrac{5\pi}{6} + i \sin \dfrac{5\pi}{6}\right)$

In Exercises 105–108, (a) express the two complex numbers in trigonometric form, and (b) use the trigonometric form to find $z_1 z_2$ and z_1/z_2.

105. $z_1 = -5, \quad z_2 = 5i$

106. $z_1 = 2\sqrt{3} - 2i, \quad z_2 = -10i$

107. $z_1 = -3(1 + i), \quad z_2 = \left(\sqrt{3} + i\right)$

108. $z_1 = 5i, \quad z_2 = 2(1 - i)$

In Exercises 109–112, use DeMoivre's Theorem to find the indicated power of the complex number. Express the result in standard form.

109. $\left[5\left(\cos \dfrac{\pi}{12} + i \sin \dfrac{\pi}{12}\right)\right]^4$

110. $\left[2\left(\cos \dfrac{4\pi}{15} + i \sin \dfrac{4\pi}{15}\right)\right]^5$

111. $(2 + 3i)^6$

112. $(1 - i)^8$

In Exercises 113–116, use DeMoivre's Theorem to find the roots of the complex number.

113. Sixth roots of: $-729i$ **114.** Fourth roots of: 256

115. Cube roots of: -1 **116.** Fourth roots of: $-1 + i$

In Exercises 117–120, find all solutions of the equation and represent the solutions on a unit circle.

117. $x^4 + 81 = 0$ **118.** $x^5 - 32 = 0$

119. $(x^3 - 1)(x^2 + 1) = 0$ **120.** $x^3 + 8i = 0$

CHAPTER 7

LINEAR MODELS AND SYSTEMS OF EQUATIONS

7.1 SOLVING SYSTEMS OF EQUATIONS ALGEBRAICALLY AND GRAPHICALLY

The Method of Substitution / **Graphical Approach to Finding Solutions** / **Applications**

The Method of Substitution

Up to this point in the text, most problems have involved either a function of one variable or a single equation in two variables. However, many problems in science, business, and engineering involve two or more equations in two or more variables. To solve such problems, you need to find solutions of a **system of equations.** Here is an example of a system of two equations in x and y.

$$2x + y = 5 \qquad \text{\textit{Equation 1}}$$
$$3x - 2y = 4 \qquad \text{\textit{Equation 2}}$$

A **solution** of this system is an ordered pair that satisfies each equation in the system. When you find the set of all solutions, you are **solving the system of equations.** For instance, the ordered pair $(2, 1)$ is a solution of this system. To check this you can substitute 2 for x and 1 for y into *each* equation, as follows.

$$2(2) + 1 = 5 \qquad \text{\textit{Equation 1 checks}}$$
$$3(2) - 2(1) = 4 \qquad \text{\textit{Equation 2 checks}}$$

There are several different ways to solve systems of equations. In this chapter you will study the three most common techniques, beginning with the **method of substitution.** This method has five basic steps, which can be labeled *solve, substitute, solve, back-substitute,* and *check.*

DISCOVERY

Use a graphing utility to graph $y = -2x + 5$ and $y = 1.5x - 2$ on the same viewing rectangle. Then use the zoom feature to find the coordinates of the point of intersection of the two lines. Do you obtain the same coordinates as are obtained algebraically in the discussion at the right?

EXAMPLE 1   Solving a System of Two Equations in Two Variables

Solve the system of equations.

$$x + y = 4 \qquad \textit{Equation 1}$$
$$x - y = 2 \qquad \textit{Equation 2}$$

SOLUTION

Solving for y in Equation 1 produces $y = 4 - x$. *Substituting* $(4 - x)$ for y in Equation 2 produces a single-variable equation that you can *solve* for x.

$$x - (4 - x) = 2 \qquad \textit{Substitute for y}$$
$$x - 4 + x = 2 \qquad \textit{Single-variable equation}$$
$$2x = 6$$
$$x = 3 \qquad \textit{Solve for x}$$

REMARK The term *back-substitution* implies that you work *backwards*. First you solve for one of the variables, and then you substitute that value *back* into one of the equations in the system to find the value of the other variable.

Finally, by *back-substituting* $x = 3$ into the equation $y = 4 - x$, you obtain $y = 1$. Thus, the solution is the ordered pair $(3, 1)$. Check this solution by substituting $x = 3$ and $y = 1$ into *both* equations in the original system.

Because many steps are required to solve a system of equations, it is very easy to make errors in arithmetic. Thus, we *strongly* suggest that you always *check your solution by substituting it into each equation in the original system.*

METHOD OF SUBSTITUTION

To solve a system of two equations in two variables, use the following steps.

1. *Solve* one of the equations for one variable in terms of the other.
2. *Substitute* the expression found in Step 1 into the other equation to obtain an equation in one variable.
3. *Solve* the equation obtained in Step 2.
4. *Back-substitute* the solution in Step 3 into the expression obtained in Step 1 to find the value of the other variable.
5. *Check* the solution to see that it satisfies *each* of the original equations.

EXAMPLE 2   Solving a System by Substitution: One-Solution Case

A total of $12,000 is invested in two funds paying 9% and 11% simple interest. If the yearly interest is $1180, how much of the $12,000 is invested at each rate?

SOLUTION

VERBAL MODEL

Investment in + Investment in = Total
9% fund   11% fund   investment

9% interest + 11% interest = Total interest

LABELS

9% fund $= x$	*(dollars)*
11% fund $= y$	*(dollars)*
Total investment $= 12{,}000$	*(dollars)*
9% interest $= 0.09x$	*(dollars)*
11% interest $= 0.11y$	*(dollars)*
Total interest $= 1180$	*(dollars)*

SYSTEM

$$x + y = 12{,}000 \qquad \textit{Equation 1}$$
$$0.09x + 0.11y = 1{,}180 \qquad \textit{Equation 2}$$

To begin, it is convenient to multiply both sides of the second equation by 100 to obtain $9x + 11y = 118{,}000$. This step eliminates the need to work with decimals.

1. Solve for x in Equation 1.

$$x = 12{,}000 - y$$

2. Substitute this expression for x into the new Equation 2.

$$9(12{,}000 - y) + 11y = 118{,}000$$

3. Solve for y.

$$108{,}000 - 9y + 11y = 118{,}000$$
$$2y = 10{,}000$$
$$y = \$5000$$

4. Back-substitute the value $y = 5000$ to solve for x.

$$x = 12{,}000 - 5000 = \$7000$$

The solution is the ordered pair (7000, 5000). Check to see that $x = 7000$ and $y = 5000$ satisfy each of the original equations.   ·········

Note that the equations in Examples 1 and 2 are linear. That is, the variables x and y occurred to the first power only. The method of substitution can also be used to solve systems in which one or both of the equations are nonlinear.

EXAMPLE 3 Solving a System by Substitution: Two-Solution Case

Solve the system of equations.

$$x^2 - x - y = 1 \qquad \text{\textit{Equation 1}}$$
$$-x + y = -1 \qquad \text{\textit{Equation 2}}$$

SOLUTION

1. Solve for y in Equation 2.

$$y = x - 1$$

2. Substitute this expression for y into Equation 1.

$$x^2 - x - (x - 1) = 1$$

3. Solve for x.

$$x^2 - 2x + 1 = 1$$
$$x^2 - 2x = 0$$
$$x(x - 2) = 0$$
$$x = 0, \qquad x = 2$$

4. Back-substitute these values of x to solve for the corresponding values of y.

For $x = 0$: $y = 0 - 1 = -1$
For $x = 2$: $y = 2 - 1 = 1$

There are two solutions: $(0, -1)$ and $(2, 1)$. Check these solutions in the original system.

EXAMPLE 4 Solving a System by Substitution: No-Solution Case

Solve the system of equations.

$$-x + y = 4 \qquad \text{\textit{Equation 1}}$$
$$x^2 + y = 3 \qquad \text{\textit{Equation 2}}$$

SOLUTION

1. In this case, you can solve for y in Equation 1.

$$y = x + 4$$

2. Substitute this expression for y into Equation 2.

$$x^2 + (x + 4) = 3$$

3. Solve for x.

$$x^2 + x + 4 = 3$$
$$x^2 + x + 1 = 0$$

$$x = \frac{-1 \pm \sqrt{1^2 - 4(1)(1)}}{2} \quad \textit{Quadratic Formula}$$

Because the discriminant is negative, the equation $x^2 + x + 1 = 0$ has no (real) solution. Hence, this system has no (real) solution.

Graphical Approach to Finding Solutions

From Examples 2, 3, and 4, you can see that a system of two equations in two unknowns can have exactly one solution, more than one solution, or no solution. In practice, you can gain insight about the location and number of solutions of a system of equations by graphing each of the equations on the same coordinate plane. The solutions of the system correspond to the **points of intersection** of the graphs. For instance, in Figure 7.1(a) the two equations graph as two lines with a *single point* of intersection. The two equations in Example 3 graph as a parabola and a line with *two points* of intersection, as shown in Figure 7.1(b). Moreover, the two equations in Example 4 graph as a line and a parabola that happen to have *no points* of intersection, as shown in Figure 7.1(c).

Sometimes the graphical approach to solving a system of equations is easier than the method of substitution, and a graphing utility is especially helpful.

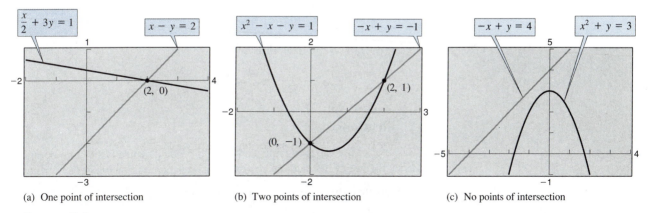

(a) One point of intersection   (b) Two points of intersection   (c) No points of intersection

FIGURE 7.1

REMARK Example 5 shows the value of a graphical approach to solving systems of equations in two variables. Notice what would have happened if you had tried only the substitution method in Example 5. By substituting $y = \ln x$ into $x + y = 1$, you obtain $x + \ln x = 1$. It would be difficult to solve this equation for x using standard algebraic techniques.

EXAMPLE 5   **Solving a System of Equations**

Solve the system of equations.

$$y = \ln x \qquad \text{\textit{Equation 1}}$$
$$x + y = 1 \qquad \text{\textit{Equation 2}}$$

SOLUTION

The graph of each equation is shown in Figure 7.2. From this sketch it is clear that there is only one point of intersection. Also, it appears that $(1, 0)$ is the solution point, and you can confirm this by checking these coordinates in *both* equations.

CHECK: Let $x = 1$ and $y = 0$.

$$0 = \ln 1 \qquad \text{\textit{Equation 1 checks}}$$
$$1 + 0 = 1 \qquad \text{\textit{Equation 2 checks}}$$

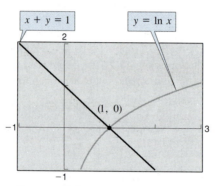

FIGURE 7.2

Applications

The total cost C of producing x units of a product typically has two components: the initial cost and the cost per unit. When enough units have been sold so that the total revenue R equals the total cost, the sales have reached the **break-even point.** The break-even point corresponds to the point of intersection of the cost and revenue curves.

EXAMPLE 6   An Application: Break-Even Analysis

A small business invests $10,000 in equipment to produce a product. Each unit of the product costs $0.65 to produce and is sold for $1.20. How many items must be sold before the business breaks even?

SOLUTION

The total cost of producing x units is

$$\overset{\substack{\text{Cost per}\\\text{unit}}}{C = 0.65x} + \overset{\substack{\text{Initial}\\\text{cost}}}{10{,}000,} \qquad\qquad \textit{Equation 1}$$

and the revenue obtained by selling x units is

$$R = \overset{\substack{\text{Price}\\\text{per unit}}}{1.2x.} \qquad\qquad \textit{Equation 2}$$

Because the break-even point occurs when $R = C$, you have

$$1.2x = 0.65x + 10{,}000 \qquad \textit{Set R equal to C}$$
$$0.55x = 10{,}000$$
$$x = \frac{10{,}000}{0.55} \approx 18{,}182 \text{ units.}$$

Note in Figure 7.3 that sales less than the break-even point correspond to an overall loss, whereas sales greater than the break-even point correspond to a profit.

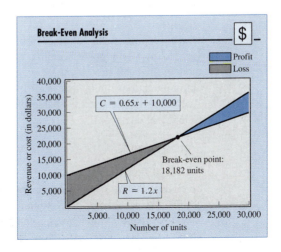

FIGURE 7.3

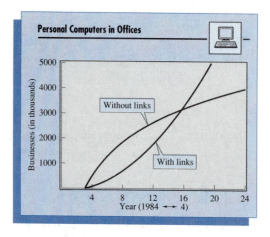

FIGURE 7.4

EXAMPLE 7   **Computers in the Office**

The number of business or government offices with personal computers (with and without communications links such as networks or modems) from 1984 to 1989 can be approximated by the models

$$y = -99.15 - 11.38t + 13.91t^2 \qquad \textit{With communications links}$$
$$y = -2202.41 + 1934.45 \ln t, \qquad \textit{Without communications links}$$

where y represents the number of businesses in thousands and t represents the year, with $t = 4$ corresponding to 1984. Sketch the graphs of these two models. In 1989 ($t = 9$), there were more offices without personal computer communications links than with personal computer communications links. According to the models, when will this change? (*Source:* Gartner Group, COMTEX database)

SOLUTION

The graphs of the two models are shown in Figure 7.4. According to the graphs, the number of offices with personal computer communications links will surpass the number without personal computer communications links in 1995 (when $t \approx 15.5$). Use the zoom and trace features of a graphing utility to verify this result.

DISCUSSION PROBLEM

POINTS OF INTERSECTION OF TWO GRAPHS

In this section, you learned that the graphs of two equations can intersect at zero, one, or more points. Use a graphing utility to graph the following systems. Which represents a system with no solution? Which represents a system with one solution? Which represents a system with two solutions?

(a) $x^2 + y^2 = 4$
$x + y = 2$

(b) $x^2 + y^2 = 4$
$x + y = \sqrt{8}$

(c) $x^2 + y^2 = 4$
$x + y = 6$

Find three other systems of equations, one with no solution, one with one solution, and one with two solutions.

WARM-UP

The following warm-up exercises involve skills that were covered in earlier sections. You will use these skills in the exercise set for this section.

In Exercises 1–4, sketch the graph of the equation.

1. $y = -\frac{1}{3}x + 6$

2. $y = 2(x - 3)$

3. $x^2 + y^2 = 4$

4. $y = 5 - (x - 3)^2$

In Exercises 5–8, perform the indicated operations and simplify.

5. $(3x + 2y) - 2(x + y)$

6. $(-10u + 3v) + 5(2u - 8v)$

7. $x^2 + (x - 3)^2 + 6x$

8. $y^2 - (y + 1)^2 + 2y$

In Exercises 9 and 10, solve the equation.

9. $3x + (x - 5) = 15 + 4$

10. $y^2 + (y - 2)^2 = 2$

SECTION 7.1 · EXERCISES

In Exercises 1–10, solve the system graphically. Check your solution algebraically.

1. $2x + y = 4$
$-x + y = 1$

2. $x - y = -5$
$x + 2y = 4$

3. $x - y = -3$
$x^2 - y = -1$

4. $3x - y = -2$
$x^3 - y = 0$

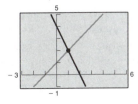
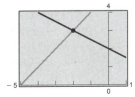
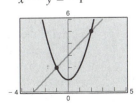
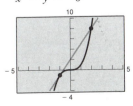

5. $x + 3y = 15$
$x^2 + y^2 = 25$

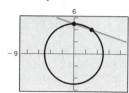

6. $x \qquad - y = 0$
$x^3 - 5x + y = 0$

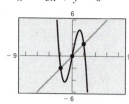

In Exercises 21–30, solve the system graphically or algebraically. Explain why you chose the method you used.

21. $y = 2x$
$y = x^2 + 1$

22. $x + y = 4$
$x^2 - y = 2$

23. $3x - 7y + 6 = 0$
$x^2 - y^2 = 4$

24. $x^2 + y^2 = 25$
$2x + y = 10$

25. $y - e^{-x} = 1$
$y - \ln x = 3$

26. $y = x^3 - 2x^2 + x - 1$
$y = -x^2 + 3x \qquad - 1$

27. $y = x^4 - 2x^2 + 1$
$y = 1 - x^2$

28. $x^2 + y = 4$
$e^x - y = 0$

29. $xy - 1 = 0$
$2x - 4y + 7 = 0$

30. $x - 2y = 1$
$y = \sqrt{x - 1}$

7. $x^2 - y = 0$
$x^2 - 4x + y = 0$

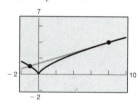

8. $y = - x^2 + 1$
$y = x^4 - 2x^2 + 1$

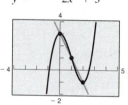

In Exercises 31–42, use a graphing utility to find all points of intersection of the graphs of the pair of equations.

31. $x + y = 4$
$x^2 + y^2 - 4x = 0$

32. $x - y + 3 = 0$
$x^2 - 4x + 7 = y$

33. $2x - y + 3 = 0$
$x^2 + y^2 - 4x = 0$

34. $3x - 2y = 0$
$x^2 - y^2 = 4$

35. $x^2 + y^2 = 25$
$(x - 8)^2 + y^2 = 41$

36. $x^2 + y^2 = 8$
$y = x^2$

37. $y = e^x$
$x - y + 1 = 0$

38. $x + 2y = 8$
$y = \log_2 x$

39. $y = \sqrt{x}$
$y = x$

40. $x - y = 3$
$x - y^2 = 1$

41. $x^2 + y^2 = 169$
$x^2 - 8y = 104$

42. $x^2 + y^2 = 4$
$2x^2 - y = 2$

9. $x - 3y = -4$
$x^2 - y^3 = 0$

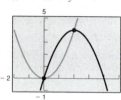

10. $y = x^3 - 3x^2 + 3$
$y = - 2x + 3$

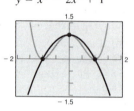

In Exercises 11–20, solve the system by the method of substitution.

11. $x - y = 0$
$5x - 3y = 10$

12. $x + 2y = 1$
$5x - 4y = -23$

13. $2x - y + 2 = 0$
$4x + y - 5 = 0$

14. $6x - 3y - 4 = 0$
$x + 2y - 4 = 0$

15. $30x - 40y - 33 = 0$
$10x + 20y - 21 = 0$

16. $1.5x + 0.8y = 2.3$
$0.3x - 0.2y = 0.1$

17. $\frac{1}{5}x + \frac{1}{2}y = 8$
$x + y = 20$

18. $\frac{1}{2}x + \frac{3}{4}y = 10$
$\frac{3}{2}x - y = 4$

19. $x - y = 0$
$2x + y = 0$

20. $x - 2y = 0$
$3x - y = 0$

Break-Even Analysis In Exercises 43–46, find the sales necessary to break even $(R = C)$ for the given cost C of x units, and the given revenue R obtained by selling x units. (Round to the nearest whole unit.)

43. $C = 8650x + 250,000, \quad R = 9950x$

44. $C = 5.5\sqrt{x} + 10,000, \quad R = 3.29x$

45. $C = 2.65x + 350,000, \quad R = 4.15x$

46. $C = 0.08x + 50,000, \quad R = 0.25x$

47. *Break-Even Point* Suppose you are setting up a small business and have invested $16,000 to produce an item that will sell for $5.95. If each unit can be produced for $3.45, how many units must be sold to break even?

48. *Break-Even Point* Suppose you are setting up a small business and have an initial investment of $5000. The unit cost of the product is $21.60, and the selling price is $34.10. How many units must be sold to break even?

49. *Investment Portfolio* A total of $25,000 is invested in two funds paying 8% and 8.5% simple interest. If the yearly interest is $2060, how much of the $25,000 is invested at each rate?

50. *Investment Portfolio* A total of $18,000 is invested in two funds paying 7.75% and 8.25% simple interest. If the yearly interest is $1455, how much of the $18,000 is invested at each rate?

51. *Choice of Two Jobs* Suppose you are offered two different jobs selling dental supplies. One company offers a straight commission of 6% of sales. The other company offers a salary of $250 per week *plus* 3% of sales. How much would you have to sell in a week in order to make the straight commission offer better?

52. *Choice of Two Jobs* Suppose you are offered two different jobs selling college textbooks. One company offers an annual salary of $20,000 *plus* a year-end bonus of 1% of your total sales. The other company offers a salary of $15,000 *plus* a year-end bonus of 2% of your total sales. How much would you have to sell in a year in order to make the second offer better?

53. *Log Volume* You are offered two different rules for estimating the number of board feet in a log that is 16 feet long. One is the *Doyle Log Rule* and is modeled by

$$V = (D - 4)^2, \qquad 5 \le D \le 40,$$

and the other is the *Scribner Log Rule* and is modeled by

$$V = 0.79D^2 - 2D - 4, \qquad 5 \le D \le 40,$$

where D is the diameter of the log and V is its volume in board feet.
  (a) Use a graphing utility to graph the log rules in the same viewing rectangle.
  (b) For what diameter trees do the two scales agree?
  (c) If you were selling large logs, which scale would you want to be used?

54. *Market Equilibrium* The supply and demand curves for a business dealing with wheat are given by

Supply: $\quad p = 1.45 + 0.00014x^2$

Demand: $\quad p = (2.388 - 0.007x)^2,$

where p is the price in dollars per bushel and x is the quantity in bushels per day. Use a graphing utility to graph the supply and demand equations and find the market equilibrium. (The *market equilibrium* is the point of intersection of the graphs for $x > 0$.)

55. *Area* What are the dimensions of a rectangular tract of land if its perimeter is 40 miles and its area is 96 square miles?

56. *Area* What are the dimensions of an isoceles right triangle with a 2-inch hypotenuse and an area of 1 square inch?

57. Find an equation of a line whose graph intersects the graph of the parabola $y = x^2$ at (a) two points, (b) one point, and (c) no points. (There is more than one correct answer for each part of the exercise.)

58. Solve the system.
  (a) $y = 2^x$
      $y = x^2$
  (a) $y = 10^x$
      $y = x^{10}$

7.2   SYSTEMS OF LINEAR EQUATIONS IN TWO VARIABLES

The Method of Elimination   /   Graphical Interpretation of Solutions   /
Applications

The Method of Elimination

In Section 7.1, you studied two methods of solving a system of equations (by substitution and by graphing). In this section, you will study a third method called the **method of elimination.** The key step in the method of elimination is to obtain, for one of the variables, coefficients that differ only in sign so that by *adding* the two equations this variable will be eliminated.

EXAMPLE 1   The Method of Elimination

Solve the system of linear equations.

$$3x + 2y = 4 \qquad \textit{Equation 1}$$
$$5x - 2y = 8 \qquad \textit{Equation 2}$$

SOLUTION

Begin by noting that the coefficients for y differ only in sign. By adding the two equations, you can eliminate y.

$$
\begin{array}{ll}
3x + 2y = 4 & \textit{Equation 1} \\
\underline{5x - 2y = 8} & \textit{Equation 2} \\
8x = 12 & \textit{Add equations}
\end{array}
$$

Therefore, $x = \frac{3}{2}$. By back-substituting this value into the first equation, you can solve for y.

$$
\begin{array}{ll}
3x + 2y = 4 & \textit{Equation 1} \\
3\left(\dfrac{3}{2}\right) + 2y = 4 & \textit{Replace x by } \tfrac{3}{2} \\
2y = 4 - \dfrac{9}{2} = -\dfrac{1}{2} & \textit{Solve for y} \\
y = -\dfrac{1}{4} &
\end{array}
$$

The solution is $(\frac{3}{2}, -\frac{1}{4})$. You can check the solution *algebraically* by substituting in the original system, or *graphically*, as shown in Figure 7.5.

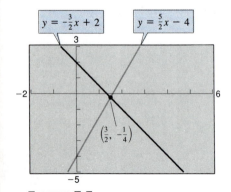

$y = -\frac{3}{2}x + 2$ $y = \frac{5}{2}x - 4$

$\left(\frac{3}{2}, -\frac{1}{4}\right)$

FIGURE 7.5

Try using the method of substitution to solve the system given in Example 1. Which method do you think is easier? Many people find that the method of elimination is more efficient.

To obtain coefficients (for one of the variables) that differ only in sign, you often need to multiply one or both of the equations by a suitable constant.

EXAMPLE 2   The Method of Elimination

Solve the system of linear equations.

$$2x - 3y = -7 \qquad \text{\textit{Equation 1}}$$
$$3x + y = -5 \qquad \text{\textit{Equation 2}}$$

SOLUTION

For this system, you can obtain coefficients that differ only in sign by multiplying the second equation by 3.

$$
\begin{array}{lll}
2x - 3y = -7 & \longrightarrow & 2x - 3y = -7 \quad \textit{Equation 1} \\
\underline{3x + y = -5} & \longrightarrow & \underline{9x + 3y = -15} \quad \textit{Multiply Equation 2 by 3} \\
& & 11x = -22 \quad \textit{Add equations}
\end{array}
$$

Thus, $x = -2$. By back-substituting this value of x into the first equation, you can solve for y.

$$
\begin{array}{ll}
2x - 3y = -7 & \textit{Equation 1} \\
2(-2) - 3y = -7 & \textit{Replace x by } -2 \\
-3y = -3 & \textit{Add 4 to both sides} \\
y = 1 & \textit{Solve for y}
\end{array}
$$

The solution is $(-2, 1)$. You can check the solution algebraically by substituting in the original system, or graphically, as shown in Figure 7.6.

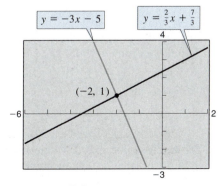

FIGURE 7.6

In Example 2, the two systems of linear equations

$$2x - 3y = -7 \quad \text{and} \quad 2x - 3y = -7$$
$$3x + y = -5 \qquad\qquad 9x + 3y = -15$$

are called **equivalent** because they have precisely the same solution set. The operations that can be performed on a system of linear equations to produce an equivalent system are (1) interchange two equations, (2) multiply an equation by a nonzero constant, and (3) add a multiple of an equation to another equation.

THE METHOD OF ELIMINATION

To use the **method of elimination** to solve a system of two linear equations in x and y, use the following steps.

1. Obtain coefficients for x (or y) that differ only in sign by multiplying all terms of one or both equations by suitably chosen nonzero constants.
2. Add the equations to eliminate one variable and solve the resulting equation.
3. Back-substitute the value obtained in Step 2 into either of the original equations and solve for the other variable.
4. Check your solution in both of the original equations.

EXAMPLE 3   The Method of Elimination

Solve the system of linear equations.

$$5x + 3y = 9 \qquad\qquad \textit{Equation 1}$$
$$2x - 4y = 14 \qquad\qquad \textit{Equation 2}$$

SOLUTION

You can obtain coefficients that differ only in sign by multiplying the first equation by 4 and multiplying the second equation by 3.

$$
\begin{array}{rcll}
5x + 3y = 9 & \longrightarrow & 20x + 12y = 36 & \textit{Multiply Equation 1 by 4} \\
2x - 4y = 14 & \longrightarrow & 6x - 12y = 42 & \textit{Multiply Equation 2 by 3} \\
\hline
& & 26x = 78 & \textit{Add equations}
\end{array}
$$

From this equation, you can see that $x = 3$. By back-substituting this value of x into the second equation, you can solve for y.

$$
\begin{array}{rcll}
2x - 4y &=& 14 & \textit{Equation 2} \\
2(3) - 4y &=& 14 & \textit{Replace x by 3} \\
-4y &=& 8 & \\
y &=& -2 & \textit{Solve for y}
\end{array}
$$

The solution is $(3, -2)$. You can check the solution algebraically by substituting in the original system, or graphically, as shown in Figure 7.7.

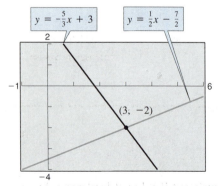

$y = -\frac{5}{3}x + 3$ $y = \frac{1}{2}x - \frac{7}{2}$

$(3, -2)$

FIGURE 7.7

Graphical Interpretation of Solutions

In Section 7.1, you saw that it is possible for a *general* system of equations to have exactly one solution, two or more solutions, or no solution. For a system of *linear* equations, this result can be strengthened. Specifically, if a system of linear equations has two different solutions, then it must have an infinite number of solutions! To see why this is true, consider the following graphical interpretations of a system of two linear equations in two variables. (Remember that the graph of a linear equation in two variables is a straight line.)

GRAPHICAL INTERPRETATION OF SOLUTIONS

For a system of two linear equations in two variables, the number of solutions is given by one of the following.

Number of Solutions	Graphical Interpretation
1. Exactly one solution	The two lines intersect at one point.
2. Infinitely many solutions	The two lines are identical.
3. No solution	The two lines are parallel.

These three possibilities are shown in Figure 7.8.

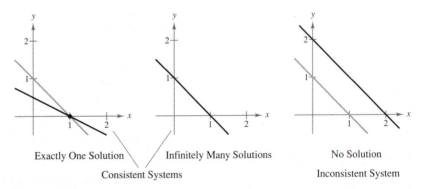

Exactly One Solution   Infinitely Many Solutions   No Solution

Consistent Systems   Inconsistent System

FIGURE 7.8

A system of linear equations is **consistent** if it has at least one solution, and it is **inconsistent** if it has no solution.

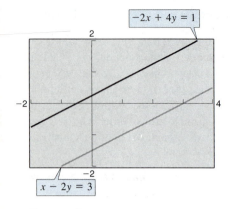

FIGURE 7.9

EXAMPLE 4   The Method of Elimination: No-Solution Case

Solve the system of linear equations.

$$x - 2y = 3 \qquad \textit{Equation 1}$$
$$-2x + 4y = 1 \qquad \textit{Equation 2}$$

SOLUTION

To obtain coefficients that differ only in sign, multiply the first equation by 2.

$x - 2y = 3$	$\longrightarrow$	$2x - 4y = 6$ *Multiply Equation 1 by 2*
$-2x + 4y = 1$	$\longrightarrow$	$-2x + 4y = 1$ *Equation 2*
		$ 0 = 7$ *False statement*

Because there are no values of x and y for which $0 = 7$, the system is inconsistent and has no solution. The lines corresponding to the two equations given in this system are shown in Figure 7.9. Note that the two lines are parallel, and therefore have no point of intersection.

EXAMPLE 5   The Method of Elimination: Many-Solutions Case

Solve the following system of linear equations.

$$2x - y = 1 \qquad \textit{Equation 1}$$
$$4x - 2y = 2 \qquad \textit{Equation 2}$$

SOLUTION

To obtain coefficients that differ only in sign, multiply the second equation by $-\frac{1}{2}$.

$2x - y = 1$	$\longrightarrow$	$2x - y = 1$ *Equation 1*
$4x - 2y = 2$	$\longrightarrow$	$-2x + y = -1$ *Multiply Equation 2 by $-\frac{1}{2}$*
		$ 0 = 0$ *Add equations*

Because the two equations turn out to be equivalent (have the same solution set), the system has infinitely many solutions. The solution set consists of all points (x, y) lying on the line $2x - y = 1$, as shown in Figure 7.10.

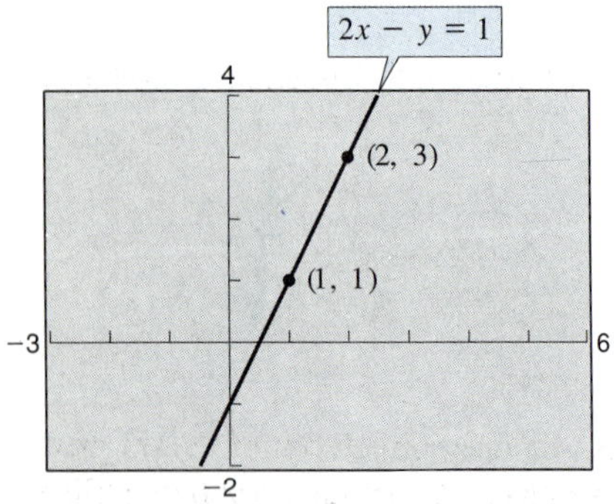

Infinite Number of Solutions

FIGURE 7.10

In Example 5, you could have reached the same conclusion by multiplying the first equation by 2 to obtain

$$4x - 2y = 2 \qquad \textit{New Equation 1}$$
$$4x - 2y = 2. \qquad \textit{Equation 2}$$

The general solution of the linear system

$$ax + by = c$$
$$dx + ey = f$$

is $x = (ce - bf)/(ae - db)$ and $y = (af - cd)/(ae - db)$. If $ae - db = 0$, then the system does not have a unique solution. A graphing utility program for solving such a system is given below. Try using this program to solve the system in Example 6.

```
Prgm1:SOLVE
:Disp "ENTER A,B,C,D,E,F"
:Input A
:Input B
:Input C
:Input D
:Input E
:Input F
:If AE-DB=0
:Goto 1
:(CE-BF)/(AE-DB)→X
:(AF-CD)/(AE-DB)→Y
:Disp X
:Disp Y
:Lbl 1
:End
```

EXAMPLE 6   **Solving a Linear System Having Decimal Coefficients**

Solve the following system of linear equations.

$$0.02x - 0.05y = -0.38 \qquad \textit{Equation 1}$$
$$0.03x + 0.04y = 1.04 \qquad \textit{Equation 2}$$

SOLUTION

Because the coefficients in this system have two decimal places, begin by multiplying each equation by 100. (This produces an equivalent system in which the coefficients are all integers.)

$$2x - 5y = -38 \qquad \textit{Revised Equation 1}$$
$$3x + 4y = 104 \qquad \textit{Revised Equation 2}$$

Now, to obtain coefficients that differ only in sign, multiply the first equation by 3 and multiply the second equation by -2.

$$
\begin{array}{rcll}
2x - 5y = -38 & \longrightarrow & 6x - 15y = -114 & \textit{Multiply Equation 1 by 3} \\
3x + 4y = 104 & \longrightarrow & \underline{-6x - 8y = -208} & \textit{Multiply Equation 2 by } -2 \\
& & {-23y} = -322 & \textit{Add equations}
\end{array}
$$

Thus, $y = \frac{-322}{-23} = 14$. Back-substituting this value into Equation 2 produces the following.

$$
\begin{array}{rll}
3x + 4y = 104 & \textit{Equation 2} \\
3x + 4(14) = 104 & \textit{Replace y by 14} \\
3x = 48 & \\
x = 16 & \textit{Solve for x}
\end{array}
$$

Therefore, the solution is (16, 14). Check this solution in each of the original equations in the system. You can also solve this system with a graphing utility by graphing the two lines

$$y = \frac{2}{5}x + \frac{38}{5}$$

$$y = -\frac{3}{4}x + 26$$

obtained by solving each equation for y. The intersection of the two lines, (16, 14), is the solution to the original system, as shown in Figure 7.11.

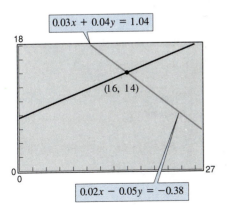

FIGURE 7.11

Applications

You have seen that systems of linear equations have many applications in science, business, health services, and government. The question that may come to mind is, How can I tell which application problems can be solved using a system of linear equations? The answer comes from the following considerations.

1. Does the problem involve more than one unknown quantity?
2. Are there two (or more) equations or conditions to be satisfied?

If one or both of these conditions occur, then the appropriate model for the problem may be a system of linear equations. Example 7 shows how to construct such a model.

EXAMPLE 7   An Application of a Linear System

An airplane flying into a headwind travels the 2000-mile flying distance between two cities in 4 hours and 24 minutes. On the return flight, the same distance is traveled in 4 hours. Find the ground speed of the plane and the speed of the wind, assuming that both remain constant.

SOLUTION

The two unknown quantities are the speeds of the wind and of the plane. If r_1 is the speed of the plane and r_2 is the speed of the wind, then

$$r_1 - r_2 = \text{speed of the plane } against \text{ the wind}$$
$$r_1 + r_2 = \text{speed of the plane } with \text{ the wind,}$$

as shown in Figure 7.12. Using the formula

$$\text{Distance} = (\text{rate})(\text{time})$$

for these two speeds, you can obtain the following equations.

$$2000 = (r_1 - r_2)\left(4 + \frac{24}{60}\right)$$
$$2000 = (r_1 + r_2)(4)$$

These two equations simplify as follows.

$$5000 = 11r_1 - 11r_2 \qquad \textit{Equation 1}$$
$$500 = r_1 + r_2 \qquad \textit{Equation 2}$$

By elimination, the solution is

$$r_1 = \frac{5250}{11} \approx 477.27 \text{ miles per hour}$$

$$r_2 = \frac{250}{11} \approx 22.73 \text{ miles per hour.}$$

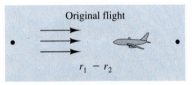
Original flight

$r_1 - r_2$

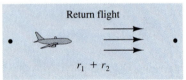
Return flight

$r_1 + r_2$

FIGURE 7.12

In a free market, the supply of and demand for many products are related to the price of the product. As the price of a product decreases, the demand (by *consumers*) increases whereas the supply (by *producers*) tends to decrease.

EXAMPLE 8   Finding the Point of Equilibrium

Suppose the demand and supply functions for a certain type of calculator are given by

$$p = 150 - 10x \qquad \text{\textit{Demand equation}}$$
$$p = 60 + 20x, \qquad \text{\textit{Supply equation}}$$

where p is the price in dollars and x represents the number of units (in millions). Find the **point of equilibrium** for this market by solving this system of equations. (The point of equilibrium is the price p and the number of units x that satisfy both the demand and supply equations.)

SOLUTION

For this system, you can use the method of substitution (because the two equations are given in a form in which p is written in terms of x). By substituting the value of p given in the second equation into the first equation, you obtain the following.

$$p = 150 - 10x \qquad \text{\textit{First equation}}$$
$$60 + 20x = 150 - 10x \qquad \text{\textit{Replace p by 60 + 20x}}$$
$$30x = 90 \qquad \text{\textit{Add 10x and subtract 60}}$$
$$x = 3 \qquad \text{\textit{Divide both sides by 30}}$$

Thus, the point of equilibrium occurs when the demand and supply are each 3 million units. (See Figure 7.13.) The price that corresponds to this x-value is obtained by back-substituting $x = 3$ into either of the original equations. For instance, back-substituting into the first equation produces

$$p = 150 - 10(3) = 150 - 30 = \$120.$$

Try back-substituting $x = 3$ into the second equation to see that you obtain the same price.

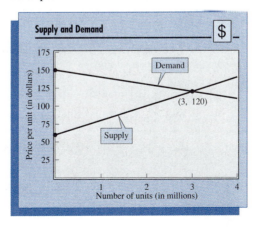

FIGURE 7.13

DISCUSSION PROBLEM

CREATING CONSISTENT AND INCONSISTENT SYSTEMS

Consider the following system of linear equations.

$$x - \ y = 4$$
$$-2x + 2y = k$$

(a) Find the value of k so that the system has an infinite number of solutions.
(b) Find one value of k so that the system has no solutions.
(c) Can the system have a unique solution? Why or why not?

WARM-UP

The following warm-up exercises involve skills that were covered in earlier sections. You will use these skills in the exercise set for this section.

In Exercises 1 and 2, sketch the graph of the equation.

1. $2x + y = 4$ **2.** $5x - 2y = 3$

In Exercises 3 and 4, find an equation of the line passing through the two points.

3. $(-1, 3), (4, 8)$ **4.** $(2, 6), (5, 1)$

In Exercises 5 and 6, determine the slope of the line.

5. $3x + 6y = 4$ **6.** $7x - 4y = 10$

In Exercises 7–10, determine whether the lines represented by the pair of equations are parallel, perpendicular, or neither.

7. $2x - 3y = -10$
   $3x + 2y = \ \ 11$

8. $\ \ 4x - 12y = 5$
   $-2x + \ \ 6y = 3$

9. $5x + \ \ y = 2$
   $3x + 2y = 1$

10. $\ \ x - 3y = 2$
   $6x + 2y = 4$

SECTION 7.2 · EXERCISES

In Exercises 1–10, solve the linear system by elimination. Label each line with the appropriate equation.

1. $2x + y = 4$
$x - y = 2$

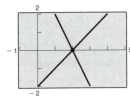

2. $x + 3y = 2$
$-x + 2y = 3$

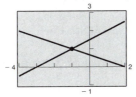

3. $x - y = 0$
$3x - 2y = -1$

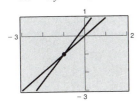

4. $2x - y = 2$
$4x + 3y = 24$

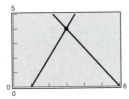

5. $x - y = 1$
$-2x + 2y = 5$

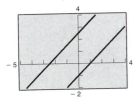

6. $3x + 2y = 2$
$6x + 4y = 14$

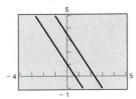

7. $3x - 2y = 6$
$-6x + 4y = -12$

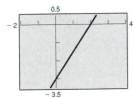

8. $x - 2y = 5$
$6x + 2y = 7$

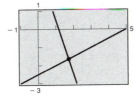

9. $9x - 3y = -1$
$3x + 6y = -5$

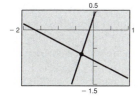

10. $5x + 3y = 18$
$2x - 7y = -1$

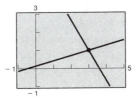

In Exercises 11–20, solve the system by elimination.

11. $x + 2y = 4$
$x - 2y = 1$

12. $3x - 5y = 2$
$2x + 5y = 13$

13. $2x + 3y = 18$
$5x - y = 11$

14. $x + 7y = 12$
$3x - 5y = 10$

15. $3x + 2y = 10$
$2x + 5y = 3$

16. $8r + 16s = 20$
$16r + 50s = 55$

17. $2u + v = 120$
$u + 2v = 120$

18. $5u + 6v = 24$
$3u + 5v = 18$

19. $6r - 5s = 3$
$10s - 12r = 5$

20. $1.8x + 1.2y = 4$
$9x + 6y = 3$

In Exercises 21–30, solve the system by elimination and verify the solution with a graphing utility.

21. $\dfrac{x}{4} + \dfrac{y}{6} = 1$
$x - y = 3$

22. $\dfrac{2}{3}x + \dfrac{1}{6}y = \dfrac{2}{3}$
$4x + y = 4$

23. $\dfrac{x + 3}{4} + \dfrac{y - 1}{3} = 1$
$2x - y = 12$

24. $\dfrac{x - 1}{2} + \dfrac{y + 2}{3} = 4$
$x - 2y = 5$

25. $2.5x - 3y = 1.5$
$10x - 12y = 6$

26. $0.02x - 0.05y = -0.19$
$0.03x + 0.04y = 0.52$

27. $0.05x - 0.03y = 0.21$
$0.07x + 0.02y = 0.16$

28. $0.2x - 0.5y = -27.8$
$0.3x + 0.4y = 68.7$

29. $4b + 3m = 3$
$3b + 11m = 13$

30. $3b + 3m = 7$
$3b + 5m = 3$

In Exercises 31 and 32, the graphs of the two equations appear to be parallel. Yet, when the system is solved algebraically, you will see that the system does have a solution. Find the solution and explain why it does not appear on the portion of the graph that is shown.

31. $200y - x = \quad 200$
$\quad\; 199y - x = -198$

32. $25x - 24y = \quad 0$
$\quad\; 13x - 12y = 120$

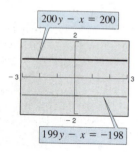

FIGURE FOR 31

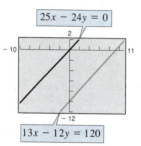

FIGURE FOR 32

33. *Airplane Speed*   An airplane flying into a headwind travels the 1800-mile flying distance between two cities in 3 hours and 36 minutes. On the return flight, the distance is traveled in 3 hours. Find the ground speed of the plane and the speed of the wind, assuming that both remain constant.

34. *Airplane Speed*   Two planes start from the same airport and fly in opposite directions. The second plane starts one-half hour after the first plane, but its speed is 50 miles per hour faster. Find the ground speed of each plane if 2 hours after the first plane starts the planes are 2000 miles apart.

35. *Acid Mixture*   Ten gallons of a 30% acid solution is obtained by mixing a 20% solution with a 50% solution. How much of each must be used?

36. *Fuel Mixture*   Five hundred gallons of 89 octane gasoline is obtained by mixing 87 octane gasoline with 92 octane gasoline. How much of each must be used?

37. *Investment Portfolio*   A total of $12,000 is invested in two corporate bonds that pay 10.5% and 12% simple interest. The annual interest is $1380. How much is invested in each bond?

38. *Investment Portfolio*   A total of $32,000 is invested in two municipal bonds that pay 5.75% and 6.25% simple interest. The annual interest is $1930. How much is invested in each bond?

39. *Ticket Sales*   Five hundred tickets were sold for a certain performance of a play. The tickets for adults and children sold for $7.50 and $4.00, respectively, and the total receipts for the performance were $3312.50. How many of each kind of ticket were sold?

40. *Shoe Sales*   Suppose you are the manager of a shoe store. On Saturday night you are going over the receipts of the previous week's sales. Two hundred and forty pairs of tennis shoes were sold. One style sold for $66.95 and the other sold for $84.95. The total receipts were $17,652. The cash register that was supposed to record the number of each type of shoe sold malfunctioned. Can you recover the information? If so, how many shoes of each type were sold?

Supply and Demand   In Exercises 41–46, use a graphing utility to graph the demand and supply equations and find the point of equilibrium.

Demand	Supply
41. $p = 50 - 0.5x$	$p = 0.125x$
42. $p = 60 - x$	$p = 10 + \frac{7}{3}x$
43. $p = 300 - x$	$p = 100 + x$
44. $p = 100 - 0.05x$	$p = 25 + 0.1x$
45. $p = 140 - 0.00002x$	$p = 80 + 0.00001x$
46. $p = 400 - 0.0002x$	$p = 225 + 0.0005x$

47. *Driving Distances*   In a trip of 300 miles, two people do the driving. One person drives three times as far as the other. Find the distance that each person drives.

48. *Truck Scheduling*   A contractor is hiring two trucking companies to haul 1600 tons of crushed stone for a highway construction project. The contracts state that one company is to haul four times as much as the other. Find the amount hauled by each.

Fitting a Line to Data   In Exercises 49–56, find the *least squares regression line*, $y = ax + b$, for the points

$(x_1, y_1), (x_2, y_2), \ldots, (x_n, y_n)$.

To find the line, solve the following system for a and b. Then use the linear regression feature of your graphing utility to confirm the result.

$$nb + \left(\sum_{i=1}^{n} x_i\right)a = \sum_{i=1}^{n} y_i$$

$$\left(\sum_{i=1}^{n} x_i\right)b + \left(\sum_{i=1}^{n} x_i^2\right)a = \sum_{i=1}^{n} x_i y_i$$

49. $5b + 10a = 20.2$
$10b + 30a = 50.1$

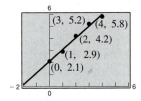

50. $5b + 10a = 11.7$
$10b + 30a = 25.6$

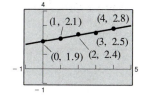

51. $7b + 21a = 35.1$
$21b + 91a = 114.2$

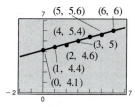

52. $6b + 15a = 23.6$
$15b + 55a = 48.8$

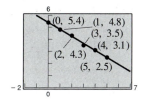

53. $(-2, 0), (0, 1), (2, 3)$

54. $(-3, 0), (-1, 1), (1, 1), (3, 2)$

55. $(0, 4), (1, 3), (1, 1), (2, 0)$

56. $(1, 0), (2, 0), (3, 0), (3, 1), (4, 1), (4, 2), (5, 2), (6, 2)$

57. *Demand Function*   A store manager wants to know the demand of a certain product as a function of the price. The daily sales for the different prices of the product are given in the following table.

Price (x)	$1.00	$1.25	$1.50
Demand (y)	450	375	330

Use the technique demonstrated in Exercises 49–56 to find the line that best fits the given data. Plot the points and graph the line. Then use the line to predict the demand when the price is $1.40.

58. *Yield*   A farmer used four test plots to determine the relationship between wheat yield in bushels per acre and the amount of fertilizer in hundreds of pounds per acre. The results are given in the following table.

Fertilizer (x)	1.0	1.5	2.0	2.5
Yield (y)	32	41	48	53

Use the technique demonstrated in Exercises 49–56 to find the line that best fits the given data. Plot the points and graph the line. Then use the line to estimate the yield for a fertilizer application of 160 pounds per acre.

In Exercises 59 and 60, find a system of linear equations having the given solution. (There is more than one correct answer.)

59. $\left(3, \frac{5}{2}\right)$ **60.** $(8, -2)$

7.3　SYSTEMS OF LINEAR EQUATIONS IN MORE THAN TWO VARIABLES

Row-Echelon Form and Back-Substitution / Gaussian Elimination / Graphical Interpretation of a Linear System in Three Variables / Nonsquare Systems / Applications

Row-Echelon Form and Back-Substitution

The method of elimination can be applied to a system of linear equations in more than two variables. The goal is to rewrite the system in a form to which back-substitution can be applied. To see how this works, consider the following two systems of linear equations.

$$\begin{aligned} x - 2y + 3z &= 9 \\ -x + 3y \phantom{{}+ 3z} &= -4 \\ 2x - 5y + 5z &= 17 \end{aligned} \qquad \begin{aligned} x - 2y + 3z &= 9 \\ y + 3z &= 5 \\ z &= 2 \end{aligned}$$

Clearly, the system on the right is easier to solve. This system is in **row-echelon form,** which means that it follows a stair-step pattern and has leading coefficients of 1. To solve such a system, use back-substitution, working from the bottom equation to the top equation.

EXAMPLE 1　Using Back-Substitution to Solve a System in Row-Echelon Form

Solve the system of linear equations.

$$\begin{aligned} x - 2y + 3z &= 9 & & \textit{Equation 1} \\ y + 3z &= 5 & & \textit{Equation 2} \\ z &= 2 & & \textit{Equation 3} \end{aligned}$$

SOLUTION

From Equation 3, you already know the value of z. To solve for y, substitute $z = 2$ into Equation 2 to obtain

$$\begin{aligned} y + 3(2) &= 5 & & \textit{Substitute } z = 2 \\ y &= -1. & & \textit{Solve for } y \end{aligned}$$

Finally, substitute $y = -1$ and $z = 2$ into Equation 1 to obtain

$$\begin{aligned} x - 2(-1) + 3(2) &= 9 & & \textit{Substitute } y = -1, z = 2 \\ x &= 1. & & \textit{Solve for } x \end{aligned}$$

The solution is $x = 1$, $y = -1$, and $z = 2$. Check this solution in the original system.

Gaussian Elimination

Two systems are **equivalent** if they have precisely the same solution set. To solve a system that is not in row-echelon form, change it to an *equivalent* system that is in row-echelon form by using the following operations. This process is called **Gaussian elimination,** after the German mathematician Carl Friedrich Gauss (1777–1855).

1. Interchange two equations.
2. Multiply one of the equations by a nonzero constant.
3. Add a multiple of one equation to another equation.

EXAMPLE 2   **Using Elimination to Solve a Linear System**

Solve the system of linear equations.

$$\begin{aligned} x - 2y + 3z &= 9 \\ -x + 3y &= -4 \\ 2x - 5y + 5z &= 17 \end{aligned}$$

SOLUTION

Work from the upper left corner, saving the x in the upper left position and eliminating the other x's from the first column.

$$\begin{aligned} x - 2y + 3z &= 9 \\ y + 3z &= 5 \quad \longleftarrow \\ 2x - 5y + 5z &= 17 \end{aligned}$$

Adding the first equation to the second equation produces a new second equation.

$$\begin{aligned} x - 2y + 3z &= 9 \\ y + 3z &= 5 \\ -y - z &= -1 \quad \longleftarrow \end{aligned}$$

Adding -2 times the first equation to the third equation produces a new third equation.

Now that all but the first x have been eliminated from the first column, go to work on the second column. (You need to eliminate y from the third equation.)

$$\begin{aligned} x - 2y + 3z &= 9 \\ y + 3z &= 5 \\ 2z &= 4 \quad \longleftarrow \end{aligned}$$

Adding the second equation to the third equation produces a new third equation.

Finally, you need a coefficient of 1 for z in the third equation.

$$\begin{aligned} x - 2y + 3z &= 9 \\ y + 3z &= 5 \\ z &= 2 \quad \longleftarrow \end{aligned}$$

Multiplying the third equation by $\frac{1}{2}$ produces a new third equation.

This is the same system solved in Example 1. As in that example, you can conclude that the solution is $x = 1$, $y = -1$, and $z = 2$. The solution can also be written as the **ordered triple** $(1, -1, 2)$. Check this solution in the original system.

If, at some stage in the elimination process, you obtain an absurdity such as $0 = 7$, you can conclude that the original system is inconsistent.

EXAMPLE 3 An Inconsistent System

Solve the system of linear equations.

$$\begin{aligned} x - 3y + z &= 1 \\ 2x - y - 2z &= 2 \\ x + 2y - 3z &= -1 \end{aligned}$$

SOLUTION

$$\begin{aligned} x - 3y + z &= 1 \\ 5y - 4z &= 0 \quad \longleftarrow \\ x + 2y - 3z &= -1 \end{aligned}$$

Adding -2 times the first equation to the second equation produces a new second equation.

$$\begin{aligned} x - 3y + z &= 1 \\ 5y - 4z &= 0 \\ 5y - 4z &= -2 \quad \longleftarrow \end{aligned}$$

Adding -1 times the first equation to the third equation produces a new third equation.

$$\begin{aligned} x - 3y + z &= 1 \\ 5y - 4z &= 0 \\ 0 &= -2 \quad \longleftarrow \end{aligned}$$

Adding -1 times the second equation to the third equation produces a new third equation.

Because the third "equation" is absurd, you can conclude that the original system is inconsistent.

As with a system of linear equations in two variables, a system of linear equations in more than two variables must have (1) exactly one solution, (2) infinitely many solutions, or (3) no solution.

There is more than one way to describe the solutions of a system with infinitely many solutions. For example, consider a solution set that is described as

$$(a, a + 1, 2a), \quad \text{where } a \text{ is any real number.}$$

This means that each real number a produces a solution of the system. A few of the infinitely many possible solutions are found by letting $a = -1, 0, 1,$ and 2 to obtain $(-1, 0, -2), (0, 1, 0), (1, 2, 2),$ and $(2, 3, 4),$ respectively. Now consider the solutions represented by the ordered triple

$$(b - 1, b, 2b - 2), \quad \text{where } b \text{ is any real number.}$$

Here again, a few possible solutions are $(-1, 0, -2)$, $(0, 1, 0)$, $(1, 2, 2)$, and $(2, 3, 4)$, found by letting $b = 0$, 1, 2, and 3, respectively. Note that both descriptions result in the same collection of solutions. Thus, when comparing descriptions of an infinite solution set, keep in mind that the way you describe the set may be different from the way another person describes the set.

EXAMPLE 4   A System with Infinitely Many Solutions

Solve the system of linear equations.

$$
\begin{aligned}
x + y - 3z &= -1 && \textit{Equation 1} \\
y - z &= 0 && \textit{Equation 2} \\
-x + 2y &= 1 && \textit{Equation 3}
\end{aligned}
$$

SOLUTION

Begin by rewriting the system in row-echelon form as follows.

$$
\begin{aligned}
x + y - 3z &= -1 \\
y - z &= 0 \\
3y - 3z &= 0 \quad \longleftarrow
\end{aligned}
$$

Adding the first equation to the third equation produces a new third equation.

$$
\begin{aligned}
x + y - 3z &= -1 \\
y - z &= 0 \\
0 &= 0 \quad \longleftarrow
\end{aligned}
$$

Adding -3 times the second equation to the third equation produces a new third equation.

This means that Equation 3 is *dependent* on Equations 1 and 2 in the sense that it gives no additional information about the variables. Thus, the original system is equivalent to the system

$$
\begin{aligned}
x + y - 3z &= -1 \\
y - z &= 0.
\end{aligned}
$$

In this last equation, solve for y in terms of z to obtain $y = z$. Back-substituting for y into the previous equation, you can find x in terms of z, as follows.

$$
\begin{aligned}
x + z - 3z &= -1 \\
x - 2z &= -1 \\
x &= 2z - 1
\end{aligned}
$$

Finally, letting $z = a$, the solutions to the given system are all of the form

$$
x = 2a - 1, \qquad y = a, \qquad z = a,
$$

where a is a real number. Therefore, every ordered triple that is in the form $(2a - 1, a, a)$, where a is a real number, is a solution of the system.

Graphical Interpretation of a Linear System in Three Variables

Solutions of equations with three variables can be pictured with a **three-dimensional coordinate system.** To construct such a system, begin with the *xy*-coordinate plane in a horizontal position. Then draw the *z*-axis as a vertical line through the origin.

Every ordered triple (x, y, z) corresponds to a point in the three-dimensional coordinate system. For instance, the points corresponding to $(-2, 5, 4)$, $(2, -5, 3)$, $/1, 6, 0)$, and $(3, 3, -2)$ are shown in Figure 7.14.

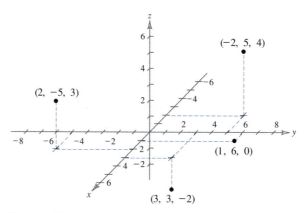

FIGURE 7.14

The **graph** of an equation in three variables consists of all points (x, y, z) that are solutions of the equation. The graph of a linear equation in three variables is a *plane*. Sketching graphs in a three-dimensional coordinate system is difficult because the sketch itself is only two-dimensional.

One technique for sketching a plane is to find the three points at which the plane intersects the axes. For instance, the plane given by

$$3x + 2y + 4z = 12$$

intersects the *x*-axis at the point $(4, 0, 0)$, the *y*-axis at the point $(0, 6, 0)$, and the *z*-axis at the point $(0, 0, 3)$.

By plotting these three points, connecting them with line segments, and shading the resulting triangular region, you can picture a portion of the graph, as shown in Figure 7.15.

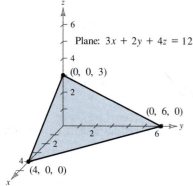

FIGURE 7.15

The graph of a system of three linear equations in three variables consists of *three* planes. When these planes intersect in a single point, the system has exactly one solution. When the three planes have no point in common, the system has no solution. When the three planes contain a common line, the system has infinitely many solutions. (See Figure 7.16.)

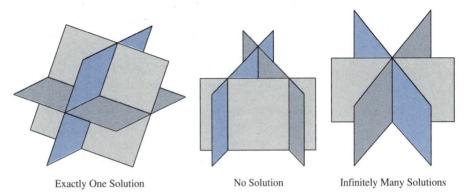

Exactly One Solution   No Solution   Infinitely Many Solutions

FIGURE 7.16

Nonsquare Systems

So far we have considered only **square** systems, for which the number of equations is equal to the number of variables. In a **nonsquare** system, the number of equations differs from the number of variables.

EXAMPLE 5   A System with Fewer Equations than Variables

Solve the system of linear equations.

$$x - 2y + z = 2 \qquad \text{\textit{Equation 1}}$$
$$2x - y - z = 1 \qquad \text{\textit{Equation 2}}$$

SOLUTION

$$x - 2y + z = 2$$
$$3y - 3z = -3 \quad \longleftarrow$$

Adding -2 times the first equation to the second equation produces a new second equation.

$$x - 2y + z = 2$$
$$y - z = -1 \quad \longleftarrow$$

Multiplying the second equation by $\frac{1}{3}$ produces a new second equation.

Solving for y in terms of z, you obtain $y = z - 1$, and back-substitution into Equation 1 yields

$$x - 2(z - 1) + z = 2$$
$$x - 2z + 2 + z = 2$$
$$x = z.$$

Finally, by letting $z = a$, you obtain the solution

$$x = a, \qquad y = a - 1, \qquad \text{and} \qquad z = a$$

where a is a real number. Thus, every ordered triple of the form $(a, a - 1, a)$, where a is a real number, is a solution of the system.

Applications

Example 6 shows how to fit a parabola through three given points in the plane. This procedure can be generalized to fit an nth degree polynomial function to $n + 1$ points in the plane. The only restriction on the procedure is that (because you are trying to fit a *function* to the points) every point must have a distinct x-coordinate.

EXAMPLE 6   Curve-Fitting

Find a quadratic function

$$f(x) = ax^2 + bx + c$$

the graph of which passes through the points $(-1, 3)$, $(1, 1)$, and $(2, 6)$.

SOLUTION

Because the graph of f passes through the points $(-1, 3)$, $(1, 1)$, and $(2, 6)$, you can write the following equations.

$$f(-1) = a(-1)^2 + b(-1) + c = 3$$
$$f(1) = a(1)^2 + b(1) + c = 1$$
$$f(2) = a(2)^2 + b(2) + c = 6$$

This produces a system of linear equations in the variables a, b, and c.

$$
\begin{aligned}
a - b + c &= 3 \\
a + b + c &= 1 \\
4a + 2b + c &= 6
\end{aligned}
$$

Using the techniques of this section, you can find the solution of this system to be $a = 2$, $b = -1$, and $c = 0$. Thus, the equation of the parabola passing through the three given points is

$$f(x) = ax^2 + bx + c = 2x^2 - x,$$

as shown in Figure 7.17. You should check that the three given points satisfy this equation.

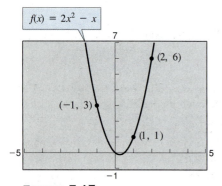

$f(x) = 2x^2 - x$

FIGURE 7.17

EXAMPLE 7   An Investment Portfolio

Suppose an investor has a portfolio totaling $450,000, and wishes to allocate this amount to the following types of investments: (1) certificates of deposit, (2) municipal bonds, (3) blue-chip stocks, and (4) growth or speculative stocks. The certificates of deposit pay 9% annually, and the municipal bonds pay 6% annually. Over a 5-year period, the investor expects the blue-chip stocks to return 10% annually, and expects the growth stocks to return 15% annually. The investor wants a combined annual return of 8%, and also wants to have only one-third of the portfolio invested in stocks. How much should be allocated to each type of investment?

SOLUTION

To solve this problem, let C, M, B, and G represent the amounts in the four types of investments. Because the total investment is $450,000, you can write

$$C + M + B + G = 450{,}000.$$

A second equation can be derived from the fact that the combined annual return should be 8%.

$$0.09C + 0.06M + 0.10B + 0.15G = 0.08(450{,}000)$$

Finally, because only one-third of the investment should be allocated to stocks, you can write

$$B + G = \frac{1}{3}(450{,}000).$$

These three equations make up the following system.

$$
\begin{aligned}
C + \quad M + \quad B + \quad G &= 450{,}000 \\
0.09C + 0.06M + 0.1B + 0.15G &= \ \ 36{,}000 \\
B + \quad G &= 150{,}000
\end{aligned}
$$

Using elimination, you can find that the system has infinitely many solutions, which can be written as follows.

$$C = -\frac{5}{3}a + 100{,}000, \qquad M = \frac{5}{3}a + 200{,}000,$$

$$B = -a + 150{,}000, \qquad G = a$$

Thus, the investor has many different options. One possible solution is to choose $a = 30{,}000$, which yields the following portfolio.

1. Certificates of deposit: $50,000
2. Municipal bonds: $250,000
3. Blue-chip stocks: $120,000
4. Growth or speculative stocks: $30,000

DISCUSSION PROBLEM
....

COMPUTER SOFTWARE

There are many computer software programs that will solve linear systems. Use your school's library, computer center, mathematics-learning center, or some other reference facility to find out about such a product. Write a short report about the capabilities of the software.

WARM-UP
....

The following warm-up exercises involve skills that were covered in earlier sections. You will use these skills in the exercise set for this section.

In Exercises 1–4, solve the system of linear equations.

1. $x + y = 25$
$\quad y = 10$

2. $2x - 3y = 4$
$\quad 6x = -12$

3. $x + y = 32$
$\quad x - y = 24$

4. $2r - s = 5$
$\quad r + 2s = 10$

In Exercises 5–8, determine whether the ordered triple is a solution of the equation.

5. $5x - 3y + 4z = 2$
$\quad (-1, -2, 1)$

6. $x - 2y + 12z = 9$
$\quad (6, 3, 2)$

7. $2x - 5y + 3z = -9$
$\quad (a - 2, a + 1, a)$

8. $-5x + y + z = 21$
$\quad (a - 4, 4a + 1, a)$

In Exercises 9 and 10, solve for x in terms of a.

9. $x + 2y - 3z = 4$
$\quad y = 1 - a, z = a$

10. $x - 3y + 5z = 4$
$\quad y = 2a + 3, z = a$

SECTION 7.3 · EXERCISES

In Exercises 1–26, solve the system of linear equations.

1. $x + y + z = 6$
$\quad 2x - y + z = 3$
$\quad 3x - z = 0$

2. $x + y + z = 2$
$\quad -x + 3y + 2z = 8$
$\quad 4x + y = 4$

3. $4x + y - 3z = 11$
$\quad 2x - 3y + 2z = 9$
$\quad x + y + z = -3$

4. $2x + 2z = 2$
$\quad 5x + 3y = 4$
$\quad 3y - 4z = 4$

5. $ 6y + 4z = -12$
$\quad 3x + 3y = 9$
$\quad 2x - 3z = 10$

6. $2x + 4y + z = -4$
$\quad 2x - 4y + 6z = 13$
$\quad 4x - 2y + z = 6$

7. $3x - 2y + 4z = 1$
$\quad x + y - 2z = 3$
$\quad 2x - 3y + 6z = 8$

8. $5x - 3y + 2z = 3$
$\quad 2x + 4y - z = 7$
$\quad x - 11y + 4z = 3$

9. $3x + 3y + 5z = 1$
$\quad 3x + 5y + 9z = 0$
$\quad 5x + 9y + 17z = 0$

10. $2x + y + 3z = 1$
$\quad 2x + 6y + 8z = 3$
$\quad 6x + 8y + 18z = 5$

11. $x + 2y - 7z = -4$
$\quad 2x + y + z = 13$
$\quad 3x + 9y - 36z = -33$

12. $2x + y - 3z = 4$
$\quad 4x + 2z = 10$
$\quad -2x + 3y - 13z = -8$

13. $x + 4z = 13$
$\quad 4x - 2y + z = 7$
$\quad 2x - 2y - 7z = -19$

14. $4x - y + 5z = 11$
$\quad x + 2y - z = 5$
$\quad 5x - 8y + 13z = 7$

15. $x - 2y + 5z = 2$
$\quad 3x + 2y - z = -2$

16. $x - 3y + 2z = 18$
$\quad 5x - 13y + 12z = 80$

17. $2x - 3y + z = -2$
$\quad -4x + 9y = 7$

18. $2x + 3y + 3z = 7$
$\quad 4x + 18y + 15z = 44$

19. $x + 3w = 4$
$\quad 2y - z - w = 0$
$\quad 3y - 2w = 1$
$\quad 2x - y + 4z = 5$

20. $x + y + z + w = 6$
$\quad 2x + 3y - w = 0$
$\quad -3x + 4y + z + 2w = 4$
$\quad x + 2y - z + w = 0$

21. $x + 4z = 1$
$\quad x + y + 10z = 10$
$\quad 2x - y + 2z = -5$

22. $3x - 2y - 6z = -4$
$\quad -3x + 2y + 6z = 1$
$\quad x - y - 5z = -3$

23. $4x + 3y + 17z = 0$
$5x + 4y + 22z = 0$
$4x + 2y + 19z = 0$

24. $2x + 3y \quad\quad = 0$
$4x + 3y - z = 0$
$8x + 3y + 3z = 0$

25. $5x + 5y - z = 0$
$10x + 5y + 2z = 0$
$5x + 15y - 9z = 0$

26. $12x + 5y + z = 0$
$12x + 4y - z = 0$

In Exercises 27–30, find the equation of the parabola $y = ax^2 + bx + c$ that passes through the given points. Verify your result with a graphing utility.

27. $(0, -4)$, $(1, 1)$, $(2, 10)$

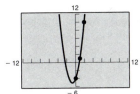

28. $(0, 5)$, $(1, 6)$, $(2, 5)$

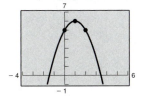

29. $(1, 0)$, $(2, -1)$, $(3, 0)$

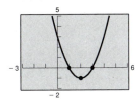

30. $(1, 2)$, $(2, 1)$, $(3, -4)$

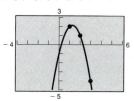

In Exercises 31–34, find the equation of the circle
$$x^2 + y^2 + Dx + Ey + F = 0$$
that passes through the given points. Verify your result with a graphing utility.

31. $(0, 0)$, $(2, -2)$, $(4, 0)$

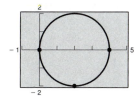

32. $(0, 0)$, $(0, 6)$, $(-3, 3)$

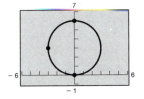

33. $(3, -1)$, $(-2, 4)$, $(6, 8)$

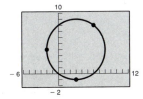

34. $(0, 0)$, $(0, 2)$, $(3, 0)$

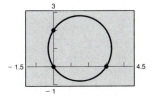

Vertical Motion   In Exercises 35–38, find the position equation
$$s = \tfrac{1}{2}at^2 + v_0 t + s_0$$
for an object that is moving vertically and is at the given positions at the specified times.

35. At $t = 1$ second, $s = 128$ feet
At $t = 2$ seconds, $s = 80$ feet
At $t = 3$ seconds, $s = 0$ feet

36. At $t = 1$ second, $s = 48$ feet
At $t = 2$ seconds, $s = 64$ feet
At $t = 3$ seconds, $s = 48$ feet

37. At $t = 1$ second, $s = 452$ feet
At $t = 3$ seconds, $s = 260$ feet
At $t = 4$ seconds, $s = 116$ feet

38. At $t = 2$ seconds, $s = 132$ feet
At $t = 3$ seconds, $s = 100$ feet
At $t = 4$ seconds, $s = 36$ feet

39. *Investments*   An inheritance of $16,000 was divided among three investments yielding a total of $990 in interest per year. The interest rates for the three investments were 5%, 6%, and 7%. Find the amount placed in each investment if the 5% and 6% investments were $3000 and $2000 less than the 7% investment, respectively.

40. *Investments*   Suppose you receive a total of $1520 a year in interest from three investments. The interest rates for the three investments are 5%, 7%, and 8%. The 5% investment is half of the 7% investment, and the 7% investment is $1500 less than the 8% investment. What is the amount of each investment?

41. *Borrowing*   A small corporation borrowed $775,000 to expand its product line. Some of the money was borrowed at 8%, some at 9%, and some at 10%. How much was borrowed at each rate if the annual interest was $67,500 and the amount borrowed at 8% was four times the amount borrowed at 10%?

42. *Borrowing*   A small corporation borrowed $800,000 to expand its product line. Some of the money was borrowed at 8%, some at 9%, and some at 10%. How much was borrowed at each rate if the annual interest was $67,000 and the amount borrowed at 8% was five times the amount borrowed at 10%?

Investment Portfolio   In Exercises 43 and 44, consider an investment with a portfolio totaling $500,000 that is to be allocated among the following types of investments: (1) certificates of deposit, (2) municipal bonds, (3) blue-chip stocks, and (4) growth or speculative stocks. How much should be allocated to each type of investment?

43. The certificates of deposit pay 10% annually, and the municipal bonds pay 8% annually. Over a 5-year period, the investor expects the blue-chip stocks to return 12% annually and the growth stocks to return 13% annually. The investor wants a combined annual return of 10% and also wants to have only one-fourth of the portfolio invested in stocks.

44. The certificates of deposit pay 9% annually, and the municipal bonds pay 5% annually. Over a 5-year period, the investor expects the blue-chip stocks to return 12% annually and the growth stocks to return 14% annually. The investor wants a combined annual return of 10% and also wants to have only one-fourth of the portfolio invested in stocks.

45. *Crop Spraying*   A mixture of 12 gallons of chemical A, 16 gallons of chemical B, and 26 gallons of chemical C is required to kill a certain destructive crop insect. Commercial spray X contains 1, 2, and 2 parts, respectively, of these chemicals. Commercial spray Y contains only chemical C. Commercial spray Z contains only chemicals A and B in equal amounts. How much of each type of commercial spray is needed to get the desired mixture?

46. *Chemistry*   A chemist needs 10 liters of a 25% acid solution. The solution is to be mixed from three solutions the concentrations of which are 10%, 20%, and 50%, respectively. How many liters of each solution should the chemist use to satisfy the following?
(a) Use as little as possible of the 50% solution.
(b) Use as much as possible of the 50% solution.
(c) Use two liters of the 50% solution.

47. *Truck Scheduling*   A small company that manufactures products A and B has an order for 15 units of product A and 16 units of product B. The company has trucks of three different sizes that can haul the products, as shown in the table.

Truck	Product A	Product B
Large	6	3
Medium	4	4
Small	0	3

How many trucks of each size are needed to deliver the order? (Give *two* possible solutions.)

48. *Electrical Networks*   Applying Kirchhoff's Laws to the electrical network in the accompanying figure, the currents I_1, I_2, and I_3 must be the solution to the system

$$I_1 - I_2 + I_3 = 0$$
$$3I_1 + 2I_2 \quad\quad = 7$$
$$\quad\quad 2I_2 + 4I_3 = 8,$$

where the current is measured in amperes. Find the currents.

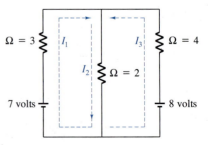

FIGURE FOR 48

49. *Pulley System* A system of pulleys that are assumed to be frictionless and without mass are loaded with 128-pound and 32-pound weights (see figure). The tensions t_1 and t_2 in the ropes, and the acceleration a of the 32-pound weight, are found by solving the system

$$
\begin{aligned}
t_1 - 2t_2 &= 0 \\
t_1 \quad\;\; - 2a &= 128 \\
t_2 + a &= 32,
\end{aligned}
$$

where t_1 and t_2 are measured in pounds and a is in feet per second squared. Solve the system.

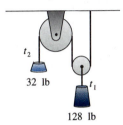

32 lb

128 lb

FIGURE FOR 49

50. *Pulley System* If the 32-pound weight is replaced by a 64-pound weight in the pulley system of Exercise 49, it is modeled by the following system of equations.

$$
\begin{aligned}
t_1 - 2t_2 &= 0 \\
t_1 \quad\;\; - 2a &= 128 \\
t_2 + a &= 64
\end{aligned}
$$

Solve the system and use your result for the acceleration to describe what (if anything) is happening in the system.

Fitting a Parabola In Exercises 51–54, find the least squares regression parabola $y = ax^2 + bx + c$ for the points

$(x_1, y_1), (x_2, y_2), \ldots, (x_n, y_n)$.

To find the parabola, solve the following system of linear equations for a, b, and c.

$$
nc + \left(\sum_{i=1}^{n} x_i\right)b + \left(\sum_{i=1}^{n} x_i^2\right)a = \sum_{i=1}^{n} y_i
$$

$$
\left(\sum_{i=1}^{n} x_i\right)c + \left(\sum_{i=1}^{n} x_i^2\right)b + \left(\sum_{i=1}^{n} x_i^3\right)a = \sum_{i=1}^{n} x_i y_i
$$

$$
\left(\sum_{i=1}^{n} x_i^2\right)c + \left(\sum_{i=1}^{n} x_i^3\right)b + \left(\sum_{i=1}^{n} x_i^4\right)a = \sum_{i=1}^{n} x_i^2 y_i
$$

51.
$$
\begin{aligned}
4c \quad\;\; + 40a &= 19 \\
40b \quad\quad &= -12 \\
40c \quad\;\; + 544a &= 160
\end{aligned}
$$

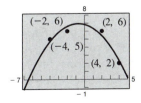

52.
$$
\begin{aligned}
5c \quad\;\; + 10a &= 8 \\
10b \quad\quad &= 12 \\
10c \quad\;\; + 34a &= 22
\end{aligned}
$$

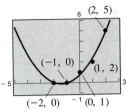

53.
$$
\begin{aligned}
4c + 9b + 29a &= 20 \\
9c + 29b + 99a &= 70 \\
29c + 99b + 353a &= 254
\end{aligned}
$$

54.
$$
\begin{aligned}
4c + 6b + 14a &= 25 \\
6c + 14b + 36a &= 21 \\
14c + 36b + 98a &= 33
\end{aligned}
$$

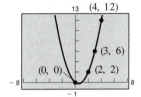

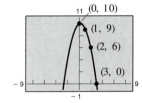

55. *Stopping Distances* In testing the new braking system on an automobile, the speed in miles per hour and the stopping distance in feet were recorded (see table).

Speed (x)	20	30	40	50	60
Stopping distance (y)	25	55	105	188	300

(a) Plot the data on the rectangular coordinate system.

(b) Fit a least squares regression parabola for the data and sketch the graph of the parabola on the coordinate system of part (a).

56. *Reproduction* A wildlife management team studied the reproduction rates of deer in five tracts of a wildlife preserve. Each tract contained 5 acres. In each tract, the number of females and the percentage of females that had offspring the following year were counted. The results are given in the following table.

Number (x)	80	100	120	140	160
Percentage (y)	80	75	68	55	30

(a) Plot the data on the rectangular coordinate system.

(b) Fit the least squares regression parabola for the data and sketch the graph of the parabola on the coordinate system of part (a).

In Exercises 57 and 58, find a system of equations having the given solution. (There is more than one correct answer for each exercise.)

57. $(4, -1, 2)$

58. $\left(-\frac{3}{2}, 4, -7\right)$

7.4   PARTIAL FRACTIONS

Introduction to Partial Fractions / Partial Fraction Decomposition

Introduction to Partial Fractions

In this section, you will study a procedure for writing a rational expression as the sum of two or more simpler rational expressions. (This procedure is useful in calculus.) For example, the rational expression $(x + 7)/(x^2 - x - 6)$ can be written as the sum of two fractions with first-degree denominators.

$$\frac{x + 7}{x^2 - x - 6} = \frac{2}{x - 3} + \frac{-1}{x + 2}$$

Each fraction on the right side of the equation is a **partial fraction,** and together they make up the **partial fraction decomposition** of the left side.

In Chapter 3, you learned that it is theoretically possible to write any polynomial as the product of linear and irreducible quadratic factors. For instance,

$$x^5 + x^4 - x - 1 = (x - 1)(x + 1)^2(x^2 + 1),$$

where $(x - 1)$ is a linear factor, $(x + 1)$ is a repeated linear factor, and $(x^2 + 1)$ is an irreducible quadratic factor.

You can use this factorization to find the partial fraction decomposition of any rational expression having $x^5 + x^4 - x - 1$ as its denominator. Specifically, if $N(x)$ is a polynomial of degree 4 or less, the partial fraction decomposition of $N(x)/(x^5 + x^4 - x - 1)$ has the form

$$\frac{N(x)}{x^5 + x^4 - x - 1} = \frac{N(x)}{(x - 1)(x + 1)^2(x^2 + 1)}$$

$$= \frac{A}{x - 1} + \frac{B}{x + 1} + \frac{C}{(x + 1)^2} + \frac{Dx + F}{x^2 + 1}.$$

Note that the factor $(x + 1)^2$ results in *two* fractions: one for $(x + 1)$ and one for $(x + 1)^2$. If $(x + 1)^3$ were a factor, you would use three fractions: one for $(x + 1)$, one for $(x + 1)^2$, and one for $(x + 1)^3$. In general, the number of fractions resulting from a repeated factor is equal to the exponent of the factor.

Partial Fraction Decomposition

The following guidelines summarize the steps used to find the partial fraction decomposition of a rational expression.

DECOMPOSITION OF $N(x)/D(x)$ INTO PARTIAL FRACTIONS

1. *Divide if improper:* If $N(x)/D(x)$ is an improper fraction, divide the denominator into the numerator to obtain

$$\frac{N(x)}{D(x)} = (\text{polynomial}) + \frac{N_1(x)}{D(x)}$$

and apply Steps 2, 3, and 4 to the proper rational expression $N_1(x)/D(x)$.

2. *Factor denominator:* Completely factor the denominator into factors of the form

$$(px + q)^m \quad \text{and} \quad (ax^2 + bx + c)^n,$$

where $(ax^2 + bx + c)$ is irreducible.

3. *Linear factors:* For *each* factor of the form $(px + q)^m$, the partial fraction decomposition must include the following sum of m fractions.

$$\frac{A_1}{(px + q)} + \frac{A_2}{(px + q)^2} + \cdots + \frac{A_m}{(px + q)^m}$$

4. *Quadratic factors:* For *each* factor of the form $(ax^2 + bx + c)^n$, the partial fraction decomposition must include the following sum of n fractions.

$$\frac{B_1x + C_1}{ax^2 + bx + c} + \frac{B_2x + C_2}{(ax^2 + bx + c)^2} + \cdots + \frac{B_nx + C_n}{(ax^2 + bx + c)^n}$$

Algebraic techniques for determining the constants in the numerators of the partial fractions are demonstrated in the examples that follow. Note that the techniques vary slightly depending on the type of factors of the denominator— linear or quadratic, distinct or repeated.

EXAMPLE 1   **Distinct Linear Factors**

Write the partial fraction decomposition for $\dfrac{x + 7}{x^2 - x - 6}$.

SOLUTION

Because $x^2 - x - 6 = (x - 3)(x + 2)$, you include one partial fraction with a constant numerator for each linear factor of the denominator and write

$$\frac{x + 7}{x^2 - x - 6} = \frac{A}{x - 3} + \frac{B}{x + 2}.$$

Multiplying both sides of this equation by the least common denominator, $(x - 3)(x + 2)$, leads to the **basic equation**

$$x + 7 = A(x + 2) + B(x - 3). \qquad \textit{Basic equation}$$

Because this equation is true for all x, you can substitute any *convenient* values of x that will help determine the constants A and B. Values of x that are especially convenient are ones that make the factors $(x + 2)$ and $(x - 3)$ equal to zero. For instance, if you let $x = -2$, then

$$-2 + 7 = A(0) + B(-5) \qquad \textit{Substitute convenient value of x}$$
$$5 = -5B$$
$$-1 = B.$$

To solve for A, let $x = 3$ to obtain

$$3 + 7 = A(5) + B(0) \qquad \textit{Substitute convenient value of x}$$
$$10 = 5A$$
$$2 = A.$$

Therefore, the decomposition is

$$\frac{x + 7}{x^2 - x - 6} = \frac{2}{x - 3} + \frac{-1}{x + 2},$$

as indicated at the beginning of this section. Check this result by combining the two partial fractions on the right side of the equation. You can check the decomposition graphically by sketching the graphs of

$$y = \frac{x + 7}{x^2 - x - 6} \quad \text{and} \quad y = \frac{2}{x - 3} - \frac{1}{x + 2}$$

on the same screen. If the decomposition is correct, the two graphs will be identical, as shown in Figure 7.18.

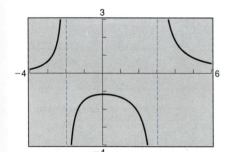

FIGURE 7.18

In Example 1, the basic equation was solved for A and B by substituting convenient values for x. The solution could also be obtained by writing the basic equation in polynomial form and *equating coefficients of like terms.* This yields a system of linear equations that can be solved by any convenient method. This approach is used in the next three examples.

EXAMPLE 2 Repeated Linear Factors

Write the partial fraction decomposition for $\dfrac{5x^2 + 20x + 6}{x^3 + 2x^2 + x}$.

SOLUTION

Because the denominator factors as

$$x^3 + 2x^2 + x = x(x^2 + 2x + 1) = x(x + 1)^2,$$

include one fraction with a constant numerator for each power of x and $(x + 1)$, and write

$$\frac{5x^2 + 20x + 6}{x(x + 1)^2} = \frac{A}{x} + \frac{B}{x + 1} + \frac{C}{(x + 1)^2}.$$

Multiplying by the least common denominator, $x(x + 1)^2$, leads to the basic equation

$$
\begin{aligned}
5x^2 + 20x + 6 &= A(x + 1)^2 + Bx(x + 1) + Cx && \textit{Basic equation}\\
&= Ax^2 + 2Ax + A + Bx^2 + Bx + Cx \\
&= (A + B)x^2 + (2A + B + C)x + A. && \textit{Polynomial form}
\end{aligned}
$$

By equating coefficients of like terms, you obtain the following system of linear equations.

$$
\begin{aligned}
5 &= A + B \\
20 &= 2A + B + C \\
6 &= A
\end{aligned}
$$

Substituting 6 for A in the first equation produces

$$
\begin{aligned}
5 &= 6 + B \\
-1 &= B.
\end{aligned}
$$

Substituting $A = 6$ and $B = -1$ in the second equation produces

$$
\begin{aligned}
20 &= 2(6) + (-1) + C \\
9 &= C.
\end{aligned}
$$

Therefore, the partial fraction decomposition is

$$\frac{5x^2 + 20x + 6}{x(x + 1)^2} = \frac{6}{x} - \frac{1}{x + 1} + \frac{9}{(x + 1)^2}.$$

Use a graphing utility to confirm this result.

EXAMPLE 3   Distinct Linear and Quadratic Factors

Write the partial fraction decomposition for $\dfrac{3x^2 + 4x + 4}{x^3 + 4x}$.

SOLUTION

Because the denominator factors as $x^3 + 4x = x(x^2 + 4)$, include one partial fraction with a constant numerator and one partial fraction with a linear numerator and write

$$\frac{3x^2 + 4x + 4}{x^3 + 4x} = \frac{A}{x} + \frac{Bx + C}{x^2 + 4}.$$

Multiplying by the least common denominator, $x(x^2 + 4)$, yields the basic equation

$$\begin{aligned}
3x^2 + 4x + 4 &= A(x^2 + 4) + (Bx + C)x && \textit{Basic equation} \\
&= Ax^2 + 4A + Bx^2 + Cx \\
&= (A + B)x^2 + Cx + 4A. && \textit{Polynomial form}
\end{aligned}$$

The resulting system of linear equations

$$3 = A + B$$
$$4 = C$$
$$4 = 4A$$

has solutions $A = 1$, $B = 2$, and $C = 4$. Therefore, the partial fraction decomposition is

$$\frac{3x^2 + 4x + 4}{x^3 + 4x} = \frac{1}{x} + \frac{2x + 4}{x^2 + 4}.$$

Use a graphing utility to confirm this result.

EXAMPLE 4   Repeated Quadratic Factors

Write the partial fraction decomposition for $\dfrac{8x^3 + 13x}{(x^2 + 2)^2}$.

SOLUTION

Include one partial fraction with a linear numerator for each power of $(x^2 + 2)$, and write

$$\frac{8x^3 + 13x}{(x^2 + 2)^2} = \frac{Ax + B}{x^2 + 2} + \frac{Cx + D}{(x^2 + 2)^2}.$$

Multiplying by the least common denominator, $(x^2 + 2)^2$, yields the basic equation

$$8x^3 + 13x = (Ax + B)(x^2 + 2) + Cx + D \qquad \textit{Basic equation}$$
$$= Ax^3 + 2Ax + Bx^2 + 2B + Cx + D$$
$$= Ax^3 + Bx^2 + (2A + C)x + (2B + D). \qquad \textit{Polynomial form}$$

By equating coefficients of like terms, you obtain the following system of linear equations.

$$8 = A$$
$$0 = B$$
$$13 = 2A + C$$
$$0 = 2B + D$$

The solution of this system is $A = 8$, $B = 0$, $C = -3$, and $D = 0$. Therefore, the partial fraction decomposition is

$$\frac{8x^3 + 13x}{(x^2 + 2)^2} = \frac{8x}{x^2 + 2} - \frac{3x}{(x^2 + 2)^2}.$$

Use a graphing utility to confirm this result.

GUIDELINES FOR SOLVING THE BASIC EQUATION

System of Linear Equations Method

1. Write the basic equation in polynomial form.
2. Equate the coefficients of like terms to obtain a system of linear equations involving A, B, C, and so on.
3. Solve the system of equations by any convenient method.

Alternative Method for Nonrepeated Linear Factors

Substitute the zeros of the distinct linear factors into the basic equation and solve for A, B, C, and so on.

Keep in mind that for *improper* rational expressions such as

$$\frac{N(x)}{D(x)} = \frac{2x^3 + x^2 - 7x + 7}{x^2 + x - 2},$$

you must first divide to obtain the form

$$\frac{N(x)}{D(x)} = (\text{polynomial}) + \frac{N_1(x)}{D(x)}.$$

The proper rational expression $N_1(x)/D(x)$ is then decomposed into its partial fractions by the usual methods.

DISCUSSION PROBLEM

································

YOU BE THE INSTRUCTOR

Suppose you are tutoring a student in algebra. In trying to find a partial fraction decomposition, your student writes the following.

$$\frac{x^2 + 1}{x(x - 1)} = \frac{A}{x} + \frac{B}{x - 1}$$

$$\frac{x^2 + 1}{x(x - 1)} = \frac{A(x - 1)}{x(x - 1)} + \frac{Bx}{x(x - 1)}$$

$$x^2 + 1 = A(x - 1) + Bx \qquad \textit{Basic equation}$$

By substituting $x = 0$ and $x = 1$ into the basic equation, your student concludes that $A = -1$ and $B = 2$. However, in checking this solution, your student obtains

$$\frac{-1}{x} + \frac{2}{x - 1} = \frac{(-1)(x - 1) + 2(x)}{x(x - 1)} = \frac{x + 1}{x(x - 1)} \neq \frac{x^2 + 1}{x(x - 1)}.$$

What has gone wrong?

WARM-UP

···················

The following warm-up exercises involve skills that were covered in earlier sections. You will use these skills in the exercise set for this section.

In Exercises 1–10, find the sum and simplify.

1. $\dfrac{2}{x} + \dfrac{3}{x + 1}$

2. $\dfrac{5}{x + 2} + \dfrac{3}{x}$

3. $\dfrac{7}{x - 2} - \dfrac{3}{2x - 1}$

4. $\dfrac{2}{x + 5} - \dfrac{5}{x + 12}$

5. $\dfrac{1}{x - 3} + \dfrac{3}{(x - 3)^2} - \dfrac{5}{(x - 3)^3}$

6. $\dfrac{-5}{x + 2} + \dfrac{4}{(x + 2)^2}$

7. $\dfrac{-3}{x} + \dfrac{3x - 1}{x^2 + 3}$

8. $\dfrac{5}{x + 1} - \dfrac{x - 6}{x^2 + 5}$

9. $\dfrac{3}{x^2 + 1} + \dfrac{x - 3}{(x^2 + 1)^2}$

10. $\dfrac{x}{x^2 + x + 1} - \dfrac{x - 1}{(x^2 + x + 1)^2}$

SECTION 7.4 · EXERCISES

In Exercises 1–32, write the partial fraction decomposition for the rational expression. Check your result algebraically *and* graphically using a graphing utility.

1. $\dfrac{1}{x^2 - 1}$

2. $\dfrac{1}{4x^2 - 9}$

3. $\dfrac{1}{x^2 + x}$

4. $\dfrac{3}{x^2 - 3x}$

5. $\dfrac{1}{2x^2 + x}$

6. $\dfrac{5}{x^2 + x - 6}$

7. $\dfrac{3}{x^2 + x - 2}$

8. $\dfrac{x + 1}{x^2 + 4x + 3}$

9. $\dfrac{5 - x}{2x^2 + x - 1}$

10. $\dfrac{3x^2 - 7x - 2}{x^3 - x}$

11. $\dfrac{x^2 + 12x + 12}{x^3 - 4x}$

12. $\dfrac{x + 2}{x(x - 4)}$

13. $\dfrac{4x^2 + 2x - 1}{x^2(x + 1)}$

14. $\dfrac{2x - 3}{(x - 1)^2}$

15. $\dfrac{x - 1}{x^3 + x^2}$

16. $\dfrac{4x^2 - 1}{2x(x + 1)^2}$

17. $\dfrac{3x}{(x - 3)^2}$

18. $\dfrac{6x^2 + 1}{x^2(x - 1)^3}$

19. $\dfrac{x^2 - 1}{x(x^2 + 1)}$

20. $\dfrac{x}{(x - 1)(x^2 + x + 1)}$

21. $\dfrac{x^2}{x^4 - 2x^2 - 8}$

22. $\dfrac{2x^2 + x + 8}{(x^2 + 4)^2}$

23. $\dfrac{x}{16x^4 - 1}$

24. $\dfrac{x^2 - 4x + 7}{(x + 1)(x^2 - 2x + 3)}$

25. $\dfrac{x^2 + x + 2}{(x^2 + 2)^2}$

26. $\dfrac{x^3}{(x + 2)^2(x - 2)^2}$

27. $\dfrac{x^2 + 5}{(x + 1)(x^2 - 2x + 3)}$

28. $\dfrac{x + 1}{x^3 + x}$

29. $\dfrac{2x^3 - 4x^2 - 15x + 5}{x^2 - 2x - 8}$

30. $\dfrac{x^3 - x + 3}{x^2 + x - 2}$

31. $\dfrac{x^4}{(x - 1)^3}$

32. $\dfrac{x^2 - x}{x^2 + x + 1}$

In Exercises 33–36, write the partial fraction decomposition for the rational expression. Check your result algebraically. Then assign a value to the constant a to check the result graphically.

33. $\dfrac{1}{a^2 - x^2}$

34. $\dfrac{1}{x(x + a)}$

35. $\dfrac{1}{x(a - x)}$

36. $\dfrac{1}{(x + 1)(a - x)}$

7.5   SYSTEMS OF INEQUALITIES

The Graph of an Inequality   /   Systems of Inequalities   /   Applications

The Graph of an Inequality

Technology Note

Most graphing utilities have the capability of shading a region that is above a graph, below a graph, or between two graphs. Learn how to use your graphing utility to graph a linear inequality in x and y. Then use your graphing utility to duplicate the result shown in Example 1.

An ordered pair (a, b) is a **solution of an inequality** in x and y if the inequality is true when a and b are substituted for x and y, respectively. The **graph** of an inequality is the collection of all solutions of the inequality. To sketch the graph of an inequality such as $3x - 2y < 6$, begin by sketching the graph of the *corresponding equation* $3x - 2y = 6$. This graph is made with a dashed line for the strict inequalities $<$ and $>$ and a solid line for the inequalities $\leq$ and $\geq$. The graph of the equation will normally separate the plane into two or more regions. In each such region, one of the following must be true.

1. *All* points in the region are solutions of the inequality.
2. *No* points in the region are solutions of the inequality.

Thus, you can determine whether the points in an entire region satisfy the inequality by simply testing *one* point in the region.

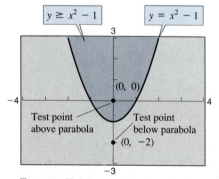

FIGURE 7.19

EXAMPLE 1   Graphing an Inequality

Sketch the graph of the inequality $y \geq x^2 - 1$.

SOLUTION

The graph of the corresponding *equation* $y = x^2 - 1$ is a parabola, as shown in Figure 7.19. By testing a point *above* the parabola $(0, 0)$ and a point *below* the parabola $(0, -2)$, you can see that the points that satisfy the inequality are those lying above (or on) the parabola.

The inequality given in Example 1 is a nonlinear inequality in two variables. This section, however, primarily discusses **linear inequalities** of the following forms.

$$ax + by < c \qquad ax + by \leq c$$
$$ax + by > c \qquad ax + by \geq c$$

The graph of each of these linear inequalities is a half-plane lying on one side of the line $ax + by = c$. The simplest linear inequalities are those corresponding to horizontal or vertical lines.

EXAMPLE 2   Graphing a Linear Inequality

Sketch the graphs of the following inequalities.

A. $x > -2$ **B.** $y \leq 3$

SOLUTION

A. The graph of the corresponding equation $x = -2$ is a vertical line. The points that satisfy the inequality $x > -2$ are those lying to the right of this line, as shown in Figure 7.20.

B. The graph of the corresponding equation $y = 3$ is a horizontal line. The points that satisfy the inequality $y \leq 3$ are those lying below (or on) this line, as shown in Figure 7.21.

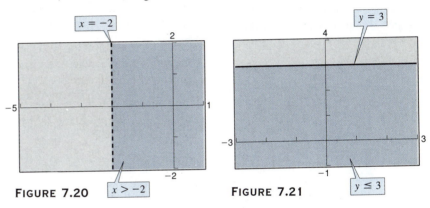

FIGURE 7.20   **FIGURE 7.21**

EXAMPLE 3   Graphing a Linear Inequality

Sketch the graph of $x - y < 2$.

SOLUTION

The graph of the corresponding equation $x - y = 2$ is a line, as shown in Figure 7.22. Because the origin $(0, 0)$ satisfies the inequality, the graph consists of the half-plane lying above the line. (Try checking a point below the line. Regardless of which point you choose, you will see that it does not satisfy the inequality.) For a linear inequality in two variables, you can sometimes simplify the graphing procedure by writing the inequality in *slope-intercept* form. For instance, by writing $x - y < 2$ in the form

$$y > x - 2,$$

you can see that the solution points lie *above* the line $x - y = 2$.

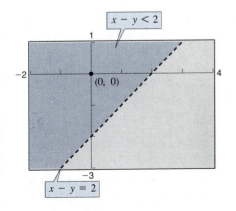

FIGURE 7.22

Systems of Inequalities

Many practical problems in business, science, and engineering involve systems of linear inequalities. Here are two examples of such systems.

$$\begin{aligned} 2x - y &\le 5 \\ x + 2y &> 2 \end{aligned} \qquad \begin{aligned} x + y &\le 12 \\ 3x - 4y &\le 15 \\ x &\ge 0 \\ y &\ge 0 \end{aligned}$$

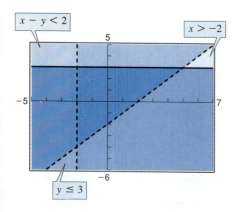

A **solution** of a system of inequalities in x and y is an ordered pair (x, y) that satisfies each inequality in the system. For instance, $(2, 4)$ is a solution of the system on the right because $x = 2$ and $y = 4$ satisfy each of the four inequalities in the system.

To graph a system of inequalities in two variables, first sketch the graph of each individual inequality (on the same coordinate system) and then find the region that is *common* to every graph in the system. For systems of linear inequalities, it is helpful to find the *vertices* of the solution region, as shown in the following example.

EXAMPLE 4 Solving a System of Inequalities

Sketch the graph (and label the vertices) of the solution set of the following system.

$$\begin{aligned} x - y &< 2 \\ x &> -2 \\ y &\le 3 \end{aligned}$$

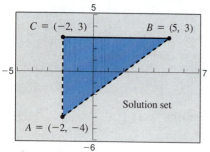

FIGURE 7.23

SOLUTION

The graphs of these inequalities were sketched in Examples 2 and 3. The triangular region common to all three graphs can be found by superimposing the graphs on the same coordinate plane, as shown in Figure 7.23. To find the vertices of the region, solve the three systems of corresponding equations obtained by taking *pairs* of equations representing the boundaries of the individual regions.

Vertex A: $(-2, -4)$	Vertex B: $(5, 3)$	Vertex C: $(-2, 3)$
Obtained by solving the system	Obtained by solving the system	Obtained by solving the system
$\begin{aligned} x - y &= 2 \\ x &= -2 \end{aligned}$	$\begin{aligned} x - y &= 2 \\ y &= 3 \end{aligned}$	$\begin{aligned} x &= -2 \\ y &= 3 \end{aligned}$

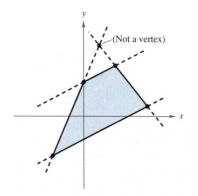

Boundary lines can intersect at a point that is not a vertex.

FIGURE 7.24

For the triangular region shown in Figure 7.23, each point of intersection of a pair of boundary lines corresponds to a vertex. With more complicated regions, two border lines can sometimes intersect at a point that is not a vertex of the region, as shown in Figure 7.24. In order to keep track of which points of intersection are actually vertices of the region, make a careful sketch of the region and refer to your sketch as you find each point of intersection.

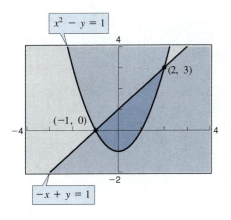

FIGURE 7.25

EXAMPLE 5   Solving a System of Inequalities

Sketch the region represented by the system.

$$x^2 - y \leq 1$$
$$-x + y \leq 1$$

SOLUTION

As shown in Figure 7.25, the points that satisfy the inequality $x^2 - y \leq 1$ are the points lying above (or on) the parabola given by $x^2 - y = 1$. The points satisfying the inequality $-x + y \leq 1$ are the points lying on or below the line given by $-x + y = 1$. To find the points of intersection of the parabola and the line, solve the following system of corresponding equations.

$$x^2 - y = 1$$
$$-x + y = 1$$

Using the method of substitution, you can find the solutions to be $(-1, 0)$ and $(2, 3)$, as shown in Figure 7.25.

When solving a system of inequalities, you should be aware that the system might have no solution. For instance, the system

$$x + y > 3$$
$$x + y < -1$$

has no solution points, because the quantity $(x + y)$ cannot be both less than -1 and greater than 3, as shown in Figure 7.26.

Another possibility is that the solution set of a system of inequalities can be unbounded. For instance, the solution set of

$$x + y < 3$$
$$x + 2y > 3$$

forms an *infinite wedge*, as shown in Figure 7.27.

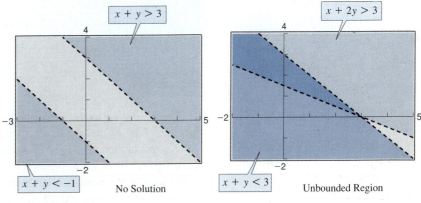

FIGURE 7.26   **FIGURE 7.27**

Applications

EXAMPLE 6   An Application of a System of Inequalities

The liquid portion of a diet is to provide at least 300 calories, 36 units of vitamin A, and 90 units of vitamin C daily. A cup of dietary drink X provides 60 calories, 12 units of vitamin A, and 10 units of vitamin C. A cup of dietary drink Y provides 60 calories, 6 units of vitamin A, and 30 units of vitamin C. Set up a system of linear inequalities that describes the minimum daily requirements for calories and vitamins. Does a liquid diet of 2 units of X and 6 units of Y meet these requirements?

SOLUTION

Let x represent the number of cups of dietary drink X and let y represent the number of cups of dietary drink Y. Then, to meet the minimum daily requirements, the following inequalities must be satisfied.

$$
\begin{aligned}
\text{For calories:} \quad & 60x + 60y \geq 300 \\
\text{For vitamin A:} \quad & 12x + 6y \geq 36 \\
\text{For vitamin C:} \quad & 10x + 30y \geq 90 \\
& x \geq 0 \\
& y \geq 0
\end{aligned}
$$

The last two inequalities are included because x and y cannot be negative. The graph of this system of inequalities is shown in Figure 7.28. Because the point (2, 6) lies inside the solution set of this system, the liquid diet meets the requirements.

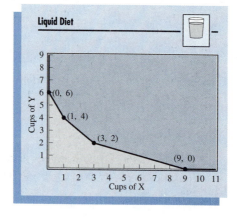

FIGURE 7.28

Example 8 in Section 7.2 discussed the *point of equilibrium* for a demand and supply function. The next example discusses two related concepts that economists call **consumer surplus** and **producer surplus.** As shown in Figure 7.29, the consumer surplus is defined to be the area of the region that lies *below* the demand curve, *above* the horizontal line passing through the equilibrium point, and to the right of the *y*-axis. Similarly, the producer surplus is defined to be the area of the region that lies *above* the supply curve, *below* the horizontal line passing through the equilibrium point, and to the right of the *y*-axis. In general terms, consumer surplus is a measure of the amount of money that consumers would have been willing to pay *above what they actually paid.* Similarly, producer surplus is a measure of the amount of money that producers would have been willing to receive *below what they actually received.*

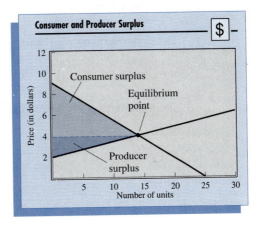

FIGURE 7.29

EXAMPLE 7   **Consumer and Producer Surplus**

Suppose that the demand and supply functions for a certain type of calculator are given by

$$p = 150 - 10x \qquad \text{\textit{Demand equation}}$$
$$p = 60 + 20x, \qquad \text{\textit{Supply equation}}$$

where *p* is the price in dollars and *x* represents the number of units (in millions). Find the consumer surplus and producer surplus for these two equations.

SOLUTION

To begin, find the point of equilibrium by solving the equation

$$60 + 20x = 150 - 10x.$$

In Example 8 of Section 7.2, the solution was found to be $x = 3$, which corresponded to an equilibrium price of $p = \$120$. Thus, the consumer surplus and producer surplus are the areas of the triangular regions given by the following sets of inequalities.

Consumer Surplus	*Producer Surplus*
$p \le 150 - 10x$	$p \ge 60 + 20x$
$p \ge 120$	$p \le 120$
$x \ge 0$	$x \ge 0$

Using Figure 7.30 and the formula for the area of a triangle, you can find that the consumer surplus is

$$\text{Consumer surplus} = \frac{1}{2}(\text{base})(\text{height}) = \frac{1}{2}(30)(3) = \$45 \text{ million}$$

and the producer surplus is

$$\text{Producer surplus} = \frac{1}{2}(\text{base})(\text{height}) = \frac{1}{2}(60)(3) = \$90 \text{ million}.$$

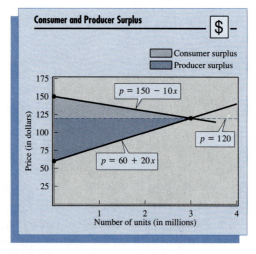

FIGURE 7.30

**DISCUSSION
PROBLEM**

. . . .

**YOU BE THE
INSTRUCTOR**

Suppose you are tutoring a student in algebra and want to construct some practice problems for your student. Write a paragraph describing how you could write a system of linear inequalities that had a given region as its solution. Then apply your procedure to find two systems of linear inequalities that have the following regions as solutions.

(a) Region with $(0, 0)$, $(0, 2)$, $(3, 0)$, and $(2, 1)$ as vertices.
(b) Region with $(0, 0)$, $(0, 3)$, $(4, 0)$, and $(3, 2)$ as vertices.

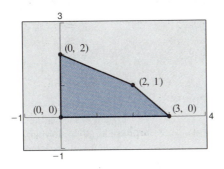
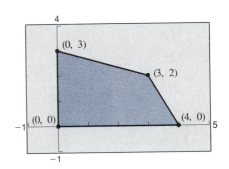

WARM-UP

. . . .

The following warm-up exercises involve skills that were covered in earlier sections. You will use these skills in the exercise set for this section.

In Exercises 1–6, identify the graph of the equation.

1. $x + y = 3$
2. $4x - y = 8$
3. $y = x^2 - 4$
4. $y = -x^2 + 1$
5. $x^2 + y^2 = 9$
6. $\dfrac{x^2}{4} + \dfrac{y^2}{9} = 1$

In Exercises 7–10, solve the system of equations.

7. $x + 2y = 3$
$4x - 7y = -3$
8. $2x - 3y = 4$
$x + 5y = 2$
9. $x^2 + y = 5$
$2x - 4y = 0$
10. $x^2 + y^2 = 13$
$x + y = 5$

SECTION 7.5 · EXERCISES

In Exercises 1–8, match the inequality with its graph. [The graphs are labeled (a), (b), (c), (d), (e), (f), (g), and (h).]

1. $x > 3$

2. $y \le 2$

3. $2x + 3y \le 6$

4. $2x - y \ge -2$

5. $x^2 + y^2 < 4$

6. $(x - 2)^2 + (y - 3)^2 > 4$

7. $xy > 2$

8. $y \le 4 - x^2$

In Exercises 9-20, sketch the graph of the inequality.

9. $x \ge 2$ 10. $x \le 4$

11. $y \ge -1$ 12. $y \le 3$

13. $y < 2 - x$ 14. $y > 2x - 4$

15. $2y - x \ge 4$ 16. $5x + 3y \ge -15$

17. $(x + 1)^2 + (y - 2)^2 < 9$ 18. $y^2 - x < 0$

19. $y \le \dfrac{1}{1 + x^2}$ 20. $y < \ln x$

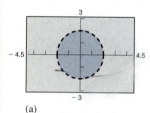

(a)

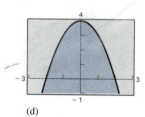

(b)

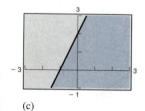

(c)

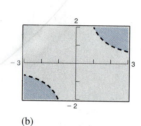

(d)

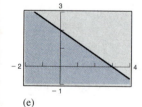

(e)

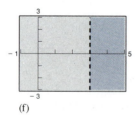

(f)

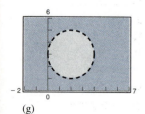

(g)

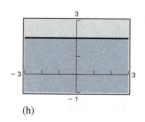

(h)

In Exercises 21–40, sketch the graph of the solution of the system of inequalities and use a graphing utility to verify your result. (*Note:* If your graphing utility has *shading* capabilities, shade the region representing the solution.)

21. $\begin{aligned} x + y &\le 1 \\ -x + y &\le 1 \\ y &\ge 0 \end{aligned}$ 22. $\begin{aligned} 3x + 2y &< 6 \\ x &> 0 \\ y &> 0 \end{aligned}$

23. $\begin{aligned} x + y &\le 5 \\ x &\ge 2 \\ y &\ge 0 \end{aligned}$ 24. $\begin{aligned} 2x + y &\ge 2 \\ x &\le 2 \\ y &\le 1 \end{aligned}$

25. $\begin{aligned} -3x + 2y &< 6 \\ x + 4y &> -2 \\ 2x + y &< 3 \end{aligned}$ 26. $\begin{aligned} x - 7y &> -36 \\ 5x + 2y &> 5 \\ 6x - 5y &> 6 \end{aligned}$

27. $\begin{aligned} 2x + y &> 2 \\ 6x + 3y &< 2 \end{aligned}$ 28. $\begin{aligned} x - 2y &< -6 \\ 5x - 3y &> -9 \end{aligned}$

29. $\begin{aligned} x &\ge 1 \\ x - 2y &\le 3 \\ 3x + 2y &\ge 9 \\ x + y &\le 6 \end{aligned}$ 30. $\begin{aligned} x - y^2 &> 0 \\ x - y &< 2 \end{aligned}$

31. $\begin{aligned} x^2 + y^2 &\le 9 \\ x^2 + y^2 &\ge 1 \end{aligned}$ 32. $\begin{aligned} x^2 + y^2 &\le 25 \\ 4x - 3y &\le 0 \end{aligned}$

33. $\begin{aligned} x &> y^2 \\ x &< y + 2 \end{aligned}$ 34. $\begin{aligned} x &< 2y - y^2 \\ 0 &< x + y \end{aligned}$

35. $\begin{aligned} y &\le \sqrt{3x} + 1 \\ y &\ge x + 1 \end{aligned}$ 36. $\begin{aligned} y &< -x^2 + 2x + 3 \\ y &> x^2 - 4x + 3 \end{aligned}$

37. $\begin{aligned} y &< x^3 - 2x + 1 \\ y &> -2x \\ x &\le 1 \end{aligned}$ 38. $\begin{aligned} y &\ge x^4 - 2x^2 + 1 \\ y &\le 1 - x^2 \end{aligned}$

39. $\begin{aligned} x^2 y &\ge 1 \\ 0 &< x \le 4 \\ y &\le 4 \end{aligned}$ 40. $\begin{aligned} y &\le e^{-x^2/2} \\ y &\ge 0 \\ -2 &\le x \le 2 \end{aligned}$

In Exercises 41–46, derive a set of inequalities to describe the region.

41. Rectangular region with vertices at $(2, 1)$, $(5, 1)$, $(5, 7)$, and $(2, 7)$

42. Parallelogram region with vertices at $(0, 0)$, $(4, 0)$, $(1, 4)$, and $(5, 4)$

43. Triangular region with vertices at $(0, 0)$, $(5, 0)$, and $(2, 3)$

44. Triangular region with vertices at $(-1, 0)$, $(1, 0)$, and $(0, 1)$

45. Sector of a circle   **46.** Sector of a circle

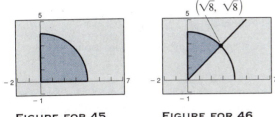

FIGURE FOR 45   **FIGURE FOR 46**

47. *Furniture Production*   A furniture company can sell all the tables and chairs it produces. Each table requires 1 hour in the assembly center and $1\frac{1}{3}$ hours in the finishing center. Each chair requires $1\frac{1}{2}$ hours in the assembly center and $1\frac{1}{2}$ hours in the finishing center. The company's assembly center is available 12 hours per day, and its finishing center is available 15 hours per day. If x is the number of tables produced per day and y is the number of chairs produced per day, find a system of inequalities describing all possible production levels. Graph the system.

48. *Computer Inventory*   A store sells two models of a certain brand of computer. Because of the demand, it is necessary to stock twice as many units of model A as units of model B. The costs to the store for the two models are $800 and $1200, respectively. The management does not want more than $20,000 in computer inventory at any one time, and it wants at least four model A computers and two model B computers in inventory at all times. Devise a system of inequalities describing all possible inventory levels, and graph the system.

49. *Investment*   A person plans to invest $20,000 in two different interest-bearing accounts. Each account is to contain at least $5000. Moreover, one account should have at least twice the amount that is in the other account. Find a system of inequalities to describe the various amounts that can be deposited in each account, and graph the system.

50. *Concert Ticket Sales*   Two types of tickets are to be sold for a concert. One type costs $15 per ticket and the other type costs $25 per ticket. The promoter of the concert must sell at least 15,000 tickets, including 8000 of the $15 tickets and 4000 of the $25 tickets. Moreover, the gross receipts must total at least $275,000 in order for the concert to be held. Find a system of inequalities describing the different numbers of tickets that can be sold, and graph the system.

51. *Diet Supplement*   A dietitian is asked to design a special diet supplement using two different foods. Each ounce of food X contains 20 units of calcium, 15 units of iron, and 10 units of vitamin B. Each ounce of food Y contains 10 units of calcium, 10 units of iron, and 20 units of vitamin B. The minimum daily requirements in the diet are 280 units of calcium, 160 units of iron, and 180 units of vitamin B. Find a system of inequalities describing the different amounts of food X and food Y that can be used in the diet, and graph the system.

52. *Diet Supplement*   A dietitian is asked to design a special diet supplement using two different foods. Each ounce of food X contains 20 units of calcium, 15 units of iron, and 10 units of vitamin B. Each ounce of food Y contains 10 units of calcium, 10 units of iron, and 20 units of vitamin B. The minimum daily requirements in the diet are 300 units of calcium, 150 units of iron, and 200 units of vitamin B. Find a system of inequalities describing the different amounts of food X and food Y that can be used in the diet, and graph the system.

Consumer and Producer Surplus   In Exercises 53–58, find the consumer surplus and producer surplus for the given pair of supply and demand equations.

Demand	Supply
53. $p = 50 - 0.5x$	$p = 0.125x$
54. $p = 60 - x$	$p = 10 + \frac{7}{3}x$
55. $p = 300 - x$	$p = 100 + x$
56. $p = 100 - 0.05x$	$p = 25 + 0.1x$
57. $p = 140 - 0.00002x$	$p = 80 + 0.00001x$
58. $p = 400 - 0.0002x$	$p = 225 + 0.0005x$

59. *Physical Fitness Facility* An indoor running track is to be constructed with a space for body-building equipment inside the track (see figure). The inside of the track must be at least 125 meters long, and the body-building space must have an area of at least 500 square meters. Find a system of inequalities describing the various sizes of the track, and graph the system.

FIGURE FOR 59

7.6   LINEAR PROGRAMMING

The Graphical Approach to Linear Programming   /   Applications

The Graphical Approach to Linear Programming

Many applications in business and economics involve a process called **optimization,** in which you are required to find the minimum cost, the maximum profit, or the minimum use of resources. In this section you will study one type of optimization problem called **linear programming.**

A two-dimensional linear programming problem consists of a linear **objective function** and a system of linear inequalities called **constraints.** The objective function gives the quantity that is to be maximized (or minimized), and the constraints determine the set of **feasible solutions.**

For example, consider a linear programming problem in which you are asked to maximize the value of

$$z = ax + by \qquad \textit{Objective function}$$

subject to a set of constraints that determine the region indicated in Figure 7.31. Because every point in the region satisfies each constraint, it is not clear how you should go about finding the point that yields a maximum value of z. Fortunately, it can be shown that if there is an optimal solution, it must occur at one of the vertices of the region. In other words, *you can find the maximum value by testing z at each of the vertices,* as illustrated in Example 1.

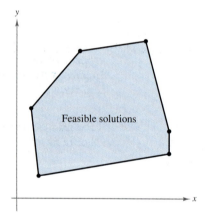

The objective function has its optimal value at one of the vertices of the region determined by the constraints.

FIGURE 7.31

OPTIMAL SOLUTION OF LINEAR PROGRAMMING PROBLEM

If a linear programming problem has a solution, it must occur at a vertex of the set of feasible solutions. If the problem has more than one solution, then at least one of them must occur at a vertex of the set of feasible solutions. In either case, the value of the objective function is unique.

EXAMPLE 1   **Solving a Linear Programming Problem**

Find the maximum value of

$$z = 3x + 2y \qquad \text{\textit{Objective function}}$$

subject to the following constraints.

$$\left.\begin{array}{r} x \geq 0 \\ y \geq 0 \\ x + 2y \leq 4 \\ x - y \leq 1 \end{array}\right\} \quad \text{\textit{Constraints}}$$

SOLUTION

The constraints form the region shown in Figure 7.32. At the four vertices of this region, the objective function has the following values.

At $(0, 0)$: $z = 3(0) + 2(0) = 0$
At $(1, 0)$: $z = 3(1) + 2(0) = 3$
At $(2, 1)$: $z = 3(2) + 2(1) = 8$ *(Maximum value of z)*
At $(0, 2)$: $z = 3(0) + 2(2) = 4$

Thus, the maximum value of z is 8, and this occurs when $x = 2$ and $y = 1$.

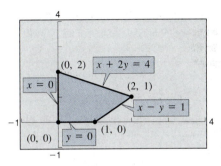

FIGURE 7.32

REMARK   In Example 1, try testing some of the *interior* points in the region. You will see that the corresponding values of z are less than 8.

To see why the maximum value of the objective function in Example 1 must occur at a vertex, consider writing the objective function in the form

$$y = -\frac{3}{2}x + \frac{z}{2},$$

where $z/2$ is the y-intercept of the objective function. This equation represents a family of lines, each of slope $-\frac{3}{2}$. Of these infinitely many lines, you want the one that has the largest z-value, while still intersecting the region determined by the constraints. In other words, of all the lines whose slope is $-\frac{3}{2}$, you want the one that has the largest y-intercept *and* intersects the given region, as shown in Figure 7.33. It should be clear that such a line will pass through one (or more) of the vertices of the region.

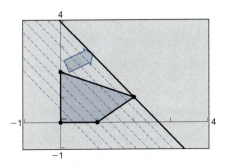

FIGURE 7.33

We outline the steps used in Example 1 as follows.

GRAPHICAL METHOD OF SOLVING A LINEAR PROGRAMMING PROBLEM

To solve a linear programming problem involving two variables by the graphical method, use the following steps.

1. Sketch the region corresponding to the system of constraints. (The points inside or on the boundary of the region are called *feasible solutions*.)
2. Find the vertices of the region.
3. Test the objective function at each of the vertices and select the values of the variables that optimize the objective function. For a bounded region, both a minimum and maximum value will exist. (For an unbounded region, *if* an optimal solution exists, it will occur at a vertex.)

These guidelines will work whether the objective function is to be maximized *or* minimized. For instance, in Example 1 the same test used to find the maximum value of z can be used to conclude that the minimum value of z is 0, and this occurs at the vertex $(0, 0)$.

EXAMPLE 2   Solving a Linear Programming Problem

Find the maximum value of the objective function

$$z = 4x + 6y, \qquad \textit{Objective function}$$

where $x \geq 0$ and $y \geq 0$, subject to the constraints

$$\left. \begin{array}{r} -x + y \leq 11 \\ x + y \leq 27 \\ 2x + 5y \leq 90. \end{array} \right\} \quad \textit{Constraints}$$

SOLUTION

The region bounded by the constraints is shown in Figure 7.34. By testing the objective function at each vertex, you obtain the following.

$$
\begin{aligned}
\text{At} \quad (0, 0): \quad & z = 4(0) + 6(0) = 0 \\
\text{At} \quad (0, 11): \quad & z = 4(0) + 6(11) = 66 \\
\text{At} \quad (5, 16): \quad & z = 4(5) + 6(16) = 116 \\
\text{At} \quad (15, 12): \quad & z = 4(15) + 6(12) = 132 \quad \textit{(Maximum value of z)} \\
\text{At} \quad (27, 0): \quad & z = 4(27) + 6(0) = 108
\end{aligned}
$$

Thus, the maximum value of z is 132, and this occurs when $x = 15$ and $y = 12$.

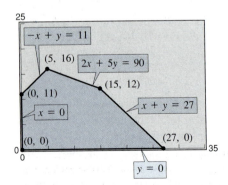

FIGURE 7.34

The next example shows that the same basic procedure can be used to solve a linear programming problem in which the objective function is to be *minimized.*

EXAMPLE 3   Minimizing an Objective Function

Find the minimum value of the objective function

$$z = 5x + 7y, \qquad \textit{Objective function}$$

where $x \geq 0$ and $y \geq 0$, subject to the constraints

$$\left. \begin{array}{r} 2x + 3y \geq 6 \\ 3x - y \leq 15 \\ -x + y \leq 4 \\ 2x + 5y \leq 27. \end{array} \right\} \quad \textit{Constraints}$$

SOLUTION

The region bounded by the constraints is shown in Figure 7.35. By testing the objective function at each vertex, you obtain the following.

At $(0, 2)$: $z = 5(0) + 7(2) = 14$ *(Maximum value of z)*
At $(0, 4)$: $z = 5(0) + 7(4) = 28$
At $(1, 5)$: $z = 5(1) + 7(5) = 40$
At $(6, 3)$: $z = 5(6) + 7(3) = 51$
At $(5, 0)$: $z = 5(5) + 7(0) = 25$
At $(3, 0)$: $z = 5(3) + 7(0) = 15$

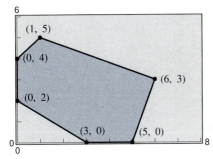

FIGURE 7.35

Thus, the minimum value of z is 14, and this occurs when $x = 0$ and $y = 2$.

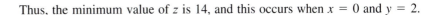

REMARK   In Example 3, note that the steps used to find the minimum value are precisely the same ones you would use to find the maximum value. In other words, once you have evaluated the objective function at the vertices of the feasible region, you simply choose the largest value as the maximum and the smallest value as the minimum.

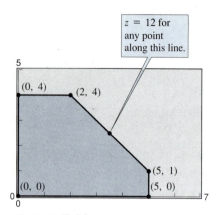

z = 12 for any point along this line.

FIGURE 7.36

In a linear programming problem, it is possible that the maximum (or minimum) value occurs at *two* different vertices. For instance, at the vertices of the region shown in Figure 7.36, the objective function

$$z = 2x + 2y \qquad \textit{Objective function}$$

has the following values.

At $(0, 0)$: $z = 2(0) + 2(0) = 0$
At $(0, 4)$: $z = 2(0) + 2(4) = 8$
At $(2, 4)$: $z = 2(2) + 2(4) = 12$ *(Maximum value of z)*
At $(5, 1)$: $z = 2(5) + 2(1) = 12$ *(Maximum value of z)*
At $(5, 0)$: $z = 2(5) + 2(0) = 10$

In this case, you can conclude that the objective function has a maximum value (of 12) not only at the vertices $(2, 4)$ and $(5, 1)$, but also at *any point on the line segment connecting these two vertices*. Note that the objective function, $y = -x + \frac{1}{2}z$, has the same slope as the line through the vertices $(2, 4)$ and $(5, 1)$.

Some linear programming problems have no optimal solution. This can occur if the region determined by the constraint is *unbounded*. Example 4 illustrates such a problem.

EXAMPLE 4   An Unbounded Region

Find the maximum value of

$$z = 4x + 2y, \qquad \textit{Objective function}$$

where $x \geq 0$ and $y \geq 0$, subject to the constraints

$$\left. \begin{array}{r} x + 2y \geq 4 \\ 3x + y \geq 7 \\ -x + 2y \leq 7. \end{array} \right\} \quad \textit{Constraints}$$

SOLUTION

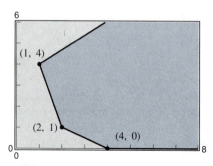

FIGURE 7.37

The region determined by the constraints is shown in Figure 7.37. For this unbounded region, there is no maximum value of z. To see this, note that the point $(x, 0)$ lies in the region for all values of $x \geq 4$. By choosing x to be large, you can obtain values of $z = 4(x) + 2(0) = 4x$ to be as large as you want. Thus, there is no maximum value of z. For this problem, there *is* a minimum value of $z = 10$, which occurs at the vertex $(2, 1)$.

Applications

EXAMPLE 5   A Business Application: Maximum Profit

A manufacturer wants to maximize the profit for two products. The first product yields a profit of $1.50 per unit, and the second product yields a profit of $2.00 per unit. Market tests and available resources have indicated the following constraints.

1. The combined production level should not exceed 1200 units per month.
2. The demand for product II is less than or equal to half of the demand of product I.
3. The production level of product I is less than or equal to 600 units plus three times the production level of product II.

SOLUTION

Let x be the number of units of product I and let y be the number of units of product II. The objective function (for the combined profit) is given by

$$P = 1.5x + 2y. \qquad \textit{Objective function}$$

The three constraints translate into the following linear inequalities.

1. $x + y \le 1200 \qquad \Rightarrow \quad x + y \le 1200$
2. $ y \le \frac{1}{2}x \qquad \Rightarrow \quad x - 2y \le 0$
3. $ x \le 3y + 600 \quad \Rightarrow \quad x - 3y \le 600$

Because neither x nor y can be negative, you also have the two additional constraints of $x \ge 0$ and $y \ge 0$. Figure 7.38 shows the region determined by the constraints. To find the maximum profit, test the value of P at the vertices of the region.

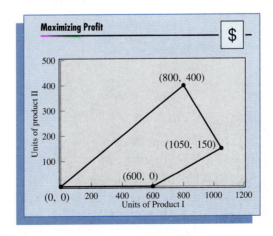

FIGURE 7.38

At $(0, 0)$: $P = 1.5(0)$ $+ 2(0)$ $=$ 0
At $(800, 400)$: $P = 1.5(800)$ $+ 2(400) = 2000$ *(Maximum profit)*
At $(1050, 150)$: $P = 1.5(1050) + 2(150) = 1875$
At $(600, 0)$: $P = 1.5(600)$ $+ 2(0)$ $=$ 900

Thus, the maximum profit is $2000, and it occurs when the monthly production consists of 800 units of product I and 400 units of product II.

━━━━━━━━━

EXAMPLE 6   An Application: Minimum Cost

In Example 6 in Section 7.5, we set up a system of linear equations for the following problem. The liquid portion of a diet is to provide at least 300 calories, 36 units of vitamin A, and 90 units of vitamin C daily. A cup of dietary drink X provides 60 calories, 12 units of vitamin A, and 10 units of vitamin C. A cup of dietary drink Y provides 60 calories, 6 units of vitamin A, and 30 units of vitamin C. Now, suppose the dietary drink X costs $0.12 per cup and drink Y costs $0.15 per cup. How many cups of each drink should be consumed each day to minimize the cost and still meet the stated daily requirements?

SOLUTION

We begin by letting x be the number of cups of dietary drink X and letting y be the number of cups of dietary drink Y. Moreover, to meet the minimum daily requirements, the following inequalities must be satisfied.

For calories: $60x + 60y \geq 300$
For vitamin A: $12x + 6y \geq 36$
For vitamin C: $10x + 30y \geq 90$ *Constraints*
$x \geq 0$
$y \geq 0$

The cost C is given by

$$C = 0.12x + 0.15y.$$ *Objective function*

The graph of the region corresponding to the constraints is shown in Figure 7.39. To determine the minimum cost, test C at each vertex of the region, as follows.

At $(0, 6)$: $C = 0.12(0) + 0.15(6) = 0.90$
At $(1, 4)$: $C = 0.12(1) + 0.15(4) = 0.72$
At $(3, 2)$: $C = 0.12(3) + 0.15(2) = 0.66$ *(Minimum value of C)*
At $(9, 0)$: $C = 0.12(9) + 0.15(0) = 1.08$

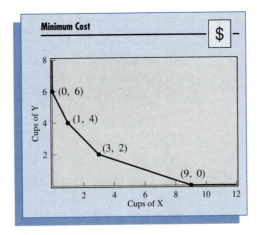

FIGURE 7.39

Thus, the minimum cost is $0.66 per day, and this occurs when three cups of drink X and two cups of drink Y are consumed each day.   ·········

EXAMPLE 7   Maximum Profit

A small computer-keyboard company makes two popular models, for which the demand is much greater than the current supply (both models have substantial back-orders). Both models take 1 hour to assemble. However, model TT1 requires only 7.5 minutes to test, whereas model TT2 requires 30 minutes to test. With the company's current facilities, there are 45,000 hours per month available for assembly, and 15,000 hours per month available for testing. The profit for model TT1 is $50.00 per unit, and the profit for model TT2 is $80.00 per unit. What is the greatest monthly profit the company can make without increasing its current facilities?

SOLUTION

To begin to solve this problem, it is helpful to organize the information in table form, as shown in Table 7.1.

TABLE 7.1

	Model TT1 (x units)	Model TT2 (y units)	Maximum hours
Assembly time per unit	1 hour	1 hour	45,000
Test time per unit	$\frac{1}{8}$ hour	$\frac{1}{2}$ hour	15,000
Profit per unit	$50.00	$80.00	—

From the third row in the table, you can see that the total profit for selling x units of model TT1 and y units of model TT2 is

$$P = 50x + 80y. \qquad \qquad \textit{Objective function}$$

This is the quantity you need to maximize. The constraints under which you are allowed to work are given by the following two inequalities.

$$x + y \leq 45,000$$

$$\frac{1}{8}x + \frac{1}{2}y \leq 15,000$$

Because $x \geq 0$ and $y \geq 0$, the possible values of x and y are those that lie in (or on the boundary of) the region shown in Figure 7.40. At the four vertices, you obtain the following profits.

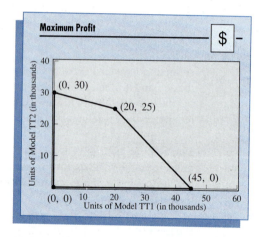

FIGURE 7.40

$$
\begin{aligned}
P &= 50(0) &+ 80(0) &= \$0 \\
P &= 50(0) &+ 80(30,000) &= \$2,400,000 \\
P &= 50(20,000) &+ 80(25,000) &= \$3,000,000 \qquad \textit{(Maximum value of P)} \\
P &= 50(45,000) &+ 80(0) &= \$2,250,000
\end{aligned}
$$

Therefore, the maximum profit can be obtained by producing 20,000 units of model TT1 and 25,000 units of model TT2.

DISCUSSION PROBLEM

CREATING A LINEAR PROGRAMMING PROBLEM

Consider the following linear programming problem.

Objective function:

$z = ax + by$

Constraints:

$x \geq 0$

$y \geq 0$

$x + 2y \leq 8$

$x + y \leq 5$

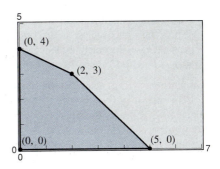

The region determined by these constraints is shown in the figure. Find, if possible, an objective function that has a *maximum* at the indicated vertex of the region.

(a) Maximum at (0, 4)

(b) Maximum at (2, 3)

(c) Maximum at (5, 0)

(d) Maximum at (0, 0)

WARM-UP

The following warm-up exercises involve skills that were covered in earlier sections. You will use these skills in the exercise set for this section.

In Exercises 1–4, sketch the graph of the linear equation.

1. $y + x = 3$

2. $y - x = 12$

3. $x = 0$

4. $y = 4$

In Exercises 5–8, solve the system of equations.

5. $x + y = 4$
$ x = 0$

6. $x + 2y = 12$
$ y = 0$

7. $x + y = 4$
$2x + 3y = 9$

8. $x + 2y = 12$
$2x + y = 9$

In Exercises 9 and 10, sketch the graph of the inequality.

9. $2x + 3y \geq 18$

10. $4x + 3y \geq 12$

SECTION 7.6 · EXERCISES

In Exercises 1–12, find the minimum and maximum values of the given objective function, subject to the indicated constraints. (For each exercise, the graph of the region determined by the constraints is provided.)

1. Objective function:
$z = 4x + 5y$
Constraints:
$x \geq 0$
$y \geq 0$
$x + y \leq 6$

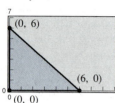

2. Objective function:
$z = 2x + 8y$
Constraints:
$x \geq 0$
$y \geq 0$
$2x + y \leq 4$

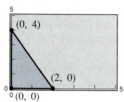

3. Objective function:
$z = 10x + 6y$
Constraints:
(See Exercise 1.)

4. Objective function:
$z = 7x + 3y$
Constraints:
(See Exercise 2.)

5. Objective function:
$z = 3x + 2y$
Constraints:
$x \geq 0$
$y \geq 0$
$x + 3y \leq 15$
$4x + y \leq 16$

6. Objective function:
$z = 4x + 3y$
Constraints:
$x \geq 0$
$2x + 3y \geq 6$
$3x - 2y \leq 9$
$x + 5y \leq 20$

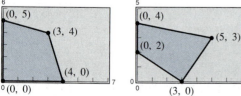

7. Objective function:
$z = 5x + 0.5y$
Constraints:
(See Exercise 5.)

8. Objective function:
$z = x + 6y$
Constraints:
(See Exercise 6.)

9. Objective function:
$z = 10x + 7y$
Constraints:
$0 \leq x \leq 60$
$0 \leq y \leq 45$
$5x + 6y \leq 420$

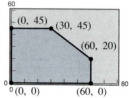

10. Objective function:
$z = 50x + 35y$
Constraints:
$x \geq 0$
$y \geq 0$
$8x + 9y \leq 7200$
$8x + 9y \geq 5400$

11. Objective function:
$z = 25x + 30y$
Constraints:
(See Exercise 9.)

12. Objective function:
$z = 16x + 18y$
Constraints:
(See Exercise 10.)

In Exercises 13–24, use a graphing utility to sketch the region determined by the constraints. Then find the minimum and maximum values of the objective function, subject to the constraints.

13. Objective function:
$z = 6x + 10y$
Constraints:
$x \geq 0$
$y \geq 0$
$2x + 5y \leq 10$

14. Objective function:
$z = 7x + 8y$
Constraints:
$x \geq 0$
$y \geq 0$
$x + \frac{1}{2}y \leq 4$

15. Objective function:
$z = 9x + 24y$
Constraints:
(See Exercise 13.)

16. Objective function:
$z = 7x + 2y$
Constraints:
(See Exercise 14.)

17. Objective function:
$z = 4x + 5y$
Constraints:
$x \geq 0$
$y \geq 0$
$4x + 3y \geq 27$
$x + y \geq 8$
$3x + 5y \geq 30$

18. Objective function:
$z = 4x + 5y$
Constraints:
$x \geq 0$
$y \geq 0$
$2x + 2y \leq 10$
$x + 2y \leq 6$

19. Objective function:
$z = 2x + 7y$
Constraints: Constraints:
(See Exercise 17.)

20. Objective function:
$z = 2x - y$
Constraints:
(See Exercise 18.)

21. Objective function:
$z = 4x + y$
Constraints:
$$x \geq 0$$
$$y \geq 0$$
$$x + 2y \leq 40$$
$$x + y \geq 30$$
$$2x + 3y \geq 72$$

22. Objective function:

Constraints:
$$x \geq 0$$
$$y \geq 0$$
$$2x + 3y \leq 60$$
$$2x + y \leq 28$$
$$4x + y \leq 48$$

23. Objective function:
$z = x + 4y$
Constraints:
(See Exercise 21.)

24. Objective function:
$z = y$
Constraints:
(See Exercise 22.)

In Exercises 25–28, maximize the objective function subject to the constraints $3x + y \leq 15$ and $4x + 3y \leq 30$, where $x \geq 0$ and $y \geq 0$.

25. $z = 2x + y$

26. $z = 5x + y$

27. $z = x + y$

28. $z = 3x + y$

In Exercises 29–32, maximize the objective function subject to the constraints $x + 4y \leq 20$, $x + y \leq 8$, and $3x + 2y \leq 21$ where $x \geq 0$ and $y \geq 0$.

29. $z = x + 5y$

30. $z = 2x + 4y$

31. $z = 4x + 5y$

32. $z = 4x + y$

33. *Maximum Profit*   A merchant plans to sell two models of home computers at costs of $250 and $400, respectively. The $250 model yields a profit of $45 and the $400 model yields a profit of $50. The merchant estimates that the total monthly demand will not exceed 250 units. Find the number of units of each model that should be stocked in order to maximize profit. Assume that the merchant does not want to invest more than $70,000 in computer inventory.

34. *Maximum Profit*   A fruit grower has 150 acres of land available to raise two crops, A and B. It takes one day to trim an acre of crop A and two days to trim an acre of crop B, and there are 240 days per year available for trimming. It takes 0.3 day to pick an acre of crop A and 0.1 day to pick an acre of crop B, and there are 30 days per year available for picking. Find the number of acres of each fruit that should be planted to maximize profit, assuming that the profit is $140 per acre for crop A and $235 per acre for crop B.

35. *Minimum Cost*   A farming cooperative mixes two brands of cattle feed. Brand X costs $25 per bag and contains 2 units of nutritional element A, 2 units of element B, and 2 units of element C. Brand Y costs $20 per bag and contains 1 unit of nutritional element A, 9 units of element B, and 3 units of element C. Find the number of bags of each brand that should be mixed to produce a mixture having a minimum cost per bag. The minimum requirements of nutrients A, B, and C are 12 units, 36 units, and 24 units, respectively.

36. *Minimum Cost*   Two gasolines, type A and type B, have octane ratings of 80 and 92, respectively. Type A costs $1.13 per gallon and type B costs $1.28 per gallon. Determine the blend of minimum cost with an octane rating of at least 90. (*Hint:* Let x be the fraction of each gallon that is type A and let y be the fraction that is type B.)

37. *Maximum Profit*   A manufacturer produces two models of bicycles. The time (in hours) required for assembling, painting, and packaging each model is as follows.

	Model A	Model B
Assembling	2	2.5
Painting	4	1
Packaging	1	0.75

The total time available for assembling, painting, and packaging is 4000 hours, 4800 hours, and 1500 hours, respectively. The profit per unit for each model is $45 (model A) and $50 (model B). How many of each type should be produced to obtain a maximum profit?

38. *Maximum Profit*   A manufacturer produces two models of bicycles. The time (in hours) required for assembling, painting, and packaging each model is as follows.

	Model A	Model B
Assembling	2.5	3
Painting	2	1
Packaging	0.75	1.25

The total time available for assembling, painting, and packaging is 4000 hours, 2500 hours, and 1500 hours, respectively. The profit per unit for each model is $50 (model A) and $52 (model B). How many of each type should be produced to obtain a maximum profit?

39. *Maximum Revenue*   An accounting firm has 900 hours of staff time and 100 hours of reviewing time available each week. It charges $2000 for an audit and $300 for a tax return. Each audit takes 100 hours of staff time and 10 hours of review time. Each tax return takes 12.5 hours of staff time and 2.5 hours of review time. What number of audits and tax returns will yield the maximum revenue?

40. *Maximum Revenue*   The accounting firm in Exercise 39 lowers its charge for an audit to $1000. What number of audits and tax returns will now yield the maximum revenue?

In Exercises 41–46, the given linear programming problem has an unusual characteristic. Sketch a graph of the solution region for the problem and describe the unusual characteristic. In each problem, the objective function is to be maximized.

41. Objective function:
$$z = 2.5x + y$$
Constraints:
$$x \geq 0$$
$$y \geq 0$$
$$3x + 5y \leq 15$$
$$5x + 2y \leq 10$$

42. Objective function:
$$z = x + y$$
Constraints:
$$x \geq 0$$
$$y \geq 0$$
$$-x + y \leq 1$$
$$-x + 2y \leq 4$$

43. Objective function:
$$z = -x + 2y$$
Constraints:
$$x \geq 0$$
$$y \geq 0$$
$$x \leq 10$$
$$x + y \leq 7$$

44. Objective function:
$$z = x + y$$
Constraints:
$$x \geq 0$$
$$y \geq 0$$
$$-x + y \leq 0$$
$$-3x + y \geq 3$$

45. Objective function:
$$z = 3x + 4y$$
Constraints:
$$x \geq 0$$
$$y \geq 0$$
$$x + y \leq 1$$
$$2x + y \leq 4$$

46. Objective function:
$$z = x + 2y$$
Constraints:
$$x \geq 0$$
$$y \geq 0$$
$$x + 2y \leq 4$$
$$2x + y \leq 4$$

In Exercises 47 and 48, determine the values of *t* such that the objective function has a maximum value at the indicated vertex.

47. Objective function:
$$z = 3x + ty$$
Constraints:
$$x \geq 0$$
$$y \geq 0$$
$$x + 3y \leq 15$$
$$4x + y \leq 16$$
(a) $(0, 5)$
(b) $(3, 4)$

48. Objective function:
$$z = 3x + ty$$
Constraints:
$$x \geq 0$$
$$y \geq 0$$
$$x + 2y \leq 4$$
$$x - y \leq 1$$
(a) $(2, 1)$
(b) $(0, 2)$

CHAPTER 7 · REVIEW EXERCISES

In Exercises 1–6, solve the system by the method of substitution. Use a graphing utility to confirm your result.

1. $x + y = 2$
$x - y = 0$

2. $2x = 3(y - 1)$
$y = x$

3. $x^2 - y^2 = 9$
$x - y = 1$

4. $x^2 + y^2 = 169$
$3x + 2y = 39$

5. $y = 2x^2$
$y = x^4 - 2x^2$

6. $x = y + 3$
$x = y^2 + 1$

In Exercises 7–12, solve the system by elimination. Use a graphing utility to confirm your result.

7. $2x - y = 2$
$6x + 8y = 39$

8. $40x + 30y = 24$
$20x - 50y = -14$

9. $3x - 2y = 0$
$3x + 2(y + 5) = 10$

10. $7x + 12y = 63$
$2x + 3y = 15$

11. $1.25x - 2y = 3.5$
$5x - 8y = 14$

12. $1.5x + 2.5y = 8.5$
$6x + 10y = 24$

In Exercises 13–16, use a graphing utility to solve the system of equations.

13. $y^2 - 2y + x = 0$
$x + y = 0$

14. $y = 2x^2 - 4x + 1$
$y = x^2 - 4x + 3$

15. $y = 2(6 - x)$
$y = 2^{x-2}$

16. $y = \ln(x - 1) - 3$
$y = 4 - \frac{1}{2}x$

17. *Break-Even Point* Suppose you are setting up a small business and have made an initial investment of $10,000. The unit cost of the product is $2.85 and the selling price is $4.95. How many units must you sell to break even? (Round your answer to the nearest whole unit.)

18. *Choice of Two Jobs* You are offered two different jobs selling personal computers. One company offers an annual salary of $22,500 plus a year-end bonus of 1.5% of your total sales. The other company offers a salary of $20,000 plus a year-end bonus of 2% of your total sales. How much would you have to sell in order to make the second offer better?

19. *Acid Mixture* One hundred gallons of a 60% acid solution are obtained by mixing a 75% solution with a 50% solution. How many gallons of each must be used to obtain the desired mixture?

20. *Cassette Tape Sales* Suppose you are the manager of a music store. At the end of the week you are going over receipts for the previous week's sales. Six hundred and fifty cassette tapes were sold. One type of cassette sold for $9.95 and another sold for $14.95. The total cassette receipts were $7717.50. The cash register that was supposed to record the number of each type of cassette sold malfunctioned. Can you recover the information? If so, how many of each type of cassette were sold?

21. *Flying Speeds* Two planes leave Pittsburgh and Philadelphia at the same time, each going to the other city. Because of the wind, one plane flies 25 miles per hour faster than the other. Find the ground speed of each plane if the cities are 275 miles apart and the planes pass one another (at different altitudes) after 40 minutes of flying time.

22. *Dimensions of a Rectangle* The perimeter of a rectangle is 480 meters and its length is 150% of its width. Find the dimensions of the rectangle.

Supply and Demand In Exercises 23 and 24, find the point of equilibrium for the pair of supply and demand equations.

Demand	Supply
23. $p = 37 - 0.0002x$	$p = 22 + 0.00001x$
24. $p = 120 - 0.0001x$	$p = 45 + 0.0002x$

In Exercises 25–34, solve the system of equations.

25.
$$\begin{aligned} x + 2y + 6z &= 4 \\ -3x + 2y - z &= -4 \\ 4x + + 2z &= 16 \end{aligned}$$

26.
$$\begin{aligned} x + 3y - z &= 13 \\ 2x - 5z &= 23 \\ 4x - y - 2z &= 14 \end{aligned}$$

27.
$$\begin{aligned} x - 2y + z &= -6 \\ 2x - 3y &= -7 \\ -x + 3y - 3z &= 11 \end{aligned}$$

28.
$$\begin{aligned} 2x + 6z &= -9 \\ 3x - 2y + 11z &= -16 \\ 3x - y + 7z &= -11 \end{aligned}$$

29.
$$\begin{aligned} 2x + 5y - 19z &= 34 \\ 3x + 8y - 31z &= 54 \end{aligned}$$

30.
$$\begin{aligned} 2x + y + z + 2w &= -1 \\ 5x - 2y + z - 3w &= 0 \\ -x + 3y + 2z + 2w &= 1 \\ 3x + 2y + 3z - 5w &= 12 \end{aligned}$$

31.
$$\begin{aligned} -x + y + 2z &= 1 \\ 2x + 3y + z &= -2 \\ 5x + 4y + 2z &= 4 \end{aligned}$$

32.
$$\begin{aligned} 2x + 3y + z &= 10 \\ 2x - 3y - 3z &= 22 \\ 4x - 2y + 3z &= -2 \end{aligned}$$

33.
$$\begin{aligned} 2x + y + 2z &= 4 \\ 2x + 2y &= 5 \\ 2x - y + 6z &= 2 \end{aligned}$$

34.
$$\begin{aligned} x + 2y + 6z &= 1 \\ 2x + 5y + 15z &= 4 \\ 3x + y + 3z &= -6 \end{aligned}$$

In Exercises 35 and 36, find the equation of the parabola $y = ax^2 + bx + c$ that passes through the given points. Use a graphing utility to verify your result.

35. $(0, -6), (1, -3), (2, 4)$

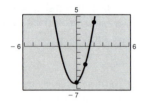

FIGURE FOR 35

36. $(-5, 0), (1, -6), (2, 14)$

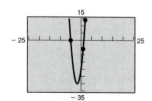

FIGURE FOR 36

In Exercises 37 and 38, find the equation of the circle
$$x^2 + y^2 + Dx + Ey + F = 0$$
that passes through the given points. Use a graphing utility to verify your result.

37. $(2, 2), (5, -1), (-1, -1)$

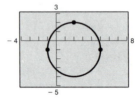

FIGURE FOR 37

38. $(4, 2), (1, 3), (-2, -6)$

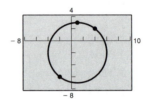

FIGURE FOR 38

39. *Crop Spraying* A mixture of 6 gallons of chemical A, 8 gallons of chemical B, and 13 gallons of chemical C is required to kill a certain destructive crop insect. Commercial spray X contains 1, 2, and 2 parts, respectively, of these chemicals. Commercial spray Y contains only chemical C. Commercial spray Z contains chemicals A, B, and C in equal amounts. How much of each type of commercial spray is needed to get the desired mixture?

40. *Investments* An inheritance of $20,000 was divided among three investments yielding $1818 in interest per year. The interest rates for the three investments were 7%, 9%, and 11%. Find the amount placed in each investment if the second and third were $3000 and $1000 less than the first, respectively.

In Exercises 41–48, write the partial fraction decomposition for the rational expression.

41. $\dfrac{4 - x}{x^2 + 6x + 8}$ **42.** $\dfrac{-x}{x^2 + 3x + 2}$

43. $\dfrac{x^2}{x^2 + 2x - 15}$ **44.** $\dfrac{9}{x^2 - 9}$

45. $\dfrac{x^2 + 2x}{x^3 - x^2 + x - 1}$ **46.** $\dfrac{4x - 2}{3(x - 1)^2}$

47. $\dfrac{3x^3 + 4x}{(x^2 + 1)^2}$ **48.** $\dfrac{4x^2}{(x - 1)(x^2 + 1)}$

In Exercises 49–56, sketch a graph of the solution set of the system of inequalities and use a graphing utility to verify your result. (*Note:* If your graphing utility has *shading* capabilities, shade the region representing the solution.)

49. $\begin{aligned} x + 2y &\le 160 \\ 3x + y &\le 180 \\ x &\ge 0 \\ y &\ge 0 \end{aligned}$ **50.** $\begin{aligned} 2x + 3y &\le 24 \\ 2x + y &\le 16 \\ x &\ge 0 \\ y &\ge 0 \end{aligned}$

51. $\begin{aligned} 3x + 2y &\ge 24 \\ x + 2y &\ge 12 \\ 2 \le x &\le 15 \\ y &\le 15 \end{aligned}$ **52.** $\begin{aligned} 2x + y &\ge 16 \\ x + 3y &\ge 18 \\ 0 \le x &\le 25 \\ 0 \le y &\le 25 \end{aligned}$

53. $\begin{aligned} y &< x + 1 \\ y &> x^2 - 1 \end{aligned}$ **54.** $\begin{aligned} y &\le 6 - 2x - x^2 \\ y &\ge x + 6 \end{aligned}$

55. $\begin{aligned} 2x - 3y &\ge 0 \\ 2x - y &\le 8 \\ y &\ge 0 \end{aligned}$ **56.** $\begin{aligned} x^2 + y^2 &\le 9 \\ (x - 3)^2 + y^2 &\le 9 \end{aligned}$

In Exercises 57 and 58, derive a set of inequalities to describe the region.

57. Parallelogram with vertices at $(1, 5)$, $(3, 1)$, $(6, 10)$, $(8, 6)$

58. Triangle with vertices at $(1, 2)$, $(6, 7)$, $(8, 1)$

In Exercises 59 and 60, determine a system of inequalities that models the description and sketch a graph of the solution of the system.

59. *Fruit Distribution*   A Pennsylvania fruit grower has 1500 bushels of apples that are to be divided between markets in Harrisburg and Philadelphia. These two markets need at least 400 bushels and 600 bushels, respectively.

60. *Inventory Costs*   A warehouse operator has 24,000 square feet of floor space in which to store two products. Each unit of product I requires 20 square feet of floor space and costs $12 per day to store. Each unit of product II requires 30 square feet of floor space and costs $8 per day to store. The total storage cost per day cannot exceed $12,400.

In Exercises 61 and 62, find the consumer surplus and producer surplus for the pair of supply and demand equations.

Demand	Supply
61. $p = 160 - 0.0001x$	$p = 70 + 0.0002x$
62. $p = 130 - 0.0002x$	$p = 30 + 0.0003x$

In Exercises 63–66, find the required optimum value of the objective function subject to the indicated constraints.

63. Maximize the objective function:

$z = 3x + 4y$

Constraints:

$\begin{aligned} x &\ge 0 \\ y &\ge 0 \\ 2x + 5y &\le 50 \\ 4x + y &\le 28 \end{aligned}$

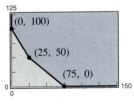

64. Minimize the objective function:

$z = 10x + 7y$

Constraints:

$\begin{aligned} x &\ge 0 \\ y &\ge 0 \\ 2x + y &\ge 100 \\ x + y &\ge 75 \end{aligned}$

65. Minimize the objective function:

$z = 1.75x + 2.25y$

Constraints:

$\begin{aligned} x &\ge 0 \\ y &\ge 0 \\ 2x + y &\ge 25 \\ 3x + 2y &\ge 45 \end{aligned}$

66. Maximize the objective function:

$z = 50x + 70y$

Constraints:

$\begin{aligned} x &\ge 0 \\ y &\ge 0 \\ x + 2y &\le 1500 \\ 5x + 2y &\le 3500 \end{aligned}$

67. *Maximum Profit*   A manufacturer produces products A and B yielding profits of $18 and $24, respectively. Each product must go through three processes with the required times per unit as shown in the following table.

Process	Hours for product A	Hours for product B	Hours available per day
I	4	2	24
II	1	2	9
III	1	1	8

Find the daily production level for each unit to maximize the profit.

68. *Maximum Revenue*   A student is working part-time as a cosmetologist to pay college expenses. The student may work no more than 21 hours per week. Haircuts cost $17 and require an average of 20 minutes; permanents cost $60 and require an average of 1 hour and 10 minutes. What combination of haircuts and/or perms will yield maximum revenue?

69. *Minimum Cost*   A pet supply company mixes two brands of dry dog food. Brand X costs $15 per bag and contains 8 units of nutritional element A, 1 unit of nutritional element B, and 2 units of nutritional element C. Brand Y costs $30 per bag and contains 2 units of nutritional element A, 1 unit of nutritional element B, and 7 units of nutritional element C. Each bag of dog food must contain at least 16 units, 5 units, and 20 units of nutritional elements A, B, and C, respectively. Find the number of bags of brands X and Y that should be mixed to produce a mixture meeting the minimum nutritional requirements and having a minimum cost per bag.

70. *Minimum Cost*   Two gasolines, type A and type B, have octane ratings of 80 and 92, respectively. Type A costs $1.25 per gallon and type B costs $1.55 per gallon. Determine the blend of minimum cost with an octane rating of at least 88. (*Hint:* Let x be the fraction of each gallon that is type A, and let y be the fraction that is type B.)

MATRICES AND DETERMINANTS

8.1 MATRICES AND SYSTEMS OF LINEAR EQUATIONS

Matrices / Elementary Row Operations / Gaussian Elimination with Back-Substitution / Gauss-Jordan Elimination

Matrices

In this section you will study a streamlined technique for solving systems of linear equations. This technique involves the use of a rectangular array of real numbers called a **matrix.**

DEFINITION OF A MATRIX

If m and n are positive integers, then an $m \times n$ **matrix** (read "m by n") is a rectangular array

$$
\begin{bmatrix}
a_{11} & a_{12} & a_{13} & \cdots & a_{1n} \\
a_{21} & a_{22} & a_{23} & \cdots & a_{2n} \\
a_{31} & a_{32} & a_{33} & \cdots & a_{3n} \\
\vdots & \vdots & \vdots & & \vdots \\
a_{m1} & a_{m2} & a_{m3} & \cdots & a_{mn}
\end{bmatrix} \Big\} \; m \text{ rows}
$$

$\underbrace{\hspace{4cm}}_{n \text{ columns}}$

in which each **entry,** a_{ij}, of the matrix is a real number. An $m \times n$ matrix has m **rows** (horizontal lines) and n **columns** (vertical lines).

REMARK   The plural of matrix is *matrices.*

The entry in the ith row and jth column is denoted by the *double subscript notation* a_{ij}. We call i the **row subscript** because it gives the position in the horizontal lines, and j the **column subscript** because it gives the position in the vertical lines. A matrix having m rows and n columns is said to be of **order** $m \times n$. If $m = n$, the matrix is **square** of order n. For a square matrix, the entries $a_{11}, a_{22}, a_{33}, \ldots$ are the **main diagonal** entries.

EXAMPLE 1   Examples of Matrices

The following matrices have the indicated orders.

A. Order: 1×4

$$\begin{bmatrix} 1 & -3 & 0 & \frac{1}{2} \end{bmatrix}$$

B. Order: 2×2

$$\begin{bmatrix} 0 & 0 \\ 0 & 0 \end{bmatrix}$$

C. Order: 2×3

$$\begin{bmatrix} -1 & 0 & 5 \\ 2 & 1 & -4 \end{bmatrix}$$

D. Order: 3×2

$$\begin{bmatrix} 5 & 0 \\ 2 & -2 \\ -7 & 4 \end{bmatrix}$$

A matrix that has only one row is a **row matrix,** and a matrix that has only one column is a **column matrix.**

A matrix derived from a system of linear equations (each written in standard form with the constant term on the right) is called the **augmented matrix** of the system. Moreover, the matrix derived from the coefficients of the system (but which does not include the constant terms) is called the **coefficient matrix** of the system. Here is an example.

REMARK   Note the use of 0 for the missing y-variable in the third equation. Also note that the fourth column of constant terms in the augmented matrix is separated by a column of vertical dots.

System	Augmented Matrix	Coefficient Matrix
$\begin{aligned} x - 4y + 3z &= 5 \\ -x + 3y - z &= -3 \\ 2x \quad\quad - 4z &= 6 \end{aligned}$	$\begin{bmatrix} 1 & -4 & 3 & \vdots & 5 \\ -1 & 3 & -1 & \vdots & -3 \\ 2 & 0 & -4 & \vdots & 6 \end{bmatrix}$	$\begin{bmatrix} 1 & -4 & 3 \\ -1 & 3 & -1 \\ 2 & 0 & -4 \end{bmatrix}$

When forming either the coefficient matrix or the augmented matrix of a system, you should begin by vertically aligning the variables in the equations and using 0's for the missing variables.

System	Line Up Variables	Form Augmented Matrix
$\begin{aligned} x + 3y &= 9 \\ -y + 4z &= -2 \\ x - 5z &= 0 \end{aligned}$	$\begin{aligned} x + 3y \quad\quad &= 9 \\ -y + 4z &= -2 \\ x \quad\quad - 5z &= 0 \end{aligned}$	$\begin{bmatrix} 1 & 3 & 0 & \vdots & 9 \\ 0 & -1 & 4 & \vdots & -2 \\ 1 & 0 & -5 & \vdots & 0 \end{bmatrix}$

Elementary Row Operations

In Section 7.3 you studied three operations that can be used on a system of linear equations to produce an equivalent system. In matrix terminology, these three operations correspond to **elementary row operations.** An elementary row operation on an augmented matrix of a given system of linear equations produces a new augmented matrix corresponding to a new (but equivalent) system of linear equations. Two matrices are **row-equivalent** if one can be obtained from the other by a sequence of elementary row operations.

ELEMENTARY ROW OPERATIONS

1. Interchange two rows.
2. Multiply a row by a nonzero constant.
3. Add a multiple of a row to another row.

Although elementary row operations are simple to perform, they involve a lot of arithmetic. Because it is easy to make a mistake, we suggest that you get in the habit of recording the elementary row operations performed in each step so that you can go back and check your work.

EXAMPLE 2   Elementary Row Operations

A. Interchange the first and second rows.

Original Matrix

$$\begin{bmatrix} 0 & 1 & 3 & 4 \\ -1 & 2 & 0 & 3 \\ 2 & -3 & 4 & 1 \end{bmatrix}$$

New Row-Equivalent Matrix

$$\begin{matrix} R_1 \\ R_2 \end{matrix} \quad \begin{bmatrix} -1 & 2 & 0 & 3 \\ 0 & 1 & 3 & 4 \\ 2 & -3 & 4 & 1 \end{bmatrix}$$

B. Multiply the first row by $\frac{1}{2}$.

Original Matrix

$$\begin{bmatrix} 2 & -4 & 6 & -2 \\ 1 & 3 & -3 & 0 \\ 5 & -2 & 1 & 2 \end{bmatrix}$$

New Row-Equivalent Matrix

$$\frac{1}{2}R_1 \longrightarrow \begin{bmatrix} 1 & -2 & 3 & -1 \\ 1 & 3 & -3 & 0 \\ 5 & -2 & 1 & 2 \end{bmatrix}$$

C. Add -2 times the first row to the third row.

Original Matrix

$$\begin{bmatrix} 1 & 2 & -4 & 3 \\ 0 & 3 & -2 & -1 \\ 2 & 1 & 5 & -2 \end{bmatrix}$$

New Row-Equivalent Matrix

$$-2R_1 + R_3 \longrightarrow \begin{bmatrix} 1 & 2 & -4 & 3 \\ 0 & 3 & -2 & -1 \\ 0 & -3 & 13 & -8 \end{bmatrix}$$

Note that we write the elementary row operation beside the row that has been changed.

In Section 7.3 you used Gaussian elimination with back-substitution to solve a system of linear equations. We now demonstrate the matrix version of Gaussian elimination. The two methods are essentially the same. The basic difference is that with matrices you do not need to keep writing the variables.

EXAMPLE 3   Using Elementary Row Operations to Solve a System

<table>
<tr><td colspan="2" align="center">*Linear System*</td><td colspan="2" align="center">*Associated Augmented Matrix*</td></tr>
</table>

Technology Note

Most graphing utilities can perform elementary row operations on matrices. Read the user's manual for your graphing utility and duplicate the elementary row operations in the example at the right. (*Hint:* If your graphing utility *does not* store the matrices obtained in temporary results, we suggest that you do this after each elementary row operation.)

$$
\begin{aligned}
x - 2y + 3z &= 9 \\
-x + 3y &= -4 \\
2x - 5y + 5z &= 17
\end{aligned}
$$

$$
\begin{bmatrix}
1 & -2 & 3 & \vdots & 9 \\
-1 & 3 & 0 & \vdots & -4 \\
2 & -5 & 5 & \vdots & 17
\end{bmatrix}
$$

Add the first equation to the second equation.

Add the first row to the second row $(R_1 + R_2)$.

$$
\begin{aligned}
x - 2y + 3z &= 9 \\
y + 3z &= 5 \\
2x - 5y + 5z &= 17
\end{aligned}
\qquad R_1 + R_2 \longrightarrow
\begin{bmatrix}
1 & -2 & 3 & \vdots & 9 \\
0 & 1 & 3 & \vdots & 5 \\
2 & -5 & 5 & \vdots & 17
\end{bmatrix}
$$

Add -2 times the first equation to the third equation.

Add -2 times the first row to the third row $(-2R_1 + R_3)$.

$$
\begin{aligned}
x - 2y + 3z &= 9 \\
y + 3z &= 5 \\
-y - z &= -1
\end{aligned}
\qquad -2R_1 + R_3 \longrightarrow
\begin{bmatrix}
1 & -2 & 3 & \vdots & 9 \\
0 & 1 & 3 & \vdots & 5 \\
0 & -1 & -1 & \vdots & -1
\end{bmatrix}
$$

Add the second equation to the third equation.

Add the second row to the third row $(R_2 + R_3)$.

$$
\begin{aligned}
x - 2y + 3z &= 9 \\
y + 3z &= 5 \\
2z &= 4
\end{aligned}
\qquad R_2 + R_3 \longrightarrow
\begin{bmatrix}
1 & -2 & 3 & \vdots & 9 \\
0 & 1 & 3 & \vdots & 5 \\
0 & 0 & 2 & \vdots & 4
\end{bmatrix}
$$

Multiply the third equation by $\frac{1}{2}$.

Multiply the third row by $\frac{1}{2}$.

$$
\begin{aligned}
x - 2y + 3z &= 9 \\
y + 3z &= 5 \\
z &= 2
\end{aligned}
\qquad \tfrac{1}{2}R_3 \longrightarrow
\begin{bmatrix}
1 & -2 & 3 & \vdots & 9 \\
0 & 1 & 3 & \vdots & 5 \\
0 & 0 & 1 & \vdots & 2
\end{bmatrix}
$$

At this point, you can use back-substitution to find that the solution is $x = 1$, $y = -1$, and $z = 2$.

The last matrix in Example 3 is in **row-echelon form.** The term *echelon* refers to the stair-step pattern formed by the nonzero elements of the matrix. To be in this form, a matrix must have the following properties.

Technology Note

Some graphing utilities, such as the TI-85, can transform a matrix to row-echelon form and to reduced row-echelon form. Read the user's manual for your graphing utility, then try to obtain the last matrix in Example 3.

DEFINITION OF ROW-ECHELON FORM AND REDUCED ROW-ECHELON FORM

A matrix in **row-echelon form** has the following properties.

1. All rows consisting entirely of zeros occur at the bottom of the matrix.
2. For each row that does not consist entirely of zeros, the first nonzero entry is 1 (a **leading 1**).
3. For two successive (nonzero) rows, the leading 1 in the higher row is farther to the left than the leading 1 in the lower row.

A matrix in *row-echelon form* is in **reduced row-echelon form** if every column that has a leading 1 has zeros in every position above and below its leading 1.

EXAMPLE 4   Row-Echelon Form

The following matrices are in row-echelon form.

A. $\begin{bmatrix} 1 & 2 & -1 & 4 \\ 0 & 1 & 0 & 3 \\ 0 & 0 & 1 & -2 \end{bmatrix}$ **B.** $\begin{bmatrix} 0 & 1 & 0 & 5 \\ 0 & 0 & 1 & 3 \\ 0 & 0 & 0 & 0 \end{bmatrix}$

C. $\begin{bmatrix} 1 & -5 & 2 & -1 & 3 \\ 0 & 0 & 1 & 3 & -2 \\ 0 & 0 & 0 & 1 & 4 \\ 0 & 0 & 0 & 0 & 1 \end{bmatrix}$ **D.** $\begin{bmatrix} 1 & 0 & 0 & -1 \\ 0 & 1 & 0 & 2 \\ 0 & 0 & 1 & 3 \\ 0 & 0 & 0 & 0 \end{bmatrix}$

The matrices in (B) and (D) also happen to be in *reduced* row-echelon form. The following matrices are not in row-echelon form.

E. $\begin{bmatrix} 1 & 2 & -3 & 4 \\ 0 & 2 & 1 & -1 \\ 0 & 0 & 1 & -3 \end{bmatrix}$ **F.** $\begin{bmatrix} 1 & 2 & -1 & 2 \\ 0 & 0 & 0 & 0 \\ 0 & 1 & 2 & -4 \end{bmatrix}$

Every matrix is row-equivalent to a matrix in row-echelon form. For instance, in Example 4, you can change the matrix in part (E) to row-echelon form by multiplying its second row by $\frac{1}{2}$.

Gaussian Elimination with Back-Substitution

Guidelines for using Gaussian elimination with back-substitution to solve a system of linear equations are summarized as follows.

GAUSSIAN ELIMINATION WITH BACK-SUBSTITUTION

1. Write the augmented matrix of the system of linear equations.
2. Use elementary row operations to rewrite the augmented matrix in row-echelon form.
3. Write the system of linear equations corresponding to the matrix in row-echelon form, and use back-substitution to find the solution.

Gaussian elimination with back-substitution works well for solving systems of linear equations with a computer. For this algorithm, the order in which the elementary row operations are performed is important. We suggest operating from *left to right by columns,* using elementary row operations to obtain zeros in all entries directly below the leading 1's.

EXAMPLE 5   Gaussian Elimination with Back-Substitution

Solve the following system.

$$\begin{aligned} y + z - 2w &= -3 \\ x + 2y - z &= 2 \\ 2x + 4y + z - 3w &= -2 \\ x - 4y - 7z - w &= -19 \end{aligned}$$

SOLUTION

The augmented matrix for this system is

$$\begin{bmatrix} 0 & 1 & 1 & -2 & \vdots & -3 \\ 1 & 2 & -1 & 0 & \vdots & 2 \\ 2 & 4 & 1 & -3 & \vdots & -2 \\ 1 & -4 & -7 & -1 & \vdots & -19 \end{bmatrix}$$

Begin by obtaining a leading 1 in the upper left corner by interchanging the first and second rows, and then proceed to obtain zeros elsewhere in the first column.

$$\begin{matrix} R_2 \\ R_1 \end{matrix} \begin{bmatrix} 1 & 2 & -1 & 0 & \vdots & 2 \\ 0 & 1 & 1 & -2 & \vdots & -3 \\ 2 & 4 & 1 & -3 & \vdots & -2 \\ 1 & -4 & -7 & -1 & \vdots & -19 \end{bmatrix}$$ *First column has leading 1 in upper left corner*

$$\begin{matrix} \\ \\ -2R_1 + R_3 \longrightarrow \\ -R_1 + R_4 \longrightarrow \end{matrix} \begin{bmatrix} 1 & 2 & -1 & 0 & \vdots & 2 \\ 0 & 1 & 1 & -2 & \vdots & -3 \\ 0 & 0 & 3 & -3 & \vdots & -6 \\ 0 & -6 & -6 & -1 & \vdots & -21 \end{bmatrix}$$ *First column has zeros below its leading 1*

Now that the first column is already in the desired form, you can change the second, third, and fourth columns as follows.

$$6R_2 + R_4 \longrightarrow \begin{bmatrix} 1 & 2 & -1 & 0 & \vdots & 2 \\ 0 & 1 & 1 & -2 & \vdots & -3 \\ 0 & 0 & 3 & -3 & \vdots & -6 \\ 0 & 0 & 0 & -13 & \vdots & -39 \end{bmatrix}$$ *Second column has zeros below its leading 1*

$$\tfrac{1}{3}R_3 \longrightarrow \begin{bmatrix} 1 & 2 & -1 & 0 & \vdots & 2 \\ 0 & 1 & 1 & -2 & \vdots & -3 \\ 0 & 0 & 1 & -1 & \vdots & -2 \\ 0 & 0 & 0 & -13 & \vdots & -39 \end{bmatrix}$$ *Third column has zeros below its leading 1*

$$-\tfrac{1}{13}R_4 \longrightarrow \begin{bmatrix} 1 & 2 & -1 & 0 & \vdots & 2 \\ 0 & 1 & 1 & -2 & \vdots & -3 \\ 0 & 0 & 1 & -1 & \vdots & -2 \\ 0 & 0 & 0 & 1 & \vdots & 3 \end{bmatrix}$$ *Fourth column has a leading 1*

The matrix is now in row-echelon form, and the corresponding system of linear equations is

$$\begin{aligned} x + 2y - z &= 2 \\ y + z - 2w &= -3 \\ z - w &= -2 \\ w &= 3. \end{aligned}$$

Using back-substitution, you can determine that the solution is $x = -1$, $y = 2$, $z = 1$, and $w = 3$. Check this solution in the original system.

When solving a system of linear equations, remember that it is possible for the system to have no solution. If, in the elimination process, you obtain a row with zeros except for the last entry, it is unnecessary to continue the elimination process. You can simply conclude that the system is inconsistent. For instance, applying Gaussian elimination to the system

$$\begin{aligned} x - y + 2z &= 4 \\ x + z &= 6 \\ 2x - 3y + 5z &= 4 \\ 3x + 2y - z &= 1 \end{aligned}$$

produces

$$\begin{bmatrix} 1 & -1 & 2 & \vdots & 4 \\ 0 & 1 & -1 & \vdots & 2 \\ 0 & 0 & 0 & \vdots & -2 \\ 0 & 5 & -7 & \vdots & -11 \end{bmatrix}.$$

Note that the third row of this matrix consists of zeros except for the last entry. This means that the original system of linear equations is *inconsistent*.

Gauss-Jordan Elimination

With Gaussian elimination, elementary row operations are applied to a matrix to obtain a (row-equivalent) row-echelon form. A second method of elimination, called **Gauss-Jordan elimination,** after Carl Friedrich Gauss and Wilhelm Jordan (1842–1899), continues the reduction process until a *reduced* row-echelon form is obtained.

EXAMPLE 6   Gauss-Jordan Elimination

Use Gauss-Jordan elimination to solve the following system.

$$\begin{aligned} x - 2y + 3z &= 9 \\ -x + 3y &= -4 \\ 2x - 5y + 5z &= 17 \end{aligned}$$

SOLUTION

In Example 3, Gaussian elimination was used to obtain the following row-echelon form.

$$\begin{bmatrix} 1 & -2 & 3 & \vdots & 9 \\ 0 & 1 & 3 & \vdots & 5 \\ 0 & 0 & 1 & \vdots & 2 \end{bmatrix}$$

Now, rather than using back-substitution, apply elementary row operations until you obtain a matrix in reduced row-echelon form, as follows.

$$2R_2 + R_1 \longrightarrow \begin{bmatrix} 1 & 0 & 9 & \vdots & 19 \\ 0 & 1 & 3 & \vdots & 5 \\ 0 & 0 & 1 & \vdots & 2 \end{bmatrix}$$

Second column has zeros above its leading 1

$$\begin{aligned} -9R_3 + R_1 &\longrightarrow \\ -3R_3 + R_2 &\longrightarrow \end{aligned} \begin{bmatrix} 1 & 0 & 0 & \vdots & 1 \\ 0 & 1 & 0 & \vdots & -1 \\ 0 & 0 & 1 & \vdots & 2 \end{bmatrix}$$

Third column has zeros above its leading 1

Now, converting back to a system of linear equations, you can write

$$\begin{aligned} x\phantom{{}+y+z} &= 1 \\ y\phantom{{}+z} &= -1 \\ z &= 2. \end{aligned}$$

The beauty of Gauss-Jordan elimination is that, from the reduced row-echelon form, you can simply read the solution.

EXAMPLE 7   A System with an Infinite Number of Solutions

Solve the system of linear equations.

$$2x + 4y - 2z = 0$$
$$3x + 5y \quad\quad = 1$$

SOLUTION

Using Gauss-Jordan elimination, the augmented matrix reduces as follows.

$$\begin{bmatrix} 2 & 4 & -2 & : & 0 \\ 3 & 5 & 0 & : & 1 \end{bmatrix} \quad \tfrac{1}{2}R_1 \longrightarrow \begin{bmatrix} 1 & 2 & -1 & : & 0 \\ 3 & 5 & 0 & : & 1 \end{bmatrix}$$

$$-3R_1 + R_2 \longrightarrow \begin{bmatrix} 1 & 2 & -1 & : & 0 \\ 0 & -1 & 3 & : & 1 \end{bmatrix}$$

$$-R_2 \longrightarrow \begin{bmatrix} 1 & 2 & -1 & : & 0 \\ 0 & 1 & -3 & : & -1 \end{bmatrix}$$

$$-2R_2 + R_1 \longrightarrow \begin{bmatrix} 1 & 0 & 5 & : & 2 \\ 0 & 1 & -3 & : & -1 \end{bmatrix}$$

The corresponding system of equations is

$$x \quad + 5z = \quad 2$$
$$y - 3z = -1.$$

Solving for x and y in terms of z, you have $x = -5z + 2$ and $y = 3z - 1$. Then, letting $z = a$, the solution set has the form

$$(-5a + 2, 3a - 1, a),$$

where a is a real number.

DISCUSSION PROBLEM

COMPARING GAUSSIAN ELIMINATION WITH GAUSS-JORDAN ELIMINATION

Solve the following system of linear equations in two ways: once with Gaussian elimination with back-substitution and once with Gauss-Jordan elimination. Then write a short paragraph describing the advantages of one method over the other.

$$3x - 2y + \quad z = -6$$
$$-x + \quad y - 2z = \quad 1$$
$$2x + 2y - 3z = -1$$

WARM-UP

The following warm-up exercises involve skills that were covered in earlier sections. You will use these skills in the exercise set for this section.

In Exercises 1–4, evaluate the given expression.

1. $2(-1) - 3(5) + 7(2)$

2. $-4(-3) + 6(7) + 8(-3)$

3. $11(\frac{1}{2}) - 7(-\frac{3}{2}) - 5(2)$

4. $\frac{2}{3}(\frac{1}{2}) + \frac{4}{3}(-\frac{1}{3})$

In Exercises 5 and 6, determine whether $x = 1$, $y = 3$, and $z = -1$ is a solution of the system of linear equations.

5. $\begin{aligned} 4x - 2y + 3z &= -5 \\ x + 3y - z &= 11 \\ -x + 2y &= 5 \end{aligned}$

6. $\begin{aligned} -x + 2y + z &= 4 \\ 2x - 3z &= 5 \\ 3x + 5y - 2z &= 21 \end{aligned}$

In Exercises 7–10, use back-substitution to solve the system of linear equations.

7. $\begin{aligned} 2x - 3y &= 4 \\ y &= 2 \end{aligned}$

8. $\begin{aligned} 5x + 4y &= 0 \\ y &= -3 \end{aligned}$

9. $\begin{aligned} x - 3y + z &= 0 \\ y - 3z &= 8 \\ z &= 2 \end{aligned}$

10. $\begin{aligned} 2x - 5y + 3z &= -2 \\ y - 4z &= 0 \\ z &= 1 \end{aligned}$

SECTION 8.1 · EXERCISES

In Exercises 1–6, determine the order of the matrix.

1. $\begin{bmatrix} 4 & -2 \\ 7 & 0 \\ 0 & 8 \end{bmatrix}$

2. $\begin{bmatrix} 5 & -3 & 8 & 7 \end{bmatrix}$

3. $\begin{bmatrix} -9 \\ 2 \\ 36 \\ 11 \\ 3 \end{bmatrix}$

4. $\begin{bmatrix} 11 & 0 & 8 & 5 & 5 \\ -3 & 7 & 15 & 0 & 10 \\ 0 & 6 & 3 & 3 & 9 \\ 12 & 4 & 16 & 9 & 0 \\ 1 & 1 & 6 & 7 & 8 \end{bmatrix}$

5. $\begin{bmatrix} 33 & 45 \\ -9 & 20 \end{bmatrix}$

6. $\begin{bmatrix} 4 \end{bmatrix}$

In Exercises 7–10, determine whether the matrix is in row-echelon form. If it is, determine if it is also in reduced row-echelon form.

7. $\begin{bmatrix} 1 & 0 & 0 & 0 \\ 0 & 1 & 1 & 5 \\ 0 & 0 & 0 & 0 \end{bmatrix}$

8. $\begin{bmatrix} 1 & 0 & 2 & 1 \\ 0 & 1 & -3 & 10 \\ 0 & 0 & 1 & 0 \end{bmatrix}$

9. $\begin{bmatrix} 2 & 0 & 4 & 0 \\ 0 & -1 & 3 & 6 \\ 0 & 0 & 1 & 5 \end{bmatrix}$

10. $\begin{bmatrix} 1 & 3 & 0 & 0 \\ 0 & 0 & 1 & 8 \\ 0 & 0 & 0 & 0 \end{bmatrix}$

In Exercises 11–14, fill in the blanks using elementary row operations to form a row-equivalent matrix.

11. $\begin{bmatrix} 1 & 4 & 3 \\ 2 & 10 & 5 \end{bmatrix}$ **12.** $\begin{bmatrix} 3 & 6 & 8 \\ 4 & -3 & 6 \end{bmatrix}$

$\begin{bmatrix} 1 & 4 & 3 \\ 0 & \blacksquare & -1 \end{bmatrix}$ $\begin{bmatrix} 1 & \blacksquare & \frac{8}{3} \\ 4 & -3 & 6 \end{bmatrix}$

13. $\begin{bmatrix} 1 & 1 & 4 & -1 \\ 3 & 8 & 10 & 3 \\ -2 & 1 & 12 & 6 \end{bmatrix}$ **14.** $\begin{bmatrix} 2 & 4 & 8 & 3 \\ 1 & -1 & -3 & 2 \\ 2 & 6 & 4 & 9 \end{bmatrix}$

$\begin{bmatrix} 1 & 1 & 4 & -1 \\ 0 & 5 & \blacksquare & \blacksquare \\ 0 & 3 & \blacksquare & \blacksquare \end{bmatrix}$ $\begin{bmatrix} 1 & \blacksquare & \blacksquare & \blacksquare \\ 1 & -1 & -3 & 2 \\ 2 & 6 & 4 & 9 \end{bmatrix}$

$\begin{bmatrix} 1 & 1 & 4 & -1 \\ 0 & 1 & \blacksquare & \blacksquare \\ 0 & 3 & 20 & 4 \end{bmatrix}$ $\begin{bmatrix} 1 & 2 & 4 & \frac{3}{2} \\ 0 & \blacksquare & -7 & \frac{1}{2} \\ 0 & 2 & \blacksquare & \blacksquare \end{bmatrix}$

15. Perform the indicated *sequence* of elementary row operations to write the matrix in reduced row-echelon form.

$\begin{bmatrix} 1 & 2 & 3 \\ 2 & -1 & -4 \\ 3 & 1 & -1 \end{bmatrix}$

(a) Add -2 times Row 1 to Row 2. (Only Row 2 should change.)

(b) Add -3 times Row 1 to Row 3. (Only Row 3 should change.)

(c) Add -1 times Row 2 to Row 3.

(d) Multiply Row 2 by $-\frac{1}{5}$.

(e) Add -2 times Row 2 to Row 1.

16. Perform the indicated *sequence* of elementary row operations to write the matrix in reduced row-echelon form.

$\begin{bmatrix} 7 & 1 \\ 0 & 2 \\ -3 & 4 \\ 4 & 1 \end{bmatrix}$

(a) Add Row 3 to Row 4. (Only Row 4 should change.)

(b) Interchange Rows 1 and 4. (Note that the first element in the matrix is now 1, and that it was obtained without introducing fractions.)

(c) Add 3 times Row 1 to Row 3.

(d) Add -7 times Row 1 to Row 4.

(e) Multiply Row 2 by $\frac{1}{2}$.

(f) Add the appropriate multiple of Row 2 to Rows 1, 3, and 4.

In Exercises 17–20, write the matrix in row-echelon form. Remember that the row-echelon form for a given matrix is not unique. (*Note:* For some exercises it may be advantageous to use the matrix capabilities of your graphing utility.)

17. $\begin{bmatrix} 1 & 1 & 0 & 5 \\ -2 & -1 & 2 & -10 \\ 3 & 6 & 7 & 14 \end{bmatrix}$

18. $\begin{bmatrix} 1 & 2 & -1 & 3 \\ 3 & 7 & -5 & 14 \\ -2 & -1 & -3 & 8 \end{bmatrix}$

19. $\begin{bmatrix} 1 & -1 & -1 & 1 \\ 5 & -4 & 1 & 8 \\ -6 & 8 & 18 & 0 \end{bmatrix}$

20. $\begin{bmatrix} 1 & -3 & 0 & -7 \\ -3 & 10 & 1 & 23 \\ 4 & -10 & 2 & -24 \end{bmatrix}$

In Exercises 21–24, write the matrix in *reduced* row-echelon form. (*Note:* For some exercises it may be advantageous to use the matrix capabilities of your graphing utility.)

21. $\begin{bmatrix} 3 & 3 & 3 \\ -1 & 0 & -4 \\ 2 & 4 & -2 \end{bmatrix}$ **22.** $\begin{bmatrix} 1 & 3 & 2 \\ 5 & 15 & 9 \\ 2 & 6 & 10 \end{bmatrix}$

23. $\begin{bmatrix} 1 & 2 & 3 & -5 \\ 1 & 2 & 4 & -9 \\ -2 & -4 & -4 & 3 \\ 4 & 8 & 11 & -14 \end{bmatrix}$ **24.** $\begin{bmatrix} 1 & -3 \\ -1 & 8 \\ 0 & 4 \\ -2 & 10 \end{bmatrix}$

In Exercises 25–28, write the system of linear equations represented by the augmented matrix.

25. $\begin{bmatrix} 4 & 3 & : & 8 \\ 1 & -2 & : & 3 \end{bmatrix}$ **26.** $\begin{bmatrix} 9 & -4 & : & 0 \\ 6 & 1 & : & -4 \end{bmatrix}$

27. $\begin{bmatrix} 1 & 0 & 2 & : & -10 \\ 0 & 3 & -1 & : & 5 \\ 4 & 2 & 0 & : & 3 \end{bmatrix}$

28. $\begin{bmatrix} 5 & 8 & 2 & 0 & : & -1 \\ -2 & 15 & 5 & 1 & : & 9 \\ 1 & 6 & -7 & 0 & : & -3 \end{bmatrix}$

In Exercises 29–32, write the system of linear equations represented by the augmented matrix. Then use back-substitution to find the solution. (Use variables x, y, and z.)

29. $\begin{bmatrix} 1 & -2 & : & 4 \\ 0 & 1 & : & -3 \end{bmatrix}$

30. $\begin{bmatrix} 1 & 5 & : & 0 \\ 0 & 1 & : & -1 \end{bmatrix}$

31. $\begin{bmatrix} 1 & -1 & 2 & : & 4 \\ 0 & 1 & -1 & : & 2 \\ 0 & 0 & 1 & : & -2 \end{bmatrix}$

32. $\begin{bmatrix} 1 & 2 & -2 & : & -1 \\ 0 & 1 & 1 & : & 9 \\ 0 & 0 & 1 & : & -3 \end{bmatrix}$

In Exercises 33–36, an augmented matrix that represents a system of linear equations (in variables x, y, and z) has been reduced using Gauss-Jordan elimination. Write the solution represented by the augmented matrix.

33. $\begin{bmatrix} 1 & 0 & : & 7 \\ 0 & 1 & : & -5 \end{bmatrix}$

34. $\begin{bmatrix} 1 & 0 & : & -2 \\ 0 & 1 & : & 4 \end{bmatrix}$

35. $\begin{bmatrix} 1 & 0 & 0 & : & -4 \\ 0 & 1 & 0 & : & -8 \\ 0 & 0 & 1 & : & 2 \end{bmatrix}$

36. $\begin{bmatrix} 1 & 0 & 0 & : & 3 \\ 0 & 1 & 0 & : & -1 \\ 0 & 0 & 1 & : & 0 \end{bmatrix}$

In Exercises 37–58, solve the system of equations. Use Gaussian elimination with back-substitution or Gauss-Jordan elimination. (*Note:* For some exercises it may be advantageous to use the matrix capabilities of your graphing utility.)

37. $x + 2y = 7$
$2x + y = 8$

38. $2x + 6y = 16$
$2x + 3y = 7$

39. $-3x + 5y = -22$
$3x + 4y = 4$
$4x - 8y = 32$

40. $x + 2y = 0$
$x + y = 6$
$3x - 2y = 8$

41. $8x - 4y = 7$
$5x + 2y = 1$

42. $2x - y = -0.1$
$3x + 2y = 1.6$

43. $-x + 2y = 1.5$
$2x - 4y = 3$

44. $x - 3y = 5$
$-2x + 6y = -10$

45. $x - 3z = -2$
$3x + y - 2z = 5$
$2x + 2y + z = 4$

46. $2x - y + 3z = 24$
$2y - z = 14$
$7x - 5y = 6$

47. $x + y - 5z = 3$
$x - 2z = 1$
$2x - y - z = 0$

48. $2x + 3z = 3$
$4x - 3y + 7z = 5$
$8x - 9y + 15z = 9$

49. $x + 2y + z = 8$
$3x + 7y + 6z = 26$

50. $4x + 12y - 7z - 20w = 22$
$3x + 9y - 5z - 28w = 30$

51. $3x + 3y + 12z = 6$
$x + y + 4z = 2$
$2x + 5y + 20z = 10$
$-x + 2y + 8z = 4$

52. $2x + 10y + 2z = 6$
$x + 5y + 2z = 6$
$x + 5y + z = 3$
$-3x - 15y - 3z = -9$

53. $2x + y - z + 2w = -6$
$3x + 4y + w = 1$
$x + 5y + 2z + 6w = -3$
$5x + 2y - z - w = 3$

54. $x + 2y + 2z + 4w = 11$
$3x + 6y + 5z + 12w = 30$

55. $x + 2y = 0$
$-x - y = 0$

56. $x + 2y = 0$
$2x + 4y = 0$

57. $x + y + z = 0$
$2x + 3y + z = 0$
$3x + 5y + z = 0$

58. $x + 2y + z + 3w = 0$
$x - y + w = 0$
$y - z + 2w = 0$

59. *Borrowing Money* A small corporation borrowed $1,500,000 to expand its product line. Some of the money was borrowed at 8%, some at 9%, and some at 12%. How much was borrowed at each rate if the annual interest was $133,000 and the amount borrowed at 8% was four times the amount borrowed at 12%?

60. *Borrowing Money* A small corporation borrowed $500,000 to expand its product line. Some of the money was borrowed at 9%, some at 10%, and some at 12%. How much was borrowed at each rate if the annual interest was $52,000 and the amount borrowed at 10% was $2\frac{1}{2}$ times the amount borrowed at 9%?

61. *Partial Fractions* Write the partial fraction decomposition for the rational expression

$$\frac{4x^2}{(x+1)^2(x-1)} = \frac{A}{x-1} + \frac{B}{x+1} + \frac{C}{(x+1)^2}.$$

62. *Electrical Network* The currents in a certain electrical network are given by the solution of the system

$I_1 - I_2 + I_3 = 0$
$2I_1 + 2I_2 = 7$
$2I_2 + 4I_3 = 8,$

where I_1, I_2, and I_3 are measured in amperes. Solve the system of equations.

In Exercises 63–66, find the specified equation that passes through the given points. Use a graphing utility to verify your result.

63. Parabola:
$$y = ax^2 + bx + c$$

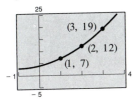

64. Parabola:
$$y = ax^2 + bx + c$$

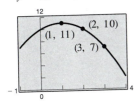

65. Circle:
$$x^2 + y^2 + Dx + Ey + F = 0$$

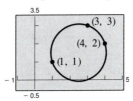

66. Cubic polynomial:
$$y = ax^3 + bx^2 + cx + d$$

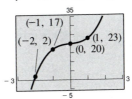

8.2   OPERATIONS WITH MATRICES

Equality of Matrices   /   Matrix Addition and Scalar Multiplication   /   Matrix Multiplication   /   Applications

Equality of Matrices

In Section 8.1 you used matrices to solve systems of linear equations. Matrices, however, can do much more than that. There is a rich mathematical theory of matrices, and its applications are numerous. This section and the next introduce some fundamentals of matrix theory. It is standard mathematical convention to represent matrices in any of the following three ways.

1. A matrix can be denoted by an uppercase letter such as A, B, or C.
2. A matrix can be denoted by a representative element enclosed in brackets, such as $[a_{ij}]$, $[b_{ij}]$, or $[c_{ij}]$.
3. A matrix can be denoted by a rectangular array of numbers such as

$$A = [a_{ij}] = \begin{bmatrix} a_{11} & a_{12} & a_{13} & \cdots & a_{1n} \\ a_{21} & a_{22} & a_{23} & \cdots & a_{2n} \\ a_{31} & a_{32} & a_{33} & \cdots & a_{3n} \\ \vdots & \vdots & \vdots & & \vdots \\ a_{m1} & a_{m2} & a_{m3} & \cdots & a_{mn} \end{bmatrix}.$$

Two matrices are **equal** if their corresponding entries are equal.

DEFINITION OF EQUALITY OF MATRICES

Two matrices $A = [a_{ij}]$ and $B = [b_{ij}]$ are **equal** if they have the same order $(m \times n)$ and

$$a_{ij} = b_{ij}$$

for $1 \leq i \leq m$ and $1 \leq j \leq n$.

EXAMPLE 1 **Equality of Matrices**

Solve for a_{11}, a_{12}, a_{21}, and a_{22} in the following matrix equation.

$$\begin{bmatrix} a_{11} & a_{12} \\ a_{21} & a_{22} \end{bmatrix} = \begin{bmatrix} 2 & -1 \\ -3 & 0 \end{bmatrix}$$

SOLUTION

Because two matrices are equal only if their corresponding entries are equal, we can conclude that

$$a_{11} = 2, \quad a_{12} = -1, \quad a_{21} = -3, \quad \text{and} \quad a_{22} = 0.$$

Matrix Addition and Scalar Multiplication

Add two matrices (of the same order) by adding their corresponding entries.

DEFINITION OF MATRIX ADDITION

If $A = [a_{ij}]$ and $B = [b_{ij}]$ are matrices of order $m \times n$, then their **sum** is the $m \times n$ matrix given by

$$A + B = [a_{ij} + b_{ij}].$$

The sum of two matrices of different orders is undefined.

Most graphing utilities can perform matrix addition and scalar multiplication. Read the user's manual for your graphing utility and duplicate the matrix operations in Examples 2 and 3. Try adding two matrices of different orders such as

$$A = \begin{bmatrix} 1 & 2 \\ 3 & 4 \end{bmatrix} \quad \text{and} \quad B = \begin{bmatrix} 5 \\ 6 \end{bmatrix}.$$

What error message does your graphing utility give?

EXAMPLE 2   Addition of Matrices

A. $\begin{bmatrix} -1 & 2 \\ 0 & 1 \end{bmatrix} + \begin{bmatrix} 1 & 3 \\ -1 & 2 \end{bmatrix} = \begin{bmatrix} -1+1 & 2+3 \\ 0-1 & 1+2 \end{bmatrix} = \begin{bmatrix} 0 & 5 \\ -1 & 3 \end{bmatrix}$

B. $\begin{bmatrix} 1 \\ -3 \\ -2 \end{bmatrix} + \begin{bmatrix} -1 \\ 3 \\ 2 \end{bmatrix} = \begin{bmatrix} 0 \\ 0 \\ 0 \end{bmatrix}$

C. The sum of

$$A = \begin{bmatrix} 2 & 1 & 0 \\ 4 & 0 & -1 \\ 3 & -2 & 2 \end{bmatrix} \quad \text{and} \quad B = \begin{bmatrix} 0 & 1 \\ -1 & 3 \\ 2 & 4 \end{bmatrix}$$

is undefined.

When working with matrices, you will usually refer to numbers as **scalars.** In this text, scalars will always be real numbers. You multiply a matrix A by a scalar c by multiplying each entry in A by c.

DEFINITION OF SCALAR MULTIPLICATION

If $A = [a_{ij}]$ is an $m \times n$ matrix and c is a scalar, then the **scalar multiple** of A by c is the $m \times n$ matrix given by

$$cA = [ca_{ij}].$$

Use $-A$ to represent the scalar product $(-1)A$. Moreover, if A and B are of the same order, then $A - B$ represents the sum of A and $(-1)B$. That is,

$$A - B = A + (-1)B. \qquad \textit{Subtraction of matrices}$$

EXAMPLE 3   Scalar Multiplication and Matrix Subtraction

For the matrices

$$A = \begin{bmatrix} 1 & 2 & 4 \\ -3 & 0 & -1 \\ 2 & 1 & 2 \end{bmatrix} \quad \text{and} \quad B = \begin{bmatrix} 2 & 0 & 0 \\ 1 & -4 & 3 \\ -1 & 3 & 2 \end{bmatrix},$$

find the following.

A. $3A$ **B.** $3A - B$

SOLUTION

A. $3A = 3\begin{bmatrix} 1 & 2 & 4 \\ -3 & 0 & -1 \\ 2 & 1 & 2 \end{bmatrix} = \begin{bmatrix} 3(1) & 3(2) & 3(4) \\ 3(-3) & 3(0) & 3(-1) \\ 3(2) & 3(1) & 3(2) \end{bmatrix} = \begin{bmatrix} 3 & 6 & 12 \\ -9 & 0 & -3 \\ 6 & 3 & 6 \end{bmatrix}$

B. $3A - B = \begin{bmatrix} 3 & 6 & 12 \\ -9 & 0 & -3 \\ 6 & 3 & 6 \end{bmatrix} - \begin{bmatrix} 2 & 0 & 0 \\ 1 & -4 & 3 \\ -1 & 3 & 2 \end{bmatrix} = \begin{bmatrix} 1 & 6 & 12 \\ -10 & 4 & -6 \\ 7 & 0 & 4 \end{bmatrix}$

It is often convenient to rewrite the scalar multiple cA by factoring c out of every entry in the matrix. For instance, in the following example, the scalar $\frac{1}{2}$ has been factored out of the matrix.

$$\begin{bmatrix} \frac{1}{2} & -\frac{3}{2} \\ \frac{5}{2} & \frac{1}{2} \end{bmatrix} = \frac{1}{2}\begin{bmatrix} 1 & -3 \\ 5 & 1 \end{bmatrix}$$

The properties of matrix addition and scalar multiplication are similar to those of addition and multiplication of real numbers, and we summarize them in the following list.

PROPERTIES OF MATRIX ADDITION AND SCALAR MULTIPLICATION

If A, B, and C are $m \times n$ matrices and c and d are scalars, then the following properties are true.

1. $A + B = B + A$ *Commutative Property of Addition*
2. $A + (B + C) = (A + B) + C$ *Associative Property of Addition*
3. $(cd)A = c(dA)$ *Associative Property of Scalar Multiplication*
4. $1A = A$ *Scalar Identity*
5. $c(A + B) = cA + cB$ *Distributive Property*
6. $(c + d)A = cA + dA$ *Distributive Property*

Note that the Associative Property of matrix addition allows you to write expressions such as $A + B + C$ without ambiguity because the same sum occurs no matter how the matrices are grouped. In other words, you obtain the same sum whether you group $A + B + C$ as $(A + B) + C$ or as $A + (B + C)$. This same reasoning applies to sums of four or more matrices.

One important property of addition of real numbers is that the number 0 is the additive identity. That is, $c + 0 = c$ for any real number c. For matrices, a similar property holds. That is, if A is an $m \times n$ matrix and O is the $m \times n$ **zero matrix** consisting entirely of zeros, then

$$A + O = A.$$

In other words, O is the **additive identity** for the set of all $m \times n$ matrices. For example, the following matrix is the additive identity for the set of all 2×3 matrices.

$$O = \begin{bmatrix} 0 & 0 & 0 \\ 0 & 0 & 0 \end{bmatrix} \qquad \textit{Zero } 2 \times 3 \textit{ matrix}$$

Similarly, the additive identity for the set of all 3×4 matrices is

$$O = \begin{bmatrix} 0 & 0 & 0 & 0 \\ 0 & 0 & 0 & 0 \\ 0 & 0 & 0 & 0 \end{bmatrix}. \qquad \textit{Zero } 3 \times 4 \textit{ matrix}$$

The algebra of real numbers and the algebra of matrices have many similarities. For example, compare the following solutions.

Real Numbers (Solve for x)	$m \times n$ Matrices (Solve for X)
$x + a = b$	$X + A = B$
$x + a + (-a) = b + (-a)$	$X + A + (-A) = B + (-A)$
$x + 0 = b - a$	$X + O = B - A$
$x = b - a$	$X = B - A$

The process of solving a matrix equation is demonstrated in Example 4.

EXAMPLE 4   Solving a Matrix Equation

Solve for X in the equation $3X + A = B$, where

$$A = \begin{bmatrix} 1 & -2 \\ 0 & 3 \end{bmatrix} \quad \text{and} \quad B = \begin{bmatrix} -3 & 4 \\ 2 & 1 \end{bmatrix}.$$

SOLUTION

Begin by solving the given equation for X to obtain

$$3X = B - A \quad \longrightarrow \quad X = \frac{1}{3}(B - A).$$

Now, using the given matrices A and B, you have

$$X = \frac{1}{3}\left(\begin{bmatrix} -3 & 4 \\ 2 & 1 \end{bmatrix} - \begin{bmatrix} 1 & -2 \\ 0 & 3 \end{bmatrix} \right) = \frac{1}{3}\begin{bmatrix} -4 & 6 \\ 2 & -2 \end{bmatrix} = \begin{bmatrix} -\frac{4}{3} & 2 \\ \frac{2}{3} & -\frac{2}{3} \end{bmatrix}.$$

Try using a graphing utility to solve this problem.

Matrix Multiplication

The third basic matrix operation is **matrix multiplication.** At first glance, the definition may seem unusual. You will see later, however, that this definition of the product of two matrices has many practical applications.

DEFINITION OF MATRIX MULTIPLICATION

If $A = [a_{ij}]$ is an $m \times n$ matrix and $B = [b_{ij}]$ is an $n \times p$ matrix, then the **product** AB is the $m \times p$ matrix

$$AB = [c_{ij}],$$

where $c_{ij} = a_{i1}b_{1j} + a_{i2}b_{2j} + a_{i3}b_{3j} + \cdots + a_{in}b_{nj}.$

This definition indicates a *row-by-column* multiplication, where the entry c_{ij} in the ith row and jth column of the product AB is obtained by multiplying the entries in the ith row of A by the corresponding entries in the jth column of B and then adding the results. The example that follows illustrates this process.

EXAMPLE 5 Finding the Product of Two Matrices

Technology Note

Most graphing utilities can perform matrix multiplication. Read the user's manual for your graphing utility and duplicate the matrix multiplication in Example 5. Then try multiplying two matrices A and B in which the number of columns of A is not equal to the number of rows of B. What error message does your graphing utility give?

Find the product AB where

$$A = \begin{bmatrix} -1 & 3 \\ 4 & -2 \\ 5 & 0 \end{bmatrix} \quad \text{and} \quad B = \begin{bmatrix} -3 & 2 \\ -4 & 1 \end{bmatrix}.$$

SOLUTION

First note that the product AB is defined because the number of columns of A is equal to the number of rows of B. Moreover, the product AB has order 3×2 and will take the form

$$\begin{bmatrix} -1 & 3 \\ 4 & -2 \\ 5 & 0 \end{bmatrix} \begin{bmatrix} -3 & 2 \\ -4 & 1 \end{bmatrix} = \begin{bmatrix} c_{11} & c_{12} \\ c_{21} & c_{22} \\ c_{31} & c_{32} \end{bmatrix}.$$

To find c_{11} (the entry in the first row and first column of the product), multiply corresponding entries in the first row of A and the first column of B. That is,

$$c_{11} = (-1)(-3) + (3)(-4) = -9$$

$$\begin{bmatrix} -1 & 3 \\ 4 & -2 \\ 5 & 0 \end{bmatrix} \begin{bmatrix} -3 & 2 \\ -4 & 1 \end{bmatrix} = \begin{bmatrix} -9 & c_{12} \\ c_{21} & c_{22} \\ c_{31} & c_{32} \end{bmatrix}.$$

Similarly, to find c_{12}, multiply corresponding entries in the first row of A and the second column of B to obtain

$$c_{12} = (-1)(2) + (3)(1) = 1$$

$$\begin{bmatrix} -1 & 3 \\ 4 & -2 \\ 5 & 0 \end{bmatrix} \begin{bmatrix} -3 & 2 \\ -4 & 1 \end{bmatrix} = \begin{bmatrix} -9 & 1 \\ c_{21} & c_{22} \\ c_{31} & c_{32} \end{bmatrix}.$$

Continuing this pattern produces the following results.

$$\begin{aligned} c_{21} &= (4)(-3) + (-2)(-4) = -4 \\ c_{22} &= (4)(2) + (-2)(1) = 6 \\ c_{31} &= (5)(-3) + (0)(-4) = -15 \\ c_{32} &= (5)(2) + (0)(1) = 10 \end{aligned}$$

Thus, the product is

$$\begin{aligned} AB &= \begin{bmatrix} -1 & 3 \\ 4 & -2 \\ 5 & 0 \end{bmatrix} \begin{bmatrix} -3 & 2 \\ -4 & 1 \end{bmatrix} \\ &= \begin{bmatrix} (-1)(-3) + (3)(-4) & (-1)(2) + (3)(1) \\ (4)(-3) + (-2)(-4) & (4)(2) + (-2)(1) \\ (5)(-3) + (0)(-4) & (5)(2) + (0)(1) \end{bmatrix} \\ &= \begin{bmatrix} -9 & 1 \\ -4 & 6 \\ -15 & 10 \end{bmatrix}. \end{aligned}$$

Be sure you understand that for the product of two matrices to be defined, the number of columns of the first matrix must equal the number of rows of the second matrix. That is, the middle two indices must be the same and the outside two indices give the order of the product, as shown in the following diagram.

$$\begin{array}{ccc} A & B & = & AB \\ m \times n & n \times p & & m \times p \end{array}$$

Equal

Order of AB

The general pattern for matrix multiplication is as follows. To obtain the entry in the ith row and the jth column of the product AB, use the ith row of A and the jth column of B.

$$\begin{bmatrix} a_{11} & a_{12} & a_{13} & \cdots & a_{1n} \\ a_{21} & a_{22} & a_{23} & \cdots & a_{2n} \\ a_{31} & a_{32} & a_{33} & \cdots & a_{3n} \\ \cdot & \cdot & \cdot & & \cdot \\ \cdot & \cdot & \cdot & & \cdot \\ a_{i1} & a_{i2} & a_{i3} & \cdots & a_{in} \\ \cdot & \cdot & \cdot & & \cdot \\ \cdot & \cdot & \cdot & & \cdot \\ a_{m1} & a_{m2} & a_{m3} & \cdots & a_{mn} \end{bmatrix} \begin{bmatrix} b_{11} & b_{12} & \cdots & b_{1j} & \cdots & b_{1p} \\ b_{21} & b_{22} & \cdots & b_{2j} & \cdots & b_{2p} \\ b_{31} & b_{32} & \cdots & b_{3j} & \cdots & b_{3p} \\ \cdot & \cdot & & \cdot & & \cdot \\ \cdot & \cdot & & \cdot & & \cdot \\ b_{n1} & b_{n2} & \cdots & b_{nj} & \cdots & b_{np} \end{bmatrix} = \begin{bmatrix} c_{11} & c_{12} & \cdots & c_{1j} & \cdots & c_{1p} \\ c_{21} & c_{22} & \cdots & c_{2j} & \cdots & c_{2p} \\ \cdot & \cdot & & \cdot & & \cdot \\ \cdot & \cdot & & \cdot & & \cdot \\ c_{i1} & c_{i2} & \cdots & c_{ij} & \cdots & c_{ip} \\ \cdot & \cdot & & \cdot & & \cdot \\ \cdot & \cdot & & \cdot & & \cdot \\ c_{m1} & c_{m2} & \cdots & c_{mj} & \cdots & c_{mp} \end{bmatrix}$$

$$a_{i1}b_{1j} + a_{i2}b_{2j} + a_{i3}b_{3j} + \cdots + a_{in}b_{nj} = c_{ij}$$

EXAMPLE 6 Matrix Multiplication

REMARK In parts (B) and (C) of Example 6, note that the two products are different. Matrix multiplication is not, in general, commutative. That is, for most matrices, $AB \neq BA$.

A. $\underset{2 \times 3}{\begin{bmatrix} 1 & 0 & 3 \\ 2 & -1 & -2 \end{bmatrix}} \underset{3 \times 3}{\begin{bmatrix} -2 & 4 & 2 \\ 1 & 0 & 0 \\ -1 & 1 & -1 \end{bmatrix}} = \underset{2 \times 3}{\begin{bmatrix} -5 & 7 & -1 \\ -3 & 6 & 6 \end{bmatrix}}$

B. $\underset{1 \times 3}{\begin{bmatrix} 1 & -2 & -3 \end{bmatrix}} \underset{3 \times 1}{\begin{bmatrix} 2 \\ -1 \\ 1 \end{bmatrix}} = \underset{1 \times 1}{\begin{bmatrix} 1 \end{bmatrix}}$

C. $\underset{3 \times 1}{\begin{bmatrix} 2 \\ -1 \\ 1 \end{bmatrix}} \underset{1 \times 3}{\begin{bmatrix} 1 & -2 & -3 \end{bmatrix}} = \underset{3 \times 3}{\begin{bmatrix} 2 & -4 & -6 \\ -1 & 2 & 3 \\ 1 & -2 & -3 \end{bmatrix}}$

D. The product AB for

$$A = \underset{3 \times 2}{\begin{bmatrix} -2 & 1 \\ 1 & -3 \\ 1 & 4 \end{bmatrix}} \quad \text{and} \quad B = \underset{3 \times 4}{\begin{bmatrix} -2 & 3 & 1 & 4 \\ 0 & 1 & -1 & 2 \\ 2 & -1 & 0 & 1 \end{bmatrix}}$$

is not defined (nor is the product BA).

Use your graphing utility to verify these products.

PROPERTIES OF MATRIX MULTIPLICATION

If A, B, and C are matrices and c is a scalar, then the following properties are true.

1. $A(BC) = (AB)C$ *Associative Property of Multiplication*
2. $A(B + C) = AB + AC$ *Left Distributive Property*
3. $(A + B)C = AC + BC$ *Right Distributive Property*
4. $c(AB) = (cA)B = A(cB)$ *Associative Property of Scalar Multiplication*

The $n \times n$ matrix that consists of 1's on its main diagonal and 0's elsewhere is the **identity matrix of order n** and is denoted by

$$I_n = \begin{bmatrix} 1 & 0 & 0 & \cdots & 0 \\ 0 & 1 & 0 & \cdots & 0 \\ 0 & 0 & 1 & \cdots & 0 \\ \vdots & \vdots & \vdots & & \vdots \\ 0 & 0 & 0 & \cdots & 1 \end{bmatrix}.$$ *Identity matrix*

Note that an identity matrix must be *square*. When the order is understood to be n, I_n is often denoted simply by I. If A is an $n \times n$ matrix, then the identity matrix has the properties

$$AI_n = A \quad \text{and} \quad I_nA = A.$$

For example,

$$\begin{bmatrix} 3 & -2 & 5 \\ 1 & 0 & 4 \\ -1 & 2 & -3 \end{bmatrix} \begin{bmatrix} 1 & 0 & 0 \\ 0 & 1 & 0 \\ 0 & 0 & 1 \end{bmatrix} = \begin{bmatrix} 3 & -2 & 5 \\ 1 & 0 & 4 \\ -1 & 2 & -3 \end{bmatrix}$$

and

$$\begin{bmatrix} 1 & 0 & 0 \\ 0 & 1 & 0 \\ 0 & 0 & 1 \end{bmatrix} \begin{bmatrix} 3 & -2 & 5 \\ 1 & 0 & 4 \\ -1 & 2 & -3 \end{bmatrix} = \begin{bmatrix} 3 & -2 & 5 \\ 1 & 0 & 4 \\ -1 & 2 & -3 \end{bmatrix}.$$

Applications

EXAMPLE 7 **An Application of Matrix Multiplication**

Two softball teams submit equipment lists to their sponsors.

	Women's Team	Men's Team
Bats	12	15
Balls	45	38
Gloves	15	17

Each bat costs $21, each ball costs $4, and each glove costs $30. Use matrices to find the total cost of equipment for each team.

SOLUTION

The equipment lists can be written in matrix form as

$$E = \begin{bmatrix} 12 & 15 \\ 45 & 38 \\ 15 & 17 \end{bmatrix},$$

and the cost per item can be written in matrix form as

$$C = [21 \quad 4 \quad 30].$$

The total cost of equipment for each team is given by the product

$$21(12) + 4(45) + 30(15) = 882 \quad \text{(Women's team)}$$

$$CE = [21 \quad 4 \quad 30] \begin{bmatrix} 12 & 15 \\ 45 & 38 \\ 15 & 17 \end{bmatrix} = [882 \quad 977].$$

$$21(15) + 4(38) + 30(17) = 977 \quad \text{(Men's team)}$$

Thus, the total cost of equipment for the women's team is $882, and the total cost of equipment for the men's team is $977.

Another useful application of matrix multiplication is in representing a system of linear equations. Note how the system

$$a_{11}x_1 + a_{12}x_2 + a_{13}x_3 = b_1$$
$$a_{21}x_1 + a_{22}x_2 + a_{23}x_3 = b_2$$
$$a_{31}x_1 + a_{32}x_2 + a_{33}x_3 = b_3$$

can be written as the matrix equation $AX = B$, where A is the *coefficient matrix* of the system, and X and B are column matrices.

$$\underset{A}{\begin{bmatrix} a_{11} & a_{12} & a_{13} \\ a_{21} & a_{22} & a_{23} \\ a_{31} & a_{32} & a_{33} \end{bmatrix}} \underset{X}{\begin{bmatrix} x_1 \\ x_2 \\ x_3 \end{bmatrix}} = \underset{B}{\begin{bmatrix} b_1 \\ b_2 \\ b_3 \end{bmatrix}}$$

EXAMPLE 8   Solving a System of Linear Equations

Solve the matrix equation $AX = B$ for X, where

Coefficient matrix   Constant matrix

$$A = \begin{bmatrix} 1 & -2 & 1 \\ 0 & 1 & 2 \\ 2 & 3 & -2 \end{bmatrix} \quad \text{and} \quad B = \begin{bmatrix} -4 \\ 4 \\ 2 \end{bmatrix}.$$

SOLUTION

As a system of linear equations, $AX = B$ is as follows.

$$\begin{aligned} x_1 - 2x_2 + x_3 &= -4 \\ x_2 + 2x_3 &= 4 \\ 2x_1 + 3x_2 - 2x_3 &= 2 \end{aligned}$$

Using Gauss-Jordan elimination on the augmented matrix of this system, you obtain

$$\begin{bmatrix} 1 & 0 & 0 & : & -1 \\ 0 & 1 & 0 & : & 2 \\ 0 & 0 & 1 & : & 1 \end{bmatrix}.$$

Thus, the solution of the system of linear equations is $x_1 = -1$, $x_2 = 2$, and $x_3 = 1$, and the solution of the matrix equation is

$$X = \begin{bmatrix} x_1 \\ x_2 \\ x_3 \end{bmatrix} = \begin{bmatrix} -1 \\ 2 \\ 1 \end{bmatrix}.$$

Use your graphing utility to verify that $AX = B$.

DISCUSSION PROBLEM

DIAGONAL MATRICES

A square matrix is called a **diagonal matrix** if each entry that is not on the main diagonal is zero. For instance,

$$A = \begin{bmatrix} -1 & 0 & 0 \\ 0 & 2 & 0 \\ 0 & 0 & 0 \end{bmatrix} \quad \text{and} \quad B = \begin{bmatrix} 2 & 0 & 0 & 0 \\ 0 & -1 & 0 & 0 \\ 0 & 0 & 3 & 0 \\ 0 & 0 & 0 & -2 \end{bmatrix}$$

are diagonal matrices. Write a paragraph describing a quick rule for multiplying a diagonal matrix by itself. Then illustrate your rule to find the matrices $A^2 = AA$ and $B^2 = BB$ for the given matrices A and B.

WARM-UP

The following warm-up exercises involve skills that were covered in earlier sections. You will use these skills in the exercise set for this section.

In Exercises 1 and 2, evaluate the expressions.

1. $-3(-\frac{5}{6}) + 10(-\frac{3}{4})4$ **2.** $-22(\frac{5}{2}) + 6(8)$

In Exercises 3 and 4, determine whether the matrices are in *reduced* row-echelon form.

3. $\begin{bmatrix} 0 & 1 & 0 & -5 \\ 1 & 0 & 3 & 2 \\ 0 & 0 & 1 & 0 \end{bmatrix}$ **4.** $\begin{bmatrix} 1 & 0 & 0 & 2 & 3 \\ 0 & 0 & 0 & 0 & 0 \\ 0 & 1 & 1 & 3 & 10 \end{bmatrix}$

In Exercises 5 and 6, write the augmented matrix for each system of linear equations.

5. $\begin{aligned} -5x + 10y &= 12 \\ 7x - 3y &= 0 \\ -x + 7y &= 25 \end{aligned}$ **6.** $\begin{aligned} 10x + 15y - 9z &= 42 \\ 6x - 5y &= 0 \end{aligned}$

In Exercises 7–10, solve the systems of linear equations represented by the augmented matrices.

7. $\begin{bmatrix} 1 & 0 & : & 0 \\ 0 & 1 & : & 2 \end{bmatrix}$ **8.** $\begin{bmatrix} 1 & 0 & -1 & : & 2 \\ 0 & 1 & 1 & : & 3 \end{bmatrix}$

9. $\begin{bmatrix} 1 & 2 & 1 & : & 0 \\ 0 & 0 & 1 & : & -1 \\ 0 & 0 & 0 & : & 0 \end{bmatrix}$ **10.** $\begin{bmatrix} 1 & -1 & 0 & : & 3 \\ 0 & 1 & -2 & : & 1 \\ 0 & 0 & 1 & : & -1 \end{bmatrix}$

SECTION 8.2 · EXERCISES

In Exercises 1–4, find x and y.

1. $\begin{bmatrix} x & -2 \\ 7 & y \end{bmatrix} = \begin{bmatrix} -4 & -2 \\ 7 & 22 \end{bmatrix}$ **2.** $\begin{bmatrix} -5 & x \\ y & 8 \end{bmatrix} = \begin{bmatrix} -5 & 13 \\ 12 & 8 \end{bmatrix}$

3. $\begin{bmatrix} 16 & 4 & 5 & 4 \\ -3 & 13 & 15 & 6 \\ 0 & 2 & 4 & 0 \end{bmatrix} = \begin{bmatrix} 16 & 4 & 2x+1 & 4 \\ -3 & 13 & 15 & 3x \\ 0 & 2 & 3y-5 & 0 \end{bmatrix}$

4. $\begin{bmatrix} x+2 & 8 & -3 \\ 1 & 2y & 2x \\ 7 & -2 & y+2 \end{bmatrix} = \begin{bmatrix} 2x+6 & 8 & -3 \\ 1 & 18 & -8 \\ 7 & -2 & 11 \end{bmatrix}$

In Exercises 5–10, find (a) $A + B$, (b) $A - B$, (c) $3A$, and (d) $3A - 2B$.

5. $A = \begin{bmatrix} 1 & -1 \\ 2 & -1 \end{bmatrix}$, $B = \begin{bmatrix} 2 & -1 \\ -1 & 8 \end{bmatrix}$

6. $A = \begin{bmatrix} 1 & 2 \\ 2 & 1 \end{bmatrix}$, $B = \begin{bmatrix} -3 & -2 \\ 4 & 2 \end{bmatrix}$

7. $A = \begin{bmatrix} 6 & -1 \\ 2 & 4 \\ -3 & 5 \end{bmatrix}$, $B = \begin{bmatrix} 1 & 4 \\ -1 & 5 \\ 1 & 10 \end{bmatrix}$

8. $A = \begin{bmatrix} 2 & 1 & 1 \\ -1 & -1 & 4 \end{bmatrix}$, $B = \begin{bmatrix} 2 & -3 & 4 \\ -3 & 1 & -2 \end{bmatrix}$

9. $A = \begin{bmatrix} 2 & 2 & -1 & 0 & 1 \\ 1 & 1 & -2 & 0 & -1 \end{bmatrix}$,

$B = \begin{bmatrix} 1 & 1 & -1 & 1 & 0 \\ -3 & 4 & 9 & -6 & -7 \end{bmatrix}$

10. $A = \begin{bmatrix} 3 \\ 2 \\ -1 \end{bmatrix}$, $B = \begin{bmatrix} -4 \\ 6 \\ 2 \end{bmatrix}$

In Exercises 11–16, find (a) AB, (b) BA, and, if possible, (c) A^2. (*Note:* $A^2 = AA$.)

11. $A = \begin{bmatrix} 1 & 2 \\ 4 & 2 \end{bmatrix}$, $B = \begin{bmatrix} 2 & -1 \\ -1 & 8 \end{bmatrix}$

12. $A = \begin{bmatrix} 2 & -1 \\ 1 & 4 \end{bmatrix}$, $B = \begin{bmatrix} 0 & 0 \\ 3 & -3 \end{bmatrix}$

13. $A = \begin{bmatrix} 3 & -1 \\ 1 & 3 \end{bmatrix}$, $B = \begin{bmatrix} 1 & -3 \\ 3 & 1 \end{bmatrix}$

14. $A = \begin{bmatrix} 1 & -1 \\ 1 & 1 \end{bmatrix}$, $B = \begin{bmatrix} 1 & 3 \\ -3 & 1 \end{bmatrix}$

15. $A = \begin{bmatrix} 1 & -1 & 7 \\ 2 & -1 & 8 \\ 3 & 1 & -1 \end{bmatrix}$, $B = \begin{bmatrix} 1 & 1 & 2 \\ 2 & 1 & 1 \\ 1 & -3 & 2 \end{bmatrix}$

16. $A = [3 \quad 2 \quad 1]$, $B = \begin{bmatrix} 2 \\ 3 \\ 0 \end{bmatrix}$

In Exercises 17–24, use a graphing utility to find AB, if possible.

17. $A = \begin{bmatrix} 2 & 1 \\ -3 & 4 \\ 1 & 6 \end{bmatrix}$, $B = \begin{bmatrix} 0 & -1 & 0 \\ 4 & 0 & 2 \\ 8 & -1 & 7 \end{bmatrix}$

18. $A = \begin{bmatrix} 0 & -1 & 0 \\ 4 & 0 & 2 \\ 8 & -1 & 7 \end{bmatrix}$, $B = \begin{bmatrix} 2 & 1 \\ -3 & 4 \\ 1 & 6 \end{bmatrix}$

19. $A = \begin{bmatrix} -1 & 3 \\ 4 & -5 \\ 0 & 2 \end{bmatrix}$, $B = \begin{bmatrix} 1 & 2 \\ 0 & 7 \end{bmatrix}$

20. $A = \begin{bmatrix} 1 & 0 & 0 \\ 0 & 4 & 0 \\ 0 & 0 & -2 \end{bmatrix}$, $D = \begin{bmatrix} 3 & 0 & 0 \\ 0 & -1 & 0 \\ 0 & 0 & 5 \end{bmatrix}$

21. $A = \begin{bmatrix} 5 & 0 & 0 \\ 0 & -8 & 0 \\ 0 & 0 & 7 \end{bmatrix}$, $B = \begin{bmatrix} \frac{1}{5} & 0 & 0 \\ 0 & -\frac{1}{8} & 0 \\ 0 & 0 & \frac{1}{2} \end{bmatrix}$

22. $A = \begin{bmatrix} 0 & 0 & 5 \\ 0 & 0 & -3 \\ 0 & 0 & 4 \end{bmatrix}$, $B = \begin{bmatrix} 6 & -11 & 4 \\ 8 & 16 & 4 \\ 0 & 0 & 0 \end{bmatrix}$

23. $A = \begin{bmatrix} 6 \\ -2 \\ 1 \\ 6 \end{bmatrix}$, $B = [10 \quad 12]$

24. $A = \begin{bmatrix} 1 & 0 & 3 & -2 & 4 \\ 6 & 13 & 8 & -17 & 10 \end{bmatrix}$, $B = \begin{bmatrix} 1 & 6 \\ 4 & 2 \end{bmatrix}$

In Exercises 25–28, solve for X given

$$A = \begin{bmatrix} -2 & -1 \\ 1 & 0 \\ 3 & -4 \end{bmatrix} \quad \text{and} \quad B = \begin{bmatrix} 0 & 3 \\ 2 & 0 \\ -4 & -1 \end{bmatrix}.$$

25. $X = 3A - 2B$

26. $2X = 2A - B$

27. $2X + 3A = B$

28. $2A + 4B = -2X$

In Exercises 29–32, find matrices A, X, and B such that the given system of linear equations can be written as the matrix equation $AX = B$. Solve the system of equations. Use a graphing utility to check your result.

29. $\begin{aligned} -x + y &= 4 \\ -2x + y &= 0 \end{aligned}$

30. $\begin{aligned} 2x + 3y &= 5 \\ x + 4y &= 10 \end{aligned}$

31. $\begin{aligned} x - 2y + 3z &= 9 \\ -x + 3y - z &= -6 \\ 2x - 5y + 5z &= 17 \end{aligned}$

32. $\begin{aligned} x + y - 3z &= -1 \\ -x + 2y &= 1 \\ -y + z &= 0 \end{aligned}$

33. If a, b, and c are real numbers such that $c \neq 0$ and $ac = bc$, then $a = b$. However, if A, B, and C are matrices such that $AC = BC$, then A is *not necessarily* equal to B. Illustrate this using the following matrices.

$$A = \begin{bmatrix} 1 & 2 & 3 \\ 0 & 5 & 4 \\ 3 & -2 & 1 \end{bmatrix}, \quad B = \begin{bmatrix} 4 & -6 & 3 \\ 5 & 4 & 4 \\ -1 & 0 & 1 \end{bmatrix},$$

$$C = \begin{bmatrix} 0 & 0 & 0 \\ 0 & 0 & 0 \\ 4 & -2 & 3 \end{bmatrix}$$

34. If a and b are real numbers such that $ab = 0$, then $a = 0$ or $b = 0$. However, if A and B are matrices such that $AB = 0$, then it is *not necessarily* true that $A = 0$ or $B = 0$. Illustrate this using the following matrices.

$$A = \begin{bmatrix} 3 & 3 \\ 4 & 4 \end{bmatrix}, \quad B = \begin{bmatrix} 1 & -1 \\ -1 & 1 \end{bmatrix}$$

35. *Factory Production*   A certain corporation has three factories, each of which manufactures two products. The number of units of product i produced at factory j in one day is represented by a_{ij} in the matrix

$$A = \begin{bmatrix} 60 & 40 & 20 \\ 30 & 90 & 60 \end{bmatrix}.$$

Find the production levels if production is increased by 20%. (*Hint:* Because an increase of 20% corresponds to 100% + 20%, multiply the given matrix by 1.2.)

36. *Factory Production*   A certain corporation has four factories, each of which manufactures two products. The number of units of product i produced at factory j in one day is represented by a_{ij} in the matrix

$$A = \begin{bmatrix} 100 & 90 & 70 & 30 \\ 40 & 20 & 60 & 60 \end{bmatrix}.$$

Find the production levels if production is increased by 10%. (*Hint:* Because an increase of 10% corresponds to 100% + 10%, multiply the given matrix by 1.1.)

37. *Crop Production*   A fruit grower raises two crops that are shipped to three outlets. The number of units of product i that are shipped to outlet j is represented by a_{ij} in the matrix

$$A = \begin{bmatrix} 100 & 75 & 75 \\ 125 & 150 & 100 \end{bmatrix}.$$

The profit per unit is represented by the matrix

$$B = [\$3.75 \quad \$7.00].$$

Find the product BA, and state what each entry of the product represents.

38. *Total Revenue*   A manufacturer produces three different models of a given product that are shipped to two different warehouses. The number of units of model i that are shipped to warehouse j is represented by a_{ij} in the matrix

$$A = \begin{bmatrix} 5{,}000 & 4{,}000 \\ 6{,}000 & 10{,}000 \\ 8{,}000 & 5{,}000 \end{bmatrix}.$$

The price per unit is represented by the matrix

$$B = [\$20.50 \quad \$26.50 \quad \$29.50].$$

Find the product BA, and state what each entry of the product represents.

39. *Inventory Levels*   A company sells five different models of computers through three retail outlets. The inventory of each model in the three outlets is given by the matrix S.

$$\begin{array}{c} \text{Model} \\ \begin{array}{ccccc} \text{A} & \text{B} & \text{C} & \text{D} & \text{E} \end{array} \\ S = \begin{bmatrix} 3 & 2 & 2 & 3 & 0 \\ 0 & 2 & 3 & 4 & 3 \\ 4 & 2 & 1 & 3 & 2 \end{bmatrix} \begin{array}{c} 1 \\ 2 \\ 3 \end{array} \right\} \text{Outlet} \end{array}$$

The wholesale and retail price for each model is given by the matrix T.

$$\begin{array}{c} \text{Price} \\ \overbrace{\begin{array}{cc} \text{Wholesale} & \text{Retail} \end{array}} \\ T = \begin{bmatrix} \$840 & \$1100 \\ \$1200 & \$1350 \\ \$1450 & \$1650 \\ \$2650 & \$3000 \\ \$3050 & \$3200 \end{bmatrix} \begin{array}{c} \text{A} \\ \text{B} \\ \text{C} \\ \text{D} \\ \text{E} \end{array} \right\} \text{Model} \end{array}$$

(a) What is the retail price of the inventory at Outlet 1?

(b) What is the wholesale price of the inventory at Outlet 3?

(c) Compute ST and interpret the result.

40. *Labor/Wage Requirements*   A company that manufactures boats has the following labor-hour and wage requirements.

Labor-hour Requirements (per boat)

	Department		
	Cutting	Assembly	Packaging
$S =$	1.0 hour	0.5 hour	0.2 hour	Small
	1.6 hours	1.0 hour	0.2 hour	Medium
	2.5 hours	2.0 hours	0.4 hour	Large

(Boat size: Small, Medium, Large)

Wage Requirements (per hour)

	Plant	
	A	B
$T =$	\$12	\$10	Cutting
	\$9	\$8	Assembly
	\$6	\$5	Packaging

(Department: Cutting, Assembly, Packaging)

(a) What is the labor cost for a medium boat at Plant B?

(b) What is the labor cost for a large boat at Plant A?

(c) Compute ST and interpret the result.

41. *Voting Preference*   The matrix

$$\begin{array}{c} \text{From} \\ \overbrace{\begin{array}{ccc} \text{R} & \text{D} & \text{I} \end{array}} \\ P = \begin{bmatrix} 0.6 & 0.1 & 0.1 \\ 0.2 & 0.7 & 0.1 \\ 0.2 & 0.2 & 0.8 \end{bmatrix} \begin{array}{c} \text{R} \\ \text{D} \\ \text{I} \end{array} \right\} \text{To} \end{array}$$

is called a stochastic matrix. Each entry p_{ij} ($i \neq j$) represents the proportion of the voting population that changes from party i to party j, and p_{ii} represents the proportion that remains loyal to the party from one election to the next. Use a graphing utility to find P^2. (This matrix gives the transition probabilities from the first election to the third.)

42. *Voting Preference*   Use a graphing utility to find P^3, P^4, P^5, P^6, P^7, and P^8 for the matrix given in Exercise 41. Can you detect a pattern as P is raised to higher and higher powers?

8.3 INVERSE MATRICES AND SYSTEMS OF LINEAR EQUATIONS

The Inverse of a Matrix / The Inverse of a 2 × 2 Matrix (Quick Method) / Systems of Linear Equations

The Inverse of a Matrix

This section further develops the algebra of matrices to include the solution of matrix equations involving matrix multiplication. To begin, consider the real number equation $ax = b$. To solve this equation for x, multiply both sides of the equation by a^{-1} (provided $a \neq 0$).

$$ax = b$$
$$(a^{-1}a)x = a^{-1}b$$
$$(1)x = a^{-1}b$$
$$x = a^{-1}b$$

The number a^{-1} is the *multiplicative inverse* of a because it has the property that $a^{-1}a = 1$. The definition of a multiplicative inverse of a matrix is similar.

DEFINITION OF AN INVERSE OF A MATRIX

Let A be a square matrix of order n. If there exists a matrix A^{-1} such that

$$AA^{-1} = I_n = A^{-1}A,$$

then A^{-1} is the **inverse** of A.

REMARK The symbol A^{-1} is read "A inverse."

If a matrix A has an inverse, then A is **invertible** (or **nonsingular**); otherwise, A is **singular**. A nonsquare matrix cannot have an inverse. To see this, note that if A is of order $m \times n$ and B is of order $n \times m$ (where $m \neq n$), then the products AB and BA are of different orders and could therefore not be equal to each other. Not all square matrices possess inverses. If, however, a matrix does possess an inverse, then that inverse is unique.

EXAMPLE 1   The Inverse of a Matrix

Show that B is the inverse of A, where

$$A = \begin{bmatrix} -1 & 2 \\ -1 & 1 \end{bmatrix} \quad \text{and} \quad B = \begin{bmatrix} 1 & -2 \\ 1 & -1 \end{bmatrix}.$$

SOLUTION

Using the definition of an inverse matrix, you can show that B is the inverse of A by showing that $AB = I = BA$ as follows.

REMARK Recall that it is not always true that $AB = BA$, even if both products are defined. However, if A and B are both square matrices and $AB = I_n$, then it can be shown that $BA = I_n$. Hence, in Example 1, you only needed to check that $AB = I_2$.

$$AB = \begin{bmatrix} -1 & 2 \\ -1 & 1 \end{bmatrix} \begin{bmatrix} 1 & -2 \\ 1 & -1 \end{bmatrix} = \begin{bmatrix} -1+2 & 2-2 \\ -1+1 & 2-1 \end{bmatrix} = \begin{bmatrix} 1 & 0 \\ 0 & 1 \end{bmatrix}$$

$$BA = \begin{bmatrix} 1 & -2 \\ 1 & -1 \end{bmatrix} \begin{bmatrix} -1 & 2 \\ -1 & 1 \end{bmatrix} = \begin{bmatrix} -1+2 & 2-2 \\ -1+1 & 2-1 \end{bmatrix} = \begin{bmatrix} 1 & 0 \\ 0 & 1 \end{bmatrix}$$

The following example shows how to use a system of equations to find the inverse.

EXAMPLE 2   Finding the Inverse of a Matrix

Find the inverse of the matrix

$$A = \begin{bmatrix} 1 & 4 \\ -1 & -3 \end{bmatrix}.$$

SOLUTION

To find the inverse of A, try to solve the matrix equation $AX = I$ for X.

$$\overset{A}{\begin{bmatrix} 1 & 4 \\ -1 & -3 \end{bmatrix}} \overset{X}{\begin{bmatrix} x_{11} & x_{12} \\ x_{21} & x_{22} \end{bmatrix}} = \overset{I}{\begin{bmatrix} 1 & 0 \\ 0 & 1 \end{bmatrix}}$$

$$\begin{bmatrix} x_{11}+4x_{21} & x_{12}+4x_{22} \\ -x_{11}-3x_{21} & -x_{12}-3x_{22} \end{bmatrix} = \begin{bmatrix} 1 & 0 \\ 0 & 1 \end{bmatrix}$$

Equating corresponding entries, you obtain the following two systems of linear equations.

$$\begin{array}{ll} x_{11} + 4x_{21} = 1 & x_{12} + 4x_{22} = 0 \\ -x_{11} - 3x_{21} = 0 & -x_{12} - 3x_{22} = 1 \end{array}$$

From the first system you find that $x_{11} = -3$ and $x_{21} = 1$, and from the second system you find that $x_{12} = -4$ and $x_{22} = 1$. Therefore, the inverse of A is

$$X = A^{-1} = \begin{bmatrix} -3 & -4 \\ 1 & 1 \end{bmatrix}.$$

Try using matrix multiplication to check this result.

EXAMPLE 3   **Finding the Inverse of a Matrix**

Use a graphing utility to find the inverse of

$$A = \begin{bmatrix} 1 & -1 & 0 \\ 1 & 0 & -1 \\ 6 & -2 & -3 \end{bmatrix}.$$

SOLUTION

$$A^{-1} = \begin{bmatrix} -2 & -3 & 1 \\ -3 & -3 & 1 \\ -2 & -4 & 1 \end{bmatrix}$$

Try using matrix multiplication to confirm this result.

REMARK   The denominator $ad - bc$ is called the **determinant** of the 2×2 matrix A.

The Inverse of a 2 × 2 Matrix (Quick Method)

If A is a 2×2 matrix given by

$$A = \begin{bmatrix} a & b \\ c & d \end{bmatrix},$$

then A is invertible if and only if $ad - bc \neq 0$. Moreover, if $ad - bc \neq 0$, then the inverse is given by

$$A^{-1} = \frac{1}{ad - bc} \begin{bmatrix} d & -b \\ -c & a \end{bmatrix}.$$

Try verifying this inverse by multiplication.

Technology Note

Graphing utilities that can perform matrix operations usually have the capability of finding the **determinant** of a matrix. It can be shown that a square matrix has an inverse if and only if its determinant is not zero. Try using this result to decide which of the following have inverses.

$$A = \begin{bmatrix} 1 & 2 \\ 3 & 4 \end{bmatrix}, \qquad B = \begin{bmatrix} -1 & 2 \\ 4 & -8 \end{bmatrix}$$

EXAMPLE 4   **Finding the Inverse of a 2 × 2 Matrix**

If possible, find the inverses of the following matrices.

A. $A = \begin{bmatrix} 3 & -1 \\ -2 & 2 \end{bmatrix}$ **B.** $B = \begin{bmatrix} -1 & 3 \\ 2 & -6 \end{bmatrix}$

SOLUTION

A. For matrix A, we apply the formula for the inverse of a 2×2 matrix to obtain $ad - bc = (3)(2) - (-1)(-2) = 4$. Because this quantity is not zero, the inverse is formed by interchanging the entries on the main diagonal and changing the signs of the other two entries as follows.

$$A^{-1} = \frac{1}{4} \begin{bmatrix} 2 & 1 \\ 2 & 3 \end{bmatrix} = \begin{bmatrix} \frac{1}{2} & \frac{1}{4} \\ \frac{1}{2} & \frac{3}{4} \end{bmatrix}$$

B. For matrix B, we have $ad - bc = (-1)(-6) - (3)(2) = 0$, which means that B is not invertible.

Systems of Linear Equations

A system of linear equations can have exactly one solution, an infinite number of solutions, or no solution. If the coefficient matrix A of a *square* system (a system that has the same number of equations as variables) is invertible, the system has a unique solution, which is given as follows.

$$AX = B \qquad\qquad \text{\textit{Original equation}}$$

$$A^{-1}AX = A^{-1}B \qquad\qquad \textit{Multiply both sides by } A^{-1} \textit{ (on the left)}$$

$$X = A^{-1}B \qquad\qquad \textit{Solution}$$

A SYSTEM OF EQUATIONS WITH A UNIQUE SOLUTION

If A is an invertible matrix, then the system of linear equations represented by $AX = B$ has a unique solution given by

$$X = A^{-1}B.$$

EXAMPLE 5   Solving a System of Equations Using an Inverse

Use an inverse matrix to solve the following system.

$$2x + 3y + z = -1$$
$$3x + 3y + z = 1$$
$$2x + 4y + z = -2$$

SOLUTION

Begin by writing the linear system in matrix form.

$$\underset{A}{\begin{bmatrix} 2 & 3 & 1 \\ 3 & 3 & 1 \\ 2 & 4 & 1 \end{bmatrix}} \underset{X}{\begin{bmatrix} x \\ y \\ z \end{bmatrix}} = \underset{B}{\begin{bmatrix} -1 \\ 1 \\ -2 \end{bmatrix}}$$

To solve the system, enter A and B into your graphing utility and compute $A^{-1}B$.

$$X = A^{-1}B = \begin{bmatrix} 2 \\ -1 \\ -2 \end{bmatrix}$$

The solution is $x = 2$, $y = -1$, and $z = -2$.

DISCUSSION PROBLEM

....

THE FACTORIZATION PRINCIPLE

For *real numbers* the Factorization Principle states that if $ab = 0$, then either $a = 0$ or $b = 0$. It is possible, however, to find nonzero matrices whose product is zero. For instance,

$$\begin{bmatrix} 2 & -1 \\ -4 & 2 \end{bmatrix} \begin{bmatrix} 3 & -3 \\ 6 & -6 \end{bmatrix} = \begin{bmatrix} 0 & 0 \\ 0 & 0 \end{bmatrix}.$$

In order to obtain a factorization principle for matrices, we must restrict ourselves to a certain type of matrix. Complete the following factorization principle for matrices by determining which type of matrix will make the statement true. Then write a short paragraph that justifies your conclusion.

"Let A and B be square matrices, each of order $n \times n$. If either A or B is ▨▨▨▨ and $AB = 0$ where 0 is the $n \times n$ zero matrix, then $A = 0$ or $B = 0$."

WARM-UP

....

The following warm-up exercises involve skills that were covered in earlier sections. You will use these skills in the exercise set for this section.

In Exercises 1–8, perform the indicated matrix operation.

1. $4 \begin{bmatrix} 1 & 6 \\ 0 & -4 \\ 12 & 2 \end{bmatrix}$

2. $\dfrac{1}{2} \begin{bmatrix} 11 & 10 & 48 \\ 1 & 0 & 16 \\ 0 & 2 & 8 \end{bmatrix}$

3. $\begin{bmatrix} 1 & -10 & 3 \\ 4 & 1 & 0 \end{bmatrix} - 2 \begin{bmatrix} 3 & -4 & 8 \\ 0 & 7 & 1 \end{bmatrix}$

4. $\begin{bmatrix} 5 & 20 \\ -7 & 15 \end{bmatrix} - 3 \begin{bmatrix} 6 & 3 \\ 4 & -2 \end{bmatrix}$

5. $\begin{bmatrix} 1 & -2 \\ -1 & 3 \end{bmatrix} \begin{bmatrix} 3 & 2 \\ 1 & 1 \end{bmatrix}$

6. $\begin{bmatrix} 1 & 0 \\ 0 & 1 \end{bmatrix} \begin{bmatrix} 6 & 5 \\ 3 & -2 \end{bmatrix}$

7. $\begin{bmatrix} 2 & 0 & 0 \\ 0 & -1 & 0 \\ 0 & 0 & 3 \end{bmatrix} \begin{bmatrix} \frac{1}{2} & 0 & 0 \\ 0 & -1 & 0 \\ 0 & 0 & \frac{1}{3} \end{bmatrix}$

8. $\begin{bmatrix} 1 & -1 & 0 \\ 1 & 0 & -1 \\ 6 & -2 & -3 \end{bmatrix} \begin{bmatrix} -2 & -3 & 1 \\ -3 & -3 & 1 \\ -2 & -4 & 1 \end{bmatrix}$

In Exercises 9 and 10, rewrite the matrices in reduced row-echelon form.

9. $\begin{bmatrix} 3 & -2 & 1 & 0 \\ 4 & -3 & 0 & 1 \end{bmatrix}$

10. $\begin{bmatrix} 1 & 1 & 2 & 1 & 0 & 0 \\ -1 & 0 & 3 & 0 & 1 & 0 \\ 1 & 2 & 8 & 0 & 0 & 1 \end{bmatrix}$

SECTION 8.3 · EXERCISES

In Exercises 1–8, show that B is the inverse of A.

1. $A = \begin{bmatrix} 2 & 1 \\ 5 & 3 \end{bmatrix}$, $B = \begin{bmatrix} 3 & -1 \\ -5 & 2 \end{bmatrix}$

2. $A = \begin{bmatrix} 1 & -1 \\ -1 & 2 \end{bmatrix}$, $B = \begin{bmatrix} 2 & 1 \\ 1 & 1 \end{bmatrix}$

3. $A = \begin{bmatrix} 1 & 2 \\ 3 & 4 \end{bmatrix}$, $B = \begin{bmatrix} -2 & 1 \\ \frac{3}{2} & -\frac{1}{2} \end{bmatrix}$

4. $A = \begin{bmatrix} 1 & -1 \\ 2 & 3 \end{bmatrix}$, $B = \begin{bmatrix} \frac{3}{5} & \frac{1}{5} \\ -\frac{2}{5} & \frac{1}{5} \end{bmatrix}$

5. $A = \begin{bmatrix} -2 & 2 & 3 \\ 1 & -1 & 0 \\ 0 & 1 & 4 \end{bmatrix}$, $B = \frac{1}{3}\begin{bmatrix} -4 & -5 & 3 \\ -4 & -8 & 3 \\ 1 & 2 & 0 \end{bmatrix}$

6. $A = \begin{bmatrix} 2 & -17 & 11 \\ -1 & 11 & -7 \\ 0 & 3 & -2 \end{bmatrix}$, $B = \begin{bmatrix} 1 & 1 & 2 \\ 2 & 4 & -3 \\ 3 & 6 & -5 \end{bmatrix}$

7. $A = \begin{bmatrix} 2 & 0 & 1 & 1 \\ 3 & 0 & 0 & 1 \\ -1 & 1 & -2 & 1 \\ 4 & -1 & 1 & 0 \end{bmatrix}$,

$B = \begin{bmatrix} -1 & 2 & -1 & -1 \\ -4 & 9 & -5 & -6 \\ 0 & 1 & -1 & -1 \\ 3 & -5 & 3 & 3 \end{bmatrix}$

8. $A = \begin{bmatrix} -1 & 1 & 0 & -1 \\ 1 & -1 & 2 & 0 \\ -1 & 1 & 2 & 0 \\ 0 & -1 & 1 & 1 \end{bmatrix}$,

$B = \frac{1}{3}\begin{bmatrix} -3 & 1 & 1 & -3 \\ -3 & -1 & 2 & -3 \\ 0 & 1 & 1 & 0 \\ -3 & -2 & 1 & 0 \end{bmatrix}$

In Exercises 9–20, find the inverse of the matrix (if it exists).

9. $\begin{bmatrix} 2 & 0 \\ 0 & 3 \end{bmatrix}$ **10.** $\begin{bmatrix} 1 & 2 \\ 3 & 7 \end{bmatrix}$

11. $\begin{bmatrix} 1 & -2 \\ 2 & -3 \end{bmatrix}$ **12.** $\begin{bmatrix} -7 & 33 \\ 4 & -19 \end{bmatrix}$

13. $\begin{bmatrix} -1 & 1 \\ -2 & 1 \end{bmatrix}$ **14.** $\begin{bmatrix} 11 & 1 \\ -1 & 0 \end{bmatrix}$

15. $\begin{bmatrix} 2 & 4 \\ 4 & 8 \end{bmatrix}$ **16.** $\begin{bmatrix} 2 & 3 \\ 1 & 4 \end{bmatrix}$

17. $\begin{bmatrix} 2 & 7 & 1 \\ -3 & -9 & 2 \end{bmatrix}$ **18.** $\begin{bmatrix} -2 & 5 \\ 6 & -15 \\ 0 & 1 \end{bmatrix}$

19. $\begin{bmatrix} 1 & 1 & 1 \\ 3 & 5 & 4 \\ 3 & 6 & 5 \end{bmatrix}$ **20.** $\begin{bmatrix} 1 & 2 & 2 \\ 3 & 7 & 9 \\ -1 & -4 & -7 \end{bmatrix}$

In Exercises 21–34, use a graphing utility to find the inverse of the matrix (if it exists).

21. $\begin{bmatrix} 1 & 2 & -1 \\ 3 & 7 & -10 \\ -5 & -7 & -15 \end{bmatrix}$ **22.** $\begin{bmatrix} 10 & 5 & -7 \\ -5 & 1 & 4 \\ 3 & 2 & -2 \end{bmatrix}$

23. $\begin{bmatrix} 1 & 1 & 2 \\ 3 & 1 & 0 \\ -2 & 0 & 3 \end{bmatrix}$ **24.** $\begin{bmatrix} 3 & 2 & 2 \\ 2 & 2 & 2 \\ -4 & 4 & 3 \end{bmatrix}$

25. $\begin{bmatrix} 0.1 & 0.2 & 0.3 \\ -0.3 & 0.2 & 0.2 \\ 0.5 & 0.4 & 0.4 \end{bmatrix}$ **26.** $\begin{bmatrix} 2 & 0 & 0 \\ 0 & 3 & 0 \\ 0 & 0 & 5 \end{bmatrix}$

27. $\begin{bmatrix} 1 & 0 & 0 \\ 3 & 4 & 0 \\ 2 & 5 & 5 \end{bmatrix}$ **28.** $\begin{bmatrix} 1 & 0 & 0 \\ 3 & 0 & 0 \\ 2 & 5 & 5 \end{bmatrix}$

29. $\begin{bmatrix} 1 & 0 & 3 & 0 \\ 0 & 2 & 0 & 4 \\ 1 & 0 & 3 & 0 \\ 0 & 2 & 0 & 4 \end{bmatrix}$

30. $\begin{bmatrix} -1 & 0 & 1 & 0 \\ 0 & 2 & 0 & -1 \\ 2 & 0 & -1 & 0 \\ 0 & -1 & 0 & 1 \end{bmatrix}$

31. $\begin{bmatrix} -8 & 0 & 0 & 0 \\ 0 & 1 & 0 & 0 \\ 0 & 0 & 4 & 0 \\ 0 & 0 & 0 & -5 \end{bmatrix}$ **32.** $\begin{bmatrix} 1 & 3 & -2 & 0 \\ 0 & 2 & 4 & 6 \\ 0 & 0 & -2 & 1 \\ 0 & 0 & 0 & 5 \end{bmatrix}$

33. $\begin{bmatrix} 1 & -2 & -1 & -2 \\ 3 & -5 & -2 & -3 \\ 2 & -5 & -2 & -5 \\ -1 & 4 & 4 & 11 \end{bmatrix}$ **34.** $\begin{bmatrix} 4 & 8 & -7 & 14 \\ 2 & 5 & -4 & 6 \\ 0 & 2 & 1 & -7 \\ 3 & 6 & -5 & 10 \end{bmatrix}$

In Exercises 35–46, use a graphing utility to solve (if possible) the system of linear equations.

35. $3x + 4y = -2$
$5x + 3y = 4$

37. $-0.4x + 0.8y = 1.6$
$2x - 4y = 5$

39. $3x + 6y = 5$
$6x + 14y = 11$

41. $4x - y + z = -5$
$2x + 2y + 3z = 10$
$5x - 2y + 6z = 1$

43. $5x - 3y + 2z = 2$
$2x + 2y - 3z = 3$
$x - 7y + 8z = -4$

45. $7x - 3y + 2w = 41$
$-2x + y - w = -13$
$4x + z - 2w = 12$
$-x + y - w = -8$

46. $2x + 5y + w = 11$
$x + 4y + 2z - 2w = -7$
$2x - 2y + 5z + w = 3$
$x - 3w = 1$

36. $18x + 12y = 13$
$30x + 24y = 23$

38. $13x - 6y = 17$
$26x - 12y = 8$

40. $3x + 2y = 1$
$2x + 10y = 6$

42. $4x - 2y + 3z = -2$
$2x + 2y + 5z = 16$
$8x - 5y - 2z = 4$

44. $2x + 3y + 5z = 4$
$3x + 5y + 9z = 7$
$5x + 9y + 17z = 13$

Circuit Analysis   In Exercises 51 and 52, consider the circuit in the figure. The currents I_1, I_2, and I_3 in amperes are given by the solution to the system of linear equations

$$2I_1 \quad\quad + 4I_3 = E_1$$
$$I_2 + 4I_3 = E_2$$
$$I_1 + I_2 - I_3 = 0,$$

where E_1 and E_2 are voltages. Find the unknown currents for the given voltages.

51. $E_1 = 14$ V, $E_2 = 28$ V   **52.** $E_1 = 10$ V, $E_2 = 10$ V

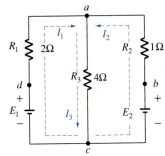

FIGURE FOR 51 AND 52

Bond Investments   In Exercises 47–50, consider a person who invests in AAA-rated bonds, A-rated bonds, and B-rated bonds. The average yield is 6.5% on AAA bonds, 7% on A bonds, and 9% on B bonds. Suppose the person invests twice as much in B bonds as in A bonds. A system of linear equations (where x, y, and z represent the amounts invested in AAA, A, and B bonds, respectively) is as follows.

$$x + y + z = \text{(total investment)}$$
$$0.065x + 0.07y + 0.09z = \text{(annual return)}$$
$$2y - z = 0$$

Find the amount invested in each type of bond, with the given total investment and annual return.

47. Total investment = $25,000
Annual return = $1900

48. Total investment = $45,000
Annual return = $3750

49. Total investment = $12,000
Annual return = $835

50. Total investment = $500,000
Annual return = $38,000

8.4   THE DETERMINANT OF A SQUARE MATRIX

The Determinant of a 2 × 2 Matrix / **Minors and Cofactors of a Square Matrix** / **The Determinant of a Square Matrix** / **Triangular Matrices**

The Determinant of a 2 × 2 Matrix

Every *square* matrix can be associated with a real number called its **determinant.** Historically, the use of determinants arose from special number patterns that occur when solving a system of linear equations. For instance, the system

$$a_1x + b_1y = c_1$$
$$a_2x + b_2y = c_2$$

has a solution given by

$$x = \frac{c_1b_2 - c_2b_1}{a_1b_2 - a_2b_1} \quad \text{and} \quad y = \frac{a_1c_2 - a_2c_1}{a_1b_2 - a_2b_1},$$

provided $a_1b_2 - a_2b_1 \neq 0$. Note that the denominator of each fraction is the same. This denominator is called the *determinant* of the coefficient matrix of the system.

Coefficient Matrix   *Determinant*

$$A = \begin{bmatrix} a_1 & b_1 \\ a_2 & b_2 \end{bmatrix} \quad\quad \det(A) = a_1b_2 - a_2b_1$$

The determinant of the matrix A can also be denoted by vertical bars on both sides of the matrix, as indicated in the following definition.

REMARK In this text, det(A) and $|A|$ are used interchangeably to represent the determinant of A. Although vertical bars are also used to denote the absolute value of a real number, the context will show which use is intended.

DEFINITION OF THE DETERMINANT OF A 2 × 2 MATRIX

The **determinant** of the matrix

$$A = \begin{bmatrix} a_1 & b_1 \\ a_2 & b_2 \end{bmatrix}$$

is given by

$$\det(A) = \begin{vmatrix} a_1 & b_1 \\ a_2 & b_2 \end{vmatrix} = a_1b_2 - a_2b_1.$$

A convenient method for remembering the formula for the determinant of a 2×2 matrix is shown in the following diagram.

$$\det(A) = \begin{vmatrix} a_1 & b_1 \\ a_2 & b_2 \end{vmatrix} = a_1 b_2 - a_2 b_1$$

Note that the determinant is given by the difference of the products of the two diagonals of the matrix.

EXAMPLE 1   The Determinant of a 2 × 2 Matrix

Find the determinant of each of the following matrices.

A. $A = \begin{bmatrix} 2 & -3 \\ 1 & 2 \end{bmatrix}$ **B.** $B = \begin{bmatrix} 2 & 1 \\ 4 & 2 \end{bmatrix}$ **C.** $C = \begin{bmatrix} 0 & 3 \\ 2 & 4 \end{bmatrix}$

SOLUTION

A. $\det(A) = \begin{vmatrix} 2 & -3 \\ 1 & 2 \end{vmatrix} = 2(2) - 1(-3) = 4 + 3 = 7$

REMARK   Notice in Example 1 that the determinant of a matrix can be positive, zero, or negative.

B. $\det(B) = \begin{vmatrix} 2 & 1 \\ 4 & 2 \end{vmatrix} = 2(2) - 4(1) = 4 - 4 = 0$

C. $\det(C) = \begin{vmatrix} 0 & 3 \\ 2 & 4 \end{vmatrix} = 0(4) - 2(3) = 0 - 6 = -6$

The determinant of a matrix of order 1×1 is defined simply as the entry of the matrix. For instance, if $A = |-2|$, then $\det(A) = -2$.

Minors and Cofactors of a Square Matrix

To define the determinant of a square matrix of order 3×3 or higher, it is convenient to introduce the concepts of **minors** and **cofactors**.

MINORS AND COFACTORS OF A SQUARE MATRIX

If A is a square matrix, then the **minor** M_{ij} of the entry a_{ij} is the determinant of the matrix obtained by deleting the ith row and jth column of A. The **cofactor** C_{ij} of the entry a_{ij} is given by

$$C_{ij} = (-1)^{i+j} M_{ij}.$$

For example, if A is a 3×3 matrix, then the minors of a_{21} and a_{22} are shown in the following diagram.

Sign Pattern for Cofactors

$$\begin{bmatrix} + & - & + \\ - & + & - \\ + & - & + \end{bmatrix}$$

3 × 3 matrix

$$\begin{bmatrix} + & - & + & - \\ - & + & - & + \\ + & - & + & - \\ - & + & - & + \end{bmatrix}$$

4 × 4 matrix

$$\begin{bmatrix} + & - & + & - & + & \cdots \\ - & + & - & + & - & \cdots \\ + & - & + & - & + & \cdots \\ - & + & - & + & - & \cdots \\ + & - & + & - & + & \cdots \\ \vdots & \vdots & \vdots & \vdots & \vdots & \end{bmatrix}$$

n × *n* matrix

Minor of a_{21}

$$\begin{bmatrix} a_{11} & a_{12} & a_{13} \\ a_{21} & a_{22} & a_{23} \\ a_{31} & a_{32} & a_{33} \end{bmatrix}, \quad M_{21} = \begin{vmatrix} a_{12} & a_{13} \\ a_{32} & a_{33} \end{vmatrix}$$

Delete Row 2 and Column 1

Minor of a_{22}

$$\begin{bmatrix} a_{11} & a_{12} & a_{13} \\ a_{21} & a_{22} & a_{23} \\ a_{31} & a_{32} & a_{33} \end{bmatrix}, \quad M_{22} = \begin{vmatrix} a_{11} & a_{13} \\ a_{31} & a_{33} \end{vmatrix}$$

Delete Row 2 and Column 2

The minors and cofactors of a matrix differ at most in sign. To obtain the cofactor C_{ij}, first find the minor M_{ij}, and then multiply by $(-1)^{i+j}$. The value of $(-1)^{i+j}$ is given by the checkerboard pattern of +'s and −'s shown at the left. Note that *odd* positions (where $i + j$ is odd) have negative signs, and *even* positions (where $i + j$ is even) have positive signs.

EXAMPLE 2   Finding the Minors and Cofactors of a Matrix

Find all the minors and cofactors of

$$A = \begin{bmatrix} 0 & 2 & 1 \\ 3 & -1 & 2 \\ 4 & 0 & 1 \end{bmatrix}.$$

SOLUTION

To find the minor M_{11}, delete the first row and first column of A and evaluate the determinant of the resulting matrix.

$$\begin{bmatrix} 0 & 2 & 1 \\ 3 & -1 & 2 \\ 4 & 0 & 1 \end{bmatrix}, \quad M_{11} = \begin{vmatrix} -1 & 2 \\ 0 & 1 \end{vmatrix} = -1(1) - 0(2) = -1$$

Similarly, to find M_{12}, delete the first row and second column.

$$\begin{bmatrix} 0 & 2 & 1 \\ 3 & -1 & 2 \\ 4 & 0 & 1 \end{bmatrix}, \quad M_{12} = \begin{vmatrix} 3 & 2 \\ 4 & 1 \end{vmatrix} = 3(1) - 4(2) = -5$$

Continuing this pattern produces the following minors.

$$M_{11} = -1 \qquad M_{12} = -5 \qquad M_{13} = 4$$
$$M_{21} = 2 \qquad M_{22} = -4 \qquad M_{23} = -8$$
$$M_{31} = 5 \qquad M_{32} = -3 \qquad M_{33} = -6$$

Now, to find the cofactors, combine the checkerboard pattern of signs with these minors to obtain the following.

$$C_{11} = -1 \qquad C_{12} = 5 \qquad C_{13} = 4$$
$$C_{21} = -2 \qquad C_{22} = -4 \qquad C_{23} = 8$$
$$C_{31} = 5 \qquad C_{32} = 3 \qquad C_{33} = -6$$

The Determinant of a Square Matrix

Having defined the minors and cofactors of a square matrix, we are ready to give a general definition of the determinant of a matrix. The definition given below is **inductive** because it uses determinants of matrices of order $n - 1$ to define the determinant of a matrix of order n.

DETERMINANT OF A SQUARE MATRIX

If A is a square matrix (of order 2×2 or greater), then the determinant of A is the sum of the entries in any row (or column) of A multiplied by their respective cofactors. For instance, expanding along the first row yields

$$|A| = a_{11} C_{11} + a_{12} C_{12} + \cdots + a_{1n} C_{1n}.$$

REMARK Try checking that for a 2×2 matrix this definition yields $|A| = a_{11}a_{22} - a_{12}a_{21}$, as previously defined.

Using this definition to evaluate a determinant is called **expanding by cofactors.** This procedure is demonstrated for a 3×3 matrix in Example 3.

EXAMPLE 3   The Determinant of a Matrix of Order 3×3

Technology Note

Most graphing utilities have the capability of evaluating the determinant of a square matrix. Try using your graphing utility to confirm the result of Example 3.

Find the determinant of

$$A = \begin{bmatrix} 0 & 2 & 1 \\ 3 & -1 & 2 \\ 4 & 0 & 1 \end{bmatrix}.$$

SOLUTION

Note that this is the same matrix that was given in Example 2. There, the cofactors of the entries in the first row were found to be

$$C_{11} = -1, \quad C_{12} = 5, \quad \text{and} \quad C_{13} = 4.$$

Therefore, by the definition of a determinant, you have the following.

$$|A| = a_{11} C_{11} + a_{12} C_{12} + a_{13} C_{13} \qquad \textit{First row expansion}$$
$$= 0(-1) + 2(5) + 1(4)$$
$$= 14$$

In Example 3 the determinant was found by expanding by the cofactors in the first row. You could have used any row or column. For instance, you could have expanded along the second row to obtain

$$|A| = a_{21} C_{21} + a_{22} C_{22} + a_{23} C_{23} \qquad \textit{Second row expansion}$$
$$= 3(-2) + (-1)(-4) + 2(8) = 14$$

or along the first column to obtain

$$|A| = a_{11} C_{11} + a_{21} C_{21} + a_{31} C_{31} \qquad \textit{First column expansion}$$
$$= 0(-1) + 3(-2) + 4(5) = 14.$$

Try some other possibilities to see that the determinant of A can be evaluated by expanding by cofactors along *any* row or column.

When expanding by cofactors you do not need to find cofactors of zero entries because zero times its cofactor is zero.

$$a_{ij} C_{ij} = (0)C_{ij} = 0$$

Thus, the row (or column) containing the most zeros is usually the best choice for expansion by cofactors. This is demonstrated in the next example.

EXAMPLE 4   The Determinant of a Matrix of Order 4

Find the determinant of

$$A = \begin{bmatrix} 1 & -2 & 3 & 0 \\ -1 & 1 & 0 & 2 \\ 0 & 2 & 0 & 3 \\ 3 & 4 & 0 & 2 \end{bmatrix}.$$

SOLUTION

Expand along the third column (because it has three entries that are zeros).

$$|A| = 3(C_{13}) + 0(C_{23}) + 0(C_{33}) + 0(C_{43})$$

Because C_{23}, C_{33}, and C_{43} have zero coefficients, you need only find the cofactor C_{13}. To do this, delete the first row and third column of A and evaluate the determinant of the resulting matrix.

$$C_{13} = (-1)^{1+3} \begin{vmatrix} -1 & 1 & 2 \\ 0 & 2 & 3 \\ 3 & 4 & 2 \end{vmatrix} = \begin{vmatrix} -1 & 1 & 2 \\ 0 & 2 & 3 \\ 3 & 4 & 2 \end{vmatrix}$$

Expanding by cofactors in the second row yields the following.

$$C_{13} = (0)(-1)^3 \begin{vmatrix} 1 & 2 \\ 4 & 2 \end{vmatrix} + (2)(-1)^4 \begin{vmatrix} -1 & 2 \\ 3 & 2 \end{vmatrix} + (3)(-1)^5 \begin{vmatrix} -1 & 1 \\ 3 & 4 \end{vmatrix} = 5$$

Thus, $|A| = 3C_{13} = 3(5) = 15$. Try using a computer or calculator to confirm this result.

There is an alternative method that is commonly used for evaluating the determinant of a 3×3 matrix A. (This method works *only* for 3×3 matrices.) To apply this method, copy the first and second columns of A to form fourth and fifth columns. The determinant of A is then obtained by adding the products of the three "downward diagonals" and subtracting the products of the three "upward diagonals," as shown in the following diagram.

Subtract these three products.

Add these three products.

Thus, the determinant of the 3×3 matrix A is given by the following.

$$|A| = a_{11}a_{22}a_{33} + a_{12}a_{23}a_{31} + a_{13}a_{21}a_{32}$$
$$- a_{31}a_{22}a_{13} - a_{32}a_{23}a_{11} - a_{33}a_{21}a_{12}$$

EXAMPLE 5   The Determinant of a 3 × 3 Matrix

Find the determinant of

$$A = \begin{bmatrix} 0 & 2 & 1 \\ 3 & -1 & 2 \\ 4 & -4 & 1 \end{bmatrix}.$$

SOLUTION

Because A is a 3×3 matrix, you can use the alternative procedure for finding $|A|$. Begin by recopying the first two columns and then computing the six diagonal products as follows.

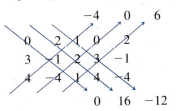

Subtract these three products.

Add these three products.

Now, by adding the lower three products and subtracting the upper three products, you can find the determinant of A to be

$$|A| = 0 + 16 - 12 - (-4) - 0 - 6 = 2.$$

REMARK   Be sure you understand that the diagonal process illustrated in Example 5 is valid *only* for matrices of order 3×3. For matrices of higher orders, another method must be used.

Triangular Matrices

Evaluating determinants of matrices of order 4 or higher *by hand* can be tedious. There is, however, an important exception: the determinant of a **triangular** matrix. A square matrix is **upper triangular** if it has all zero entries below its main diagonal and **lower triangular** if it has all zero entries above its main diagonal. A matrix that is both upper and lower triangular is **diagonal.** That is, a diagonal matrix is one in which all entries above and below the main diagonal are zero.

Upper Triangular Matrix

$$\begin{bmatrix} a_{11} & a_{12} & a_{13} & \cdots & a_{1n} \\ 0 & a_{22} & a_{23} & \cdots & a_{2n} \\ 0 & 0 & a_{33} & \cdots & a_{3n} \\ \vdots & \vdots & \vdots & & \vdots \\ 0 & 0 & 0 & \cdots & a_{nn} \end{bmatrix}$$

Lower Triangular Matrix

$$\begin{bmatrix} a_{11} & 0 & 0 & \cdots & 0 \\ a_{21} & a_{22} & 0 & \cdots & 0 \\ a_{31} & a_{32} & a_{33} & \cdots & 0 \\ \vdots & \vdots & \vdots & & \vdots \\ a_{n1} & a_{n2} & a_{n3} & \cdots & a_{nn} \end{bmatrix}$$

D I S C O V E R Y

The formula for the determinant of a triangular matrix (discussed at the right) is only one of many properties of matrices. You can use a computer or calculator to discover other properties. For instance, how is $|cA|$ related to $|A|$? How are $|A|$ and $|B|$ related to $|A + B|$?

To find the determinant of a triangular matrix, simply form the product of the entries on the main diagonal. It is easy to see that this procedure is valid for triangular matrices of order 2×2 or 3×3. For instance, the determinant of

$$A = \begin{bmatrix} 2 & 3 & -1 \\ 0 & -1 & 2 \\ 0 & 0 & 3 \end{bmatrix}$$

can be found by expanding by the third row to obtain

$$|A| = 0 \begin{vmatrix} 3 & -1 \\ -1 & 2 \end{vmatrix} - 0 \begin{vmatrix} 2 & -1 \\ 0 & 2 \end{vmatrix} + 3 \begin{vmatrix} 2 & 3 \\ 0 & -1 \end{vmatrix} = 3(2)(-1) = -6,$$

which is the product of the entries on the main diagonal.

EXAMPLE 6 **The Determinant of a Triangular Matrix**

Find the determinant of each of the following matrices.

A. $A = \begin{bmatrix} 2 & 0 & 0 & 0 \\ 4 & -2 & 0 & 0 \\ -5 & 6 & 1 & 0 \\ 1 & 5 & 3 & 3 \end{bmatrix}$

B. $B = \begin{bmatrix} -1 & 0 & 0 & 0 & 0 \\ 0 & 3 & 0 & 0 & 0 \\ 0 & 0 & 2 & 0 & 0 \\ 0 & 0 & 0 & 4 & 0 \\ 0 & 0 & 0 & 0 & -2 \end{bmatrix}$

SOLUTION

A. The determinant of this triangular matrix is given by

$$|A| = (2)(-2)(1)(3) = -12.$$

B. The determinant of this diagonal matrix is given by

$$|B| = (-1)(3)(2)(4)(-2) = 48.$$

DISCUSSION PROBLEM

A MATRIX WITH A DETERMINANT OF ZERO

Let a, b, and c be *any* real numbers. Write a paragraph explaining why the determinant of the following matrix is zero (regardless of the values of a, b, and c).

$$A = \begin{bmatrix} a & b & c \\ 1 & 2 & 3 \\ 2 & 4 & 6 \end{bmatrix}$$

WARM-UP

The following warm-up exercises involve skills that were covered in earlier sections. You will use these skills in the exercise set for this section.

In Exercises 1–4, perform the indicated matrix operations.

1. $\begin{bmatrix} 1 & -2 \\ 0 & 3 \end{bmatrix} + \begin{bmatrix} 2 & 7 \\ 4 & -3 \end{bmatrix}$ **2.** $\begin{bmatrix} -2 & 5 \\ 3 & -2 \end{bmatrix} - \begin{bmatrix} 0 & -3 \\ 1 & 2 \end{bmatrix}$

3. $3\begin{bmatrix} 3 & -4 & 2 \\ 1 & 0 & -1 \\ 0 & 1 & -2 \end{bmatrix}$ **4.** $4\begin{bmatrix} 0 & 2 & 3 \\ -1 & 2 & 3 \\ -2 & 1 & -2 \end{bmatrix}$

In Exercises 5–10, perform the indicated arithmetic operations.

5. $[(1)(3) + (-3)(2)] - [(1)(4) + (3)(5)]$

6. $[(4)(4) + (-1)(-3)] - [(-1)(2) + (-2)(7)]$

7. $\dfrac{4(7) - 1(-2)}{(-5)(-2) - 3(4)}$ **8.** $\dfrac{3(6) - 2(7)}{6(-5) - 2(1)}$

9. $-5(-1)^2[6(-2) - 7(-3)]$ **10.** $4(-1)^3[3(6) - 2(7)]$

SECTION 8.4 · EXERCISES

In Exercises 1–24, find the determinant of the matrix.

1. $[5]$

2. $[-8]$

3. $\begin{bmatrix} 2 & 1 \\ 3 & 4 \end{bmatrix}$

4. $\begin{bmatrix} -3 & 1 \\ 5 & 2 \end{bmatrix}$

5. $\begin{bmatrix} 5 & 2 \\ -6 & 3 \end{bmatrix}$

6. $\begin{bmatrix} 2 & -2 \\ 4 & 3 \end{bmatrix}$

7. $\begin{bmatrix} -7 & 6 \\ \frac{1}{2} & 3 \end{bmatrix}$

8. $\begin{bmatrix} 4 & -3 \\ 0 & 0 \end{bmatrix}$

9. $\begin{bmatrix} 2 & 6 \\ 0 & 3 \end{bmatrix}$

10. $\begin{bmatrix} 2 & -3 \\ -6 & 9 \end{bmatrix}$

11. $\begin{bmatrix} 2 & -1 & 0 \\ 4 & 2 & 1 \\ 4 & 2 & 1 \end{bmatrix}$

12. $\begin{bmatrix} -2 & 2 & 3 \\ 1 & -1 & 0 \\ 0 & 1 & 4 \end{bmatrix}$

13. $\begin{bmatrix} 0.3 & 0.2 & 0.2 \\ 0.2 & 0.2 & 0.2 \\ -0.4 & 0.4 & 0.3 \end{bmatrix}$

14. $\begin{bmatrix} 0.1 & 0.2 & 0.3 \\ -0.3 & 0.2 & 0.2 \\ 0.5 & 0.4 & 0.4 \end{bmatrix}$

15. $\begin{bmatrix} 1 & 4 & -2 \\ 3 & 6 & -6 \\ -2 & 1 & 4 \end{bmatrix}$

16. $\begin{bmatrix} 2 & 3 & 1 \\ 0 & 5 & -2 \\ 0 & 0 & -2 \end{bmatrix}$

17. $\begin{bmatrix} 6 & 3 & -7 \\ 0 & 0 & 0 \\ 4 & -6 & 3 \end{bmatrix}$

18. $\begin{bmatrix} 1 & 1 & 2 \\ 3 & 1 & 0 \\ -2 & 0 & 3 \end{bmatrix}$

19. $\begin{bmatrix} -1 & 2 & -5 \\ 0 & 3 & 4 \\ 0 & 0 & 3 \end{bmatrix}$

20. $\begin{bmatrix} 1 & 0 & 0 \\ -4 & -1 & 0 \\ 5 & 1 & 5 \end{bmatrix}$

21. $\begin{bmatrix} -1 & 0 & 0 & 0 \\ 2 & 3 & 0 & 0 \\ -4 & 5 & 3 & 0 \\ 1 & 0 & 2 & 2 \end{bmatrix}$

22. $\begin{bmatrix} -2 & 0 & 0 & 0 & 0 \\ 0 & 3 & 0 & 0 & 0 \\ 0 & 0 & -1 & 0 & 0 \\ 0 & 0 & 0 & 2 & 0 \\ 0 & 0 & 0 & 0 & -4 \end{bmatrix}$

23. $\begin{bmatrix} x & y & 1 \\ -2 & -2 & 1 \\ 1 & 5 & 1 \end{bmatrix}$

24. $\begin{bmatrix} 3 - \lambda & 2 \\ 4 & 1 - \lambda \end{bmatrix}$

In Exercises 25–28, find (a) all minors and (b) all cofactors for the given matrix.

25. $\begin{bmatrix} 3 & 4 \\ 2 & -5 \end{bmatrix}$

26. $\begin{bmatrix} 11 & 0 \\ -3 & 2 \end{bmatrix}$

27. $\begin{bmatrix} 3 & -2 & 8 \\ 3 & 2 & -6 \\ -1 & 3 & 6 \end{bmatrix}$

28. $\begin{bmatrix} -2 & 9 & 4 \\ 7 & -6 & 0 \\ 6 & 7 & -6 \end{bmatrix}$

In Exercises 29–34, find the determinant of the matrix by the method of expansion by cofactors. Expand using the indicated row or column.

29. $\begin{bmatrix} -3 & 2 & 1 \\ 4 & 5 & 6 \\ 2 & -3 & 1 \end{bmatrix}$

   (a) Row 1
   (b) Column 2

30. $\begin{bmatrix} -3 & 4 & 2 \\ 6 & 3 & 1 \\ 4 & -7 & -8 \end{bmatrix}$

   (a) Row 2
   (b) Column 3

31. $\begin{bmatrix} 5 & 0 & -3 \\ 0 & 12 & 4 \\ 1 & 6 & 3 \end{bmatrix}$

   (a) Row 2
   (b) Column 2

32. $\begin{bmatrix} 10 & -5 & 5 \\ 30 & 0 & 10 \\ 0 & 10 & 1 \end{bmatrix}$

   (a) Row 3
   (b) Column 1

33. $\begin{bmatrix} 6 & 0 & -3 & 5 \\ 4 & 13 & 6 & -8 \\ -1 & 0 & 7 & 4 \\ 8 & 6 & 0 & 2 \end{bmatrix}$

   (a) Row 2
   (b) Column 2

34. $\begin{bmatrix} 10 & 8 & 3 & -7 \\ 4 & 0 & 5 & -6 \\ 0 & 3 & 2 & 7 \\ 1 & 0 & -3 & 2 \end{bmatrix}$

   (a) Row 3
   (b) Column 1

In Exercises 35–44, find the determinant of the matrix.

35. $\begin{bmatrix} 1 & 4 & -2 \\ 3 & 2 & 0 \\ -1 & 4 & 3 \end{bmatrix}$

36. $\begin{bmatrix} 2 & -1 & 3 \\ 1 & 4 & 4 \\ 1 & 0 & 2 \end{bmatrix}$

37. $\begin{bmatrix} 2 & 4 & 6 \\ 0 & 3 & 1 \\ 0 & 0 & -5 \end{bmatrix}$

38. $\begin{bmatrix} -3 & 0 & 0 \\ 7 & 11 & 0 \\ 1 & 2 & 2 \end{bmatrix}$

39. $\begin{bmatrix} 3 & 6 & -5 & 4 \\ -2 & 0 & 6 & 0 \\ 1 & 1 & 2 & 2 \\ 0 & 3 & -1 & -1 \end{bmatrix}$

40. $\begin{bmatrix} 2 & 6 & 6 & 2 \\ 2 & 7 & 3 & 6 \\ 1 & 5 & 0 & 1 \\ 3 & 7 & 0 & 7 \end{bmatrix}$

41. $\begin{bmatrix} 5 & 3 & 0 & 6 \\ 4 & 6 & 4 & 12 \\ 0 & 2 & -3 & 4 \\ 0 & 1 & -2 & 2 \end{bmatrix}$ **42.** $\begin{bmatrix} 1 & 4 & 3 & 2 \\ -5 & 6 & 2 & 1 \\ 0 & 0 & 0 & 0 \\ 3 & -2 & 1 & 5 \end{bmatrix}$

43. $\begin{bmatrix} 3 & 2 & 4 & -1 & 5 \\ -2 & 0 & 1 & 3 & 2 \\ 1 & 0 & 0 & 4 & 0 \\ 6 & 0 & 2 & -1 & 0 \\ 3 & 0 & 5 & 1 & 0 \end{bmatrix}$

44. $\begin{bmatrix} 5 & 2 & 0 & 0 & -2 \\ 0 & 1 & 4 & 3 & 2 \\ 0 & 0 & 2 & 6 & 3 \\ 0 & 0 & 3 & 4 & 1 \\ 0 & 0 & 0 & 0 & 2 \end{bmatrix}$

In Exercises 45–60, use a graphing utility to evaluate the determinant.

45. $\begin{vmatrix} 1 & 2 & 5 \\ 1 & 4 & 2 \\ 0 & 3 & -4 \end{vmatrix}$ **46.** $\begin{vmatrix} 1 & 7 & -3 \\ 1 & 3 & 1 \\ 4 & 8 & 1 \end{vmatrix}$

47. $\begin{vmatrix} 3 & -1 & -3 \\ -1 & -4 & -2 \\ 3 & -1 & -1 \end{vmatrix}$ **48.** $\begin{vmatrix} 4 & 3 & -2 \\ 5 & 4 & 1 \\ -2 & 3 & 4 \end{vmatrix}$

49. $\begin{vmatrix} 3 & 8 & -7 \\ 0 & -5 & 4 \\ 8 & 1 & 6 \end{vmatrix}$ **50.** $\begin{vmatrix} 5 & -8 & 0 \\ 9 & 7 & 4 \\ -8 & 7 & 1 \end{vmatrix}$

51. $\begin{vmatrix} 2 & -1 & 3 \\ 1 & 2 & -1 \\ 3 & -4 & 7 \end{vmatrix}$ **52.** $\begin{vmatrix} 2 & 0 & 1 \\ 4 & -4 & 0 \\ -1 & 5 & 2 \end{vmatrix}$

53. $\begin{vmatrix} 7 & 0 & -14 \\ -2 & 5 & 4 \\ -6 & 2 & 12 \end{vmatrix}$ **54.** $\begin{vmatrix} 3 & 0 & 0 \\ -2 & 5 & 0 \\ 12 & 5 & 7 \end{vmatrix}$

55. $\begin{vmatrix} 4 & -8 & 5 & 0 \\ 8 & -5 & 3 & 0 \\ 8 & 5 & 2 & 0 \\ 1 & 7 & -5 & 1 \end{vmatrix}$ **56.** $\begin{vmatrix} 4 & -7 & 9 & 1 \\ 6 & 2 & 7 & 0 \\ 3 & 6 & -3 & 3 \\ 0 & 7 & 4 & -1 \end{vmatrix}$

57. $\begin{vmatrix} 0 & -3 & 8 & 2 \\ 8 & 1 & -1 & 6 \\ -4 & 6 & 0 & 9 \\ -7 & 0 & 0 & 14 \end{vmatrix}$ **58.** $\begin{vmatrix} 1 & -1 & 8 & 4 \\ 2 & 6 & 0 & -4 \\ 2 & 0 & 2 & 6 \\ 0 & 2 & 8 & 0 \end{vmatrix}$

59. $\begin{vmatrix} 3 & -2 & 4 & 3 & 1 \\ -1 & 0 & 2 & 1 & 0 \\ 5 & -1 & 0 & 3 & 2 \\ 4 & 7 & -8 & 0 & 0 \\ 1 & 2 & 3 & 0 & 2 \end{vmatrix}$ **60.** $\begin{vmatrix} 4 & 2 & -1 & 0 & 3 \\ 0 & 1 & 1 & 2 & -3 \\ 0 & 0 & -2 & 8 & 12 \\ 0 & 0 & 0 & 5 & 13 \\ 0 & 0 & 0 & 0 & 3 \end{vmatrix}$

In Exercises 61 and 62, solve for x.

61. $\begin{vmatrix} x - 1 & 2 \\ 3 & x - 2 \end{vmatrix} = 0$ **62.** $\begin{vmatrix} x - 2 & -1 \\ -3 & x \end{vmatrix} = 0$

In Exercises 63–68, evaluate the determinant of the matrix where the entries are functions. Determinants of this type occur in calculus.

63. $\begin{bmatrix} 4u & -1 \\ -1 & 2v \end{bmatrix}$ **64.** $\begin{bmatrix} 3x^2 & -3y^2 \\ 1 & 1 \end{bmatrix}$

65. $\begin{bmatrix} e^{2x} & e^{3x} \\ 2e^{2x} & 3e^{3x} \end{bmatrix}$ **66.** $\begin{bmatrix} e^{-x} & xe^{-x} \\ -e^{-x} & (1 - x)e^{-x} \end{bmatrix}$

67. $\begin{bmatrix} x & \ln x \\ 1 & \frac{1}{x} \end{bmatrix}$ **68.** $\begin{bmatrix} x & x \ln x \\ 1 & 1 + \ln x \end{bmatrix}$

In Exercises 69–73, find (a) $|A|$, (b) $|B|$, (c) AB, and (d) $|AB|$.

69. $A = \begin{bmatrix} -1 & 0 \\ 0 & 3 \end{bmatrix}$, $B = \begin{bmatrix} 2 & 0 \\ 0 & -1 \end{bmatrix}$

70. $A = \begin{bmatrix} -2 & 1 \\ 4 & -2 \end{bmatrix}$, $B = \begin{bmatrix} 1 & 2 \\ 0 & -1 \end{bmatrix}$

71. $A = \begin{bmatrix} -1 & 2 & 1 \\ 1 & 0 & 1 \\ 0 & 1 & 0 \end{bmatrix}$ $B = \begin{bmatrix} -1 & 0 & 0 \\ 0 & 2 & 0 \\ 0 & 0 & 3 \end{bmatrix}$

72. $A = \begin{bmatrix} 2 & 0 & 1 \\ 1 & -1 & 2 \\ 3 & 1 & 0 \end{bmatrix}$ $B = \begin{bmatrix} 2 & -1 & 4 \\ 0 & 1 & 3 \\ 3 & -2 & 1 \end{bmatrix}$

73. Find the square matrices A and B to demonstrate that
$$|A + B| \neq |A| + |B|.$$

74. If the matrix B is obtained from matrix A by adding a nonzero multiple of any row of A to another row of A, then $|B| = |A|$. Demonstrate this property of determinants for the matrix
$$A = \begin{bmatrix} 1 & 10 & -6 \\ 2 & -3 & 4 \\ 7 & 6 & 3 \end{bmatrix}.$$

8.5 APPLICATIONS OF MATRICES AND DETERMINANTS

Cramer's Rule / Area of Triangle / Lines in the Plane / Cryptography

Cramer's Rule

So far, you have studied several methods for solving a system of linear equations: substitution, elimination (with equations), elimination (with matrices), and use of inverse matrices. In this section, you will study one more method, **Cramer's Rule,** named after Gabriel Cramer (1704–1752). This rule uses determinants to write the solution of a system of linear equations. To see how Cramer's Rule works, let's take another look at the solution described at the beginning of Section 8.4. There, we pointed out that the system

$$a_1x + b_1y = c_1$$
$$a_2x + b_2y = c_2$$

has a solution given by

$$x = \frac{c_1b_2 - c_2b_1}{a_1b_2 - a_2b_1}$$

and

$$y = \frac{a_1c_2 - a_2c_1}{a_1b_2 - a_2b_1},$$

provided $a_1b_2 - a_2b_1 \neq 0$. Each numerator and denominator in this solution can be expressed as a determinant, as follows.

$$x = \frac{c_1b_2 - c_2b_1}{a_1b_2 - a_2b_1} = \frac{\begin{vmatrix} c_1 & b_1 \\ c_2 & b_2 \end{vmatrix}}{\begin{vmatrix} a_1 & b_1 \\ a_2 & b_2 \end{vmatrix}}, \qquad y = \frac{a_1c_2 - a_2c_1}{a_1b_2 - a_2b_1} = \frac{\begin{vmatrix} a_1 & c_1 \\ a_2 & c_2 \end{vmatrix}}{\begin{vmatrix} a_1 & b_1 \\ a_2 & b_2 \end{vmatrix}}$$

Relative to the original system, the denominator for x and y is simply the determinant of the *coefficient* matrix of the system. This determinant is denoted by D. The numerators for x and y are denoted by D_x and D_y, respectively. They are formed by using the column of constants as replacements for the coefficients of x and y, as follows.

Coefficient Matrix	D	D_x	D_y
$\begin{bmatrix} a_1 & b_1 \\ a_2 & b_2 \end{bmatrix}$	$\begin{vmatrix} a_1 & b_1 \\ a_2 & b_2 \end{vmatrix}$	$\begin{vmatrix} c_1 & b_1 \\ c_2 & b_2 \end{vmatrix}$	$\begin{vmatrix} a_1 & c_1 \\ a_2 & c_2 \end{vmatrix}$

EXAMPLE 1 Using Cramer's Rule for a 2 × 2 System

Use Cramer's Rule to solve the following system of linear equations.

$$4x - 2y = 10$$
$$3x - 5y = 11$$

SOLUTION

Begin by finding the determinant of the coefficient matrix.

$$D = \begin{vmatrix} 4 & -2 \\ 3 & -5 \end{vmatrix} = -14$$

Because this determinant is not zero, you can apply Cramer's Rule to find the solution, as follows.

$$x = \frac{D_x}{D} = \frac{\begin{vmatrix} 10 & -2 \\ 11 & -5 \end{vmatrix}}{-14} = \frac{-28}{-14} = 2$$

$$y = \frac{D_y}{D} = \frac{\begin{vmatrix} 4 & 10 \\ 3 & 11 \end{vmatrix}}{-14} = \frac{14}{-14} = -1$$

Therefore, the solution is $x = 2$ and $y = -1$. Try checking this solution in the original system of equations.

Cramer's Rule generalizes easily to systems of n equations in n variables. The value of each variable is given as the quotient of two determinants. The denominator is the determinant of the coefficient matrix, and the numerator is the determinant of the matrix formed by replacing the column corresponding to the variable (being solved for) with the column representing the constants. For example, the solution for x_3 in the system

$$a_{11}x_1 + a_{12}x_2 + a_{13}x_3 = b_1$$
$$a_{21}x_1 + a_{22}x_2 + a_{23}x_3 = b_2$$
$$a_{31}x_1 + a_{32}x_2 + a_{33}x_3 = b_3$$

is given by

$$x_3 = \frac{|A_3|}{|A|} = \frac{\begin{vmatrix} a_{11} & a_{12} & b_1 \\ a_{21} & a_{22} & b_2 \\ a_{31} & a_{32} & b_3 \end{vmatrix}}{\begin{vmatrix} a_{11} & a_{12} & a_{13} \\ a_{21} & a_{22} & a_{23} \\ a_{31} & a_{32} & a_{33} \end{vmatrix}}.$$

CRAMER'S RULE

If a system of n linear equations in n variables has a coefficient matrix A with a *nonzero* determinant $|A|$, then the solution of the system is given by

$$x_1 = \frac{|A_1|}{|A|}, \quad x_2 = \frac{|A_2|}{|A|}, \quad \ldots, \quad x_n = \frac{|A_n|}{|A|},$$

where the ith column of A_i is the column of constants in the system of equations. If the coefficient matrix is zero, then the system has either no solution *or* infinitely many solutions.

EXAMPLE 2   Using Cramer's Rule for a 3 × 3 System

Use Cramer's Rule to solve the following system of linear equations.

$$\begin{aligned}
-x + 2y - 3z &= 1 \\
2x \quad\quad + z &= 0 \\
3x - 4y + 4z &= 2
\end{aligned}$$

SOLUTION

Begin by finding the determinant of the coefficient matrix.

$$D = \begin{vmatrix} -1 & 2 & -3 \\ 2 & 0 & 1 \\ 3 & -4 & 4 \end{vmatrix} = 10$$

Because this determinant is not zero, you can apply Cramer's Rule.

$$x = \frac{D_x}{D} = \frac{\begin{vmatrix} 1 & 2 & -3 \\ 0 & 0 & 1 \\ 2 & -4 & 4 \end{vmatrix}}{10} = \frac{8}{10} = \frac{4}{5}$$

REMARK   When using Cramer's Rule, remember that the method *does not* apply if the determinant of the coefficient matrix is zero.

$$y = \frac{D_y}{D} = \frac{\begin{vmatrix} -1 & 1 & -3 \\ 2 & 0 & 1 \\ 3 & 2 & 4 \end{vmatrix}}{10} = \frac{-15}{10} = -\frac{3}{2}$$

$$z = \frac{D_z}{D} = \frac{\begin{vmatrix} -1 & 2 & 1 \\ 2 & 0 & 0 \\ 3 & -4 & 2 \end{vmatrix}}{10} = \frac{-16}{10} = -\frac{8}{5}$$

Therefore, the solution is $\left(\frac{4}{5}, -\frac{3}{2}, -\frac{8}{5}\right)$. Check this solution in the original system of equations.

REMARK   To see the benefit of the "determinant formula for area," you should try finding the area of the triangle in Example 3 using the standard formula: area = $\frac{1}{2}$(base)(height).

Area of a Triangle

Throughout this chapter you have studied several applications of matrices and determinants that involve systems of linear equations. We now look at some *other* types of applications involving determinants and matrices.

The first gives a formula for finding the area of a triangle whose vertices are given by three points in a rectangular coordinate system.

AREA OF A TRIANGLE

The area of a triangle with vertices (x_1, y_1), (x_2, y_2), and (x_3, y_3) is given by

$$\text{Area} = \pm \frac{1}{2} \begin{vmatrix} x_1 & y_1 & 1 \\ x_2 & y_2 & 1 \\ x_3 & y_3 & 1 \end{vmatrix},$$

where the symbol ($\pm$) indicates that the appropriate sign should be chosen to yield a positive area.

EXAMPLE 3   Finding the Area of a Triangle

Find the area of the triangle whose vertices are $(1, 0)$, $(2, 2)$, and $(4, 3)$, as shown in Figure 8.1.

SOLUTION

Choose $(x_1, y_1) = (1, 0)$, $(x_2, y_2) = (2, 2)$, and $(x_3, y_3) = (4, 3)$. To find the area of the triangle, evaluate the "area determinant" as follows.

$$\begin{vmatrix} x_1 & y_1 & 1 \\ x_2 & y_2 & 1 \\ x_3 & y_3 & 1 \end{vmatrix} = \begin{vmatrix} 1 & 0 & 1 \\ 2 & 2 & 1 \\ 4 & 3 & 1 \end{vmatrix} = -3$$

Using this value, you can conclude that the area of the triangle is

$$\text{Area} = -\frac{1}{2} \begin{vmatrix} 1 & 0 & 1 \\ 2 & 2 & 1 \\ 4 & 3 & 1 \end{vmatrix} = -\frac{1}{2}(-3) = \frac{3}{2}.$$

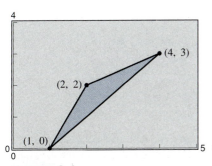

FIGURE 8.1

Lines in the Plane

Suppose that the three points in Example 3 had been on the same line. What would have happened had you applied the area formula to three such points? The answer is that the determinant would have been zero. Consider for instance, the three collinear points $(0, 1)$, $(2, 2)$, and $(4, 3)$, as shown in Figure 8.2. The area of the "triangle" that has these three points as vertices is

$$\frac{1}{2}\begin{vmatrix} 0 & 1 & 1 \\ 2 & 2 & 1 \\ 4 & 3 & 1 \end{vmatrix} = \frac{1}{2}\left(0\begin{vmatrix} 2 & 1 \\ 3 & 1 \end{vmatrix} - 1\begin{vmatrix} 2 & 1 \\ 4 & 1 \end{vmatrix} + 1\begin{vmatrix} 2 & 2 \\ 4 & 3 \end{vmatrix}\right)$$

$$= \frac{1}{2}[0(-1) - 1(-2) + 1(-2)] = 0.$$

This result is generalized as follows.

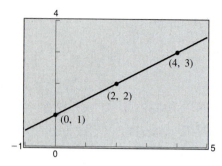

FIGURE 8.2

TEST FOR COLLINEAR POINTS

Three points (x_1, y_1), (x_2, y_2), and (x_3, y_3) are collinear (lie on the same line) if and only if

$$\begin{vmatrix} x_1 & y_1 & 1 \\ x_2 & y_2 & 1 \\ x_3 & y_3 & 1 \end{vmatrix} = 0.$$

EXAMPLE 4   Testing for Collinear Points

Determine whether the points $(-2, -2)$, $(1, 1)$, and $(7, 5)$ lie on the same line. (See Figure 8.3.)

SOLUTION

Letting $(x_1, y_1) = (-2, -2)$, $(x_2, y_2) = (1, 1)$, and $(x_3, y_3) = (7, 5)$, you have

$$\begin{vmatrix} x_1 & y_1 & 1 \\ x_2 & y_2 & 1 \\ x_3 & y_3 & 1 \end{vmatrix} = \begin{vmatrix} -2 & -2 & 1 \\ 1 & 1 & 1 \\ 7 & 5 & 1 \end{vmatrix}$$

$$= -2\begin{vmatrix} 1 & 1 \\ 5 & 1 \end{vmatrix} - (-2)\begin{vmatrix} 1 & 1 \\ 7 & 1 \end{vmatrix} + 1\begin{vmatrix} 1 & 1 \\ 7 & 5 \end{vmatrix}$$

$$= -2(-4) - (-2)(-6) + 1(-2)$$

$$= -6.$$

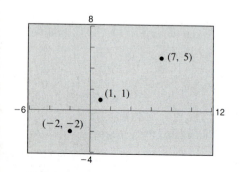

FIGURE 8.3

Because the value of this determinant is *not* zero, you can conclude that the three points do not lie on the same line.

The test for collinear points can be adapted to another use. That is, if you are given two points in a rectangular coordinate system, then you can find the equation of the line passing through the two points as follows.

TWO-POINT FORM OF THE EQUATION OF A LINE

An equation of the line passing through the distinct points (x_1, y_1) and (x_2, y_2) is given by

$$\begin{vmatrix} x & y & 1 \\ x_1 & y_1 & 1 \\ x_2 & y_2 & 1 \end{vmatrix} = 0.$$

EXAMPLE 5   Finding an Equation of a Line

Find an equation of the line passing through the two points $(2, 4)$ and $(-1, 3)$, as shown in Figure 8.4.

SOLUTION

Applying the determinant formula for the equation of the line passing through these two points produces

$$\begin{vmatrix} x & y & 1 \\ 2 & 4 & 1 \\ -1 & 3 & 1 \end{vmatrix} = 0.$$

To evaluate this determinant, expand by cofactors along the first row to obtain the following.

$$x\begin{vmatrix} 4 & 1 \\ 3 & 1 \end{vmatrix} - y\begin{vmatrix} 2 & 1 \\ -1 & 1 \end{vmatrix} + 1\begin{vmatrix} 2 & 4 \\ -1 & 3 \end{vmatrix} = x - 3y + 10 = 0$$

Therefore, an equation of the line is

$$x - 3y + 10 = 0.$$

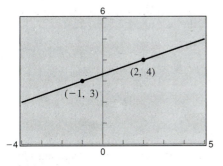

FIGURE 8.4

Cryptography

A **cryptogram** is a message written according to a secret code. (The Greek word "kryptos" means "hidden.") We describe next a method for using matrix multiplication to **encode** and **decode** messages.

Begin by assigning a number to each letter in the alphabet (with 0 assigned to a blank space) as follows.

0 = _	9 = I	18 = R
1 = A	10 = J	19 = S
2 = B	11 = K	20 = T
3 = C	12 = L	21 = U
4 = D	13 = M	22 = V
5 = E	14 = N	23 = W
6 = F	15 = O	24 = X
7 = G	16 = P	25 = Y
8 = H	17 = Q	26 = Z

Then the message is converted to numbers and partitioned into **uncoded row matrices,** each having n entries, as demonstrated in Example 6.

EXAMPLE 6   Forming Uncoded Row Matrices

Write the uncoded row matrices of order 1×3 for the message MEET ME MONDAY.

SOLUTION

Partitioning the message (including blank spaces, but ignoring other punctuation) into groups of three produces the following uncoded row matrices.

$$[13 \quad 5 \quad 5] \, [20 \quad 0 \quad 13] \, [5 \quad 0 \quad 13] \, [15 \quad 14 \quad 4] \, [1 \quad 25 \quad 0]$$
$$\text{M} \quad \text{E} \quad \text{E} \quad \text{T} \quad \_ \quad \text{M} \quad \text{E} \quad \_ \quad \text{M} \quad \text{O} \quad \text{N} \quad \text{D} \quad \text{A} \quad \text{Y} \quad \_$$

Note that a blank space is used to fill out the last uncoded row matrix.

To encode a message, choose an $n \times n$ invertible matrix A and multiply the uncoded row matrices (on the right) by A to obtain **coded row matrices.** This process is demonstrated in Example 7.

EXAMPLE 7   Encoding a Message

Use the following matrix to encode the message MEET ME MONDAY.

$$A = \begin{bmatrix} 1 & -2 & 2 \\ -1 & 1 & 3 \\ 1 & -1 & -4 \end{bmatrix}$$

SOLUTION

The coded row matrices are obtained by multiplying each of the uncoded row matrices found in Example 6 by the matrix A as follows.

Uncoded Row Matrix	Encoding Matrix A	Coded Row Matrix

$$[13 \quad 5 \quad 5] \begin{bmatrix} 1 & -2 & 2 \\ -1 & 1 & 3 \\ 1 & -1 & -4 \end{bmatrix} = [13 \quad -26 \quad 21]$$

$$[20 \quad 0 \quad 13] \begin{bmatrix} 1 & -2 & 2 \\ -1 & 1 & 3 \\ 1 & -1 & -4 \end{bmatrix} = [33 \quad -53 \quad -12]$$

$$[5 \quad 0 \quad 13] \begin{bmatrix} 1 & -2 & 2 \\ -1 & 1 & 3 \\ 1 & -1 & -4 \end{bmatrix} = [18 \quad -23 \quad -42]$$

$$[15 \quad 14 \quad 4] \begin{bmatrix} 1 & -2 & 2 \\ -1 & 1 & 3 \\ 1 & -1 & -4 \end{bmatrix} = [5 \quad -20 \quad 56]$$

$$[1 \quad 25 \quad 0] \begin{bmatrix} 1 & -2 & 2 \\ -1 & 1 & 3 \\ 1 & -1 & -4 \end{bmatrix} = [-24 \quad 23 \quad 77]$$

Thus, the sequence of coded row matrices is

$$[13 \,-26 \, 21][33 \,-53 \,-12][18 \,-23 \,-42][5 \,-20 \, 56][-24 \, 23 \, 77].$$

Finally, removing the matrix notation produces the following cryptogram.

$$13 \,-26 \, 21 \, 33 \,-53 \,-12 \, 18 \,-23 \,-42 \, 5 \,-20 \, 56 \,-24 \, 23 \, 77$$

For those who do not know the matrix A, decoding the cryptogram found in Example 7 is difficult. But for an authorized receiver who knows the matrix A, decoding is simple. The receiver need only multiply the coded row matrices by A^{-1} to retrieve the uncoded row matrices.

EXAMPLE 8   **Decoding a Message**

Use the inverse of the matrix

$$A = \begin{bmatrix} 1 & -2 & 2 \\ -1 & 1 & 3 \\ 1 & -1 & -4 \end{bmatrix}$$

to decode the cryptogram

13 −26 21 33 −53 −12 18 −23 −42 5 −20 56 −24 23 77.

SOLUTION

Begin by finding A^{-1}.

$$A^{-1} = \begin{bmatrix} -1 & -10 & -8 \\ -1 & -6 & -5 \\ 0 & -1 & -1 \end{bmatrix}$$

Now, to decode the message, partition the message into groups of three to form the coded row matrices

$$[13 \ -26 \ 21][33 \ -53 \ -12][18 \ -23 \ -42][5 \ -20 \ 56][-24 \ 23 \ 77].$$

Then, multiply each coded row matrix by A^{-1} (on the right) to obtain the decoded row matrices.

Coded Row Matrix	Decoding Matrix A^{-1}	Decoded Row Matrix
$[13 \quad -26 \quad -21]$	$\begin{bmatrix} -1 & -10 & -8 \\ -1 & -6 & -5 \\ 0 & -1 & -1 \end{bmatrix}$	$= [13 \ 5 \ 5]$
$[33 \quad -53 \quad -12]$	$\begin{bmatrix} -1 & -10 & -8 \\ -1 & -6 & -5 \\ 0 & -1 & -1 \end{bmatrix}$	$= [20 \ 0 \ 13]$
$[18 \quad -23 \quad -42]$	$\begin{bmatrix} -1 & -10 & -8 \\ -1 & -6 & -5 \\ 0 & -1 & -1 \end{bmatrix}$	$= [5 \ 0 \ 13]$
$[5 \quad -20 \quad 56]$	$\begin{bmatrix} -1 & -10 & -8 \\ -1 & -6 & -5 \\ 0 & -1 & -1 \end{bmatrix}$	$= [15 \ 14 \ 4]$
$[-24 \quad 23 \quad 77]$	$\begin{bmatrix} -1 & -10 & -8 \\ -1 & -6 & -5 \\ 0 & -1 & -1 \end{bmatrix}$	$= [1 \ 25 \ 0]$

Thus, the sequence of decoded row matrices is

$$[13 \quad 5 \quad 5][20 \quad 0 \quad 13][5 \quad 0 \quad 13][15 \quad 14 \quad 4][1 \quad 25 \quad 0]$$

and the message is as follows.

$$[13 \quad 5 \quad 5][20 \quad 0 \quad 13] [5 \quad 0 \quad 13][15 \quad 14 \quad 4] [1 \quad 25 \quad 0]$$
$$\text{M} \quad \text{E} \quad \text{E} \quad \text{T} \; \_ \; \text{M} \; \text{E} \; \_ \; \text{M} \; \text{O} \; \text{N} \; \text{D} \; \text{A} \; \text{Y} \; \_$$

DISCUSSION PROBLEM

....

COMPARING TECHNIQUES

In Chapters 7 and 8, you have looked at several techniques for solving a system of linear equations.

(a) Elimination method using equations (Section 7.3)

(b) Gaussian elimination with back-substitution (Section 8.1)

(c) Gauss-Jordan elimination (Section 8.1)

(d) Inverse matrix method (Section 8.3)

(e) Cramer's Rule (Section 8.5)

Write a short paper describing the advantages and disadvantages of each method.

WARM-UP

....

The following warm-up exercises involve skills that were covered in earlier sections. You will use these skills in the exercise set for this section.

In Exercises 1–4, solve the system of equations using Gaussian elimination with back-substitution or Gauss-Jordan elimination.

1. $\begin{aligned} x - 3y &= -2 \\ x + y &= 2 \end{aligned}$ **2.** $\begin{aligned} -x + 3y &= 5 \\ 4x - y &= 2 \end{aligned}$

3. $\begin{aligned} x + 2y - z &= 7 \\ -y - z &= 4 \\ 4x \quad\;\; - z &= 16 \end{aligned}$ **4.** $\begin{aligned} 3x \quad\;\; + 6z &= 0 \\ -2x + y \quad\;\; &= 5 \\ y + 2z &= 3 \end{aligned}$

In Exercises 5–10, evaluate the determinant.

5. $\begin{vmatrix} 10 & 8 \\ -6 & -4 \end{vmatrix}$ **6.** $\begin{vmatrix} -7 & 14 \\ 2 & 3 \end{vmatrix}$

7. $\begin{vmatrix} 1 & 0 & -2 \\ 0 & 1 & 0 \\ -2 & 0 & 1 \end{vmatrix}$ **8.** $\begin{vmatrix} 0 & 3 & 1 \\ 5 & -2 & 1 \\ 1 & 6 & 1 \end{vmatrix}$

9. $\begin{vmatrix} 0 & -2 & 1 & 0 \\ -2 & 0 & 5 & -1 \\ 1 & 5 & 0 & 2 \\ 0 & -1 & 2 & 0 \end{vmatrix}$ **10.** $\begin{vmatrix} 1 & 0 & -2 & 1 \\ 0 & 2 & 5 & 1 \\ 3 & -3 & 2 & 1 \\ 0 & 0 & 4 & 1 \end{vmatrix}$

SECTION 8.5 · EXERCISES

In Exercises 1–10, use Cramer's Rule to solve (if possible) the system of equations.

1. $x + 2y = 5$
$-x + y = 1$

2. $2x - y = -10$
$3x + 2y = -1$

3. $3x + 4y = -2$
$5x + 3y = 4$

4. $18x + 12y = 13$
$30x + 24y = 23$

5. $20x + 8y = 11$
$12x - 24y = 21$

6. $13x - 6y = 17$
$26x - 12y = 8$

7. $-0.4x + 0.8y = 1.6$
$2x - 4y = 5$

8. $-0.4x + 0.8y = 1.6$
$0.2x + 0.3y = 2.2$

9. $3x + 6y = 5$
$6x + 14y = 11$

10. $3x + 2y = 1$
$2x + 10y = 6$

In Exercises 11–20, use Cramer's Rule to solve (if possible) the system. Use a graphing utility to evaluate the determinants.

11. $4x - y + z = -5$
$2x + 2y + 3z = 10$
$5x - 2y + 6z = 1$

12. $4x - 2y + 3z = -2$
$2x + 2y + 5z = 16$
$8x - 5y - 2z = 4$

13. $3x + 4y + 4z = 11$
$4x - 4y + 6z = 11$
$6x - 6y = 3$

14. $14x - 21y - 7z = 10$
$-4x + 2y - 2z = 4$
$56x - 21y + 7z = 5$

15. $3x + 3y + 5z = 1$
$3x + 5y + 9z = 2$
$5x + 9y + 17z = 4$

16. $2x + 3y + 5z = 4$
$3x + 5y + 9z = 7$
$5x + 9y + 17z = 13$

17. $5x - 3y + 2z = 2$
$2x + 2y - 3z = 3$
$x - 7y + 8z = -4$

18. $3x + 2y + 5z = 4$
$4x - 3y - 4z = 1$
$-8x + 2y + 3z = 0$

19. $7x - 3y + 2w = 41$
$-2x + y - w = -13$
$4x + z - 2w = 12$
$-x + y - w = -8$

20. $2x + 5y + w = 11$
$x + 4y + 2z - 2w = -7$
$2x - 2y + 5z + w = 3$
$x - 3w = 1$

21. *Circuit Analysis*   Consider the circuit in the figure. The currents I_1, I_2, and I_3 in amperes are given by the solution to the system of linear equations

$-4I_1 + 10I_3 = 5$
$5I_2 + 10I_3 = 70$
$I_1 - I_2 + I_3 = 0.$

Use Cramer's Rule to find the three currents.

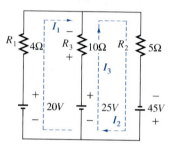

FIGURE FOR 21

22. *Circuit Analysis*   Consider the circuit in the figure. The currents I_1, I_2, and I_3 in amperes are given by the solution to the system of linear equations

$4I_1 + 8I_3 = 2$
$2I_2 + 8I_3 = 6$
$I_1 + I_2 - I_3 = 0.$

Use Cramer's Rule to find the three currents.

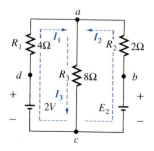

FIGURE FOR 22

23. *Maximum Social Security Contribution* The maximum Social Security contributions for an employee between 1981 and 1989 are shown in the figure. (The figure shows the amount contributed by the *employee*. This amount is matched by the employer.) The least squares regression line $y = a + bt$ for these data is found by solving the system

$9a + 45b = 24.983$

$45a + 285b = 137.012,$

where y is the contribution in thousands of dollars and t is the calendar year, with $t = 1$ corresponding to 1981. Use Cramer's Rule to solve this system, and use the result to approximate the maximum Social Security contribution in 1992. (*Source:* U.S. Social Security Administration)

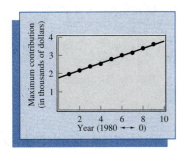

FIGURE FOR 23

24. *Pulley System* A system of pulleys that is assumed frictionless and without mass is loaded with 192-pound and 64-pound weights (see figure). The tensions t_1 and t_2 in the ropes and the acceleration a of the 64-pound weight are found by solving the system

$t_1 - 2t_2 \qquad = \quad 0$

$t_1 \qquad - 3a = 192$

$\qquad t_2 + 2a = \quad 64,$

where t_1 and t_2 are measured in pounds and a is in feet per second squared. Use Cramer's Rule to find the acceleration a of the system.

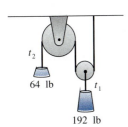

FIGURE FOR 24

In Exercises 25–34, use a determinant to find the area of the triangle with given vertices.

25. $(0, 0), (3, 1), (1, 5)$

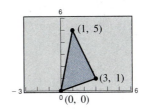

26. $(0, 0), (5, -2), (4, 5)$

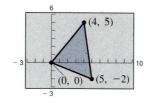

27. $(-2, -3), (2, -3), (0, 4)$

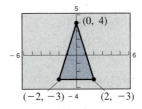

28. $(-2, 1), (3, -1), (1, 6)$

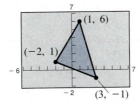

29. $\left(0, \frac{1}{2}\right), \left(\frac{5}{2}, 0\right), (4, 3)$

30. $(-4, -5), (6, -1), (6, 10)$

31. $(-2, 4), (2, 3), (-1, 5)$

32. $(0, -2), (-1, 4), (3, 5)$

33. $(-3, 5), (2, 6), (3, -5)$

34. $(-2, 4), (1, 5), (3, -2)$

35. *Area of a Region* A large region of forest has been infested with gypsy moths. The region is roughly triangular, as shown in the figure. From the northernmost vertex *A* of the region, the distance to vertex *B* is 25 miles south and 10 miles east, and the distance to vertex *C* is 20 miles south and 28 miles east. Approximate the number of square miles in this region.

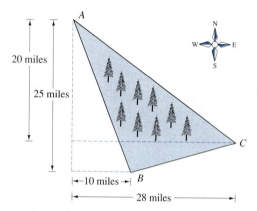

FIGURE FOR 35

36. *Area of a Region* Suppose you purchased a triangular tract of land, as shown in the figure. To estimate the number of square feet in the tract, you start at one vertex and walk 65 feet east and 50 feet north to the second vertex. Then, from the second vertex you walk 85 feet west and 30 feet north to the third vertex. How many square feet are there in the tract of land?

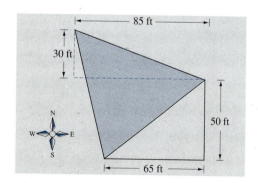

FIGURE FOR 36

In Exercises 37–42, use a determinant to ascertain if the points are collinear.

37. $(3, -1)$, $(0, -3)$, $(12, 5)$

38. $(-3, -5)$, $(6, 1)$, $(10, 2)$

39. $\left(2, -\frac{1}{2}\right)$, $(-4, 4)$, $(6, -3)$

40. $(0, 1)$, $(4, -2)$, $(-8, 7)$

41. $(0, 2)$, $(1, 2.4)$, $(-1, 1.6)$

42. $(2, 3)$, $(3, 3.5)$, $(-1, 2)$

In Exercises 43–48, use a determinant to find an equation of the line through the given points (x_1, y_1) and (x_2, y_2).

43. $(0, 0)$, $(5, 3)$ **44.** $(0, 0)$, $(-2, 2)$

45. $(-4, 3)$, $(2, 1)$ **46.** $(10, 7)$, $(-2, -7)$

47. $\left(-\frac{1}{2}, 3\right)$, $\left(\frac{5}{2}, 1\right)$ **48.** $\left(\frac{2}{3}, 4\right)$, $(6, 12)$

In Exercises 49–52, write a cryptogram for each message using the matrix

$$A = \begin{bmatrix} 1 & 2 & 2 \\ 3 & 7 & 9 \\ -1 & -4 & -7 \end{bmatrix}.$$

49. LANDING SUCCESSFUL

50. BEAM ME UP SCOTTY

51. HAPPY BIRTHDAY

52. OPERATION OVERLOAD

In Exercises 53 and 54, decode the cryptogram by using the inverse of matrix *A* from Exercises 49–52.

53. 20 17 −15 −12 −56 −104 1 −25 −65
62 143 181

54. 13 −9 −59 61 112 106 −17 −73 −131
11 24 29 65 144 172

CHAPTER 8 · REVIEW EXERCISES

In Exercises 1–8, solve the system of equations. (*Note:* For some exercises it may be advantageous to use the matrix capabilities of your graphing utility.)

1. $\begin{aligned} 5x + 4y &= 2 \\ -x + y &= -22 \end{aligned}$

2. $\begin{aligned} 2x - 5y &= 2 \\ 3x - 7y &= 1 \end{aligned}$

3. $\begin{aligned} 0.2x - 0.1y &= 0.07 \\ 0.4x - 0.5y &= -0.01 \end{aligned}$

4. $\begin{aligned} 2x + y &= 0.3 \\ 3x - y &= -1.3 \end{aligned}$

5. $\begin{aligned} 2x + 3y + 3z &= 3 \\ 6x + 6y + 12z &= 13 \\ 12x + 9y - z &= 2 \end{aligned}$

6. $\begin{aligned} 4x + 4y + 4z &= 5 \\ 4x - 2y - 8z &= 1 \\ 5x + 3y + 8z &= 6 \end{aligned}$

7. $\begin{aligned} x + 2y + + w &= 3 \\ -3y + 3z &= 0 \\ 4x + 4y + z + 2w &= 0 \\ 2x + z &= 3 \end{aligned}$

8. $\begin{aligned} 3x + 21y - 29z &= -1 \\ 2x + 15y - 21z &= 0 \end{aligned}$

In Exercises 9–16, perform the indicated matrix operations (if possible).

9. $\begin{bmatrix} 2 & 1 & 0 \\ 0 & 5 & -4 \end{bmatrix} - 3\begin{bmatrix} 5 & 3 & -6 \\ 0 & -2 & 5 \end{bmatrix}$

10. $-2\begin{bmatrix} 1 & 2 \\ 5 & -4 \\ 6 & 0 \end{bmatrix} + 8\begin{bmatrix} 7 & 1 \\ 1 & 2 \\ 1 & 4 \end{bmatrix}$

11. $\begin{bmatrix} 1 & 2 \\ 5 & -4 \\ 6 & 0 \end{bmatrix}\begin{bmatrix} 6 & -2 & 8 \\ 4 & 0 & 0 \end{bmatrix}$

12. $\begin{bmatrix} 1 & 5 & 6 \\ 2 & -4 & 0 \end{bmatrix}\begin{bmatrix} 6 & -2 & 8 \\ 4 & 0 & 0 \end{bmatrix}$

13. $\begin{bmatrix} 1 & 5 & 6 \\ 2 & -4 & 0 \end{bmatrix}\begin{bmatrix} 6 & 4 \\ -2 & 0 \\ 8 & 0 \end{bmatrix}$

14. $\begin{bmatrix} 4 \\ 6 \end{bmatrix}\begin{bmatrix} 6 & -2 \end{bmatrix}$

15. $\begin{bmatrix} 1 & 3 & 2 \\ 0 & 2 & -4 \\ 0 & 0 & 3 \end{bmatrix}\begin{bmatrix} 4 & -3 & 2 \\ 0 & 3 & -1 \\ 0 & 0 & 2 \end{bmatrix}$

16. $\begin{bmatrix} 2 & 1 \\ 6 & 0 \end{bmatrix}\left(\begin{bmatrix} 4 & 2 \\ -3 & 1 \end{bmatrix} + \begin{bmatrix} -2 & 4 \\ 0 & 4 \end{bmatrix}\right)$

In Exercises 17–20, solve for X given

$$A = \begin{bmatrix} -4 & 0 \\ 1 & -5 \\ -3 & 2 \end{bmatrix} \text{ and } B = \begin{bmatrix} 1 & 2 \\ -2 & 1 \\ 4 & 4 \end{bmatrix}.$$

17. $X = 3A - 2B$

18. $6X = 4A + 3B$

19. $3X + 2A = B$

20. $2A - 5B = 3X$

21. Write the system of linear equations represented by the matrix equation

$$\begin{bmatrix} 5 & 4 \\ -1 & 1 \end{bmatrix}\begin{bmatrix} x \\ y \end{bmatrix} = \begin{bmatrix} 2 \\ -22 \end{bmatrix}.$$

22. Write the matrix equation $AX = B$ for the following system of linear equations.

$$\begin{aligned} 2x + 3y + z &= 10 \\ 2x - 3y - 3z &= 22 \\ 4x - 2y + 3z &= -2 \end{aligned}$$

In Exercises 23–26, find the inverse of the matrix (if it exists).

23. $\begin{bmatrix} 2 & 6 \\ 3 & -6 \end{bmatrix}$

24. $\begin{bmatrix} 3 & -10 \\ 4 & 2 \end{bmatrix}$

25. $\begin{bmatrix} 2 & 0 & 3 \\ -1 & 1 & 1 \\ 2 & -2 & 1 \end{bmatrix}$

26. $\begin{bmatrix} 1 & 4 & 6 \\ 2 & -3 & 1 \\ -1 & 18 & 16 \end{bmatrix}$

In Exercises 27–30, evaluate the determinant.

27. $\begin{vmatrix} 50 & -30 \\ 10 & 5 \end{vmatrix}$

28. $\begin{vmatrix} 8 & 5 \\ 2 & -4 \end{vmatrix}$

29. $\begin{vmatrix} 3 & 0 & -4 & 0 \\ 0 & 8 & 1 & 2 \\ 6 & 1 & 8 & 2 \\ 0 & 3 & -4 & 1 \end{vmatrix}$

30. $\begin{vmatrix} -5 & 6 & 0 & 0 \\ 0 & 1 & -1 & 2 \\ -3 & 4 & -5 & 1 \\ 1 & 6 & 0 & 3 \end{vmatrix}$

In Exercises 31–38, solve (if possible) the system of linear equations using (a) the inverse of the coefficient matrix and (b) Cramer's Rule. (*Note:* For some exercises it may be advantageous to use the matrix capabilities of your graphing utility.)

31. $x + 2y = -1$
$3x + 4y = -5$

32. $x + 3y = 23$
$-6x + 2y = -18$

33. $-3x - 3y - 4z = 2$
$y + z = -1$
$4x + 3y + 4z = -1$

34. $x - 3y - 2z = 8$
$-2x + 7y + 3z = -19$
$x - y - 3z = 3$

35. $x + 3y + 2z = 2$
$-2x - 5y - z = 10$
$2x + 4y = -12$

36. $2x + 4y = -12$
$3x + 4y - 2z = -14$
$-x + y + 2z = -6$

37. $2x + 3y - 4z = 1$
$x - y + 2z = -4$
$3x + 7y - 10z = 0$

38. $-x + y + z = 6$
$4x - 3y + z = 20$
$2z - y + 3z = 8$

In Exercises 39–42, use a determinant to find the area of the triangle with the given vertices.

39. $(1,0), (5,0), (5,8)$

40. $(-4,0), (4,0), (0,6)$

41. $(1,2), (4,-5), (3,2)$

42. $\left(\frac{3}{2},1\right), \left(4,-\frac{1}{2}\right), (4,2)$

In Exercises 43–46, use a determinant to find an equation of the line through the given points.

43. $(-4,0), (4,4)$

44. $(2,5), (6,-1)$

45. $\left(-\frac{5}{2},3\right), \left(\frac{7}{2},1\right)$

46. $(-0.8,0.2), (0.7,3.2)$

47. *Mixture Problem* A florist wants to arrange a dozen flowers consisting of two varieties—carnations and roses. Carnations cost $0.75 each and roses cost $1.50 each. How many of each should the florist use in order for the arrangement to cost $12.00?

48. *Mixture Problem* One hundred gallons of a 60% acid solution are obtained by mixing a 75% solution with a 50% solution. How many gallons of each must be used to obtain the desired mixture?

49. *Fitting a Line to Data* Find the least squares regression line $y = ax + b$ for the points
$(x_1,y_1), (x_2,y_2), \ldots, (x_n,y_n)$.
To find the line, solve the following system of linear equations for a and b.
$5b + 10a = 17.8$
$10b + 30a = 45.7$

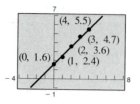

FIGURE FOR 49

50. *Fitting a Parabola to Data* Find the least squares regression parabola $y = ax^2 + bx + c$ for the points
$(x_1,y_1), (x_2,y_2), \ldots, (x_n,y_n)$.
To find the parabola, solve the following system of linear equations for a, b, and c.
$5c + 10a = 9.1$
$10b = 8.0$
$10c + 34a = 19.8$

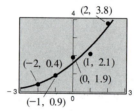

FIGURE FOR 50

51. *Fitting a Parabola to Three Points* Find an equation of the parabola $y = ax^2 + bx + c$ passing through the points $(-1, 2)$, $(0, 3)$, and $(1, 6)$.

52. *Break-Even Point* A small business invests $25,000 in equipment to produce a product. Each unit of the product costs $3.75 to produce and is sold for $5.25. How many items must be sold before the business breaks even?

53. If A is a 3×3 matrix and $|A| = 2$, what is the value of $|4A|$? Give the reason for your answer.

54. Verify that
$$\begin{vmatrix} a_{11} & a_{12} & a_{13} \\ a_{21} & a_{22} & a_{23} \\ a_{31} + c_1 & a_{32} + c_2 & a_{33} + c_3 \end{vmatrix}$$
$$= \begin{vmatrix} a_{11} & a_{12} & a_{13} \\ a_{21} & a_{22} & a_{23} \\ a_{31} & a_{32} & a_{33} \end{vmatrix} + \begin{vmatrix} a_{11} & a_{12} & a_{13} \\ a_{21} & a_{22} & a_{23} \\ c_1 & c_2 & c_3 \end{vmatrix}.$$

CUMULATIVE TEST FOR CHAPTERS 6–8

Take this test as you would take a test in class. After you are done, check your work against the answers in the back of the book.

1. Add and simplify: $\dfrac{1 + \cos \beta}{\sin \beta} + \dfrac{\sin \beta}{1 + \cos \beta}$

2. Find all the solutions to the equation $2 \cos^2 x - \cos x = 0$ in the interval $[0, 2\pi)$.

3. Given $\sin x = \frac{2}{3}$, find the exact value of $\sin(2x)$, where x is a first quadrant angle.

4. Evaluate $\cos 105° = \cos(135° - 30°)$ *without* the aid of a calculator.

5. Find the remaining sides and angles of each triangle.

   (a)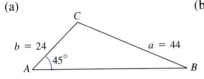   (b)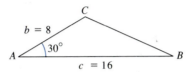

6. In a (square) baseball diamond with 90-foot sides, the pitcher's mound is 60 feet from home plate. How far is a runner from the pitcher's mound when the runner is halfway from third base to home?

7. Write the complex number $z = 2(1 - i)$ in trigonometric form and find z^3.

8. Solve the system by the method of substitution.

$$y = 3 - x^2$$
$$2(y - 2) = x - 1$$

9. Use a graphing utility to solve the linear system graphically.

$$x + 3y = -1$$
$$2x + 4y = 0$$

10. Use Gauss-Jordan elimination to solve the linear system.

$$x + 3y - 2z = -7$$
$$-2x + y - z = -5$$
$$4x + y + z = 3$$

11. Find the value of a such that the system is inconsistent.

$$ax - 8y = 9$$
$$3x + 4y = 0$$

12. Sketch a graph of the solution of the system of inequalities.

$$3x + 4y \geq 16$$
$$3x - 4y \leq 8$$
$$y \leq 4$$

13. Maximize the objective function $z = 3x + 2y$ subject to the following constraints:

$$x \geq 0$$
$$y \geq 0$$
$$x + 4y \leq 20$$
$$2x + y \leq 12$$

14. Find $2A - B$ given the following matrices.

$$A = \begin{bmatrix} 6 & -1 \\ 2 & 4 \\ -3 & 5 \end{bmatrix}, \quad B = \begin{bmatrix} 1 & 4 \\ -1 & 5 \\ 1 & 10 \end{bmatrix}$$

15. Find AB, if possible.

$$A = \begin{bmatrix} 4 & -3 \\ 2 & 1 \\ 5 & 0 \end{bmatrix}, \quad B = \begin{bmatrix} 3 & -2 \\ 1 & -3 \end{bmatrix}$$

16. Find the inverse (if it exists) of the matrix.

$$A = \begin{bmatrix} 1 & 2 & -1 \\ 3 & 7 & -10 \\ -5 & -7 & -15 \end{bmatrix}$$

17. Evaluate the determinant.

$$\begin{vmatrix} 1 & 1 & 1 \\ 2 & -1 & -2 \\ 1 & -2 & -1 \end{vmatrix}$$

18. Use determinants to find the area of the triangle with vertices $(0, 0)$, $(6, 2)$, and $(8, 10)$.

CHAPTER 9

SEQUENCES, PROBABILITY, AND STATISTICS

9.1 SEQUENCES AND SUMMATION NOTATION

Sequences / Factorial Notation / Summation Notation / The Sum of an Infinite Sequence / Application

Sequences

A sequence is a *function* whose domain is the set of positive integers. Sequences are usually represented, however, by subscript notation, rather than the standard function notation. For instance, the terms of the sequence

$$f(1), f(2), f(3), f(4), \ldots, f(n), \ldots$$

are usually written as

$$a_1, a_2, a_3, a_4, \ldots, a_n, \ldots.$$

Note that subscripts make up the domain of the sequence, and they serve to identify the location of a term within the sequence. For instance, a_4 is the fourth term of the sequence and a_n is called the **nth term** of the sequence. The entire sequence is sometimes denoted by the short form $\{a_n\}$.

DEFINITION OF A SEQUENCE

An **infinite sequence** $\{a_n\}$ is a function whose domain is the set of positive integers. The function values

$$a_1, a_2, a_3, a_4, \ldots, a_n, \ldots$$

are called the **terms** of the sequence. If the domain of the function consists of the first n positive integers only, then the sequence is called a **finite sequence.**

On occasion it is convenient to begin subscripting a sequence with 0 instead of 1 so that the terms of the sequence become

$$a_0, a_1, a_2, a_3, a_4, \ldots, a_n, \ldots.$$

In such cases, a_n is still called the nth term of the sequence, even though it occupies the $(n + 1)$ position in the sequence.

EXAMPLE 1   Finding Terms in a Sequence

A. The first four terms of the sequence whose nth term is $a_n = 3n - 2$ are:

$$a_1 = 3(1) - 2 = 1$$
$$a_2 = 3(2) - 2 = 4$$
$$a_3 = 3(3) - 2 = 7$$
$$a_4 = 3(4) - 2 = 10.$$

B. The first four terms of the sequence whose nth term is $a_n = 3 + (-1)^n$ are:

$$a_1 = 3 + (-1)^1 = 3 - 1 = 2$$
$$a_2 = 3 + (-1)^2 = 3 + 1 = 4$$
$$a_3 = 3 + (-1)^3 = 3 - 1 = 2$$
$$a_4 = 3 + (-1)^4 = 3 + 1 = 4.$$

Technology Note _____

Try using your graphing utility to graph the sequences in Example 2. Try to duplicate the screens shown in the figure.

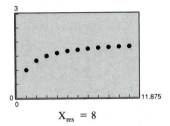

$X_{res} = 8$

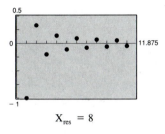

$X_{res} = 8$

EXAMPLE 2   Finding Terms in a Sequence

A. The first four terms of the sequence whose nth term is $a_n = \dfrac{2n}{1+n}$ are:

$$a_1 = \frac{2(1)}{1+1} = \frac{2}{2} = 1$$

$$a_2 = \frac{2(2)}{1+2} = \frac{4}{3}$$

$$a_3 = \frac{2(3)}{1+3} = \frac{6}{4} = \frac{3}{2}$$

$$a_4 = \frac{2(4)}{1+4} = \frac{8}{5}.$$

B. The first four terms of the sequence whose nth term is $a_n = \dfrac{(-1)^n}{2n-1}$ are:

$$a_1 = \frac{(-1)^1}{2(1)-1} = \frac{-1}{2-1} = -1$$

$$a_2 = \frac{(-1)^2}{2(2)-1} = \frac{1}{4-1} = \frac{1}{3}$$

$$a_3 = \frac{(-1)^3}{2(3)-1} = \frac{-1}{6-1} = -\frac{1}{5}$$

$$a_4 = \frac{(-1)^4}{2(4)-1} = \frac{1}{8-1} = \frac{1}{7}.$$

REMARK   Try finding the first four terms of the sequence whose nth term is

$$a_n = \frac{(-1)^{n+1}}{2n-1}.$$

How do they differ from the first four terms of the sequence in Example 2(B)?

It is important to realize that simply listing the first few terms is not sufficient to define a sequence—the nth term *must be given*. To see this, consider the following sequences, both of which have the same first three terms.

$$\frac{1}{2}, \frac{1}{4}, \frac{1}{8}, \frac{1}{16}, \ldots, \frac{1}{2^n}, \ldots$$

$$\frac{1}{2}, \frac{1}{4}, \frac{1}{8}, \frac{1}{15}, \ldots, \frac{6}{(n+1)(n^2-n+6)}, \ldots$$

When given the first few terms of a sequence, you can, however, be asked to find the *apparent* nth term.

EXAMPLE 3 Finding the Apparent *n*th Term of a Sequence

Find the apparent *n*th term of each sequence.

A. 1, 3, 5, 7, . . .   **B.** 2, 5, 10, 17, . . .   **C.** $\dfrac{2}{1}, \dfrac{3}{2}, \dfrac{4}{3}, \dfrac{5}{4}, \dots$

SOLUTION

A.   *n:* 1 2 3 4 . . . *n*
           ↓ ↓ ↓ ↓ . . . ↓
   Terms: 1 3 5 7 . . . a_n

Apparent pattern: Each term is 1 less than twice *n*, which implies that $a_n = 2n - 1$.

B.   *n:* 1 2 3 4 . . . *n*
           ↓ ↓ ↓ ↓ . . . ↓
   Terms: 2 5 10 17 . . . a_n

Apparent pattern: Each term is 1 more than the square of *n*, which implies that $a_n = n^2 + 1$.

C.   *n:* 1 2 3 4 . . . *n*
           ↓ ↓ ↓ ↓ . . . ↓
   Terms: $\dfrac{2}{1}$ $\dfrac{3}{2}$ $\dfrac{4}{3}$ $\dfrac{5}{2}$ . . . a_n

Apparent pattern: Each term has a denominator of *n* and a numerator that is 1 more than *n*, which implies that $a_n = (n + 1)/n$.

Factorial Notation

Some very important sequences in mathematics involve terms that are defined with special types of products called **factorials.**

> **DEFINITION OF FACTORIAL**
>
> If *n* is a positive integer, then **_n_ factorial** is defined by
>
> $$n! = 1 \cdot 2 \cdot 3 \cdot 4 \cdots (n - 1) \cdot n.$$
>
> As a special case, zero factorial is defined to be $0! = 1$.

Technology Note _____

Most graphing utilities have the ability to compute $n!$. Learn how to use this feature of your graphing utility, and then use it to confirm the results given here. How large a value of $n!$ will your graphing utility allow you to compute?

Here are some values of $n!$ for the first several nonnegative integers.

$$0! = 1 \qquad\qquad 3! = 1 \cdot 2 \cdot 3 = 6$$
$$1! = 1 \qquad\qquad 4! = 1 \cdot 2 \cdot 3 \cdot 4 = 24$$
$$2! = 1 \cdot 2 = 2 \qquad 5! = 1 \cdot 2 \cdot 3 \cdot 4 \cdot 5 = 120$$

The value of n does not have to be very large before the value of $n!$ becomes huge. For instance, $10! = 3,628,800$. Many calculators have a factorial key, denoted by $\boxed{x!}$.

Factorials follow the same conventions for order of operations as do exponents. For instance,

$$2n! = 2(n!) = 2(1 \cdot 2 \cdot 3 \cdot 4 \cdots n),$$

whereas $(2n)! = 1 \cdot 2 \cdot 3 \cdot 4 \cdots n \cdots (2n - 1)(2n)$.

EXAMPLE 4   **Finding Terms of a Sequence Involving Factorials**

List the first five terms of the sequence whose nth term is $a_n = 2^n/n!$. Begin with $n = 0$.

SOLUTION

$$a_0 = \frac{2^0}{0!} = \frac{1}{1} = 1 \qquad a_3 = \frac{2^3}{3!} = \frac{8}{6} = \frac{4}{3}$$

$$a_1 = \frac{2^1}{1!} = \frac{2}{1} = 2 \qquad a_4 = \frac{2^4}{4!} = \frac{16}{24} = \frac{2}{3}$$

$$a_2 = \frac{2^2}{2!} = \frac{4}{2} = 2$$

When working with fractions involving factorials, you will often find that the fractions can be reduced. Here are two examples.

$$\frac{n!}{(n-1)!} = \frac{1 \cdot 2 \cdot 3 \cdots (n-1) \cdot n}{1 \cdot 2 \cdot 3 \cdots (n-1)} = n$$

$$\frac{8!}{2! \cdot 6!} = \frac{1 \cdot 2 \cdot 3 \cdot 4 \cdot 5 \cdot 6 \cdot 7 \cdot 8}{1 \cdot 2 \cdot 1 \cdot 2 \cdot 3 \cdot 4 \cdot 5 \cdot 6} = \frac{7 \cdot 8}{2} = 28$$

Summation Notation

There is a convenient notation for the sum of the terms of a finite sequence. It is called **summation notation** or **sigma notation** because it involves the use of the uppercase Greek letter sigma, written as Σ.

DEFINITION OF SUMMATION NOTATION

The sum of the first n terms of a sequence is represented by

$$\sum_{i=1}^{n} a_i = a_1 + a_2 + a_3 + a_4 + \cdots + a_n,$$

where i is the **index of summation,** n is the **upper limit of summation,** and 1 is the **lower limit of summation.**

EXAMPLE 5 Summation Notation for Sums

REMARK In Example 5, note that the lower limit of a summation does not have to be 1. Also note that the index does not have to be the letter i. For instance, in part (B), the letter k is the index.

A. $\displaystyle\sum_{i=1}^{5} 3i = 3(1) + 3(2) + 3(3) + 3(4) + 3(5)$

$\qquad\qquad = 3(1 + 2 + 3 + 4 + 5)$

$\qquad\qquad = 3(15) = 45$

B. $\displaystyle\sum_{k=3}^{6} (1 + k^2) = (1 + 3^2) + (1 + 4^2) + (1 + 5^2) + (1 + 6^2)$

$\qquad\qquad\qquad = 10 + 17 + 26 + 37 = 90$

C. $\displaystyle\sum_{i=0}^{8} \frac{1}{i!} = \frac{1}{0!} + \frac{1}{1!} + \frac{1}{2!} + \frac{1}{3!} + \frac{1}{4!} + \frac{1}{5!} + \frac{1}{6!} + \frac{1}{7!} + \frac{1}{8!}$

$\qquad\quad = 1 + 1 + \frac{1}{2} + \frac{1}{6} + \frac{1}{24} + \frac{1}{120} + \frac{1}{720} + \frac{1}{5040} + \frac{1}{40,320}$

$\qquad\quad \approx 2.71828$

Technology Note

Some graphing utilities allow you to evaluate summations. For instance, on the TI-85 you can verify Example 5 (B) by going to the MATH MISC menu and entering

sum seq$(1 + x^2, x, 3, 6, 1)$

Did you obtain the correct answer? Verify part C of Example 5.

For this summation, note that the sum is very close to the irrational number $e \approx 2.718281828$. It can be shown that, as more terms of the sequence whose nth term is $1/n!$ are added, the sum becomes closer and closer to e.

PROPERTIES OF SUMS

1. $\displaystyle\sum_{i=1}^{n} ca_i = c\sum_{i=1}^{n} a_i$, c is any constant

2. $\displaystyle\sum_{i=1}^{n} (a_i + b_i) = \sum_{i=1}^{n} a_i + \sum_{i=1}^{n} b_i$

3. $\displaystyle\sum_{i=1}^{n} (a_i - b_i) = \sum_{i=1}^{n} a_i - \sum_{i=1}^{n} b_i$

Variations in the upper and lower limits of summation can produce quite different-looking summation notations for *the same sum*. For example, consider the following two sums.

$$\sum_{i=1}^{5} 3(2^i) = 3 \sum_{i=1}^{5} 2^i = 3(2^1 + 2^2 + 2^3 + 2^4 + 2^5)$$

$$\sum_{i=0}^{4} 3(2^{i+1}) = 3 \sum_{i=0}^{4} 2^{i+1} = 3(2^1 + 2^2 + 2^3 + 2^4 + 2^5)$$

The Sum of an Infinite Sequence

It is easy to see that the sum of the terms of any *finite* sequence must be a finite number. However, it may not be as easy to convince yourself that the sum of all the terms of an *infinite* sequence can also be a finite number. For instance, it can be shown that the sum of all of the terms of the sequence whose nth term is $1/2^n$ (with n beginning at 0) is

$$\sum_{n=0}^{\infty} \frac{1}{2^n} = 1 + \frac{1}{2} + \frac{1}{4} + \frac{1}{8} + \frac{1}{16} + \frac{1}{32} + \cdots = 2.$$

(See Figure 9.1.) This type of summation is called an **infinite series.**

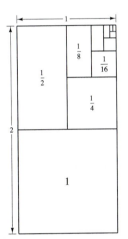

The total area is 2.

FIGURE 9.1

EXAMPLE 6 Finding the Sum of an Infinite Sequence

Find the sum of the infinite sequence.

$$\frac{3}{10^1}, \frac{3}{10^2}, \frac{3}{10^3}, \frac{3}{10^4}, \frac{3}{10^5}, \cdots$$

SOLUTION

$$\sum_{n=1}^{\infty} \frac{3}{10^n} = \frac{3}{10^1} + \frac{3}{10^2} + \frac{3}{10^3} + \frac{3}{10^4} + \frac{3}{10^5} + \cdots$$

$$= 0.3 + 0.03 + 0.003 + 0.0003 + 0.00003 + 0.000003 + \cdots$$

$$= 0.33333 \ldots$$

$$= \frac{1}{3}$$

Application

EXAMPLE 7   Population of the United States

From 1950 to 1989, the resident population of the United States can be approximated by the model

$$a_n = \sqrt{22{,}926 + 902.5n + 2.01n^2}, \qquad n = 0, 1, \ldots, 39,$$

where a_n is the population in millions and n represents the year, with $n = 0$ corresponding to 1950. (*Source:* U.S. Bureau of Census) Find the last five terms of this finite sequence.

SOLUTION

The last five terms of this finite sequence are as follows.

$$a_{35} = \sqrt{22{,}926 + 902.5(35) + 2.01(35^2)} \approx 238.7 \qquad \textit{1985 population}$$
$$a_{36} = \sqrt{22{,}926 + 902.5(36) + 2.01(36^2)} \approx 240.9 \qquad \textit{1986 population}$$
$$a_{37} = \sqrt{22{,}926 + 902.5(37) + 2.01(37^2)} \approx 243.0 \qquad \textit{1987 population}$$
$$a_{38} = \sqrt{22{,}926 + 902.5(38) + 2.01(38^2)} \approx 245.2 \qquad \textit{1988 population}$$
$$a_{39} = \sqrt{22{,}926 + 902.5(39) + 2.01(39^2)} \approx 247.3 \qquad \textit{1989 population}$$

The bar graph in Figure 9.2 graphically represents the population given by this sequence for the entire 40-year period from 1950 to 1989.

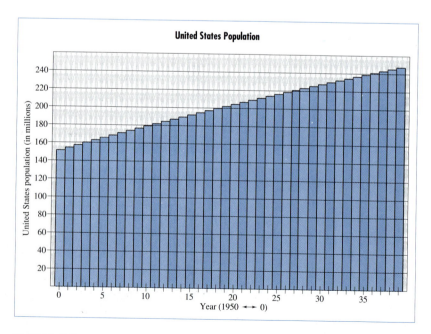

FIGURE 9.2

DISCUSSION PROBLEM

....

FINDING THE NTH TERM OF A SEQUENCE

Consider the following sequence.

$$a_1 = 1$$

$$a_2 = 1 \cdot 3$$

$$a_3 = 1 \cdot 3 \cdot 5$$

$$a_4 = 1 \cdot 3 \cdot 5 \cdot 7$$

$$\vdots$$

$$a_n = 1 \cdot 3 \cdot 5 \cdots (2n - 1)$$

Is it true that the nth term of this sequence can be written as

$$a_n = \frac{(2n)!}{2^n n!}?$$

Write a paragraph that justifies your answer.

WARM-UP

....

The following warm-up exercises involve skills that were covered in earlier sections. You will use these skills in the exercise set for this section.

In Exercises 1 and 2, find the required value of the function.

1. $f(n) = \dfrac{2n}{n^2 + 1}$, $f(2)$ **2.** $f(n) = \dfrac{4}{3(n + 1)}$, $f(3)$

In Exercises 3–6, factor the expression.

3. $4n^2 - 1$ **4.** $4n^2 - 8n + 3$

5. $n^2 - 3n + 2$ **6.** $n^2 + 3n + 2$

In Exercises 7–10, perform the indicated operations and/or simplify.

7. $\left(\dfrac{2}{3}\right)\left(\dfrac{3}{4}\right)\left(\dfrac{4}{5}\right)\left(\dfrac{5}{6}\right)$ **8.** $\dfrac{2 \cdot 4 \cdot 6 \cdot 8}{2^4}$

9. $\dfrac{1}{2 \cdot 2} + \dfrac{1}{2 \cdot 3} + \dfrac{1}{2 \cdot 4}$ **10.** $\dfrac{1}{1 \cdot 2} + \dfrac{1}{2 \cdot 3} + \dfrac{1}{3 \cdot 4}$

SECTION 9.1 · EXERCISES

In Exercises 1–18, write the first five terms of the indicated sequence. (Assume n begins with 1.)

1. $a_n = 2n + 1$

2. $a_n = 4n - 3$

3. $a_n = 2^n$

4. $a_n = (\frac{1}{2})^n$

5. $a_n = (-2)^n$

6. $a_n = (-\frac{1}{2})^n$

7. $a_n = \dfrac{1 + (-1)^n}{n}$

8. $a_n = \dfrac{n}{n + 1}$

9. $a_n = 3 - \dfrac{1}{2^n}$

10. $a_n = \dfrac{3^n}{4^n}$

11. $a_n = \dfrac{1}{n^{3/2}}$

12. $a_n = \dfrac{3n^2 - n + 4}{2n^2 + 1}$

13. $a_n = \dfrac{3^n}{n!}$

14. $a_n = \dfrac{n!}{n}$

15. $a_n = \dfrac{(-1)^n}{n^2}$

16. $a_n = (-1)^n \left(\dfrac{n}{n + 1} \right)$

17. $a_1 = 3$ and $a_{k+1} = 2(a_k - 1)$

18. $a_1 = 4$ and $a_{k+1} = \left(\dfrac{k + 1}{2} \right) a_k$

In Exercises 19–24, simplify the ratio of factorials.

19. $\dfrac{4!}{6!}$

20. $\dfrac{25!}{23!}$

21. $\dfrac{(n + 1)!}{n!}$

22. $\dfrac{(n + 2)!}{n!}$

23. $\dfrac{(2n - 1)!}{(2n + 1)!}$

24. $\dfrac{(2n + 2)!}{(2n)!}$

In Exercises 25–36, write an expression for the *most apparent* nth term of the sequence. (Assume n begins with 1.)

25. $1, 4, 7, 10, 13, \ldots$

26. $3, 7, 11, 15, 19, \ldots$

27. $0, 3, 8, 15, 24, \ldots$

28. $1, \frac{1}{4}, \frac{1}{9}, \frac{1}{16}, \frac{1}{25}, \ldots$

29. $\frac{1}{2}, \frac{-1}{4}, \frac{1}{8}, \frac{-1}{16}, \ldots$

30. $\frac{1}{3}, \frac{2}{9}, \frac{4}{27}, \frac{8}{81}, \ldots$

31. $1 + \frac{1}{1}, 1 + \frac{1}{2}, 1 + \frac{1}{3}, 1 + \frac{1}{4}, 1 + \frac{1}{5}, \ldots$

32. $1 + \frac{1}{2}, 1 + \frac{3}{4}, 1 + \frac{7}{8}, 1 + \frac{15}{16}, 1 + \frac{31}{32}, \ldots$

33. $1, \frac{1}{2}, \frac{1}{6}, \frac{1}{24}, \frac{1}{120}, \ldots$

34. $2, -4, 6, -8, 10, \ldots$

35. $1, -1, 1, -1, 1, \ldots$

36. $1, 2, \dfrac{2^2}{2}, \dfrac{2^3}{6}, \dfrac{2^4}{24}, \dfrac{2^5}{120}, \ldots$

In Exercises 37–50, find the sum.

37. $\displaystyle\sum_{i=1}^{5} (2i + 1)$

38. $\displaystyle\sum_{i=1}^{6} (3i - 1)$

39. $\displaystyle\sum_{k=1}^{4} 10$

40. $\displaystyle\sum_{k=1}^{5} 6$

41. $\displaystyle\sum_{i=0}^{4} i^2$

42. $\displaystyle\sum_{i=0}^{5} 3i^2$

43. $\displaystyle\sum_{k=0}^{3} \dfrac{1}{k^2 + 1}$

44. $\displaystyle\sum_{j=3}^{5} \dfrac{1}{j}$

45. $\displaystyle\sum_{i=1}^{4} [(i - 1)^2 + (i + 1)^3]$

46. $\displaystyle\sum_{k=2}^{5} (k + 1)(k - 3)$

47. $\displaystyle\sum_{i=1}^{4} (9 + 2i)$

48. $\displaystyle\sum_{j=0}^{4} (-2)^j$

49. $\displaystyle\sum_{k=0}^{4} \dfrac{(-1)^k}{k + 1}$

50. $\displaystyle\sum_{k=0}^{4} \dfrac{(-1)^k}{k!}$

In Exercises 51–60, use sigma notation to write the given sum.

51. $\dfrac{1}{3(1)} + \dfrac{1}{3(2)} + \dfrac{1}{3(3)} + \cdots + \dfrac{1}{3(9)}$

52. $\dfrac{5}{1 + 1} + \dfrac{5}{1 + 2} + \dfrac{5}{1 + 3} + \cdots + \dfrac{5}{1 + 15}$

53. $[2(\frac{1}{8}) + 3] + [2(\frac{2}{8}) + 3] + \cdots + [2(\frac{8}{8}) + 3]$

54. $[1 - (\frac{1}{6})^2] + [1 - (\frac{2}{6})^2] + \cdots + [1 - (\frac{6}{6})^2]$

55. $3 - 9 + 27 - 81 + 243 - 729$

56. $1 - \frac{1}{2} + \frac{1}{4} - \frac{1}{8} + \cdots - \frac{1}{128}$

57. $\dfrac{1}{1^2} - \dfrac{1}{2^2} + \dfrac{1}{3^2} - \dfrac{1}{4^2} + \cdots - \dfrac{1}{20^2}$

58. $\dfrac{1}{1 \cdot 3} + \dfrac{1}{2 \cdot 4} + \dfrac{1}{3 \cdot 5} + \cdots + \dfrac{1}{10 \cdot 12}$

59. $\frac{1}{4} + \frac{3}{8} + \frac{7}{16} + \frac{15}{32} + \frac{31}{64}$

60. $\frac{1}{2} + \frac{2}{4} + \frac{6}{8} + \frac{24}{16} + \frac{120}{32} + \frac{720}{64}$

61. *Compound Interest* A deposit of $5000 is made in an account that earns 8% interest compounded quarterly. The balance in the account after n quarters is given by

$$A_n = 5000 \left(1 + \dfrac{0.08}{4} \right)^n, \quad n = 1, 2, 3, \ldots.$$

(a) Compute the first eight terms of this sequence.

(b) Find the balance in this account after 10 years by computing the 40th term of the sequence.

62. *Compound Interest*   A deposit of $100 is made *each* month in an account that earns 12% interest compounded monthly. The balance in the account after *n* months is given by

$$A_n = 100(101)[(1.01)^n - 1], \quad n = 1, 2, 3, \ldots.$$

(a) Compute the first six terms of this sequence.

(b) Find the balance after 5 years by computing the 60th term of the sequence.

(c) Find the balance after 20 years by computing the 240th term of the sequence.

63. *Hospital Costs*   The average cost of a day in a hospital from 1980 to 1987 is given by the model

$$a_n = 242.67 + 42.67n, \quad n = 0, 1, 2, \ldots, 7,$$

where a_n is the average cost in dollars and *n* is the year, with $n = 0$ corresponding to 1980. (*Source:* American Hospital Association) Find the terms of this finite sequence and construct a bar graph that represents the sequence.

64. *Federal Debt*   It took more than 200 years for the U.S. to accumulate a $1 trillion debt. Then it took just 8 years to get to $3 trillion. (*Source:* Treasury Department) The federal debt during the decade of the 1980s is approximated by the model

$$a_n = 0.1\sqrt{82 + 9n^2}, \quad n = 0, 1, 2, \ldots, 10,$$

where a_n is the debt in trillions and *n* is the year, with $n = 0$ corresponding to 1980. Find the terms of this finite sequence and construct a bar graph that represents the sequence.

65. *Corporate Dividends*   The dividends declared per share of common stock of Ameritech Corporation for the years 1985 through 1990 are shown in the figure. (*Source:* Ameritech 1990 Annual Report) These dividends can be approximated by the model

$$a_n = 0.20n + 1.17, \quad n = 5, 6, 7, 8, 9, 10,$$

where a_n is the dividend in dollars and *n* is the year, with $n = 5$ corresponding to 1985. Approximate the sum of the dividends per share of common stock for the years 1985 through 1990 by evaluating

$$\sum_{n=5}^{10} (0.20n + 1.17).$$

Compare this sum with the result of adding the dividends as shown in the figure.

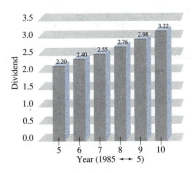

FIGURE FOR 65

66. *Total Revenue*   From 1980 through 1989, the total annual sales for MCI Communications can be approximated by the model

$$a_n = 48.217n^2 + 228.1n + 311.28, \quad n = 0, 1, \ldots, 9,$$

where a_n is the annual sales (in millions of dollars) and *n* is the year, with $n = 0$ corresponding to 1980. (*Source:* MCI Communications) Find the total revenue from 1980 through 1989 by evaluating the sum

$$\sum_{n=0}^{9} (48.217n^2 + 228.1n + 311.28).$$

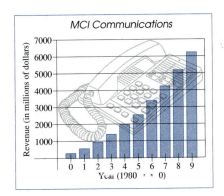

FIGURE FOR 66

In Exercises 67 and 68, use the following definition of the arithmetic mean $\bar{x}$ of a set of *n* measurements $x_1, x_2, \ldots, x_n$.

$$\bar{x} = \frac{1}{n} \sum_{i=1}^{n} x_i$$

67. Prove that $\displaystyle\sum_{i=1}^{n} (x_i - \bar{x}) = 0.$

68. Prove that $\displaystyle\sum_{i=1}^{n} (x_i - \bar{x})^2 = \sum_{i=1}^{n} x_i^2 - \frac{1}{n}\left(\sum_{i=1}^{n} x_i\right)^2.$

69. *Fibonacci Sequence* In the study of the progeny of rabbits, Fibonacci (c. 1175–c. 1250) encountered the now famous sequence bearing his name. The sequence is defined recursively by

$a_{n+2} = a_n + a_{n+1}$ where $a_1 = a_2 = 1.$

(a) Write out the first 12 terms of the sequence.

(b) Write out the first 10 terms of the sequence defined by

$$b_n = \frac{a_{n+1}}{a_n}, \qquad \text{for } n > 1.$$

(c) Using the definition of part (b), show that

$$b_n = 1 + \frac{1}{b_{n-1}}.$$

9.2   ARITHMETIC SEQUENCES

Arithmetic Sequences   /   The Sum of an Arithmetic Sequence   /   Arithmetic Mean   /   Applications

Arithmetic Sequences

A sequence whose consecutive terms have a common difference is called an **arithmetic sequence.**

DEFINITION OF AN ARITHMETIC SEQUENCE

A sequence is **arithmetic** if the differences between consecutive terms are the same. Thus, the sequence

   $a_1, a_2, a_3, a_4, \ldots, a_n, \ldots$

is arithmetic if there is a number d such that

   $a_2 - a_1 = d, \qquad a_3 - a_2 = d, \qquad a_4 - a_3 = d,$

and so on. The number d is the **common difference** of the arithmetic sequence.

EXAMPLE 1 Examples of Arithmetic Sequences

A. The sequence whose nth term is $4n + 3$ is arithmetic. For this sequence, the common difference between consecutive terms is 4.

   $7, 11, 15, 19, \ldots, 4n + 3, \ldots$

   $11 - 7 = 4$

B. The sequence whose nth term is $7 - 5n$ is arithmetic. For this sequence, the common difference between consecutive terms is -5.

$$2, -3, -8, -13, \ldots, 7 - 5n, \ldots$$
$$-3 - 2 = -5$$

C. The sequence whose nth term is $\frac{1}{4}(n + 3)$ is arithmetic. For this sequence, the common difference between consecutive terms is $\frac{1}{4}$.

$$1, \frac{5}{4}, \frac{3}{2}, \frac{7}{4}, \ldots, \frac{n + 3}{4}, \ldots$$
$$\frac{5}{4} - 1 = \frac{1}{4}$$

THE NTH TERM OF AN ARITHMETIC SEQUENCE

The nth term of an arithmetic sequence has the form

$$a_n = dn + c,$$

where d is the common difference between consecutive terms of the sequence and $c = a_1 - d$. An alternative form of the nth term of an arithmetic sequence is $a_n = a_1 + (n - 1)d$.

EXAMPLE 2   **Finding the nth Term of an Arithmetic Sequence**

Find a formula for the nth term of the arithmetic sequence whose common difference is 3 and whose first term is 2.

SOLUTION

Because the common difference is $d = 3$, you can write the following.

$$a_n = dn + c \qquad \textit{Formula for nth term}$$
$$a_n = 3n + c \qquad \textit{Substitute } d = 3$$

Using $a_1 = 2$, it follows that $2 = 3(1) + c$, which implies that $c = -1$. Thus, a formula for the nth term is

$$a_n = 3n - 1.$$

The sequence has the following form.

$$2, 5, 8, 11, 14, \ldots, 3n - 1, \ldots$$

EXAMPLE 3 Finding the *n*th Term of an Arithmetic Sequence

Find a formula for the *n*th term of the arithmetic sequence whose common difference is 5 and whose *second* term is 12. What is the 18th term of this sequence?

SOLUTION

Because the common difference is $d = 5$, you can write the following.

$$a_n = dn + c \qquad \textit{Formula for nth term}$$
$$a_n = 5n + c \qquad \textit{Substitute } d = 5$$

Using $a_2 = 12$, it follows that $12 = 5(2) + c$, which implies that $c = 2$. Thus, a formula for the *n*th term is

$$a_n = 5n + 2.$$

The sequence therefore has the form

$$7, 12, 17, 22, 27, \ldots, 5n + 2, \ldots,$$

and the 18th term of the sequence is $a_{18} = 5(18) + 2 = 92$.

If you know the *n*th term of an arithmetic sequence *and* you know the common difference of the sequence, you can find the $(n + 1)$th term by using the following **recursion formula.**

$$a_{n+1} = a_n + d$$

With such a formula, you can find any term of an arithmetic sequence, *provided* you know the previous term. For example, if you know the first term, you can find the second term. Then, knowing the second term, you can find the third term, and so on.

EXAMPLE 4 Using a Recursion Formula

Find the ninth term of the arithmetic sequence whose first two terms are 2 and 9.

SOLUTION

For this sequence you can find that $a_1 = 2$ and $a_2 = 9$ so that the common difference is $d = 9 - 2 = 7$. There are two ways to find the ninth term. One way is simply to write out the first nine terms (by repeatedly adding 7).

$$2, 9, 16, 23, 30, 37, 44, 51, 58$$

Another way to find the ninth term is first to find a formula for the nth term. Because the first term is 2, it follows that $c = a_1 - d = 2 - 7 = -5$. Therefore, a formula for the nth term of the sequence is $a_n = 7n - 5$, which implies that the ninth term is

$$a_9 = 7(9) - 5 = 58.$$

EXAMPLE 5   **Finding the nth Term of an Arithmetic Sequence**

The fourth term of an arithmetic sequence is 20, and the 13th term is 65. Write the first several terms of this sequence.

SOLUTION

To obtain the 13th term from the fourth term, you can add the common difference d to the fourth term nine times. That is

$$a_{13} = a_4 + 9d.$$

Because $a_4 = 20$ and $a_{13} = 65$, it follows that $65 = 20 + 9d$, which implies that $d = 5$. Now, from the formula for the nth term of an arithmetic sequence, you can write

$$a_n = dn + c$$
$$a_4 = 5(4) + c$$
$$20 = 20 + c$$
$$0 = c.$$

Thus, $a_n = dn = 5n$ and the first several terms of the sequence are as follows.

$$\begin{array}{cccccccccccccc} 1 & 2 & 3 & 4 & 5 & 6 & 7 & 8 & 9 & 10 & 11 & 12 & 13 \\ 5, & 10, & 15, & 20, & 25, & 30, & 35, & 40, & 45, & 50, & 55, & 60, & 65, & \ldots \end{array}$$

The Sum of an Arithmetic Sequence

The following result gives a formula for finding the *sum* of a finite arithmetic sequence.

THE SUM OF A FINITE ARITHMETIC SEQUENCE

The sum of a finite arithmetic sequence with n terms is

$$S = \frac{n}{2}(a_1 + a_n).$$

PROOF

Begin by generating the terms of the arithmetic sequence in two ways. In the first way, repeatedly add d to the first term to obtain

$$S = a_1 + a_2 + a_3 + \cdots + a_{n-2} + a_{n-1} + a_n$$
$$= a_1 + [a_1 + d] + [a_1 + 2d] + \cdots + [a_1 + (n-1)d].$$

REMARK Be sure you see that this formula only works for *arithmetic* sequences.

In the second way, repeatedly subtract d from the nth term to obtain

$$S = a_n + a_{n-1} + a_{n-2} + \cdots + a_3 + a_2 + a_1$$
$$= a_n + [a_n - d] + [a_n - 2d] + \cdots + [a_n - (n-1)d].$$

If you add these two versions of S, the multiples of d subtract out and you obtain

$$\overbrace{2S = (a_1 + a_n) + (a_1 + a_n) + (a_1 + a_n) + \cdots + (a_1 + a_n)}^{n \text{ terms}}$$
$$= n(a_1 + a_n).$$

Thus, you have

$$S = \frac{n}{2}(a_1 + a_n).$$

EXAMPLE 6   Finding the Sum of an Arithmetic Sequence

Find the following sum.

$$1 + 3 + 5 + 7 + 9 + 11 + 13 + 15 + 17 + 19$$

SOLUTION

To begin, notice that the sequence is arithmetic (with a common difference of 2). Moreover, the sequence has 10 terms. Thus, the sum of the sequence is

$$S = 1 + 3 + 5 + 7 + 9 + 11 + 13 + 15 + 17 + 19$$

$$= \frac{n}{2}(a_1 + a_n)$$

$$= \frac{10}{2}(1 + 19)$$

$$= 5(20)$$

$$= 100.$$

EXAMPLE 7   Finding the Sum of an Arithmetic Sequence

Find the sum of the integers from 1 to 100.

SOLUTION

The integers from 1 to 100 form an arithmetic sequence

$$1, 2, 3, 4, 5, 6, \ldots, 99, 100$$

that has 100 terms. Thus, you can use the formula for the sum of an arithmetic sequence, as follows.

$$S = 1 + 2 + 3 + 4 + 5 + 6 + \cdots + 99 + 100$$

$$= \frac{n}{2}(a_1 + a_n)$$

$$= \frac{100}{2}(1 + 100)$$

$$= 50(101)$$

$$= 5050$$

EXAMPLE 8   **Finding the Sum of an Arithmetic Sequence**

Find the sum of the first 150 terms of the arithmetic sequence

$$5, \ 16, \ 27, \ 38, \ 49, \ \ldots .$$

SOLUTION

For this arithmetic sequence, you have $a_1 = 5$ and $d = 16 - 5 = 11$. Thus, $c = a_1 - d = 5 - 11 = -6$, and the nth term is

$$a_n = 11n - 6.$$

Therefore, $a_{150} = 11(150) - 6 = 1644$, and the sum of the first 150 terms is as follows.

$$S = \frac{n}{2}(a_1 + a_n)$$

$$= \frac{150}{2}(5 + 1644)$$

$$= 75(1649)$$

$$= 123{,}675$$

EXAMPLE 9   **Finding the Sum of an Arithmetic Sequence**

Verify the formula

$$S = 1 + 3 + 5 + \cdots + (2n - 1) = n^2.$$

SOLUTION

Using the formula for the sum of a finite arithmetic sequence, you can write the following.

$$S = 1 + 3 + 5 + \cdots + (2n - 1) \qquad a_n = 2n - 1$$

$$= \frac{n}{2}(a_1 + a_n)$$

$$= \frac{n}{2}[1 + (2n - 1)]$$

$$= \frac{n}{2}(2n)$$

$$= n^2$$

Arithmetic Mean

Recall that $(a + b)/2$ is the midpoint between the two numbers a and b on the real number line. As a result, the terms

$$a, \frac{a + b}{2}, b$$

have a common difference. The number $(a + b)/2$ is the **arithmetic mean** of the numbers a and b. You can generalize this concept by finding k numbers m_1, $m_2, m_3, \ldots, m_k$ between a and b such that the terms

$$a, m_1, m_2, m_3, \ldots, m_k, b$$

have a common difference. This process is referred to as **inserting k arithmetic means** between a and b.

EXAMPLE 10   Inserting Arithmetic Means Between Two Numbers

Insert three arithmetic means between 4 and 15.

SOLUTION

You need to find three numbers m_1, m_2, and m_3 such that the terms

$$4, m_1, m_2, m_3, 15$$

have a common difference. In this case you have $a_1 = 4$, $n = 5$, and $a_5 = 15$. Therefore, using the alternative form of a_n, you have

$$a_5 = 15 = a_1 + (n - 1)d = 4 + 4d.$$

Because $15 = 4 + 4d$, you find that $d = \frac{11}{4}$, and the three arithmetic means are as follows.

$$m_1 = a_1 + d = 4 + \frac{11}{4} = \frac{27}{4}$$

$$m_2 = m_1 + d = \frac{27}{4} + \frac{11}{4} = \frac{38}{4}$$

$$m_3 = m_2 + d = \frac{38}{4} + \frac{11}{4} = \frac{49}{4}$$

Applications

EXAMPLE 11 Seating Capacity

An auditorium has 20 rows of seats. There are 20 seats in the first row, 21 seats in the second row, 22 seats in the third row, and so on. (See Figure 9.3.) How many seats are there in all 20 rows?

SOLUTION

The number of seats in the rows forms an arithmetic sequence in which the common difference is $d = 1$. Because $c = a_1 - d = 20 - 1 = 19$, you can determine that the formula for the nth term in the sequence is $a_n = n + 19$. Therefore, the 20th term in the sequence is $a_{20} = 20 + 19 = 39$, and the total number of seats is

$$S = 20 + 21 + 22 + \cdots + 39$$

$$= \frac{n}{2}(a_1 + a_{20})$$

$$= \frac{20}{2}(20 + 39)$$

$$= 10(59)$$

$$= 590.$$

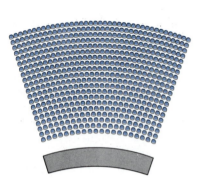

FIGURE 9.3

EXAMPLE 12 Total Sales

A small business sells $10,000 worth of products during its first year. The owner of the business has set a goal of increasing annual sales by $7500 each year for 9 years. Assuming that this goal is met, find the total sales during the first 10 years this business is in operation.

SOLUTION

The annual sales form an arithmetic sequence in which $a_1 = 10,000$ and $d = 7500$. Thus, $c = a_1 - d = 10,000 - 7500 = 2500$, and the nth term of the sequence is

$$a_n = 7500n + 2500.$$

This implies that the tenth term of the sequence is $a_{10} = 7500(10) + 2500 = 77,500$. Therefore, the total sales for the first 10 years is as follows.

$$S = \frac{n}{2}(a_1 + a_{10})$$

$$= \frac{10}{2}(10,000 + 77,500)$$

$$= 5(87,500)$$

$$= \$437,500$$

**DISCUSSION
PROBLEM**

**ARITHMETIC
SEQUENCES**

The first five terms of each of the indicated sequences are 1, 3, 5, 7, 9.

(a) $a_n = 2n - 1$ (b) $b_n = \sqrt{4n^2 - 4n + 1}$ (c) $c_n = \dfrac{2n^2 + n - 1}{n + 1}$

Does this fact *alone* mean that each sequence is arithmetic? Is each sequence arithmetic? Write a paragraph justifying your answer. Here is another sequence whose first five terms are 1, 3, 5, 7, 9. Is it arithmetic?

$$d_n = n^6 - 15n^5 + 85n^4 - 225n^3 + 274n^2 - 118n - 1$$

WARM-UP

The following warm-up exercises involve skills that were covered in earlier sections. You will use these skills in the exercise set for this section.

In Exercises 1 and 2, find the sum.

1. $\displaystyle\sum_{i=1}^{6} (2i - 1)$ **2.** $\displaystyle\sum_{i=1}^{10} (4i + 2)$

In Exercises 3 and 4, find the distance between the two real numbers.

3. $\frac{5}{2}, 8$ **4.** $\frac{4}{3}, \frac{14}{3}$

In Exercises 5 and 6, find the required value of the function.

5. $f(n) = 10 + (n - 1)4, \quad f(3)$ **6.** $f(n) = 1 + (n - 1)\frac{1}{3}, \quad f(10)$

In Exercises 7–10, evaluate the expression.

7. $\frac{11}{2}(1 + 25)$ **8.** $\frac{16}{2}(4 + 16)$

9. $\frac{20}{2}[2(5) + (12 - 1)3]$ **10.** $\frac{8}{2}[2(-3) + (15 - 1)5]$

SECTION 9.2 · EXERCISES

In Exercises 1–10, determine whether the sequence is arithmetic. If it is, find the common difference.

1. $4, 7, 10, 13, 16, \ldots$
2. $10, 8, 6, 4, 2, \ldots$
3. $1, 2, 4, 8, 16, \ldots$
4. $3, \frac{5}{2}, 2, \frac{3}{2}, 1, \ldots$
5. $\frac{9}{4}, 2, \frac{7}{4}, \frac{3}{2}, \frac{5}{4}, \ldots$
6. $-12, -8, -4, 0, 4, \ldots$
7. $\frac{1}{3}, \frac{2}{3}, \frac{4}{3}, \frac{8}{3}, \frac{16}{3}, \ldots$
8. $\ln 1, \ln 2, \ln 3, \ln 4, \ln 5, \ldots$
9. $5.3, 5.7, 6.1, 6.5, 6.9, \ldots$
10. $1^2, 2^2, 3^2, 4^2, 5^2, \ldots$

In Exercises 11–18, write the first five terms of the specified sequence. Determine whether the sequence is arithmetic, and if it is, find the common difference.

11. $a_n = 5 + 3n$
12. $a_n = (2^n)n$
13. $a_n = \dfrac{1}{n + 1}$
14. $a_n = 1 + (n - 1)4$
15. $a_n = 100 - 3n$
16. $a_n = 2^{n-1}$
17. $a_n = 3 + \dfrac{(-1)^n 2}{n}$
18. $a_n = (-1)^n$

In Exercises 19–30, find a formula for a_n for the given arithmetic sequence.

19. $a_1 = 1, d = 3$
20. $a_1 = 15, d = 4$
21. $a_1 = 100, d = -8$
22. $a_1 = 0, d = -\frac{2}{3}$
23. $a_1 = x, d = 2x$
24. $a_1 = -y, d = 5y$
25. $4, \frac{3}{2}, -1, -\frac{7}{2}, \ldots$
26. $10, 5, 0, -5, -10, \ldots$
27. $a_1 = 5, a_4 = 15$
28. $a_1 = -4, a_5 = 16$
29. $a_3 = 94, a_6 = 85$
30. $a_5 = 190, a_{10} = 115$

In Exercises 31–40, write the first five terms of the arithmetic sequence.

31. $a_1 = 5, d = 6$
32. $a_1 = 5, d = -\frac{3}{4}$
33. $a_1 = -2.6, d = -0.4$
34. $a_1 = 16.5, d = 0.25$
35. $a_1 = \frac{3}{2}, a_{k+1} = a_k - \frac{1}{4}$
36. $a_1 = 6, a_{k+1} = a_k + 12$
37. $a_1 = 2, a_{12} = 46$
38. $a_4 = 16, a_{10} = 46$
39. $a_8 = 26, a_{12} = 42$
40. $a_3 = 19, a_{15} = -1.7$

In Exercises 41–48, find the sum of the first n terms of the arithmetic sequence.

41. $8, 20, 32, 44, \ldots, \quad n = 10$
42. $2, 8, 14, 20, \ldots, \quad n = 25$
43. $-6, -2, 2, 6, \ldots, \quad n = 50$
44. $0.5, 0.9, 1.3, 1.7, \ldots, \quad n = 10$
45. $40, 37, 34, 31, \ldots, \quad n = 10$
46. $1.50, 1.45, 1.40, 1.35, \ldots, \quad n = 20$
47. $a_1 = 100, a_{25} = 220, n = 25$
48. $a_1 = 15, a_{100} = 307, n = 100$

In Exercises 49–60, find the indicated sum.

49. $\displaystyle\sum_{n=1}^{50} n$
50. $\displaystyle\sum_{n=1}^{100} 2n$
51. $\displaystyle\sum_{n=1}^{100} 5n$
52. $\displaystyle\sum_{n=51}^{100} 7n$
53. $\displaystyle\sum_{n=11}^{30} n - \sum_{n=1}^{10} n$
54. $\displaystyle\sum_{n=51}^{100} n - \sum_{n=1}^{50} n$
55. $\displaystyle\sum_{n=1}^{500} (n + 3)$
56. $\displaystyle\sum_{n=1}^{250} (1000 - n)$
57. $\displaystyle\sum_{n=1}^{20} (2n + 5)$
58. $\displaystyle\sum_{n=1}^{100} \frac{n + 4}{2}$
59. $\displaystyle\sum_{n=0}^{50} (1000 - 5n)$
60. $\displaystyle\sum_{n=0}^{100} \frac{8 - 3n}{16}$

In Exercises 61–64, insert *k* arithmetic means between the given pair of numbers.

61. 5, 17, $k = 2$

62. 24, 56, $k = 3$

63. 3, 6, $k = 3$

64. 2, 5, $k = 4$

65. Find the sum of the first 100 odd integers.

66. Find the sum of the integers from -10 to 50.

67. *Job Offer*   A person accepts a position with a company and will receive a salary of $27,500 for the first year. The person is guaranteed a raise of $1500 per year for the first 5 years.
(a) Determine the person's salary during the sixth year of employment.
(b) Determine the person's total compensation from the company through 6 full years of employment.

68. *Job Offer*   A person accepts a position with a company and will receive a salary of $32,800 for the first year. The person is guaranteed a raise of $1750 per year for the first 5 years.
(a) Determine the person's salary during the sixth year of employment.
(b) Determine the person's total compensation from the company through 6 full years of employment.

69. *Seating Capacity*   Determine the seating capacity of an auditorium with 30 rows of seats if there are 20 seats in the first row, 24 seats in the second row, 28 seats in the third row, and so on.

70. *Seating Capacity*   Determine the seating capacity of an auditorium with 36 rows of seats if there are 15 seats in the first row, 18 seats in the second row, 21 seats in the third row, and so on.

71. *Brick Pattern*   A brick patio is in roughly the shape of a trapezoid (see figure). The patio has 20 rows of bricks. The first row has 14 bricks, and the 20th row has 33 bricks. How many bricks are in the patio?

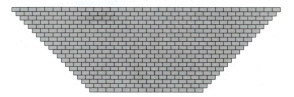

FIGURE FOR 71

72. *Falling Object*   An object (with negligible air resistance) is dropped from a plane. During the first second of fall, the object falls 4.9 meters; during the next second, it falls 14.7 meters; during the third second, it falls 24.5 meters; during the fourth second, it falls 34.3 meters. If this arithmetic pattern continues, how many meters will the object have fallen in 10 seconds?

73. Prove that the sum of the first *n* positive integers,
$$S = 1 + 2 + 3 + \cdots + n,$$
is given by
$$S = \frac{n(n + 1)}{2}.$$

9.3 GEOMETRIC SEQUENCES

**Geometric Sequences / The Sum of a Finite Geometric Sequence /
Application**

Geometric Sequences

In Section 9.2, you learned that a sequence whose consecutive terms have a common *difference* is an arithmetic sequence. In this section you will study another important type of sequence called a **geometric sequence.** Consecutive terms of a geometric sequence have a common *ratio*, as indicated in the following definition.

DEFINITION OF A GEOMETRIC SEQUENCE

A sequence is **geometric** if the ratios of consecutive terms are the same. Thus, the sequence

$$a_1, a_2, a_3, a_4, \ldots, a_n, \ldots$$

is geometric if there is a number r, $r \neq 0$, such that

$$\frac{a_2}{a_1} = r, \qquad \frac{a_3}{a_2} = r, \qquad \frac{a_4}{a_3} = r,$$

and so on. The number r is the **common ratio** of the geometric sequence.

EXAMPLE 1   **Examples of Geometric Sequences**

A. The sequence whose nth term is 2^n is geometric. For this sequence, the common ratio between consecutive terms is 2.

$$2, 4, 8, 16, \ldots, 2^n, \ldots$$

$$\frac{4}{2} = 2$$

B. The sequence whose nth term is $4(3^n)$ is geometric. For this sequence, the common ratio between consecutive terms is 3.

$$12, 36, 108, 324, \ldots, 4(3^n), \ldots$$

$$\frac{36}{12} = 3$$

c. The sequence whose nth term is $(-\frac{1}{3})^n$ is geometric. For this sequence, the common ratio between consecutive terms is $-\frac{1}{3}$.

$$-\frac{1}{3}, \frac{1}{9}, -\frac{1}{27}, \frac{1}{81}, \ldots, \left(-\frac{1}{3}\right)^n, \ldots$$

$$\underbrace{\frac{1/9}{-1/3} = -\frac{1}{3}}$$

In Example 1, notice that each of the geometric sequences has an nth term that is of the form ar^n, where the common ratio of the sequence is r.

REMARK If you know the nth term of a geometric sequence, the $(n + 1)$th term can be found by multiplying by r. That is,

$$a_{n+1} = ra_n.$$

THE NTH TERM OF A GEOMETRIC SEQUENCE

The nth term of a geometric sequence has the form

$$a_n = a_1 r^{n-1},$$

where r is the common ratio of consecutive terms of the sequence. Thus, every geometric sequence can be written in the following form.

$$a_1, \quad a_2, \quad a_3, \quad a_4, \quad a_5, \quad \ldots, \quad a_n, \quad \ldots$$

$$\downarrow \quad \downarrow \quad \downarrow \quad \downarrow \quad \downarrow \quad \ldots, \quad \downarrow \quad \ldots$$

$$a_1, \quad a_1 r, \quad a_1 r^2, \quad a_1 r^3 \quad a_1 r^4, \quad \ldots, \quad a_1 r^{n-1}$$

EXAMPLE 2   Finding the Terms of a Geometric Sequence

Write the first five terms of the geometric sequence whose first term is $a_1 = 3$ and whose common ratio is $r = 2$.

SOLUTION

Starting with 3, repeatedly multiply by 2 to obtain the following.

$$a_1 = 3$$
$$a_2 = 3(2^1) = 6$$
$$a_3 = 3(2^2) = 12$$
$$a_4 = 3(2^3) = 24$$
$$a_5 = 3(2^4) = 48$$

EXAMPLE 3 Finding a Term of a Geometric Sequence

Find the 12th term of the geometric sequence 5, 15, 45, ...

SOLUTION

The common ratio of this sequence is $r = \frac{15}{5} = 3$. Therefore, because the first term is $a_1 = 5$, you can determine the 12th term ($n = 12$) to be

$$a_{12} = a_1 r^{n-1} = 5(3)^{11} = 5(177,147) = 885,735.$$

EXAMPLE 4 Finding a Term of a Geometric Sequence

Find the 15th term of the geometric sequence whose first term is 20 and whose common ratio is 1.05.

SOLUTION

You can obtain the 15th term by multiplying the first term by r 14 times. Thus, because $a_1 = 20$ and $r = 1.05$, the 15th term ($n = 15$) is

$$a_{15} = a_1 r^{n-1} = 20(1.05)^{14} \approx 39.599.$$

EXAMPLE 5 Finding a Term of a Geometric Sequence

The 4th term of a geometric sequence is 125, and the 10th term is $\frac{125}{64}$. Find the 14th term.

SOLUTION

To obtain the 10th term from the 4th term, you can multiply the 4th term by r^6. That is, $a_{10} = a_4 r^6$. Because $a_{10} = \frac{125}{64}$ and $a_4 = 125$, you obtain the following.

$$\frac{125}{64} = 125r^6$$

$$\frac{1}{64} = r^6$$

$$\frac{1}{2} = r$$

Now, you can obtain the 14th term by multiplying the 10th term by r^4. That is,

$$a_{14} = a_{10}r^4 = \frac{125}{64}\left(\frac{1}{2}\right)^4 = \frac{125}{1024}.$$

The Sum of a Finite Geometric Sequence

THE SUM OF A FINITE GEOMETRIC SEQUENCE

The sum of the finite geometric sequence

$$a_1, a_1 r, a_1 r^2, a_1 r^3, a_1 r^4, \ldots, a_1 r^{n-1}$$

with common ratio $r \neq 1$ is given by

$$S = a_1 \left(\frac{1 - r^n}{1 - r} \right).$$

EXAMPLE 6   **Finding the Sum of a Finite Geometric Sequence**

$$\sum_{n=1}^{8} (0.3)^n = (0.3) + (0.3)^2 + (0.3)^3 + \cdots + (0.3)^8$$

$$= a_1 \left(\frac{1 - r^n}{1 - r} \right)$$

$$= (0.3) \left(\frac{1 - (0.3)^8}{1 - 0.3} \right)$$

$$\approx 0.429$$

When using the formula for the sum of a finite geometric sequence, be careful to check that the index begins at $i = 1$. If the index begins at $i = 0$, you must adjust the formula for the nth partial sum, as demonstrated in the following example.

EXAMPLE 7   **Finding the Sum of a Finite Geometric Sequence**

Find the sum

$$\sum_{i=0}^{10} 10 \left(-\frac{1}{2} \right)^i.$$

SOLUTION

By writing out a few terms you have

$$\sum_{i=0}^{10} 10 \left(-\frac{1}{2} \right)^i = 10 - 10 \left(\frac{1}{2} \right) + 10 \left(\frac{1}{2} \right)^2 - \cdots + 10 \left(\frac{1}{2} \right)^{10}.$$

You can see that the *first* term is $a_1 = 10$ and $r = -\frac{1}{2}$. Moreover, by starting with $i = 0$ and ending with $i = 10$, you are adding $n = 11$ terms, which means that the sum is

$$\sum_{i=0}^{10} 10 \left(-\frac{1}{2} \right)^i = a_1 \left(\frac{1 - r^n}{1 - r} \right) = 10 \left(\frac{1 - (-\frac{1}{2})^{11}}{1 - (-\frac{1}{2})} \right) \approx 6.670.$$

The formula for the sum of a *finite* geometric sequence can, depending on the value of r, be extended to find the sum of an *infinite* geometric sequence. Specifically, if the common ratio r has the property that $|r| < 1$, then it can be shown that r^n becomes arbitrarily close to zero as n increases without bound. Consequently,

$$S \rightarrow a_1 \left(\frac{1 - 0}{1 - r} \right) = \frac{a_1}{1 - r} \quad \text{as} \quad n \rightarrow \infty.$$

SUM OF AN INFINITE GEOMETRIC SEQUENCE

If $|r| < 1$, then the infinite geometric sequence

$$a_1, a_1 r, a_1 r^2, a_1 r^3, \ldots, a_1 r^{n-1}, \ldots$$

has the sum

$$S = \sum_{n=1}^{\infty} a_1 r^{n-1} = \frac{a_1}{1 - r}.$$

REMARK   The summation $a_1 + a_1 r + a_1 r^2 + a_1 r^3 + \cdots$ is called an **infinite geometric series**.

EXAMPLE 8   Finding the Sum of an Infinite Geometric Sequence

Find the sum of the following infinite geometric sequence.

$$4, 4(0.6), 4(0.6)^2, 4(0.6)^3, \ldots, 4(0.6)^{n-1}, \ldots$$

SOLUTION

Because $a_1 = 4$, $r = 0.6$, and $|r| < 1$, you have

$$S = \sum_{n=1}^{\infty} 4(0.6)^{n-1}$$

$$= \frac{a_1}{1 - r}$$

$$= \frac{4}{1 - 0.6}$$

$$= 10.$$

Figure 9.4 shows the graph of

$$f(x) = \frac{4}{1 - 0.6} - \frac{4(0.6)^x}{1 - 0.6} = 10 - 10(0.6)^x,$$

which has a horizontal asymptote of $y = 10$.

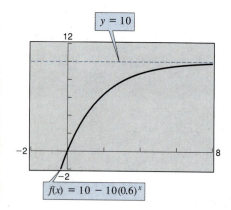

$y = 10$

$f(x) = 10 - 10(0.6)^x$

FIGURE 9.4

Application

EXAMPLE 9   **An Application: Compound Interest**

A deposit of $50 is made the first day of each month in a savings account that pays 12% compounded monthly. What is the balance at the end of 2 years?

SOLUTION

The formula for compound interest is

$$A = P\left(1 + \frac{r}{n}\right)^{tn},$$

where A is the balance of the account, P is the initial deposit, r is the annual percentage rate, n is the number of compoundings per year, and t is the time (in years). To find the balance in the account after 24 months, it is helpful to consider each of the 24 deposits separately. For example, the first deposit will gain interest for a full 24 months, and its balance will be

$$A_{24} = 50\left(1 + \frac{0.12}{12}\right)^{24} = 50(1.01)^{24}.$$

The second deposit will gain interest for 23 months, and its balance will be

$$A_{23} = 50\left(1 + \frac{0.12}{12}\right)^{23} = 50(1.01)^{23}.$$

The last (24th) deposit will gain interest for only 1 month, and its balance will be

REMARK   This type of investment account is an *annuity*.

$$A_1 = 50\left(1 + \frac{0.12}{12}\right)^{1} = 50(1.01).$$

Finally, the total balance in the account will be the sum of the balances of the 24 deposits.

$$S = A_1 + A_2 + A_3 + \cdots + A_{23} + A_{24}$$
$$= 50(1.01) + 50(1.01)^2 + \cdots + 50(1.01)^{23} + 50(1.01)^{24}$$

Using the formula for the sum of a geometric sequence, with $A_1 = 50(1.01)$ and $r = 1.01$, you have

$$S = 50(1.01)\left(\frac{1 - (1.01)^{24}}{1 - 1.01}\right) = \$1362.16.$$

DISCUSSION PROBLEM

········

COMPARING TWO SEQUENCES

The first several terms of the sequences whose nth terms are $a_n = (0.99)^{n-1}$ and $b_n = 1 - (n - 1)(0.01)$ are almost the same.

(a) $a_1 = 1$
$a_2 = 0.99$
$a_3 = (0.99)^2 \approx 0.98$
$a_4 = (0.99)^3 \approx 0.97$
$a_5 = (0.99)^4 \approx 0.96$

(b) $b_1 = 1$
$b_2 = 1 - 0.01 = 0.99$
$b_3 = 1 - 2(0.01) = 0.98$
$b_4 = 1 - 3(0.01) = 0.97$
$b_5 = 1 - 4(0.01) = 0.96$

Yet the two sequences have very basic differences. Write a paragraph describing some of the differences between the two sequences.

WARM-UP

··············

The following warm-up exercises involve skills that were covered in earlier sections. You will use these skills in the exercise set for this section.

In Exercises 1–4, evaluate the expression.

1. $\left(\frac{4}{5}\right)^3$ **2.** $\left(\frac{3}{4}\right)^2$

3. 2^{-4} **4.** $\dfrac{5}{3^4}$

In Exercises 5–10, simplify the expression.

5. $(2n)(3n^2)$ **6.** $n(3n)^3$

7. $\dfrac{4n^5}{n^2}$ **8.** $\dfrac{(2n)^3}{8n}$

9. $[2(3)^{-4}]^n$ **10.** $3(4^2)^{-n}$

SECTION 9.3 · EXERCISES

In Exercises 1–10, determine whether the sequence is geometric. If it is, find its common ratio.

1. 5, 15, 45, 135, . . .   **2.** 3, 12, 48, 192, . . .

3. 3, 12, 21, 30, . . .   **4.** 1, −2, 4, −8, . . .

5. $1, -\frac{1}{2}, \frac{1}{4}, -\frac{1}{8}, \ldots$ **6.** 5, 1, 0.2, 0.04, . . .

7. $\frac{1}{2}, \frac{2}{3}, \frac{3}{4}, \frac{4}{5}, \ldots$ **8.** $9, -6, 4, -\frac{8}{3}, \ldots$

9. $1, \frac{1}{2}, \frac{1}{3}, \frac{1}{4}, \ldots$ **10.** $\frac{1}{5}, \frac{2}{3}, \frac{3}{9}, \frac{4}{11}, \ldots$

In Exercises 11–20, write the first five terms of the geometric sequence.

11. $a_1 = 2, r = 3$ **12.** $a_1 = 6, r = 2$

13. $a_1 = 1, r = \frac{1}{2}$ **14.** $a_1 = 1, r = \frac{1}{3}$

15. $a_1 = 5, r = -\frac{1}{10}$ **16.** $a_1 = 6, r = -\frac{1}{4}$

17. $a_1 = 1, r = e$ **18.** $a_1 = 2, r = \sqrt{3}$

19. $a_1 = 3, r = \dfrac{x}{2}$ **20.** $a_1 = 5, r = 2x$

In Exercises 21–32, find the nth term of the geometric sequence.

21. $a_1 = 4$, $r = \frac{1}{2}$, $n = 10$

22. $a_1 = 5$, $r = \frac{3}{2}$, $n = 8$

23. $a_1 = 6$, $r = -\frac{1}{3}$, $n = 12$

24. $a_1 = 8$, $r = \sqrt{5}$, $n = 9$

25. $a_1 = 100$, $r = e^x$, $n = 9$

26. $a_1 = 1$, $r = -\dfrac{x}{3}$, $n = 7$

27. $a_1 = 500$, $r = 1.02$, $n = 40$

28. $a_1 = 1000$, $r = 1.005$, $n = 60$

29. $a_1 = 16$, $a_4 = \frac{27}{4}$, $n = 3$

30. $a_2 = 3$, $a_5 = \frac{3}{64}$, $n = 1$

31. $a_2 = -18$, $a_5 = \frac{2}{3}$, $n = 6$

32. $a_3 = \frac{16}{3}$, $a_5 = \frac{64}{27}$, $n = 7$

33. *Compound Interest*   A principal of $1000 is invested at 10% interest. Find the amount after 10 years if the interest is compounded (a) annually, (b) semiannually, (c) quarterly, (d) monthly, and (e) daily.

34. *Compound Interest*   A principal of $2500 is invested at 12% interest. Find the amount after 20 years if the interest is compounded (a) annually, (b) semiannually, (c) quarterly, (d) monthly, and (e) daily.

35. *Depreciation*   A company buys a machine for $135,000 that depreciates at the rate of 30% per year. (In other words, at the end of each year the depreciated value is 70% of what it was at the beginning of the year.) Find the depreciated value of the machine after 5 full years.

36. *Population Growth*   A city of 250,000 people is growing at the rate of 1.3% per year. Estimate the population of the city 30 years from now.

In Exercises 37–46, find the indicated sum.

37. $\displaystyle\sum_{n=1}^{9} 2^{n-1}$

38. $\displaystyle\sum_{n=1}^{9} (-2)^{n-1}$

39. $\displaystyle\sum_{i=1}^{7} 64\left(-\tfrac{1}{2}\right)^{i-1}$

40. $\displaystyle\sum_{i=1}^{6} 32\left(\tfrac{1}{4}\right)^{i-1}$

41. $\displaystyle\sum_{i=1}^{10} 8\left(\tfrac{-1}{4}\right)^{i-1}$

42. $\displaystyle\sum_{i=1}^{10} 5\left(\tfrac{-1}{3}\right)^{i-1}$

43. $\displaystyle\sum_{n=0}^{20} 3\left(\tfrac{3}{2}\right)^{n}$

44. $\displaystyle\sum_{n=0}^{15} 2\left(\tfrac{4}{3}\right)^{n}$

45. $\displaystyle\sum_{n=0}^{5} 300(1.06)^{n}$

46. $\displaystyle\sum_{n=0}^{6} 500(1.04)^{n}$

47. *Annuities*   A deposit of $100 is made at the beginning of each month for 5 years in an account that pays 10%, compounded monthly. What is the balance A in the account at the end of 5 years?

$$A = 100\left(1 + \frac{0.10}{12}\right)^{1} + \cdots + 100\left(1 + \frac{0.10}{12}\right)^{60}$$

48. *Annuities*   A deposit of $50 is made at the beginning of each month for 5 years in an account that pays 12%, compounded monthly. What is the balance A in the account at the end of 5 years?

$$A = 50\left(1 + \frac{0.12}{12}\right)^{1} + \cdots + 50\left(1 + \frac{0.12}{12}\right)^{60}$$

49. *Annuities*   A deposit of P dollars is made at the beginning of each month in an account at an annual interest rate r, compounded monthly. The balance A after t years is

$$A = P\left(1 + \frac{r}{12}\right) + P\left(1 + \frac{r}{12}\right)^{2} + \cdots$$
$$+ P\left(1 + \frac{r}{12}\right)^{12t}.$$

Show that the balance is given by

$$A = P\left[\left(1 + \frac{r}{12}\right)^{12t} - 1\right]\left(1 + \frac{12}{r}\right).$$

50. *Annuities*   A deposit of P dollars is made at the beginning of each month in an account at an annual interest rate r, compounded continuously. The balance A after t years is

$$A = Pe^{r/12} + Pe^{2r/12} + \cdots + Pe^{12tr/12}.$$

Show that the balance is given by

$$A = \frac{Pe^{r/12}(e^{rt} - 1)}{e^{r/12} - 1}.$$

Annuities   In Exercises 51–54, consider making monthly deposits of P dollars into a savings account at an annual interest rate r. Use the results of Exercises 49 and 50 to find the balance A after t years if the interest is compounded (a) monthly and (b) continuously.

51. $P = \$50$, $r = 7\%$, $t = 20$ years

52. $P = \$75$, $r = 9\%$, $t = 25$ years

53. $P = \$100$, $r = 10\%$, $t = 40$ years

54. $P = \$20$, $r = 6\%$, $t = 50$ years

55. *Annuities* Consider an initial deposit of P dollars into an account at an annual interest rate r, compounded monthly. At the end of each month a withdrawal of W dollars will occur and the account will be depleted in t years. The magnitude of the initial deposit is given by

$$P = W\left(1 + \frac{r}{12}\right)^{-1} + W\left(1 + \frac{r}{12}\right)^{-2} + \cdots$$
$$+ W\left(1 + \frac{r}{12}\right)^{-12t}.$$

Show that the initial deposit is given by

$$P = W\left(\frac{12}{r}\right)\left[1 - \left(1 + \frac{r}{12}\right)^{-12t}\right].$$

56. *Annuities* Determine the amount that is required in an individual retirement account for an individual who retires at age 65 and wants an income from the account of $2000 per month for 20 years. Use the result of Exercise 55 and assume the account earns 9% compounded monthly.

57. *Profit* The annual profit for the H. J. Heinz Company from 1980 through 1989 can be approximated by the model

$$a_n = 167.5e^{0.12n}, \qquad n = 0, 1, 2, \ldots, 9,$$

where a_n is the annual profit in millions of dollars and n represents the year, with $n = 0$ corresponding to 1980 (see figure). Use the formula for the sum of a geometric sequence to approximate the total profit earned during this 10-year period. (*Source:* H. J. Heinz Company)

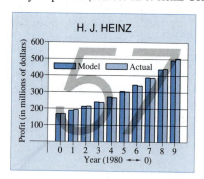

FIGURE FOR 57

58. *Would You Take This Job?* Suppose you went to work at a company that pays $0.01 for the first day, $0.02 for the second day, $0.04 for the third day, and so on. If the daily wage keeps doubling, determine what your total income would be for working (a) 29 days, (b) 30 days, and (c) 31 days.

59. *Salary* You accept a job with a salary of $30,000 for the first year. Suppose that during the next 39 years, you receive a 5% raise each year. What would be your total compensation over the 40-year period?

60. *Area* The sides of a square are 16 inches in length. The square is divided into nine smaller squares and the center square is shaded (see figure). Each of the eight unshaded squares is then divided into nine smaller squares and the center square of each is shaded. If this process is repeated four more times, determine the area of the shaded region.

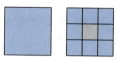

FIGURE FOR 60

In Exercises 61–70, find the sum of the infinite geometric sequence.

61. $\displaystyle\sum_{n=0}^{\infty} \left(\tfrac{1}{2}\right)^n = 1 + \tfrac{1}{2} + \tfrac{1}{4} + \tfrac{1}{8} + \cdots$

62. $\displaystyle\sum_{n=0}^{\infty} 2\left(\tfrac{2}{3}\right)^n = 2 + \tfrac{4}{3} + \tfrac{8}{9} + \tfrac{16}{27} + \cdots$

63. $\displaystyle\sum_{n=0}^{\infty} \left(-\tfrac{1}{2}\right)^n = 1 - \tfrac{1}{2} + \tfrac{1}{4} - \tfrac{1}{8} + \cdots$

64. $\displaystyle\sum_{n=0}^{\infty} 2\left(-\tfrac{2}{3}\right)^n = 2 - \tfrac{4}{3} + \tfrac{8}{9} - \tfrac{16}{27} + \cdots$

65. $\displaystyle\sum_{n=0}^{\infty} 4\left(\tfrac{1}{4}\right)^n = 4 + 1 + \tfrac{1}{4} + \tfrac{1}{16} + \cdots$

66. $\displaystyle\sum_{n=0}^{\infty} \left(\tfrac{1}{10}\right)^n = 1 + 0.1 + 0.01 + 0.001 + \cdots$

67. $8 + 6 + \tfrac{9}{2} + \tfrac{27}{8} + \cdots$

68. $3 - 1 + \tfrac{1}{3} - \tfrac{1}{9} + \cdots$

69. $4 - 2 + 1 - \tfrac{1}{2} + \cdots$

70. $2 + \sqrt{2} + 1 + \dfrac{1}{\sqrt{2}} + \cdots$

In Exercises 71 and 72, use a graphing utility to graph the function. Identify the horizontal asymptote of the graph and determine its relationship to the given series.

Function	Series
71. $f(x) = 6\left[\dfrac{1 - (0.5)^x}{1 - (0.5)}\right]$	$\displaystyle\sum_{n=0}^{\infty} 6\left(\tfrac{1}{2}\right)^n$
72. $f(x) = 2\left[\dfrac{1 - (0.8)^x}{1 - (0.8)}\right]$	$\displaystyle\sum_{n=0}^{\infty} 2\left(\tfrac{4}{5}\right)^n$

73. *Distance*   A ball is dropped from a height of 16 feet. Each time it drops h feet, it rebounds $0.81h$ feet. Find the total distance traveled by the ball.

74. *Time*   The ball in Exercise 73 takes the following times for each fall.

$s_1 = -16t^2 + 16,$ $s_1 = 0$ if $t = 1$

$s_2 = -16t^2 + 16(0.81),$ $s_2 = 0$ if $t = 0.9$

$s_3 = -16t^2 + 16(0.81)^2,$ $s_3 = 0$ if $t = (0.9)^2$

$s_4 = -16t^2 + 16(0.81)^3,$ $s_4 = 0$ if $t = (0.9)^3$

$\vdots$ $\vdots$

$s_n = -16t^2 + 16(0.81)^{n-1},$ $s_n = 0$ if $t = (0.9)^{n-1}$

Beginning with s_2, the ball takes the same amount of time to bounce up as it does to fall, and thus the total time elapsed before it comes to rest is

$$t = 1 + 2 \sum_{n=1}^{\infty} (0.9)^n.$$

Find this total.

9.4   MATHEMATICAL INDUCTION

Mathematical Induction   /   Sums of Powers of Integers   /   Pattern Recognition   /   Finite Differences

Mathematical Induction

In this section you will study a form of mathematical proof, called the principle of **mathematical induction.** It is important that you clearly see the logical need for it, so let's take a closer took at a problem that was discussed earlier.

$S_1 = 1 = 1^2$

$S_2 = 1 + 3 = 2^2$

$S_3 = 1 + 3 + 5 = 3^2$

$S_4 = 1 + 3 + 5 + 7 = 4^2$

$S_5 = 1 + 3 + 5 + 7 + 9 = 5^2$

Judging from the pattern formed by these first five sums, it appears that the sum of the first n odd integers is

$$S_n = 1 + 3 + 5 + 7 + 9 + \cdots + (2n - 1) = n^2.$$

Although this particular formula is valid, it is important for you to see that recognizing a pattern and then simply *jumping to the conclusion* that the pattern must be true for all values of n is *not* a logically valid method of proof. There are many examples in which a pattern appears to be developing for small values of n and then at some point the pattern fails. One of the most famous cases of this was the conjecture by the French mathematician Pierre de Fermat (1601–1655), who speculated that all numbers of the form

$$F_n = 2^{2^n} + 1, \quad n = 0, 1, 2, \ldots$$

are prime. For $n = 0, 1, 2, 3,$ and 4, the conjecture is true.

$$F_0 = 3, \ F_1 = 5, \ F_2 = 17, \ F_3 = 257, \ F_4 = 65{,}537$$

The size of the next Fermat number ($F_5 = 4{,}294{,}967{,}297$) is so great that it was difficult for Fermat to determine whether it was prime or not. However, another well-known mathematician, Leonhard Euler (1707–1783), later found the factorization

$$F_5 = 4{,}294{,}967{,}297 = 641(6{,}700{,}417),$$

which proved that F_5 is not prime, and therefore Fermat's conjecture was false.

Just because a rule, pattern, or formula seems to work for several values of n, you cannot simply decide that it is valid for all values of n without going through a *legitimate proof.*

REMARK It is important to recognize that both parts of the Principle of Mathematical Induction are necessary.

THE PRINCIPLE OF MATHEMATICAL INDUCTION

Let P_n be a statement involving the positive integer n. If

1. P_1 is true, and
2. the truth of P_k implies the truth of P_{k+1}, for every positive integer k,

then P_n must be true for all positive integers n.

To apply the Principle of Mathematical Induction, you need to be able to determine the statement P_{k+1} for a given statement P_k.

EXAMPLE 1 A Preliminary Example

Determine the statement P_{k+1} for each of the following statements P_k.

A. P_k is the statement: $\quad S_k = \dfrac{k^2(k+1)^2}{4}$.

B. P_k is the statement:

$$S_k = 1 + 5 + 9 + \cdots + [4(k-1) - 3] + (4k - 3).$$

C. P_k is the statement: $\quad 3^k \geq 2k + 1$.

SOLUTION

A. Substituting $k + 1$ for k, you obtain the statement P_{k+1}.

$$P_{k+1}: \quad S_{k+1} = \frac{(k+1)^2(k+1+1)^2}{4} \qquad \textit{Replace k by k + 1}$$

$$= \frac{(k+1)^2(k+2)^2}{4} \qquad \textit{Simplify}$$

B. In this case you have

$$P_{k+1}: \quad S_{k+1} = 1 + 5 + 9 + \cdots + (4[(k + 1) - 1] - 3)$$
$$+ [4(k + 1) - 3]$$
$$= 1 + 5 + 9 + \cdots + (4k - 3) + (4k + 1).$$

C. Replacing k by $k + 1$ in the statement $3^k \geq 2k + 1$, you have

$$P_{k+1}: \quad 3^{k+1} \geq 2(k + 1) + 1$$
$$3^{k+1} \geq 2k + 3.$$

EXAMPLE 2   **Using Mathematical Induction**

Use mathematical induction to prove the following formula.

$$S_n = 1 + 3 + 5 + 7 + \cdots + (2n - 1) = n^2$$

SOLUTION

Mathematical induction consists of two distinct parts. First, you must show that the formula is true when $n = 1$.

REMARK When using mathematical induction to prove a *summation* formula (like the one in Example 2), it is helpful to think of S_{k+1} as $S_{k+1} = S_k + a_{k+1}$, where a_{k+1} is the $(k + 1)$th term of the original sum.

1. When $n = 1$, the formula is valid, because

$$S_1 = 1 = 1^2.$$

The second part of mathematical induction has two steps. The first step is to assume that the formula is valid for *some* integer k. The second step is to use this assumption to prove that the formula is valid for the next integer, $k + 1$.

2. Assuming that the formula

$$S_k = 1 + 3 + 5 + 7 + \cdots + (2k - 1) = k^2$$

is true, you must show that the formula $S_{k+1} = (k + 1)^2$ is true.

$$S_{k+1} = 1 + 3 + 5 + 7 + \cdots + (2k - 1) + [2(k + 1) - 1]$$
$$= [1 + 3 + 5 + 7 + \cdots + (2k - 1)] + (2k + 2 - 1)$$
$$= S_k + (2k + 1)$$
$$= k^2 + 2k + 1 \qquad \textit{Because } S_k = k^2 \textit{ from above}$$
$$= (k + 1)^2$$

Combining the results of parts (1) and (2), you can conclude by mathematical induction that the formula is valid for *all* positive integer values of n.

A well-known illustration used to explain why the principle of mathematical induction works is the unending line of dominoes shown in Figure 9.5. If the line actually contains infinitely many dominoes, then it is clear that you could not knock the entire line down by knocking down only *one domino* at a time. However, suppose it were true that each domino would knock down the next one as it fell. Then you could knock them all down simply by pushing the first one and starting a chain reaction. Mathematical induction works in the same way. If the truth of P_k implies the truth of P_{k+1} and if P_1 is true, then the chain reaction proceeds as follows:

P_1 implies P_2

P_2 implies P_3

P_3 implies P_4

and so on.

The first domino knocks over the second which knocks over
the third which knocks over the fourth and so on.

FIGURE 9.5

It occasionally happens that a statement involving natural numbers is not true for the first $k - 1$ positive integers but is true for all values of $n \geq k$. In these instances, you can use a slight variation of the principle of mathematical induction in which you verify P_k rather than P_1. This variation is called the **extended principle of mathematical induction.** To see the validity of this, note from Figure 9.5 that all but the first $k - 1$ dominoes can be knocked down by knocking over the kth domino. This suggests that you can prove a statement P_n to be true for $n \geq k$ by showing that P_k is true and that P_k implies P_{k+1}.

EXAMPLE 3   Using Mathematical Induction

Use mathematical induction to prove the following formula.

$$S_n = 1^2 + 2^2 + 3^2 + 4^2 + \cdots + n^2 = \frac{n(n+1)(2n+1)}{6}$$

SOLUTION

1. When $n = 1$, the formula is valid, because

$$S_1 = 1^2 = \frac{1(1+1)[2(1)+1]}{6} = \frac{1(2)(3)}{6}.$$

2. Assuming that

$$S_k = 1^2 + 2^2 + 3^2 + 4^2 + \cdots + k^2 = \frac{k(k+1)(2k+1)}{6},$$

you must show that

$$S_{k+1} = \frac{(k+1)(k+2)(2k+3)}{6}.$$

To do this, write the following.

$$\begin{aligned}
S_{k+1} &= S_k + a_{k+1}\\
&= (1^2 + 2^2 + 3^2 + 4^2 + \cdots + k^2) + (k+1)^2\\
&= \frac{k(k+1)(2k+1)}{6} + (k+1)^2 \qquad \textit{By assumption}\\
&= \frac{k(k+1)(2k+1) + 6(k+1)^2}{6}\\
&= \frac{(k+1)[k(2k+1) + 6(k+1)]}{6}\\
&= \frac{(k+1)(2k^2 + 7k + 6)}{6}\\
&= \frac{(k+1)(k+2)(2k+3)}{6}
\end{aligned}$$

Combining the results of parts (1) and (2), you can conclude by mathematical induction that the formula is valid for *all* $n \geq 1$.

Sums of Powers of Integers

The formula in Example 3 is one of a collection of useful summation formulas.

SUMS OF POWERS OF INTEGERS

1. $\displaystyle\sum_{i=1}^{n} i = 1 + 2 + 3 + 4 + \cdots + n = \frac{n(n+1)}{2}$

2. $\displaystyle\sum_{i=1}^{n} i^2 = 1^2 + 2^2 + 3^2 + 4^2 + \cdots + n^2 = \frac{n(n+1)(2n+1)}{6}$

3. $\displaystyle\sum_{i=1}^{n} i^3 = 1^3 + 2^3 + 3^3 + 4^3 + \cdots + n^3 = \frac{n^2(n+1)^2}{4}$

4. $\displaystyle\sum_{i=1}^{n} i^4 = 1^4 + 2^4 + 3^4 + 4^4 + \cdots + n^4$

   $= \dfrac{n(n+1)(2n+1)(3n^2+3n-1)}{30}$

5. $\displaystyle\sum_{i=1}^{n} i^5 = 1^5 + 2^5 + 3^5 + 4^5 + \cdots + n^5$

   $= \dfrac{n^2(n+1)^2(2n^2+2n-1)}{12}$

EXAMPLE 4   **Finding a Sum of Powers of Integers**

Find the following sum.

$$1^3 + 2^3 + 3^3 + 4^3 + 5^3 + 6^3 + 7^3$$

SOLUTION

Using the formula for the sum of the cubes of the first n positive integers produces the following.

$$1^3 + 2^3 + 3^3 + 4^3 + 5^3 + 6^3 + 7^3 = \frac{7^2(7+1)^2}{4}$$

$$= \frac{49(64)}{4}$$

$$= 784$$

Check this sum by adding the numbers 1, 8, 27, 64, 125, 216, and 343.

Pattern Recognition

Although choosing a formula on the basis of a few observations does *not* guarantee the validity of the formula, pattern recognition *is* important. Once you have a pattern or formula that you think works, you can try using mathematical induction to prove your formula.

FINDING A FORMULA FOR THE *N*TH TERM OF A SEQUENCE

1. Calculate the first several terms of the sequence. (It is often a good idea to write the terms in both simplified and factored form.)
2. Try to find a recognizable pattern from these terms, and write down a formula for the *n*th term of the sequence. (This is your hypothesis. You might try computing one or two more terms in the sequence to test your hypothesis.)
3. Use mathematical induction to attempt to prove your hypothesis.

EXAMPLE 5 **Finding a Formula for a Finite Sum**

Find a formula for the following finite sum.

$$\frac{1}{1 \cdot 2} + \frac{1}{2 \cdot 3} + \frac{1}{3 \cdot 4} + \frac{1}{4 \cdot 5} + \cdots + \frac{1}{n(n + 1)}$$

SOLUTION

Begin by writing out the sums of the first few terms.

$$S_1 = \frac{1}{1 \cdot 2} = \frac{1}{2} = \frac{1}{1 + 1}$$

$$S_2 = \frac{1}{1 \cdot 2} + \frac{1}{2 \cdot 3} = \frac{4}{6} = \frac{2}{3} = \frac{2}{2 + 1}$$

$$S_3 = \frac{1}{1 \cdot 2} + \frac{1}{2 \cdot 3} + \frac{1}{3 \cdot 4} = \frac{9}{12} = \frac{3}{4} = \frac{3}{3 + 1}$$

$$S_4 = \frac{1}{1 \cdot 2} + \frac{1}{2 \cdot 3} + \frac{1}{3 \cdot 4} + \frac{1}{4 \cdot 5} = \frac{48}{60} = \frac{4}{5} = \frac{4}{4 + 1}$$

Now, from this sequence of four sums, it appears that the formula for the *k*th sum is

$$S_k = \frac{1}{1 \cdot 2} + \frac{1}{2 \cdot 3} + \frac{1}{3 \cdot 4} + \frac{1}{4 \cdot 5} + \cdots + \frac{1}{k(k + 1)} = \frac{k}{k + 1}.$$

To prove the validity of this formula, use mathematical induction, as follows. Note that you have already verified the formula for $n = 1$, so you can begin by assuming that the formula is valid for $n = k$ and try to show that it is valid for $n = k + 1$.

$$S_{k+1} = \left[\frac{1}{1 \cdot 2} + \frac{1}{2 \cdot 3} + \frac{1}{3 \cdot 4} + \cdots + \frac{1}{k(k+1)}\right] + \frac{1}{(k+1)(k+2)}$$

$$= \frac{k}{k+1} + \frac{1}{(k+1)(k+2)} \qquad \textit{By assumption}$$

$$= \frac{k(k+2) + 1}{(k+1)(k+2)}$$

$$= \frac{k^2 + 2k + 1}{(k+1)(k+2)}$$

$$= \frac{(k+1)^2}{(k+1)(k+2)}$$

$$= \frac{k+1}{k+2}$$

Thus, the formula is valid.

EXAMPLE 6 Proving an Inequality by Mathematical Induction

Prove that $n < 2^n$ for all positive integers n.

SOLUTION

1. For $n = 1$, the formula is true, because

$$1 < 2^1.$$

2. Assuming that

$$k < 2^k,$$

you need to show that $k + 1 < 2^{k+1}$. You have

$$2^{k+1} = 2(2^k) > 2(k) = 2k. \qquad \textit{By assumption}$$

Because $2k = k + k > k + 1$ for all $k > 1$, it follows that

$$2^{k+1} > 2k > k + 1$$

or

$$k + 1 < 2^{k+1}.$$

Hence, $n < 2^n$ for all integers $n \geq 1$.

Finite Differences

The **first differences** of a sequence are found by subtracting consecutive terms. The **second differences** are found by subtracting consecutive first differences. The first and second differences of the sequence 3, 5, 8, 12, 17, 23, . . . , are as follows.

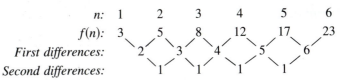

For this sequence, the second differences are all the same number. When this happens, the sequence has a quadratic model. (If the first differences are all the same number, then the sequence has a linear model. That is, it is arithmetic.)

EXAMPLE 7   Finding a Quadratic Model

Find a quadratic model for the sequence whose first three terms are 3, 5, and 8.

SOLUTION

You know the model has the form

$$f(n) = an^2 + bn + c.$$

By substituting 1, 2, and 3 for n, you can obtain a system of three linear equations in three variables.

$f(1) = a(1^2) + b(1) + c = 3$ *Substitute 1 for n*

$f(2) = a(2^2) + b(2) + c = 5$ *Substitute 2 for n*

$f(3) = a(3^2) + b(3) + c = 8$ *Substitute 3 for n*

You now have a system of three equations in a, b, and c.

$a + b + c = 3$ *Equation 1*

$4a + 2b + c = 5$ *Equation 2*

$9a + 3b + c = 8$ *Equation 3*

Using the techniques discussed in Chapter 7, you can find the solution to be $a = \frac{1}{2}$, $b = \frac{1}{2}$, and $c = 2$. Thus, a quadratic model is

$$f(n) = \frac{1}{2}n^2 + \frac{1}{2}n + 2.$$

Try checking the values of $f(1)$, $f(2)$, and $f(3)$.

Notice that we used the fact that the second differences are all the same to conclude that the function in Example 7 *has* a quadratic model. To *find* the model, we used the first three values of the function to create a system of three linear equations.

DISCUSSION PROBLEM

···················

THE SUM OF THE ANGLES OF A REGULAR POLYGON

A *regular n*-sided polygon is a polygon that has *n* equal sides and *n* equal angles. For instance, an equilateral triangle is a regular three-sided polygon. Each angle of an equilateral triangle measures 60° and the sum of all three angles is 180°. Similarly, the sum of the four angles of a regular four-sided polygon (a square) is 360°. From the following four regular polygons (see figure), find a formula for the sum of the angles of a regular *n*-sided polygon. Write a paragraph describing how you could *prove* that your formula is valid. Do you think that mathematical induction would be an appropriate technique to prove the validity of your formula?

(a) Equilateral   (b) Square   (c) Regular   (d) Regular
    Triangle   (360°)   Pentagon   Hexagon
    (180°)   (540°)   (720°)

WARM-UP

················

The following warm-up exercises involve skills that were covered in earlier sections. You will use these skills in the exercise set for this section.

In Exercises 1–4, find the required sum.

1. $\displaystyle\sum_{k=3}^{6} (2k - 3)$ **2.** $\displaystyle\sum_{j=1}^{5} (j^2 - j)$

3. $\displaystyle\sum_{k=2}^{5} \frac{1}{k}$ **4.** $\displaystyle\sum_{i=1}^{2} \left(1 + \frac{1}{i}\right)$

In Exercises 5–10, simplify the expression.

5. $\dfrac{2(k + 1) + 3}{5}$ **6.** $\dfrac{3(k + 1) - 2}{6}$

7. $2 \cdot 2^{2(k+1)}$ **8.** $\dfrac{3^{2k}}{3^{2(k+1)}}$

9. $\dfrac{k + 1}{k^2 + k}$ **10.** $\dfrac{\sqrt{32}}{\sqrt{50}}$

SECTION 9.4 · EXERCISES

In Exercises 1–4, find P_{k+1} for the given P_k.

1. $P_k = \dfrac{5}{k(k+1)}$

2. $P_k = \dfrac{1}{(k+1)(k+3)}$

3. $P_k = \dfrac{k^2(k+1)^2}{4}$

4. $P_k = \dfrac{k}{2}(3k-1)$

In Exercises 5–18, use mathematical induction to prove the given formula for every positive integer n.

5. $2 + 4 + 6 + 8 + \cdots + 2n = n(n+1)$

6. $3 + 7 + 11 + 15 + \cdots + (4n-1) = n(2n+1)$

7. $2 + 7 + 12 + 17 + \cdots + (5n-3) = \dfrac{n}{2}(5n-1)$

8. $1 + 4 + 7 + 10 + \cdots + (3n-2) = \dfrac{n}{2}(3n-1)$

9. $1 + 2 + 2^2 + 2^3 + \cdots + 2^{n-1} = 2^n - 1$

10. $2(1 + 3 + 3^2 + 3^3 + \cdots + 3^{n-1}) = 3^n - 1$

11. $1 + 2 + 3 + 4 + \cdots + n = \dfrac{n(n+1)}{2}$

12. $1^2 + 2^2 + 3^2 + 4^2 + \cdots + n^2 = \dfrac{n(n+1)(2n+1)}{6}$

13. $1^3 + 2^3 + 3^3 + 4^3 + \cdots + n^3 = \dfrac{n^2(n+1)^2}{4}$

14. $\left(1 + \dfrac{1}{1}\right)\left(1 + \dfrac{1}{2}\right)\left(1 + \dfrac{1}{3}\right) \cdots \left(1 + \dfrac{1}{n}\right) = n + 1$

15. $\displaystyle\sum_{i=1}^{n} i^5 = \dfrac{n^2(n+1)^2(2n^2 + 2n - 1)}{12}$

16. $\displaystyle\sum_{i=1}^{n} i^4 = \dfrac{n(n+1)(2n+1)(3n^2 + 3n - 1)}{30}$

17. $\displaystyle\sum_{i=1}^{n} i(i+1) = \dfrac{n(n+1)(n+2)}{3}$

18. $\displaystyle\sum_{i=1}^{n} \dfrac{1}{(2i-1)(2i+1)} = \dfrac{n}{2n+1}$

In Exercises 19–28, find the indicated sum using the formulas for the sums of powers of integers.

19. $\displaystyle\sum_{n=1}^{20} n$ **20.** $\displaystyle\sum_{n=1}^{50} n$

21. $\displaystyle\sum_{n=1}^{6} n^2$ **22.** $\displaystyle\sum_{n=1}^{10} n^2$

23. $\displaystyle\sum_{n=1}^{5} n^3$ **24.** $\displaystyle\sum_{n=1}^{8} n^3$

25. $\displaystyle\sum_{n=1}^{6} n^4$ **26.** $\displaystyle\sum_{n=1}^{4} n^5$

27. $\displaystyle\sum_{n=1}^{6} (n^2 - n)$ **28.** $\displaystyle\sum_{n=1}^{10} (n^3 - n^2)$

In Exercises 29–34, find a formula for the sum of the first n terms of the sequence.

29. $3, 7, 11, 15, \ldots$

30. $25, 22, 19, 16, \ldots$

31. $1, \dfrac{9}{10}, \dfrac{81}{100}, \dfrac{729}{1000}, \ldots$

32. $3, -\dfrac{9}{2}, \dfrac{27}{4}, -\dfrac{81}{8}, \ldots$

33. $\dfrac{1}{4}, \dfrac{1}{12}, \dfrac{1}{24}, \dfrac{1}{40}, \ldots, \dfrac{1}{2n(n-1)}, \ldots$

34. $\dfrac{1}{2 \cdot 3}, \dfrac{1}{3 \cdot 4}, \dfrac{1}{4 \cdot 5}, \dfrac{1}{5 \cdot 6}, \ldots, \dfrac{1}{(n+1)(n+2)}, \ldots$

In Exercises 35–38, use mathematical induction to prove the given inequality for the indicated integer values of n.

35. $\left(\dfrac{4}{3}\right)^n > n, \quad n \geq 7$

36. $\dfrac{1}{\sqrt{1}} + \dfrac{1}{\sqrt{2}} + \dfrac{1}{\sqrt{3}} + \cdots + \dfrac{1}{\sqrt{n}} > \sqrt{n}, \quad n \geq 2$

37. $n! > 2^n, \quad n \geq 4$

38. $\left(\dfrac{x}{y}\right)^{n+1} < \left(\dfrac{x}{y}\right)^n$, if $n \geq 1$ and $0 < x < y$.

In Exercises 39–46, use mathematical induction to prove the given property for all positive integers n.

39. $(ab)^n = a^n b^n$ **40.** $\left(\dfrac{a}{b}\right)^n = \dfrac{a^n}{b^n}$

41. If $x_1 \neq 0$, $x_2 \neq 0$, . . . , $x_n \neq 0$, then
$(x_1 x_2 x_3 \cdots x_n)^{-1} = x_1^{-1} x_2^{-1} x_3^{-1} \cdots x_n^{-1}$.

42. If $x_1 > 0$, $x_2 > 0$, . . . , $x_n > 0$, then
$\ln(x_1 x_2 x_3 \cdots x_n) = \ln x_1 + \ln x_2 + \cdots + \ln x_n$.

43. Generalized Distributive Law:
$x(y_1 + y_2 + \cdots + y_n) = xy_1 + xy_2 + \cdots + xy_n$

44. $(a + bi)^n$ and $(a - bi)^n$ are complex conjugates for all $n \geq 1$.

45. A factor of $(n^3 + 3n^2 + 2n)$ is 3

46. A factor of $(2^{2n-1} + 3^{2n-1})$ is 5

In Exercises 47–50, write the first five terms of the sequence.

47. $a_0 = 1$
$a_n = a_{n-1} + 2$

48. $a_1 = 10$
$a_n = 4a_{n-1}$

49. $a_0 = 4$
$a_1 = 2$
$a_n = a_{n-1} - a_{n-2}$

50. $a_0 = 0$
$a_1 = 2$
$a_n = a_{n-1} + 2a_{n-2}$

In Exercises 51–60, write the first five terms of the sequence. Then calculate the first and second differences of the sequence. Does the sequence have a linear or quadratic model?

51. $f(1) = 0$
$a_n = a_{n-1} + 3$

52. $f(1) = 2$
$a_n = n - a_{n-1}$

53. $f(1) = 3$
$a_n = a_{n-1} - n$

54. $f(2) = -3$
$a_n = -2a_{n-1}$

55. $a_0 = 0$
$a_n = a_{n-1} + n$

56. $a_0 = 2$
$a_n = (a_{n-1})^2$

57. $f(1) = 2$
$a_n = a_{n-1} + 2$

58. $f(1) = 0$
$a_n = a_{n-1} + 2n$

59. $a_0 = 1$
$a_n = a_{n-1} + n^2$

60. $a_0 = 0$
$a_n = a_{n-1} - 1$

In Exercises 61–64, find a quadratic model for the sequence with the indicated terms.

61. $a_0 = 3$, $a_1 = 3$, $a_4 = 15$

62. $a_0 = 7$, $a_1 = 6$, $a_3 = 10$

63. $a_0 = -3$, $a_2 = 1$, $a_4 = 9$

64. $a_0 = 3$, $a_2 = 0$, $a_6 = 36$

9.5 THE BINOMIAL THEOREM

Binomial Coefficients / Pascal's Triangle / Binomial Expansions

Binomial Coefficients

Recall that a **binomial** is a polynomial that has two terms. In this section, you will study a formula that gives a quick method of raising a binomial to a power. To begin, look at the expansions of $(x + y)^n$ for several values of n.

$$(x + y)^0 = 1$$
$$(x + y)^1 = x + y$$
$$(x + y)^2 = x^2 + 2xy + y^2$$
$$(x + y)^3 = x^3 + 3x^2y + 3xy^2 + y^3$$
$$(x + y)^4 = x^4 + 4x^3y + 6x^2y^2 + 4xy^3 + y^4$$
$$(x + y)^5 = x^5 + 5x^4y + 10x^3y^2 + 10x^2y^3 + 5xy^4 + y^5$$

There are several observations you can make about these expansions of $(x + y)^n$.

1. In each expansion, there are $n + 1$ terms.
2. In each expansion, x and y have symmetrical roles. The powers of x decrease by 1 in successive terms, whereas the powers of y increase by 1.
3. The sum of the powers of each term in a binomial expansion is n. For example, in the expansion of $(x + y)^5$, the sum of the powers of each term is 5, as follows.

$$4 + 1 = 5 \qquad 3 + 2 = 5$$
$$(x + y)^5 = x^5 + 5\overbrace{x^4 y^1} + 10\overbrace{x^3 y^2} + 10x^2 y^3 + 5xy^4 + y^5$$

4. The first term is x^n, the last term is y^n, and each of these terms has a coefficient of 1.
5. The coefficients increase and then decrease in a symmetrical pattern. For $(x + y)^5$, the pattern is

$$1 \quad 5 \quad 10 \quad 10 \quad 5 \quad 1.$$

The most difficult part of a binomial expansion is finding the coefficients of the interior terms. To find these **binomial coefficients,** you can use a well-known theorem called the **Binomial Theorem.**

REMARK   The symbol $\binom{n}{m}$ is often used in place of $_nC_m$ to denote binomial coefficients.

THE BINOMIAL THEOREM

In the expansion of $(x + y)^n$,

$$(x + y)^n = x^n + nx^{n-1}y + \cdots + {}_nC_m x^{n-m}y^m + \cdots + nxy^{n-1} + y^n,$$

the coefficient of $x^{n-m}y^m$ is given by

$$_nC_m = \frac{n!}{(n - m)!\,m!}.$$

Technology Note

Most graphing utilities have the ability to compute binomial coefficients. Learn how to use this feature of your graphing utility, and then use it to confirm the results given in Example 1.

EXAMPLE 1   **Finding Binomial Coefficients**

Find the following binomial coefficients.

A. $_8C_2$ **B.** $_{10}C_3$ **C.** $_7C_0$ **D.** $_6C_6$

SOLUTION

Note in parts (A) and (B) how the numerator is factored to cancel part of the denominator.

A. $_8C_2 = \dfrac{8!}{6!2!} = \dfrac{(8 \cdot 7) \cdot 6!}{6! \cdot 2!} = \dfrac{8 \cdot 7}{2 \cdot 1} = 28$

B. $_{10}C_3 = \dfrac{10!}{7!3!} = \dfrac{(10 \cdot 9 \cdot 8) \cdot 7!}{7! \cdot 3!} = \dfrac{10 \cdot 9 \cdot 8}{3 \cdot 2 \cdot 1} = 120$

C. $_7C_0 = \dfrac{7!}{7! \cdot 0!} = 1$

D. $_6C_6 = \dfrac{6!}{0! \cdot 6!} = 1$

Some calculators have a built-in binomial coefficient key. The TI-81 uses the nCr symbol in the math menu. Try verifying the above computations with your calculator.

When $m \neq 0$ or $m \neq n$, as in parts (A) and (B) above, there is a simple pattern for evaluating binomial coefficients.

$$_8C_2 = \frac{\overbrace{8 \cdot 7}^{2 \text{ factors}}}{\underbrace{2 \cdot 1}_{2 \text{ factorial}}} \quad \text{and} \quad _{10}C_3 = \frac{\overbrace{10 \cdot 9 \cdot 8}^{3 \text{ factors}}}{\underbrace{3 \cdot 2 \cdot 1}_{3 \text{ factorial}}}$$

In general, you have the following.

$$_nC_m = \frac{\overbrace{n(n-1)(n-2)\cdots}^{m \text{ factors}}}{m!}, \qquad 0 < m < n$$

EXAMPLE 2 Finding Binomial Coefficients

Find the following binomial coefficients.

A. $_7C_3$ **B.** $_7C_4$ **C.** $_{12}C_1$ **D.** $_{12}C_{11}$

SOLUTION

A. $_7C_3 = \dfrac{7 \cdot 6 \cdot 5}{3 \cdot 2 \cdot 1} = 35$

B. $_7C_4 = \dfrac{7 \cdot 6 \cdot 5 \cdot 4}{4 \cdot 3 \cdot 2 \cdot 1} = 35$

C. $_{12}C_1 = \dfrac{12}{1} = 12$

D. $_{12}C_{11} = \dfrac{12!}{1!11!} = \dfrac{(12) \cdot 11!}{1! \cdot 11!} = \dfrac{12}{1} = 12$

It is not a coincidence that the results in parts (A) and (B), and similarly in parts (C) and (D), are the same. In general it is true that $_nC_m = {_nC_{n-m}}$. For instance,

$$_6C_0 = {_6C_6} = 1, \quad _6C_1 = {_6C_5} = 6, \quad \text{and} \quad _6C_2 = {_6C_4} = 15.$$

This shows the symmetric property of binomial coefficients that was identified earlier. Notice how this property is used in Example 3.

EXAMPLE 3   **Finding a Binomial Coefficient**

Find the binomial coefficient $_{12}C_{10}$.

SOLUTION

Rather than calculating the binomial coefficient

$$_{12}C_{10} = \frac{12 \cdot 11 \cdot 10 \cdot 9 \cdot 8 \cdot 7 \cdot 6 \cdot 5 \cdot 4 \cdot 3}{10 \cdot 9 \cdot 8 \cdot 7 \cdot 6 \cdot 5 \cdot 4 \cdot 3 \cdot 2 \cdot 1},$$

you can use the fact that $_{12}C_{10} = {}_{12}C_2$ as follows.

$$_{12}C_{10} = {}_{12}C_2 = \frac{12 \cdot 11}{2 \cdot 1} = 66$$

Pascal's Triangle

There is a convenient way to remember the pattern for binomial coefficients. By arranging the coefficients in a triangular pattern, you obtain the following array, which is called **Pascal's Triangle.** This triangle is named after the famous French mathematician Blaise Pascal (1623–1662).

```
                    1
                 1     1
              1     2     1
           1     3     3     1
        1     4     6     4     1
     1     5    10    10     5     1
  1     6    15    20    15     6     1
1     7    21    35    35    21     7     1
```

The first and last numbers in each row of Pascal's Triangle are 1. Every other number in each row is formed by adding the two numbers immediately above the number. For example, the two numbers above 35 are 15 and 20.

$$15 \quad 20$$
$$\searrow \swarrow$$
$$35 \qquad\qquad\qquad 15 + 20 = 35$$

Pascal noticed that numbers in this triangle are precisely the same numbers that are the coefficients of binomial expansions, as follows.

$$(x + y)^0 = 1$$
$$(x + y)^1 = 1x + 1y$$
$$(x + y)^2 = 1x^2 + 2xy + 1y^2$$
$$(x + y)^3 = 1x^3 + 3x^2y + 3xy^2 + 1y^3$$
$$(x + y)^4 = 1x^4 + 4x^3y + 6x^2y^2 + 4xy^3 + 1y^4$$
$$(x + y)^5 = 1x^5 + 5x^4y + 10x^3y^2 + 10x^2y^3 + 5xy^4 + 1y^5$$
$$(x + y)^6 = 1x^6 + 6x^5y + 15x^4y^2 + 20x^3y^3 + 15x^2y^4 + 6xy^5 + 1y^6$$
$$(x + y)^7 = 1x^7 + 7x^6y + 21x^5y^2 + 35x^4y^3 + 35x^3y^4 + 21x^2y^5 + 7xy^6 + 1y^7$$

Because the top row in Pascal's Triangle corresponds to the binomial expansion $(x + y)^0 = 1$, it is called the **zero row.** Similarly, the next row corresponds to the binomial expansion $(x + y)^1 = 1x + 1y$, and it is called the **first row.** In general, the **nth row** in Pascal's Triangle gives the coefficients of $(x + y)^n$.

EXAMPLE 4   Using Pascal's Triangle

Use Pascal's Triangle to find the following binomial coefficients.

$$_8C_0, \; _8C_1, \; _8C_2, \; _8C_3, \; _8C_4, \; _8C_5, \; _8C_6, \; _8C_7, \; _8C_8$$

SOLUTION

These nine binomial coefficients represent the eighth row of Pascal's Triangle. Thus, using the seventh row of the triangle, you can calculate the numbers in the eighth row, as follows

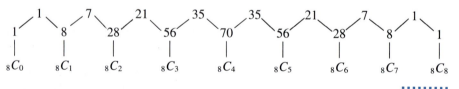

Binomial Expansions

EXAMPLE 5   Expanding a Binomial

Write the expansion of $(x + 1)^3$.

SOLUTION

The binomial coefficients from the third row of Pascal's Triangle are 1, 3, 3, 1. Therefore, the expansion is as follows.

$$(x + 1)^3 = (1)x^3 + (3)x^2(1) + (3)x(1^2) + (1)(1^3)$$
$$= x^3 + 3x^2 + 3x + 1$$

To expand binomials representing *differences,* rather than sums, you must alternate signs. Here are two examples.

$$(x - 1)^3 = x^3 - 3x^2 + 3x - 1$$
$$(x - 1)^4 = x^4 - 4x^3 + 6x^2 - 4x + 1$$

EXAMPLE 6   Expanding a Binomial

Write the expansion of $(x - 2)^3$.

SOLUTION

The binomial coefficients from the third row of Pascal's Triangle are 1, 3, 3, 1. Therefore, the expansion is as follows.

$$(x - 1)^3 = (1)x^3 - (3)x^2(2) + (3)x(2^2) - (1)(2^3)$$
$$= x^3 - 6x^2 + 12x - 8$$

EXAMPLE 7   Expanding a Binomial

Write the expansion for the following expression.

$$(x + 3)^4$$

SOLUTION

The binomial coefficients from the fourth row of Pascal's Triangle are 1, 4, 6, 4, 1. Therefore, the expansion is as follows.

$$(x + 3)^4 = (1)x^4 + (4)x^3(3) + (6)x^2(3^2) + (4)x(3^3) + (1)(3^4)$$
$$= x^4 + 12x^3 + 54x^2 + 108x + 81$$

EXAMPLE 8   Expanding a Binomial

Write the expansion of $(x - 2y)^4$.

SOLUTION

The binomial coefficients from the fourth row of Pascal's Triangle are 1, 4, 6, 4, 1. Therefore, the expansion is as follows.

$$(x - 2y)^4 = (1)x^4 - (4)x^3(2y) + (6)x^2(2y)^2 - (4)x(2y)^3 + (1)(2y)^4$$
$$= x^4 - 8x^3y + 24x^2y^2 - 32xy^3 + 16y^4$$

EXAMPLE 9   Finding a Specified Term in a Binomial Expansion

Find the sixth term in the expansion of $(3a + 2b)^{12}$.

SOLUTION

Using the Binomial Theorem, let $x = 3a$ and $y = 2b$ and note that in the *sixth* term the exponent of y is $m = 5$ and the exponent of x is $n - m = 12 - 5 = 7$. Consequently, the sixth term of the expansion is

$$_{12}C_5 x^7 y^5 = \frac{12 \cdot 11 \cdot 10 \cdot 9 \cdot 8}{5!} (3a)^7 (2b)^5.$$

DISCUSSION PROBLEM

THE ROWS OF PASCAL'S TRIANGLE

By adding the terms in each of the rows of Pascal's Triangle, you obtain the following.

Row 0: $1 = 1$
Row 1: $1 + 1 = 2$
Row 2: $1 + 2 + 1 = 4$
Row 3: $1 + 3 + 3 + 1 = 8$
Row 4: $1 + 4 + 6 + 4 + 1 = 16$

Can you find a pattern for this sequence? Use this pattern to find the sum of the terms in the 10th row of Pascal's Triangle. Then check your answer by actually adding the terms of the 10th row.

WARM-UP

The following warm-up exercises involve skills that were covered in earlier sections. You will use these skills in the exercise set for this section.

In Exercises 1–6, perform the indicated operations and/or simplify.

1. $5x^2(x^3 + 3)$ **2.** $(x + 5)(x^2 - 3)$ **3.** $(x + 4)^2$
4. $(2x - 3)^2$ **5.** $x^2 y(3xy^{-2})$ **6.** $(-2z)^5$

In Exercises 7–10, evaluate the expression.

7. $5!$ **8.** $\dfrac{8!}{5!}$ **9.** $\dfrac{10!}{7!}$ **10.** $\dfrac{6!}{3!3!}$

SECTION 9.5 · EXERCISES

In Exercises 1–10, evaluate $_n C_m$. If your graphing utility has a binomial coefficient key (also called combinations function), use it to verify your answer.

1. $_5 C_3$ **2.** $_8 C_6$

3. $_{12} C_0$ **4.** $_{20} C_{20}$

5. $_{20} C_{15}$ **6.** $_{12} C_5$

7. $_{100} C_{98}$ **8.** $_{10} C_4$

9. $_{100} C_2$ **10.** $_{10} C_6$

In Exercises 11–30, use the Binomial Theorem to expand and simplify the expression.

11. $(x + 1)^4$ **12.** $(x + 1)^6$

13. $(a + 2)^3$ **14.** $(a + 3)^4$

15. $(y - 2)^4$ **16.** $(y - 2)^5$

17. $(x + y)^5$ **18.** $(x + y)^6$

19. $(r + 3s)^6$ **20.** $(x + 2y)^4$

21. $(x - y)^5$ **22.** $(2x - y)^5$

23. $(1 - 2x)^3$ **24.** $(5 - 3y)^3$

25. $(x^2 + 5)^4$ **26.** $(x^2 + y^2)^6$

27. $\left(\dfrac{1}{x} + y\right)^5$ **28.** $\left(\dfrac{1}{x} + 2y\right)^6$

29. $2(x - 3)^4 + 5(x - 3)^2$

30. $3(x + 1)^5 - 4(x + 1)^3$

In Exercises 31–36, use the Binomial Theorem to expand the complex number. Simplify your answer by recalling that $i^2 = -1$.

31. $(1 + i)^4$ **32.** $(2 - i)^5$

33. $(2 - 3i)^6$ **34.** $(5 + \sqrt{-9})^3$

35. $\left(\dfrac{-1}{2} + \dfrac{\sqrt{3}}{2}i\right)^3$ **36.** $(5 - \sqrt{3}i)^4$

In Exercises 37–40, expand the binomial using Pascal's Triangle to determine the coefficients.

37. $(2t - s)^5$ **38.** $(x + 2y)^5$

39. $(3 - 2z)^4$ **40.** $(3y + 2)^5$

In Exercises 41–48, find the coefficient a of the given term in the expansion of the binomial.

Binomial	Term
41. $(x + 3)^{12}$	ax^5
42. $(x^2 + 3)^{12}$	ax^8
43. $(x - 2y)^{10}$	$ax^8 y^2$
44. $(4x - y)^{10}$	$ax^2 y^8$
45. $(3x - 2y)^9$	$ax^4 y^5$
46. $(2x - 3y)^8$	$ax^6 y^2$
47. $(x^2 + 1)^{10}$	ax^8
48. $(z^2 - 1)^{12}$	az^6

Probability In Exercises 49–52, consider n independent trials of an experiment where each trial has two possible outcomes called success and failure. The probability of a success on each trial is p and the probability of a failure is $q = 1 - p$. In this context the term

$$_n C_k p^k q^{n-k}$$

of the expansion of $(p + q)^n$ gives the probability of k successes in the n trials of the experiment. Expand the given binomial and find the required probability.

49. A fair coin is tossed seven times. Find the probability of obtaining four heads.

$$\left(\frac{1}{2} + \frac{1}{2}\right)^7$$

50. The probability of a baseball player getting a hit on any given time at bat is $\frac{1}{4}$. Find the probability that the player gets three hits in the next 10 times at bat.

$$\left(\frac{1}{4} + \frac{3}{4}\right)^{10}$$

51. The probability of a salesperson making a sale with any one customer is $\frac{1}{3}$. If the salesperson makes eight contacts on a given day, find the probability that he or she makes four sales.

$$\left(\frac{1}{3} + \frac{2}{3}\right)^8$$

52. Find the probability that the salesperson in Exercise 51 makes four sales in eight contacts if the probability of a sale with any one customer is $\frac{1}{2}$.

$$\left(\frac{1}{2} + \frac{1}{2}\right)^8$$

In Exercises 53–56, use the Binomial Theorem to approximate the given quantity accurate to three decimal places. For example, in Exercise 53 you have

$(1.02)^8 = (1 + 0.02)^8 = 1 + 8(0.02) + 28(0.02)^2 + \cdots$

53. $(1.02)^8$ **54.** $(2.005)^{10}$
55. $(2.99)^{12}$ **56.** $(1.98)^9$

In Exercises 57–60, use a graphing utility to graph f and g in the same viewing rectangle. What is the relationship between the two graphs? Use the Binomial Theorem to write the polynomial function g in standard form.

57. $f(x) = -x^2 + 3x + 2$
   $g(x) = f(x - 2)$
58. $f(x) = 2x^2 - 4x + 1$
   $g(x) = f(x + 3)$
59. $f(x) = x^3 - 4x$
   $g(x) = f(x + 4)$
60. $f(x) = -x^4 + 4x^2 - 1$
   $g(x) = f(x - 3)$

61. *Life Insurance*   The average amount of life insurance per household (for households that carry life insurance) from 1970 through 1988 can be approximated by the model

$f(t) = 0.2187t^2 + 0.6715t + 26.67, \qquad 0 \le t \le 18.$

In this model $f(t)$ represents the amount of life insurance (in thousands of dollars) and t represents the year, with $t = 0$ corresponding to 1970 (see figure). You want to adjust this model so that $t = 0$ corresponds to 1980 rather than 1970. To do this, you shift the graph of f 10 units *to the left* and obtain

$g(t) = f(t + 10) = 0.2187(t + 10)^2 + 0.6715(t + 10)$
$+ 26.67.$

Write this new polynomial function in standard form. (*Source:* American Council of Life Insurance)

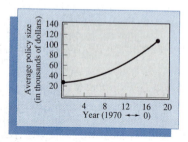

FIGURE FOR 61

62. *Health Maintenance Organizations*   The number of people enrolled in health maintenance organizations (HMOs) in the United States from 1976 through 1988 can be approximated by the model

$f(t) = 215t^2 - 470t + 6700, \qquad 0 \le t \le 12.$

In this model, $f(t)$ represents the number of people (in thousands) and t represents the year, with $t = 0$ corresponding to 1976 (see figure). You want to adjust this model so that $t = 0$ corresponds to 1980 rather than 1976. To do this, shift the graph of f four units *to the left* and obtain

$g(t) = f(t + 4) = 215(t + 4)^2 - 470(t + 4) + 6700.$

Write this new polynomial function in standard form. (*Source:* Group Health Insurance Association of America)

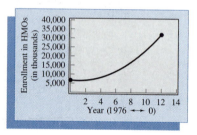

FIGURE FOR 62

In Exercises 63–66, prove the given property for all integers m and n where $0 \le m \le n$.

63. $_nC_m = {}_nC_{n-m}$
64. $_nC_0 - {}_nC_1 + {}_nC_2 - \cdots \pm {}_nC_n = 0$
65. $_{n+1}C_m = {}_nC_m + {}_nC_{m-1}$
66. $_{2n}C_n = ({}_nC_0)^2 + ({}_nC_1)^2 + ({}_nC_2)^2 + \cdots + ({}_nC_n)^2$

67. Prove that the sum of the numbers in the nth row of Pascal's Triangle is 2^n. [*Hint:* Consider $2^n = (1 + 1)^n$.]
68. Use a graphing utility to graph the following functions in the same viewing rectangle. Compare the graphs as you add terms. Which two functions have identical graphs and why?
   (a) $f(x) = (1 - x)^3$
   (b) $g(x) = 1 - 3x$
   (c) $h(x) = 1 - 3x + 3x^2$
   (d) $p(x) = 1 - 3x + 3x^2 - x^3$

9.6   COUNTING PRINCIPLES, PERMUTATIONS, COMBINATIONS

**Simple Counting Problems   /   Counting Principles   /
Permutations   /   Combinations**

Simple Counting Problems

In this section and in Section 9.7, we give a brief introduction to some of the basic counting principles and their application to probability. In Section 9.7, you will see that much of the probability has to do with counting the number of ways an event can occur. Examples 1, 2, and 3 describe some simple counting problems.

EXAMPLE 1   A Random Number Generator

A random number generator (on a computer) selects an integer from 1 to 40. Find the number of ways the following events can occur.

A. An even integer is selected.

B. A number less than 10 is selected.

C. A prime number is selected.

SOLUTION

A. Since half of the numbers between 1 and 40 are even, this event can occur in 20 different ways.

B. The integers between 1 and 40 that are less than 10 are given in the following set.

$$\{1, 2, 3, 4, 5, 6, 7, 8, 9\}$$

Since this set has nine members, there are nine different ways this event can happen.

C. The prime numbers between 1 and 40 are given in the following set.

$$\{2, 3, 5, 7, 11, 13, 17, 19, 23, 29, 31, 37\}$$

Since this set has 12 members, there are 12 different ways this event can happen.

EXAMPLE 2   Selecting Pairs of Numbers at Random

Eight pieces of paper are numbered from 1 to 8 and placed in a box. One piece of paper is drawn from the box, its number is written down, and the piece of paper is replaced in the box. Then a second piece of paper is drawn from the box, and its number is written down. Finally, the two numbers are added together. How many different ways can a total of 12 be obtained?

SOLUTION

To solve this problem, you count the different ways that a total of 12 can be obtained using two numbers between 1 and 8.

First number + Second number = 12

After considering the various possibilities, you see that this equation can be solved in the following five ways.

First Number	Second Number
4	8
5	7
6	6
7	5
8	4

Thus, a total of 12 can be obtained in five different ways.

Solving counting problems can be tricky. Often, seemingly minor changes in the statement of a problem can affect the answer. For instance, compare the counting problem in the next example with that given in Example 2.

EXAMPLE 3   Selecting Pairs of Numbers at Random

Eight pieces of paper are numbered from 1 to 8 and placed in a box. Two pieces of paper are drawn from the box, the number on each piece of paper is written down, and the two are totaled. How many different ways can a total of 12 be obtained?

SOLUTION

To solve this problem, you count the different ways that a total of 12 can be obtained *using two different numbers* between 1 and 8.

First Number	Second Number
4	8
5	7
7	5
8	4

Thus, a total of 12 can be obtained in four different ways.

REMARK   The difference between the counting problems in Examples 2 and 3 is that the random selection in Example 2 occurs **with replacement,** whereas the random selection in Example 3 occurs **without replacement,** which eliminates the possibility of choosing two 6's.

Counting Principles

In the first three examples, we looked at simple counting problems in which we can *list* each possible way that an event can occur. When it is possible, this is always the best way to solve a counting problem. However, some events can occur in so many different ways that it is not feasible to write out the entire list. In such cases, we must rely on formulas and counting principles. The most important of these is the **Fundamental Counting Principle.**

REMARK   The Fundamental Counting Principle can be extended to three or more events. For instance, the number of ways that three events E_1, E_2, and E_3 can occur is $m_1 \cdot m_2 \cdot m_3$.

FUNDAMENTAL COUNTING PRINCIPLE

Let E_1 and E_2 be two events. The first event E_1 can occur in m_1 different ways. After E_1 has occurred, E_2 can occur in m_2 different ways. The number of ways that the two events can occur is

$$m_1 \cdot m_2.$$

EXAMPLE 4   Applying the Fundamental Counting Principle

How many different pairs of letters from the English alphabet (with replacement) are possible? (Disregard the difference between uppercase and lowercase letters.)

SOLUTION

This experiment has two events. The first event is the choice of the first letter, and the second event is the choice of the second letter. Because the English alphabet contains 26 letters, it follows that each event can occur in 26 ways.

Two-letter "words"

26   26

Thus, using the Fundamental Counting Principle, it follows that the number of two-letter "words" is

$$26 \cdot 26 = 676.$$

EXAMPLE 5   Applying the Fundamental Counting Principle

Telephone numbers in the United States have ten digits. The first three are the *area code* and the next seven are the *local telephone number*. How many different telephone numbers are possible within each area code? (Note that a local telephone number cannot begin with 0 or 1.)

SOLUTION

Since the first digit cannot be 0 or 1, there are only eight choices for the first digit. For each of the other six digits, there are ten choices.

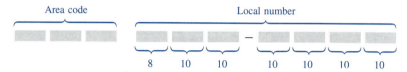

Area code   Local number

8   10   10   10   10   10   10

Thus, using the Fundamental Counting Principle, the number of local telephone numbers that are possible within each area code is

$$8 \cdot 10 \cdot 10 \cdot 10 \cdot 10 \cdot 10 \cdot 10 = 8,000,000.$$

Permutations

One important application of the Fundamental Counting Principle is in determining the number of ways that *n* elements can be arranged (in order). We call an ordering of *n* elements a **permutation** of the elements.

DEFINITION OF PERMUTATION

A **permutation** of *n* different elements is an ordering of the elements such that one element is first, one is second, one is third, and so on.

EXAMPLE 6   Listing Permutations

Write the different permutations of the letters A, B, and C.

SOLUTION

These three letters can be arranged in the following six different ways.

A, B, C,   B, A, C,   C, A, B
A, C, B,   B, C, A   C, B, A

Thus, you see that these three letters have six different permutations.

In Example 6, you were able to list the different permutations of three letters. However, you could also have used the Fundamental Counting Principle. To do this, you could reason that there are three choices for the first letter, two choices for the second, and only one choice for the third, as follows.

Permutations of three letters

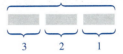

3   2   1

Thus, the number of permutations of three letters is

$3 \cdot 2 \cdot 1 = 3! = 6.$

EXAMPLE 7   Finding the Number of Permutations of n Elements

How many different permutations are possible for the letters A, B, C, D, E, and F?

SOLUTION

There are too many different permutations to list, so you use the following reasoning.

1st position:	Any of the *six* letters
2nd position:	Any of the remaining *five* letters
3rd position:	Any of the remaining *four* letters
4th position:	Any of the remaining *three* letters
5th position:	Any of the remaining *two* letters
6th position:	The *one* remaining letter

Thus, the number of choices for the six positions is as follows.

Permutations of six letters

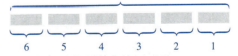

6   5   4   3   2   1

Using the Fundamental Counting Principle, you find that the total number of permutations of the six letters is

$6 \cdot 5 \cdot 4 \cdot 3 \cdot 2 \cdot 1 = 6! = 720.$

The results obtained in Examples 6 and 7 can be generalized to conclude that the number of permutations of n different elements is $n!$.

NUMBER OF PERMUTATIONS OF N ELEMENTS

The number of permutations of n elements is given by

$$n \cdot (n - 1) \cdots 4 \cdot 3 \cdot 2 \cdot 1 = n!.$$

In other words, there are $n!$ different ways that n elements can be ordered.

EXAMPLE 8   Finding the Number of Permutations

Suppose that you are a supervisor for 11 different employees. One of your responsibilities is to perform an annual evaluation for each employee, and then rank the 11 different performances. (In other words, one employee must be ranked as having the best performance, one must be ranked second, and so on.) How many different rankings are possible?

SOLUTION

Since there are 11 different employees, you have 11 choices for first ranking. After choosing the first ranking, you can choose any of the remaining 10 for second ranking, and so on.

Rankings of 11 employees

    11   10   9   8   7   6   5   4   3   2   1

Thus, the number of different rankings is

$$11! = 39,916,800.$$

Occasionally, you may be interested in ordering a *subset* of a collection of elements rather than the entire collection. For example, you might want to choose (and order) m elements out of a collection of n elements. Such an ordering is a **permutation of n elements taken m at a time.**

EXAMPLE 9   Permutations of n Elements Taken m at a Time

Eight horses are running in a race. In how many different ways can these horses come in first, second, and third? (Assume that there are no ties.)

SOLUTION

You have the following possibilities.

Win (1st position):   *Eight* choices
Place (2nd position):   *Seven* choices
Show (3rd position):   *Six* choices

Using the Fundamental Counting Principle, multiply these three numbers together to obtain the following.

Different orders of horses

8   7   6

Thus, there are $8 \cdot 7 \cdot 6 = 336$ different orders.

The result of Example 9 can be generalized as follows.

PERMUTATIONS OF *N* ELEMENTS TAKEN *M* AT A TIME

The number of permutations of *n* elements taken *m* at a time is

$$_nP_m = \frac{n!}{(n-m)!} = n(n-1)(n-2) \cdots (n-m+1).$$

Using this formula, you can rework Example 9 to find that the number of permutations of eight horses taken three at a time is

$$_8P_3 = \frac{8!}{(8-3)!} = \frac{8!}{5!} = \frac{8 \cdot 7 \cdot 6 \cdot 5!}{5!} = 8 \cdot 7 \cdot 6 = 336,$$

which is the same answer you obtained in the solution of Example 9.

Remember that for permutations, order is important. Thus, if you are looking at the possible permutations of the letters A, B, C, and D taken three at a time, the permutations (A, B, D) and (B, A, D) would be different (since the *order* of the elements is different).

Suppose, however, that you are asked to find the possible permutations of the letters A, A, B, and C. The total number of permutations of the four letters would be $_4P_4 = 4!$. However, not all of these arrangements would be *distin guishable* because there are two A's in the list. To find the number of distinguishable permutations, you can use the following formula.

DISTINGUISHABLE PERMUTATIONS

Suppose a set of *n* objects has n_1 of one kind of object, n_2 of a second kind, n_3 of a third kind, and so on, with $n = n_1 + n_2 + n_3 + \cdots + n_k$. Then the number of **distinguishable permutations** of the *n* objects is

$$\frac{n!}{n_1! \cdot n_2! \cdot n_3! \cdots n_k!}.$$

EXAMPLE 10   Distinguishable Permutations

In how many distinguishable ways can the letters in BANANA be written?

SOLUTION

This word has six letters, of which three are A's, two are N's, and one is a B. Thus, the number of distinguishable ways the letter can be written is

$$\frac{6!}{3! \cdot 2! \cdot 1!} = \frac{6 \cdot 5 \cdot 4 \cdot 3!}{3! \cdot 2!} = 60.$$

The 60 different "words" are as follows.

AAABNN	AAANBN	AAANNB	AABANN	AABNAN	AABNNA
AANABN	AANANB	AANBAN	AANBNA	AANNAB	AANNBA
ABAANN	ABANAN	ABANNA	ABNAAN	ABNANA	ABNNAA
ANAABN	ANAANB	ANABAN	ANABNA	ANANAB	ANANBA
ANBAAN	ANBANA	ANBNAA	ANNAAB	ANNABA	ANNBAA
BAAANN	BAANAN	BAANNA	BANAAN	BANANA	BANNAA
BNAAAN	BNAANA	BNANAA	BNNAAA	NAAABN	NAAANB
NAABAN	NAABNA	NAANAB	NAANBA	NABAAN	NABANA
NABNAA	NANAAB	NANABA	NANBAA	NBAAAN	NBAANA
NBANAA	NBNAAA	NNAAAB	NNAABA	NNABAA	NNBAAA

Combinations

When counting the numbers of possible permutations of a set of elements, *order* is important. As a final topic in this section, we look at a method of selecting subsets of a larger set in which order is *not important.* Such subsets are **combinations of *n* elements taken *m* at a time.** For instance, the combinations

{A, B, C}   and   {B, A, C}

are equivalent because both sets contain the same three elements, and the order in which the elements are listed is *not important.* Hence, we would count only one of the two sets. A common example of how a combination occurs is a card game in which the player is free to reorder the cards after they have been dealt.

EXAMPLE 11   Combination of *n* Elements Taken *m* at a Time

In how many different ways can three letters be chosen from the letters A, B, C, D, and E? (The order of the three letters is not important.)

SOLUTION

The following subsets represent the different combinations of three letters that can be chosen from five letters.

{A, B, C}   {A, B, D}
{A, B, E}   {A, C, D}
{A, C, E}   {A, D, E}
{B, C, D}   {B, C, E}
{B, D, E}   {C, D, E}

From this list, you conclude that there are ten different ways that three letters can be chosen from five letters. ■■■■■■■■■

The formula for the number of *combinations* of n elements taken m at a time is as follows.

COMBINATIONS OF *N* ELEMENTS TAKEN *M* AT A TIME

The number of combinations of n elements taken m at a time is

$$_nC_m = \frac{n!}{(n-m)!m!}.$$

Technology Note _____

Most graphing utilities allow you to evaluate combinations of n elements taken r at a time. For instance, on the TI-81, you can verify that $_5C_3 = 10$ by entering 5, going to the MATH PRB menu and selecting $_nC_r$, and finally entering 3.

Note that the formula for $_nC_m$ is the same as the one given for binomial coefficients. To see how this formula is used, solve the counting problem given in Example 11. In that problem, you must find the number of combinations of five elements taken three at a time. Thus, $n = 5$, $m = 3$, and the number of combinations is

$$_5C_3 = \frac{5!}{2!3!} = \frac{5 \cdot 4 \cdot 3}{3 \cdot 2 \cdot 1} = 10,$$

which is the same answer you obtained in Example 11.

EXAMPLE 12   Combinations of *n* Elements Taken *m* at a Time

A standard poker hand consists of five cards dealt from a deck of 52. How many different poker hands are possible? (After the cards are dealt, the player may reorder them, and therefore order is not important.)

SOLUTION

Use the formula for the number of combinations of 52 elements taken five at a time, as follows.

$$_{52}C_5 = \frac{52!}{47!5!} = \frac{52 \cdot 51 \cdot 50 \cdot 49 \cdot 48}{5 \cdot 4 \cdot 3 \cdot 2 \cdot 1} = 2{,}598{,}960 \text{ different hands}$$

■■■■■■■■■

EXAMPLE 13   Combinations and the Fundamental Counting Principle

The traveling squad for a college basketball team consists of two centers, five forwards, and four guards. In how many ways can the coach select a starting team of one center, two forwards, and two guards?

SOLUTION

The number of ways to select one center is

$$_2C_1 = \frac{2!}{1!(1!)} = 2.$$

The number of ways to select two forwards from among five is

$$_5C_2 = \frac{5!}{3!(2!)} = 10.$$

The number of ways to select two guards from among four is

$$_4C_2 = \frac{4!}{2!(2!)} = 6.$$

Therefore, the total number of ways to select a starting team is

$$_2C_1 \cdot {_5C_2} \cdot {_4C_2} = 2 \cdot 10 \cdot 6 = 120.$$

DISCUSSION PROBLEM

YOU BE THE INSTRUCTOR

Suppose you are teaching an algebra class and are writing a test for this chapter. Create two word problems that you think are appropriate for the test. One of the word problems should deal with permutations and the other should deal with combinations. (Assume that your students will have only five minutes to solve each problem.)

WARP-UP

The following warm-up exercises involve skills that were covered in earlier sections. You will use these skills in the exercise set for this section.

In Exercises 1–4, evaluate the expression.

1. $13 \cdot 8^2 \cdot 2^3$

2. $10^2 \cdot 9^3 \cdot 4$

3. $\dfrac{12!}{2!(7!)(3!)}$

4. $\dfrac{25!}{22!}$

In Exercises 5 and 6, find the binomial coefficient.

5. $_{12}C_7$

6. $_{25}C_{22}$

In Exercises 7–10, simplify the expression.

7. $\dfrac{n!}{(n-4)!}$

8. $\dfrac{(2n)!}{4(2n-3)!}$

9. $\dfrac{2 \cdot 4 \cdot 6 \cdot 8 \cdots (2n)}{2^n}$

10. $\dfrac{3 \cdot 6 \cdot 9 \cdot 12 \cdots (3n)}{3^n}$

SECTION 9.6 · EXERCISES

1. A bag contains 10 marbles numbered 1 through 10. A marble is selected, its number is recorded, and the marble is *replaced* in the bag. Then a second marble is drawn and its number recorded. Finally, the recorded numbers are added. How many different ways can a sum of 8 be obtained?

2. A bag contains 10 marbles numbered 1 through 10. Two marbles are selected and their numbers are recorded. Then the recorded numbers are added. How many different ways can a sum of 8 be obtained?

3. *Job Applicants* A small college needs two additional faculty members: a chemist and a statistician. In how many ways can these positions be filled if there are three applicants for the chemistry position and four for the position in statistics?

4. *Computer Systems* A customer in a computer store can choose one of three monitors, one of two keyboards, and one of four computers. If all the choices are compatible, how many different systems could be chosen?

5. *Toboggan Ride* Four people are lining up for a ride on a toboggan, but only two of the four are willing to take the first position. With that constraint, in how many ways can the four people be seated on the toboggan?

6. *Course Schedule* A college student is preparing a course schedule for the next semester. The student may select one of two mathematics courses, one of three science courses, and one of five courses from the social sciences and humanities. How many schedules are possible?

7. *License Plate Numbers* In a certain state the automobile license plates consist of two letters followed by a four-digit number. How many distinct license plate numbers can be formed?

8. *License Plate Numbers* In a certain state the automobile license plates consist of two letters followed by a four-digit number. To avoid confusion between "O" and "zero" and "I" and "one," the letters "O" and "I" are not used. How many distinct license plate numbers can be formed?

9. *True-False Exam* In how many ways can a six-question true-false exam be answered? (Assume that no questions are omitted.)

10. *Multiple Choice* In how many ways can a 10-question multiple choice exam be answered if there are four choices of answers for each question? (Assume that no questions are omitted.)

11. *Three-Digit Numbers* How many three-digit numbers can be formed under the following conditions?
(a) The leading digit cannot be zero.
(b) The leading digit cannot be zero and no repetition of digits is allowed.
(c) The leading digit cannot be zero and the number must be a multiple of 5.
(d) The number is at least 400.

12. *Four-Digit Numbers* How many four-digit numbers can be formed under the following conditions?
(a) The leading digit cannot be zero.
(b) The leading digit cannot be zero and no repetition of digits is allowed.
(c) The leading digit cannot be zero and the number must be less than 5000.
(d) The leading digit cannot be zero and the number must be even.

13. *Combination Lock* A combination lock will open when the right choice of three numbers (from 1 to 40, inclusive) is selected. How many different lock combinations are possible?

14. *Combination Lock* A combination lock will open when the right choice of three numbers (from 1 to 50, inclusive) is selected. How many different lock combinations are possible?

15. *Concert Seats* Three couples have reserved seats in a given row for a concert. In how many different ways can they be seated if
(a) there are no seating restrictions?
(b) the two members of each couple wish to sit together?

16. *Single File* In how many orders can three girls and two boys walk through a doorway single-file if
(a) there are no restrictions?
(b) the boys go before the girls?
(c) the girls go before the boys?

In Exercises 17–26, evaluate $_nP_m$.

17. $_4P_4$ 18. $_5P_5$
19. $_8P_3$ 20. $_{20}P_2$
21. $_{20}P_5$ 22. $_{100}P_1$
23. $_{100}P_2$ 24. $_{10}P_2$
25. $_5P_4$ 26. $_7P_4$

27. Write all the permutations of the letters A, B, C, and D.

28. Write all the permutations of the letters A, B, C, and D if the letters B and C must remain between the letters A and D.

29. *Posing for a Photograph* In how many ways can five children line up in one row to have their picture taken?

30. *Riding in a Car* In how many ways can six people sit in a six-passenger car?

31. *Choosing Officers* From a pool of 12 candidates, the offices of president, vice-president, secretary, and treasurer will be filled. In how many different ways can the offices be filled, if each of the 12 candidates can hold any office?

32. *Assembly Line Production* There are four processes involved in assembling a certain product, and these can be performed in any order. The management wants to test each order to determine which is the least time-consuming. How many different orders will have to be tested?

In Exercises 33–38, find the number of distinguishable permutations of the given group of letters.

33. A, A, G, E, E, E, M
34. B, B, B, T, T, T, T, T
35. A, A, Y, Y, Y, Y, X, X, X
36. K, K, M, M, M, L, L, N, N
37. A, L, G, E, B, R, A
38. M, I, S, S, I, S, S, I, P, P, I

39. Write all the possible selections of two letters that can be formed from the letters A, B, C, D, E, and F. (The order of the two letters is not important.)

40. Write all the possible selections of three letters that can be formed from the letters A, B, C, D, E, and F. (The order of the three letters is not important.)

41. *Forming an Experimental Group* In order to conduct a certain experiment, four students are randomly selected from a class of 20. How many different groups of four students are possible?

42. *Test Questions* A student may answer any 10 questions from a total of 12 questions on an exam. In how many different ways can the student select the questions?

43. *Lottery Choices* There are 40 numbers in a particular state lottery. In how many ways can a player select six of the numbers? (The order of selection is not important.)

44. *Lottery Choices* There are 50 numbers in a particular state lottery. In how many ways can a player select six of the numbers? (The order of selection is not important.)

45. *Number of Subsets* How many subsets of four elements can be formed from a set of 100 elements?

46. *Number of Subsets* How many subsets of five elements can be formed from a set of 80 elements?

47. *Forming a Committee* A committee composed of three graduate students and two undergraduate students is to be selected from a group of eight graduates and five undergraduates. How many different committees can be formed?

48. *Defective Units* A shipment of 12 microwave ovens contains three defective units. In how many ways can a vending company purchase four of these units and receive (a) all good units, (b) two good units, and (c) at least two good units?

49. *Job Applicants* An employer interviews eight people for four openings in the company. Three of the eight people are women. If all eight are qualified, in how many ways could the employer fill the four positions if (a) the selection is random and (b) exactly two are women?

50. *Poker Hand* Five cards are selected from an ordinary deck of 52 playing cards. In how many ways can you get a full house? (A full house consists of three of one kind and two of another. For example, A-A-A-5-5 and K-K-K-10-10 are full houses.)

51. *Forming a Committee* Four people are to be selected at random from a group of four couples. In how many ways can this be done, given the following conditions?
(a) There are no restrictions.
(b) There is to be at least one couple in the group of four.
(c) The selection must include one member from each couple.

52. *Interpersonal Relationships* The complexity of the interpersonal relationships increases dramatically as the size of a group increases. Determine the number of two-person relationships in a group of people of size (a) 3, (b) 8, (c) 12, and (d) 20.

In Exercises 53–56, find the number of diagonals of the given polygon. (A line segment connecting any two non-adjacent vertices is called a *diagonal* of the polygon.)

53. Pentagon **54.** Hexagon

55. Octagon **56.** Decagon (10 sides)

In Exercises 57 and 58, solve for n.

57. $14 \cdot {}_nP_3 = {}_{n+2}P_4$ **58.** ${}_nP_5 = 18 \, {}_{n-2}P_4$

In Exercises 59–63, prove the identity.

59. ${}_nP_{n-1} = {}_nP_n$ **60.** ${}_nP_1 = {}_nC_1$

61. ${}_nC_{n-1} = {}_nC_1$ **62.** ${}_nC_n = {}_nC_0$

63. ${}_nC_m = \dfrac{{}_nP_m}{m!}$

9.7   PROBABILITY

Sample Spaces / **The Probability of an Event** / **Mutually Exclusive Events** / **Independent Events** / **The Complement of an Event**

Sample Spaces

As a member of a complex society, you are used to living with varying amounts of uncertainty. For example, you may be questioning the likelihood of getting a good job after graduation, of winning a state lottery, of having an accident on your next trip home, or of any of several other possibilities.

In assigning measurements to uncertainties in everyday life, we often use ambiguous terminology, such as *fairly certain, probable,* or *highly unlikely.* In mathematics, we attempt to remove this ambiguity by assigning a number to the likelihood of the occurrence of an event. We call this measurement the **probability** that the event will occur. For example, if we toss a fair coin, we say that the probability that it will land heads up is one-half, or 50%.

In the study of probability, any happening whose result is uncertain is an **experiment.** The various possible results of the experiment are **outcomes,** and the collection of all possible outcomes of an experiment is the **sample space** of the experiment. Finally, any subcollection of a sample space is an **event.** In this section we will deal only with sample spaces in which each outcome is equally likely, such as flipping a fair coin or tossing a fair die.

EXAMPLE 1   Finding the Sample Space

An experiment consists of tossing a six-sided die.

A. What is the sample space?
B. Describe the event corresponding to a number greater than 2 turning up.

SOLUTION

A. The sample space consists of six outcomes, which you represent by the numbers 1 through 6. That is,

$$S = \{1, 2, 3, 4, 5, 6\}.$$

Note that each of the outcomes in the sample space is equally likely (assuming the die is balanced).

B. The *event* corresponding to a number greater than 2 turning up is the following subset of S.

$$A = \{3, 4, 5, 6\}$$

To describe sample spaces in such a way that each outcome is equally likely, we must sometimes distinguish between various outcomes in ways that appear artificial. The next example illustrates such a situation.

EXAMPLE 2   Finding the Sample Space

Find the sample spaces for the following.

A. One coin is tossed.
B. Two coins are tossed.
C. Three coins are tossed.

SOLUTION

A. Since the coin will land either heads up (denoted by H) or tails up (denoted by T), the sample space is

$$S = \{H, T\}.$$

B. Since either coin can land heads up or tails up, the possible outcomes are as follows.

$HH =$ heads up on both coins
$HT =$ heads up on first coin and tails up on second coin
$TH =$ tails up on first coin and heads up on second coin
$TT =$ tails up on both coins

Thus, the sample space is

$$S = \{HH, HT, TH, TT\}.$$

Note that you must distinguish between the two cases HT and TH, even though these two outcomes appear to be similar.

C. Following the notation of part (B), the sample space is

$$S = \{HHH, HHT, HTH, HTT, THH, THT, TTH, TTT\}.$$

The Probability of an Event

To calculate the probability of an event, you count the number of outcomes in the event and in the sample space. The *number of outcomes* in event E is denoted by $n(E)$, and the number of outcomes in the sample space S is denoted by $n(S)$.

THE PROBABILITY OF AN EVENT

If an event E has $n(E)$ equally likely outcomes and its sample space S has $n(S)$ equally likely outcomes, then the **probability** of event E is

$$P(E) = \frac{n(E)}{n(S)}.$$

Because the number of outcomes in an event must be less than or equal to the number of outcomes in the sample space, you can see that the probability of an event must be a number between 0 and 1. That is, for any event E, it must be true that $0 \le P(E) \le 1$.

PROPERTIES OF THE PROBABILITY OF AN EVENT

Let E be an event that is a subset of a finite sample space S.

1. $0 \le P(E) \le 1$
2. If $P(E) = 0$, then the event E *cannot occur*, and E is an **impossible event.**
3. If $P(E) = 1$, then the event E *must occur*, and E is a **certain event.**

EXAMPLE 3   Finding the Probability of an Event

Find the probabilities of the following events.

A. Two coins are tossed. Both coins land heads up.
B. A card is drawn from a standard deck of playing cards. The card drawn is an ace.

SOLUTION

A. Following the procedure in Example 2(B), let

$$E = \{HH\} \quad \text{and} \quad S = \{HH, HT, TH, TT\}.$$

The probability of getting two heads is

$$P(E) = \frac{n(E)}{n(S)} = \frac{1}{4}.$$

B. Since there are 52 cards in a standard deck of playing cards and there are four aces (one in each suit), the probability of drawing an ace is

$$P(E) = \frac{n(E)}{n(S)} = \frac{4}{52} = \frac{1}{13}.$$

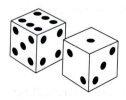

FIGURE 9.6

EXAMPLE 4   Finding the Probability of an Event

Two six-sided dice are tossed. What is the probability that the total of the two dice is 7? (See Figure 9.6.)

SOLUTION

Since there are six possible outcomes on each die, you use the Fundamental Counting Principle to conclude that there are

$6 \cdot 6 = 36$ different outcomes

when two dice are tossed. To find the probability of rolling a total of 7, you must first count the number of ways this can occur.

	Total of 7					
First die	1	2	3	4	5	6
Second die	6	5	4	3	2	1

Thus, a total of 7 can be rolled in six ways, which means that the probability of rolling a 7 is

$$P(E) = \frac{n(E)}{n(S)} = \frac{6}{36} = \frac{1}{6}.$$

You could have written out each sample space in Examples 3 and 4 and simply counted the outcomes in the desired events. For larger sample spaces, however, you must make more use of the counting principles discussed in the previous section.

EXAMPLE 5   Finding the Probability of an Event

Twelve-sided dice can be constructed (in the shape of regular dodecahedrons) so that each of the numbers from 1 to 6 appears twice on each die, as shown in Figure 9.7. Prove that these dice can be used in any game requiring ordinary six-sided dice without changing the probability of different outcomes.

FIGURE 9.7

SOLUTION

For an ordinary six-sided die, each of the numbers 1, 2, 3, 4, 5, and 6 occurs only once, so the probability of any particular number coming up is

$$P(E) = \frac{n(E)}{n(S)} = \frac{1}{6}.$$

For one of the twelve-sided dice, each number occurs twice, so the probability of any particular number coming up is

$$P(E) = \frac{n(E)}{n(S)} = \frac{2}{12} = \frac{1}{6}.$$

Thus, the twelve-sided dice can be used in place of the six-sided dice without changing the probabilities.

EXAMPLE 6   The Probability of Winning a Lottery

A state lottery is set up so that each player chooses six different numbers from 1 to 40. If these six numbers match the six numbers drawn by the lottery commission, the player wins (or shares) the top prize. What is the probability of winning the top prize in this game?

SOLUTION

Since the order of the numbers is not important, use the formula for the number of combinations of 40 elements taken six at a time to determine the size of the sample space.

$$n(S) = {}_{40}C_6 = \frac{40 \cdot 39 \cdot 38 \cdot 37 \cdot 36 \cdot 35}{6 \cdot 5 \cdot 4 \cdot 3 \cdot 2 \cdot 1} = 3{,}838{,}380$$

If a person buys only one ticket, the probability of winning is

$$P(E) = \frac{n(E)}{n(S)} = \frac{1}{3{,}838{,}380}.$$

EXAMPLE 7   Random Selection

The total number of colleges and universities in the United States in 1987 is shown in Figure 9.8. (*Source:* U.S. National Center for Education Statistics) Suppose one institution is selected at random. What is the probability that the institution is in one of three southern regions?

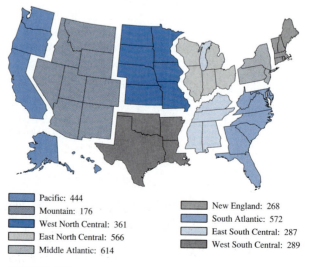

Pacific: 444
Mountain: 176
West North Central: 361
East North Central: 566
Middle Atlantic: 614

New England: 268
South Atlantic: 572
East South Central: 287
West South Central: 289

FIGURE 9.8

SOLUTION

Begin by finding the total number of colleges and universities.

Total $= 268 + 614 + 566 + 361 + 572 + 287 + 289 + 176 + 444 = 3577$

Since there are $572 + 287 + 289 = 1148$ colleges and universities in the three southern regions, the probability that the institution is from one of these regions is

$$P(E) = \frac{n(E)}{n(S)} = \frac{1148}{3577} \approx 0.321.$$

Mutually Exclusive Events

Two events A and B (from the same sample space) are **mutually exclusive** if A and B have no outcomes in common. In the terminology of sets, we say that the **intersection of A and B** is the empty set, which implies that

$P(A \cap B) = 0.$

For instance, if two dice are tossed, the event A of rolling a total of six and the event B of rolling a total of nine are mutually exclusive. To find the probability that one or the other of two mutually exclusive events will occur, *add* their individual probabilities.

PROBABILITY OF THE UNION OF TWO EVENTS

If A and B are events in the same sample space, then the probability of A or B occurring is given by

$$P(A \cup B) = P(A) + P(B) - P(A \cap B).$$

If A and B are mutually exclusive, then $P(A \cap B) = 0$ and it follows that

$$P(A \cup B) = P(A) + P(B).$$

EXAMPLE 8 **Finding the Probability of the Union of Two Events**

One card is selected from a standard deck of 52 playing cards. What is the probability that the card is either a heart or a face card?

SOLUTION

Since the deck has 13 hearts, the probability of selecting a heart (event A) is

$$P(A) = \frac{13}{52}.$$ *Selecting a heart*

Similarly, since the deck has 12 face cards, the probability of selecting a face card (event B) is

$$P(B) = \frac{12}{52}.$$ *Selecting a face card*

Now, because three of the cards are hearts and face cards (see Figure 9.9), it follows that

$$P(A \cap B) = \frac{3}{52}.$$

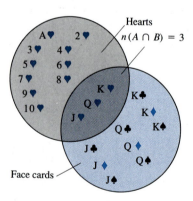

FIGURE 9.9

Finally, applying the formula for the probability of the union of two events, you conclude that the probability of selecting a heart or a face card is

$$P(A \cup B) = P(A) + P(B) - P(A \cap B)$$
$$= \frac{13}{52} + \frac{12}{52} - \frac{3}{52}$$
$$= \frac{22}{52} \approx 0.423.$$

Independent Events

Two events are **independent** if the occurrence of one has no effect on the occurrence of the other. To find the probability that two independent events will occur, *multiply* the probabilities of each. For instance, rolling a total of 12 with two six-sided dice has no effect on the outcome for future rolls of the dice.

PROBABILITY OF INDEPENDENT EVENTS

If A and B are independent events, then the probability that both A and B will occur is

$$P(A \text{ and } B) = P(A) \cdot P(B).$$

EXAMPLE 9   Probability of Independent Events

A random number generator on a computer selects three integers from 1 to 20. What is the probability that all three numbers are less than or equal to 5?

SOLUTION

If the random number generator is truly random, then you can conclude that the selection of any given number will not affect the selection of the next number. This means that the three choices represent independent events. Furthermore, since the probability of selecting a number from 1 to 5 is

$$P(A) = \frac{5}{20} = \frac{1}{4},$$

you can conclude that the probability of selecting all three numbers less than or equal to 5 is

$$P(A) \cdot P(A) \cdot P(A) = \left(\frac{1}{4}\right)\left(\frac{1}{4}\right)\left(\frac{1}{4}\right) = \frac{1}{64}.$$

EXAMPLE 10   Probability of Independent Events

In 1988, 54.2% of the population of the United States was 30 years old or older. Suppose that in a survey, 10 people were chosen at random from the population. What is the probability that all 10 were 30 years old or older?

SOLUTION

Let A represent choosing a person who is 30 years old or older. Since the probability of choosing a person who is 30 years old or older is 0.542, you conclude that the probability that all 10 people are 30 years old or older is

$$P(A)^{10} = (0.542)^{10} \approx 0.0022.$$

The Complement of an Event

The **complement of an event** A is the collection of all outcomes in the sample space that are not in A. We denote the complement of event A by A'. Since $P(A \text{ or } A') = 1$ and since A and A' are mutually exclusive, we have $P(A) + P(A') = 1$. Therefore, the probability of A' is given by

$$P(A') = 1 - P(A).$$

For instance, if the probability of *winning* a certain game is

$$P(A) = \frac{1}{4},$$

then the probability of *losing* the game is

$$P(A') = 1 - \frac{1}{4} = \frac{3}{4}.$$

EXAMPLE 11   **Finding the Probability of the Complement of an Event**

A manufacturer has determined that a certain machine averages one faulty unit for every 1000 it produces. What is the probability that an order of 200 units will have one or more faulty units?

SOLUTION

To solve this problem as stated, you would need to find the probability of having exactly one faulty unit, exactly two faulty units, exactly three faulty units, and so on. However, using complements, you can simply find the probability that all units are perfect and then subtract this value from 1. Since the probability that any given unit is perfect is 999/1000, the probability that all 200 units are perfect is

$$P(A) = \left(\frac{999}{1000}\right)^{200} \approx 0.8186.$$

Therefore, the probability that at least one unit is faulty is

$$P(A') = 1 - P(A) \approx 0.1814.$$

DISCUSSION PROBLEM

····················

AN EXPERIMENT IN PROBABILITY

In this section you have been finding probabilities from a *theoretical* point of view. Another way to find probabilities is from an *experimental* point of view. For instance, suppose you want to find the probability of obtaining a given total when two six-sided dice are tossed. The following BASIC program simulates the tossing of a pair of dice 5000 times.

```
 10 RANDOMIZE
 20 DIM TALLY(12)
 30 FOR I=1 TO 5000
 40 ROLLONE=INT(6*RND)+1
 50 ROLLTWO=INT(6*RND)+1
 60 DICETOTAL=ROLLONE+ROLLTWO
 70 TALLY(DICETOTAL)=TALLY(DICETOTAL)+1
 80 NEXT
 90 FOR I=2 TO 12
100 PRINT "TOTAL OF",I,"OCCURRED",TALLY(I),
    "TIMES"
110 NEXT
120 END
```

When you run this program, the printout is as follows.

```
TOTAL OF  2 OCCURRED 139 TIMES
TOTAL OF  3 OCCURRED 264 TIMES
TOTAL OF  4 OCCURRED 443 TIMES
TOTAL OF  5 OCCURRED 553 TIMES
TOTAL OF  6 OCCURRED 691 TIMES
TOTAL OF  7 OCCURRED 810 TIMES
TOTAL OF  8 OCCURRED 715 TIMES
TOTAL OF  9 OCCURRED 557 TIMES
TOTAL OF 10 OCCURRED 398 TIMES
TOTAL OF 11 OCCURRED 270 TIMES
TOTAL OF 12 OCCURRED 160 TIMES
```

In Example 4 you found that the theoretical probability of tossing a total of 7 on a pair of dice is $\frac{1}{6} \approx 0.167$. From this experiment, you find that the experimental probability of tossing a total of 7 on a pair of dice is $810/5000 \approx 0.162$. Try this experiment on a computer that has the BASIC language. By increasing the number of trials from 5000 to 10,000, does your experimental result get closer to the theoretical result?

WARM-UP

The following warm-up exercises involve skills that were covered in earlier sections. You will use these skills in the exercise set for this section.

In Exercises 1–8, evaluate the expression.

1. $\frac{1}{4} + \frac{5}{8} - \frac{5}{16}$ **2.** $\frac{4}{15} + \frac{3}{5} - \frac{1}{3}$

3. $\frac{5 \cdot 4}{5!}$ **4.** $\frac{5!22!}{27!}$

5. $\frac{4!8!}{12!}$ **6.** $\frac{9 \cdot 8 \cdot 7 \cdot 6 \cdot 5}{9!}$

7. $\frac{_5C_3}{_{10}C_3}$ **8.** $\frac{_{10}C_2 \cdot {}_{10}C_2}{_{20}C_4}$

In Exercises 9 and 10, evaluate the expression. (Round to three decimal places.)

9. $\left(\frac{99}{100}\right)^{100}$ **10.** $1 - \left(\frac{89}{100}\right)^{50}$

SECTION 9.7 · EXERCISES

In Exercises 1–6, determine the sample space for the given experiment.

1. A coin and a die are tossed.

2. A die is tossed twice and the sum of the points is recorded.

3. A taste tester has to rank three varieties of yogurt, A, B, and C, according to preference.

4. Two marbles are selected from a sack containing two red marbles, two blue marbles, and one black marble. The color of each marble is recorded.

5. Two county supervisors are selected from five supervisors, A, B, C, D, and E, to study a recycling plan.

6. A salesperson makes a presentation about a product in three homes per day. In each home there may be a sale (denote by *S*) or there may be no sale (denote by *F*).

Heads or Tails   In Exercises 7–10, find the required probability in the experiment of tossing a coin three times. Use the sample space $S = \{HHH, HHT, HTH, HTT, THH, THT, TTH, TTT\}$.

7. The probability of getting exactly one tail.

8. The probability of getting a head on the first toss.

9. The probability of getting at least one head.

10. The probability of getting at least two heads.

Drawing a Card   In Exercises 11–14, find the required probability in the experiment of selecting one card from a standard deck of 52 playing cards. (A face card is a J, Q, or K.)

11. The probability of getting a face card.

12. The probability of not getting a face card.

13. The probability of getting a black card that is not a face card.

14. The probability that the card will be a 6 or less.

Tossing a Die   In Exercises 15–20, find the required probability in the experiment of tossing a six-sided die twice.

15. The probability that the sum is 4.

16. The probability that the sum is less than 11.

17. The probability that the sum is at least 7.

18. The probability that the total is 2, 3, or 12.

19. The probability that the sum is odd and no more than 7.

20. The probability that the sum is odd or a prime.

Drawing Marbles In Exercises 21–24, find the required probability in the experiment of drawing two marbles (the first is *not* replaced before the second is drawn) from a bag containing one green, two yellow, and three red marbles.

21. The probability of drawing two red marbles.

22. The probability of drawing two yellow marbles.

23. The probability of drawing neither yellow marble.

24. The probability of drawing marbles of different colors.

In Exercises 25 and 26, you are given the probability that an event *will* happen. Find the probability that the event *will not* happen.

25. $p = 0.7$ 26. $p = 0.36$

In Exercises 27 and 28, you are given the probability that an event *will not* happen. Find the probability that the event *will* happen.

27. $p = 0.15$ 28. $p = 0.84$

29. *Alumni Association* The alumni office of a college is sending a survey to selected members of the class of 1990. Of the 1254 people who graduated that year, 672 were women, 124 of whom went on to graduate school. Of the 582 male graduates, 198 went on to graduate school. If an alumni member is selected at random, what is the probability that the person is (a) female, (b) male, and (c) female and did not attend graduate school?

30. *Post High School Education* In a high school graduating class of 72 students, 28 are on the honor roll. Of these 28, 18 are going on to college. Of the other 44 students, 12 are going on to college. If a student is selected at random from the class, what is the probability that the person chosen is (a) going to college, (b) not going to college, and (c) on the honor roll but not going to college?

31. *Winning an Election* Taylor, Moore, and Jenkins are candidates for public office. It is estimated that Moore and Jenkins have about the same probability of winning, and Taylor is believed to be twice as likely to win as either of the others. Find the probability of each candidate winning the election.

32. *Winning an Election* Three people have been nominated for president of a college class. From a small poll, it is estimated that the probability of the first candidate winning the election is 0.37, and the probability of the second candidate winning the election is 0.44. What is the probability that the third candidate will win?

33. *Preparing for a Test* An instructor gives her class a list of 20 study problems, from which she will select 10 to be answered on an exam. If a given student knows how to solve 15 of the problems, find the probability that the student will be able to answer (a) all 10 questions on the exam, (b) exactly 8 questions on the exam, and (c) at least 9 questions on the exam.

34. *Preparing for a Test* An instructor gives his class a list of eight study problems, from which he will select five to be answered on an exam. If a given student knows how to solve six of the problems, find the probability that the student will be able to answer (a) all five questions on the exam, (b) exactly four questions on the exam, and (c) at least four questions on the exam.

35. *Letter Mix-Up* Four letters and envelopes are addressed to four different people. If the letters are randomly inserted into the envelopes, what is the probability that (a) exactly one will be inserted in the correct envelope and (b) at least one will be inserted in the correct envelope?

36. *Payroll Mix-Up* Five paychecks and envelopes are addressed to five different people. If the paychecks are randomly inserted into the envelopes, what is the probability that (a) exactly one will be inserted in the correct envelope and (b) at least one will be inserted in the correct envelope?

37. *Game Show* On a game show you are given five digits to arrange in the proper order to give the price of a car. If you are correct, you win the car. What is the probability of winning, given the following conditions?
(a) You guess the position of each digit.
(b) You know the first digit, but must guess the remaining four.

38. *Game Show* On a game show you are given four digits to arrange in the proper order to give the price of a car. If you are correct, you win the car. What is the probability of winning, given the following conditions?
(a) You guess the position of each digit.
(b) You know the first digit, but must guess the remaining three.

39. *Drawing Cards from a Deck* Two cards are selected at random from an ordinary deck of 52 playing cards. Find the probability that two aces are selected, given the following conditions.
(a) The cards are drawn in sequence, with the first card being replaced and the deck reshuffled prior to the second drawing.
(b) The two cards are drawn consecutively, without replacement.

40. *Poker Hand*   Five cards are drawn from an ordinary deck of 52 playing cards. What is the probability of getting a full house?

41. *Defective Units*   A shipment of 12 microwave ovens contains three defective units. A vending company has ordered four of these 12 units, and since each is identically packaged, the selection will be random. What is the probability that (a) all four units are good, (b) exactly two units are good, and (c) at least two units are good?

42. *Defective Units*   A shipment of 20 compact disc players contains four defective units. A retail outlet has ordered five of these units. What is the probability that (a) all five players are good, (b) exactly four of the players are good, and (c) at least one of the players is defective?

43. *Random Number Generator*   Two integers (between 1 and 30 inclusive) are chosen by a random number generator on a computer. What is the probability that (a) the numbers are both even, (b) one number is even and one is odd, (c) both numbers are less than 10, and (d) the same number is chosen twice?

44. *Random Number Generator*   Two integers (between 1 and 40 inclusive) are chosen by a random number generator on a computer. What is the probability that (a) the numbers are both even, (b) one number is even and one is odd, (c) both numbers are no more than 30, and (d) the same number is chosen twice?

45. *Backup System*   A space vehicle has an independent backup system for one of its communication networks. The probability that either system will function satisfactorily for the duration of a flight is 0.985. What is the probability that during a given flight (a) both systems function satisfactorily, (b) at least one system functions satisfactorily, and (c) both systems fail?

46. *Backup Vehicle*   A fire company keeps two rescue vehicles to serve the community. Because of the demand on the company's time and the chance of mechanical failure, the probability that a specific vehicle is available when needed is 90%. If the availability of one vehicle is *independent* of the other, find the probability that (a) both vehicles are available at a given time, (b) neither vehicle is available at a given time, and (c) at least one vehicle is available at a given time.

47. *Making a Sale*   A sales representative makes sales at a rate of approximately one sale for every four calls. If, on a given day, the representative contacts five potential clients, what is the probability that a sale will be made with (a) all five contacts, (b) none of the contacts, and (c) at least one contact?

48. *Making a Sale*   A sales representative makes sales at a rate of approximately one sale for every three calls. If, on a given day, the representative contacts four potential clients, what is the probability that a sale will be made with (a) all four contacts, (b) none of the contacts, and (c) at least one contact?

49. *A Boy or a Girl?*   Assume that the probability of the birth of a child of a particular sex is 50%. In a family with four children, what is the probability that (a) all the children are boys, (b) all the children are the same sex, and (c) there is at least one boy?

50. *A Boy or a Girl?*   Assume that the probability of the birth of a child of a particular sex is 50%. In a family with six children, what is the probability that (a) all the children are girls, (b) all the children are the same sex, and (c) there is at least one girl?

51. *Is That Cash or Charge?*   According to a survey by *USA Today,* the method used by Christmas shoppers to pay for gifts is as shown in the pie chart. Suppose two Christmas shoppers are chosen at random. What is the probability that both shoppers paid for their gifts only in cash?

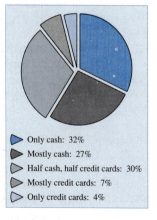

Only cash: 32%
Mostly cash: 27%
Half cash, half credit cards: 30%
Mostly credit cards: 7%
Only credit cards: 4%

FIGURE FOR 51

52. *Flexible Work Hours*   In a survey by Robert Hall International, people were asked if they would prefer to work flexible hours—even if it meant slower career advancement—so they could spend more time with their families. The results of the survey are shown in the figure. Suppose three people from the survey were chosen at random. What is the probability that all three people would prefer flexible work hours?

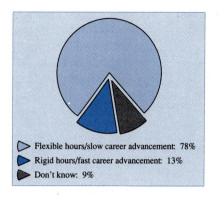

▷ Flexible hours/slow career advancement: 78%

▶ Rigid hours/fast career advancement: 13%

▶ Don't know: 9%

FIGURE FOR 52

9.8   EXPLORING DATA: MEASURES OF CENTRAL TENDENCY

Mean, Median, and Mode   /   Working with Organized Data   /   Choosing a Measure of Central Tendency

Mean, Median, and Mode

In many real-life situations, it is helpful to describe data by a single number that is most representative of the entire collection of numbers. Such a number is called a **measure of central tendency.** The most commonly used measures are as follows.

1. The **mean,** or **average,** of n numbers is the sum of the numbers divided by n.
2. The **median** of n numbers is the middle number when the numbers are written in order. If n is even, the median is the average of the two middle numbers.
3. The **mode** of n numbers is the number that occurs most frequently. If two numbers tie for most frequent occurrence, the collection has two modes and is called **bimodal.**

EXAMPLE 1   Finding Measures of Central Tendency

Find the mean, median, and mode of the following numbers.

5, 8, 3, 10, 8, 7, 9, 7, 5, 6, 7, 8, 4, 7, 6,
6, 5, 9,   4, 6, 7, 4, 8, 7, 9, 5, 4, 6, 7, 8

SOLUTION

The collection has 30 numbers. Thus, the mean is

$$\text{Mean} = \frac{5 + 8 + 3 + \cdots + 7 + 8}{30} = \frac{195}{30} = 6.5.$$

To find the median, order the numbers as follows.

3, 4, 4, 4, 4, 5, 5, 5, 5, 6, 6, 6, 6, 6, 7,
7, 7, 7, 7, 7, 7, 8, 8, 8, 8, 8, 9, 9, 9, 10

Because the two middle numbers are 7's, the median is 7. To find the mode of the numbers, construct a frequency distribution, as shown in Figure 9.10. From the distribution, you can see that the mode is 7.

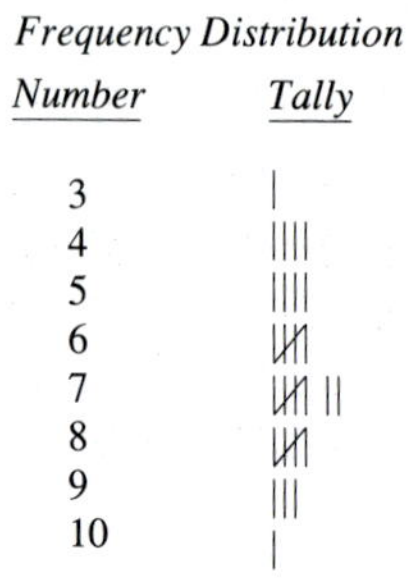

Frequency Distribution

Number	Tally
3	|
4	||||
5	||||
6	Ⅲ|
7	Ⅲ ||
8	Ⅲ
9	|||
10	|

FIGURE 9.10

In Example 1, all three measures of central tendency are about the same. If this always happened, there would be no need for different types of measures. The next example shows how different the values of the three measures can be.

EXAMPLE 2   **Comparing Measures of Central Tendency**

You are interviewing for a job. The person who is interviewing you tells you that the average income of the 25 employees is $60,849. The actual annual incomes of the 25 employees are shown below. What are the mean, median, and mode of the incomes? Was the person telling you the truth?

$17,305, $478,320, $45,678, $18,980, $17,408,
$25,676, $28,906, $12,500, $24,540, $33,450,
$12,500, $33,855, $37,450, $20,432, $28,956,
$34,983, $36,540, $250,921, $36,853, $16,430,
$32,654, $98,213, $48,980, $94,024, $35,671

Technology Note _____

Statistical calculators have built-in programs for calculating the mean of a collection of numbers. Try using your calculator to find the mean shown in Example 2. Then use your calculator to sort the incomes shown in Example 2.

SOLUTION

The mean of the incomes is

$$\text{Mean} = \frac{17,305 + 478,320 + 45,678 + 18,980 + \cdots + 35,671}{25}$$

$$= \frac{1,521,225}{25}$$

$$= \$60,849.$$

To find the median, order the incomes as follows.

$12,500, $12,500, $16,430, $17,305, $17,408,
$18,980, $20,432, $24,540, $25,676, $28,906,
$28,956, $32,654, $33,450, $33,855, $34,983,
$35,671, $36,540, $36,853, $37,450, $45,678,
$48,980, $94,024, $98,213, $250,921, $478,320

From this list, you can see that the median (the middle number) is $33,450. From the same list, you can see that $12,500 is the only income that occurs more than once. Thus, the mode is $12,500. Technically, the person was telling the truth because the average is (generally) defined to be the mean. However, of the three measures of central tendency

Mean: $60,849   *Median:* $33,450   *Mode:* $12,500

it seems clear that the median is the most representative. The mean is inflated by the two highest salaries.

Working with Organized Data

In Examples 1 and 2, you were asked to find the mean, median, and mode of "raw data." Suppose, however, that the data were already organized in a frequency distribution or histogram. To find (or approximate) the measures of central tendency in such cases, follow the procedure demonstrated in the next two examples.

EXAMPLE 3   Working with Frequency Distributions

A pair of six-sided dice is tossed 500 times.* The frequency of the totals is shown in the following frequency distribution. What are the mean, median, and mode of the totals?

Total	2	3	4	5	6	7	8	9	10	11	12
Frequency	17	22	46	50	63	87	64	65	47	23	16

Technology Note _____

Most computers and calculators that have statistical programs are capable of working with grouped data. Try entering the data given in Example 3 into your calculator. Then use the calculator to find the mean of the data.

SOLUTION

The mean of the numbers is

$$\text{Mean} = \frac{17(2) + 22(3) + 46(4) + \cdots + 16(12)}{500}$$

$$= \frac{3533}{500}$$

$$= 7.066.$$

Both the median and mode of the totals are 7. A histogram of the data is shown in Figure 9.11. This type of histogram is called *bell-shaped*. For such distributions, the mean, median, and mode are approximately the same—they all lie near the center of the histogram.

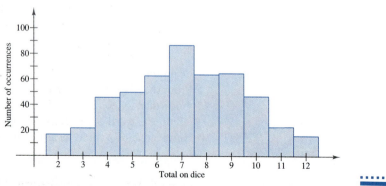

FIGURE 9.11

* The totals given in the table were simulated by a computer with a random number generator. Such simulations are called **"Monte Carlo"** simulations.

EXAMPLE 4   **Working with Grouped Data**

This histogram in Figure 9.12 represents the heights of 78 women who were chosen at random from a college freshman class. Approximate the mean height of the women.

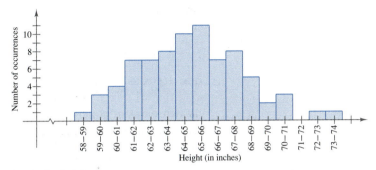

FIGURE 9.12

SOLUTION

Notice that the histogram groups the heights into 1-inch intervals. Because you are not given the exact heights, you cannot find the exact mean. You can, however, approximate the mean by using the midpoint of each interval as its representative height.

Height Interval	Midpoint, h	Frequency, f	hf
58–59	58.5	1	(58.5)(1) = 58.5
59–60	59.5	3	(59.5)(3) = 178.5
60–61	60.5	4	(60.5)(4) = 242.0
61–62	61.5	7	(61.5)(7) = 430.5
62–63	62.5	7	(62.5)(7) = 437.5
63–64	63.5	8	(63.5)(8) = 508.0
64–65	64.5	10	(64.5)(10) = 645.0
65–66	65.5	11	(65.5)(11) = 720.5
66–67	66.5	7	(66.5)(7) = 465.5
67–68	67.5	8	(67.5)(8) = 540.0
68–69	68.5	5	(68.5)(5) = 342.5
69–70	69.5	2	(69.5)(2) = 139.0
70–71	70.5	3	(70.5)(3) = 211.5
71–72	71.5	0	(71.5)(0) = 0.0
72–73	72.5	1	(72.5)(1) = 72.5
73–74	73.5	1	(73.5)(1) = 73.5
			Total 5065.0

From this table, you can approximate the mean height to be

$$\text{Mean} = \frac{5065}{78} \approx 64.9.$$

Choosing a Measure of Central Tendency

Which of the three measures of central tendency is the most representative? The answer is that it depends on the distribution of the data *and* the way in which you plan to use the data.

For instance, in Example 2, the mean salary of $60,849 does not seem very representative to a potential employee. To a city income tax collector who wants to estimate 1% of the total income of the 25 employees, however, the mean is precisely the right measure.

EXAMPLE 5   Choosing a Measure of Central Tendency

Which measure of central tendency is the most representative of the data given in the following frequency distribution?

A. Number	Tally		B. Number	Tally		C. Number	Tally
1	7		1	9		1	6
2	20		2	8		2	1
3	15		3	7		3	2
4	11		4	6		4	3
5	8		5	5		5	5
6	3		6	6		6	5
7	2		7	7		7	4
8	0		8	8		8	3
9	15		9	9		9	0

SOLUTION

A. For these data, the mean is 4.23, the median is 3, and the mode is 2. Of these, the mode is probably the most representative.

B. For these data, the mean and median are each 5 and the modes are 1 and 9 (the distribution is bimodal). Of these, the mean or median is the most representative.

C. For these data, the mean is 4.59, the median is 5, and the mode is 1. Of these, the mean or median is the most representative.

DISCUSSION PROBLEM

COMPARING MEASURES OF CENTRAL TENDENCY

Find collections of numbers that have the following properties. If this is not possible, explain why it is not.
(a) Mean = 6,   Median = 4,   Mode = 4
(b) Mean = 6,   Median = 6,   Mode = 4
(c) Mean = 6,   Median = 4,   Mode = 6

WARM-UP

The following warm-up exercises involve skills that were covered in earlier sections. You will use these skills in the exercise set for this section.

1. The scores for a 25-point quiz were 22, 17, 21, 25, 23, 23, 21, 13, 19, 23, 22, 22, 23, 19. Use a line plot to organize these data.

2. The numbers represent the grade-point averages of 20 college freshmen. Use a stem-and-leaf plot to organize these data.

2.5	1.7	2.5	3.5	1.8
2.0	2.6	2.9	2.3	2.2
3.0	2.4	4.0	2.4	2.8
3.2	3.5	1.5	2.6	3.7

3. The table shows crude oil prices (in constant 1982 cents per million Btu's) from 1981 through 1990. Construct a line graph for these data. (*Source:* U.S. Energy Information Administration)

Year	1981	1982	1983	1984	1985
Price	547.8	491.7	434.6	414.3	374.5

Year	1986	1987	1988	1989	1990
Price	189.5	226.1	178.8	216.5	262.6

4. The data below show the per capita consumption (in pounds) of selected meats in 1990. Construct a bar graph for these data. (*Source:* U.S. Department of Agriculture)

Beef	64.0	*Pork*	46.3	*Fish*	15.5
Chicken	49.3	*Turkey*	14.4		

In Exercises 5–8, evaluate the sum.

5. $\displaystyle\sum_{k=1}^{10} 4$

6. $\displaystyle\sum_{k=1}^{10} 4k$

7. $\displaystyle\sum_{j=1}^{8} 2j$

8. $\displaystyle\sum_{j=1}^{12} 3j$

In Exercises 9 and 10, use the properties of summation to simplify the expression.

9. $\displaystyle\sum_{i=1}^{6} (i - a)$

10. $\displaystyle\sum_{i=1}^{5} (i - a)^2$

SECTION 9.8 · EXERCISES

In Exercises 1–6, find the mean, median, and mode of the set of measurements.

1. 5, 12, 7, 14, 8, 9, 7
2. 30, 37, 32, 39, 33, 34, 32
3. 5, 12, 7, 24, 8, 9, 7
4. 20, 37, 32, 39, 33, 34, 32
5. 5, 12, 7, 14, 9, 7
6. 30, 37, 32, 39, 34, 32

7. Compare your answers for Exercises 1 and 3 with those for Exercises 2 and 4. Which of the measures of central tendency is sensitive to extreme measurements? Explain your reasoning.

8. (a) Add 6 to each measurement in Exercise 1 and calculate the mean, median, and mode of the revised measurements. How are the measures of central tendency changed?
   (b) If a constant k is added to each measurement in a set of data, how will the measures of central tendency change? Use the properties of summation to prove the result for the mean.

9. *Electric Bills*   A person had the following monthly bills for electricity. What are the mean and median of this collection of bills?

January	$67.92	February	$59.84
March	$52.00	April	$52.50
May	$57.99	June	$65.35
July	$81.76	August	$74.98
September	$87.82	October	$83.18
November	$65.35	December	$57.00

10. *Car Rental*   A car rental company kept the following record of the number of miles driven by a car that was rented. What are the mean, median, and mode of these data?

Monday	410	Tuesday	260
Wednesday	320	Thursday	320
Friday	460	Saturday	150

11. *Six-Child Families*   A study was done on families having six children. The table gives the number of families in the study with the indicated number of girls. Determine the mean, median, and mode of these data.

Number of girls	0	1	2	3	4	5	6
Frequency	1	24	45	54	50	19	7

12. *Baseball*   A baseball fan examined the records of a favorite baseball player's performance during his last 50 games. The numbers of games in which the player had 0, 1, 2, 3, and 4 hits are recorded in the table.

Number of hits	0	1	2	3	4
Frequency	14	26	7	2	1

(a) Determine the average number of hits per game.
(b) Determine the player's batting average if he had 200 at bats during the 50-game series.

13. *Tire Wear*   A tire company tested a new tire design on 100 cars. The useful tread lives (in thousands of miles) for the cars are given in the table. Approximate the average number of miles in the tread life of these tires.

Miles	26–28	28–30	30–32
Frequency	1	7	5
Miles	32–34	34–36	36–38
Frequency	11	19	25

Miles	38–40	40–42	42–44
Frequency	12	11	5
Miles	44–46	46–48	
Frequency	3	1	

14. *Health* The following frequency distributions give the numbers of deaths (in thousands) attributed to heart disease and accidents by age group in the United States. (*Source:* U.S. National Center for Health Statistics)

Age	15–24	25–34	35–44
Heart disease	0.9	3.5	11.8
Accidents	16.7	16.6	11.9
Age	45–54	55–64	65–74
Heart disease	30.9	81.4	165.8
Accidents	7.5	7.6	8.8

(a) Approximate the mean age of death by heart disease.

(b) Approximate the mean age of death by accidents.

15. *Test Scores* A professor records the following scores for a 100-point exam.

99, 64, 80, 77, 59, 72, 87, 79, 92, 88, 90, 42, 20, 89, 42, 100, 98, 84, 78, 91

Which measure of central tendency best describes these test scores?

16. *Shoe Sales* A salesman sold eight pairs of a certain style of men's shoes. The sizes of the eight pairs were as follows: $10\frac{1}{2}$, 8, 12, $10\frac{1}{2}$, 10, $9\frac{1}{2}$, 11, and $10\frac{1}{2}$. Which measure of central tendency best describes the typical shoe size for these data?

17. *Hourly Wage* The median hourly wage at a particular company is $9.18. If there are 254 employees who earn hourly wages, is it true that at least 127 employees earn at least $9.18 per hour?

9.9   EXPLORING DATA: MEASURES OF DISPERSION

Variance and Standard Deviation / Computing Standard Deviation / Box-and-Whisker Plots

Variance and Standard Deviation

In Section 9.8, you saw that very different sets of numbers can have the same mean. In this section, you will study two **measures of dispersion,** which give you an idea of how much the numbers in the set differ from the mean of the set. These two measures are called the *variance* of the set and the *standard deviation* of the set.

DEFINITIONS OF VARIANCE AND STANDARD DEVIATION

Consider a set of numbers $\{x_1, x_2, \cdots, x_n\}$ with a mean of $\bar{x}$. The **variance** of the set is

$$v = \frac{(x_1 - \bar{x})^2 + (x_2 - \bar{x})^2 + \cdots + (x_n - \bar{x})^2}{n},$$

and the **standard deviation** of the set is

$$\sigma = \sqrt{v}$$

(σ is the lowercase Greek letter *sigma*).

The standard deviation of a set is a measure of how much a typical number in the set differs from the mean. The greater the standard deviation, the more the numbers in the set *vary* from the mean. For instance, each of the following sets has a mean of 5.

$$\{5, 5, 5, 5\}, \quad \{4, 4, 6, 6\}, \quad \text{and} \quad \{3, 3, 7, 7\}$$

The standard deviations of the sets are 0, 1, and 2.

$$\sigma_1 = \sqrt{\frac{(5-5)^2 + (5-5)^2 + (5-5)^2 + (5-5)^2}{4}} = 0$$

$$\sigma_2 = \sqrt{\frac{(4-5)^2 + (4-5)^2 + (6-5)^2 + (6-5)^2}{4}} = 1$$

$$\sigma_3 = \sqrt{\frac{(3-5)^2 + (3-5)^2 + (7-5)^2 + (7-5)^2}{4}} = 2$$

EXAMPLE 1   Estimations of Standard Deviation

Consider the three sets of data represented by the following bar graphs. Which set has the smallest standard deviation? Which has the largest?

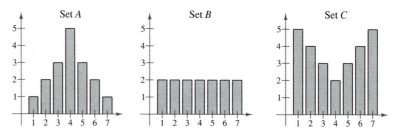

SOLUTION

Of the three sets, the numbers in set *A* are grouped most closely to the center and the numbers in set *C* are the most dispersed. Thus, set *A* has the smallest standard deviation and set *C* has the largest standard deviation.

EXAMPLE 2   Finding Standard Deviation

Find the standard deviation of each set shown in Example 1.

SOLUTION

Because of the symmetry of each bar graph, you can conclude that each has a mean of $\bar{x} = 4$. The standard deviation of set *A* is

$$\sigma = \sqrt{\frac{(-3)^2 + 2(-2)^2 + 3(-1)^2 + 5(0)^2 + 3(1)^2 + 2(2)^2 + (3)^2}{17}}$$

$$\approx 1.53.$$

The standard deviation of set B is

$$\sigma = \sqrt{\frac{2(-3)^2 + 2(-2)^2 + 2(-1)^2 + 2(0)^2 + 2(1)^2 + 2(2)^2 + 2(3)^2}{14}}$$

$$= 2.$$

The standard deviation of set C is

$$\sigma = \sqrt{\frac{5(-3)^2 + 4(-2)^2 + 3(-1)^2 + 2(0)^2 + 3(1)^2 + 4(2)^2 + 5(3)^2}{26}}$$

$$\approx 2.22.$$

These values confirm the results of Example 1. That is, set A has the smallest standard deviation and set C has the largest.   ■■■■■■■■

Computing Standard Deviation

Most computations in statistics are performed with calculators and computers that are programmed with formulas for the mean, variance, and standard deviation. The following two examples, however, show how to organize computations that are done by hand.

EXAMPLE 3   Using a Table

Find the standard deviation of the following set of numbers.

5, 6, 6, 7, 7, 8, 8, 8, 9, 10

SOLUTION

Begin by finding the mean of the set.

$$\bar{x} = \frac{5 + 6 + 6 + 7 + 7 + 8 + 8 + 8 + 9 + 10}{10} = \frac{74}{10} = 7.4$$

To find the variance, you can use the following tabular format.

x	$x - \bar{x}$	$(x - \bar{x})^2$
5	−2.4	5.76
6	−1.4	1.96
6	−1.4	1.96
7	−0.4	0.16
7	−0.4	0.16
8	0.6	0.36
8	0.6	0.36
8	0.6	0.36
9	1.6	2.56
10	2.6	6.76
		20.4

← *Sum of squared differences*

Technology Note _____

If you have access to a computer or calculator with a standard deviation program, try using it to obtain the result given in Example 3. If you do this, the program will probably output two versions of the standard deviation. In one, the sum of the squared differences is divided by n, and in the other it is divided by $n - 1$. In this text, we always divide by n.

To find the variance, divide the sum of the squared differences by 10 (the number of entries in the set).

$$v = \frac{20.4}{10} = 2.04$$

Finally, the standard deviation is the square root of the variance.

$$\sigma = \sqrt{2.04} \approx 1.43$$

Because the mean of a set is often a rounded decimal, the technique shown in Example 3 can be cumbersome *and* it can involve round-off error. The following alternative formula provides a more efficient way to compute the standard deviation.

ALTERNATIVE FORMULA FOR STANDARD DEVIATION

The standard deviation of $\{x_1, x_2, \cdots, x_n\}$ is

$$\sigma = \sqrt{\frac{x_1{}^2 + x_2{}^2 + \cdots + x_n{}^2}{n} - \bar{x}^2}.$$

Because of messy computations, this formula is difficult to verify. Conceptually, however, the process is straightforward. It consists of showing that the expressions

$$\sqrt{\frac{(x_1 - \bar{x})^2 + (x_2 - \bar{x})^2 + \cdots + (x_n - \bar{x})^2}{n}}$$

and

$$\sqrt{\frac{x_1{}^2 + x_2{}^2 + \cdots + x_n{}^2}{n} - \bar{x}^2}$$

are equivalent. Try verifying this equivalence for the set $\{x_1, x_2, x_3\}$ with $\bar{x} = (x_1 + x_2 + x_3)/3$.

EXAMPLE 4   Using the Alternative Formula

Use the alternative formula for standard deviation to find the standard deviation of the set given in Example 3.

5, 6, 6, 7, 7, 8, 8, 8, 9, 10

SOLUTION

From Example 3, you know the mean is 7.4. Thus, the standard deviation is

$$\sigma = \sqrt{\frac{5^2 + 2(6^2) + 2(7^2) + 3(8^2) + 9^2 + 10^2}{10} - (7.4)^2}$$

$$= \sqrt{\frac{568}{10} - 54.76}$$

$$= \sqrt{2.04}$$

$$\approx 1.43.$$

Note that this result agrees with that obtained in Example 3.   ⬛⬛⬛⬛⬛⬛

A well-known theorem in statistics, called *Chebychev's Theorem,* states that at least

$$1 - \frac{1}{k^2}$$

of the numbers in a distribution must lie within k standard deviations of the mean. Thus, 75% of the numbers in a collection must lie within two standard deviations of the mean, and at least 88.9% of the numbers must lie within three standard deviations of the mean. For most distributions, these percentages are low. For instance, in all three distributions shown in Example 1, 100% of the numbers lie within two standard deviations of the mean.

EXAMPLE 5   Describing a Distribution

The following table shows the number of dentists (per 100,000 people) in each state and the District of Columbia. Find the mean and standard deviation of the numbers. What percent of the numbers lie within two standard deviations of the mean? (*Source:* American Dental Association)

AK	66	AL	40	AR	39	AZ	51	CA	62
CO	69	CT	80	DC	94	DE	44	FL	50
GA	46	HI	80	IA	55	ID	53	IL	61
IN	47	KS	51	KY	53	LA	45	MA	74
MD	68	ME	47	MI	62	MN	67	MO	53
MS	37	MT	62	NC	42	ND	47	NE	63
NH	59	NJ	77	NM	45	NV	49	NY	73
OH	55	OK	47	OR	70	PA	61	RI	56
SC	41	SD	49	TN	53	TX	47	UT	66
VA	54	VT	57	WA	68	WI	65	WV	43
WY	52								

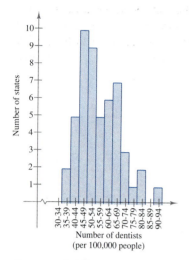

FIGURE 9.13

SOLUTION

Begin by entering the numbers into a computer or calculator that has a standard deviation program. After running the program, you should obtain

$$\bar{x} \approx 56.76 \quad \text{and} \quad \sigma \approx 12.14.$$

The interval that contains all numbers that lie within two standard deviations of the mean is

$$[56.76 - 2(12.14), 56.76 + 2(12.14)] \quad \text{or} \quad [32.48, 81.04].$$

From the histogram in Figure 9.13, you can see that all but one of the numbers (98%) lie in this interval—that number corresponds to the number of dentists (per 100,000 people) in Washington, D.C.

Box-and-Whisker Plots

Standard deviation is the measure of dispersion that is associated with the mean. **Quartiles** are measures of dispersion that are associated with the median.

DEFINITION OF QUARTILES

Consider an ordered set of numbers whose median is m. The **lower quartile** is the median of the numbers that occur before m. The **upper quartile** is the median of the numbers that occur after m.

EXAMPLE 6   **Finding Quartiles of a Set**

Find the lower and upper quartiles for the following set.

34, 14, 24, 16, 12, 18, 20, 24, 16, 26, 13, 27

SOLUTION

Begin by ordering the set.

12, 13, 14, 16, 16, 18, 20, 24, 24, 26, 27, 34

1st 25%   2nd 25%   3rd 25%   4th 25%

The median of the entire set is 19. The median of the six numbers that are less than 19 is 15. Thus, the lower quartile is 15. The median of the six numbers that are greater than 19 is 25. Thus, the upper quartile is 25.

Quartiles are represented graphically by a **box-and-whisker plot,** as shown in Figure 9.14. In the plot, notice that five numbers are listed: the smallest number, the lower quartile, the median, the upper quartile, and the largest number. Also, notice that the numbers are spaced proportionally, as though they were on a real number line.

12   15   19   25   34

FIGURE 9.14

The next example shows how to find quartiles when the number of elements in a set is not divisible by 4.

EXAMPLE 7   Sketching Box-and-Whisker Plots

Sketch a box-and-whisker plot for each of the following sets.

A. 27, 28, 30, 42, 45, 50, 50, 61, 62, 64, 66
B. 82, 82, 83, 85, 87, 89, 90, 94, 95, 95, 96, 98, 99
C. 11, 13, 13, 15, 17, 18, 20, 24, 24, 27

SOLUTION

A. This set has 11 numbers. The median is 50 (the sixth number). The lower quartile is 30 (the median of the first five numbers). The upper quartile is 62 (the median of the last five numbers). A box-and-whisker plot for the data is shown below.

27   30   50   62   66

B. This set has 13 numbers. The median is 90 (the seventh number). The lower quartile is 84 (the median of the first six numbers). The upper quartile is 95.5 (the median of the last six numbers). A box-and-whisker plot for the data is shown below.

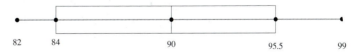

82   84   90   95.5   99

C. This set has 10 numbers. The median is 17.5 (the average of the fifth and sixth numbers). The lower quartile is 13 (the median of the first five numbers). The upper quartile is 24 (the median of the last five numbers). A box-and-whisker plot for the data is shown below.

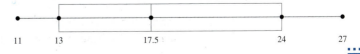

11   13   17.5   24   27

DISCUSSION
PROBLEM
. . . .

ESTIMATING
STANDARD
DEVIATION

Estimate the standard deviations of the data represented by the bar graphs.
Then, calculate the standard deviations. Compare your calculated results with
your estimates.

(a)

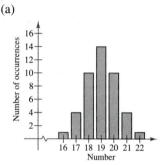

(b)

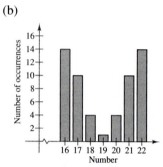

WARM-UP
. . . .

The following warm-up exercises involve skills that were covered in earlier sections. You
will use these skills in the exercise set for this section.

In Exercises 1–4, find the mean of the measurements.

1. 14, 15, 13, 21, 15, 14, 26, 13

2. 0.4, 3.8, 8.7, 2.6, 7.2, 5.6, 1.3

3. $34.95, $31.19, $43.96, $27.49, $38.21

4. 77°, 68°, 64°, 65°, 73°, 81°, 79°

In Exercises 5–10, evaluate the expression.

5. $(15 - 9)^2 + (4 - 9)^2 + (9 - 9)^2 + (8 - 9)^2$

6. $(4 - 9.4)^2 + (9 - 9.4)^2 + (13 - 9.4)^2 + (6 - 9.4)^2 + (15 - 9.4)^2$

7. $\dfrac{7^2 + 1^2 + 9^2 + 10^2 + 13^2 + 5^2}{6} - 7^2$

8. $\dfrac{22^2 + 20^2 + 10^2 + 18^2 + 30^2}{5} - 20^2$

9. $\sqrt{\dfrac{(11 - 13.75)^2 + (18 - 13.75)^2 + (12 - 13.75)^2 + (14 - 13.75)^2}{4}}$

10. $\sqrt{\dfrac{11^2 + 18^2 + 12^2 + 14^2}{4} - (13.75)^2}$

SECTION 9.9 · EXERCISES

In Exercises 1–8, find the mean, variance, and standard deviation of the numbers.

1. 4, 10, 8, 2

2. 3, 15, 6, 9, 2

3. 0, 1, 1, 2, 2, 2, 3, 3, 4

4. 2, 2, 2, 2, 2, 2

5. 1, 2, 3, 4, 5, 6, 7

6. 1, 1, 1, 5, 5, 5

7. 49, 62, 40, 29, 32, 70

8. 1.5, 0.4, 2.1, 0.7, 0.8

In Exercises 9–14, use the alternative formula to find the standard deviation of the numbers.

9. 2, 4, 6, 6, 13, 5

10. 10, 25, 50, 26, 15, 33, 29, 4

11. 246, 336, 473, 167, 219, 359

12. 6.0, 9.1, 4.4, 8.7, 10.4

13. 8.1, 6.9, 3.7, 4.2, 6.1

14. 9.0, 7.5, 3.3, 7.4, 6.0

15. Without calculating the standard deviation, explain why the set {4, 4, 20, 20} has a standard deviation of 8.

16. If the standard deviation of a set of numbers is 0, what does that imply about the set?

In Exercises 17 and 18, the line plots of sets of data are given. Determine the mean and standard deviation of each set.

17. (a)

(b)

(c)

(d)

18. (a)

(b)

(c)

(d)

19. *Test Scores* An instructor adds five points to each student's exam score. Will this change the mean or standard deviation of the exam scores? Explain.

20. Consider the four sets of data represented by the histograms. Order the sets from the smallest to the largest variance.

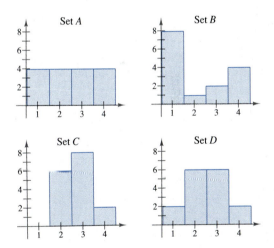

21. *Test Scores* The scores of a mathematics exam given to 600 science and engineering students at a college had a mean and standard deviation equal to 235 and 28, respectively. Use Chebychev's Theorem to determine the intervals containing at least $\frac{3}{4}$ and at least $\frac{8}{9}$ of the scores. How would the intervals change if the standard deviation were 16?

22. *Precipitation*   The following data represent the annual precipitation (in inches) at Erie, Pennsylvania for the years 1960 through 1989. Use a computer or calculator to find the mean, variance, and standard deviation of the data. What percent of the data lies within two standard deviations of the mean? (*Source:* National Oceanic and Atmospheric Administration)

27.41, 36.50, 36.90, 28.11, 36.47,
38.41, 37.74, 37.78, 34.33, 36.58,
41.50, 34.06, 43.55, 38.04, 41.83,
43.03, 43.85, 61.70, 35.04, 55.31,
47.04, 41.97, 41.56, 46.25, 37.79,
45.87, 47.30, 44.86, 38.87, 41.88

In Exercises 23–26, sketch a box-and-whisker plot for the data.

23. 23, 15, 14, 23, 13, 14, 13, 20, 12

24. 11, 10, 11, 14, 17, 16, 14, 11, 8, 14, 20

25. 46, 48, 48, 50, 52, 47, 51, 47, 49, 53

26. 25, 20, 22, 28, 24, 28, 25, 19, 27, 29, 28, 21

27. *Product Lifetime*   A company has redesigned a product in an attempt to increase the lifetime of the product. The two sets of data give the lifetimes (in months) of samples of 20 units before and 20 units after the design change. Create a box-and-whisker plot for each set of data, and then comment on the differences between the plots.

Original Design
15.1, 78.3, 56.3, 68.9, 30.6,
27.2, 12.5, 42.7, 72.7, 20.2,
53.0, 13.5, 11.0, 18.4, 85.2,
10.8, 38.3, 85.1, 10.0, 12.6

New Design
55.8, 71.5, 25.6, 19.0, 23.1,
37.2, 60.0, 35.3, 18.9, 80.5,
46.7, 31.1, 67.9, 23.5, 99.5,
54.0, 23.2, 45.5, 24.8, 87.8

CHAPTER 9 · REVIEW EXERCISES

In Exercises 1–4, use sigma notation to write the given sum.

1. $\dfrac{1}{2(1)} + \dfrac{1}{2(2)} + \dfrac{1}{2(3)} + \cdots + \dfrac{1}{2(20)}$

2. $2(1^2) + 2(2^2) + 2(3^2) + \cdots + 2(9^2)$

3. $\frac{1}{2} + \frac{2}{3} + \frac{3}{4} + \cdots + \frac{9}{10}$

4. $1 - \frac{1}{3} + \frac{1}{9} - \frac{1}{27} + \cdots$

In Exercises 5–18, find the sum.

5. $\displaystyle\sum_{i=1}^{6} 5$

6. $\displaystyle\sum_{k=2}^{5} 4k$

7. $\displaystyle\sum_{j=3}^{10} (2j - 3)$

8. $\displaystyle\sum_{j=1}^{8} (20 - 3j)$

9. $\displaystyle\sum_{i=0}^{6} 2^i$

10. $\displaystyle\sum_{i=0}^{4} 3^i$

11. $\displaystyle\sum_{i=0}^{\infty} \left(\frac{7}{8}\right)^i$

12. $\displaystyle\sum_{i=0}^{\infty} \left(\frac{1}{3}\right)^i$

13. $\displaystyle\sum_{k=0}^{\infty} 4\left(\frac{2}{3}\right)^k$

14. $\displaystyle\sum_{k=0}^{\infty} 1.3\left(\frac{1}{10}\right)^k$

15. $\displaystyle\sum_{k=1}^{11} \left(\frac{2}{3}k + 4\right)$

16. $\displaystyle\sum_{k=1}^{25} \left(\frac{3k + 1}{4}\right)$

17. $\displaystyle\sum_{n=0}^{10} (n^2 + 3)$

18. $\displaystyle\sum_{n=1}^{100} \left(\frac{1}{n} - \frac{1}{n + 1}\right)$

In Exercises 19–22, write the first five terms of the arithmetic sequence.

19. $a_1 = 3,\ d = 4$

20. $a_1 = 8,\ d = -2$

21. $a_4 = 10,\ a_{10} = 28$

22. $a_2 = 14,\ a_6 = 22$

In Exercises 23 and 24, write an expression for the *n*th term of the specified arithmetic sequence and find the sum of the first 20 terms of the sequence.

23. $a_1 = 100,\ d = -3$

24. $a_1 = 10,\ a_3 = 28$

25. Find the sum of the first 100 positive multiples of 5.

26. Find the sum of the integers from 20 to 80 (inclusive).

In Exercises 27–30, write the first five terms of the geometric sequence.

27. $a_1 = 4, r = -\frac{1}{4}$ **28.** $a_1 = 2, r = 2$

29. $a_1 = 9, a_3 = 4$ **30.** $a_1 = 2, a_3 = 12$

In Exercises 31 and 32, write an expression for the nth term of the specified geometric sequence and find the sum of the first 20 terms of the sequence.

31. $a_1 = 16, a_2 = -8$ **32.** $a_1 = 100, r = 1.05$

33. *Depreciation* A company buys a machine for $120,000. During the next 5 years it will depreciate at the rate of 30% per year. (That is, at the end of each year the depreciated value will be 70% of what it was at the beginning of the year.)
   (a) Find the formula for the nth term of a geometric sequence that gives the value of the machine t full years after it was purchased.
   (b) Find the depreciated value of the machine at the end of 5 full years.

34. *Total Compensation* Suppose you accept a job that pays a salary of $32,000 the first year and that you will receive a 5.5% raise each year for each of the next 39 years. What will your total salary be over the 40-year period?

35. *Compound Interest* A deposit of $200 is made at the beginning of each month for 2 years into an account that pays 6%, compounded monthly. What is the balance in the account at the end of 2 years?

36. *Compound Interest* A deposit of $100 is made at the beginning of each month for 10 years into an account that pays 6.5%, compounded monthly. What is the balance in the account at the end of 10 years?

In Exercises 37–40, use mathematical induction to prove the given formula for every positive integer n.

37. $1 + 4 + \cdots + (3n - 2) = \dfrac{n}{2}(3n - 1)$

38. $1 + \dfrac{3}{2} + 2 + \dfrac{5}{2} + \cdots + \dfrac{1}{2}(n + 1) = \dfrac{n}{4}(n + 3)$

39. $\displaystyle\sum_{i=0}^{n-1} ar^i = \dfrac{a(1 - r^n)}{1 - r}$

40. $\displaystyle\sum_{k=0}^{n-1} (a + kd) = \dfrac{n}{2}[2a + (n - 1)d]$

In Exercises 41–44, evaluate $_nC_m$. If your graphing utility has a binomial coefficient key (also called the combinations function), use it to verify your answer.

41. $_6C_4$ **42.** $_{10}C_7$

43. $_{25}C_5$ **44.** $_{12}C_3$

In Exercises 45–50, use the Binomial Theorem to expand the binomial. Simplify your result. (Remember that $i = \sqrt{-1}$.)

45. $\left(\dfrac{x}{2} + y\right)^4$ **46.** $(a - 3b)^5$

47. $\left(\dfrac{2}{x} - 3x\right)^6$ **48.** $(3x + y^2)^7$

49. $(5 + 2i)^4$ **50.** $(4 - 5i)^3$

51. *Interpersonal Relationships* The complexity of interpersonal relationships increases dramatically as the size of a group increases. Determine the number of different two-person relationships in a family of (a) 2, (b) 4, and (c) 6.

52. *Morse Code* In Morse Code, all characters are transmitted using a sequence of dits and dahs. How many different characters can be formed by a sequence of three dits and dahs? (These can be repeated. For example, dit-dit-dit represents the letter s.)

53. *Amateur Radio* A Novice Amateur Radio license consists of two letters, one digit, and then three more letters. How many different licenses can be issued if no restrictions are placed on the letters or digits?

54. *Connecting Points with Lines* How many different straight-line segments are determined by (a) five noncollinear points and (b) ten noncollinear points?

55. *Matching Socks*   A man has five pairs of socks (no two pairs are the same color). If he randomly selects two socks from the drawer, what is the probability that he gets a matched pair?

56. *Bookshelf Order*   A child carries a five-volume set of books to a bookshelf. The child is not able to read, and hence cannot distinguish one volume from another. What is the probability that the books are shelved in the correct order?

57. *Roll of the Dice*   Are the chances of rolling a 3 with one die the same as the chances of rolling a total of 6 with two dice? If not, which has the higher probability?

58. *Roll of the Dice*   A die is rolled six times. What is the probability that each side will appear exactly once?

59. *Tossing a Coin*   Find the probability of obtaining at least one tail when a coin is tossed five times.

60. *Parental Independence*   In the *Marriott Seniors' Attitudes Survey,* senior citizens were asked if they would live with their children when they reached the point of not being able to live alone. The results are shown in the figure. Suppose three senior citizens who could not live alone are randomly selected. What is the probability that all three are not living with their children?

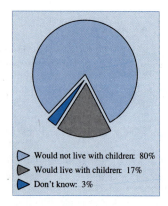

> Would not live with children: 80%
> Would live with children: 17%
> Don't know: 3%

FIGURE FOR 60

61. *Card Game*   Five cards are drawn from an ordinary deck of 52 playing cards. Find the probability of getting two pairs. (For example, the hand could be A-A-5-5-Q or 4-4-7-7-K.)

62. *Birthday Problem*
  (a) What is the probability that, in a group of 10 people, at least two have the same birthday? (Assume there are 365 different birthdays in a year.)
  (b) How large must the group be before the probability that two have the same birthday is at least 50%?

In Exercises 63 and 64, the line plots of sets of data are given. Determine the mean and standard deviation of each set.

63. (a)    (b)

(c) 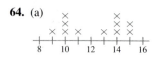   (d)

64. (a)    (b)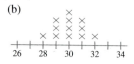

(c) 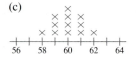   (d)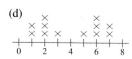

In Exercises 65–68, find the mean, median, variance, and standard deviation of the data.

65. 14, 12, 6, 4, 11, 6, 18, 14, 6, 16

66. 5, 18, 8, 11, 12, 10, 18, 20, 15

67. 104.0, 143.9, 167.1, 182.4, 194.3, 206.0

68. 228.1, 298.6, 309.1, 335.2, 356.6, 393.8

Automotive Batteries   In Exercises 69 and 70, use the following data, which represent the lifetimes (in months) of a sample of 45 automotive batteries.

> 35, 48, 38, 41, 42, 40, 48, 50, 55, 52, 55, 52, 47, 37, 57, 46, 42, 53, 44, 59, 62, 56, 56, 47, 47, 56, 56, 48, 63, 51, 52, 56, 36, 59, 53, 64, 62, 46, 44, 51, 46, 58, 54, 46, 56

69. Determine the mean, median, variance, and standard deviation of the data.

70. Organize the data graphically using a box-and-whisker plot.

CONICS, PARAMETRIC EQUATIONS, AND POLAR COORDINATES

10.1   INTRODUCTION TO CONICS: PARABOLAS

Conics / Parabolas / Application

Conics

Conic sections were discovered during the classical Greek period (600 to 300 B.C.). This early Greek study was concerned largely with the geometrical properties of conics. It was not until the early 17th century that the broad applicability of conics became apparent, and they then played a prominent role in the early development of calculus.

A **conic section** (or simply **conic**) is the intersection of a plane and a double-napped cone. Notice in Figure 10.1 that in the formation of the four basic conics, the intersecting plane does not pass through the vertex of the cone. When the plane does pass through the vertex, the resulting figure is a **degenerate conic,** as shown in Figure 10.2.

Circle   Ellipse   Parabola   Hyperbola

FIGURE 10.1

Point   Line   Two intersecting lines

FIGURE 10.2

Parabolas

There are several ways to begin a study of conics. You could define conics as the intersections of planes and cones, as the Greeks did, or you could define them algebraically in terms of the general second-degree equation

$$Ax^2 + Bxy + Cy^2 + Dx + Ey + F = 0.$$

However, we will use a third approach, in which each of the conics is defined as a collection of points satisfying a certain geometric property. For example, in Section 7 of the Prerequisites chapter, you saw how the definition of a circle as *the collection of all points* (x, y) *that are equidistant from a fixed point* (h, k) led easily to the standard equation of a circle,

$$(x - h)^2 + (y - k)^2 = r^2.$$

In this and the following two sections, we give similar definitions for the other three types of conics.

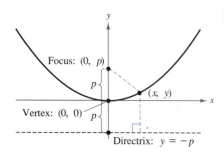

FIGURE 10.3

DEFINITION OF A PARABOLA

A **parabola** is the set of all points (x, y) that are equidistant from a fixed line (**directrix**) and a fixed point (**focus**) not on the line.

The midpoint between the focus and the directrix is the **vertex,** and the line passing through the focus and the vertex is the **axis** of the parabola. Note in Figure 10.3 that a parabola is symmetric with respect to its axis.

REMARK Be sure you understand that the term *parabola* is a technical term used in mathematics and does not simply refer to any U-shaped curve.

STANDARD EQUATION OF A PARABOLA

The **standard form** of the equation of a parabola with vertex at (h, k) is as follows.

$$(x - h)^2 = 4p(y - k), \qquad p \neq 0 \quad \textit{Vertical axis, directrix: } y = k - p$$
$$(y - k)^2 = 4p(x - h), \qquad p \neq 0 \quad \textit{Horizontal axis, directrix: } x = h - p$$

The focus lies on the axis p units (*directed distance*) from the vertex. If the vertex is at the origin $(0, 0)$, the equation takes one of the following forms.

$$x^2 = 4py \qquad\qquad \textit{Vertical axis}$$
$$y^2 = 4px \qquad\qquad \textit{Horizontal axis}$$

PROOF

We prove only the case for which the directrix is parallel to the x-axis and the focus lies above the vertex, as shown in Figure 10.4(a). If (x, y) is any point on the parabola, then by definition it is equidistant from the focus $(h, k + p)$ and the directrix $y = k - p$, and we have

$$\sqrt{(x - h)^2 + [y - (k + p)]^2} = y - (k - p)$$
$$(x - h)^2 + [y - (k + p)]^2 = [y - (k - p)]^2$$
$$(x - h)^2 + y^2 - 2y(k + p) + (k + p)^2 = y^2 - 2y(k - p) + (k - p)^2$$
$$(x - h)^2 - 2py + 2pk = 2py - 2pk$$
$$(x - h)^2 = 4p(y - k).$$

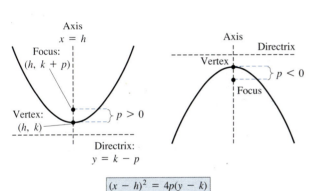

(a) Vertical axis: $p > 0$ (b) Vertical axis: $p < 0$

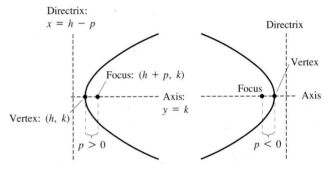

(c) Horizontal axis: $p > 0$ (d) Horizontal axis: $p < 0$

Parabolic Orientations

FIGURE 10.4

REMARK   By expanding the standard equation in Example 1, you obtain the more common quadratic form $y = \frac{1}{12}(x^2 - 4x + 16)$.

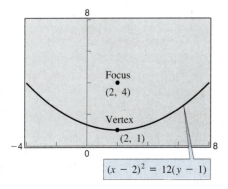

FIGURE 10.5

EXAMPLE 1   Finding the Standard Equation of a Parabola

Find the standard form of the equation of the parabola with vertex $(2, 1)$ and focus $(2, 4)$.

SOLUTION

Because the axis of the parabola is vertical, consider the equation

$$(x - h)^2 = 4p(y - k),$$

where $h = 2$, $k = 1$, and $p = 4 - 1 = 3$. Thus, the standard form is

$$(x - 2)^2 = 12(y - 1).$$

The graph of this parabola is shown in Figure 10.5.

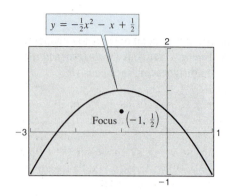

FIGURE 10.6

EXAMPLE 2   Finding the Focus of a Parabola

Find the focus of the parabola given by

$$y = -\frac{1}{2}x^2 - x + \frac{1}{2}.$$

SOLUTION

To find the focus, convert to standard form by completing the square.

$y = -\dfrac{1}{2}x^2 - x + \dfrac{1}{2}$	*Original equation*
$-2y = x^2 + 2x - 1$	*Multiply by −2*
$1 - 2y = x^2 + 2x$	*Group terms*
$2 - 2y = x^2 + 2x + 1$	*Add 1 to both sides*
$-2(y - 1) = (x + 1)^2$	*Standard form*

Comparing this equation with $(x - h)^2 = 4p(y - k)$, you can conclude that $h = -1$, $k = 1$, and $p = -\frac{1}{2}$. Because p is negative, the parabola opens downward, as shown in Figure 10.6. Therefore, the focus of the parabola is

$$(h, k + p) = \left(-1, \frac{1}{2}\right). \qquad \textit{Focus}$$

EXAMPLE 3   Vertex at the Origin

Find the standard equation of the parabola with vertex at the origin and focus at (2, 0).

SOLUTION

The axis of the parabola is horizontal, passing through the origin and (2, 0), as shown in Figure 10.7. Thus, the standard form is

$$y^2 = 4px,$$

where $h = k = 0$ and $p = 2$. Therefore, the equation is

$$y^2 = 8x.$$

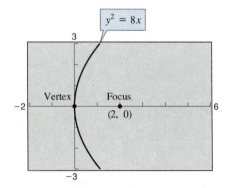

FIGURE 10.7

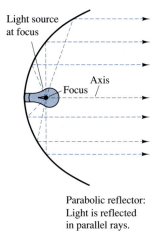

Parabolic reflector:
Light is reflected
in parallel rays.

FIGURE 10.8

Application

A line segment that passes through the focus of a parabola and has endpoints on the parabola is called a **focal chord.** The specific focal chord perpendicular to the axis of the parabola is called the **latus rectum.**

Parabolas occur in a wide variety of applications. For instance, a parabolic reflector can be formed by revolving a parabola about its axis. The resulting surface has the property that all incoming rays parallel to the axis are reflected through the focus of the parabola; this is the principle behind the construction of the parabolic mirrors used in reflecting telescopes. Conversely, the light rays emanating from the focus of a parabolic reflector used in a flashlight are all reflected parallel to one another, as shown in Figure 10.8.

A line is **tangent** to a parabola at a point on the parabola if the line intersects, but does not cross, the parabola at that point. Tangent lines to parabolas have special properties related to the use of parabolas in constructing reflective surfaces.

REFLECTIVE PROPERTY OF A PARABOLA

The tangent line to a parabola at a point P makes equal angles with the following two lines (see figure).

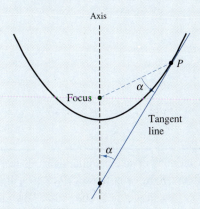

1. The line passing through P and the focus.
2. The axis of the parabola.

*Technology Note*_____

In Example 4, use a graphing utility to
graph the parabola and the tangent
line on the same screen. Then zoom-in
to the point (1, 1). You should be able
to zoom-in close enough so that the
parabola and the tangent line are
indistinguishable. Why?

EXAMPLE 4   Finding the Tangent Line at a Point on a Parabola

Find the equation of the tangent line to the parabola given by $y = x^2$ at the
point (1, 1).

SOLUTION

For this parabola, $p = \frac{1}{4}$ and the focus is $(0, \frac{1}{4})$, as shown in Figure 10.9. You
can find the y-intercept $(0, b)$ of the tangent line by equating the lengths of the
two sides of the isosceles triangle

$$d_1 = \frac{1}{4} - b$$

and

$$d_2 = \sqrt{(1 - 0)^2 + (1 - \tfrac{1}{4})^2} = \frac{5}{4},$$

as shown in Figure 10.9. Setting $d_1 = d_2$ produces

$$\frac{1}{4} - b = \frac{5}{4}$$
$$b = -1.$$

Thus, the slope of the tangent line is

$$m = \frac{1 - (-1)}{1 - 0} = 2,$$

and its slope-intercept equation is

$$y = 2x - 1.$$

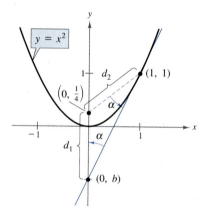

FIGURE 10.9

DISCUSSION
P R O B L E M
•••••••••••••••••••

TELEVISION
ANTENNA
DISHES

Cross sections of television antenna dishes are parabolic in shape. Write a paragraph describing why these dishes are parabolic.

WARM-UP

The following warm-up exercises involve skills that were covered in earlier sections. You will use these skills in the exercise set for this section.

In Exercises 1–4, expand and simplify the expression.

1. $(x - 5)^2 - 20$ **2.** $(x + 3)^2 - 1$ **3.** $10 - (x + 4)^2$ **4.** $4 - (x - 2)^2$

In Exercises 5–8, complete the square for the quadratic expression.

5. $x^2 + 6x + 8$ **6.** $x^2 - 10x + 21$ **7.** $-x^2 + 2x + 1$ **8.** $-2x^2 + 4x - 2$

In Exercises 9 and 10, find an equation of the line passing through the given point with the specified slope.

9. $m = -\frac{2}{3}$, $(1, 6)$ **10.** $m = \frac{3}{4}$, $(3, -2)$

SECTION 10.1 · EXERCISES

In Exercises 1–6, match the equation with its graph, and describe the given viewing rectangle. [The graphs are labeled (a), (b), (c), (d), (e), and (f).]

1. $y^2 = 4x$ **2.** $x^2 = -2y$
3. $x^2 = 8y$ **4.** $y^2 = -12x$
5. $(y - 1)^2 = 4(x - 2)$ **6.** $(x + 3)^2 = -2(y - 2)$

(a)

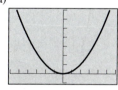

(b)

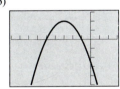

(c)

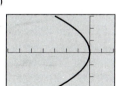

(d)

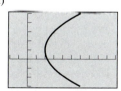

(e)

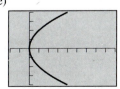

(f)

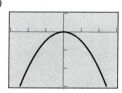

In Exercises 7–22, find the vertex, focus, and directrix of the parabola, and sketch its graph.

7. $y = 4x^2$ **8.** $y = 2x^2$
9. $y^2 = -6x$ **10.** $y^2 = 3x$
11. $x^2 + 8y = 0$ **12.** $x + y^2 = 0$
13. $(x - 1)^2 + 8(y + 2) = 0$
14. $(x + 3) + (y - 2)^2 = 0$
15. $\left(y + \frac{1}{2}\right)^2 = 2(x - 5)$
16. $\left(x + \frac{1}{2}\right)^2 = 4(y - 3)$
17. $y = \frac{1}{4}(x^2 - 2x + 5)$
18. $4x - y^2 - 2y - 33 = 0$
19. $y^2 - 4y - 4x = 0$
20. $y^2 + 6y + 8x + 25 = 0$
21. $y^2 + 4y + 8x - 12 = 0$
22. $x^2 + 4x + 4y - 4 = 0$

In Exercises 23–26, find the vertex, focus, and directrix of the parabola, and use a graphing utility to graph the parabola.

23. $y = -\frac{1}{6}(x^2 + 4x - 2)$
24. $x^2 - 2x + 8y + 9 = 0$
25. $y^2 + x + y = 0$
26. $y^2 - 4x - 4 = 0$

In Exercises 27–36, find an equation of the specified parabola.

27. Vertex: (0, 0)
  Focus: $(0, -\frac{3}{2})$

28. Vertex: (0, 0)
  Focus: (−2, 0)

29. Vertex: (0, 0)
  Directrix: $x = 3$

30. Vertex: (3, 2)
  Focus: (1, 2)

31. Vertex: (0, 4)
  Directrix: $y = 2$

32. Vertex: (0, 0)
  Directrix: $y = 4$

33. Axis: Parallel to y-axis
  Passes through the points: (0, 3), (3, 4), (4, 11)

34. Axis: Parallel to x-axis
  Passes through the points: (4, −2), (0, 0), (3, −3)

35.

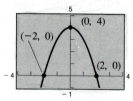

36.

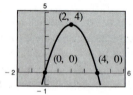

37. *Satellite Antenna*   The receiver in a parabolic television dish antenna is 3 feet from the vertex and is located at the focus (see figure). Find an equation of a cross section of the reflector. (Assume that the dish is directed upward and the vertex is at the origin.)

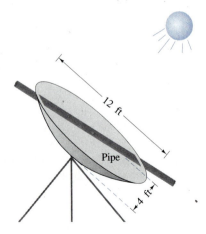

FIGURE FOR 37

38. *Solar Collector*   A solar collector for heating water is constructed with a sheet of stainless steel that is formed into the shape of a parabola (see figure). The water flows through a pipe that passes through the focus of the parabola. At what distance is the pipe from the vertex?

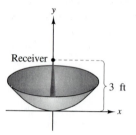

FIGURE FOR 38

39. *Suspension Bridge*   Each cable of a suspension bridge is suspended (in the shape of a parabola) between two towers that are 400 feet apart and 50 feet above the roadway (see figure). The cables touch the roadway midway between the towers.
  (a) Find an equation for the parabolic shape of each cable.
  (b) Find the length of the vertical supporting cable when $x = 100$.

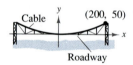

FIGURE FOR 39

40. *Beam Deflection*   A simply supported beam that is 100 feet long has a load concentrated at its center (see figure). The deflection of the beam at its center is 3 inches. If the shape of the deflected beam is parabolic, find the equation of the parabola. (Assume the vertex is at the origin.)

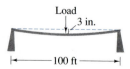

FIGURE FOR 40

41. *Escape Velocity* A satellite in a 100-mile-high circular orbit around the earth has a velocity of approximately 17,500 miles per hour. If this velocity is multiplied by $\sqrt{2}$, the satellite will have the minimum velocity necessary to escape the earth's gravity and it will follow a parabolic path with the center of the earth as the focus (see figure).
(a) Find the escape velocity of the satellite.
(b) Find the equation of its path (assume that the radius of the earth is 4000 miles).

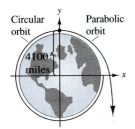

FIGURE FOR 41

42. *Highway Design* Highway engineers design a parabolic curve for an entrance ramp from a straight street to an interstate highway (see figure). Find an equation of the parabola.

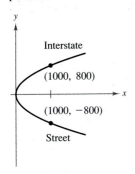

FIGURE FOR 42

Projectile Motion In Exercises 43–45, consider the path of a projectile projected horizontally with a velocity of v feet per second at a height of s feet, where the model for the path is given by

$$y = -\frac{16}{v^2}x^2 + s.$$

In this model, air resistance is disregarded and y is the height (in feet) of the projectile t seconds after its release.

43. A ball is thrown horizontally from the top of a 75-foot tower with a velocity of 32 feet per second.
(a) Find the equation of the parabolic path.
(b) How far does the ball travel horizontally before striking the ground?

44. A ball is thrown horizontally from the top of a 100-foot tower with a velocity of 32 feet per second.
(a) Find the equation of the parabolic path.
(b) How far does the ball travel horizontally before striking the ground?

45. A bomber flying due east at 550 miles per hour at an altitude of 42,000 feet releases a bomb. Determine how far the bomb travels horizontally before striking the ground.

46. Find the equation of the tangent line to the parabola $y = ax^2$ at $x = x_0$. Prove that the x-intercept of this tangent line is $(x_0/2, 0)$.

In Exercises 47–50, find an equation of the tangent line to the parabola at the given point and find the x-intercept of the line.

47. $y = \frac{1}{2}x^2$, $(4, 8)$ **48.** $y = \frac{1}{2}x^2$, $\left(-3, \frac{9}{2}\right)$
49. $y = -2x^2$, $(-1, -2)$ **50.** $y = -2x^2$, $(3, -18)$

10.2 ELLIPSES

Ellipses / Application / Eccentricity

Ellipses

The second type of conic is called an **ellipse,** and is defined as follows.

DEFINITION OF AN ELLIPSE

An **ellipse** is the set of all points (x, y) the sum of whose distances from two distinct fixed points **(foci)** is constant.

$d_1 + d_2$ is constant.

The line through the foci intersects the ellipse at two points **(vertices).** The chord joining the vertices is the **major axis,** and its midpoint is the **center** of the ellipse. The chord perpendicular to the major axis at the center is the **minor axis** of the ellipse.

You can visualize the definition of an ellipse by imagining two thumbtacks placed at the foci, as shown in Figure 10.10. If the ends of a fixed length of string are fastened to the thumbtacks and the string is drawn taut with a pencil, the path traced by the pencil will be an ellipse.

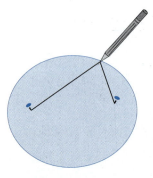

FIGURE 10.10

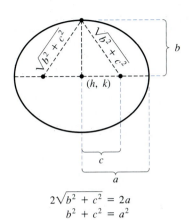

$2\sqrt{b^2 + c^2} = 2a$
$b^2 + c^2 = a^2$

FIGURE 10.11

To derive the standard form of the equation of an ellipse, consider the ellipse in Figure 10.11 with the following points.

Center: (h, k)

Vertices: $(h \pm a, k)$

Foci: $(h \pm c, k)$

The sum of the distances from any point on the ellipse to the two foci is constant. At a vertex, this constant sum is

$$(a + c) + (a - c) = 2a, \qquad \text{\textit{Length of major axis}}$$

or simply the length of the major axis. Now, if you let (x, y) be *any* point on the ellipse, the sum of the distances between (x, y) and the two foci must also be $2a$. That is,

$$\sqrt{[x - (h - c)]^2 + (y - k)^2} + \sqrt{[x - (h + c)]^2 + (y - k)^2} = 2a.$$

Finally, in Figure 10.11, you can see that $b^2 = a^2 - c^2$, which implies that the equation of the ellipse is

$$b^2(x - h)^2 + a^2(y - k)^2 = a^2 b^2$$

$$\frac{(x - h)^2}{a^2} + \frac{(y - k)^2}{b^2} = 1.$$

Had you chosen a vertical major axis, you would have obtained a similar equation. Both results are summarized as follows.

STANDARD EQUATION OF AN ELLIPSE

The standard form of the equation of an ellipse, with center (h, k) and major and minor axes of lengths $2a$ and $2b$, where $0 < b < a$, is

$$\frac{(x - h)^2}{a^2} + \frac{(y - k)^2}{b^2} = 1 \qquad \text{\textit{Major axis is horizontal}}$$

$$\frac{(x - h)^2}{b^2} + \frac{(y - k)^2}{a^2} = 1. \qquad \text{\textit{Major axis is vertical}}$$

The foci lie on the major axis, c units from the center, with $c^2 = a^2 - b^2$. If the center is at the origin $(0, 0)$, the equation takes one of the following forms.

$$\frac{x^2}{a^2} + \frac{y^2}{b^2} = 1 \qquad \text{\textit{Major axis is horizontal}}$$

$$\frac{x^2}{b^2} + \frac{y^2}{a^2} = 1 \qquad \text{\textit{Major axis is vertical}}$$

Figure 10.12 shows both the vertical and horizontal orientations for an ellipse.

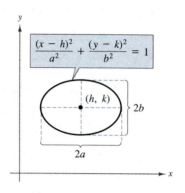 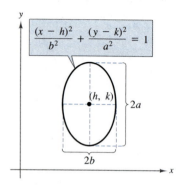

FIGURE 10.12

EXAMPLE 1   **Finding the Standard Equation of an Ellipse**

Find the standard form of the equation of the ellipse having foci at $(0, 1)$ and $(4, 1)$ and a major axis of length 6, as shown in Figure 10.13.

SOLUTION

Because the foci occur at $(0, 1)$ and $(4, 1)$, the center of the ellipse is $(2, 1)$. This implies that the distance from the center to one of the foci is $c = 2$, and since $2a = 6$ you know that $a = 3$. Now, using $c^2 = a^2 - b^2$, you have

$$b = \sqrt{a^2 - c^2} = \sqrt{9 - 4} = \sqrt{5}.$$

Because the major axis is horizontal, the standard equation is

$$\frac{(x - 2)^2}{9} + \frac{(y - 1)^2}{5} = 1.$$

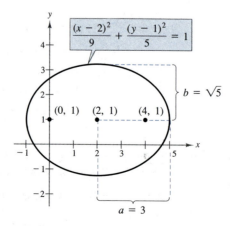

FIGURE 10.13

Technology Note

Most graphing utilities have a "parametric mode." Try using the parametric mode to graph the ellipse given by $x = 2 + 3 \cos t$ and $y = 1 + \sqrt{4} \sin t$. How does the result compare with the graph given in Figure 10.13? (*Hint:* Set your graphing utility to radian mode before graphing the ellipse.)

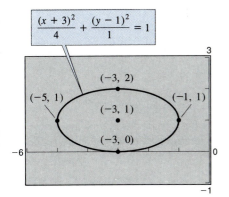

FIGURE 10.14

EXAMPLE 2 Writing an Equation in Standard Form

Sketch the graph of the ellipse whose equation is

$$x^2 + 4y^2 + 6x - 8y + 9 = 0.$$

SOLUTION

Begin by writing the given equation is standard form.

$x^2 + 4y^2 + 6x - 8y + 9 = 0$	*Original equation*
$(x^2 + 6x + \quad) + (4y^2 - 8y + \quad) = -9$	*Group terms*
$(x^2 + 6x + \quad) + 4(y^2 - 2y + \quad) = -9$	*Factor 4 out of y-term*
$(x^2 + 6x + 9) + 4(y^2 - 2y + 1) = -9 + 9 + 4$	*Add 9 and 4 to both sides*
$(x + 3)^2 + 4(y - 1)^2 = 4$	*Completed square form*
$\dfrac{(x + 3)^2}{4} + \dfrac{(y - 1)^2}{1} = 1$	*Standard form*

From the standard form, you can see that the center occurs at $(h, k) = (-3, 1)$. Because the denominator of the x-term is $a^2 = 2^2$, locate the endpoints of the major axis two units to the right and left of the center. Similarly, because the denominator of the y-term is $b^2 = 1^2$, locate the endpoints of the minor axis one unit up and down from the center. The graph of this ellipse is shown in Figure 10.14.

EXAMPLE 3 Analyzing an Ellipse

Find the center, vertices, and foci of the ellipse given by

$$4x^2 + y^2 - 8x + 4y - 8 = 0.$$

SOLUTION

By completing the square, you can write the given equation in standard form.

$$4x^2 + y^2 - 8x + 4y - 8 = 0$$
$$4(x^2 - 2x + 1) + (y^2 + 4y + 4) = 8 + 4 + 4$$
$$4(x - 1)^2 + (y + 2)^2 = 16$$
$$\frac{(x - 1)^2}{4} + \frac{(y + 2)^2}{16} = 1$$

Thus, the major axis is vertical, where $h = 1$, $k = -2$, $a = 4$, $b = 2$, and $c = \sqrt{16 - 4} = 2\sqrt{3}$. Therefore, you have the following.

Center: $(1, -2)$	Vertices: $(1, -6)$	Foci: $\left(1, -2 - 2\sqrt{3}\right)$
	$(1, 2)$	$\left(1, -2 + 2\sqrt{3}\right)$

The graph of the ellipse is shown in Figure 10.15.

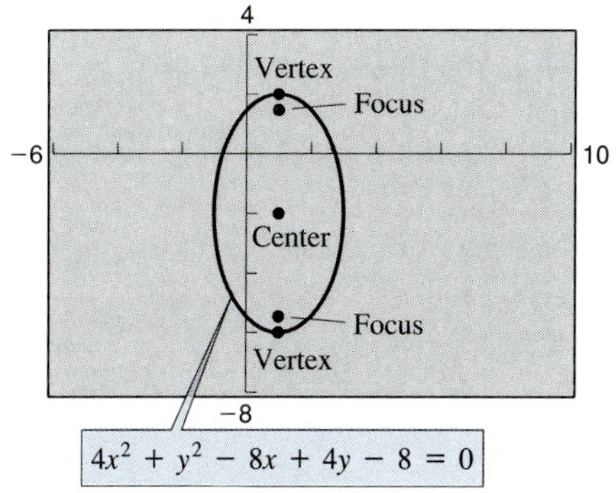

FIGURE 10.15

Technology Note

You can also use a graphing utility to graph an ellipse by graphing the upper and lower portions on the same viewing rectangle. For example, to graph the ellipse in Example 3, first solve for y

$$\frac{(x-1)^2}{4} + \frac{(y+2)^2}{16} = 1$$

$$(y+2)^2 = 16\left(1 - \frac{(x-1)^2}{4}\right)$$

$$y + 2 = \pm 4\sqrt{1 - \frac{(x-1)^2}{4}}$$

$$y = -2 \pm 4\sqrt{1 - \frac{(x-1)^2}{4}}$$

If you then graph the two equations

$$y_1 = -2 + 4\sqrt{1 - \frac{(x-1)^2}{4}}$$

$$y_2 = -2 - 4\sqrt{1 - \frac{(x-1)^2}{4}}$$

you will obtain the ellipse in Figure 10.15.

REMARK If the constant term in the equation in Example 3 had been $F \geq 8$, you would have obtained one of the following degenerate cases.

1. Single point: $\dfrac{(x-1)^2}{4} + \dfrac{(y+2)^2}{16} = 0$ $F = 8$

2. No solution points: $\dfrac{(x-1)^2}{4} + \dfrac{(y+2)^2}{16} < 0$ $F > 8$

Application

Ellipses have many practical and aesthetic uses. For instance, machine gears, supporting arches, and acoustical designs often involve elliptical shapes. The orbits of satellites and planets are also ellipses. Example 4 investigates the elliptical orbit of the moon about the earth.

EXAMPLE 4   An Application Involving an Elliptical Orbit

The moon travels about the earth in an elliptical orbit with the earth at one focus, as shown in Figure 10.16. The major and minor axes of the orbit have lengths of 768,806 kilometers and 767,746 kilometers, respectively. Find the greatest and least distances (the apogee and perigee) from the earth's center to the moon's center.

SOLUTION

Because $2a = 768,806$ and $2b = 767,746$, you have $a = 384,403$, $b = 383,873$, and

$$c = \sqrt{a^2 - b^2} \approx 20,179.$$

Therefore, the greatest distance between the center of the earth and the center of the moon is

$$a + c \approx 404,582 \text{ km},$$

and the least distance is

$$a - c \approx 364,224 \text{ km}.$$

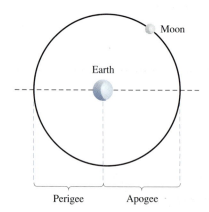

FIGURE 10.16

Eccentricity

One of the reasons it was difficult for early astronomers to detect that the orbits of the planets are ellipses is that the foci of the planetary orbits are relatively close to their centers, thus making the orbits nearly circular. To measure the ovalness of an ellipse, we use the concept of **eccentricity.**

DEFINITION OF ECCENTRICITY

The **eccentricity** e of an ellipse is given by the ratio

$$e = \frac{c}{a}.$$

To see how this ratio is used to describe the shape of an ellipse, note that since the foci of an ellipse are located along the major axis between the vertices and the center, it follows that

$$0 < c < a.$$

For an ellipse that is nearly circular, the foci are close to the center and the ratio c/a is small, as shown in Figure 10.17(a). On the other hand, for an elongated ellipse, the foci are close to the vertices, and the ratio c/a is close to 1, as shown in Figure 10.17(b).

REMARK   Note that $0 < e < 1$ for every ellipse.

The orbit of the moon has an eccentricity of $e = 0.0549$, and the eccentricities of the nine planetary orbits are as follows.

Mercury:	$e = 0.2056$	Saturn:	$e = 0.0543$
Venus:	$e = 0.0068$	Uranus:	$e = 0.0460$
Earth:	$e = 0.0167$	Neptune:	$e = 0.0082$
Mars:	$e = 0.0934$	Pluto:	$e = 0.2481$
Jupiter:	$e - 0.0484$		

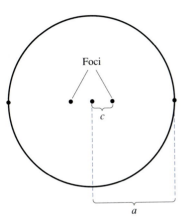

(a) $\dfrac{c}{a}$ is small.

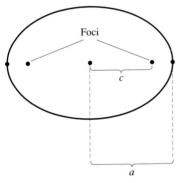

(b) $\dfrac{c}{a}$ is close to 1.

FIGURE 10.17

DISCUSSION PROBLEM

····················

IS A CIRCLE AN ELLIPSE?

If $a = b = r$, then the equation

$$\frac{x^2}{a^2} + \frac{y^2}{b^2} = 1$$

can be rewritten as

$$x^2 + y^2 = r^2.$$

Does this imply that a circle is an ellipse? Why or why not?

WARP-UP

WARM-UP

The following warm-up exercises involve skills that were covered in earlier sections. You will use these skills in the exercise set for this section.

In Exercises 1–4, sketch a graph of the equation.

1. $x^2 = 9y$ **2.** $y^2 = 9x$ **3.** $y^2 = -9x$ **4.** $x^2 = -9y$

In Exercises 5–8, find the unknown in the equation $c^2 = a^2 - b^2$. (Assume a, b, and c are positive.)

5. $a = 13, b = 5$ **6.** $a = \sqrt{10}, c = 3$ **7.** $b = 6, c = 8$ **8.** $a = 7, b = 5$

In Exercises 9 and 10, simplify the compound fraction.

9. $\dfrac{x^2}{\frac{1}{4}} + \dfrac{y^2}{\frac{1}{3}}$ **10.** $\dfrac{(x-1)^2}{\frac{4}{9}} + \dfrac{(y+2)^2}{\frac{1}{9}}$

SECTION 10.2 · EXERCISES

In Exercises 1–6, match the equation with its graph, and describe the given viewing rectangle. [The graphs are labeled (a), (b), (c), (d), (e), and (f).]

1. $\dfrac{x^2}{1} + \dfrac{y^2}{9} = 1$ **2.** $\dfrac{x^2}{9} + \dfrac{y^2}{1} = 1$

3. $\dfrac{x^2}{9} + \dfrac{y^2}{4} = 1$ **4.** $\dfrac{y^2}{9} + \dfrac{x^2}{9} = 1$

5. $\dfrac{(x-2)^2}{16} + \dfrac{(y+1)^2}{4} = 1$ **6.** $\dfrac{(x+2)^2}{4} + \dfrac{(y+2)^2}{25} = 1$

(e)

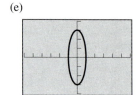

(f)

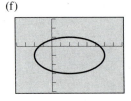

(a)

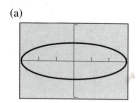

(b)

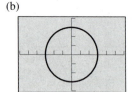

(c)

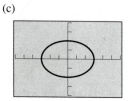

(d)

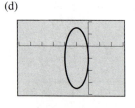

In Exercises 7–22, find the center, foci, vertices, and eccentricity of the ellipse and sketch its graph.

7. $\dfrac{x^2}{25} + \dfrac{y^2}{16} = 1$ **8.** $\dfrac{x^2}{144} + \dfrac{y^2}{169} = 1$

9. $\dfrac{x^2}{16} + \dfrac{y^2}{25} = 1$ **10.** $\dfrac{x^2}{169} + \dfrac{y^2}{144} = 1$

11. $\dfrac{x^2}{9} + \dfrac{y^2}{5} = 1$ **12.** $\dfrac{x^2}{28} + \dfrac{y^2}{64} = 1$

13. $x^2 + 4y^2 = 4$ **14.** $5x^2 + 3y^2 = 15$

15. $3x^2 + 2y^2 = 6$ **16.** $5x^2 + 7y^2 = 70$

17. $\dfrac{(x-1)^2}{9} + \dfrac{(y-5)^2}{25} = 1$

18. $(x+2)^2 + \dfrac{(y+4)^2}{\frac{1}{4}} = 1$

19. $9x^2 + 4y^2 + 36x - 24y + 36 = 0$
20. $9x^2 + 4y^2 - 36x + 8y + 31 = 0$
21. $16x^2 + 25y^2 - 32x + 50y + \blacksquare = 0$
22. $9x^2 + 25y^2 - 36x - 50y + 61 = 0$

In Exercises 23–28, find the center, foci, and vertices of the ellipse. Use a graphing utility to graph the ellipse. (Explain how you used the utility to obtain the graph.)

23. $4x^2 + y^2 = 1$
24. $16x^2 + 25y^2 = 1$
25. $12x^2 + 20y^2 - 12x + 40y - 37 = 0$
26. $36x^2 + 9y^2 + 48x - 36y + 43 = 0$
27. $x^2 + 2y^2 - 3x + 4y + 0.25 = 0$
28. $2x^2 + y^2 + 4.8x - 6.4y + 3.12 = 0$

In Exercises 29–36, find an equation of the specified ellipse.

29. Vertices: $(\pm 6, 0)$
    Foci: $(\pm 5, 0)$

30. Vertices: $(0, \pm 8)$
    Foci: $(0, \pm 4)$

31. Vertices: $(0, \pm 2)$
    Minor axis of length 2

32. Vertices: $(0, 2)$, $(4, 2)$
    Minor axis of length 2

33. Foci: $(0, 0)$, $(0, 8)$
    Major axis of length 16

34. Vertices: $(0, \pm 5)$
    Solution point: $(4, 2)$

35. Center: $(3, 2)$, $a = 3c$
    Foci: $(1, 2)$, $(5, 2)$

36. Vertices: $(\pm 5, 0)$
    Eccentricity: $\frac{3}{5}$

37. *Fireplace Arch*   A fireplace arch is to be constructed in the shape of a semi-ellipse. The opening is to have a height of 2 feet at the center and a width of 5 feet along the base (see figure). The contractor draws the outline of the ellipse by the method shown in Figure 10.10. Where should the tacks be placed and what should be the length of the piece of string?

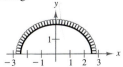

FIGURE FOR 37

38. *Mountain Tunnel*   A semi-elliptical arch over a tunnel for a road through a mountain has a major axis of 100 feet, and its height at the center is 30 feet (see figure). Determine the height of the arch 5 feet from the edge of the tunnel.

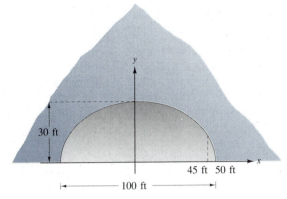

FIGURE FOR 38

39. Sketch a graph of the ellipse that consists of all points (x, y) such that the sum of the distances between (x, y) and two fixed points is 16 units and the foci are located at the centers of the two sets of concentric circles in the figure.

40. A line segment through a focus with endpoints on the ellipse and perpendicular to the major axis is called a **latus rectum** of the ellipse. Therefore, an ellipse has two latus recta. Knowing the length of the latus recta is helpful in sketching an ellipse because it yields other points on the curve (see figure). Show that the length of each latus rectum is $2b^2/a$.

FIGURE FOR 39

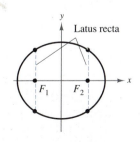

FIGURE FOR 40

In Exercises 41–44, sketch the graph of the ellipse, making use of the latus recta (see Exercise 40).

41. $\dfrac{x^2}{4} + \dfrac{y^2}{1} = 1$ **42.** $\dfrac{x^2}{9} + \dfrac{y^2}{16} = 1$

43. $9x^2 + 4y^2 = 36$ **44.** $5x^2 + 3y^2 = 15$

45. *Orbit of the Earth*   The earth moves in an elliptical orbit with the sun at one of the foci (see figure). The length of half of the major axis is 92.957×10^6 miles and the eccentricity is 0.017. Find the smallest distance (*perihelion*) and the greatest distance (*aphelion*) of the earth from the sun.

46. *Orbit of Pluto*   The planet Pluto moves in an elliptical orbit with the sun at one of the foci (see figure). The length of half of the major axis is 3.666×10^9 miles and the eccentricity is 0.248. Find the smallest distance and the greatest distance of Pluto from the sun.

47. *Orbit of Saturn*   The planet Saturn moves in an elliptical orbit with the sun at one of the foci (see figure). The smallest distance and the greatest distance of the planet from the sun are 1.3495×10^9 kilometers and 1.5045×10^9 kilometers, respectively. Find the eccentricity of the orbit.

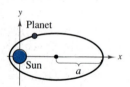

FIGURE FOR 45–47

48. *Satellite Orbit*   If the apogee and the perigee (see Example 4) of an elliptical orbit of an earth satellite are given by A and P, respectively, show that the eccentricity of the orbit is given by

$$e = \frac{A - P}{A + P}.$$

49. *Sputnik I*   The first artificial satellite to orbit the earth was Sputnik I (launched by the Soviet Union in 1957). Its highest point above the earth's surface was 583 miles, and its lowest point was 132 miles. Assume that the center of the earth is the focus of the elliptical orbit and the radius of the earth is 4000 miles. Find the eccentricity of the orbit.

50. *Explorer 18*   On November 26, 1963, the United States launched Explorer 18. Its low and high points over the surface of the earth were 119 miles and 122,000 miles, respectively. Find the eccentricity of its elliptical orbit.

51. Show that the equation of an ellipse can be written as

$$\frac{(x - h)^2}{a^2} + \frac{(y - k)^2}{a^2(1 - e^2)} = 1.$$

Note that as e approaches zero the ellipse approaches a circle of radius a.

10.3 HYPERBOLAS

Hyperbolas / Asymptotes of a Hyperbola / Applications

Hyperbolas

The definition of a hyperbola parallels that of an ellipse. The difference is that for an ellipse the *sum* of the distances between the foci and a point on the ellipse is fixed, while for a hyperbola the *difference* of these distances is fixed.

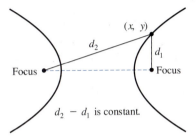

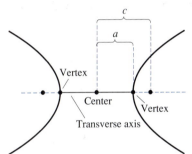

FIGURE 10.18

DEFINITION OF A HYPERBOLA

A **hyperbola** is the set of all points (x, y) the difference of whose distances from two distinct fixed points (foci) is constant (see Figure 10.18).

Every hyperbola has two disconnected parts (**branches**). The line through the two foci intersects a hyperbola at two points (**vertices**). The line segment connecting the vertices is the **transverse axis,** and the midpoint of the transverse axis is the **center** of the hyperbola.

The development of the standard form of the equation of a hyperbola is similar to that of an ellipse, and we list the following result without proof.

STANDARD EQUATION OF A HYPERBOLA

The standard form of the equation of a hyperbola with center at (h, k) is

$$\frac{(x - h)^2}{a^2} - \frac{(y - k)^2}{b^2} = 1 \qquad \textit{Transverse axis is horizontal}$$

$$\frac{(y - k)^2}{a^2} - \frac{(x - h)^2}{b^2} = 1. \qquad \textit{Transverse axis is vertical}$$

The vertices are a units from the center, and the foci are c units from the center. Moreover, $b^2 = c^2 - a^2$. If the center of the hyperbola is at the origin $(0, 0)$, the equation takes one of the following forms.

$$\frac{x^2}{a^2} - \frac{y^2}{b^2} = 1 \qquad \textit{Transverse axis is horizontal}$$

$$\frac{y^2}{a^2} - \frac{x^2}{b^2} = 1 \qquad \textit{Transverse axis is vertical}$$

Figure 10.19 shows both the horizontal and vertical orientations of a hyperbola.

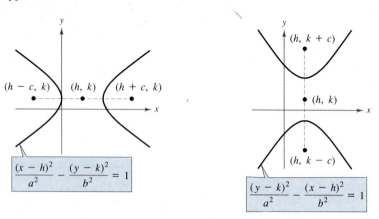

FIGURE 10.19   Standard Equations of Hyperbolas

EXAMPLE 1   Finding the Standard Equation of a Hyperbola

Find the standard form of the equation of the hyperbola with foci at $(-1, 2)$ and $(5, 2)$ and vertices at $(0, 2)$ and $(4, 2)$.

SOLUTION

By the Midpoint Formula, the center of the hyperbola occurs at the point $(2, 2)$. Furthermore, $c = 3$ and $a = 2$, and it follows that

$$b^2 = 3^2 - 2^2 = 9 - 4 = 5.$$

Thus, the equation of the hyperbola is

$$\frac{(x - 2)^2}{4} - \frac{(y - 2)^2}{5} = 1.$$

Figure 10.20 shows the graph of the hyperbola.

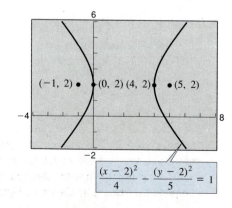

FIGURE 10.20

Technology Note

You can use a graphing utility to graph a hyperbola in two ways. For example, to graph the hyperbola in Example 1, you can use the following equations in parametric mode.

$X_{1T} = 2 + 2 \sec t = 2 + 2/\cos t$

$Y_{1T} = 2 + \sqrt{5} \tan t$

Note how this method is based on the trigonometric identity $\tan^2 t + 1 = \sec^2 t$. Alternatively, you can solve for y as follows.

$$\frac{(x - 2)^2}{4} - \frac{(y - 2)^2}{5} = 1$$

$$(y - 2)^2 = 5\left(\frac{(x - 2)^2}{4} - 1\right)$$

$$y - 2 = \pm\sqrt{5}\sqrt{\frac{(x - 2)^2}{4} - 1}$$

$$y = 2 \pm \sqrt{5}\sqrt{\frac{(x - 2)^2}{4} - 1}$$

If you graph the two equations

$$y_1 = 2 + \sqrt{5}\sqrt{\frac{(x - 2)^2}{4} - 1}$$

$$y_2 = 2 - \sqrt{5}\sqrt{\frac{(x - 2)^2}{4} - 1}$$

you will obtain the hyperbola in Figure 10.20. Which method of graphing do you prefer?

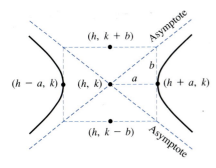

FIGURE 10.21

Asymptotes of a Hyperbola

An important aid in sketching the graph of a hyperbola is the determination of its **asymptotes,** as shown in Figure 10.21. Each hyperbola has two asymptotes that intersect at the center of the hyperbola. The asymptotes pass through the vertices of a rectangle of dimensions $2a$ by $2b$, with its center at (h, k). The line segment of length $2b$ joining $(h, k + b)$ and $(h, k - b)$ is referred to as the **conjugate axis** of the hyperbola. The following result identifies the equations for the asymptotes.

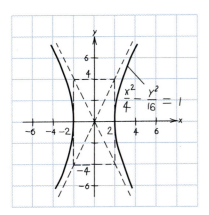

FIGURE 10.22

ASYMPTOTES OF A HYPERBOLA

For a *horizontal* transverse axis, the equations of the asymptotes are

$$y = k + \frac{b}{a}(x - h) \quad \text{and} \quad y = k - \frac{b}{a}(x - h).$$

For a *vertical* transverse axis, the equations of the asymptotes are

$$y = k + \frac{a}{b}(x - h) \quad \text{and} \quad y = k - \frac{a}{b}(x - h).$$

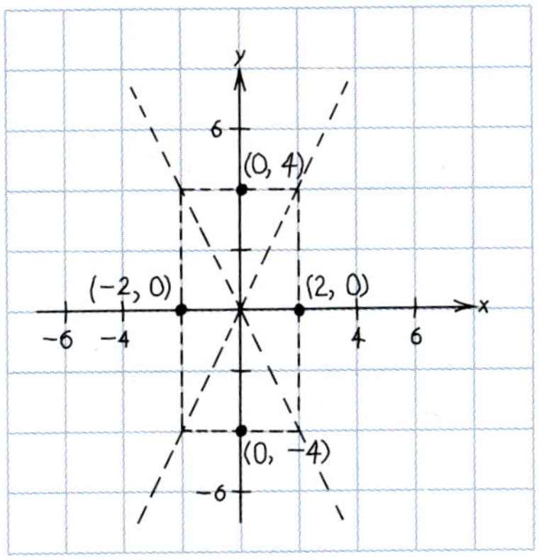

EXAMPLE 2   **Using Asymptotes to Sketch a Hyperbola**

Sketch the hyperbola whose equation is $4x^2 - y^2 = 16$.

SOLUTION

$$4x^2 - y^2 = 16 \qquad \text{\textit{Original equation}}$$

$$\frac{4x^2}{16} - \frac{y^2}{16} = \frac{16}{16}$$

$$\frac{x^2}{2^2} - \frac{y^2}{4^2} = 1 \qquad \text{\textit{Standard form}}$$

From this, it follows that the transverse axis is horizontal and the vertices occur at $(-2, 0)$ and $(2, 0)$. Moreover, the ends of the conjugate axis occur at $(0, -4)$ and $(0, 4)$, and you can sketch the rectangle shown in Figure 10.22. Finally, by drawing the asymptotes through the corners of this rectangle, you can complete the sketch, as shown in Figure 10.23.

FIGURE 10.23

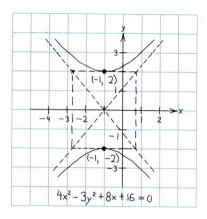

FIGURE 10.24

REMARK If the constant term F in the equation in Example 3 had been $F = -4$ instead of 16, you would have obtained the following degenerate case (two intersecting lines).

$$\frac{y^2}{4} - \frac{(x + 1)^2}{3} = 0$$

EXAMPLE 3 Finding the Asymptotes of a Hyperbola

Sketch the hyperbola given by $4x^2 - 3y^2 + 8x + 16 = 0$ and find the equations of its asymptotes.

SOLUTION

$$4x^2 - 3y^2 + 8x + 16 = 0 \qquad \textit{Original equation}$$
$$4(x^2 + 2x) - 3y^2 = -16$$
$$-4(x^2 + 2x + 1) + 3y^2 = 16 - 4$$
$$-4(x + 1)^2 + 3y^2 = 12$$
$$\frac{y^2}{4} - \frac{(x + 1)^2}{3} = 1 \qquad \textit{Standard form}$$

From this equation it follows that the hyperbola is centered at $(-1, 0)$ and has vertices at $(-1, 2)$ and $(-1, -2)$, and that the ends of the conjugate axis occur at $\left(-1 - \sqrt{3}, 0\right)$ and $\left(-1 + \sqrt{3}, 0\right)$. To sketch the graph of the hyperbola, draw a rectangle through the vertices and the ends of the conjugate axis. The asymptotes are the lines passing through the corners of the rectangle, as shown in Figure 10.24. Finally, using $a = 2$ and $b = \sqrt{3}$, it follows that the equations of the asymptotes are

$$y = \frac{2}{\sqrt{3}}(x + 1) \quad \text{and} \quad y = -\frac{2}{\sqrt{3}}(x + 1).$$

EXAMPLE 4 Using Asymptotes to Find the Standard Equation

Find the standard form of the equation of the hyperbola having vertices at $(3, -5)$ and $(3, 1)$ and asymptotes $y = 2x - 8$ and $y = -2x + 4$, as shown in Figure 10.25.

SOLUTION

By the Midpoint Formula, the center of the hyperbola is at $(3, -2)$. Furthermore, the hyperbola has a vertical transverse axis with $a = 3$. From the given equation, you can determine the slopes of the asymptotes to be

$$m_1 = 2 = \frac{a}{b} \quad \text{and} \quad m_2 = -2 = -\frac{a}{b},$$

and since $a = 3$, you can conclude that $b = \frac{3}{2}$. Thus, the standard equation is

$$\frac{(y + 2)^2}{9} - \frac{(x - 3)^2}{\frac{9}{4}} = 1.$$

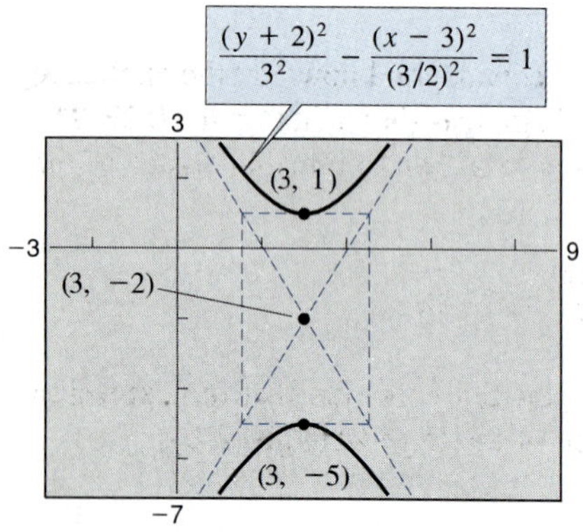

FIGURE 10.25

As with ellipses, the **eccentricity** of a hyperbola is $e = c/a$, and because $c > a$ it follows that $e > 1$. If the eccentricity is large, the branches of the hyperbola are nearly flat. If the eccentricity is close to 1, the branches of the hyperbola are more pointed, as shown in Figure 10.26.

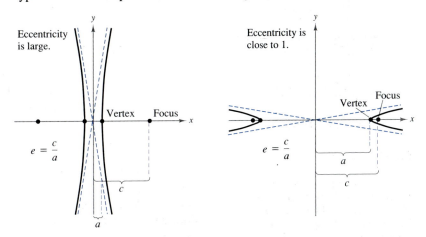

FIGURE 10.26

Applications

The following application was developed during World War II. It shows how the properties of hyperbolas can be used in radar and other detection systems.

EXAMPLE 5   **An Application Involving Hyperbolas**

Two microphones, 1 mile apart, record an explosion. Microphone A received the sound 2 seconds before microphone B. Where did the explosion occur?

SOLUTION

Assuming sound travels at 1100 feet per second, you know that the explosion took place 2200 feet further from B than from A, as shown in Figure 10.27. The locus of all points that are 2200 feet closer to A than to B is one branch of the hyperbola $(x^2/a^2) - (y^2/b^2) = 1$, where

$$c = \frac{5280}{2} = 2640 \quad \text{and} \quad a = \frac{2200}{2} = 1100.$$

Thus, $b^2 = c^2 - a^2 = 5{,}759{,}600$ and you can conclude that the explosion occurred somewhere on the right branch of the hyperbola given by

$$\frac{x^2}{1{,}210{,}000} - \frac{y^2}{5{,}759{,}600} = 1.$$

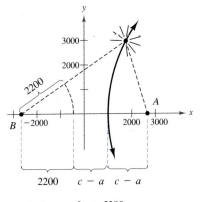

$$2c = 5280$$
$$2200 + 2(c - a) = 5280$$

FIGURE 10.27

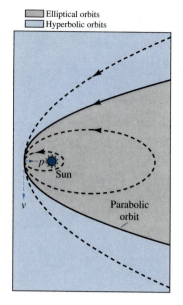

Elliptical orbits
Hyperbolic orbits

Sun

Parabolic
orbit

FIGURE 10.28

In Example 5, you were able to determine only the hyperbola on which the explosion occurred, but not the exact location of the explosion. If, however, you had received the sound from a third position C, then two other hyperbolas would have been determined. The exact location of the explosion would have been the point where these three hyperbolas intersected.

Another interesting application of conic sections involves the orbits of comets in our solar system. Of the 610 comets identified prior to 1970, 245 having elliptical orbits, 295 have parabolic orbits, and 70 have hyperbolic orbits. The center of the sun is a focus of each of these orbits, and each orbit has a vertex at the point where the comet is closest to the sun, as shown in Figure 10.28. Undoubtedly there have been many comets with parabolic or hyperbolic orbits that were not identified. We only get to see such comets *once*. Comets with elliptical orbits, such as Halley's comet, are the only ones that remain in our solar system.

If p is the distance between the vertex and the focus in meters, and v is the velocity of the comet at the vertex in meters per second, then the type of orbit is determined as follows.

1. Ellipse: $v < \sqrt{2GM/p}$
2. Parabola: $v = \sqrt{2GM/p}$
3. Hyperbola: $v > \sqrt{2GM/p}$

In each of these equations, $M \approx 1.991 \times 10^{30}$ kilograms (the mass of the sun) and $G \approx 6.67 \times 10^{-11}$ cubic meters per gram-second squared.

We conclude this section with a procedure for classifying a conic using the coefficients in the general form of its equation.

CLASSIFYING A CONIC FROM ITS GENERAL EQUATION

The graph of $Ax^2 + Cy^2 + Dx + Ey + F = 0$ is one of the following (except in degenerate cases).

1. Circle: $A = C$
2. Parabola: $AC = 0$ *$A = 0$ or $C = 0$, but not both.*
3. Ellipse: $AC > 0$ *A and C have like signs.*
4. Hyperbola: $AC < 0$ *A and C have unlike signs.*

EXAMPLE 6   Classifying Conics from General Equations

A. For the general equation

$$4x^2 - 9x + y - 5 = 0,$$

you have $AC = 4(0) = 0$. Thus, the graph is a parabola.

B. For the general equation

$$4x^2 - y^2 + 8x - 6y + 4 = 0,$$

you have $AC = 4(-1) < 0$. Thus, the graph is a hyperbola.

C. For the general equation

$$2x^2 + 4y^2 - 4x + 12y = 0,$$

you have $AC = (2)(4) > 0$. Thus, the graph is an ellipse.

DISCUSSION PROBLEM

HYPERBOLAS IN APPLICATIONS

At the beginning of Section 10.1, we mentioned that each type of conic section can be formed by the intersection of a plane and a double-napped cone. Three examples of how such intersections can occur in physical situations are shown below.

Identify the cone and hyperbola (or portion of a hyperbola) in each of the three situations. Can you think of other examples of physical situations in which hyperbolas are formed?

WARM-UP

The following warm-up exercises involve skills that were covered in earlier sections. You will use these skills in the exercise set for this section.

In Exercises 1 and 2, find the distance between the two points.

1. $(4, 1), (10, 6)$ **2.** $(-1, 5), (3, -2)$

In Exercises 3–6, sketch the graph of the lines on the same set of coordinate axes.

3. $y = \pm\frac{1}{2}x$ **4.** $y = 3 \pm \frac{1}{2}x$

5. $y = 3 \pm \frac{1}{2}(x - 4)$ **6.** $y = \pm\frac{1}{2}(x - 4)$

In Exercises 7–10, identify the graph of the equation.

7. $x^2 + 4y = 4$ **8.** $x^2 + 4y^2 = 4$

9. $4x^2 + 4y^2 = 4$ **10.** $x + 4y^2 = 4$

SECTION 10.3 · EXERCISES

In Exercises 1–6, match the equation with its graph, and describe the given viewing rectangle. [The graphs are labeled (a), (b), (c), (d), (e), and (f).]

1. $\dfrac{x^2}{9} - \dfrac{y^2}{4} = 1$ **2.** $\dfrac{y^2}{9} - \dfrac{x^2}{4} = 1$

3. $\dfrac{y^2}{1} - \dfrac{x^2}{16} = 1$ **4.** $\dfrac{y^2}{16} - \dfrac{x^2}{1} = 1$

5. $\dfrac{(x-2)^2}{9} - \dfrac{y^2}{4} = 1$ **6.** $\dfrac{(x+1)^2}{16} - \dfrac{(y-3)^2}{9} = 1$

(a)

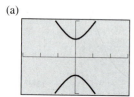

(b)

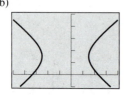

(c)

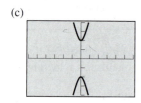

(d)

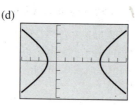

(e)

(f)

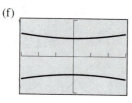

In Exercises 7–22, find the center, foci, and vertices of the hyperbola and sketch its graph using asymptotes as an aid.

7. $x^2 - y^2 = 1$ **8.** $\dfrac{x^2}{9} - \dfrac{y^2}{16} = 1$

9. $y^2 - \dfrac{x^2}{4} = 1$ **10.** $\dfrac{y^2}{9} - \dfrac{x^2}{1} = 1$

11. $\dfrac{y^2}{25} - \dfrac{x^2}{144} = 1$ **12.** $\dfrac{x^2}{36} - \dfrac{y^2}{4} = 1$

13. $5y^2 = 4x^2 + 20$ **14.** $7x^2 - 3y^2 = 21$

15. $\dfrac{(x-1)^2}{4} - \dfrac{(y+2)^2}{1} = 1$ **16.** $\dfrac{(x+1)^2}{144} - \dfrac{(y-4)^2}{25} = 1$

17. $(y+6)^2 - (x-2)^2 = 1$

18. $\dfrac{(y-1)^2}{\frac{1}{4}} - \dfrac{(x+3)^2}{\frac{1}{9}} = 1$

19. $9x^2 - y^2 - 36x - 6y + 18 = 0$

20. $x^2 - 9y^2 + 36y - 72 = 0$

21. $x^2 - 9y^2 + 2x - 54y - 80 = 0$

22. $16y^2 - x^2 + 2x + 64y + 63 = 0$

In Exercises 23–28, find the center, foci, and vertices of the hyperbola and sketch a graph of the hyperbola and its asymptotes with the aid of a graphing utility.

23. $2x^2 - 3y^2 = 6$ **24.** $3y^2 = 5x^2 + 15$

25. $9y^2 - x^2 + 2x + 54y + 62 = 0$

26. $9x^2 - y^2 + 54x + 10y + 55 = 0$

27. $3x^2 - 2y^2 - 6x - 12y - 27 = 0$

28. $3y^2 - x^2 + 6x - 12y = 0$

In Exercises 29–38, find an equation of the specified hyperbola.

29. Vertices: $(0, \pm 2)$
Foci: $(0, \pm 4)$

30. Vertices: $(0, \pm 3)$
Asymptotes: $y = \pm 3x$

31. Vertices: $(\pm 1, 0)$
Asymptotes: $y = \pm 3x$

32. Vertices: $(0, \pm 3)$
Solution point: $(-2, 5)$

33. Vertices: $(2, 0), (6, 0)$
Foci: $(0, 0), (8, 0)$

34. Vertices: $(4, 1), (4, 9)$
Foci: $(4, 0), (4, 10)$

35. Vertices: $(2, \pm 3)$
Solution point: $(0, 5)$

36. Vertices: $(\pm 2, 1)$
Solution point: $(4, 3)$

37. Vertices: $(0, 2), (6, 2)$
Asymptotes: $y = \frac{2}{3}x$
$y = 4 - \frac{2}{3}x$

38. Vertices: $(3, 0), (3, 4)$
Asymptotes: $y = \frac{2}{3}x$
$y = 4 - \frac{2}{3}x$

39. *Sound Location* Three listening stations located at $(4400, 0)$, $(4400, 1100)$, and $(-4400, 0)$ monitor an explosion. If the latter two stations detect the explosion 1 second and 5 seconds after the first, respectively, determine the coordinates of the explosion. (Assume that the coordinate system is measured in feet and that sound travels at 1100 feet per second.)

40. *LORAN*   Long-distance radio navigation for aircraft and ships uses synchronized pulses transmitted by widely separated transmitting stations. These pulses travel at the speed of light (186,000 miles per second). The difference in the times of arrival of these pulses at an aircraft or ship is constant on a hyperbola having the transmitting stations as foci. Assume that two stations, 300 miles apart, are positioned on the rectangular coordinate system at points with coordinates $(-150, 0)$ and $(150, 0)$ and that a ship is traveling on a path with coordinates $(x, 75)$ (see figure). Find the x-coordinate of the position of the ship if the time difference between the pulses from the transmitting stations is 1000 microseconds (0.001 second).

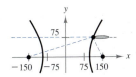

FIGURE FOR 40

41. *Hyperbolic Mirror*   A hyperbolic mirror (used in some telescopes) has the property that a light ray directed at the focus will be reflected to the other focus (see figure). The focus of a hyperbolic mirror has coordinates $(12, 0)$. Find the vertex of the mirror if its mount has coordinates $(12, 12)$.

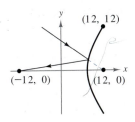

FIGURE FOR 41

42. Sketch a graph of the hyperbola that consists of all points (x, y) such that the difference of the distances between (x, y) and two fixed points is 10 units. The foci are located at the centers of the two sets of concentric circles in the figure.

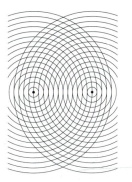

FIGURE FOR 42

In Exercises 43–50, classify the graph of each equation as a circle, a parabola, an ellipse, or a hyperbola.

43. $x^2 + y^2 - 6x + 4y + 9 = 0$

44. $x^2 + 4y^2 - 6x + 16y + 21 = 0$

45. $4x^2 - y^2 - 4x - 3 = 0$

46. $y^2 - 4y - 4x = 0$

47. $4x^2 + 3y^2 + 8x - 24y + 51 = 0$

48. $4y^2 - 2x^2 - 4y - 8x - 15 = 0$

49. $25x^2 - 10x - 200y - 119 = 0$

50. $4x^2 + 4y^2 - 16y + 15 = 0$

10.4 ROTATION AND SYSTEMS OF QUADRATIC EQUATIONS

Rotation / Invariants Under Rotation / Systems of Quadratic Equations

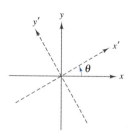

Rotated: x'-axis
y'-axis

FIGURE 10.29

Rotation

In Section 10.3 you saw that the equation of a conic with its axis parallel to one of the coordinate axes has a standard form that can be written in the general form

$$Ax^2 + Cy^2 + Dx + Ey + F = 0. \qquad \textit{Horizontal or vertical axis}$$

In this section you will study the equations of conics whose axes are rotated so that they are not parallel to either the x-axis or the y-axis. The general equation for such conics contains an *xy-term*.

$$Ax^2 + Bxy + Cy^2 + Dx + Ey + F = 0 \qquad \textit{Equation in xy-plane}$$

To eliminate this xy-term, you can use a procedure called **rotation of axes.** The objective is to rotate the x- and y-axes until they are parallel to the axes of the conic. The rotated axes are denoted as the x'-axis and the y'-axis, as shown in Figure 10.29. The equation of the conic in the new $x'y'$-plane will have the form

$$A'(x')^2 + C'(y')^2 + D'x' + E'y' + F' = 0. \qquad \textit{Equation in x'y'-plane}$$

Because this equation has no xy-term, you can obtain a standard form by completing the square.

The following result identifies how much to rotate the axes to eliminate the xy-term and also the equations for determining the new coefficients A', C', D', E', and F'.

ROTATION OF AXES TO ELIMINATE AN XY-TERM

The general second-degree equation $Ax^2 + Bxy + Cy^2 + Dx + Ey + F = 0$ can be rewritten as

$$A'(x')^2 + C'(y')^2 + D'x' + E'y' + F' = 0$$

by rotating the coordinate axes through an angle θ, where

$$\cot 2\theta = \frac{A - C}{B}.$$

The coefficients of the new equation are obtained by making the substitutions

$$x = x' \cos \theta - y' \sin \theta \quad \text{and} \quad y = x' \sin \theta + y' \cos \theta.$$

PROOF

You need to discover how the coordinates in the xy-system are related to the coordinates in the $x'y'$-system. To do this, choose a point $P = (x, y)$ in the original system and attempt to find its coordinates (x', y') in the rotated system. In either system, the distance r between the point P and the origin is the same; thus, the equations for x, y, x', and y' are those given in Figure 10.30. Using the formulas for the sine and cosine of the difference of two angles, you have the following.

$$\begin{aligned}
x' &= r\cos(\alpha - \theta) \\
&= r(\cos \alpha \cos \theta + \sin \alpha \sin \theta) \\
&= r\cos \alpha \cos \theta + r \sin \alpha \sin \theta \\
&= x \cos \theta + y \sin \theta \\
y' &= r\sin(\alpha - \theta) \\
&= r(\sin \alpha \cos \theta - \cos \alpha \sin \theta) \\
&= r\sin \alpha \cos \theta - r \cos \alpha \sin \theta \\
&= y \cos \theta - x \sin \theta
\end{aligned}$$

Solving this system for x and y yields

$$\begin{aligned}
x &= x' \cos \theta - y' \sin \theta \\
y &= x' \sin \theta + y' \cos \theta.
\end{aligned}$$

Finally, by substituting these values for x and y into the original equation and collecting terms, you obtain

$$\begin{aligned}
A' &= A \cos^2 \theta + B \cos \theta \sin \theta + C \sin^2 \theta \\
C' &= A \sin^2 \theta - B \cos \theta \sin \theta + C \cos^2 \theta \\
D' &= D \cos \theta + E \sin \theta \\
E' &= -D \sin \theta + E \cos \theta \\
F' &= F.
\end{aligned}$$

Now, in order to eliminate the $x'y'$-term, you must select θ so that $B' = 0$, as follows.

$$\begin{aligned}
B' &= 2(C - A) \sin \theta \cos \theta + B(\cos^2 \theta - \sin^2 \theta) \\
&= (C - A) \sin 2\theta + B \cos 2\theta \\
&= B(\sin 2\theta)\left(\frac{C - A}{B} + \cot 2\theta\right) = 0, \quad \sin 2\theta \neq 0
\end{aligned}$$

If $B = 0$, no rotation is necessary, because the xy-term is not present in the original equation. If $B \neq 0$, the only way to make $B' = 0$ is to let

$$\cot 2\theta = \frac{A - C}{B}, \quad B \neq 0.$$

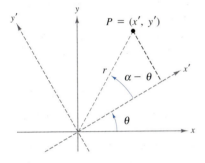

Rotated: $x' = r \cos(\alpha - \theta)$
$\qquad\quad y' = r \sin(\alpha - \theta)$

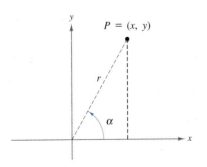

Original: $x = r \cos \alpha$
$\qquad\quad\; y = r \sin \alpha$

FIGURE 10.30

EXAMPLE 1   Rotation of Axes for a Hyperbola

Write the equation $xy - 1 = 0$ in standard form.

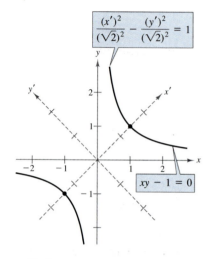

$$\frac{(x')^2}{(\sqrt{2})^2} - \frac{(y')^2}{(\sqrt{2})^2} = 1$$

$xy - 1 = 0$

Vertices:
In $x'y'$-system: $(\sqrt{2},\ 0),\ (-\sqrt{2},\ 0)$
In xy-system: $(1,\ 1),\ (-1,\ -1)$

FIGURE 10.31

SOLUTION

Because $A = 0$, $B = 1$, and $C = 0$, you have

$$\cot 2\theta = \frac{A - C}{B} = 0 \rightarrow 2\theta = \frac{\pi}{2} \rightarrow \theta = \frac{\pi}{4},$$

which implies that

$$x = x' \cos \frac{\pi}{4} - y' \sin \frac{\pi}{4}$$

$$= x'\left(\frac{\sqrt{2}}{2}\right) - y'\left(\frac{\sqrt{2}}{2}\right) = \frac{x' - y'}{\sqrt{2}}$$

and

$$y = x' \sin \frac{\pi}{4} + y' \cos \frac{\pi}{4}$$

$$= x'\left(\frac{\sqrt{2}}{2}\right) + y'\left(\frac{\sqrt{2}}{2}\right) = \frac{x' + y'}{\sqrt{2}}.$$

The equation in the $x'y'$-system is obtained by substituting these expressions into the equation $xy - 1 = 0$.

$$\left(\frac{x' - y'}{\sqrt{2}}\right)\left(\frac{x' + y'}{\sqrt{2}}\right) - 1 = 0$$

$$\frac{(x')^2 - (y')^2}{2} - 1 = 0$$

$$\frac{(x')^2}{(\sqrt{2})^2} - \frac{(y')^2}{(\sqrt{2})^2} = 1 \qquad \textit{Standard form}$$

This is the equation of a hyperbola centered at the origin with vertices at $\left(\pm\sqrt{2},\ 0\right)$ in the $x'y'$-system, as shown in Figure 10.31. To find the coordinates of the vertices in the xy-system, substitute the coordinates $\left(\pm\sqrt{2},\ 0\right)$ into the equations

$$x = \frac{x' - y'}{\sqrt{2}} \quad \text{and} \quad y = \frac{x' + y'}{\sqrt{2}}.$$

This substitution yields the vertices $(1, 1)$ and $(-1, -1)$ in the xy-system. Note also that the asymptotes of the hyperbola have equations $y' = \pm x'$, which correspond to the original x- and y-axes.

REMARK   Remember that the substitutions

$x = x' \cos \theta - y' \sin \theta$

and

$y = x' \sin \theta + y' \cos \theta$

were developed to eliminate the $x'y'$-term in the rotated system. You can use this as a check on your work. In other words, if your final equation contains an $x'y'$-term, you know that you have made a mistake.

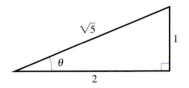

FIGURE 10.32

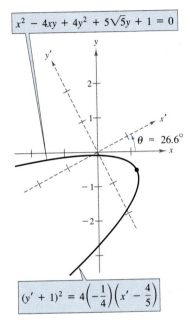

$$(y' + 1)^2 = 4\left(-\frac{1}{4}\right)\left(x' - \frac{4}{5}\right)$$

Vertex:

In $x'y'$-system: $\left(\frac{4}{5}, -1\right)$

In xy-system: $\left(\frac{13}{5\sqrt{5}}, -\frac{6}{5\sqrt{5}}\right)$

FIGURE 10.33

EXAMPLE 2 Rotation of Axes for a Parabola

Sketch the graph of $x^2 - 4xy + 4y^2 + 5\sqrt{5}\,y + 1 = 0$.

SOLUTION

Because $A = 1$, $B = -4$, and $C = 4$, you have

$$\cot 2\theta = \frac{A - C}{B} = \frac{1 - 4}{-4} = \frac{3}{4}.$$

Using the identity $\cot 2\theta = (\cot^2 \theta - 1)/(2 \cot \theta)$ produces

$$\cot 2\theta = \frac{3}{4} = \frac{\cot^2 \theta - 1}{2 \cot \theta},$$

from which you obtain the equation

$$4 \cot^2 \theta - 4 = 6 \cot \theta$$
$$4 \cot^2 \theta - 6 \cot \theta - 4 = 0$$
$$(2 \cot \theta - 4)(2 \cot \theta + 1) = 0.$$

Considering $0 < \theta < \pi/2$, you have $2 \cot \theta = 4$. Thus,

$$\cot \theta = 2 \rightarrow \theta \approx 26.6°.$$

From the triangle in Figure 10.32 you obtain $\sin \theta = 1/\sqrt{5}$ and $\cos \theta = 2/\sqrt{5}$. Consequently, you should use the substitutions

$$x = x' \cos \theta - y' \sin \theta = x'\left(\frac{2}{\sqrt{5}}\right) - y'\left(\frac{1}{\sqrt{5}}\right) = \frac{2x' - y'}{\sqrt{5}}$$

$$y = x' \sin \theta + y' \cos \theta = x'\left(\frac{1}{\sqrt{5}}\right) + y'\left(\frac{2}{\sqrt{5}}\right) = \frac{x' + 2y'}{\sqrt{5}}.$$

Substituting these expressions into the original equation, you have

$$x^2 - 4xy + 4y^2 + 5\sqrt{5}\,y + 1 = 0$$

$$\left(\frac{2x' - y'}{\sqrt{5}}\right)^2 - 4\left(\frac{2x' - y'}{\sqrt{5}}\right)\left(\frac{x' + 2y'}{\sqrt{5}}\right) + 4\left(\frac{x' + 2y'}{\sqrt{5}}\right)^2 + 5\sqrt{5}\left(\frac{x' + 2y'}{\sqrt{5}}\right) + 1 = 0,$$

which simplifies as follows.

$$5(y')^2 + 5x' + 10y' + 1 = 0$$

$$5(y' + 1)^2 = -5x' + 4 \qquad \textit{Complete the square}$$

$$(y' + 1)^2 = (-1)\left(x' - \frac{4}{5}\right) \qquad \textit{Standard form}$$

The graph of this equation is a parabola with its vertex at $(\frac{4}{5}, -1)$. Its axis is parallel to the x'-axis in the $x'y'$-system, as shown in Figure 10.33.

Invariants Under Rotation

In the rotation of axes theorem listed at the beginning of this section, note that the constant term $F' = F$ is the same in both equations, and is said to be **invariant under rotation.** The next theorem lists some other rotation invariants.

ROTATION INVARIANTS

The rotation of coordinate axes through an angle θ that transforms the equation $Ax^2 + Bxy + Cy^2 + Dx + Ey + F = 0$ into the form

$$A'(x')^2 + C'(y')^2 + D'x' + E'y' + F' = 0$$

has the following rotation invariants.

1. $F = F'$
2. $A + C = A' + C'$
3. $B^2 - 4AC = (B')^2 - 4A'C'$

You can use these results to classify the graph of a second-degree equation *with* an *xy*-term in much the same way as was done for second-degree equations *without* an *xy*-term. Note that, because $B' = 0$, the invariant $B^2 - 4AC$ reduces to

$$B^2 - 4AC = -4A'C'. \qquad \textit{Discriminant}$$

This quantity is the **discriminant** of the equation

$$Ax^2 + Bxy + Cy^2 + Dx + Ey + F = 0.$$

Now, from the classification procedure given in Section 10.3, you know that the sign of $A'C'$ determines the type of graph for the equation

$$A'(x')^2 + C'(y')^2 + D'x' + E'y' + F' = 0.$$

Consequently, the sign of $B^2 - 4AC$ will determine the type of graph for the original equation, as shown in the following classification.

CLASSIFICATION OF CONICS BY THE DISCRIMINANT

The graph of the equation $Ax^2 + Bxy + Cy^2 + Dx + Ey + F = 0$ is, except in degenerate cases, determined by its discriminant, as follows.

1. Ellipse or circle: $B^2 - 4AC < 0$
2. Parabola: $B^2 - 4AC = 0$
3. Hyperbola: $B^2 - 4AC > 0$

EXAMPLE 3   Using the Discriminant

Classify the graph of each of the following equations.

A. $4xy - 9 = 0$
B. $2x^2 - 3xy + 2y^2 - 2x = 0$
C. $x^2 - 6xy + 9y^2 - 2y + 1 = 0$
D. $3x^2 + 8xy + 4y^2 - 7 = 0$

SOLUTION

A. Because $B^2 - 4AC = 16 - 0 > 0$, the graph is a hyperbola.
B. Because $B^2 - 4AC = 9 - 16 < 0$, the graph is a circle or an ellipse.
C. Because $B^2 - 4AC = 36 - 36 = 0$, the graph is a parabola.
D. Because $B^2 - 4AC = 64 - 48 > 0$, the graph is a hyperbola.

Systems of Quadratic Equations

To find the points of intersection of two conics, you can use elimination or substitution, as demonstrated in Examples 4 and 5.

EXAMPLE 4   Solving a Quadratic System by Elimination

Solve the system of quadratic equations.

$$x^2 + y^2 - 16x + 39 = 0 \qquad \text{\textit{Equation 1}}$$
$$x^2 - y^2 \quad 9 = 0 \qquad \text{\textit{Equation 2}}$$

SOLUTION

You can eliminate the y^2-term by adding the two equations. The resulting equation can then be solved for x.

$$2x^2 - 16x + 30 = 0$$
$$2(x - 3)(x - 5) = 0$$

There are two real solutions: $x = 3$ and $x = 5$. The corresponding y-values are $y = 0$ and $y = \pm 4$. Thus the graphs of the equations have three points of intersection: $(3, 0)$, $(5, 4)$, and $(5, -4)$, as shown in Figure 10.34.

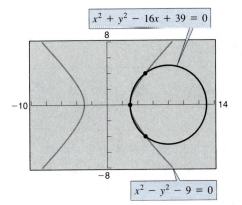

$x^2 + y^2 - 16x + 39 = 0$

$x^2 - y^2 - 9 = 0$

FIGURE 10.34

EXAMPLE 5   Solving a Quadratic System by Substitution

Find the points of intersection of the graphs of the system.

$$x^2 + 4y^2 - 4x - 8y + 4 = 0 \qquad \textit{Equation 1}$$
$$x^2 + 4y - 4 = 0 \qquad \textit{Equation 2}$$

SOLUTION

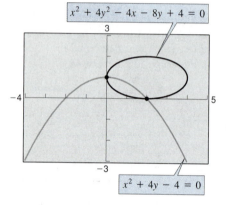

Because the second equation has no y^2-term, solve that equation for y to obtain $y = 1 - \frac{1}{4}x^2$. Next, substitute this expression for y in the first equation and solve for x.

$$x^2 + 4y^2 - 4x - 8y + 4 = 0$$
$$x^2 + 4(1 - \tfrac{1}{4}x^2)^2 - 4x - 8(1 - \tfrac{1}{4}x^2) + 4 = 0$$
$$x^2 + 4 - 2x^2 + \tfrac{1}{4}x^4 - 4x - 8 + 2x^2 + 4 = 0$$
$$\tfrac{1}{4}x^4 + x^2 - 4x = 0$$
$$x^4 + 4x^2 - 16x = 0$$
$$x(x - 2)(x^2 + 2x + 8) = 0$$

The last equation has only two real solutions, $x = 0$ and $x = 2$. The corresponding values of y are $y = 1$ and $y = 0$. Thus the original system of equations has two solutions: $(0, 1)$ and $(2, 0)$, as shown in Figure 10.35.

FIGURE 10.35

DISCUSSION PROBLEM

CLASSIFYING A GRAPH AS A HYPERBOLA

The graph of the rational function

$$y = \frac{1}{x}$$

was discussed in Section 3.5. Write a short paragraph describing how you could use the techniques in this section to show that the graph of this function is a hyperbola.

WARM-UP

The following warm-up exercises involve skills that were covered in earlier sections. You will use these skills in the exercise set for this section.

In Exercises 1–6, match the equation with its graph. [The graphs are labeled (a), (b), (c), (d), (e), and (f).]

1. $\dfrac{x^2}{1} + \dfrac{y^2}{4} = 1$ **2.** $x^2 = 4y$

3. $\dfrac{x^2}{1} - \dfrac{y^2}{4} = 1$ **4.** $(x - 1)^2 + y^2 = 4$

5. $y^2 = -4x$ **6.** $x^2 - 5y^2 = -5$

(a)   (b)   (c)

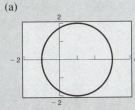   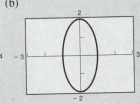   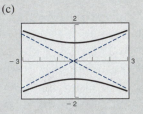

(d)   (e)   (f)

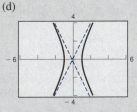   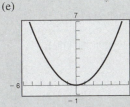   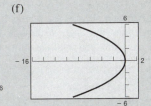

In Exercises 7 and 8, evaluate the trigonometric functions and rewrite the expression.

7. $x \cos \dfrac{\pi}{3} - y \sin \dfrac{\pi}{3}$

8. $x \sin\left(-\dfrac{\pi}{6}\right) + y \cos\left(-\dfrac{\pi}{6}\right)$

In Exercises 9 and 10, expand the expression.

9. $\left(\dfrac{2x - 3y}{\sqrt{13}}\right)^2$

10. $\left(\dfrac{x - \sqrt{2}y}{\sqrt{3}}\right)^2$

SECTION 10.4 · EXERCISES

In Exercises 1–12, rotate the axes to eliminate the xy-term. Sketch the graph of the resulting equation, showing both sets of axes.

1. $xy + 1 = 0$
2. $xy - 4 = 0$
3. $x^2 - 10xy + y^2 + 1 = 0$
4. $xy + x - 2y + 3 = 0$
5. $xy - 2y - 4x = 0$
6. $13x^2 + 6\sqrt{3}\,xy + 7y^2 - 16 = 0$
7. $5x^2 - 2xy + 5y^2 - 12 = 0$
8. $2x^2 - 3xy - 2y^2 + 10 = 0$
9. $3x^2 - 2\sqrt{3}\,xy + y^2 + 2x + 2\sqrt{3}\,y = 0$
10. $16x^2 - 24xy + 9y^2 - 60x - 80y + 100 = 0$
11. $9x^2 + 24xy + 16y^2 + 90x - 130y = 0$
12. $9x^2 + 24xy + 16y^2 + 80x - 60y = 0$

In Exercises 13–18, use a graphing utility to graph the conic. Determine the angle θ through which the axes are rotated. Explain how you used the utility to obtain the graph.

13. $x^2 + xy + y^2 = 10$
14. $x^2 - 4xy + 2y^2 = 6$
15. $17x^2 + 32xy - 7y^2 = 75$
16. $40x^2 + 36xy + 25y^2 = 52$
17. $32x^2 + 50xy + 7y^2 = 52$
18. $4x^2 - 12xy + 9y^2 + \left(4\sqrt{13} - 12\right)x$
    $\qquad - \left(6\sqrt{13} + 8\right)y = 91$

In Exercises 19–26, use the discriminant to determine whether the graph of the equation is a parabola, an ellipse, or a hyperbola.

19. $16x^2 - 24xy + 9y^2 - 30x - 40y = 0$
20. $x^2 - 4xy - 2y^2 - 6 = 0$
21. $13x^2 - 8xy + 7y^2 - 45 = 0$
22. $2x^2 + 4xy + 5y^2 + 3x - 4y - 20 = 0$
23. $x^2 - 6xy - 5y^2 + 4x - 22 = 0$
24. $36x^2 - 60xy + 25y^2 + 9y = 0$
25. $x^2 + 4xy + 4y^2 - 5x - y - 3 = 0$
26. $x^2 + xy + 4y^2 + x + y = 0$

In Exercises 27–34, use a graphing utility to graph the equations and find any points of intersection of the graphs by the method of elimination.

27. $-x^2 + y^2 + 4x - 6y + 4 = 0$
    $x^2 + y^2 - 4x - 6y + 12 = 0$
28. $-x^2 - y^2 - 8x + 20y - 7 = 0$
    $x^2 + 9y^2 + 8x + 4y + 7 = 0$
29. $-4x^2 - y^2 - 32x + 24y - 64 = 0$
    $4x^2 + y^2 + 56x - 24y + 304 = 0$
30. $\quad x^2 - 4y^2 - 20x - 64y - 172 = 0$
    $16x^2 + 4y^2 - 320x + 64y + 1600 = 0$
31. $x^2 - y^2 - 12x + 12y - 36 = 0$
    $x^2 + y^2 - 12x - 12y + 36 = 0$
32. $x^2 + 4y^2 - 2x - 8y + 1 = 0$
    $-x^2 + 2x - 4y - 1 = 0$
33. $-16x^2 - y^2 + 24y - 80 = 0$
    $16x^2 + 25y^2 - 400 = 0$
34. $16x^2 - y^2 + 16y - 128 = 0$
    $y^2 - 48x - 16y - 32 = 0$

In Exercises 35–40, use a graphing utility to graph the equations and find any points of intersection of the graphs by the method of substitution.

35. $x^2 + y^2 - 25 = 0$
    $9x - 4y^2 = 0$
36. $4x^2 + 9y^2 - 36y = 0$
    $x^2 + 9y - 27 = 0$
37. $x^2 + 2y^2 - 4x + 6y - 5 = 0$
    $x + y + 5 = 0$
38. $x^2 + 2y^2 - 4x + 6y - 5 = 0$
    $x^2 - 4x - y + 4 = 0$
39. $xy + x - 2y + 3 = 0$
    $x^2 + 4y^2 - 9 = 0$
40. $5x^2 - 2xy + 5y^2 - 12 = 0$
    $x + y - 1 = 0$

41. Show that the equation $x^2 + y^2 = r^2$ is invariant under rotation of axes.

10.5 PLANE CURVES AND PARAMETRIC EQUATIONS
Plane Curves / Sketching a Plane Curve / Eliminating the Parameter /
Finding Parametric Equations for a Graph

Plane Curves

Up to this point, you have been representing a graph by a single equation involving the *two* variables x and y. In this section, you will study situations in which it is useful to introduce a *third* variable to represent a curve in the plane.

To see the usefulness of this procedure, consider the path followed by an object that is propelled into the air at an angle of 45°. If the initial velocity of the object is 48 feet per second, it can be shown that it follows the parabolic path given by

$$y = -\frac{x^2}{72} + x, \qquad \text{\textit{Rectangular equation}}$$

as shown in Figure 10.36. However, this equation does not tell the whole story.

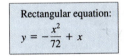

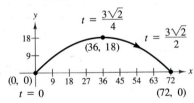

Curvilinear Motion:
two variables for position
one variable for time

FIGURE 10.36

Although it does tell us *where* the object has been, it doesn't tell us *when* the object was at a given point (x, y) on the path. To determine this time, we introduce a third variable t, which we call a **parameter.** It is possible to write both x and y as functions of t to obtain the **parametric equations**

$$x = 24\sqrt{2}\,t \quad \text{and} \quad y = -16t^2 + 24\sqrt{2}\,t. \quad \textit{Parametric equations}$$

From this set of equations we can determine that at time $t = 0$, the object is at the point $(0, 0)$. Similarly, at time $t = 1$, the object is at the point $(24\sqrt{2},\ 24\sqrt{2} - 16)$, and so on.

For this particular motion problem, x and y are continuous functions of t, and we call the resulting path a **plane curve.** (Recall that a *continuous function* is one whose graph can be traced without lifting the pencil from the paper.)

DEFINITION OF A PLANE CURVE

If f and g are continuous functions of t on an interval I, then the set of ordered pairs $(f(t), g(t))$ is a **plane curve** C. The equations $x = f(t)$ and $y = g(t)$ are **parametric equations** for C, and t is the **parameter.**

Sketching a Plane Curve

One way to sketch a curve represented by a pair of parametric equations is to plot points in the xy plane. Each set of coordinates (x, y) is determined from a value chosen for the parameter t. By plotting the resulting points in the order of *increasing* values of t, you trace the curve in a specific direction. This is called the **orientation** of the curve.

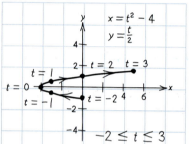

FIGURE 10.37

EXAMPLE 1   Sketching a Curve

Sketch the curve described by the parametric equations

$$x = t^2 - 4 \quad \text{and} \quad y = \frac{t}{2}, \qquad -2 \le t \le 3.$$

SOLUTION

Using values of t in the given interval, the parametric equations yield the points (x, y) shown in the table.

t	-2	-1	0	1	2	3
x	0	-3	-4	-3	0	5
y	-1	$-\frac{1}{2}$	0	$\frac{1}{2}$	1	$\frac{3}{2}$

By plotting these points in the order of increasing t, we obtain the curve shown in Figure 10.37. Note that the arrows on the curve indicate its orientation as t increases from -2 to 3.

The graph shown in Figure 10.37 does not define y as a function of x. This points out one benefit of parametric equations—they can be used to represent graphs that are more general than graphs of functions.

DISCOVERY

Set your graphing utility to parametric mode and enter the two functions

$X_{1T} = T^2 - 4$

$Y_{1T} = T/2$

Use the range settings $-2 \le T \le 3$, Tstep $= .1$, $-6 \le x \le 10$, and $-4 \le y \le 4$. You should obtain the same curve as in Figure 10.37. Trace along the curve and observe how the t values determine the x and y coordinates of the point. What is the orientation of the curve? What happens if you change Tstep to .01?

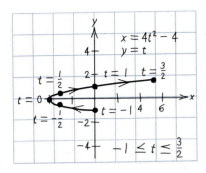

FIGURE 10.38

Two different sets of parametric equations can have the same graph. For example, the set of parametric equations

$$x = 4t^2 - 4 \quad \text{and} \quad y = t, \qquad -1 \le t \le \frac{3}{2}$$

has the same graph as the set given in Example 1. However, by comparing the values of t in Figures 10.37 and 10.38, you can see that this second graph is traced out more *rapidly* (considering t as time) than the first graph. Thus, in applications, different parametric representations can be used to represent various *speeds* at which objects travel along a given path.

Another way to display a curve represented by a pair of parametric equations is to use a graphing utility. When you do this, be sure to set the graphing utility to parametric mode.

EXAMPLE 2   **Using a Graphing Utility in Parametric Mode**

Use a graphing utility to graph the curves represented by the parametric equations. For which curve is y a function of x?

A. $x = \cos^3 t$ **B.** $x = t$
    $y = \sin^3 t$ $y = t^3$

SOLUTION

Begin by setting the graphing utility to parametric *and* radian mode. When choosing a viewing rectangle, you must not only set minimum and maximum values of x and y, you must also set minimum and maximum values of t.

A. Enter the parametric equations for x and y. The curve is shown in Figure 10.39(a). From the graph, you can see that y is *not* a function of x.
B. Enter the parametric equations for x and y. The curve is shown in Figure 10.39(b). From the graph, you can see that y *is* a function of x.

Technology Note

When you use a graphing utility to graph the curve represented by a set of parametric equations, be sure to set the utility to parametric mode. If the parameter involves an angle, you must also set the angle measure to the proper mode (usually radians).

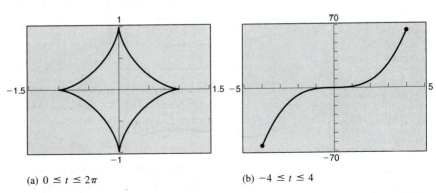

(a) $0 \le t \le 2\pi$ (b) $-4 \le t \le 4$

FIGURE 10.39

Eliminating the Parameter

Many curves that are represented by a set of parametric equations have graphs that can also be represented by a rectangular equation (in x and y). The process of finding the rectangular equation is called **eliminating the parameter.** Here is an example.

$$
\begin{array}{cccc}
\text{Parametric} & \text{Solve for } t \text{ in} & \text{Substitute into} & \text{Rectangular} \\
\text{equations} \rightarrow & \text{one equation} \rightarrow & \text{other equation} \rightarrow & \text{equation} \\
x = t^2 - 4 & t = 2y & x = (2y)^2 - 4 & x = 4y^2 - 4 \\
y = \dfrac{t}{2} & & &
\end{array}
$$

After eliminating the parameter, you can recognize that the curve is a parabola with a horizontal axis and vertex at $(-4, 0)$.

Converting equations from parametric to rectangular form can change the ranges of x and y. In such cases, you should restrict x and y in the rectangular equation so that the graph of this equation matches the graph of the parametric equations.

EXAMPLE 3 **Adjusting the Domain After Eliminating the Parameter**

Identify the curve represented by the equations

$$x = \frac{1}{\sqrt{t+1}} \quad \text{and} \quad y = \frac{t}{t+1}.$$

SOLUTION

Solving for t in the equation for x produces

$$x = \frac{1}{\sqrt{t+1}} \quad \text{or} \quad x^2 = \frac{1}{t+1},$$

which implies that $t = (1 - x^2)/x^2$. Substituting into the equation for y, we obtain

$$y = \frac{t}{t+1} = \frac{\dfrac{1-x^2}{x^2}}{\dfrac{1-x^2}{x^2} + 1} = 1 - x^2.$$

From the rectangular equation, you can recognize the curve to be a parabola that opens downward and has its vertex at $(0, 1)$. The rectangular equation is defined for all values of x. From the parametric equation for x, however, you can see that the curve is defined only when $-1 < t$. Thus, you should restrict the domain of x to positive values, as shown in Figure 10.40.

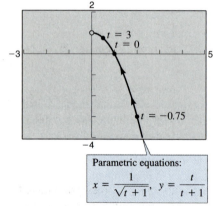

Parametric equations:
$$x = \frac{1}{\sqrt{t+1}}, \quad y = \frac{t}{t+1}$$

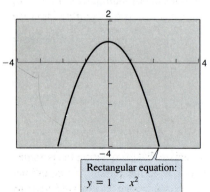

Rectangular equation:
$$y = 1 - x^2$$

FIGURE 10.40

It is not necessary for the parameter in a set of parametric equations to represent time. For instance, in the set of parametric equations in Example 4, *angle* is used as the parameter.

EXAMPLE 4   Using a Trigonometric Identity to Eliminate a Parameter

Identify the curve represented by

$$x = 3 \cos \theta \quad \text{and} \quad y = 4 \sin \theta, \qquad 0 \le \theta \le 2\pi.$$

SOLUTION

Begin by solving for $\cos \theta$ and $\sin \theta$ in the given equations.

$$\cos \theta = \frac{x}{3} \quad \text{and} \quad \sin \theta = \frac{y}{4} \qquad \textit{Solve for } \cos \theta \textit{ and } \sin \theta$$

You can use the identity $\cos^2 \theta + \sin^2 \theta = 1$ to form an equation involving only x and y.

$$\cos^2 \theta + \sin^2 \theta = 1 \qquad \textit{Trigonometric Identity}$$

$$\cos^2 \theta + \sin^2 \theta = \left(\frac{x}{3}\right)^2 + \left(\frac{y}{4}\right)^2 = 1 \qquad \textit{Substitute}$$

$$\frac{x^2}{9} + \frac{y^2}{16} = 1 \qquad \textit{Rectangular equation}$$

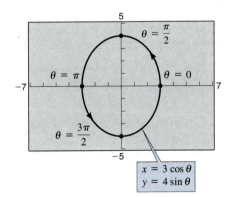

$x = 3 \cos \theta$
$y = 4 \sin \theta$

FIGURE 10.41

From the rectangular equation, you can see that the graph is an ellipse centered at $(0, 0)$, with vertices at $(0, 4)$ and $(0, -4)$, and minor axis of length $2b = 6$, as shown in Figure 10.41. Note that the ellipse is traced out *counterclockwise* as θ varies from 0 to 2π.

In Examples 3 and 4, it is important to realize that eliminating the parameter is primarily an aid to identifying the curve. If the parametric equations represent the path of a moving object, the graph alone is not sufficient to describe the object's motion. You still need the parametric equations to determine the *position, direction,* and *speed* at a given time.

Finding Parametric Equations for a Graph

How can we determine a set of parametric equations for a given graph or a given physical description? From the discussion following Example 1, we know that such a representation is not unique. This is further demonstrated in the following example, in which we find two different parametric representations for a graph.

EXAMPLE 5   Finding Parametric Equations for a Given Graph

Find a set of parametric equations to represent the graph of $y = 1 - x^2$, using the following parameters.

A. $t = x$ **B.** $t = 1 - x$

SOLUTION

A. Letting $t = x$, you obtain the parametric equations

$$x = t \quad \text{and} \quad y = 1 - x^2 = 1 - t^2.$$

B. Letting $t = 1 - x$, you obtain

$$x = 1 - t \quad \text{and} \quad y = 1 - (1 - t)^2 = 2t - t^2.$$

In Figure 10.42, note how the resulting curve is oriented by the increasing values of t. For part (A), the curve would have the opposite orientation.

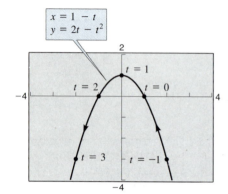

Graph of $y = 1 - x^2$

FIGURE 10.42

EXAMPLE 6   Parametric Equations for a Cycloid

Determine the **cycloid** traced out by a point P on the circumference of a circle of radius a as the circle rolls along a straight line in a plane.

SOLUTION

Use the parameter θ, where θ is the measure of the circle's rotation. Assume that the point $P = (x, y)$ begins at the origin. When $\theta = 0$, P is at the origin; when $\theta = \pi$, P is at a maximum point $(\pi a, 2a)$; and when $\theta = 2\pi$, P is back on the x-axis at $(2\pi a, 0)$. In Figure 10.43, you can see that $\angle APC = 180° - \theta$.

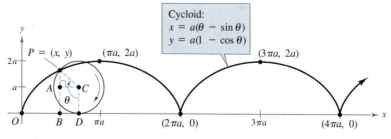

FIGURE 10.43

Hence, you have

$$\sin \theta = \sin(180° - \theta) = \sin(\angle APC) = \frac{AC}{a} = \frac{BD}{a}$$

$$\cos \theta = -\cos(180° - \theta) = -\cos(\angle APC) = \frac{AP}{-a},$$

which implies that $AP = -a \cos \theta$ and $BD = a \sin \theta$. Now, because the circle rolls along the x-axis, we know that $OD = \overset{\frown}{PD} = a\theta$. Furthermore, because $BA = DC = a$, we have

$$x = OD - BD = a\theta - a \sin \theta$$
$$y = BA + AP = a - a \cos \theta.$$

Therefore, the parametric equations are

$$x = a(\theta - \sin \theta) \quad \text{and} \quad y = a(1 - \cos \theta).$$

DISCUSSION PROBLEM

CHANGING THE ORIENTATION OF A CURVE

The **orientation** of a curve refers to the direction in which the curve is traced as the values of the parameter increase. For instance, as t increases, the circle given by

$$x = \cos t \quad \text{and} \quad y = \sin t$$

is traced out *counterclockwise*. Find a parametric representation for which the circle is traced out *clockwise*.

WARM-UP

The following warm-up exercises involve skills that were covered in earlier sections. You will use these skills in the exercise set for this section.

In Exercises 1–6, sketch the graph of the equation.

1. $y = -\frac{1}{4}x^2$ **2.** $y = 4 - \frac{1}{4}(x - 2)^2$

3. $16x^2 + y^2 = 16$ **4.** $-16x^2 + y^2 = 16$

5. $x + y = 4$ **6.** $x^2 + y^2 = 16$

In Exercises 7–10, simplify the expression.

7. $10 \sin^2 \theta + 10 \cos^2 \theta$ **8.** $5 \sec^2 \theta - 5$

9. $\sec^4 x - \tan^4 x$ **10.** $\dfrac{\sin 2\theta}{4 \cos \theta}$

SECTION 10.5 · EXERCISES

In Exercises 1–20, sketch the curve represented by the parametric equations (indicate the direction of the curve). Use a graphing utility to confirm your result. Then eliminate the parameter and write a rectangular equation whose graph represents the curve.

1. $x = t$
$y = -2t$

2. $x = t$
$y = \frac{1}{2}t$

3. $x = 3t - 1$
$y = 2t + 1$

4. $x = 3 - 2t$
$y = 2 + 3t$

5. $x = \frac{1}{4}t$
$y = t^2$

6. $x = t$
$y = t^3$

7. $x = t + 1$
$y = t^2$

8. $x = \sqrt{t}$
$y = 1 - t$

9. $x = t^3$
$y = \dfrac{t}{2}$

10. $x = t - 1$
$y = \dfrac{t}{t - 1}$

11. $x = 3 \cos \theta$
$y = 3 \sin \theta$

12. $x = 4 \sin 2\theta$
$y = 2 \cos 2\theta$

13. $x = \cos \theta$
$y = 2 \sin^2 \theta$

14. $x = \sec \theta$
$y = \cos \theta$

15. $x = 4 + 2 \cos \theta$
$y = -1 + 4 \sin \theta$

16. $x = 4 \sec \theta$
$y = 3 \tan \theta$

17. $x = e^{-t}$
$y = e^{3t}$

18. $x = e^{2t}$
$y = e^{t}$

19. $x = t^3$
$y = 3 \ln t$

20. $x = \ln t$
$y = t^2$

In Exercises 21 and 22, describe how the curves differ from each other.

21. (a) $x = t$
$y = 2t + 1$

(b) $x = \cos \theta$
$y = 2 \cos \theta + 1$

(c) $x = e^{-t}$
$y = 2e^{-t} + 1$

(d) $x = e^{t}$
$y = 2e^{t} + 1$

22. (a) $x = t$
$y = t^2 - 1$

(b) $x = t^2$
$y = t^4 - 1$

(c) $x = \sin t$
$y = \sin^2 t - 1$

(d) $x = e^{t}$
$y = e^{2t} - 1$

In Exercises 23–26, eliminate the parameter and obtain the standard form of the rectangular equation.

23. Line through (x_1, y_1) and (x_2, y_2):
$x = x_1 + t(x_2 - x_1)$
$y = y_1 + t(y_2 - y_1)$

24. Circle:
$x = h + r \cos \theta$
$y = k + r \sin \theta$

25. Ellipse:
$x = h + a \cos \theta$
$y = k + b \sin \theta$

26. Hyperbola:
$x = h + a \sec \theta$
$y = k + b \tan \theta$

In Exercises 27–34, use the results of Exercises 23–26 to find a set of parametric equations for the line or conic.

27. Line: Passes through $(0, 0)$ and $(5, -2)$
28. Line: Passes through $(1, 4)$ and $(5, -2)$
29. Circle: Center: $(2, 1)$
Radius: 4
30. Circle: Center: $(-3, 1)$
Radius: 3
31. Ellipse: Vertices: $(\pm 5, 0)$
Foci: $(\pm 4, 0)$
32. Ellipse: Vertices: $(4, 7), (4, -3)$
Foci: $(4, 5), (4, -1)$
33. Hyperbola: Vertices: $(\pm 4, 0)$
Foci: $(\pm 5, 0)$
34. Hyperbola: Vertices: $(0, \pm 1)$
Foci: $(0, \pm 5)$

In Exercises 35 and 36, find two different sets of parametric equations for the rectangular equation.

35. $y = x^3$ **36.** $y = x^2$

In Exercises 37–42, use a graphing utility to graph the curve represented by the parametric equations.

37. Cycloid: $x = 2(\theta - \sin \theta)$
$\quad\quad\quad\quad\quad\; y = 2(1 - \cos \theta)$

38. Cycloid: $x = \theta + \sin \theta$
$\quad\quad\quad\quad\quad\; y = 1 - \cos \theta$

39. Prolate cycloid: $x = \theta - \frac{3}{2} \sin \theta$
$\quad\quad\quad\quad\quad\quad\quad\; y = 1 - \frac{3}{2} \cos \theta$

40. Curtate cycloid: $x = 2\theta - \sin \theta$
$\quad\quad\quad\quad\quad\quad\quad\; y = 2 - \cos \theta$

41. Witch of Agnesi: $x = 2 \cot \theta$
$\quad\quad\quad\quad\quad\quad\quad\; y = 2 \sin^2 \theta$

42. Folium of Descartes: $x = \dfrac{3t}{1 + t^3}$
$$y = \dfrac{3t^2}{1 + t^3}$$

In Exercises 43–46, match the parametric equation with the correct graph and describe the viewing rectangle. [The graphs are labeled (a), (b), (c), and (d).]

43. Lissajous curve: $x = 4 \cos \theta$
$\quad\quad\quad\quad\quad\quad\; y = 2 \sin 2\theta$

44. Evolute of ellipse: $x = \cos^3 \theta$
$\quad\quad\quad\quad\quad\quad\quad\; y = 2 \sin^3 \theta$

45. Involute of circle: $x = \cos \theta + \theta \sin \theta$
$\quad\quad\quad\quad\quad\quad\quad\; y = \sin \theta - \theta \cos \theta$

46. Serpentine curve: $x = \cot \theta$
$\quad\quad\quad\quad\quad\quad\quad\; y = 4 \sin \theta \cos \theta$

(a)
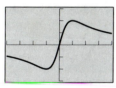

(b)
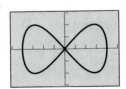

(c)
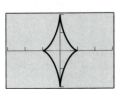

(d)
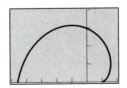

47. A wheel of radius a rolls along a straight line without slipping (see figure). Find the parametric equations for the curve generated by a point P that is b units from the center of the wheel. This curve is called a *curtate cycloid* when $b < a$.

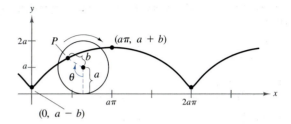

FIGURE FOR 47

48. A wheel of radius 1 rolls around the outside of a circle of radius 2 without slipping (see figure). Show that the parametric equations for the curve generated by a point on the rolling wheel are

$x = 3 \cos \theta - \cos 3\theta$ and $y = 3 \sin \theta - \sin 3\theta$.

This curve is called an *epicycloid.*

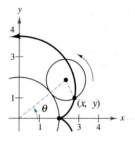

FIGURE FOR 48

10.6 POLAR COORDINATES
....
Introduction / **Coordinate Conversion** / **Equation Conversion**

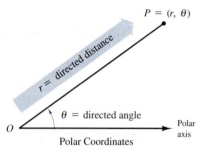

FIGURE 10.44

Introduction

To form the **polar coordinate system** in the plane, fix a point O, called the **pole** (or **origin**), and construct from O an initial ray called the **polar axis,** as shown in Figure 10.44. Each point P in the plane is assigned **polar coordinates** (r, θ) as follows.

1. $r = $ *directed distance* from O to P
2. $\theta = $ *directed angle*, counterclockwise from polar axis to segment $\overline{OP}$

   In the polar coordinate system, it is convenient to locate points with respect to a grid of concentric circles intersected by **radial lines** through the pole. This procedure is shown in Example 1.

EXAMPLE 1   Plotting Points in the Polar Coordinate System
....

A. The point $(r, \theta) = (2, \pi/3)$ lies at the intersection of a circle of radius $r = 2$ and the terminal side of the angle $\theta = \pi/3$, as shown in Figure 10.45(a).
B. The point $(r, \theta) = (3, -\pi/6)$ lies in the fourth quadrant, 3 units from the pole. Note that negative angles are measured *clockwise,* as shown in Figure 10.45(b).
C. The point $(r, \theta) = (3, 11\pi/6)$ coincides with the point $(3, -\pi/6)$, as shown in Figure 10.45(c).

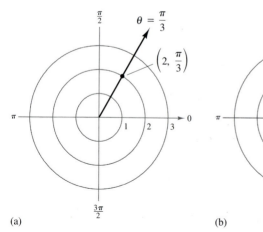

(a)

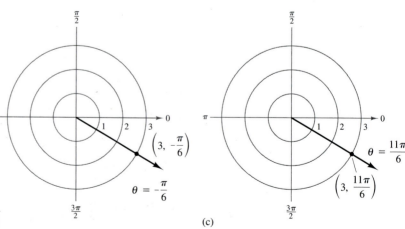

(b)   (c)

FIGURE 10.45

On rectangular coordinates, each point (x, y) has a unique representation. This is not true for polar coordinates. For instance, the coordinates (r, θ) and $(r, 2\pi + \theta)$ represent the same point. Another way to obtain multiple representations of a point is to use negative values of r. Because r is a *directed distance*, the coordinates (r, θ) and $(-r, \theta + \pi)$ represent the same point. In general, the point (r, θ) can be represented as

$$(r, \theta \pm 2n\pi) \quad \text{or} \quad (-r, \theta \pm (2n + 1)\pi),$$

where n is any integer. Moreover, the pole is represented by $(0, \theta)$, where θ is any angle.

EXAMPLE 2   Multiple Representation of Points

Plot the point $(3, -3\pi/4)$ and find three additional polar representations of this point, using $-2\pi < \theta < 2\pi$.

SOLUTION

The point is shown in Figure 10.46. Three other representations are as follows.

$$\left(3, \frac{-3\pi}{4} + 2\pi\right) = \left(3, \frac{5\pi}{4}\right) \qquad \textit{Add } 2\pi \textit{ to } \theta$$

$$\left(-3, \frac{-3\pi}{4} - \pi\right) = \left(-3, \frac{-7\pi}{4}\right) \qquad \textit{Replace } r \textit{ with } -r; \\ \textit{subtract } \pi \textit{ from } \theta$$

$$\left(-3, \frac{-3\pi}{4} + \pi\right) = \left(-3, \frac{\pi}{4}\right) \qquad \textit{Replace } r \textit{ with } -r; \textit{ add } \pi \textit{ to } \theta$$

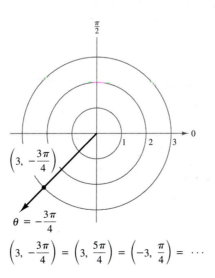

$$\left(3, -\frac{3\pi}{4}\right) = \left(3, \frac{5\pi}{4}\right) = \left(-3, \frac{\pi}{4}\right) = \cdots$$

FIGURE 10.46

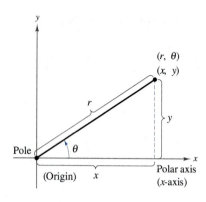

Relating Polar and Rectangular Coordinates

FIGURE 10.47

Coordinate Conversion

To establish the relationship between polar and rectangular coordinates, let the polar axis coincide with the positive x-axis and the pole with the origin, as shown in Figure 10.47. Because (x, y) lies on a circle of radius r, it follows that $r^2 = x^2 + y^2$. Moreover, for $r > 0$, the definitions of the trigonometric functions imply that

$$\tan \theta = \frac{y}{x}, \quad \cos \theta = \frac{x}{r}, \quad \text{and} \quad \sin \theta = \frac{y}{r}.$$

If $r < 0$, you can show that the same relationships hold. For example, consider the point (r, θ), where $r < 0$. Then, because $(-r, \theta + \pi)$ represents the same point and $-r > 0$, you have $-\sin \theta = \sin(\theta + \pi) = -y/r$, which implies that $\sin \theta = y/r$.

These relationships allow you to convert *coordinates* or *equations* from one system to the other, as indicated in the following rule.

COORDINATE CONVERSION

The polar coordinates (r, θ) are related to the rectangular coordinates (x, y) as follows.

$$x = r \cos \theta \quad \text{and} \quad \tan \theta = \frac{y}{x}$$

$$y = r \sin \theta \qquad\qquad r^2 = x^2 + y^2$$

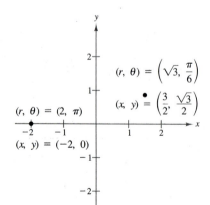

FIGURE 10.48

EXAMPLE 3 Polar-to-Rectangular Conversion

A. For the point $(r, \theta) = (2, \pi)$, you can write the following.

$$x = r \cos \theta = 2 \cos \pi = -2$$
$$y = r \sin \theta = 2 \sin \pi = 0$$

The rectangular coordinates are $(x, y) = (-2, 0)$, as shown in Figure 10.48.

B. For the point $(r, \theta) = (\sqrt{3}, \pi/6)$, you can write the following.

$$x = \sqrt{3} \cos \frac{\pi}{6} = \sqrt{3}\left(\frac{\sqrt{3}}{2}\right) = \frac{3}{2}$$

$$y = \sqrt{3} \sin \frac{\pi}{6} = \sqrt{3}\left(\frac{1}{2}\right) = \frac{\sqrt{3}}{2}$$

The rectangular coordinates are $(x, y) = (3/2, \sqrt{3}/2)$, as shown in Figure 10.48.

EXAMPLE 4 **Rectangular-to-Polar Conversion**

A. For the second-quadrant point $(x, y) = (-1, 1)$, you can write

$$\tan \theta = \frac{y}{x} = -1 \quad \longrightarrow \quad \theta = \frac{3\pi}{4}.$$

Because θ lies in the same quadrant as (x, y), use positive r.

$$r = \sqrt{x^2 + y^2} = \sqrt{(-1)^2 + (1)^2} = \sqrt{2}$$

Thus, *one* set of polar coordinates is $(r, \theta) = (\sqrt{2}, 3\pi/4)$, as shown in Figure 10.49(a).

B. Because the point $(x, y) = (0, 2)$ lies on the positive y-axis, choose $\theta = \pi/2$ and $r = 2$. Thus, one set of polar coordinates is $(r, \theta) = (2, \pi/2)$, as shown in Figure 10.49(b).

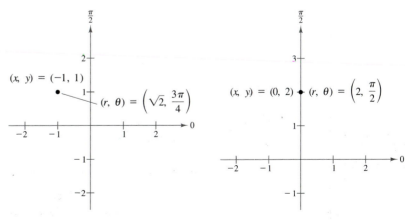

(a)　　　　　　　　　　　　　　(b)

FIGURE 10.49

Equation Conversion

By comparing Examples 3 and 4, you can see that point conversion from the polar to the rectangular system is straightforward, whereas point conversion from the rectangular to the polar system is more involved. For equations, the opposite is true. To convert a rectangular equation to polar form, simply replace x by $r \cos \theta$ and y by $r \sin \theta$. For instance, the rectangular equation $y = x^2$ can be written in polar form as follows.

$$\underbrace{y = x^2}_{\substack{\text{Rectangular} \\ \text{equation}}} \longrightarrow \underbrace{r \sin \theta = (r \cos \theta)^2}_{\text{Polar equation}}$$

On the other hand, converting a polar equation to rectangular form can require considerable ingenuity.

EXAMPLE 5   Converting Polar Equations to Rectangular Form

Convert each polar equation to rectangular form. Then identify the graph.

A. $r = 2$ **B.** $\theta = \dfrac{\pi}{3}$ **C.** $r = \sec \theta$

SOLUTION

A. The graph of the polar equation $r = 2$ consists of all points that are two units from the pole. In other words, this graph is a circle centered at the origin and having a radius of 2, as shown in Figure 10.50(a). You can confirm this by converting to rectangular coordinates, using the relationship $r^2 = x^2 + y^2$.

$$\underbrace{r = 2}_{\text{Polar equation}} \longrightarrow r^2 = 2^2 \longrightarrow \underbrace{x^2 + y^2 = 2^2}_{\text{Rectangular equation}}$$

B. The graph of the polar equation $\theta = \pi/3$ consists of all points on the line that makes an angle of $\pi/3$ with the positive x-axis, as shown in Figure 10.50(b). To convert to rectangular form, you can make use of the relationship $\tan \theta = y/x$.

$$\underbrace{\theta = \frac{\pi}{3}}_{\text{Polar equation}} \longrightarrow \tan \theta = \sqrt{3} \longrightarrow \underbrace{y = \sqrt{3}x}_{\text{Rectangular equation}}$$

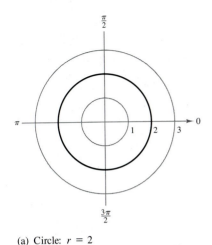

(a) Circle: $r = 2$

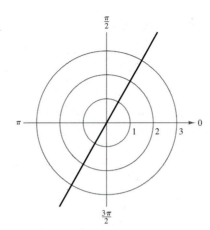

(b) Radial line: $\theta = \dfrac{\pi}{3}$

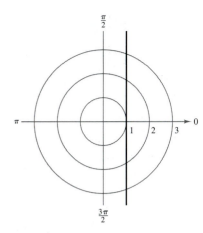

(c) Vertical line: $r = \sec \theta$

FIGURE 10.50

c. The graph of the polar equation $r = \sec \theta$ is not evident by simple inspection. You can convert to rectangular form by using the relationship $r \cos \theta = x$.

$$\underbrace{r = \sec \theta}_{\text{Polar equation}} \longrightarrow r \cos \theta = 1 \longrightarrow \underbrace{x = 1}_{\text{Rectangular equation}}$$

Now, you can see that the graph is a vertical line, as shown in Figure 10.50(c).

Curve sketching by converting to rectangular form is not always convenient. In the next section, you will learn other techniques for sketching polar equations.

DISCUSSION PROBLEM

SIMPLIFYING A POLAR EQUATION

In the discussion before Example 5, we showed how to convert the rectangular equation $y = x^2$ to the polar equation

$$r \sin \theta = (r \cos \theta)^2.$$

Simplifying this equation by dividing both sides by r produces

$$\sin \theta = r \cos^2 \theta \quad \text{or} \quad r = \frac{\sin \theta}{\cos^2 \theta}.$$

By simplifying in this way, however, you risk the possibility of losing the pole $(0, \theta)$ as a solution point. For the equation shown above, division by r does not change the set of solution points, because the pole is a solution point of the simplified equation. Find a polar equation for which division by r *would* change the set of solution points.

WARM-UP

The following warm-up exercises involve skills that were covered in earlier sections. You will use these skills in the exercise set for this section.

In Exercises 1 and 2, find a positive angle coterminal with the given angle.

1. $\dfrac{11\pi}{4}$
2. $-\dfrac{5\pi}{6}$

In Exercises 3 and 4, find the sine and cosine of the angle in standard position with terminal side passing through the given point.

3. $(2, 1)$
4. $(4, -3)$

In Exercises 5 and 6, find the magnitude (in radians) of an angle in standard position with terminal side passing through the given point.

5. $(-4, 4)$
6. $(3, 2)$

In Exercises 7 and 8, evaluate the trigonometric function without the aid of a calculator.

7. $\sin \dfrac{4\pi}{3}$
8. $\cos \dfrac{3\pi}{4}$

In Exercises 9 and 10, use a calculator to evaluate the trigonometric function.

9. $\cos \dfrac{3\pi}{5}$
10. $\sin 1.34$

SECTION 10.6 · EXERCISES

In Exercises 1–10, plot the point given in polar coordinates and find the corresponding rectangular coordinates for the point.

1. $\left(4, \dfrac{3\pi}{6}\right)$
2. $\left(4, \dfrac{3\pi}{2}\right)$

3. $\left(-1, \dfrac{5\pi}{4}\right)$
4. $(0, -\pi)$

5. $\left(4, -\dfrac{\pi}{3}\right)$
6. $\left(-1, -\dfrac{3\pi}{4}\right)$

7. $\left(0, -\dfrac{7\pi}{6}\right)$
8. $\left(\dfrac{3}{2}, \dfrac{5\pi}{2}\right)$

9. $(\sqrt{2}, 2.36)$
10. $(-3, -1.57)$

In Exercises 11–20, the rectangular coordinates of a point are given. Plot the point and find *two* sets of polar coordinates for the point for $0 \le \theta < 2\pi$.

11. $(1, 1)$
12. $(0, -5)$
13. $(-6, 0)$
14. $(-3, -3)$
15. $(-3, 4)$
16. $(3, -1)$
17. $(-\sqrt{3}, -\sqrt{3})$
18. $(-2, 0)$
19. $(4, 6)$
20. $(5, 12)$

In Exercises 21–34, convert the rectangular equation to polar form.

21. $x^2 + y^2 = 9$

22. $x^2 + y^2 = a^2$

23. $x^2 + y^2 - 2ax = 0$

24. $x^2 + y^2 - 2ay = 0$

25. $y = 4$

26. $y = b$

27. $x = 10$

28. $x = a$

29. $3x - y + 2 = 0$

30. $4x + 7y - 2 = 0$

31. $xy = 4$

32. $y = x$

33. $(x^2 + y^2)^2 - 9(x^2 - y^2) = 0$

34. $y^2 - 8x - 16 = 0$

In Exercises 35–44, convert the polar equation to rectangular form.

35. $r = 4 \sin \theta$

36. $r = 4 \cos \theta$

37. $\theta = \dfrac{\pi}{6}$

38. $r = 4$

39. $r = 2 \csc \theta$

40. $r^2 = \sin 2\theta$

41. $r = 2 \sin 3\theta$

42. $r = \dfrac{1}{1 - \cos \theta}$

43. $r = \dfrac{6}{2 - 3 \sin \theta}$

44. $r = \dfrac{6}{2 \cos \theta - 3 \sin \theta}$

In Exercises 45–50, convert the polar equation to rectangular form and sketch its graph.

45. $r = 3$

46. $r = 8$

47. $\theta = \dfrac{\pi}{4}$

48. $\theta = \dfrac{5\pi}{6}$

49. $r = 3 \sec \theta$

50. $r = 2 \csc \theta$

51. Show that the distance between the points (r_1, θ_1) and (r_2, θ_2) is given by
$$\sqrt{r_1^2 + r_2^2 - 2r_1 r_2 \cos(\theta_1 - \theta_2)}.$$

52. Choose two points in the polar coordinate system. Use the formula given in Exercise 51 to find the distance between the two points. Then choose different polar coordinate representations of the same two points and apply the distance formula again. Discuss your result.

53. Convert the polar equation
$$r = 2(h \cos \theta + k \sin \theta)$$
to rectangular form and verify that it is the equation of a circle. Find the radius and the rectangular coordinates of the center of the circle.

54. Convert the polar equation $r = \cos \theta + 3 \sin \theta$ to rectangular form and identify the graph.

10.7 GRAPHS OF POLAR EQUATIONS

Introduction   /   Using a Graphing Utility   /   Symmetry   /   Maximum
r-Values   /   Special Polar Graphs

Introduction

In previous chapters, you spent a lot of time learning how to sketch graphs in rectangular coordinates. You began with the basic point-plotting method. Then you used sketching aids such as a graphing utility, symmetry, intercepts, asymptotes, periods, and shifts to further investigate the nature of the graph. In this section, you will approach curve sketching in the polar coordinate system in a similar way.

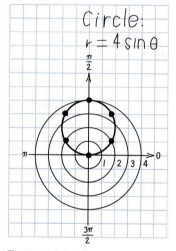

FIGURE 10.51

EXAMPLE 1   Graphing a Polar Equation by Point Plotting

Sketch the graph of the polar equation $r = 4 \sin \theta$.

SOLUTION

The sine function is periodic, so you can get a full range of *r*-values by considering values of θ in the interval $0 \le \theta \le 2\pi$, shown in the table.

θ	0	$\dfrac{\pi}{6}$	$\dfrac{\pi}{3}$	$\dfrac{\pi}{2}$	$\dfrac{2\pi}{3}$	$\dfrac{5\pi}{6}$	π	$\dfrac{7\pi}{6}$	$\dfrac{3\pi}{2}$	$\dfrac{11\pi}{6}$	2π
$4 \sin \theta$	0	2	$2\sqrt{3}$	4	$2\sqrt{3}$	2	0	-2	-4	-2	0

Plot these points as shown in Figure 10.51. In the figure, it appears that the graph is a circle of radius 2 whose center is at the point $(x, y) = (0, 2)$.

Using a Graphing Utility

Some graphing utilities have a polar-coordinate graphing mode. If your graphing utility *doesn't* have such a mode, but *does* have a parametric mode, you can use the following conversion to graph a polar equation.

POLAR EQUATIONS IN PARAMETRIC FORM

The graph of the polar equation $r = f(\theta)$ can be written in parametric form, using *t* as a parameter, as follows.

$$x = f(t) \cos t \quad \text{and} \quad y = f(t) \sin t$$

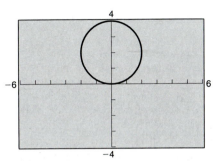

FIGURE 10.52

EXAMPLE 2   **Graphing a Polar Equation with a Graphing Utility**

Use a graphing utility to graph $r = 4 \sin \theta$.

SOLUTION

Begin by writing the equation in parametric form.

$$x = 4 \sin t \cos t$$
$$y = 4 \sin t \sin t$$

Set the graphing utility to parametric and radian mode. Use a viewing rectangle of $-6 \le x \le 6$ and $-4 \le y \le 4$, and let the parameter t vary from 0 to π. The graph is shown in Figure 10.52. In the figure, it appears that the graph is a circle, which reinforces the result obtained in Example 1.

REMARK   You can confirm that the graph given in Examples 1 and 2 is a circle by rewriting the polar equation in rectangular form. If you do this, you will find that the rectangular equation is $x^2 + y^2 = 4y$. By completing the square, this equation can be written as $x^2 + (y - 2)^2 = 2^2$, which you know is the equation of a circle.

Symmetry

In Figure 10.52, the entire graph of the polar equation was traced as t increased from 0 to π. Try resetting your graphing utility so that t varies from 0 to 2π. Then, use the trace feature to follow the curve through increasing values of t, and you will see that the curve is traced out *twice*.

In Figure 10.52, you should also notice that the graph is *symmetric with respect to the line $\theta = \pi/2$*. Symmetry with respect to the line $\theta = \pi/2$ is one of three important types of symmetry to consider in polar curve sketching. (See Figure 10.53.)

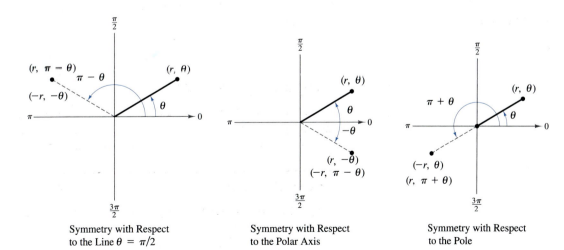

Symmetry with Respect to the Line $\theta = \pi/2$

Symmetry with Respect to the Polar Axis

Symmetry with Respect to the Pole

FIGURE 10.53

TEST FOR SYMMETRY IN POLAR COORDINATES

The graph of a polar equation is symmetric with respect to the following if the given substitution yields an equivalent equation.

1. The line $\theta = \pi/2$:   replace (r, θ) by $(r, \pi - \theta)$ or $(-r, -\theta)$.
2. The polar axis: replace (r, θ) by $(r, -\theta)$ or $(-r, \pi - \theta)$.
3. The pole: replace (r, θ) by $(r, \pi + \theta)$ or $(-r, \theta)$.

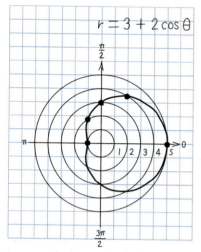

FIGURE 10.54

EXAMPLE 3   Finding Symmetry of the Graph of a Polar Equation

Sketch the graph of $r = 3 + 2 \cos \theta$, and discuss its symmetry.

SOLUTION

Replacing (r, θ) by $(r, -\theta)$ produces

$$r = 3 + 2 \cos(-\theta) = 3 + 2 \cos \theta.$$

Thus, you conclude that the curve is symmetric with respect to the polar axis. This conclusion can be confirmed by sketching the graph of the polar equation, as shown in Figure 10.54. A set of parametric equations for the graph is

$$x = (3 + 2 \cos t)(\cos t)$$
$$y = (3 + 2 \cos t)(\sin t).$$

By letting t vary from 0 to 2π, you can obtain the graph shown in Figure 10.54.

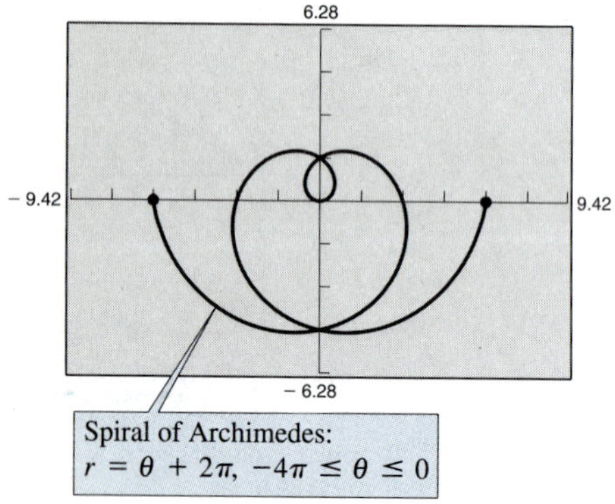

The three tests given for symmetry in polar coordinates are sufficient to guarantee symmetry, but they are not necessary. For instance, Figure 10.55 shows the graph of $r = \theta + 2\pi$ to be symmetric with respect to the line $\theta = \pi/2$. Yet the test fails to indicate symmetry because neither of the following replacements yields an equivalent equation.

Original Equation	Replacement	New Equation
$r = \theta + 2\pi$	(r, θ) by $(-r, -\theta)$	$-r = -\theta + 2\pi$
$r = \theta + 2\pi$	(r, θ) by $(r, \pi - \theta)$	$r = -\theta + 3\pi$

The equations discussed in Examples 1–3 are of the form

$$r = 4 \sin \theta = f(\sin \theta) \quad \text{and} \quad r = 3 + 2 \cos \theta = g(\cos \theta).$$

The graph of the first equation is symmetric with respect to the line $\theta = \pi/2$, and the graph of the second equation is symmetric with respect to the polar axis. This observation can be generalized to yield the following *quick test for symmetry.*

Spiral of Archimedes:
$r = \theta + 2\pi, \ -4\pi \le \theta \le 0$

FIGURE 10.55

1. The graph of $r = f(\sin \theta)$ is symmetric with respect to the line $\theta = \pi/2$.
2. The graph of $r = g(\cos \theta)$ is symmetric with respect to the polar axis.

Maximum *r*-Values

The graph of a polar equation has a maximum *r*-value if there is a point on the graph that is farthest from the pole. For instance, in Example 1, the maximum value of r for $r = 4 \sin \theta$ is 4. This value of r occurs when $\theta = \pi/2$.

EXAMPLE 4   Finding a Maximum *r*-Value of a Polar Graph

Find the maximum value of r for the graph of $r = 1 - 2 \cos \theta$.

SOLUTION

Because the polar equation is of the form $r = 1 - 2 \cos \theta = g(\cos \theta)$, you know that the graph is symmetric with respect to the polar axis. To sketch a graph of the equation, write it in parametric form.

$$x = (1 - 2 \cos t)(\cos t)$$
$$y = (1 - 2 \cos t)(\sin t)$$

Using a graphing utility and letting t vary from 0 to 2π, you can display the graph shown in Figure 10.56. Finally, using the trace feature of the graphing utility, you can conclude that the maximum value of r (the r-value of the point that is farthest from the pole) is 3. This value of r occurs when $t = \pi$ (or when $\theta = \pi$).

REMARK   Note how the negative *r*-values determine the *inner loop* of the graph in Figure 10.56. This type of graph is called a **limaçon.**

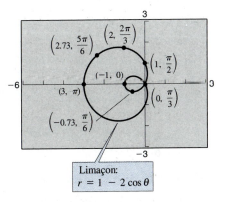

Limaçon:
$r = 1 - 2 \cos \theta$

FIGURE 10.56

Some curves reach their maximum *r*-values at more than one point.

EXAMPLE 5   Finding a Maximum *r*-Value of a Polar Graph

Find the maximum *r*-value of the graph of $r = 2 \cos 3\theta$.

SOLUTION

Because $2 \cos 3\theta = 2 \cos(-3\theta)$, you can conclude that the graph is symmetric with respect to the polar axis. A set of parametric equations for the graph is given by

$$x = 2 \cos 3t \cos t$$
$$y = 2 \cos 3t \sin t.$$

REMARK The graph shown in Figure 10.57 is called a **rose curve,** and each of the loops on the graph is called a *petal* of the rose curve.

By letting *t* vary from 0 to π, you can use a graphing utility to display the graph shown in Figure 10.57. The maximum *r*-value for this graph is 2. This value occurs at *three* different points on the graph: $(r, \theta) = (2, 0)$, $(-2, \pi/3)$, and $(2, 2\pi/3)$.

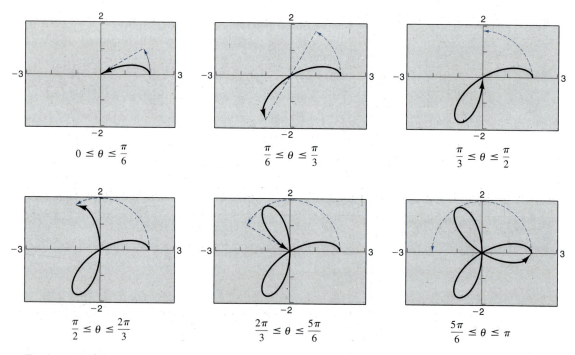

$$0 \leq \theta \leq \frac{\pi}{6} \qquad \frac{\pi}{6} \leq \theta \leq \frac{\pi}{3} \qquad \frac{\pi}{3} \leq \theta \leq \frac{\pi}{2}$$

$$\frac{\pi}{2} \leq \theta \leq \frac{2\pi}{3} \qquad \frac{2\pi}{3} \leq \theta \leq \frac{5\pi}{6} \qquad \frac{5\pi}{6} \leq \theta \leq \pi$$

FIGURE 10.57

Special Polar Graphs

Several important types of graphs have equations that are simpler in polar form than in rectangular form. For example, the circle $r = 4 \sin \theta$ in Example 1 has the more complicated rectangular equation $x^2 + (y - 2)^2 = 4$. The following list gives several other types of graphs that have simple polar equations.

Limaçons

$r = a \pm b \cos \theta$
$r = a \pm b \sin \theta$
$(0 < a, 0 < b)$

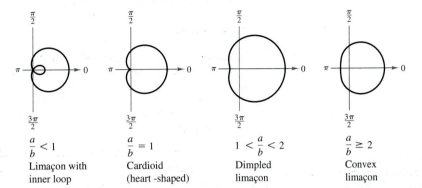

$\dfrac{a}{b} < 1$	$\dfrac{a}{b} = 1$	$1 < \dfrac{a}{b} < 2$	$\dfrac{a}{b} \geq 2$
Limaçon with inner loop	Cardioid (heart-shaped)	Dimpled limaçon	Convex limaçon

Rose Curves

n petals if n is odd
$2n$ petals if n is even
$(n \geq 2)$

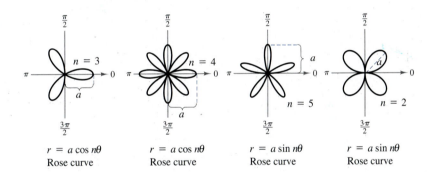

$r = a \cos n\theta$	$r = a \cos n\theta$	$r = a \sin n\theta$	$r = a \sin n\theta$
Rose curve	Rose curve	Rose curve	Rose curve

Circles and Lemniscates

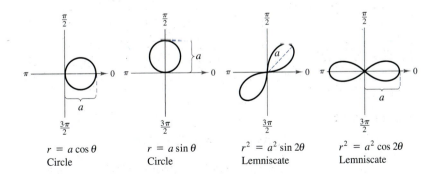

$r = a \cos \theta$	$r = a \sin \theta$	$r^2 = a^2 \sin 2\theta$	$r^2 = a^2 \cos 2\theta$
Circle	Circle	Lemniscate	Lemniscate

EXAMPLE 6   Sketching a Rose Curve

Sketch the graph of $r = 3 \cos 2\theta$.

SOLUTION

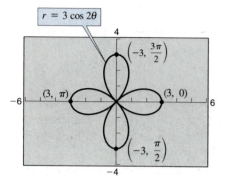

$r = 3 \cos 2\theta$

FIGURE 10.58

Begin with an analysis of the basic features of the graph.

Type of curve: Rose curve with $2n = 4$ petals

Symmetry: With respect to polar axis and the line $\theta = \dfrac{\pi}{2}$

Maximum value of r: $|r| = 3$ when $\theta = 0, \dfrac{\pi}{2}, \pi, \dfrac{3\pi}{2}$

Parametric equations: $x = 3 \cos 2t \cos t$
$y = 3 \cos 2t \sin t$

Using a graphing utility and letting t vary from 0 to 2π, you can graph the rose curve shown in Figure 10.58.

EXAMPLE 7   Sketching a Lemniscate

Sketch the graph of $r^2 = 9 \sin 2\theta$.

SOLUTION

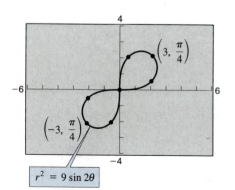

$r^2 = 9 \sin 2\theta$

FIGURE 10.59

Begin with an analysis of the basic features of the graph.

Type of curve: Lemniscate

Symmetry: With respect to the pole

Maximum value of r: $|r| = 3$ when $\theta = \dfrac{\pi}{4}$

Two sets of parametric equations: $x = 3\sqrt{\sin(2t)}\,(\cos t)$
$y = 3\sqrt{\sin(2t)}\,(\sin t)$
$x = -3\sqrt{\sin(2t)}\,(\cos t)$
$y = -3\sqrt{\sin(2t)}\,(\sin t)$

Using a graphing utility and letting t vary from 0 to $\pi/2$, you can graph the lemniscate shown in Figure 10.59.

DISCUSSION PROBLEM .... **DIFFERENT TYPES OF LIMAÇONS**	There are four different types of limaçons in the table showing special polar graphs. Explain the differences among the four types. Then classify each of the following as one of the special types and sketch its graph. (a) $r = 3 + 3 \sin \theta$ (b) $r = 4 + 3 \sin \theta$ (c) $r = 2 + 3 \sin \theta$ (d) $r = 3 + \sin \theta$

WARM-UP
....

The following warm-up exercises involve skills that were covered in earlier sections. You will use these skills in the exercise set for this section.

In Exercises 1–4, determine the amplitude and period of the function.

1. $y = 5 \sin 4x$ **2.** $y = 3 \cos 2\pi x$

3. $y = -5 \cos \dfrac{5x}{2}$ **4.** $y = -\dfrac{1}{2} \sin \dfrac{x}{2}$

In Exercises 5–8, sketch the graph of the function through two periods.

5. $y = 2 \sin x$ **6.** $y = 3 \cos x$

7. $y = 4 \cos 2x$ **8.** $y = 2 \sin \pi x$

In Exercises 9 and 10, use the sum and difference identities to simplify the trigonometric expressions.

9. $\sin\left(x - \dfrac{\pi}{6}\right)$ **10.** $\sin\left(x + \dfrac{\pi}{4}\right)$

SECTION 10.7 · EXERCISES

In Exercises 1–6, test for symmetry with respect to $\theta = \pi/2$, the polar axis, and the pole.

1. $r = 10 + 6 \cos \theta$ **2.** $r = 16 \cos 3\theta$

3. $r = \dfrac{2}{1 + \sin \theta}$ **4.** $r = 6 \sin \theta$

5. $r = 4 \sec \theta \csc \theta$ **6.** $r^2 = 25 \sin 2\theta$

In Exercises 7–10, find the maximum value of $|r|$.

7. $r = 5 \cos 3\theta$

8. $r = -2 \cos \theta$

9. $r = 10(1 - \sin \theta)$

10. $r = 6 + 12 \cos \theta$

In Exercises 11–30, sketch the graph of the polar equation.

11. $r = 5$ **12.** $r = 2$

13. $\theta = \dfrac{\pi}{6}$ **14.** $\theta = -\dfrac{\pi}{4}$

15. $r = 3 \sin \theta$ **16.** $r = 3(1 - \cos \theta)$

17. $r = 4(1 + \sin \theta)$ **18.** $r = 3 - 2 \cos \theta$

19. $r = 4 + 3 \cos \theta$ **20.** $r = 2 + 4 \sin \theta$

21. $r = 3 - 4 \cos \theta$ **22.** $r = 2 \cos 3\theta$

23. $r = 3 \sin 2\theta$ **24.** $r = 2 \sec \theta$

25. $r = \dfrac{3}{\sin \theta - 2 \cos \theta}$ **26.** $r = \dfrac{6}{2 \sin \theta - 3 \cos \theta}$

27. $r^2 = 4 \cos 2\theta$ **28.** $r^2 = 4 \sin \theta$

29. $r = \dfrac{\theta}{2}$ **30.** $r = \theta$

In Exercises 31–42, use a graphing utility to graph the polar equation.

31. $r = 6 \cos \theta$ **32.** $r = \dfrac{\theta}{2}$

33. $r = 3(2 - \sin \theta)$ **34.** $r = \cos 2\theta$

35. $r = 4 \sin \theta \cos^2 \theta$ **36.** $r = 3 \cos 2\theta \sec \theta$

37. $r = 2 \csc \theta + 5$ **38.** $r = 2 \cos(3\theta - 2)$

39. $r = 2 - \sec \theta$ **40.** $r = 2 + \csc \theta$

41. $r = \dfrac{2}{\theta}$ **42.** $r = 2 \cos 2\theta \sec \theta$

In Exercises 43–50, use a graphing utility to graph the polar equation. Find the interval for θ over which the graph is traced only once.

43. $r = 3 - 4 \cos \theta$ **44.** $r = 2(1 - 2 \sin \theta)$

45. $r = 2 + \sin \theta$ **46.** $r = 4 + 3 \cos \theta$

47. $r = 2 \cos\left(\dfrac{3\theta}{2}\right)$ **48.** $r = 3 \sin\left(\dfrac{5\theta}{2}\right)$

49. $r^2 = 4 \sin 2\theta$ **50.** $r^2 = 3 \cos 4\theta$

In Exercises 51 and 52, convert the polar equation to rectangular form and show that the indicated line is an asymptote of the graph.

Polar Equation	*Asymptote*
51. $r = 2 - \sec \theta$	$x = -1$
52. $r = 2 + \csc \theta$	$y = 1$

53. The graph of $r = f(\theta)$ is rotated about the pole through an angle ϕ. Show that the equation for the rotated graph is $r = f(\theta - \phi)$.

54. Consider the graph of $r = f(\sin \theta)$.
    (a) Show that if the graph is rotated counterclockwise $\pi/2$ radians about the pole, then the equation for the rotated graph is $r = f(-\cos \theta)$.
    (b) Show that if the graph is rotated counterclockwise π radians about the pole, then the equation for the rotated graph is $r = f(-\sin \theta)$.
    (c) Show that if the graph is rotated counterclockwise $3\pi/2$ radians about the pole, then the equation for the rotated graph is $r = f(\cos \theta)$.

In Exercises 55–58, use the results of Exercises 53 and 54.

55. Write an equation for the rose curve $r = 2 - \sin \theta$ after it has been rotated by the given amount.
    (a) $\dfrac{\pi}{4}$ (b) $\dfrac{\pi}{2}$ (c) π (d) $\dfrac{3\pi}{2}$

56. Write an equation for the rose curve $r = 2 \sin 2\theta$ after it has been rotated by the given amount.
    (a) $\dfrac{\pi}{6}$ (b) $\dfrac{\pi}{2}$ (c) $\dfrac{2\pi}{3}$ (d) π

57. Sketch the graph of each equation.
    (a) $r = 1 - \sin \theta$ (b) $r = 1 - \sin\left(\theta - \dfrac{\pi}{4}\right)$

58. Sketch the graph of each equation.
    (a) $r = 3 \sec \theta$ (b) $r = 3 \sec\left(\theta - \dfrac{\pi}{4}\right)$
    (c) $r = 3 \sec\left(\theta + \dfrac{\pi}{3}\right)$ (d) $r = 3 \sec\left(\theta - \dfrac{\pi}{2}\right)$

10.8 POLAR EQUATIONS OF CONICS

Alternative Definition of Conics / *Polar Equations of Conics* / Application

Alternative Definition of Conics

In Sections 10.2 and 10.3, you saw that the rectangular equations of ellipses and hyperbolas take simple forms when the origin lies at their *center*. There are, however, many important applications of conics in which it is more convenient to use one of the *foci* as the origin of the coordinate system. For example, the sun lies at the focus of the earth's orbit. Similarly, the light source of a parabolic reflector lies at its focus. In this section you will see that polar equations of conics take simple forms if one of the foci lies at the pole.

ALTERNATIVE DEFINITION OF CONIC

The locus of a point in the plane that moves so that its distance from a fixed point (focus) is in constant ratio to its distance from a fixed line (directrix) is a **conic.** The constant ratio is the **eccentricity** of the conic and is denoted by e. Moreover, the conic is an **ellipse** if $e < 1$, a **parabola** if $e = 1$, and a **hyperbola** if $e > 1$.

In Figure 10.60, note that for each type of conic, the pole corresponds to the fixed point (focus) given in the definition.

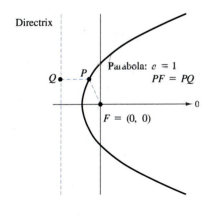

Parabola: $e = 1$
$PF = PQ$
$F = (0, 0)$

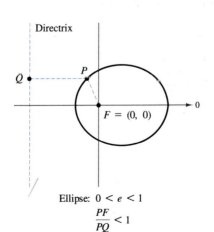

Ellipse: $0 < e < 1$
$$\frac{PF}{PQ} < 1$$
$F = (0, 0)$

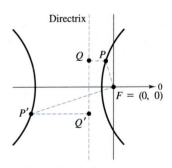

Hyperbola: $e > 1$
$$\frac{PF}{PQ} = \frac{P'F}{P'Q'} > 1$$
$F = (0, 0)$

FIGURE 10.60

Polar Equations of Conics

POLAR EQUATIONS OF CONICS

The graph of a polar equation of the form

1. $r = \dfrac{ep}{1 \pm e \cos \theta}$

2. $r = \dfrac{ep}{1 \pm e \sin \theta}$

is a conic, where $e > 0$ is the eccentricity and $|p|$ is the distance
between the focus (pole) and the directrix.

PROOF

We give a proof for $r = ep/(1 + e \cos \theta)$ with $p > 0$. In Figure 10.61
consider a vertical directrix, p units to the right of the focus $F = (0, 0)$. If
$P = (r, \theta)$ is a point on the graph of $r = ep/(1 + e \cos \theta)$, then the distance
between P and the directrix is

$$PQ = |p - x| = |p - r \cos \theta|$$

$$= \left| p - \left(\frac{ep}{1 + e \cos \theta} \right) \cos \theta \right|$$

$$= \left| p \left(1 - \frac{e \cos \theta}{1 + e \cos \theta} \right) \right|$$

$$= \left| \frac{p}{1 + e \cos \theta} \right|$$

$$= \left| \frac{r}{e} \right|.$$

Because the distance between P and the pole is simply $PF = |r|$, the ratio of
PF to PQ is

$$\frac{PF}{PQ} = \frac{|r|}{\left| \dfrac{r}{e} \right|} = |e| = e,$$

and, by definition, the graph of the equation must be a conic. ⋯⋯⋯⋯

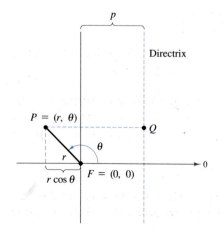

FIGURE 10.61

By completing the proofs of the other three cases, you can see that the equations

$$r = \frac{ep}{1 \pm e \cos \theta}$$ *Vertical directrix*

correspond to conics with vertical directrices and the equations

$$r = \frac{ep}{1 \pm e \sin \theta}$$ *Horizontal directrix*

correspond to conics with horizontal directrices. The converse is also true. That is, any conic with a focus at the pole and having a horizontal or vertical directrix can be represented by one of the given equations.

EXAMPLE 1   Determining a Conic from Its Equation

Identify the conic given by

$$r = \frac{15}{3 - 2 \cos \theta}$$

and sketch its graph.

SOLUTION

To determine the type of conic, rewrite the equation as

$$r = \frac{15}{3 - 2 \cos \theta} = \frac{5}{1 - \frac{2}{3} \cos \theta}.$$

From this form you can conclude that the graph is an ellipse with $e = \frac{2}{3}$. To graph the ellipse, use a graphing utility and the following parametric equations.

$$x = \frac{15 \cos t}{3 - 2 \cos t} \quad \text{and} \quad y = \frac{15 \sin t}{3 - 2 \cos t}$$

Letting t vary from 0 to 2π produces the graph shown in Figure 10.62. Notice that the graph has symmetry with respect to the polar axis.   ·········

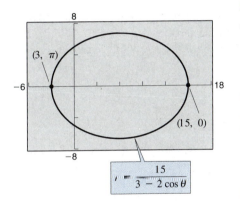

FIGURE 10.62

For the ellipse in Figure 10.62, the major axis is horizontal and the vertices lie at $(15, 0)$ and $(3, \pi)$. Thus, the length of the *major* axis is $2a = 18$. To find the length of the *minor* axis, you can use the equations $e = c/a$ and $b^2 = a^2 - c^2$ to conclude that

$$b^2 = a^2 - c^2 = a^2 - (ea)^2 = a^2(1 - e^2).   \quad \textit{Ellipse}$$

Because $e = \frac{2}{3}$, it follows that $b^2 = 9^2[1 - (2/3)^2] = 45$, which implies that $b = \sqrt{45} = 3\sqrt{5}$. Thus, the length of the minor axis is $2b = 6\sqrt{5}$. A similar analysis for hyperbolas yields

$$b^2 = c^2 - a^2 = (ea)^2 - a^2 = a^2(e^2 - 1).   \quad \textit{Hyperbola}$$

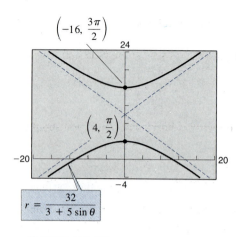

$\left(-16, \dfrac{3\pi}{2}\right)$

$\left(4, \dfrac{\pi}{2}\right)$

$r = \dfrac{32}{3 + 5 \sin \theta}$

FIGURE 10.63

EXAMPLE 2 Sketching a Conic from Its Polar Equation

Identify the conic given by

$$r = \frac{32}{3 + 5 \sin \theta}$$

and sketch its graph.

SOLUTION

To determine the type of conic, rewrite the equation as

$$r = \frac{\frac{32}{3}}{1 + \frac{5}{3} \sin \theta}.$$

Because $e = \frac{5}{3} > 1$, the graph is a hyperbola. The transverse axis of the hyperbola lies on the line $\theta = \pi/2$, and the vertices occur at $(4, \pi/2)$ and $(-16, 3\pi/2)$. Because the length of the transverse axis is 12, it follows that $a = 6$. To find b, you can write

$$b^2 = a^2(e^2 - 1) = 6^2\left[\left(\frac{5}{3}\right)^2 - 1\right] = 64.$$

Therefore, $b = 8$. Finally, use a and b to determine the asymptotes of the hyperbola and obtain the graph shown in Figure 10.63.

In the next example you are asked to find a polar equation for a specified conic. To do this, let p be the distance between the pole and the directrix. With this interpretation of p, we suggest the following guidelines for finding a polar equation for a conic.

1. Horizontal directrix above the pole: $\quad r = \dfrac{ep}{1 + e \sin \theta}$

2. Horizontal directrix below the pole: $\quad r = \dfrac{ep}{1 - e \sin \theta}$

3. Vertical directrix to the right of the pole: $\quad r = \dfrac{ep}{1 + e \cos \theta}$

4. Vertical directrix to the left of the pole: $\quad r = \dfrac{ep}{1 - e \cos \theta}$

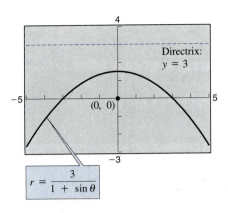

Directrix:
y = 3

(0, 0)

$$r = \frac{3}{1 + \sin \theta}$$

FIGURE 10.64

EXAMPLE 3   Finding the Polar Equation for a Conic
....

Find the polar equation for the parabola whose focus is the pole and whose directrix is the line $y = 3$.

SOLUTION

In Figure 10.64, you can see that the directrix is horizontal. Thus, you should choose an equation of the form

$$r = \frac{ep}{1 + e \sin \theta}.$$

Moreover, because the eccentricity of a parabola is $e = 1$ and the distance between the pole and the directrix is $p = 3$, you obtain

$$r = \frac{3}{1 + \sin \theta}.$$

....

Application

Kepler's Laws (listed below), named after the German astronomer Johannes Kepler (1571–1630), can be used to describe the orbits of the planets about the sun.

1. Each planet moves in an elliptical orbit with the sun as a focus.
2. A ray from the sun to the planet sweeps out equal areas of the ellipse in equal times.
3. The square of the period is proportional* to the cube of the mean distance between the planet and the sun.

Although Kepler simply stated these laws on the basis of observation, they were later validated by Isaac Newton (1642–1727). In fact, Newton was able to show that each law can be deduced from a set of universal laws of motion and gravitation which govern the movement of all heavenly bodies, including comets and satellites. This is illustrated in the next example involving the comet named after the English mathematician and physicist Edmund Halley (1656–1742).

---

*Using Earth as a reference with a period of 1 year and a distance of 1 astronomical unit, the proportionality constant is 1. For example, because Mars has a mean distance to the sun of $d = 1.523$ AU, its period P is given by $d^3 = P^2$. Thus, the period for Mars is $P = 1.88$ years.

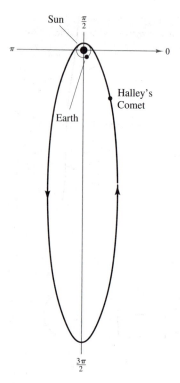

FIGURE 10.65

EXAMPLE 4   Halley's Comet

Halley's comet has an elliptical orbit with an eccentricity of $e \approx 0.97$. The length of the major axis of the orbit is approximately 36.18 astronomical units. (An astronomical unit is defined as the mean distance between the earth and the sun, 93 million miles.) Find a polar equation for the orbit. How close does Halley's comet come to the sun?

SOLUTION

Using a vertical axis, as shown in Figure 10.65, choose an equation of the form $r = ep/(1 + e \sin \theta)$. Because the vertices of the ellipse occur when $\theta = \pi/2$ and $\theta = 3\pi/2$, you can determine the length of the major axis to be the sum of the r-values of the vertices. That is,

$$2a = \frac{0.97p}{1 + 0.97} + \frac{0.97p}{1 - 0.97} \approx 32.83p \approx 36.18.$$

Thus, $p \approx 1.102$ and $ep \approx (0.97)(1.102) \approx 1.069$. Using this value in the equation, you have

$$r = \frac{1.069}{1 + 0.97 \sin \theta},$$

where r is measured in astronomical units. To find the closest point to the sun (the focus), you can write

$$c = ea \approx (0.97)(18.09) \approx 17.55.$$

Because c is the distance between the focus and the center, the closest point is

$$a - c \approx 18.09 - 17.55 \approx 0.54 \ AU \approx 50,000,000 \text{ miles.}$$

DISCUSSION PROBLEM

COMETS IN OUR SOLAR SYSTEM

Halley's comet is not the only spectacular comet that is periodically visible to viewers on earth. Use your school library to find information about another comet, and write a paragraph describing some of its characteristics.

WARM-UP

The following warm-up exercises involve skills that were covered in earlier sections. You will use these skills in the exercise set for this section.

In Exercises 1 and 2, plot the point given in polar coordinates and find the corresponding rectangular coordinates for the point.

1. $\left(-3, \dfrac{3\pi}{4}\right)$ **2.** $\left(4, -\dfrac{2\pi}{3}\right)$

In Exercises 3 and 4, plot the point given in rectangular coordinates and find two sets of polar coordinates for the point where $0 \le \theta < 2\pi$.

3. $(0, -3)$ **4.** $(-5, 12)$

In Exercises 5 and 6, convert the rectangular equation to polar form.

5. $x^2 + y^2 = 25$ **6.** $x^2y = 4$

In Exercises 7 and 8, convert the polar equation to rectangular form.

7. $r \sin \theta = -4$ **8.** $r = 4 \cos \theta$

In Exercises 9 and 10, identify and sketch the graph of the polar equation.

9. $r = 1 - \sin \theta$ **10.** $r = 1 + 2 \cos \theta$

SECTION 10.8 · EXERCISES

In Exercises 1–6, match the polar equation with the correct graph. [The graphs are labeled (a), (b), (c), (d), (e), and (f).]

1. $r = \dfrac{6}{1 - \cos \theta}$

2. $r = \dfrac{2}{2 - \cos \theta}$

3. $r = \dfrac{3}{1 - 2 \sin \theta}$

4. $r = \dfrac{2}{1 + \sin \theta}$

5. $r = \dfrac{6}{2 - \sin \theta}$

6. $r = \dfrac{2}{2 + 3 \cos \theta}$

(c)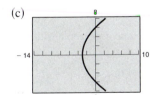

(d)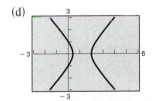

(a)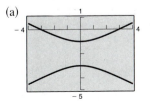

(b)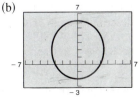

(e)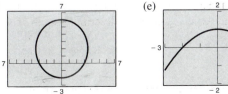

(f)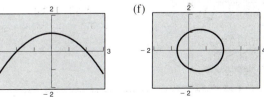

In Exercises 7–18, identify and sketch the graph of the polar equation. Use a graphing utility to confirm your sketch.

7. $r = \dfrac{2}{1 - \cos \theta}$

8. $r = \dfrac{4}{1 + \sin \theta}$

9. $r = \dfrac{5}{1 + \sin \theta}$

10. $r = \dfrac{6}{1 + \cos \theta}$

11. $r = \dfrac{2}{2 - \cos \theta}$

12. $r = \dfrac{3}{3 + \sin \theta}$

13. $r = \dfrac{4}{2 + \sin \theta}$

14. $r = \dfrac{6}{3 - 2 \cos \theta}$

15. $r = \dfrac{3}{2 + 4 \sin \theta}$

16. $r = \dfrac{5}{-1 + 2 \cos \theta}$

17. $r = \dfrac{3}{2 - 6 \cos \theta}$

18. $r = \dfrac{3}{2 + 6 \sin \theta}$

Conic	Vertex or Vertices
25. Parabola	$\left(1, -\dfrac{\pi}{2}\right)$
26. Parabola	$(4, 0)$
27. Parabola	$(5, \pi)$
28. Parabola	$\left(10, \dfrac{\pi}{2}\right)$
29. Ellipse	$(2, 0), (8, \pi)$
30. Ellipse	$\left(2, \dfrac{\pi}{2}\right), \left(4, \dfrac{3\pi}{2}\right)$
31. Ellipse	$(20, 0), (4, \pi)$
32. Hyperbola	$(2, 0), (10, 0)$
33. Hyperbola	$\left(1, \dfrac{3\pi}{2}\right), \left(9, \dfrac{3\pi}{2}\right)$
34. Hyperbola	$\left(4, \dfrac{\pi}{2}\right), \left(-1, \dfrac{3\pi}{2}\right)$

35. Show that the polar equation of the ellipse
$$\frac{x^2}{a^2} + \frac{y^2}{b^2} = 1 \quad \text{is} \quad r^2 = \frac{b^2}{1 - e^2 \cos^2 \theta}.$$

36. Show that the polar equation of the hyperbola
$$\frac{x^2}{a^2} - \frac{y^2}{b^2} = 1 \quad \text{is} \quad r^2 = \frac{-b^2}{1 - e^2 \cos^2 \theta}.$$

In Exercises 19–34, find a polar equation of the conic with its focus at the pole.

Conic	Eccentricity	Directrix
19. Parabola	$e = 1$	$x = -1$
20. Parabola	$e = 1$	$y = -2$
21. Ellipse	$e = \frac{1}{2}$	$y = 1$
22. Ellipse	$e = \frac{3}{4}$	$y = -2$
23. Hyperbola	$e = 2$	$x = 1$
24. Hyperbola	$e = \frac{3}{2}$	$x = -1$

In Exercises 37–42, use the results of Exercises 35 and 36 to write the polar form of the equation of the conic.

37. $\dfrac{x^2}{169} + \dfrac{y^2}{144} = 1$

38. $\dfrac{x^2}{25} + \dfrac{y^2}{16} = 1$

39. $\dfrac{x^2}{9} - \dfrac{y^2}{16} = 1$

40. $\dfrac{x^2}{36} - \dfrac{y^2}{4} = 1$

41. Hyperbola   One focus: $(5, 0)$
Vertices: $(4, 0), (4, \pi)$

42. Ellipse   One focus: $(4, 0)$
Vertices: $(5, 0), (5, \pi)$

In Exercises 43 and 44, sketch the graph of the rotated conic.

43. $r = \dfrac{2}{1 - \cos(\theta - \pi/4)}$ (See Exercise 7)

44. $r = \dfrac{4}{1 + \sin(\theta - \pi/3)}$ (See Exercise 8)

45. *Orbits of Planets*   The planets travel in elliptical orbits with the sun as a focus. Assume that the focus is at the pole, the major axis lies on the polar axis, and the length of the major axis is $2a$ (see figure). Show that the polar equation of the orbit is given by

$$r = \frac{(1 - e^2)a}{1 - e \cos \theta},$$

where e is the eccentricity.

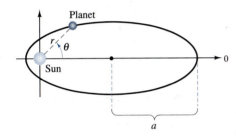

FIGURE FOR 45

46. *Orbits of Planets*   Use the result of Exercise 45 to show that the minimum distance (*perihelion distance*) from the sun to the planet is $r = a(1 - e)$ and the maximum distance (*aphelion distance*) is $r = a(1 + e)$.

Orbits of Planets   In Exercises 47 and 48, use the results of Exercises 45 and 46 to find the polar equation of the planet and the perihelion and aphelion distances.

47. Earth $a = 92.957 \times 10^6$ miles
$e = 0.0167$

48. Pluto $a = 3.666 \times 10^9$ miles
$e = 0.2481$

49. *Satellite Tracking*   A satellite in a 100-mile-high circular orbit around the earth has a velocity of approximately 17,500 miles per hour. If this velocity is multiplied by $\sqrt{2}$, then the satellite will have the minimum velocity necessary to escape the earth's gravity and it will follow a parabolic path with the center of the earth as the focus (see figure). Find a polar equation of the parabolic path of the satellite (assume the radius of the earth is 4000 miles).

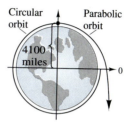

FIGURE FOR 49

50. *Explorer 18*   On November 26, 1963, the United States launched Explorer 18. Its low and high points over the surface of the earth were 119 miles and 122,000 miles, respectively (see figure). The center of the earth is the focus of the orbit. Find the polar equation for the orbit and find the distance between the surface of the earth and the satellite when $\theta = 60°$.

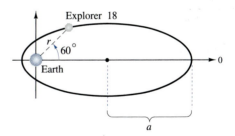

FIGURE FOR 50

CHAPTER 10 · REVIEW EXERCISES

In Exercises 1–12, identify the conic and sketch its graph.

1. $4x - y^2 = 0$ **2.** $8y + x^2 = 0$

3. $x^2 - 6x + 2y + 9 = 0$

4. $y^2 - 12y - 8x + 20 = 0$

5. $x^2 + y^2 - 2x - 4y + 5 = 0$

6. $16x^2 + 16y^2 - 16x + 24y - 3 = 0$

7. $4x^2 + y^2 = 16$ **8.** $2x^2 + 6y^2 = 18$

9. $x^2 + 9y^2 + 10x - 18y + 25 = 0$

10. $4x^2 + y^2 - 16x + 15 = 0$

11. $5y^2 - 4x^2 = 20$

12. $x^2 - 9y^2 + 10x + 18y + 7 = 0$

In Exercises 13 and 14, consider the general form of a conic whose axes are not parallel to either the x-axis or the y-axis (note the xy term). Use the quadratic formula to solve for y, and use a graphing utility to graph the equation. Identify the conic.

13. $x^2 - 10xy + y^2 + 1 = 0$

14. $40x^2 + 36xy + 25y^2 - 52 = 0$

In Exercises 15–18, find an equation of the specified parabola.

15. Vertex: $(4, 2)$
Focus: $(4, 0)$

16. Vertex: $(2, 0)$
Focus: $(0, 0)$

17. Vertex: $(0, 2)$
Passes through $(-1, 0)$
Horizontal axis

18. Vertex: $(2, 2)$
Directrix: $y = 0$

In Exercises 19–22, find an equation of the specified ellipse.

19. Vertices: $(-3, 0), (7, 0)$
Foci: $(0, 0), (4, 0)$

20. Vertices: $(2, 0), (2, 4)$
Foci: $(2, 1), (2, 3)$

21. Vertices: $(0, \pm 6)$
Passes through $(2, 2)$

22. Vertices: $(0, 1), (4, 1)$
Minor axis endpoints:
$(2, 0), (2, 2)$

In Exercises 23–26, find an equation of the specified hyperbola.

23. Vertices: $(0, \pm 1)$
Foci: $(0, \pm 3)$

24. Vertices: $(2, 2), (-2, 2)$
Foci: $(4, 2), (-4, 2)$

25. Foci: $(0, 0), (8, 0)$
Asymptotes: $y = \pm 2(x - 4)$

26. Foci: $(3, \pm 2)$
Asymptotes: $y = \pm 2(x - 3)$

27. *Satellite Antenna*   A cross section of a large parabolic antenna (see figure) is given by

$$y = \frac{x^2}{200}, \qquad 0 \le x \le 100.$$

The receiving and transmitting equipment is positioned at the focus. Find the coordinates of the focus.

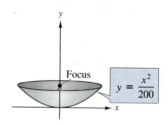

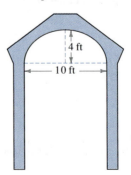

FIGURE FOR 27

28. *Semi-elliptical Archway*   A semi-elliptical archway is to be formed over the entrance to an estate. The arch is to be set on pillars that are 10 feet apart and is to have a height (atop the pillars) of 4 feet (see figure). Where should the foci be placed in order to sketch the semi-elliptical arch?

FIGURE FOR 28

In Exercises 29–40, sketch the curve represented by the parametric equations. Use a graphing utility to confirm your sketch. If possible, write the corresponding rectangular equation by eliminating the parameter.

29. $x = 2t$
$y = 4t$

30. $x = t^2$
$y = \sqrt{t}$

31. $x = 1 + 4t$
$y = 2 - 3t$

32. $x = t + 4$
$y = t^2$

33. $x = \dfrac{1}{t}$
$y = t^2$

34. $x = \dfrac{1}{t}$
$y = 2t + 3$

35. $x = 6 \cos \theta$
$y = 6 \sin \theta$

36. $x = 3 + 3 \cos \theta$
$y = 2 + 5 \sin \theta$

37. $x = \cos^3 \theta$
$y = 4 \sin^3 \theta$

38. $x = \sec \theta$
$y = \tan \theta$

39. $x = e^t$
$y = e^{-t}$

40. $x = 2\theta - \sin \theta$
$y = 2 - \cos \theta$

In Exercises 41–56, identify and sketch the graph of the polar equation. Use a graphing utility to confirm your sketch.

41. $r = 4$

42. $\theta = \dfrac{\pi}{12}$

43. $r = 4 \sin 2\theta$

44. $r = 2\theta$

45. $r = -2(1 + \cos \theta)$

46. $r = 3 - 4 \cos \theta$

47. $r = 4 - 3 \cos \theta$

48. $r = \cos 5\theta$

49. $r = -3 \cos 3\theta$

50. $r^2 = \cos 2\theta$

51. $r^2 = 4 \sin 2\theta$

52. $r = 3 \csc \theta$

53. $r = \dfrac{3}{\cos\left(\theta - \dfrac{\pi}{4}\right)}$

54. $r = \dfrac{4}{5 - 3 \cos \theta}$

55. $r = \dfrac{2}{1 - \sin \theta}$

56. $r = \dfrac{1}{1 + 2 \sin \theta}$

In Exercises 57–62, convert the polar equation to rectangular form.

57. $r = 3 \cos \theta$

58. $r = 4 \sec\left(\theta - \dfrac{\pi}{3}\right)$

59. $r = \dfrac{2}{1 + \sin \theta}$

60. $r = \dfrac{1}{2 - \cos \theta}$

61. $r^2 = \cos 2\theta$

62. $r = 10$

In Exercises 63 and 64, convert the rectangular equation to polar form.

63. $(x^2 + y^2)^2 = ax^2y$

64. $x^2 + y^2 - 4x = 0$

In Exercises 65–70, find a polar equation for the line or conic.

65. Circle   Center: $\left(5, \dfrac{\pi}{2}\right)$
Solution point: $(0, 0)$

66. Line   Solution point: $(0, 0)$
Slope: $\sqrt{3}$

67. Parabola   Vertex: $(2, \pi)$
Focus: $(0, 0)$

68. Parabola   Vertex: $\left(2, \dfrac{\pi}{2}\right)$
Focus: $(0, 0)$

69. Ellipse   Vertices: $(5, 0), (1, \pi)$
One focus: $(0, 0)$

70. Hyberbola   Vertices: $(1, 0), (7, 0)$
One focus: $(0, 0)$

71. Find a parametric representation of the ellipse with center at $(-3, 4)$, major axis horizontal and eight units in length, and minor axis six units in length.

72. Find a parametric representation of the hyperbola with vertices $(0, \pm 4)$ and foci at $(0, \pm 5)$.

73. Show that the Cartesian equation of a cycloid is
$$x = a \arccos \frac{a - y}{a} \pm \sqrt{2ay - y^2}.$$

74. The *involute of a circle* is described by the endpoint P of a string that is held taut as it is unwound from a spool (see figure). The spool does not rotate. Show that a parametric representation of the involute of a circle is given by
$$x = r(\cos \theta + \theta \sin \theta) \quad \text{and} \quad y = r(\sin \theta - \theta \cos \theta).$$

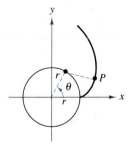

FIGURE FOR 74

75. Sketch the graph of the involute of a circle given by the parametric equations
$$x = 2(\cos \theta + \theta \sin \theta)$$
$$y = 2(\sin \theta - \theta \cos \theta).$$

VECTORS IN THE PLANE AND IN SPACE

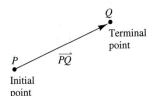

FIGURE 11.1

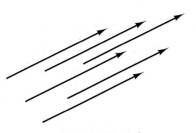

Equivalent Directed
Line Segments

FIGURE 11.2

11.1 VECTORS IN THE PLANE

Vectors in the Plane / **Component Form of a Vector** / **Vector Operations** / **Unit Vectors** / **Direction Angles** / **Applications of Vectors**

Vectors in the Plane

Many quantities in geometry and physics, such as area, time, and temperature, can be represented by a single real number. Other quantities, such as force and velocity, involve both *magnitude* and *direction* and cannot be completely characterized by a single real number. To represent such a quantity, we use a **directed line segment,** as shown in Figure 11.1. The directed line segment $\overrightarrow{PQ}$ has **initial point** P and **terminal point** Q, and we denote its **length** by $\|PQ\|$.

Two directed line segments that have the same length (or magnitude) and direction are called **equivalent.** For example, the directed line segments in Figure 11.2 are all equivalent. The set of all directed line segments that are equivalent to a given directed line segment $\overrightarrow{PQ}$ is a **vector v in the plane,** and we write $\mathbf{v} = \overrightarrow{PQ}$. Vectors are denoted by lowercase, boldface letters such as $\mathbf{u}$, $\mathbf{v}$, and $\mathbf{w}$.

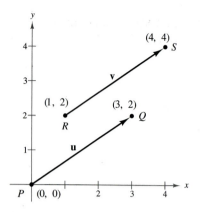

FIGURE 11.3

EXAMPLE 1   **Vector Representation by Directed Line Segments**

The vector **u** is represented by the directed line segment from $P = (0, 0)$ to $Q = (3, 2)$, and the vector **v** is represented by the directed line segment from $R = (1, 2)$ to $S = (4, 4)$, as shown in Figure 11.3. Show that **u** = **v**.

SOLUTION

From the distance formula, you can see that $\vec{PQ}$ and $\vec{RS}$ have the *same length*.

$$\|\vec{PQ}\| = \sqrt{(3 - 0)^2 + (2 - 0)^2} = \sqrt{13}$$
$$\|\vec{RS}\| = \sqrt{(4 - 1)^2 + (4 - 2)^2} = \sqrt{13}$$

Moreover, both line segments have the *same direction* because they are both directed toward the upper right on lines having a slope of $\frac{2}{3}$. Thus, $\vec{PQ}$ and $\vec{RS}$ have the same length and direction, and you can conclude that **u** = **v**.

Component Form of a Vector

The directed line segment whose initial point is the origin is often the most convenient representative of a set of equivalent directed line segments. This representative of the vector **v** is in **standard position.**

A vector whose initial point is at the origin $(0, 0)$ can be uniquely represented by the coordinates of its terminal point (v_1, v_2). We call this the **component form of a vector v** and write

$$\mathbf{v} = \langle v_1, v_2 \rangle.$$

The coordinates v_1 and v_2 are the **components** of **v**. If both the initial point and the terminal point lie at the origin, then **v** is the **zero vector** and is denoted by $\mathbf{0} = \langle 0, 0 \rangle$. To convert directed line segments to component form, use the following procedure.

COMPONENT FORM OF A VECTOR

The component form of the vector with initial point $P = (p_1, p_2)$ and terminal point $Q = (q_1, q_2)$ is

$$\vec{PQ} = \langle q_1 - p_1, q_2 - p_2 \rangle = \langle v_1, v_2 \rangle = \mathbf{v}.$$

The **length** (or magnitude) of **v** is given by

$$\|\mathbf{v}\| = \sqrt{(q_1 - p_1)^2 + (q_2 - p_2)^2} = \sqrt{v_1^2 + v_2^2}.$$

If $\|\mathbf{v}\| = 1$, then **v** is a **unit vector.** Moreover, $\|\mathbf{v}\| = 0$ if and only if **v** is the **zero vector 0.**

Two vectors $\mathbf{u} = \langle u_1, u_2 \rangle$ and $\mathbf{v} = \langle v_1, v_2 \rangle$ are **equal** if and only if $u_1 = v_1$ and $u_2 = v_2$. For instance, in Example 1, the vector $\mathbf{u}$ from $P = (0, 0)$ to $Q = (3, 2)$ is

$$\mathbf{u} = \overrightarrow{PQ} = \langle 3 - 0, 2 - 0 \rangle = \langle 3, 2 \rangle,$$

and the vector $\mathbf{v}$ from $R = (1, 2)$ to $S = (4, 4)$ is

$$\mathbf{v} = \overrightarrow{RS} = \langle 4 - 1, 4 - 2 \rangle = \langle 3, 2 \rangle,$$

which shows that $\mathbf{u}$ and $\mathbf{v}$ are equal.

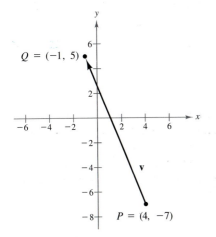

Component form of **v**:
$$\mathbf{v} = \langle -5, 12 \rangle$$

FIGURE 11.4

EXAMPLE 2   Finding the Component Form and Length of a Vector

Find the component form and length of the vector $\mathbf{v}$ that has initial point $(4, -7)$ and terminal point $(-1, 5)$.

SOLUTION

Let $P = (4, -7) = (p_1, p_2)$ and $Q = (-1, 5) = (q_1, q_2)$. The components of $\mathbf{v} = \langle v_1, v_2 \rangle$ are given by

$$v_1 = q_1 - p_1 = -1 - 4 = -5$$
$$v_2 = q_2 - p_2 = 5 - (-7) = 12.$$

Thus, $\mathbf{v} = \langle -5, 12 \rangle$, and the length of $\mathbf{v}$ is

$$\|\mathbf{v}\| = \sqrt{(-5)^2 + 12^2} = \sqrt{169} = 13,$$

as shown in Figure 11.4.

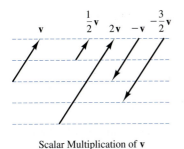

Scalar Multiplication of **v**

FIGURE 11.5

Vector Operations

The two basic vector operations are **scalar multiplication** and **vector addition.** (In this text, we use the term **scalar** to mean a real number.) Geometrically, the product of a vector $\mathbf{v}$ and a scalar k is the vector that is $|k|$ times as long as $\mathbf{v}$. If k is positive, then $k\mathbf{v}$ has the same direction as $\mathbf{v}$, and if k is negative, then $k\mathbf{v}$ has the opposite direction of $\mathbf{v}$, as shown in Figure 11.5.

To add two vectors geometrically, position them (without changing length or direction) so that the initial point of one coincides with the terminal point of the other. The sum **u** + **v** is formed by joining the initial point of the second vector **v** with the terminal point of the first vector **u**, as shown in Figure 11.6. Because the vector **u** + **v** is the diagonal of a parallelogram having **u** and **v** as its adjacent sides, we call this the **parallelogram law** for vector addition.

Vector addition and scalar multiplication can also be defined using components of vectors.

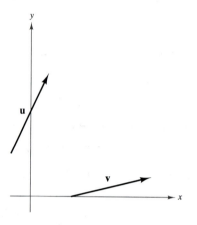

To find **u** + **v**,

move the initial point of **v** to the terminal point of **u**.

FIGURE 11.6

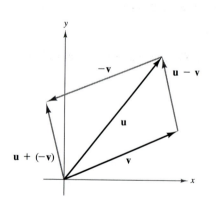

Vector Subtraction

FIGURE 11.7

DEFINITION OF VECTOR ADDITION AND SCALAR MULTIPLICATION

Let $\mathbf{u} = \langle u_1, u_2 \rangle$ and $\mathbf{v} = \langle v_1, v_2 \rangle$ be vectors and let k be a scalar (a real number). Then the **sum** of **u** and **v** is the vector

$$\mathbf{u} + \mathbf{v} = \langle u_1 + v_1, u_2 + v_2 \rangle, \qquad \textit{Sum}$$

and the **scalar multiple** of k times **u** is the vector

$$k\mathbf{u} = k\langle u_1, u_2 \rangle = \langle ku_1, ku_2 \rangle. \qquad \textit{Scalar multiple}$$

The **negative** of $\mathbf{v} = \langle v_1, v_2 \rangle$ is

$$-\mathbf{v} = (-1)\mathbf{v} = \langle -v_1, -v_2 \rangle, \qquad \textit{Negative}$$

and the **difference** of **u** and **v** is

$$\mathbf{u} - \mathbf{v} = \mathbf{u} + (-\mathbf{v}) = \langle u_1 - v_1, u_2 - v_2 \rangle. \qquad \textit{Difference}$$

To represent **u** − **v** graphically, we use directed line segments with the *same* initial points. The difference **u** − **v** is the vector from the terminal point of **v** to the terminal point of **u**, as shown in Figure 11.7.

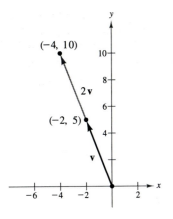

(a)

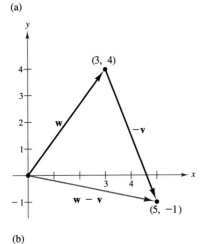

(b)

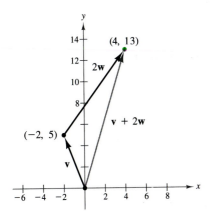

(c)

FIGURE 11.8

EXAMPLE 3   Vector Operations

Let $\mathbf{v} = \langle -2, 5 \rangle$ and $\mathbf{w} = \langle 3, 4 \rangle$, and find the following vectors.

A. $2\mathbf{v}$
B. $\mathbf{w} - \mathbf{v}$
C. $\mathbf{v} + 2\mathbf{w}$

SOLUTION

Sketches of the vectors are shown in Figure 11.8.

A. Because $\mathbf{v} = \langle -2, 5 \rangle$, you have

$$2\mathbf{v} = \langle 2(-2), 2(5) \rangle = \langle -4, 10 \rangle.$$

B. The difference of $\mathbf{w}$ and $\mathbf{v}$ is given by

$$\mathbf{w} - \mathbf{v} = \langle 3 - (-2), 4 - 5 \rangle = \langle 5, -1 \rangle.$$

C. Because $2\mathbf{w} = \langle 6, 8 \rangle$, it follows that

$$\begin{aligned} \mathbf{v} + 2\mathbf{w} &= \langle -2, 5 \rangle + \langle 6, 8 \rangle \\ &= \langle -2 + 6, 5 + 8 \rangle \\ &= \langle 4, 13 \rangle. \end{aligned}$$

Vector addition and scalar multiplication share many of the properties of ordinary arithmetic.

PROPERTIES OF VECTOR ADDITION AND SCALAR MULTIPLICATION

Let $\mathbf{u}$, $\mathbf{v}$, and $\mathbf{w}$ be vectors, and let c and d be scalars. Then the following properties are true.

1. $\mathbf{u} + \mathbf{v} = \mathbf{v} + \mathbf{u}$
2. $(\mathbf{u} + \mathbf{v}) + \mathbf{w} = \mathbf{u} + (\mathbf{v} + \mathbf{w})$
3. $\mathbf{u} + \mathbf{0} = \mathbf{u}$
4. $\mathbf{u} + (-\mathbf{u}) = \mathbf{0}$
5. $c(d\mathbf{u}) = (cd)\mathbf{u}$
6. $(c + d)\mathbf{u} = c\mathbf{u} + d\mathbf{u}$
7. $c(\mathbf{u} + \mathbf{v}) = c\mathbf{u} + c\mathbf{v}$
8. $1(\mathbf{u}) = \mathbf{u}$, $0(\mathbf{u}) = \mathbf{0}$
9. $\|c\mathbf{v}\| = |c| \, \|\mathbf{v}\|$

REMARK Property 9 can be stated as follows: The length of the vector $c\mathbf{v}$ is the absolute value of c times the length of $\mathbf{v}$.

Unit Vectors

In many applications of vectors, it is useful to find a unit vector that has the same direction as a given nonzero vector $\mathbf{v}$. To do this, divide $\mathbf{v}$ by its length to obtain

$$\mathbf{u} = \frac{\mathbf{v}}{\|\mathbf{v}\|} = \left(\frac{1}{\|\mathbf{v}\|}\right)\mathbf{v}. \qquad \textit{Unit vector}$$

Note that $\mathbf{u}$ is a scalar multiple of $\mathbf{v}$. The vector $\mathbf{u}$ has length 1 and the same direction as $\mathbf{v}$. We call $\mathbf{u}$ a **unit vector in the direction of v.**

EXAMPLE 4   Finding a Unit Vector

Find a unit vector in the direction of $\mathbf{v} = \langle -2, 5 \rangle$ and verify that the result has length 1.

SOLUTION

The unit vector in the direction of $\mathbf{v}$ is

$$\frac{\mathbf{v}}{\|\mathbf{v}\|} = \frac{\langle -2, 5 \rangle}{\sqrt{(-2)^2 + (5)^2}} = \frac{1}{\sqrt{29}}\langle -2, 5 \rangle = \left\langle \frac{-2}{\sqrt{29}}, \frac{5}{\sqrt{29}} \right\rangle.$$

This vector has length 1 because

$$\sqrt{\left(\frac{-2}{\sqrt{29}}\right)^2 + \left(\frac{5}{\sqrt{29}}\right)^2} = \sqrt{\frac{4}{29} + \frac{25}{29}} = \sqrt{\frac{29}{29}} = 1.$$

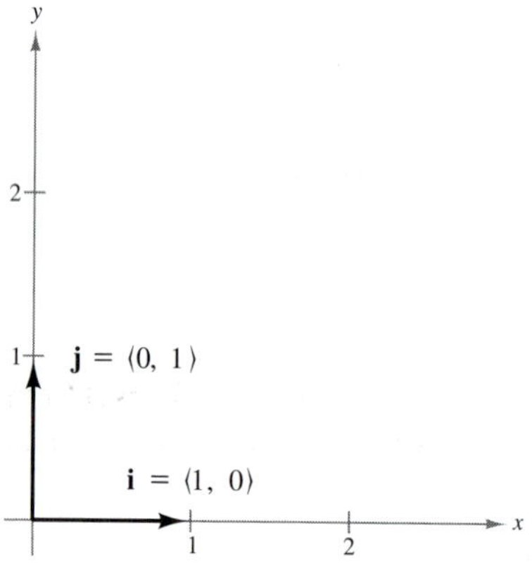

The unit vectors $\langle 1, 0 \rangle$ and $\langle 0, 1 \rangle$ are the **standard unit vectors** and are denoted by

$$\mathbf{i} = \langle 1, 0 \rangle \quad \text{and} \quad \mathbf{j} = \langle 0, 1 \rangle,$$

as shown in Figure 11.9. (Note that the lowercase letter $\mathbf{i}$ is written in boldface to distinguish it from the imaginary number $i = \sqrt{-1}$.) These vectors can be used to represent any vector $\mathbf{v} = \langle v_1, v_2 \rangle$ as follows.

$$\mathbf{v} = \langle v_1, v_2 \rangle = v_1 \langle 1, 0 \rangle + v_2 \langle 0, 1 \rangle = v_1 \mathbf{i} + v_2 \mathbf{j}$$

The scalars v_1 and v_2 are the **horizontal and vertical components of v,** respectively. The vector sum $v_1 \mathbf{i} + v_2 \mathbf{j}$ is a **linear combination** of the vectors $\mathbf{i}$ and $\mathbf{j}$. Any vector in the plane can be expressed as a linear combination of the standard unit vectors $\mathbf{i}$ and $\mathbf{j}$.

EXAMPLE 5   Representing a Vector as a Linear Combination of Unit Vectors

Let $\mathbf{u}$ be the vector with initial point $(2, -5)$ and terminal point $(-1, 3)$. Write $\mathbf{u}$ as a linear combination of the standard unit vectors $\mathbf{i}$ and $\mathbf{j}$.

Standard Unit Vectors $\mathbf{i}$ and $\mathbf{j}$

FIGURE 11.9

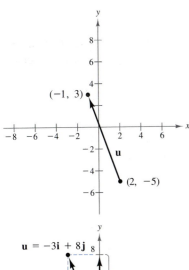

SOLUTION

$$\begin{aligned}
\mathbf{u} &= \langle -1 - 2, \, 3 - (-5) \rangle \\
&= \langle -3, \, 8 \rangle \\
&= -3\mathbf{i} + 8\mathbf{j}
\end{aligned}$$

This result is shown graphically in Figure 11.10.

EXAMPLE 6   Vector Operations

Let $\mathbf{u} = -3\mathbf{i} + 8\mathbf{j}$ and $\mathbf{v} = 2\mathbf{i} - \mathbf{j}$. Find $2\mathbf{u} - 3\mathbf{v}$.

SOLUTION

$$\begin{aligned}
2\mathbf{u} - 3\mathbf{v} &= 2(-3\mathbf{i} + 8\mathbf{j}) - 3(2\mathbf{i} - \mathbf{j}) \\
&= -6\mathbf{i} + 16\mathbf{j} - 6\mathbf{i} + 3\mathbf{j} \\
&= -12\mathbf{i} + 19\mathbf{j}
\end{aligned}$$

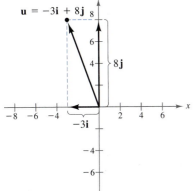

Direction Angles

If $\mathbf{u}$ is a *unit vector* such that θ is the angle (measured counterclockwise) from the positive x-axis to $\mathbf{u}$, then the terminal point of $\mathbf{u}$ lies on the unit circle and you have

$$\mathbf{u} = \langle \cos \theta, \, \sin \theta \rangle = (\cos \theta)\mathbf{i} + (\sin \theta)\mathbf{j},$$

FIGURE 11.10

as shown in Figure 11.11. We call θ the **direction angle** of the vector $\mathbf{u}$.

Suppose that $\mathbf{u}$ is a unit vector with direction angle θ. If $\mathbf{v}$ is any vector that makes an angle θ with the positive x-axis, then it has the same direction as $\mathbf{u}$ and you can write

$$\mathbf{v} = \|\mathbf{v}\|\langle \cos \theta, \, \sin \theta \rangle = \|\mathbf{v}\|(\cos \theta)\,\mathbf{i} + \|\mathbf{v}\|(\sin \theta)\,\mathbf{j}.$$

For instance, the vector $\mathbf{v}$ of length 3 making an angle of $30°$ with the positive x-axis is given by

$$\mathbf{v} = 3(\cos 30°)\,\mathbf{i} + 3(\sin 30°)\,\mathbf{j} = \frac{3\sqrt{3}}{2}\mathbf{i} + \frac{3}{2}\mathbf{j},$$

where $\|\mathbf{v}\| = 3$.

Because $\mathbf{v} = a\mathbf{i} + b\mathbf{j} = \|\mathbf{v}\|(\cos \theta)\,\mathbf{i} + \|\mathbf{v}\|(\sin \theta)\,\mathbf{j}$, it follows that the direction angle θ for $\mathbf{v}$ is determined from

$$\tan \theta = \frac{\sin \theta}{\cos \theta} = \frac{\|\mathbf{v}\| \sin \theta}{\|\mathbf{v}\| \cos \theta} = \frac{b}{a}.$$

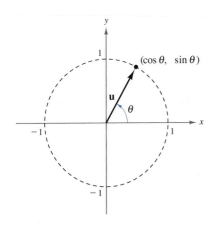

FIGURE 11.11

EXAMPLE 7   Finding Direction Angles of Vectors

Find the direction angles of the vectors.

A. $\mathbf{u} = 3\mathbf{i} + 3\mathbf{j}$
B. $\mathbf{v} = 3\mathbf{i} - 4\mathbf{j}$

SOLUTION

A. The direction angle is given by

$$\tan\theta = \frac{b}{a} = \frac{3}{3} = 1.$$

Therefore, $\theta = 45°$, as shown in Figure 11.12.

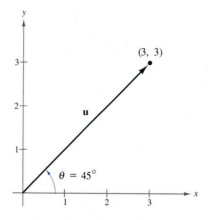

FIGURE 11.12

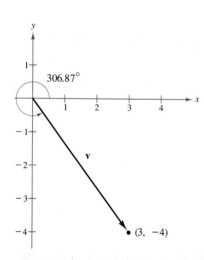

FIGURE 11.13

B. The direction angle is given by

$$\tan\theta = \frac{b}{a} = \frac{-4}{3}.$$

Moreover, because $\mathbf{v} = 3\mathbf{i} - 4\mathbf{j}$ lies in Quadrant IV, θ lies in Quadrant IV and its reference angle is

$$\theta' = \left| \arctan\left(-\frac{4}{3}\right) \right| \approx |-53.13°| = 53.13°.$$

Therefore, $\theta \approx 360° - 53.13° = 306.87°$, as shown in Figure 11.13.

Applications of Vectors

Many applications of vectors involve the use of triangles and trigonometry in their solutions.

EXAMPLE 8   **Finding Component Form, Given Magnitude and Direction**

Find the component form of the vector that represents the velocity of an airplane descending at a speed of 100 miles per hour at an angle 30° below horizontal, as shown in Figure 11.14.

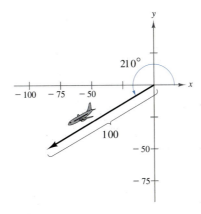

FIGURE 11.14

SOLUTION

The velocity vector **v** has a magnitude of 100 and a direction angle of $\theta = 210°$. Hence, the component form of **v** is

$$\mathbf{v} = \|\mathbf{v}\| \, (\cos \theta) \, \mathbf{i} + \|\mathbf{v}\| \, (\sin \theta) \, \mathbf{j}$$
$$= 100(\cos 210°) \, \mathbf{i} + 100(\sin 210°) \, \mathbf{j}$$
$$= 100\left(\frac{-\sqrt{3}}{2}\right) \mathbf{i} + 100\left(\frac{-1}{2}\right) \mathbf{j}$$
$$= -50\sqrt{3} \, \mathbf{i} - 50\mathbf{j}$$
$$= \langle -50\sqrt{3}, -50 \rangle.$$

You should check to see that $\|\mathbf{v}\| = 100$.

For an object to be in *equilibrium*, it must be at rest and the sum of all force vectors acting on the object must be the zero vector.

EXAMPLE 9 An Equilibrium Problem
·················

A person is standing on a tightrope, as shown in Figure 11.15. The total weight of the person and the balancing pole is 200 pounds. Find the force (or tension) on each end of the tightrope.

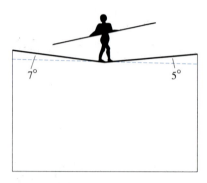

FIGURE 11.15

SOLUTION

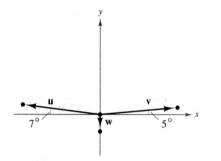

FIGURE 11.16

Begin by sketching a diagram in which all force vectors are in standard position (with initial points at the origin), as shown in Figure 11.16. In the diagram,

$$\mathbf{w} = -200\mathbf{j} \qquad \text{\textit{Weight of person}}$$

represents the force of the person. (Note that this force is acting straight down.) The forces corresponding to the left and right ends of the rope are as follows.

$$\mathbf{u} = -\cos 7° \, \|\mathbf{u}\| \, \mathbf{i} + \sin 7° \, \|\mathbf{u}\| \, \mathbf{j} \qquad \text{\textit{Left end of rope}}$$
$$\mathbf{v} = \cos 5° \, \|\mathbf{v}\| \, \mathbf{i} + \sin 5° \, \|\mathbf{v}\| \, \mathbf{j} \qquad \text{\textit{Right end of rope}}$$

Because all three forces must sum to zero, you can write the following.

$$\begin{aligned}
\mathbf{0} &= \mathbf{u} + \mathbf{v} + \mathbf{w} \\
&= -\cos 7° \, \|\mathbf{u}\| \, \mathbf{i} + \sin 7° \, \|\mathbf{u}\| \, \mathbf{j} + \cos 5° \, \|\mathbf{v}\| \, \mathbf{i} + \sin 5° \, \|\mathbf{v}\| \, \mathbf{j} - 200\mathbf{j} \\
&= (-\cos 7° \, \|\mathbf{u}\| + \cos 5° \, \|\mathbf{v}\|) \, \mathbf{i} + (\sin 7° \, \|\mathbf{u}\| + \sin 5° \, \|\mathbf{v}\| - 200) \, \mathbf{j} \\
&= 0\mathbf{i} + 0\mathbf{j}
\end{aligned}$$

Because two vectors are equal only if their corresponding components are equal, you can write the following system of equations.

$$\begin{aligned}
-\cos 7° \, \|\mathbf{u}\| + \cos 5° \, \|\mathbf{v}\| &= 0 \\
\sin 7° \, \|\mathbf{u}\| + \sin 5° \, \|\mathbf{v}\| &= 200
\end{aligned}$$

The solution of this system is

$$\|\mathbf{u}\| \approx 958.3 \text{ pounds} \quad \text{and} \quad \|\mathbf{v}\| \approx 954.8 \text{ pounds.}$$

These two quantities represent the forces acting on the left and right ends of the rope. Are you surprised at the amounts of these forces?

**DISCUSSION
PROBLEM**

....

**SOLVING AN
EQUILIBRIUM
PROBLEM**

In Example 9, suppose the tightrope was deflected with different angles, as shown below. Would the tension at each end of the rope be greater or less than the tensions found in Example 9? Explain your reasoning.

(a)

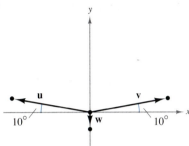

(b)

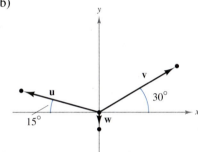

WARM-UP

....

The following warm-up exercises involve skills that were covered in earlier sections. You will use these skills in the exercise set for this section.

In Exercises 1 and 2, find the distance between the points.

1. $(-2, 6), (5, -15)$ **2.** $(0, 0), (-3, -7)$

In Exercises 3 and 4, find an equation of the line through the two points.

3. $(3, 1), (-2, 4)$ **4.** $(-2, -3), (4, 5)$

In Exercises 5 and 6, find an angle θ $(0 \le \theta \le 360°)$ whose vertex is at the origin and whose terminal side passes through the given point.

5. $(-2, 5)$ **6.** $(4, -3)$

In Exercises 7–10, find the sine and cosine of the angle θ.

7. $\theta = 30°$ **8.** $\theta = 120°$ **9.** $\theta = 300°$ **10.** $\theta = 210°$

SECTION 11.1 · EXERCISES

In Exercises 1–6, use the figure to sketch a graph of the specified vector.

1. −**u**   **2.** 3**v**
3. **u** + **v**   **4.** **u** + 2**v**
5. **u** − **v**   **6.** **v** − $\frac{1}{2}$**u**

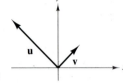

FIGURE FOR 1–6

In Exercises 7–16, find the component form and the magnitude of the vector **v**.

7.
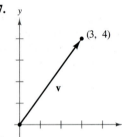

8.
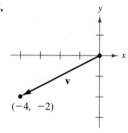

9.
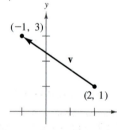

10.
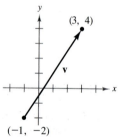

11.
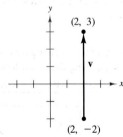

12.
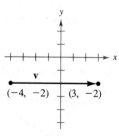

13. Initial point: $(-1, 5)$
 Terminal point: $(15, 2)$
14. Initial point: $(1, 11)$
 Terminal point: $(9, 3)$
15. Initial point: $(-3, -5)$
 Terminal point: $(5, -1)$
16. Initial point: $(-3, 11)$
 Terminal point: $(9, 40)$

In Exercises 17–26, find (a) **u** + **v**, (b) **u** − **v**, and (c) 2**u** − 3**v**.

17. **u** = $\langle 1, 2 \rangle$,   **v** = $\langle 3, 1 \rangle$
18. **u** = $\langle 2, 3 \rangle$,   **v** = $\langle 4, 0 \rangle$
19. **u** = $\langle -2, 3 \rangle$,   **v** = $\langle -2, 1 \rangle$
20. **u** = $\langle 0, 1 \rangle$,   **v** = $\langle 0, -1 \rangle$
21. **u** = $\langle 4, -2 \rangle$,   **v** = $\langle 0, 0 \rangle$
22. **u** = $\langle 0, 0 \rangle$,   **v** = $\langle 2, 1 \rangle$
23. **u** = **i** + **j**,   **v** = 2**i** − 3**j**
24. **u** = 2**i** − **j**,   **v** = −**i** + **j**
25. **u** = 2**i**,   **v** = **j**
26. **u** = 3**j**,   **v** = 2**i**

In Exercises 27–30, find the magnitude and direction angle of the vector **v**.

27. **v** = 5⟨cos 30°, sin 30°⟩
28. **v** = 8⟨cos 135°, sin 135°⟩
29. **v** = 6**i** − 6**j**
30. **v** = −2**i** + 5**j**

In Exercises 31–38, sketch **v** and find its component form. (Assume θ is measured counterclockwise from the x-axis to the vector.)

31. $\|\mathbf{v}\| = 3$, $\theta = 0°$
32. $\|\mathbf{v}\| = 1$, $\theta = 45°$
33. $\|\mathbf{v}\| = 1$, $\theta = 150°$
34. $\|\mathbf{v}\| = \frac{5}{2}$, $\theta = 45°$
35. $\|\mathbf{v}\| = 3\sqrt{2}$, $\theta = 150°$
36. $\|\mathbf{v}\| = 8$, $\theta = 90°$
37. $\|\mathbf{v}\| = 2$,   **v** in the direction **i** + 3**j**
38. $\|\mathbf{v}\| = 3$,   **v** in the direction 3**i** + 4**j**

In Exercises 39–44, find the component form of **v** and sketch the specified vector operations geometrically, where

$$\mathbf{u} = 2\mathbf{i} - \mathbf{j} \quad \text{and} \quad \mathbf{w} = \mathbf{i} + 2\mathbf{j}.$$

39. **v** = $\frac{3}{2}$**u**
40. **v** = **u** + **w**
41. **v** = **u** + 2**w**
42. **v** = −**u** + **w**
43. **v** = $\frac{1}{2}$(3**u** + **w**)
44. **v** = **u** − 2**w**

In Exercises 45–48, find the component form of the sum of the vectors **u** and **v** with direction angles $\theta_{\mathbf{u}}$ and $\theta_{\mathbf{v}}$, respectively.

45. $\|\mathbf{u}\| = 5$, $\theta_{\mathbf{u}} = 0°$
   $\|\mathbf{v}\| = 5$, $\theta_{\mathbf{v}} = 90°$

46. $\|\mathbf{u}\| = 2$, $\theta_{\mathbf{u}} = 30°$
   $\|\mathbf{v}\| = 2$, $\theta_{\mathbf{v}} = 90°$

47. $\|\mathbf{u}\| = 20$, $\theta_{\mathbf{u}} = 45°$
   $\|\mathbf{v}\| = 50$, $\theta_{\mathbf{v}} = 180°$

48. $\|\mathbf{u}\| = 35$, $\theta_{\mathbf{u}} = 25°$
   $\|\mathbf{v}\| = 50$, $\theta_{\mathbf{v}} = 120°$

In Exercises 49–52, find a unit vector in the direction of the given vector.

49. $\mathbf{v} = 4\mathbf{i} - 3\mathbf{j}$

50. $\mathbf{v} = \mathbf{i} + \mathbf{j}$

51. $\mathbf{v} = 2\mathbf{j}$

52. $\mathbf{v} = \mathbf{i} - 2\mathbf{j}$

In Exercises 53–56, use the Law of Cosines to find the angle α between the given vectors. (*Assume* $0° \le \alpha \le 180°$.)

53. $\mathbf{v} = \mathbf{i} + \mathbf{j}$, $\mathbf{w} = 2(\mathbf{i} - \mathbf{j})$

54. $\mathbf{v} = 3\mathbf{i} + \mathbf{j}$, $\mathbf{w} = 2\mathbf{i} - \mathbf{j}$

55. $\mathbf{v} = \mathbf{i} + \mathbf{j}$, $\mathbf{w} = 3\mathbf{i} - \mathbf{j}$

56. $\mathbf{v} = \mathbf{i} + 2\mathbf{j}$, $\mathbf{w} = 2\mathbf{i} - \mathbf{j}$

In Exercises 57 and 58, find the angle between the forces, given the magnitude of their resultant (vector sum). (*Hint:* Write one force as a vector in the direction of the positive x-axis and the other as a vector at an angle θ with the positive x-axis.)

57. Force one: 45 pounds
   Force two: 60 pounds
   Resultant force: 90 pounds

58. Force one: 3000 pounds
   Force two: 1000 pounds
   Resultant force: 3750 pounds

59. *Resultant Force*   Forces with magnitudes of 35 pounds and 50 pounds act on a hook (see figure). The angle between the two forces is 30°. Find the direction and magnitude of the resultant (vector sum) of these two forces.

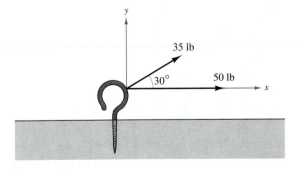

FIGURE FOR 59

60. *Resultant Force*   Forces with magnitudes of 500 pounds and 200 pounds act on a machine part at angles of 30° and −45°, respectively, with the x-axis (see figure). Find the direction and magnitude of the resultant (vector sum) of these forces.

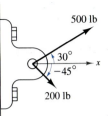

FIGURE FOR 60

61. *Resultant Force*   Three forces with magnitudes of 75 pounds, 100 pounds, and 125 pounds act on an object at angles of 30°, 45°, and 120°, respectively, with the positive x-axis. Find the direction and magnitude of the resultant of these forces.

62. *Resultant Force*   Three forces with magnitudes of 70 pounds, 40 pounds, and 60 pounds act on an object at angles of −30°, 45°, and 135°, respectively, with the positive x-axis. Find the direction and magnitude of the resultant of these forces.

63. *Horizontal and Vertical Components of Velocity*   A ball is thrown with an initial velocity of 80 feet per second, at an angle of 50° with the horizontal (see figure). Find the vertical and horizontal components of the velocity.

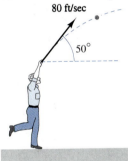

FIGURE FOR 63

64. *Horizontal and Vertical Components of Velocity*   A gun with a muzzle velocity of 1200 feet per second is fired at an angle of 6° with the horizontal. Find the vertical and horizontal components of the velocity.

Cable Tension   In Exercises 65 and 66, use the figure to determine the tension in each cable supporting the given load.

65.

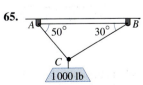

66.

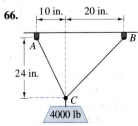

67. *Barge Towing* A loaded barge is being towed by two tugboats, and the magnitude of the resultant is 6000 pounds directed along the axis of the barge (see figure). Find the tension in the tow lines if they each make a 20° angle with the axis of the barge.

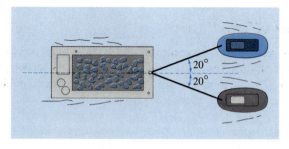

FIGURE FOR 67

68. *Shared Load* To carry a 100-pound cylindrical weight, two people lift on the ends of short ropes that are tied to an eyelet on the top center of the cylinder. Find the tension in the ropes if they each make a 30° angle with the vertical (see figure).

FIGURE FOR 68

69. *Navigation* An airplane is flying in the direction S 32° E, with an air speed of 540 miles per hour. Because of the wind, the plane's ground speed and direction are 500 miles per hour and S 40° E, respectively (see figure). Find the direction and speed of the wind.

70. *Navigation* An airplane's velocity with respect to the air is 580 miles per hour, and it is headed N 58° W. The wind, at the altitude of the plane, is from the southwest and has a velocity of 60 miles per hour (see figure). What is the true direction of the plane, and what is its speed with respect to the ground?

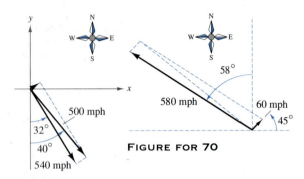

FIGURE FOR 70

FIGURE FOR 69

71. *Work* A heavy implement is pulled 10 feet across a floor, using a force of 85 pounds. Find the work done if the direction of the force is 60° above the horizontal (see figure). (Use the formula for work, $W = FD$, where F is the component of the force in the direction of motion and D is the distance.)

72. *Tether Ball* A tether ball weighing 1 pound is pulled outward from the pole by a horizontal force **u** until the rope makes a 30° angle with the pole (see figure). Determine the resulting tension in the rope and the magnitude of **u**.

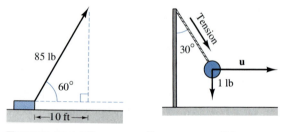

FIGURE FOR 71   **FIGURE FOR 72**

73. *Technology* Enter the following program in a graphing utility. (This program is for the TI-81; other versions are listed in the appendix.) Use $-6 \le x \le 6$ and $-2 \le y \le 6$, and enter $A = 5$, $B = 2$, $C = -4$, and $D = 3$. Explain what the program does.

```
:Disp "ENTER, (A, B)"        :Line (0, 0, A, B)
:Disp "ENTER, A"             :Line (0, 0, C, D)
:Input A                     :A+C→E
:Disp "ENTER, B"             :B+D→F
:Input B                     :Line (0, 0, E, F)
:Disp "ENTER, (C, D)"        :Line (A, B, E, F)
:Disp "ENTER, C"             :Line (C, D, E, F)
:Input C                     :Pause
:Disp "ENTER, D"             :ClrDraw
:Input D                     :End
```

11.2 THE DOT PRODUCT OF TWO VECTORS

The Dot Product of Two Vectors / Angle Between Two Vectors / Finding Vector Components / Work

The Dot Product of Two Vectors

So far you have studied two vector operations—vector addition and multiplication by a scalar—each of which yields another vector. In this section you will study a third vector operation, the **dot product.** This product yields a scalar, rather than a vector.

DEFINITION OF DOT PRODUCT

The **dot product** of $\mathbf{u} = \langle u_1, u_2 \rangle$ and $\mathbf{v} = \langle v_1, v_2 \rangle$ is

$$\mathbf{u} \cdot \mathbf{v} = u_1 v_1 + u_2 v_2.$$

PROPERTIES OF THE DOT PRODUCT

Let $\mathbf{u}$, $\mathbf{v}$, and $\mathbf{w}$ be vectors in the plane and let c be a scalar.

1. $\mathbf{u} \cdot \mathbf{v} = \mathbf{v} \cdot \mathbf{u}$ 2. $\mathbf{0} \cdot \mathbf{v} = 0$
3. $\mathbf{u} \cdot (\mathbf{v} + \mathbf{w}) = \mathbf{u} \cdot \mathbf{v} + \mathbf{u} \cdot \mathbf{w}$ 4. $\mathbf{v} \cdot \mathbf{v} = \|\mathbf{v}\|^2$
5. $c(\mathbf{u} \cdot \mathbf{v}) = c\mathbf{u} \cdot \mathbf{v} = \mathbf{u} \cdot c\mathbf{v}$

PROOF

To prove the first property, let $\mathbf{u} = \langle u_1, u_2 \rangle$ and $\mathbf{v} = \langle v_1, v_2 \rangle$. Then

$$\mathbf{u} \cdot \mathbf{v} = u_1 v_1 + u_2 v_2 = v_1 u_1 + v_2 u_2 = \mathbf{v} \cdot \mathbf{u}.$$

For the fourth property, let $\mathbf{v} = \langle v_1, v_2 \rangle$. Then

$$\begin{aligned} \mathbf{v} \cdot \mathbf{v} &= v_1{}^2 + v_2{}^2 \\ &= \left(\sqrt{v_1{}^2 + v_2{}^2} \right)^2 \\ &= \|\mathbf{v}\|^2. \end{aligned}$$

Proofs of the other properties are left to you.

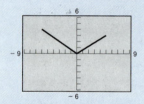
EXAMPLE 1   Finding Dot Products

Find the following dot products.

A. $\langle 4, 5 \rangle \cdot \langle 2, 3 \rangle$ **B.** $\langle 2, -1 \rangle \cdot \langle 1, 2 \rangle$ **C.** $(-5\mathbf{i} - 2\mathbf{j}) \cdot (3\mathbf{i} + 2\mathbf{j})$

SOLUTION

A. $\langle 4, 5 \rangle \cdot \langle 2, 3 \rangle = 4(2) + 5(3) = 8 + 15 = 23$
B. $\langle 2, -1 \rangle \cdot \langle 1, 2 \rangle = 2(1) + (-1)(2) = 2 - 2 = 0$
C. $(-5\mathbf{i} - 2\mathbf{j}) \cdot (3\mathbf{i} + 2\mathbf{j}) = (-5)(3) + (-2)(2)$
$$= -15 - 4$$
$$= -19$$

REMARK In Example 1, be sure you see that the dot product of two vectors is a scalar (a real number), not a vector. Moreover, notice that the dot product can be positive, zero, or negative.

EXAMPLE 2   Using Properties of Dot Products

Let $\mathbf{u} = \langle -1, 3 \rangle$, $\mathbf{v} = \langle 2, -4 \rangle$, and $\mathbf{w} = \langle 1, -2 \rangle$. Find the following products.

A. $(\mathbf{u} \cdot \mathbf{v}) \, \mathbf{w}$ **B.** $\mathbf{u} \cdot (2\mathbf{v})$

SOLUTION

Begin by finding the dot product of $\mathbf{u}$ and $\mathbf{v}$.

$$\mathbf{u} \cdot \mathbf{v} = \langle -1, 3 \rangle \cdot \langle 2, -4 \rangle = (-1)(2) + 3(-4) = -14$$

A. $(\mathbf{u} \cdot \mathbf{v}) \, \mathbf{w} = -14\langle 1, -2 \rangle = \langle -14, 28 \rangle$
B. $\mathbf{u} \cdot (2\mathbf{v}) = 2(\mathbf{u} \cdot \mathbf{v}) = 2(-14) = -28$

Notice that the first product is a vector, whereas the second is a scalar. Can you see why?

EXAMPLE 3   Dot Product and Length

The dot product of $\mathbf{u}$ with itself is 5. What is the length of $\mathbf{u}$?

SOLUTION

Because $\|\mathbf{u}\|^2 = \mathbf{u} \cdot \mathbf{u} = 5$, it follows that

$$\|\mathbf{u}\| = \sqrt{\mathbf{u} \cdot \mathbf{u}} = \sqrt{5}.$$

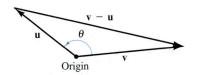

Angle Between Two Vectors

FIGURE 11.17

Angle Between Two Vectors

The **angle between two nonzero vectors** is the angle θ, $0 \leq \theta \leq \pi$, between their respective standard position vectors, as shown in Figure 11.17. This angle can be found using the dot product. (Note that we do not define the angle between the zero vector and another vector.)

> **ANGLE BETWEEN TWO VECTORS**
>
> If θ is the angle between two nonzero vectors **u** and **v**, then
> $$\cos \theta = \frac{\mathbf{u} \cdot \mathbf{v}}{\|\mathbf{u}\| \|\mathbf{v}\|}.$$

PROOF

Consider the triangle determined by vectors **u**, **v**, and **v** − **u**, as shown in Figure 11.17. By the Law of Cosines, you can write

$$\|\mathbf{v} - \mathbf{u}\|^2 = \|\mathbf{u}\|^2 + \|\mathbf{v}\|^2 - 2\|\mathbf{u}\| \|\mathbf{v}\| \cos \theta$$
$$(\mathbf{v} - \mathbf{u}) \cdot (\mathbf{v} - \mathbf{u}) = \|\mathbf{u}\|^2 + \|\mathbf{v}\|^2 - 2\|\mathbf{u}\| \|\mathbf{v}\| \cos \theta$$
$$(\mathbf{v} - \mathbf{u}) \cdot \mathbf{v} - (\mathbf{v} - \mathbf{u}) \cdot \mathbf{u} = \|\mathbf{u}\|^2 + \|\mathbf{v}\|^2 - 2\|\mathbf{u}\| \|\mathbf{v}\| \cos \theta$$
$$\mathbf{v} \cdot \mathbf{v} - \mathbf{u} \cdot \mathbf{v} - \mathbf{v} \cdot \mathbf{u} + \mathbf{u} \cdot \mathbf{u} = \|\mathbf{u}\|^2 + \|\mathbf{v}\|^2 - 2\|\mathbf{u}\| \|\mathbf{v}\| \cos \theta$$
$$\|\mathbf{v}\|^2 - 2\mathbf{u} \cdot \mathbf{v} + \|\mathbf{u}\|^2 = \|\mathbf{u}\|^2 + \|\mathbf{v}\|^2 - 2\|\mathbf{u}\| \|\mathbf{v}\| \cos \theta$$
$$-2\mathbf{u} \cdot \mathbf{v} = -2\|\mathbf{u}\| \|\mathbf{v}\| \cos \theta$$
$$\cos \theta = \frac{\mathbf{u} \cdot \mathbf{v}}{\|\mathbf{u}\| \|\mathbf{v}\|}.$$

EXAMPLE 4   Finding the Angle Between Two Vectors

Find the angle between $\mathbf{u} = \langle 4, 3 \rangle$ and $\mathbf{v} = \langle 3, 5 \rangle$.

SOLUTION

$$\cos \theta = \frac{\mathbf{u} \cdot \mathbf{v}}{\|\mathbf{u}\| \|\mathbf{v}\|} = \frac{\langle 4, 3 \rangle \cdot \langle 3, 5 \rangle}{\|\langle 4, 3 \rangle\| \|\langle 3, 5 \rangle\|} = \frac{27}{5\sqrt{34}}$$

This implies that the angle between the two vectors is

$$\theta = \arccos \frac{27}{5\sqrt{34}} \approx 22.2°,$$

as shown in Figure 11.18.

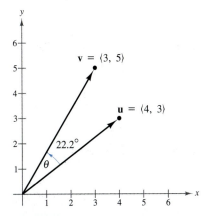

FIGURE 11.18

Rewriting the expression for the angle between two vectors in the form

$$\mathbf{u} \cdot \mathbf{v} = \|\mathbf{u}\| \|\mathbf{v}\| \cos \theta \qquad \textit{Alternative form of dot product}$$

produces an alternative way to calculate the dot product. From this form, you can see that because $\|\mathbf{u}\|$ and $\|\mathbf{v}\|$ are always positive, $\mathbf{u} \cdot \mathbf{v}$ and $\cos \theta$ will always have the same sign. Figure 11.19 shows the five possible orientations of two vectors.

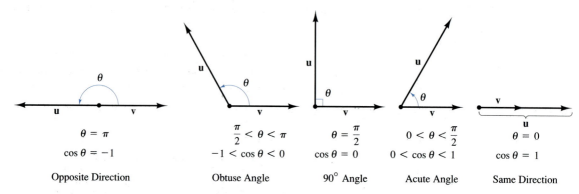

$\theta = \pi$	$\dfrac{\pi}{2} < \theta < \pi$	$\theta = \dfrac{\pi}{2}$	$0 < \theta < \dfrac{\pi}{2}$	$\theta = 0$
$\cos \theta = -1$	$-1 < \cos \theta < 0$	$\cos \theta = 0$	$0 < \cos \theta < 1$	$\cos \theta = 1$
Opposite Direction	Obtuse Angle	90° Angle	Acute Angle	Same Direction

FIGURE 11.19

DEFINITION OF ORTHOGONAL VECTORS

The vectors $\mathbf{u}$ and $\mathbf{v}$ are **orthogonal** if $\mathbf{u} \cdot \mathbf{v} = 0$.

The terms "orthogonal" and "perpendicular" mean essentially the same thing—meeting at right angles. By definition, however, the zero vector is orthogonal to every vector $\mathbf{u}$, because $\mathbf{0} \cdot \mathbf{u} = 0$.

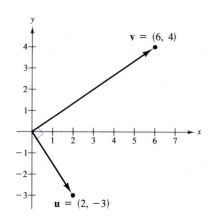

FIGURE 11.20

EXAMPLE 5   **Determining Orthogonal Vectors**

Are the vectors $\mathbf{u} = \langle 2, -3 \rangle$ and $\mathbf{v} = \langle 6, 4 \rangle$ orthogonal?

SOLUTION

Begin by finding the dot product of the two vectors.

$$
\begin{aligned}
\mathbf{u} \cdot \mathbf{v} &= \langle 2, -3 \rangle \cdot \langle 6, 4 \rangle \\
&= 2(6) + (-3)(4) \\
&= 0
\end{aligned}
$$

Because the dot product is 0, the two vectors are orthogonal, as shown in Figure 11.20.

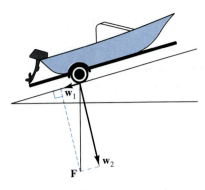

FIGURE 11.21

Finding Vector Components

You have already seen applications in which two vectors are added to produce a resultant vector. Many applications in physics and engineering pose the reverse problem—decomposing a given vector into the sum of two **vector components.**

Consider a boat on an inclined ramp, as shown in Figure 11.21. The force **F** due to gravity pulls the boat *down* the ramp and *against* the ramp. These two orthogonal forces, $\mathbf{w}_1$ and $\mathbf{w}_2$, are vector components of **F**. That is,

$$\mathbf{F} = \mathbf{w}_1 + \mathbf{w}_2. \qquad \textit{Vector components of } \mathbf{F}$$

The negative of component $\mathbf{w}_1$ represents the force needed to keep the boat from rolling down the ramp, whereas $\mathbf{w}_2$ represents the force that the tires must withstand against the ramp. A procedure for finding $\mathbf{w}_1$ and $\mathbf{w}_2$ is shown below.

DEFINITION OF VECTOR COMPONENTS

Let **u** and **v** be nonzero vectors such that

$$\mathbf{u} = \mathbf{w}_1 + \mathbf{w}_2,$$

where $\mathbf{w}_1$ and $\mathbf{w}_2$ are orthogonal and $\mathbf{w}_1$ is parallel to **v**, as shown in Figure 11.22. The vectors $\mathbf{w}_1$ and $\mathbf{w}_2$ are called **vector components** of **u**. The vector $\mathbf{w}_1$ is the **projection** of **u** onto **v** and is denoted by

$$\mathbf{w}_1 = \text{proj}_\mathbf{v}\,\mathbf{u}.$$

The vector $\mathbf{w}_2$ is given by $\mathbf{w}_2 = \mathbf{u} - \mathbf{w}_1$.

From this definition, you can see that it is easy to find the component $\mathbf{w}_2$ once you have found the projection of **u** onto **v**. To find the projection, you can use the dot product, as shown below.

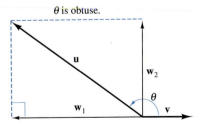

$\mathbf{w}_1 = \text{proj}_\mathbf{v}\,\mathbf{u} = $ projection of **u** onto **v**.
$\mathbf{w}_2 = $ vector component of
   u orthogonal to **v**.

FIGURE 11.22

PROJECTION OF U ONTO V

Let **u** and **v** be nonzero vectors. The projection of **u** onto **v** is

$$\text{proj}_\mathbf{v}\,\mathbf{u} = \left(\frac{\mathbf{u} \cdot \mathbf{v}}{\|\mathbf{v}\|^2}\right)\mathbf{v}.$$

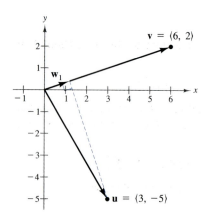

FIGURE 11.23

EXAMPLE 6   Decomposing a Vector into Components

Find the projection of $\mathbf{u} = \langle 3, -5 \rangle$ onto $\mathbf{v} = \langle 6, 2 \rangle$. Then write $\mathbf{u}$ as the sum of two orthogonal vectors, one of which is $\mathrm{proj}_\mathbf{v}\, \mathbf{u}$.

SOLUTION

The projection of $\mathbf{u}$ onto $\mathbf{v}$ is

$$\mathbf{w}_1 = \mathrm{proj}_\mathbf{v}\, \mathbf{u} = \left(\frac{\mathbf{u} \cdot \mathbf{v}}{\|\mathbf{v}\|^2} \right) \mathbf{v} = \left(\frac{8}{40} \right) \langle 6, 2 \rangle = \left\langle \frac{6}{5}, \frac{2}{5} \right\rangle,$$

as shown in Figure 11.23. The other component, $\mathbf{w}_2$, is

$$\mathbf{w}_2 = \mathbf{u} - \mathbf{w}_1 = \langle 3, -5 \rangle - \left\langle \frac{6}{5}, \frac{2}{5} \right\rangle = \left\langle \frac{9}{5}, -\frac{27}{5} \right\rangle.$$

Thus, $\mathbf{u} = \mathbf{w}_1 + \mathbf{w}_2 = \left\langle \frac{6}{5}, \frac{2}{5} \right\rangle + \left\langle \frac{9}{5}, -\frac{27}{5} \right\rangle = \langle 3, -5 \rangle.$

EXAMPLE 7   Finding a Force

A 600-pound boat sits on a ramp inclined at 30°, as shown in Figure 11.24. What force is required to keep the boat from rolling down the ramp?

SOLUTION

Because the force due to gravity is vertical and downward, you can represent the gravitational force by the vector

$$\mathbf{F} = -600\mathbf{j}. \qquad \textit{Force due to gravity}$$

To find the force required to keep the boat from rolling down the ramp, project $\mathbf{F}$ onto a unit vector $\mathbf{v}$ in the direction of the ramp, as follows.

$$\mathbf{v} = \cos 30°\, \mathbf{i} + \sin 30°\, \mathbf{j} = \frac{\sqrt{3}}{2}\mathbf{i} + \frac{1}{2}\mathbf{j} \qquad \textit{Unit vector along ramp}$$

Therefore, the projection of $\mathbf{F}$ onto $\mathbf{v}$ is given by

$$\begin{aligned} \mathbf{w}_1 = \mathrm{proj}_\mathbf{v}\, \mathbf{F} &= \left(\frac{\mathbf{F} \cdot \mathbf{v}}{\|\mathbf{v}\|^2} \right) \mathbf{v} \\ &= (\mathbf{F} \cdot \mathbf{v})\, \mathbf{v} = (-600)\left(\frac{1}{2} \right) \mathbf{v} \\ &= -300\left(\frac{\sqrt{3}}{2}\mathbf{i} + \frac{1}{2}\mathbf{j} \right). \end{aligned}$$

The magnitude of this force is 300, and therefore a force of 300 pounds is required to keep the boat from rolling down the ramp.

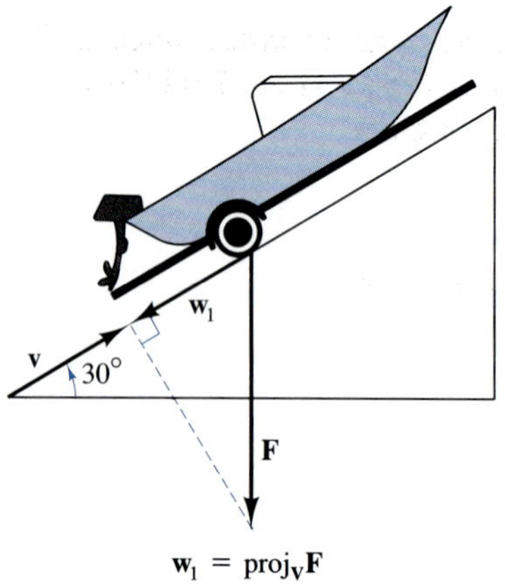

FIGURE 11.24

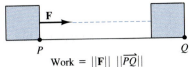

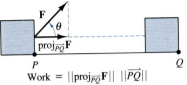

Work = $\| \mathbf{F} \| \| \overrightarrow{PQ} \|$

(a) Force acts along the line of motion.

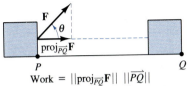

Work = $\| \text{proj}_{\overrightarrow{PQ}} \mathbf{F} \| \| \overrightarrow{PQ} \|$

(b) Force acts at angle θ with the line of motion.

FIGURE 11.25

Work

The work W done by a constant force $\mathbf{F}$ acting along the line of motion of an object is given by

$$W = (\text{magnitude of force})(\text{distance}) = \| \mathbf{F} \| \| \overrightarrow{PQ} \|,$$

as shown in Figure 11.25(a). If the constant force $\mathbf{F}$ is not directed along the line of motion, then you can see from Figure 11.25(b) that the work W done by the force is

$$W = \| \text{proj}_{\overrightarrow{PQ}} \mathbf{F} \| \| \overrightarrow{PQ} \| = (\cos \theta) \| \mathbf{F} \| \| \overrightarrow{PQ} \| = \mathbf{F} \cdot \overrightarrow{PQ}.$$

This notion of work is summarized in the following definition.

DEFINITION OF WORK

The **work** W done by a constant force $\mathbf{F}$ as its point of application moves along the vector $\overrightarrow{PQ}$ is given by either of the following.

1. $W = \| \text{proj}_{\overrightarrow{PQ}} \mathbf{F} \| \| \overrightarrow{PQ} \|$ *Projection form*

2. $W = \mathbf{F} \cdot \overrightarrow{PQ}$ *Dot product form*

EXAMPLE 8   Finding Work

To close a sliding door, a person pulls on a rope with a constant force of 50 pounds at a constant angle of 60°, as shown in Figure 11.26. Find the work done in moving the door 12 feet to its closed position.

SOLUTION

Using a projection, you can calculate the work as follows.

$$\begin{aligned} W &= \| \text{proj}_{\overrightarrow{PQ}} \mathbf{F} \| \| \overrightarrow{PQ} \| \\ &= (\cos 60°) \| \mathbf{F} \| \| \overrightarrow{PQ} \| \\ &= \frac{1}{2}(50)(12) \\ &= 300 \text{ ft-lb} \end{aligned}$$

Thus, the work done is 300 foot-pounds.

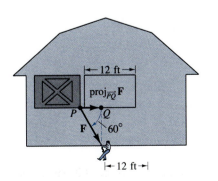

FIGURE 11.26

**DISCUSSION
PROBLEM**
....

**THE SIGN OF
THE DOT
PRODUCT**

In this section, you were given the alternative form of the dot product of two vectors.

$$\mathbf{u} \cdot \mathbf{v} = \|\mathbf{u}\| \|\mathbf{v}\| \cos \theta$$

Use this form to determine the sign of the dot product of **u** and **v** for the given angle. Explain your reasoning.

(a)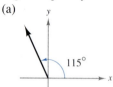

(b)

WARM-UP
....

The following warm-up exercises involve skills that were covered in earlier sections. You will use these skills in the exercise set for this section.

In Exercises 1–4, find (a) $\mathbf{u} + 2\mathbf{v}$, and (b) $\|\mathbf{u}\|$.

1. $\mathbf{u} = \langle 6, -3 \rangle$
    $\mathbf{v} = \langle -10, -1 \rangle$

2. $\mathbf{u} = \langle \frac{3}{8}, \frac{4}{5} \rangle$
    $\mathbf{v} = \langle \frac{5}{2}, -\frac{1}{10} \rangle$

3. $\mathbf{u} = 4\mathbf{i} - 16\mathbf{j}$
    $\mathbf{v} = -5\mathbf{i} + 10\mathbf{j}$

4. $\mathbf{u} = 0.5\mathbf{i} + 1.4\mathbf{j}$
    $\mathbf{v} = 4.1\mathbf{i} - 1.8\mathbf{j}$

In Exercises 5–8, find the values of θ in the interval $0 \le \theta < 2\pi$ that satisfy the equation. If the exact value is not known, round the result to two decimal places.

5. $\cos \theta = -\frac{1}{2}$

6. $\cos \theta = 0$

7. $\cos \theta = 0.5403$

8. $\cos \theta = -0.9689$

In Exercises 9 and 10, find a unit vector (a) in the direction of **u** and (b) in the direction opposite that of **u**.

9. $\mathbf{u} = \langle 120, -50 \rangle$

10. $\mathbf{u} = \langle \frac{4}{5}, \frac{1}{3} \rangle$

SECTION 11.2 · EXERCISES
....

In Exercises 1–6, find the dot product of **u** and **v**.

1. $\mathbf{u} = \langle 6, 2 \rangle$
    $\mathbf{v} = \langle 2, -3 \rangle$

2. $\mathbf{u} = \langle 4, 10 \rangle$
    $\mathbf{v} = \langle -3, 1 \rangle$

5. $\mathbf{u} = 4\mathbf{i} - 2\mathbf{j}$
    $\mathbf{v} = \mathbf{i} + 3\mathbf{j}$

6. $\mathbf{u} = 2\mathbf{i} + 5\mathbf{j}$
    $\mathbf{v} = 9\mathbf{i} - 3\mathbf{j}$

3. $\mathbf{u} = \langle 3, -3 \rangle$
    $\mathbf{v} = \langle 0, 5 \rangle$

4. $\mathbf{u} = \mathbf{j}$
    $\mathbf{v} = \mathbf{j}$

In Exercises 7–10, use the vectors $\mathbf{u} = \langle 2, 2 \rangle$ and $\mathbf{v} = \langle -3, 4 \rangle$ to find the indicated quantity. State whether the result is a vector or a scalar.

7. $\mathbf{u} \cdot \mathbf{u}$ **8.** $\|\mathbf{u}\|^2$

9. $(\mathbf{u} \cdot \mathbf{v}) \mathbf{v}$ **10.** $\mathbf{u} \cdot (2\mathbf{v})$

In Exercises 11 and 12, find $\mathbf{u} \cdot \mathbf{v}$, where θ is the angle between $\mathbf{u}$ and $\mathbf{v}$.

11. $\|\mathbf{u}\| = 4$, $\|\mathbf{v}\| = 10$, $\theta = \dfrac{2\pi}{3}$

12. $\|\mathbf{u}\| = 100$, $\|\mathbf{v}\| = 250$, $\theta = \dfrac{\pi}{6}$

13. *Revenue*   The vector $\mathbf{u} = \langle 1245, 2600 \rangle$ gives the number of units of two products produced by a company. The vector $\mathbf{v} = \langle 12.20, 8.50 \rangle$ gives the price (in dollars) of each unit, respectively. Find the dot product, $\mathbf{u} \cdot \mathbf{v}$, and explain what information it gives.

14. Repeat Exercise 13 after increasing the prices by 5%.

In Exercises 15–22, find the angle θ between the given vectors.

15. $\mathbf{u} = \langle 1, 0 \rangle$ **16.** $\mathbf{u} = \langle 4, 4 \rangle$
    $\mathbf{v} = \langle 0, -2 \rangle$ $\mathbf{v} = \langle 2, 0 \rangle$

17. $\mathbf{u} = \langle 3, 4 \rangle$ **18.** $\mathbf{u} = \langle -2, 5 \rangle$
    $\mathbf{v} = \langle 5, 3 \rangle$ $\mathbf{v} = \langle 1, 4 \rangle$

19. $\mathbf{u} = 2\mathbf{i} + 3\mathbf{j}$ **20.** $\mathbf{u} = \mathbf{i} - 6\mathbf{j}$
    $\mathbf{v} = -2\mathbf{i} + 2\mathbf{j}$ $\mathbf{v} = 2\mathbf{i} - 12\mathbf{j}$

21. $\mathbf{u} = \cos\left(\dfrac{\pi}{3}\right)\mathbf{i} + \sin\left(\dfrac{\pi}{3}\right)\mathbf{j}$

    $\mathbf{v} = \cos\left(\dfrac{3\pi}{4}\right)\mathbf{i} + \sin\left(\dfrac{3\pi}{4}\right)\mathbf{j}$

22. $\mathbf{u} = \cos\left(\dfrac{\pi}{4}\right)\mathbf{i} + \sin\left(\dfrac{\pi}{4}\right)\mathbf{j}$

    $\mathbf{v} = \cos\left(\dfrac{\pi}{2}\right)\mathbf{i} + \sin\left(\dfrac{\pi}{2}\right)\mathbf{j}$

In Exercises 23–26, use a graphing utility to sketch the vectors and find the degree measure of the angle between the vectors.

23. $\mathbf{u} = 3\mathbf{i} + 4\mathbf{j}$ **24.** $\mathbf{u} = -6\mathbf{i} - 3\mathbf{j}$
    $\mathbf{v} = -7\mathbf{i} + 5\mathbf{j}$ $\mathbf{v} = -8\mathbf{i} + 4\mathbf{j}$

25. $\mathbf{u} = 5\mathbf{i} + 5\mathbf{j}$ **26.** $\mathbf{u} = 2\mathbf{i} - 3\mathbf{j}$
    $\mathbf{v} = -6\mathbf{i} + 6\mathbf{j}$ $\mathbf{v} = 4\mathbf{i} + 3\mathbf{j}$

In Exercises 27–32, determine whether $\mathbf{u}$ and $\mathbf{v}$ are orthogonal, parallel, or neither.

27. $\mathbf{u} = \langle 4, 6 \rangle$ **28.** $\mathbf{u} = \langle 4, 14 \rangle$
    $\mathbf{v} = \langle 3, -2 \rangle$ $\mathbf{v} = \langle 6, 21 \rangle$

29. $\mathbf{u} = \langle -12, 30 \rangle$ **30.** $\mathbf{u} = \langle 15, 45 \rangle$
    $\mathbf{v} = \langle \frac{1}{2}, -\frac{5}{4} \rangle$ $\mathbf{v} = \langle -5, 12 \rangle$

31. $\mathbf{u} = -\frac{1}{4}(3\mathbf{i} - \mathbf{j})$ **32.** $\mathbf{u} = \cos 20° \, \mathbf{i} + \sin 20° \, \mathbf{j}$
    $\mathbf{v} = 5\mathbf{i} + 6\mathbf{j}$ $\mathbf{v} = \sin 70° \, \mathbf{i} + \cos 70° \, \mathbf{j}$

In Exercises 33–38, find the projection of $\mathbf{u}$ onto $\mathbf{v}$. Then find the vector component of $\mathbf{u}$ orthogonal to $\mathbf{v}$.

33. $\mathbf{u} = \langle 3, 4 \rangle$ **34.** $\mathbf{u} = \langle 4, -2 \rangle$
    $\mathbf{v} = \langle 8, 2 \rangle$ $\mathbf{v} = \langle 10, 2 \rangle$

35. $\mathbf{u} = \langle 4, 2 \rangle$ **36.** $\mathbf{u} = \langle 0, 3 \rangle$
    $\mathbf{v} = \langle 1, -2 \rangle$ $\mathbf{v} = \langle 2, 15 \rangle$

37. $\mathbf{u} = \langle 2, 1 \rangle$ **38.** $\mathbf{u} = \langle -5, -1 \rangle$
    $\mathbf{v} = \langle 0, 8 \rangle$ $\mathbf{v} = \langle -1, 1 \rangle$

39. *Braking Load*   A truck with a gross weight of 26,000 pounds is parked on a 10° slope (see figure). Assume the only force to overcome is that due to gravity.
    (a) Find the force required to keep the truck from rolling down the hill.
    (b) Find the force perpendicular to the hill.

40. *Work*   An object is pulled 20 feet across a floor using a force of 45 pounds. Find the work done if the direction of the force is 30° above the horizontal (see figure).

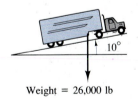

Weight = 26,000 lb

FIGURE FOR 39

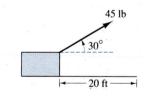

45 lb

30°

20 ft

FIGURE FOR 40

41. *Work*   A tractor pulls a log 2500 feet and the tension in the cable connecting the tractor and log is approximately 3600 pounds. Approximate the work done if the direction of the force is 35° above the horizontal.

42. Prove Property 2 of the dot product: $\mathbf{0} \cdot \mathbf{v} = 0$.

43. Prove Property 3 of the dot product: $\mathbf{u} \cdot (\mathbf{v} + \mathbf{w}) = \mathbf{u} \cdot \mathbf{v} + \mathbf{u} \cdot \mathbf{w}$.

44. Prove Property 5 of the dot product: $c(\mathbf{u} \cdot \mathbf{v}) = c\mathbf{u} \cdot \mathbf{v} = \mathbf{u} \cdot c\mathbf{v}$.

11.3 THE THREE-DIMENSIONAL COORDINATE SYSTEM

The Three-Dimensional Coordinate System / The Distance and Midpoint
Formulas / The Equation of a Sphere

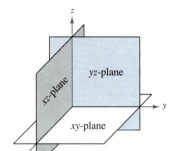

FIGURE 11.27

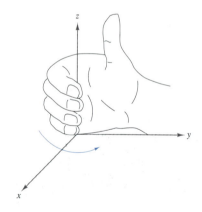

FIGURE 11.28

The Three-Dimensional Coordinate System

Recall that the Cartesian plane is determined by two perpendicular number lines called the x-axis and the y-axis. These axes together with their point of intersection (the origin) provide a two-dimensional coordinate system for identifying points in a plane. To identify a point in space, you must introduce a third dimension to the model. The geometry of this three-dimensional model is called **solid analytic geometry.**

You can construct a **three-dimensional coordinate system** by passing a z-axis perpendicular to both the x- and y-axes at the origin. Figure 11.27 shows the positive portion of each coordinate axis. Taken as pairs, the axes determine three **coordinate planes:** the **xy-plane,** the **xz-plane,** and the **yz-plane.** These three coordinate planes separate the three-dimensional coordinate system into eight **octants.** The first octant is the one for which all three coordinates are positive. In this three-dimensional system, a point P in space is determined by an ordered triple (x, y, z), where x, y, and z are as follows.

x = directed distance from yz-plane to P
y = directed distance from xz-plane to P
z = directed distance from xy-plane to P

A three-dimensional coordinate system can have either a **left-handed** or a **right-handed** orientation. In this text, we work exclusively with right-handed systems, as shown in Figure 11.28.

EXAMPLE 1   Plotting Points in Space

Plot the following points in space.

A. $(2, -3, 3)$ **B.** $(-2, 6, 2)$ **C.** $(1, 4, 0)$ **D.** $(2, 2, -3)$

SOLUTION

To plot the point $(2, -3, 3)$, notice that $x = 2$, $y = -3$, and $z = 3$. To help visualize the point, locate the point $(2, -3)$ in the xy-plane (denoted by a cross in Figure 11.29). The point $(2, -3, 3)$ lies 3 units above the cross. The other three points are also shown in Figure 11.29.

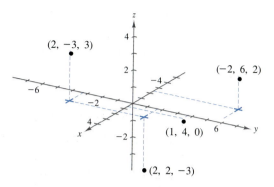

FIGURE 11.29

The Distance and Midpoint Formulas

Many of the formulas established for the two-dimensional coordinate system can be extended to three dimensions. For example, to find the distance between two points in space, you can use the Pythagorean Theorem twice, as shown in Figure 11.30. By doing this, you will obtain the formula for the distance between two points in space.

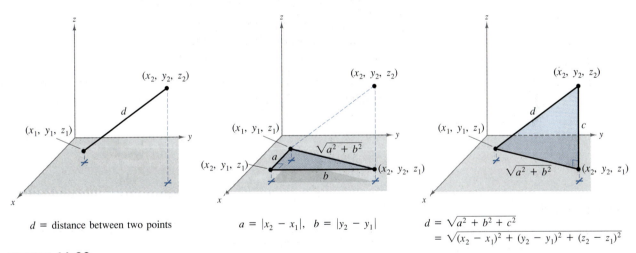

FIGURE 11.30

DISTANCE FORMULA IN SPACE

The distance between the points (x_1, y_1, z_1) and (x_2, y_2, z_2) is

$$d = \sqrt{(x_2 - x_1)^2 + (y_2 - y_1)^2 + (z_2 - z_1)^2}.$$

EXAMPLE 2   Finding the Distance Between Two Points

Find the distance between $(1, 0, 2)$ and $(2, 4, -3)$.

SOLUTION

$$
\begin{aligned}
d &= \sqrt{(x_2 - x_1)^2 + (y_2 - y_1)^2 + (z_2 - z_1)^2} \\
&= \sqrt{(2 - 1)^2 + (4 - 0)^2 + (-3 - 2)^2} \qquad \textit{Substitute} \\
&= \sqrt{1 + 16 + 25} \qquad \textit{Simplify} \\
&= \sqrt{42} \qquad \textit{Simplify}
\end{aligned}
$$

Notice the similarity between the Distance Formulas in the plane and in space, and the similarity between the standard equation of a circle and a sphere. The Midpoint Formulas in the plane and in space are also similar.

MIDPOINT FORMULA IN SPACE

The midpoint of the line segment joining the points (x_1, y_1, z_1) and (x_2, y_2, z_2) is

$$
\left(\frac{x_1 + x_2}{2}, \frac{y_1 + y_2}{2}, \frac{z_1 + z_2}{2} \right).
$$

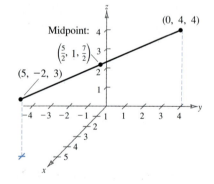

FIGURE 11.31

EXAMPLE 3   Using the Midpoint Formula

Find the midpoint of the line segment joining $(5, -2, 3)$ and $(0, 4, 4)$.

SOLUTION

Using the Midpoint Formula, the midpoint is

$$
\left(\frac{5 + 0}{2}, \frac{-2 + 4}{2}, \frac{3 + 4}{2} \right) = \left(\frac{5}{2}, 1, \frac{7}{2} \right),
$$

as shown in Figure 11.31.

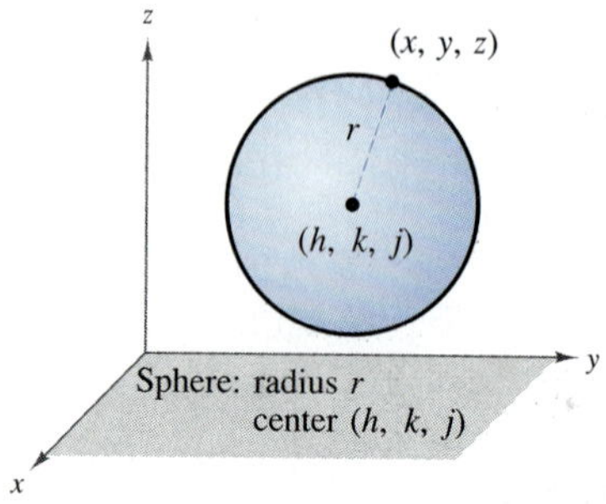

The Equation of a Sphere

A **sphere** with center at (h, k, j) and radius r is defined to be the set of all points (x, y, z) such that the distance between (x, y, z) and (h, k, j) is r, as shown in Figure 11.32. Using the Distance Formula, this condition can be written as

$$
\sqrt{(x - h)^2 + (y - k)^2 + (z - j)^2} = r.
$$

By squaring both sides of this equation, you obtain the standard equation of a sphere.

FIGURE 11.32

The **standard equation of a sphere** whose center is (h, k, j) and whose radius is r is

$$(x - h)^2 + (y - k)^2 + (z - j)^2 = r^2.$$

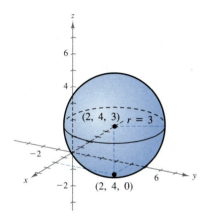

FIGURE 11.33

EXAMPLE 4   **Finding the Equation of a Sphere**

Find the standard equation for the sphere whose center is $(2, 4, 3)$ and whose radius is 3. Does this sphere intersect the xy-plane?

SOLUTION

$$\begin{aligned}(x - h)^2 + (y - k)^2 + (z - j)^2 &= r^2 \qquad \textit{Standard equation}\\ (x - 2)^2 + (y - 4)^2 + (z - 3)^2 &= 3^2 \qquad \textit{Substitute}\end{aligned}$$

From the graph shown in Figure 11.33, you can see that the center of the sphere lies 3 units above the xy-plane. Because the sphere has a radius of 3, you can conclude that it does intersect the xy-plane—at the point $(2, 4, 0)$.

EXAMPLE 5   **Finding the Equation of a Sphere**

Find the equation of the sphere that has the points $(3, -2, 6)$ and $(-1, 4, 2)$ as endpoints of a diameter.

SOLUTION

By the Midpoint Rule, the center of the sphere is

$$\left(\frac{3 - 1}{2}, \frac{-2 + 4}{2}, \frac{6 + 2}{2}\right) = (1, 1, 4).$$

By the Distance Formula, the radius is

$$r = \sqrt{(3 - 1)^2 + (-2 - 1)^2 + (6 - 4)^2} = \sqrt{17}.$$

Therefore, the standard equation of the sphere is

$$(x - 1)^2 + (y - 1)^2 + (z - 4)^2 = 17.$$

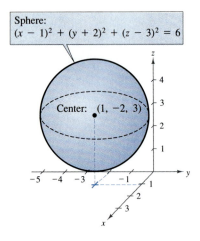

Sphere:
$(x - 1)^2 + (y + 2)^2 + (z - 3)^2 = 6$

Center: $(1, -2, 3)$

FIGURE 11.34

EXAMPLE 6 Finding the Center and Radius of a Sphere

Find the center and radius of the sphere whose equation is

$$x^2 + y^2 + z^2 - 2x + 4y - 6z + 8 = 0.$$

SOLUTION

You can obtain the standard equation of this sphere by completing the square, as follows.

$$x^2 + y^2 + z^2 - 2x + 4y - 6z + 8 = 0$$
$$(x^2 - 2x + \ ) + (y^2 + 4y + \ ) + (z^2 - 6z + \ ) = -8$$
$$(x^2 - 2x + 1) + (y^2 + 4y + 4) + (z^2 - 6z + 9) = -8 + 1 + 4 + 9$$
$$(x - 1)^2 + (y + 2)^2 + (z - 3)^2 = 6$$

Therefore, the center of the sphere is $(1, -2, 3)$, and its radius is $\sqrt{6}$, as shown in Figure 11.34.

Note in Example 6 that the points satisfying the equation of the sphere are "surface points," not "interior points." In general, the collection of points satisfying an equation involving x, y, and z is called a **surface in space.**

Finding the intersection of a surface with one of the three coordinate planes (or with a plane parallel to one of the three coordinate planes) helps visualize the surface. Such an intersection is called a **trace** of the surface. For example, the xy-trace of a surface consists of all points that are common to both the surface *and* the xy-plane. Similarly, the xz-trace of a surface consists of all points that are common to both the surface and the xz-plane.

EXAMPLE 7 Finding a Trace of a Surface

Sketch the xy-trace of the sphere whose equation is

$$(x - 3)^2 + (y - 2)^2 + (z + 4)^2 = 5^2.$$

SOLUTION

To find the xy-trace of this surface, use the fact that every point in the xy-plane has a z-coordinate of zero. This means that if you substitute $z = 0$ into the given equation, the resulting equation will represent the intersection of the surface with the xy-plane.

$$(x - 3)^2 + (y - 2)^2 + (0 + 4)^2 = 25$$
$$(x - 3)^2 + (y - 2)^2 + 16 = 25$$
$$(x - 3)^2 + (y - 2)^2 = 9$$
$$(x - 3)^2 + (y - 2)^2 = 3^2$$

From this form, you can see that the xy-trace is a circle of radius 3, as shown in Figure 11.35.

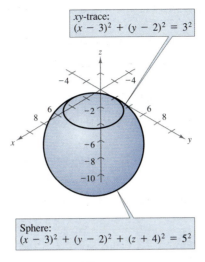

xy-trace:
$(x - 3)^2 + (y - 2)^2 = 3^2$

Sphere:
$(x - 3)^2 + (y - 2)^2 + (z + 4)^2 = 5^2$

FIGURE 11.35

DISCUSSION PROBLEM

COMPARING TWO AND THREE DIMENSIONS

In this section, you saw similarities between formulas in two-dimensional coordinate geometry and three-dimensional coordinate geometry. In two-dimensional coordinate geometry, the graph of the equation

$$ax + by + c = 0$$

is a line. In three-dimensional coordinate geometry, what is the graph of equation

$$ax + by + cz = 0?$$

Is it a line? Explain your reasoning.

WARM-UP

The following warm-up exercises involve skills that were covered in earlier sections. You will use these skills in the exercise set for this section.

In Exercises 1–6, find the distance between A and B and the midpoint of the line segment joining A and B.

1. $A(0, 0)$, $B(5, 12)$ **2.** $A(-4, 1)$, $B(3, 8)$

3. $A(1, 6)$, $B(6, -1)$ **4.** $A(0.5, 0.8)$, $B(-1.2, -2.4)$

5. $A\left(\frac{1}{2}, 1\right)$, $B\left(\frac{5}{2}, \frac{1}{2}\right)$ **6.** $A\left(\frac{4}{3}, \frac{3}{2}\right)$, $B\left(\frac{2}{3}, 1\right)$

In Exercises 7–10, find the standard equation of the circle satisfying the given conditions.

7. Center: $(4, -5)$; Radius: 4 **8.** Center: $(-1, 3)$; Radius: 1

9. Endpoints of a diameter: $(1, 4)$, $(5, 2)$ **10.** Endpoints of a diameter: $(-1, 0)$, $(0, 6)$

SECTION 11.3 · EXERCISES

In Exercises 1–4, plot the points on the same three-dimensional coordinate system.

1. (a) $(1, 2, 4)$
(b) $(1, -2, 1)$

2. (a) $(3, -1, 4)$
(b) $(2, 4, -3)$

3. (a) $(5, -3, 4)$
(b) $(5, -3, -4)$

4. (a) $(2, 0, 5)$
(b) $(0, 3, -5)$

In Exercises 5–8, find the lengths of the triangle with the indicated vertices, and determine whether the triangle is a right triangle, an isosceles triangle, or neither.

5. $(0, 0, 0)$
$(3, 3, 2)$
$(3, -6, 2)$

6. $(3, 1, 2)$
$(5, -1, 1)$
$(1, 3, 1)$

7. $(2, -1, 0)$
$(6, 0, 3)$
$(0, 2, 3)$

8. $(6, 0, 0)$
$(0, 3, 0)$
$(0, 0, -2)$

In Exercises 9–12, find the coordinates of the midpoint of the line segment joining the given points.

9. $(3, -6, 10)$, $(-3, 2, 2)$

10. $(2, -2, -8)$, $(5, 6, 18)$

11. $(4, -2, 5)$, $(-4, 2, 8)$

12. $(-3, 5, 7)$, $(-6, 4, 10)$

In Exercises 13–18, find the standard equation of the sphere.

13. Center: $(0, 4, 3)$; Radius: 4

14. Center: $(1, -2, 3)$; Radius: 5

15. Center: $(-3, 7, 5)$; Diameter: 10

16. Center: $(0, 5, -9)$; Diameter: 6

17. Endpoints of a diameter: $(3, 0, 0)$, $(0, 0, 6)$

18. Endpoints of a diameter: $(2, -2, 2)$, $(-1, 4, 6)$

In Exercises 19–24, find the center and radius of the sphere.

19. $x^2 + y^2 + z^2 - 4x + 2y - 6z + 10 = 0$

20. $x^2 + y^2 + z^2 - 6x + 4y + 9 = 0$

21. $x^2 + y^2 + z^2 + 4x - 8z + 19 = 0$

22. $x^2 + y^2 + z^2 - 8y - 6z + 13 = 0$

23. $9x^2 + 9y^2 + 9z^2 - 18x - 6y - 72z + 73 = 0$

24. $2x^2 + 2y^2 + 2z^2 - 2x - 6y - 4z + 5 = 0$

In Exercises 25 and 26, sketch the graph of the equation and sketch the specified traces.

25. $x^2 + y^2 + z^2 = 16$
(a) yz-trace   (b) xy-trace

26. $x^2 + y^2 + (z - 3)^2 = 9$
(a) yz-trace   (b) xz-trace

In Exercises 27–30, use a graphing utility to graph the sphere. (*Hint:* Solve for z and graph the two resulting expressions in x and y.)

27. $x^2 + y^2 + z^2 - 16 = 0$

28. $x^2 + y^2 + z^2 - 4y - 4 = 0$

29. $(x - 3)^2 + (y - 4)^2 + (z - 5)^2 = 4$

30. $x^2 + y^2 + z^2 + 6y - 8z + 21 = 0$

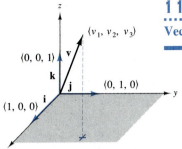

Standard Unit Vectors in Space

FIGURE 11.36

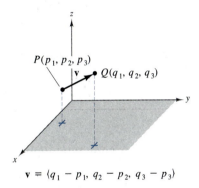

$$\mathbf{v} = \langle q_1 - p_1, q_2 - p_2, q_3 - p_3 \rangle$$

FIGURE 11.37

11.4 VECTORS IN SPACE

Vectors in Space / Parallel Vectors / Application

Vectors in Space

Physical forces and velocities are not confined to the plane, so it is natural to extend the concept of vectors from two-dimensional space to three-dimensional space. In space, vectors are denoted by ordered triples

$$\mathbf{v} = \langle v_1, v_2, v_3 \rangle. \qquad \textit{Component form}$$

The **zero vector** is denoted by $\mathbf{0} = \langle 0, 0, 0 \rangle$. Using the unit vectors $\mathbf{i} = \langle 1, 0, 0 \rangle$, $\mathbf{j} = \langle 0, 1, 0 \rangle$, and $\mathbf{k} = \langle 0, 0, 1 \rangle$ in the direction of the positive z-axis, the **standard unit vector notation** for $\mathbf{v}$ is

$$\mathbf{v} = v_1 \mathbf{i} + v_2 \mathbf{j} + v_3 \mathbf{k}, \qquad \textit{Unit vector form}$$

as shown in Figure 11.36. If $\mathbf{v}$ is represented by the directed line segment from $P(p_1, p_2, p_3)$ to $Q(q_1, q_2, q_3)$, as shown in Figure 11.37, the component form of $\mathbf{v}$ is given by subtracting the coordinates of the initial point from the coordinates of the terminal point, as follows.

$$\mathbf{v} = \langle v_1, v_2, v_3 \rangle = \langle q_1 - p_1, q_2 - p_2, q_3 - p_3 \rangle$$

VECTORS IN SPACE

1. *Equality of vectors:* Two vectors are equal if and only if their corresponding components are equal.
2. *Length of a vector:* The length of $\mathbf{u} = \langle u_1, u_2, u_3 \rangle$ is

$$\|\mathbf{u}\| = \sqrt{u_1{}^2 + u_2{}^2 + u_3{}^2}.$$

3. *Unit vector:* A unit vector $\mathbf{v}$ in the direction of $\mathbf{u}$ is given by

$$\mathbf{v} = \frac{\mathbf{u}}{\|\mathbf{u}\|}, \qquad \mathbf{u} \neq \mathbf{0}.$$

4. *Vector addition:* The sum of the vectors $\mathbf{u} = \langle u_1, u_2, u_3 \rangle$ and $\mathbf{v} = \langle v_1, v_2, v_3 \rangle$ is

$$\mathbf{v} + \mathbf{u} = \langle v_1 + u_1, v_2 + u_2, v_3 + u_3 \rangle.$$

5. *Scalar multiplication:* The scalar multiple of the real number c and the vector $\mathbf{u} = \langle u_1, u_2, u_3 \rangle$ is

$$c \mathbf{u} = \langle cu_1, cu_2, cu_3 \rangle.$$

6. *Dot product:* The dot product of the vectors $\mathbf{u} = \langle u_1, u_2, u_3 \rangle$ and $\mathbf{v} = \langle v_1, v_2, v_3 \rangle$ is

$$\mathbf{u} \cdot \mathbf{v} = u_1 v_1 + u_2 v_2 + u_3 v_3.$$

REMARK Note how similar these definitions are to the corresponding definitions for vectors in the plane. The properties of vector operations discussed in Section 11.1 and 11.2 are also valid for vectors in space.

EXAMPLE 1   Finding the Component Form of a Vector

Find the component form and length of the vector **v** having initial point (3, 4, 2) and terminal point (3, 6, 4). Then find a unit vector in the direction of **v**.

SOLUTION

The component form of **v** is

$$\mathbf{v} = \langle 3 - 3, 6 - 4, 4 - 2 \rangle = \langle 0, 2, 2 \rangle,$$

which implies that its length is

$$\|\mathbf{v}\| = \sqrt{0^2 + 2^2 + 2^2} = \sqrt{8} = 2\sqrt{2}.$$

The unit vector in the direction of **v** is

$$\mathbf{u} = \frac{\mathbf{v}}{\|\mathbf{v}\|} = \frac{1}{2\sqrt{2}} \langle 0, 2, 2 \rangle = \left\langle 0, \frac{1}{\sqrt{2}}, \frac{1}{\sqrt{2}} \right\rangle.$$

Technology Note _____

Some graphing utilities have the capability to perform vector operations, such as the dot product. For example, on the TI-85, you can use the VECTR menu to verify the dot product in Example 2 as follows.

dot ([0, 3, −2], [4, −2, 3])

EXAMPLE 2   Finding the Dot Product of Two Vectors

Find the following dot product.

$$\langle 0, 3, -2 \rangle \cdot \langle 4, -2, 3 \rangle$$

SOLUTION

$$\begin{aligned}
\langle 0, 3, -2 \rangle \cdot \langle 4, -2, 3 \rangle &= 0(4) + 3(-2) + (-2)(3) \\
&= 0 - 6 - 6 \\
&= -12
\end{aligned}$$

EXAMPLE 3   Standard Unit Vector Notation

Write the vector $\mathbf{v} = 2\mathbf{j} - 6\mathbf{k}$ in component form.

SOLUTION

Because **i** is missing, its component is 0 and

$$\mathbf{v} = 2\mathbf{j} - 6\mathbf{k} = \langle 0, 2, -6 \rangle.$$

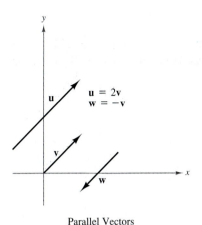

y

$\mathbf{u} = 2\mathbf{v}$
$\mathbf{w} = -\mathbf{v}$

$\mathbf{u}$

$\mathbf{v}$

$\mathbf{w}$

x

Parallel Vectors

FIGURE 11.38

Parallel Vectors

Recall from the definition of scalar multiplication that positive scalar multiples of a nonzero vector **v** have the same direction as **v**, whereas negative multiples have the direction opposite that of **v**. In general, two nonzero vectors **u** and **v** are **parallel** if there is some scalar c such that $\mathbf{u} = c\,\mathbf{v}$. For example, in Figure 11.38, the vectors **u**, **v**, and **w** are parallel because $\mathbf{u} = 2\mathbf{v}$ and $\mathbf{w} = -\mathbf{v}$.

EXAMPLE 4   Parallel Vectors

Vector **w** has initial point $(1, -2, 0)$ and terminal point $(3, 2, 1)$. Which of the following vectors is parallel to **w**?

A. $\mathbf{u} = \langle 4, 8, 2 \rangle$ **B.** $\mathbf{v} = \langle 4, 8, 4 \rangle$

SOLUTION

Begin by writing **w** in component form.

$$\mathbf{w} = \langle 3 - 1, 2 + 2, 1 - 0 \rangle = \langle 2, 4, 1 \rangle$$

A. Because

$$\begin{aligned}
\mathbf{u} &= \langle 4, 8, 2 \rangle \\
&= 2\langle 2, 4, 1 \rangle \\
&= 2\mathbf{w},
\end{aligned}$$

you can conclude that **u** *is* parallel to **w**.

B. In this case, you need to find a scalar c such that

$$\langle 4, 8, 4 \rangle = c\langle 2, 4, 1 \rangle.$$

However, equating corresponding components produces $c = 2$ for the first two components and $c = 4$ for the third. Hence, the equation has no solution, and the vectors are *not* parallel.

You can use vectors to determine whether three points are collinear (lie on the same line). The points P, Q, and R are collinear if and only if the vectors $\overrightarrow{PQ}$ and $\overrightarrow{PR}$ are parallel.

EXAMPLE 5   Using Vectors to Determine Collinear Points

Determine whether the following points lie on the same line.

$$P(2, -1, 4), \quad Q(5, 4, 6), \quad \text{and} \quad R(-4, -11, 0)$$

SOLUTION

The component forms of $\overrightarrow{PQ}$ and $\overrightarrow{PR}$ are

$$\overrightarrow{PQ} = \langle 5 - 2, 4 + 1, 6 - 4 \rangle = \langle 3, 5, 2 \rangle$$

and

$$\overrightarrow{PR} = \langle -4 - 2, -11 + 1, 0 - 4 \rangle = \langle -6, -10, -4 \rangle.$$

Because $\overrightarrow{PR} = -2\overrightarrow{PQ}$, you can conclude that they are parallel. Therefore, the points P, Q, and R lie on the same line.

EXAMPLE 6   **Finding the Terminal Point of a Vector**

The initial point of the vector $\mathbf{v} = 4\mathbf{i} + 2\mathbf{j} - \mathbf{k}$ is $P(3, -1, 6)$. What is the terminal point of this vector?

SOLUTION

Using the component form of the vector whose initial point is P and whose terminal point is Q, you can write

$$\begin{aligned}
\overrightarrow{PQ} &= \langle q_1 - p_1, q_2 - p_2, q_3 - p_3 \rangle \\
&= \langle q_1 - 3, q_2 + 1, q_3 - 6 \rangle \\
&= \langle 4, 2, -1 \rangle.
\end{aligned}$$

This implies that

$$q_1 - 3 = 4, \quad q_2 + 1 = 2, \quad \text{and} \quad q_3 - 6 = -1.$$

The solutions of these three equations are $q_1 = 7$, $q_2 = 1$, and $q_3 = 5$. Thus, the terminal point is $Q(7, 1, 5)$.

Application

In Section 11.1, you saw how to use vectors to solve an equilibrium problem in a plane. The next example shows how to use vectors to solve an equilibrium problem in space.

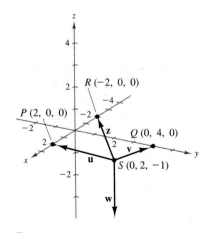

FIGURE 11.39

EXAMPLE 7   **Solving an Equilibrium Problem**

A weight of 480 pounds is supported by three ropes. As shown in Figure 11.39, the weight is located at $S(0, 2, -1)$. The ropes are tied to the points $P(2, 0, 0)$, $Q(0, 4, 0)$, and $R(-2, 0, 0)$. Find the force (or tension) on each rope.

SOLUTION

The (downward) force of the weight is represented by the vector $\mathbf{w} = \langle 0, 0, -480 \rangle$. The force vectors corresponding to the ropes are as follows.

$$\mathbf{u} = \|\mathbf{u}\| \frac{\overrightarrow{SP}}{\|\overrightarrow{SP}\|} = \|\mathbf{u}\| \frac{\langle 2 - 0, 0 - 2, 0 + 1 \rangle}{3} = \|\mathbf{u}\| \left\langle \frac{2}{3}, -\frac{2}{3}, \frac{1}{3} \right\rangle$$

$$\mathbf{v} = \|\mathbf{v}\| \frac{\overrightarrow{SQ}}{\|\overrightarrow{SQ}\|} = \|\mathbf{v}\| \frac{\langle 0 - 0, 4 - 2, 0 + 1 \rangle}{\sqrt{5}} = \|\mathbf{v}\| \left\langle 0, \frac{2}{\sqrt{5}}, \frac{1}{\sqrt{5}} \right\rangle$$

$$\mathbf{z} = \|\mathbf{z}\| \frac{\overrightarrow{SR}}{\|\overrightarrow{SR}\|} = \|\mathbf{z}\| \frac{\langle -2 - 0, 0 - 2, 0 + 1 \rangle}{3} = \|\mathbf{z}\| \left\langle -\frac{2}{3}, -\frac{2}{3}, \frac{1}{3} \right\rangle$$

For the system to be in equilibrium, it must be true that

$$\mathbf{u} + \mathbf{v} + \mathbf{z} + \mathbf{w} = \mathbf{0} \quad \text{or} \quad \mathbf{u} + \mathbf{v} + \mathbf{z} = -\mathbf{w}.$$

This yields the following system of linear equations.

$$\frac{2}{3}\|\mathbf{u}\| \qquad\qquad - \frac{2}{3}\|\mathbf{z}\| = 0$$

$$-\frac{2}{3}\|\mathbf{u}\| + \frac{2}{\sqrt{5}}\|\mathbf{v}\| - \frac{2}{3}\|\mathbf{z}\| = 0$$

$$\frac{1}{3}\|\mathbf{u}\| + \frac{1}{\sqrt{5}}\|\mathbf{v}\| + \frac{1}{3}\|\mathbf{z}\| = 480$$

The solution of this system is

$$\|\mathbf{u}\| \approx 360.0, \quad \|\mathbf{v}\| = 536.7, \quad \text{and} \quad \|\mathbf{z}\| \approx 360.0.$$

Thus, the rope attached at point P has about 360 pounds of tension, the rope attached at point Q has about 536.7 pounds of tension, and the rope attached at point R has about 360 pounds of tension.

DISCUSSION PROBLEM

PROPERTIES OF VECTORS IN SPACE

The properties of vectors in the plane that you studied earlier in the text also apply to vectors in space. Choose one of the following properties. Then show how the properties of real numbers can be used to prove the property for vectors in space.

1. $\mathbf{u} + \mathbf{v} = \mathbf{v} + \mathbf{u}$
2. $(\mathbf{u} + \mathbf{v}) + \mathbf{w} = \mathbf{u} + (\mathbf{v} + \mathbf{w})$
3. $\mathbf{u} + \mathbf{0} = \mathbf{u}$
4. $\mathbf{u} + (-\mathbf{u}) = \mathbf{0}$
5. $c(d\mathbf{u}) = (cd)\mathbf{u}$
6. $(c + d)\mathbf{u} = c\mathbf{u} + d\mathbf{u}$
7. $c(\mathbf{u} + \mathbf{v}) = c\mathbf{u} + c\mathbf{v}$
8. $1(\mathbf{u}) = \mathbf{u}$
9. $0(\mathbf{u}) = \mathbf{0}$
10. $\|c\mathbf{u}\| = |c|\,\|\mathbf{u}\|$

WARM-UP

The following warm-up exercises involve skills that were covered in earlier sections. You will use these skills in the exercise set for this section.

In Exercises 1–4, find the component forms of the vectors $\overrightarrow{AB}$, $\overrightarrow{AC}$, and $\overrightarrow{AB} + \overrightarrow{AC}$. Sketch the three vectors.

1. $A(0, 0)$
   $B(1, -2)$
   $C(2, 4)$

2. $A(0, 0)$
   $B(-4, -2)$
   $C(-2, 3)$

3. $A(6, 4)$
   $B(0, -1)$
   $C(-2, 4)$

4. $A(-3, -5)$
   $B(4, -1)$
   $C(1, 3)$

In Exercises 5 and 6, find a unit vector in the direction of **u**.

5. $\mathbf{u} = \langle 5, -12 \rangle$

6. $\mathbf{u} = \langle 3 \cos 38°, 3 \sin 38° \rangle$

In Exercises 7–10, find (a) $\mathbf{u} \cdot \mathbf{v}$ and (b) the angle between **u** and **v**.

7. $\mathbf{u} = \langle 4, 0 \rangle$
   $\mathbf{v} = \langle 2, -2 \rangle$

8. $\mathbf{u} = \langle 3, 4 \rangle$
   $\mathbf{v} = \langle -3, 4 \rangle$

9. $\mathbf{u} = \langle 3, 4 \rangle$
   $\mathbf{v} = \langle 4, -3 \rangle$

10. $\mathbf{u} = \langle -2, 8 \rangle$
   $\mathbf{v} = \langle 3, -12 \rangle$

SECTION 11.4 · EXERCISES

In Exercises 1–4, (a) find the component form of the vector **v** and (b) sketch the vector with its initial point at the origin.

1.

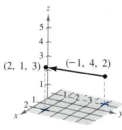

2.

3.

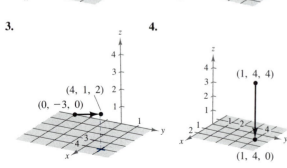

4.

In Exercises 5 and 6, write the component form of **v**. Then write the component forms of (a) a vector parallel to **v** and (b) a vector in the opposite direction of **v**. (There are many correct answers.)

5. Initial point: $(-1, -2, 1)$
   Terminal point: $(3, 2, 5)$

6. Initial point: $(-4, 5, 5)$
   Terminal point: $(4, 0, 0)$

In Exercises 7–12, find the vector **z**, given $\mathbf{u} = \langle -1, 3, 2 \rangle$, $\mathbf{v} = \langle 1, -2, -2 \rangle$, and $\mathbf{w} = \langle 5, 0, -5 \rangle$.

7. $\mathbf{z} = \mathbf{u} - 2\mathbf{v}$

8. $\mathbf{z} = 3\mathbf{u} - \mathbf{v} + \mathbf{w}$

9. $\mathbf{z} = 2\mathbf{u} + 8\mathbf{v} - \mathbf{w}$

10. $\mathbf{z} = -7\mathbf{u} + \mathbf{v} - \frac{1}{5}\mathbf{w}$

11. $2\mathbf{z} - 4\mathbf{u} = \mathbf{w}$

12. $\mathbf{u} + \mathbf{v} - 2\mathbf{w} + \mathbf{z} = 0$

In Exercises 13 and 14, find the length of **v**.

13. $\mathbf{v} = \langle 4, 1, 4 \rangle$

14. $\mathbf{v} = 4\mathbf{i} - 3\mathbf{j} - 7\mathbf{k}$

In Exercises 15 and 16, find a unit vector in the direction of **u**.

15. $\mathbf{u} = 8\mathbf{i} + 3\mathbf{j} - \mathbf{k}$

16. $\mathbf{u} = -3\mathbf{i} + 5\mathbf{j} + 10\mathbf{k}$

In Exercises 17–20, find the dot product of **u** and **v**.

17. $\mathbf{u} = \langle 3, -3, 5 \rangle$
$\mathbf{v} = \langle 0, 5, 3 \rangle$

18. $\mathbf{u} = \langle 4, 4, -1 \rangle$
$\mathbf{v} = \langle 2, -5, -8 \rangle$

19. $\mathbf{u} = 4\mathbf{i} - 2\mathbf{j} + \mathbf{k}$
$\mathbf{v} = \mathbf{i} + 3\mathbf{j} - \mathbf{k}$

20. $\mathbf{u} = 2\mathbf{i} + 5\mathbf{j} - 3\mathbf{k}$
$\mathbf{v} = 9\mathbf{i} - 3\mathbf{j} + \mathbf{k}$

In Exercises 21–24, find the angle θ between the given vectors.

21. $\mathbf{u} = \langle 0, 2, 2 \rangle$
$\mathbf{v} = \langle 3, 0, -4 \rangle$

22. $\mathbf{u} = \langle 4, 1, 2 \rangle$
$\mathbf{v} = \langle 2, -4, 1 \rangle$

23. $\mathbf{u} = 10\mathbf{i} + 40\mathbf{j}$
$\mathbf{v} = -3\mathbf{j} + 8\mathbf{k}$

24. $\mathbf{u} = \mathbf{i} - 6\mathbf{j} + 2\mathbf{k}$
$\mathbf{v} = 2\mathbf{i} - 4\mathbf{j} - 3\mathbf{k}$

In Exercises 25–28, determine whether **u** and **v** are orthogonal, parallel, or neither.

25. $\mathbf{u} = \langle -12, 6, 15 \rangle$
$\mathbf{v} = \langle 8, -4, -10 \rangle$

26. $\mathbf{u} = \langle 6, -3, 3 \rangle$
$\mathbf{v} = \langle 1, 5, 3 \rangle$

27. $\mathbf{v} = \frac{3}{4}\mathbf{i} - \frac{1}{2}\mathbf{j} + 2\mathbf{k}$
$\mathbf{v} = 4\mathbf{i} + 10\mathbf{j} + \mathbf{k}$

28. $\mathbf{u} = -7\mathbf{i} - 14\mathbf{j} + 21\mathbf{k}$
$\mathbf{v} = \mathbf{i} + 2\mathbf{j} - \mathbf{k}$

In Exercises 29 and 30, find the projection $\mathbf{w}_1$ of **u** onto **v**. Then write **u** as the sum of $\mathbf{w}_1$ and $\mathbf{w}_2$, where $\mathbf{w}_2$ is orthogonal to $\mathbf{w}_1$.

29. $\mathbf{u} = \langle 2, 1, 0 \rangle$
$\mathbf{v} = \langle 0, 8, 3 \rangle$

30. $\mathbf{u} = \langle -5, -1, 2 \rangle$
$\mathbf{v} = \langle -1, 1, 1 \rangle$

In Exercises 31–34, use vectors to determine whether the given points lie in a straight line.

31. $(1, 3, 2)$
$(-1, 2, 5)$
$(3, 4, -1)$

32. $(0, 4, 4)$
$(-1, 5, 6)$
$(-2, 6, 7)$

33. $(5, 4, 1)$
$(7, 3, -1)$
$(4, 5, 3)$

34. $(-2, 7, 4)$
$(\,4, 8, 1)$
$(0, 6, 7)$

In Exercises 35 and 36, the vector **v** and its initial point are given. Find the terminal point.

35. $\mathbf{v} = \langle 2, -4, 7 \rangle$
Initial point: $(1, 5, 0)$

36. $\mathbf{v} = \left\langle 4, \frac{3}{2}, -\frac{1}{4} \right\rangle$
Initial point: $\left(-2, 1, -\frac{3}{2} \right)$

In Exercises 37 and 38, write the component form of **v**.

37. $\|\mathbf{v}\| = 4$, **v** lies in the yz-plane and makes an angle of $45°$ with the positive y-axis.

38. $\|\mathbf{v}\| = 10$, **v** lies in the xz-plane and makes an angle of $60°$ with the positive z-axis.

39. The initial and terminal points of the vector **v** are (x_1, y_1, z_1) and (x, y, z), respectively. Describe the set of all points (x, y, z) such that $\|\mathbf{v}\| = 9$.

40. *Light Installation* The lights in an auditorium are 30-pound disks with a radius of 24 inches. Each disk is supported by three equally spaced 60-inch wires from the ceiling (see figure). Find the tension in each wire.

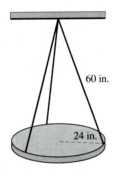

60 in.

24 in.

FIGURE FOR 40

Work In Exercises 41 and 42, find the work done in moving a particle from P to Q if the force is given by **F**.

41. $P(0, 0, 0)$, $Q(10, 5, 4)$, $\mathbf{F} = \langle 3, 2, 7 \rangle$

42. $P(2, 4, 0)$, $Q(-4, 4, 10)$, $\mathbf{F} = -6\mathbf{i} + 2\mathbf{j} + 6\mathbf{k}$

11.5 THE CROSS PRODUCT OF TWO VECTORS
....

The Cross Product / Geometric Properties of the Cross Product / The Triple Scalar Product

The Cross Product

Many applications in physics, engineering, and geometry involve finding a vector in space that is orthogonal to two given vectors. In this section you will study a product that will yield such a vector. It is called the **cross product,** and it is most conveniently defined and calculated using the standard unit vector form.

DEFINITION OF CROSS PRODUCT OF TWO VECTORS IN SPACE

Let $\mathbf{u} = u_1\,\mathbf{i} + u_2\,\mathbf{j} + u_3\,\mathbf{k}$ and $\mathbf{v} = v_1\,\mathbf{i} + v_2\,\mathbf{j} + v_3\,\mathbf{k}$ be vectors in space. The **cross product** of $\mathbf{u}$ and $\mathbf{v}$ is the vector

$$\mathbf{u} \times \mathbf{v} = (u_2 v_3 - u_3 v_2)\,\mathbf{i} - (u_1 v_3 - u_3 v_1)\,\mathbf{j} + (u_1 v_2 - u_2 v_1)\,\mathbf{k}.$$

REMARK   Be sure you see that this definition applies only to three-dimensional vectors. The cross product is not defined for two-dimensional vectors. ▪▪▪▪▪▪

A convenient way to calculate $\mathbf{u} \times \mathbf{v}$ is to use the following *determinant form* with cofactor expansion. (This 3×3 determinant form is used simply to help remember the formula for the cross product—it is technically not a determinant, because the entries of the corresponding matrix are not all real numbers.)

$$\mathbf{u} \times \mathbf{v} = \begin{vmatrix} \mathbf{i} & \mathbf{j} & \mathbf{k} \\ u_1 & u_2 & u_3 \\ v_1 & v_2 & v_3 \end{vmatrix} \quad \begin{matrix} \\ \leftarrow \text{Put } \mathbf{u} \text{ in Row 2.} \\ \leftarrow \text{Put } \mathbf{v} \text{ in Row 3.} \end{matrix}$$

$$= \begin{vmatrix} u_2 & u_3 \\ v_2 & v_3 \end{vmatrix} \mathbf{i} - \begin{vmatrix} u_1 & u_3 \\ v_1 & v_3 \end{vmatrix} \mathbf{j} + \begin{vmatrix} u_1 & u_2 \\ v_1 & v_2 \end{vmatrix} \mathbf{k}$$

$$= (u_2 v_3 - u_3 v_2)\,\mathbf{i} - (u_1 v_3 - u_3 v_1)\,\mathbf{j} + (u_1 v_2 - u_2 v_1)\,\mathbf{k}$$

Note the minus sign in front of the **j**-component.

Technology Note

Some graphing utilities have the capability to perform vector operations, such as the cross product. For example, on the TI-85, you can use the VECTR menu to verify the cross product in Example 1 as follows.

cross ([1, 2, 1], [3, 1, 2])

EXAMPLE 1   Finding Cross Products

Given $\mathbf{u} = \mathbf{i} + 2\mathbf{j} + \mathbf{k}$ and $\mathbf{v} = 3\mathbf{i} + \mathbf{j} + 2\mathbf{k}$, find the following.

A. $\mathbf{u} \times \mathbf{v}$ **B.** $\mathbf{v} \times \mathbf{u}$ **C.** $\mathbf{v} \times \mathbf{v}$

SOLUTION

A. $\mathbf{u} \times \mathbf{v} = \begin{vmatrix} \mathbf{i} & \mathbf{j} & \mathbf{k} \\ 1 & 2 & 1 \\ 3 & 1 & 2 \end{vmatrix} = \begin{vmatrix} 2 & 1 \\ 1 & 2 \end{vmatrix} \mathbf{i} - \begin{vmatrix} 1 & 1 \\ 3 & 2 \end{vmatrix} \mathbf{j} + \begin{vmatrix} 1 & 2 \\ 3 & 1 \end{vmatrix} \mathbf{k}$

$= (4 - 1)\,\mathbf{i} - (2 - 3)\,\mathbf{j} + (1 - 6)\,\mathbf{k}$

$= 3\mathbf{i} + \mathbf{j} - 5\mathbf{k}$

B. $\mathbf{v} \times \mathbf{u} = \begin{vmatrix} \mathbf{i} & \mathbf{j} & \mathbf{k} \\ 3 & 1 & 2 \\ 1 & 2 & 1 \end{vmatrix} = \begin{vmatrix} 1 & 2 \\ 2 & 1 \end{vmatrix} \mathbf{i} - \begin{vmatrix} 3 & 2 \\ 1 & 1 \end{vmatrix} \mathbf{j} + \begin{vmatrix} 3 & 1 \\ 1 & 2 \end{vmatrix} \mathbf{k}$

$= (1 - 4)\,\mathbf{i} - (3 - 2)\,\mathbf{j} + (6 - 1)\,\mathbf{k}$

$= -3\mathbf{i} - \mathbf{j} + 5\mathbf{k}$

Note that this result is the negative of that in part **a**.

C. $\mathbf{v} \times \mathbf{v} = \begin{vmatrix} \mathbf{i} & \mathbf{j} & \mathbf{k} \\ 3 & 1 & 2 \\ 3 & 1 & 2 \end{vmatrix} = \mathbf{0}$

The results obtained in Example 1 suggest some interesting *algebraic* properties of the cross product. For instance

$$\mathbf{u} \times \mathbf{v} = -(\mathbf{v} \times \mathbf{u}) \quad \text{and} \quad \mathbf{v} \times \mathbf{v} = \mathbf{0}.$$

These properties, and several others, are summarized in the following list.

ALGEBRAIC PROPERTIES OF THE CROSS PRODUCT

Let $\mathbf{u}$, $\mathbf{v}$, and $\mathbf{w}$ be vectors in space and let c be a scalar.

1. $\mathbf{u} \times \mathbf{v} = -(\mathbf{v} \times \mathbf{u})$
2. $\mathbf{u} \times (\mathbf{v} + \mathbf{w}) = (\mathbf{u} \times \mathbf{v}) + (\mathbf{u} \times \mathbf{w})$
3. $c(\mathbf{u} \times \mathbf{v}) = (c\mathbf{u}) \times \mathbf{v} = \mathbf{u} \times (c\mathbf{v})$
4. $\mathbf{u} \times \mathbf{0} = \mathbf{0} \times \mathbf{u} = \mathbf{0}$
5. $\mathbf{u} \times \mathbf{u} = \mathbf{0}$
6. $\mathbf{u} \cdot (\mathbf{v} \times \mathbf{w}) = (\mathbf{u} \times \mathbf{v}) \cdot \mathbf{w}$

Geometric Properties of the Cross Product

The first property listed on the previous page indicates that the cross product is *not commutative*. In particular, this property indicates that the vectors $\mathbf{u} \times \mathbf{v}$ and $\mathbf{v} \times \mathbf{u}$ have equal lengths but opposite directions. The following list gives some other *geometric* properties of the cross product of two vectors.

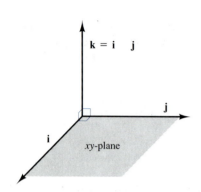

$\mathbf{k} = \mathbf{i} \quad \mathbf{j}$

$\mathbf{j}$

$\mathbf{i}$ *xy*-plane

Right-handed Systems

FIGURE 11.40

GEOMETRIC PROPERTIES OF THE CROSS PRODUCT

Let $\mathbf{u}$ and $\mathbf{v}$ be nonzero vectors in space, and let θ be the angle between $\mathbf{u}$ and $\mathbf{v}$.

1. $\mathbf{u} \times \mathbf{v}$ is orthogonal to both $\mathbf{u}$ and $\mathbf{v}$.
2. $\|\mathbf{u} \times \mathbf{v}\| = \|\mathbf{u}\| \, \|\mathbf{v}\| \sin \theta$.
3. $\mathbf{u} \times \mathbf{v} = \mathbf{0}$ if and only if $\mathbf{u}$ and $\mathbf{v}$ are scalar multiples.
4. $\|\mathbf{u} \times \mathbf{v}\|$ = area of parallelogram having $\mathbf{u}$ and $\mathbf{v}$ as sides.

Both $\mathbf{u} \times \mathbf{v}$ and $\mathbf{v} \times \mathbf{u}$ are perpendicular to the plane determined by $\mathbf{u}$ and $\mathbf{v}$. One way to remember the orientation of the vectors $\mathbf{u}$, $\mathbf{v}$, and $\mathbf{u} \times \mathbf{v}$ is to compare them with the unit vectors $\mathbf{i}$, $\mathbf{j}$, and $\mathbf{k} = \mathbf{i} \times \mathbf{j}$, as shown in Figure 11.40. The three vectors $\mathbf{u}$, $\mathbf{v}$, and $\mathbf{u} \times \mathbf{v}$ form a *right-handed system*.

EXAMPLE 2   Using the Cross Product

Find a unit vector that is orthogonal to both

$$\mathbf{u} = 3\mathbf{i} - 4\mathbf{j} + \mathbf{k} \quad \text{and} \quad \mathbf{v} = -3\mathbf{i} + 6\mathbf{j}.$$

SOLUTION

The cross product $\mathbf{u} \times \mathbf{v}$, as shown in Figure 11.41, is orthogonal to both $\mathbf{u}$ and $\mathbf{v}$.

$$\mathbf{u} \times \mathbf{v} = \begin{vmatrix} \mathbf{i} & \mathbf{j} & \mathbf{k} \\ 3 & -4 & 1 \\ -3 & 6 & 0 \end{vmatrix} = -6\mathbf{i} - 3\mathbf{j} + 6\mathbf{k}$$

Because

$$\|\mathbf{u} \times \mathbf{v}\| = \sqrt{(-6)^2 + (-3)^2 + 6^2} = \sqrt{81} = 9,$$

a unit vector orthogonal to both $\mathbf{u}$ and $\mathbf{v}$ is

$$\frac{\mathbf{u} \times \mathbf{v}}{\|\mathbf{u} \times \mathbf{v}\|} = -\frac{2}{3}\mathbf{i} - \frac{1}{3}\mathbf{j} + \frac{2}{3}\mathbf{k}.$$

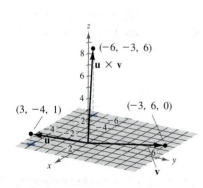

FIGURE 11.41

In Example 2, note that you could have used the cross product $\mathbf{v} \times \mathbf{u}$ to form a unit vector that is orthogonal to both $\mathbf{u}$ and $\mathbf{v}$. With that choice, you would have obtained the negative of the unit vector found in the example.

The fourth geometric property of cross products states that $\|\mathbf{u} \times \mathbf{v}\|$ is the area of the parallelogram that has $\mathbf{u}$ and $\mathbf{v}$ as adjacent sides. A simple example of this is given by the unit square with adjacent sides of $\mathbf{i}$ and $\mathbf{j}$. Because

$$\mathbf{i} \times \mathbf{j} = \mathbf{k}$$

and $\|\mathbf{k}\| = 1$, it follows that the square has an area of 1. This geometric property of the cross product is illustrated further in the next example.

EXAMPLE 3 Geometric Application of the Cross Product

Show that the quadrilateral with vertices at the following points is a parallelogram. Then find the area of the parallelogram.

$$A = (5, 2, 0) \qquad B = (2, 6, 1)$$
$$C = (2, 4, 7) \qquad D = (5, 0, 6)$$

SOLUTION

From Figure 11.42 you can see that the sides of the quadrilateral correspond to the following four vectors.

$$\vec{AB} = -3\mathbf{i} + 4\mathbf{j} + \mathbf{k}$$
$$\vec{CD} = 3\mathbf{i} - 4\mathbf{j} - \mathbf{k} = -\vec{AB}$$
$$\vec{AD} = 0\mathbf{i} - 2\mathbf{j} + 6\mathbf{k}$$
$$\vec{CB} = 0\mathbf{i} + 2\mathbf{j} - 6\mathbf{k} = -\vec{AD}$$

Because $\vec{AB}$ is parallel to $\vec{CD}$ and $\vec{AD}$ is parallel to $\vec{CB}$, it follows that the quadrilateral is a parallelogram with $\vec{AB}$ and $\vec{AD}$ as adjacent sides. Moreover, because

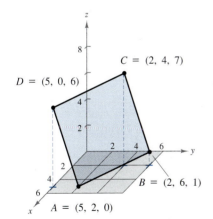

FIGURE 11.42

$$\vec{AB} \times \vec{AD} = \begin{vmatrix} \mathbf{i} & \mathbf{j} & \mathbf{k} \\ -3 & 4 & 1 \\ 0 & -2 & 6 \end{vmatrix} = 26\mathbf{i} + 18\mathbf{j} + 6\mathbf{k},$$

the area of the parallelogram is

$$\|\vec{AB} \times \vec{AD}\| = \sqrt{1036} \approx 32.19.$$

Is the parallelogram a rectangle? You can tell whether it is by finding the angle between the vectors $\vec{AB}$ and $\vec{AD}$.

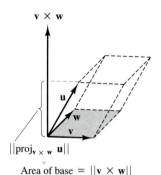

v × w

u

w

v

$\|\text{proj}_{\mathbf{v} \times \mathbf{w}} \mathbf{u}\|$

Area of base $= \|\mathbf{v} \times \mathbf{w}\|$
Volume of
parallelepiped $= |\mathbf{u} \cdot (\mathbf{v} \times \mathbf{w})|$

FIGURE 11.43

The Triple Scalar Product

For vectors **u**, **v**, and **w** in space, the dot product of **u** and **v** × **w** is called the **triple scalar product** of **u**, **v**, and **w**.

THE TRIPLE SCALAR PRODUCT

The **triple scalar product** of **u**, **v**, and **w** is given by

$$\mathbf{u} \cdot (\mathbf{v} \times \mathbf{w}) = \begin{vmatrix} u_1 & u_2 & u_3 \\ v_1 & v_2 & v_3 \\ w_1 & w_2 & w_3 \end{vmatrix}.$$

If the vectors **u**, **v**, and **w** do not lie in the same plane, then the triple scalar product **u** · (**v** × **w**) can be used to determine the volume of the parallelepiped with **u**, **v**, and **w** as adjacent edges, as shown in Figure 11.43.

GEOMETRIC PROPERTY OF TRIPLE SCALAR PRODUCT

The volume V of a parallelepiped with vectors **u**, **v**, and **w** as adjacent edges is given by

$$V = |\mathbf{u} \cdot (\mathbf{v} \times \mathbf{w})|.$$

EXAMPLE 4   Volume by the Triple Scalar Product

Find the volume of the parallelepiped having $\mathbf{u} = 3\mathbf{i} - 5\mathbf{j} + \mathbf{k}$, $\mathbf{v} = 2\mathbf{j} - 2\mathbf{k}$, and $\mathbf{w} = 3\mathbf{i} + \mathbf{j} + \mathbf{k}$ as adjacent edges, as shown in Figure 11.44.

SOLUTION

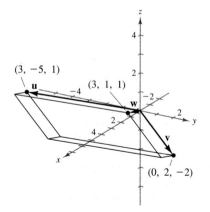

$(3, -5, 1)$
u
$(3, 1, 1)$
w
v
$(0, 2, -2)$

FIGURE 11.44

The volume of the parallelepiped is

$$V = |\mathbf{u} \cdot (\mathbf{v} \times \mathbf{w})|$$
$$= \begin{vmatrix} 3 & -5 & 1 \\ 0 & 2 & -2 \\ 3 & 1 & 1 \end{vmatrix}$$
$$= 3\begin{vmatrix} 2 & -2 \\ 1 & 1 \end{vmatrix} - (-5)\begin{vmatrix} 0 & -2 \\ 3 & 1 \end{vmatrix} + (1)\begin{vmatrix} 0 & 2 \\ 3 & 1 \end{vmatrix}$$
$$= 3(4) + 5(6) + 1(-6)$$
$$= 36.$$

DISCUSSION PROBLEM

....

COPLANAR VECTORS

Consider three vectors **u**, **v**, and **w** with the same initial point. It can be shown that they are coplanar if and only if

$$\mathbf{u} \cdot (\mathbf{v} \times \mathbf{w}) = \begin{vmatrix} u_1 & u_2 & u_3 \\ v_1 & v_2 & v_3 \\ w_1 & w_2 & w_3 \end{vmatrix} = 0.$$

Explain why this test for coplanar vectors is valid.

WARM-UP

....

The following warm-up exercises involve skills that were covered in earlier sections. You will use these skills in the exercise set for this section.

In Exercises 1–4, find (a) $\mathbf{u} \cdot \mathbf{v}$, (b) $\|\mathbf{u}\|^2$, and (c) $(\mathbf{u} \cdot \mathbf{v})\,\mathbf{v}$.

1. $\mathbf{u} = \langle 0, 8 \rangle$
   $\mathbf{v} = \langle 6, 8 \rangle$

2. $\mathbf{u} = \langle 25, 15 \rangle$
   $\mathbf{v} = \langle -3, 5 \rangle$

3. $\mathbf{u} = \langle 12, -6, 16 \rangle$
   $\mathbf{v} = \langle 9, -\frac{9}{2}, 12 \rangle$

4. $\mathbf{u} = 4\mathbf{i} + 3\mathbf{k}$
   $\mathbf{v} = -2\mathbf{i} + 6\mathbf{j}$

In Exercises 5–7, find any values of the constant k so the vectors **u** and **v** are orthogonal.

5. $\mathbf{u} = \langle 15, k \rangle$
   $\mathbf{v} = \langle 6, 12 \rangle$

6. $\mathbf{u} = \langle k, 9 \rangle$
   $\mathbf{v} = \langle -4, 8 \rangle$

7. $\mathbf{u} = \langle -2, k, 7 \rangle$
   $\mathbf{v} = \langle 8, k, 1 \rangle$

In Exercises 8–10, evaluate the determinant.

8. $\begin{vmatrix} 10 & 4 \\ -6 & 2 \end{vmatrix}$

9. $\begin{vmatrix} 4 & -7 \\ -2 & 1 \end{vmatrix}$

10. $\begin{vmatrix} 8 & 0 & 2 \\ -3 & -2 & 1 \\ 0 & 4 & 1 \end{vmatrix}$

SECTION 11.5 · EXERCISES

In Exercises 1–4, find the cross product of the given unit vectors and sketch your result.

1. $\mathbf{i} \times \mathbf{j}$ **2.** $\mathbf{j} \times \mathbf{k}$

3. $\mathbf{i} \times \mathbf{k}$ **4.** $\mathbf{k} \times \mathbf{i}$

In Exercises 5–16, find $\mathbf{u} \times \mathbf{v}$ and show that it is orthogonal to both **u** and **v**.

5. $\mathbf{u} = \langle 1, -4, 0 \rangle$
   $\mathbf{v} = \langle 2, 6, 0 \rangle$

6. $\mathbf{u} = \langle -3, 2, 3 \rangle$
   $\mathbf{v} = \langle 0, 1, 0 \rangle$

7. $\mathbf{u} = \langle 7, -5, 2 \rangle$
   $\mathbf{v} = \langle -1, 4, -1 \rangle$

8. $\mathbf{u} = \langle -5, 5, 11 \rangle$
   $\mathbf{v} = \langle 2, 2, 3 \rangle$

9. $\mathbf{u} = \langle 2, 4, 3 \rangle$
   $\mathbf{v} = \langle 0, -2, 1 \rangle$

10. $\mathbf{u} = \langle 4, -2, 6 \rangle$
   $\mathbf{v} = \langle -1, 5, 7 \rangle$

11. $\mathbf{u} = 6\mathbf{i} + 2\mathbf{j} + \mathbf{k}$
   $\mathbf{v} = \mathbf{i} + 3\mathbf{j} - 2\mathbf{k}$

12. $\mathbf{u} = 6\mathbf{k}$
   $\mathbf{v} = -\mathbf{i} + 3\mathbf{j} + \mathbf{k}$

13. $\mathbf{u} = \mathbf{i} + \frac{3}{2}\mathbf{j} - \frac{5}{2}\mathbf{k}$
   $\mathbf{v} = \frac{1}{2}\mathbf{i} - \frac{3}{4}\mathbf{j} + \frac{1}{4}\mathbf{k}$

14. $\mathbf{u} = \frac{2}{3}\mathbf{i}$
   $\mathbf{v} = \frac{1}{3}\mathbf{i} + 3\mathbf{k}$

15. $\mathbf{u} = 6\mathbf{i} - 5\mathbf{j} + \mathbf{k}$
   $\mathbf{v} = \frac{1}{3}\mathbf{i} - \frac{1}{3}\mathbf{j} + \frac{2}{3}\mathbf{k}$

16. $\mathbf{u} = -\mathbf{i} + \mathbf{k}$
   $\mathbf{v} = \mathbf{j} - 2\mathbf{k}$

In Exercises 17–22, find a unit vector orthogonal to **u** and **v**.

17. $\mathbf{u} = 3\mathbf{i} + \mathbf{j}$
    $\mathbf{v} = \mathbf{j} + \mathbf{k}$

18. $\mathbf{u} = \mathbf{i} + 2\mathbf{j}$
    $\mathbf{v} = \mathbf{i} - 3\mathbf{j}$

19. $\mathbf{u} = -2\mathbf{i} + \mathbf{j} + 3\mathbf{k}$
    $\mathbf{v} = \mathbf{i} + 4\mathbf{j} + 6\mathbf{k}$

20. $\mathbf{u} = 7\mathbf{i} - 14\mathbf{j} + 5\mathbf{k}$
    $\mathbf{v} = 14\mathbf{i} + 28\mathbf{j} - 15\mathbf{k}$

21. $\mathbf{u} = \mathbf{i} + \mathbf{j} - \mathbf{k}$
    $\mathbf{v} = \mathbf{i} + \mathbf{j} + \mathbf{k}$

22. $\mathbf{u} = \mathbf{i} - 2\mathbf{j} + 2\mathbf{k}$
    $\mathbf{v} = 2\mathbf{i} - \mathbf{j} - 2\mathbf{k}$

In Exercises 23–28, find the area of the parallelogram that has the vectors as adjacent sides.

23. $\mathbf{u} = \mathbf{k}$
    $\mathbf{v} = \mathbf{i} + \mathbf{k}$

24. $\mathbf{u} = \mathbf{i} + 2\mathbf{j} + 2\mathbf{k}$
    $\mathbf{v} = \mathbf{i} + \mathbf{k}$

25. $\mathbf{u} = 3\mathbf{i} + 4\mathbf{j} + 6\mathbf{k}$
    $\mathbf{v} = 2\mathbf{i} - \mathbf{j} + 5\mathbf{k}$

26. $\mathbf{u} = \langle -2, 3, 2 \rangle$
    $\mathbf{v} = \langle 1, 2, 4 \rangle$

27. $\mathbf{u} = \langle 2, 2, -3 \rangle$
    $\mathbf{v} = \langle 0, 2, 3 \rangle$

28. $\mathbf{u} = \langle 4, -3, 2 \rangle$
    $\mathbf{v} = \langle 5, 0, 1 \rangle$

In Exercises 29 and 30, verify that the points are the vertices of a parallelogram and find its area.

29. $A(2, -1, 4)$, $B(3, 1, 2)$, $C(0, 5, 6)$, $D(-1, 3, 8)$

30. $A(3, 5, 0)$, $B(-1, 8, 5)$, $C(1, 3, 11)$, $D(5, 0, 6)$

In Exercises 31–34, find the area of the triangle with the given vertices. (The area of the triangle having **u** and **v** as adjacent sides is $\frac{1}{2}\|\mathbf{u} \times \mathbf{v}\|$.)

31. $(0, 0, 0)$, $(4, -2, 6)$, $(-4, 0, 3)$

32. $(1, -4, 3)$, $(2, 0, 2)$, $(-2, 2, 0)$

33. $(2, 3, -5)$, $(-2, -2, 0)$, $(3, 0, 6)$

34. $(2, 4, 0)$, $(-2, -4, 0)$, $(0, 0, 4)$

In Exercises 35 and 36, find $\mathbf{u} \cdot (\mathbf{v} \times \mathbf{w})$.

35. $\mathbf{u} = \langle 2, 3, 3 \rangle$, $\mathbf{v} = \langle 4, 4, 0 \rangle$, $\mathbf{w} = \langle 0, 0, 4 \rangle$

36. $\mathbf{u} = \langle 20, 10, 10 \rangle$, $\mathbf{v} = \langle 1, 4, 4 \rangle$, $\mathbf{w} = \langle 0, 2, 2 \rangle$

In Exercises 37 and 38, use the triple scalar product to find the volume of the parallelepiped having adjacent edges **u**, **v**, and **w**.

37. $\mathbf{u} = 2\mathbf{i}$, $\mathbf{v} = 2\mathbf{j}$, $\mathbf{w} = 2\mathbf{i} + 2\mathbf{j} + \mathbf{k}$

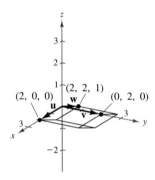

FIGURE FOR 37

38. $\mathbf{u} = \langle 1, 1, 3 \rangle$, $\mathbf{v} = \langle 0, 3, 3 \rangle$,
    $\mathbf{w} = \langle 3, 0, 3 \rangle$

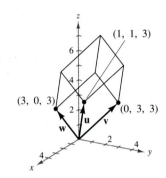

FIGURE FOR 38

In Exercises 39 and 40, find the volume of the parallelepiped with the given vertices.

39. $(0, 0, 0)$, $(4, 0, 0)$, $(4, -2, 3)$, $(0, -2, 3)$, $(4, 5, 3)$, $(0, 5, 3)$, $(0, 3, 6)$, $(4, 3, 6)$

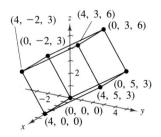

FIGURE FOR 39

40. $(3, 0, 0), (4, 1, 2), (3, -1, 4), (2, -2, 2), (-1, 5, 4), (0, 6, 6),$
$(-1, 4, 8), (-2, 3, 6)$

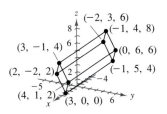

FIGURE FOR 40

In Exercises 41 and 42, prove the property of the cross product where $\mathbf{u} = \langle u_1, u_2, u_3 \rangle$ and $\mathbf{v} = \langle v_1, v_2, v_3 \rangle$.

41. $\mathbf{u} \times \mathbf{u} = \mathbf{0}$

42. $\mathbf{u} \times \mathbf{v}$ is orthogonal to both $\mathbf{u}$ and $\mathbf{v}$.

43. Consider the vectors $\mathbf{u} = \langle \cos \beta, \sin \beta, 0 \rangle$ and $\mathbf{v} = \langle \cos \alpha, \sin \alpha, 0 \rangle$ where $\alpha > \beta$. Find the cross product of the vectors and use the result to prove the identity

$$\sin(\alpha - \beta) = \sin \alpha \cos \beta - \cos \alpha \sin \beta.$$

11.6   EQUATIONS OF LINES AND PLANES IN SPACE
....

Lines in Space / Planes in Space / Sketching Planes in Space / Distance Between a Point and a Plane

Lines in Space

In the plane, *slope* is used to determine an equation of a line. In space, it is more convenient to use *vectors* to determine the equation of a line. In Figure 11.45, consider the line L through the point $P = (x_1, y_1, z_1)$ and parallel to the vector $\mathbf{v} = \langle a, b, c \rangle$. The vector $\mathbf{v}$ is the **direction vector** for the line L, and a, b, and c are the **direction numbers.** One way of describing the line L is to say that it consists of all points $Q = (x, y, z)$ for which the vector $\overrightarrow{PQ}$ is parallel to $\mathbf{v}$. This means that $\overrightarrow{PQ}$ is a scalar multiple of $\mathbf{v}$, and you can write $\overrightarrow{PQ} = t\mathbf{v}$, where t is a scalar.

$$\overrightarrow{PQ} = \langle x - x_1, y - y_1, z - z_1 \rangle = \langle at, bt, ct \rangle = t\mathbf{v}$$

By equating corresponding components, you can obtain the **parametric equations** of a line in space.

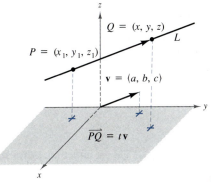

Line has direction vector $\mathbf{v}$.

FIGURE 11.45

PARAMETRIC EQUATIONS OF A LINE IN SPACE

A line L parallel to the vector $\mathbf{v} = \langle a, b, c \rangle$ and passing through the point $P = (x_1, y_1, z_1)$ is represented by the **parametric equations**

$$x = x_1 + at, \qquad y = y_1 + bt, \qquad z = z_1 + ct.$$

If the direction numbers a, b, and c are all nonzero, you can eliminate the parameter t to obtain the **symmetric equations** of a line.

$$\frac{x - x_1}{a} = \frac{y - y_1}{b} = \frac{z - z_1}{c} \qquad \textit{Symmetric equations}$$

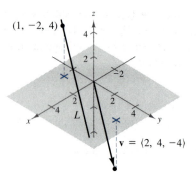

(1, −2, 4)

$\mathbf{v} = \langle 2, 4, -4 \rangle$

FIGURE 11.46

EXAMPLE 1 **Finding Parametric and Symmetric Equations**

Find parametric and symmetric equations of the line L that passes through the point $(1, -2, 4)$ and is parallel to $\mathbf{v} = \langle 2, 4, -4 \rangle$.

SOLUTION

To find a set of parametric equations of the line, use the coordinates $x_1 = 1$, $y_1 = -2$, and $z_1 = 4$ and direction numbers $a = 2$, $b = 4$, and $c = -4$ (see Figure 11.46).

$$x = 1 + 2t, \qquad y = -2 + 4t, \qquad z = 4 - 4t \qquad \textit{Parametric equations}$$

Because a, b, and c are all nonzero, a set of symmetric equations is

$$\frac{x - 1}{2} = \frac{y + 2}{4} = \frac{z - 4}{-4}. \qquad \textit{Symmetric equations}$$

Neither the parametric equations nor the symmetric equations of a given line are unique. For instance, in Example 1, by letting $t = 1$ in the parametric equations you would obtain the point $(3, 2, 0)$. Using this point with the direction numbers $a = 2$, $b = 4$, and $c = -4$ produces the parametric equations

$$x = 3 + 2t, \qquad y = 2 + 4t, \qquad z = -4t.$$

EXAMPLE 2 **Parametric Equations of a Line Through Two Points**

Find a set of parametric equations of the line that passes through the points $(-2, 1, 0)$ and $(1, 3, 5)$.

SOLUTION

Begin by letting $P = (-2, 1, 0)$ and $Q = (1, 3, 5)$. Then a direction vector for the line passing through P and Q is given by

$$\mathbf{v} = \overrightarrow{PQ} = \langle 1 - (-2), 3 - 1, 5 - 0 \rangle = \langle 3, 2, 5 \rangle = \langle a, b, c \rangle.$$

Using the direction numbers $a = 3$, $b = 2$, and $c = 5$, with the point $P = (-2, 1, 0)$, you can obtain the parametric equations

$$x = -2 + 3t, \qquad y = 1 + 2t, \qquad z = 5t.$$

Planes in Space

You have seen how an equation of a line in space can be obtained from a point on the line and a vector *parallel* to it. You will now see that an equation of a plane in space can be obtained from a point in the plane and a vector *normal* (perpendicular) to it.

Consider the plane containing the point $P = (x_1, y_1, z_1)$ and having a nonzero normal vector $\mathbf{n} = \langle a, b, c \rangle$, as shown in Figure 11.47. This plane consists of all points $Q = (x, y, z)$ for which the vector $\overrightarrow{PQ}$ is orthogonal to $\mathbf{n}$. Using the dot product, you can write the following.

$$\mathbf{n} \cdot \overrightarrow{PQ} = 0$$
$$\langle a, b, c \rangle \cdot \langle x - x_1, y - y_1, z - z_1 \rangle = 0$$
$$a(x - x_1) + b(y - y_1) + c(z - z_1) = 0$$

The third equation of the plane is said to be in **standard form.**

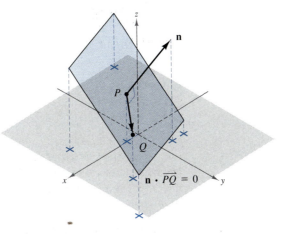

FIGURE 11.47

STANDARD EQUATION OF A PLANE IN SPACE

The plane containing the point (x_1, y_1, z_1) and having a normal vector $\mathbf{n} = \langle a, b, c \rangle$ can be represented, in **standard form,** by the equation

$$a(x - x_1) + b(y - y_1) + c(z - z_1) = 0.$$

By regrouping terms, you obtain the **general form** of the equation of a plane in space,

$$ax + by + cz + d = 0. \qquad \textit{General form of equation of a plane}$$

Given the general form of the equation of a plane, it is easy to find a vector normal to the plane. Simply use the coefficients of x, y, and z and write $\mathbf{n} = \langle a, b, c \rangle$.

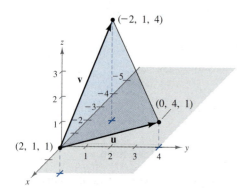

Plane determined by **u** and **v**.

FIGURE 11.48

EXAMPLE 3 **Finding an Equation of a Plane in Three-Space**
....

Find the general equation of the plane containing the points $(2, 1, 1)$, $(0, 4, 1)$, and $(-2, 1, 4)$.

SOLUTION

To find the equation of the plane, you need a point in the plane and a vector that is normal to the plane. There are three choices for the point, but no normal vector is given. To obtain a normal vector, use the cross product of vectors **u** and **v** extending from the point $(2, 1, 1)$ to the points $(0, 4, 1)$ and $(-2, 1, 4)$, as shown in Figure 11.48. The component forms of **u** and **v** are

$$\mathbf{u} = \langle 0 - 2, 4 - 1, 1 - 1 \rangle = \langle -2, 3, 0 \rangle$$
$$\mathbf{v} = \langle -2 - 2, 1 - 1, 4 - 1 \rangle = \langle -4, 0, 3 \rangle,$$

and it follows that

$$\mathbf{n} = \mathbf{u} \times \mathbf{v} = \begin{vmatrix} \mathbf{i} & \mathbf{j} & \mathbf{k} \\ -2 & 3 & 0 \\ -4 & 0 & 3 \end{vmatrix} = 9\mathbf{i} + 6\mathbf{j} + 12\mathbf{k} = \langle a, b, c \rangle$$

is normal to the given plane. Using the direction numbers for **n** and the point $(x_1, y_1, z_1) = (2, 1, 1)$, you can determine an equation of the plane to be

$$a(x - x_1) + b(y - y_1) + c(z - z_1) = 0$$
$$9(x - 2) + 6(y - 1) + 12(z - 1) = 0 \qquad \textit{Standard form}$$
$$9x + 6y + 12z - 36 = 0$$
$$3x + 2y + 4z - 12 = 0. \qquad \textit{General form}$$

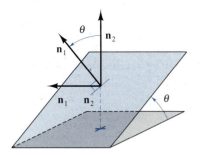

Angle Between Two Planes

FIGURE 11.49

REMARK In Example 3, check to see that each of the three points satisfies the equation $3x + 2y + 4z - 12 = 0$.

Two distinct planes in three-space either are parallel or intersect in a line. If they intersect, you can determine the angle between them from the angle between their normal vectors, as shown in Figure 11.49. Specifically, if vectors $\mathbf{n}_1$ and $\mathbf{n}_2$ are normal to two intersecting planes, then the angle θ between the normal vectors is equal to the angle between the two planes and is given by

$$\cos \theta = \frac{|\mathbf{n}_1 \cdot \mathbf{n}_2|}{\|\mathbf{n}_1\| \|\mathbf{n}_2\|}. \qquad \textit{Angle between two planes}$$

Consequently, two planes with normal vectors $\mathbf{n}_1$ and $\mathbf{n}_2$ are

1. *perpendicular* if $\mathbf{n}_1 \cdot \mathbf{n}_2 = 0$
2. *parallel* if $\mathbf{n}_1$ is a scalar multiple of $\mathbf{n}_2$.

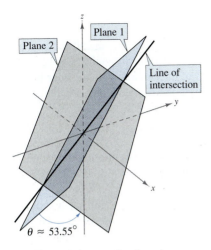

The angle between the planes is approximately 53.55°.

FIGURE 11.50

EXAMPLE 4   **Finding the Line of Intersection of Two Planes**

Find the angle between the two planes given by

$$x - 2y + z = 0 \qquad\qquad \text{Equation for Plane 1}$$
$$2x + 3y - 2z = 0 \qquad\qquad \text{Equation for Plane 2}$$

and find parametric equations of their line of intersection (see Figure 11.50).

SOLUTION

The normal vectors for the planes are $\mathbf{n}_1 = \langle 1, -2, 1 \rangle$ and $\mathbf{n}_2 = \langle 2, 3, -2 \rangle$. Consequently, the angle between the two planes is determined as follows.

$$\cos \theta = \frac{|\mathbf{n}_1 \cdot \mathbf{n}_2|}{\|\mathbf{n}_1\| \, \|\mathbf{n}_2\|} = \frac{|-6|}{\sqrt{6}\sqrt{17}} = \frac{6}{\sqrt{102}} \approx 0.59409$$

This implies that the angle between the two planes is $\theta \approx 53.55°$. You can find the line of intersection of the two planes by simultaneously solving the two linear equations representing the planes. One way to do this is to multiply the first equation by -2 and add the result to the second equation.

$$
\begin{array}{lll}
x - 2y + z = 0 & \to & -2x + 4y - 2z = 0 \\
2x + 3y - 2z = 0 & & \underline{2x + 3y - 2z = 0} \\
& & 7y - 4z = 0 \quad \to \quad y = \dfrac{4z}{7}
\end{array}
$$

Substituting $y = 4z/7$ back into one of the original equations, you can determine that $x = z/7$. Finally, by letting $t = z/7$, you obtain the parametric equations

$$x = x_1 + at = t, \qquad y = y_1 + bt = 4t, \qquad z = z_1 + ct = 7t,$$

which indicate that $a = 1, b = 4$, and $c = 7$ are direction numbers for the line of intersection.

Note that the direction numbers in Example 4 can be obtained from the cross product of the two normal vectors as follows.

$$
\mathbf{n}_1 \times \mathbf{n}_2 = \begin{vmatrix} \mathbf{i} & \mathbf{j} & \mathbf{k} \\ 1 & -2 & 1 \\ 2 & 3 & -2 \end{vmatrix}
$$

$$
= \begin{vmatrix} -2 & 1 \\ 3 & -2 \end{vmatrix} \mathbf{i} - \begin{vmatrix} 1 & -2 \\ 2 & -2 \end{vmatrix} \mathbf{j} + \begin{vmatrix} 1 & -2 \\ 2 & 3 \end{vmatrix} \mathbf{k}
$$

$$
= \mathbf{i} + 4\mathbf{j} + 7\mathbf{k}
$$

This means that the line of intersection of the two planes is parallel to the cross product of their normal vectors.

Sketching Planes in Space

If a plane in space intersects one of the coordinate planes, we call the line of intersection the **trace** of the given plane in the coordinate plane. To sketch a plane in space, it is helpful to find its points of intersection with the coordinate axes and its traces in the coordinate planes. For example, consider the plane given by

$$3x + 2y + 4z = 12. \qquad \text{\textit{Equation of plane}}$$

You can find the xy-trace by letting $z = 0$ and sketching the line

$$3x + 2y = 12 \qquad \text{\textit{xy-trace}}$$

in the xy-plane. This line intersects the x-axis at $(4, 0, 0)$ and the y-axis at $(0, 6, 0)$. In Figure 11.51, this process is continued by finding the yz-trace and the xz-trace, and then shading in the triangular region lying in the first octant.

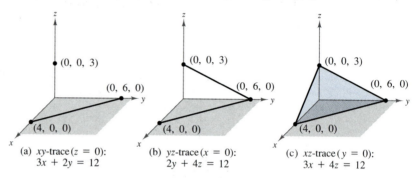

(a) xy-trace $(z = 0)$:
$3x + 2y = 12$

(b) yz-trace $(x = 0)$:
$2y + 4z = 12$

(c) xz-trace $(y = 0)$:
$3x + 4z = 12$

Traces of the Plane: $3x + 2y + 4z = 12$

FIGURE 11.51

If the equation of a plane has a missing variable such as $2x + z = 1$, then the plane must be *parallel to the axis* represented by the missing variable, as shown in Figure 11.52. If two variables are missing from the equation of a plane, then it is *parallel to the coordinate plane* represented by the missing variables, as shown in Figure 11.53.

[Figure 11.52 — left margin]

$(0, 0, 1)$

$\left(\frac{1}{2}, 0, 0\right)$

Plane: $2x + z = 1$

FIGURE 11.52

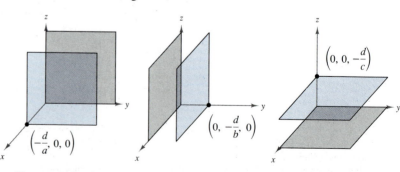

FIGURE 11.53

Plane $ax + d = 0$
is parallel to yz-plane.

$\left(-\frac{d}{a}, 0, 0\right)$

Plane $by + d = 0$
is parallel to xz-plane.

$\left(0, -\frac{d}{b}, 0\right)$

Plane $cz + d = 0$
is parallel to xy-plane.

$\left(0, 0, -\frac{d}{c}\right)$

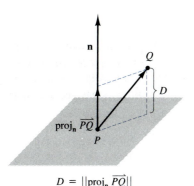

$$D = \|\text{proj}_\mathbf{n}\ \overrightarrow{PQ}\|$$

FIGURE 11.54

Distance Between a Point and a Plane

The distance D between a point Q and a plane is the length of the shortest line segment connecting Q to the plane, as shown in Figure 11.54. If P is *any* point in the plane, you can find this distance by projecting the vector $\overrightarrow{PQ}$ onto the normal vector $\mathbf{n}$. The length of this projection is the desired distance.

DISTANCE BETWEEN A POINT AND A PLANE

The distance between a plane and a point Q (not in the plane) is

$$D = \|\text{proj}_\mathbf{n}\ \overrightarrow{PQ}\| = \frac{|\overrightarrow{PQ} \cdot \mathbf{n}|}{\|\mathbf{n}\|},$$

where P is a point in the plane and $\mathbf{n}$ is normal to the plane.

To find a point in the plane given by $ax + by + cz + d = 0$, where $a \neq 0$, let $y = 0$ and $z = 0$. Then, from the equation $ax + d = 0$, you can conclude that the point $(-d/a, 0, 0)$ lies in the plane.

EXAMPLE 5   Finding the Distance Between a Point and a Plane

Find the distance between the point $Q = (1, 5, -4)$ and the plane given by $3x - y + 2z = 6$.

SOLUTION

You know that $\mathbf{n} = \langle 3, -1, 2 \rangle$ is normal to the given plane. To find a point in the plane, let $y = 0$ and $z = 0$, and obtain the point $P = (2, 0, 0)$. The vector from P to Q is given by

$$\overrightarrow{PQ} = \langle 1 - 2, 5 - 0, -4 - 0 \rangle$$
$$= \langle -1, 5, -4 \rangle.$$

The formula for the distance between a point and a plane produces

$$D = \frac{|\overrightarrow{PQ} \cdot \mathbf{n}|}{\|\mathbf{n}\|} = \frac{|\langle -1, 5, -4 \rangle \cdot \langle 3, -1, 2 \rangle|}{\sqrt{9 + 1 + 4}}$$
$$= \frac{|-3 - 5 - 8|}{\sqrt{14}}$$
$$= \frac{16}{\sqrt{14}}.$$

REMARK   The choice of the point P in Example 5 is arbitrary. Try choosing a different point to verify that you obtain the same distance.

WARM-UP

The following warm-up exercises involve skills that were covered in earlier sections. You will use these skills in the exercise set for this section.

In Exercises 1–4, find $\mathbf{u} \cdot \mathbf{v}$ and $\mathbf{u} \times \mathbf{v}$.

1. $\mathbf{u} = \langle 2, 10, -6 \rangle$
   $\mathbf{v} = \langle 3, -4, -1 \rangle$

2. $\mathbf{u} = \langle \frac{2}{3}, -\frac{5}{3}, 2 \rangle$
   $\mathbf{v} = \langle 3, -6, -\frac{1}{2} \rangle$

3. $\mathbf{u} = 20\mathbf{i} - 15\mathbf{j} + 50\mathbf{k}$
   $\mathbf{v} = -2\mathbf{i} + \mathbf{j} + 2\mathbf{k}$

4. $\mathbf{u} = \mathbf{i} - 3\mathbf{j}$
   $\mathbf{v} = \mathbf{j} - 7\mathbf{k}$

In Exercises 5–8, find the angle θ between the vectors $\mathbf{u}$ and $\mathbf{v}$.

5. $\mathbf{u} = \langle 3, -3, 1 \rangle$
   $\mathbf{v} = \langle 2, 2, -1 \rangle$

6. $\mathbf{u} = \langle -8, -12, 20 \rangle$
   $\mathbf{v} = \langle 12, 18, 30 \rangle$

7. $\mathbf{u} = 5\mathbf{i} - 3\mathbf{j} + \mathbf{k}$
   $\mathbf{v} = -2\mathbf{i} - \mathbf{j} + 7\mathbf{k}$

8. $\mathbf{u} = 2\mathbf{i} - 5\mathbf{j}$
   $\mathbf{v} = \mathbf{k}$

In Exercises 9 and 10, find a unit vector in the direction of $\mathbf{v}$.

9. $\mathbf{v} = \langle 6, -2, 3 \rangle$ **10.** $\mathbf{v} = \langle \frac{3}{4}, 1, \frac{1}{2} \rangle$

SECTION 11.6 · EXERCISES

In Exercises 1–6, find sets of (a) parametric equations and (b) symmetric equations of the line through the point and parallel to the specified vector or line. (For each line, express the direction numbers as integers.)

Point	Parallel to
1. $(0, 0, 0)$	$\mathbf{v} = \langle -2, 4, 1 \rangle$
2. $(0, 0, 0)$	$\mathbf{v} = \langle 3, -7, -10 \rangle$
3. $(-4, 1, 0)$	$\mathbf{v} = \frac{1}{2}\mathbf{i} + \frac{4}{3}\mathbf{j} - \mathbf{k}$
4. $(5, 0, 10)$	$\mathbf{v} = 4\mathbf{i} + 3\mathbf{k}$
5. $(2, -3, 5)$	$x = 5 + 2t$
	$y = 7 - 3t$
	$z = -2 + t$
6. $(10, -18, 36)$	$\dfrac{x - 2}{3} = \dfrac{y + 2}{-3} = z - 5$

In Exercises 7 and 8, find sets of (a) parametric equations and (b) symmetric equations of the line through the two points. (For each line, express the direction numbers as integers.)

7. $(3, -5, -4), \left(-\frac{3}{2}, \frac{3}{2}, 2 \right)$ **8.** $(4, 2, -2), (8, 0, -1)$

In Exercises 9 and 10, find a set of parametric equations of the line.

9. The line passes through the point $(-3, 8, 15)$ and is parallel to the xz-plane and the yz-plane.

10. The line passes through the point $(4, -3, 8)$ and is perpendicular to the plane given by $3x + 2y - z = 6$.

In Exercises 11 and 12, determine which of the points lie on the line L.

11. The line L passes through the point $(-4, -1, 7)$ and is parallel to the vector $\mathbf{v} = 3\mathbf{i} - \mathbf{j}$.

    (a) $(-4, -1, 0)$ (b) $(-1, -2, 7)$

    (c) $(-10, 1, 7)$ (d) $(4, 1, -7)$

12. The line L passes through the points $(3, 1, 0)$ and $(2, 4, -2)$.

    (a) $(4, -2, 2)$ (b) $\left(\frac{3}{2}, -\frac{5}{2}, -\frac{3}{4}\right)$

    (c) $(3, 1, 1)$ (d) $\left(\frac{3}{2}, \frac{11}{2}, -3\right)$

In Exercises 13 and 14, find the point of intersection of the lines and the angle between the lines.

13. $x = 6 + t$ $x = 7 + 5s$
    $y = -5 - 2t$ $y = 5 + 2s$
    $z = 1 + 3t$ $z = -8 + 3s$

14. $x = 3 - 2t$ $x = 3 - s$
    $y = 8 + t$ $y = 15 + 4s$
    $z = 12 + 3t$ $z = 3 - 3s$

In Exercises 15 and 16, use a graphing utility to graph the line.

15. $x = 2t$ **16.** $x = 5 - 2t$
    $y = 2 + t$ $y = 1 + t$
    $z = 1 + \frac{1}{2}t$ $z = 5 - \frac{1}{2}t$

In Exercises 17–22, find an equation of the plane passing through the point and perpendicular to the specified vector or line.

Point	Perpendicular to
17. $(3, 4, -2)$	$\mathbf{n} = \mathbf{j}$
18. $(2, 3, 5)$	$\mathbf{n} = \mathbf{k}$
19. $(5, 6, 3)$	$\mathbf{n} = -2\mathbf{i} + \mathbf{j} - 2\mathbf{k}$
20. $(0, 0, 0)$	$\mathbf{n} = -3\mathbf{j} + 5\mathbf{k}$
21. $(2, 0, 0)$	$x = 3 - t, y = 2 - 2t, z = 4 + t$
22. $(1, 3, 1)$	$\dfrac{x - 2}{3} = y + 1 = \dfrac{z + 2}{-2}$

In Exercises 23–26, find an equation of the plane passing through the three points.

23. $(0, 0, 0)$, $(1, 1, 0)$, $(0, 3, 3)$

24. $(0, 0, 0)$, $(2, 1, 3)$, $(-2, 1, 3)$

25. $(4, -1, 3)$, $(2, 5, 1)$, $(-1, 2, 1)$

26. $(5, 1, -6)$, $(2, 3, 10)$, $(1, -2, 4)$

In Exercises 27–30, find an equation of the plane.

27. The plane passes through the point $(2, 5, 3)$ and is parallel to the xz-plane.

28. The plane passes through the point $(2, 5, 3)$ and is parallel to the xy-plane.

29. The plane contains the x-axis and makes an angle of $45°$ with the positive y-axis.

30. The plane passes through the points $(4, 0, 0)$ and $(0, 2, 0)$ and is perpendicular to the plane $x + 2y + 2z = 4$.

In Exercises 31–34, find the angle between the two planes and find parametric equations of their line of intersection.

31. $x \qquad + z = 4$ **32.** $x - 3y - z = 3$
    $x - y \qquad = 1$ $2x + y - z = 10$

33. $x + y + 2z = 4$ **34.** $x - 3y + z = 8$
    $x - 2y + z = 10$ $2y - z = -1$

In Exercises 35–40, mark the intercepts and sketch a graph of the plane.

35. $2x + 3y + 4z = 12$ **36.** $x + 2y + 3z = 6$

37. $2x - y + 4z = 4$ **38.** $x + y - 2z = 4$

39. $x + z = 3$ **40.** $y + 2z = 4$

In Exercises 41–44, use a graphing utility to graph the plane.

41. $3x + 2y - z = 6$ **42.** $x - 3y = 3$

43. $x + 2y - 6z = 8$ **44.** $3x - 4y - z = -12$

In Exercises 45 and 46, find the distance between the point and the plane.

45. $(0, 0, 0)$ **46.** $(2, -3, 3)$
    $3x + 2y + z = 12$ $2x - y + 3z = 4$

47. (a) Describe and find an equation for the surface generated by all points (x, y, z) that are two units from the point $(4, -1, 1)$.

(b) Describe and find an equation for the surface generated by all points (x, y, z) that are two units from the plane $4x - 3y + z = 10$.

48. Consider the two nonzero vectors **u** and **v**. Describe the geometric figure generated by the terminal points of the following vectors, where s and t represent all real numbers.

(a) $t\mathbf{v}$ (b) $\mathbf{u} + t\mathbf{v}$ (c) $s\mathbf{u} + t\mathbf{v}$

49. *Mechanical Design* A chute at the top of the grain elevator of a combine funnels grain into a bin (see figure). Use the specified measurements to find the angle between two adjacent sides.

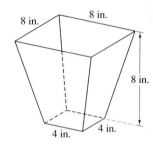

FIGURE FOR 49

CHAPTER 11 · REVIEW EXERCISES

In Exercises 1 and 2, find the component forms of the vectors $\overrightarrow{AB}$, $\overrightarrow{AC}$, and $\overrightarrow{AB} + \overrightarrow{AC}$. Sketch the three vectors.

1. $A(2, 1)$, $B(5, 0)$, $C(4, 3)$ **2.** $A(4, 6)$, $B(0, 5)$, $C(-2, 0)$

In Exercises 3–6, find the component form of the vector **v** satisfying the given conditions.

3. Initial point: $(0, 10)$ **4.** Initial point: $(1, 5)$
Terminal point: $(7, 3)$ Terminal point: $(15, 9)$

5. $\|\mathbf{v}\| = 8$, $\theta = 120°$ **6.** $\|\mathbf{v}\| = \frac{1}{2}$, $\theta = 225°$

In Exercises 7 and 8, sketch **v** and find its component form. (Assume θ is measured counterclockwise from the x-axis to the vector.)

7. $\|\mathbf{v}\| = 3$, $\theta = 135°$ **8.** $\|\mathbf{v}\| = 5$, $\theta = 300°$

In Exercises 9–12, find the component form of the specified vector and sketch its graph, given that $\mathbf{u} = 6\mathbf{i} - 5\mathbf{j}$ and $\mathbf{v} = 10\mathbf{i} + 3\mathbf{j}$.

9. $\dfrac{1}{\|\mathbf{u}\|}\mathbf{u}$ **10.** $3\mathbf{v}$

11. $4\mathbf{u} - 5\mathbf{v}$ **12.** $\frac{1}{2}\mathbf{v}$

In Exercises 13 and 14, find (a) $\mathbf{u} + \mathbf{v}$, (b) $\mathbf{u} - \mathbf{v}$, and (c) $2\mathbf{u} - 3\mathbf{v}$.

13. $\mathbf{u} = \frac{7}{2}\mathbf{i} - \mathbf{j}$, $\mathbf{v} = -\mathbf{i} + 2\mathbf{j}$ **14.** $\mathbf{u} = -3\mathbf{i} + 5\mathbf{j}$, $\mathbf{v} = 10\mathbf{i}$

In Exercises 15 and 16, find the length and direction angle of the vector **v**.

15. $\mathbf{v} = 2\mathbf{i} - 2\sqrt{3}\,\mathbf{j}$ **16.** $\mathbf{v} = -3\mathbf{i} - 3\mathbf{j}$

17. *Resultant Force* Find the direction and magnitude of the resultant of the three forces shown in the figure.

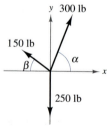

$\tan \beta = \frac{3}{4}$ $\tan \alpha = \frac{12}{5}$

FIGURE FOR 17

18. *Resultant Force* Forces with magnitudes of 85 pounds and 50 pounds act on a single point. Find the magnitude of the resultant if the angle between the forces is 15°.

19. *Rope Tension* A 100-pound weight is supported by two ropes, as shown in the figure. Find the tension in each rope.

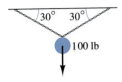

FIGURE FOR 19

20. *Braking Force* A 500-pound motorcycle is headed up a hill inclined at 12°. What force is required to keep the motorcycle from rolling back down the hill when stopped at a red light?

21. *Navigation* An airplane has an air speed of 450 miles per hour at a bearing of N 30° E. If the wind velocity is 20 miles per hour from the west, find the ground speed and the direction of the plane.

22. *Angle Between Forces* Forces of 60 pounds and 100 pounds have a resultant force of 125 pounds. Find the angle between the two forces.

In Exercises 23 and 24, find the standard form of the equation of the sphere that has the given points as the endpoints of a diameter.

23. $(0, 0, 4)$, $(4, 6, 0)$ **24.** $(3, 4, 2)$, $(5, 8, 1)$

In Exercises 25 and 26, find a unit vector in the direction of $\overrightarrow{PQ}$.

25. $P(7, -4, 3)$ **26.** $P(0, 3, -1)$
     $Q(-3, 2, 10)$ $Q(5, -8, 6)$

In Exercises 27 and 28, determine if the vectors are orthogonal, parallel, or neither.

27. $\langle 39, -12, 21 \rangle$ **28.** $\langle 8, 5, -8 \rangle$
     $\langle -26, 8, -14 \rangle$ $\langle -2, 4, \frac{1}{2} \rangle$

In Exercises 29–32, find the angle θ between the vectors **u** and **v**.

29. $\mathbf{u} = 32 \left[\cos\left(\dfrac{7\pi}{4}\right) \mathbf{i} + \sin\left(\dfrac{7\pi}{4}\right) \mathbf{j} \right]$

$\mathbf{v} = 14 \left[\cos\left(\dfrac{5\pi}{6}\right) \mathbf{i} + \sin\left(\dfrac{5\pi}{6}\right) \mathbf{j} \right]$

30. $\mathbf{u} = \langle -6, -3, 18 \rangle$
     $\mathbf{v} = \langle 4, 2, -12 \rangle$

31. $\mathbf{u} = \langle 2\sqrt{2}, -4, 4 \rangle$ **32.** $\mathbf{u} = \langle 3, 1, -1 \rangle$
     $\mathbf{v} = \langle -\sqrt{2}, 1, 2 \rangle$ $\mathbf{v} = \langle 4, 5, 2 \rangle$

In Exercises 33–36, find $\text{proj}_\mathbf{v}\, \mathbf{u}$.

33. $\mathbf{u} = \langle -4, 3 \rangle$, $\mathbf{v} = \langle -8, -2 \rangle$
34. $\mathbf{u} = \langle 5, 6 \rangle$, $\mathbf{v} = \langle 10, 0 \rangle$
35. $\mathbf{u} = \langle 2, 7, 4 \rangle$, $\mathbf{v} = \langle 1, -1, 0 \rangle$
36. $\mathbf{u} = \langle -3, 5, 1 \rangle$, $\mathbf{v} = \langle -5, 2, 6 \rangle$

In Exercises 37 and 38, find $\mathbf{u} \times \mathbf{v}$.

37. $\mathbf{u} = \langle -2, 8, 2 \rangle$, $\mathbf{v} = \langle 1, 1, -1 \rangle$
38. $\mathbf{u} = \langle 20, 15, 5 \rangle$, $\mathbf{v} = \langle 5, -3, 0 \rangle$

In Exercises 39–43, let $\mathbf{u} = \langle 1, -2, -1 \rangle$, $\mathbf{v} = \langle 2, 4, 0 \rangle$, and $\mathbf{w} = \langle 3, 4, 5 \rangle$.

39. Show that $\mathbf{u} \cdot \mathbf{u} = \|\mathbf{u}\|^2$.
40. Determine a unit vector perpendicular to the vectors **v** and **w**.
41. Show that $\mathbf{u} \times \mathbf{v} = -(\mathbf{v} \times \mathbf{u})$.
42. Show that $\mathbf{u} \times (\mathbf{v} + \mathbf{w}) = (\mathbf{u} \times \mathbf{v}) + (\mathbf{u} + \mathbf{w})$.
43. *Volume* Find the volume of the solid whose edges are **u**, **v**, and **w**.

44. *Volume* Use the triple scalar product to find the volume of the parallelepiped with vertices $(3, 2, -1)$, $(2, 5, 1)$, $(3, 2, 8)$, $(4, -1, 6)$, $(-5, 4, 2)$, $(-6, 7, 4)$, $(-5, 4, 11)$, and $(-4, 1, 9)$.

45. *Machine Design*   A component used to channel grain into the end of an elevator has the shape and dimensions shown in the figure. In fabricating the part, it is necessary to know the angle θ between the adjacent sides. Find the angle.

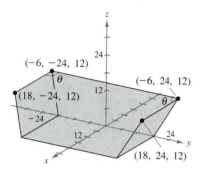

FIGURE FOR 45

46. *Cable Tension*   In a manufacturing process, an electric hoist lifts 200-pound ingots (see figure). Find the tension in the supporting cables.

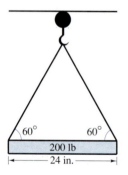

FIGURE FOR 46

In Exercises 47–50, find a set of parametric equations for the given line.

47. The line passes through the points $(-1, 3, 5)$ and $(3, 6, -1)$.

48. The line passes through the points $(0, -10, 3)$ and $(5, 10, 0)$.

49. The line passes through $(3, 1, 2)$ and is parallel to the line given by $x = y = z$.

50. The line is the intersection of the planes given by
$$2\,x - y - 2z = 4$$
$$x - y - z = 8.$$

In Exercises 51–54, find an equation of the plane.

51. The plane passes through the points $(0, 0, 0)$, $(5, 0, 2)$, and $(2, 3, 8)$.

52. The plane passes through the points $(-1, 3, 4)$, $(4, -2, 2)$, and $(2, 8, 6)$.

53. The plane passes through the point $(3, 1, 2)$ and is orthogonal to the line given by $x = y = z$.

54. The plane passes through the point $(5, 3, 2)$ and is parallel to the xy-coordinate plane.

In Exercises 55 and 56, find the distance from the point to the plane.

55. $(2, 3, 10)$
$2x - 20y + 6z = 6$

56. $(0, 0, 0)$
$x - 10y + 3z = 2$

CUMULATIVE TEST FOR CHAPTERS 9–11

Take this test as you would take a test in class. After you are done, check your work against the answers in the back of the book.

1. Find the sum of the first 20 terms of the arithmetic sequence.

   8, 12, 16, 20, . . .

2. Write the first 5 terms of the geometric sequence in which $a_1 = 54$ and $r = -\frac{1}{3}$.

3. Find the sum of the infinite series: $\displaystyle\sum_{i=0}^{\infty} 3(\tfrac{1}{2})^n$.

4. Use mathematical induction to prove the formula

   $3 + 7 + 11 + 15 + \cdots + (4n - 1) = n(2n + 1)$.

5. Use the Binomial Theorem to expand and simplify $(z - 3)^4$.

6. You accept a job with a salary of \$32,000 for the first year. Suppose that during the next 9 years you receive a 5% raise each year. What would your total compensation be over the 10-year period?

7. A personnel manager has ten applicants to fill three positions in a corporation. In how many ways can this be done, assuming all the applicants are qualified for any of the three positions?

8. On a game show, the digits 3, 4, and 5 are given to you to arrange in the proper order to show the price of an appliance. If you are correct, you win the appliance. What is the probability of winning if
   (a) you have no idea of the price of the appliance?
   (b) you know that the price of the appliance is at least \$400?

9. Find the mean, median, and standard deviation of the numbers 21, 27, 23, 25, 30, and 19.

10. Sketch the graph of each of the following equations.

    (a) $6x - y^2 = 0$ (b) $\dfrac{(x - 2)^2}{4} + \dfrac{(y + 1)^2}{9} = 1$

11. Find an equation of the hyperbola with foci $(0, 0)$ and $(0, 4)$ and asymptotes $y = \pm\frac{1}{2}x + 2$.

12. Find a set of parametric equations of the line passing through the points $(2, -3)$ and $(6, 4)$. (The answer is not unique.)

13. Sketch the curve represented by the parametric equations $x = 3 + 3 \cos \theta$ and $y = 2 + 2 \sin \theta$. Write the corresponding rectangular equation by eliminating the parameter.

14. Convert the polar equation $r = 6 \cos \theta$ to rectangular form and sketch its graph.

15. Use a graphing utility to graph the polar equation $r = 2 \cos 2\theta \sec \theta$. Identify any asymptotes of the graph and write the equation of any asymptotes in rectangular form.

16. Find a polar equation of the parabola with focus at the origin and vertex at $(1, \pi)$.

17. Find the component form of $\mathbf{u}$ if $\|\mathbf{u}\| = 3$ and $\theta = 30°$. (Assume the angle is measured counterclockwise from the x-axis to the vector.)

18. A projectile is fired with an initial velocity of 120 feet per second at an angle of $15°$ with the horizontal. Find the vertical and horizontal components of the velocity accurate to two decimal places.

19. Find an equation of the sphere for which the endpoints of a diameter are $(0, 0, 0)$ and $(4, 4, 8)$.

20. Find $\mathbf{u} \cdot \mathbf{v}$ and $\mathbf{u} \times \mathbf{v}$ if $\mathbf{u} = \langle -3, 4, 1 \rangle$ and $\mathbf{v} = \langle 5, 0, 2 \rangle$.

21. Find a set of parametric equations of the line passing through the points $(-2, 3, 0)$ and $(5, 8, 25)$.

22. Find an equation of the plane passing through the points $(0, 0, 0)$, $(-2, 3, 0)$, and $(5, 8, 25)$.

CHAPTER 12

LIMITS AND AN INTRODUCTION TO CALCULUS

12.1 INTRODUCTION TO LIMITS

The Limit Concept / Definition of Limit / Limits That Fail to Exist / Properties of Limits

The Limit Concept

The notion of a limit is a *fundamental* concept of calculus. In this chapter, you will learn how to evaluate limits and how they are used in the two basic problems of calculus: "the tangent line problem" and "the area problem."

EXAMPLE 1   Finding a Rectangle of Maximum Area

You are given 24 inches of wire and are asked to form a rectangle whose area is as large as possible. What dimensions should the rectangle have?

SOLUTION

Let w represent the width of the rectangle and let l represent the length of the rectangle, as shown in Figure 12.1. Because $2w + 2l = 24$, it follows that $l = 12 - w$, and the area of the rectangle is

$$A = wl = w(12 - w) = 12w - w^2.$$

Using this model for area, you can experiment with different values of w to see how to obtain a maximum area. After trying several values, it appears that the maximum area occurs when $w = 6$.

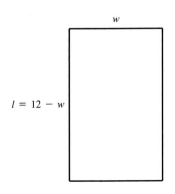

FIGURE 12.1

Width, w	5.0	5.5	5.9	6.0	6.1	6.5	7.0
Area, A	35.0	35.75	35.99	36.0	35.99	35.75	35.0

In limit terminology, we say that the limit of A as w approaches 6 is 36. This is written as

$$\lim_{w \to 6} A = 36.$$

■ ■ ■ ■ ■ ■ ■ ■ ■

Definition of Limit

Some people think that a limit is a quantity that can only be approached and cannot actually be reached. Some limits are like that, but may are not.

EXAMPLE 2   Finding a Limit That Can Be Reached

Use a table to estimate the limit: $\lim\limits_{x \to 2}(3x - 2)$.

SOLUTION

Let $f(x) = 3x - 2$. Then construct a table that shows values of $f(x)$ when x is close to 2.

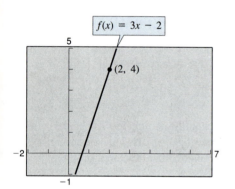

$f(x) = 3x - 2$

(2, 4)

FIGURE 12.2

x	1.9	1.99	1.999	2.0	2.001	2.01	2.1
$f(x)$	3.7	3.97	3.997	?	4.003	4.03	4.3

From the table, it appears that the closer x gets to 2, the closer $f(x)$ gets to 4. Thus, you can estimate the limit to be 4. For this particular function, you can obtain the limit simply by substituting 2 for x to obtain

$$\lim_{x \to 2}(3x - 2) = 3(2) - 2 = 4.$$

Figure 12.2 adds further support to this conclusion.

■ ■ ■ ■ ■ ■ ■ ■ ■

In Figure 12.2, note that the graph of $f(x) = 3x - 2$ is continuous. That is, it has no gaps or jumps. For graphs that are not continuous, finding a limit can be more difficult.

EXAMPLE 3   Finding a Limit That Cannot Be Reached

Use a table to estimate the limit: $\lim\limits_{x \to 1} \dfrac{x^3 - x^2 + x - 1}{x - 1}$.

SOLUTION

Let $f(x) = (x^3 - x^2 + x - 1)/(x - 1)$. Then construct a table that shows values of $f(x)$ when x is close to 1.

x	0.9	0.99	0.999	1.0	1.001	1.01	1.1
$f(x)$	1.81	1.980	1.998	?	2.002	2.020	2.21

From the table, it appears that the limit is 2. The graph in Figure 12.3 supports this conclusion. In this case, notice that you cannot obtain the limit simply by evaluating $f(x)$ when $x = 1$.

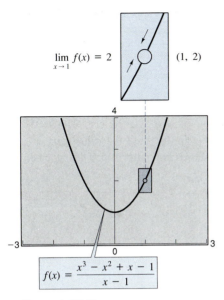

$$\lim_{x \to 1} f(x) = 2 \qquad (1, 2)$$

$$f(x) = \frac{x^3 - x^2 + x - 1}{x - 1}$$

FIGURE 12.3

DEFINITION OF LIMIT

If $f(x)$ becomes arbitrarily close to a unique number L as x approaches c from either side, then the **limit** of $f(x)$ as x approaches c is L. This is written as

$$\lim_{x \to c} f(x) = L.$$

EXAMPLE 4 Using a Calculator to Find a Limit

Use a table to estimate the limit: $\lim\limits_{x \to 0} \dfrac{x}{\sqrt{x + 1} - 1}$.

SOLUTION

Let $f(x) = x/(\sqrt{x + 1} - 1)$. Then use a calculator to construct a table that shows values of $f(x)$ when x is close to 0.

x	-0.01	-0.001	-0.0001	0	0.0001	0.001	0.01
$f(x)$	1.995	1.9995	1.9999	?	2.0001	2.0005	2.005

From the table, it appears that the limit is 2. If you have access to a graphing utility, try using its *zoom* and *trace* features to graphically verify that the limit is 2 (see Figure 12.4).

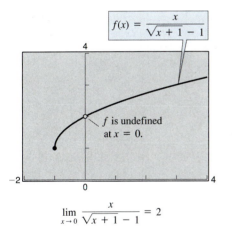

$$f(x) = \frac{x}{\sqrt{x + 1} - 1}$$

f is undefined at $x = 0$.

$$\lim_{x \to 0} \frac{x}{\sqrt{x + 1} - 1} = 2$$

FIGURE 12.4

In Example 4, note that $f(x)$ has a limit when $x \to 0$ although the function is not defined when $x = 0$. This often happens, and it is important to realize that *the existence or nonexistence of $f(x)$ when $x = c$ has no bearing on the existence of the limit of $f(x)$ as x approaches c.*

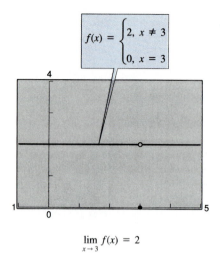

$$f(x) = \begin{cases} 2, & x \neq 3 \\ 0, & x = 3 \end{cases}$$

$$\lim_{x \to 3} f(x) = 2$$

FIGURE 12.5

EXAMPLE 5   Using a Graph to Find a Limit

Find the limit of $f(x)$ as x approaches 3, where f is defined by

$$f(x) = \begin{cases} 2, & x \neq 3 \\ 0, & x = 3. \end{cases}$$

SOLUTION

Because $f(x) = 2$ for all x other than $x = 3$ and because the value of $f(3)$ is immaterial, it follows that the limit is 2, as shown in Figure 12.5. Thus, you can write

$$\lim_{x \to 3} f(x) = 2.$$

Limits That Fail to Exist

In the next three examples you will examine some functions for which limits do not exist.

EXAMPLE 6   Comparing Left and Right Behavior

Show that the following limit does not exist.

$$\lim_{x \to 0} \frac{|x|}{x}$$

SOLUTION

Consider the graph of the function $f(x) = |x|/x$. From Figure 12.6, you can see that for positive x-values,

$$\frac{|x|}{x} = 1, \qquad x > 0$$

and for negative x-values

$$\frac{|x|}{x} = -1, \qquad x < 0.$$

This means that no matter how close x gets to 0, there will be both positive and negative x-values that yield $f(x) = 1$ and $f(x) = -1$. This implies that the limit does not exist.

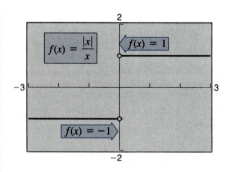

$$\lim_{x \to 0} f(x) \text{ does not exist.}$$

FIGURE 12.6

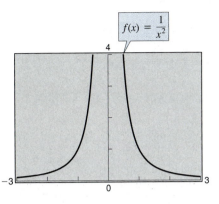

$$f(x) = \frac{1}{x^2}$$

$$\lim_{x \to 0} f(x) \text{ does not exist.}$$

FIGURE 12.7

EXAMPLE 7   Unbounded Behavior

Discuss the existence of the limit

$$\lim_{x \to 0} \frac{1}{x^2}.$$

SOLUTION

Let $f(x) = 1/x^2$. In Figure 12.7, note that as x approaches 0 from either the right or the left, $f(x)$ increases without bound. This means that by choosing x close enough to 0, you can force $f(x)$ to be as large as you want. For instance, $f(x)$ will be larger than 100 if you choose x that is within $\frac{1}{10}$ of 0. That is,

$$0 < |x| < \frac{1}{10} \quad \rightarrow \quad f(x) = \frac{1}{x^2} > 100.$$

Similarly, you can force $f(x)$ to be larger than 1,000,000, as follows.

$$0 < |x| < \frac{1}{1000} \quad \rightarrow \quad f(x) = \frac{1}{x^2} > 1,000,000$$

Because $f(x)$ is not approaching a real number L as x approaches 0, you can conclude that the limit does not exist.

$$f(x) = \sin\left(\frac{1}{x}\right)$$

$$\lim_{x \to 0} f(x) \text{ does not exist.}$$

FIGURE 12.8

EXAMPLE 8   Oscillating Behavior

Discuss the existence of the limit

$$\lim_{x \to 0} \sin\left(\frac{1}{x}\right).$$

SOLUTION

Let $f(x) = \sin(1/x)$. In Figure 12.8, you can see that as x approaches 0, $f(x)$ oscillates between -1 and 1. Therefore, the limit does not exist because no matter how close you are to 0, it is possible to choose values of x_1 and x_2 such that $\sin(1/x_1) = 1$ and $\sin(1/x_2) = -1$. as indicated in the table below.

x	$\dfrac{2}{\pi}$	$\dfrac{2}{3\pi}$	$\dfrac{2}{5\pi}$	$\dfrac{2}{7\pi}$	$\dfrac{2}{9\pi}$	$\dfrac{2}{11\pi}$	$x \to 0$
$\sin\left(\dfrac{1}{x}\right)$	1	-1	1	-1	1	-1	Limit does not exist.

Examples 6, 7, and 8 show three of the most common types of behavior associated with the nonexistence of limits.

CONDITIONS UNDER WHICH LIMITS DO NOT EXIST

The limit of $f(x)$ as $x \to c$ does not exist if any one of the following conditions is true.

1. $f(x)$ approaches a different number from the right side of c than it approaches from the left side of c.
2. $f(x)$ increases or decreases without bound as x approaches c.
3. $f(x)$ oscillates between two fixed values as x approaches c.

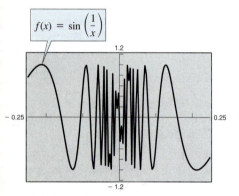

$$f(x) = \sin\left(\frac{1}{x}\right)$$

FIGURE 12.9

A graphing utility can help you discover the behavior of a function near the x-value at which you are trying to evaluate a limit. When you use a graphing utility, however, you should realize that you can't always trust the pictures that graphing utilities draw. For instance, if you use a graphing utility to sketch the graph of the function in Example 8 over an interval containing 0, you will most likely obtain an incorrect graph—such as that shown in Figure 12.9. The reason that a graphing utility can't show the correct graph is that the graph has infinitely many oscillations over any interval that contains 0.

Properties of Limits

You have seen that sometimes the limit of $f(x)$ as $x \to c$ is simply $f(c)$. In such cases, we say that the limit can be evaluated by *direct substitution*. That is,

$$\lim_{x \to c} f(x) = f(c). \qquad \text{\textit{Substitute c for x}}$$

There are many "well-behaved" functions that have this property—some of the basic ones are included in the following list.

PROPERTIES OF LIMITS

Let b and c be real numbers and let n be a positive integer.

1. $\lim_{x \to c} b = b$

2. $\lim_{x \to c} x = c$

3. $\lim_{x \to c} x^n = c^n$

4. $\lim_{x \to c} \sqrt[n]{x} = \sqrt[n]{c}$,   for n even and $c \geq 0$

Trigonometric functions could also have been included in this list. For instance,

$$\lim_{x \to \pi} \sin x = \sin \pi = 0 \quad \text{and} \quad \lim_{x \to 0} \cos x = \cos 0 = 1.$$

By combining the properties of limits with the following operations, you can find limits for a wide variety of functions.

OPERATIONS WITH LIMITS

Let b and c be real numbers, let n be a positive integer, and let f and g be functions with the following limits.

$$\lim_{x \to c} f(x) = L \quad \text{and} \quad \lim_{x \to c} g(x) = K$$

1. Scalar multiple: $\lim_{x \to c}[b\,f(x)] = bL$

2. Sum or difference: $\lim_{x \to c}[f(x) \pm g(x)] = L \pm K$

3. Product: $\lim_{x \to c}[f(x)g(x)] = LK$

4. Quotient: $\lim_{x \to c}\dfrac{f(x)}{g(x)} = \dfrac{L}{K},$ provided $K \neq 0$

5. Power: $\lim_{x \to c}[f(x)]^n = L^n$

EXAMPLE 9   **Finding Limits by Direct Substitution**

Find the following limits.

A. $\lim_{x \to 4} x^2$

B. $\lim_{x \to 4} 5$

C. $\lim_{x \to 4}(x^2 + 2x - 5)$

D. $\lim_{x \to 4} \dfrac{x^2 + 2x - 5}{x - 1}$

SOLUTION

You can use the Properties of Limits and direct substitution to evaluate each limit.

A. $\lim_{x \to 4} x^2 = (4)^2 = 16$

B. $\lim_{x \to 4} 5 = 5$

C. $\lim_{x \to 4}(x^2 + 2x - 5) = (4)^2 + 2(4) - 5 = 16 + 8 - 5 = 19$

D. $\lim_{x \to 4} \dfrac{x^2 + 2x - 5}{x - 1} = \dfrac{(4)^2 + 2(4) - 5}{4 - 1} = \dfrac{16 + 8 - 5}{3} = \dfrac{19}{3}$

When evaluating limits, remember that there are several ways to solve most problems. Often, a problem can be solved *numerically, graphically,* or *analytically.* For instance, the limits in Example 1–4 were found numerically (by constructing a table). The limit in Example 5 was found graphically, and the limits in Example 9 were found analytically.

DISCUSSION
PROBLEM
· ·
GRAPHS WITH
GAPS

Match each graph with one of the following functions.

$$f(x) = x + 1 \qquad g(x) = \frac{x^2 - 1}{x - 1} \qquad h(x) = \frac{x^3 - 2x^2 - x + 2}{x^2 - 3x + 2}$$

Find the limit of each function as x approaches 1 and as x approaches 2. What conclusion can you make?

(a)   (b)   (c)

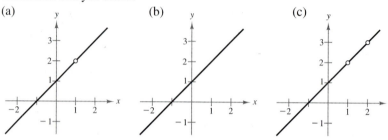

Use a graphing utility to graph each function. Does the graphing utility distinguish between the graphs? Explain your answer.

WARM-UP
· · · · · · · · · · · · · · · · · ·

The following warm-up exercises involve skills that were covered in earlier sections. You will use these skills in the exercise set for this section.

In Exercises 1–4, evaluate (if possible) the function at the specified value of the independent variable and simplify the results.

1. $f(x) = x^2 - 2x + 2$

(a) $f(-1)$
(b) $f(c)$
(c) $f(x + h)$

2. $f(x) = \begin{cases} 2x + 1, & x < 0 \\ 2x + 2, & x \geq 0 \end{cases}$

(a) $f(-1)$
(b) $f(2)$
(c) $f(t^2 + 1)$

3. $f(x) = 2x^2 - 3x + 1$

$$\frac{f(2 + h) - f(2)}{h}$$

4. $f(x) = \frac{3}{x}$

$$\frac{f(1 + h) - f(1)}{h}$$

In Exercises 5–8, find the domain and range of the function, and sketch its graph.

5. $g(x) = \frac{4}{x}$

6. $f(x) = \sqrt{16 - x^2}$

7. $f(x) = |x - 4|$

8. $f(x) = \frac{|x|}{x}$

In Exercises 9 and 10, determine whether y is a function of x.

9. $4x^2 + 9y^2 = 36$

10. $x^2y + 4x = y$

SECTION 12.1 · EXERCISES

In Exercises 1–4, complete the table to find the limit. Can the limit be reached? Explain your reasoning.

1. $\lim_{x \to 3}(3 - 2x)$

x	2.9	2.99	2.999	3.0	3.001	3.01	3.1
$f(x)$				?			

2. $\lim_{x \to 4}(\frac{1}{2}x^2 - 2x + 3)$

x	3.9	3.99	3.999	4.0	4.001	4.01	4.1
$f(x)$				?			

3. $\lim_{x \to 2} \dfrac{x - 2}{x^2 - 4}$

x	1.9	1.99	1.999	2.0	2.001	2.01	2.1
$f(x)$				?			

4. $\lim_{x \to -1} \dfrac{x + 1}{x^2 - x - 2}$

x	−1.1	−1.01	−1.001	−1.0
$f(x)$				?
x	−0.999	−0.99	−0.9	
$f(x)$				

In Exercises 5–10, use a calculator to complete the table and estimate the limit. Use a graphing utility to confirm your result.

5. $\lim_{x \to 1} \dfrac{x - 1}{x^2 + 2x - 3}$

x	0.9	0.99	0.999	1.0	1.001	1.01	1.1
$f(x)$				?			

6. $\lim_{x \to -3} \dfrac{x + 3}{x^2 + x - 6}$

x	−3.1	−3.01	−3.001	−3.0
$f(x)$				?
x	−2.999	−2.99	−2.9	
$f(x)$				

7. $\lim_{x \to -4} \dfrac{\frac{x}{x + 2} - 2}{x + 4}$

x	−4.1	−4.01	−4.001	−4.0
$f(x)$				?
x	−3.999	−3.99	−3.9	
$f(x)$				

8. $\lim_{x \to 2} \dfrac{\frac{1}{x + 2} - \frac{1}{4}}{x - 2}$

x	1.9	1.99	1.999	2.0	2.001	2.01	2.1
$f(x)$				?			

9. $\lim_{x \to 0} \dfrac{\sin x}{x}$

x	−0.1	−0.01	−0.001
$f(x)$			
x	0	0.001	0.01	0.1
$f(x)$	?			

10. $\lim_{x \to 0} \dfrac{\cos x - 1}{x}$

x	−0.1	−0.01	−0.001
$f(x)$			
x	0	0.001	0.01	0.1
$f(x)$	?			

In Exercises 11–16, use the graph to find the limit (if it exists). If the limit does not exist, state the reason.

11. $\lim_{x \to -1} \sin \dfrac{\pi x}{2}$

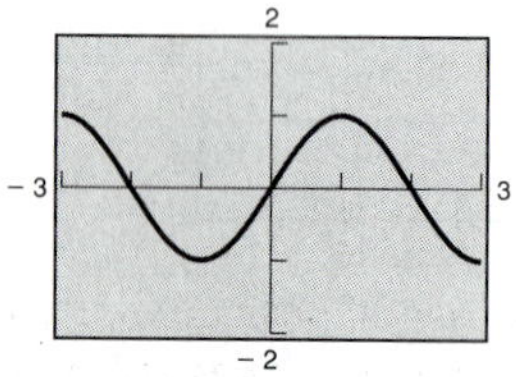

12. $\lim_{x \to 2} \dfrac{3x^2 - 12}{x - 2}$

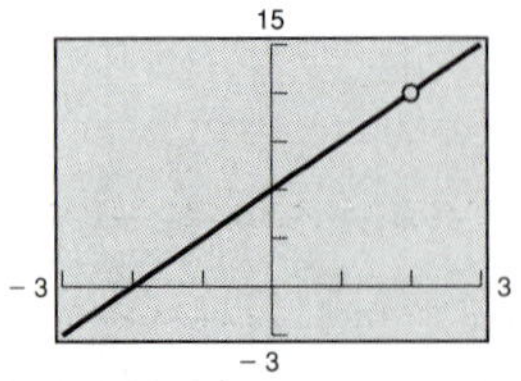

13. $\lim_{x \to -2} \dfrac{|x + 2|}{x + 2}$

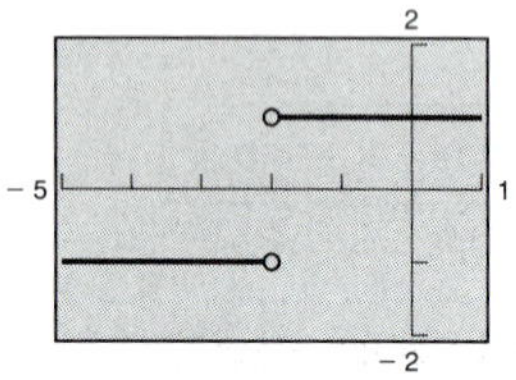

14. $\lim_{x \to 1} \dfrac{1}{x - 1}$

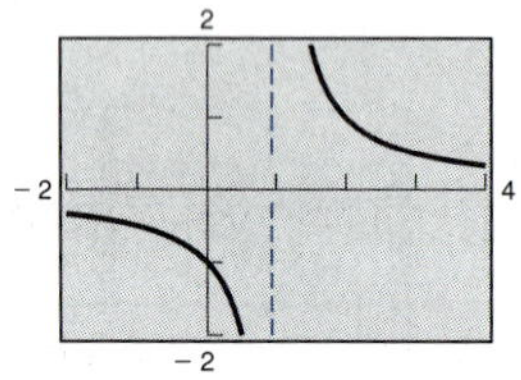

15. $\lim_{x \to 0} 2 \cos \dfrac{\pi}{x}$

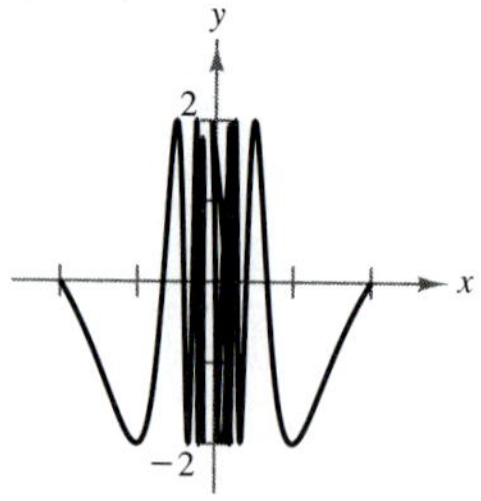

16. $\lim_{x \to \pi/3} \sec x$

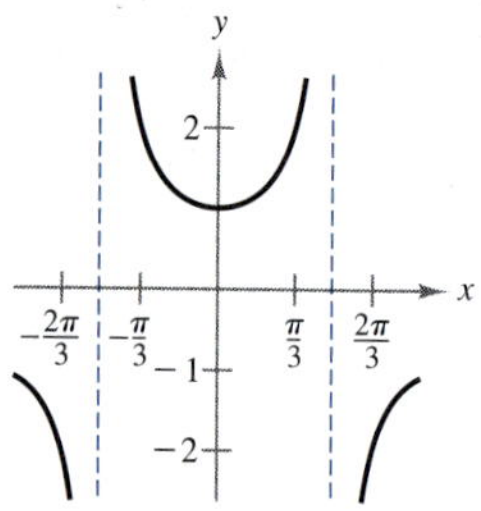

In Exercises 17–28, find the limit by direct substitution.

17. $\lim_{x \to 2} (x^2 + 3x - 4)$

18. $\lim_{x \to 2} \sqrt[3]{10x + 7}$

19. $\lim_{x \to 3} \dfrac{12}{x}$

20. $\lim_{x \to -5} \dfrac{4}{x + 2}$

21. $\lim_{x \to -1} \dfrac{x^2 - 1}{x}$

22. $\lim_{x \to 8} \dfrac{\sqrt{x + 1}}{x - 4}$

23. $\lim_{x \to 2} e^x$

24. $\lim_{x \to e} \ln x$

25. $\lim_{x \to \pi} \sin 2x$

26. $\lim_{x \to 1/2} \dfrac{\tan \pi x}{2}$

27. $\lim_{x \to 1/2} \arcsin x$

28. $\lim_{x \to 1} \arccos \dfrac{x}{2}$

In Exercises 29–31, find the limit (if it exists) of $f(x)$ as $x \to 2$.

29. $f(x) = \begin{cases} 2x + 1, & x < 2 \\ 2x + 2, & x \geq 2 \end{cases}$

30. $f(x) = \begin{cases} x^2 - 4, & x < 2 \\ x - 2, & x \geq 2 \end{cases}$

31. $f(x) = \begin{cases} x^2 - x, & x < 2 \\ x^2 - 2, & x \geq 2 \end{cases}$

32. If $\lim_{x \to c} f(x) = 4$ and $\lim_{x \to c} g(x) = 5$, find

(a) $\lim_{x \to c} [-2g(x)]$

(b) $\lim_{x \to c} [f(x) + g(x)]$

(c) $\lim_{x \to c} \dfrac{f(x)}{g(x)}$

(d) $\lim_{x \to c} \sqrt{f(x)}$.

33. If $\lim_{x \to c} f(x) = \frac{3}{2}$ and $\lim_{x \to c} g(x) = -\frac{1}{2}$, find

(a) $\lim_{x \to c} [f(x) + g(x)]^2$

(b) $\lim_{x \to c} [6f(x)g(x)]$

(c) $\lim_{x \to c} \dfrac{5g(x)}{4f(x)}$

(d) $\lim_{x \to c} \dfrac{1}{\sqrt{f(x)}}$.

34. Use a graphing utility to sketch the graph of the function

$$f(x) = \frac{x - 4}{\sqrt{x} - 2}.$$

Does the limit of $f(x)$ exist as $x \to 4$? What is the domain of the function? Can you determine the domain of a function solely by analyzing the graph drawn by a graphing utility? Write a short paragraph about the importance of examining a function analytically as well as graphically.

12.2 TECHNIQUES FOR EVALUATING LIMITS

Limits of Polynomial and Rational Functions / Cancellation Technique / Rationalizing Technique / Using Technology / One-Sided Limits / A Limit from Calculus

Limits of Polynomial and Rational Functions

In Section 12.1, you saw how direct substitution and operations with limits can be used to evaluate limits of certain well-behaved functions, such as polynomial functions and rational functions with nonzero denominators. This result is summarized as follows.

LIMITS OF POLYNOMIAL AND RATIONAL FUNCTIONS

1. If p is a polynomial function and c is a real number, then
$$\lim_{x \to c} p(x) = p(c).$$

2. If r is a rational function given by $r(x) = p(x)/q(x)$, and c is a real number such that $q(c) \neq 0$, then
$$\lim_{x \to c} r(x) = r(c) = \frac{p(c)}{q(c)}, \qquad q(c) \neq 0.$$

EXAMPLE 1   Evaluating Limits by Direct Substitution

Find the following limits.

A. $\displaystyle\lim_{x \to -1} (x^2 + x - 6)$ **B.** $\displaystyle\lim_{x \to -1} \frac{x^2 + x - 6}{x + 3}$

SOLUTION

The first function is a polynomial function and the second is a rational function (with a nonzero denominator at $x = -1$). Thus, you can evaluate the limits by direct substitution.

A. $\displaystyle\lim_{x \to -1} (x^2 + x - 6) = (-1)^2 + (-1) - 6 = -6$

B. $\displaystyle\lim_{x \to -1} \frac{x^2 + x - 6}{x + 3} = \frac{-6}{-1 + 3} = -3$

Cancellation Technique

In Example 1(B), suppose you were asked to find the limit as $x \to -3$.

$$\lim_{x \to -3} \frac{x^2 + x - 6}{x + 3}$$

Direct substitution would fail because -3 is a zero of the denominator. When a table is constructed, however, it appears that the limit of the function as $x \to -3$ is -5.

x	-3.01	-3.001	-3.0001	-3	-2.9999	-2.999	-2.99
$\dfrac{x^2 + x - 6}{x + 3}$	-5.01	-5.001	-5.0001	?	-4.9999	-4.999	-4.99

Another way to find the limit of this function is to factor the numerator and cancel common factors, as shown in Example 2.

EXAMPLE 2   Cancellation Technique

Find the following limit.

$$\lim_{x \to -3} \frac{x^2 + x - 6}{x + 3}$$

SOLUTION

Begin by factoring the numerator and canceling any common factors.

$$\lim_{x \to -3} \frac{x^2 + x - 6}{x + 3} = \lim_{x \to -3} \frac{(x - 2)(x + 3)}{x + 3} \qquad \textit{Factor numerator}$$

$$= \lim_{x \to -3} \frac{(x - 2)(x + 3)}{x + 3} \qquad \textit{Cancel common factor}$$

$$= \lim_{x \to -3} (x - 2) \qquad \textit{Simplify}$$

$$= -5 \qquad \textit{Direct substitution}$$

This procedure for evaluating a limit is called the *cancellation technique*. The validity of the procedure stems from the fact that if two functions agree at all but a single number c, they must have identical limit behavior at $x = c$. In Example 2, the functions

$$f(x) = \frac{x^2 + x - 6}{x + 3} \quad \text{and} \quad g(x) = x - 2$$

agree at all values of x other than $x = -3$. Hence, you can use $g(x)$ to find the limit of $f(x)$.

D I S C O V E R Y

Use a graphing utility to compare the graphs of

$$y = \frac{x^2 + x - 6}{x + 3}$$

and

$$y = x - 2.$$

Can you zoom-in close enough to the point $(-3, -5)$ so that the utility distinguishes between the two graphs? What can you conclude?

The cancellation technique should be applied only when direct substitution produces 0 in both the numerator *and* the denominator. The resulting fraction, $\frac{0}{0}$, has no meaning as a real number. It is called an **indeterminate form** because you cannot, from the form alone, determine the limit. When you encounter this form by direct substitution into a rational function, you can conclude that the numerator and denominator must have a common factor. After factoring and canceling, you should try direct substitution again.

EXAMPLE 3   Cancellation Technique

Find the following limit.

$$\lim_{x \to 1} \frac{x - 1}{x^3 - x^2 + x - 1}$$

SOLUTION

Begin by substituting $x = 1$ in the numerator and denominator.

$$1 - 1 = 0 \qquad \text{\textit{Numerator is 0 when } x = 1}$$
$$(1)^3 - (1)^2 + 1 - 1 = 0 \qquad \text{\textit{Denominator is 0 when } x = 1}$$

Because both the numerator and denominator are zero when $x = 1$, direct substitution will not yield the limit. To find the limit, you should factor the numerator and denominator, cancel any common factors, and then try direct substitution again.

$$\lim_{x \to 1} \frac{x - 1}{x^3 - x^2 + x - 1} = \lim_{x \to 1} \frac{x - 1}{(x - 1)(x^2 + 1)} \qquad \text{\textit{Factor}}$$

$$= \lim_{x \to 1} \frac{\cancel{x - 1}}{\cancel{(x - 1)}(x^2 + 1)} \qquad \text{\textit{Cancel}}$$

$$= \lim_{x \to 1} \frac{1}{x^2 + 1} \qquad \text{\textit{Simplify}}$$

$$= \frac{1}{(1)^2 + 1} \qquad \text{\textit{Substitute}}$$

$$= \frac{1}{2} \qquad \text{\textit{Simplify}}$$

This result is shown graphically in Figure 12.10.

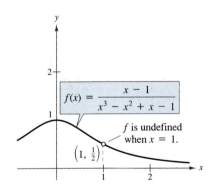

$$f(x) = \frac{x - 1}{x^3 - x^2 + x - 1}$$

f is undefined when $x = 1$.

$\left(1, \frac{1}{2}\right)$

FIGURE 12.10

REMARK   In Example 3, the factorization of the denominator can be obtained by dividing by $(x - 1)$ or by grouping as follows.

$$x^3 - x^2 + x - 1 = x^2(x - 1) + (x - 1)$$
$$= (x - 1)(x^2 + 1)$$

Rationalizing Technique

EXAMPLE 4 **Rationalizing Technique**

Find the following limit.

$$\lim_{x \to 0} \frac{\sqrt{x + 1} - 1}{x}$$

SOLUTION

By direct substitution, you obtain the indeterminate form $\frac{0}{0}$.

$$\lim_{x \to 0} \frac{\sqrt{x + 1} - 1}{x} = \frac{0}{0} \qquad \textit{Indeterminate form}$$

In this case, you can rewrite the fraction by rationalizing the numerator.

$$\frac{\sqrt{x + 1} - 1}{x} = \left(\frac{\sqrt{x + 1} - 1}{x}\right)\left(\frac{\sqrt{x + 1} + 1}{\sqrt{x + 1} + 1}\right)$$

$$= \frac{(x + 1) - 1}{x(\sqrt{x + 1} + 1)} \qquad \textit{Multiply}$$

$$= \frac{x}{x(\sqrt{x + 1} + 1)} \qquad \textit{Simplify}$$

$$= \frac{\cancel{x}}{\cancel{x}(\sqrt{x + 1} + 1)} \qquad \textit{Cancel}$$

$$= \frac{1}{\sqrt{x + 1} + 1}, \qquad x \neq 0 \qquad \textit{Simplify}$$

Now you can evaluate the limit by direct substitution.

$$\lim_{x \to 0} \frac{\sqrt{x + 1} - 1}{x} = \lim_{x \to 0} \frac{1}{\sqrt{x + 1} + 1} = \frac{1}{1 + 1} = \frac{1}{2}$$

You can reinforce your conclusion that the limit is $\frac{1}{2}$ by constructing a table, or by sketching a graph, as shown in Figure 12.11.

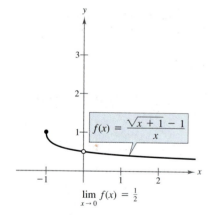

$$f(x) = \frac{\sqrt{x + 1} - 1}{x}$$

$$\lim_{x \to 0} f(x) = \frac{1}{2}$$

FIGURE 12.11

x	-0.1	-0.01	-0.001	0	0.001	0.01	0.1
$f(x)$	0.5132	0.5013	0.5001	?	0.4999	0.4988	0.4881

REMARK The rationalization technique for evaluating limits is based on multiplication by a convenient form of 1. In Example 4, the convenient form is

$$1 = \frac{\sqrt{x + 1} + 1}{\sqrt{x + 1} + 1}.$$

Using Technology

The cancellation and rationalization techniques work well for finding the limit of a rational function or for finding the limit of a function involving a radical. To find limits of nonalgebraic functions, you often need to use more sophisticated analytic techniques. For instance, how would you find the following limits?

$$\lim_{x \to 0}(1 + x)^{1/x} \quad \text{and} \quad \lim_{x \to 0} \frac{\sin x}{x}$$

Both of these limits are important in calculus. The next two examples show how technology can be used to approximate limits.

EXAMPLE 5   Approximating a Limit

Approximate the limit $\lim_{x \to 0}(1 + x)^{1/x}$.

SOLUTION

To approximate the limit graphically, enter the function

$$f(x) = (1 + x)^{1/x}$$

in a graphing utility, as shown in Figure 12.12. Using the *zoom* and *trace* features of the utility, choose two points on the graph of f, such as

$(-0.001645, 2.7205)$ and $(0.001645, 2.7160)$.

Because the x-coordinates of these two points are equidistant from 0, you can approximate the limit as the average of the y-coordinates. That is,

$$\lim_{x \to 0}(1 + x)^{1/x} \approx \frac{2.7205 + 2.7160}{2} = 2.71825.$$

Using analytic means, it can be shown that this approximation is accurate to four decimal places. The actual limit is $e \approx 2.71828$.

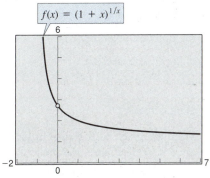

FIGURE 12.12

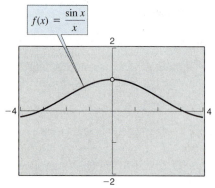

FIGURE 12.13

EXAMPLE 6   Approximating a Limit

Approximate the limit $\lim\limits_{x \to 0} \dfrac{\sin x}{x}$.

SOLUTION

Notice that direct substitution does not work because it produces the indeterminate form $\frac{0}{0}$. To approximate the limit, use the procedure described in Example 5. Begin by sketching the graph of $f(x) = (\sin x)/x$, as shown in Figure 12.13. Then use the *zoom* and *trace* features of the utility to choose a point on each side of 0, such as

$$(-0.001234, 0.9999998) \quad \text{and} \quad (0.001234, 0.9999998).$$

Finally, use interpolation to approximate the limit as the average of the *y*-coordinates of these two points.

$$\lim_{x \to 0} \frac{\sin x}{x} \approx 0.9999998$$

By use of analytic means, it can be shown that this limit is exactly 1.

One-Sided Limits

In Section 12.1, you saw that one way in which a limit can fail to exist is when a function approaches a different value from the left side of c than it approaches from the right side of c. This type of behavior can be described more concisely with the concept of a **one-sided limit.**

$$\lim_{x \to c^-} f(x) = L \qquad\qquad \textit{Limit from the left}$$
$$\lim_{x \to c^+} f(x) = L \qquad\qquad \textit{Limit from the right}$$

EXAMPLE 7   Evaluating One-Sided Limits

Find the limit as $x \to 0$ from the left and the limit as $x \to 0$ from the right for the function given by

$$f(x) = \frac{|\,2x\,|}{x}.$$

SOLUTION

From the graph of f, shown in Figure 12.14, you can see that $f(x) = -2$ for all $x < 0$. Therefore, the limit from the left is

$$\lim_{x \to 0^-} \frac{|\,2x\,|}{x} = -2. \qquad\qquad \textit{Limit from the left}$$

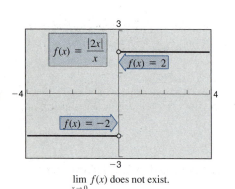

$\lim\limits_{x \to 0} f(x)$ does not exist.

FIGURE 12.14

Because $f(x) = 2$ for all $x > 0$, the limit from the right is

$$\lim_{x \to 0^+} \frac{|2x|}{x} = 2.$$ *Limit from the right*

In Example 7, note that the function approaches different limits from the left and from the right. In such cases, the limit of $f(x)$ as $x \to c$ does not exist. For the limit of a function to exist as $x \to c$, it must be true that both one-sided limits exist and are equal.

EXISTENCE OF A LIMIT

If f is a function and c and L are real numbers, then

$$\lim_{x \to c} f(x) = L$$

if and only if both the left and right limits are equal to L.

EXAMPLE 8   Evaluating One-Sided Limits

Find the limit of $f(x)$ as x approaches 1.

$$f(x) = \begin{cases} 4 - x, & x < 1 \\ 4x - x^2, & x > 1 \end{cases}$$

SOLUTION

Remember that you are concerned about the value of f *near* $x = 1$ rather than *at* $x = 1$. Thus, for $x < 1$, $f(x)$ is given by $4 - x$, and you can use direct substitution to obtain

$$\lim_{x \to 1^-} f(x) = \lim_{x \to 1^-} (4 - x) = 4 - 1 = 3.$$

For $x > 1$, $f(x)$ is given by $4x - x^2$, and you can use direct substitution to obtain

$$\lim_{x \to 1^+} f(x) = \lim_{x \to 1^+} (4x - x^2) = 4 - 1 = 3.$$

Because the one-sided limits both exist and are equal to 3, it follows that

$$\lim_{x \to 1} f(x) = 3.$$

The graph in Figure 12.15 confirms this conclusion.

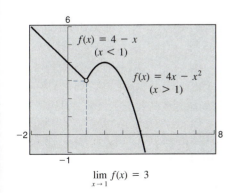

$f(x) = 4 - x$
$(x < 1)$

$f(x) = 4x - x^2$
$(x > 1)$

$\lim_{x \to 1} f(x) = 3$

FIGURE 12.15

EXAMPLE 9 **Comparing Limits from the Left and Right**

An overnight delivery service charges $8 for the first pound and $2 for each additional pound. Let x represent the weight of a parcel and let $f(x)$ represent the shipping cost.

$$f(x) = \begin{cases} 8, & 0 < x \le 1 \\ 10, & 1 < x \le 2 \\ 12, & 2 < x \le 3 \end{cases}$$

Show that the limit of $f(x)$ as $x \to 2$ does not exist.

SOLUTION

The graph of f is shown in Figure 12.16. The limit of $f(x)$ as x approaches 2 from the left is

$$\lim_{x \to 2^-} f(x) = 10,$$

whereas the limit of $f(x)$ as x approaches 2 from the right is

$$\lim_{x \to 2^+} f(x) = 12.$$

Because these one-sided limits are not equal, the limit of $f(x)$ as $x \to 2$ does not exist.

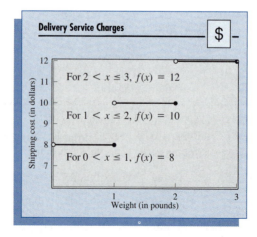

FIGURE 12.16

A Limit from Calculus

In the next section, you will study an important type of limit from calculus—the limit of a *difference quotient*.

$$\lim_{h \to 0} \frac{f(x + h) - f(x)}{h}$$

Direct substitution into the difference quotient always produces the indeterminate form $\frac{0}{0}$.

EXAMPLE 10 Evaluating a Limit from Calculus

For the function $f(x) = x^2 - 1$, find the following limit.

$$\lim_{h \to 0} \frac{f(3 + h) - f(3)}{h}$$

SOLUTION

Direct substitution produces an indeterminate form.

$$\lim_{h \to 0} \frac{f(3 + h) - f(3)}{h} = \lim_{h \to 0} \frac{[(3 + h)^2 - 1] - [(3)^2 - 1]}{h}$$

$$= \lim_{h \to 0} \frac{9 + 6h + h^2 - 1 - 9 + 1}{h}$$

$$= \lim_{h \to 0} \frac{6h + h^2}{h}$$

$$= \frac{0}{0}$$

By factoring and canceling, you can obtain the following.

$$\lim_{h \to 0} \frac{f(3 + h) - f(3)}{h} = \lim_{h \to 0} \frac{h(6 + h)}{h} = \lim_{h \to 0}(6 + h) = 6$$

Thus, the limit is 6.

DISCUSSION PROBLEM

LIMITS OF RATIONAL FUNCTIONS

Consider the limit of the rational function $p(x)/q(x)$, where p and q are polynomial functions. What conclusion can you make if direct substitution produces the given expression? Explain your reasoning.

(a) $\lim_{x \to c} \dfrac{p(x)}{q(x)} = \dfrac{0}{1}$ (b) $\lim_{x \to c} \dfrac{p(x)}{q(x)} = \dfrac{1}{1}$

(c) $\lim_{x \to c} \dfrac{p(x)}{q(x)} = \dfrac{1}{0}$ (d) $\lim_{x \to c} \dfrac{p(x)}{q(x)} = \dfrac{0}{0}$

WARM-UP

The following warm-up exercises involve skills that were covered in earlier sections. You will use these skills in the exercise set for this section.

In Exercises 1–4, completely factor the polynomial.

1. $x^3 - 36x$

2. $10x^2 - 13x - 3$

3. $s^3 + 27$

4. $y^3 - y^2 + y - 1$

In Exercises 5–8, write the fraction in reduced form.

5. $\dfrac{x^2 - 36}{6 - x}$

6. $\dfrac{x^2 - 4}{x^3 + x^2 - 6x}$

7. $\dfrac{x - 3}{2x^3 + 4x^2 - 18x + 36}$

8. $\dfrac{x^3 + 6x^2 - x - 6}{x + 6}$

In Exercises 9 and 10, rationalize the numerator.

9. $\dfrac{\sqrt{x + 1} - 2}{x - 3}$

10. $\dfrac{\sqrt{3 + x} - \sqrt{3}}{x}$

SECTION 12.2 · EXERCISES

In Exercises 1–4, use the graph to determine each limit (if it exists). Then identify another function that agrees with the given function at all but one point.

1. $g(x) = \dfrac{-2x^2 + x}{x}$

2. $h(x) = \dfrac{x^2 - 3x}{x}$

3. $g(x) = \dfrac{x^3 - x}{x - 1}$

4. $f(x) = \dfrac{x^2 - 1}{x + 1}$

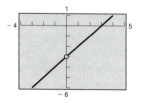

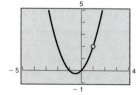

(a) $\lim\limits_{x \to 0} g(x)$

(b) $\lim\limits_{x \to -1} g(x)$

(c) $\lim\limits_{x \to -2} g(x)$

(a) $\lim\limits_{x \to -2} h(x)$

(b) $\lim\limits_{x \to 0} h(x)$

(c) $\lim\limits_{x \to 3} h(x)$

(a) $\lim\limits_{x \to 1} g(x)$

(b) $\lim\limits_{x \to -1} g(x)$

(c) $\lim\limits_{x \to 0} g(x)$

(a) $\lim\limits_{x \to 1} f(x)$

(b) $\lim\limits_{x \to 2} f(x)$

(c) $\lim\limits_{x \to -1} f(x)$

In Exercises 5–10, find the limit by direct substitution.

5. $\lim\limits_{x \to 5} 5(x^2 - 1)$

6. $\lim\limits_{x \to 10} (x^2 - 5x - 25)$

7. $\lim\limits_{x \to -3} \ln e^x$

8. $\lim\limits_{x \to -1} e^{x^2 - 1}$

9. $\lim\limits_{\theta \to 1} \sin \dfrac{2\pi\theta}{3}$

10. $\lim\limits_{t \to \pi/2} \cot t$

In Exercises 11–22, find the limit (if it exists).

11. $\lim\limits_{x \to 7} \dfrac{x - 7}{x^2 - 49}$

12. $\lim\limits_{x \to 3} \dfrac{3 - x}{x^2 - 9}$

13. $\lim\limits_{x \to -1} \dfrac{1 - 2x - 3x^2}{1 + x}$

14. $\lim\limits_{t \to -2} \dfrac{t^3 + 8}{t + 2}$

15. $\lim\limits_{y \to 0} \dfrac{\sqrt{3 + y} - \sqrt{3}}{y}$

16. $\lim\limits_{z \to 0} \dfrac{\sqrt{6 - z} - \sqrt{6}}{z}$

17. $\lim\limits_{x \to -3} \dfrac{\sqrt{x + 7} - 2}{x + 3}$

18. $\lim\limits_{x \to 2} \dfrac{4 - \sqrt{18 - x}}{x - 2}$

19. $\lim\limits_{x \to 0} \dfrac{\dfrac{1}{x + 1} - 1}{x}$

20. $\lim\limits_{x \to 0} \dfrac{\dfrac{1}{5 - x} - \dfrac{1}{5}}{x}$

21. $\lim\limits_{x \to 0} \dfrac{\sec x}{\tan x}$

22. $\lim\limits_{x \to \pi/2} \dfrac{1 - \sin x}{\cos x}$

In Exercises 23–28, graph the function. Determine the limit (if it exists) by evaluating the corresponding one-sided limits.

23. $\lim\limits_{x \to 3} \dfrac{|x - 3|}{x - 3}$

24. $\lim\limits_{x \to 3} |x - 3|$

25. $\lim\limits_{x \to 1} \dfrac{1}{x^2 + 1}$

26. $\lim\limits_{x \to 1} \dfrac{1}{x^2 - 1}$

27. $\lim\limits_{x \to 2} f(x)$

$f(x) = \begin{cases} x - 1, & x \le 2 \\ 2x - 3, & x > 2 \end{cases}$

28. $\lim\limits_{x \to 1} f(x)$

$f(x) = \begin{cases} 4 - x^2, & x \le 1 \\ 3 - x, & x > 1 \end{cases}$

In Exercises 29–31, use a graphing utility to estimate the limit. Construct a table to reinforce your conclusion. Then find the limit by analytic methods.

29. $\lim\limits_{x \to 0} \dfrac{\sqrt{1 - x} - 1}{x}$

30. $\lim\limits_{x \to 0} \dfrac{[1/(3 + x)] - (1/3)}{x}$

31. $\lim\limits_{x \to 0} \dfrac{[1/(2x - 5)] + (1/5)}{x}$

In Exercises 32–34, find the limit of $[f(x + h) - f(x)]/h$ as $h \to 0$.

32. $f(x) = 3x - 1$

33. $f(x) = x^2 - 3x$

34. $f(x) = 4 - 2x - x^2$

In Exercises 35–42, use a graphing utility to approximate the limit. Write an approximation that you believe is accurate to within three decimal places.

35. $\lim\limits_{x \to 0^+} (x \ln x)$

36. $\lim\limits_{x \to 0^+} (x^2 \ln x)$

37. $\lim\limits_{x \to 0} \dfrac{\sin 2x}{x}$

38. $\lim\limits_{x \to 0} \dfrac{\sin 3x}{x}$

39. $\lim\limits_{x \to 0} \dfrac{\tan x}{x}$

40. $\lim\limits_{x \to 0} \dfrac{1 - \cos 2x}{x}$

41. $\lim\limits_{x \to 1} \dfrac{1 - \sqrt[3]{x}}{1 - x}$

42. $\lim\limits_{x \to 0} (1 + 2x)^{1/x}$

In Exercises 43–45, state which limit can be evaluated using direct substitution. Then evaluate or approximate each limit.

43. (a) $\lim\limits_{x \to 0} x^2 \sin x^2$

(b) $\lim\limits_{x \to 0} \dfrac{\sin x^2}{x^2}$

44. (a) $\lim\limits_{x \to 0} \dfrac{x}{\cos x}$

(b) $\lim\limits_{x \to 0} \dfrac{1 - \cos x}{x}$

45. (a) $\lim\limits_{x \to 0} (1 + x^2)^x$

(b) $\lim\limits_{x \to 0} (1 + x^2)^{1/x}$

Velocity of a Free-Falling Object   In Exercises 46 and 47, use the position function $s(t) = -16t^2 + 128$ that gives the height (in feet) of a free-falling object. The velocity at time $t = a$ seconds is

$$\lim\limits_{t \to a} \dfrac{s(a) - s(t)}{a - t}.$$

46. Find the velocity when $t = 1$ second.

47. Find the velocity when $t = 2$ seconds.

12.3 THE TANGENT LINE PROBLEM

Tangent Line to a Graph / Slope of a Graph / Slope and the Limit Process / The Derivative of a Function

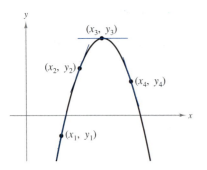

The slope of a graph changes from one point to another.

FIGURE 12.17

Tangent Line to a Graph

Calculus is a branch of mathematics that studies rates of change of functions. If you go on to take a course in calculus, you will learn that rates of change have many applications in real life.

Earlier in the text, you learned how the slope of a line indicates the rate at which a line rises or falls. For a line, this rate (or slope) is the same at every point on the line. For graphs other than lines, the rate at which the graph rises or falls changes from point to point. For instance, in Figure 12.17, the parabola is rising more quickly at the point (x_1, y_1) than it is at the point (x_2, y_2). At the vertex (x_3, y_3), the graph levels off, and at the point (x_4, y_4), the graph is falling.

To determine the rate at which a graph rises or falls at a *single point,* you can find the slope of the tangent line at the point. In simple terms, the **tangent line** to a graph of a function f at a point $P(x_1, y_1)$ is the line that best approximates the graph at the point. Figure 12.18 shows other examples of tangent lines.

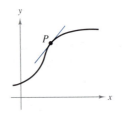 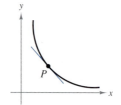 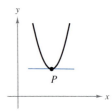

Tangent Line to a Graph at a Point

FIGURE 12.18

From geometry, you know that a line is tangent to a circle if the line intersects the circle at only one point. Tangent lines to noncircular graphs can intersect the graph at more than one point. For instance, in the first graph above, if the tangent line were extended, it would intersect the graph at a point other than the point of tangency.

Slope of a Graph

Because a tangent line approximates the graph at a point, the problem of finding the slope of a graph at a point becomes one of finding the slope of the tangent line at the point.

EXAMPLE 1   **Approximating the Slope of a Graph**

Use the graph in Figure 12.19 to approximate the slope of the graph of $f(x) = x^2$ at the point $(1, 1)$.

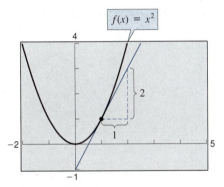

FIGURE 12.19

SOLUTION

From the graph of $f(x) = x^2$, you can see that the tangent line at $(1, 1)$ rises approximately 2 units for each unit change in x. Thus, the slope of the tangent line at $(1, 1)$ is given by

$$\text{Slope} = \frac{\text{change in } y}{\text{change in } x} \approx \frac{2}{1} = 2.$$

Because the tangent line at the point $(1, 1)$ has a slope of about 2, you can conclude that the graph has a slope of about 2 at the point $(1, 1)$.

When visually approximating the slope of a graph, note that the scales on the horizontal and vertical axes may differ. When this happens (as it frequently does in applications), the slope of the tangent line is distorted, and you must be careful to account for the difference in scales.

EXAMPLE 2   Approximating the Slope of a Graph

Figure 12.20 graphically depicts the average daily temperature (in degrees Fahrenheit) in Dallas, Texas. Estimate the slope of this graph at the indicated point and give a physical interpretation of the result.

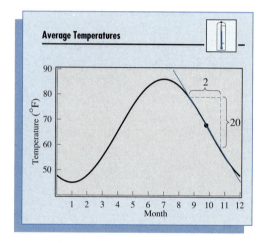

FIGURE 12.20

SOLUTION

From the graph, you can see that the tangent line at the given point falls approximately 20 units for each 2-unit change in x. Thus, you can estimate the slope at the given point to be

$$\text{Slope} = \frac{\text{change in } y}{\text{change in } x} \approx \frac{-20}{2} = -10 \text{ degrees per month.}$$

This means that you can expect the average daily temperatures in November to be about 10 degrees lower than the corresponding temperatures in October.

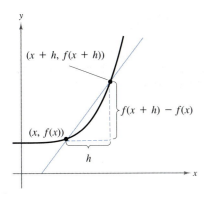

The Secant Line Through
$(x, f(x))$ and $(x + h, f(x + h))$

FIGURE 12.21

Slope and the Limit Process

In Examples 1 and 2, you approximated the slope of a graph at a point by making a careful graph and then "eyeballing" the tangent line at the point of tangency. A more precise method of approximating tangent lines makes use of a **secant line** through the point of tangency and a second point on the graph, as shown in Figure 12.21. If $(x, f(x))$ is the point of tangency and $(x + h, f(x + h))$ is a second point on the graph of f, the slope of the secant line through the two points is

$$m_{\text{sec}} = \frac{f(x + h) - f(x)}{h}. \qquad \textit{Slope of secant line}$$

The right side of this equation is called the **difference quotient.** The denominator h is the **change in x,** and the numerator is the **change in y.** The beauty of this procedure is that you obtain better and better approximations to the slope of the tangent line by choosing the second point closer and closer to the point of tangency, as shown in Figure 12.22.

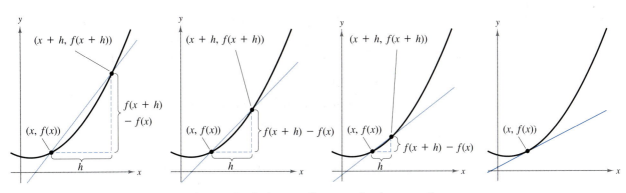

FIGURE 12.22

As h approaches 0, the secant line approaches the tangent line.

Using the limit process, you can find the *exact* slope of the tangent line at $(x, f(x))$.

DEFINITION OF THE SLOPE OF A GRAPH

The **slope** m of the graph of f at the point $(x, f(x))$ is equal to the slope of its tangent line at $(x, f(x))$, and is given by

$$m = \lim_{h \to 0} m_{\text{sec}} = \lim_{h \to 0} \frac{f(x + h) - f(x)}{h},$$

provided this limit exists.

EXAMPLE 3   **Finding the Slope of a Graph**

Find the slope of the graph of $f(x) = x^2$ at the point $(-2, 4)$.

SOLUTION

Find an expression that represents the slope of a secant line at the point $(-2, 4)$.

$$m_{\sec} = \frac{f(-2 + h) - f(-2)}{h} \qquad \textit{Set up difference quotient}$$

$$= \frac{(-2 + h)^2 - (-2)^2}{h} \qquad \textit{Use } f(x) = x^2$$

$$= \frac{4 - 4h + h^2 - 4}{h} \qquad \textit{Expand terms}$$

$$= \frac{-4h + h^2}{h} \qquad \textit{Simplify}$$

$$= \frac{\cancel{h}(-4 + h)}{\cancel{h}} \qquad \textit{Factor and cancel}$$

$$= -4 + h, \qquad h \neq 0 \qquad \textit{Simplify}$$

Next, take the limit of $m_{\sec}$ as $h \to 0$.

$$m = \lim_{h \to 0} m_{\sec} = \lim_{h \to 0}(-4 + h) = -4$$

Thus, the graph of f has a slope of -4 at the point $(-2, 4)$ (see Figure 12.23).

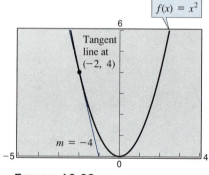

FIGURE 12.23

EXAMPLE 4   **Finding the Slope of a Graph**

Find the slope of $f(x) = -2x + 4$.

SOLUTION

You know from your study of linear functions that the line given by $f(x) = -2x + 4$ has a slope of -2, as shown in Figure 12.24. This conclusion is consistent with that obtained by the limit definition of slope.

$$m = \lim_{h \to 0} \frac{f(x + h) - f(x)}{h}$$

$$= \lim_{h \to 0} \frac{[-2(x + h) + 4] - [-2x + 4]}{h}$$

$$= \lim_{h \to 0} \frac{-2x - 2h + 4 + 2x - 4}{h}$$

$$= \lim_{h \to 0} \frac{-2h}{h}$$

$$= -2.$$

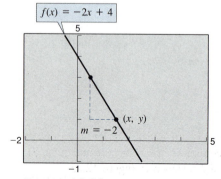

FIGURE 12.24

It is important that you see the difference between the ways in which the difference quotients were set up in Examples 3 and 4. In Example 3, you were finding the slope of a graph at a specific point $(c, f(c))$. To find the slope, you can use the following form of a difference quotient.

$$m = \lim_{h \to 0} \frac{f(c + h) - f(c)}{h} \qquad \textit{Slope at specific point}$$

In Example 4, however, you were finding a formula for the slope at *any* point on the graph. In such cases, you should use x, rather than c, in the difference quotient.

$$m = \lim_{h \to 0} \frac{f(x + h) - f(x)}{h} \qquad \textit{Formula for slope}$$

Except for linear functions, this form will always produce a function of x, which then can be evaluated to find the slope at any desired point.

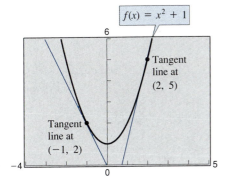

$f(x) = x^2 + 1$

Tangent line at $(2, 5)$

Tangent line at $(-1, 2)$

FIGURE 12.25

Technology Note

Try verifying the result in Example 5 by graphing the three functions

$y_1 = x^2 + 1$

$y_2 = -2x$

$y_3 = 4x - 3$

on the same viewing rectangle. Some graphing utilities even have the capability to automatically graph the tangent line to a curve at a given point. If you have such a graphing utility, try verifying Example 5.

EXAMPLE 5   Finding a Formula for the Slope of a Graph

Find a formula for the slope of the graph of $f(x) = x^2 + 1$. What is the slope at the points $(-1, 2)$ and $(2, 5)$?

SOLUTION

$$m_{\text{sec}} = \frac{f(x + h) - f(x)}{h} \qquad \textit{Set up difference quotient}$$

$$= \frac{[(x + h)^2 + 1] - [x^2 + 1]}{h} \qquad \textit{Use } f(x) = x^2 + 1$$

$$= \frac{x^2 + 2xh + h^2 + 1 - x^2 - 1}{h} \qquad \textit{Expand terms}$$

$$= \frac{2xh + h^2}{h} \qquad \textit{Simplify}$$

$$= \frac{h(2x + h)}{h} \qquad \textit{Factor and cancel}$$

$$= 2x + h, \quad h \neq 0 \qquad \textit{Simplify}$$

Next, take the limit of m_{sec} as $h \to 0$.

$$m = \lim_{h \to 0} m_{\text{sec}} = \lim_{h \to 0} (2x + h) = 2x$$

Using the formula $m = 2x$ for the slope at $(x, f(x))$, you can find the slope at the specified points. At $(-1, 2)$, the slope is $m = 2(-1) = -2$, and at $(2, 5)$, the slope is $m = 2(2) = 4$. The graph of f is shown in Figure 12.25.

The Derivative of a Function

In Example 5, you started with the function $f(x) = x^2 + 1$, and used the limit process to derive another function, $m = 2x$, that represents the slope of the graph of f at the point $(x, f(x))$. This derived function is called the **derivative** of f at x. It is denoted by $f'(x)$, which is read as "f prime of x."

DEFINITION OF THE DERIVATIVE

The **derivative** of f at x is given by

$$f'(x) = \lim_{h \to 0} \frac{f(x + h) - f(x)}{h},$$

provided this limit exists.

Remember that the derivative, $f'(x)$, is a formula for the slope of the tangent line to the graph of f at the point $(x, f(x))$.

EXAMPLE 6   Finding a Derivative

Find the derivative of $f(x) = 3x^2 - 2x$.

SOLUTION

$$
\begin{aligned}
f'(x) &= \lim_{h \to 0} \frac{f(x + h) - f(x)}{h} \\
&= \lim_{h \to 0} \frac{[3(x + h)^2 - 2(x + h)] - [3x^2 - 2x]}{h} \\
&= \lim_{h \to 0} \frac{3x^2 + 6xh + 3h^2 - 2x - 2h - 3x^2 + 2x}{h} \\
&= \lim_{h \to 0} \frac{6xh + 3h^2 - 2h}{h} \\
&= \lim_{h \to 0} \frac{h(6x + 3h - 2)}{h} \\
&= \lim_{h \to 0} (6x + 3h - 2) \\
&= 6x - 2
\end{aligned}
$$

Thus, the derivative of $f(x) = 3x^2 - 2x$ is

$$f'(x) = 6x - 2.$$

EXAMPLE 7   Using the Derivative

Find $f'(x)$ for $f(x) = \sqrt{x}$. Then find the slope of the graph of f at the points $(1, 1)$ and $(4, 2)$.

SOLUTION

Use the procedure for rationalizing numerators, as discussed in Section 12.2.

$$
\begin{aligned}
f'(x) &= \lim_{h \to 0} \frac{f(x + h) - f(x)}{h} \\
&= \lim_{h \to 0} \frac{\sqrt{x + h} - \sqrt{x}}{h} \\
&= \lim_{h \to 0} \left(\frac{\sqrt{x + h} - \sqrt{x}}{h} \right) \left(\frac{\sqrt{x + h} + \sqrt{x}}{\sqrt{x + h} + \sqrt{x}} \right) \\
&= \lim_{h \to 0} \frac{(x + h) - x}{h(\sqrt{x + h} + \sqrt{x})} \\
&= \lim_{h \to 0} \frac{h}{h(\sqrt{x + h} + \sqrt{x})} \\
&= \lim_{h \to 0} \frac{1}{\sqrt{x + h} + \sqrt{x}} \\
&= \frac{1}{2\sqrt{x}}
\end{aligned}
$$

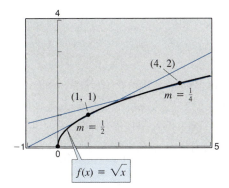

(4, 2)

$m = \frac{1}{4}$

(1, 1)

$m = \frac{1}{2}$

$f(x) = \sqrt{x}$

FIGURE 12.26

At the point $(1, 1)$, the slope is $f'(1) = \frac{1}{2}$. At the point $(4, 2)$, the slope is $f'(4) = \frac{1}{4}$. The graph of f is shown in Figure 12.26.

DISCUSSION PROBLEM

USING A DERIVATIVE TO FIND SLOPE

In many applications, it is convenient to use a variable other than x as the independent variable. Complete the following limit process to find the derivative of $f(t) = 3/t$. Then use the result to find the slope of the graph of $f(t) = 3/t$ at the point $(6, \frac{1}{2})$.

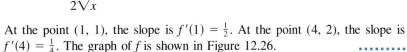

$$
\begin{aligned}
f'(t) &= \lim_{h \to 0} \frac{f(t + h) - f(t)}{h} \\
&= \lim_{h \to 0} \frac{\dfrac{3}{t + h} - \dfrac{3}{t}}{h} \\
&= \lim_{h \to 0} \frac{\dfrac{3t - 3(t + h)}{t(t + h)}}{h} \\
&= \cdots
\end{aligned}
$$

WARM-UP

The following warm-up exercises involve skills that were covered in earlier sections. You will use these skills in the exercise set for this section.

In Exercises 1–4, determine the slope of the line through the points.

1. $(-1, -2), (4, 8)$

2. $\left(\frac{2}{3}, -2\right), \left(\frac{10}{3}, -2\right)$

3. $\left(-1, \frac{3}{5}\right), \left(\frac{12}{5}, 0\right)$

4. $(0.4, 7.3), (5.8, -0.9)$

In Exercises 5–8, find the general form of the equation of the line through the given point with the specified slope.

5. $(3, 9), m = \frac{3}{4}$

6. $(-2, 0), m$ is undefined

7. $\left(4, \frac{7}{8}\right), m = 0$

8. $\left(0, \frac{15}{2}\right), m = -\frac{3}{2}$

In Exercises 9 and 10, find the limit (if it exists).

9. $\lim\limits_{h \to 0} \dfrac{3(x + h) - 3x}{h}$

10. $\lim\limits_{h \to 0} \dfrac{(x + h)^2 - x^2}{h}$

SECTION 12.3 · EXERCISES

In Exercises 1–4, estimate the slope of the curve at the point (x, y).

1.

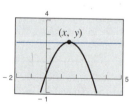

2.

3.

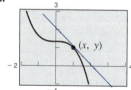

4.

In Exercises 5–8, sketch the graph of the function. Then sketch the tangent line at the point $(1, f(1))$. Use your sketch to estimate the slope of the tangent line.

5. $f(x) = x^2 - 2$

6. $f(x) = x^2 - 2x + 1$

7. $f(x) = \sqrt{2 - x}$

8. $f(x) = \dfrac{3}{2 - x}$

9. The figure graphically depicts the per capita debt, where $t = 0$ corresponds to 1950. Estimate the slope of the graph when $t = 30$ and give an interpretation of the result.

FIGURE FOR 9

10. The figure graphically depicts the number of fish in a lake t months after it was stocked. Estimate the slope of the graph when $t = 4$ and give an interpretation of the result.

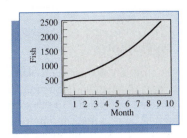

FIGURE FOR 10

In Exercises 11–20, use the limit process to find the slope of the graph of the function at the specified point.

11. $g(x) = 4 - 3x$, $(1, 1)$ 12. $h(x) = 2x + 5$, $(-1, 3)$
13. $f(x) = x^2 - 3$, $(2, 1)$ 14. $g(x) = x^2 - 2x$, $(3, 3)$
15. $f(x) = 9 - x^2$, $(2, 5)$ 16. $f(x) = 10x - 2x^2$, $(3, 12)$
17. $g(x) = \dfrac{4}{x}$, $(2, 2)$ 18. $g(x) = \dfrac{1}{x - 2}$, $\left(4, \dfrac{1}{2}\right)$
19. $h(x) = \sqrt{x}$, $(9, 3)$ 20. $h(x) = \sqrt{x + 10}$, $(-1, 3)$

In Exercises 21–24, find a formula for the slope of the graph. Use the formula to find the slope at the two points.

21. $g(x) = 4 - x^2$
   (a) $(0, 4)$ (b) $(-1, 3)$
22. $g(x) = x^3$
   (a) $(1, 1)$ (b) $(-2, -8)$
23. $g(x) = \dfrac{1}{x + 4}$
   (a) $\left(0, \dfrac{1}{4}\right)$ (b) $\left(-2, \dfrac{1}{2}\right)$
24. $g(x) = \sqrt{x - 1}$
   (a) $(5, 2)$ (b) $(10, 3)$

In Exercises 25–30, find the derivative of the function.

25. $f(x) = 5$
26. $f(x) = -4x + 2$
27. $g(x) = 6 - \frac{2}{3}x$
28. $f(x) = x^2 - 2x + 3$
29. $f(x) = \dfrac{1}{x^2}$
30. $h(s) = \dfrac{1}{\sqrt{s + 1}}$

In Exercises 31–34, find the slope of the graph of f at the indicated point. Use the result to find an equation of the tangent line to the graph at that point. Then verify your answer by sketching both the graph of f and the tangent line.

31. $f(x) = x^2 - 1$, $(2, 3)$ 32. $f(x) = x^3 - x$, $(2, 6)$
33. $f(x) = \sqrt{x + 1}$, $(3, 2)$ 34. $f(x) = 2x + \dfrac{4}{x}$, $(2, 6)$

In Exercises 35–38, use a graphing utility to graph f over the interval $[-2, 2]$. Then complete the table by graphically estimating the slope of the graph at the indicated points. Finally, evaluate the slope analytically and compare your results with those obtained graphically.

x	-2	-1.5	-1	-0.5	0	0.5	1	1.5	2
$f(x)$									
$f'(x)$									

35. $f(x) = \frac{1}{2}x^2$ 36. $f(x) = \sqrt{x + 3}$
37. $f(x) = \dfrac{1}{4}x^3$ 38. $f(x) = \dfrac{x^2 - 4}{x + 4}$

In Exercises 39–44, find the derivative of f. Use the derivative to determine any points on the graph of f where the tangent line is horizontal. Verify your results by using a graphing utility to graph f.

39. $f(x) = 9 - x^2$ 40. $f(x) = x^2 - 4x + 3$
41. $f(x) = x^3 + 3$ 42. $f(x) = \frac{1}{3}(x^3 - 3x)$
43. $f(x) = 3x^3 - 9x$ 44. $f(x) = 3x^4 + 4x^3$

In Exercises 45–48, use your knowledge of the graph of the function and the geometric interpretation of the derivative to match the function with the graph of its *derivative*. It is not necessary to find the derivative of the function. [The graphs are labeled (a), (b), (c), and (d).]

45. $f(x) = \frac{1}{2}x$ 46. $f(x) = x^2$
47. $f(x) = \sqrt{x}$ 48. $f(x) = \dfrac{1}{x}$

(a)
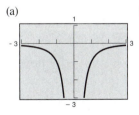

(b)
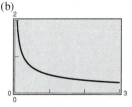

(c)
(d)
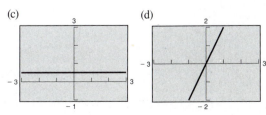

12.4 LIMITS AT INFINITY AND LIMITS OF SEQUENCES

Limits at Infinity and Horizontal Asymptotes / Limits of Sequences

Limits at Infinity and Horizontal Asymptotes

As pointed out at the beginning of this chapter, there are two basic problems in calculus: finding tangent lines and finding the area of a region. In Section 12.3, you saw how limits can be used to solve the tangent line problem. In this section and the next, you will see how a different type of limit, a *limit at infinity,* can be used to solve the area problem. To get an idea of what is meant by a limit at infinity, consider the function

$$f(x) = \frac{x + 1}{2x}.$$

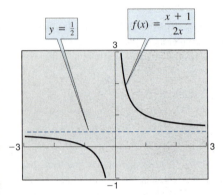

$y = \frac{1}{2}$ $f(x) = \frac{x + 1}{2x}$

FIGURE 12.27

The graph of f is shown in Figure 12.27. From earlier work, you know that $y = \frac{1}{2}$ is a horizontal asymptote of the graph of this function. Using limit notation, this can be written as follows.

$$\lim_{x \to -\infty} f(x) = \frac{1}{2} \qquad \textit{Horizontal asymptote to the left}$$

$$\lim_{x \to \infty} f(x) = \frac{1}{2} \qquad \textit{Horizontal asymptote to the right}$$

These limits mean that the value of $f(x)$ gets arbitrarily close to $\frac{1}{2}$ as x decreases or increases without bound.

LIMITS AT INFINITY

If f is a function and L_1 and L_2 are real numbers, the statements

$$\lim_{x \to -\infty} f(x) = L_1$$

and

$$\lim_{x \to \infty} f(x) = L_2$$

denote the **limits at infinity.** The first is read as *the limit of $f(x)$ as x approaches $-\infty$ is L_1,* and the second is read as *the limit of $f(x)$ as x approaches ∞ is L_2.*

To help evaluate limits at infinity, you can use the following.

LIMITS AT INFINITY

If r is a positive real number, then

$$\lim_{x \to \infty} \frac{1}{x^r} = 0. \qquad \textit{Limit toward the right}$$

Furthermore, if x^r is defined when $x < 0$, then

$$\lim_{x \to -\infty} \frac{1}{x^r} = 0. \qquad \textit{Limit toward the left}$$

Limits at infinity share many of the properties of limits listed in Section 12.1. Some of these are demonstrated in the next example.

EXAMPLE 1   Evaluating a Limit at Infinity

Find the following limit.

$$\lim_{x \to \infty}\left(4 - \frac{3}{x^2}\right)$$

SOLUTION

Try to identify the properties of limits that are used in the following steps.

$$\lim_{x \to \infty}\left(4 - \frac{3}{x^2}\right) = \lim_{x \to \infty} 4 - \lim_{x \to \infty} \frac{3}{x^2}$$

$$= \lim_{x \to \infty} 4 - 3\left[\lim_{x \to \infty} \frac{1}{x^2}\right]$$

$$= 4 - 3(0)$$

$$= 4$$

Thus, the limit of $f(x) = 4 - (3/x^2)$ as x approaches ∞ is 4. This conclusion is confirmed graphically in Figure 12.28. Note that the line $y = 4$ is a horizontal asymptote to the right.

In Figure 12.28, it appears that the line $y = 4$ is also a horizontal asymptote *to the left*. You can verify this by showing that

$$\lim_{x \to -\infty}\left(4 - \frac{3}{x^2}\right) = 4.$$

The graph of a rational function need not have a horizontal asymptote. If it does, however, it must approach the same asymptote to the left and to the right.

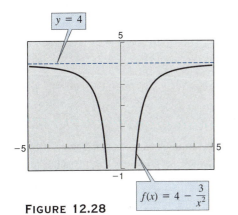

$y = 4$

$$f(x) = 4 - \frac{3}{x^2}$$

FIGURE 12.28

EXAMPLE 2   Comparing Limits at Infinity

Find the limit as $x \to \infty$ for the following functions.

A. $f(x) = \dfrac{-2x + 3}{3x^2 + 1}$ **B.** $f(x) = \dfrac{-2x^2 + 3}{3x^2 + 1}$ **C.** $f(x) = \dfrac{-2x^3 + 3}{3x^2 + 1}$

SOLUTION

In each case, you can begin by dividing both the numerator and denominator by x^2, the highest power of x in the denominator.

A. $\displaystyle\lim_{x \to \infty} \frac{-2x + 3}{3x^2 + 1} = \lim_{x \to \infty} \frac{-\dfrac{2}{x} + \dfrac{3}{x^2}}{3 + \dfrac{1}{x^2}}$

$\qquad\qquad = \dfrac{-0 + 0}{3 + 0}$

$\qquad\qquad = 0$

B. $\displaystyle\lim_{x \to \infty} \frac{-2x^2 + 3}{3x^2 + 1} = \lim_{x \to \infty} \frac{-2 + \dfrac{3}{x^2}}{3 + \dfrac{1}{x^2}}$

$\qquad\qquad = \dfrac{-2 + 0}{3 + 0}$

$\qquad\qquad = -\dfrac{2}{3}$

C. $\displaystyle\lim_{x \to \infty} \frac{-2x^3 + 3}{3x^2 + 1} = \lim_{x \to \infty} \frac{-2x + \dfrac{3}{x^2}}{3 + \dfrac{1}{x^2}}$

In this case, you can conclude that the limit does not exist because the numerator increases without bound as the denominator approaches 3.

In Example 2, observe that when the degree of the numerator is less than the degree of the denominator, the limit is 0. When the degrees of the numerator and denominator are equal, the limit is the ratio of the coefficients of the highest-powered terms. When the degree of the numerator is greater than the degree of the denominator, the limit does not exist.

This result seems reasonable when you realize that for large values of x, the highest-powered term of a polynomial is the most "influential" term. That is, a polynomial tends to behave as its highest-powered term as x approaches positive or negative infinity.

LIMITS AT INFINITY FOR RATIONAL FUNCTIONS

For the rational function $f(x) = p(x)/q(x)$, where

$$p(x) = a_n x^n + \cdots + a_0 \quad \text{and} \quad q(x) = b_m x^m + \cdots + b_0,$$

the limit as x approaches positive or negative infinity is as follows.

$$\lim_{x \to \pm\infty} f(x) = \begin{cases} 0, & n < m(\textit{degree of } p(x)) < (\textit{degree of } q(x)). \\ \dfrac{a_n}{b_m}, & n = m(\textit{degree of } p(x)) = (\textit{degree of } q(x)). \end{cases}$$

If $n > m$, the limit does not exist.

EXAMPLE 3   Finding the Average Cost

You are manufacturing a product that costs $0.50 per unit to produce. Your initial investment is $5000, which implies that the total cost of producing x units is $C = 0.5x + 5000$. The average cost per unit is given by

$$\overline{C} = \frac{C}{x} = \frac{0.5x + 5000}{x}.$$

Find the average cost per unit when $x = 1000$, 10,000, and 100,000. What is the limit of $\overline{C}$ when x approaches infinity?

SOLUTION

When $x = 1000$, the average cost per unit it

$$\overline{C} = \frac{0.5(1000) + 5000}{1000} = \$5.50.$$

When $x = 10,000$, the average cost per unit is

$$\overline{C} = \frac{0.5(10,000) + 5000}{10,000} = \$1.00.$$

When $x = 100,000$, the average cost per unit is

$$\overline{C} = \frac{0.5(100,000) + 5000}{100,000} = \$0.55.$$

As x approaches infinity, the limit of $\overline{C}$ is

$$\lim_{x \to \infty} \frac{0.5x + 5000}{x} = \$0.50.$$

The graph of $\overline{C}$ is shown in Figure 12.29.

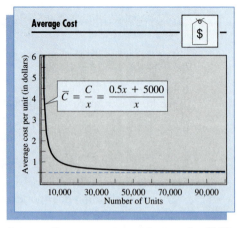

As $x \to \infty$, the average cost per unit approaches \$0.50.

FIGURE 12.29

Limits of Sequences

Limits of sequences have many of the same properties as limits of functions. For instance, consider the sequence whose nth term is $a_n = 1/2^n$.

$$\frac{1}{2}, \frac{1}{4}, \frac{1}{8}, \frac{1}{16}, \frac{1}{32}, \cdots$$

As n increases without bound, the terms of this sequence get closer and closer to 0, and the sequence is said to **converge** to 0. Using limit notation, you can write

$$\lim_{n \to \infty} \frac{1}{2^n} = 0.$$

The following shows how limits of functions of x can be used to evaluate the limit of a sequence.

LIMIT OF A SEQUENCE

Let f be a function of a real variable, such that

$$\lim_{x \to \infty} f(x) = L.$$

If $\{a_n\}$ is a sequence such that $f(n) = a_n$ for every positive integer n, then

$$\lim_{x \to \infty} a_n = L.$$

A sequence that does not converge is said to **diverge.** For instance, the sequence $1, -1, 1, -1, 1, \ldots$ diverges.

EXAMPLE 4   Finding the Limit of a Sequence

Find the limit of the following sequences.

A. $a_n = \dfrac{2n + 1}{n + 4}$ **B.** $b_n = \dfrac{2n + 1}{n^2 + 4}$

SOLUTION

A. $\displaystyle\lim_{n\to\infty} \frac{2n + 1}{n + 4} = 2$ $\dfrac{3}{5}, \dfrac{5}{6}, \dfrac{7}{7}, \dfrac{9}{8}, \dfrac{11}{9}, \dfrac{13}{10}, \ldots, \to 2$

B. $\displaystyle\lim_{n\to\infty} \frac{2n + 1}{n^2 + 4} = 0$ $\dfrac{3}{5}, \dfrac{5}{8}, \dfrac{7}{13}, \dfrac{9}{20}, \dfrac{11}{29}, \dfrac{13}{40}, \ldots, \to 0$

In the next section, you will encounter limits of sequences like that shown in Example 5. A strategy for evaluating such limits is to begin by writing the nth term in standard rational function form. Then you can determine the limit by comparing the degrees of the numerator and denominator.

EXAMPLE 5   Finding the Limit of a Sequence

Find the limit of the sequence whose nth term is

$$a_n = \frac{8}{n^3}\left[\frac{n(n + 1)(2n + 1)}{6}\right].$$

SOLUTION

Begin by writing the nth term in standard rational function form—as the ratio of two polynomials.

$$a_n = \frac{8}{n^3}\left[\frac{n(n + 1)(2n + 1)}{6}\right] \qquad \textit{Original nth term}$$

$$= \frac{8(n)(n + 1)(2n + 1)}{6n^3} \qquad \textit{Multiply fractions}$$

$$= \frac{8n^3 + 12n^2 + 4n}{3n^3} \qquad \textit{Standard rational form}$$

From this form, you can see that the degree of the numerator is equal to the degree of the denominator. Thus, the limit of the sequence is the ratio of the coefficients of the highest-powered terms.

$$\lim_{n\to\infty} \frac{8n^3 + 12n^2 + 4n}{3n^3} = \frac{8}{3}$$

After obtaining a limit analytically, you can check your result numerically by constructing a table. For instance, the following table shows that a_n gets closer and closer to $\frac{8}{3} \approx 2.667$ as $n \to \infty$.

n	1	10	100	1000	10,000
a_n	8	3.08	2.707	2.671	2.667

DISCUSSION PROBLEM

. . . .

COMPARING RATES OF CONVERGENCE

In the table above, the values of a_n are approaching their limit of $\frac{8}{3}$ rather slowly. (The first term to be accurate to three decimal places is $a_{4801} \approx 2.667$.) Each of the following sequences converges to 0. Which converges most rapidly? Which converges most slowly?

(a) $a_n = \dfrac{1}{n}$ (b) $b_n = \dfrac{1}{n^2}$ (c) $c_n = \dfrac{1}{2^n}$ (d) $d_n = \dfrac{1}{n!}$

WARM-UP

. . . .

The following warm-up exercises involve skills that were covered in earlier sections. You will use these skills in the exercise set for this section.

In Exercises 1–6, determine any vertical and horizontal asymptotes of the rational function.

1. $f(x) = \dfrac{3x}{x^2 + 1}$ **2.** $f(x) = \dfrac{2}{x - 2}$

3. $h(x) = \dfrac{-4x}{2x - 3}$ **4.** $h(x) = \dfrac{5x^3 - 3x}{x^3 + 1}$

5. $g(x) = 4 - \dfrac{3}{x}$ **6.** $f(x) = \dfrac{3x^3}{x - 2}$

In Exercises 7–10, write the first five terms of the sequence. (Assume n begins with 1.)

7. $a_n = \dfrac{2n}{n + 1}$ **8.** $a_n = \dfrac{2^n}{3^n}$

9. $a_n = \left(-\dfrac{2}{3}\right)^n$ **10.** $a_n = \dfrac{n!}{(n + 1)!}$

SECTION 12.4 · EXERCISES

In Exercises 1–14, find the limit (if it exists).

1. $\lim\limits_{x \to \infty} \dfrac{2}{x^2}$

2. $\lim\limits_{x \to \infty} \dfrac{4}{2x + 3}$

3. $\lim\limits_{x \to \infty} \dfrac{2 + x}{2 - x}$

4. $\lim\limits_{x \to \infty} \dfrac{1 - 5x}{1 + 2x}$

5. $\lim\limits_{x \to -\infty} \dfrac{4x - 3}{2x + 1}$

6. $\lim\limits_{x \to -\infty} \dfrac{x^2 + 1}{x^2}$

7. $\lim\limits_{t \to \infty} \dfrac{t^2}{t + 2}$

8. $\lim\limits_{y \to \infty} \dfrac{4y^4}{y^2 + 1}$

9. $\lim\limits_{x \to -\infty} \left(-2 + \dfrac{2}{x} \right)$

10. $\lim\limits_{x \to \infty} \dfrac{2x^2 - 6}{(x - 1)^2}$

11. $\lim\limits_{x \to -\infty} \dfrac{x}{(x + 1)^2}$

12. $\lim\limits_{x \to \infty} \left[3 + \dfrac{2x^2}{(x + 5)^2} \right]$

13. $\lim\limits_{t \to \infty} \left(\dfrac{1}{3t^2} - \dfrac{5t}{t + 2} \right)$

14. $\lim\limits_{x \to \infty} \left[\dfrac{x}{2x + 1} + \dfrac{3x^2}{(x - 3)^2} \right]$

In Exercises 15–18, you are given the nth term of a sequence. Find the limit of the sequence (if it exists).

15. $a_n = \dfrac{1}{n}$

16. $a_n = \dfrac{3}{n^2}$

17. $a_n = \dfrac{5n}{n - 5}$

18. $a_n = \dfrac{1 - n^2}{n - 1}$

In Exercises 19–22, match the function with its graph by using limits at infinity. [The graphs are labeled (a), (b), (c), and (d).]

(a)

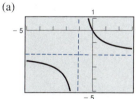

(b)

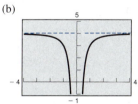

(c)

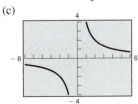

(d)

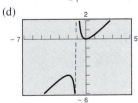

19. $f(x) = \dfrac{4}{x}$

20. $f(x) = \dfrac{-2x}{x + 1}$

21. $f(x) = \dfrac{x^2}{x + 1}$

22. $f(x) = 4 - \dfrac{1}{x^2}$

In Exercises 23–26, use a calculator to complete the table and estimate the limit as x approaches infinity. Use a graphing utility to graph the function.

x	10^0	10^1	10^2	10^3	10^4	10^5	10^6
$f(x)$							

23. $f(x) = x - \sqrt{x^2 + 2}$

24. $f(x) = 3x - \sqrt{9x^2 + 1}$

25. $f(x) = 3\left(2x - \sqrt{4x^2 + x} \right)$

26. $f(x) = 4\left(4x - \sqrt{16x^2 - x} \right)$

27. *Average Cost* The cost function for a certain product is given by

$$C = 1.35x + 4570,$$

where C is measured in dollars and x is the number of units produced.

(a) Find the average cost per unit when $x = 100$ and when $x = 1000$.

(b) Determine the limit of the average cost function as x approaches infinity.

28. *Learning Curve* Psychologists have developed mathematical models to predict performance as a function of the number of trials n for a certain task. One such model is

$$P = \dfrac{b + \theta a(n - 1)}{1 + \theta(n - 1)},$$

where P is the percentage of correct responses after n trials and a, b, and θ are constants depending on the actual learning situation. Find the limit of P as n approaches infinity.

In Exercises 29–40, write the first five terms of the sequence and find the limit of the sequence (if it exists). Assume n begins with 1.

29. $a_n = \dfrac{n + 2}{2n - 1}$

30. $a_n = \dfrac{3n - 1}{n + 2}$

31. $a_n = \dfrac{n + 1}{n^2 + 1}$

32. $a_n = \dfrac{2n^2}{n^4 + n + 1}$

33. $a_n = \dfrac{n^2}{5n + 2}$

34. $a_n = 2n - 5$

35. $a_n = \dfrac{(n + 1)!}{n!}$

36. $a_n = \dfrac{(2n - 1)!}{(2n + 1)!}$

37. $a_n = 2[5 + (n - 1)^3]$

38. $a_n = \dfrac{(-1)^{n+1}}{n^2}$

39. $a_n = (-1)^{n-1}\left(\dfrac{n}{2n + 1}\right)$

40. $a_n = (-1)^{n+1}\left(\dfrac{2n - 1}{n + 4}\right)$

In Exercises 41–46, find the limit of the sequence and complete the table.

n	10^0	10^1	10^2	10^3	10^4	10^5	10^6
a_n							

41. $a_n = \dfrac{1}{n}\left(n + \dfrac{1}{n}\left[\dfrac{n(n + 1)}{2}\right]\right)$

42. $a_n = \dfrac{4}{n}\left(n + \dfrac{4}{n}\left[\dfrac{n(n + 1)}{2}\right]\right)$

43. $a_n = 6 - \dfrac{4}{n^2}\left[\dfrac{n(n + 1)}{2}\right]$

44. $a_n = \dfrac{16}{n^4}\left[\dfrac{n^2(n + 1)^2}{4}\right]$

45. $a_n = \dfrac{16}{n^3}\left[\dfrac{n(n + 1)(2n + 1)}{6}\right]$

46. $a_n = \dfrac{n(n + 1)}{n^2} - \dfrac{1}{n^4}\left[\dfrac{n(n + 1)}{2}\right]^2$

In Exercises 47–50, use a calculator to evaluate the nth term of the sequence for "large" values of n. Based on the results, decide whether the sequence converges or diverges. If it converges, estimate its limit.

47. $a_n = 4(\tfrac{2}{3})^n$

48. $a_n = 3(\tfrac{3}{2})^n$

49. $a_n = \dfrac{3[1 - (1.5)^n]}{1 - 1.5}$

50. $a_n = \dfrac{3[1 - (0.5)^n]}{1 - 0.5}$

51. Use the results of Exercises 47–50 to make an inference about the values of r for which the sequence

$$a_n = \dfrac{a(1 - r^n)}{1 - r}$$

will converge. Explain your answer.

12.5   THE AREA PROBLEM
....
Limits of Summations   /   The Area Problem

Limits of Summations

Earlier in the text, you used the concept of a limit to obtain a formula for the sum S of an infinite geometric sequence

$$S = a_1 + a_1 r + a_1 r^2 + \cdots = \sum_{i=1}^{\infty} a_1 r^{i-1} = \frac{a_1}{1-r}, \quad |r| < 1.$$

Using limit notation, this sum can be written as

$$S = \lim_{n \to \infty} \sum_{i=1}^{n} a_1 r^{i-1} = \lim_{n \to \infty} \frac{a_1(1-r^n)}{1-r} = \frac{a_1}{1-r}.$$

The following summation formulas and properties are used to evaluate finite and infinite summations.

SUMMATION FORMULAS AND PROPERTIES

1. $\displaystyle\sum_{i=1}^{n} c = cn$
2. $\displaystyle\sum_{i=1}^{n} i = \frac{n(n+1)}{2}$

3. $\displaystyle\sum_{i=1}^{n} i^2 = \frac{n(n+1)(2n+1)}{6}$
4. $\displaystyle\sum_{i=1}^{n} i^3 = \frac{n^2(n+1)^2}{4}$

5. $\displaystyle\sum_{i=1}^{n} (a_i \pm b_i) = \sum_{i=1}^{n} a_i \pm \sum_{i=1}^{n} b_i$
6. $\displaystyle\sum_{i=1}^{n} k a_i = k \sum_{i=1}^{n} a_i$

Technology Note _____

Some graphing utilities have the capability to compute summations. For example, on the TI-85, you can use the MATH MISC menu to verify the summation in Example 1 as follows.

   sum seq(N, N, 1, 200, 1)

EXAMPLE 1   Evaluating a Summation
....

Using the second summation formula with $n = 200$, you can write

$$\sum_{i=1}^{200} i = 1 + 2 + 3 + 4 + \cdots + 200$$

$$= \frac{n(n+1)}{2} = \frac{200(201)}{2} = 20{,}100.$$

EXAMPLE 2   Evaluating a Summation

Evaluate the summation

$$S = \sum_{i=1}^{n} \frac{i + 2}{n^2} = \frac{3}{n^2} + \frac{4}{n^2} + \frac{5}{n^2} + \cdots + \frac{n + 2}{n^2}$$

for $n = 10, 100, 1000,$ and $10,000.$

SOLUTION

Begin by applying summation formulas and properties to simplify S. In the second line of the solution, note that $1/n^2$ can be factored out of the sum because n is considered to be constant. You could not factor i out of the summation because i is the (variable) index of summation.

$$S = \sum_{i=1}^{n} \frac{i + 2}{n^2} \qquad \textit{Original form of summation}$$

$$= \frac{1}{n^2} \sum_{i=1}^{n} (i + 2) \qquad \textit{Factor constant } 1/n^2 \textit{ out of sum}$$

$$= \frac{1}{n^2} \left(\sum_{i=1}^{n} i + \sum_{i=1}^{n} 2 \right) \qquad \textit{Write as two sums}$$

$$= \frac{1}{n^2} \left[\frac{n(n + 1)}{2} + 2n \right] \qquad \textit{Apply Formulas 1 and 2}$$

$$= \frac{1}{n^2} \left[\frac{n^2 + 5n}{2} \right] \qquad \textit{Add fractions}$$

$$= \frac{n + 5}{2n} \qquad \textit{Simplify}$$

Now you can evaluate the sum by substituting the appropriate values of n as shown in the following table.

n	10	100	1000	10,000
$\sum_{i=1}^{n} \dfrac{i + 2}{n^2} = \dfrac{n + 5}{2n}$	0.75	0.525	0.5025	0.50025

In Example 2, note that the sum appears to approach a limit as n increases. To find the limit of $(n + 5)/2n$ as n approaches infinity, you can write

$$\lim_{n \to \infty} \frac{n + 5}{2n} = \frac{1}{2}.$$

Be sure you notice the strategy used in Example 2. Rather than separately evaluating the sums

$$\sum_{i=1}^{10} \frac{i+2}{n^2}, \qquad \sum_{i=1}^{100} \frac{i+2}{n^2}, \qquad \sum_{i=1}^{1000} \frac{i+2}{n^2}, \qquad \sum_{i=1}^{10,000} \frac{i+2}{n^2},$$

it was more efficient to first convert to rational form.

$$S = \underbrace{\sum_{i=1}^{n} \frac{i+2}{n^2}}_{\text{Summation form}} = \underbrace{\frac{n+5}{2n}}_{\text{Rational form}}$$

With this form, each sum can be evaluated by simply substituting appropriate values of n into the rational form.

EXAMPLE 3   Finding the Limit of a Summation

Let $S(n)$ be defined by

$$S(n) = \sum_{i=1}^{n} \left(1 + \frac{i}{n}\right)^2 \left(\frac{1}{n}\right),$$

and find the limit of $S(n)$ as $n \to \infty$.

SOLUTION

Begin by using the summation formulas and properties to rewrite the summation in rational form.

$$S(n) = \sum_{i=1}^{n} \left(1 + \frac{i}{n}\right)^2 \left(\frac{1}{n}\right)$$

$$= \sum_{i=1}^{n} \left(\frac{n^2 + 2ni + i^2}{n^2}\right)\left(\frac{1}{n}\right)$$

$$= \frac{1}{n^3} \sum_{i=1}^{n} (n^2 + 2ni + i^2)$$

$$= \frac{1}{n^3} \left(\sum_{i=1}^{n} n^2 + \sum_{i=1}^{n} 2ni + \sum_{i=1}^{n} i^2\right)$$

$$= \frac{1}{n^3} \left(n^3 + 2n\left[\frac{n(n+1)}{2}\right] + \frac{n(n+1)(2n+1)}{6}\right)$$

$$= \frac{14n^3 + 9n^2 + n}{6n^3}$$

In this rational form, you can now find the limit as $n \to \infty$.

$$\lim_{n\to\infty} S(n) = \lim_{n\to\infty} \frac{14n^3 + 9n^2 + n}{6n^3} = \frac{14}{6} = \frac{7}{3}$$

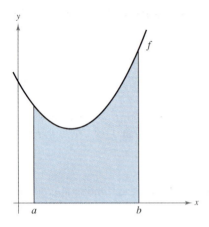

Region Bounded by the Graph
of a Function, the x-axis, $x = a$,
and $x = b$

FIGURE 12.30

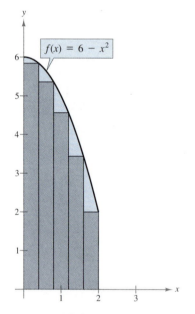

$f(x) = 6 - x^2$

FIGURE 12.31

The Area Problem

You now have the tools needed to solve the second basic problem of calculus: the area problem. The problem is to find the *area* of the region R bounded by the graph of a nonnegative, continuous function f, the x-axis, and the vertical lines $x = a$ and $x = b$, as shown in Figure 12.30.

If the region R is a square, triangle, trapezoid, or semicircle, you can find its area by using a geometric formula. For more general regions, however, you must use a different approach—one that involves the limit of a summation. The basic strategy is to use a collection of rectangles that approximates the region R, as illustrated in Example 4.

EXAMPLE 4   Approximating the Area of a Region

Use the five rectangles in Figure 12.31 to approximate the area of the region bounded by the graph of $f(x) = 6 - x^2$, the x-axis, and the lines $x = 0$ and $x = 2$.

SOLUTION

The right endpoints of the five intervals are $\frac{2}{5}i$, where $i = 1, 2, 3, 4, 5$. The width of each rectangle is $\frac{2}{5}$, and the height of each rectangle can be obtained by evaluating f at the right endpoint of each interval. The five intervals are

$$\left[0, \frac{2}{5}\right], \quad \left[\frac{2}{5}, \frac{4}{5}\right], \quad \left[\frac{4}{5}, \frac{6}{5}\right], \quad \left[\frac{6}{5}, \frac{8}{5}\right], \quad \left[\frac{8}{5}, \frac{10}{5}\right].$$

The sum of the areas of the five rectangles is

$$\sum_{i=1}^{5} f\left(\overbrace{\frac{2i}{5}}^{\text{Height}}\right)\left(\overbrace{\frac{2}{5}}^{\text{Width}}\right) = \sum_{i=1}^{5}\left[6 - \left(\frac{2i}{5}\right)^2\right]\left(\frac{2}{5}\right) = \frac{212}{25} = 8.48.$$

Thus, you can approximate the area of R to be 8.48.

By increasing the number of rectangles used in Example 4, you can obtain closer and closer approximations of the area of the region. For instance, using 25 rectangles of width $\frac{2}{25}$ each, you can approximate the area to be $A \approx 9.17$. The following table shows even better approximations.

n	5	25	100	1000	5000
Approximate area	8.48	9.17	9.29	9.33	9.33

The *exact* area of the region R is given by the limit of the sum of n rectangles as $n \to \infty$.

AREA OF A PLANE REGION

Let f be continuous and nonnegative on the interval $[a, b]$. The **area** A of the region bounded by the graph of f, the x-axis, and the vertical lines $x = a$ and $x = b$ is

$$A = \lim_{n \to \infty} \sum_{i=1}^{n} f\left[a + \frac{(b - a)i}{n}\right]\left(\frac{b - a}{n}\right).$$

EXAMPLE 5   Finding the Area of a Region

Find the area of the region bounded by the graph of $f(x) = x^2$ and the x-axis between $x = 0$ and $x = 1$.

SOLUTION

Begin by finding the dimensions of the rectangles.

$$\textit{Width:} \quad \frac{b - a}{n} = \frac{1}{n} \qquad \textit{Height:} \quad f\left[a + \frac{(b - a)i}{n}\right] = f\left(\frac{i}{n}\right) = \frac{i^2}{n^2}$$

Next, approximate the area as the sum of the areas of n rectangles.

$$\begin{aligned}
A &\approx \sum_{i=1}^{n} f\left[a + \frac{(b - a)i}{n}\right]\left(\frac{b - a}{n}\right) \\
&= \sum_{i=1}^{n} \left(\frac{i^2}{n^2}\right)\left(\frac{1}{n}\right) \\
&= \sum_{i=1}^{n} \frac{i^2}{n^3} \\
&= \frac{1}{n^3} \sum_{i=1}^{n} i^2 \\
&= \frac{1}{n^3}\left[\frac{n(n + 1)(2n + 1)}{6}\right] \\
&= \frac{2n^3 + 3n^2 + n}{6n^3}
\end{aligned}$$

Finally, find the exact area by taking the limit as $n \to \infty$.

$$A = \lim_{n \to \infty} \frac{2n^3 + 3n^2 + n}{6n^3} = \frac{1}{3}$$

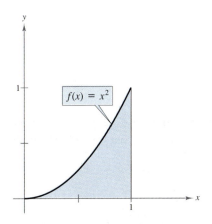

FIGURE 12.32

Figure 12.32 shows the region R.

EXAMPLE 6   Finding the Area of a Region

Find the area of the region bounded by the graph of

$$f(x) = 3x - x^2$$

and the x-axis between $x = 1$ and $x = 2$.

SOLUTION

Begin by finding the dimensions of the rectangles.

$$\text{Width:} \quad \frac{b - a}{n} = \frac{1}{n}$$

$$\text{Height:} \quad f\left[a + \frac{(b - a)i}{n}\right] = f\left(1 + \frac{i}{n}\right)$$

$$= 3\left(1 + \frac{i}{n}\right) - \left(1 + \frac{i}{n}\right)^2$$

$$= 3 + \frac{3i}{n} - \left(1 + \frac{2i}{n} + \frac{i^2}{n^2}\right)$$

$$= 2 + \frac{i}{n} - \frac{i^2}{n^2}$$

Next, approximate the area as the sum of the areas of n rectangles.

$$A \approx \sum_{i=1}^{n} f\left[a + \frac{(b - a)i}{n}\right]\left(\frac{b - a}{n}\right)$$

$$= \sum_{i=1}^{n}\left(2 + \frac{i}{n} - \frac{i^2}{n^2}\right)\left(\frac{1}{n}\right)$$

$$= \frac{1}{n}\sum_{i=1}^{n} 2 + \frac{1}{n^2}\sum_{i=1}^{n} i - \frac{1}{n^3}\sum_{i=1}^{n} i^2$$

$$= \frac{1}{n}(2n) + \frac{1}{n^2}\left[\frac{n(n + 1)}{2}\right] - \frac{1}{n^3}\left[\frac{n(n + 1)(2n + 1)}{6}\right]$$

$$= 2 + \frac{n^2 + n}{2n^2} - \frac{2n^3 + 3n^2 + n}{6n^3}$$

$$= 2 + \frac{1}{2} + \frac{1}{2n} - \frac{1}{3} - \frac{1}{2n} - \frac{1}{6n^2}$$

$$= \frac{13}{6} - \frac{1}{6n^2}$$

Finally, find the exact area by taking the limit as $n \to \infty$.

$$A = \lim_{n \to \infty}\left(\frac{13}{6} - \frac{1}{6n^2}\right) = \frac{13}{6}$$

Figure 12.33 shows the region R.

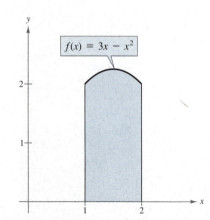

$f(x) = 3x - x^2$

FIGURE 12.33

DISCUSSION PROBLEM

·······················

COMPARING GEOMETRY AND CALCULUS

Show how to use a formula from geometry to find the area of the region bounded by the graph of

$$f(x) = 2 - x$$

and the x-axis between $x = 0$ and $x = 2$. Then show how the same area can be found using the limit procedure described in this section.

WARM-UP

·················

The following warm-up exercises involve skills that were covered in earlier sections. You will use these skills in the exercise set for this section.

In Exercises 1–4, use sigma notation to write the sum.

1. 4, 5, 6, 7, 8, 9

2. 2, 6, 12, 20, 30, 42

3. $3, 2, \frac{5}{3}, \frac{3}{2}, \frac{7}{5}, \frac{4}{3}$

4. $\frac{1}{2}, 2, \frac{9}{2}, 8, \frac{25}{2}, 18$

In Exercises 5–8, find the limit (if it exists).

5. $\displaystyle\lim_{x \to \infty} \frac{2x - 3}{x + 1}$

6. $\displaystyle\lim_{x \to \infty} \frac{2x - 3}{x^2 + 1}$

7. $\displaystyle\lim_{x \to \infty} \frac{2x^2 - 3}{x + 1}$

8. $\displaystyle\lim_{n \to \infty} \frac{9}{n^2}\left(\frac{2n^2 + 3n + 1}{2}\right)$

In Exercises 9 and 10, simplify the expression.

9. $\dfrac{4}{n}(n) - \dfrac{20^2}{n}\left[\dfrac{n(n + 1)}{2}\right]$

10. $\dfrac{16}{n^3}\left[\dfrac{n(n + 1)(2n + 1)}{6}\right]$

SECTION 12.5 · EXERCISES

In Exercises 1–6, find the sum using the summation formulas and properties.

1. $\displaystyle\sum_{i=1}^{60} i$

2. $\displaystyle\sum_{i=1}^{30} i^2$

3. $\displaystyle\sum_{k=1}^{20} k^3$

4. $\displaystyle\sum_{k=1}^{50} (2k + 1)$

5. $\displaystyle\sum_{j=1}^{25} (j^2 + j)$

6. $\displaystyle\sum_{j=1}^{10} (j^3 - 3j^2)$

In Exercises 7–18, use summation formulas and properties to rewrite the sum as a function $S(n)$. Use the function to complete the following table. Then find $\lim_{n\to\infty} S(n)$.

n	10^0	10^1	10^2	10^3
$S(n)$				

7. $\displaystyle\sum_{i=1}^{n} \frac{4i^2}{n^3}$

8. $\displaystyle\sum_{i=1}^{n} \frac{i}{n^2}$

9. $\displaystyle\sum_{i=1}^{n} \frac{3}{n^3}(1 + i^2)$

10. $\displaystyle\sum_{i=1}^{n} \frac{i^3}{n^4}$

11. $\displaystyle\sum_{i=1}^{n} \frac{2i + 3}{n^2}$

12. $\displaystyle\sum_{i=1}^{n} \frac{24i^2}{n^3}$

13. $\displaystyle\sum_{i=1}^{n} \left(\frac{i^2}{n^3} + \frac{2}{n}\right)\left(\frac{1}{n}\right)$

14. $\displaystyle\sum_{i=1}^{n} \left[3 - 2\left(\frac{i}{n}\right)\right]\left(\frac{1}{n}\right)$

15. $\displaystyle\sum_{i=1}^{n} \left[1 - \left(\frac{i}{n}\right)^2\right]\left(\frac{1}{n}\right)$

16. $\displaystyle\sum_{i=1}^{n} \left(\frac{4}{n} + \frac{2i}{n^2}\right)\left(\frac{2i}{n}\right)$

17. $\displaystyle\sum_{i=1}^{n} \left(\frac{4i^2}{n^2} - \frac{i}{n}\right)\left(\frac{1}{n}\right)$

18. $\displaystyle\sum_{i=1}^{n} \left[4 - \left(\frac{3i}{n}\right)^2\right]\left(\frac{3i}{n^2}\right)$

In Exercises 19–24, approximate the area of the region using the indicated number of rectangles of equal width.

19. $f(x) = x + 1$

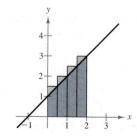

20. $f(x) = 4 - x$

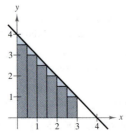

21. $f(x) = 4 - x^2$

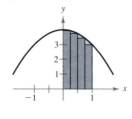

22. $f(x) = x^2 + 1$

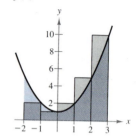

23. $f(x) = \frac{1}{4}x^3$

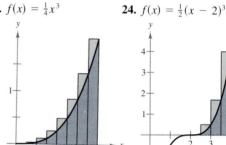

24. $f(x) = \frac{1}{2}(x - 2)^3$

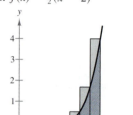

In Exercises 25–30, complete the table showing the approximate area of the region in the graph using n rectangles of equal width.

n	4	8	20	50
Approximate area				

25. $f(x) = -\frac{1}{2}x + 4$

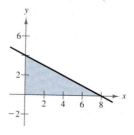

26. $f(x) = 2x + 1$

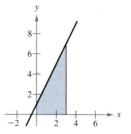

27. $f(x) = \frac{1}{4}x^2$

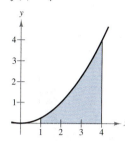

28. $f(x) = 9 - x^2$

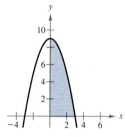

29. $f(x) = \frac{1}{9}x^3$

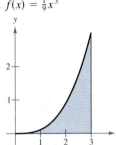

30. $f(x) = 3 - \frac{1}{4}x^3$

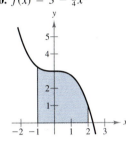

In Exercises 31–39, use the limit process to find the area of the region between the graph of the function and the x-axis over the specified interval.

31. $f(x) = 3x$, $[0, 2]$

32. $f(x) = \frac{1}{3}x + 1$, $[0, 3]$

33. $f(x) = 4 - \frac{1}{2}x$, $[2, 8]$

34. $f(x) = x + 2$, $[-2, 2]$

35. $f(x) = 3 - x^2$, $[-1, 1]$

36. $f(x) = \frac{1}{4}x^2 + 1$, $[1, 4]$

37. $g(x) = 8 - x^3$, $[1, 2]$

38. $g(x) = 4x - x^3$, $[0, 2]$

39. $f(x) = \frac{1}{4}(x^2 + 4x)$, $[1, 4]$

40. The boundaries of a parcel of land are two edges modeled by the coordinate axes and a stream modeled by the equation

$$y = (-3.0 \times 10^{-6})x^3 + 0.002x^2 - 1.05x + 400.$$

Use a graphing utility to graph the equation and find the area of the property, assuming all distances are measured in feet.

CHAPTER 12 · REVIEW EXERCISES

In Exercises 1 and 2, use a calculator to complete the table and use the result to estimate the limit.

1. $\lim\limits_{x \to 2} \dfrac{x - 2}{3x^2 - 4x - 4}$

x	1.9	1.99	1.999	2.0	2.001	2.01	2.1
$f(x)$				?			

2. $\lim\limits_{x \to 3} \dfrac{[x/(x + 1)] - (3/4)}{x - 3}$

x	2.9	2.99	2.999	3.0	3.001	3.01	3.1
$f(x)$				?			

In Exercises 3–10, find the limit (if it exists).

3. $\lim\limits_{x \to 3}(5x - 4)$

4. $\lim\limits_{x \to -2}(5 - 2x - x^2)$

5. $\lim\limits_{x \to 5} \dfrac{x - 5}{x^2 + 5x - 50}$

6. $\lim\limits_{x \to 4} \dfrac{x^3 - 64}{x^2 - 16}$

7. $\lim\limits_{u \to 0} \dfrac{\sqrt{4 + u} - 2}{u}$

8. $\lim\limits_{v \to 2} \dfrac{\sqrt{v + 7} - 3}{v}$

9. $\lim\limits_{x \to -1} \dfrac{[1/(x + 2)] - 1}{x + 1}$

10. $\lim\limits_{x \to 6^-} \dfrac{|x - 6|}{x - 6}$

In Exercises 11 and 12, find $\lim\limits_{h \to 0} \dfrac{f(x + h) - f(x)}{h}$

11. $f(x) = 3x - x^2$

12. $f(x) = x^3 + 5$

In Exercises 13–16, find a formula for the slope of the graph. Use the formula to find the slope at the two points.

13. $g(x) = x^2 - 4x$
   (a) $(0, 0)$ (b) $(5, 5)$

14. $g(x) = \sqrt{x}$
   (a) $(1, 1)$ (b) $(4, 2)$

15. $f(x) = \dfrac{4}{x - 6}$
   (a) $(7, 4)$ (b) $(8, 2)$

16. $f(x) = \dfrac{1}{4}x^4$
   (a) $(-2, 4)$ (b) $(1, \frac{1}{4})$

In Exercises 17–22, find the derivative of the function.

17. $g(x) = -4$

18. $f(x) = 3x$

19. $h(x) = 5 - \frac{1}{2}x$

20. $f(x) = \frac{1}{2}x^2 + 3$

21. $f(t) = \sqrt{t + 5}$

22. $g(s) = \dfrac{4}{s + 5}$

In Exercises 23–28, find the limit (if it exists).

23. $\displaystyle\lim_{x \to -\infty} \frac{4x}{2x - 3}$

24. $\displaystyle\lim_{x \to \infty} \frac{7x}{14x + 2}$

25. $\displaystyle\lim_{x \to \infty} \frac{x^2 + 5x}{x^2 - 25}$

26. $\displaystyle\lim_{x \to \infty} \frac{x^2}{2x + 3}$

27. $\displaystyle\lim_{x \to \infty} \left(4 - \frac{7}{x^3} \right)$

28. $\displaystyle\lim_{x \to \infty} (x - 2)^{-3}$

In Exercises 29 and 30, find the limit of the sequence.

29. $a_n = \dfrac{1}{2n^2}[3 - 2n(n + 1)]$

30. $a_n = \left(\dfrac{2}{n}\right)\left[n + \dfrac{2}{n}\left(\dfrac{n(n - 1)}{2} - n \right) \right]$

In Exercises 31–36, approximate the area of the region using the indicated number of rectangles of equal width. Then use the limit process to find the exact area.

31. $y = 10 - x$

32. $y = 2x - 6$

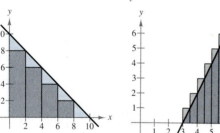

33. $y = x^2 + 4$

34. $y = 8(x - x^2)$

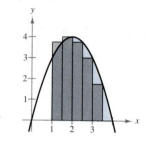

35. $y = 2(x^2 - x^3)$

36. $y = 4 - (x - 2)^2$

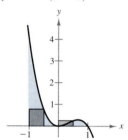

APPENDIX A

GRAPHING UTILITIES

Introduction

In Section 1.1, you studied the point-plotting method for sketching the graph of an equation. One of the disadvantages of the point-plotting method is that in order to get a good idea about the shape of a graph, you need to plot *many* points. With only a few points, you could badly misrepresent the graph. For instance, consider the equation

$$y = \frac{1}{30}x(39 - 10x^2 + x^4).$$

Suppose you plotted only five points: $(-3, -3)$, $(-1, -1)$, $(0, 0)$, $(1, 1)$, and $(3, 3)$, as shown in Figure A.1. From these five points, you might assume that the graph of the equation is a straight line. That, however, is not correct. By plotting several more points you can see that the actual graph is not straight at all! (See Figure A.2.)

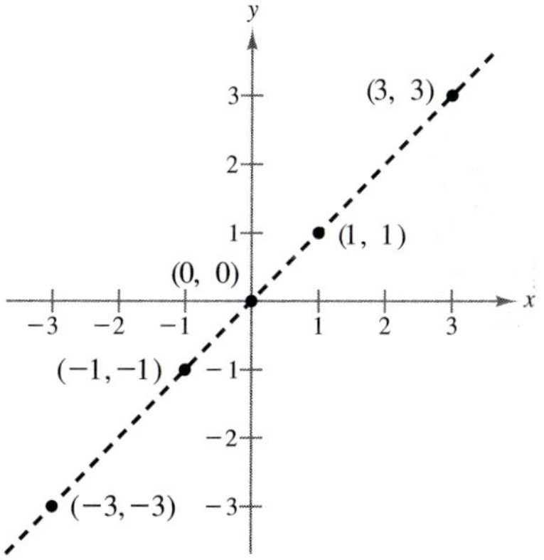

FIGURE A.1

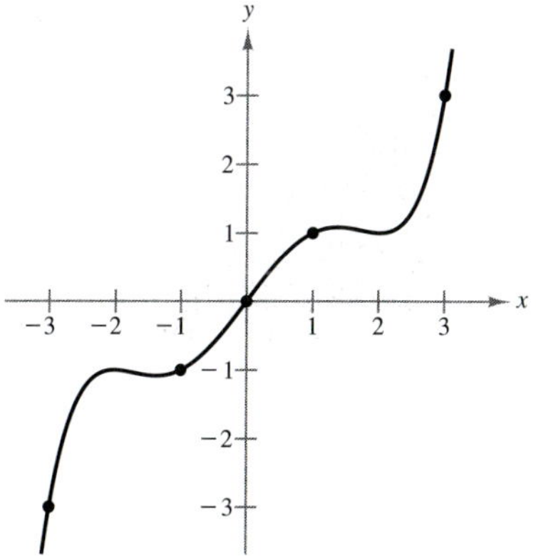

FIGURE A.2

Thus, the point-plotting method leaves us with a dilemma. On the one hand, the method can be very inaccurate if only a few points are plotted. But, on the other hand, it is very time consuming to plot a dozen (or more) points.

Technology can help us solve this dilemma. Plotting several (even several hundred) points in a rectangular coordinate system is something that a graphing utility can do easily.

The point-plotting method is the method used by *all* graphing packages for computers and *all* graphing calculators. Each computer or calculator screen is made up of a grid of hundreds or thousands of small areas called **pixels.** Screens that have many pixels per inch are said to have a higher **resolution** than screens that don't have as many. For instance, the screen shown in Figure A.3(a) has a higher resolution than the screen shown in Figure A.3(b). Note that the "graph" of the line on the first screen looks more like a line than the "graph" on the second screen.

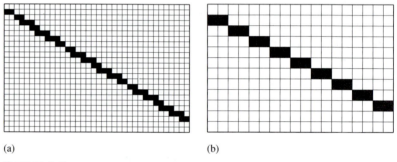

(a)   (b)

FIGURE A.3

Screens on most graphing calculators have 48 pixels per inch. Screens on computer monitors typically have between 32 and 100 pixels per inch.

EXAMPLE 1   Using Pixels to Sketch a Graph

Use the grid shown in Figure A.4 to sketch a graph of $y = \frac{1}{2}x^2$. Each pixel on the grid must be either on (shaded black) or off (unshaded).

SOLUTION

To shade the grid, we use the following rule. If a pixel contains a plotted point of the graph, then it will be "on"; otherwise, the pixel will be "off." Using this rule, the graph of the curve looks like that shown in Figure A.5.

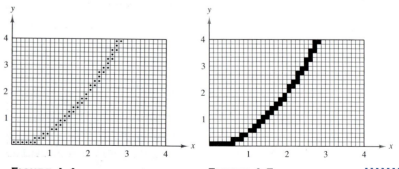

FIGURE A.4   **FIGURE A.5**

Basic Graphing

There are many different types of graphing utilities—graphing calculators and software packages for computers. The procedures used to draw a graph are similar with most of these utilities.

BASIC GRAPHING STEPS FOR A GRAPHING UTILITY

To draw the graph of an equation involving x and y with a graphing utility, use the following steps.

1. Rewrite the equation so that y is isolated on the left side of the equation.
2. Set the boundaries of the viewing rectangle by entering the minimum and maximum x-values and the minimum and maximum y-values.
3. Enter the equation in the form $y =$ (expression involving x). Read the user's guide that accompanies your graphing utility to see how the equation should be entered.
4. Activate the graphing utility.

EXAMPLE 2 **Sketching the Graph of an Equation**

Sketch the graph of $2y + x^3 = 4x$.

SOLUTION

To begin, solve the given equation for y in terms of x.

$$2y + x^3 = 4x \qquad \textit{Given equation}$$

$$2y = -x^3 + 4x \qquad \textit{Subtract } x^3 \textit{ from both sides}$$

$$y = -\frac{1}{2}x^3 + 2x \qquad \textit{Divide both sides by 2}$$

Set the viewing rectangle so that $-10 \le x \le 10$ and $-10 \le y \le 10$. (On some graphing utilities, this is the default setting.) Next, enter the equation into the graphing utility.

$$Y = -X \wedge 3/2 + 2 * X$$

Finally, activate the graphing utility. The display screen should look like that shown in Figure A.6.

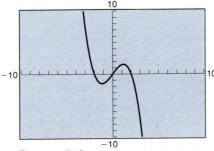

FIGURE A.6

In Figure A.6, notice that the calculator screen does not label the tick marks on the *x*-axis or the *y*-axis. To see what the tick marks represent, check the values in the utility's "range."

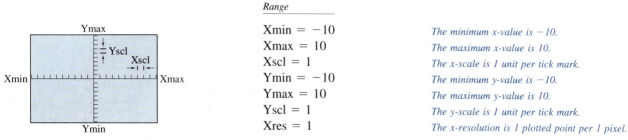

Range

Xmin = -10		*The minimum x-value is -10.*
Xmax = 10		*The maximum x-value is 10.*
Xscl = 1		*The x-scale is 1 unit per tick mark.*
Ymin = -10		*The minimum y-value is -10.*
Ymax = 10		*The maximum y-value is 10.*
Yscl = 1		*The y-scale is 1 unit per tick mark.*
Xres = 1		*The x-resolution is 1 plotted point per 1 pixel.*

FIGURE A.7

These settings are summarized visually in Figure A.7.

EXAMPLE 3 **Graphing an Equation Involving Absolute Value**

Sketch the graph of $y = |x - 3|$.

SOLUTION

This equation is already written so that *y* is isolated on the left side of the equation, so you can enter the equation as follows.

$$Y = \text{abs}(X - 3)$$

After activating the graphing utility, its screen should look like the one shown in Figure A.8.

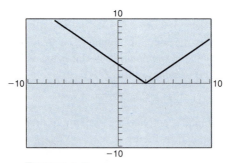

FIGURE A.8

Special Features

In order to be able to use your graphing calculator to its best advantage, you must be able to determine a proper viewing rectangle and use the zoom feature. The next two examples show how this is done.

EXAMPLE 4   **Determining a Viewing Rectangle**

Sketch the graph of $y = x^2 + 12$.

SOLUTION

Begin as usual by entering the equation.

$Y = X \wedge 2 + 12$

Activate the graphing utility. If you used a viewing rectangle in which $-10 \leq x \leq 10$ and $-10 \leq y \leq 10$, then no part of the graph will appear on the screen, as shown in Figure A.9(a). The reason for this is that the lowest point on the graph of $y = x^2 + 12$ occurs at the point (0, 12). With the viewing rectangle in Figure A.9(a), the largest y-value is 10. In other words, none of the graph is visible on a screen whose y-values range between -10 and 10.

To be able to see the graph, change Ymax = 10 to Ymax = 30, Yscl = 1 to Yscl = 5. Now activate the graphing utility and you will obtain the graph shown in Figure A.9(b). On this graph, note that each tick mark on the y-axis represents 5 units because you changed the y-scale to 5. Also note that the highest point on the y-axis is now 30 because you changed the maximum value of y to 30.

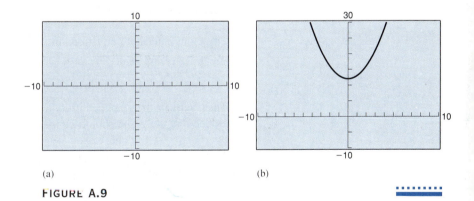

(a)   (b)

FIGURE A.9

EXAMPLE 5   **Using the Zoom Feature**

Sketch the graph of $y = x^3 - x^2 - x$. How many x-intercepts does this graph have?

SOLUTION

Begin by drawing the graph on a "standard" viewing rectangle as shown in Figure A.10(a). From the display screen, it is clear that the graph has at least one intercept (just to the left of $x = 2$), but it is difficult to determine whether

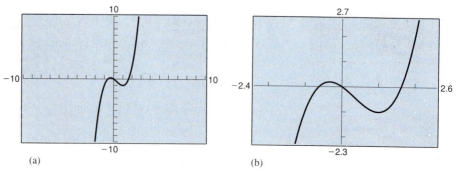

FIGURE A.10   (a)   (b)

the graph has other intercepts. To obtain a better view of the graph near $x = -1$, you can use the zoom feature of the graphing utility. The redrawn screen is shown in Figure A.10(b). From this screen you can tell that the graph has three x-intercepts whose x-coordinates are approximately -0.6, 0, and 1.6.

EXAMPLE 6   **Sketching More Than One Graph on the Same Screen**

Sketch the graphs of $y = -\sqrt{36 - x^2}$ and $y = \sqrt{36 - x^2}$ on the same screen.

SOLUTION

To begin, enter both equations in the graphing utility.

$$Y = \sqrt{(36 - X \wedge 2)}$$
$$Y = -\sqrt{(36 - X \wedge 2)}$$

Then, activate the graphing utility to obtain the graph shown in Figure A.11(a). Notice that the graph should be the upper and lower parts of the circle given by $x^2 + y^2 = 6^2$. The reason it doesn't look like a circle is that, with the standard settings, the tick marks on the x-axis are farther apart than the tick marks on the y-axis. To correct this, change the viewing rectangle so that $-15 \le x \le 15$. The redrawn screen is shown in Figure A.11(b). Notice that in this screen the graph appears to be more circular.

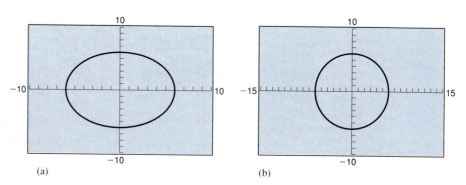

(a)   (b)

FIGURE A.11

DISCUSSION PROBLEM
∙∙∙∙∙∙∙∙∙∙∙∙∙∙∙∙∙∙∙∙∙∙∙

A MISLEADING GRAPH

Sketch the graph of $y = x^2 - 12x$, using $-10 \leq x \leq 10$ and $-10 \leq y \leq 10$. The graph appears to be a straight line, as shown in the figure. However, this is misleading because the screen doesn't show an important portion of the graph. Can you find a range setting that reveals a better view of this graph?

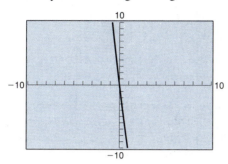

WARM-UP
∙∙∙∙∙∙∙∙∙∙∙∙∙∙∙∙∙∙

In Exercises 1–10, solve for y in terms of x.

1. $3x + y = 4$ **2.** $x - y = 0$

3. $2x + 3y = 2$ **4.** $4x - 5y = -2$

5. $3x + 4y - 5 = 0$ **6.** $-2x - 3y + 6 = 0$

7. $x^2 + y - 4 = 0$ **8.** $-2x^2 + 3y + 2 = 0$

9. $x^2 + y^2 = 4$ **10.** $x^2 - y^2 = 9$

APPENDIX A · EXERCISES
∙∙∙

In Exercises 1–20, use a graphing utility to sketch the graph of the equation. Use a setting on each graph of $-10 \leq x \leq 10$ and $-10 \leq y \leq 10$.

```
RANGE
Xmin=-10
Xmax=10
Xscl=1
Ymin=-10
Ymax=10
Yscl=1
Xres=1
```

1. $y = x - 5$ **2.** $y = -x + 4$

3. $y = -\frac{1}{2}x + 3$ **4.** $y = \frac{2}{3}x + 1$

5. $2x - 3y = 4$ **6.** $x + 2y = 3$

7. $y = \frac{1}{2}x^2 - 1$ **8.** $y = -x^2 + 6$

9. $y = x^2 - 4x - 5$ **10.** $y = x^2 - 3x + 2$

11. $y = -x^2 + 2x + 1$ **12.** $y = -x^2 + 4x - 1$

13. $2y = x^2 + 2x - 3$ **14.** $3y = -x^2 - 4x + 5$

15. $y = |x + 5|$ **16.** $y = \frac{1}{2}|x - 6|$

17. $y = \sqrt{x^2 + 1}$ **18.** $y = 2\sqrt{x^2 + 2} - 4$

19. $y = \frac{1}{5}(-x^3 + 16x)$ **20.** $y = \frac{1}{8}(x^3 + 8x^2)$

In Exercises 21–30, use a graphing utility to match the equation with its graph. [The graphs are labeled (a), (b), (c), (d), (e), (f), (g), (h), (i), and (j).]

21. $y = x$ **22.** $y = -x$

23. $y = x^2$ **24.** $y = -x^2$

25. $y = x^3$

26. $y = -x^3$

27. $y = |x|$

28. $y = -|x|$

29. $y = \sqrt{x}$

30. $y = -\sqrt{x}$

(a)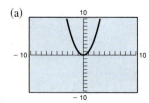

(b)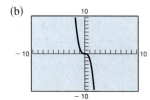

(c)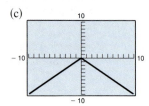

(d)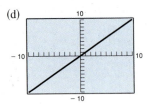

(e)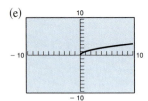

(f)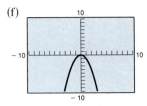

(g)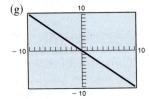

(h)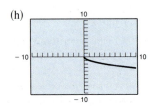

(i)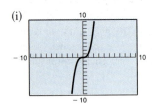

(j)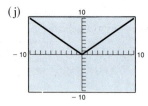

In Exercises 31–34, use a graphing utility to sketch the graph of the equation. Use the indicated setting.

31. $y = -2x^2 + 12x + 14$ **32.** $y = -x^2 + 5x + 6$

```
RANGE
Xmin=-5
Xmax=10
Xscl=1
Ymin=-5
Ymax=35
Yscl=5
Xres=1
```

```
RANGE
Xmin=-8
Xmax=4
Xscl=1
Ymin=-5
Ymax=15
Yscl=5
Xres=1
```

33. $y = x^3 + 6x^2$ **34.** $y = -x^3 + 16x$

```
RANGE
Xmin=-10
Xmax=5
Xscl=1
Ymin=-4
Ymax=36
Yscl=3
Xres=1
```

```
RANGE
Xmin=-6
Xmax=6
Xscl=1
Ymin=-25
Ymax=25
Yscl=5
Xres=1
```

In Exercises 35–38, find a setting on a graphing utility so that the graph of the equation agrees with the graph shown.

35. $y = -x^2 - 4x + 20$ **36.** $y = x^2 + 12x - 8$

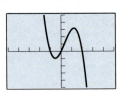

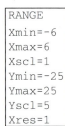

37. $y = -x^3 + x^2 + 2x$ **38.** $y = x^3 + 3x^2 - 2x$

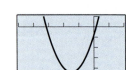

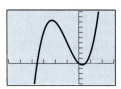

In Exercises 39–42, use a graphing utility to find the number of x-intercepts of the equation.

39. $y = \frac{1}{8}(4x^2 - 32x + 65)$

40. $y = \frac{1}{4}(-4x^2 + 16x - 15)$

41. $y = 4x^3 - 20x^2 - 4x + 61$

42. $y = \frac{1}{4}(2x^3 + 6x^2 - 4x + 1)$

In Exercises 43–46, use a graphing utility to sketch the graphs of the equations on the same screen. Using a "square setting," what geometrical shape is bounded by the graphs?

43. $y = |x| - 4$
$y = -|x| + 4$

44. $y = x + |x| - 4$
$y = x - |x| + 4$

45. $y = -\sqrt{25 - x^2}$
$y = \sqrt{25 - x^2}$

46. $y = 6$
$y = -\sqrt{3}x - 4$
$y = \sqrt{3}x - 4$

Ever Been Married? In Exercises 47–50, use the following models, which relate ages to the percentages of American males and females who have never been married.

$$y = \frac{0.36 - 0.0056x}{1 - 0.0817x + 0.00226x^2}, \quad \begin{array}{l} \text{Males} \\ 20 \le x \le 50 \end{array}$$

$$y = \frac{100}{8.944 - 0.886x + 0.249x^2}, \quad \begin{array}{l} \text{Females} \\ 20 \le x \le 50 \end{array}$$

In these models, y is the percent of the population (in decimal form) who have never been married and x is the age of the person. (*Source:* U.S. Bureau of the Census)

47. Use a graphing utility to sketch the graph of both equations giving the percentages of American males and females who have never been married. Use the following range settings.

```
RANGE
Xmin=20
Xmax=50
Xscl=5
Ymin=0
Ymax=1
Yscl=0.1
Xres=1
```

48. Write a short paragraph describing the relationship between the two graphs that were plotted in Exercise 47.

49. Suppose an American male is chosen at random from the population. If the person is 25 years old, what is the probability that he has never been married?

50. Suppose an American female is chosen at random from the population. If the person is 25 years old, what is the probability that she has never been married?

Earnings and Dividends In Exercises 51–54, use the following model, which approximates the relationship between dividends per share and earnings per share for the Pall Corporation between 1982 and 1989.

$$y = -0.166 + 0.502x - 0.0953x^2, \quad 0.25 \le x \le 2$$

In this model, y is the dividends per share (in dollars) and x is the earnings per share (in dollars). (*Source:* Standard ASE Stock Reports)

51. Use a graphing utility to sketch the graph of the model that gives the dividend per share in terms of the earnings per share. Use the following range settings.

```
RANGE
Xmin=0
Xmax=2
Xscl=0.25
Ymin=0
Ymax=0.5
Yscl=0.1
Xres=1
```

52. According to the given model, what size dividend would the Pall Corporation pay if the earnings per share were $1.30?

53. Use a trace feature on your graphing utility to estimate the earnings per share that would produce a dividend per share of $0.25. The choices are labeled (a), (b), (c), and (d). (Find the y-value that is as close to 0.25 as possible. The x-value that is displayed will then be the approximate earnings per share that would produce a dividend per share of $0.25.)
(a) $1.00   (b) $1.03   (c) $1.06   (d) $1.09

54. The **Payout Ratio** for a stock is the ratio of the dividend per share to earnings per share. Use the model to find the payout ratio for an earnings per share of (a) $0.75, (b) $1.00, and (c) $1.25.

....

PROGRAMS

Programs for the TI-81 graphing calculator are given in several sections in the text. This appendix contains translations of these text programs for several other graphics calculators from Texas Instruments, Casio, and Sharp. Similar programs can be written for other brands and models of graphics calculators.

Enter a program in your calculator, then refer to the text discussion and apply the program as appropriate. Section references are provided to help you locate the text discussion of the programs and their use.

To illustrate the power and versatility of programmable calculators, a variety of types of programs are presented, including a simulation program (A Graph Reflecting Program), a tutorial program (A Program for Practice), and programs for solving an equation or system of equations.

Functions   (Section 1.3)

The program, found in the Technology Note, can be used to evaluate a function at a real value of x.

For a TI-81 program, see page 84.

Casio 6300 Program

EVALUATE:
Lbl 1:
"X="?→X:
"F(X)="◢Prog 0◢
Goto 1

To use this program, write the function as Prog 0.

Casio 7700 Program

EVALUATE
Lbl 1
"X="?→X
"F(X)=":f₁◢
Goto 1

To use this program, enter a function in f_1.

Sharp EL 9200, 9300 Program

```
Evaluate
-----------REAL
Goto top
Label eqtn
Y=f(X)
Return
Label top
Input X
Gosub equation
Print Y
Goto top
End
```

TI-85 Program

```
 PROGRAM:
:Lbl A
:Input "Enter x ",x
:Disp y1
:Goto A
```

To use this program, enter a function in $y1$.

Shifting, Reflecting, and Stretching Graphs   (Section 1.5)

The program, shown in Example 5, A Program for Practice, will sketch a graph of the function $y = R(x + H)^2 + V$ where $R = \pm 1$, H is an integer between -6 and 6, and V is an integer between -3 and 3. This program gives you practice working with reflections, horizontal shifts, and vertical shifts.

For a TI-81 program, see page 109.

Casio 6300 Program

```
R(X+H)²+V:
-6+Int (12Ran#)→H:
-3+Int (6Ran#)→V:
Ran#<.5⇒-1→R:1→R:
Range -9,9,1,-6,6,1:
Graph Y=R(X+H)²+V◢
"Y=R(X+H)²+V"◢
"R="◢R◢
"H="◢H◢
"V="◢V
```

Casio 7700 Program

```
R(X+H)²+V
-6+Int (12Ran#)→H
-3+Int (6Ran#)→V
Ran#<.5⇒-1→R:1→R
Range -9,9,1,-6,6,1
Graph Y=R(X+H)²+V◢
"Y=R(X+H)²+V"
"R=":R◢
"H=":H◢
"V=":V
```

Sharp EL 9200, 9300 Program

```
Parabola
-----------REAL
H=int (random*12)-6
V=int (random*6)-3
S=(random*2)-1
R=S/abs S
Range -9,9,1,-6,6,1
Graph R(X+H)²+V
Wait
Print "Y-R(X+H)²+V
Print R
Print H
Print V
End
```

Pressing Enter after the graph will display the coefficients.

TI-85 Program

```
PROGRAM:
:rand→H
:-6+int(12H)→H
:rand→V
:-3+int(6V)→V
:rand→R
:If R<.5
:-1→R
:If R>.49
:1→R
:y1=R(x+H)²+V
:-9→xMin
:9→xMax
:1→xScl
:-6→yMin
:6→yMax
:1→yScl
:DispG
:Pause
:Disp "Y=R(X+H)²+V"
:Disp "R=",R
:Disp "H=",H
:Disp "V=",V
```

Inverse Functions   (Section 1.7)

The program, shown in Example 5, A Graph Reflecting Program, graphs a function f *and* its reflection in the line $y = x$.

For a TI-81 program, see page 126.

Casio 6300 Program

REFLECTION:
"-A TO A"◢
"A="?→A:
Range -A,A,1,-2A+3,2A+3,1:
-A→B:
Lbl 1:
B→X:
Prog 0:
Ans→Y:
Plot B,Y:
B+A÷24→B:
B≤A⇒Goto 1:
-A→B:
Lbl 2:
B→X:
Prog 0:
Ans→Y:
Plot Y,B:
B+A÷24→B:
B≤A⇒Goto 2:Graph Y=X

To use this program, write the function as Prog 0 and set a viewing rectangle.

Casio 7700 Program

REFLECTION
"GRAPH -A TO A"
"A="?→A
Range -A,A,1,-2A+3,2A+3,1
Graph Y=f_1
-A→B
Lbl 1
B→X
Plot f_1,B
B+A÷32→B
B≤A⇒Goto 1:Graph Y=X

To use this program, enter the function in f_1 and set a viewing rectangle.

Sharp EL 9200, 9300 Program

```
Reflection
-----------REAL
Goto top
Label eqtn
Y=X^3+X+1
Return
Label rng
xmin=-10
xmax=10
xstp=(xmax-xmin)/10
ymin=2xmin/3
ymax=2xmax/3
ystp=xstp
Range xmin,xmax,xstp,ymin,ymax,ystp
Return
Label top
Gosub rng
Graph X
step=(xmax-xmin)/(94*2)
X=xmin
Label 1
Gosub eqtn
Plot X,Y
Plot Y,X
X=X+step
If X<=xmax Goto 1
End
```

TI-85 Program

```
PROGRAM:
:63xMin/127→yMin
:63xMax/127→yMax
:xScl→yScl
:y2=x
:DispG
:(xMax-xMin)/126→I
:xMin→x
:Lbl A
:PtOn(y1,x)
:x+I→x
:If x>xMax
:Stop
:Goto A
```

Solving Equations Algebraically   (Section 2.4)

The program, shown in a Technology Note, will display solutions to quadratic equations or the words "No Real Solution." To use the program, write the equation in standard form and then enter the values of a, b, and c.

For a TI-81 program, see page 167.

Casio 6300 Program

QUADRATICS:
"AX²+BX+C=0"◢
"A="?→A:
"B="?→B:
"C="?→C:
B²-4AC→D:
D<0⇒Goto 1:
"X="◢(-B+√D)÷2A◢
"OR X="◢(-B-√D)÷2A:
Goto 2:
Lbl 1:
"NO REAL SOLUTION"
Lbl 2

Casio 7700 Program

QUADRATICS
"AX²+BX+C=0"
"A="?→A
"B="?→B
"C="?→C
B²-4AC→D
D<0⇒Goto 1
"X=":(-B+√D)÷2A◢
"OR X=":(-B-√D)÷2A
Goto 2
Lbl 1
"NO REAL SOLUTION"
Lbl 2

Sharp EL 9200, 9300 Program

```
Quadratic
----------COMPLEX
Input A
Input B
Input C
D=B²-4AC
x1=(-B+√ D)/(2A)
x2=(-B-√ D)/(2A)
Print x1
Print x2
X=x1
Y=x2
End
```

This program is written in the program's complex mode, so both real and complex answers are given. The answers are also stored under variables X and Y so they can be used in the calculator mode.

TI-85 Program

```
PROGRAM:
:Input "ENTER A ",A
:Input "ENTER B ",B
:Input "ENTER C ",C
:B²-4*A*C→D
:Disp (-B+√D)/(2A)
:Disp (-B-√D)/(2A)
```

Solutions to quadratic equations are also available directly by using the TI-85 POLY function.

Solving Inequalities Algebraically and Graphically (Section 2.5)

The program presented in the Discussion Problem, A Computer Experiment, evaluates an expression at several values. A visual display is printed indicating if the value is negative, zero, or positive.

Casio 6300 Program

SIGN:
"Left="?→A:
"Right="?→B:
Range A,B,1,-2,2,1:
(B-A)÷24→C:
Lbl 1:
A→X:
Prog 0:
Ans→Y:
Y=0⇒Plot A,0:Goto 2:
Goto 3:
Lbl 2:
Plot A,Y÷Abs Y:
Lbl 3:
A+C→A:
A≤B+1⇒Goto 1

To evaluate $x^2 - 9$, enter the expression as Prog 0. Then run the program. Press G↔T to view the graph.

Casio 7700 Program

SIGN
"GRAPH A TO B"
"A="?→A
"B="?→B
Range A,B,1,-2,2,1
(B-A)÷32→C
Lbl 1
A→X
f_1=0⇒Plot A,0:Goto 2
Goto 3
Lbl 2
Plot A,f_1÷Abs f_1
Lbl 3
A+C→A
A≤B+1⇒Goto 1

To evaluate $x^2 - 9$, enter the expression in f_1. Then run the program. Press G↔T to view the graph.

Radian and Degree Measure (Section 5.1)

The program, found in Discussion Problem, An Angle-Drawing Program, can be used to draw several different angles in either radian or degree mode.

For a TI-81 program, see page 350.

Casio 6300 Program

```
ANGLE:
"MODE:"◢
"0 = RADIAN"◢
"1 = DEGREE"?→M:
"ANGLE="?→T:
M=1⇒πT÷180→T:
Rad:
Range -1.5,1.5,1,-1,1,1:
T=0⇒Goto 2:
T÷Abs T→F:
0→N:
Lbl 1:
.25+.04N→R:
Plot Rcos (FN), Rsin (FN):
N+.2→N:
N≤AbsT⇒Goto 1:
Plot 0,0:
Plot cos T,sin T:
Line:
Lbl 2
```

Casio 7700 Program

```
ANGLE
"MODE:"
"0 FOR RADIAN"
"1 FOR DEGREE"
?→M
"ANGLE="?→T
M=1⇒πT÷180→T
Rad
Range -1.5,1.5,1,-1,1,1,0,T,.15
Graph r=.25+.04Abs θ
Plot 0,0
Plot cos T,sin T
Line
```

Press Mode Shift to change to polar mode when starting to write this program.

Sharp EL 9200, 9300 Program

```
Angle
-----------REAL
m=sin⁻¹ 1/(π/2)
Input angle
Print " 0 for rad        1 for deg
Input d
If d=0 Goto 1
angle=angle*π/180
Label 1
Range -2.25,2.25,1.5,-1.5,1.5,1.5
s=angle/abs angle
loops=ipart (s*angle/(2π))+1
θ=0
ostep=angle/(30*loops)
r=1/loops
rstep=(1-r)/(30*loops)
xo=r
yo=0
n=0
Label top
r=r+rstep
θ=θ+ostep
x=r*cos (mθ)
y=r*sin (mθ)
Line xo,yo,x,y
xo=x
yo=y
n=n+1
If n<30*loops Goto top
Line 0,0,1.5x,1.5y
End
```

TI-85 Program

```
PROGRAM:
:Disp "ENTER MODE"
:Disp "0 RADIAN"
:Disp "1 DEGREE"
:Input M
:Input "ENTER ANGLE ",T
:If M==1
:πT/180→T
:Radian
:ClDrw
:FnOff
:Param
:-1.5→xMin
:1.5→xMax
:1→xScl
:-1→yMin
:1→yMax
:1→yScl
:0→tMin
:abs T→tMax
:.15→tStep
:cos T→A
:sin T→B
:xt1=(.25+.04t)cos t
:yt1=sign T*(.25+.04t)sin t
:DispG
:Line(0,0,A,B)
:Pause
:Func
```

Graphs of Sine and Cosine Functions   (Section 5.5)

The program, presented in the Discussion Problem, A Sine Show, simultaneously draws the unit circle and the corresponding points on the sine curve. After the circle and sine curve are drawn, you can connect points on the unit circle with their corresponding points on the sine curve by repeatedly pressing Enter.

For a TI-81 program, see page 394.

Casio 6300 Program

"SINESHOW"◢
Rad:
Range -2.25,π÷2,3,-1.19,
    1.19,1:
0→T:
Lbl 1:
Plot -1.25+cos T,sin T:
Plot T÷4,sin T:
T+.15→T:
T≤6.3⇒Goto 1:
Plot -1.25,-1.19
Plot -1.25,1.19:
Line:
0→N:
Lbl 2:
N+1→N:
Nπ÷6.5→T:
-1.25+cos T→A:
sin T→B:
T÷4→C:
Plot A,B:
Plot C,B:
Line◢
N<12⇒Goto 2

Casio 7700 Program

SINESHOW
Rad
Range -2.25,π÷2,3,-1.19,
    1.19,1,0,6.3,.15
Graph(X,Y)=(-1.25+cos T,sin
    T):Graph(X,Y)=(T÷4,sin T)
Plot -1.25,-1.19
Plot -1.25,1.19
Line
0→N
Lbl 1
N+1→N
Nπ÷6.5→T
-1.25+cos T→A
sin T→B
T÷4→C
Plot A,B
Plot C,B
Line◢
N<12⇒Goto 1

Press Mode Shift X to change to parametric mode when starting to write this program.

Sharp EL 9200, 9300 Program

```
SinShow
-----------REAL
m=sin⁻¹ 1/(π/2)
Range -2.25,π/2,3,-1.19,119,1
step=π/15
θ=0
xco=-0.25
xso-0
yo=0
Label 1
θ=θ+step
xc=cos (mθ)-1.25
xs=θ/4
y=sin (mθ)
Line xco,yo,xc,y
Line xso,yo,xs,y
xco=xc
xso=xs
yo=y
If θ<2π Goto 1
θ=0
Label 2
θ=θ+step
xc=cos (mθ)-1.25
xs=θ/4
y=sin (mθ)
Line xc,y,xs,y
If θ<2π Goto 2
End
```

TI-85 Program

```
 PROGRAM:
:Radian
:ClDrw:FnOff
:Param:SimulG
: -2.25→xMin
:π/2→xMax
:3→xScl
: -1.1→yMin
:1.1→yMax
:1→yScl
:0→tMin
:6.3→tMax
:.15→tStep
:xt1= -1.25+cos t
:yt1=sin t
:xt2=t/4
:yt2=sin t
:Line( -1.25,1.1, -1.25,1.1)
:DispG
:0→N
:Lbl A
:N+1→N
:N*π/6.5→T
: -1.25+cos T→A
:sin T→B
:T/4→C
:Line(A,B,C,B)
:Pause
:If N==12
:Goto B
:Goto A
:Lbl B
:Pause:Func:SeqG:Disp
```

Systems of Linear Equations in Two Variables (Section 7.2)

The program, discussed in a Technology Note, will solve a system of two linear equations.

For a TI-81 program, see page 539.

Casio 6300 Program

```
SOLVE:
"A="?→A:
"B="?→B:
"C="?→C:
"D="?→D:
"E="?→E:
"F="?→F:
AE-DB=0⇒Goto 1:
"X="◢(CE-BF)÷(AE-DB)◢
"Y="◢(AF-CD)÷(AE-DB):
Goto 2:
Lbl 1:
"NO UNIQUE SOLUTION":
Lbl 2
```

Casio 7700 Program

```
SOLVE
"ENTER A,B,C,D,E,F"
"A="?→A
"B="?→B
"C="?→C
"D="?→D
"E="?→E
"F="?→F
AE-DB=0⇒Goto 1
"X=":(CE-BF)÷(AE-DB)◢
"Y=":(AF-CD)÷(AE-DB)
Goto 2
Lbl 1
"NO UNIQUE SOLUTION"
Lbl 2
```

Sharp EL 9200, 9300 Program

```
Solve
----------REAL
Input A
Input B
Input C
Input D
Input E
Input F
If A*E-D*B=0 Goto 1
X=(C*E-B*F)/(A*E-D*B)
Y=(A*F-C*D)/(A*E-D*B)
Print X
Print Y
End
Label 1
Print "No solution
End
```

TI-85 Program

```
PROGRAM:
:Disp "ax+by=c"
:Input "Enter a ",A
:Input "Enter b ",B
:Input "Enter c ",C
:Disp "dx+ey=f"
:Input "Enter d ",D
:Input "Enter e ",E
:Input "Enter f ",F
:If A*E-D*B=0
:Goto A
:(C*E-B*F)/(A*E-D*B)→X
:(A*F-C*D)/(A*E-D*B)→Y
:Disp X
:Disp Y
:Lbl A
```

Equations must be entered in the form: $Ax + By = C$; $Dx + Ey = F$. Uppercase letters are used so that the values can be accessed in the calculation mode of the calculator.

Sequences and Summation Notation   (Section 9.1)

The program, from a Discovery note, finds the sum of a finite sequence. For a TI-81 program, see page 660.

Casio 6300 Program

```
SUM:
"LOWER LIMIT"?→M:
"UPPER LIMIT"?→N:
0→S:
M→X:
Lbl 1:
Prog 0:
Ans+S→S:
X+1→X:
X≤N⇒Goto 1:
"SUM="◢S
```

To use this program, store the formula for the nth term as Prog 0.

Casio 7700 Program

```
SUM
"LOWER LIMIT"?→M
"UPPER LIMIT"?→N
0→S
M→X
Lbl 1
S+f₁→S
X+1→X
X≤N⇒Goto 1
"SUM=":S
```

To use this program, store the formula for the nth term in f_1.

Sharp EL 9200, 9300 Program

```
Sum
----------REAL
Goto 1
Label eqtn
y=√(22926+902.5x+2.01x²
Return
Label 1
Print "Enter start
Input m
Print "Enter end
Input n
If n<m Goto 1
sum=0
x=m
Label 2
Gosub eqtn
sum=sum+y
x=x+1
If x<=n Goto 2
Print sum
End
```

To use this program, store the formula for the nth term into the subroutine 'eqtn' at the beginning of the program.

TI-85 Program

```
PROGRAM:
:Input "ENTER M ",M
:Input "ENTER N ",N
:0→S
:M→x
:Lbl A
:S+y1→S
:If x=N
:Goto B
:x+1→x
:Goto A
:Lbl B
:Disp S
```

To use this program, store the formula for the nth term as $y1$.

Vectors in the Plane   (Section 11.1)

The program is presented in Exercise 73. You are asked to explain what the program does.

For TI-81 program, see page 838.

Casio 6300 Program

```
"ENTER (A,B)"◢
"A="?→A:
"B="?→B:
"ENTER (C,D)"◢
"C="?→C:
"D="?→D:
Range -6,6,1,-2,6,1:
Plot 0,0:
Plot A,B:
Line:
Plot 0,0:
Plot C,D:
Line◢
A+C→E:
B+D→F:
Plot 0,0:
Plot E,F:
Line:
Plot A,B:
Plot E,F:
Line:
Plot C,D:
Plot E,F:
Line:
Lbl 1
```

Casio 7700 Program

```
"ENTER (A,B)"
"A="?→A
"B="?→B
"ENTER (C,D)"
"C="?→C
"D="?→D
Range -6,6,1,-2,6,1
Plot 0,0
Plot A,B
Line
Plot 0,0
Plot C,D
Line◢
A+C→E
B+D→F
Plot 0,0
Plot E,F
Line
Plot A,B
Plot E,F
Line
Plot C,D
Plot E,F
Line
Lbl 1
```

Sharp EL 9200, 9300 Program

```
Input a
Input b
Input c
Input d
Range -6,6,1,-2,6,1
Line 0,0,a,b
Line 0,0,c,d
a=a+c
f=b+d
Line 0,0,e,f
Line a,b,e,f
Line c,d,e,f
End
```

TI-85 Program

```
PROGRAM:
:Disp "ENTER (A,B)"
:Input "ENTER A ",A
:Input "ENTER B ",B
:Disp "ENTER (C,D)"
:Input "ENTER C ",C
:Input "ENTER D ",D
:Line(0,0,A,B)
:Line(0,0,C,D)
:A+C→E
:B+D→F
:Line(0,0,E,F)
:Line(A,B,E,F)
:Line(C,D,E,F)
:Pause
:ClDrw:Disp
```

The Dot Product of Two Vectors   (Section 11.2)

The program, presented in a Discovery note, sketches two vectors and finds the angle between the vectors.

For a TI-81 program, see page 840.

Casio 6300 Program

```
"ENTER (A,B)"◢
"A="?→A:
"B="?→B:
"ENTER (C,D)"◢
"C"=?→C:
"D"=?→D:
Range -6,6,1,-9,9,1:
Plot 0,0:
Plot A,B:
Line:
Plot 0,0:
Plot C,D:
Line ◢
AC+BD→E:
√(A²+B²)→U:
√(C²+D²)→V:
cos⁻¹(E/UV)→T:
"ANGLE="◢
T
```

Casio 7700 Program

```
"ENTER (A,B)"
"A="?→A
"B="?→B
"ENTER (C,D)"
"C"=?→C
"D"=?→D
Range -6,6,1,-9,9,1
Plot 0,0
Plot A,B
Line
Plot 0,0
Plot C,D
Line ◢
AC+BD→E
√(A²+B²)→U
√(C²+D²)→V
cos⁻¹(E÷UV)→T
"ANGLE="
T
```

Sharp EL 9200, 9300 Program

```
Input a
Input b
Input c
Input d
Range -6,6,1,-9,9,1
Line 0,0,a,b
Line 0,0,c,d
Wait
e=a*c+b*d
u=√(a²+b²)
v=√(c²+d²)
t=cos⁻¹(e/(u*v))
Print t
End
```

TI-85 Program

```
:Disp "ENTER (A,B)"
:Input "ENTER A ",A
:Input "ENTER B ",B
:Disp "ENTER (C,D)"
:Input "ENTER C ",C
:Input "ENTER D ",D
:Line (0,0,A,B)
:Line (0,0,C,D)
:Pause
:A*C+B*D→E
:√(A²+B²)→U
:√(C²+D²)→V
:cos⁻¹ (E/U*V))→T
:Disp T
:ClDrw
```

FRACTAL PROGRAMS

Fractal Gasket

```
:0→Xmin
:100→Xmax
:10→Xscl
:0→Ymin
:100→Ymax
:10→Yscl
:0→A
:0→B
:50→C
:100→D
:100→E
:0→F
:ClrDraw
```

```
:0→L
:PT-On(A,B)
:PT-On(C,D)
:PT-On(E,F)
:Disp "START(X,Y)"
:Input G
:Input H
:Lbl 1
:L+1→L
:Int(3Rand+1)→T
:If T=1
:G+(A-G)/2→G
:If T=1
```

```
:H+(B-H)/2→H
:If T=2
:G+(C-G)/2→G
:If T=2
:H+(D-H)/2→H
:If T=3
:G+(E-G)/2→G
:If T=3
:H+(F-H)/2→H
:PT-On(G,H)
:If L<1000
:Goto 1
:End
```

Any point (G, H) may be entered. Depending on the choice of (G, H) the first few points may lie outside the Fractal Gasket.

Fractal Dragon

```
:0→Xmin
:10→Xmax
:1→Xscl
:0→Ymin
:10→Ymax
:1→Yscl
:ClrDraw
:0→X
:0→Y
:1→N
:Lbl 1
:Rand→P
:If P<.5
:Goto 2
:If P>.5
```

```
:Goto 3
:Lbl 4
:AX+BY+C→R
:DX+EY+F→S
:3.2R+2.5→U
:4.25S+6.5→V
:PT-On(U,V)
:R→X
:S→Y
:N+1→N
:If N<3000
:Goto 1
:End
:Lbl 2
:.5→A
```

```
:.5→B
:0→C
:-.5→D
:.5→E
:0→F
:Goto 4
:Lbl 3
:-.5→A
:.5→B
:2→C
:-.5→D
:-.5→E
:0→F
:Goto 4
:End
```

Fractal Fern

```
:0→Xmin              :-.0436→A            :.16→S
:50→Xmax             :-.0436→B            :0→A
:10→Xscl             :1→K                 :0→B
:0→Ymin              :If Z<.005           :0→K
:50→Ymax             :Goto 2              :Goto 5
:10→Yscl             :If Z<.1025          :Lbl 3
:ClrDraw             :Goto 3              :.3→R
:25→A                :If Z<.2             :.34→S
:25→B                :Goto 4              :.8552→A
:0→J                 :Lbl 5               :.8552→B
:Disp "START(X, Y)"  :RCcos A-SDsinB→E    :1.6→K
:Input C             :RCsinA+SDcosB+K→F   :Goto 5
:Input D             :E→C                 :Lbl 4
:Lbl 1               :F→D                 :.3→R
:8(A+D)-1→U          :J+1→J               :.37→S
:16(B+C)+20→V        :If J<10000          :2.0944→A
:PT-On(U,V)          :Goto 1              :-.8552→B
:Rand→Z              :End                 :.44→K
:.85→R               :Lbl 2               :Goto 5
:.85→S               :0→R
```

Any point (C, D) may be entered. Depending on the choice of (C, D) the first few points may lie outside the Fractal Fern. *Note:* Calculator should be in *radian* mode.

ANSWERS
Warm-Ups, Odd-Numbered Exercises, and Cumulative Tests

PREREQUISITES

Section 1 (page 7)

1. (a) 5, 1   **(b)** $-9, 5, 0, 1$ **(c)** $-9, -\frac{7}{2}, 5, \frac{2}{3}, 0, 1$
(d) $\sqrt{2}$
3. (a) $12, 1, \sqrt{4}$ **(b)** $12, -13, 1, \sqrt{4}$
(c) $12, -13, 1, \sqrt{4}, \frac{3}{2}$ **(d)** $\sqrt{6}$
5. (a) $\frac{6}{3}$ **(b)** $\frac{6}{3}$ **(c)** $-\frac{1}{3}, \frac{6}{3}, -7.5$ **(d)** $-\pi, \frac{1}{2}\sqrt{2}$
7. $\frac{3}{2} < 7$ **9.** $-4 > -8$

11. $\frac{5}{6} > \frac{2}{3}$

13. $-1 \le x \le 3$, bounded   **15.** $10 < x$, unbounded
17. $(-\infty, 5]$ **19.** $(-\infty, 0)$

21. $[4, \infty)$ **23.** $(-2, 2)$

25. $[-1, 0)$

27. $(-\infty, 0)$ **29.** $(-\infty, 25]$
31. $[30, \infty)$ **33.** $[0.035, 0.06]$ **35.** -1 **37.** -9
39. 3.75 **41.** $|-3| > -|-3|$ **43.** $-5 = -|5|$
45. $-|-2| = -|2|$ **47.** $\frac{5}{2}$ **49.** 51 **51.** 14.99
53. $|x - 5| \le 3$ **55.** $|z - \frac{3}{2}| > 1$ **57.** $|y| \ge 6$
59. $\frac{127}{90}, \frac{584}{413}, \frac{7071}{5000}, \sqrt{2}, \frac{47}{33}$ **61.** 0.625 **63.** $0.123123\ldots$
65. False. The reciprocal of 2 is $\frac{1}{2}$, which is not an integer.
67. True

Section 2 (page 16)

1. $7x, 4$ **3.** $x^2, -4x, 8$ **5.** $4x^3, x, -5$
7. (a) -10 **(b)** -6 **9. (a)** 14 **(b)** 2
11. (a) Division by 0 is undefined.   **(b)** 0

13. Commutative (addition)   **15.** Inverse (multiplication)
17. Distributive Property   **19.** Identity (multiplication)
21. Associative and Commutative (multiplication)   **23.** 0
25. Division by 0 is undefined.   **27.** 2 **29.** 6
31. $\frac{1}{2}$ **33.** $\frac{3}{8}$ **35.** $\frac{3}{10}$ **37.** $\frac{1}{12}$ **39.** 48
41. -2.57 **43.** 1.56 **45.** 0.33 **47.** 1.11
49. 125 **51.** 729 **53.** 5184 **55.** $\frac{16}{3}$ **57.** -24
59. 5 **61.** $-125z^3$ **63.** $24y^{10}$ **65.** $3x^2$ **67.** $\frac{7}{x}$
69. $\frac{4}{3}(x + y)^2$ **71.** 1 **73.** $-2x^3$ **75.** $\frac{10}{x}$
77. 3^{3n} **79.** 5.75×10^7 **81.** 8.99×10^{-5}
83. $524{,}000{,}000$ **85.** 0.00000000048
87. (a) 954.448 **(b)** 3.077×10^{10}

Section 3 (page 24)

1. $9^{1/2} = 3$ **3.** $\sqrt{196} = 14$ **5.** $(-216)^{1/3} = -6$
7. $\sqrt[3]{27^2} = 9$ **9.** $81^{3/4} = 27$ **11.** 3 **13.** 2
15. 6 **17.** 3 **19.** $\frac{1}{2}$ **21.** -125 **23.** 4 **25.** $\frac{1}{8}$
27. $\frac{27}{8}$ **29.** -4 **31.** $2\sqrt{2}$ **33.** 3×10^{-2}
35. $6|x|\sqrt{2x}$ **37.** $2x\sqrt[3]{2x^2}$ **39.** $\frac{5|x|\sqrt{3}}{y^2}$ **41.** $\frac{\sqrt{3}}{3}$
43. $4\sqrt[3]{4}$ **45.** $\frac{x(5 + \sqrt{3})}{11}$ **47.** $3(\sqrt{6} - \sqrt{5})$
49. $\frac{2}{\sqrt{2}}$ **51.** $\frac{2}{3(\sqrt{5} - \sqrt{3})}$ **53.** $-\frac{1}{2(\sqrt{7} + 3)}$
55. $3^{1/2} = \sqrt{3}$ **57.** $(x + 1)^{2/3} = \sqrt[3]{(x + 1)^2}$
59. $2\sqrt[4]{2}$ **61.** $\sqrt[8]{2x}$ **63.** $2\sqrt{x}$ **65.** $34\sqrt{2}$
67. $4\sqrt{y}$ **69.** $\frac{1}{x^3}$ **71.** $\frac{9y^{3/2}}{x^{2/3}}$ **73.** $\frac{3y^2}{4z^{4/3}}$ **75.** $x^{1/4}$
77. $c^{1/2}$ **79.** 7.550 **81.** 2.236 **83.** 9.137
85. $\sqrt{5} + \sqrt{3} > \sqrt{5 + 3}$ **87.** $5 > \sqrt{3^2 + 2^2}$
89. $\sqrt{3} \cdot \sqrt[3]{3} > \sqrt[8]{3}$ **91.** 1

Section 4 (*page 29*)

	Degree	*Leading Coefficient*
1.	2	2
3.	5	1
5.	5	4

7. Polynomial: $-3x^3 + 2x + 8$

9. Not a polynomial because of the operation of division.

11. Polynomial: $-y^4 + y^3 + y^2$ **13.** $-2x - 10$

15. $3x^3 - 2x + 2$ **17.** $8x^3 + 29x^2 + 11$

19. $12z + 8$ **21.** $3x^3 - 6x^2 + 3x$ **23.** $-15z^2 + 5z$

25. $30x^3 + 12x^2$ **27.** $x^2 + 7x + 12$

29. $6x^2 - 7x - 5$ **31.** $x^2 + 12x + 36$

33. $4x^2 - 20xy + 25y^2$

35. $x^2 + 2xy + y^2 - 6x - 6y + 9$ **37.** $x^2 - 100$

39. $x^2 - 4y^2$ **41.** $m^2 - n^2 - 6m + 9$ **43.** $4r^4 - 25$

45. $x^3 + 3x^2 + 3x + 1$ **47.** $8x^3 - 12x^2y + 6xy^2 - y^3$

49. $x - y$ **51.** $16x^6 - 24x^3 + 9$

53. $x^4 - x^3 + 5x^2 - 9x - 36$ **55.** $x^4 + x^2 + 1$

57. $2x^2 + 2x$ **59.** $x^3 + 4x^2 - 5x - 20$ **61.** $m + n$

Section 5 (*page 35*)

1. $3(x + 2)$ **3.** $2x(x^2 - 3)$ **5.** $(x - 1)(x + 5)$

7. $(x + 6)(x - 6)$ **9.** $(4y + 3)(4y - 3)$

11. $(x + 1)(x - 3)$ **13.** $(x - 2)^2$ **15.** $(2t + 1)^2$

17. $(5y - 1)^2$ **19.** $(x + 2)(x - 1)$

21. $(s - 3)(s - 2)$ **23.** $(y + 5)(y - 4)$

25. $(x - 20)(x - 10)$ **27.** $(3x - 2)(x - 1)$

29. $(3z + 1)(3z - 2)$ **31.** $(5x + 1)(x + 5)$

33. $(x - 2)(x^2 + 2x + 4)$ **35.** $(y + 4)(y^2 - 4y + 16)$

37. $(2t - 1)(4t^2 + 2t + 1)$ **39.** $(x - 1)(x^2 + 2)$

41. $(2x - 1)(x^2 - 3)$ **43.** $(3 + x)(2 - x^3)$

45. $(x + 2)(3x + 4)$ **47.** $(2x - 1)(3x + 2)$

49. $(3x - 1)(5x - 2)$

51.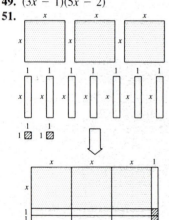

53.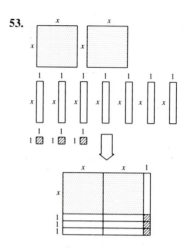

55. $x(x + 3)(x - 3)$ **57.** $x^2(x - 4)$ **59.** $(x - 1)^2$

61. $(1 - 2x)^2$ **63.** $2x(x + 1)(x - 2)$

65. $(9x + 1)(x + 1)$ **67.** $(3x + 1)(x^2 + 5)$

69. $x(x - 4)(x^2 + 1)$ **71.** $-x(x + 10)$

73. $(x + 1)^2(x - 1)^2$ **75.** $2(t - 2)(t^2 + 2t + 4)$

77. $(2x - 1)(6x - 1)$ **79.** $-(x + 1)(x - 3)(x + 9)$

Section 6 (*page 41*)

1. All real numbers   **3.** All nonnegative real numbers

5. All real numbers x such that $x \neq 2$

7. All real numbers x such that $x \neq 0$ and $x \neq 4$

9. All real numbers x such that $x \geq -1$ **11.** $3x, \, x \neq 0$

13. $x - 2, \, x \neq 2$ **15.** $x + 2, \, x \neq -2$ **17.** $\dfrac{3x}{2}, \, x \neq 0$

19. $\dfrac{3y}{y + 1}, \, y \neq 0$ **21.** $-\dfrac{1}{2}, \, x \neq 5$ **23.** $\dfrac{x(x + 3)}{x - 2}, \, x \neq -2$

25. $\dfrac{y - 4}{y + 6}, \, y \neq 3$ **27.** $-(x^2 + 1), \, x \neq 2$ **29.** $z - 2$

31. $\dfrac{1}{5(x - 2)}, \, x \neq 1$ **33.** $\dfrac{r + 1}{r}, \, r \neq 1$

35. $\dfrac{2(y^2 + 2y + 4)}{y^2(y - 3)}, \, y \neq 2$ **37.** $\dfrac{3}{2}$

39. $x(x + 1)$, $x \neq 0, -1$ **41.** $\dfrac{x + 5}{x - 1}$ **43.** $\dfrac{6x + 13}{x + 3}$

45. $-\dfrac{2}{x - 2}$ **47.** $\dfrac{x - 4}{(x + 2)(x - 2)(x - 1)}$

49. $\dfrac{2 - x}{x^2 + 1}$, $x \neq 0$ **51.** $\dfrac{2x^2 - 1}{(x^2 - 1)^{1/2}}$ **53.** $\dfrac{2(x - 3)}{(x - 2)^{4/3}}$

55. $\dfrac{1}{2}$ **57.** $\dfrac{1}{x}$, $x \neq -1$ **59.** $-\dfrac{2x + h}{x^2(x + h)^2}$, $h \neq 0$

61. $\dfrac{a + 2b}{(ab)^2}$ **63.** $\dfrac{2x - 1}{2x}$, $x > 0$ **65.** $-\dfrac{1}{t^2\sqrt{t^2 + 1}}$

67. $\dfrac{1}{\sqrt{x + 2} + \sqrt{x}}$

Section 7   (page 48)

1.

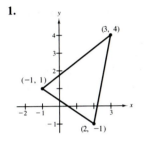

3.

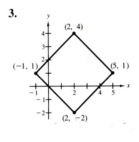

5. 8   **7.** 5   **9.** $a = 4$, $b = 3$, $c = 5$

11. $a = 10$, $b = 3$, $c = \sqrt{109}$

13. (a)   (b) 10   (c) $(5, 4)$

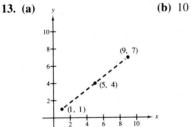

15. (a)   (b) 17   (c) $\left(0, \dfrac{5}{2}\right)$

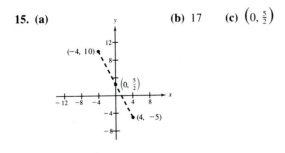

17. (a)   (b) $2\sqrt{10}$ (c) $(2, 3)$

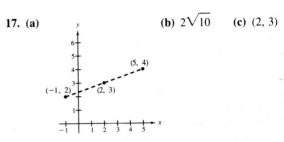

19. (a)   (b) $\dfrac{\sqrt{82}}{3}$ (c) $\left(-1, \dfrac{7}{6}\right)$

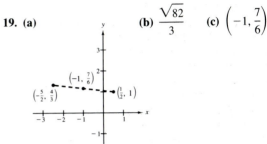

21. (a)   (b) $\sqrt{110.97}$ (c) $(1.25, 3.6)$

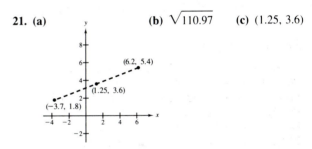

23. (a)   (b) $6\sqrt{277}$ (c) $(6, -45)$

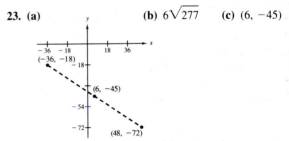

25. \$630,000   **27.** $\left(\sqrt{5}\right)^2 + \left(\sqrt{45}\right)^2 = \left(\sqrt{50}\right)^2$

29. All sides have a length of $\sqrt{5}$.   **31.** $x = 6, -4$

33. $y = \pm 15$ **35.** $3x - 2y - 1 = 0$

37. Quadrant IV   **39.** Quadrant I   **41.** Quadrant II

43. Quadrant III or IV   **45.** Quadrant I or III

47. Quadrant III

49.

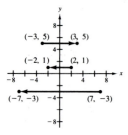

The points are reflected through the y-axis.

51. $x^2 + y^2 = 9$ **53.** $(x - 2)^2 + (y + 1)^2 = 16$
55. $(x + 1)^2 + (y - 2)^2 = 5$
57. $(x - 3)^2 + (y - 4)^2 = 25$

Section 8   (*page 55*)

1. 15

3. 81 and 85

5. Stems	Leaves
7	0 5 5 5 7 7 8 8 8
8	1 1 1 1 2 3 4 5 5 5 5 7 8 9 9 9
9	0 2 9
10	0 0

7. Stems	Leaves
6	18 68 71 94
7	16 19 25 25 42 57 58 76 84 88
8	13 28 35 41 46 59 61 62 66 81 83 89 91 92 92
9	05 06 15 18 25 25 26 28 41 44 83 90 92
10	04 10 62 95
11	51 78 86
12	23
13	
14	
15	
16	26

9.

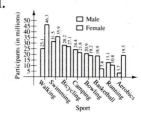

11.

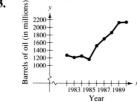

13.

CHAPTER 1

Section 1.1   *(page 64)*

(page 64)

WARM-UP   **1.** $14x - 42$ **2.** $-17s$ **3.** $-24y^7$

4. $\dfrac{2a}{3b^5}$ **5.** $5x^2\sqrt{6}$ **6.** $\dfrac{1}{2}$ **7.** $2x^2(x - 3)$

8. $2(t + 3)(t^2 + 3t - 2)$ **9.** $(3z + 10)(2z - 5)$

10. $(2s + 5)(2s - 5)$

---

1. (a) Yes   **(b)** Yes   **3. (a)** No   **(b)** Yes

5. (a) Yes   **(b)** Yes   **7.** 2   **9.** 4

11.

x	-4	-2	0	2	4
y	11	7	3	-1	-5
(x, y)	$(-4, 11)$	$(-2, 7)$	$(0, 3)$	$(2, -1)$	$(4, -5)$

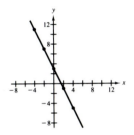

13. c

```
Xmin=-2
Xmax=6
Xscl=1
Ymin=-1
Ymax=5
Yscl=1
```

16. a

```
Xmin=-1
Xmax=8
Xscl=1
Ymin=-1
Ymax=4
Yscl=1
```

14. f

```
Xmin=-4
Xmax=2
Xscl=1
Ymin=-2
Ymax=4
Yscl=1
```

17. e

```
Xmin=-2
Xmax=2
Xscl=1
Ymin=-2
Ymax=2
Yscl=1
```

15. d

```
Xmin=-3
Xmax=3
Xscl=1
Ymin=-1
Ymax=3
Yscl=1
```

18. b

```
Xmin=-4
Xmax=4
Xscl=1
Ymin=-3
Ymax=3
Yscl=1
```

19.

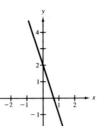

21.

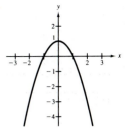

23.

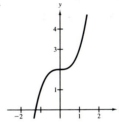

25.

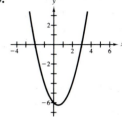

27.

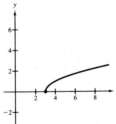

29.

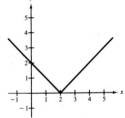

31.

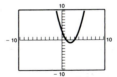

The graph intersects the x-axis twice.
The graph intersects the y-axis once.

33.

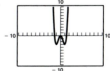

The graph intersects the x-axis three times.
The graph intersects the y-axis once.

35.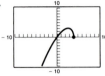

The graph intersects the x-axis twice.
The graph intersects the y-axis once.

37.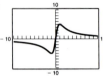

The graph intersects the x-axis once.
The graph intersects the y-axis once.

39.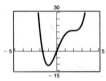

The graph intersects the x-axis twice.
The graph intersects the y-axis once.

41.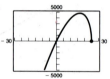

The graph intersects the x-axis twice.
The graph intersects the y-axis once.

43.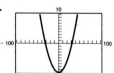

The graph intersects the x-axis twice.
The graph intersects the y-axis once.

45. $y = \sqrt{64 - x^2}, \; y = -\sqrt{64 - x^2}$

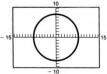

47. $y = \sqrt{6(12 - x^2)}, \; y = -\sqrt{6(12 - x^2)}$

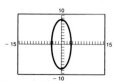

49.
```
Xmin=-1
Xmax=100
Xscl=5
Ymin=-1
Ymax=40
Yscl=2
```

51.
```
Xmin=-1
Xmax=20
Xscl=1
Ymin=-25
Ymax=150
Yscl=10
```

53.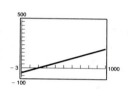

The person would select the first viewing rectangle.

55. (a)

x	-1	0	1
y	-1	0	1

(b)

x	-1	$-\frac{3}{4}$	$-\frac{1}{2}$	$-\frac{1}{4}$	0	$\frac{1}{4}$	$\frac{1}{2}$	$\frac{3}{4}$	1
y	-1	-0.91	-0.79	-0.63	0	0.63	0.79	0.91	1

(c)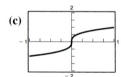

57.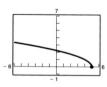

(a) $y = 1.73$
(b) $x = -4$

59.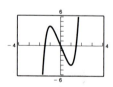

(a) $y = 2.47$
(b) $x = -1.65, 1$

61. (b)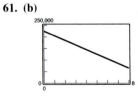

63.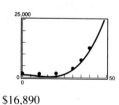

$16,890

65.

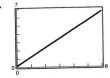

67. There are an unlimited number of correct answers, one of which is $a = 1$ and $b = -5$.

Section 1.2   *(page 77)*

WARM-UP   **1.** $-\frac{9}{2}$ **2.** $-\frac{13}{3}$ **3.** $-\frac{5}{4}$ **4.** $\frac{1}{2}$
5. $y = \frac{2}{3}x - \frac{5}{3}$ **6.** $y = -2x$ **7.** $y = 3x - 1$
8. $y = \frac{2}{3}x + 5$ **9.** $y = -2x + 7$ **10.** $y = x + 3$

1. $\frac{6}{5}$ **3.** 0   **5.** -3

7. $m = -3$ $m = 1$ **9.** $m = 2$

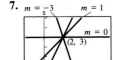

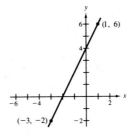

11. m is undefined.   **13.** $m = \frac{4}{3}$

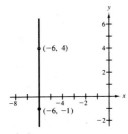

15. $(0, 1), (3, 1), (-1, 1)$ **17.** $(6, -5), (7, -4), (8, -3)$
19. Perpendicular   **21.** Parallel   **23.** Collinear
25. $16,667 \text{ ft} \approx 3.16 \text{ mi}$

27. $m = 5$; Intercept: $(0, 3)$

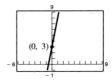

29. m is undefined. There is no y-intercept.

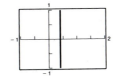

31. $3x + 5y - 10 = 0$ **33.** $x + 2y - 3 = 0$
35. $x + 8 = 0$ **37.** $3x - y - 2 = 0$
39. $8x + 6y - 47 = 0$
41. $x - 6 = 0$
43.

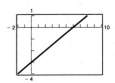

45.   **47.**

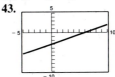

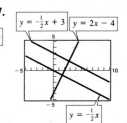

49.

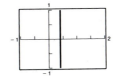

a and b are the x- and y-intercepts of the graph of the line.
51. $3x + 2y - 6 = 0$
53. (a) $2x - y - 3 = 0$ **(b)** $x + 2y - 4 = 0$
55. (a) $y = 0$ **(b)** $x + 1 = 0$ **57.** $V = 125t + 2540$
59. $V = 2000t + 20,400$ **61.** b   **62.** c   **63.** a
64. d   **65.** $F = \frac{9}{5}C + 32$
67. (a) $S = 2200t + 28,500$ **(b)** \$39,500

69. (a) $V = -175t + 875$ **(b)**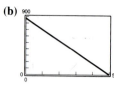
(c) \$525 **(d)** 3.86 yr

71. $W = 0.07S + 2500$
73. (a) $C = 16.75t + 36,500$ **(b)** $R = 37t$
(c) $t \approx 1802$ hr
75. The model approximates the daily cost y of producing x units of a product.

Section 1.3 (page 90)

WARM-UP **1.** -73 **2.** 13 **3.** $2(x + 2)$
4. $-8(x - 2)$ **5.** $y = \frac{7}{5} - \frac{2}{5}x$ **6.** $y = \pm x$
7. $5xy\sqrt{2x}$ **8.** $3z\sqrt{2z^2 - 1}$ **9.** $\frac{x - y}{x + y}$
10. $\frac{x(x - 3)}{2}$

1. (a) Function
(b) Not a function because the element 1 in A is matched with two elements, -2 and 1, in B.
(c) Function
(d) Not a function because not all elements of A are matched with an element in B.
3. Not a function **5.** Function **7.** Function
9. Not a function
11. (a) $6 - 4(3) = -6$ **(b)** $6 - 4(-7) = 34$
(c) $6 - 4(t) = 6 - 4t$
(d) $6 - 4(c + 1) = 2(1 - 2c)$
13. (a) $\frac{1}{4 + 1} = \frac{1}{5}$ **(b)** $\frac{1}{0 + 1} = 1$ **(c)** $\frac{1}{4x + 1}$
(d) $\frac{1}{(x + h) + 1}$
15. (a) -1 **(b)** -9 **(c)** $2x - 5$ **(d)** $-\frac{5}{2}$
17. (a) 0 **(b)** 3 **(c)** $x^2 + 2x$ **(d)** -0.75
19. (a) 1 **(b)** -1 **(c)** 1 **(d)** $\frac{|x - 1|}{x - 1}$
21. (a) -1 **(b)** 2 **(c)** 4 **(d)** 6
23. All real numbers **25.** $y \geq 10$
27. All real numbers except $t = 0$
29. All real numbers except $x = 0, -2$
31. $(-2, 4), (-1, 1), (0, 0), (1, 1), (2, 4)$
33. $(-2, 0), (-1, 1), (0, \sqrt{2}), (1, \sqrt{3}), (2, 2)$
35. (a) 1056.250 **(b)** 1470.084
37. (a) 0.118 **(b)** 21.277 **39.** $g(x) = cx^2, c = -2$

41. $r(x) = \frac{c}{x}, c = 32$ **43.** $2, h \neq 0$
45. $3xh + 3x^2 + h^2, h \neq 0$ **47.** $A = \frac{C^2}{4\pi}$
49. $A = \frac{x^2}{x - 1}, x > 1$
51. $V = 4x^2(27 - x), 0 < x < 27$
53. $h = \sqrt{d^2 - 2000^2}, 0 \leq d$
55. (a) $C = 12.30x + 98,000$ **(b)** $R = 17.98x$
(c) $P = 5.68x - 98,000$

Section 1.4 (page 101)

WARM-UP **1.** 2 **2.** 0 **3.** $-\frac{3}{x}$ **4.** $x^2 + 3$
5. x-intercept: $(-\frac{12}{5}, 0)$
y-intercept: $(0, 6)$
6. x-intercept: $(5, 0)$
y-intercept: $(0, 3)$
7. All real numbers except $x = 4$
8. All real numbers except $x = 4, 5$ **9.** $t \leq \frac{5}{3}$
10. All real numbers

1. 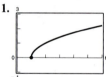 **3.**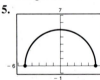

Domain: $[1, \infty)$ Domain: $(-\infty, -2], [2, \infty)$
Range: $[0, \infty)$ Range: $[0, \infty)$
5. **7.** Function

```
Xmin=-3
Xmax=3
Xscl=1
Ymin=-1
Ymax=4
Yscl=1
```

Domain: $[-5, 5]$
Range: $[0, 5]$
9. Not a function **11.** Function

```
Xmin=-1
Xmax=5
Xscl=1
Ymin=-3
Ymax=3
Yscl=1
```

```
Xmin=-5
Xmax=5
Xscl=1
Ymin=-4
Ymax=4
Yscl=1
```

13. b **15.** a

17.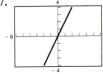
Increasing on $(-\infty, \infty)$

19.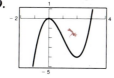
Increasing on $(-\infty, 0)$, $(2, \infty)$
Decreasing on $(0, 2)$

21.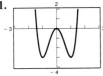
Increasing on $(-1, 0)$, $(1, \infty)$
Decreasing on $(-\infty, -1)$, $(0, 1)$

23.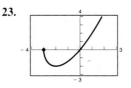
Increasing on $(-2, \infty)$
Decreasing on $(-3, -2)$

25. Relative minimum: $(3, -9)$
27. Relative minimum: $(1, -7)$
Relative maximum: $(-2, 20)$
29. Relative minimum: $(0.33, -0.38)$
31. (b)

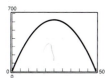

33.
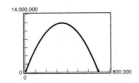

(c) Maximum area:
625 sq ft, 25 ft × 25 ft

350,000 units

35.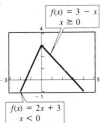
$f(x) = 3 - x$, $x \ge 0$
$f(x) = 2x + 3$, $x < 0$

37.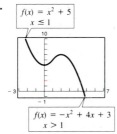
$f(x) = x^2 + 5$, $x \le 1$
$f(x) = -x^2 + 4x + 3$, $x > 1$

39.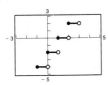

41. (a) $C = 0.65 + 0.42[\![t]\!]$
(b)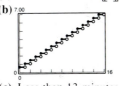
(c) Less than 13 minutes

43. (a) $(-5, 6)$ **(b)** $(-5, -6)$
45. (a) $\left(\frac{3}{2}, -2\right)$ **(b)** $\left(\frac{3}{2}, 2\right)$

47.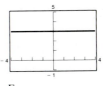
Even

49.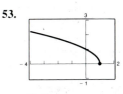
Neither even nor odd

51.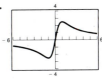
Odd

53.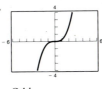
Neither even nor odd

55.
Odd

57. Even

59. Odd
61. Neither even nor odd
63. $(-\infty, 4]$
65. $[-1, 1]$

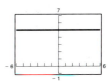

67. $f(x) > 0$ for all x
69. $h = (4x - x^2) - 3$

71. $h = 4x - 2x^2$ **73.** $L = (4 - y^2) - (y + 2)$

75.
Interval	Intake Pipe	Drain Pipe 1	Drain Pipe 2
[0, 5]	Open	Closed	Closed
[5, 10]	Open	Open	Closed
[10, 20]	Closed	Closed	Closed
[20, 30]	Closed	Closed	Open
[30, 40]	Open	Open	Open
[40, 45]	Open	Closed	Open
[45, 50]	Open	Open	Open
[50, 60]	Open	Open	Closed

Section 1.5   *(page 112)*

WARM-UP

1.

Domain: $[-4, 4]$
Range: $[0, 4]$

2.

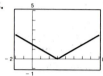

Domain: $(-\infty, \infty)$
Range: $[0, \infty)$

3.

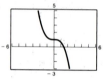

Domain: $(-\infty, \infty)$
Range: $[-1, \infty)$

4.

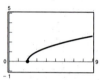

Domain: $(-\infty, \infty)$
Range: $(-\infty, \infty)$

5.

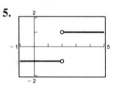

Domain: all real numbers
except $x = 2$; Range: $\{-1, 1\}$

6.

Domain: $[2, \infty)$
Range: $[0, \infty)$

7. Odd   **8.** Even   **9.** Even
10. Neither even nor odd

1.

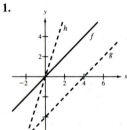

3.

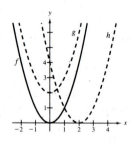

5.

7.

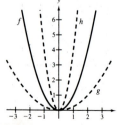

9. $g(x) = (x - 1)^2 + 1$
$h(x) = -(x - 2)^2 + 4$

11. $g(x) = (x - 3)^2 - 2$
$h(x) = -x^2 + 3$

13.

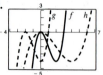

$g(x)$ is shifted two units to the left of $f(x)$.
$h(x)$ is a vertical shrink of $f(x)$.

15.

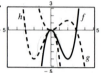

$g(x)$ is a vertical shrink of $f(x)$ and a reflection across
the x-axis.
$h(x)$ is a reflection across the y-axis.

17. $y = -(x^3 - 3x^2) + 1$
19. y is $f(x)$ shifted up two units.
21. y is $f(x)$ shifted right two units.
23. y is a vertical stretch of $f(x)$ by $\sqrt{2}$.
25. y is $f(x)$ shifted down one unit.
27. y is $f(x)$ shifted right one unit.
29. y is $f(x)$ reflected in the y-axis.

31.

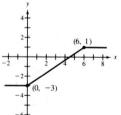

33.

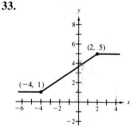

35.

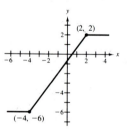

37. A reflection in the x-axis followed by a vertical shift of
four units upward.
39. A horizontal shift of two units to the left and a vertical
shrink.
41. A horizontal stretch and a vertical shift of two units
upward.

43. (a)

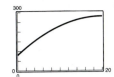

(b) $P(x) = -2420 + 20x - 0.5x^2$
Vertical shift
(c) $P(x) = 80 + 0.2x - 0.00005x^2$
Horizontal stretch

45.

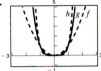

The three even functions are nonnegative. As the exponents increase, the graphs become flatter in the interval $(-1, 1)$.

47.

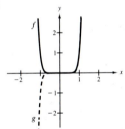

49.

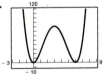

51.

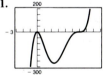

Section 1.6   *(page 120)*

WARM-UP   **1.** $\dfrac{1}{x(1 - x)}$ **2.** $-\dfrac{12}{(x + 3)(x - 3)}$

3. $\dfrac{3x - 2}{x(x - 2)}$ **4.** $\dfrac{4x - 5}{3(x - 5)}$ **5.** $\sqrt{\dfrac{x - 1}{x + 1}}$

6. $\dfrac{x + 1}{x(x + 2)}$ **7.** $5(x - 2)$ **8.** $\dfrac{x + 1}{(x - 2)(x + 3)}$

9. $\dfrac{1 + 5x}{3x - 1}$ **10.** $\dfrac{x + 4}{4x}$

1. (a) $2x$ **(b)** 2 **(c)** $x^2 - 1$ **(d)** $\dfrac{x + 1}{x - 1}, x \neq 1$

3. (a) $x^2 + 5 + \sqrt{1 - x}$ **(b)** $x^2 + 5 - \sqrt{1 - x}$
  (c) $(x^2 + 5)\sqrt{1 - x}$ **(d)** $\dfrac{x^2 + 5}{\sqrt{1 - x}}, x < 1$

5. (a) $\dfrac{x + 1}{x^2}$ **(b)** $\dfrac{x - 1}{x^2}$ **(c)** $\dfrac{1}{x^3}$ **(d)** $x, x \neq 0$

7. 9 **9.** $4t^2 - 2t + 5$ **11.** 0 **13.** 26 **15.** $\frac{3}{5}$

17.

19.

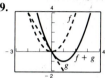

21.

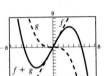

23. $T = \frac{3}{4}x + \frac{1}{15}x^2$

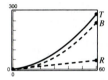

$0 \leq x \leq 2, f$
$x > 5, g$

25. (a) $(x - 1)^2$ **(b)** $x^2 - 1$ **(c)** x^4
27. (a) $20 - 3x$ **(b)** $-3x$ **(c)** $9x + 20$
29. (a) $\sqrt{x^2 + 4}$ **(b)** $x + 4$
31. (a) $x - \frac{8}{3}$ **(b)** $x - 8$
33. (a) $|x + 6|$ **(b)** $|x| + 6$ **35. (a)** 3 **(b)** 0
37. (a) 0 **(b)** 4 **39.** $f(x) = x^2, g(x) = 2x + 1$
41. $f(x) = \sqrt[3]{x}, g(x) = x^2 - 4$
43. $f(x) = \dfrac{1}{x}, g(x) = x + 2$
45. $f(x) = x^2 + 2x, g(x) = x + 4$
47. (a) $x \geq 0$ **(b)** All real numbers
  (c) All real numbers
49. (a) All real numbers except $x = \pm 1$
  (b) All real numbers
  (c) All real numbers except $x = -2, 0$
51. (a) $(A \circ r)(t) = 0.36\pi t^2$
    $A \circ r$ represents the area of the circle at time t.
  (b)

4.2 sec

53. (a) $(C \circ x)(t) = 3000t + 750$

$C \circ x$ represents the cost after t production hours.

(b)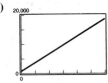

4.75 hr

55. $R(x) = x - 1200$, $D(x) = 0.85x$

$(D \circ R)(x) = 0.85(x - 1200)$

$(D \circ R)(18,400) = \$14,620$

$(R \circ D)(x) = 0.85x - 1200$

$(R \circ D)(18,400) = \$14,440$

57. Odd

Section 1.7 (page 130)

WARM-UP **1.** All real numbers **2.** $[-1, \infty)$

3. All real numbers except $x = 0, 2$

4. All real numbers except $x = -\frac{5}{3}$ **5.** x **6.** x

7. x **8.** x **9.** $x = \frac{3}{2}y + 3$ **10.** $x = \frac{y^3}{2} + 2$

1. $f^{-1}(x) = \frac{1}{8}x$ **3.** $f^{-1}(x) = x - 10$

5. $f^{-1}(x) = x^3$

7. (a) $(f \circ g)(x) = f\left(\frac{x}{2}\right) = 2\left(\frac{x}{2}\right) = x$

$(g \circ f)(x) = g(2x) = \frac{2x}{2} = x$

(b)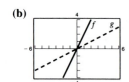

9. (a) $(f \circ g)(x) = f\left(\frac{x - 1}{5}\right) = \frac{5x + 1 - 1}{5} = x$

$(g \circ f)(x) = g(5x + 1) = 5\left(\frac{x - 1}{5}\right) + 1 = x$

(b)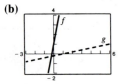

11. (a) $(f \circ g)(x) = f\left(\sqrt[3]{x}\right) = \left(\sqrt[3]{x}\right)^3 = x$

$(g \circ f)(x) = g(x^3) = \sqrt[3]{x^3} = x$

(b)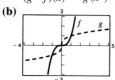

13. (a) $(f \circ g)(x) = f(x^2 + 4)$, $x \geq 0$

$= \sqrt{(x^2 + 4)} - 4 = x$

$(g \circ f)(x) = g\left(\sqrt{x - 4}\right)$

$= \left(\sqrt{x - 4}\right)^2 + 4 = x$

(b)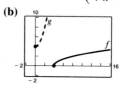

15. (a) $(f \circ g)(x) = f\left(\sqrt{9 - x}\right)$, $x \leq 9$

$= 9 - \left(\sqrt{9 - x}\right)^2 = x$

$(g \circ f)(x) = g(9 - x^2)$

$= \sqrt{9 - (9 - x)^2} = x$

(b)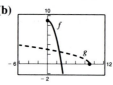

17. One-to-one **19.** Not one-to-one

21. Not one-to-one **23.** $f^{-1}(x) = \frac{x + 3}{2}$

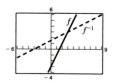

25. $f^{-1}(x) = \sqrt[5]{x}$ **27.** $f^{-1}(x) = x^2$, $x \geq 0$

 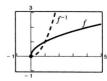

29. $f^{-1}(x) = \sqrt{4 - x^2}, 0 \le x \le 2$ **31.** $f^{-1}(x) = x^3 + 1$

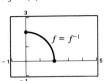

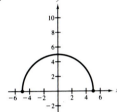

33. Not one-to-one   **35.** $g^{-1}(x) = 8x$

37. $f^{-1}(x) = \sqrt{x} - 3, x \ge 0$ **39.** $h^{-1}(x) = \dfrac{1}{x}$

41. $f^{-1}(x) = \dfrac{x^2 - 3}{2}, x \ge 0$

43. Not one-to-one   **45.** $f^{-1}(x) = -\sqrt{25 - x}$

47. $f^{-1}(x) = \sqrt{x} + 3, x \ge 0$ **49.** $f^{-1}(x) = x - 3, x \ge 0$

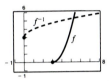

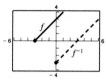

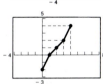

51.
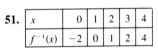

x	0	1	2	3	4
$f^{-1}(x)$	-2	0	1	2	4

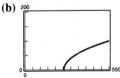

53. 32   **55.** 600   **57.** $\dfrac{x + 1}{2}$ **59.** $\dfrac{x + 1}{2}$

61. (a) $y = \sqrt{\dfrac{x - 254.50}{0.03}}$

   y: percentage load; x: exhaust temperature

(b)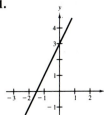

63. False   **65.** True

Chapter 1 Review Exercises   (*page 132*)

1.

3.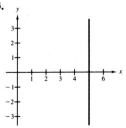

5.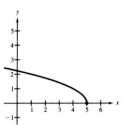

7.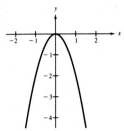

9.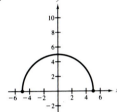

11.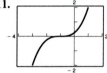

The graph intersects the x-axis once.
The graph intersects the y-axis once.

13.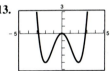

The graph intersects the x-axis three times.
The graph intersects the y-axis once.

15.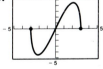

The graph intersects the x-axis three times.
The graph intersects the y-axis once.

17. 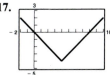   **19.**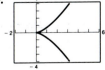

The graph intersects the x-axis twice.
The graph intersects the y-axis once.

21. $t = \frac{7}{3}$ **23.** $t = 3$ **25.** $x = 0$

27. $5x - 12y + 2 = 0$ **29.** $2x - 7y + 2 = 0$

31. $3x - 2y - 10 = 0$ **33.** $2x + 3y - 6 = 0$

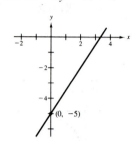

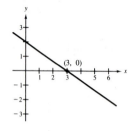

35. (a) $5x - 4y - 23 = 0$ **(b)** $4x + 5y - 2 = 0$

37. \$210,000

39. (a) 5   **(b)** 17   **(c)** $t^4 + 1$ **(d)** $-x^2 - 1$

41. (a) -14 **(b)** $-5x^2 - 30x - 39$ **(c)** -30
    (d) $-10x - 5t, t \neq 0$

43. (a)

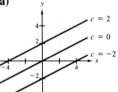

(b)

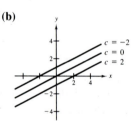

(c)

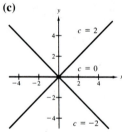

45. $[-5, 5]$ **47.** All real numbers except $s = 3$

49. All real numbers except $x = -2, 3$

51. (a) 16 ft/sec   **(b)** 1.5 sec   **(c)** -16 ft/sec

53. $A = x(12 - x), (0, 6]$ **55. (a)**

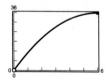

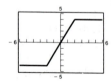

Constant: $(-\infty, -2), (2, \infty)$
Increasing: $(-2, 2)$
(b) Minimum: -4 on $(-\infty, -2)$
Maximum: 4 on $(2, \infty)$
(c) Odd

57. (a)

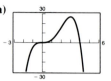

Increasing: $(-\infty, 0), (0, 3)$
Decreasing: $(3, \infty)$
(b) Maximum: $(3, 27)$ **(c)** Neither even nor odd

59. (b) $t = 4.92$ hr, $d = 64.02$ mi

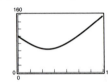

61. (a) $f^{-1}(x) = 2x + 6$
(b)

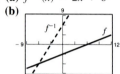

(c) $f^{-1}(f(x)) = f^{-1}(\frac{1}{2}x - 3)$
$= 2(\frac{1}{2}x - 3) + 6 = x$
$f(f^{-1}(x)) = f(2x + 6)$
$= \frac{1}{2}(2x + 6) - 3 = x$

63. (a) $f^{-1}(x) = x^2 - 1, x \geq 0$
(b)

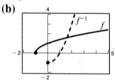

(c) $f^{-1}(f(x)) = f^{-1}(\sqrt{x + 1}), x \geq 0$
$= (\sqrt{x + 1})^2 - 1 = x$
$f(f^{-1}(x)) = f(x^2 - 1)$
$= \sqrt{(x^2 - 1) + 1} = x$

65. (a) $f^{-1}(x) = \sqrt{x + 5}, x \geq -5$
(b)

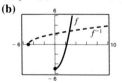

(c) $f^{-1}(f(x)) = f^{-1}(x^2 - 5), x \geq -5$
$= \sqrt{(x^2 - 5) + 5} = x$
$f(f^{-1}(x)) = f(\sqrt{x + 5}), x \geq 0$
$= (\sqrt{x + 5})^2 - 5 = x$

67. $x \geq 4$, $f^{-1}(x) = \sqrt{\dfrac{x}{2}} + 4$

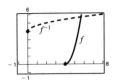

69. $x \geq 2$, $f^{-1}(x) = \sqrt{x^2 + 4}$, $x \geq 0$

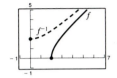

71. -7 **73.** 5   **75.** 23   **77.** 9

Chapter 2

Section 2.1   *(page 143)*

WARM-UP   **1.** $-3x - 10$ **2.** $5x - 12$ **3.** x

4. $x + 26$ **5.** $\dfrac{8x}{15}$ **6.** $\dfrac{3x}{4}$ **7.** $-\dfrac{1}{x(x + 1)}$

8. $\dfrac{5}{x}$ **9.** $\dfrac{7x - 8}{x(x - 2)}$ **10.** $-\dfrac{2}{x^2 - 1}$

1. Identity   **3.** Conditional   **5.** Identity
7. Identity   **9.** Conditional
11. (a) No   (b) No   (c) Yes   (d) No
13. (a) No   (b) No   (c) Yes   (d) Yes
15. 5   **17.** -4 **19.** 3   **21.** 9   **23.** -26
25. -4 **27.** $-\dfrac{6}{5}$ **29.** No solution   **31.** 10
33. 4   **35.** 3   **37.** 5   **39.** No solution   **41.** $\dfrac{11}{6}$
43. $\dfrac{5}{3}$ **45.** No solution   **47.** All real numbers
49. $\dfrac{1}{3 - a}$, $a \neq 3$ **51.** $\dfrac{1 + 4b}{2 + a}$, $a \neq -2$
53. $x \approx 138.889$ **55.** $x \approx 62.372$
57. One equation of this form is $x + 4 = 3x$.
59. 15 ft $\times$ $22\frac{1}{2}$ ft   **61.** 3 hr   **63.** $\frac{1}{3}$ hr
65. (a) 3.8 hr, 3.2 hr   (b) 1.1 hr   (c) 25.6 mi
67. $66\frac{2}{3}$ mph   **69.** 57.1 ft   **71.** 42 ft   **73.** 97
75. You have \$4500 in the 10.5% fund and \$7500 in the 13% fund.
77. $r = 11.43\%$ **79.** 32.1 gal
81. At most, the company can manufacture 8823 units.

83. $\dfrac{2A}{b}$ **85.** $\dfrac{l}{PT}$ **87.** $\dfrac{2A - ah}{h}$

89. $\dfrac{3V + \pi h^3}{3\pi h^2}$ **91.** $\dfrac{Fr^2}{\alpha m_1}$ **93.** $\dfrac{S - a}{S - L}$ **95.** 6 ft

Section 2.2   *(page 155)*

WARM-UP   **1.** $\frac{3}{2}$ **2.** -9 **3.** $\frac{40}{3}$ **4.** $\frac{250}{7}$ **5.** 2
6. -4 **7.**   **8.**

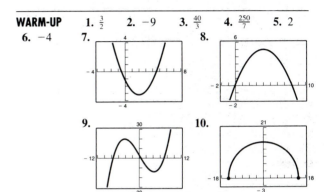

9.   **10.**

1. $(5, 0)$, $(0, -5)$ **3.** $(-2, 0)$, $(1, 0)$, $(0, -2)$
5. $(0, 0)$, $(-2, 0)$ **7.** $(1, 0)$, $(0, \frac{1}{2})$
9.   **11.** $(-1.5, 0)$

13.   **15.** $13x - 89 = 0$

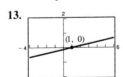

17. $\dfrac{2x}{3} + \dfrac{1}{x} - 10 = 0$

19. $\dfrac{3}{x + 2} - \dfrac{4}{x - 2} - 5 = 0$ **21.** $\dfrac{15}{4}$ **23.** 6
25. $-\frac{5}{8}$ **27.** 2.172, 7.828   **29.** -1.379
31. $\frac{1}{2}$, -3, 3   **33.** -0.717, 2.107   **35.** -1.333
37. $(1, 1)$ **39.** $(1.449, 1.898)$, $(-3.449, -7.898)$
41. $(-2, 8)$, $(1.333, 8)$ **43.** $(-1, 0)$, $(2, 9)$
45. $(0, 0)$, $(2, 8)$, $(-2, 8)$
47. (a) 6.46   (b) 6.41, yes
49. (a) 1.00   (b) 1.01, yes

51. (a) $T = \dfrac{x}{63} + \dfrac{280 - x}{54}$

(b)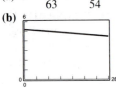

$0 < x < 280$

(c) 164.5 mi

53. (a) $A = 0.33(55 - x) + x$

(b)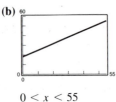

$0 < x < 55$

(c) 22.2 gal

55. $S = 8000 + \frac{1}{2}x$

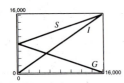

(a) Decreases

(b) Increases

57. $7600.00

59.

61. 0, 0, 0, 0, 0, 0

63. $0.5i$, $-0.25 + 0.5i$, $-0.1875 + 0.25i$, $-0.0273 + 0.4063i$, $-0.1643 + 0.4778i$, $-0.2013 + 0.3430i$

65. 1, 2, 5, 26, 677, 458, 330 **67.** 8, 8, 8

Section 2.3 (page 165)

WARM-UP **1.** $2\sqrt{3}$ **2.** $10\sqrt{5}$ **3.** $\sqrt{5}$

4. $-6\sqrt{3}$ **5.** 12 **6.** 48 **7.** $\dfrac{\sqrt{3}}{3}$ **8.** $\sqrt{2}$

9. $-\dfrac{1}{2} \pm \dfrac{\sqrt{5}}{2}$ **10.** $-1 \pm \sqrt{2}$

1. i, -1, $-i$, 1, i, -1, $-i$, 1, i, -1, $-i$, 1, i, -1, $-i$, 1
3. $a = -10$, $b = 6$ **5.** $a = 6$, $b = 5$ **7.** $4 + 3i$
9. $2 - 3\sqrt{3}i$ **11.** $5\sqrt{3}i$ **13.** $-1 - 6i$ **15.** $-5i$
17. 8 **19.** $11 - i$ **21.** 4 **23.** $3 - 3\sqrt{2}i$
25. $\frac{1}{6} + \frac{7}{6}i$ **27.** $5 - 3i$, 34 **29.** $-2 + \sqrt{5}i$, 9
31. $-2\sqrt{3}$ **33.** -10 **35.** $5 + i$ **37.** $12 + 30i$
39. 24 **41.** $-9 + 40i$ **43.** -10 **45.** $\frac{16}{41} + \frac{20}{41}i$
47. $\frac{3}{5} + \frac{4}{5}i$ **49.** $-7 - 6i$ **51.** $-\frac{5}{4} - \frac{5}{4}i$
53. $\frac{35}{29} + \frac{595}{29}i$

55.

57.

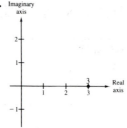

Section 2.4 (page 178)

WARM-UP **1.** $\dfrac{\sqrt{14}}{10}$ **2.** $4\sqrt{2}$ **3.** 14 **4.** $\dfrac{\sqrt{10}}{4}$

5. $x(3x + 7)$ **6.** $(2x + 5)(2x - 5)$
7. $-(x - 7)(x - 15)$ **8.** $(x - 2)(x + 9)$
9. $(5x - 1)(2x + 3)$ **10.** $(6x - 1)(x - 12)$

1. 0, $-\frac{1}{2}$ **3.** 4, -2 **5.** -5 **7.** 3, $-\frac{1}{2}$
9. 2, -6 **11.** $-a$ **13.** $\pm 2\sqrt{3} \approx \pm 3.46$
15. $12 + 3\sqrt{2} \approx 16.24$, $12 - 3\sqrt{2} \approx 7.76$
17. $-2 + 2\sqrt{3} \approx 1.46$, $-2 - 2\sqrt{3} \approx -5.46$ **19.** 2
21. $-3 \pm \sqrt{7}$ **23.** $1 \pm \dfrac{\sqrt{6}}{3}$ **25.** $2 \pm 2\sqrt{3}$
27. $\frac{1}{2}$, -1 **29.** $\frac{1}{4}$, $-\frac{3}{4}$ **31.** $1 \pm \sqrt{3}$
33. $-7 \pm \sqrt{5}$ **35.** $-4 \pm 2\sqrt{5}$ **37.** $\dfrac{2}{3} \pm \dfrac{\sqrt{7}}{3}$
39. $-\dfrac{1}{3} \pm \dfrac{\sqrt{11}}{6}$ **41.** $-\dfrac{1}{2} \pm \sqrt{2}$ **43.** $\dfrac{2}{7}$
45. $-\dfrac{8}{5} \pm \dfrac{\sqrt{3}}{5}$ **47.** $6 \pm \sqrt{11}$ **49.** 3, -1, 0
51. -3, 0 **53.** 3, 1, -1 **55.** ± 3, ± 1
57. $-\frac{1}{5}$, $-\frac{1}{3}$ **59.** 1, $-\frac{125}{8}$ **61.** 50 **63.** -16
65. 0 **67.** 36 **69.** $\frac{101}{4}$ **71.** 4, -5
73. $\dfrac{-3 \pm \sqrt{21}}{6}$ **75.** 1, -3 **77.** 3, -2
79. $\sqrt{3}$, -3 **81.** ± 1 **83.** $\pm\_$ **85.** 0.25
87. 2, -5 **89.** 0, 4 **91.** 2, -1.50 **93.** -1
95. 10, -1 **97.** ± 1.04 **99.** 16.76
101. $\dfrac{\sqrt{1821}}{4} \approx 10.67$ sec
103. 35 ft by 20 ft or 15 ft by $\frac{140}{3}$ ft

105. (a) $C = 0.45x^2 - 1.65x + 50.75$, $10 \le x \le 25$

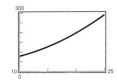

(b) If $C = 150$, then $x = 16.797°$

(c) If the temperature is increased from $10°$ to $20°$, then C increases from 79.25 to 197.75, a factor of 2.5.

107. 400 mi/hr

109. (a)

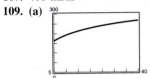

(b) $211.6°$ **(c)** 24.725 psi

111. 500 units

Section 2.5   *(page 191)*

WARM-UP   **1.** $-\frac{1}{2}$ **2.** $-\frac{1}{6}$ **3.** -3 **4.** -6
5. $x \ge 0$ **6.** $-3 < z < 10$ **7.** $P \le 2$
8. $W \ge 200$ **9.** 2, 7   **10.** 0, 1

1. c   **2.** h   **3.** f   **4.** e   **5.** g   **6.** a
7. b   **8.** d
9. $x < 3$ **11.** $x > -4$

13. $x \ge 12$ **15.** $x < -\frac{1}{2}$

17. $-1 < x < 3$ **19.** $-\frac{9}{2} < x < \frac{15}{2}$

21. $-\frac{3}{4} < x < -\frac{1}{4}$ **23.** $-5 < x < 5$

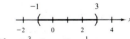

25. $x < -6$, $x > 6$ **27.** $16 \le x \le 24$

29. $x \le 16$, $x \ge 24$ **31.** $4 < x < 5$

 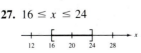

33. No solution   **35.** $x > 0.50$ **37.** $x \le 5.45$
39. $-3 < x < 2.67$ **41.** $x \le -7$, $x \ge 13$
43. $x \le -14.50$, $x \ge -5.50$ **45.** $|x| \le 2$
47. $|x - 9| \ge 3$ **49.** $|x - 12| \le 10$
51. $|x + 3| > 5$ **53.** $[-3, 3]$ **55.** $(-\infty, -5]$, $[1, \infty)$
57. $(-3, 2)$ **59.** $(-\infty, -1)$, $(1, \infty)$ **61.** $(-3, 1)$
63. $(-\infty, 0)$, $\left(0, \frac{3}{2}\right)$ **65.** $[-2, 0]$, $[2, \infty)$
67. $(-\infty, -1)$, $(0, 1)$ **69.** $(-\infty, -1)$, $(4, \infty)$
71. $(5, 15)$ **73.** $\left(-5, -\frac{3}{2}\right)$, $(-1, \infty)$ **75.** $(1, 2)$, $(0, 1)$
77. $[-2, \infty)$ **79.** $(-0.13, 25.13)$
81. $(-14, -2)$, $(6, \infty)$ **83.** $(-\infty, -2)$, $(-1, 1)$, $(3, \infty)$
85. $x \le 3$ **87.** $x \le \frac{7}{2}$ **89.** $[-2, 2]$
91. $(-\infty, 3]$, $[4, \infty)$ **93.** $[-4, 3]$ **95.** $(-3, 0]$, $(3, \infty)$
97. $x \ge 36$ units
99. (a) $C = 1.45x + 150$
(b) $R = 2.95x$
(c) $P = 1.50x - 150$
(d) $134 \le x \le 233$

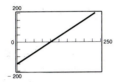

(e) $143 \le x \le 243$
101. 1991   **103.** $65.8 \le h \le 71.2$
105. (a) 10 sec   **(b)** 4 sec $< t <$ 6 sec
107. (a) $R = \dfrac{R_1 R_2}{R_1 + R_2}$ **(b)** $R = \dfrac{2R_1}{R_1 + 2}$
(c) $R_1 \ge 2$

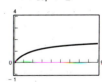

Section 2.6   *(page 202)*

WARM-UP
1.   **2.**

3.

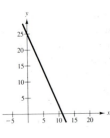

4.

5. $y = -\frac{5}{8}x + 5$ **6.** $y = 2x - 2$ **7.** $y = x - 4$

8. $y = -\frac{12}{11}x + \frac{75}{11}$ **9.** $y = -\frac{4}{5}x + 8$

10. $y = \frac{3}{2}x + 2$

11. $y = \frac{16}{35}x + \frac{39}{35}$ **13.** $y = \frac{30}{7}x$

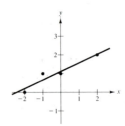

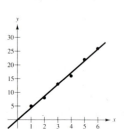

15. $y = -2.179x + 22.964$

17. $y = 2.378x + 23.546$

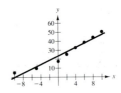

1.

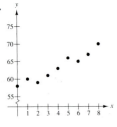

3. $y = 57.49 + 1.43x$; 71.8

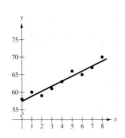

5.

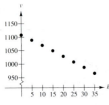

7. $v = 1117.3 - 4.1h$; 1006.6

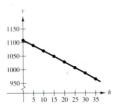

9. (a)

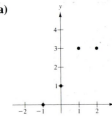

(b) $y = 1.1x + 1.2$; 0.7

(c) $y = \frac{11}{10}x + \frac{6}{5}$; 0.7

19. $S = 384.1 + 21.2x$; \$95,784.10

$r = 0.996$

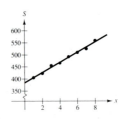

Chapter 2 Review Exercises   (*page 204*)

   1. Identity

   3. (a) No   **(b)** Yes   **(c)** Yes   **(d)** No

   5. 20   **7.** $\frac{1}{5}$ **9.** $\frac{4}{3}, -\frac{1}{2}$ **11.** $-4 \pm 3\sqrt{2}$

13. $6 \pm \sqrt{6}$ **15.** 0, 1, 2   **17.** 2, 6   **19.** 5

21. $\dfrac{25}{4}$ **23.** $-124, 126$ **25.** $-4, \dfrac{-10 \pm \sqrt{95}}{5}$

27. $-5, 15$ **29.** $-6.464, 0.464$ **31.** 1.944

33. ±0.707 35. 1, 3 37. $r = \sqrt{\dfrac{3V}{\pi h}}$

39. $p = \dfrac{k}{3\pi r^2 L}$ 41. $-14 + 5i$ 43. $40 + 65i$

45. $1 - 6i$ 47. $f(x) = 6x^4 + 13x^3 + 7x^2 - x - 1$

49. $f(x) = 3x^4 - 14x^3 + 17x^2 - 42x + 24$

51. $\left(-\frac{5}{3}, \infty\right)$ 53. $(-\infty, 3), (5, \infty)$ 55. $(1, 3)$

57. $(-\infty, 0], [3, \infty)$ 59. $(-\infty, 17.143)$ 61. $(4, \infty)$

63. $[5, \infty)$ 65. September: \$325,000; October: \$364,000

67. 2.86 quarts 69. $1000\sqrt{2} \approx 1414.21$ ft

71. 4 farmers 73. 143,202 units 75. $x \geq 36$

77. $L \geq \dfrac{32}{\pi^2} \approx 3.24$ ft

CUMULATIVE TEST for Prerequisites, Chapters 1 and 2 (page 207)

1. $-\dfrac{437}{45}$ 2. $\dfrac{4x^3}{15y^5}$ 3. $2x^2|y|\sqrt{6y}$

4. $x^3 - x^2 - 5x + 6$ 5. $-x(x + 1)(6x - 1)$

6. $\dfrac{s - 1}{(s + 1)(s + 3)}$ 7. $\dfrac{1}{2(2x + 1)\sqrt{x}}$

8. (a) (b)

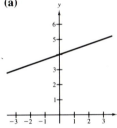

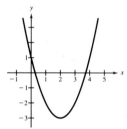

(c) (d)

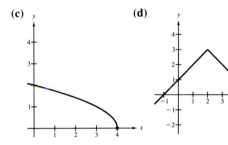

9. $C = 3$ 10. $(1, 11), (7, 7), (10, 5)$

11. $2x - y + 2 = 0$

12. (a) $\dfrac{3}{2}$ (b) Undefined (c) $\dfrac{t}{t - 1}$ (d) $\dfrac{s + 2}{s}$

13. $-\infty < x < 3, 0 < y$ 14. $A = \dfrac{\sqrt{3}}{4}s^2$

15. For some values of x, there is more than one corresponding value of y.

16. (a) Horizontal stretch by a factor of 2
   (b) Vertical shrink by a factor of $\frac{1}{2}$
   (c) Vertical translation of two units upward
   (d) Horizontal translation of two units to the left

17. $(f \circ g)(x) = \sqrt{x^2 + 3}$ 18. $h^{-1}(x) = \frac{1}{5}(x + 2)$

19. (a) 7 (b) 5 (c) $\dfrac{-3 \pm \sqrt{3}}{3}$ (d) 6

20. 7.1 21. $-1 < x < 5$ 22. $-2 < x < 3$

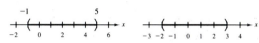

23. \$8000 at 7.5%, \$4000 at 9% 24. 1.8 gal

25. $n = 9$

CHAPTER 3

Section 3.1 (page 216)

WARM-UP 1. $\frac{1}{2}, -6$ 2. $-\frac{3}{5}, 3$ 3. $\frac{3}{2}, -1$

4. -10 5. $3 \pm \sqrt{5}$ 6. $-2 \pm \sqrt{3}$

7. $4 \pm \dfrac{\sqrt{14}}{2}$ 8. $-5 \pm \dfrac{\sqrt{3}}{3}$ 9. $-\dfrac{3}{2} \pm \dfrac{\sqrt{5}}{2}$

10. $-\dfrac{3}{2} \pm \dfrac{\sqrt{21}}{2}$

1. f

```
Xmin=-1
Xmax=6
Xscl=1
Ymin=-1
Ymax=5
Yscl=1
```

2. d

```
Xmin=-8
Xmax=1
Xscl=1
Ymin=-1
Ymax=6
Yscl=1
```

3. c

```
Xmin=-4
Xmax=4
Xscl=1
Ymin=-5
Ymax=2
Yscl=1
```

4. e

```
Xmin=-4
Xmax=4
Xscl=1
Ymin=-1
Ymax=6
Yscl=1
```

5. b

```
Xmin=-3
Xmax=5
Xscl=1
Ymin=-2
Ymax=5
Yscl=1
```

6. a

```
Xmin=-5
Xmax=1
Xscl=1
Ymin=-3
Ymax=3
Yscl=1
```

7.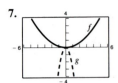

9.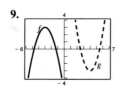

11. $f(x) = (x - 2)^2$ **13.** $f(x) = -(x + 2)^2 + 4$

15. $f(x) = -2(x + 3)^2 + 3$

17. Intercepts: $(\pm\sqrt{5}, 0), (0, -5)$
   Vertex: $(0, -5)$

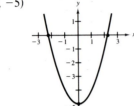

19. Intercepts: $(-5 \pm \sqrt{6}, 0), (0, 19)$
   Vertex: $(-5, -6)$

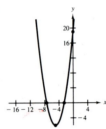

21. Intercepts: $(4, 0), (0, 16)$
   Vertex: $(4, 0)$

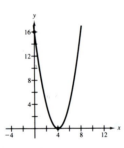

23. Intercepts: $(1 \pm \sqrt{6}, 0), (0, 5)$
   Vertex: $(1, 6)$

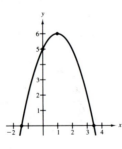

25. Intercepts: $(0, 21)$
   Vertex: $(\frac{1}{2}, 20)$

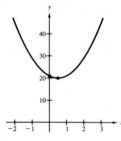

27. Intercepts: $(1, 0), (-3, 0), (0, 3)$
   Vertex: $(-1, 4)$

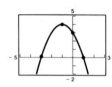

29. Intercepts: $(3.29, 0), (4.71, 0), (0, 31)$
   Vertex: $(4, -1)$

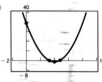

31. $f(x) = -\frac{1}{2}(x - 3)^2 + 4$ **33.** $f(x) = \frac{3}{4}(x - 5)^2 + 12$

35. $f(x) = x^2 - 2x - 3, g(x) = -x^2 + 2x + 3$
   (The answer is not unique.)

37. $f(x) = x^2 - 10x, g(x) = -x^2 + 10x$
   (The answer is not unique.)

39. $f(x) = 2x^2 + 7x + 3, g(x) = -2x^2 - 7x - 3$
   (The answer is not unique.)

41. 55, 55

43. $A = x(50 - x), 0 < x < 50$

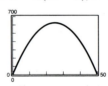

   25 ft × 25 ft

45. 25 ft × $33\frac{1}{3}$ ft   **47.** 4500 units

49. (a) No. The function is still increasing.
   (b) 4024.5, 11.0

51. (a)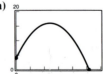

   (b) 4 ft
   (c) 16 ft
   (d) $12 + 8\sqrt{3} \approx 25.86$ ft

53. (a)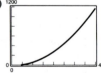

   (b) 166.7 board feet
   (c) 26.6 in.

Section 3.2   (page 228)

WARM-UP   **1.** $(3x - 2)(4x + 5)$ **2.** $x(5x - 6)^2$
3. $z^2(12z + 5)(z + 1)$ **4.** $(y + 5)(y^2 - 5y + 25)$
5. $(x + 3)(x + 2)(x - 2)$ **6.** $(x + 2)(x^2 + 3)$
7. No real solution   **8.** $3 \pm \sqrt{5}$ **9.** $-\frac{1}{2} \pm \sqrt{3}$
10. ± 3

1. (a) $f(x)$ is the graph of y shifted right two units.
   (b) $f(x)$ is the graph of y shifted down two units.
   (c) $f(x)$ is the graph of y shifted right two units and
      down two units.
   (d) $f(x)$ is the mirror image (with respect to the x-axis)
      of the graph of y with a vertical shrink of $\frac{1}{2}$.

3. e   **4.** c

```
Xmin=-3        Xmin=-2
Xmax=5         Xmax=4
Xscl=1         Xscl=1
Ymin=-1        Ymin=-2
Ymax=6         Ymax=3
Yscl=1         Yscl=1
```

5. b   **6.** f   **7.** a

```
Xmin=-5     Xmin=-3     Xmin=-3
Xmax=1      Xmax=3      Xmax=3
Xscl=1      Xscl=1      Xscl=1
Ymin=-12    Ymin=-6     Ymin=-3
Ymax=3      Ymax=8      Ymax=2
Yscl=3      Yscl=2      Yscl=1
```

8. g   **9.** d   **10.** h

```
Xmin=-4     Xmin=-3     Xmin=-3
Xmax=4      Xmax=2      Xmax=3
Xscl=1      Xscl=1      Xscl=1
Ymin=-2     Ymin=-2     Ymin=-4
Ymax=5      Ymax=4      Ymax=4
Yscl=1      Yscl=1      Yscl=1
```

11.   **13.**

15. Rises to the left. Rises to the right.
17. Falls to the left. Falls to the right.
19. Rises to the left. Falls to the right.   **21.** ± 5 **23.** 3
25. $1, -2$ **27.** $2 \pm \sqrt{3}$ **29.** $2, 0$ **31.** ± 1
33. $\pm \sqrt{5}$ **35.** No real zeros   **37.** $f(x) = x^2 - 10x$
39. $f(x) = x^2 + 4x - 12$ **41.** $f(x) = x^3 + 5x^2 + 6x$
43. $f(x) = x^4 - 4x^3 - 9x^2 + 36x$
45. $f(x) = x^2 - 2x - 2$

47. 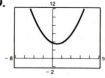   **49.**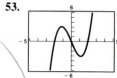

51.   **53.**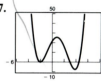

55. 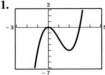   **57.**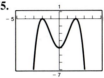

59. 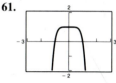   **61.**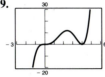

63. $[-1, 0], [1, 2], [2, 3]$ **65.** $[-2, -1], [0, 1]$
67. (b) Domain: $0 < x < 6$ **69.**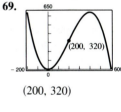
   (c)

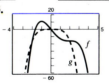

$(200, 320)$

Maximum when $x = 2$

Section 3.3   (page 244)

WARM-UP   **1.** $x^3 - x^2 + 2x + 3$
2. $2x^3 + 4x^2 - 6x - 4$ **3.** $x^4 - 2x^3 + 4x^2 - 2x - 7$
4. $2x^4 + 12x^3 - 3x^2 - 18x - 5$ **5.** $(x - 3)(x - 1)$
6. $2x(2x - 3)(x - 1)$ **7.** $x^3 - 7x^2 + 12x$
8. $x^2 + 5x - 6$ **9.** $x^3 + x^2 - 7x - 3$
10. $x^4 - 3x^3 - 5x^2 + 9x - 2$

1. $2x + 4$ **3.** $x^2 - 3x + 1$ **5.** $x^3 + 3x^2 - 1$
7. $7 - \dfrac{11}{x + 2}$ **9.** $3x + 5 - \dfrac{2x - 3}{2x^2 + 1}$

11. $x^2 + 2x + 4 + \dfrac{2x - 11}{x^2 - 2x + 3}$

13. $2x - \dfrac{17x - 5}{x^2 - 2x + 1}$ **15.** $3x^2 - 2x + 5$

17. $4x^2 - 9$ **19.** $-x^2 + 10x - 25$

21. $5x^2 + 14x + 56 + \dfrac{232}{x - 4}$

23. $10x^3 + 10x^2 + 60x + 360 + \dfrac{1360}{x - 6}$

25. $x^2 - 8x + 64$

27. $-3x^3 - 6x^2 - 12x - 24 - \dfrac{48}{x - 2}$

29. $-x^2 + 3x - 6 + \dfrac{11}{x + 1}$ **31.** $4x^2 + 14x - 30$

33. $(x - 2)(x + 3)(x - 1)$; Zeros: $x = 2, x = -3, x = 1$

35. $(2x - 1)(x - 5)(x - 2)$; Zeros: $x = \frac{1}{2}, x = 5, x = 2$

37. $\left(x + \sqrt{3}\right)\left(x - \sqrt{3}\right)(x + 2)$; Zeros: $x = -\sqrt{3},$
$x = \sqrt{3}, x = -2$

39. $(x - 4)(x^2 + 3x - 2) + 3, f(4) = 3$

41. $\left(x - \sqrt{2}\right)\left[x^2 + \left(3 + \sqrt{2}\right)x + 3\sqrt{2}\right] - 8,$
$f(\sqrt{2}) = -8$

43. (a) 1   (b) 4   (c) 4   (d) 1954

45. (a) 97   (b) $-\frac{5}{3}$ (c) 17   (d) -199

47. One negative zero   **49.** No real zeros

51. One positive zero   **53.** One or three positive zeros

55. Two or no positive zeros   **57.** $\pm 1, \pm 2, \pm 4$

59. $\pm 1, \pm 3, \pm \frac{1}{2}, \pm \frac{3}{2}, \pm \frac{1}{4}, \pm \frac{3}{4}$

61. (a) $\pm 1, \pm 2, \pm 4, \pm 8, \pm \frac{1}{2}$

(b) 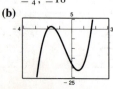   (c) $-\frac{1}{2}, 1, 2, 4$

63. (a) $\pm 1, \pm 3, \pm \frac{1}{2}, \pm \frac{3}{2}, \pm \frac{1}{4}, \pm \frac{3}{4}, \pm \frac{1}{8}, \pm \frac{3}{8}, \pm \frac{1}{16}, \pm \frac{3}{16},$
$\pm \frac{1}{32}, \pm \frac{3}{32},$

(b) 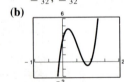   (c) $1, \frac{3}{4}, -\frac{1}{8}$

65. (a) $\pm 1, \pm 2, \pm 3, \pm 6, \pm 9, \pm \frac{1}{2}, \pm \frac{3}{2}, \pm \frac{9}{2}, \pm \frac{1}{4}, \pm \frac{3}{4},$
$\pm \frac{9}{4}, \pm 18$

(b)

(c) $-2, \dfrac{1}{8} \pm \dfrac{\sqrt{145}}{8}$

67. (a) Upper bound   (b) Lower bound   (c) Neither

69. (a) Neither   (b) Lower bound   (c) Upper bound

71. $1, 2, 3$ **73.** $1, -1, 4$ **75.** $-1, -10$ **77.** $1, 2$

79. $\frac{1}{2}, -1$ **81.** $1, -\frac{1}{2}$ **83.** $-\frac{3}{4}$ **85.** $\pm 1, \pm \sqrt{2}$

87. $\pm 2, \pm \frac{3}{2}$ **89.** $\pm 1, \frac{1}{4}$ **91.** d   **92.** a

93. b   **94.** c

95. (a) ± 1 **97.** (a) $\pm 1, \pm 3$

(b) 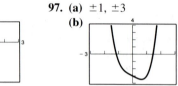   (b)

(c) 0.68   (c) $-1.16, 1.45$

99. 3.77 in. $\times$ 7.77 in. $\times$ 0.614 in.   **101.** \$384,356

103. 4000   **105.** 0.53   **107.** 1.56

Section 3.4   *(page 253)*

WARM-UP   **1.** $4 - \sqrt{29}\,i, 4 + \sqrt{29}\,i$
2. $-5 - 12i, -5 + 12i$ **3.** $-1 + 4\sqrt{2}\,i, -1 - 4\sqrt{2}\,i$
4. $6 + \frac{1}{2}i, 6 - \frac{1}{2}i$ **5.** $-13 + 9i$ **6.** $12 + 16i$
7. $26 + 22i$ **8.** 29   **9.** i **10.** $-9 + 46i$

1. $\pm 5i$; $(x + 5i)(x - 5i)$

3. $2 \pm \sqrt{3}$; $\left(x - 2 - \sqrt{3}\right)\left(x - 2 + \sqrt{3}\right)$

5. $\pm 3, \pm 3i$; $(x + 3)(x - 3)(x + 3i)(x - 3i)$

7. $1 \pm i$; $(z - 1 + i)(z - 1 - i)$

9. $-5, 4 \pm 3i$; $(t + 5)(t - 4 + 3i)(t - 4 - 3i)$

11. $-\frac{3}{4}, 1 \pm \frac{1}{2}i$; $(4x + 3)(2x - 2 + i)(2x - 2 - i)$

13. $-\frac{1}{5}, 1 \pm \sqrt{5}\,i$
$(5x + 1)\left(x - 1 + \sqrt{5}\,i\right)\left(x - 1 - \sqrt{5}\,i\right)$

15. $\pm i, \pm 3i$; $(x + i)(x - i)(x + 3i)(x - 3i)$

17. $2, \pm 2i$; $(x - 2)^2(x + 2i)(x - 2i)$

19. $-2, -\frac{1}{2}, \pm i$; $(x + 2)(2x + 1)(x + i)(x - i)$

21. $x^3 - x^2 + 25x - 25$ **23.** $x^3 - 10x^2 + 33x - 34$

25. $x^4 + 37x^2 + 36$ **27.** $x^4 + 8x^3 + 9x^2 - 10x + 100$

29. $16x^4 + 36x^3 + 16x^2 + x - 30$

31. (a) $(x^2 + 9)(x^2 - 3)$

(b) $(x^2 + 9)\left(x + \sqrt{3}\right)\left(x - \sqrt{3}\right)$

(c) $(x + 3i)(x - 3i)\left(x + \sqrt{3}\right)\left(x - \sqrt{3}\right)$

33. (a) $(x^2 - 2x - 2)(x^2 - 2x + 3)$
(b) $\left(x - 1 + \sqrt{3}\right)\left(x - 1 - \sqrt{3}\right)(x^2 - 2x + 3)$
(c) $\left(x - 1 + \sqrt{3}\right)\left(x - 1 - \sqrt{3}\right)\left(x - 1 + \sqrt{2}\,i\right)$
$\left(x - 1 - \sqrt{2}\,i\right)$

35. $\pm 5i, -\frac{3}{2}$ **37.** $\pm 2i, 1, -\frac{1}{2}$ **39.** $-3 \pm i, \frac{1}{4}$

41. $2, -3 \pm \sqrt{2}\,i$ **43.** $\frac{3}{4}, \frac{1}{2} \pm \frac{\sqrt{5}}{2}i$

45. No; setting $h = 64$ and solving the resulting equation yields imaginary roots.

47. $x^2 + b$

Section 3.5   *(page 262)*

WARM-UP   **1.** $(x - 5)(x + 2)$ **2.** $(x - 5)(x - 2)$
3. $x(x + 1)(x + 3)$ **4.** $(x^2 - 2)(x - 4)$

5.
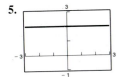

6.

7.
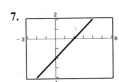

8.
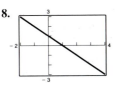

9. $x + 9 + \dfrac{42}{x - 4}$ **10.** $x + 1 + \dfrac{2}{x + 4}$

1. (a)

x	$f(x)$		x	$f(x)$		x	$f(x)$
1	-1		3	1		3	1
1.5	-2		2.5	2		5	0.3333
1.9	-10		2.1	10		10	0.1250
1.99	-100		2.01	100		100	0.0102
1.999	-1000		2.001	1000		1000	0.0010

(b) Vertical asymptote: $x = 2$; Horizontal asymptote: $y = 0$
(c) Domain: all real numbers except $x = 2$

3. (a)

x	$f(x)$		x	$f(x)$		x	$f(x)$
1	3		3	9		3	9
1.5	9		2.5	15		5	5
1.9	57		2.1	63		10	3.75
1.99	597		2.01	603		100	3.0612
1.999	5997		2.001	6003		1000	3.0060

(b) Vertical asymptote: $x = 2$; Horizontal asymptote: $y = 3$ as $x \to +\infty$ and $y = -3$ as $x \to -\infty$
(c) Domain: all real numbers except $x = 2$

5. (a)

x	$f(x)$		x	$f(x)$		x	$f(x)$
1	-1		3	5.4		3	5.4
1.5	-3.8571		2.5	8.33		5	3.5714
1.9	-27.77		2.1	32.27		10	3.1250
1.99	-297.75		2.01	302.25		100	3.0012
1.999	-2997.75		2.001	3002.25		1000	3.0000

(b) Vertical asymptote: $x = \pm 2$; Horizontal asymptote: $y = 3$
(c) Domain: all real numbers except $x = \pm 2$

7. Domain: all $x \neq 0$
Vertical asymptote: $x = 0$
Horizontal asymptote: $y = 0$

9. Domain: all $x \neq 2$
Vertical asymptote: $x = 2$
Horizontal asymptote: $y = -1$

11. Domain: all $x \neq \pm 1$
Vertical asymptote: $x = \pm 1$
Slant asymptote: $y = x$

13. Domain: all reals
Horizontal asymptote: $y = 3$

15. f   **16.** e   **17.** a

```
Xmin=-5        Xmin=-1        Xmin=-3
Xmax=3         Xmax=8         Xmax=3
Xscl=1         Xscl=1         Xscl=1
Ymin=-3        Ymin=-3        Ymin=-1
Ymax=3         Ymax=3         Ymax=3
Yscl=1         Yscl=1         Yscl=1
```

18. b   **19.** c   **20.** d

```
Xmin=-3        Xmin=-2        Xmin=-4
Xmax=3         Xmax=4         Xmax=2
Xscl=1         Xscl=1         Xscl=1
Ymin=-5        Ymin=-1        Ymin=-3
Ymax=1         Ymax=3         Ymax=1
Yscl=1         Yscl=1         Yscl=1
```

21. (a) Domain of f: $(-\infty, -2)$, $(-2, \infty)$
Domain of g: $(-\infty, \infty)$
(b) No vertical asymptote
(c)

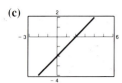

(d) Some viewing rectangles will show the difference in domains. For instance, on the TI-81, use the standard viewing rectangle.

23. (a) Domain of f: $(-\infty, 0)$, $(0, 3)$, $(3, \infty)$
Domain of g: $(-\infty, 0)$, $(0, \infty)$
(b) $x = 0$
(c)

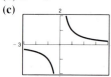

(d) Some viewing rectangles will show the difference in domains. For instance, on the TI-81, use the following range:

```
Xmin=-4.5
Xmax=5
Xscl=1
Ymin=-2
Ymax=2
Yscl=1
```

25. (a) $\$28\frac{1}{3}$ million   (b) $\$170$ million
(c) $\$765$ million   (d) No
27. (a) 167, 250, 400   (b) 750
29. (a)

n	1	2	3	4	5	6	7	8	9	10
P	0.50	0.74	0.82	0.86	0.89	0.91	0.92	0.93	0.94	0.95

(b) 100%

Section 3.6   (page 272)

WARM-UP   **1.** $(-3, 0)$, $(0, 6)$ **2.** $(-3, 0)$, $(0, \frac{12}{5})$
3. Odd   **4.** Even
5. Domain: all real numbers $x \neq 8$
Vertical asymptote: $x = 8$
Horizontal asymptote: $y = 0$
6. Domain: all real numbers $x \neq -\frac{1}{4}$
Vertical asymptote: $x = -\frac{1}{4}$
Horizontal asymptote: $y = \frac{3}{4}$
7. Domain: all real numbers $x \neq \pm 3$
Vertical asymptotes: $x = \pm 3$
Horizontal asymptote: $y = 2$

8. Domain: all real numbers $x \neq 0$
Vertical asymptote: $x = 0$
Horizontal asymptote: $y = 4$
9. $2x + \dfrac{7}{2} + \dfrac{23}{2(2x - 1)}$ **10.** $x + \dfrac{1}{x^2}$

1.

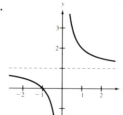

3.

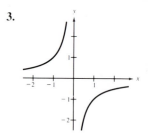

5.

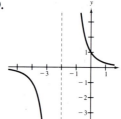

7.

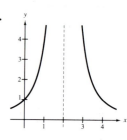

9.

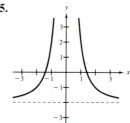

11.

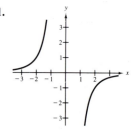

13.

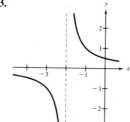

15.

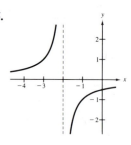

17.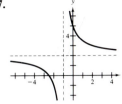

19.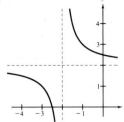

37.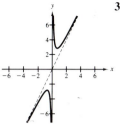

39.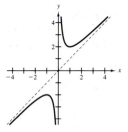

21.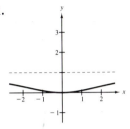

23.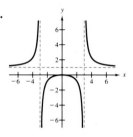

41.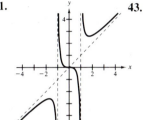

43.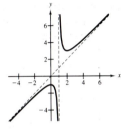

25.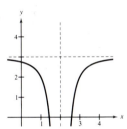

27.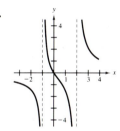

45.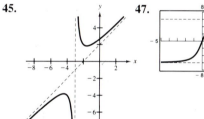

47.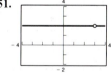

29.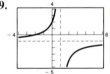

Domain:
all real numbers $x \neq 1$

Range:
all real numbers $y \neq -1$

31.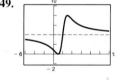

Domain:
all real numbers $t \neq 0$

Range:
all real numbers $y \neq 3$

49.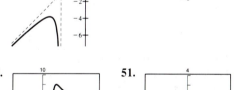

51.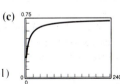

The fraction is not
reduced to lowest terms.

55. (b) $0 \leq x \leq 240$

(c)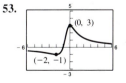

Increases at a lower
rate approaching 75%

33.

Domain:
all real numbers

Range: $0 < y \leq 4$

35.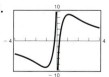

Domain:
all real numbers $x \neq 0$

Range: $-\infty < y < \infty$

53.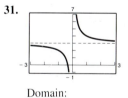

Relative minimum: $(-2, -1)$
Relative maximum: $(0, 3)$

57. (b) $x > 2$

(c)

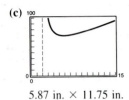

5.87 in. $\times$ 11.75 in.

Chapter 3 Review Exercises *(page 275)*

1. Intercept: $(0, \frac{13}{4})$
   Vertex: $(-\frac{3}{2}, 1)$

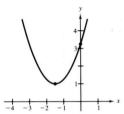

3. Intercepts: $\left(\dfrac{-5 \pm \sqrt{41}}{2}, 0\right), \left(0, -\dfrac{4}{3}\right)$

   Vertex: $\left(-\dfrac{5}{2}, -\dfrac{41}{12}\right)$

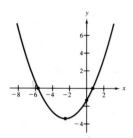

5. $f(x) = (x - 1)^2 - 4$ **7.** Minimum: -1
9. Maximum: 9   **11.** Maximum: 3
13. Minimum: $-\frac{41}{4}$ **15.** $(3, \frac{3}{2})$ **17.** 4500 units
19. $1029   **21.** Falls to the left. Falls to the right.
23. Rises to the left. Rises to the right.

25.

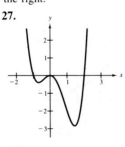

27.

29.

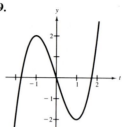

31.

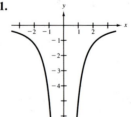

33. $8x + 5 + \dfrac{2}{3x - 2}$ **35.** $x^2 - 2$

37. $x^2 - 3x + 2 - \dfrac{1}{x^2 + 2}$

39. $0.25x^3 - 3.5x^2 - 7x - 14 - \dfrac{28}{x - 2}$

41. $2x^2 - (3 - 4i)x + (1 - 2i)$
43. (a) No   **(b)** Yes   **(c)** Yes   **(d)** No
45. (a) No   **(b)** Yes   **(c)** Yes   **(d)** No
47. (a) 580   **(b)** 0   **49. (a)** -421 **(b)** 96
51. Two or no positive zeros. One negative zero.
53. $\pm 1, \pm 3, \pm 5, \pm 15, \pm \frac{1}{2}, \pm \frac{3}{2}, \pm \frac{5}{2}, \pm \frac{15}{2}, \pm \frac{1}{4}, \pm \frac{3}{4}, \pm \frac{5}{4},$
   $\pm \frac{15}{4}$
55. $1, \frac{3}{4}$ **57.** $\frac{5}{6}, \pm 2i$ **59.** $-1, \frac{3}{2}, 3, \frac{2}{3}$
61. (a) ± 1 **63. (a)** $\pm 1, \pm 2, \pm 5, \pm 10$
   (b)   **(b)**

(c) $-1.40, 0.47$ **(c)** 3.26
65. 11.30
67. Domain: all reals such that $x \neq -3$
   Vertical asymptote: $x = -3$
   Horizontal asymptote: $y = 0$
69. Domain: all reals such that $x \neq \pm 2$
   Vertical asymptotes: $x = 2, x = -2$
   Horizontal asymptote: $y = 1$
71.   **73.**

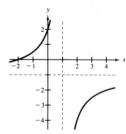

75.

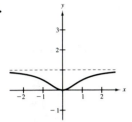

77.

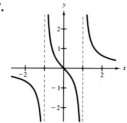

79.

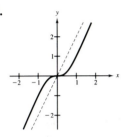

81.

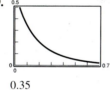

83.

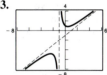

85. As x increases, the cost approaches the horizontal asymptote, $\overline{C} = 0.5$.

87. **(a)** $176 million
  (b) $528 million
  (c) $1584 millior
  (d) Not possible, C approaches infinity

89.

0.35

CHAPTER 4

Section 4.1   (page 289)

WARM-UP   **1.** 5^x **2.** 3^{2x} **3.** 4^{3x} **4.** 10^x
5. 4^{2x} **6.** 4^{10x} **7.** $\left(\frac{3}{2}\right)^x$ **8.** 4^{3x}
9. 2^{-x} **10.** 2^x

1. 946.852   **3.** 747.258   **5.** 0.006   **7.** 673.639
9. 0.472

11. g

```
Xmin=-3
Xmax=3
Xscl=1
Ymin=-1
Ymax=3
Yscl=1
```

12. e

```
Xmin=-3
Xmax=3
Xscl=1
Ymin=-5
Ymax=1
Yscl=1
```

13. b

```
Xmin=-3
Xmax=3
Xscl=1
Ymin=-1
Ymax=3
Yscl=1
```

14. h

```
Xmin=-3
Xmax=3
Xscl=1
Ymin=-5
Ymax=1
Yscl=1
```

15. d

```
Xmin=-3
Xmax=3
Xscl=1
Ymin=-5
Ymax=1
Yscl=1
```

16. a

```
Xmin=-3
Xmax=3
Xscl=1
Ymin=0
Ymax=4
Yscl=1
```

17. f

```
Xmin=-1
Xmax=5
Xscl=1
Ymin=-5
Ymax=1
Yscl=1
```

18. c

```
Xmin=-2
Xmax=4
Xscl=1
Ymin=-1
Ymax=3
Yscl=1
```

19.

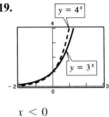

$y = 4^x$

$y = 3^x$

$x < 0$

21.

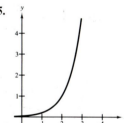

23.

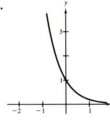

25.

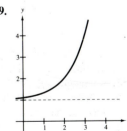

27.

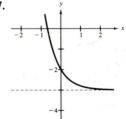

29.

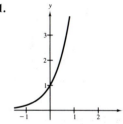

31.

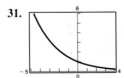

33.

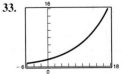

35.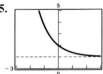

37.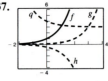

39. (a)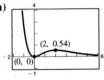

Increasing: (0, 2)
Decreasing: $(-\infty, 0)$, $(2, \infty)$
Relative minimum: (0, 0)
Relative maximum: (2, 0.54)

(b)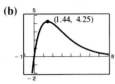

Increasing: $(-\infty, 1.44)$
Decreasing: $(1.44, \infty)$
Relative maximum:
(1.44, 4.25)

41.

n	1	2	4
A	\$7764.62	\$8017.84	\$8155.09

n	12	365	Continuous compounding
A	\$8250.97	\$8298.66	\$8300.29

43.

n	1	2	4
A	\$24,115.73	\$25,714.29	\$26,602.23

n	12	365	Continuous compounding
A	\$27,231.38	\$27,547.07	\$27,557.94

45.

t	1	10	20
P	\$91,393.12	\$40,656.97	\$16,529.89

t	30	40	50
P	\$6720.55	\$2732.37	\$1110.90

47.

t	1	10	20
P	\$90,521.24	\$36,940.70	\$13,646.15

t	30	40	50
P	\$5040.98	\$1862.17	\$687.90

49. \$222,822.57
51. (a) \$472.70
(b) \$298.29; The graph of p is a decreasing function
with domain (0, 1725) and range (0, 499.50).

53. (a) 100
(b) 300
(c) 900

55. (a) 25 units
(b) 16.30 units
(c)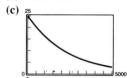

57. (a)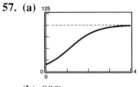

(b) 80%
(c) 1550

59.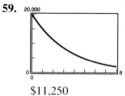

\$11,250

Section 4.2   (page 298)

WARM-UP   **1.** 3   **2.** 0   **3.** -1 **4.** 1   **5.** 7.389
6. 0.368   **7.** Graph is shifted two units to the left.
8. Graph is reflected about the x-axis.
9. Graph is shifted down one unit.
10. Graph is reflected about the y-axis.

1. 4   **3.** $\frac{1}{2}$ **5.** 0   **7.** -2 **9.** 3
11. 2   **13.** $\log_5 125 = 3$ **15.** $\log_{81} 3 = \frac{1}{4}$
17. $\log_6 \frac{1}{36} = -2$ **19.** $\ln 20.0855 \approx 3$ **21.** 2.538
23. -0.319 **25.** 2.913

27.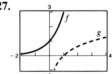

29.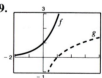

31. d
32. e
33. a

Xmin=-1
Xmax=4
Xscl=1
Ymin=-1
Ymax=3
Yscl=1

Xmin=-1
Xmax=4
Xscl=1
Ymin=-2
Ymax=2
Yscl=1

Xmin=-3
Xmax=3
Xscl=1
Ymin=-2
Ymax=2
Yscl=1

34. c

```
Xmin=-1
Xmax=4
Xscl=1
Ymin=-2
Ymax=2
Yscl=1
```

35. f

```
Xmin=-3
Xmax=2
Xscl=1
Ymin=-2
Ymax=2
Yscl=1
```

36. b

```
Xmin=-4
Xmax=1
Xscl=1
Ymin=-2
Ymax=2
Yscl=1
```

37. Domain: $(0, \infty)$
Vertical asymptote: $x = 0$
Intercept: $(1, 0)$

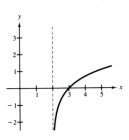

39. Domain: $(-2, \infty)$
Vertical asymptote: $x = -2$
Intercepts: $(-1, 0)$, $(0, -\log_3 2)$

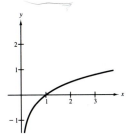

41. Domain: $(2, \infty)$
Vertical asymptote: $x = 2$
Intercept: $(3, 0)$

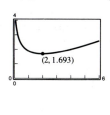

43. Increasing: $(2, \infty)$
Decreasing: $(0, 2)$
Relative minimum: $(2, 1.693)$

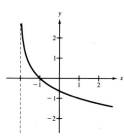
(2, 1.693)

45. (a)

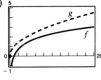

$\sqrt{x}$ is increasing more rapidly than $\ln x$.

(b)

$\sqrt[4]{x}$ is increasing more rapidly than $\ln x$.

47. (a) 80   (b) 68.1   (c) 62.3

49.

r	0.005	0.010	0.015	0.020	0.025	0.030
t	138.6 yr	69.3 yr	46.2 yr	34.7 yr	27.7 yr	23.1 yr

51.

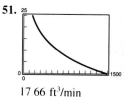

17 66 ft³/min

53. 20 yr

55. \$280,178.40

57. 21,357 ft·lb

59. (a)

x	1	5	10	10^2	10^4	10^6
$f(x)$	0	0.322	0.230	0.046	0.00092	0.0000138

(b) 0

Section 4.3   (page 306)

WARM-UP   **1.** 2   **2.** -5 **3.** -2 **4.** -3

5. e^5 **6.** $\dfrac{1}{e}$ **7.** e^6 **8.** 1   **9.** x^{-2}

10. $x^{1/2}$

1. $\dfrac{\log_{10} 5}{\log_{10} 3}$ **3.** $\dfrac{\log_{10} x}{\log_{10} 2}$ **5.** $\dfrac{\ln 5}{\ln 3}$ **7.** $\dfrac{\ln x}{\ln 2}$

9. 1.771   **11.** -2.000 **13.** -0.417 **15.** 2.633

17. $\log_{10} 5 + \log_{10} x$ **19.** $\log_{10} 5 - \log_{10} x$

21. $4 \log_8 x$ **23.** $\frac{1}{2} \ln z$ **25.** $\ln x + \ln y + \ln z$

27. $\frac{1}{2} \ln(a - 1)$

29. $\ln z + 2 \ln(z - 1)$ **31.** $\frac{1}{3} \ln x - \frac{1}{3} \ln y$

33. $4 \ln x + \frac{1}{2} \ln y - 5 \ln z$

35. $2 \log_b x - 2 \log_b y - 3 \log_b z$ **37.** $\ln 2x$

39. $\log_4 \dfrac{z}{y}$ **41.** $\log_2(x + 4)^2$ **43.** $\log_3 \sqrt[3]{5x}$

45. $\ln \dfrac{x}{(x + 1)^3}$ **47.** $\log_3 \dfrac{x - 2}{x + 2}$

49. $\ln \dfrac{x}{(x^2 - 4)^2}$ **51.** $\ln \sqrt[3]{\dfrac{x(x + 3)^2}{x^2 - 1}}$

53. $\ln \dfrac{\sqrt[3]{y(y + 4)^2}}{y - 1}$ **55.** $\ln \dfrac{9}{\sqrt{x^2 + 1}}$ **57.** 0.9208

59. 1.8957   **61.** -0.1781 **63.** 0.9136   **65.** 2.0365

67. 2   **69.** 2.4   **71.** 4.5   **73.** $\frac{3}{2}$

75. $\frac{1}{2} + \frac{1}{2} \log_7 10$ **77.** $-3 - \log_5 2$ **79.** $6 + \ln 5$

81. $\beta = 10(\log_{10} I + 16)$, 60 db

83.

f and h have the same graphs.

Section 4.4 (page 316)

WARM-UP   **1.** $\dfrac{\ln 3}{\ln 2}$ **2.** $1 + \dfrac{2}{\ln 4}$ **3.** $\dfrac{e}{2}$ **4.** $2e$

5. $2 \pm i$ **6.** $\frac{1}{2}, 1$ **7.** $2x$ **8.** $3x$ **9.** $2x$

10. $-x^2$

---

1. 2   **3.** -2 **5.** 3   **7.** 64   **9.** $\frac{1}{10}$ **11.** x^2

13. $5x + 2$ **15.** x^2 **17.** $\ln 10 \approx 2.303$ **19.** 0

21. $\dfrac{\ln 12}{3} \approx 0.828$ **23.** $\ln \dfrac{5}{3} \approx 0.511$ **25.** $\ln 5 \approx 1.609$

27. $2 \ln 75 \approx 8.6350$ **29.** $\log_{10} 42 \approx 1.623$

31. $\dfrac{\ln 80}{2 \ln 3} \approx 1.994$ **33.** 2

35. $\dfrac{\ln 2}{12 \ln(1 + 0.10/12)} \approx 6.960$

37. 3.847   **39.** 12.207

41. 0.059   **43.** 21.330   **45.** $e^{-3} \approx 0.050$

47. $\dfrac{e^{2.4}}{2} \approx 5.512$ **49.** $e^2 - 2 \approx 5.389$

51. $1 + \sqrt{1 + e} \approx 2.928$ **53.** 103

55. $\dfrac{-1 + \sqrt{17}}{2} \approx 1.562$ **57.** 4   **59.** No solution

61. 33.115   **63.** 14.369   **65.** 8.2 yr   **67.** 12.9 yr

69. (a) 1426 units   (b) 1498 units

71. (a)

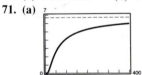

(b) $V = 6.7$

The asymptote represents the limiting yield per acre.

(c) 29.33 yr

73. (a)

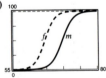

(b) $y = 0,\ y = 100$

(c) Males: 69.71 in.; Females: 64.51 in.

Section 4.5 (page 325)

WARM-UP

1.

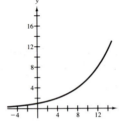

2.

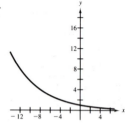

3.

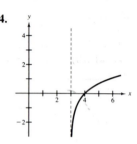

4.

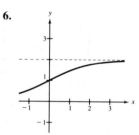

5.

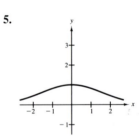

6.

7. $\dfrac{1}{2} \ln \dfrac{7}{3} \approx 0.424$ **8.** $\dfrac{\ln(0.001)}{-0.2} \approx 34.539$

9. $\frac{1}{5} e^{7/2} \approx 6.623$ **10.** $\frac{1}{2} e^2 \approx 3.695$

	Initial Investment	Annual % Rate	Effective Yield	Time to Double	Amount After 10 Years
1.	$1000	12%	12.75%	5.78 yr	$3320.12
3.	$750	8.94%	9.35%	$7\frac{3}{4}$ yr	$1833.67
5.	$500	9.5%	9.97%	7.30 yr	$1292.85
7.	$6392.79	11%	11.63%	6.30 yr	$19,205.00
9.	$5000	8%	8.33%	8.66 yr	$11,127.70

11. $112,087.09

13. (a) 6.642 yr   (b) 6.330 yr   (c) 6.302 yr

     (d) 6.301 yr

15.

r	2%	4%	6%	8%	10%	12%
t	54.93	27.47	18.31	13.73	10.99	9.16

17.

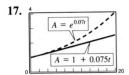

Continuous compounding at 7% grows faster.

Isotope	Half-life (Years)	Initial Quantity	Amount After 1000 Years	Amount After 10,000 Years
19. ^{226}Ra	1620	10 g	6.52 g	0.14 g
21. ^{14}C	5730	6.70 g	5.95 g	2 g
23. ^{230}Pu	24,360	2.16 g	2.1 g	1.63 g

25. $\frac{1}{4}\ln 10 \approx 0.5756$ **27.** $\frac{1}{4}\ln\frac{1}{4} \approx -0.3466$

29. 105,300, 2013 **31.** $k = 0.0137$, 3288

33. $y = 4.22e^{0.0430t}$, 9.97 million **35.** 3.15 hr

37. 95.8% **39.** $9281

41. (a) $S(t) = 100(1 - e^{-0.1625t})$ (b) 55,625

43.

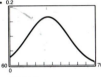

64.5 in

45. (a)

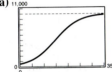

$p = 10,000$. This represents the carrying capacity of the lake for this species of fish.

(b) 1252 fish (c) 7.8 months

47. (a) $S = 10(1 - e^{-0.0575x})$ (b) 3314

49. (a) 7.91 (b) 7.68

51. (a) 20 (b) 70 (c) 95 (d) 120

53. 95% **55.** 4.64 **57.** 1.58×10^{-6} moles per liter

59. 10^7

61. (a)

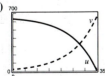

(b) Interest payments, 27.7 yr

63. 7:30 A.M.

Section 4.6   *(page 335)*

WARM-UP

1.

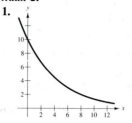

2.

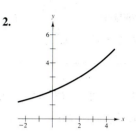

3.

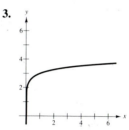

4.

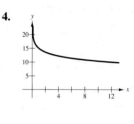

5.

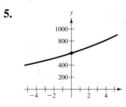

6.

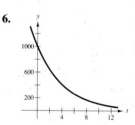

7. $y = -0.7x + 2.6$

8. $y = \frac{2}{3}x + 4.75$

9. $y = 3.9x + 9.9$

10. $y = -x + 8.4$

1. $y = 2.152 + 2.704 \ln x$ **3.** $y = 3.114e^{0.2935x}$

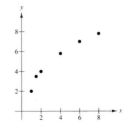

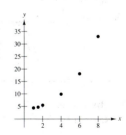

5. $y = -0.788x + 8.257$

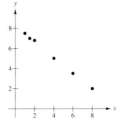

7. $y = 3.807e^{0.2667x}$

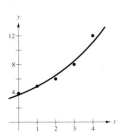

9. $y = 8.463e^{-0.2517x}$

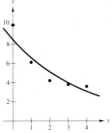

11. $y = 2.083 + 1.257 \ln x$

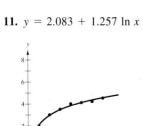

13. $y = 9.826 - 4.097 \ln x$

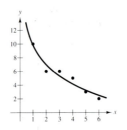

15. $y = 1,985x^{0.760}$

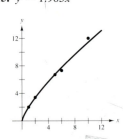

17. $y = 16.103x^{-3.174}$

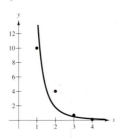

19. $y = 4.754(6.774)^x$

21. $y = 19.752 + 3.541 \ln t$

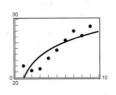

23. $y = 5.088x^{0.645}$

Chapter 4 Review Exercises (*page 337*)

1. d

```
Xmin=-4
Xmax=3
Xscl=1
Ymin=-1
Ymax=4
Yscl=1
```

2. f

```
Xmin=-3
Xmax=4
Xscl=1
Ymin=-1
Ymax=4
Yscl=1
```

3. a

```
Xmin=-4
Xmax=4
Xscl=1
Ymin=-5
Ymax=1
Yscl=1
```

4. b

```
Xmin=-4
Xmax=4
Xscl=1
Ymin=0
Ymax=6
Yscl=1
```

5. c

```
Xmin=-3
Xmax=6
Xscl=1
Ymin=-3
Ymax=3
Yscl=1
```

6. e

```
Xmin=-1
Xmax=7
Xscl=1
Ymin=-4
Ymax=3
Yscl=1
```

7.

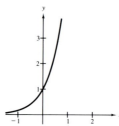

9.

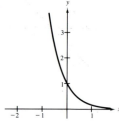

11.

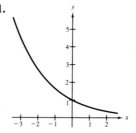

13.

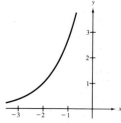

15.

n	1	2	4
A	\$9499.28	\$9738.91	\$9867.22

n	12	365	Continuous compounding
A	\$9956.20	\$10,000.27	\$10,001.78

17.

t	1	10	20
P	\$184,623.27	\$89,865.79	\$40,379.30

t	30	40	50
P	\$18,143.59	\$8152.44	\$3663.13

19. \$1,069,047.14   **21.** 229.2 units per milliliter

23.

Speed	50	55	60	65	70
Miles per gallon	28	26.4	24.8	23.4	22.0

25.

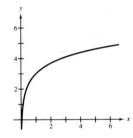

27.

29.

31. $\log_4 64 = 3$

33. 3   **35.** -2 **37.** 7   **39.** 0   **41.** 1.585
43. 2.132   **45.** $1 + 2 \log_5 x$
47. $\log_{10} 5 + \frac{1}{2} \log_{10} y - 2 \log_{10} x$
49. $\ln(x^2 + 1) + \ln(x - 1)$
51. $\log_2 5x$ **53.** $\ln \dfrac{\sqrt{|2x - 1|}}{(x + 1)^2}$ **55.** $\ln \dfrac{3\sqrt[3]{4 - x^2}}{x}$
57. False   **59.** False   **61.** 1.6542   **63.** 0.2823
65. 27.16 mi   **67.** $\ln 12 \approx 2.485$
69. $-\dfrac{\ln 44}{5} \approx -0.757$ **71.** $\ln 2 \approx 0.693$, $\ln 5 \approx 1.609$
73. $\frac{1}{3} e^{8.2} \approx 1213.650$ **75.** $3e^2 \approx 22.167$
77. $y = 2e^{0.1014t}$ **79.** $y = 4e^{-0.4159t}$
81. (a) 1151 units   **(b)** 1325 units
83. (a) 8.94%   **(b)** \$1834.37   **(c)** 9.36%
85. $10^{-3.5}$

87. $y = 234.684e^{-0.134x}$

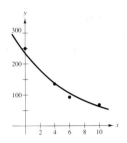

CHAPTER 5

Section 5.1   (*page 350*)

WARM-UP   **1.** 45   **2.** 70   **3.** $\dfrac{\pi}{6}$ **4.** $\dfrac{\pi}{3}$ **5.** $\dfrac{\pi}{4}$
6. $\dfrac{4\pi}{3}$ **7.** $\dfrac{\pi}{9}$ **8.** $\dfrac{11\pi}{6}$ **9.** 45   **10.** 45

1. (a) Quadrant I   **3. (a)** Quadrant IV
(b) Quadrant III   **(b)** Quadrant III
5. (a) Quadrant II
(b) Quadrant IV
7. (a)   **(b)**

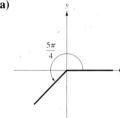

9. (a)   **(b)**

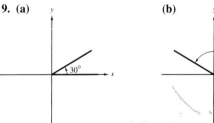

11. (a) $\dfrac{19\pi}{9}$, $-\dfrac{17\pi}{9}$ **(b)** $\dfrac{10\pi}{3}$, $-\dfrac{2\pi}{3}$
13. (a) $396°$, $-324°$ **15. (a)** $660°$, $-60°$
(b) $315°$, $-405°$ **(b)** $20°$, $-340°$
17. (a) Complement: $\dfrac{\pi}{6}$; Supplement: $\dfrac{2\pi}{3}$
(b) Complement: none; Supplement: $\dfrac{\pi}{4}$

19. (a) Complement: 72°; Supplement: 162°
   (b) Complement: none; Supplement: 65°
21. (a) 270°   **23. (a)** 420°
   (b) 210°   **(b)** −66°
25. (a) $\dfrac{\pi}{6}$ **27. (a)** $-\dfrac{\pi}{9}$
   (b) $\dfrac{5\pi}{6}$ **(b)** $-\dfrac{4\pi}{3}$
29. (a) 2.007   **31. (a)** 9.285
   (b) 1.525   **(b)** 0.009
33. (a) 25.714°   **35. (a)** −756°
   (b) 81.818°   **(b)** 275.020°
37. (a) 245.167°
   (b) 2.2°
39. (a) 240° 36′   **41. (a)** 143° 14′ 22″
   (b) −145° 48′   **(b)** −205° 7′ 8″
43. $\frac{4}{15}$ rad   **45.** 1.724 rad   **47.** 15π in. $\approx$ 47.12 in.
49. 12 m   **51.** 591.72 mi   **53.** 1141 mi   **55.** 4.655°
57. $\frac{1}{4}$ rad $\approx$ 14.32°
59. (a) 560.2 rev/min
   (b) 3520 rad/min
61. (a) Angular speed of motor pulley = 3400π rad/min
       Angular speed of saw arbor = 1700π rad/min
   (b) 850 rev/min
63. (a) 80π rad/sec
   (b) 78.54 ft/sec

Section 5.2   (*page 361*)

WARM-UP   **1.** $-\dfrac{\sqrt{3}}{3}$ **2.** −1   **3.** $\dfrac{2\pi}{3}$ **4.** $\dfrac{7\pi}{4}$

5. $\dfrac{\pi}{6}$ **6.** $\dfrac{3\pi}{4}$ **7.** 60°   **8.** −270°   **9.** 2π

10. π

---

1. $\left(\dfrac{\sqrt{2}}{2}, \dfrac{\sqrt{2}}{2}\right)$ **3.** $\left(-\dfrac{\sqrt{3}}{2}, \dfrac{1}{2}\right)$ **5.** $\left(-\dfrac{1}{2}, -\dfrac{\sqrt{3}}{2}\right)$

7. $(0, -1)$

9. $\sin\dfrac{\pi}{4} = \dfrac{\sqrt{2}}{2}$ **11.** $\sin-\dfrac{5\pi}{4} = \dfrac{\sqrt{2}}{2}$

   $\cos\dfrac{\pi}{4} = \dfrac{\sqrt{2}}{2}$ $\cos-\dfrac{5\pi}{4} = -\dfrac{\sqrt{2}}{2}$

   $\tan\dfrac{\pi}{4} = 1$ $\tan-\dfrac{5\pi}{4} = -1$

13. $\sin\dfrac{11\pi}{6} = -\dfrac{1}{2}$ **15.** $\sin\dfrac{4\pi}{3} = -\dfrac{\sqrt{3}}{2}$

   $\cos\dfrac{11\pi}{6} = \dfrac{\sqrt{3}}{2}$ $\cos\dfrac{4\pi}{3} = -\dfrac{1}{2}$

   $\tan\dfrac{11\pi}{6} = -\dfrac{\sqrt{3}}{3}$ $\tan\dfrac{4\pi}{3} = \sqrt{3}$

17. $\sin\dfrac{3\pi}{4} = \dfrac{\sqrt{2}}{2}$ **19.** $\sin\dfrac{\pi}{2} = 1$

   $\cos\dfrac{3\pi}{4} = -\dfrac{\sqrt{2}}{2}$ $\cos\dfrac{\pi}{2} = 0$

   $\tan\dfrac{3\pi}{4} = -1$ $\tan\dfrac{\pi}{2}$ is undefined

   $\csc\dfrac{3\pi}{4} = \sqrt{2}$ $\csc\dfrac{\pi}{2} = 1$

   $\sec\dfrac{3\pi}{4} = -\sqrt{2}$ $\sec\dfrac{\pi}{2}$ is undefined

   $\cot\dfrac{3\pi}{4} = -1$ $\cot\dfrac{\pi}{2} = 0$

21. $\sin\left(-\dfrac{4\pi}{3}\right) = \dfrac{\sqrt{3}}{2}$

   $\cos\left(-\dfrac{4\pi}{3}\right) = -\dfrac{1}{2}$

   $\tan\left(-\dfrac{4\pi}{3}\right) = -\sqrt{3}$

   $\csc\left(-\dfrac{4\pi}{3}\right) = \dfrac{2\sqrt{3}}{3}$

   $\sec\left(-\dfrac{4\pi}{3}\right) = -2$

   $\sin\left(-\dfrac{4\pi}{3}\right) = -\dfrac{\sqrt{3}}{3}$

23. $\sin\pi = 0$ **25.** $\cos\dfrac{2\pi}{3} = -\dfrac{1}{2}$

27. $\cos\dfrac{7\pi}{6} = -\dfrac{\sqrt{3}}{2}$ **29.** $\sin\dfrac{7\pi}{4} = -\dfrac{\sqrt{2}}{2}$

31. (a) $-\frac{1}{3}$ **(b)** −3   **33. (a)** $-\frac{7}{8}$ **(b)** $-\frac{8}{7}$
35. (a) $\frac{4}{5}$ **(b)** $-\frac{4}{5}$ **37.** 0.7071   **39.** 0.8271
41. −2.6746   **43.** 1.3940   **45. (a)** −1   **(b)** −0.4
47. (a) 0.25, 2.9   **(b)** 1.8, 4.5
49. (a) 0.2500 ft   **(b)** 0.0177 ft   **(c)** −0.2475 ft
51. 0.794

Section 5.3   (*page 370*)

WARM-UP   **1.** $2\sqrt{5}$ **2.** $3\sqrt{10}$ **3.** 10
4. $3\sqrt{2}$ **5.** 1.24   **6.** 317.55   **7.** 63.13
8. 133.57   **9.** 2,785,714.29   **10.** 28.80

1. $\sin\theta = \dfrac{1}{2}$

$\cos\theta = \dfrac{\sqrt{3}}{2}$

$\tan\theta = \dfrac{\sqrt{3}}{3}$

$\csc\theta = 2$

$\sec\theta = \dfrac{2\sqrt{3}}{3}$

$\cot\theta = \sqrt{3}$

3. $\sin\theta = \dfrac{3}{5}$

$\cos\theta = \dfrac{4}{5}$

$\tan\theta = \dfrac{3}{4}$

$\csc\theta = \dfrac{5}{3}$

$\sec\theta = \dfrac{5}{4}$

$\cot\theta = \dfrac{4}{3}$

5. $\sin\theta = \dfrac{\sqrt{161}}{15}$

$\cos\theta = \dfrac{8}{15}$

$\tan\theta = \dfrac{\sqrt{161}}{8}$

$\csc\theta = \dfrac{15\sqrt{161}}{161}$

$\sec\theta = \dfrac{15}{8}$

$\cot\theta = \dfrac{8\sqrt{161}}{161}$

7. $\sin\theta = \dfrac{15}{17}$

$\cos\theta = \dfrac{8}{17}$

$\tan\theta = \dfrac{15}{8}$

$\csc\theta = \dfrac{17}{15}$

$\sec\theta = \dfrac{17}{8}$

$\cot\theta = \dfrac{8}{15}$

9. $\cos\theta = \dfrac{\sqrt{5}}{3}$

$\tan\theta = \dfrac{2\sqrt{5}}{5}$

$\csc\theta = \dfrac{3}{2}$

$\sec\theta = \dfrac{3\sqrt{5}}{5}$

$\cot\theta = \dfrac{\sqrt{5}}{2}$

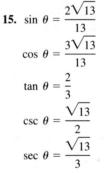

11. $\sin\theta = \dfrac{\sqrt{3}}{2}$

$\cos\theta = \dfrac{1}{2}$

$\tan\theta = \sqrt{3}$

$\csc\theta = \dfrac{2\sqrt{3}}{3}$

$\cot\theta = \dfrac{\sqrt{3}}{3}$

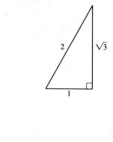

13. $\sin\theta = \dfrac{3\sqrt{10}}{10}$

$\cos\theta = \dfrac{\sqrt{10}}{10}$

$\csc\theta = \dfrac{\sqrt{10}}{3}$

$\sec\theta = \sqrt{10}$

$\cot\theta = \dfrac{1}{3}$

15. $\sin\theta = \dfrac{2\sqrt{13}}{13}$

$\cos\theta = \dfrac{3\sqrt{13}}{13}$

$\tan\theta = \dfrac{2}{3}$

$\csc\theta = \dfrac{\sqrt{13}}{2}$

$\sec\theta = \dfrac{\sqrt{13}}{3}$

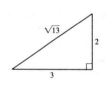

17. (a) $\sqrt{3}$

(b) $\dfrac{1}{2}$

(c) $\dfrac{\sqrt{3}}{2}$

(d) $\dfrac{\sqrt{3}}{3}$

19. (a) $\dfrac{1}{3}$

(b) $\dfrac{2\sqrt{2}}{3}$

(c) $\dfrac{\sqrt{2}}{4}$

(d) 3

21. (a) $\dfrac{1}{2}$

(b) 1

23. (a) 1

(b) $\dfrac{\sqrt{2}}{2}$

25. (a) 0.1736
(b) 0.1736

27. (a) 0.2815
(b) 3.5523

29. (a) 1.3499
(b) 1.3432

31. (a) 5.0273
(b) 0.1989

33. (a) 1.1884   **35.** (a) $30° = \dfrac{\pi}{6}$

(b) 1.1884   (b) $30° = \dfrac{\pi}{6}$

37. (a) $60° = \dfrac{\pi}{3}$ **39.** (a) $60° = \dfrac{\pi}{3}$

(b) $45° = \dfrac{\pi}{4}$ (b) $45° = \dfrac{\pi}{4}$

41. (a) $55° \approx 0.96$ **43.** (a) $50° \approx 0.873$
(b) $89° \approx 1.55$ (b) $25° \approx 0.436$

45. 57.74   **47.** 14.43   **49.** 15.56

51. 9.19   **53.** 15 ft   **55.** 19.32 ft   **57.** 2145.10 ft

59. $(x_1, y_1) = \left(10\sqrt{3}, 10\right)$, $(x_2, y_2) = \left(10, 10\sqrt{3}\right)$

61. $\sin 25° \approx 0.42$, $\cos 25° \approx 0.91$, $\tan 25° \approx 0.47$,
$\csc 25° \approx 2.37$, $\sec 25° \approx 1.10$, $\cot 25° \approx 2.14$

63. True, $\csc x = \dfrac{1}{\sin x}$ **65.** False, $\dfrac{\sqrt{2}}{2} + \dfrac{\sqrt{2}}{2} \neq 1$

67. False, $1.7321 \neq \sin 2°$ **69.** (a) 29.73 ft
(b) 82.34°
(c) 29.74 ft
(d) The difference is due to rounding.

Section 5.4   (page 382)

WARM-UP   **1.** $\sin 30° = \dfrac{1}{2}$ **2.** $\tan 45° = 1$

3. $\cos \dfrac{\pi}{4} = \dfrac{\sqrt{2}}{2}$ **4.** $\cot \dfrac{\pi}{3} = \dfrac{\sqrt{3}}{3}$

5. $\sec \dfrac{\pi}{6} = \dfrac{2\sqrt{3}}{3}$ **6.** $\csc \dfrac{\pi}{4} = \sqrt{2}$

7. $\sin \theta = \dfrac{3\sqrt{13}}{13}$ **8.** $\sin \theta = \dfrac{\sqrt{5}}{3}$

$\cos \theta = \dfrac{2\sqrt{13}}{13}$ $\tan \theta = \dfrac{\sqrt{5}}{2}$

$\csc \theta = \dfrac{\sqrt{13}}{3}$ $\csc \theta = \dfrac{3\sqrt{5}}{5}$

$\sec \theta = \dfrac{\sqrt{13}}{2}$ $\sec \theta = \dfrac{3}{2}$

$\cot \theta = \dfrac{2}{3}$ $\cot \theta = \dfrac{2\sqrt{5}}{5}$

9. $\cos \theta = \dfrac{2\sqrt{6}}{5}$ **10.** $\sin \theta = \dfrac{2\sqrt{2}}{3}$

$\tan \theta = \dfrac{\sqrt{6}}{12}$ $\cos \theta = \dfrac{1}{3}$

$\csc \theta = 5$ $\tan \theta = 2\sqrt{2}$

$\sec \theta = \dfrac{5\sqrt{6}}{12}$ $\csc \theta = \dfrac{3\sqrt{2}}{4}$

$\cot \theta = 2\sqrt{6}$ $\cot \theta = \dfrac{\sqrt{2}}{4}$

1. (a) $\sin \theta = \frac{4}{5}$ (b) $\sin \theta = -\frac{15}{17}$

$\cos \theta = \frac{3}{5}$ $\cos \theta = \frac{8}{17}$

$\tan \theta = \frac{4}{3}$ $\tan \theta = -\frac{15}{8}$

$\csc \theta = \frac{5}{4}$ $\csc \theta = -\frac{17}{15}$

$\sec \theta = \frac{5}{3}$ $\sec \theta = \frac{17}{8}$

$\cot \theta = \frac{3}{4}$ $\cot \theta = -\frac{8}{15}$

3. (a) $\sin \theta = \dfrac{1}{2}$ (b) $\sin \theta = -\dfrac{\sqrt{2}}{2}$

$\cos \theta = -\dfrac{\sqrt{3}}{2}$ $\cos \theta = -\dfrac{\sqrt{2}}{2}$

$\tan \theta = -\dfrac{\sqrt{3}}{3}$ $\tan \theta = 1$

$\csc \theta = 2$ $\csc \theta = -\sqrt{2}$

$\sec \theta = -\dfrac{2\sqrt{3}}{3}$ $\sec \theta = -\sqrt{2}$

$\cot \theta = -\sqrt{3}$ $\cot \theta = 1$

5. (a) $\sin \theta = \frac{24}{25}$ (b) $\sin \theta = -\frac{24}{25}$

$\cos \theta = \frac{7}{25}$ $\cos \theta = \frac{7}{25}$

$\tan \theta = \frac{24}{7}$ $\tan \theta = -\frac{24}{7}$

$\csc \theta = \frac{25}{24}$ $\csc \theta = -\frac{25}{24}$

$\sec \theta = \frac{25}{7}$ $\sec \theta = \frac{25}{7}$

$\cot \theta = \frac{7}{24}$ $\cot \theta = -\frac{7}{24}$

7. (a) $\sin \theta = \dfrac{5\sqrt{29}}{29}$ (b) $\sin \theta = -\dfrac{5\sqrt{34}}{34}$

$\cos \theta = -\dfrac{2\sqrt{29}}{29}$ $\cos \theta = \dfrac{3\sqrt{34}}{34}$

$\tan \theta = -\dfrac{5}{2}$ $\tan \theta = -\dfrac{5}{3}$

$\csc \theta = \dfrac{\sqrt{29}}{5}$ $\csc \theta = -\dfrac{\sqrt{34}}{5}$

$\sec \theta = -\dfrac{\sqrt{29}}{2}$ $\sec \theta = \dfrac{\sqrt{34}}{3}$

$\cot \theta = -\dfrac{2}{5}$ $\cot \theta = -\dfrac{3}{5}$

9. (a) $c_1 = 5$
$b_2 = 12$
$c_2 = 15$

(b) $\sin \alpha_1 = \dfrac{a_1}{c_1} = \dfrac{3}{5} = \dfrac{a_2}{c_2} = \sin \alpha_2$

$\cos \alpha_1 = \dfrac{b_1}{c_1} = \dfrac{4}{5} = \dfrac{b_2}{c_2} = \cos \alpha_2$

$\tan \alpha_1 = \dfrac{a_1}{b_1} = \dfrac{3}{4} = \dfrac{a_2}{b_2} = \tan \alpha_2$

$\csc \alpha_1 = \dfrac{c_1}{a_1} = \dfrac{5}{3} = \dfrac{c_2}{a_2} = \csc \alpha_2$

$\sec \alpha_1 = \dfrac{c_1}{b_1} = \dfrac{5}{4} = \dfrac{c_2}{b_2} = \sec \alpha_2$

$\cot \alpha_1 = \dfrac{b_1}{a_1} = \dfrac{4}{3} = \dfrac{b_2}{a_2} = \cot \alpha_2$

11. (a) $b_1 = \sqrt{3}$

$a_2 = \dfrac{5\sqrt{3}}{3}$

$c_2 = \dfrac{10\sqrt{3}}{3}$

(b) $\sin \alpha_1 = \dfrac{a_1}{c_1} = \dfrac{1}{2} = \dfrac{a_2}{c_2} = \sin \alpha_2$

$\cos \alpha_1 = \dfrac{b_1}{c_1} = \dfrac{\sqrt{3}}{2} = \dfrac{b_2}{c_2} = \cos \alpha_2$

$\tan \alpha_1 = \dfrac{a_1}{b_1} = \dfrac{\sqrt{3}}{3} = \dfrac{a_2}{b_2} = \tan \alpha_2$

$\csc \alpha_1 = \dfrac{c_1}{a_1} = 2 = \dfrac{c_2}{a_2} = \csc \alpha_2$

$\sec \alpha_1 = \dfrac{c_1}{b_1} = \dfrac{2\sqrt{3}}{3} = \dfrac{c_2}{b_2} = \sec \alpha_2$

$\cot \alpha_1 = \dfrac{b_1}{a_1} = \sqrt{3} = \dfrac{b_2}{a_2} = \cot \alpha_2$

13. (a) Quadrant III
(b) Quadrant II

15. (a) Quadrant II
(b) Quadrant IV

17. $\sin \theta = \frac{3}{5}$
$\cos \theta = -\frac{4}{5}$
$\tan \theta = -\frac{3}{4}$
$\csc \theta = \frac{5}{3}$
$\sec \theta = -\frac{5}{4}$
$\cot \theta = -\frac{4}{3}$

19. $\sin \theta = -\frac{15}{17}$
$\cos \theta = \frac{8}{17}$
$\tan \theta = -\frac{15}{8}$
$\csc \theta = -\frac{17}{15}$
$\sec \theta = \frac{17}{8}$
$\cot \theta = -\frac{8}{15}$

21. $\sin \theta = \dfrac{\sqrt{3}}{2}$

$\cos \theta = -\dfrac{1}{2}$

$\tan \theta = -\sqrt{3}$

$\csc \theta = \dfrac{2\sqrt{3}}{3}$

$\sec \theta = -2$

$\cot \theta = -\dfrac{\sqrt{3}}{3}$

23. $\sin \theta = 0$
$\cos \theta = -1$
$\tan \theta = 0$
$\csc \theta$ is undefined.
$\sec \theta = -1$
$\cot \theta$ is undefined.

25. $\sin \theta = -\dfrac{2\sqrt{5}}{5}$

$\cos \theta = -\dfrac{\sqrt{5}}{5}$

$\tan \theta = 2$

$\csc \theta = \dfrac{-\sqrt{5}}{2}$

$\sec \theta = -\sqrt{5}$

$\cot \theta = \dfrac{1}{2}$

27. (a) $\theta' = 23°$

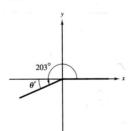

(b) $\theta' = 53°$

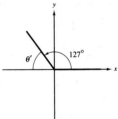

29. (a) $\theta' = 65°$

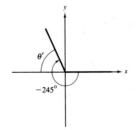

(b) $\theta' = 72°$

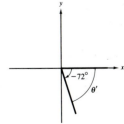

31. (a) $\theta' = \dfrac{\pi}{3}$ (b) $\theta' = \dfrac{\pi}{6}$

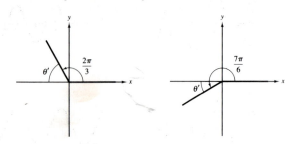

33. (a) $\theta' = 3.5 - \pi$ (b) $\theta' = 2\pi - 5.8$

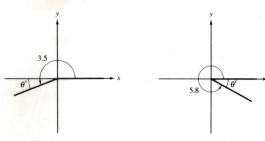

35. (a) $\sin(225°) = -\dfrac{\sqrt{2}}{2}$ (b) $\sin(-225°) = \dfrac{\sqrt{2}}{2}$

$\cos(225°) = -\dfrac{\sqrt{2}}{2}$. $\cos(-225°) = -\dfrac{\sqrt{2}}{2}$

$\tan(225°) = 1$ $\tan(-225°) = -1$

37. (a) $\sin(750°) = \dfrac{1}{2}$ (b) $\sin(510°) = \dfrac{1}{2}$

$\cos(750°) = \dfrac{\sqrt{3}}{2}$ $\cos(510°) = -\dfrac{\sqrt{3}}{2}$

$\tan(750°) = \dfrac{\sqrt{3}}{3}$ $\tan(510°) = -\dfrac{\sqrt{3}}{3}$

39. (a) $\sin\dfrac{4\pi}{3} = -\dfrac{\sqrt{3}}{2}$ (b) $\sin\dfrac{2\pi}{3} = \dfrac{\sqrt{3}}{2}$

$\cos\dfrac{4\pi}{3} = -\dfrac{1}{2}$ $\cos\dfrac{2\pi}{3} = -\dfrac{1}{2}$

$\tan\dfrac{4\pi}{3} = \sqrt{3}$ $\tan\dfrac{2\pi}{3} = -\sqrt{3}$

41. (a) $\sin\left(-\dfrac{\pi}{6}\right) = -\dfrac{1}{2}$ (b) $\sin\dfrac{5\pi}{6} = \dfrac{1}{2}$

$\cos\left(-\dfrac{\pi}{6}\right) = \dfrac{\sqrt{3}}{2}$ $\cos\dfrac{5\pi}{6} = -\dfrac{\sqrt{3}}{2}$

$\tan\left(-\dfrac{\pi}{6}\right) = -\dfrac{\sqrt{3}}{3}$ $\tan\dfrac{5\pi}{6} = -\dfrac{\sqrt{3}}{3}$

43. (a) $\sin\dfrac{11\pi}{4} = \dfrac{\sqrt{2}}{2}$ (b) $\sin\left(-\dfrac{13\pi}{6}\right) = -\dfrac{1}{2}$

$\cos\dfrac{11\pi}{4} = -\dfrac{\sqrt{2}}{2}$ $\cos\left(-\dfrac{13\pi}{6}\right) = \dfrac{\sqrt{3}}{2}$

$\tan\dfrac{11\pi}{4} = -1$ $\tan\left(-\dfrac{13\pi}{6}\right) = -\dfrac{\sqrt{3}}{3}$

45. (a) 0.1736   **47.** (a) -0.3420
(b) 5.7588   (b) -0.3420

49. (a) 1.7321   **51.** (a) 0.3640
(b) 1.7321   (b) 0.3640

53. (a) $30° = \dfrac{\pi}{6}$, $150° = \dfrac{5\pi}{6}$

(b) $210° = \dfrac{7\pi}{6}$, $330° = \dfrac{11\pi}{6}$

55. (a) $60° = \dfrac{\pi}{3}$, $120° = \dfrac{2\pi}{3}$

(b) $135° = \dfrac{3\pi}{4}$, $315° = \dfrac{7\pi}{4}$

57. (a) $45° = \dfrac{\pi}{4}$, $225° = \dfrac{5\pi}{4}$ **59.** (a) 54.99°, 125.01°

(b) $150° = \dfrac{5\pi}{6}$, $330° = \dfrac{11\pi}{6}$ (b) 195.00°, 345.00°

61. (a) 0.175, 6.109   **63.** (a) 0.873, 4.014
(b) 2.201, 4.083   (b) 1.693, 4.835

65. $\dfrac{4}{5}$ **67.** $-\sqrt{3}$

69. $\sin^2\theta + \cos^2\theta = 1$
$\sin^2 2 + \cos^2 2 = 1$

71. (a) 25.2°F   **73.** (a) 10 mi
(b) 65.1°F   (b) 5.18 mi
(c) 50.8°F   (c) 5 mi

Section 5.5   (*page 395*)

1. Period: π
Amplitude: 2

```
Xmin=0
Xmax=3π
Xscl=π/2
Ymin=-3
Ymax=3
Yscl=1
```

3. Period: 4π
Amplitude: $\frac{3}{2}$

```
Xmin=0
Xmax=4π
Xscl=π
Ymin=-2
Ymax=2
Yscl=1
```

5. Period: 2
Amplitude: $\frac{1}{2}$

```
Xmin=0
Xmax=6
Xscl=1
Ymin=-2
Ymax=2
Yscl=0.5
```

7. Period: $\dfrac{\pi}{5}$; Amplitude: 3   **9.** Period: $\dfrac{1}{2}$; Amplitude: 3

11. *Shift* the graph of f π units to the right to obtain the graph of g.   **13.** *Reflect* the graph of f about the x-axis to obtain the graph of g.   **15.** *Shift* the graph of f two units up to obtain the graph of g.

17.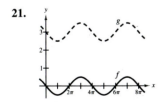

19.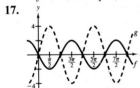

21.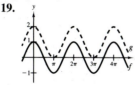

23.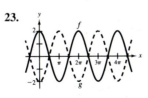

25.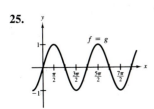

27.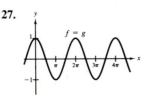

29.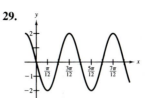

31.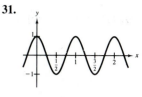

33.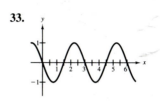

35.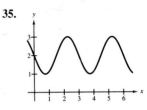

37.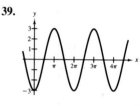

39.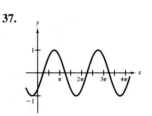

41.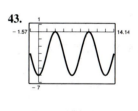

43.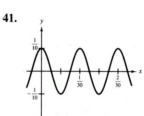

45.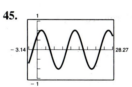

47.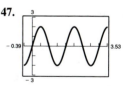

49.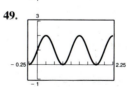

51.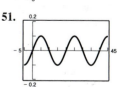

53.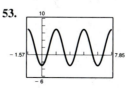

55. $-\dfrac{5\pi}{6}, -\dfrac{\pi}{6}, \dfrac{7\pi}{6}, \dfrac{11\pi}{6}$ **57.** $-\dfrac{7\pi}{4}, -\dfrac{\pi}{4}, \dfrac{\pi}{4}, \dfrac{7\pi}{4}$

59. $f(x) = 2\cos x + 2$ **61.** $y = 2\sin 4x$

63. $y = \cos\left(2x + \dfrac{\pi}{2}\right)$

65.

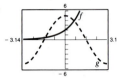

$(-1.480, 0.455), (0.672, 3.914)$

67. (a) 6
  (b) 10 cycles/min
  (c)

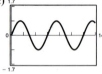

69. (a) $\frac{1}{440}$
  (b) 440
  (c)

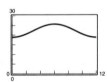

71.

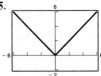

Maximum sales: June
Minimum sales: December

73. $g(x) = 2f(x)$

75. $g(x) = f(x - \pi)$

Section 5.6   (page 407)

WARM-UP

1. $f(x) = -1: \dfrac{3\pi}{2}$

  $f(x) = 0: 0, \pi, 2\pi$

  $f(x) = 1: \dfrac{\pi}{2}$

2. $f(x) = -1: \pi$

  $f(x) = 0: \dfrac{\pi}{2}, \dfrac{3\pi}{2}$

  $f(x) = 1: 0, 2\pi$

3. $f(x) = -1: \dfrac{3\pi}{4}, \dfrac{7\pi}{4}$

  $f(x) = 0: 0, \dfrac{\pi}{2}, \pi, \dfrac{3\pi}{2}, 2\pi$

  $f(x) = 1: \dfrac{\pi}{4}, \dfrac{5\pi}{4}$

4. $f(x) = -1: 2\pi$

  $f(x) = 0: \pi$

5.

6.

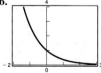

7.

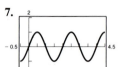

8.

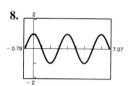

9. $0, \dfrac{\sqrt{3}\pi}{12}, \dfrac{\sqrt{2}\pi}{8}, \dfrac{\pi}{6}, 0$

10. $0, \dfrac{3 + \pi}{6}, \dfrac{2\sqrt{2} + \pi}{4}, \dfrac{3\sqrt{3} + 2\pi}{6}, \dfrac{\pi + 2}{2}$

1. c

```
Xmin=-π/2
Xmax=3π/2
Xscl=π/4
Ymin=-3
Ymax=3
Yscl=1
```

2. g

```
Xmin=-π/3
Xmax=π
Xscl=π/6
Ymin=-2
Ymax=2
Yscl=1
```

3. e

```
Xmin=-2π
Xmax=2π
Xscl=π/2
Ymin=-3
Ymax=3
Yscl=1
```

4. a

```
Xmin=-π
Xmax=5π
Xscl=π
Ymin=-5
Ymax=5
Yscl=1
```

5. d

```
Xmin=-1
Xmax=4
Xscl=1
Ymin=-2
Ymax=2
Yscl=1
```

6. h

```
Xmin=-1
Xmax=3
Xscl=0.5
Ymin=-2
Ymax=2
Yscl=1
```

7. b

```
Xmin=-π
Xmax=2π
Xscl=π/2
Ymin=-3
Ymax=3
Yscl=1
```

8. f

```
Xmin=-0.5
Xmax=2
Xscl=0.25
Ymin=-4
Ymax=4
Yscl=1
```

9.

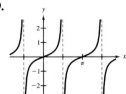

11.

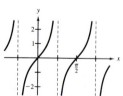

13.

15.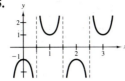

17.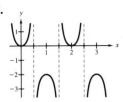

19.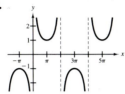

21.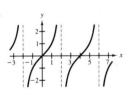

23.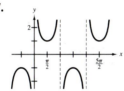

25.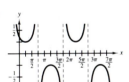

27.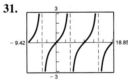

29.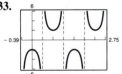

31.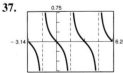

33.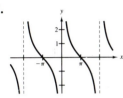

35.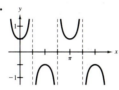

37.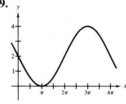

39.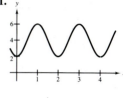

41. $-\dfrac{7\pi}{4}, -\dfrac{3\pi}{4}, \dfrac{\pi}{4}, \dfrac{5\pi}{4}$

43. $-\dfrac{4\pi}{3}, -\dfrac{2\pi}{3}, \dfrac{2\pi}{3}, \dfrac{4\pi}{3}$

45. Even

47. (a)

(b) $(0.524, 2.618)$

(c) As x approaches π, f approaches 0 and g increases without bound. g is the reciprocal of f.

49.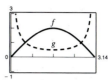

51.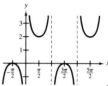

53.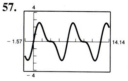

55.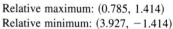

Period: 2π

Relative maximum: $(0.785, 1.414)$

Relative minimum: $(3.927, -1.414)$

57.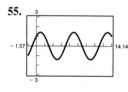

Period: 2π

Relative maximum: $(1.047, 2.598)$

Relative minimum: $(5.236, -2.598)$

59.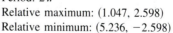

Period: 4π

Relative maxima: $(0, 0), (6.2831, 2)$

Relative minima: $(2.636, -1.125), (9.930, -1.125)$

61.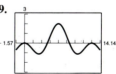

63.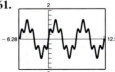

65.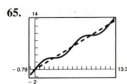

67.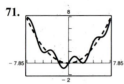

69.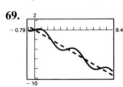

71.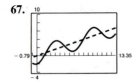

73.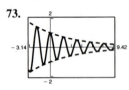

The functional values approach 0 as x increases without bound.

75.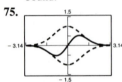

The functional values approach 0 as x increases without bound.

77.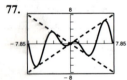

The functional values approach 0 as x approaches 0.

79.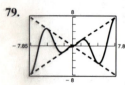

The functional values approach 0 as x approaches 0.

81.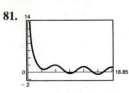

y increases without bound.

83.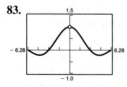

$g(x)$ approaches the value of 1.

85.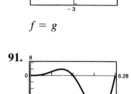

$f(x)$ oscillates between -1 and 1.

87.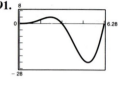

$f = g$

89.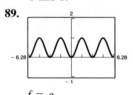

$f = g$

91.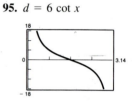

93.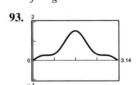

95. $d = 6 \cot x$

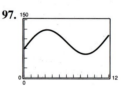

97.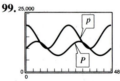

99.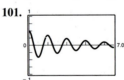

We can explain the cycles of this predator-prey population by noting the cause and effect pattern.

101.

Section 5.7 (page 418)

WARM-UP	**1.** -1	**2.** -1	**3.** -1	**4.** $\dfrac{\sqrt{2}}{2}$
5. 0	**6.** $\dfrac{\pi}{6}$	**7.** π	**8.** $\dfrac{\pi}{4}$	**9.** 0 **10.** $-\dfrac{\pi}{4}$

1. $\dfrac{\pi}{6}$ **3.** $\dfrac{\pi}{3}$ **5.** $\dfrac{\pi}{6}$ **7.** $\dfrac{5\pi}{6}$ **9.** $-\dfrac{\pi}{3}$ **11.** $\dfrac{2\pi}{3}$

13. $\dfrac{\pi}{3}$ **15.** 0 **17.** 1.29 **19.** -0.85 **21.** -1.11

23. 0.32 **25.** 1.99 **27.** 0.74 **29.**

31. 0.3 **33.** -0.1 **35.** 0

37. $\dfrac{3}{5}$ **39.** $\dfrac{\sqrt{5}}{5}$ **41.** $\dfrac{12}{13}$

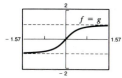

43. $\dfrac{\sqrt{34}}{5}$

45. $y = \pm 1$ **47.** $\dfrac{1}{x}$

49. $\sqrt{1 - 4x^2}$ **51.** $\sqrt{1 - x^2}$ **53.** $\dfrac{\sqrt{9 - x^2}}{x}$

55. $\dfrac{\sqrt{2}}{\sqrt{2 - x^2}}$ **57.** $\dfrac{9}{\sqrt{x^2 + 81}}$ **59.** $\dfrac{|x - 1|}{\sqrt{x^2 - 2x + 10}}$

61. **63.**

65. $f(t) = 3\sqrt{2} \sin\left(2t + \dfrac{\pi}{4}\right)$

67. (a)

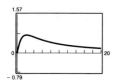

(b) 2.236 ft

(c) $\beta = 0$; As the distance from the picture increases without bound, the angle β approaches 0.

69. (a) 14.5°
   (b) 30.0°

71. $y = \operatorname{arccot} x$ if and only if $\cot y = x$, where $0 < y < \pi$.

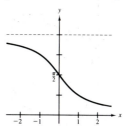

73. $y = \operatorname{arccsc} x$ if and only if $\csc y = x$, where $-\dfrac{\pi}{2} \le y < 0$ or $0 < y \le \dfrac{\pi}{2}$.

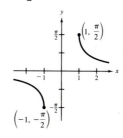

Section 5.8 (page 430)

WARM-UP **1.** 8.45 **2.** 78.99 **3.** 1.06 **4.** 1.24
5. 4.88 **6.** 34.14 **7.** 4, π **8.** $\frac{1}{2}$, 2 **9.** 3, $\frac{2}{3}$
10. 0.2, 8π

1. $a \approx 3.64$ **3.** $a \approx 8.26$
  $c \approx 10.64$ $c \approx 25.38$
  $B = 70°$ $A = 19°$
5. $a \approx 91.34$ **7.** $c \approx 11.66$
  $b \approx 420.70$ $A \approx 30.96°$
  $B = 77° \, 45'$ $B \approx 59.04°$
9. $a \approx 49.48$
  $A \approx 72.08°$
  $B \approx 17.92°$

11. 2.56 in.   **13.** 121.2 ft   **15.** 15.4 ft   **17.** 56.3°
19. 12.68°   **21.** 5099 ft   **23.** 19.9 ft
25. 508 mi north; 650 mi east   **27.** N 56.3° W
29. (a) N 58° E   **(b)** 68.8 yd   **31.** 1657 ft
33. 17,054 ft $\approx$ 3.23 mi   **35.** 29.389 in.
37. $y = \sqrt{3}r$ **39.** $a \approx 7$, $c \approx 12.2$
41. (a) 4   **43. (a)** $\frac{1}{16}$
   (b) 4   **(b)** 60
   (c) $\frac{1}{16}$ **(c)** $\frac{1}{120}$
45. $\omega = 528\pi$

Chapter 5 Review Exercises   (*page 435*)

1. $\dfrac{3\pi}{4}$, $-\dfrac{5\pi}{4}$

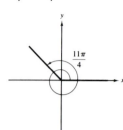

3. $250°$, $-470°$

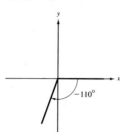

5. $135.28°$ **7.** $5.38°$ **9.** $135°\ 16'\ 12''$ **11.** $-85°\ 9'$

13. $128.57°$ **15.** $-200.54°$ **17.** 8.3776

19. -0.5890 **21.** $72°$ **23.** $\dfrac{\pi}{5}$

25. $\sin\theta = \dfrac{4}{5}$

$\cos\theta = \dfrac{3}{5}$

$\tan\theta = \dfrac{4}{3}$

$\csc\theta = \dfrac{5}{4}$

$\sec\theta = \dfrac{5}{3}$

$\cot\theta = \dfrac{3}{4}$

27. $\sin\theta = \dfrac{2\sqrt{53}}{53}$

$\cos\theta = -\dfrac{7\sqrt{53}}{53}$

$\tan\theta = -\dfrac{2}{7}$

$\csc\theta = \dfrac{\sqrt{53}}{2}$

$\sec\theta = -\dfrac{\sqrt{53}}{7}$

$\cot\theta = -\dfrac{7}{2}$

29. $\sin\theta = -\dfrac{3\sqrt{13}}{13}$

$\cos\theta = -\dfrac{2\sqrt{13}}{13}$

$\tan\theta = \dfrac{3}{2}$

$\csc\theta = -\dfrac{\sqrt{13}}{3}$

$\sec\theta = -\dfrac{\sqrt{13}}{2}$

$\cot\theta = \dfrac{2}{3}$

31. $\sin\theta = -\dfrac{\sqrt{11}}{6}$

$\cos\theta = \dfrac{5}{6}$

$\tan\theta = -\dfrac{\sqrt{11}}{5}$

$\csc\theta = -\dfrac{6\sqrt{11}}{11}$

$\cot\theta = -\dfrac{5\sqrt{11}}{11}$

33. $\cos\theta = -\dfrac{\sqrt{55}}{8}$

$\tan\theta = -\dfrac{3\sqrt{55}}{55}$

$\csc\theta = \dfrac{8}{3}$

$\sec\theta = -\dfrac{8\sqrt{55}}{55}$

$\cot\theta = -\dfrac{\sqrt{55}}{3}$

35. $\sqrt{3}$ **37.** $-\dfrac{\sqrt{3}}{2}$ **39.** $-\dfrac{\sqrt{2}}{2}$

41. 0.65 **43.** 3.24 **45.** $135° = \dfrac{3\pi}{4}$, $225° = \dfrac{5\pi}{4}$

47. $210° = \dfrac{7\pi}{6}$, $330° = \dfrac{11\pi}{6}$

49. $57° \approx 0.9948$, $123° \approx 2.1467$

51. $165° \approx 2.8798$, $195° \approx 3.4034$

53. $\dfrac{\sqrt{-x^2 + 2x}}{-x^2 + 2x}$ **55.** $\dfrac{2\sqrt{4 - 2x^2}}{4 - x^2}$

57.

59.

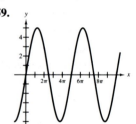

61.

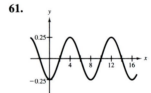

63.

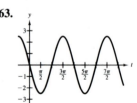

65.

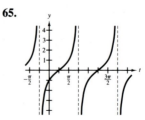

67.

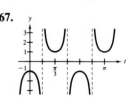

69.

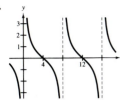

71.

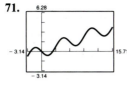

73.

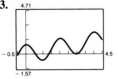

75.

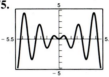

77.

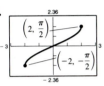

$\left(2, \dfrac{\pi}{2}\right)$ $\left(-2, -\dfrac{\pi}{2}\right)$

79.

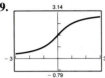

81.

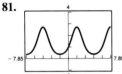

Periodic
Relative maximum: (1.571, 2.718)
Relative minimum: (4.712, 0.368)

83.

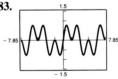

Periodic
Relative maxima: (0.615, 0.770), (2.526, 0.770), (4.712, 0)
Relative minima: (1.571, 0), (3.757, −0.770),
(5.668, −0.770)

85. 7.66 m   **87.** 1.33 mi   **89.** 268.8 ft

91. (a)

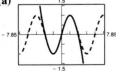

(b)

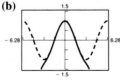

(c) $\sin x = x - \dfrac{x^3}{3!} + \dfrac{x^5}{5!} - \dfrac{x^7}{7!} + \dfrac{x^9}{9!}$

$\cos x = 1 - \dfrac{x^2}{2!} + \dfrac{x^4}{4!} - \dfrac{x^6}{6!} + \dfrac{x^8}{8!}$

The accuracy increases with additional terms.

CUMULATIVE TEST for Chapters 3–5 (*page 438*)

1. Vertex: $\left(\frac{3}{2}, 2\right)$ **2.** $f(x) = -\frac{1}{4}x^2 + 6$

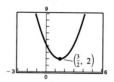

$\left(\frac{3}{2}, 2\right)$

3. Upward to the left;
Downward to the right
4. $f(x) = 2x^3 + 3x^2 - 18x + 8$

5.

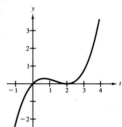

6. $3x - 2 - \dfrac{3x - 2}{2x^2 + 1}$ **7.** $3x^2 + 6x + 7 + \dfrac{18}{x - 2}$

8. $-\frac{4}{3}, \frac{3}{2}, 4$ **9.** 2.4

10. (a)

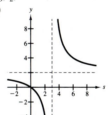

(b)

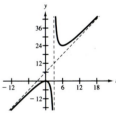

11. (a)

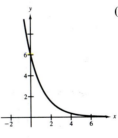

(b)

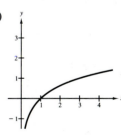

12. 3   **13.** $\ln\left(\dfrac{x^2}{\sqrt{x + 5}}\right)$ **14. (a)** 1.24
     (b) 12.80

15. $16,302.05

16. $80°$

17. $-\dfrac{2\pi}{3}$

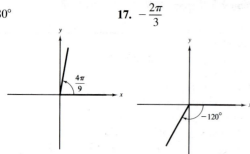

18. $\sin\theta = \frac{5}{13}$ $\csc\theta = \frac{13}{5}$
$\cos\theta = \frac{12}{13}$ $\sec\theta = \frac{13}{12}$
$\tan\theta = \frac{5}{12}$ $\cot\theta = \frac{12}{5}$

19. $\sin t = \dfrac{\sqrt{5}}{3}$, $\tan t = -\dfrac{\sqrt{5}}{2}$

20. -0.9490 **21.** $61.93°,\ 298.07°$

22. (a)   **(b)**

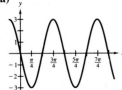

(c)   **(d)**

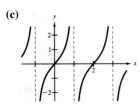

23. (a) The graph of g is 10 units above the graph of f.
(b) The period of f is 2π and the period of g is 4.
(c) The graph of g is $\pi/4$ units to the left of the graph of f.
(d) The graph of g is a reflection in the x-axis of the graph of f.

24. $y = 3\sin\!\left(2x - \dfrac{\pi}{2}\right)$ **25. (a)**

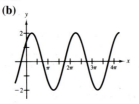

(b) 2π
(c) $x \approx 1.9$
(d) $f(2.7) \approx 2.8$

26. (a)

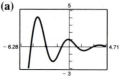

(b) No
(c) $t \approx 0.75$

27. (a) $\dfrac{\pi}{6}$ **(b)** $\dfrac{\pi}{3}$ **28.** $\sqrt{1 - 4x^2}$

29. 382 ft

CHAPTER 6
Section 6.1   (page 447)

WARM-UP

1. $\sin\theta = \dfrac{3\sqrt{13}}{13}$
$\cos\theta = \dfrac{2\sqrt{13}}{13}$
$\csc\theta = \dfrac{\sqrt{13}}{3}$
$\sec\theta = \dfrac{\sqrt{13}}{2}$
$\cot\theta = \dfrac{2}{3}$

2. $\sin\theta = \dfrac{2\sqrt{2}}{3}$
$\cos\theta = \dfrac{1}{3}$
$\tan\theta = 2\sqrt{2}$
$\csc\theta = \dfrac{3\sqrt{2}}{4}$
$\cot\theta = \dfrac{\sqrt{2}}{4}$

3. $\sin\theta = -\dfrac{3\sqrt{58}}{58}$
$\cos\theta = \dfrac{7\sqrt{58}}{58}$
$\tan\theta = -\dfrac{3}{7}$
$\csc\theta = -\dfrac{\sqrt{58}}{3}$
$\sec\theta = \dfrac{\sqrt{58}}{7}$
$\cot\theta = -\dfrac{7}{3}$

4. $\sin\theta = \dfrac{\sqrt{5}}{5}$
$\cos\theta = -\dfrac{2\sqrt{5}}{5}$
$\tan\theta = -\dfrac{1}{2}$
$\csc\theta = \sqrt{5}$
$\sec\theta = -\dfrac{\sqrt{5}}{2}$
$\cot\theta = -2$

5. $\dfrac{1}{2}$ **6.** $\dfrac{5}{4}$ **7.** $\dfrac{\sqrt{73}}{8}$ **8.** $\dfrac{2}{3}$ **9.** $\dfrac{x^2 + x + 16}{4(x + 1)}$

10. $\dfrac{8x - 2}{1 - x^2}$

1. $\tan x = \dfrac{\sqrt{3}}{3}$

$\csc x = 2$

$\sec x = \dfrac{2\sqrt{3}}{3}$

$\cot x = \sqrt{3}$

3. $\cos \theta = \dfrac{\sqrt{2}}{2}$

$\tan \theta = -1$

$\csc \theta = -\sqrt{2}$

$\cot \theta = -1$

5. $\sin x = \dfrac{2}{3}$

$\cos x = -\dfrac{\sqrt{5}}{3}$

$\csc x = \dfrac{3}{2}$

$\sec x = -\dfrac{3\sqrt{5}}{5}$

$\cot x = -\dfrac{\sqrt{5}}{2}$

7. $\sin \theta = -\dfrac{2\sqrt{5}}{5}$

$\cos \theta = -\dfrac{\sqrt{5}}{5}$

$\csc \theta = -\dfrac{\sqrt{5}}{2}$

$\sec \theta = -\sqrt{5}$

$\cot \theta = \dfrac{1}{2}$

9. $\cos \theta = 0$; $\tan \theta$ is undefined. $\csc \theta = -1$; $\sec \theta$ is undefined.

11. d   **12.** e   **13.** a   **14.** f   **15.** b   **16.** c

17. b   **18.** c   **19.** e   **20.** a   **21.** f   **22.** d

23. $\sec \phi$ **25.** $\sin \beta$ **27.** $\cos x$ **29.** 1

31. $-\tan x$ **33.** $\tan x$ **35.** $1 + \sin y$ **37.** $\sin^2 x$

39. $\sin^2 x \tan^2 x$ **41.** $\sec^4 x$ **43.** $\sin^2 x - \cos^2 x$

45. $1 + 2 \sin x \cos x$ **47.** $\tan^2 x$ **49.** $2 \csc^2 x$

51. $2 \sec x$ **53.** $1 + \cos y$ **55.** $3(\sec x + \tan x)$

57. Identity   **59.** Not an identity

61. (a) $0 \leq x \leq \dfrac{\pi}{2}, \dfrac{3\pi}{2} \leq x \leq 2\pi$ (b) $\dfrac{\pi}{2} \leq x \leq \dfrac{3\pi}{2}$

63. $5 \cos \theta$ **65.** $3 \tan \theta$ **67.** $5 \sec \theta$ **69.** $\cos \theta$

71. $27 \sec^3 \theta$ **73.** $0 \leq \theta < \dfrac{\pi}{2}, \dfrac{3\pi}{2} < \theta \leq 2\pi$

75. $\ln |\cot \theta|$ **77.** False, $\dfrac{\sin k\theta}{\cos k\theta} = \tan k\theta$ **79.** True

81. (a) $\csc^2 132° - \cot^2 132° \approx 1.8107 - 0.8107 = 1$

(b) $\csc^2 \dfrac{2\pi}{7} - \cot^2 \dfrac{2\pi}{7} \approx 1.6360 - 0.6360 = 1$

83. (a) $\cos(90° - 80°) = \sin 80° \approx 0.9848$

(b) $\cos\left(\dfrac{\pi}{2} - 0.8\right) = \sin 0.8 \approx 0.7174$

85. $\cos \theta = \pm\sqrt{1 - \sin^2 \theta}$

$\tan \theta = \pm \dfrac{\sin \theta}{\sqrt{1 - \sin^2 \theta}}$

$\csc \theta = \dfrac{1}{\sin \theta}$

$\sec \theta = \pm \dfrac{1}{\sqrt{1 - \sin^2 \theta}}$

$\cot \theta = \pm \dfrac{\sqrt{1 - \sin^2 \theta}}{\sin \theta}$

Section 6.2   (*page 455*)

WARM-UP

1. (a) $x^2(1 - y^2)$ **2.** (a) $x^2(1 + y^2)$

(b) $\sin^4 x$ (b) 1

3. (a) $(x^2 + 1)(x^2 - 1)$

(b) $\sec^2 x(\tan^2 x - 1)$

4. (a) $(z + 1)(z^2 - z + 1)$

(b) $(\tan x + 1)(\tan^2 x - \tan x + 1)$

5. (a) $(x - 1)(x^2 + 1)$ **6.** (a) $(x^2 - 1)^2$

(b) $(\cot x - 1) \csc^2 x$ (b) $\cos^4 x$

7. (a) $\dfrac{y^2 - x^2}{x}$ **8.** (a) $\dfrac{x^2 - 1}{x^2}$

(b) $\tan x$ (b) $\sin^2 x$

9. (a) $\dfrac{y^2 + (1 + z)^2}{y(1 + z)}$ **10.** (a) $\dfrac{y(1 + y) - z^2}{z(1 + y)}$

(b) $2 \csc x$ (b) $\dfrac{\tan x - 1}{\sec x(1 + \tan x)}$

Exercises 1–50 are proofs.

51. $\sin \theta = \pm\sqrt{1 - \cos^2 \theta}; \dfrac{7\pi}{4}$

53. $\sqrt{\tan^2 x} = |\tan x|; \dfrac{3\pi}{4}$ **55.** $\mu = \tan \theta$

Section 6.3   (*page 466*)

WARM-UP

1. $\dfrac{2\pi}{3}, \dfrac{4\pi}{3}$ **2.** $\dfrac{\pi}{3}, \dfrac{2\pi}{3}$ **3.** $\dfrac{\pi}{4}, \dfrac{7\pi}{4}$ **4.** $\dfrac{7\pi}{4}, \dfrac{5\pi}{4}$

5. $\dfrac{\pi}{3}, \dfrac{4\pi}{3}$ **6.** $\dfrac{3\pi}{4}, \dfrac{7\pi}{4}$ **7.** $\dfrac{15}{8}$ **8.** $-3, \dfrac{5}{2}$

9. $\dfrac{2 \pm \sqrt{14}}{2}$ **10.** $-1, 3$

Exercises 1–6 are proofs.

7. $\dfrac{2\pi}{3} + 2n\pi, \dfrac{4\pi}{3} + 2n\pi$ **9.** $\dfrac{\pi}{3} + 2n\pi, \dfrac{2\pi}{3} + 2n\pi$

11. $\dfrac{\pi}{4} + \dfrac{n\pi}{2}$ **13.** $\dfrac{\pi}{6} + n\pi, \dfrac{5\pi}{6} + n\pi$ **15.** $n\pi, \dfrac{\pi}{4} + n\pi$

17. $n\pi, \dfrac{3\pi}{2} + 2n\pi$ **19.** $\dfrac{\pi}{3} + n\pi, \dfrac{2\pi}{3} + n\pi$

21. $\dfrac{\pi}{3}, \dfrac{5\pi}{3}$ **23.** $\dfrac{7\pi}{6}, \dfrac{3\pi}{2}, \dfrac{11\pi}{6}$ **25.** $\dfrac{\pi}{6}, \dfrac{5\pi}{6}, \dfrac{7\pi}{6}, \dfrac{11\pi}{6}$

27. No solution   **29.** $\dfrac{2\pi}{3}, \dfrac{5\pi}{6}, \dfrac{5\pi}{3}, \dfrac{11\pi}{6}$

31. $\dfrac{\pi}{2}$ **33.** $\dfrac{2}{3}, \dfrac{3}{2}$; 0.8411, 5.4421

35. $4 \pm \sqrt{3}$; 1.1555, 1.3981, 4.2971, 4.5397

37. 1.107, 4.249   **39.** 1.0472, 5.2360

41. 0, 1.895   **43.** 0, 2.678, 3.142, 5.820, 6.283

45. 0.9828, 1.7682, 4.1244, 4.9098

47. 0.3398, 0.8481, 2.2935, 2.8018   **49.** 0.4271, 2.7145

51. 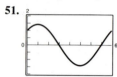   **53.** 0.7391, approximate the zeros of $f(x) = \cos x - x$.

Maximum: (0.785, 1.414)
Minimum: (3.927, −1.414)

55.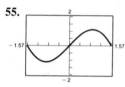

1.122 square units

57. 0.04, 0.43, 0.83   **59.** 37°, 53°   **61.** 1   **63.** $\sqrt{2}$

Section 6.4   (page 474)

WARM-UP

1. $\dfrac{\sqrt{10}}{10}$ **2.** $\dfrac{-5\sqrt{34}}{34}$ **3.** $-\dfrac{\sqrt{7}}{4}$ **4.** $\dfrac{2\sqrt{2}}{3}$

5. $\dfrac{\pi}{4}, \dfrac{3\pi}{4}$ **6.** $\dfrac{\pi}{2}, \dfrac{3\pi}{2}$ **7.** $\tan^3 x$ **8.** $\cot^2 x$

9. $\sec x$ **10.** $1 - \tan^2 x$

1. $\sin 75° = \dfrac{\sqrt{2}}{4}(1 + \sqrt{3})$

$\cos 75° = \dfrac{\sqrt{2}}{4}(\sqrt{3} - 1)$

$\tan 75° = \sqrt{3} + 2$

3. $\sin 105° = \dfrac{\sqrt{2}}{4}(\sqrt{3} + 1)$

$\cos 105° = \dfrac{\sqrt{2}}{4}(1 - \sqrt{3})$

$\tan 105° = -2 - \sqrt{3}$

5. $\sin 195° = \dfrac{\sqrt{2}}{4}(1 - \sqrt{3})$

$\cos 195° = -\dfrac{\sqrt{2}}{4}(\sqrt{3} + 1)$

$\tan 195° = 2 - \sqrt{3}$

7. $\sin \dfrac{11\pi}{12} = \dfrac{\sqrt{2}}{4}(\sqrt{3} - 1)$

$\cos \dfrac{11\pi}{12} = -\dfrac{\sqrt{2}}{4}(\sqrt{3} + 1)$

$\tan \dfrac{11\pi}{12} = -2 + \sqrt{3}$

9. $\sin \dfrac{17\pi}{12} = -\dfrac{\sqrt{2}}{4}(\sqrt{3} + 1)$ **11.** $\cos 40°$

$\cos \dfrac{17\pi}{12} = \dfrac{\sqrt{2}}{4}(1 - \sqrt{3})$

$\tan \dfrac{17\pi}{12} = 2 + \sqrt{3}$

13. $\sin 200°$ **15.** $\tan 239°$ **17.** $\sin 1.8$ **19.** $\tan 3x$

21. $\frac{33}{65}$ **23.** $-\frac{56}{65}$ **25.** $-\frac{3}{5}$ **27.** $\frac{44}{125}$

Exercises 29–46 are proofs.

47. (a) $\sqrt{2} \sin\left(\theta + \dfrac{\pi}{4}\right)$ **49.** (a) $13 \sin(3\theta + 0.3948)$
          (b) $13 \cos(3\theta - 1.1760)$

(b) $\sqrt{2} \cos\left(\theta - \dfrac{\pi}{4}\right)$

51. $\sqrt{2} \sin \theta + \sqrt{2} \cos \theta$ **53.** 1

55. $\dfrac{\pi}{2}$ **57.** $\dfrac{\pi}{4}, \dfrac{7\pi}{4}$ **59.** 3.927, 5.498

Section 6.5   (page 485)

WARM-UP   **1.** $\sin x(2 + \cos x)$

2. $(\cos x - 2)(\cos x + 1)$ **3.** $0, \dfrac{\pi}{2}, \pi, \dfrac{3\pi}{2}$

4. $\dfrac{\pi}{4}, \dfrac{3\pi}{4}, \dfrac{5\pi}{4}, \dfrac{7\pi}{4}$ **5.** π **6.** 0   **7.** $\dfrac{2 - \sqrt{2}}{\ell}$

8. $\frac{3}{4}$ **9.** $\tan 3x$ **10.** $\cos x(1 - 4 \sin^2 x)$

1.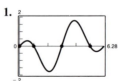

$0, \dfrac{\pi}{3}, \pi, \dfrac{5\pi}{3}$

3.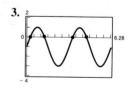

$\dfrac{\pi}{12}, \dfrac{5\pi}{12}, \dfrac{13\pi}{12}, \dfrac{17\pi}{12}$

5.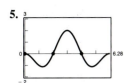

$$0, \frac{2\pi}{3}, \frac{4\pi}{3}$$

7.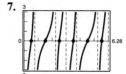

$$\frac{\pi}{2}, \frac{\pi}{6}, \frac{5\pi}{6}, \frac{7\pi}{6}, \frac{3\pi}{2}, \frac{11\pi}{6}$$

9.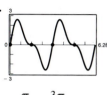

$$0, \frac{\pi}{2}, \pi, \frac{3\pi}{2}$$

11. $f(x) = 3 \sin 2x$

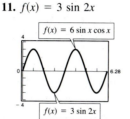

Relative maxima: $(0.785, 3)$, $(3.927, 3)$
Relative minima: $(2.356, -3)$, $(5.498, -3)$

13. $g(x) = 4 \cos 2x$

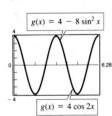

Relative maxima: $(0, 4)$, $(3.142, 4)$
Relative minima: $(1.571, -4)$, $(4.712, -4)$

15. $\sin 2u = \frac{24}{25}$
$\cos 2u = -\frac{7}{25}$
$\tan 2u = -\frac{24}{7}$

17. $\sin 2u = \frac{4}{5}$
$\cos 2u = \frac{3}{5}$
$\tan 2u = \frac{4}{3}$

19. $\sin 2u = -\frac{4\sqrt{21}}{25}$
$\cos 2u = -\frac{17}{25}$
$\tan 2u = \frac{4\sqrt{21}}{17}$

21. $\frac{1}{8}(3 + 4 \cos 2x + \cos 4x)$ **23.** $\frac{1}{8}(1 - \cos 4x)$
25. $\frac{1}{16}(1 + \cos 2x)(1 - \cos 4x)$
27. $\sin 105° = \frac{1}{2}\sqrt{2 + \sqrt{3}}$
$\cos 105° = -\frac{1}{2}\sqrt{2 - \sqrt{3}}$
$\tan 105° = -2 - \sqrt{3}$
29. $\sin 112° \, 30' = \frac{1}{2}\sqrt{2 + \sqrt{2}}$
$\cos 112° \, 30' = -\frac{1}{2}\sqrt{2 - \sqrt{2}}$
$\tan 112° \, 30' = -1 - \sqrt{2}$

31. $\sin \frac{\pi}{8} = \frac{1}{2}\sqrt{2 - \sqrt{2}}$
$\cos \frac{\pi}{8} = \frac{1}{2}\sqrt{2 + \sqrt{2}}$
$\tan \frac{\pi}{8} = \sqrt{2} - 1$

33. $\sin \frac{u}{2} = \frac{5\sqrt{26}}{26}$
$\cos \frac{u}{2} = \frac{\sqrt{26}}{26}$
$\tan \frac{u}{2} = 5$

35. $\sin \frac{u}{2} = \sqrt{\frac{89 - 8\sqrt{89}}{178}}$
$\cos \frac{u}{2} = -\sqrt{\frac{89 + 8\sqrt{89}}{178}}$
$\tan \frac{u}{2} = \frac{8 - \sqrt{89}}{5}$

37. $\sin \frac{u}{2} = \frac{3\sqrt{10}}{10}$
$\cos \frac{u}{2} = -\frac{\sqrt{10}}{10}$
$\tan \frac{u}{2} = -3$

39. $|\sin 3x|$ **41.** $-|\tan 4x|$
43. π

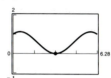

45. $\frac{\pi}{3}, \pi, \frac{5\pi}{3}$

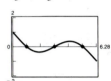

47. $3\left(\sin \frac{\pi}{2} + \sin 0\right)$ **49.** $\frac{1}{2}(\sin 8\theta + \sin 2\theta)$
51. $\frac{5}{2}(\cos 8\beta + \cos 2\beta)$ **53.** $\frac{1}{2}(\cos 2y - \cos 2x)$
55. $\frac{1}{2}(\sin 2\theta + \sin 2\pi)$ **57.** $2 \sin 45° \cos 15°$
59. $-2 \sin \frac{\pi}{2} \sin \frac{\pi}{4}$ **61.** $2 \cos 4x \cos 2x$
63. $2 \cos \alpha \sin \beta$ **65.** $2 \cos(\phi + \pi) \cos \pi$
67. $0, \frac{\pi}{4}, \frac{\pi}{2}, \frac{3\pi}{4}, \pi, \frac{5\pi}{4}, \frac{3\pi}{2}, \frac{7\pi}{4}$ **69.** $\frac{\pi}{6}, \frac{5\pi}{6}$

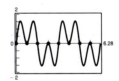 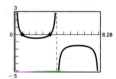

Exercises 71–87 are proofs.

89.

91. $2x\sqrt{1 - x^2}$ **93.** Proof
95. (a) $A = 100 \sin \frac{\theta}{2} \cos \frac{\theta}{2}$
(b) $A = 50 \sin \theta$; The area is maximum when $\theta = \frac{\pi}{2}$.

Section 6.6 (page 496)

WARM-UP   **1.** $b = 3\sqrt{3}$, $A = 30°$, $B = 60°$
2. $c = 5\sqrt{2}$, $A = 45°$, $B = 45°$
3. $a = 8$, $A \approx 28.07°$, $B \approx 61.93°$
4. $b \approx 8.33$, $c \approx 11.21$, $B = 48°$
5. $a \approx 22.69$, $c \approx 23.04$, $A = 80°$
6. $a \approx 45.73$, $b \approx 142.86$, $A = 17° \ 45'$
7. 8.48   **8.** 12.94   **9.** 2.25   **10.** 91.06

1. $C = 105°$, $b \approx 14.14$, $c \approx 19.32$
3. $C = 110°$, $b \approx 22.44$, $c \approx 24.35$
5. $B \approx 21.55°$, $C \approx 122.45°$, $c \approx 11.49$
7. $B = 10°$, $b \approx 69.46$, $c \approx 136.81$
9. $B = 42° \ 4'$, $a \approx 22.05$, $b \approx 14.88$
11. $A \approx 10° \ 11'$, $C \approx 154° \ 19'$, $c \approx 11.03$
13. $A \approx 25.57°$, $B \approx 9.43°$, $a \approx 10.5$
15. $B \approx 18° \ 13'$, $C \approx 51° \ 32'$, $c \approx 40.06$
17. No solution
19. Two solutions: $B \approx 70.4°$, $C \approx 51.6°$, $c \approx 4.16$;
    $B \approx 109.6°$, $C \approx 12.4°$, $c \approx 1.14$
21. No solution
23. (a) $b \le 5$, $b = \dfrac{5}{\sin 36°}$
    (b) $5 < b < \dfrac{5}{\sin 36°}$
    (c) $b > \dfrac{5}{\sin 36°}$
25. 10.4   **27.** 1675.2   **29.** 474.9   **31.** 6 ft
33. 77 yd   **35.** 5 mi   **37.** 26.1 mi, 15.9 mi
39. 4.55 mi   **41.** No solution

Section 6.7 (page 504)

WARM-UP   **1.** $2\sqrt{13}$ **2.** $3\sqrt{5}$ **3.** $4\sqrt{10}$
4. $3\sqrt{13}$ **5.** 20   **6.** 48
7. $a \approx 4.62$, $c \approx 26.20$, $B = 70°$
8. $a \approx 34.20$, $b \approx 93.97$, $B = 70°$ **9.** No solution
10. $a \approx 15.09$, $B \approx 18.97°$, $C \approx 131.03°$

1. $A \approx 27.7°$, $B \approx 40.5°$, $C \approx 111.8°$
3. $B \approx 23.8°$, $C \approx 126.2°$, $a \approx 12.4$
5. $A \approx 36.9°$, $B \approx 53.1°$, $C \approx 90°$
7. $A \approx 92.9°$, $B \approx 43.55°$, $C \approx 43.55°$
9. $a \approx 11.79$, $B \approx 12.7°$, $C \approx 47.3°$
11. $A \approx 158° \ 36'$, $C \approx 12° \ 38'$, $b \approx 10.4$
13. $A = 27° \ 10'$, $B = 27° \ 10'$, $c \approx 56.9$

	a	b	c	d	θ	ϕ
15.	4	6	9.67	3.23	30°	150°
17.	10	14	20	13.86	68.2°	111.8°
19.	10	11.57	18	12	67.1°	112.9°

21. 16.25   **23.** 54   **25.** 96.82   **27.** 97,979.6 ft^2
29. N 52° 37′ E, S 64° 40′ E   **31.** 43.3 mi
33. 1344 ft   **35.** 114.95°
37. $\overline{PQ} \approx 9.4$ ft, $\overline{QS} \approx 5.0$ ft, $\overline{RS} \approx 12.8$ ft
39. (a) N 58.3° W   **41.** (a) 63.7 ft
    (b) S 81.6° W   (b) 47.6 ft
43. 3.26 ft
45. (a) 570.60   (b) 5910.68   (c) 177.09

Section 6.8 (page 517)

WARM-UP   **1.** $3\sqrt[3]{2}$ **2.** $2\sqrt{2}$
3. $5\sqrt{2}(\cos 135° + i \sin 135°)$
4. $3(\cos 270° + i \sin 270°)$
5. $12(\cos 180° + i \sin 180°)$ **6.** $12(\cos 0° + i \sin 0°)$
7. $\cos \dfrac{3\pi}{4} + i \sin \dfrac{3\pi}{4}$ **8.** $\cos \dfrac{11\pi}{12} + i \sin \dfrac{11\pi}{12}$
9. $2\left(\cos \dfrac{\pi}{2} + i \sin \dfrac{\pi}{2}\right)$ **10.** $\dfrac{2}{3}(\cos 45° + i \sin 45°)$

1. 5

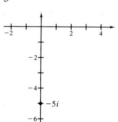

3. $4\sqrt{2}$

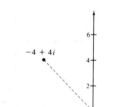

$-4 + 4i$

5. $\sqrt{85}$

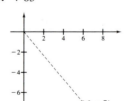

$6 - 7i$

7. $4\left(\cos \dfrac{\pi}{2} + i \sin \dfrac{\pi}{2}\right)$
9. $3\sqrt{2}\left(\cos \dfrac{5\pi}{4} + i \sin \dfrac{5\pi}{4}\right)$

11. $3\sqrt{2}\left(\cos\dfrac{7\pi}{4} + i\sin\dfrac{7\pi}{4}\right)$ **13.** $2\left(\cos\dfrac{\pi}{6} + i\sin\dfrac{\pi}{6}\right)$

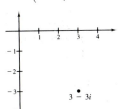

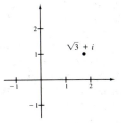

15. $4\left(\cos\dfrac{4\pi}{3} + i\sin\dfrac{4\pi}{3}\right)$ **17.** $6\left(\cos\dfrac{\pi}{2} + i\sin\dfrac{\pi}{2}\right)$

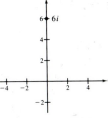

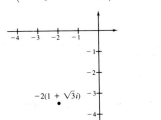

19. $\sqrt{65}(\cos 2.62 + i\sin 2.62)$ **21.** $7(\cos 0 + i\sin 0)$

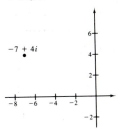

23. $\sqrt{37}(\cos 1.41 + i\sin 1.41)$

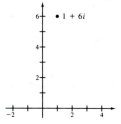

25. $\sqrt{10}(\cos 3.46 + i\sin 3.46)$

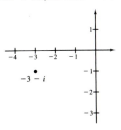

27. $-\sqrt{3} + i$ **29.** $\dfrac{3}{4} - \dfrac{3\sqrt{3}}{4}i$

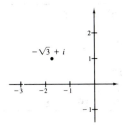

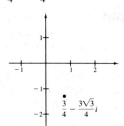

31. $-\dfrac{15\sqrt{2}}{8} + \dfrac{15\sqrt{2}}{8}i$ **33.** $-4i$

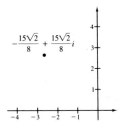

35. $2.8408 + 0.9643i$

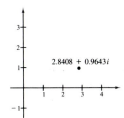

37. $12\left(\cos\dfrac{\pi}{2} + i\sin\dfrac{\pi}{2}\right)$

39. $\dfrac{10}{9}(\cos 200° + i\sin 200°)$

41. $0.27(\cos 150° + i\sin 150°)$

43. $\frac{1}{2}(\cos 80° + i\sin 80°)$

45. $\cos\dfrac{2\pi}{3} + i\sin\dfrac{2\pi}{3}$

47. $4[\cos(-58°) + i\sin(-58°)]$

49. (a) $2\sqrt{2}(\cos 45° + i\sin 45°)$,
$\sqrt{2}[\cos(-45°) + i\sin(-45°)]$
(b) $4(\cos 0° + i\sin 0°) = 4$
(c) 4

51. (a) $2[\cos(-90°) + i\sin(-90°)]$,
$\sqrt{2}(\cos 45° + i\sin 45°)$
(b) $2\sqrt{2}[\cos(-45°) + i\sin(-45°)] = 2 - 2i$
(c) $-2i - 2i^2 = -2i + 2 = 2 - 2i$

53. (a) $5(\cos 0° + i\sin 0°)$, $\sqrt{13}(\cos 56.31° + i\sin 56.31°)$
(b) $\dfrac{5}{\sqrt{13}}[\cos(-56.31°) + i\sin(-56.31°)]$
$\approx 0.7692 - 1.154i$
(c) $\frac{10}{13} - \frac{15}{13}i \approx 0.7692 - 1.154i$

55. Proof

57. (a) r^2
  (b) $\cos 2\theta + i \sin 2\theta$

59.

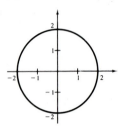

61. $-4 - 4i$ **63.** $-32i$ **65.** $-128\sqrt{3} - 128i$

67. $\dfrac{125}{2} + \dfrac{125\sqrt{3}}{2}i$ **69.** i **71.** $608.02 + 144.69i$

73. (a) $3(\cos 60° + i \sin 60°)$
  $3(\cos 240° + i \sin 240°)$

  (b)

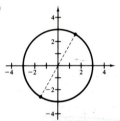

  (c) $\dfrac{3}{2} + \dfrac{3\sqrt{3}}{2}i, \ -\dfrac{3}{2} - \dfrac{3\sqrt{3}}{2}i$

75. (a) $2\left(\cos \dfrac{\pi}{3} + i \sin \dfrac{\pi}{3}\right)$
  $2\left(\cos \dfrac{5\pi}{6} + i \sin \dfrac{5\pi}{6}\right)$
  $2\left(\cos \dfrac{4\pi}{3} + i \sin \dfrac{4\pi}{3}\right)$
  $2\left(\cos \dfrac{11\pi}{6} + i \sin \dfrac{11\pi}{6}\right)$

  (b)

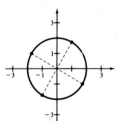

  (c) $1 + \sqrt{3}i, \ -\sqrt{3} + i, \ -1 - \sqrt{3}i, \ \sqrt{3} - i$

77. (a) $5\left(\cos \dfrac{3\pi}{4} + i \sin \dfrac{3\pi}{4}\right)$
  $5\left(\cos \dfrac{7\pi}{4} + i \sin \dfrac{7\pi}{4}\right)$

  (b)

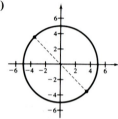

  (c) $-\dfrac{5\sqrt{2}}{2} + \dfrac{5\sqrt{2}}{2}i, \ \dfrac{5\sqrt{2}}{2} - \dfrac{5\sqrt{2}}{2}i$

79. (a) $5\left(\cos \dfrac{4\pi}{9} + i \sin \dfrac{4\pi}{9}\right)$
  $5\left(\cos \dfrac{10\pi}{9} + i \sin \dfrac{10\pi}{9}\right)$
  $5\left(\cos \dfrac{16\pi}{9} + i \sin \dfrac{16\pi}{9}\right)$

  (b)

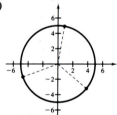

  (c) $0.8682 + 4.924i, \ -4.698 - 1.710i, \ 3.830 - 3.214i$

81. (a) $2(\cos 0 + i \sin 0)$
  $2\left(\cos \dfrac{2\pi}{3} + i \sin \dfrac{2\pi}{3}\right)$
  $2\left(\cos \dfrac{4\pi}{3} + i \sin \dfrac{4\pi}{3}\right)$

  (b)

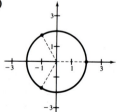

  (c) $2, \ -1 + \sqrt{3}i, \ -1 - \sqrt{3}i$

83. (a) $\cos 0 + i \sin 0$

$\cos \dfrac{2\pi}{5} + i \sin \dfrac{2\pi}{5}$

$\cos \dfrac{4\pi}{5} + i \sin \dfrac{4\pi}{5}$

$\cos \dfrac{6\pi}{5} + i \sin \dfrac{6\pi}{5}$

$\cos \dfrac{8\pi}{5} + i \sin \dfrac{8\pi}{5}$

(b)

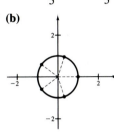

(c) $1,\ 0.3090 + 0.9511i,\ -0.8090 + 0.5878i,$
$-0.8090 - 0.5878i,\ 0.3090 - 0.9511i$

85. $\cos \dfrac{\pi}{8} + i \sin \dfrac{\pi}{8}$

$\cos \dfrac{5\pi}{8} + i \sin \dfrac{5\pi}{8}$

$\cos \dfrac{9\pi}{8} + i \sin \dfrac{9\pi}{8}$

$\cos \dfrac{13\pi}{8} + i \sin \dfrac{13\pi}{8}$

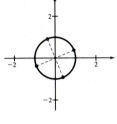

87. $3\left(\cos \dfrac{\pi}{5} + i \sin \dfrac{\pi}{5}\right)$

$3\left(\cos \dfrac{3\pi}{5} + i \sin \dfrac{3\pi}{5}\right)$

$3(\cos \pi + i \sin \pi)$

$3\left(\cos \dfrac{7\pi}{5} + i \sin \dfrac{7\pi}{5}\right)$

$3\left(\cos \dfrac{9\pi}{5} + i \sin \dfrac{9\pi}{5}\right)$

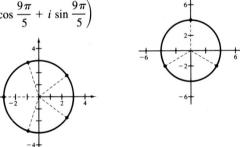

89. $4\left(\cos \dfrac{\pi}{2} + i \sin \dfrac{\pi}{2}\right)$

$4\left(\cos \dfrac{7\pi}{6} + i \sin \dfrac{7\pi}{6}\right)$

$4\left(\cos \dfrac{11\pi}{6} + i \sin \dfrac{11\pi}{6}\right)$

91. $\sqrt[6]{2}(\cos 105° + i \sin 105°)$

$\sqrt[6]{2}(\cos 225° + i \sin 225°)$

$\sqrt[6]{2}(\cos 345° + i \sin 345°)$

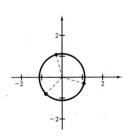

Chapter 6 Review Exercises (*page 519*)

1. $\sin^2 x$ **3.** $1 + \cot \alpha$ **5.** 1   **7.** 1
9. $\cos^2 2x$

Exercises 11–35 are proofs.

37. $\dfrac{\sqrt{2}}{4}\left(\sqrt{3} + 1\right)$ **39.** $-\dfrac{1}{2}\sqrt{2 + \sqrt{2}}$

41. $-\dfrac{3}{52}\left(5 + 4\sqrt{7}\right)$ **43.** $\dfrac{1}{52}\left(36 + 5\sqrt{7}\right)$

45. $\dfrac{1}{4}\sqrt{2\left(4 - \sqrt{7}\right)}$

47. False; If $\dfrac{\pi}{2} < \theta < \pi$, then $\cos \dfrac{\theta}{2} > 0$.   **49.** True

51. $0,\ \pi$

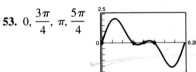

53. $0,\ \dfrac{3\pi}{4},\ \pi,\ \dfrac{5\pi}{4}$

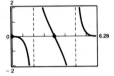

55. $0,\ \dfrac{\pi}{2},\ \pi$

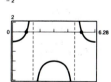

57. $\dfrac{\pi}{3},\ \dfrac{5\pi}{3}$

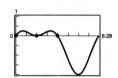

59. $\dfrac{\pi}{4},\ \dfrac{5\pi}{4}$

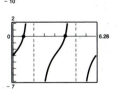

61. $2\cos \dfrac{5\theta}{2}\cos \dfrac{\theta}{2}$

63. $\dfrac{1}{2}(\cos \alpha - \cos 5\alpha)$ **65.** $8x^2 - 1$ **67.** Proof

69. (a) $y = \frac{1}{2}\sqrt{10}\ \sin(8t - \arctan\frac{1}{3})$

(b)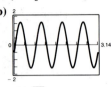

(c) $\frac{1}{2}\sqrt{10}$

(d) $\dfrac{4}{\pi}$

71. $A \approx 29.7°,\ B \approx 52.4°,\ C \approx 97.9°$

73. $C \approx 110°,\ b \approx 20.4,\ c \approx 22.6$

75. $A \approx 35°,\ C \approx 35°,\ b \approx 6.6$ **77.** No solution

79. $A \approx 25.9°,\ C \approx 39.1°,\ c \approx 10.1$

81. $B \approx 31.2°,\ C \approx 133.8°,\ c \approx 13.9$
$\qquad B \approx 148.8°,\ C \approx 16.2°,\ c \approx 5.39$

83. $A \approx 9.9°,\ C \approx 20.1°,\ b \approx 29.1$

85. $A \approx 40.9°,\ C \approx 114.1°,\ c \approx 8.6$
$\qquad A \approx 139.1°,\ C \approx 15.9°,\ c \approx 2.6$

87. 9.798   **89.** 9.08   **91.** 31 ft   **93.** 31.1 m

95. 1135 mi

97. $5\sqrt{2}(\cos 315° + i \sin 315°)$

99. $13(\cos 67.38° + i \sin 67.38°)$

101. $-50 - 50\sqrt{3}i$ **103.** 13

105. (a) $z_1 = 5(\cos \pi + i \sin \pi)$

$\qquad z_2 = 5\left(\cos \dfrac{\pi}{2} + i \sin \dfrac{\pi}{2}\right)$

(b) $z_1 z_2 = 25\left(\cos \dfrac{3\pi}{2} + i \sin \dfrac{3\pi}{2}\right)$

$\qquad \dfrac{z_1}{z_2} = \cos \dfrac{\pi}{2} + i \sin \dfrac{\pi}{2}$

107. (a) $z_1 = 3\sqrt{2}\left(\cos \dfrac{5\pi}{4} + i \sin \dfrac{5\pi}{4}\right)$

$\qquad z_2 = 2\left(\cos \dfrac{\pi}{6} + i \sin \dfrac{\pi}{6}\right)$

(b) $z_1 z_2 = 6\sqrt{2}\left(\cos \dfrac{17\pi}{12} + i \sin \dfrac{17\pi}{12}\right)$

$\qquad \dfrac{z_1}{z_2} = \dfrac{3\sqrt{2}}{2}\left(\cos \dfrac{13\pi}{12} + i \sin \dfrac{13\pi}{12}\right)$

109. $\dfrac{625}{2} + \dfrac{625\sqrt{3}}{2}i$ **111.** $2035 - 828i$

113. $3\left(\cos \dfrac{\pi}{4} + i \sin \dfrac{\pi}{4}\right)$ $3\left(\cos \dfrac{5\pi}{4} + i \sin \dfrac{5\pi}{4}\right)$

$\qquad 3\left(\cos \dfrac{7\pi}{12} + i \sin \dfrac{7\pi}{12}\right)$ $3\left(\cos \dfrac{19\pi}{12} + i \sin \dfrac{19\pi}{12}\right)$

$\qquad 3\left(\cos \dfrac{11\pi}{12} + i \sin \dfrac{11\pi}{12}\right)$ $3\left(\cos \dfrac{23\pi}{12} + i \sin \dfrac{23\pi}{12}\right)$

115. $\cos \dfrac{\pi}{3} + i \sin \dfrac{\pi}{3} = \dfrac{1}{2} + \dfrac{\sqrt{3}}{2}i$

$\qquad \cos \pi + i \sin \pi = -1$

$\qquad \cos \dfrac{5\pi}{3} + i \sin \dfrac{5\pi}{3} = \dfrac{1}{2} - \dfrac{\sqrt{3}}{2}i$

117. $3\left(\cos \dfrac{\pi}{4} + i \sin \dfrac{\pi}{4}\right) = \dfrac{3\sqrt{2}}{2} + \dfrac{3\sqrt{2}}{2}i$

$\qquad 3\left(\cos \dfrac{3\pi}{4} + i \sin \dfrac{3\pi}{4}\right) = -\dfrac{3\sqrt{2}}{2} + \dfrac{3\sqrt{2}}{2}i$

$\qquad 3\left(\cos \dfrac{5\pi}{4} + i \sin \dfrac{5\pi}{4}\right) = -\dfrac{3\sqrt{2}}{2} - \dfrac{3\sqrt{2}}{2}i$

$\qquad 3\left(\cos \dfrac{7\pi}{4} + i \sin \dfrac{7\pi}{4}\right) = \dfrac{3\sqrt{2}}{2} - \dfrac{3\sqrt{2}}{2}i$

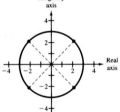

119. $\cos 0 + i \sin 0 = 1$

$\qquad \cos \dfrac{\pi}{2} + i \sin \dfrac{\pi}{2} = i$

$\qquad \cos \dfrac{2\pi}{3} + i \sin \dfrac{2\pi}{3} = -\dfrac{1}{2} + \dfrac{\sqrt{3}}{2}i$

$\qquad \cos \dfrac{4\pi}{3} + i \sin \dfrac{4\pi}{3} = -\dfrac{1}{2} - \dfrac{\sqrt{3}}{2}i$

$\qquad \cos \dfrac{3\pi}{2} + i \sin \dfrac{3\pi}{2} = -i$

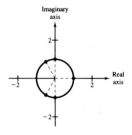

Section 7.1 *(page 531)*

WARM-UP

1.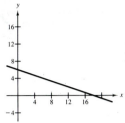

2.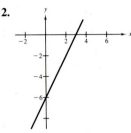

3.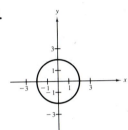

4.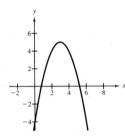

5. x **6.** $-37v$ **7.** $2x^2 + 9$ **8.** -1
9. $x = 6$ **10.** $y = 1$

1. $(1, 2)$ **3.** $(-1, 2), (2, 5)$ **5.** $(0, 5), (3, 4)$
7. $(0, 0), (2, 4)$ **9.** $(-1, 1), (8, 4)$ **11.** $(5, 5)$
13. $\left(\frac{1}{2}, 3\right)$ **15.** $(1.5, 0.3)$ **17.** $\left(\frac{20}{3}, \frac{40}{3}\right)$ **19.** $(0, 0)$
21. $(1, 2)$ **23.** $\left(\frac{29}{10}, \frac{21}{10}\right), (-2, 0)$ **25.** $(0.28, 1.75)$
27. $(-1, 0), (0, 1), (1, 0)$ **29.** $\left(\frac{1}{2}, 2\right), \left(-4, -\frac{1}{4}\right)$

31.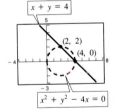

$(2, 2), (4, 0)$

33.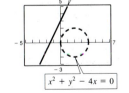

No points of intersection

35.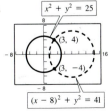

$(3, \pm 4)$

37.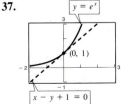

$(0, 1)$

39.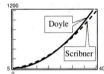

$(0, 0), (1, 1)$

41.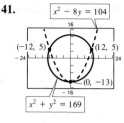

$(0, -13), (\pm 12, 5)$

43. 193 units   **45.** 233,334 units   **47.** 6400 units
49. \$13,000 at 8%, \$12,000 at 8.5%   **51.** \$8333.33
53. **(a)**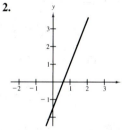

(b) 24.7 in.   **(c)** Doyle's Rule   **55.** 8 mi $\times$ 12 mi
57. **(a)** $y = 2x$ **(b)** $y = 0$ **(c)** $y = x - 2$

Section 7.2 *(page 542)*

WARM-UP
1.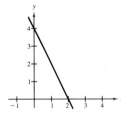

2.

3. $x - y + 4 = 0$ **4.** $5x + 3y - 28 = 0$ **5.** $-\frac{1}{2}$
6. $\frac{7}{4}$ **7.** Perpendicular   **8.** Parallel
9. Neither parallel nor perpendicular   **10.** Perpendicular

1. $(2, 0)$ **3.** $(-1, -1)$ **5.** Inconsistent
7. $(2a, 3a - 3)$ **9.** $\left(-\frac{1}{3}, -\frac{2}{3}\right)$ **11.** $\left(\frac{5}{2}, \frac{3}{4}\right)$
13. $(3, 4)$ **15.** $(4, -1)$ **17.** $(40, 40)$
19. Inconsistent   **21.** $\left(\frac{18}{5}, \frac{3}{5}\right)$ **23.** $(5, -2)$
25. $\left(a, \frac{5}{6}a - \frac{1}{2}\right)$ **27.** $\left(\frac{90}{31}, -\frac{67}{31}\right)$ **29.** $\left(-\frac{6}{35}, \frac{43}{35}\right)$
31. $(79,400, 398)$; The scale is incorrect for seeing the point
of intersection.
33. 550 mph, 50 mph
35. $\frac{20}{3}$ gal of 20% solution, $\frac{10}{3}$ gal of 50% solution
37. \$4000 at 10.5%, \$8000 at 12%
39. 375 adults, 125 children

41.

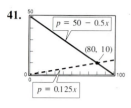

$(80, 10)$

43.

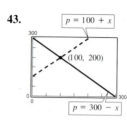

$(100, 200)$

45.

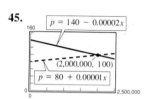

$(2,000,000, 100)$

49. $y = 0.97x + 2.10$

53. $y = \frac{3}{4}x + \frac{4}{3}$

57.

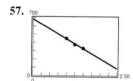

$y = -240x + 685$

349 units

Section 7.3   (*page 555*)

WARM-UP   **1.** $(15, 10)$ **2.** $\left(-2, -\frac{8}{3}\right)$ **3.** $(28, 4)$
4. $(4, 3)$ **5.** Not a solution   **6.** Not a solution
7. Solution   **8.** Solution   **9.** $5a + 2$
10. $a + 13$

1. $(1, 2, 3)$ **3.** $(2, -3, -2)$ **5.** $(5, -2, 0)$
7. Inconsistent   **9.** $\left(1, -\frac{3}{2}, \frac{1}{2}\right)$
11. $(-3a + 10, 5a - 7, a)$ **13.** $\left(13 - 4a, \frac{45}{2} - \frac{15}{2}a, a\right)$
15. $(-a, 2a - 1, a)$ **17.** $\left(\frac{1}{2} - \frac{3}{2}a, 1 - \frac{2}{3}a, a\right)$
19. $(1, 1, 1, 1)$ **21.** Inconsistent   **23.** $(0, 0, 0)$
25. $\left(-\frac{3}{5}a, \frac{4}{5}a, a\right)$ **27.** $y = 2x^2 + 3x - 4$
29. $y = x^2 - 4x + 3$ **31.** $x^2 + y^2 - 4x = 0$
33. $x^2 + y^2 - 6x - 8y = 0$
35. $a = -32, v_0 = 0, s_0 = 144$
37. $a = -32, v_0 = -32, s_0 = 500$
39. \$4000 at 5%, \$5000 at 6%, \$7000 at 7%
41. \$300,000 at 8%, \$400,000 at 9%, \$75,000 at 10%

43.

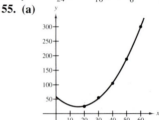

$(100, 200)$

47. 75 mi, 225 mi

51. $y = 0.318x + 4.061$

55. $y = -2x + 4$

59. $x + 2y = 8$
$x + 4y = 13$

43. $250,000 - \frac{1}{2}s$ in certificates of deposit, $125,000 + \frac{1}{2}s$ in municipal bonds, $125,000 - s$ in blue-chip stocks, s in growth stocks
45. 20 gal of spray X, 18 gal of spray Y, 16 gal of spray Z
47. 4 medium *or* 2 large, 1 medium 2 small
49. $t_1 = 96$ lb, $t_2 = 48$ lb, $a = -16$ ft/sec^2
51. $y = -\frac{5}{24}x^2 - \frac{3}{10}x + \frac{41}{6}$ **53.** $y = x^2 - x$
55. (a)

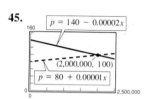

(b) $y = 0.141x^2 - 4.427x + 58.400$
57. $2x + 3y + z = 7$
$\quad x - 2y + 4z = 14$
$\quad x - 2y + 2z = 10$

Section 7.4   (*page 564*)

WARM-UP   **1.** $\dfrac{5x + 2}{x(x + 1)}$ **2.** $\dfrac{2(4x + 3)}{x(x + 2)}$
3. $\dfrac{11x - 1}{(x - 2)(2x - 1)}$ **4.** $-\dfrac{3x + 1}{(x + 5)(x + 12)}$
5. $\dfrac{x^2 - 3x - 5}{(x - 3)^3}$ **6.** $-\dfrac{5x + 6}{(x + 2)^2}$ **7.** $-\dfrac{x + 9}{x(x^2 + 3)}$
8. $\dfrac{4x^2 + 5x + 31}{(x + 1)(x^2 + 5)}$ **9.** $\dfrac{x(3x + 1)}{(x^2 + 1)^2}$ **10.** $\dfrac{x^3 + x^2 + 1}{(x^2 + x + 1)^2}$

1. $\dfrac{1}{2}\left(\dfrac{1}{x - 1} - \dfrac{1}{x + 1}\right)$ **3.** $\dfrac{1}{x} - \dfrac{1}{x + 1}$
5. $\dfrac{1}{x} - \dfrac{2}{2x + 1}$ **7.** $\dfrac{1}{x - 1} - \dfrac{1}{x + 2}$
9. $\dfrac{3}{2x - 1} - \dfrac{2}{x + 1}$ **11.** $-\dfrac{3}{x} - \dfrac{1}{x + 2} + \dfrac{5}{x - 2}$
13. $\dfrac{3}{x} - \dfrac{1}{x^2} + \dfrac{1}{x + 1}$ **15.** $\dfrac{2}{x} - \dfrac{1}{x^2} - \dfrac{2}{x + 1}$
17. $\dfrac{3}{x - 3} + \dfrac{9}{(x - 3)^2}$ **19.** $-\dfrac{1}{x} + \dfrac{2x}{x^2 + 1}$
21. $\dfrac{1}{3(x^2 + 2)} - \dfrac{1}{6(x + 2)} + \dfrac{1}{6(x - 2)}$
23. $\dfrac{1}{8(2x + 1)} + \dfrac{1}{8(2x - 1)} - \dfrac{x}{2(4x^2 + 1)}$

25. $\dfrac{1}{x^2 + 2} + \dfrac{x}{(x^2 + 2)^2}$ **27.** $\dfrac{1}{x + 1} + \dfrac{2}{x^2 - 2x + 3}$

29. $2x + \dfrac{1}{2}\left(\dfrac{3}{x - 4} - \dfrac{1}{x + 2}\right)$

31. $x + 3 + \dfrac{6}{x - 1} + \dfrac{4}{(x - 1)^2} + \dfrac{1}{(x - 1)^3}$

33. $\dfrac{1}{2a}\left(\dfrac{1}{a + x} + \dfrac{1}{a - x}\right)$ **35.** $\dfrac{1}{a}\left(\dfrac{1}{x} + \dfrac{1}{a - x}\right)$

Section 7.5   (page 573)

WARM-UP   **1.** Line   **2.** Line   **3.** Parabola
4. Parabola   **5.** Circle   **6.** Ellipse   **7.** (1, 1)
8. (2, 0)   **9.** $(2, 1), \left(-\frac{5}{2}, -\frac{5}{4}\right)$ **10.** (2, 3), (3, 2)

1. f   **2.** h   **3.** e   **4.** c   **5.** a   **6.** g
7. b   **8.** d

9.

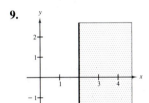

11.

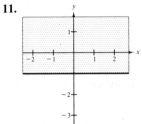

13.

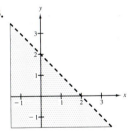

15.

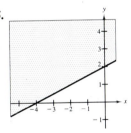

17.

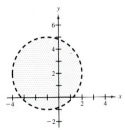

19.

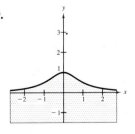

21.

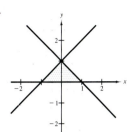

23.

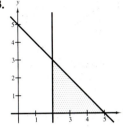

25.

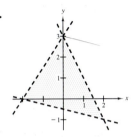

27.

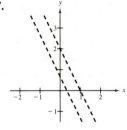

29.

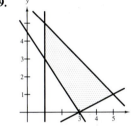

31.

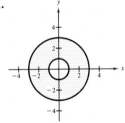

33.

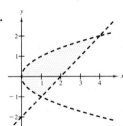

35.

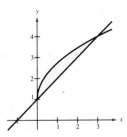

37.

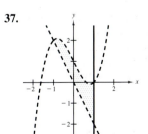

39.

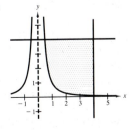

41. $2 \leq x \leq 5$
$1 \leq y \leq 7$
43. $y \leq \frac{3}{2}x$
$y \leq -x + 5$
$y \geq 0$
45. $x^2 + y^2 \leq 16$
$x \geq 0$
$y \geq 0$
47. $x + \frac{3}{2}y \leq 12$
$\frac{4}{3}x + \frac{3}{2}y \leq 15$
$x \geq 0$
$y \geq 0$

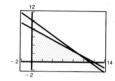

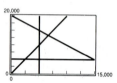

49. $x + y \leq 20,000$
$y \geq 2x$
$x \geq 5,000$
$y \geq 5,000$

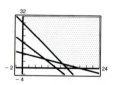

51. $20x + 10y \geq 280$
$15x + 10y \geq 160$
$10x + 20y \geq 180$
$x \geq 0$
$y \geq 0$

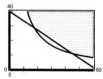

53. Consumer surplus: 1600; Producer surplus: 400
55. Consumer surplus: 5000; Producer surplus: 5000
57. Consumer surplus: 40,000,000; Producer surplus: 20,000,000
59. $xy \geq 500$
$2x + \pi y \geq 125$
$x \geq 0$
$y \geq 0$

Section 7.6   (page 585)

WARM-UP

1.

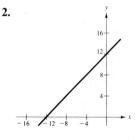

2.

3.

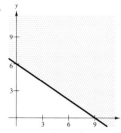

4.

5. (0, 4)   **6.** (12, 0)   **7.** (3, 1)   **8.** (2, 5)

9.

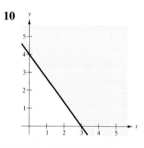

10

1. Minimum at (0, 0): 0; Maximum at (0, 6): 30
3. Minimum at (0, 0): 0; Maximum at (6, 0): 60
5. Minimum at (0, 0): 0; Maximum at (3, 4): 17
7. Minimum at (0, 0): 0; Maximum at (4, 0): 20
9. Minimum at (0, 0): 0; Maximum at (60, 20): 740
11. Minimum at (0, 0): 0; Maximum at any point on the line segment connecting (60, 20) and (30, 45): 2100
13.

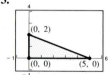

Minimum at (0, 0): 0; Maximum at (5,0): 30
15.

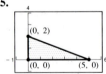

Minimum at (0, 0): 0; Maximum at (0, 2): 48
17.

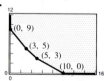

Minimum at (5, 3): 35; No maximum

19.

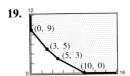

Minimum at $(10, 0)$: 20; No maximum

21.

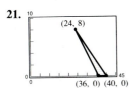

Minimum at $(24, 8)$: 104; Maximum at $(40, 0)$: 160

23.

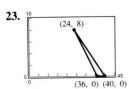

Minimum at $(36, 0)$: 36; Maximum at $(24, 8)$: 56

25. Maximum at $(3, 6)$: 12   **27.** Maximum at $(0, 10)$: 10
29. Maximum at $(0, 5)$: 25   **31.** Maximum at $(4, 4)$: 36
33. 200 units of the $250 model, 50 units of the $400 model
35. 3 bags of Brand X, 6 bags of Brand Y
37. 750 units of Model A, 1000 units of Model B,
     Maximum profit: $83,750
39. 8 audits, 8 tax returns, Maximum revenue: $18,400
41.

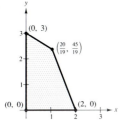

z is maximum at any point on the line segment between $(2, 0)$ and $\left(\frac{20}{19}, \frac{45}{19}\right)$.

43.

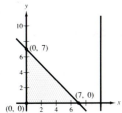

The constraint $x \le 10$ is extraneous.
Maximum at $(0, 7)$: 14

45.

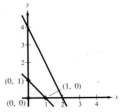

The constraint $2x + y \le 4$ is extraneous.
Maximum at $(0, 1)$: 4
47. **(a)** $t > 9$ **(b)** $\frac{3}{4} < t < 9$

Chapter 7 Review Exercises *(page 589)*

1. $(1, 1)$ **3.** $(5, 4)$ **5.** $(0, 0), (2, 8), (-2, 8)$
7. $\left(\frac{5}{2}, 3\right)$ **9.** $(0, 0)$ **11.** $\left(\frac{14}{5} + \frac{8}{5}a, a\right)$

13.

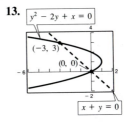

15.

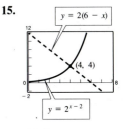

$(0, 0), (-3, 3)$ $(4, 4)$

17. 4762 units
19. 40 gal of the 75% solution, 60 gal of the 50% solution
21. Pittsburgh to Philadelphia: 218.75 mph; Philadelphia to
     Pittsburgh: 193.75 mph
23. $\left(\dfrac{500{,}000}{7}, \dfrac{159}{7}\right)$ **25.** $(4.8, 4.4, -1.6)$
27. $(3a + 4, 2a + 5, a)$ **29.** $(-3a + 2, 5a + 6, a)$
31. $(2, -3, 3)$ **33.** $\left(-2a + \frac{3}{2}, 2a + 1, a\right)$
35. $y = 2x^2 + x - 6$
37. $x^2 + y^2 - 4x + 2y - 4 = 0$
39. 10 gal of spray X, 5 gal of spray Y, 12 gal of spray Z
41. $\dfrac{3}{x + 2} - \dfrac{4}{x + 4}$ **43.** $1 - \dfrac{25}{8(x + 5)} + \dfrac{9}{8(x - 3)}$
45. $\dfrac{1}{2}\left(\dfrac{3}{x - 1} - \dfrac{x - 3}{x^2 + 1}\right)$ **47.** $\dfrac{3x}{x^2 + 1} + \dfrac{x}{(x^2 + 1)^2}$
49.

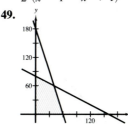

51.

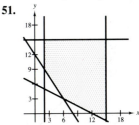

53.

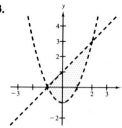

55.

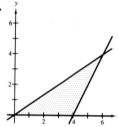

57. $-x + y \leq 4$
$2x + y \leq 22$
$-x + y \geq -2$
$-2x + y \geq 7$

59. $x + y \leq 1500$
$x \geq 400$
$y \geq 600$

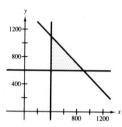

61. Consumer surplus: 4,500,000; Producer surplus: 9,000,000

63. Maximum at (5, 8): 47

65. Minimum at (15, 0): 26.25

67. 5 units of A, 2 units of B, Maximum profit: $138

69. 3 bags of Brand X, 2 bags of Brand Y

CHAPTER 8

Section 8.1   (page 602)

WARM-UP   **1.** -3 **2.** 30   **3.** 6   **4.** $-\frac{1}{9}$
5. Solution   **6.** Not a solution   **7.** (5, 2)
8. $\left(\frac{12}{5}, -3\right)$ **9.** (40, 14, 2)   **10.** $\left(\frac{15}{2}, 4, 1\right)$

---

1. 3×2 **3.** 5×1 **5.** 2×2
7. Reduced row-echelon form
9. Not in row-echelon form
11. $\begin{bmatrix} 1 & 4 & 3 \\ 0 & 2 & -1 \end{bmatrix}$

13. $\begin{bmatrix} 1 & 1 & 4 & -1 \\ 0 & 5 & -2 & 6 \\ 0 & 3 & 20 & 4 \end{bmatrix}, \begin{bmatrix} 1 & 1 & 4 & -1 \\ 0 & 1 & -\frac{2}{5} & \frac{6}{5} \\ 0 & 3 & 20 & 4 \end{bmatrix}$

15. (a) $\begin{bmatrix} 1 & 2 & 3 \\ 0 & -5 & -10 \\ 3 & 1 & -1 \end{bmatrix}$ **(b)** $\begin{bmatrix} 1 & 2 & 3 \\ 0 & -5 & -10 \\ 0 & -5 & -10 \end{bmatrix}$

(c) $\begin{bmatrix} 1 & 2 & 3 \\ 0 & -5 & -10 \\ 0 & 0 & 0 \end{bmatrix}$ **(d)** $\begin{bmatrix} 1 & 2 & 3 \\ 0 & 1 & 2 \\ 0 & 0 & 0 \end{bmatrix}$

(e) $\begin{bmatrix} 1 & 0 & -1 \\ 0 & 1 & 2 \\ 0 & 0 & 0 \end{bmatrix}$

17. $\begin{bmatrix} 1 & 1 & 0 & 5 \\ 0 & 1 & 2 & 0 \\ 0 & 0 & 1 & -1 \end{bmatrix}$ **19.** $\begin{bmatrix} 1 & -1 & -1 & 1 \\ 0 & 1 & 6 & 3 \\ 0 & 0 & 0 & 0 \end{bmatrix}$

21. $\begin{bmatrix} 1 & 0 & 0 \\ 0 & 1 & 0 \\ 0 & 0 & 1 \end{bmatrix}$ **23.** $\begin{bmatrix} 1 & 2 & 0 & 0 \\ 0 & 0 & 1 & 0 \\ 0 & 0 & 0 & 1 \\ 0 & 0 & 0 & 0 \end{bmatrix}$

25. $4x + 3y = 8$
$x - 2y = 3$

27. $x \qquad + 2z = -10$
$3y - z = 5$
$4x + 2y \qquad = 3$

29. $x - 2y = 4$
$y = -3$
$(-2, -3)$

31. $x - y + 2z = 4$
$y - z = 2$
$z = -2$
$(8, 0, -2)$

33. (7, -5)   **35.** (-4, -8, 2)   **37.** (3, 2)
39. (4, -2)   **41.** $\left(\frac{1}{2}, -\frac{3}{4}\right)$ **43.** Inconsistent
45. (4, -3, 2)   **47.** $(2a + 1, 3a + 2, a)$
49. $(5a + 4, -3a + 2, a)$ **51.** $(0, 2 - 4a, a)$
53. $(1, 0, 4, -2)$ **55.** (0, 0)   **57.** $(-2a, a, a)$
59. $800,000 at 8%, $500,000 at 9%, $200,000 at 12%
61. $\dfrac{1}{x-1} + \dfrac{3}{x+1} - \dfrac{2}{(x+1)^2}$ **63.** $y = x^2 + 2x + 4$
65. $x^2 + y^2 - 5x - 3y + 6 = 0$

Section 8.2   (page 616)

WARM-UP   **1.** -27.5 **2.** -7
3. Not in reduced row-echelon form
4. Not in reduced row-echelon form
5. $\begin{bmatrix} -5 & 10 & : & 12 \\ 7 & -3 & : & 0 \\ -1 & 7 & : & 25 \end{bmatrix}$

6. $\begin{bmatrix} 10 & 15 & -9 & : & 42 \\ 6 & -5 & 0 & : & 0 \end{bmatrix}$

7. $(0, 2)$ **8.** $(2 + a, 3 - a, a)$

9. $(1 - 2a, a, -1)$ **10.** $(2, -1, -1)$

---

1. $x = -4, y = 22$ **3.** $x = 2, y = 3$

5. (a) $\begin{bmatrix} 3 & -2 \\ 1 & 7 \end{bmatrix}$ **(b)** $\begin{bmatrix} -1 & 0 \\ 3 & -9 \end{bmatrix}$ **(c)** $\begin{bmatrix} 3 & -3 \\ 6 & -3 \end{bmatrix}$

(d) $\begin{bmatrix} -1 & -1 \\ 8 & -19 \end{bmatrix}$

7. (a) $\begin{bmatrix} 7 & 3 \\ 1 & 9 \\ -2 & 15 \end{bmatrix}$ **(b)** $\begin{bmatrix} 5 & -5 \\ 3 & -1 \\ -4 & -5 \end{bmatrix}$

(c) $\begin{bmatrix} 18 & -3 \\ 6 & 12 \\ -9 & 15 \end{bmatrix}$ **(d)** $\begin{bmatrix} 16 & -11 \\ 8 & 2 \\ -11 & -5 \end{bmatrix}$

9. (a) $\begin{bmatrix} 3 & 3 & -2 & 1 & 1 \\ -2 & 5 & 7 & -6 & -8 \end{bmatrix}$

(b) $\begin{bmatrix} 1 & 1 & 0 & -1 & 1 \\ 4 & -3 & -11 & 6 & 6 \end{bmatrix}$

(c) $\begin{bmatrix} 6 & 6 & -3 & 0 & 3 \\ 3 & 3 & -6 & 0 & -3 \end{bmatrix}$

(d) $\begin{bmatrix} 4 & 4 & -1 & -2 & 3 \\ 9 & -5 & -24 & 12 & 11 \end{bmatrix}$

11. (a) $\begin{bmatrix} 0 & 15 \\ 6 & 12 \end{bmatrix}$ **(b)** $\begin{bmatrix} -2 & 2 \\ 31 & 14 \end{bmatrix}$ **(c)** $\begin{bmatrix} 9 & 6 \\ 12 & 12 \end{bmatrix}$

13. (a) $\begin{bmatrix} 0 & -10 \\ 10 & 0 \end{bmatrix}$ **(b)** $\begin{bmatrix} 0 & -10 \\ 10 & 0 \end{bmatrix}$ **(c)** $\begin{bmatrix} 8 & -6 \\ 6 & 8 \end{bmatrix}$

15. (a) $\begin{bmatrix} 6 & -21 & 15 \\ 8 & -23 & 19 \\ 4 & 7 & 5 \end{bmatrix}$ **(b)** $\begin{bmatrix} 9 & 0 & 13 \\ 7 & -2 & 21 \\ 1 & 4 & -19 \end{bmatrix}$

(c) $\begin{bmatrix} 20 & 7 & -8 \\ 24 & 7 & -2 \\ 2 & -5 & 30 \end{bmatrix}$

17. Not possible

19. $\begin{bmatrix} -1 & 19 \\ 4 & -27 \\ 0 & 14 \end{bmatrix}$ **21.** $\begin{bmatrix} 1 & 0 & 0 \\ 0 & 1 & 0 \\ 0 & 0 & \frac{7}{2} \end{bmatrix}$

23. $\begin{bmatrix} 60 & 72 \\ -20 & -24 \\ 10 & 12 \\ 60 & 72 \end{bmatrix}$ **25.** $\begin{bmatrix} -6 & -9 \\ -1 & 0 \\ 17 & -10 \end{bmatrix}$

27. $\begin{bmatrix} 3 & 3 \\ -\frac{1}{2} & 0 \\ -\frac{13}{2} & \frac{11}{2} \end{bmatrix}$

29. $A = \begin{bmatrix} -1 & 1 \\ -2 & 1 \end{bmatrix}$, $X = \begin{bmatrix} x \\ y \end{bmatrix}$, $B = \begin{bmatrix} 4 \\ 0 \end{bmatrix}$, $x = 4, y = 8$

31. $A = \begin{bmatrix} 1 & -2 & 3 \\ -1 & 3 & -1 \\ 2 & -5 & 5 \end{bmatrix}$, $X = \begin{bmatrix} x \\ y \\ z \end{bmatrix}$, $B = \begin{bmatrix} 9 \\ -6 \\ 17 \end{bmatrix}$,

$x = 1, y = -1, z = 2$

33. $AC = BC = \begin{bmatrix} 12 & -6 & 9 \\ 16 & -8 & 12 \\ 4 & -2 & 3 \end{bmatrix}$ **35.** $\begin{bmatrix} 72 & 48 & 24 \\ 36 & 108 & 72 \end{bmatrix}$

37. $AB = [\$1250 \quad \$1331.25 \quad \$981.25]$; The entries represent the profit from the two products at each of the three outlets.

39. (a) $\$18,300$ **(b)** $\$21,260$

(c) $\begin{bmatrix} \$15,770 & \$18,300 \\ \$26,500 & \$29,250 \\ \$21,260 & \$24,150 \end{bmatrix}$

The entries are the wholesale and retail prices of the inventory at each outlet.

41. $\begin{bmatrix} 0.40 & 0.15 & 0.15 \\ 0.28 & 0.53 & 0.17 \\ 0.32 & 0.32 & 0.68 \end{bmatrix}$

Section 8.3 (page 623)

WARM-UP
1. $\begin{bmatrix} 4 & 24 \\ 0 & -16 \\ 48 & 8 \end{bmatrix}$ **2.** $\begin{bmatrix} \frac{11}{2} & 5 & 24 \\ \frac{1}{2} & 0 & 8 \\ 0 & 1 & 4 \end{bmatrix}$

3. $\begin{bmatrix} -5 & -2 & -13 \\ 4 & -13 & -2 \end{bmatrix}$ **4.** $\begin{bmatrix} -13 & 11 \\ -19 & 21 \end{bmatrix}$

5. $\begin{bmatrix} 1 & 0 \\ 0 & 1 \end{bmatrix}$ **6.** $\begin{bmatrix} 6 & 5 \\ 3 & -2 \end{bmatrix}$ **7.** $\begin{bmatrix} 1 & 0 & 0 \\ 0 & 1 & 0 \\ 0 & 0 & 1 \end{bmatrix}$

8. $\begin{bmatrix} 1 & 0 & 0 \\ 0 & 1 & 0 \\ 0 & 0 & 1 \end{bmatrix}$ **9.** $\begin{bmatrix} 1 & 0 & 3 & -2 \\ 0 & 1 & 4 & -3 \end{bmatrix}$

10. $\begin{bmatrix} 1 & 0 & 0 & -6 & -4 & 3 \\ 0 & 1 & 0 & 11 & 6 & -5 \\ 0 & 0 & 1 & -2 & -1 & 1 \end{bmatrix}$

Exercises 1–8 are proofs.

9. $\begin{bmatrix} \frac{1}{2} & 0 \\ 0 & \frac{1}{3} \end{bmatrix}$ 11. $\begin{bmatrix} -3 & 2 \\ -2 & 1 \end{bmatrix}$ 13. $\begin{bmatrix} 1 & -1 \\ 2 & -1 \end{bmatrix}$

15. Does not exist   17. Does not exist

19. $\begin{bmatrix} 1 & 1 & -1 \\ -3 & 2 & -1 \\ 3 & -3 & 2 \end{bmatrix}$ 21. $\begin{bmatrix} -175 & 37 & -13 \\ 95 & -20 & 7 \\ 14 & -3 & 1 \end{bmatrix}$

23. $\frac{1}{2}\begin{bmatrix} -3 & 3 & 2 \\ 9 & -7 & -6 \\ -2 & 2 & 2 \end{bmatrix}$ 25. $\frac{5}{11}\begin{bmatrix} 0 & -4 & 2 \\ -22 & 11 & 11 \\ 22 & -6 & -8 \end{bmatrix}$

27. $\begin{bmatrix} 1 & 0 & 0 \\ -0.75 & 0.25 & 0 \\ 0.35 & -0.25 & 0.2 \end{bmatrix}$ 29. Does not exist

31. $\begin{bmatrix} -\frac{1}{8} & 0 & 0 & 0 \\ 0 & 1 & 0 & 0 \\ 0 & 0 & \frac{1}{4} & 0 \\ 0 & 0 & 0 & -\frac{1}{5} \end{bmatrix}$ 33. $\begin{bmatrix} -24 & 7 & 1 & -2 \\ -10 & 3 & 0 & -1 \\ -29 & 7 & 3 & -2 \\ 12 & -3 & -1 & 1 \end{bmatrix}$

35. $(2, -2)$ 37. Inconsistent   39. $\left(\frac{2}{3}, \frac{1}{2}\right)$

41. $(-1, 3, 2)$ 43. $\left(\frac{5}{16}a + \frac{13}{16}, \frac{19}{16}a + \frac{11}{16}, a\right)$

45. $(5, 0, -2, 3)$

47. \$10,000 in AAA rated bonds, \$5000 in A rated bonds, \$10,000 in B rated bonds

49. \$9000 in AAA rated bonds, \$1000 in A rated bonds, \$2000 in B rated bonds

51. $I_1 = -3$ amps, $I_2 = 8$ amps, $I_3 = 5$ amps

Section 8.4   (page 633)

WARM-UP   1. $\begin{bmatrix} 3 & 5 \\ 4 & 0 \end{bmatrix}$ 2. $\begin{bmatrix} -2 & 8 \\ 2 & -4 \end{bmatrix}$

3. $\begin{bmatrix} 9 & -12 & 6 \\ 3 & 0 & -3 \\ 0 & 3 & -6 \end{bmatrix}$ 4. $\begin{bmatrix} 0 & 8 & 12 \\ -4 & 8 & 12 \\ -8 & 4 & -8 \end{bmatrix}$ 5. -22

6. 35   7. -15 8. $-\frac{1}{8}$ 9. -45 10. -16

1. 5   3. 5   5. 27   7. -24 9. 6   11. 0

13. -0.002 15. 0   17. 0   19. -9 21. -18

23. $-7x + 3y - 8$

25. (a) $M_{11} = -5, M_{12} = 2, M_{21} = 4, M_{22} = 3$

(b) $C_{11} = -5, C_{12} = -2, C_{21} = -4, C_{22} = 3$

27. (a) $M_{11} = 30, M_{12} = 12, M_{13} = 11, M_{21} = -36,$
$M_{22} = 26, M_{23} = 7, M_{31} = -4, M_{32} = -42,$
$M_{33} = 12$

(b) $C_{11} = 30, C_{12} = -12, C_{13} = 11, C_{21} = 36,$
$C_{22} = 26, C_{23} = -7, C_{31} = -4, C_{32} = 42, C_{33} = 12$

29. -75 31. 96   33. 170   35. -58 37. -30

39. -108 41. 0   43. 412

45. 1   47. -26 49. -126 51. 0   53. 0

55. 236   57. 7441   59. 410   61. $x = -1, x = 4$

63. $8uv - 1$ 65. e^{5x} 67. $1 - \ln x$

69. (a) -3 (b) -2 (c) $\begin{bmatrix} -2 & 0 \\ 0 & -3 \end{bmatrix}$ (d) 6

71. (a) 2   (b) -6 (c) $\begin{bmatrix} 1 & 4 & 3 \\ -1 & 0 & 3 \\ 0 & 2 & 0 \end{bmatrix}$ (d) -12

Section 8.5   (page 645)

WARM-UP   1. $(1, 1)$ 2. $(1, 2)$ 3. $(3, 0, -4)$
4. $(-2, 1, 1)$ 5. 8   6. -49 7. -3 8. 20
9. 9   10. 35

1. $(1, 2)$ 3. $(2, -2)$ 5. $\left(\frac{3}{4}, -\frac{1}{2}\right)$

7. Cramer's Rule does not apply.   9. $\left(\frac{2}{3}, \frac{1}{2}\right)$ 11. -1

13. 1   15. 0   17. Cramer's Rule does not apply.

19. 5

21. $I_1 = \frac{125}{22}$ amps
$I_2 = \frac{93}{11}$ amps
$I_3 = \frac{61}{22}$ amps

23. $y = 1.768 + 0.202t$; Maximum contribution is about \$4200.   25. 7   27. 14   29. $\frac{33}{8}$ 31. $\frac{5}{2}$

33. 28   35. 250 mi²   37. Collinear

39. Not collinear   41. Collinear   43. $3x - 5y = 0$

45. $x + 3y - 5 = 0$ 47. $2x + 3y - 8 = 0$

49. 1 -25 -65 17 15 -9 -12 -62 -119 27 51 48 43 67 48 57 111 117

51. -5 -41 -87 91 207 257 11 -5 -41 40 80 84 76 177 227

53. SEND PLANES

Chapter 8 Review Exercises   (page 649)

1. $(10, -12)$ 3. $(0.6, 0.5)$ 5. $\left(\frac{1}{2}, -\frac{1}{3}, 1\right)$

7. Inconsistent   9. $\begin{bmatrix} -13 & -8 & 18 \\ 0 & 11 & -19 \end{bmatrix}$

11. $\begin{bmatrix} 14 & -2 & 8 \\ 14 & -10 & 40 \\ 36 & -12 & 48 \end{bmatrix}$ 13. $\begin{bmatrix} 44 & 4 \\ 20 & 8 \end{bmatrix}$

15. $\begin{bmatrix} 4 & 6 & 3 \\ 0 & 6 & -10 \\ 0 & 0 & 6 \end{bmatrix}$ 17. $\begin{bmatrix} -14 & -4 \\ 7 & -17 \\ -17 & -2 \end{bmatrix}$

19. $\frac{1}{3}\begin{bmatrix} 9 & 2 \\ -4 & 11 \\ 10 & 0 \end{bmatrix}$

21. $5x + 4y = 2$
    $-x + y = -22$

23. $\begin{bmatrix} \frac{1}{5} & \frac{1}{5} \\ \frac{1}{10} & -\frac{1}{15} \end{bmatrix}$ **25.** $\begin{bmatrix} \frac{1}{2} & -1 & -\frac{1}{2} \\ \frac{1}{2} & -\frac{2}{3} & -\frac{5}{6} \\ 0 & \frac{2}{3} & \frac{1}{3} \end{bmatrix}$

27. 550   **29.** 279   **31.** $(-3, 1)$ **33.** $(1, 1, -2)$
35. $(2, -4, 6)$ **37.** Inconsistent   **39.** 16   **41.** 7
43. $x - 2y + 4 = 0$ **45.** $2x + 6y - 13 = 0$
47. 8 carnations, 4 roses   **49.** $y = 1.01x + 1.54$
51. $y = x^2 + 2x + 3$
53. 128; Each of the three rows is multiplied by 4.

CUMULATIVE TEST for Chapters 6–8   (*page 651*)

1. $2 \csc \beta$
2. $\dfrac{\pi}{3}, \dfrac{\pi}{2}, \dfrac{3\pi}{2}, \dfrac{5\pi}{3}$ **3.** $\dfrac{4\sqrt{5}}{9}$ **4.** $\dfrac{\sqrt{2}}{4}\left(1 - \sqrt{3}\right)$
5. (a) $c \approx 57.57, B \approx 22.69°, C \approx 112.31°$
    (b) $a \approx 9.91, B \approx 23.79°, C \approx 126.21°$
6. 42.5 ft
7. $z = 2\sqrt{2}\,(\cos 315° + i \sin 315°)$
    $z^3 = 16\sqrt{2}\,(\cos 225° + i \sin 225°) = -16 - 16i$
8. $(1, 2), \left(-\frac{3}{2}, \frac{3}{4}\right)$ **9.** $(2, -1)$
10. $(1, -2, 1)$ **11.** $a = -6$
12.

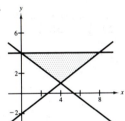

13. Maximum at $(4, 4)$: $z = 20$
14. $\begin{bmatrix} 11 & -6 \\ 5 & 3 \\ -7 & 0 \end{bmatrix}$ **15.** $\begin{bmatrix} 9 & 1 \\ 7 & -7 \\ 15 & -10 \end{bmatrix}$
16. $\begin{bmatrix} -175 & 37 & -13 \\ 95 & -20 & 7 \\ 14 & -3 & 1 \end{bmatrix}$ **17.** -6 **18.** 22

CHAPTER 9

Section 9.1   (*page 661*)

WARM-UP   **1.** $\frac{4}{5}$ **2.** $\frac{1}{3}$ **3.** $(2n + 1)(2n - 1)$
4. $(2n - 1)(2n - 3)$ **5.** $(n - 1)(n - 2)$
6. $(n + 1)(n + 2)$ **7.** $\frac{1}{3}$ **8.** 24   **9.** $\frac{13}{24}$
10. $\frac{3}{4}$

---

1. 3, 5, 7, 9, 11   **3.** 2, 4, 8, 16, 32
5. $-2, 4, -8, 16, -32$ **7.** $0, 1, 0, \frac{1}{2}, 0$
9. $\dfrac{5}{2}, \dfrac{11}{4}, \dfrac{23}{8}, \dfrac{47}{16}, \dfrac{95}{32}$ **11.** $1, \dfrac{1}{2^{3/2}}, \dfrac{1}{3^{3/2}}, \dfrac{1}{4^{3/2}}, \dfrac{1}{5^{3/2}}$
13. $3, \frac{9}{2}, \frac{9}{2}, \frac{27}{8}, \frac{81}{40}$ **15.** $-1, \frac{1}{4}, -\frac{1}{9}, \frac{1}{16}, -\frac{1}{25}$
17. 3, 4, 6, 10, 18   **19.** $\frac{1}{30}$ **21.** $n + 1$
23. $\dfrac{1}{2n(2n + 1)}$ **25.** $a_n = 3n - 2$ **27.** $a_n = n^2 - 1$
29. $a_n = \dfrac{(-1)^{n+1}}{2^n}$ **31.** $a_n = 1 + \dfrac{1}{n}$ **33.** $a_n = \dfrac{1}{n!}$
35. $a_n = (-1)^{n+1}$ **37.** 35   **39.** 40   **41.** 30
43. $\dfrac{9}{5}$ **45.** 238   **47.** 56   **49.** $\dfrac{47}{60}$ **51.** $\displaystyle\sum_{i=1}^{9} \dfrac{1}{3i}$
53. $\displaystyle\sum_{i=1}^{8} \left[2\left(\dfrac{i}{8}\right) + 3\right]$ **55.** $\displaystyle\sum_{i=1}^{6} (-1)^{i+1}3^i$
57. $\displaystyle\sum_{i=1}^{20} \dfrac{(-1)^{i+1}}{i^2}$ **59.** $\displaystyle\sum_{i=1}^{5} \dfrac{2^i - 1}{2^{i+1}}$
61. (a) $A_1 = \$5100.00, A_2 = \$5202.00, A_3 = \$5306.04,$
        $A_4 = \$5412.16, A_5 = \$5520.40, A_6 = \$5630.81,$
        $A_7 = \$5743.43, A_8 = \$5858.30$
    (b) $\$11,040.20$
63. (a) $a_0 = 242.67, a_1 = 285.34, a_2 = 328.01,$
        $a_3 = 370.68, a_4 = 413.35, a_5 = 456.02,$
        $a_6 = 498.69, a_7 = 541.36$
    (b)

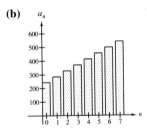

65. 16.02
67. Proof
69. (a) 1, 1, 2, 3, 5, 8, 13, 21, 34, 55, 89, 144
    (b) $2, \frac{3}{2}, \frac{5}{3}, \frac{8}{5}, \frac{13}{8}, \frac{21}{13}, \frac{34}{21}, \frac{55}{34}, \frac{89}{55}, \frac{144}{89}$

Section 9.2   (*page 673*)

WARM-UP   **1.** 36   **2.** 240   **3.** $\frac{11}{2}$ **4.** $\frac{10}{3}$ **5.** 18
6. 4   **7.** 143   **8.** 160   **9.** 430   **10.** 256

---

1. Arithmetic sequence, $d = 3$
3. Not an arithmetic sequence
5. Arithmetic sequence, $d = -\frac{1}{4}$
7. Not an arithmetic sequence
9. Arithmetic sequence, $d = 0.4$

11. 8, 11, 14, 17, 20; Arithmetic sequence, $d = 3$
13. $\frac{1}{2}, \frac{1}{3}, \frac{1}{4}, \frac{1}{5}, \frac{1}{6}$; Not an arithmetic sequence
15. 97, 94, 91, 88, 85; Arithmetic sequence, $d = -3$
17. $1, 4, \frac{7}{3}, \frac{7}{2}, \frac{13}{5}$; Not an arithmetic sequence
19. $a_n = 1 + (n - 1)3$ **21.** $a_n = 100 + (n - 1)(-8)$
23. $a_n = x + (n - 1)(2x)$ **25.** $a_n = 4 + (n - 1)\left(-\frac{5}{2}\right)$
27. $a_n = 5 + (n - 1)\left(\frac{10}{3}\right)$ **29.** $a_n = 100 + (n - 1)(-3)$
31. 5, 11, 17, 23, 29, . . .
33. $-2.6, -3.0, -3.4, -3.8, -4.2, \ldots$
35. $\frac{3}{2}, \frac{5}{4}, 1, \frac{3}{4}, \frac{1}{2}, \ldots$ **37.** 2, 6, 10, 14, 18, . . .
39. $-2, 2, 6, 10, 14, \ldots$ **41.** 620   **43.** 4600
45. 265   **47.** 4000   **49.** 1275   **51.** 25,250
53. 355   **55.** 126,750   **57.** 520   **59.** 44,625
61. 9, 13   **63.** $\frac{15}{4}, \frac{9}{2}, \frac{21}{4}$ **65.** 10,000
67. (a) $35,000   **(b)** $187,500   **69.** 2340
71. 470 bricks

Section 9.3   (*page 682*)

WARM-UP   **1.** $\frac{64}{125}$ **2.** $\frac{9}{16}$ **3.** $\frac{1}{16}$ **4.** $\frac{5}{81}$

5. $6n^3$ **6.** $27n^4$ **7.** $4n^3$ **8.** n^2 **9.** $\dfrac{2^n}{81^n}$

10. $\dfrac{3}{16^n}$

1. Geometric sequence, $r = 3$
3. Not a geometric sequence
5. Geometric sequence, $r = -\frac{1}{2}$
7. Not a geometric sequence
9. Not a geometric sequence   **11.** 2, 6, 18, 54, 162, . . .
13. $1, \frac{1}{2}, \frac{1}{4}, \frac{1}{8}, \frac{1}{16}, \ldots$ **15.** $5, -\frac{1}{2}, \frac{1}{20}, -\frac{1}{200}, \frac{1}{2000}, \ldots$
17. $1, e, e^2, e^3, e^4$ **19.** $3, \dfrac{3x}{2}, \dfrac{3x^2}{4}, \dfrac{3x^3}{8}, \dfrac{3x^4}{16}, \ldots$
21. $\left(\dfrac{1}{2}\right)^7$ **23.** $-\dfrac{2}{3^{10}}$ **25.** $100e^{8x}$ **27.** $500(1.02)^{39}$
29. 9   **31.** $-\frac{2}{9}$
33. (a) $2593.74   **(b)** $2653.30   **(c)** $2685.06
(d) $2707.04   **(e)** $2717.91
35. $22,689.45   **37.** 511   **39.** 43   **41.** 6.4
43. 29,921.31   **45.** 2092.60   **47.** $7808.24
49. Proof   **51. (a)** $26,198.27   **(b)** $26,263.88
53. (a) $637,678.02   **(b)** $645,861.43
57. $3048.1 million   **59.** $3,623,993.23   **61.** 2
63. $\frac{2}{3}$ **65.** $\frac{16}{3}$ **67.** 32   **69.** $\frac{8}{3}$

71.

73. 152.42 ft

The horizontal asymptote of the graph of f is the sum of the series.

Section 9.4   (*page 694*)

WARM-UP   **1.** 24   **2.** 40   **3.** $\frac{77}{60}$ **4.** $\frac{7}{2}$
5. $\dfrac{2k + 5}{5}$ **6.** $\dfrac{3k + 1}{6}$ **7.** $8 \cdot 2^{2k} = 2^{2k+3}$ **8.** $\frac{1}{9}$
9. $\dfrac{1}{k}$ **10.** $\dfrac{4}{5}$

1. $\dfrac{5}{(k + 1)(k + 2)}$ **3.** $\dfrac{(k + 1)^2(k + 2)^2}{4}$

Exercises 5–17 are proofs.

19. 210   **21.** 91   **23.** 225   **25.** 2275   **27.** 70
29. $n(2n + 1)$ **31.** $10[1 - (0.9)^n]$ **33.** $\dfrac{n}{2(n + 1)}$

Exercises 35–45 are proofs.

47. 1, 3, 5, 7, 9   **49.** 4, 2, -2, -4, -2
51. 0   3   6   9   12   **53.** 3   1 -2 -6 -11
    3   3   3   3 -2 -3 -4 -5
    0   0   0 -1 -1 -1
Linear   Quadratic
55. 0   1   3   6   10   **57.** 2   4   6   8   10
    1   2   3   4   2   2   2   2
    1   1   1   0   0   0
Quadratic   Linear
59. 1   2   6   15   31
    1   4   9   16
    3   5   7
Neither linear nor quadratic
61. $f(n) = n^2 - n + 3$ **63.** $f(n) = \frac{1}{2}n^2 + n - 3$

Section 9.5   (*page 702*)

WARM-UP   **1.** $5x^5 + 15x^2$ **2.** $x^3 + 5x^2 - 3x - 15$
3. $x^2 + 8x + 16$ **4.** $4x^2 - 12x + 9$ **5.** $\dfrac{3x^3}{y}$
6. $-32z^5$ **7.** 120   **8.** 336   **9.** 720   **10.** 20

1. 10   **3.** 1   **5.** 15,504   **7.** 4950   **9.** 4950
11. $x^4 + 4x^3 + 6x^2 + 4x + 1$
13. $a^3 + 6a^2 + 12a + 8$
15. $y^4 - 8y^3 + 24y^2 - 32y + 16$
17. $x^5 + 5x^4y + 10x^3y^2 + 10x^2y^3 + 5xy^4 + y^5$
19. $r^6 + 18r^5s + 135r^4s^2 + 540r^3s^3 + 1215r^2s^4 + 1458rs^5 + 729s^6$
21. $x^5 - 5x^4y + 10x^3y^2 - 10x^2y^3 + 5xy^4 - y^5$
23. $1 - 6x + 12x^2 - 8x^3$

25. $x^8 + 20x^6 + 150x^4 + 500x^2 + 625$

27. $\dfrac{1}{x^5} + \dfrac{5y}{x^4} + \dfrac{10y^2}{x^3} + \dfrac{10y^3}{x^2} + \dfrac{5y^4}{x} + y^5$

29. $2x^4 - 24x^3 + 113x^2 - 246x + 207$ **31.** -4

33. $2035 + 828i$ **35.** 1

37. $32t^5 - 80t^4s + 80t^3s^2 - 40t^2s^3 + 10ts^4 - s^5$

39. $81 - 216z + 216z^2 - 96z^3 + 16z^4$ **41.** 1,732,104

43. 180   **45.** $-326,592$ **47.** 210   **49.** $\dfrac{35}{128} \approx 0.273$

51. $\dfrac{1120}{6561} \approx 0.171$ **53.** 1.172   **55.** 510,568.785

57.

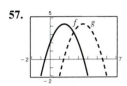

Shift two units to the right.
$g(x) = -x^2 + 7x - 8$

59.

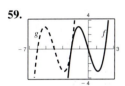

Shift four units to the left.
$g(x) = x^3 + 12x^2 + 44x + 48$

61. $g(t) = 0.2187t^2 + 5.0455t + 55.255$

Section 9.6   *(page 715)*

WARM-UP   **1.** 6656   **2.** 291,600   **3.** 7960
4. 13,800   **5.** 792   **6.** 2300
7. $n(n-1)(n-2)(n-3)$ **8.** $n(n-1)(2n-1)$
9. $n!$ **10.** $n!$

1. 7   **3.** 12   **5.** 12   **7.** 6,760,000   **9.** 64
11. (a) 900   (b) 648   (c) 180   (d) 600
13. 64,000   **15.** (a) 720   (b) 48   **17.** 24
19. 336   **21.** 1,860,480   **23.** 9900   **25.** 120
27. ABCD, ABDC, ACBD, ACDB, ADBC, ADCB, BACD,
BADC, CABD, CADB, DABC, DACB, BCAD, BDAC,
CBAD, CDAB, DBAC, DCAB, BCDA, BDCA, CBDA,
CDBA, DBCA, DCBA
29. 120   **31.** 11,880   **33.** 420   **35.** 1260
37. 2520
39. AB, AC, AD, AE, AF, BC, BD, BE, BF, CD, CE, CF,
DE, DF, EF

41. 4845   **43.** 3,838,380   **45.** 3,921,225   **47.** 560
49. (a) 70   (b) 30   **51.** (a) 70   (b) 54   (c) 16
53. 5   **55.** 20   **57.** $n = 5$ or $n = 6$

Section 9.7   *(page 728)*

WARM-UP   **1.** $\dfrac{9}{16}$ **2.** $\dfrac{8}{15}$ **3.** $\dfrac{1}{6}$ **4.** $\dfrac{1}{80,730}$ **5.** $\dfrac{1}{495}$
6. $\dfrac{1}{24}$ **7.** $\dfrac{1}{12}$ **8.** $\dfrac{135}{323}$ **9.** 0.366   **10.** 0.997

1. $\{(h, 1), (h, 2), (h, 3), (h, 4), (h, 5), (h, 6), (t, 1), (t, 2),$
$(t, 3), (t, 4), (t, 5), (t, 6)\}$
3. $\{ABC, ACB, BAC, BCA, CAB, CBA\}$
5. $\{AB, AC, AD, AE, BC, BD, BE, CD, CE, DE\}$
7. $\dfrac{3}{8}$ **9.** $\dfrac{7}{8}$ **11.** $\dfrac{3}{13}$ **13.** $\dfrac{5}{13}$ **15.** $\dfrac{1}{12}$ **17.** $\dfrac{7}{12}$
19. $\dfrac{1}{3}$ **21.** $\dfrac{1}{5}$ **23.** $\dfrac{2}{5}$ **25.** 0.3   **27.** 0.85
29. (a) $\dfrac{112}{209}$ (b) $\dfrac{97}{209}$ (c) $\dfrac{274}{627}$
31. $P(\{\text{Taylor wins}\}) = 0.50, P(\{\text{Moore wins}\}) = 0.25$
$P(\{\text{Jenkins wins}\}) = 0.25$
33. (a) $\dfrac{21}{1292} \approx 0.016$ (b) $\dfrac{225}{646} \approx 0.348$ (c) $\dfrac{49}{323} \approx 0.152$
35. (a) $\dfrac{1}{3}$ (b) $\dfrac{5}{8}$ **37.** (a) $\dfrac{1}{120}$ (b) $\dfrac{1}{24}$
39. (a) $\dfrac{1}{169}$ (b) $\dfrac{1}{221}$ **41.** (a) $\dfrac{14}{55}$ (b) $\dfrac{12}{55}$ (c) $\dfrac{54}{55}$
43. (a) $\dfrac{1}{4}$ (b) $\dfrac{1}{2}$ (c) $\dfrac{9}{100}$ (d) $\dfrac{1}{30}$
45. (a) 0.9702   (b) 0.9998   (c) 0.0002
47. (a) $\dfrac{1}{1024}$ (b) $\dfrac{243}{1024}$ (c) $\dfrac{781}{1024}$
49. (a) $\dfrac{1}{16}$ (b) $\dfrac{1}{8}$ (c) $\dfrac{15}{16}$ **51.** 0.1024

Section 9.8   *(page 737)*

WARM-UP
1.

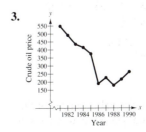

2.

Stems	Leaves
1.	5 7 8
2.	0 2 3 4 4 5 5 6 6 8 9
3.	0 2 5 5 7
4.	0

3.

4.

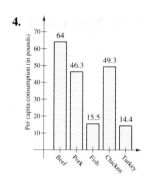

5. 40   **6.** 220   **7.** 72   **8.** 234   **9.** 21 − 6a
10. $5a^2 - 30a + 55$

---

1. Mean: 8.86; Median: 8; Mode: 7   **3.** Mean: 10.29;
Median: 8; Mode: 7   **5.** Mean: 9; Median: 8;
Mode: 7   **7.** The mean is sensitive to extreme
values.   **9.** Mean: $67.14; Median: $65.35
11. Mean: 3.07; Median: 3; Mode: 3   **13.** Mean: 36,540 mi
15. Mean: 76.6; Median: 82; Mode: 42. The median gives the
most representative description.
17. Yes.

Section 9.9   (*page 746*)

WARM-UP   **1.** 16.38   **2.** 4.23   **3.** $35.16
4. 72.43°   **5.** 62   **6.** 85.2   **7.** 21.83   **8.** 41.6
9. 2.68   **10.** 2.68

---

1. $\bar{x} = 6$, $v = 10$, $s = 3.16$ **3.** $\bar{x} = 2$, $v = \frac{4}{3}$, $s = 1.15$
5. $\bar{x} = 4$, $v = 4$, $s = 2$ **7.** $\bar{x} = 47$, $v = 226$, $s = 15.03$
9. 3.42   **11.** 101.55   **13.** 1.65
15. $\bar{x} = 12$ and $|x_i - 12| = 8$ for all x_i.
17. (a) $\bar{x} = 12$, $s = 2.83$ **(b)** $\bar{x} = 20$, $s = 2.83$
(c) $\bar{x} = 12$, $s = 1.41$ **(d)** $\bar{x} = 9$, $s = 1.41$
19. It will increase the mean by 5, but the standard deviation
will not change.
21. With $\bar{x} = 235$ and $s = 28$:
At least 75% of the scores in [179, 291].
At least 89.9% of the scores in [151, 319].
With $\bar{x} = 235$ and $s = 16$:
At least 75% of the scores in [203, 267]. At
least 89.9% of the scores in [187, 283].

23.   **25.**

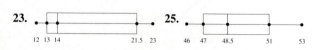

12 13 14   21.5 23   46 47 48.5   51 53

---

27. *Original Design*   *New Design*

10 13.05 28.9   62.6   85.2   18.9 24.15 41.35   63.95   99.5

From the plots, you can see that the lifetime of the
product increased. (The median increased by over
1 year)

Chapter 9 Review Exercises   (*page 748*)

1. $\sum_{k=1}^{20} \frac{1}{2k}$ **3.** $\sum_{k=1}^{9} \frac{k}{k+1}$ **5.** 30   **7.** 80
9. 127   **11.** 8   **13.** 12   **15.** 88   **17.** 418
19. 3, 7, 11, 15, 19   **21.** 1, 4, 7, 10, 13
23. $a_n = 100 + (n-1)(-3)$, 1430   **25.** 25,250
27. 9, 6, 4, $\frac{8}{3}$, $\frac{16}{9}$ or 9, -6, 4, $-\frac{8}{3}$, $\frac{16}{9}$
31. $a_n = 16\left(-\frac{1}{2}\right)^{n-1}$, 10.67
33. (a) $a_t = 120,000(0.7)^t$ **(b)** $20,168.50
35. $5111.82

Exercises 37 and 39 are proofs.

41. 15   **43.** 53,130
45. $\dfrac{x^4}{16} + \dfrac{x^3y}{2} + \dfrac{3x^2y^2}{2} + 2xy^3 + y^4$
47. $\dfrac{64}{x^6} + \dfrac{576}{x^4} + \dfrac{2160}{x^2} - 4320 + 4860x^2 - 2916x^4$
$+ 729x^6$
49. $41 + 840i$ **51. (a)** 1   **(b)** 6   **(c)** 15
53. 118,813,750   **55.** $\frac{1}{9}$
57. $P(\{3\}) = \frac{1}{6}$
$P(\{(1, 5), (5, 1), (2, 4), (4, 2), (3, 3)\}) = \frac{5}{36}$
59. $\dfrac{31}{32}$ **61.** $\dfrac{123,552}{2,598,960} \approx 0.0475$
63. (a) $\bar{x} = 4$, $s = 1.79$
(b) $\bar{x} = 10$, $s = 1.79$
(c) $\bar{x} = 8$, $s = 0.89$
(d) $\bar{x} = 20$, $s = 0.89$
65. Mean: 10.7; Median: 11.5; $v = 21.61$; $s = 4.65$
67. Mean: 166.28; Median: 174.75; $v = 1167.2$; $s = 34.16$
69. Mean: 50.36; Median: 51; $v = 54.54$; $s = 7.39$

CHAPTER 10

Section 10.1   (*page 757*)

WARM-UP   **1.** $x^2 - 10x + 5$ **2.** $x^2 + 6x + 8$
3. $-x^2 - 8x - 6$ **4.** $-x^2 + 4x$ **5.** $(x + 3)^2 - 1$
6. $(x - 5)^2 - 4$ **7.** $2 - (x - 1)^2$ **8.** $-2(x - 1)^2$
9. $2x + 3y - 20 = 0$ **10.** $3x - 4y - 17 = 0$

1. e   **2.** f   **3.** a   **4.** c   **5.** d   **6.** b

7. Vertex: $(0, 0)$; Focus: $\left(0, \frac{1}{16}\right)$; Directrix: $y = -\frac{1}{16}$

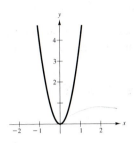

9. Vertex: $(0, 0)$; Focus $\left(-\frac{3}{2}, 0\right)$; Directrix: $x = \frac{3}{2}$

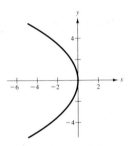

11. Vertex: $(0, 0)$; Focus: $(0, -2)$; Directrix: $y = 2$

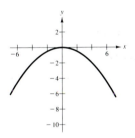

13. Vertex: $(1, -2)$; Focus: $(1, -4)$; Directrix: $y = 0$

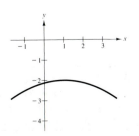

15. Vertex: $\left(5, -\frac{1}{2}\right)$; Focus: $\left(\frac{11}{2}, -\frac{1}{2}\right)$; Directrix: $x = \frac{9}{2}$

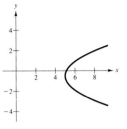

17. Vertex: $(1, 1)$; Focus: $(1, 2)$; Directrix: $y = 0$

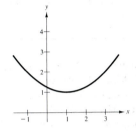

19. Vertex: $(-1, 2)$; Focus: $(0, 2)$; Directrix: $x = -2$

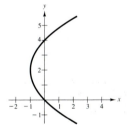

21. Vertex: $(2, -2)$; Focus: $(0, -2)$; Directrix: $x = 4$

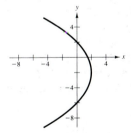

23. Vertex: $(-2, 1)$; Focus: $\left(-2, -\frac{1}{2}\right)$; Directrix: $y = \frac{5}{2}$

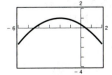

25. Vertex: $\left(\frac{1}{4}, -\frac{1}{2}\right)$; Focus: $\left(0, -\frac{1}{2}\right)$; Directrix: $x = \frac{1}{2}$

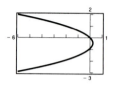

27. $x^2 = -6y$ **29.** $y^2 = -12x$
31. $x^2 - 8y + 32 = 0$ **33.** $5x^2 - 14x - 3y + 9 = 0$
35. $x^2 + y - 4 = 0$ **37.** $x^2 = 12y$
39. (a) $y = \frac{1}{800}x^2$ (b) $\frac{25}{2}$ ft
41. (a) 24,749 mph   (b) $x^2 = -16,400(y - 4100)$
43. (a) $y = -\frac{1}{64}x^2 + 75$ (b) 69.3 ft   **45.** 41,329 ft
47. $y = 4x - 8$; $(2, 0)$ **49.** $y = 4x + 2$; $\left(-\frac{1}{2}, 0\right)$

Section 10.2   (*page 766*)

WARM-UP

1.

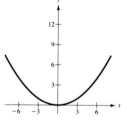

2.

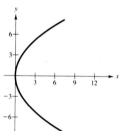

3.

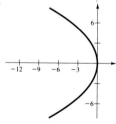

4.

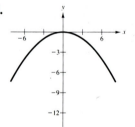

5. $c = 12$ **6.** $b = 1$ **7.** $a = 10$ **8.** $c = 2\sqrt{6}$
9. $4x^2 + 3y^2$ **10.** $\dfrac{9(x-1)^2}{4} + 9(y+2)^2$

1. e   **3.** c   **5.** f
7. Center: $(0, 0)$; Foci: $(\pm 3, 0)$;
   Vertices: $(\pm 5, 0)$: $e = \frac{3}{5}$

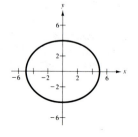

9. Center: $(0, 0)$; Foci: $(0, \pm 3)$; Vertices: $(0, \pm 5)$: $e = \frac{3}{5}$

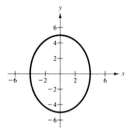

11. Center: $(0, 0)$; Foci: $(\pm 2, 0)$; Vertices: $(\pm 3, 0)$: $e = \frac{2}{3}$

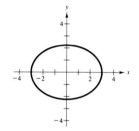

13. Center: $(0, 0)$; Foci: $\left(\pm\sqrt{3}, 0\right)$; Vertices: $(\pm 2, 0)$:
   $e = \dfrac{\sqrt{3}}{2}$

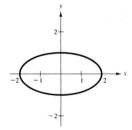

15. Center: $(0, 0)$; Foci: $(0, \pm 1)$; Vertices: $\left(0, \pm\sqrt{3}\right)$;
   $e = \dfrac{1}{\sqrt{3}} = \dfrac{\sqrt{3}}{3}$

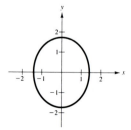

17. Center: (1, 5); Foci: (1, 9), (1, 1); Vertices: (1, 10); (1, 0); $e = \frac{4}{5}$

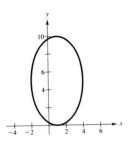

19. Center: (−2, 3); Foci: $\left(-2, 3 \pm \sqrt{5}\right)$;

Vertices: (−2, 6), (−2, 0); $e = \dfrac{\sqrt{5}}{3}$

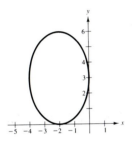

21. Center: (1, −1); Foci: $\left(\frac{7}{4}, -1\right)$, $\left(\frac{1}{4}, -1\right)$; Vertices: $\left(\frac{9}{4}, -1\right)$, $\left(-\frac{1}{4}, -1\right)$; $e = \frac{3}{5}$

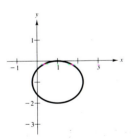

23. Center: (0, 0); Foci: $\left(0, \dfrac{\pm\sqrt{3}}{2}\right)$; Vertices: (0, ±1)

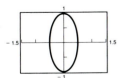

25. Center: $\left(\frac{1}{2}, -1\right)$; Foci: $\left(\frac{1}{2} \pm \sqrt{2}, -1\right)$; Vertices: $\left(\frac{1}{2} \pm \sqrt{5}, -1\right)$

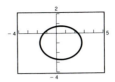

27. Center: $\left(\frac{3}{2}, -1\right)$; Foci: $\left(\frac{3}{2} \pm \sqrt{2}, -1\right)$; Vertices: $\left(-\frac{1}{2}, -1\right)$, $\left(\frac{7}{2}, -1\right)$

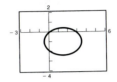

29. $\dfrac{x^2}{36} + \dfrac{y^2}{11} = 1$ **31.** $\dfrac{y^2}{4} + x^2 = 1$

33. $\dfrac{(y-4)^2}{64} + \dfrac{x^2}{48} = 1$ **35.** $\dfrac{(x-3)^2}{36} + \dfrac{(y-2)^2}{32} = 1$

37. Place tacks 1.5 ft from center. Length of string: $2a = 5$ ft

39.   **41.**

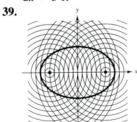

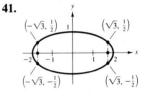

43.

45. Least distance: $a - c \approx 91.377$ million miles; Greatest distance: $a + c \approx 94.537$ million miles

47. $e = 0.0543$ **49.** $e = 0.052$

Section 10.3   (page 775)

WARM-UP   **1.** $\sqrt{61}$ **2.** $\sqrt{65}$

3.

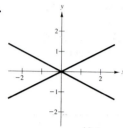

4.

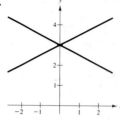

5.

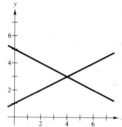

6.

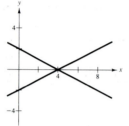

7. Parabola   **8.** Ellipse   **9.** Circle   **10.** Parabola

1. e   **3.** f   **5.** d
7. Center: $(0, 0)$; Vertices: $(\pm 1, 0)$; Foci: $(\pm\sqrt{2}, 0)$;
Asymptotes: $y = \pm x$

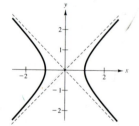

9. Center: $(0, 0)$; Vertices: $(0, \pm 1)$; Foci: $(0, \pm\sqrt{5})$;
Asymptotes: $y = \pm\frac{1}{2}x$

11. Center: $(0, 0)$; Vertices: $(0, \pm 5)$; Foci: $(0, \pm 13)$;
Asymptotes: $y = \pm\frac{5}{12}x$

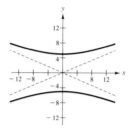

13. Center: $(0, 0)$; Vertices: $(0, \pm 2)$; Foci: $(0, \pm 3)$;
Asymptotes: $y = \pm\dfrac{2}{\sqrt{5}}x$

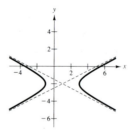

15. Center: $(1, -2)$; Vertices: $(-1, -2)$, $(3, -2)$;
Foci: $(1 \pm \sqrt{5}, -2)$;
Asymptotes: $y = -2 \pm \frac{1}{2}(x - 1)$

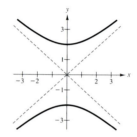

17. Center: $(2, -6)$; Vertices: $(2, -5)$, $(2, -7)$; Foci:
$(2, -6 \pm \sqrt{2})$; Asymptotes: $y = -6 \pm (x - 2)$

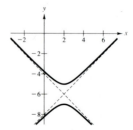

19. Center: $(2, -3)$; Vertices: $(1, -3)$, $(3, -3)$; Foci: $(2 \pm \sqrt{10}, -3)$; Asymptotes: $y = -3 \pm 3(x - 2)$

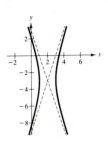

21. Degenerate hyperbola is two intersecting lines

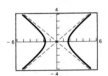

23. Center: $(0, 0)$; Vertices: $(\pm\sqrt{3}, 0)$; Foci: $(\pm\sqrt{5}, 0)$; Asymptotes: $y = \pm\sqrt{\frac{2}{3}}x$

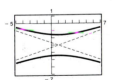

25. Center: $(1, -3)$; Vertices: $(1, -3 \pm \sqrt{2})$; Foci: $(1, -3 \pm 2\sqrt{5})$; Asymptotes: $y = -3 \pm \frac{1}{3}(x - 1)$

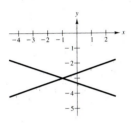

27. Center: $(1, -3)$; Vertices: $(1 \pm 2, -3)$; Foci: $(1 \pm \sqrt{10}, -3)$; Asymptotes: $y = -3 \pm \frac{\sqrt{6}}{2}(x - 1)$

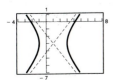

29. $\dfrac{y^2}{4} - \dfrac{x^2}{12} = 1$ **31.** $x^2 - \dfrac{y^2}{9} = 1$

33. $\dfrac{(x - 4)^2}{4} - \dfrac{y^2}{12} = 1$ **35.** $\dfrac{y^2}{9} - \dfrac{(x - 2)^2}{9/4} = 1$

37. $\dfrac{(x - 3)^2}{9} - \dfrac{(y - 2)^2}{4} = 1$ **39.** $(4400, -4290)$

41. $\left(\sqrt{216 - 72\sqrt{5}}, 0\right) \approx (7.42, 0)$ **43.** Circle

45. Hyperbola   **47.** Ellipse   **49.** Parabola

Section 10.4   (*page 785*)

WARM-UP   **1.** b   **2.** e   **3.** d   **4.** a   **5.** f

6. c   **7.** $\dfrac{1}{2}x - \dfrac{\sqrt{3}}{2}y$ **8.** $-\dfrac{1}{2}x + \dfrac{\sqrt{3}}{2}y$

9. $\dfrac{4x^2 - 12xy + 9y^2}{13}$ **10.** $\dfrac{x^2 - 2\sqrt{2}xy + 2y^2}{3}$

1. $\dfrac{(y')^2}{2} - \dfrac{(x')^2}{2} = 1$ **3.** $\dfrac{(x')^2}{1/4} - \dfrac{(y')^2}{1/6} = 1$

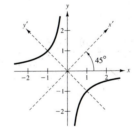   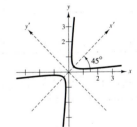

5. $\dfrac{(x' - 3\sqrt{2})^2}{16} - \dfrac{(y' - \sqrt{2})^2}{16} = 1$

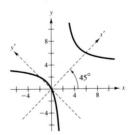

7. $\dfrac{(x')^2}{3} + \dfrac{(y')^2}{2} = 1$ **9.** $4(y')^2 + 4x' = 0,\ x' = -(y')^2$

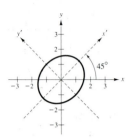
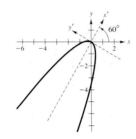

11. $(x' - 1)^2 = 4\left(\tfrac{3}{2}\right)\left(y' + \tfrac{1}{6}\right)$

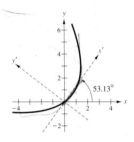

13.

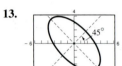

15.

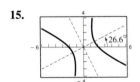

17.

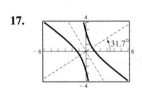

19. Parabola   **21.** Ellipse or circle
23. Hyperbola   **25.** Parabola
27.

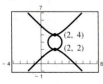

29.

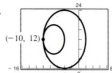

31.

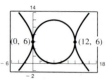

33.

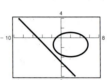

35.

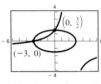

37. No points of intersection   **39.**

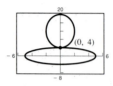

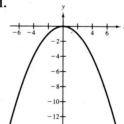

Section 10.5   (page 793)

WARM-UP

1.

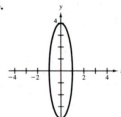

2.

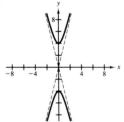

3.

4.

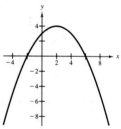

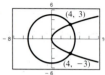

5.

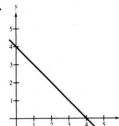

6.

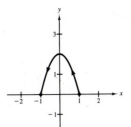

7. 10   **8.** $5 \tan^2 \theta$

9. $\sec^2 x + \tan^2 x$ **10.** $\frac{1}{2} \sin \theta$

13.

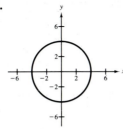

$y = 2 - 2x^2, \; -1 \le x \le 1$

15.

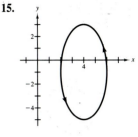

$\dfrac{(x-4)^2}{4} + \dfrac{(y+1)^2}{16} = 1$

1.

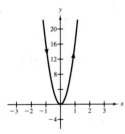

$y = -2x$

3.

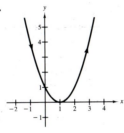

$2x - 3y + 5 = 0$

17.

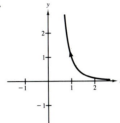

$\sqrt[3]{y} = \dfrac{1}{x}, \; y = \dfrac{1}{x^3}, \; x > 0, \; y > 0$

19.

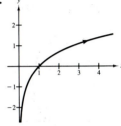

$y = \ln x$

5.

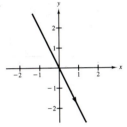

$y = 16x^2$

7.

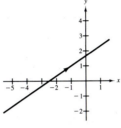

$y = (x-1)^2$

9.

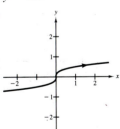

$y = \frac{1}{2}\sqrt[3]{x}$

11.

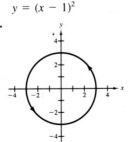

$x^2 + y^2 = 9$

21. Each curve represents a portion of the line $y = 2x + 1$.

	Domain	Orientation
(a)	$-\infty < x < \infty$	Up
(b)	$-1 \le x \le 1$	Oscillates
(c)	$0 < x < \infty$	Down
(d)	$0 < x < \infty$	Up

23. $y - y_1 = \dfrac{y_2 - y_1}{x_2 - x_1}(x - x_1)$

25. $\dfrac{(x-h)^2}{a^2} + \dfrac{(y-k)^2}{b^2} = 1$

27. $x = 5t$
$y = -2t$
Solution not unique

29. $x = 2 + 4 \cos \theta$
$y = 1 + 4 \sin \theta$
Solution not unique

31. $x = 5 \cos \theta$
$y = 3 \sin \theta$
Solution not unique

33. $x = 4 \sec \theta$
$y = 3 \tan \theta$
Solution not unique

35. *Examples*

$x = t, \; y = t^3$
$x = \sqrt[3]{t}, \; y = t$
$x = \tan t, \; y = \tan^3 t$

37.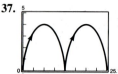

39.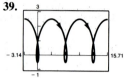

41.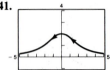

43. b

45. d

47. $x = a\theta - b\sin\theta$ and $y = a - b\cos\theta$.

Section 10.6 (page 802)

WARM-UP

1. $\dfrac{3\pi}{4}$ **2.** $\dfrac{7\pi}{6}$ **3.** $\sin\theta = \dfrac{\sqrt{5}}{5}$, $\cos\theta = \dfrac{2\sqrt{5}}{5}$

4. $\sin\theta = -\dfrac{3}{5}$, $\cos\theta = \dfrac{4}{5}$ **5.** $\dfrac{3\pi}{4}$ **6.** 0.5880

7. $-\dfrac{\sqrt{3}}{2}$ **8.** $-\dfrac{\sqrt{2}}{2}$ **9.** -0.3090 **10.** 0.9735

1.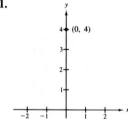

$(0, 4)$

3.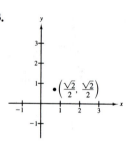

$\left(\dfrac{\sqrt{2}}{2}, \dfrac{\sqrt{2}}{2}\right)$

5.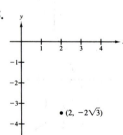

$\left(2, -2\sqrt{3}\right)$

7.

$(0, 0)$

9.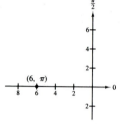

$(-1.004, 0.996)$

11.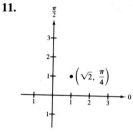

$\left(\sqrt{2}, \dfrac{\pi}{4}\right), \left(-\sqrt{2}, \dfrac{5\pi}{4}\right)$

13.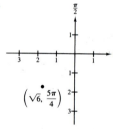

$(6, \pi), (-6, 0)$

15.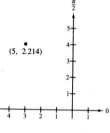

$(5, 2.214), (-5, 5.356)$

17.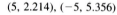

$\left(\sqrt{6}, \dfrac{5\pi}{4}\right), \left(-\sqrt{6}, \dfrac{\pi}{4}\right)$

19.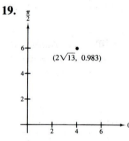

$\left(2\sqrt{13}, 0.983\right), \left(-2\sqrt{13}, 4.124\right)$

21. $r = 3$ **23.** $r = 2a\cos\theta$ **25.** $r = 4\csc\theta$

27. $r = 10\sec\theta$ **29.** $r = \dfrac{-2}{3\cos\theta - \sin\theta}$

31. $r^2 = 4\sec\theta\csc\theta = 8\csc 2\theta$

33. $r^2 = 9\cos 2\theta$ **35.** $x^2 + y^2 - 4y = 0$

37. $\sqrt{3}x - 3y = 0$ **39.** $y = 2$

41. $(x^2 + y^2)^2 = 6x^2y - 2y^3$

43. $4x^2 - 5y^2 - 36y - 36 = 0$

45. $x^2 + y^2 = 9$

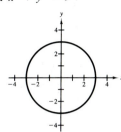

47. $x - y = 0$

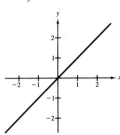

49. $x - 3 = 0$

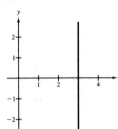

51. Proof
53. $(x - h)^2 + (y - k)^2 = h^2 + k^2$
Center: (h, k)
Radius: $\sqrt{h^2 + k^2}$

9. Maximum: $|r| = 20$ when $\theta = \dfrac{3\pi}{2}$

Zero: $r = 0$ when $\theta = \dfrac{\pi}{2}$

11.
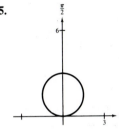

13.
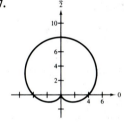

15.
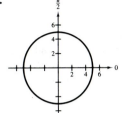

17.
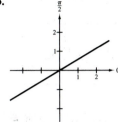

Section 10.7   (*page 811*)

WARM-UP   **1.** Amplitude: 5; Period: $\pi/2$
2. Amplitude: 3; Period: 1
3. Amplitude: 5; Period: $\frac{4}{5}$ **4.** Amplitude: $\frac{1}{2}$; Period: 4π

5.

6.
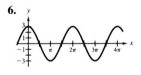

7.
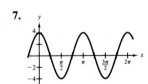

8.
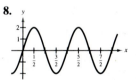

9. $\dfrac{1}{2}\left(\sqrt{3}\sin x - \cos x\right)$ **10.** $\dfrac{\sqrt{2}}{2}(\cos x + \sin x)$

1. Polar axis   **3.** $\theta = \dfrac{\pi}{2}$ **5.** $\theta = \dfrac{\pi}{2}$, polar axis, pole

7. Maximum: $|r| = 5$ when $\theta = 0, \dfrac{\pi}{3}, \dfrac{2\pi}{3}$

19.
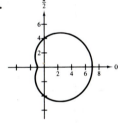

21.
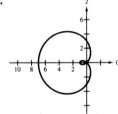

23.
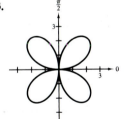

25.
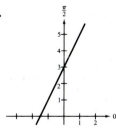

27.
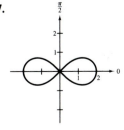

29.
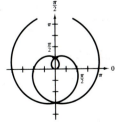

31.

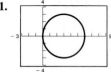

33.

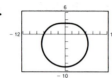

35.

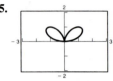

37.

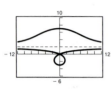

39.

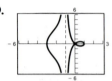

41.

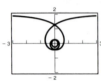

43.

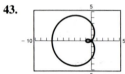

45.

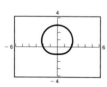

47.

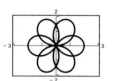

49.

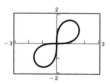

51. $y = \pm \left| \dfrac{x}{x+1} \right| \sqrt{3 - 2x - x^2}$ **53.** Proof

55. (a) $r = 2 - \dfrac{\sqrt{2}}{2}(\sin \theta - \cos \theta)$

(b) $r = 2 + \cos \theta$

(c) $r = 2 + \sin \theta$

(d) $r = 2 - \cos \theta$

57. (a)

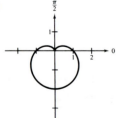

(b)

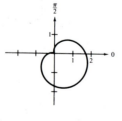

Section 10.8   (*page 819*)

WARM-UP

1. $\left(\dfrac{3\sqrt{2}}{2}, -\dfrac{3\sqrt{2}}{2} \right)$ **2.** $\left(-2, -2\sqrt{3} \right)$

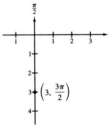

3. $\left(3, \dfrac{3\pi}{2} \right), \left(-3, \dfrac{\pi}{2} \right)$ **4.** $(13, 1.9656), (-13, 5.1072)$

5. $r = 5$ **6.** $r^3 = 4 \sec^2 \theta \csc \theta$

7. $y = -4$ **8.** $x^2 + y^2 - 4x = 0$

9.

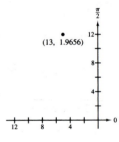

10.

1. c   **3.** a   **5.** b

7.

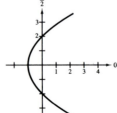

9.

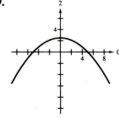

11.

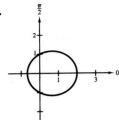

13.

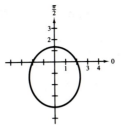

15.

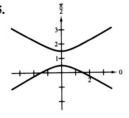

17.

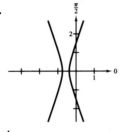

19. $r = \dfrac{1}{1 - \cos \theta}$ **21.** $r = \dfrac{1}{2 + \sin \theta}$

23. $r = \dfrac{2}{1 + 2 \cos \theta}$ **25.** $r = \dfrac{2}{1 - \sin \theta}$

27. $r = \dfrac{10}{1 - \cos \theta}$ **29.** $r = \dfrac{16}{5 + 3 \cos \theta}$

31. $r = \dfrac{20}{3 - 2 \cos \theta}$ **33.** $r = \dfrac{9}{4 - 5 \sin \theta}$

35. Proof **37.** $r^2 = \dfrac{24{,}336}{169 - 25 \cos^2 \theta}$

39. $r^2 = \dfrac{144}{25 \cos^2 \theta - 9}$ **41.** $r^2 = \dfrac{144}{25 \cos^2 \theta - 16}$

43.

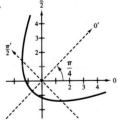

45. Proof

47. $r = \dfrac{9.2931 \times 10^7}{1 - 0.0167 \cos \theta}$, 9.1405×10^7, 9.4509×10^7

49. $r = \dfrac{8200}{1 + \sin \theta}$

Chapter 10 Review Exercises (*page 822*)

1. Parabola

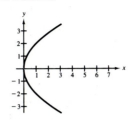

3. Parabola

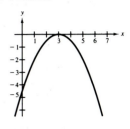

5. Degenerate circle

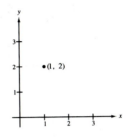

7. Ellipse

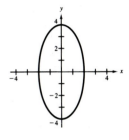

9. Ellipse

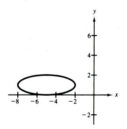

11. Hyperbola

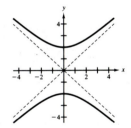

13. Hyperbola
$$y = 5x + \sqrt{24x^2 - 1}$$
$$y = 5x - \sqrt{24x^2 - 1}$$

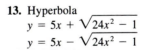

15. $(x - 4)^2 = -8(y - 2)$ **17.** $(y - 2)^2 = -4x$

19. $\dfrac{(x - 2)^2}{25} + \dfrac{y^2}{21} = 1$ **21.** $\dfrac{2x^2}{9} + \dfrac{y^2}{36} = 1$

23. $\dfrac{y^2}{1} - \dfrac{x^2}{8} = 1$ **25.** $\dfrac{5(x - 4)^2}{16} - \dfrac{5y^2}{64} = 1$

27. $(0, 50)$

29. $2x - y = 0$

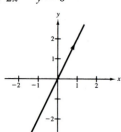

31. $3x + 4y - 11 = 0$

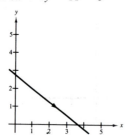

33. $y = \dfrac{1}{x^2}$

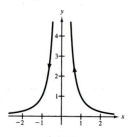

35. $x^2 + y^2 = 36$

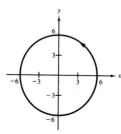

37. $x^{2/3} + \left(\dfrac{y}{4}\right)^{2/3} = 1$

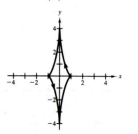

39. $xy = 1,\ x > 0,\ y > 0$

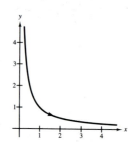

41. Circle

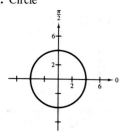

43. Rose curve

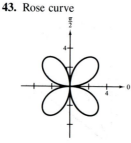

45. Cardioid

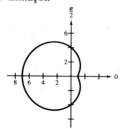

47. Limaçon

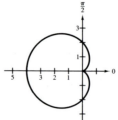

49. Rose curve

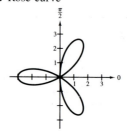

51. Lemniscate

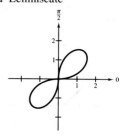

53. Line

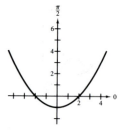

55. Parabola

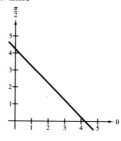

57. $x^2 + y^2 - 3x = 0$ **59.** $x^2 + 4y - 4 = 0$

61. $(x^2 + y^2)^2 - x^2 + y^2 = 0$ **63.** $r = a \cos^2 \theta \sin \theta$

65. $r = 10 \sin \theta$ **67.** $r = \dfrac{4}{1 - \cos \theta}$

69. $r = \dfrac{5}{3 - 2 \cos \theta}$

71. $x = -3 + 4 \cos \theta,\ y = 4 + 3 \sin \theta$

73. Proof

75.

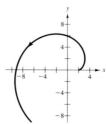

CHAPTER 11

Section 11.1   (*page 835*)

WARM-UP   **1.** $7\sqrt{10}$ **2.** $\sqrt{58}$

3. $3x + 5y - 14 = 0$ **4.** $4x - 3y - 1 = 0$

5. $111.8°$ **6.** $323.1°$ **7.** $\dfrac{1}{2}, \dfrac{\sqrt{3}}{2}$ **8.** $\dfrac{\sqrt{3}}{2}, -\dfrac{1}{2}$

9. $-\dfrac{\sqrt{3}}{2}, \dfrac{1}{2}$ **10.** $-\dfrac{1}{2}, -\dfrac{\sqrt{3}}{2}$

1.

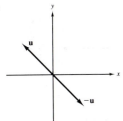

3.

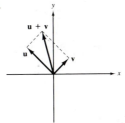

5.

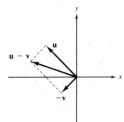

7. $\mathbf{v} = \langle 3, 4 \rangle, \|\mathbf{v}\| = 5$ **9.** $\mathbf{v} = \langle -3, 2 \rangle, \|\mathbf{v}\| = \sqrt{13}$

11. $\mathbf{v} = \langle 0, 5 \rangle, \|\mathbf{v}\| = 5$

13. $\mathbf{v} = \langle 16, -3 \rangle, \|\mathbf{v}\| = \sqrt{265}$

15. $\mathbf{v} = \langle 8, 4 \rangle, \|\mathbf{v}\| = 4\sqrt{5}$

17. (a) $\langle 4, 3 \rangle$ **19.** (a) $\langle -4, 4 \rangle$
    (b) $\langle -2, 1 \rangle$ (b) $\langle 0, 2 \rangle$
    (c) $\langle -7, 1 \rangle$ (c) $\langle 2, 3 \rangle$

21. (a) $\langle 4, -2 \rangle$ **23.** (a) $3\mathbf{i} - 2\mathbf{j}$
    (b) $\langle 4, -2 \rangle$ (b) $-\mathbf{i} + 4\mathbf{j}$
    (c) $\langle 8, -4 \rangle$ (c) $-4\mathbf{i} + 11\mathbf{j}$

25. (a) $2\mathbf{i} + \mathbf{j}$ **27.** $\|\mathbf{v}\| = 5, \theta = 30°$
    (b) $2\mathbf{i} - \mathbf{j}$
    (c) $4\mathbf{i} - 3\mathbf{j}$

29. $\|\mathbf{v}\| = 6\sqrt{2}, \theta = 315°$

31. $\mathbf{v} = \langle 3, 0 \rangle$

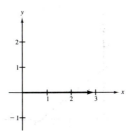

33. $\mathbf{v} = \left\langle -\dfrac{\sqrt{3}}{2}, \dfrac{1}{2} \right\rangle$ **35.** $\mathbf{v} = \left\langle -\dfrac{3\sqrt{6}}{2}, \dfrac{3\sqrt{2}}{2} \right\rangle$

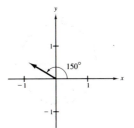   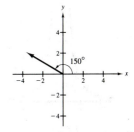

37. $\mathbf{v} = \left\langle \dfrac{\sqrt{10}}{5}, \dfrac{3\sqrt{10}}{5} \right\rangle$ **39.** $\mathbf{v} = \left\langle 3, -\dfrac{3}{2} \right\rangle$

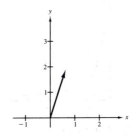   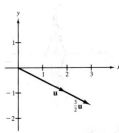

41. $\mathbf{v} = \langle 4, 3 \rangle$ **43.** $\mathbf{v} = \langle \frac{7}{2}, -\frac{1}{2} \rangle$

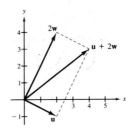   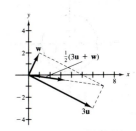

45. $\langle 5, 5 \rangle$ **47.** $\langle (10\sqrt{2} - 50), 10\sqrt{2} \rangle$ **49.** $\frac{4}{5}\mathbf{i} - \frac{3}{5}\mathbf{j}$

51. $\mathbf{j}$ **53.** $90°$ **55.** $63.4°$ **57.** $62.7°$

59. $12.3°, 82.2$ lb   **61.** $71.3°, 228.5$ lb

63. Horizontal component: $80 \cos 50° \approx 51.42$ ft/sec;
Vertical component: $80 \sin 50° \approx 61.28$ ft/sec
65. $T_{AC} \approx 879.4$ lb, $T_{BC} \approx 652.7$ lb
67. 3192.5 lb   **69.** N 25.2° E, 82.8 mph
71. 425 ft-lb

Section 11.2   (page 846)

WARM-UP   **1. (a)** $\langle -14, -5 \rangle$ **(b)** $3\sqrt{5}$
2. (a) $\langle \frac{43}{8}, \frac{3}{5} \rangle$ **(b)** $\frac{1}{40}\sqrt{1249} \approx 0.88$
3. (a) $-6\mathbf{i} + 4\mathbf{j} - 18\mathbf{k}$ **(b)** $4\sqrt{26} \approx 20.4$
4. (a) $8.7\mathbf{i} - 2.2\mathbf{j}$ **(b)** $\sqrt{12.45} \approx 3.53$
5. $\dfrac{2\pi}{5}, \dfrac{4\pi}{3}$ **6.** $\dfrac{\pi}{2}, \dfrac{3\pi}{2}$ **7.** 1.00, 5.28
8. 2.89, 3.39
9. $\dfrac{1}{\sqrt{569}}\langle 12, -5, 20 \rangle, -\dfrac{1}{\sqrt{569}}\langle 12, -5, 20 \rangle$
10. $\dfrac{1}{\sqrt{394}}\langle 12, 5, -15 \rangle, -\dfrac{1}{\sqrt{394}}\langle 12, 5, -15 \rangle$

1. 6   **3.** -15 **5.** -2
7. 8, scalar   **9.** $\langle -6, 8 \rangle$, vector   **11.** -20
13. \$37,289; It is the total revenue if all the units are sold at the given price.
15. 90°   **17.** $\approx 22.17°$ **19.** $\approx 78.69°$
21. $\dfrac{5\pi}{12}$ or 75°
23. $\approx 91.33°$ **25.** 90°

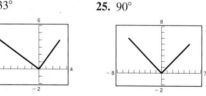

27. Orthogonal   **29.** Parallel
31. Neither   **33.** $\mathbf{w}_1 = \langle \frac{64}{17}, \frac{16}{17} \rangle$; $\mathbf{w}_2 = \langle -\frac{13}{17}, \frac{52}{17} \rangle$
35. $\mathbf{w}_1 = \langle 0 \rangle$; $\mathbf{w}_2 = \langle 4, 2 \rangle$ **37.** $\mathbf{w}_1 = \langle 0, 1 \rangle$; $\mathbf{w}_2 = \langle 2, 0 \rangle$
39. (a) ≈ 4514.9 lb   **(b)** $\approx 25,605$ lb
41. $\approx 7.37 \times 10^6$ ft-lb

Section 11.3   (page 853)

WARM-UP   **1.** 13, $\left(\frac{5}{2}, 6 \right)$ **2.** $7\sqrt{2}, \left(-\frac{1}{2}, \frac{9}{2} \right)$
3. $\sqrt{74}, \left(\frac{7}{2}, \frac{5}{2} \right)$ **4.** $\sqrt{13.13} \approx 3.62, (-0.35, -0.8)$
5. $\frac{1}{2}\sqrt{17} \approx 2.06, \left(\frac{3}{2}, \frac{3}{4} \right)$ **6.** $\frac{5}{6}, \left(1, \frac{5}{4} \right)$
7. $(x - 4)^2 + (y + 5)^2 = 16$

8. $(x + 1)^2 + (y - 3)^2 = 1$
9. $(x - 3)^2 + (y - 3)^2 = 5$
10. $\left(x + \frac{1}{2} \right)^2 + (y - 3)^2 = \frac{37}{4}$

1.

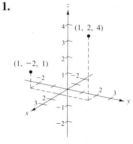

3.

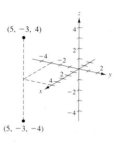

5. $AB = \sqrt{22}$ **7.** $AB = \sqrt{26}$
    $BC = 9$ $BC = \sqrt{40}$
    $AC = 7$ $AC = \sqrt{22}$
    Neither   Neither
9. $(0, -2, 6)$ **11.** $\left(0, 0, \frac{13}{2} \right)$
13. $x^2 + (y - 4)^2 + (z - 3)^2 = 16$
15. $(x + 3)^2 + (y - 7)^2 + (z - 5)^2 = 25$
17. $\left(x - \frac{3}{2} \right)^2 + y^2 + (z - 3)^2 = \frac{45}{4}$
19. Center: $(2, -1, 3)$; Radius: 2
21. Center: $(-2, 0, 4)$; Radius: 1
23. Center: $\left(1, \frac{1}{3}, 4 \right)$; Radius: 3
25.

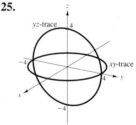

27.

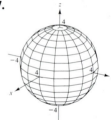

29.

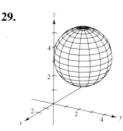

Section 11.4   (page 860)

WARM-UP

1. $\overrightarrow{AB} = \langle 1, -2 \rangle$
$\overrightarrow{AC} = \langle 2, 4 \rangle$
$\overrightarrow{AB} + \overrightarrow{AC} = \langle 3, 2 \rangle$

2. $\overrightarrow{AB} = \langle -4, -2 \rangle$
$\overrightarrow{AC} = \langle -2, 3 \rangle$
$\overrightarrow{AB} + \overrightarrow{AC} = \langle -6, 1 \rangle$

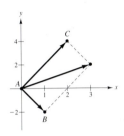   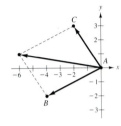

3. $\overrightarrow{AB} = \langle -6, -5 \rangle$
$\overrightarrow{AC} = \langle -8, 0 \rangle$
$\overrightarrow{AB} + \overrightarrow{AC} = \langle -14, -5 \rangle$

4. $\overrightarrow{AB} = \langle 7, 4 \rangle$
$\overrightarrow{AC} = \langle 4, 8 \rangle$
$\overrightarrow{AB} + \overrightarrow{AC} = \langle 11, 12 \rangle$

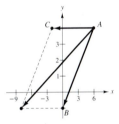   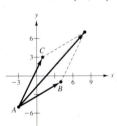

5. $\left\langle \frac{5}{13}, -\frac{12}{13} \right\rangle$ **6.** $\langle \cos 38°, \sin 38° \rangle$
7. $8, 45°$ **8.** $7, \approx 73.74°$
9. $0, 90°$ **10.** $-102, 180°$

1. $\mathbf{v} = \langle -2, 3, 1 \rangle$ **3.** $\mathbf{v} = \langle 4, 4, 2 \rangle$

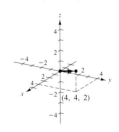

5. $\mathbf{v} = \langle 4, 4, 4 \rangle; c\langle 1, 1, 1 \rangle, c > 0; c\langle 1, 1, 1 \rangle, c < 0$
7. $\mathbf{z} = \langle -3, 7, 6 \rangle$ **9.** $\mathbf{z} = \langle 1, -10, -7 \rangle$ **11.** $\mathbf{z} = \left\langle \frac{1}{2}, 6, \frac{3}{2} \right\rangle$
13. $\sqrt{33}$ **15.** $\dfrac{1}{\sqrt{74}}(8\mathbf{i} + 3\mathbf{j} - \mathbf{k})$ **17.** 0
19. -3 **21.** $\approx 124.45°$ **23.** $\approx 109.92°$
25. Parallel   **27.** Orthogonal

29. $\mathbf{w}_1 = \left\langle 0, \frac{64}{73}, \frac{24}{73} \right\rangle; \mathbf{w}_2 = \left\langle 2, \frac{9}{73}, -\frac{24}{73} \right\rangle$
31. Collinear   **33.** Not collinear
35. $\langle 3, 1, 7 \rangle$ **37.** $\langle 0, 2\sqrt{2}, 2\sqrt{2} \rangle$
39. Sphere: $(x - x_1)^2 + (y - y_1)^2 + (z - z_1)^2 = 81$
41. 68 (work units)

Section 11.5   (page 867)

WARM-UP

1. (a) 64   (b) 64   (c) $\langle 384, 512 \rangle$
2. (a) 0   (b) 850   (c) $\langle 0, 0 \rangle$
3. (a) 327   (b) 436   (c) $\left\langle 2943, -\frac{2943}{2}, 3924 \right\rangle$
4. (a) -8 (b) 25   (c) $16\mathbf{i} - 48\mathbf{j}$
5. $k = -\frac{15}{2}$ **6.** $k = 18$ **7.** $k = \pm 3$
8. 44   **9.** -10 **10.** -72

1. $\mathbf{i} \times \mathbf{j} = \mathbf{k}$ **3.** $\mathbf{i} \times \mathbf{k} = -\mathbf{j}$

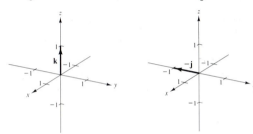

5. $\langle 0, 0, 14 \rangle$ **7.** $\langle -3, 5, 23 \rangle$ **9.** $\langle 10, -2, -4 \rangle$
11. $-7\mathbf{i} + 13\mathbf{j} + 16\mathbf{k}$ **13.** $-\frac{3}{2}\mathbf{i} - \frac{3}{2}\mathbf{j} - \frac{3}{2}\mathbf{k}$
15. $-3\mathbf{i} - \frac{11}{3}\mathbf{j} - \frac{1}{3}\mathbf{k}$
17. $\dfrac{1}{\sqrt{19}}(\mathbf{i} - 3\mathbf{j} + 3\mathbf{k})$
19. $\dfrac{1}{\sqrt{342}}(-6\mathbf{i} + 15\mathbf{j} - 9\mathbf{k})$
21. $\dfrac{1}{2\sqrt{2}}(2\mathbf{i} - 2\mathbf{j})$ **23.** 1   **25.** $\sqrt{806}$ **27.** 14
29. $\overrightarrow{AB} = \langle 1, 2, -2 \rangle$ is parallel to $\overrightarrow{DC} = \langle 1, 2, -2 \rangle$.
$\overrightarrow{AD} = \langle -3, 4, 4 \rangle$ is parallel to $\overrightarrow{BC} = \langle -3, 4, 4 \rangle$. Area is
$\| \overrightarrow{AB} \times \overrightarrow{AD} \| = 6\sqrt{10}$.
31. $\sqrt{349}$ sq. units   **33.** $\frac{1}{2}\sqrt{4290}$ sq. units
35. -16 **37.** 4 cu. units   **39.** 84 cu. units

41. $\begin{vmatrix} \mathbf{i} & \mathbf{j} & \mathbf{k} \\ u_1 & u_2 & u_3 \\ u_1 & u_2 & u_3 \end{vmatrix} = 0\mathbf{i} - 0\mathbf{j} + 0\mathbf{k} = \mathbf{0}$

43. $\begin{vmatrix} \mathbf{i} & \mathbf{j} & \mathbf{k} \\ \cos \beta & \sin \beta & 0 \\ \cos \alpha & \sin \alpha & 0 \end{vmatrix} = (\sin \alpha \cos \beta - \cos \alpha \sin \beta)\mathbf{k}$

Section 11.6   (page 876)

WARM-UP   **1.** $-28, -34\mathbf{i} - 16\mathbf{j} - 38\mathbf{k}$
2. $11, \frac{77}{6}\mathbf{i} + \frac{19}{3}\mathbf{j} + \mathbf{k}$ **3.** $45, -80\mathbf{i} - 140\mathbf{j} - 10\mathbf{k}$
4. $-3, 21\mathbf{i} + 7\mathbf{j} + \mathbf{k}$ **5.** $\approx 94.39°$
6. $\approx 71.59°$ **7.** $90°$ **8.** $90°$
9. $\left\langle \frac{6}{7}, -\frac{2}{7}, \frac{3}{7} \right\rangle$ **10.** $\left\langle \frac{3}{\sqrt{29}}, \frac{4}{\sqrt{29}}, \frac{2}{\sqrt{29}} \right\rangle$

1. (a) $x = -2t, y = 4t, z = t$
  (b) $-\dfrac{x}{2} = \dfrac{y}{4} = z$
3. (a) $x = -4 + 3t, y = 1 + 8t, z = -6t$
  (b) $\dfrac{x+4}{3} = \dfrac{y-1}{8} = -\dfrac{z}{6}$
5. (a) $x = 2 + 2t, y = -3 - 3t, z = 5 + t$
  (b) $\dfrac{x-2}{2} = -\dfrac{y+3}{3} = z - 5$
7. (a) $x = 3 + 9t, y = -5 - 13t, z = -4 - 12t$
  (b) $\dfrac{x-3}{9} = -\dfrac{y+5}{13} = -\dfrac{z+4}{12}$
9. $x = -3, y = 8, z = 15 + t$
11. b, c **13.** $(2, 3, -11); \approx 64.31°$
15.

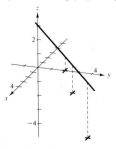

17. $y - 4 = 0$ **19.** $-2x + y - 2z + 10 = 0$
21. $-x - 2y + z + 2 = 0$ **23.** $x - y + z = 0$
25. $x - y - 4z + 7 = 0$ **27.** $y - 5 = 0$
29. $y - z = 0$ **31.** $60°; x = t, y = -1 + t, z = 4 - t$
33. $\approx 80.41°; x = 16 + 5t, y = t, z = -6 - 3t$
35.   **37.**

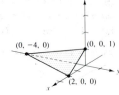

39.

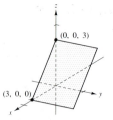

41.   **43.**

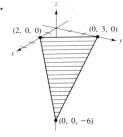   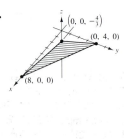

45. $\dfrac{12}{\sqrt{14}}$

47. (a) Sphere: $(x - 4)^2 + (y + 1)^2 + (z - 1)^2 = 4$
  (b) 2 planes: $4x - 3y + z = \left(10 \pm 2\sqrt{26}\right)$
49. $\approx 93.37°$

Chapter 11   Review Exercises (page 878)

1. $\overrightarrow{AB} = \langle 3, -1 \rangle$
  $\overrightarrow{AC} = \langle 2, 2 \rangle$
  $\overrightarrow{AB} + \overrightarrow{AC} = \langle 5, 1 \rangle$

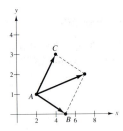

3. $\langle 7, -7 \rangle$ **5.** $\langle 8 \cos 120°, 8 \sin 120° \rangle$
7. $\mathbf{v} = 3\langle \cos 135°, \sin 135° \rangle$

9. $\left\langle \dfrac{6}{\sqrt{61}}, -\dfrac{5}{\sqrt{61}} \right\rangle$ **11.** $\langle -26, -35 \rangle$

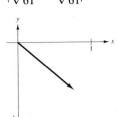

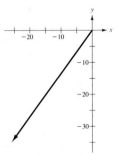

13. (a) $\frac{5}{2}\mathbf{i} + \mathbf{j}$ **(b)** $\frac{9}{2}\mathbf{i} - 3\mathbf{j}$ **(c)** $10\mathbf{i} - 8\mathbf{j}$
15. $\|\mathbf{v}\| = 4$, $\theta = 300°$
17. $92.3°$, 117.0 **19.** 100 lb
21. 460.3 mph; N $32.2°$ E
23. $(x - 2)^2 + (y - 3)^2 + (z - 2)^2 = 17$
25. $\dfrac{1}{\sqrt{185}}\langle -10, 6, 7 \rangle$ **27.** Parallel   **29.** $\dfrac{11\pi}{12}$ or $165°$
31. $90°$ **33.** $-\frac{13}{17}\langle 4, 1 \rangle$ **35.** $-\frac{5}{2}\langle 1, -1, 0 \rangle$
37. $\langle -10, 0, -10 \rangle$ **39.** $\mathbf{u} \cdot \mathbf{u} = 6 = \|\mathbf{u}\|^2$
41. $\mathbf{u} \times \mathbf{v} = \langle 4, -2, 8 \rangle = -(\mathbf{v} \times \mathbf{u})$ **43.** $.44$ cu. units
45. $\approx 78.46°$
47. $x = -1 + 4t$, $y = 3 + 3t$, $z = 5 - 6t$
49. $x = 3 + t$, $y = 1 + t$, $z = 2 + t$
51. $-2x - 12y + 5z = 0$
53. $x + y + z - 6 = 0$
55. $\dfrac{1}{\sqrt{110}}$

CUMULATIVE TEST for Chapters 9–11   (*page 881*)

1. 920 **2.** 54, -18, 6, -1, $\frac{2}{3}$ **3.** 6 **4.** Proof
5. $z^4 - 12z^3 + 54z^2 - 108z + 81$ **6.** $\$402,492.56$
7. 120 **8. (a)** $\frac{1}{6}$ **(b)** $\frac{1}{4}$
9. Mean: 24.17
  Median: 24
  Standard deviation: 3.67
10. (a)   **(b)**

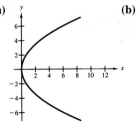

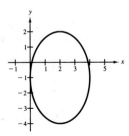

11. $\dfrac{(y - 2)^2}{4/5} - \dfrac{x^2}{16/5} = 1$
12. $x = t$
  $y = \dfrac{7}{4}t - \dfrac{13}{2}$
13. $\dfrac{(x - 3)^2}{9} + \dfrac{(y - 2)^2}{4} = 1$ **14.** $x^2 + y^2 - 6x = 0$

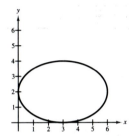

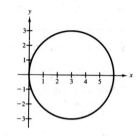

15. Vertical asymptote: $x = -2$ **16.** $r = \dfrac{2}{1 - \cos\theta}$

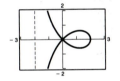

17. $\left\langle \dfrac{3\sqrt{3}}{2}, \dfrac{3}{2} \right\rangle$ **18.** $\langle 115.91, 31.06 \rangle$
19. $(x - 2)^2 + (y - 2)^2 + (z - 4)^2 = 24$
20. $\mathbf{u} \cdot \mathbf{v} = -13$; $\mathbf{u} \times \mathbf{v} = 8\mathbf{i} + 11\mathbf{j} - 20\mathbf{k}$
21. $x = -2 + 7t$, $y = 3 + 5t$, $z = 25t$
22. $75x + 50y - 31z = 0$

CHAPTER 12
Section 12.1   (*page 890*)

WARM-UP
1. (a) 5 **(b)** $c^2 - 2c + 2$
    (c) $x^2 + 2xh + h^2 - 2x - 2h + 2$
2. (a) -1 **(b)** 6 **(c)** $2(t^2 + 2)$
3. $2h + 5$, $h \neq 0$ **4.** $-\dfrac{3}{1 + h}$, $h \neq 0$

5. Domain: $(-\infty, 0) \cup (0, \infty)$
Range: $(-\infty, 0) \cup (0, \infty)$

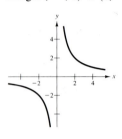

6. Domain: $[-4, 4]$
Range: $[0, 4]$

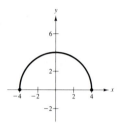

7. Domain: all real numbers
Range: $[0, \infty)$

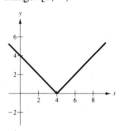

8. Domain: $(-\infty, 0) \cup (0, \infty)$
Range: $\{-1, 1\}$

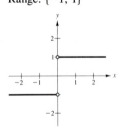

9. Not a function   **10.** Function

1. $\lim_{x \to 3}(3 - 2x) = -3$

x	2.9	2.99	2.999	3.0	3.001	3.01	3.1
$f(x)$	-2.8	-2.98	-2.998	-3	-3.002	-3.02	-3.2

3. $\lim_{x \to 2} \dfrac{x - 2}{x^2 - 4} = \dfrac{1}{4}$

x	1.9	1.99	1.999	2.0	2.001	2.01	2.1
$f(x)$	0.2564	0.2506	0.2501	?	0.2499	0.2494	0.2439

5. $\lim_{x \to 1} \dfrac{x - 1}{x^2 + 2x - 3} = \dfrac{1}{4}$

x	0.9	0.99	0.999	1.0	1.001	1.01	1.1
$f(x)$	0.2564	0.2506	0.2501	?	0.2499	0.2464	0.2439

7. $\lim_{x \to -4} \dfrac{[x/(x + 2)] - 2}{x + 4} = \dfrac{1}{2}$

x	-4.1	-4.01	-4.001	-4.0	-3.999	-3.99	-3.9
$f(x)$	0.4762	0.4975	0.4998	?	0.5003	0.5025	0.5263

9. $\lim_{x \to 0} \dfrac{\sin x}{x} = 1$

x	-0.1	-0.01	-0.001	0	0.001	0.01	0.1
$f(x)$	0.9983	0.99998	0.9999998	?	0.9999998	0.99998	0.9983

11. -1
13. The limit does not exist because $f(x)$ approaches different values from the left of $x = -2$ and the right of $x = -2$.
15. The limit does not exist because $f(x)$ oscillates between 2 and -2.
17. 6   **19.** 4   **21.** 0   **23.** $e^2 \approx 7.389$
25. 0   **27.** $\arcsin \frac{1}{2} \approx 0.524$
29. The limit does not exist.   **31.** 2
33. (a) 1   **(b)** $-\dfrac{9}{2}$ **(c)** $-\dfrac{5}{12}$ **(d)** $\dfrac{1}{\sqrt{\frac{3}{2}}}$

Section 12.2   (page 902)

WARM-UP
1. $x(x + 6)(x - 6)$ **2.** $(5x + 1)(2x - 3)$
3. $(s + 3)(s^2 - 3s + 9)$ **4.** $(y - 1)(y^2 + 1)$
5. $-(x + 6)$ **6.** $\dfrac{x + 2}{x(x + 3)}$ **7.** $\dfrac{1}{2(x + 2)(x + 3)}$
8. $(x + 1)(x - 1)$ **9.** $\dfrac{1}{\sqrt{x + 1} + 2}$
10. $\dfrac{1}{\sqrt{3 + x} + \sqrt{3}}$

1. (a) 1   **(b)** 3   **(c)** 5, $g_2(x) = -2x + 1$
3. (a) 2   **(b)** 0   **(c)** 0, $g_2(x) = x(x + 1)$ **5.** 120
7. -3 **9.** 0.8660   **11.** $\dfrac{1}{14}$ **13.** 4   **15.** $\dfrac{1}{2\sqrt{3}}$
17. $\dfrac{1}{4}$ **19.** -1 **21.** Does not exist.

23. Does not exist.

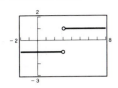

25. Limit is $\frac{1}{2}$.

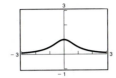

27. Limit is 1.

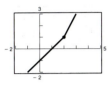

29. $-\frac{1}{2}$ **31.** $-\frac{2}{25}$ **33.** $2x - 3$ **35.** 0   **37.** 2
39. 1   **41.** 0.333
43. (a) 0, direct substitution   **(b)** 1
45. (a) 1, direct substitution   **(b)** 1
47. -64 ft/sec

Section 12.3   (*page 912*)

WARM-UP
1. 2   **2.** 0   **3.** $-\frac{3}{17}$ **4.** ≈ -1.52
5. $3x - 4y + 27 = 0$ **6.** $x = -2$ **7.** $y = \frac{7}{8}$
8. $3x + 2y - 15 = 0$ **9.** 3   **10.** $2x$

1. 0   **3.** $\frac{1}{2}$
5. $m = 2$

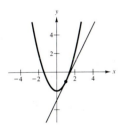

7. $m = -\frac{1}{2}$

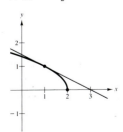

9. 500; From 1979 to 1980, per capita debt increased by about \$500.
11. -3 **13.** 4   **15.** -4
17. -1 **19.** $\frac{1}{6}$ **21.** $-2x$;   **(a)** 0   **(b)** 2
23. $-\dfrac{1}{(x + 4)^2}$;   **(a)** $-\dfrac{1}{16}$ **(b)** $-\dfrac{1}{4}$ **25.** 0
27. $-\dfrac{2}{3}$ **29.** $-\dfrac{2}{x^3}$ **31.** 4; $4x - y - 5 = 0$
33. $\dfrac{1}{4}$; $x - 4y + 5 = 0$

35.

x	-2	-1.5	-1	-0.5	0	0.5	1	1.5	2
$f(x)$	2	1.125	0.5	0.125	0	0.125	0.5	1.125	2
$f'(x)$	-2	-1.5	-1	-0.5	0	0.5	1	1.5	2

37.

x	-2	-1.5	-1	-0.5	0	0.5	1	1.5	2
$f(x)$	-2	-0.844	-0.25	-0.031	0	0.031	0.25	0.844	2
$f'(x)$	3	1.688	0.75	0.188	0	0.188	0.75	1.688	3

39. $-2x$; $(0, 9)$ **41.** $3x^2$; $(0, 3)$
43. $9x^2 - 9$; $(-1, 6)$, $(1, -6)$ **45.** c   **47.** b

Section 12.4   (*page 920*)

WARM-UP
1. Horizontal asymptote: $y = 0$
2. Vertical asymptote: $x = 2$
   Horizontal asymptote: $y = 0$
3. Vertical asymptote: $x = \frac{3}{2}$
   Horizontal asymptote: $y = -2$
4. Vertical asymptote: $x = -1$
   Horizontal asymptote: $y = 5$
5. Vertical asymptote: $x = 0$
   Horizontal asymptote: $y = 4$
6. Vertical asymptote: $x = 2$
7. $1, \frac{4}{3}, \frac{3}{2}, \frac{8}{5}, \frac{5}{3}$ **8.** $\frac{2}{3}, \frac{4}{9}, \frac{8}{27}, \frac{16}{81}, \frac{32}{243}$
9. $-\frac{2}{3}, \frac{4}{9}, -\frac{8}{27}, \frac{16}{81}, -\frac{32}{243}$ **10.** $\frac{1}{2}, \frac{1}{3}, \frac{1}{4}, \frac{1}{5}, \frac{1}{6}$

1. 0   **3.** -1 **5.** 2   **7.** Does not exist   **9.** -2
11. 0   **13.** -5 **15.** 0   **17.** 5   **19.** c   **21.** d

23. Limit is 0.

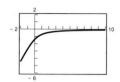

x	10^0	10^1	10^2	10^3
$f(x)$	-0.7321	-0.0995	-0.00999	-0.001

x	10^4	10^5	10^6	
$f(x)$	-1.0×10^{-4}	-1.0×10^{-5}	-1.0×10^{-6}	

25. Limit is $-\frac{3}{4}$

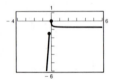

x	10^0	10^1	10^2	10^3
$f(x)$	-0.7082	-0.7454	-0.7495	-0.74995

x	10^4	10^5	10^6	
$f(x)$	-0.749995	-0.7499994	-0.75	

27. (a) \$47.05; \$5.92   **(b)** \$1.35   **29.** $3, \frac{4}{3}, 1, \frac{6}{7}, \frac{7}{9}; \frac{1}{2}$
31. $1, \frac{3}{5}, \frac{2}{5}, \frac{5}{17}, \frac{3}{13}; 0$ **33.** $\frac{1}{7}, \frac{1}{3}, \frac{9}{17}, \frac{8}{11}, \frac{25}{27}$; Does not exist.
35. $2, 3, 4, 5, 6$; Does not exist.
37. $10, 12, 26, 64, 138$; Does not exist.
39. $\frac{1}{3}, -\frac{2}{5}, \frac{3}{7}, -\frac{4}{9}, \frac{5}{11}$; Does not exist.
41. $\frac{3}{2}$

n	10^0	10^1	10^2	10^3	10^4	10^5	10^6
a_n	2	1.55	1.505	1.5005	1.50005	1.500005	1.5000005

43. 4

n	10^0	10^1	10^2	10^3	10^4	10^5	10^6
a_n	2	3.8	3.98	3.998	3.9998	3.99998	3.999998

45. $\frac{16}{3}$

n	10^0	10^1	10^2	10^3	10^4	10^5	10^6
a_n	16	6.16	5.4136	5.3413	5.3341	5.33341	5.333341

47. Converges; 0   **49.** Diverges
51. $|r| < 1$

Section 12.5   *(page 929)*

WARM-UP

1. $\displaystyle\sum_{i=1}^{6} (i + 3)$ **2.** $\displaystyle\sum_{i=1}^{6} i(i + 1)$ **3.** $\displaystyle\sum_{i=1}^{6} \frac{i + 2}{i}$

4. $\displaystyle\sum_{i=1}^{6} \frac{i^2}{2}$ **5.** 2   **6.** 0   **7.** Does not exist.   **8.** 9

9. $4 - 200(n + 1)$ **10.** $\dfrac{16n^2 + 24n + 8}{3n^2}$

1. 1830   **3.** 44,100   **5.** 5850
7. $S(n) = \dfrac{4n^3 + 6n^2 + 2n}{3n^3}$

n	10^0	10^1	10^2	10^3
$S(n)$	4	1.54	1.353	1.335

Limit: $\dfrac{4}{3}$

9. $S(n) = \dfrac{3n}{n^3} + \dfrac{6n^3 + 9n^2 + 3n}{6n^3}$

n	10^0	10^1	10^2	10^3
$S(n)$	6	1.185	1.0154	1.0015

Limit: 1

11. $S(n) = \dfrac{n^2 + n}{n^2} + \dfrac{3n}{n^2}$

n	10^0	10^1	10^2	10^3
$S(n)$	5	1.4	1.04	1.004

Limit: 1

13. $S(n) = \dfrac{2n^3 + 3n^2 + n}{6n^4} + \dfrac{2n}{n^2}$

n	10^0	10^1	10^2	10^3
$S(n)$	3	0.2385	0.02338	0.00233

Limit: 0

15. $S(n) = 1 - \dfrac{2n^3 + 3n^2 + n}{6n^3}$

n	10^0	10^1	10^2	10^3
$S(n)$	0	0.615	0.66165	0.66617

Limit: $\dfrac{2}{3}$

17. $S(n) = \dfrac{4n^3 + 6n^2 + 2n}{3n^3} - \dfrac{n^2 + n}{2n^2}$

n	10^0	10^1	10^2	10^3
$S(n)$	3	0.99	0.8484	0.8348

Limit: $\dfrac{5}{6}$

19. 4.5   **21.** 3.53   **23.** 1.266

25.

n	4	8	20	50
Approximate Area	12	14	15.2	15.68

27.

n	4	8	20	50
Approximate Area	6.727	5.971	5.534	5.363

29.

n	4	8	20	50
Approximate Area	3	2.597	2.369	2.280

31. 6   **33.** 9   **35.** $\frac{16}{3}$ **37.** $\frac{17}{4}$ **39.** $\frac{51}{4}$

Chapter 12   Review Exercises (*page 931*)

1. $\displaystyle\lim_{x \to 2} \dfrac{x - 2}{3x^2 - 4x - 4} = \dfrac{1}{8}$

x	1.9	1.99	1.999	2.0	2.001	2.01	2.1
$f(x)$	0.1299	0.1255	0.1250	?	0.1250	0.1245	0.1205

3. 11   **5.** $\frac{1}{15}$ **7.** $\frac{1}{4}$ **9.** -1 **11.** $3 - 2x$

13. $m = 2x - 4$;
    (a) At $(0, 0)$, $m = -4$.   **(b)** At $(5, 5)$, $m = 6$.

15. $m = -\dfrac{4}{(x - 6)^2}$;
    (a) At $(7, 4)$, $m = -4$.   **(b)** At $(8, 2)$, $m = -1$.

17. $g'(x) = 0$ **19.** $h'(x) = -\dfrac{1}{2}$

21. $f'(t) = \dfrac{1}{2\sqrt{t + 5}}$

23. 2   **25.** 1   **27.** 4   **29.** -1

31. Approximation: 40; Exact area: 50

33. Approximation: 15.875; Exact area: 15

35. Approximation: 0.5; Exact area: $\frac{4}{3}$

APPENDIX A   (*page A7*)

WARM-UP   **1.** $y = 4 - 3x$ **2.** $y = x$
3. $y = \frac{2}{3}(1 - x)$ **4.** $y = \frac{2}{3}(2x + 1)$
5. $y = \frac{1}{4}(5 - 3x)$ **6.** $y = \frac{2}{3}(-x + 3)$
7. $y = 4 - x^2$ **8.** $y = \frac{2}{3}(x^2 - 1)$
9. $y = \pm\sqrt{4 - x^2}$ **10.** $y = \pm\sqrt{x^2 - 9}$

1.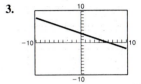

3.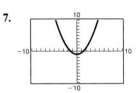

5.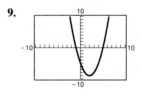

7.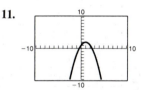

9.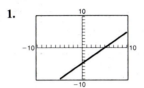

11.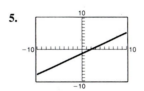

13.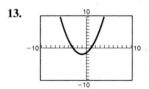

15.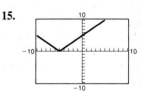

17.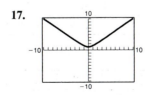

19.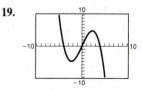

21. (d)   **23.** (a)   **25.** (i)   **27.** (j)   **29.** (e)   **47.** 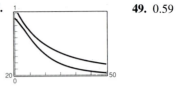   **49.** 0.59

31. 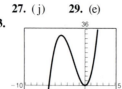   **33.**

35.
```
Xmin=-10
Xmax=10
Xscl=1
Ymin=-12
Ymax=30
Yscl=6
Xres=1
```

37.
```
Xmin=-10
Xmax=10
Xscl=1
Ymin=-10
Ymax=10
Yscl=1
Xres=1
```

51. 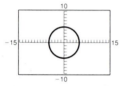   **53.** (b)

39. No intercepts   **41.** Three *x*-intercepts

43. Square   **45.** Circle

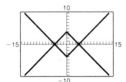 

INDEX OF APPLICATIONS

Chemistry and Physics Applications

Construction Applications

Consumer Applications

INDEX